D0811844

Space Satellite Handbook

Third Edition

Gulf Publishing Company
Houston, London, Paris, Zurich, Tokyo

Space Satellite Handbook

Third Edition

Anthony R. Curtis, Editor

Space Satellite Handbook

Third Edition

10 9 8 7 6 5 4 3 2 1

Gulf Publishing Company
Book Division
P.O. Box 2608 □ Houston, Texas 77252-2608

Printed in the United States of America.

Library of Congress Cataloging-in-Publication Data

Curtis, Anthony R., 1940–
Space satellite handbook / Anthony R. Curtis.
p. cm.
Includes index.
ISBN 0-88415-192-1
1. Artificial satellites—Registers. 2. Artificial satellites—Handbooks, manuals, etc. 3. Artificial satellites—History. I. Title.
TL796.6.E2C867 1994
629.46—dc20

93-37563
CIP

Table of Contents

Preface

An eagle-eyed satellite looks down through mountains of shifting sands to find shadows of a lost civilization buried in an ancient city beneath the desert.

★Hubble Space Telescope may be bleary-eyed, but it still searches distant galaxies, farther away than ever seen before, to answer age-old questions.

★Seven American satellites are blasted to space from one B-52 bomber.

★The Cold War ends and Russia finds its fleet of satellites used to communicate with KGB spies around the globe is rusting away.

★Shuttle astronauts corral a dead satellite in its useless orbit and kick-start it on the way to a new working life in a proper orbit.

★A flotilla of communications and science satellites is operated for decades by hobbyists for fun and public service.

★A red-hot space race, brewing in the Middle East between Israel and Iraq, is squashed by the Persian Gulf War.

★A motorist steers through a maze of Chicago streets, following a course drawn on a dashboard map as his car receives data directly from a squadron of orbiting satellites.

Did you hear about these extraordinary moments, mishaps and milestones? Countless exciting events are not reported by most news media and go unnoticed by the public.

If you're like me, when TV plays a ten-second sound bite of a space shuttle rumbling upward from Cape Canaveral, a newspaper prints a grainy photo of a mighty booster lifting off a launch pad, or a magazine serves up an intriguing blurb about a new satellite, your inner eye turns skyward and you yearn for the rest of the story.

Hearing that an old satellite has fallen silent and an extraordinary new one has been launched, you surf the TV channels, hoping for a scrap of information, but you find precious little to satisfy that hunger to know more.

What's happening up there? The answers are in your hands right now. This SPACE SATELLITE HANDBOOK stretches a canvas of satellite history and paints it with a kaleidoscope of fascinating stories and little-known facts about the more than 22,000 artificial, man-made satellites and bits and pieces of space junk ever sent to orbit.

First up, a snap course in space satellite origins. Then we'll dig right into the considerable uses for these brilliant machines—communicating, broadcasting, rescuing lost persons, forecasting tomorrow's weather, observing the mark of man on land, sea and air, navigating, spying, eavesdropping, defending, testing the effects of weightlessness on monkeys and frogs, scrutinizing the fissure in the ozone mantle, watching for earthquakes and erupting volcanoes, tracking wild animal migrations, listening for the myriad chirps and chortles of energetic particles flying across the Universe, and visiting planets far away across our Solar System.

Finally, and most importantly, a detailed reference list of *every satellite ever orbited* starts on page 190. If it's been in space, it's here in the SPACE SATELLITE HANDBOOK.

The author is grateful to National Aeronautics and Space Administration, Goddard Space Flight Center, Johnson Space Center, Jet Propulsion Laboratory and other NASA centers, the U.S. Space Command, Smithsonian Astrophysical Observatory, the European Space Agency and Arianespace, spacefaring governments around the world, satellite owners, commercial launch companies and other helpful sources.

He would be grateful to hear of satellite developments around the globe. If you have suggestions for future editions of this book, please address the author in care of Gulf Publishing Co., Post Office Box 2608, Houston, TX 77252-2608.

In the meantime, check chapter one for tips on spotting satellites from your backyard and keep an eye out for the Hubble telescope, space station Mir, an orbiter in the American space shuttle fleet, one of the massive radar-imaging battlestars and the rest of the more than 7,000 man-made points of light racing across the sky!

—Anthony R. Curtis

Satellites And The Space Age

The Space Age roared to life on a Friday in October 1957 when Russia's Old Number Seven rocket lobbed a thermometer, a silver-zinc battery and a radio transmitter in a 23-in. polished aluminum ball from the surface of Earth to a place so high above the atmosphere it couldn't come all the way down again.

The 184-lb. ball was "Sputnik," mankind's first artificial Earth satellite. The satellite was pressurized with nitrogen circulated by a cooling fan. The thermometer changed the frequency of the two radio beacons. Attached to the outside of the satellite were two eight-ft. and two ten-ft. radio antenna whips.

For three weeks, as it twirled around the world every 96 minutes in a globe-girdling orbit 588 miles above our heads, Sputnik beep-beeped its visionary message of a future above the ocean of air.

After 92 days, Sputnik burned as it fell from orbit into the atmosphere January 4, 1958.

Three centuries ago. The English mathematician Isaac Newton, working on his new theory of gravitation in 1687, uncovered the theoretical possibility of orbiting an artificial satellite of Earth.

However, it wasn't until the early 20th century that scientists figured out how to send a satellite high enough and fast enough to place it in orbit around Earth. Russian thinker Konstantin Tsiolkovsky originated theories and American tinkerer Robert Goddard conducted experiments which confirmed a satellite might be carried to space on a rocket.

World War II research between 1943 and 1946 indicated the rockets then available were too weak to boost a satellite to Earth orbit. Engineers returned to their drawing boards.

After the war, far-reaching research on rockets for upper-atmosphere research and military missiles was extensive. By 1954, engineers knew they would be able to launch a satellite to Earth orbit sooner or later. That year, the International Geophysical Year (IGY) Committee asked nations to launch science satellites for space exploration. The USSR and the U.S. announced plans for small IGY satellites in 1955.

Missiles. A weapons race started in the 1950s when the U.S. and USSR built the first hydrogen bombs. America decided to use airplanes to carry its H-bombs, but Russia turned to a new means of transporting the heavy nuclear weapons to targets—intercontinental ballistic missiles or ICBMs.

First Satellites Launched By Spacefaring Nations

1957 Oct 4	**USSR**
satellite:	Sputnik 1
rocket:	Old Number Seven
launch site:	Baikonur Cosmodrome
1958 Jan 31	**USA**
satellite:	Explorer 1
rocket:	Jupiter-C
launch site:	Cape Canaveral
1965 Nov 26	**France**
satellite:	Asterix 1
rocket:	Diamant
launch site:	Algeria
1970 Feb 11	**Japan**
satellite:	Ohsumi
rocket:	Lambda 4S-5
launch site:	Kagoshima
1970 Apr 24	**China**
satellite:	Mao 1
rocket:	Long March-1
launch site:	Inner Mongolia
1971 Oct 28	**Great Britain**
satellite:	Black Knight 1
rocket:	Black Arrow
launch site:	Woomera Australia
1979 Dec 24	**Europe**
satellite:	CAT
rocket:	Ariane
launch site:	Kourou, French Guiana
1980 Jul 18	**India**
satellite:	Rohini 1
rocket:	Satellite Launch Vehicle
launch site:	Sriharikota Island
1988 Sep 19	**Israel**
satellite:	Horizon 1
rocket:	Shavit
launch site:	Negev Desert
1989 Dec 5	**Iraq**
satellite:	rocket 3rd stage
rocket:	three-stage rocket
launch site:	Al-Anbar

Russian engineers, headed by Sergei Korolev, designed the first ICBM rocket in the mid-1950s. Known officially as R-7, insiders thought of it as Old Number Seven.

Tests on parts of the ICBM started in 1956. The next year, the first long-range flight-test rocket exploded on the launch pad in May, but momentum didn't falter. Eight test firings were made in 1957, including two rockets which flew 4,000 miles that August.

Race to orbit. The idea dawned on Soviet engineers that Old Number Seven not only could blast a warhead thousands of miles to impact on Earth, but it also could carry a payload at such great speed and altitude it would be above the atmosphere and orbiting Earth. An artificial Earth satellite could be created.

The first satellite built to ride Old Number Seven was completed in June 1957. Its official Russian name was Iskustvennyi Sputnik Zemli, or "Fellow Traveler of the Earth." The short name Sputnik stuck and today is applied generally to all Russian space satellites.

Members of the Korolev design team referred to their first Sputnik as the "Preliminary Satellite," or PS for short. Sometimes the technicians would honor Korolev by reversing the letters and using his first two initials, SP for Sergei Pavlovich.

Just 24 days after the first successful 4,000-mi. ICBM test, Soviet Premier Nikita Khrushchev on August 27 gave the okay to launch a satellite-carrying R-7 toward space.

Khrushchev wanted to launch the beach-ball-sized satellite on September 17, the 100th anniversary of the birth of Konstantin Tsiolkolvsky, known as the father of cosmonautics. However, technical problems pushed the blast-off into October.

The satellite was attached to its rocket at Baikonur Cosmodrome near Tyuratam on October 2 and covered with a nosecone. The radio was tested in the assembly building and then the booster was transported outside to the launch pad on a rail car.

Hydraulic lifts erected the rocket onto the pad October 3. Four arms supported the rocket against the wind and a white cloth shaded the Sun's heat from the nosecone. Compressed air was blown from a hose onto the nosecone to keep the satellite inside from overheating.

Last-minute work delayed blastoff into the evening of October 4. Finally, a bugler sounded several notes on the launch pad and, at 10:28 p.m. Moscow time, the cluster of rockets forming R-7 was ignited. Thrust built up and the booster was released to climb into the night sky. Strangely, Soviet secrecy prevailed and no photographs recorded the momentous event.

A nail-biting 95 minutes after Sputnik rode Old Number Seven into history, the satellite flew over the cosmodrome and its radio signal confirmed it was in orbit. When the official TASS news agency telegraphed the news around the globe at 5:58 p.m., New York time, October 4, everyone knew the Space Age had dawned.

Launch of Sputnik was the most significant event since explosion of the atom bomb in 1945. The new satellite stirred the political stew as it raced around Earth in an orbit ranging from 142 to 588 miles altitude, transmitting on frequencies of 20.005 and 40.002 megahertz.

Sputnik's one-watt transmitter generated sufficient strength to be received easily by amateur radio operators around the world. Its signals were received first in the U.S. at 8:07 p.m. Eastern time October 4. Thousands of hams recorded its cricket-chirp sound as the satellite's silver-zinc battery lasted beyond its 14-day design life to October 22.

Those times. It was an intense period in the Cold War, not even a year since Khrushchev had boasted, "We will bury you," and a mere six weeks after the USSR had demonstrated its intercontinental ballistic missile.

Sputnik horrified the American ego, sparking a comeback rally in U.S. schools and colleges. Teachers and scientists fondly recall the end of the 1950s as a golden era with U.S. President Dwight D. "Ike" Eisenhower marshaling the intellectual elite to resuscitate math and science education, but others remember America was mortified by what a writer called that "humiliating beep-beep in the high heavens."

Each beep was an "outer-space raspberry to a decade of American pretensions that the American way of life was a gilt-edged guarantee of our national superiority," according to Claire Booth Luce, a member of Congress and wife of the publisher of *Time* magazine.

Sen. Henry M. Jackson of Washington reeled under the "devastating blow to the prestige of the United States as the leader in the scientific and technical world."

Labor leader Walter Reuther declared Sputnik a "bloodless Pearl Harbor."

When reporters questioned the President October 9 about Sputnik, Eisenhower called it "one small ball in the air. I wouldn't believe that at this moment you have to fear the intelligence aspects of this." Admitting Sputnik's 184-lb. weight "astonished our scientists," Eisenhower conceded the USSR had "a great psychological advantage throughout the world."

More sputniks. Not only was the USSR's first satellite launch attempt successful, but a second try 30 days later was too. The 1,119-lb. Sputnik 2 was launched November 3 carrying the live dog Laika on a life-support system.

The capsule remained attached to the converted ICBM. The dog captured hearts around the world as life slipped away from Laika a few days into her journey. Later, Sputnik burned in the atmosphere April 14, 1958.

Sputnik 2 forced the U.S. administration to action. The President made a radio-TV broadcast November 7, naming the first White House science adviser. Later, Eisenhower approved $1 billion for the first direct federal aid to education "to meet the pressing demands of national security in the years ahead."

Vanguard. The U.S. Army and Navy had ballistic-missile projects underway. Each wanted to be the first American service to orbit a satellite.

Two months after the launch of Sputnik 1, Eisenhower gave the Navy a chance to send the first American satellite to orbit aboard that service's Vanguard rocket. The Navy achieved only a spectacular failure of its booster on December 6, 1957. In fact, fearing it might fail, the Navy had referred to the rocket before launch as Test Vehicle No. 3 (TV-3).

With world-wide news media tuned in, Vanguard lost thrust two seconds after launch, a mere four feet off the launch pad. It fell back to the pad and exploded. The tiny six-in. satellite popped out of the flames and rolled away, transmitting its radio signal on the ground. Newspapers called it Kaputnik and Stayputnik.

The Navy tried again to beat the Army on January 25, 1958, but that Vanguard rocket fizzled within 14 seconds of ignition.

Explorer 1. The U.S. Army's Redstone Arsenal at Huntsville, Alabama, was a vast base where America's rocket pioneers later would design the gargantuan boosters which carried astronauts to the Moon in 1969.

In the late 1950s, those pioneers also had been working on a response to the launch of Sputnik by the Soviet Union.

Worried about American prestige lost in the Navy fizzles, Washington officials asked the Army to try to send a satellite to orbit.

The Army did the job January 31, 1958, launching its Explorer-1 satellite from Cape Canaveral on a Jupiter-C rocket, a modified version of the Redstone ballistic missile.

Ground Control Stations

All kinds of satellites—weather, observation, navigation, communications—must stay in touch with someone on the ground. The work of satellites orbiting hundreds or thousands of miles above our planet is directed by human operators in "ground control stations" on Earth.

Ground control stations are located all around the globe. To tell an unmanned spacecraft what to do, controllers send coded radio signals to the satellite and receive similar signals from it.

For instance, a newly-orbiting satellite can be told to unfold its electrical solar panels and communication antennas and turn on science instruments. Ground controllers can instruct an observation satellite to look elsewhere. A weather satellite might be told to stare at a hurricane. The radio frequency used by a communications satellite to beam down a TV movie can be changed. A spysat can be ordered to modify its orbit to fly its sensors over specific ground targets.

NASA controls its constellation of TDRS communication satellites from a ground station in White Sands, New Mexico. Many military satellites are operated from ground control stations in Colorado and California.

NASA's deep-space interplanetary probes operate in far-flung regions of the Solar System—Magellan at Venus, Galileo approaching Jupiter, Ulysses crossing the Sun, the twin Voyagers at the far outer edge of the System. All are controlled from NASA's Jet Propulsion Laboratory in Pasadena, California, through three giant satellite dish antennas in California, Spain and Australia.

The 31-lb. Explorer satellite had been designed by James Van Allen at the Jet Propulsion Laboratory in California. Liftoff took place from Launch Complex 26 at what now is known as Cape Canaveral Air Force Station. Today, the site is home to the U.S. Air Force Space Museum.

Explorer-1's batteries died five months after launch, but by then the 31-pound satellite had discovered intriguing belts of intense radiation circling planet Earth. The belts were named for James Van Allen. On March 31, 1970, Explorer-1 plunged into Earth's atmosphere and burned.

Sputnik's wake-up call and Explorer's discovery of the Van Allen radiation belts spurred further work on artificial Earth satellites.

The Army did have embarrassments. Its second try, Explorer-2, failed to reach orbit March 5, 1958. The rocket fourth stage didn't ignite.

The Navy finally reached space-launch success when it sent the tiny 3-lb. satellite Vanguard-1 to orbit from the Cape on March 17, 1958. And the Army's Explorer-3 made it to orbit March 26,

1958, carrying instruments to measure cosmic rays, meteorites and temperature.

Sputnik 3. Meanwhile, the third time was not a charm for the USSR. A rocket failed to boost a large geophysical observatory to orbit February 3, 1958. But, the fourth try, carrying another geophysical observatory, was successful May 15. Sputnik-3 was solar powered and weighed 2,925 lbs.

Explorer 4. The U.S. Army launched Explorer-4 from the Cape July 26, 1958. James Van Allen, analyzing data from Explorers 1, 3 and 4, discovered belts of radiation trapped in the magnetic field surrounding planet Earth.

NASA. Despite resistance from the Department of Defense which wanted to keep space research entirely military, legislation was enacted in the U.S. on July 29, 1958—the National Aeronautics and Space Act of 1958—enabling the birth of the civilian National Aeronautics and Space Administration (NASA). On October 1, NASA was founded. It included the old National Advisory Committee for Aeronautics (NACA).

A crash program—the National Defense Education Act (NDEA)—started that September, granted money to graduate students and helped local school districts pay for math and science teachers, school equipment and buildings.

The National Science Foundation budget nearly tripled from 1958 to 1959, then doubled again by 1962. NSF persuaded scientists to write new text books for high schools, then held summer sessions for high school science teachers.

Explorer-6. The earliest TV pictures of Earth's cloud cover were sent down by Explorer 6 after it was launched August 7, 1959.

TIROS-1. The prototype weather satellite, Television Infra-Red Orbital Satellite 1 (TIROS-1), was launched April 4, 1960. It sent down 22,952 cloud photos in 77 days.

Echo-1. A 100-ft. aluminum-coated balloon was launched August 12, 1960. Radio and TV signals from New Jersey were bounced off the balloon to France.

Man in space. In the 1960s, U.S. and USSR launches were highlighted by manned flights to Earth orbit in small capsules, manned flights to the Moon, unmanned probes to Venus, Mars and the Moon, and launch of the first amateur radio satellites.

Yuri Gagarin was launched to Earth orbit in the USSR's Vostok-1 April 12, 1961, followed by Alan Shepard on a U.S. suborbital trip to space in Mercury-3 on May 5, 1961. OSCAR 1, the first amateur communications satellite, was launched December 12, 1961.

Third nation in space. On December 15, 1964, Italy became the third nation to have one of its satellites sent to space as an Italian launch team helped fire an American Scout rocket from a platform in the Indian Ocean off East Africa, carrying Italy's San Marco-1 satellite to orbit.

Italy's launch site, known as the San Marco Equatorial Range, is a pair of platforms, similar to rigs used in offshore oil drilling. They are anchored three miles offshore in Kenya's Ngwana Bay, 90 miles north of Kenya's port city, Mombasa.

Operated by the University of Rome, the launch pad was established in 1964 and named for the patron saint of navigators. Launches from San Marco have been so popular with the Italian public that Pope Paul VI blessed them.

The location close to the equator gave satellite controllers access to an equatorial orbit. Still in use, San Marco Equatorial Range has been the site of ten launches since 1964, all successful, all with U.S. Scout rockets.

The 1964 San Marco-1 launch was the first firing of a NASA rocket by foreign technicians. Since then, Italy has been the most-active foreign user of NASA's Scout. The four-stage solid-fuel Scout can lift a 475-lb. payload to a 300-mi.-high orbit. Over the years, Scout has been one of NASA's most reliable boosters.

The first satellite, San Marco-1, stayed in orbit until September 13, 1965, when it fell into the atmosphere and burned.

Spacefaring nations. Launching its own satellite to orbit shows a nation has a rocket powerful enough for space launches. The first nine countries able to launch their own satellites to orbit were the USSR, U.S., France, Japan, China, Great Britain, India, Israel and Iraq.

Brazil is expected to be tenth, but North Korea, South Africa, Pakistan, Germany, Australia and others probably have the technology.

The Italians had used an American rocket for their 1964 San Marco launch. Today they still don't have their own space rocket, but they do participate vigorously in the European Space Agency and use its powerful Ariane rockets.

Since 1957, more than 4,100 satellites have been launched. Nearly 3,000 of those have been launched by the Russians since Sputnik-1 in 1957. Altogether, the U.S. has launched more than 1,200 satellites.

America's Explorer-1 satellite followed Sputnik in January 1958. On March 17, 1958, Vanguard-1 became the second American satellite in orbit. Today, Vanguard-1 is the oldest satellite still in orbit from any nation.

French space age. Asterix-1 was a small satellite launched from North Africa which initiated France into an exclusive club of nations. It was the third nation with a rocket powerful enough to boost a satellite into Earth orbit.

The French launch pad was outside of the adobe village of Hammaguir, 900 miles into the Sahara Desert from Algiers.

Camels, goats and sheep foraging among brittle weeds were startled November 26, 1965, when French technicians wearing burnooses blew charges, exploding bolts holding down the silvery, needle-like Diamant single-stage rocket.

Exhaust pouring from the rocket's one large nozzle pushed the small satellite to space. France's network of receiving stations tracked the 19-lb. Asterix, named for a red-whiskered Celtic barbarian in French comics.

Old-fashioned chemical batteries powering the radio beacon in Asterix lasted only two days. The satellite fell from orbit to burn up in the atmosphere March 25, 1967.

Shortly after the Asterix launch, another French satellite, known as FR-1, rode a U.S. Scout rocket to orbit December 6 from Cape Canaveral.

Since then France has orbited more than 50 satellites, as a key partner in the European Space Agency and on its own.

East Asia. Japan launched its first satellite, Ohsumi-1, to orbit February 11, 1970, making it the fourth nation with a powerful space rocket.

Later that year, China launched its first satellite to Earth orbit April 24, 1970, on a Long March rocket. The satellite was named Mao-1.

Europe. Great Britain launched its first satellite, Black Knight-1, on a Black Arrow rocket from Woomera, Australia, October 28, 1971.

The first Ariane rocket launched by the European Space Agency from Kourou, French Guiana, carried ESA's first satellite, referred to as CAT, to orbit December 24, 1979.

South Asia. India became the seventh nation to demonstrate a space rocket when it launched its first satellite, Rohini 1, on a Satellite Launch Vehicle (SLV) rocket July 18, 1980.

Middle East. Israel fired its first satellite, Horizon-1 (Ofek-1) to space September 19, 1988.

Onlookers saw the three-stage silver Shavit rocket rise on a column of flame from a military launch pad south of Tel Aviv in the Negev Desert.

It arched high in the noon Middle Eastern sky where boosters separated and fell into the Mediterranean Sea. Shavit is Hebrew for comet.

Horizon-1 was a gray, 6.6-lb. satellite, beeping its radio beacon from a low elliptical orbit until it fell into the atmosphere and burned after 4 months.

During its short life in space, the satellite's orbit ranged from an apogee (high point) of 620 miles above Earth to a perigee (low point) of 155 miles. Horizon-1 took 90 minutes to circle the globe.

Another satellite, Horizon-2 (Ofek-2), was launched April 3, 1990.

Satellite Launches Do Fail

Many satellite launch attempts by nations around the globe have failed over the years.

For instance, a Russian Proton rocket exploded on its way to orbit in 1988, destroying three satellites—Cosmos-1917, Cosmos-1918 and Cosmos-1919. The accident sent large chunks of debris hurtling to Earth. Debris from the exploded space booster landed off Australia. A piece hit central Siberia. A third, smaller chunk remained in orbit. The shards crashing to Earth caused neither personal injury nor property damage.

A year earlier, in 1987, a 10-ton satellite, Cosmos-1871, failed to reach orbit and fell harmlessly into the South Pacific Ocean 1,500 miles east of New Zealand near Antarctica, without causing damage or injuries.

The Shavit rocket, a converted Jericho II medium-range ballistic missile, was test fired in 1988 on a short flight to a splashdown in the Mediterranean Sea. As a missile weapon, the rocket could carry an atomic bomb 900 miles to a target on Earth.

Israel had created a space agency in 1983 to work on spy satellites. The Horizon-1 and Horizon-2 test flights probably were preliminary to later launch of Israeli-built spysats to monitor Arab enemies.

Israel had been relying on pilotless drone aircraft and manned reconnaissance aircraft to gather intelligence above teeming Middle Eastern cities. Since the Horizon test flights went well, the Israelis are likely to blast a spysat to space for a two-year assignment in orbit. Such a low-flying photography satellite might pass over key Arab cities up to 16 times a day.

The Jerusalem government probably purchased sophisticated American optical gear for use in spy satellites. With its own electronic eye staring down through the dusty Middle Eastern sky at Arab states, Israel no longer would be fully dependent on the U.S. for satellite intelligence.

Jerusalem said it also is developing science satellites to be launched on European Ariane rockets. Israel plans to launch Amos-1 and Amos-2 communications satellites on Ariane rockets. Israeli amateur radio operators are building a hamsat for launch on a Russian rocket in 1995.

Regional space race. For a time, Israel and Iraq seemed to be pitted against each other in a Middle Eastern space race. However, the 1991 Persian Gulf War may have ended Iraq's ability to participate for many years.

Iraq launched a 48-ton three-stage rocket December 5, 1989, from the Al-Anbar Space Research Center 50 miles west of Baghdad on a six-orbit flight.

It was not a separate satellite, but the third stage of a rocket which swept around the Earth for 6 revolutions before falling out of orbit.

A U.S. nuclear-attack-warning satellite spotted the fiery Iraqi rocket exhaust as it blasted off. North American Aerospace Defense Command tracked the third stage around the globe.

The 75-ft. rocket was similar to a U.S. Scout rocket used to send small satellites to low orbits. The Iraqi booster may have been a modified version of Argentina's Condor ballistic missile.

The launch was the first time Iraq exposed its secret space research. Such a ballistic missile could carry a nuclear warhead 1,240 miles.

Iraq already had a 600-mi. ballistic missile built from the USSR's Scud design. Scuds were launched against Iranian cities in the eight-year Persian Gulf war in the 1980s and against Israel and Saudi Arabia in the 1991 Persian Gulf War.

Satellite Applications

The first satellites were used mostly to take the measure of the new space environment and blaze a path for communications, weather, and navigation satellites and manned spaceflight.

The first communications satellites—known as SCORE, ECHO, TELSTAR, RELAY, and SYNCOM—were launched between 1958 and 1963. The first weather satellite, TIROS, and the first navigation satellite, TRANSIT, were in 1960.

The U.S. military launched its first spysat, SAMOS, in 1961. Since then, lots of military reconnaissance satellites, including a model known as Big Bird, have been outfitted with elaborate photographic equipment and high-tech electronic snooping gear and sent to orbit on secret missions.

Following the original Sputnik in 1957, more than 22,000 useful payloads and useless chunks of space debris have been placed in orbit.

The majority of satellites have been built by Russia and the United States, but the countries of Western Europe in the European Space Agency, Japan, China, India, Canada, Israel, Brazil and others are actively engaged in satellite development.

Satellites are part of daily life, used for communications, weather forecasting, navigation, observing land, sea and air, other scientific research, and military reconnaissance. Hundreds of men and women have lived and worked aboard manned satellites—space shuttles and space stations—in Earth orbit.

What's inside? Most satellites have internal sensors to measure voltages, currents, temperatures and other information on the health of equipment. The measurement data is encoded into a telemetry signal and relayed by transmitter to ground controllers on Earth.

Communications satellites have additional radio transmitters and receivers to send many signals to and from Earth. Observation satellites often have optical and infrared cameras and radar for altitude measurements. Navigation satellites have highly-stable radio receivers and transmitters.

Communications satellites relayed military TV from Somalia in 1992 and later years, while U.S. spysats were moved into position over the Balkans to stare down at the former Yugoslavia.

Many satellites have attitude-control equipment to maintain desired orbits, to point sensors at their targets in space or on the ground, to point radio antennas at ground stations, and to keep solar-power generators pointing at the Sun.

Satellites usually are powered by the Sun, but nuclear-powered electricity generators sometimes are built into spacecraft.

Satellite Orbits

Most people think of satellites as flying east-west, circling the globe above the equator. Many do fly in such low equatorial orbits, but polar-orbiting satellites travel a north-south path across the poles.

Equatorial and polar satellites mostly fly at low altitudes between 100 and 1,000 miles. To an observer on the ground, they seem to climb up from the horizon, pass overhead, and fall below the horizon several times a day. Stationary satellites are in equatorial orbits 22,300 miles above Earth. They are so far above the planet's surface they seem to stand still.

Polar satellites look down on Earth's entire surface, passing above the North and South poles several times a day. As the satellite loops around the globe, Earth seems to rotate under the orbit.

Observation of the ground is improved if the surface always is illuminated at the same Sun angle when viewed from the satellite. Many weather, resource and reconnaissance satellites are in so-called Sun-synchronous polar orbits.

Examples include LANDSAT, Nimbus, GEOS, TIROS, NOAA and Navy Oscar satellites. Geodetic-survey and navigation satellites usually are in almost-perfectly-circular polar orbits.

The first polar-orbiting U.S. weather satellite was Nimbus-1 launched in 1964. The first U.S. Earth-resources observation satellite, Landsat-1, was sent to polar orbit in 1972.

Stationary satellites, on the other hand, orbit at altitudes around 22,300 miles. Back in 1945, science-fiction author Arthur C. Clarke imagined communications satellites in stationary orbits where they would travel around the world at the same speed the globe is spinning, making them hang stationary over one spot on Earth's surface.

A satellite on its way to stationary orbit is fired to a special equatorial orbit known as a geostationary transfer orbit (GTO). In this highly-elliptical orbit, the satellite swings out 22,300 miles and back in to an altitude of 100 miles above Earth. At an assigned time and place, a "kick motor" attached to the satellite pushes it on out to a circular orbit at 22,300 miles altitude.

The stationary-orbit region of space is referred to as the Clarke Belt. Satellites there are said to be synchronous, geostationary or geosynchronous.

A synchronous satellite has a high, fixed vantage point from which it can look down continuously on a large portion of Earth. That makes stationary satellites ideal for pinpoint broadcasting and for monitoring continent-wide weather patterns and environmental conditions.

Scores of communications and weather satellites, operated by many countries and international organizations, occupy positions assigned to every nation in the Clarke Belt.

Many popular television satellites, sending programs to backyard dish antennas, are in stationary orbits.

The U.S. communications satellite Syncom-3 was the first launched successfully to stationary orbit. Other well-known ones have included Inmarsat and TDRS.

North American cloud photos seen in television weathercasts usually have been made by American satellites known as Geostationary Operational Environmental Satellites (GOES).

Altitudes

Earth is ensnared today in a thick spider web of orbits. Satellites with different assignments fly at different altitudes:

Altitude Miles	Satellite Types
100-300	shuttles, space stations, spysats, navsats, hamsats
300-600	weather sats, photo sats
600-1,200	spysats, military comsats, hamsats
3,000-6,000	science sats
6,000-12,000	navsats
22,300 (stationary)	communications, broadcast, weather
250-50,000 (elliptical)	early-warning, Molniya broadcast, communications, spysats, hamsats

U.S. space shuttles are manned satellites of Earth. They usually fly at altitudes around 200 miles above Earth. Rarely, they fly near 400 miles altitude. Russia's orbiting space station Mir is another manned satellite. It maintains its own orbit just above 200 miles altitude.

Russian and American navigation satellites guide ships at sea from 100 to 300 miles altitude.

Photo-intelligence (PHOTOINT) satellites shoot clear pictures and infrared images of installations on the ground from the same altitude range. Radar images of targets on Earth's oceans are recorded at these altitudes.

Civilian photography satellites, such as the American Landsat and the French Spot, orbit at altitudes ranging from 300 to 600 miles. American NOAA and Russian Meteor weather satellites are at these same altitudes.

Spysats and military communications satellites dominate space from 600 to 1,200 miles altitude. The spysats gather electronic intelligence (ELINT), signal intelligence (SIGINT) and radar intelligence (RADINT). Hamsats also operate there.

Science research satellites do much of their work at altitudes between 3,000 and 6,000 miles above Earth. Their findings are radioed to Earth as telemetry data.

From 6,000 to 12,000 miles altitude, navigation satellites operate. Best known are the U.S. global-positioning system (GPS) and Russia's equivalent GLONASS satellites.

As described above, the so-called Clarke Belt is the region of space 22,300 miles above Earth where satellites seem stationary above the rotating Earth. Best known occupants of the Clarke Belt are the many domestic and international TV broadcast, weather reporting and communications satellites.

Inmarsats are examples among many non-broadcast communications satellites in the Clarke Belt. Europe's Meteosat and America's GOES are weather satellites in stationary orbits. At the same altitudes, but less-well known, are satellites such as NASA's TDRS (Tracking and Data Relay

Satellites), Russia's similar Satellite Data Relay Network (SDRN), U.S. Milstar military communications satellites, the Pentagon's Defense Satellite Communications System (DSCS) and the U.S. Navy's Fltsatcom and Ultra High Frequency (UHF) communications satellites.

The Russians have a series of communications satellites, known as Molniya, in long elliptical orbits which repeatedly carry them out beyond 22,000 miles before they swing back to within 1,000 miles of Earth's surface. Molniyas relay television broadcasts as well as man-in-space and military communications.

Other satellites in such elliptical orbits range from as close as 250 miles out to 60,000 miles. They include intelligence-agency communications craft and early-warning satellites which would report a launch of nuclear weapons.

Falling Satellites Burn

The old U.S. space station Skylab, at 77.5 tons and 118 feet long, was one of the largest satellites ever sent to Earth orbit. It provided quite a spectacle as it died July 11, 1979.

Skylab blazed through the atmosphere, sprinkling debris into the Indian Ocean and onto the Australian Out Back.

Some 26 tons survived reentry heat. Thousands of pieces, weighing up to two tons, hit water and land along a 3,600-mi.-long and 100-mi.-wide path. No one was hurt and no damage was done.

Spaceports

Rockets can be fired in any direction from a "pad" at a launch site. However, human neighbors dictate the direction of launch.

From Cape Canaveral in Florida, rockets mostly are launched east over the Atlantic Ocean, but sometimes northeast along the Atlantic Coast. Rockets are fired west and north over the Pacific Ocean into polar orbits from Vandenberg Air Force Base in California.

Russia's Baikonur Cosmodrome is not in Russia, but in its southern neighbor, Kzakhstan. Rockets launched from Baikonur carry satellites to equatorial orbits. Russia's Northern Cosmodrome is inside Russia at Plesetsk. Boosters from Plesetsk carry satellites to polar orbits.

The European Space Agency's launch site is in South America near the equator at Kourou, French Guiana. Rockets are fired east and north over the Atlantic to equatorial orbits.

Air launches. Satellites don't have to be launched from the ground to reach polar orbit. Since 1990, the U.S. has fired small satellites to polar orbit on 50-ft. Pegasus rockets carried aloft by an Air Force B-52 bomber.

For the trip to space, a winged Pegasus is strapped under the airplane's wing, ferried seven miles above Earth and dropped. Pegasus then ignites and blasts on up to a 350-mi.-high polar orbit where the satellite payload is released.

Pegsat was a communications satellite with science experiments launched by Pegasus in 1990. Another Pegasus carried seven small communications satellites to orbit in 1991. A third Pegasus ferried to space in 1993 a Brazilian science satellite to study rainforests.

Backyard Satellite Watching

Numerous satellites can be seen by naked eye from the ground at night after the Sun has set and our sky is dark, but satellites overhead are still in full sunlight. They look like points of light racing across the sky to disappear suddenly into Earth's shadow. They can be separated from high-flying airplanes because satellites don't display navigation lights or strobes and they don't change direction, make sounds or leave jet contrails.

Only large objects in low orbits can be seen by naked eye from the ground. Numerous TV satellites stare down at us constantly from stationary orbits, but are too high to be seen by naked eye from the ground.

Russia's 100-ft.-long orbiting Mir space station is the brightest man-made object in the sky. Sunlight reflects off its large surface area.

Hubble Space Telescope and space shuttles are big bright objects, readily seen by southern U.S. viewers from Los Angeles to Atlanta. Launched from the Kennedy Space Center at 28.4 degrees north latitude, their most efficient orbits are confined to 28.4 degrees north and south of the equator. Only the occasional shuttle mission cruises as far north as Chicago. Hubble follows that path, but at a higher altitude.

How to do it. Satellites can be spotted just by sitting out in the backyard at night and looking up. Most are dim and only visible directly overhead for a moment on a clear night.

Some spent rockets and worn-out satellites tumble slowly, seeming to wink off and on as they traverse the sky. They can be discerned from high-altitude airplanes which display navigation lights.

To see a satellite, go to a very dark spot and take the time to let your eyes become acclimated to the darkness. Near a city, light pollution washes out all but the brightest space objects.

Determine the direction north from your position. Don't stare at the horizon. Atmospheric haze obscures even the brightest satellite until it is 20 degrees above the horizon.

When a satellite climbs up the western sky in the evening, the Sun is below the horizon, but behind the satellite. Looking west, we have a very hard time seeing the unlit side of the satellite.

A satellite low on the western horizon in the evening, or low on the eastern horizon in the morning, can be very dim until it climbs higher in the sky, revealing some of its sunlit side.

Satellites in north-south polar orbits tend to be spy satellites, except for the occasional Earth-resource civilian observation satellite.

Spysats usually fly low orbits for better photo detail, but that makes them easy targets for spotters.

You won't be able to see a small satellite—say, the size of a beach ball—200 miles overhead with the unaided eye; binoculars are required.

An inexpensive pair rated at 7x35 will help you find many dim satellites. A more expensive pair rated 7x50 will collect even more light. The higher second number, indicating light gathering power, is desired, not the first number, magnification.

By the way, you will find a telescope too cumbersome. Its high power and narrow field of view make following a fast-moving satellite with a telescope difficult.

Spotting is easier with prior knowledge of the direction from which a satellite will rise, the direction in which it will set, and the satellite's zenith or highest point in the sky.

Knowing the satellite's track will reveal the general area of the sky in which to look. Use a personal computer to calculate a satellite's orbit, then scan its path across the sky. There are many satellite-tracking software programs for Macintosh and IBM-compatibles.

Accurate forecasting depends on current data, known as Keplarians. Use a computer modem to download Keplarian data from a public and private computer bulletin board system (bbs).

One source is Celestial BBS, Fairborn, Ohio, at (513) 427-0674. Another is the Dallas Remote Imaging Group at (214) 394-7438. The Radio Amateur Satellite Corp. (AMSAT) P.O. Box 27, Washington, D.C. 20044, is another source.

Keps grow old. If the information is a month old, use a modem to obtain fresh data from a bbs. You will need to know the latitude and longitude of your location within ten miles, so call city hall.

In the end, have patience, trust your prediction program and you will find that knowing what you're seeing in the sky can be electrifying.

How Solar Power Works

The majority of satellites use solar cells to convert the Sun's energy to electricity to power spacecraft control systems, communications gear, science instruments and steering equipment.

Solar-powered satellites either have their outside skins covered with photovoltaic cells or they sprout large wings covered with photovoltaic cells. The photovoltaic material extracts electricity from sunlight. That electricity is sent to the satellite's various electrical and electronic systems and to a storage battery which holds it for times when the satellite is in Earth's shadow.

Electricity can be generated from solar heat as well as sunlight. A dish-shaped solar heat collector attached to a satellite could heat a fluid to boiling. The resulting steam would spin the blades of a turbine to generate electricity, just like a coal-burning power plant on the ground. Such a device might electrify a space station someday.

Giant power satellites with several square miles of photovoltaic cells could collect sunlight and beam the power down to Earth.

A power satellite would convert sunlight to elecricity, then to microwave radio energy in a beam aimed at a receiving station on Earth's surface. The ground station would convert the microwave energy back into electricity.

Space Junk

During the decades of the Space Age, the region near our planet above Earth's atmosphere has become a graveyard for thousands of dead satellites and spent rockets, and myriad pieces of junk.

These days U.S. Space Command radar tracks more than 7,000 objects above Earth. Along with 2,000 useful payloads, the radar follows 5,000 pieces of junk big enough to be seen from Earth.

That junk is many things: old rockets, dead satellites, empty fuel tanks, a dozen screws from a 1984 shuttle flight, a screwdriver which floated away from a cosmonaut spacewalking outside Mir station, camera lens caps, even paint flakes.

The growing cloud is about half from Russia and half from the United States, which would like to clean up the debris before the junk damages satellites and threatens manned flights.

Unseen threats. Paint flakes are very small, of course, and the government doesn't really know how many tiny objects are up there. Space Command only tracks 7,000 pieces down to 4 inches—the size of a softball.

NASA suspects there may be 100,000 objects

from one-tenth inch to four inches in size. That does not include billions of paint flakes and clouds of burned solid-rocket propellant racing around Earth at orbital velocity.

Those unseen bits could whack a satellite or shuttle with the wallop of a 400-pound safe zooming along at 60 mph.

France Litters Australia

A large chunk of a French Globus satellite, with two cylinders and two parachutes, was found in 1989 on a farm 1,100 miles northwest of Brisbane in a remote section of northeastern Australia. A police helicopter from the nearby town of Conclurry retrieved the wreckage.

A one-tenth-inch piece of debris traveling at 6.2 miles per second packs the punch of a bowling ball at 60 mph. To compound the problem, collisions spew out untold scores of new particles, adding to the debris count.

To see what's really up there requires a radar capable of detecting objects as small as one centimeter—like spotting a dime at 360 miles.

Solar cleaner. The peak of sunspot activity on the Sun in the early 1990s had some beneficial effect in cleaning out space near Earth.

Our planet's atmosphere expanded as increased ultraviolet light from the Sun heated it, causing drag on thousands of small, man-made objects in orbit at altitudes of 400 miles and below.

The drag slowed the objects causing many of them to fall lower and burn in the atmosphere.

Unfortunately, much junk remains in Earth orbit. Getting rid of a lot of it would reduce the likelihood of collisions and make it safer for men and machines in orbit.

It also would remove some of the light-scattering debris which interferes with astronomers' telescopes.

As if there isn't enough stuff up there...a Houston rocket company has been thinking about launching a mausoleum satellite to space. A Florida funeral home would send remains to the orbiting tomb. The mausoleum would house 10,000 tubes of remains. The funeral home claims to have 3,500 people signed up at $4,000 per tube.

Burning trash. Pieces of spaceflight junk frequently re-enter Earth's atmosphere and burn.

Infrequently, a chunk makes it all the way to the surface of our planet. Such hot debris usually falls onto the seventy percent of Earth's surface which is ocean.

There have been spectacular examples of large scraps making it all the way down to the surface:

★Debris and radioactivity from the USSR's Cosmos-954 satellite rained on 20,000 square miles of Canadian wilderness January 24, 1978.

★Pieces of the U.S. space station Skylab came down in remote western Australia July 11, 1979.

★The USSR's Cosmos-1402 satellite came down in two parts in 1983. The four-ton main body fell into the Indian Ocean in January and its 110-lb. nuclear-reactor electricity generator dropped into the South Atlantic in February.

★Cosmos-1871 would have been the heaviest satellite ever placed in orbit around Earth's poles, but the ten-ton Russian spysat reached only an insufficient 90 miles altitude after blast-off August 1, 1987. Cosmos-1871 fell harmlessly into the South Pacific 1,500 miles east of New Zealand near Antarctica August 10, 1987. It was the first time the USSR announced beforehand one of its satellites was dropping.

★On December 2, 1989, Solar Max, America's large Solar Maximum Mission science satellite, fell into the Indian Ocean.

Communications Satellites

One of the oldest and most important uses for satellites has been relaying human communications around the globe by radio. The first radio relays in the sky were launched between 1958 and 1963. Known as Score, Echo, Telstar, Relay and Syncom, they blazed a path for today's sophisticated communications satellites which relay many telephone, data and television signals simultaneously.

In 1945, author Arthur C. Clarke outlined the possibility of a global communication system using satellites in stationary orbit. He assumed it would involve manned space stations. Today, Clarke lives in Sri Lanka.

In April 1955, John R. Pierce of Bell Laboratories published a paper laying out technical requirements for a satellite communication system.

On October 4, 1957, the USSR launched Sputnik 1, the first artificial satellite in orbit. Sputnik 2 was launched November 3, 1957.

1958. Three months later, on January 31, 1958, the U.S. launched Explorer-1, America's first satellite in orbit.

Later that same year, the Score satellite was launched on December 18. Its transmitter was used to broadcast from space to Earth a pre-recorded message, taped on the ground before launch by President Dwight D. Eisenhower.

1960. On April 4, 1960, the U.S. launched TIROS-1, the first weather satellite in orbit. However, the next month, NASA failed on May 13 to launch Echo A-10, a 100-ft. balloon intended to be a passive communications-reflector satellite.

Three months later, the United States successfully launched Echo-1 on August 12. The 100-foot aluminum-coated Mylar balloon was a passive communications satellite. The aluminum coating acted as a mirror to radio and TV signals from a transmitter in New Jersey. Those signals bounced off the balloon to a receiver in France.

That same year, the United States launched Courier-1B on October 4, the first successful active communications satellite. It had the first store-and-forward message bulletin board equipment in space.

The next year, on April 12, 1961, Yuri Gagarin became the first man in space, flying the USSR's Vostok capsule. A month later, Alan Shepard flew on May 5 in his Mercury capsule. The suborbital flight made him the first American in space, but he was not in orbit.

1961. On October 21, 1961, the U.S. blasted a satellite known as Midas-4 to orbit carrying the West Ford passive dipoles for a passive communications-reflector experiment. The tiny dipoles failed to disperse after release in space. Radio signals were to have been bounced off the metal needles, reflecting signals back to Earth. A second West Ford passive dipoles try was made in 1963 and was successful.

Also in 1961, OSCAR-1, the first amateur radio satellite, was launched December 12.

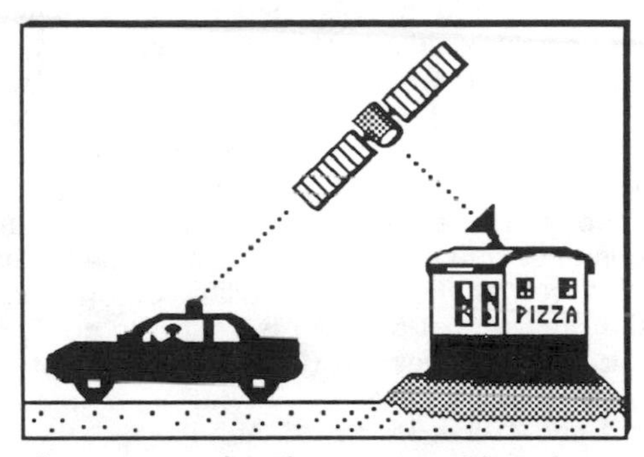

A communications satellite is a radio relay station in the sky. Signals are transmitted up to the satellite which receives them and retransmits them down to Earth.

1962. Mercury astronaut John Glenn became the first American in orbit on February 20, 1962. Not to be left behind by America, the USSR launched Cosmos-4, its first weather satellite and photo-reconnaissance spysat, April 26.

Just ten weeks later, the world became a smaller place on July 10, 1962, when NASA launched Telstar-1.

Continents previously had been linked only by copper cables. TV film footage had been sent across oceans by airplane. Long-distance telephone calls had been special occasions.

Telstar was the first active real-time communications satellite. Owned by AT&T Bell Laboratories, the "bird" was the first to receive, amplify and immediately retransmit telephone and television signals from Earth—the forerunner of modern communications satellites.

The 3-ft., 170-lb. orbiting sphere made possible live television across the Atlantic, letting TV viewers see world events as they happened.

The first telephone call through Telstar was between U.S. Vice President Lyndon Johnson and AT&T Chairman Fred Kappel. The next day, French singer Yves Montand crooned "live" for an American audience. A few weeks later, U.S. President John Kennedy addressed Europeans on live television.

Prestigious Echoes

Many times during the early decades of the Space Age, America seemed to be playing catch-up with the Russians. For instance, entering the 1960s, America was looking for ways to recover prestige lost when the USSR launched unmanned Sputniks and the Korabl Sputnik carrying two live dogs which parachuted back to Earth.

The world was anticipating a manned flight and, as we came to know, a Russian satellite capsule called Vostok would carry Yuri Gagarin in 1961.

But, in 1960 before the Gagarin flight, U.S. President Dwight Eisenhower itched for some swift, spectacular, low-cost space achievements. One of those was to be a big balloon named Echo.

Echo A-10, an immense aluminum-coated balloon to be inflated in orbit, was set to be the first passive communications satellite. Unfortunately for the presidential plan, NASA failed to get Echo A-10 to orbit in a May 13, 1960, launch attempt.

Eisenhower's luck was better three months later. Echo-1 was launched August 12, 1960. In orbit, the aluminum-coated Mylar balloon was inflated to a diameter near 100 feet.

Echo-1 was the 49th satellite ever sent successfully from Earth to orbit by any country. It also was the world's first successful passive communications satellite. Radio signals beamed toward the satellite from the ground bounced off the highly-reflective exterior of Echo-1 to other listening posts on the ground.

Echo-1 excited ordinary folks on the ground when it turned out to be brighter than anything else in the sky, except for the Sun and Moon. Sighting it by naked eye was popular around the world for several weeks, until Echo-1 descended to burn in the atmosphere May 24, 1968.

Four years later, another 100-ft. Mylar balloon became the first-ever space-cooperation project between Americans and Russians. Echo-2 went up January 25, 1964, and came down June 7, 1969. It was the 740th satellite ever orbited.

Telstar-1 sparked a musical score for electric guitar. "Telstar" was recorded by The Tornadoes and hit No.1 on pop charts.

Telstar-1 succumbed to space radiation seven months after its launch. It's deaf and dumb, but it continues to loop around the globe every 2 hours and 40 minutes.

On August 31, 1962, President John Kennedy signed legislation creating the Communications Satellite Corp. It was to establish a worldwide satellite network promoting world peace and understanding.

Canada's Alouette-1 satellite was launched September 29 to test the effect of the ionosphere on communications.

Then, on December 13, Relay-1, an active real-time communications satellite built by RCA was launched to a polar orbit ranging from 800 to 4,500 miles above Earth. It offered 12 telephone channels and one television channel. Relay-1 linked Europe, Brazil and the United States.

1963. NASA tried to test Arthur C. Clarke's vision of linking the entire globe with three satellites in stationary orbits. They were referred to as Synchronous Communications Satellite, or Syncom for short. Electronics failed in Syncom-1 when it was launched February 14, 1963.

NASA launched another active real-time communications satellite, AT&T's Telstar-2, into an elliptical orbit on May 7.

Two days later, on May 9, a new set of the West Ford metal-needle dipoles dispersed successfully in space to become passive communications reflectors.

NASA finally was able to place one of its Syncom satellites in orbit July 26, 1963. Syncom-2, built by Hughes, was in an orbit inclined 33 degrees to the equator. U.S. President John F. Kennedy used it to speak by telephone with Nigeria's prime minister.

1964. RCA's Relay-2, launched to a polar orbit January 21, 1964, was another active real-time communications satellite. It added 12 telephone channels and a television channel.

Four days later, Echo-2 was launched on January 25. It was another 100-foot aluminum-coated Mylar balloon used as a passive radio reflector. Signaling a brief thaw in the Cold War, Echo-2 was America's first joint space project with the USSR.

The third Syncom was to be the world's first satellite in stationary orbit. Syncom-3, launched August 19, 1964, reached its position 22,244 miles above the equator in the Clarke Belt at 0.10 degrees on September 23.

Unfortunately, NASA's test of Arthur C. Clarke's three-satellite hypothesis didn't work out because Syncom-1 failed and Syncom-2 wasn't in a stationary orbit. However, satellite history was recorded two months later in October when TV pictures of the 1964 Olympic Games in Japan were sent across the Pacific via Syncom-3 to America. Those pictures were bounced on across the Atlantic to a European audience via Relay-1.

1965. NASA launched an active real-time comsat known as Les-1 on February 11, 1965.

A month later, the first active real-time amateur radio communications satellite, OSCAR-3, was launched on March 9.

The big news in 1965 was a satellite called Early Bird, launched to stationary orbit April 6. It was the first commercial communications satellite. Also known as Intelsat-1, Early Bird started the global space communications network. It was built by Hughes and could carry 240 telephone calls or one television program between Europe and America from its position over the Atlantic at 325 degrees East longitude.

Today, Intelsat is a commercial cooperative of 125 member nations owning and operating a globe-spanning system of many communications satellites used worldwide by 180 countries, territories and dependencies for their international and domestic communications. Intelsat is the major provider of transoceanic telephone and television services. It also offers international video, teleconferencing, facsimile, data and telex.

Meanwhile, back in 1965, the USSR was staying in the game. It launched its first active real-time communications satellite, Molniya-1A, on April 23 to a long elliptical orbit. One Molniya in a series could cover an area on Earth's surface for eight hours. Three synchronized Molniyas could cover the same area for a full 24-hour day. Satellites in the Molniya-1 series also carried TV cameras to snap cloud photos.

The American's launched Les-2 on May 6.

The USSR launched Molniya-1B October 14, 1965. It was the USSR's second active real-time communications satellite.

At the end of the year, the first high-altitude amateur radio communications satellite, OSCAR-4, was launched December 21.

1966. The U.S. military lofted eight IDCSP communications satellites in one launch to a high orbit 21,000 miles above Earth June 16.

An American series of five Applied Technology Satellites (ATS) were to test communications techniques from stationary orbit, as well as meteorology and navigation satellite systems. ATS-1 was launched December 7, 1966, to a position over the Pacific where it relayed medical, scientific and cultural information to Australia, New Zealand, Fiji, Tonga, Samoa, The Solomon and Gilbert islands, and Papua-New Guinea.

1967. Four satellites of the Intelsat-2 class were launched. The second reached stationary orbit over the Pacific January 11, 1967, after the first failed to reach orbit in 1966. Also in 1967, a third was launched over the Atlantic and a fourth over the Pacific. They linked multiple ground stations in northern and southern hemispheres.

The USSR orbited its first communications satellites for contact with clandestine spy missions in 1967.

That same year, an American TV program known as "Our World" used five satellites on June 25 to bring together artists around the globe, such as the Beatles in London, Marc Chagall and Joan Miro in Paris, and Van Cliburn and Leonard Bernstein in New York.

America's third Applied Technology Satellite (ATS-3) was launched November 5, 1967, to relay medical data between offshore oil platforms and shore stations and between hospitals and mobile health centers in rural areas.

1968. Eight Intelsat-3 communication satellites were launched between 1968 and 1970.

The Vietnam War became the "first television war" when pictures of violence and horror startled home viewers night after night in the late 1960s, igniting a peace movement.

1969. Canada's research into the effect of the ionosphere on communications, started in 1962 with Alouette satellites, was continued with ISIS-1 launched January 30, 1969, and ISIS-2 launched two years later in 1971.

The U.S. military launched a stationary-orbit communications satellite known as TACsat on February 9, 1969.

A British military stationary-orbit satellite called Skynet was launched November 22, 1969.

America's fifth Applied Technology Satellite (ATS-5) was launched on August 12 to relay television and signals.

1970. A North Atlantic Treaty Organization (NATO) stationary-orbit military communications satellite, NATO-A, was launched March 20.

The USSR countered with Cosmos-336, a military communications satellite fired to a 900-mi.-high orbit April 25, 1970.

1971. Seven fourth-generation Intelsats were launched from 1971 to 1975. The first Intelsat-4 was launched January 25, 1971.

The U.S. military launched SDS, a communications satellite, into a Molniya elliptical orbit on March 21, 1971. SDS's long path through space took it from a low point about 250 miles above Earth to a high point 21,000 miles from Earth.

The U.S. also launched a new stationary-orbit military communications satellite in 1971. It was DSCS-1 fired to space on November 3.

The first satellite in the USSR's Molniya-2 series was launched November 24, 1971. The satellites relayed radio and television programs, phone calls, telegrams, and facsimiles of weather charts and newspaper pages. Molniya-2 satellites carried the Moscow-Washington Hotline.

Moscow-Washington Hotline

In 1962, America discovered the USSR had placed in Cuba missiles capable of reaching U.S. cities. U.S. President John F. Kennedy ordered U.S. Navy ships to turn away vessels delivering missiles. The crisis brought America and Russia to the brink of nuclear war. War seemed at hand for a week until the Soviets removed the missiles.

After the affair ended, an editor wrote an open letter in newspapers asking superpower leaders to communicate directly. Kennedy liked the idea and a hotline to Moscow was hooked up June 20, 1963.

The hotline is a direct link between the heads of state in Russia and the U.S., bypassing slower diplomatic channels when either nation is planning military action in some part of the world.

The hotline is not a red telephone as depicted in movies. Instead, the White House and Kremlin communicate via teleprinter. The only telephone is one technicians use to keep the hotline in repair.

Today's satellite hotline connection wasn't always there. Early hotline communications went by land—until a Finnish farmer plowed through the cable one day. Satellite links started in 1978.

Duplicate transmissions guarantee a message actually will arrive in a crisis. Each written hotline message takes three routes to the other nation. It is sent simultaneously through an American satellite and a Russian satellite and via land cable.

For example, a teleprinter message from the White House travels from Washington 40 miles north to Fort Detrick at Frederick, Maryland. From there, the message is transmitted to one of Russia's Molniya communications satellites. Molniya relays it to a Russian Earth station south of Moscow which sends it to the Kremlin.

Fax was added to the circuit in 1986 letting leaders supplement texts with maps and charts. The link between Russian and American presidents was modernized again in 1990.

To make sure the line is open, Fort Detrick technicians check it every hour. Another station on the American end is at Etam, West Virginia.

The hotline has been used only about two dozen times over the years. For example, it was used during the Arab-Israeli Six-Day War in 1967. In another instance, President Ronald Reagan used the hotline as he sent troops to invade Grenada.

1972. The world's first commercial domestic television satellite in stationary orbit was Canada's Anik-A1 launched November 9. Also known as Telesat-1, it could relay twelve TV channels.

1973. Anik-A2, or Telesat-2, was launched to stationary orbit April 20, 1973.

1974. The first satellite for commercial domestic communications within the United States was Western Union's Westar-1, launched to stationary orbit April 13. Westar-2 was launched in 1974 and Westar-3 was launched in 1979.

America's sixth Applied Technology Satellite (ATS-6) was launched May 30 to relay educational television, air traffic control communications, teleconferences, and rural medical data.

The first satellite in the USSR's Molniya-3 series was launched November 21, 1974.

Symphonie-1 was an experimental French-German communications satellite launched to stationary orbit December 19, 1974. It could carry four color TV programs or 1,600 telephone calls. Symphonie-2 was launched in 1975.

The first experimental USSR communications satellites in stationary orbit were Molniya-1S and Cosmos-637, launched in 1974.

1975. The first non-experimental USSR communications satellite, operational in stationary orbit over the equator, was known as Raduga, or Statsionar. It was launched December 22, 1975. Statsionar means stationary. The Raduga series transmitted television, telephone and telegraph signals as well as facsimiles of newspaper pages to the remote cities of Khabarovsk and Novosibirsk. Radugas were similar to Russia's Ghorizont communications satellites.

Anik-A3, or Telesat-3, was launched to stationary orbit May 7, 1975.

Six Intelsat-4A communications satellites were launched between 1975 and 1978.

Satcom-1, the first of RCA's domestic communications satellite series was launched over North America on December 13, 1975. Satcoms carry television, telephone calls and computer data.

1976. Canada's Communications Technology Satellite (CTS) was launched January 17 to stationary orbit. Later renamed Hermes, it was used for medical and educational communications.

The first maritime satellite-communications system was inaugurated February 19, 1976, with the launch to stationary orbit of Marisat-1. Marisat -2 and -3 were launched the same year. The satellites above the Atlantic, Pacific and Indian oceans relayed voice, teletype, fax and data for commercial shipping fleets and the U.S. Navy.

Comstar was the name for four U.S. domestic

communications satellites owned by Comsat General Corp. and leased to the AT&T and GTE telephone companies. Comstar-D1 to -D4 were launched to stationary orbit between 1976-1981.

Indonesia's first satellite, Palapa-A1, was launched to stationary orbit July 8, 1976. Palapa-A2 was launched in 1977. The name Palapa stood for unity, suggesting the unification of the diverse Indonesian nation by satellite communications.

In September 1976, Taylor Howard of San Andreas, California, became the first American to receive satellite TV signals on a home system.

Ekran was a new series of USSR broadcast satellites started in 1976. Ekran-1 was launched October 26, 1976, and positioned at 99 degrees East longitude. The Ekran series were Statsionar satellites which carried domestic television broadcasts for the USSR and its close neighbors. Statsionar means stationary.

Atlanta's local TV station WTBS became Superstation TBS in December to send its programs via satellite to cable-TV systems across America. Later, CNN was spun off.

1977. Japan launched its first stationary-orbit satellite on January 23, 1977. Lofting Kiku-2 on its own N-1 rocket made Japan the third nation able to launch satellites to the Clarke Belt. Kiku-2 also was called Engineering Test Satellite (ETS).

Italy's experimental communications satellite, Sirio, was launched August 25, 1977.

European Space Agency's first Orbital Test Satellite (OTS-1) was destroyed September 14 when its rocket exploded 54 seconds into flight from Cape Canaveral.

The predecessor of the Family Channel was created to feed religious and entertainment programs to local cable systems via satellite.

Japan's experimental domestic communications satellite Sakura was launched to stationary orbit December 15, 1977, from the United States. Sakura also was known simply as Communications satellite (CS).

1978. The U.S. military launched a new series of stationary-orbit communications satellites in 1978. The Navy's first Fltsatcom was fired to space on February 9.

Japan's experimental direct-to-home broadcast satellite, Yuri, was launched from Cape Canaveral April 8. Yuri also was known as Broadcasting Satellite Experiment (BSE).

European Space Agency's Orbital Test Satellite (OTS-2) made it to stationary orbit May 12 to relay television and telephone calls.

The first two USSR amateur radio communications satellites, Radiosputniks RS-1 and RS-2, were launched October 26, 1978.

Chicago's local station WGN became a superstation by providing its programs via satellite to cable-TV systems across America.

Canada's Anik-B, also known as Telesat-4, was launched December 15, 1978.

The USSR launched the first Ghorizont communications satellite to an elliptical orbit December 19, 1978. Later Ghorizonts were sent to stationary orbits. Ghorizonts were part of the Eastern Bloc's Intersputnik system.

Needle Dipoles

Bouncing radio signals off of space objects has been popular over the decades of the Space Age. First, there was Earth's natural satellite, the Moon. Then 100-ft. aluminum-coated Mylar balloons. Then clusters of metal needles fired to orbit and released in space.

The United States called the project West Ford. The needles were cut to appropriate lengths to act as radio dipole antennas. They reflected signals sent up from Earth.

An October 21, 1961, launch of West Ford dipoles from a satellite named Midas-4 failed when the needles didn't disperse, but a second try was successful 19 months later on May 9, 1963.

1979. Both of Japan's experimental satellites named Ayame were lost enroute to stationary orbit in February 1979. They also were known as Experimental Communications Satellite (ECS).

C-Span was started in 1979 by the U.S. cable-TV industry to televise via satellite meetings of the U.S. House of Representatives and Senate.

The Soviet Union lofted Ghorizont-2 to stationary orbit July 5, 1979. A Statsionar spacecraft, it was Europe's first television-broadcast satellite.

ESPN started sending sports, American Movie Classics (AMC) started feeding old movies and Nickelodeon started providing entertainment programs to cable systems via satellite.

Neiman Marcus featured a $36,000 home satellite TV system on the cover of its 1979 Christmas catalog.

1980. Cable News Network (CNN) became the first global satellite TV broadcaster.

Three satellites operated by Satellite Business Systems relayed phone calls, computer data and teleconferences for corporations. SBS-1 was launched November 16, 1980. SBS-2 was launched in 1981. SBS-3 was launched in 1982.

Nine Intelsat-5 satellites were launched between 1980-1984, adding maritime communications.

NASA's TDRS Satellites

When NASA needed its own satellite network to communicate with orbiting shuttles and research satellites, it designed the Tracking and Data Relay Satellite System (TDRSS).

Built by TRW, the satellites are owned and operated by Space Communications Company (Spacecom) which rents them to NASA. Spacecom also leases some TDRS C-band and Ku-band transponders to commercial interests.

Six of the 2.5-ton TDRS satellites have been launched to relay voices of astronauts laboring in space shuttles and millions of data bits from orbiting observatories. One TDRS was lost in the Challenger disaster.

All five TDRS satellites working in orbit transmit to a ground station at White Sands, New Mexico. Just one TDRS can relay all of the information in a 24-volume encyclopedia in five seconds.

TDRS-1 was hauled to space in 1983 in the maiden voyage of shuttle Challenger. At that time, it was the largest civilian communications satellite ever launched. It had four S-band, twelve C-band and two Ku-band transponders.

After TDRS-2 was destroyed in 1986 in the Challenger disaster, NASA returned to space in 1988 with shuttle Discovery carrying TDRS-3.

Six months later, Discovery carried TDRS-4 in 1989. TDRS-5 was ferried to space in shuttle Atlantis in 1991. TDRS-6 went to orbit in shuttle Endeavour in 1993. The space agency plans to send TDRS-7 to space in 1995.

Each 50-ft. TDRS has two arrays of solar cells and seven antennas. Each radio channel can transmit more than 300 million bits of information per second to White Sands.

Besides relaying astronaut chatter and shuttle data from orbit, the TDRS network relays science data to White Sands from unmanned science satellites and observatories such as Hubble Space Telescope, Compton Gamma Ray Observatory and the Upper Atmosphere Research Satellite.

Prior to TDRS, NASA had some twenty ground stations around the globe. Controllers could communicate with a shuttle only during fifteen percent of its orbit. With TDRS satellites in stationary orbits over the equator above the Atlantic and Pacific oceans, controllers can communicate with a shuttle across 85 to 100 percent of its orbit. TDRS communications are managed by NASA's Goddard Space Flight Center, Greenbelt, Maryland.

USA Network started feeding entertainment programs to cable systems via satellite in 1980.

1981. Japan launched another experimental communications satellite, Kiku-3, to stationary orbit on an N-2 rocket February 11. Kiku-3 also was called Engineering Test Satellite (ETS).

Apple was the nickname for a communications satellite for India named Ariane Passenger Payload Experiment. Apple was developed by the Indian Space Research Organization (ISRO) and launched June 19, 1981, by the European Space Agency to stationary orbit. It had two C-band transponders for educational telecasts and communication among portable data terminals.

Iskra-1, a USSR amateur radio satellite built by students at the Ordzhjonikidzé Aviation Institute, was launched July 10, 1981. It had a transponder, telemetry beacon, ground-command radio and message bbs. Iskra is "spark" in Russian.

Meanwhile, British amateur radio operator students at the University of Surrey launched the first hamsat with a television camera October 6, 1981. UoSAT-OSCAR-9 also had a magnetometer, radiation and particles detectors, synthesized radio voice with 150-word vocabulary, and message bbs.

Six amateur radio satellites from the USSR were launched on one rocket December 17, 1981, from the Northern Cosmodrome at Plesetsk. Radiosputniks 3-8 were the largest clutch of amateur radio satellites ever orbited at one time.

A maritime communications satellite named Marecs-A was built by the European Space Agency and launched to stationary orbit December 20 for the International Maritime Satellite Organization (Inmarsat). Marecs-B2 was launched in 1985. Today, Inmarsat serves 59 countries with mobile telephone, telex, fax (facsimile) and data transmissions via several stationary satellites to some 9,000 ships and shore stations.

MusicTV (MTV) began sending music videos to local cable systems via satellite in 1981.

1982. Western Union's second-generation Westar-4 was launched to stationary orbit February 25, 1982. It relayed color TV programs, fax and data, and could relay 28,000 telephone calls. Westar-5 was launched in 1982. Westar-6 was ferried to space in 1984 in a U.S. shuttle. It had to be retrieved from space later after it failed to reach stationary orbit. A refurbished Westar-6 was relaunched in 1990 as ASIAsat-1.

Canada's Anik-D1, also known as Telesat-5, was launched to stationary orbit August 26, 1982. Anik-D2, also known as Telesat-8, was launched by shuttle in 1984. Both were used for international telephone and television relays.

CNN created a separate Headline News satellite channel for cable systems and The Weather Channel started providing meteorological news to local cable systems via satellite.

Japan launched another experimental comsat, Kiku-4, to a very low orbit on an N-1 rocket September 3, 1982. Kiku-4 also was called Engineering Test Satellite (ETS).

Three Canadian satellites in the Anik-C series carried domestic communications. Anik-C3 or Telesat-6 was launched to stationary orbit by the first U.S. commercial shuttle flight November 11, 1982. Anik-C2 or Telesat-7 was launched by shuttle in 1983 and Anik-C1 or Telesat-9 was launched by shuttle in 1985.

1983. NASA's Tracking and Data Relay Satellite (TDRS-1) was carried to orbit by shuttle April 4, 1983. It was maneuvered over the Atlantic to 41 degrees West longitude. TDRS was the largest civilian communications satellite ever launched and the first able to operate in three frequency bands at one time.

Built in the U.S. for India's Space Research Organization (ISRO), INSAT-A was launched April 10, 1982. Its attitude control broke down and the satellite couldn't be used. Better luck accompanied INSAT-B when it rode a U.S. shuttle to stationary transfer orbit August 30, 1983. It was maneuvered over the equator to 74 degrees East longitude. INSAT-B not only relayed television programs into rural communities, but collected weather data from ground platforms strewn across the South Asian nation.

Europe's first operational, continental comsat was European Communications Satellite (ECS-1) launched to stationary orbit on June 16, 1983. ECS-2 was launched in 1984.

Two days later, Indonesia's stationary satellite Palapa-B1 was carried to transfer orbit by a U.S. space shuttle on June 18. It had twice as many transponders as Palapa-A launched in 1976. Palapa-B2 failed to reach stationary orbit in 1984.

Hughes series of Galaxies satellites started with the launch of Galaxy-1 on June 28, 1983. Galaxy-2 was launched in September and Galaxy-3 was launched in 1984. The 24 transponders in each carried television programs to cable systems.

AT&T launched its Telstar-3 series of domestic comsats between 1983 and 1985. The first, Telstar-301, was launched to stationary orbit July 28, 1983. Telstar-302 was launched in 1984 and Telstar-303 was launched in 1985. They carried 24 color TV programs and high-speed data across the continental United States and into Hawaii, Alaska and Puerto Rico.

The Nashville Network (TNN) was formed to send country-and-western music and entertainment programs to local cable systems via satellite.

1984. The world's first non-experimental, operational, direct-to-home broadcast satellite was Japan's BS-2A, or Yuri-2A, launched January 23, 1984. Yuri-2A was an improved version of Yuri-1 which had been launched in 1978.

People's Republic of China launched its first stationary communications satellite April 8 on a Long March rocket from Xichang Launch Center in southwestern China. The satellite, known as STW-1, China-15 or PRC-15, could carry two TV programs or 300 telephone calls. STW-2 was a backup satellite launched from Xichang in 1986.

The domestic American television, telephone and data satellite, Spacenet-1, was launched to stationary orbit May 23, 1984. RCA built the satellite for GTE which rented out its transponders. Spacenet-2 was launched in November 1984.

The French space agency, Centre National D'Etudes Spatiales (CNES), modified an ECS framework into a new domestic communications satellite, Telecom-1, and launched it to stationary orbit August 4. It relayed civilian and military phone calls, television, telegraph and data signals.

U.S. President Ronald Reagan signed a law in October 1984 legalizing private reception of unscrambled satellite TV programming.

Lifetime began sending women's news and entertainment programs, and Arts & Entertainment Network (A&E) started sending programs, to local cable systems via satellite in 1984.

1985. Brasilsat-1 was a satellite built in Canada for the Brazilian government. It was launched to stationary orbit February 8 to relay TV, telex, data and telephone calls for Brazilians. Brasilsat-2 was launched in 1986.

Not to be outdone in space, the Russians lofted four communications relays like NASA's TDRS satellites to stationary orbits in 1984 and 1985. The two launched in 1985 were Cosmos-1629 on February 21 and Cosmos-1700 on October 25.

Alongside Brasilsat-1 atop its European rocket on February 8 was the Arab League's Arabsat-1. A U.S. shuttle launched Arabsat-2 in June 1985.

Arab League Satellites

The Arab Satellite Communication Organisation, an Arab League group based in Riyadh, Saudi Arabia, commissions the building and launching of Arabsat communications satellites.

Arabsats relay telephone conversations and broadcast radio and television programs to the 21 states in the Arab League, as well as central and southern Europe, India and South Asia.

Arabsat-1A and Arabsat-1B were launched in 1985. Arabsat-1C was launched in 1992. A new generation is expected to be launched in 1995.

The Intelsat-5A series of satellites was launched with the first on March 22, 1985. It also was tenth in the Intelsat-5 series.

Discovery Channel began feeding programs via satellite to cable systems in 1985.

GTE had launched its Spacenet communications satellite series in 1984. A different series of GTE domestic American television, telephone and data satellites was launched in 1985 when GStar-1 was launched to stationary orbit May 8. RCA again built the satellite for GTE. GStar-2 was launched in 1986. GTE had four Spacenets and three GStars in its network. Each Spacenet had eighteen C-band and six Ku-band transponders while each GStar had sixteen Ku-band transponders.

Mexico's Morelos-1 was launched by U.S. shuttle on June 17, 1985. Hughes built the satellite for the Secretariat of Communications to relay telephone, TV, fax and data transmissions across Mexico. Another shuttle ferried Morelos-2 to a transfer orbit in November 1985.

A private stationary-orbit satellite, ASC-1, was launched by U.S. shuttle August 27, 1985. RCA built the satellite for the American Satellite Company to carry business communications. ASC-1 was the first satellite to have commands encrypted to prevent unauthorized access. ASC-2 was launched in 1991.

Aboard the same shuttle which took ASC-1 to space August 27 was Australia's Aussat-1. Among the usual TV, telephone, fax and data transmissions, Aussat's fifteen Ku-band transponders beamed direct-to-home telecasts into the remote Out Back, relayed air traffic control communications and aided mariners at sea. Aussat-1 was positioned over the equator north of Papua-New Guinea. Aussat-2 was launched from a different shuttle flight in November 1985.

1986. Home Box Office Inc. (HBO), a company transmitting television movies via satellite, announced on January 15, 1986, it would be the first programmer to scramble its satellite TV signal full time.

Shuttle Challenger exploded during liftoff January 28, killing seven crew members. The loss left the United States without a launch vehicle for satellites for nearly three years.

The USSR launched its Mir space station to a low Earth orbit on February 19, 1986.

QVC started selling merchandise via satellite directly to consumers in 1986.

John MacDougall of Ocala, Florida—Captain Midnight—illegally interrupted an HBO broadcast for more than four minutes April 27 to display a message vowing not to pay for the movie service.

Scrambling & Captain Midnight

Home Box Office Inc. (HBO), and others transmitting television movies via satellite to local cable systems, lobbied Congress in the early 1980s to outlaw home reception of their programs.

U.S. President Ronald Reagan signed a law in October 1984 legalizing private reception of unscrambled satellite TV programming, but outlawing reception of scrambled programming.

HBO announced on January 15, 1986, it would scramble its satellite TV signals full time. Other American satellite programmers added scramblers.

Meanwhile, video hacker John MacDougall of Ocala, Florida was working part-time at a satellite uplink station. Embittered by the scrambling, he vowed not to pay for movie service.

As Captain Midnight, MacDougall interrupted an HBO broadcast of the film "Falcon and the Snowman" on April 27, 1986. For more than four minutes, he transmitted his protest message against paying for movies to the satellite Galaxy-1, overpowering HBO's uplink.

"Falcon and the Snowman" is about military spy satellites. Captain Midnight was a hero of a 1960s television series. MacDougall was fined $5000.

Today, most popular satellite channels are scrambled. Broadcasters use a technique from General Instruments Inc. known as VC II+. Modern satellite TVRO receivers include descramblers (decoders). When home viewers subscribe to coded channels, signals are unscrambled.

European satellite broadcasters have a variety of different scrambling (coding) methods known as Eurocrypt-S, Eurocrypt-M, D-MAC, D2-MAC and HD-MAC for high definition television (HDTV). Swedish telecasters are developing a digital HDTV coding system to be known as HD-Divine.

1987. Pope John Paul II led Roman Catholics on five continents in prayers for peace in 35 languages during a live broadcast via 23 communications satellites June 6, 1987.

On October 26, the first CNN World Report from satellite telecaster Cable News Network compiled news from thirty nations into the first regularly-scheduled newscast with reports from anywhere on the globe.

France and West Germany agreed in 1980 to build identical direct-to-home television broadcast satellites. Germany's TVsat-1 was launched November 21, 1987. France's TDF-1 was launched October 28, 1988. TVsat-2 and Scandinavia's Tele-X, basically the same as TVsat, were launched in 1989. TDF-2 was launched in 1990.

1988. Panamsat, sometimes known as Alpha Lyracom Inc., is a small satellite company trying to compete with the giant Intelsat. Its first communications satellite, Panamsat-1 or PAS-1, was launched June 15 on a European rocket to provide low-cost communications between Latin America, Europe and the United States.

With the launch of shuttle Discovery September 29, America regained its satellite-launching capability.

Turner Network Television (TNT) was formed in 1988 to feed movies and programs to local cable systems via satellite.

Luxembourg's SES changed satellite broadcasting over Europe on December 11 with the launch of Astra-1A.

Amateur radio satellite UO-11 guided Russian and Canadian skiers across the North Pole.

1989. Tele-X was a satellite sponsored by Sweden, Norway and Finland and much like France's TDF-1 and German's TVsat-1. Scandinavia's Tele-X was launched to stationary orbit April 2, 1989.

Japan's industrial company C. Itoh operates a series of stationary-orbit satellites known as JCsat. JCsat-1 was launched March 6, 1989. JCsat-2 was launched in 1990.

Another large Japanese industrial concern, Mitsubishi, launched its series of Superbird communications satellites in 1989. Superbird-A was lofted to stationary orbit June 5. TV crews return news coverage to studios via Superbirds. Superbird-B1 was launched in 1992.

The European Space Agency's experimental communications satellite, Olympus, was launched July 12, 1989. It provided direct-to-home TV broadcasts to Italy as well as the usual relay of business communications.

The Intelsat-6 series was launched in 1989 with the blast off of Intelsat-6A on October 27. An Intelsat-6 had 50 transponders providing 40,000 telephone circuits and two television channels.

1990. Palapa-B2 and Westar-6, which had been retrieved from space after they failed to reach stationary orbit in 1984, were relaunched in 1990. Palapa-B2R was launched April 13.

A refurbished Westar-6 was launched by China as ASIAsat-1 or Asia No. 1. The first commercial satellite for Asian countries, ASIAsat-1 was launched April 7, 1990, on China's Long March rocket. Half of its 24 channels covered northern Asia and the other half covered southern Asia.

Pegasus, the first new unmanned American space rocket since the 1960s, was launched April 5, 1990, from an airplane seven miles above Earth. It carried Pegsat, a miniature U.S. Navy communications satellite to a low polar orbit.

News agencies around the globe distributed their information via satellite in the 1990s. For instance, England's BBC Monitoring Service subscribed to Russia's ITAR-TASS news agency via a data link through the satellite Intelsat-601. The satellite replaced a system of wire cables, allowing BBC to receive TASS information from Moscow in English and Russian.

1991. The Persian Gulf War of 1990-91 was the "first live TV war" broadcast around the world by many satellites which relayed television news broadcasts and military communications.

Arthur C. Clarke, the English writer who proposed communications satellites in 1945, said satellite networks meld residents of Earth into a global family, whether they like it or not.

ITALsat-1 was Italy's experimental stationary-orbit satellite launched January 15, 1991.

A pair of identical American Satellite Co. stationary-orbit satellites were built in the early 1980s to link businesses from space. ASC-1 was launched in 1985. ASC-2 was scheduled for shuttle launch in 1987, but the 1986 Challenger disaster forced NASA to ban commercial payloads from shuttles. ASC-2 was mothballed. Later, American Satellite Co. merged with Contel, then Contel ASC merged with GTE Spacenet Corp. As demand for satellite communications services increased, ASC-2 was refurbished. ASC-2 became the ninth Spacenet satellite in space April 13, 1991. It is known as ASC-2/Spacenet-5. There is a great demand for satellite transponders to relay news pictures around the globe. Many Spacenet users are news media organizations. Spacenet is the largest satellite news relayer in the U.S. Much of the news relayed to U.S. viewers each day passes through Spacenet transponders.

Alaska's Aurora-2 stationary-orbit satellite was launched May 29, 1991, to carry broadcasts, telephone calls, and maritime and emergency communications. Aurora-2 is owned by GE and Anchorage telephone company Alascom Inc.

1992. Satellites had relayed TV broadcasts for years, showing millions starving in Somalia, energizing an international movement and a 1992 U.N. rescue mission to the African nation.

Satellites also relayed pictures of the horrors of war in the Balkans in the 1990s, kindling frustration when the United Nations found no way to ease the pain in the former Yugoslavia.

Milstar is the latest generation of American military communications satellites. The secret satellites enable communications among all U.S. military services. Ten slots in the Clarke Belt have been set aside for Milstars, while federal funding has been approved for six.

1993. NASA's Advanced Communications Technology Satellite was launched July 17. ACTS was furnished with three times the capacity of satellites of comparable size. While most satellites have one antenna broadcasting one broad beam, ACTS has multiple antennas which send out individual narrow beams. Those spot beams let smaller, less-expensive ground stations receive communications relayed through ACTS. The master ground station for ACTS is at NASA's Lewis Research Center at Cleveland, Ohio.

A Georgia advertising marketer proposed an advertising billboard orbiting 180 miles above Earth with letters, or corporate logo, large enough to be seen by naked eye from the ground. The mile-long rectangle would be an inflatable skin of reflective plastic over a tube skeleton. From Earth, the billboard might seem as large as a full Moon as it floated around the globe for a month, slowly sinking into the atmosphere where it would burn. Members of the U.S. House of Representatives did a slow burn when they heard of the idea. They proposed outlawing "the placement of images or objects in outer space that are visible from Earth, for purposes of marketing."

What's Inside?

Signals transmitted from the ground to a satellite are uplinks. Signals transmitted from a satellite to the ground are downlinks.

Communications satellites are outfitted with radio receivers, amplifiers, transmitters and multiplexing computers to relay many telephone, data and television signals simultaneously.

Like most satellites, communications satellites have internal sensors measuring voltages, currents and temperatures. That data is encoded into telemetry signals transmitted to ground controllers.

Many communications satellites have attitude-control equipment to maintain desired orbits, to point radio antennas at ground stations, and to keep solar-power generators pointing at the Sun. Most are powered by the Sun.

Control. Communications satellites must stay in touch so their work can be directed by human operators in ground control stations.

For instance, the radio frequency used by a broadcast satellite to beam down a TV movie can be changed.

Controllers talk with their birds by sending coded radio signals to them and receiving similar signals from them. For instance, NASA's constellation of TDRS communication satellites are controlled from a ground station in White Sands, New Mexico.

Orbits. Some small commercial, government and amateur communications satellites are in low Earth orbits (LEO) at 100 to 1,000-mi. altitudes.

Larger communications satellites, such as Russia's Molniya series, are in long elliptical orbits which repeatedly take them out beyond 22,000 miles before they swing back to within 1,000 miles of Earth's surface.

Major satellites are in high orbits from where they seem to hang stationary in the sky. Actually, they whirl around the globe at the same speed as Earth rotates, girdling the equator at 22,300 miles altitude. They can spritz radio signals over an entire hemisphere.

As mentioned earlier, science-fiction author Arthur C. Clarke imagined communications satellites in such orbits as far back as 1945. Today, that region of space is known as the Clarke Belt and satellites there are said to be geostationary, synchronous or geosynchronous.

From the high vantage point of the Clarke Belt, a geostationary satellite looks down continuously on a large portion of Earth. Scores of television satellites broadcast programs to backyard dish antennas from there.

Station keeping. A stationary satellite always beams its signal onto the same area of the Earth. The surface area covered by the satellite's signal is its footprint. As the satellite orbits Earth, its footprint on the planet should not change.

No orbit is perfect, however, and any satellite will drift around the sky. A controlled satellite drifts only a small distance.

The drift is predictable, mostly back and forth in a north-south direction. If drift is not controlled, the radio footprint moves north and south on the surface of the Earth and ground stations can lose contact with the satellite. To maintain good contact with the ground, a satellite must remain in its exact assigned spot in the sky.

A satellite carries a small amount of rocket fuel. To maneuver the satellite around the sky, ground controllers send commands by radio directing it to expel fuel through its tiny thrusters to hold an exact location 22,300 miles above the equator. On command, the small thrusters push the satellite back into its assigned spot.

Saving fuel. Over several years, a satellite can use a lot of fuel to hold its position. Eventually, fuel runs out and the uncontrollable satellite become useless.

The communications satellite company Comsat found a way to extend the useful life of a satellite in orbit without flying more fuel to it from the ground. The company perfected a maneuver in which a satellite uses less fuel to hold its position.

The maneuver changes the angle of the satellite as it looks down on the Earth. It still drifts north-south, but its radio footprint stays over the same Earth-surface area. Ground stations maintain constant contact with the satellite.

As the satellite drifts north and south like clockwork, ground antennas can track the small motion. Contact is not lost. Less fuel is used from the satellite to keep it on station. The result is an extension of the useful life of a satellite.

Comsat demonstrated the maneuver's efficiency with its Comstar satellite which had been using 37 lbs. of fuel a year to keep itself on station. After the maneuver, Comstar used only three lbs.

Narrow beams. Once, telephone calls were relayed from point to point through a network of wires. Later, microwave radio signals carried additional calls. Then communications satellites brought a big leap in the number of calls which could be handled at one time.

Today we also have telephone transmission cables woven from hair-like strands of glass called fiber optics.

AT&T and other communications networks pushed fiber optics to a point where the new technology offered capacity as great as satellites. Some observers said the satellite communications business might subside as terrestrial fiber optics networks were built up.

Satellite owners, who make a lot of money from their orbiting birds, were not about to quit in the face of competition. Instead, they improved the next satellites to meet the challenge.

North American satellites had been using one antenna to send down a beam broad enough to cover the entire United States. Now more antennas appear on some satellites to narrow the beams of signals heading down to Earth.

Satellites sprouting several antennas can beam several signals to individual parts of the country. Instead of one giant radio footprint covering the continent, many small, overlapping footprints can blanket the states.

Tightly focusing a radio beam concentrates the strength of a communications signal arriving on Earth. That means smaller receiving antennas will be required, making satellite communications cheaper for broadcasters, governments, educational institutions and corporations.

Where a 45-ft. to 100-ft. dish antenna had been necessary, a 5-ft. to 10-ft. antenna would suffice.

Eventually, it will be possible to send several streams of different information down each beam, increasing the capacity of the satellite.

Broadcast Satellites

The impact of that first Telstar in 1962 and all of its many successors can be seen any evening on the nightly news. Today there are more than 100 commercial TV and telecommunications satellites in geostationary orbit. Other communications satellites, such as Russia's Molniya series, are in non-stationary orbits. Many more communications satellites of all types are scheduled for launch in the next few years.

Broadcast satellites have been used to speed video and audio across vast land masses and oceans from terrestrial network to terrestrial broadcast station and from terrestrial programmer to terrestrial cable system. Today, they also are used to broadcast directly from terrestrial stations and programmers to homes on the ground.

Not long after the first primitive Telstars and Early Birds went into service, Vietnam War film of dead bodies and jungle combat started appearing in American homes during the dinner hour.

Asian satellite uplink stations were in Hong Kong, Thailand and the Philippines, not close to American journalists in Saigon. And those satellite uplinks had to be booked in advance. Sending video home via satellite was expensive, costing about $3,000 for 10 minutes.

Only stories judged extremely important were sent back to the United States via satellite. Most film was flown on commercial flights to the United States.

As technology improved and more satellites were launched, the price of feeding news back to headquarters via satellite dropped. Networks began lugging satellite-dish trucks around on giant cargo planes.

Then CBS News came up with a collapsible satellite dish antenna in a packing crate. Its correspondents were seen talking live with news anchors in New York from anywhere in the world. Other networks followed suit.

Today, correspondents reporting from odd locales converse live with studio anchors in TV stations everywhere at any time.

The innovation of the 1990s is the small satellite dish antenna, assembled like petals of a flower, atop a remote television-studio-in-a-suitcase. It allows roving correspondents to broadcast live from the most remote war zones and disaster areas.

Many satellites relayed television news broadcasts from the Persian Gulf War in 1991, from the U.N. rescue mission in Somalia in 1992 and from several war fronts in the former Yugoslavia in 1993.

Satellite Television Broadcasting Frequency Bands

L-band	1000-2000 MHz
S-band	1700-3000 MHz
C-band	3700-4200 MHz
Ku1-band	10.9-11.75 GHz
Ku2-band	11.75-12.5 GHz (DBS)
Ku3-band	12.5-12.75 GHz
Ka-band	18.0-20.0 GHz

1000 MHz equals 1 GHz

North American satellites. The world's first commercial domestic television satellite in stationary orbit was Canada's Anik-A1 launched November 9, 1972. It also was called Telesat-1.

It was stationed over the equator at 104 degrees West longitude. Anik's first transmissions, in 1973, revolutionized Canadian communications.

In particular, lives of the Eskimo people in the vast territory north of the tree line were changed forever when they saw television for the first time. The Inuit, as they call themselves, were captivated by ice-hockey games and soap operas.

The Inuit Broadcasting Co. started delivering programs to tiny settlements above the Arctic Circle by satellite in 1983, including favorites such as *Whale Hunt* and *Old Places For Walrus.*

The first satellite for commercial domestic communications within the United States was Western Union's Westar-1, launched in 1974.

TVRO. Equipment used to receive television programs in the home from a stationary satellite is known as Television Receive Only (TVRO). A TVRO setup cannot transmit to a satellite.

Broadcast satellites are designed to relay commercial networks and are not intended for direct home viewing. Even so, millions of TVRO systems are in use in private homes around the world, particularly in North America and Europe. Several million TVRO dishes have been installed outside North American homes.

Most North American TVRO reception is in the C-Band, but the use of Ku-band is increasing.

A TVRO user's backyard C-band dish antenna can be identified by its large diameter of seven to twelve feet. Ku-band antennas are much smaller.

A transponder is a signal repeater inside a satellite. It receives a signal from the ground and retransmits it to the ground. One transponder can carry a TV channel and several radio channels or many telephone voice and data channels.

The typical C-band satellite above North America houses 24 transponders.

Older satellites are less powerful. Their transponders each transmit 5 to 8 watts of output power toward the ground. Newer satellite transponders, such as Satcom-C4 and Telstar-401, transmit 16 to 20 watts.

Two dozen C-Band broadcast satellites arc across the North American sky from 69 degrees West longitude to 143 degrees West.

Some of the satellites above North America are:

Satcom-C5 at 139 degrees, also known as Aurora-2, is near the end of its working life and is used mostly for non-video purposes such as the Digital Audio Transmission System (DATS).

Satcom-C1 at 137 degrees.

Satcom-C4 at 135 degrees is a powerful communications satellite launched in 1992.

Galaxy-1 is at 133 degrees, moved from 134.

Satcom-C3 at 131 degrees is a powerful communications satellite launched in 1992.

ASC-1 at 128 degrees has 18 C-band and 6 Ku-band transponders.

Gstar-4 at 125 degrees offers Ku-band only.

Galaxy-5 at 125 degrees was the first of a new generation of high-powered satellites.

Telstar-303 at 123 degrees, moved from 125.

Spacenet-1 was over North America at 120 degrees, but moved to Asia at 155 degrees East.

Morelos-1 at 113.5 degrees is Mexico's.

Morelos-2 at 116.8 degrees also is Mexico's.

North American C-band/Ku-Band Downlink Frequencies In MHz

Channel 1	3720/11730
Channel 2	3740/11743
Channel 3	3760/11791
Channel 4	3780/11804
Channel 5	3800/11852
Channel 6	3820/11865
Channel 7	3840/11913
Channel 8	3860/11926
Channel 9	3880/11974
Channel 10	3900/11987
Channel 11	3920/12035
Channel 12	3940/12048
Channel 13	3960/12096
Channel 14	3980/12109
Channel 15	4000/12157
Channel 16	4020/12170
Channel 17	4040
Channel 18	4060
Channel 19	4080
Channel 20	4100
Channel 21	4120
Channel 22	4140
Channel 23	4160
Channel 24	4180

Ku-Band frequencies are a 16-channel system. Half-spacing is used for a 32-channel format.

Anik-E1 at 111.1 degrees is Canada's.

Anik-E2 at 107.3 degrees also is Canada's.

Gstar-2 at 105 degrees is Ku-band only.

Spacenet-4 at 101.5 degrees was launched in 1991, but already three of its six Ku-band transponders are unusable.

Galaxy-4 orbits 22,300 miles above the eastern Pacific Ocean at 99 degrees west longitude. Launched in 1993 to work thirteen years in space, Galaxy-4 is a twin to Galaxy-7 launched just eight months earlier. They are similar to a satellite over Europe, Astra-1C, launched just six weeks before Galaxy-4. Both Galaxy-4 and Galaxy-7 are high-power, C-band and Ku-band satellites. Each has 24 transponders in each band. Galaxy-4 relays broadcast television and radio programs, business television, distance-learning services and data transmissions across the United States. Many TVRO homeowners have added extra-low-noise amplifiers to receive Ku signals.

SBS-6 at 99 degrees is Ku-band only.

Galaxy-6 at 99 degrees took up position when Westar 4 ran out of fuel there in 1991.

Telstar-301 is at 96 degrees.

Galaxy-3 is at 93.5 degrees.

Galaxy-7 at 91 degrees is a high-powered satellite with 24 transponders in the C-band transmitting 16 watts each, and 24 transponders in the Ku-band transmitting 50 watts each.

SBS-4 at 91 degrees is Ku-band only.

Spacenet-3 is at 87 degrees.

Satcom-K1 at 85 degrees is Ku-band only.

Telstar-302 is at 85 degrees.

Anik-D2 at 82 degrees is a Canadian satellite which replaced Satcom-F4 when that satellite was withdrawn from service; now all those video services have moved on to Satcom-C4 and Anik-D2 has been returned to Telecom Canada.

Satcom-K2 at 81 degrees is Ku-band only.

Galaxy-2 is at 74 degrees.

Satcom-2R is at 72 degrees.

Spacenet-2 is at 69 degrees West.

Intelsat-513 at 53 degrees relays signals from Europe to North America. Australia uses the satellite to relay TV programs from London to Los Angeles from where they are retransmitted to Australia on Intelsat-508 over the Pacific.

Panamsat-1 at 45 degrees relays C-band across the Atlantic to the Caribbean and Latin America and Ku-band to North America and Europe.

Intelsat-601 at 27.5 degrees is a European satellite which also carries NTSC transmissions to North America.

Intelsat-K at 21.5 degrees is a Ku-band only satellite launched in 1992 to relay signals across the Atlantic. It has 32 high-powered TV channels which can be directed into North America as far as the Midwest, into Europe as far east as Greece, and into some areas of South America. The beams are on 11.45-11.7 GHz with 11.7-11.95 GHz added for transmissions to the Americas.

Telstar-401 is a new high power satellite, built by America's Martin Marietta Astro Space company for launch on a European Ariane rocket. PBS will use digital compression techniques to send 8 to 10 TV channels through each Telstar-401 transponder.

What's on? Here is a sampling of programs seen on satellite channels by TVRO homeowners:

ABC network, Alaska Satellite TV, Arab Network of America, Arts and Entertainment, Asia Network in Korean, Australian feeds, Automobile Satellite TV Network, British Broadcasting Corporation (BBC), BBC News, BBC World Service Television, BBC "Breakfast News", BBC "Nine O'Clock News", Brightside Network, C-SPAN, Cable News Network (CNN), CNN news, CNN Headline News, CNN In-Flight News, CNN International, CNN's Airport Channel, California's Cal-Span, Canadian Broadcasting Corporation (CBC), CBC Atlantic, CBC French, CBC Newsworld, CBC Pacific, CBC World News, Canadian House of Commons in English and French, Canadian stations and programs, Canal Sur Spanish programs, car manufacturer networks, Caribbean Satellite Network, CBS network, China Central Television, Cinemax, Cinemax 2, college sports feeds, Comedy Central, Comedy Channel, Courtroom TV, Dallas local TV station, Deep Dish TV, Denver TV stations, Deutsche Welle from Germany, Digital Music Express, Discovery Channel, Disney Channel, ESPN sports, Family Channel, Fire and Emergency TV Network, Fox Network, Galavision Spanish programs, Germany's Deutsche Welle, Great American Broadcasting, Home Box Office (HBO), HBO, HBO 2, HBO 3, Home Shopping Network, Italy's RAI, ITN World News from Great Britain, Jade Channel in Chinese, Keystone Inspirational Network, Learning Channel, Lifetime, Los Angeles local TV station, MBC Arabic channel, Meteomedia Canadian weather, Mind Extension University, MTV, NASA Select TV, NBC network, Nebraska Educational Television, New England News Channel, New York local TV station, sports feeds, news feeds, NHK's TV Japan, Nickelodeon, North American Chinese TV, Nostalgia Television, Peru's Canal Sur, PrimeStar DBS pay channels, pro sports feeds, Public Broadcasting System (PBS network), QVC home shopping network, RAI Italy, RAI Europlus, religious channels, Reuters TV, Science Fiction Channel, SCOLA news from TV stations around the world, shopping networks, Showtime, Showtime 2, space shuttle missions live video, Spanish programs, sports feeds, TBS, Telemundo Spanish programs, Tribune Entertainment Company, Turner Broadcasting, Turner Network Television (TNT), TNT Latin America, TV Japan, TV Northern Canada for Native Americans, TVN pay-per-view movies, U.S. Congress on C-SPAN, U.S. Information Agency Worldnet, USA network, VH-1, VisNews/Reuters TV, Warner Brothers, Worldvision, WTBS Atlanta, and many more.

European broadcast sats. In the 1970s, the Soviet Union's Ghorizont was the first stationary TV broadcasting satellite over Europe.

Ghorizont satellites operate primarily in the C-Band, but carry experimental Ku-band transponders which are used infrequently.

For instance, Ghorizont-7 had six C-band transponders and one Ku-band transponder. Ghorizont-15 had six C-band transponders plus two Ku-band transponders. The C-band transponders relay television, telephone and telegraph signals interchangeably.

Satellite Downlink Frequency Differences In Western Europe

West European satellite builders segment the Ku-band for different uses.

Band	Frequency
Ku1	10.90-11.75 GHz
Commercial network relays, like U.S. C-band. Not intended for home viewing.	
Ku2	11.75-12.50 GHz
DBS band, intended for home reception.	
Ku3	12.50-12.75 GHz
Telecom band, not intended for home viewing.	

Eutelsat. Three experimental satellites led to commercial comsats over Europe.

Canada, Europe and the United States flew a test spacecraft known as CTS (Communications Technology Satellite). It was launched from Cape Canaveral January 17, 1976, to stationary orbit at 116 degrees West longitude. CTS later was renamed Hermes. It was used for medical, educational and business communications.

Italy built an experimental communications satellite known as Sirio. It was launched by the United States from Cape Canaveral on August 25, 1977. Its stationary orbit was over the equator at 15 degrees East longitude.

European Space Agency's first experimental communications satellite was called Orbital Test Satellite. OTS-1 was destroyed September 14, 1977, when its rocket exploded 54 seconds into flight from Cape Canaveral. OTS-2 made it to orbit May 12, 1978, and was positioned over the equator at 10 degrees East longitude from where it relayed television, telephone calls and data.

The first non-experimental, operational Western European comsat was European Communications Satellite (ECS-1) launched June 16, 1983, to a stationary orbit at 10 degrees East longitude. Its fourteen Ku-band transponders carried television programs, video conferences, telephone, data, fax, telex and computer links.

ECS-2 was launched in 1984 and positioned at 7 degrees East longitude.

The first European broadcast satellites were built and operated by the cooperative international organization Eutelsat, based in Paris and composed of the public telephone and telegraph (PTTs) and Telecom administrations of 33 European countries.

The growing membership of Eutelsat includes Austria, Belgium, Cyprus, Denmark, Finland, France, Germany, Iceland, Ireland, Italy, Liechtenstein, Luxembourg, Malta, Monaco, the Netherlands, Portugal, San Marino, Spain, Sweden, Switzerland, Turkey, Vatican City and elements of the former Yugoslavia.

Eutelsats carry Eurovision television programs from the European Broadcasting Union (EBU).

Astra. Eutelsat wanted individual countries to use high-power DBS satellites to deliver only two to five channels of television to homes. However, in 1988 a satellite named Astra controverted Eutelsat's plan and changed the TV picture for Western Europe forever.

A project of Luxembourg's Societe Europeenne des Satellites (SES, the European Satellite Society), Astra-1A was launched December 11, 1988. It used the Ku1 band, offered only medium-power signals, and boosted the number of channels to sixteen.

It was the first satellite capable of beaming sixteen channels directly to homes in the most densely populated areas of Europe. A second satellite, Astra-1B, was launched in 1991 adding sixteen more channels. A third, Astra-1C, was launched in 1993.

The competition. In 1977, France and Germany started planning their own five-channel DBS satellites, known as TDF. However, in the 1980s the long-time monopolies on broadcasting by European governments were beginning to fade with the advent of cable TV and the prospect of continent-wide satellite broadcasts. Luxembourg broadcasters took a financial gamble to push through innovations, starting the Astra project in 1983. The satellite was manufactured by RCA.

Just a month before Astra-1A was launched, a European Space Agency Ariane rocket had boosted TDF-1, France's competing direct-to-home television broadcasting satellite, to space. Where TDF-1 had only five channels, Astra-1A had 16.

Today, French broadcasters do not use Astra because France wants its Telecom-2A satellite at 8 degrees West used for broadcasts direct to homes.

Astra coverage. Astra signals covered 85 percent of the homes in Western Europe and could be seen in North Africa. Its movies, news, sports and other TV programs paid for by advertising were received in 55 million homes with 2-ft. diameter, umbrella-sized dish antennas. The programs were seen by 25 million more viewers through cable systems.

The major markets where viewers needed only small antennas to receive Astra were Austria, Belgium, Denmark, France, Germany, Great Britain, Italy, the Netherlands, Switzerland and Luxembourg, a tiny European country surrounded by Belgium, France and Germany. Viewers in other nations needed larger dish antennas.

Astra-1A, -1B and -1C are in orbit at 19.2 degrees East. Astra-1D was to fly in 1994, and Astra-1E in 1995. Each of the later satellites will have 18 transponders with 85-watt transmitters. Astra-1A had sixteen channels and 45-watt transmitters.

Astra-1A, -1B and -1C provide a total of 48 channels of direct-to-home (DTH) programming. Astra-1D will be a backup standing by in orbit. Astra-1E will be used for digital broadcasting.

Because Astras are at the same spot in the sky, viewers tune in all channels without antenna rotors. Pay-per-view channels use a variety of scrambling techniques, mostly Eurocrypt for D2-MAC channels and Videocrypt for PAL channels.

Television programs on the many channels in Astra satellites are a cosmopolitan blend in numerous languages. Here are samples:

Danish: TV3-Denmark and FilmNet.
Dutch: Eurosport and RTL.
English: Children's Channel, CNN, Eurosport, FilmNet, JSTV, MTV, Sky Movies Plus, Sky News, Sky One, Sky Sports, The Movie Channel, TV Asia/Sky Gold/Adult Channel, TV1000, TV3-Norway, TV3-Sweden, VOX and UK-Gold which broadcasts classics from the archives of the BBC and Thames Television. Sky News, Sky One entertainment, Sky Movies Plus, the Movie Channel, Sky Movies Gold, and Sky Sports are programs from the company British Sky Broadcasting.
German: Children's Channel, N3, DSF Sports, Eurosport, Premiere, Pro-7, RTL, Teleclub, Vox and 3-SAT, German public broadcaster ARD, SAT-1, and n-tv News. Started in 1992 on Germany's Kopernikus-1 satellite, n-tv from Berlin was the first all-news channel in German. On Astra, it also relays BBC World Service news in the original English.
Japanese: Children's Channel and JSTV.
Norwegian: TV3-Norway and FilmNet.
Spanish: Cinemania and Documania.
Swedish: TV3-Sweden, FilmNet and TV1000.

Also turning up on Astra may be Disney Channel, Games Channel, Super Channel, Cartoon Channel, Discovery Channel, Nickelodeon, Bravo, and Family Channel from International Family Entertainment controlled by American evangelist Pat Robertson.

Other Europeans. Astra is not the only KU-band satellite over Europe. Others include:

Eutelsat-1-F3 at 25 degrees feeding programs.
Eutelsat-1-F4 at 25 degrees feeding programs.
Eutelsat-1-F5 at 21.5 degrees.
Eutelsat-2-F1 at 13 degrees with many European programs.
Eutelsat-2-F2 at 10 degrees with Italian, Spanish and Turkish programs.
Eutelsat-2-F3 at 16 degrees with Spanish and Middle Eastern programs.
Eutelsat-2-F4 at 7 degrees with Eastern European and European Broadcasting Union programs.
Ghorizont-4 at 14 degrees relaying Reuters TV.
Intelsat-512 at 1 degree West with Norwegian channels.
Intelsat-515 at 18 degrees with Norwegian channels.
Intelsat-601 at 27.5 degrees with European programs, BBC and Discovery.
Intelsat-602 at 63 degrees East with Italy and Iran programs.
Intelsat-604 at 60 degrees with four Turkish transponders.
Intelsat-K at 21.5 degrees carrying trans-Atlantic program feeds.
Kopernikus DFS-1 at 33.5 degrees with German broadcasters.
Kopernikus DFS-2 at 28.5 degrees with German broadcasters.
Kopernikus DFS-3 at 23.5 degrees with German programs.
PAS-1 at 45 degrees carrying mostly program feeds in PAL and NTSC.

Eutelsat programs. Eutelsat-2 satellites transmit higher power than Eutelsat-1 satellites. Eutelsat-2 satellites are nearly as powerful as Astra satellites. Some of the programs relayed by various Eutelsat-2s are:

Eutelsat-1-F2 carries twelve TV programs. In addition, two transponders carry high-speed digital communications across Europe for businesses and governments. Services include teleconferencing, remote printing, facsimile (fax), telex, electronic mail (E-mail), and computer information transfers.

Eutelsat-2-F1 located at 13 degrees East: ARD in German, BBC World Service, Der Kabelkanal in German, Euronews with multi-lingual sound, Eurosport, FilmNet-Belgium with digital sound, FilmNet-Holland with digital sound, France Telecom, Maxat news feeds from London, Middle Eastern Broadcast Centre (MBC) programs in Arabic, Vatican Radio, RTL 2 in German, Super Channel Radio and Super Channel's Far Eastern programming including Europe-China Satellite TV and China News Europe with reports from China, Taiwan and Hong Kong in Cantonese, Mandarin and English.

Also TV5 Europe, The U.S. Information Agency's

World Net, the pornography channel Red Hot Dutch from Denmark, VisEurope with coded news feeds, the German music television channel Viva, Voice of America radio to Europe, Turkish State Television's TRT International news bulletins in English and German, and world news from Germany's international radio broadcaster, Deutsche Welle, which took over the American TV station RIAS in Berlin.

Also the European Broadcasting Union all-news station Euronews, which has no news anchors or presenters. Euronews footage, funded by ten public broadcasters, is accompanied by soundtracks in various languages on different audio subcarriers: Arabic at 7.74 MHz, English at 7.02 MHz, French at 7.20 MHz, German at 6.65 MHz, Italian at 7.56 MHz, and Spanish at 7.38 MHz.

Eutelsat-2-F2 at 10 degrees East relays Spain's TVE, three private Turkish stations, and two of Italy's RAI programs.

Eutelsat-2-F3 at 16 degrees East carries Croatia's HTV, Hungary's Duna TV, Polsat and Polish TV channels, the Portuguese RTP International service which also transmits to Africa on Russia's Ghorizont-12 satellite at 40 degrees East, Turkish channels, and Middle Eastern broadcasters reaching Arab speakers in Europe including TV7-Tunisia, the Egyptian Space Channel with audio subcarriers carrying the Egyptian Radio General Program on 7.02 MHz, the Voice of the Arabs on 7.20, and Middle East Radio on 7.38 MHz, and Morocco's RTM. Tunisian Radio is heard sometimes on a subcarriers at 7.02 and 7.20 MHz in English, German and French.

Eutelsat-2-F4 at 7 degrees East carries Greece's ET-1, Cyprus' PIK, Turkey's Kanal-6, Serbia's RTV from Belgrade, the "Vatican View" program from Vatican City, and four European Broadcasting Union (EBU) programs.

Eutelsat-2-F5 is at 36 degrees East.

Eutelsat-2-F6 to be alongside Eutelsat-2-F1 at 13 degrees East will add sixteen high-power TV channels, carrying only entertainment services during prime-time hours. Eutelsat labels its position at 13 degrees East as Europe's principal slot for direct-to-home reception and for cable television feeds.

Telecom band. A few satellites use Europe's 12.5-12.75 GHz Telecom band:

Eutelsat-2-F1 at 13 degrees.
Eutelsat-2-F2 at 10 degrees.
Eutelsat-2-F3 at 16 degrees.
Kopernikus DFS-1 from Germany at 33.5 degrees East.
Kopernikus DFS-2 from Germany at 28.5 degrees.
Kopernikus DFS-3 from Germany at 23.5 degrees.
Telecom-1C from France at 3 degrees West.
Telecom-2A from France at 8 degrees West.
Telecom-2B from France at 5 degrees West.

France wants to use its Telecom-2A at 8 degrees West to broadcast direct to homes, so French broadcasters don't use Astra. Telecom-1C is used for feeds in PAL, D2-MAC, and B-MAC.

Programs relayed by Telecom-2A include Canal Jimmy, Canal Plus, Cine Cinefil, Cine Cinemas, MCM, Planete Cable, and the French edition of Eurosport known as TV-Sport.

Intelsat programs. Intelsat is a commercial cooperative of 125 nations owning and operating a globe-spanning system of many communications satellites. Intelsats are used by 180 countries, territories and dependencies for both domestic and international communications.

The growing membership of Intelsat includes Afghanistan, Algeria, Angola, Argentina, Australia, Austria, the Bahamas, Bangladesh, Barbados, Belgium, Benin, Bolivia, Brazil, Burkina Faso, Cameroon, Canada, Cape Verde, Central African Republic, Chad, Chile, the People's Republic of China, Colombia, the Congo, Costa Rica, Côte d'Ivoire, Cyprus, Denmark, Dominican Republic, Ecuador, Egypt, El Salvador, Ethiopia, Fiji, Finland, France.

Also Gabon, Germany, Ghana, Great Britain, Greece, Guatemala, Guinea, Haiti, Honduras, Iceland, India, Indonesia, Iran, Iraq, Ireland, Israel, Italy, Jamaica, Japan, Jordan, Kenya, Kuwait, Lebanon, Libya, Liechtenstein, Luxembourg, Madagascar, Malawi, Malaysia, Mali, Mauritania, Mauritius, Mexico, Monaco, Morocco, Mozambique, Nepal, the Netherlands, New Zealand, Nicaragua, Niger, Nigeria, Norway, Oman, Pakistan, Panama, Papua-New Guinea, Paraguay, Peru, the Philippines, Portugal.

Also Qatar, Romania, Rwanda, Saudi Arabia, Senegal, Singapore, Somalia, South Africa, South Korea, Spain, Sri Lanka, Sudan, Swaziland, Sweden, Switzerland, Syria, Tanzania, Thailand, Togo, Trinidad and Tobago, Tunisia, Turkey, Uganda, United Arab Emirates, the United States, Uruguay, Vatican City, Venezuela, Vietnam, Yemen, elements of the former Yugoslavia, Zaire, Zambia and Zimbabwe.

By 1991, Intelsat's relay capability had grown to a total of 454 transponders in 15 satellites. The total included 370 C-band, 81 Ku-band and three L-band transponders aboard thirteen Intelsat-5 and Intelsat-5A satellites and two Intelsat-7s.

Some of the hundreds of programs relayed by various Intelsat satellites are:

Intelsat-601 at 27.5 degrees West has distributed CNN, Discovery, British Parliament telecasts, Children's Channel, Country Music Television Europe and BBC World Service Television. CNN, Discovery and Children's Channel moved to Astra while BBC World Service and Country Music Television Europe moved to Eutelsat.

Intelsat-603 was fired to a bad orbit in 1990. Astronauts from the U.S. space shuttle Endeavour captured the satellite in 1992 and relaunched it to its proper orbit at 34.5 degrees West from where it relays C-band feeds to the Canary Islands and Greenland and distributes the BBC on Ku-band.

Intelsat-K at 21.5 degrees West carries signals across the Atlantic Ocean, such as "Good Morning America" for London's Sky News and Super Channel from CNBC. Intelsat-K was the international organization's first Ku-band-only satellite. It has 32 high-power transponders which can be pointed at North America as far west as the Midwest, at Europe as far east as Greece, and at parts of South America.

Panamsats. Once known as Alpha Lyracom Inc., Panamsat is a small commercial satellite company trying to compete with the giant Intelsat.

Its first communications satellite, PAS-1, was launched in 1988 to 45 degrees West to relay programs between North America and Europe. It uses several channels for news feeds and relays the American Galavision programming to Europe.

Panamsat plans PAS-2 and PAS-3 located near PAS-1. PAS-2 will be held in reserve in orbit as a backup spare. PAS-3 will be stationed at 43.5 degrees so it can reach farther southeast into the Middle East and farther north into Scandinavia.

Panamsat will cover the Pacific with PAS-4 at 166 degrees East and PAS-5 at 168 degrees East. They also are called Orb-X by Panamsat. Each has fourteen C-band and Ku-band transponders.

PAS-6 and PAS-7 eventually will be above the Indian Ocean at 68 degrees and 72 degrees.

Russian satellites. The Soviet Union launched its first communications satellite, Molniya-1A, to an elliptical orbit on April 23, 1965. Molniya-1B was launched later that year on October 14.

Over the years, Molniya satellites have not been in stationary orbit like ordinary Western broadcast satellites, but in elliptical orbits covering high latitudes beyond the reach of common stationary satellites which are too low on the horizon.

Today, Molniyas serve various communications purposes, including relaying broadcasts of Double-2, Russia's second national television program, to eastern Siberia at 3.875 GHz.

The USSR's Ghorizont, also referred to as Statsionar, was launched December 19, 1978, the first stationary television satellite to serve Europe.

The Soviets concentrated on C-band transmissions, while other European satellite broadcasters developed use of the Ku-band. Even so, Ku-band transponders were built into Ghorizont satellites.

Far North TV

Sweden's TV channels known as Television-1 and Television-2 are uplinked from Oslo, Norway, to the satellite Intelsat-515 at 18 degrees West for downlink to North Sea oil platforms and to a cable system in remote Spitsbergen on the Norwegian archipelago in the Arctic Ocean.

Intelsat-515 also carries the Norwegian public broadcaster NRK, Norwegian terrestrial commercial station TV-2 and private TV-Norge.

Norway has programs on five different satellites, leading 80 percent of satellite TVRO owners in Norway to have motorized dish antennas so they can point to various positions in the sky.

Ghorizonts are over Europe at 14 degrees West and 11 degrees West. Each has a transponder at 11.525 GHz. The 14-degree satellite relays Reuters TV (Visnews). The 11-degree Ghorizont relays CNN with Russian narration and audio programs.

Most antennas in use for TVRO in Europe are small Ku-band dishes. They usually are too small for C-band reception. However, C-band signals at 3.675 GHz from Russian satellites at 14 degrees West and at 40.5 degrees East are many times stronger than ordinary C-band signals. Sensitive C-band receivers have made possible reception of those strong Russian C-band signals with dish antennas as small as five feet in diameter.

Near the Ghorizonts at 14 degrees West is the Russian satellite ZSSRD at 16 degrees West. Its two Ku-band transponders relay signals from Russia's orbiting Mir space station. Video is in clear SECAM at 10.835 GHz which is below frequencies received by most TVROs. Data beamed from ZSSRD at 11.375 GHz looks like flashing lines on a TV screen.

Intersputnik. While Western nations formed Intelsat, the East Bloc formed a similar space cooperative known as Intersputnik.

Intersputnik launched satellites and sent cosmonauts to visit orbiting Soviet space stations.

In 1987, before the break up of the Soviet Union, Intersputnik members were Afghanistan, Bulgaria, Cuba, Czechoslovakia, the German Democratic Republic (East Germany), Hungary, Laos, Mongolia, North Korea, Poland, Romania, the USSR, Vietnam and Yemen.

Later, Czechoslovakia separated into two nations, East Germany was unified with West Germany, and the USSR devolved into the Commonwealth of Independent States (CIS) aligning Russia, Armenia, Belarus, Kazakhstan, Kyrgyzstan, Moldova, Tajikistan, Turkmenistan, Ukraine, and Uzbekistan.

Comsat Capabilities

Specifications on some communications satellites, including numbers of transponders, television channels and telephone voice circuits.

Most satellite transponders carry television, telephone and telegraph signals interchangeably. Intelsat is one of very few satellite owners who distinguish the number of voice channels from the number of broadband transponders.

Satellite	Transp	TV Ch	Voice
Arabsat-F2	25	25	
ASIAsat-1	24	24	
Eutelsat-1-F1	12	12	
Eutelsat-1-F2	14	12	2 data channels
Ghorizont-7	6	7	
Ghorizont-15	8	8	
Intelsat-4A	20	2	6,000
Intelsat-5	29	2	12,000
Intelsat-5A	30	2	15,000
Intelsat-5A (IBS)	30	2	18,000
Intelsat-6	48	2	35,000
Intelsat-7A (launch 1995)		3	22,500
Palapa-B1	24	24	

Asian comsats. Today, three dozen stationary satellites float 22,300 miles above the Pacific Ocean and the Asian continent. Half a dozen are Intelsats and half a dozen are Russian satellites.

Intelsat-501 at 91.5 degrees East covers many countries in Asia and the Pacific.

Intelsat-505 at 66 degrees East relays USIA WorldNet, Deutsche Welle, and Chinese and Malaysian channels.

Intelsat-508, at 180 degrees East, uses C-band to relay CNN, ESPN, Australia's Nine Network and Network 10, America's ABC, CBS and NBC networks, Japan's NHK, RFO-France to Tahiti, USIA WorldNet, Deutsche Welle, and WTN news feeds. Ku-band transponders carry Fuji TV and Turner Broadcasting.

Intelsat-602 at 63 degrees distributes two Thailand television channels. It also carries Iran's TV1 and TV2 and the U.S. Armed Forces Radio and Television Service (AFRTS) on Ku-band.

Intelsat-604 at 60 degrees East relays four channels of Turkish State Television's TRT International broadcasts to 57 million Turkish speakers in six Moslem states of the former USSR in central Asia using Ku-band transponders.

Intelsat-7 will be located at 174 degrees East.

Russia's Ghorizonts are spotted across the Eastern sky at 80, 90, 96.5 and 103 degrees East. Each Ghorizont has six C-Band transponders. The 80, 90 and 96.5-degree satellites have Ku-Band transponders at 11.525 GHz.

A group of non-resident Indian businessmen has started Asian TV Network, a Hindi-language channel for the South Asian subcontinent, via Ghorizont 19 at 96.5 degrees East.

Russia's Ekran at 99 degrees East is a powerful broadcast satellite relaying relay Russia's Orbita-3 program on 714 MHz. That frequency is between channels 51 and 52 in Europe and channels 53 and 54 in Japan.

China's DFH2-A1 at 87.5 degrees East and DFH2-A2 at 110.5 degrees have four C-band transponders each.

ASIAsat. The first commercial satellite designed for Asian countries, was Asia No. 1, or ASIAsat-1, ferried to orbit in 1990 by one of China's Long March rockets. Located at 105.5 degrees East, ASIAsat houses 24 C-Band transponders. Twelve transmit NTSC signals in a North Beam covering northern Asia, and twelve transmit PAL signals in a South Beam covering southern Asia.

ASIAsat had the first pan-Asian satellite broadcaster, Star-TV, which went on the air in August 1991. Owned by Hong Kong's Hutchvision, its programs include BBC World Service Television, Burma's Myanmar-TV, Cantonese TV, CCTV-1, Mandarin TV, Mongolian Television, MTV Asia, Pakistan's PTV-2, Star TV entertainment, Star-Sports, and Zee TV in the Hindi-language for South Asia.

ASIAsat-2 was to be a General Electric satellite made in the United States and launched in 1995 on a Chinese Long March rocket from Xichang Space Centre. It will house 40 C-band transponders and 9 Ku-band transponders with a footprint stretching from Tokyo to Australia to Berlin.

The Republic of China (Taiwan) and the People's Republic of China (mainland China) bought into the ASIAsat joint venture company in Hong Kong to enable launch of ASIAsat-2 and ASIAsat-3. The PRC's Great Wall Industry Corp was the biggest investor. Taiwan's Chinese Engineering Development Trust (CEDT), owned by the ruling Nationalist Party, will have use of 10 per cent of the satellite's frequencies.

Apstar-1 and **Apstar-2** satellites will house 24 transponders each after they are launched on Chinese Long March rockets for Hong Kong's APT Satellite.

Korea's first two commercial communications satellites, Koreasat-1 and Koreasat-2, are scheduled for launch on an American Delta-2 rocket from Cape Canaveral in 1995. Built in the U.S. by General Electric for Korea Telecom, the satellites will be named Mugunghwa, after Korea's national flower, Rose of Sharon. The ground control center is in Yongin, Kyonggi Province, 30 miles south of Seoul.

Japanese comsats. Japan's electronics industry made that nation a pioneer in direct-to-home satellite broadcasting. The world's first non-experimental, operational direct broadcast satellite (DBS) was Japan's BS-2A launched to stationary orbit in 1984. BS-2A also was known as Yuri-2A.

The DBS satellites broadcast directly to dish antennas on homes in Japan. BS-3A and BS-3B, at 110 degrees East, each carry three television channels and one data channel.

Space Communications Corp. (SCC), part of Japan's industrial giant Mitsubishi, operates a series of high-power Japanese communications satellites known as Superbird, providing telephone, telegraph and telex communications.

Superbird A was launched in 1989. TV crews roaming the Japanese islands transmit news coverage to studios via Superbird satellites.

Superbird B-1 launched to 162 degrees East in 1992. The satellite has 19 Ku-band transponders and 10 Ka-band transponders. It can be received on 20-24 cm dish antennas.

Superbird carries six channels, including CNN, MTV and BBC World Service Television's 24-hour news and information channel.

SCC's main rival, Japan Communications Satellites, part of the industrial firm C. Itoh, operates JCsat communications satellites.

JCsat-1 at 150 degrees East and JCsat-2 at 154 degrees East. Each offers 32 Ku-band transponders.

Another Japanese communications satellite, CS-3a at 132 degrees, carries 2 C-Band and 10 Ka-Band transponders.

Despite its high-tech industry, the large Japanese communications and broadcast satellites have been built in the United States and launched to orbit on European and American rockets. JCsats were built by Hughes Aircraft Corp. Superbirds have been built by Ford Aerospace, now Space Systems/Loral. NHK's BS-series were built by General Electric's Astro Space Division.

Chinese comsats. Satellites bouncing television programs around other Asian nations also drench the largest population on Earth in the once forbidden nation of China with a flood of Western influence.

The People's Republic builds, launches and operates its own rockets and satellites. China first launched an experimental communications satellite in 1984, then an operational comsat in 1986.

Southeast Asia. For many years, Indonesia was the only Southeast Asian nation with a satellite in orbit. Then Thailand became a player and Malaysia also wanted in on the game.

Palapa is the name for domestic Indonesian communication satellites. Palapas also relay communications for Malaysia, the Philippines, Thailand and Australia's Topaz network.

Indonesia's Palapa-B2P satellite at 113 degrees East delivers television to Malaysia, the Philippines, Thailand and Indonesia. It competes with Star-TV which uses the ASIAsat satellite. Hong Kong Telecom International programs CNN, ESPN and HBO on Palapa-B2P.

Australia Television International started broadcasting to Asia via Palapa-B2P in 1993. Programs for thirty Asian and Pacific countries offer international news, science and technology reports, socio-cultural and classroom education broadcasts. Vulgar language and nude scenes are carved out of movies so Asian Moslems aren't offended.

Indonesia's Palapa B2R at 108 degrees East feeds Brunei, Malaysia, Papua-New Guinea, the Philippines, Singapore, Thailand and Indonesia.

The Friendly Islands

A trifling strand of pearls lost in the Pacific, known mostly for coconuts, bananas and delicious tapioca pudding made from cassava roots, suddenly has become a communications satellite powerhouse in the 1990s.

Tonga Islands, also known as Friendly Islands, and not to be confused with Tongaland in Africa, is an archipelago of 150 islands discovered in 1616 in the southwest Pacific Ocean by Dutch navigator Jakob Lemaire. It was a British protectorate in 1900 and achieved independence as an island kingdom of 100,000 souls in 1970.

Today, Tonga is a powerhouse among spacefaring nations because every country, no matter how small, is allocated satellite positions along the Clarke Belt above the equator. Satellite operators, Unicom and Rimsat, use Tonga's allocations for Tongasats.

Tongasat-1 is a used Ghorizont satellite, purchased from Russia and moved to one of Tonga's slots in the Clarke Belt where it covers a vast region from Iran to Hawaii.

Unicom Tongasats would cover the Pacific from Asia to the western United States with 12 C-band and six Ku-band transponders. The first would be at 138 degrees East; the second at 170.75 degrees. Others would be at 130 and 134 degrees.

Rimsat would buy seven new satellites from Informocosmos, a consortium of Russian satellite manufacturers, and use Russian rockets, launched from the Baikonur Cosmodrome in Kazakhstan, to blast them to Tonga's positions above the equator.

One fly in the ointment: the Friendly Islands weren't so friendly in 1993 after Indonesia tried to claim Tonga's 170.75 slot for a Palapa satellite.

India. Indians have only 25 million television sets, yet many viewers would like alternatives to government programs. Local businessmen have tried to meet the demand by starting satellite-television distribution services. Alternative satellite TV now is seen by three million persons, mostly in cities, while government TV still reaches a wider audience.

To get into the alternative satellite television business, an entrepreneur buys a dish antenna, converter boxes for receivers and miles of cable. He points his antenna at a Western nation's stationary satellite in the sky. After connecting business and home subscribers to his antenna, he sends Western news, films and music through the cable.

India's South Asian Sats

There are 844 million people in India. A large majority has not been exposed to education. The South Asian nation uses orbiting satellites to spread knowledge, education, science and culture to rural areas beyond mainstream Indian city life.

First, India used the American experimental communications satellite ATS-6 in 1975 to carry education, science and technology to rural towns.

That success convinced the Indian government of the need for its own communications satellites. To get started quickly, India decided to buy a first-generation satellite from a foreign builder. The second-generation would be made in India. All four satellites in the first-generation series, known as India National Satellite or INSAT, were bought from Ford Aerospace of the United States.

INSAT-1A was launched on an American rocket from Cape Canaveral in 1982. Its attitude control broke down and the satellite couldn't be used. That led to the launch of INSAT-1B which rode a U.S. shuttle to space in 1983.

INSAT-1C also was to have been ferried to orbit by U.S. shuttle, but was switched to an unmanned European rocket after the 1986 Challenger disaster. An Ariane boosted it to space in 1988. It was expected to work ten years, but part of its electrical system broke down in orbit, leaving only half of its electronics working. INSAT-1D, was launched in 1990 from Cape Canaveral.

Second-generation INSAT-2A and INSAT-2B, built by the Indian Space Research Organization and similar to their predecessors, were launched in 1992 and 1993 on European rockets. ISRO also fabricated INSAT-2C for launch in 1994, INSAT-2D in 1996, and INSAT-2E in 1997. A third-generation satellite, INSAT-3, is being designed.

INSATs are dual-purpose, stationary-orbit satellites relaying television programs into rural communities and telephone calls and data communications across the large country. They also collect weather data from 110 automatic ground stations strewn across the South Asian nation and record weather photos in visible light and infrared light every half hour.

They use C-band to relay national and regional Indian channels, including Tamil Nadu, Karnataka and Maharashtra programs. They beam high-power signals down to more than 100,000 low-cost TVRO receivers in rural areas and to disaster warning devices across the nation.

INSATs relay telephone calls and data between 31 ground stations and even monitor the weather.

The success of the INSAT-1 and INSAT-2 series boosted India's plan for a rural satellite network, called Gramsat, or Village Satellite, disseminating information on health, environment, family planning and literacy. INSATs also provide disaster warning and distress alert systems.

Thailand's first two communications satellites relay television programs and link telephone, telex, fax and data across the Southeast Asian nation.

Thaicom-1 and Thaicom-2, from their positions above the equator at 101 degrees east, also could tie in Thailand's three neighboring Indo-Chinese countries, Vietnam, Cambodia and Laos.

The first two Thaicoms are Hughes HS-376 communications satellites carried to stationary orbit on a European Ariane rocket. The HS-376s are expected to work 15 years in orbit. Two new satellites would be sent to replace them in orbit at that time.

The Transport and Communications Ministry gave the local Shinawatra Computer and Communications Co. Ltd., the right to operate Thailand's first national satellite system for 30 years.

In yet another dispute over slots in the Clarke Belt, ASIAsat tried to use one of Thailand's positions over the equator. The Hong Kong consortium wanted to station ASIAsat-2 near 105 degrees East where Thailand wanted to place Thaicom-1.

Malaysia wants to launch a multi-purpose satellite for communications, land survey and petroleum exploration, reminiscent of India's INSATs which combine meteorology with communications. Malaysia has been paying $30 million a year to Indonesia to rent the services of a Palapa satellite.

Pacific comsats. Australia, New Zealand and Tonga are the big names in southern Pacific Ocean communications satellite coverage.

Australia placed Aussat-A1, Aussat-A2, and Aussat-A3 at 156, 160, and 164 degrees East, to beam television, telephone and data signals across the continent, to the Outback, to New Zealand, and to the Pacific islands.

Each Aussat has 15 Ku-Band transponders. Even Australian air traffic controllers use Aussats.

U.S. space shuttles ferried Aussat-1 to orbit in 1985 and Aussat-2 three months later in 1985. Aussat-3 was launched by a European Ariane rocket in 1987—during the 975-day period when U.S. shuttles were grounded following the catastrophic loss of Challenger in 1986.

Optus Communications, a Sydney, Australia, company, asked the Chinese government's Great Wall Industrial Corporation to launch two replacement Aussats made in the U.S. by Hughes. The launch by China was approved by the U.S. government in 1988. They were the first American satellites launched on a non-Western rocket.

Optus was to be a new generation of satellites replacing Aussats as they grew old and ran out of the fuel needed to hold their sky positions.

Aussat-B1 was renamed Optus-B1 and assigned to replace the out-of-fuel Aussat-A1. China used one of its Long March rockets to launch Optus-B1 in August 1992 to a position over the equator at

160 degrees East. Optus-B1 has 15 Ku-band transponders, some of which will be used to distribute a six-channel subscription-TV service.

Unfortunately, the December 1992 launch of Optus-B2 on a Long March rocket from China's Xichang Space Center in southwest Sichuan Province went awry. A fireball ignited 48 seconds into the flight. Scraps on the ground after the explosion made it seem something happened to the satellite's protective nosecone four miles up on the way to space. The satellite would have been buffeted by a 1,000 mph slipstream which would have destroyed it.

Meanwhile, Australia Television International broadcasts to thirty Asian and Pacific nations via Indonesia's Palapa-B2P satellite.

New Zealand has a broadcast communication satellite service on the drawing board. It would cover the western Pacific and East Asia.

The Pacific is a rapidly developing market for coverage by a variety of new communications and broadcast satellites.

Panamsat competes with Intelsat. Its PAS-1 was launched in 1988 to relay programs between North America and Europe. Panamsat will cover the Pacific with PAS-4 at 166 degrees and PAS-5 at 168 degrees East.

Panamsat refers to PAS-4 and PAS-5 as Orb-X. Each has fourteen C-band and Ku-band transponders.

Later, Panamsat will position PAS-6 and PAS-7 above the Indian Ocean at 68 and 72 degrees.

Pacificom-1 is a high-power satellite to be operated by the American conglomerate TRW at 172 degrees East. Pacificom has eleven Ku-band DBS transponders and eight C-band transponders. Its footprint would encompass Australia, Asia and the West Coast of the United States.

Pacstar-1, at 167.5 degrees East, would cover East Asia, Pacific islands and U.S. West Coast.

Latin America comsats. Latin American broadcasts can be received from satellites such as Intelsat-513, Intelsat-601, Intelsat-K, Hispasat and Panamsat. They relay programs into South America from Europe, while Hispanic telecasts from North America are feed south via Spacenet-2.

Going the other direction, from South America to North America, is a Spanish and Portuguese satellite service on Spacenet-2, transponder 22. Known as Canal SUR, which stands for Sistema Unida de Retransmision, it repeats broadcasts from local stations in Argentina, Brazil, Chile, Colombia, Ecuador, and Peru.

Mexico and Brazil have communications satellites and Argentina wants one of its own.

Mexico's Morelos satellites were launched in 1985 by U.S. space shuttles. Mexico is replacing the Morelos with satellites known as Solidaridad-1 and Solidaridad-2, each with three times the capacity of a Morelos.

They relay television, telephone, data, facsimile (fax) and business network services in C-band and Ku-band as well as mobile communications in the L-band between 1,000-2,000 MHz.

Brazil's two Brasilsats were launched in 1985 and 1986 on Europe's Ariane rockets to stationary orbits over the equator at 70 degrees West.

Brasilsat-S1 and Brasilsat-S2, manufactured by Canada's Spar Aerospace, are C-band satellites linking the Amazon jungles with television, telephone and data links between Brazilian cities.

Brazil is replacing the pair with Brasilsat-B1 and Brasilsat-B2, manufactured by Hughes with greater capacity than the earlier Brasilsats.

Argentina is planning its first commercial communications satellite to be located in the Argentine slot in the Clarke Belt.

Middle Eastern comsats. Television pictures relayed out of the region by satellites have made the rest of the world more familiar with the ancient Middle East.

Arab League's Arabsats, and Turkey's telecasts to its emigrants in Europe and the new independent Asian countries, lead the Middle East's use of communications satellites in the 1990s.

Persian Gulf military action in 1991 was the "first live TV war." Pictures from Baghdad, Iraq, and from the United Nations coalition front were broadcast live around the world by many satellites.

Arabsat, the Arab Space Communications Corporation, was formed by the Arab League in 1976. Its two large satellites first linked cities across the Arabian desert in 1985 with television, telephone and data communications.

The growing membership of Arabsat includes Algeria, Bahrain, Djibouti, Iraq, Jordan, Kuwait, Lebanon, Libya, Mauritania, Morocco, Oman, the Palestine Liberation Organization, Qatar, Saudi Arabia, Somalia, Sudan, Syria, Tunisia, Yemen and the United Arab Emirates.

Arabsat-1A was launched in February 1985 on a European Ariane rocket. Four months later, Arabsat-1B was ferried to orbit in a June 1985 U.S. space shuttle flight.

Saudi Arabian Prince Sultan Salman Abdel Aziz Al-Saud, the first Arab in space and the first royalty in orbit, was among the shuttle crew for seven days in space.

When Aerospatiale, the French manufacturer of all three Arabsats, completed Arabsat-1C in 1986, the satellite was stored away for future use as needed. By 1992, launch of Arabsat-1C had become necessary. The old Arabsats, -1A and -1B, had been allowed to slide into inclined orbits to extend their useful lives.

Arabsat-1C was taken out of storage and launched in 1992 on a European Ariane rocket to its orbital station.

Arabsat-1A was retired, leaving the Arab League with 25 C-Band transponders each in Arabsat-1B and Arabsat-1C, as well as a strong S-Band transponder near 2600 MHz.

Arabsat-1B continued drifting eastward, beyond 43 degrees East, where it was used to relay programs from Mauritania, Morocco, Oman and Saudi Arabia. One transponder was used for Inter-Arab news broadcasts from Tunis.

Arabsat-1C is the premier Middle Eastern satellite today. From its post over the equator at 26 degrees East, Arabsat-1C's coverage of central Europe, the Mediterranean, India and South Asia link the Arab world with Europe and Asia.

Here is a sample of programs carried into the Middle East by the Arabsat system:

CNN rents a transponder to relay news to the Mediterranean and the Arab world. Radio Orient is relayed. Turner Broadcasting System (TBS) sends its Cartoon Channel and Turner Network Television (TNT) movies to the Middle East. Saudi Arabia uses more than three C-band transponders and an S-band transponder to relay domestic TV programs. An Italian broadcaster uses a transponder. The Arab States Broadcast Union exchanges daily news programs and transmits Arab regional TV programs. India programs more than a dozen transponders. Arab States use ten transponders for telephone service. Jordan, Mauritania, Morocco and Oman use C-band channels to relay domestic programs. Egypt uses an S-band transponder for its Egyptian Space Channel. The Kuwait Satellite Channel broadcasts in Arabic, except for an English news bulletin.

Middle East Broadcasting Centre (MBC) telecasts in Arabic from London are backed by Saudis. Even though MBC's chairman is a brother-in-law of King Fahd, the channel once was banned in Saudi Arabia because it used bare-headed female announcers.

New satellites and more channels are needed. Arabsat has ordered two high-power, high-capacity HS-601 satellites from the U.S. manufacturer Hughes Communications to be launched for Arabsat after 1995. Each would have eighteen C-band channels, twelve Ku-band transponders and two S-band channels.

V-SAT is a complete Ku-band ground station in a briefcase for communications. A user of the new generation of Arabsats will be able to operate from a V-SAT (very-small aperture terminal).

Turkey's satellites, Turksat-1 and Turksat-2, are intended to improve communications from Istanbul across the rough terrain of that nation's eastern and southeastern regions.

The French firm Aerospatiale built the sixteen-channel satellites to be launched on European Ariane rockets. They beam twelve channels of Turkish television and radio programs across the country, and four channels of telephone, telex, fax, and data transmissions.

Israel's communications satellites, Amos-1 and Amos-2, housing several Ku-Band transponders, are to be launched on European Ariane rockets in the 1990s.

African comsats. Satellite broadcasting is expanding across the African continent, especially into South Africa.

Somalia's massive problems in the late 1980s and early 1990s were shown to the world as satellite television broadcasts out of Africa revealed the starving millions. The pictures started an international movement and led to a 1992 United Nations rescue mission to the African nation.

The equator bisects Africa from Gabon to Kenya, leaving the great bulk of the continent north of the equator.

Astra and other European satellites orbiting above the equator intend to broadcast signals northward into Europe, but those signals also can be received southward into South Africa.

Intelsat-601 at 27.5 degrees West is another example. European broadcasters use Intelsat-601 for Ku-band transmissions into Europe and C-band for transmissions into Africa.

Broadcasters using Intelsat-601 to reach into Africa include: BBC World Service Television, BOP-TV from the South African homeland Bophutaswana, Canal France International, CNN International, M-Net South Africa, USIA WorldNet and Deutsche Welle.

Intelsat-602 at 63 degrees East carries SABC and M-Net from South Africa.

Intelsat-605 at 24.5 degrees West is used by South African Broadcasting Corporation's external services using the name Channel Africa. Television services and Radio RSA broadcast news and educational programs in English and French and are rebroadcast by other African stations.

Intelsat-505 at 66 degrees East carries BOP-TV from the South African homeland Bophutaswana, Canal France International, South Africa's SAIS, USIA WorldNet, Deutsche Welle, and Zaire television.

Ghorizont-12, the Russian communications satellite at 40 degrees East, relays to Africa the French channels TV5-Europe and TV5-Afrique.

Portugal's RTP International also is relayed via Ghorizont-12. The satellite is in an inclined orbit, forcing ground stations to track its figure-eight drift pattern.

Nigeria wants its own satellite. The Canadian manufacturer Spar was training Nigerians in satellite communications techniques for the Nigerian Transport and Communications Ministry, according to The Republic newspaper of Lagos.

DBS Satellites

Rural homeowners with rotatable Yagi antennas on towers always had trouble receiving even two or three local TV broadcast stations.

After broadcast satellites were launched to relay programs from networks to stations, rural homeowners discovered they could get better reception and dozens of channels by picking up satellite signals with big TVRO dish antennas in their backyards.

Since broadcast satellites existed to relay signals across professional networks, home TVRO was a secondary use. Still, programmers wanted to make money from homeowners so they scrambled their signals. For a fee, TVRO users could have their signals unscrambled.

As demand grew for satellite signals in the home, broadcasters and satellite manufacturers decided they could increase profits by intentionally feeding programs directly into homes.

DBS. A new class of satellites, known as direct broadcast satellites (DBS), was built to beam high-powered signals directly to homes.

To receive older broadcast satellites, TVRO viewers had needed motorized dish antennas of 7-ft. to 12-ft. diameter. Neighbors sometimes saw those big backyard dishes as eyesores.

By comparison, DBS users need dish antennas only one foot in diameter—so small they can be hidden from view—and they need not move.

DBS differs from TVRO in that DBS is intended for home reception. Direct broadcast satellites constitute a new industry for delivery of television programs to individual homes. Based in space, DBS differs from ordinary terrestrial over-the-air broadcast stations and from wired cable television (CATV) systems on the ground.

DBS satellites have very large footprints to cover as many subscribers' homes as possible.

Japanese first. Japan's state-of-the-art consumer electronics industry led that nation to become a DBS pioneer. The world's first direct-to-home broadcast satellite (DBS) was Japan's BS-2A (Yuri-2A) launched to stationary orbit in 1984.

The BS-series of DBS satellites have been operated since then by the state public service television network Nippon Hoso Kyokai (NHK), broadcasting directly to dish antennas on home roofs and balconies in the Japanese archipelago.

BS-3A and BS-3B, at 110 degrees East, each carry three TV channels and one data channel. DBS programs are provided by NHK and the commercial network JSB. BS-3B broadcasts the new high definition television (HDTV) using the Japanese Hi-Vision system.

European DBS. Several DBS satellites are in orbit above Europe:

European Space Agency's Olympus is at 18.8 degrees West. Rome's RAISAT and Canadian Broadcasting Corporation's (CBC) news in English and French are seen on Olympus.

France's TDF-1 and **TDF-2** are at 19 degrees.

Germany's TV-SAT 2 is at 19.2 degrees.

Great Britain's Marco Polo-1 and Marco Polo-2 DBS satellites were at 31 degrees. They lost money and were closed down, sold and moved.

Norway's Thor is at 0.8 degrees West. There is strong DBS competition above Scandinavia between Sweden and Norway. It came to a head when Great Britain decided to sell its Marco Polo-2 satellite. Swedish Space Corporation and Norwegian Telecom bid on Marco Polo-2. The Norwegians prevailed, moved the satellite to 0.8 degrees West, and renamed it Thor. Thor offers five TV channels to Scandinavian viewers on the ground. Norwegian Telecom may buy two more DBS satellites and place them at the same position in the sky to increase the number channels available to fifteen. CNN, Filmnet Movies and Eurosport are seen on Thor.

Sweden's Tele-X, Thor's Scandinavian competitor at 5 degrees East, is owned mostly by South Africa's M-Net, owner of Filmnet Movie Channel which is seen on Tele-X along with TV5-Nordic and the Stockholm newspaper "Svenska Dagbladet" business channel Executive Television. To increase channels for Scandinavia, Filmnet may buy another satellite, such as Britain's Marco Polo-1, to place beside Tele-X. Radio Sweden's satellite schedule is transmitted over the TV4 transponder audio subcarrier 7.38 MHz. The Swedish news agency TT sends two minute newscasts every hour to community radio stations across Sweden, using the TV5-Nordic transponder audio subcarrier at 7.56 MHz.

Spain's Hispasat-1A at 30 degrees and **Hispasat-1B** hurt Astra's Spanish-language DBS business. Hispasat-1A, launched in 1992, transmits five DBS TV channels and Hispasat-1B, launched in 1993, offers five. Hispasats are aimed at Spain, the Canary Islands and South America. Spanish channels include Antenna Tres, Canal Plus Espana and Tele Cinco. They also relay phone calls and data.

European DBS has a future. Current DBS satellites have whetted European appetites for more channels. Viewers want more than five programs as well as programs from many countries.

While smaller might seem better, European viewers apparently think a two-ft. dish antenna receiving 16 to 48 channels is more desirable than a one-ft. dish receiving three stations.

France is trying to fill the gap with its medium-powered Telecom broadcast satellites and letting its older TDF-1 and TDF-2 DBS satellites run down. Similarly, Germany is concentrating more on its Kopernikus broadcast satellite, rather than the older TV-SAT.

European satellite builders are going to introduce a new generation of satellites offering more channels at higher transmitter power for greater coverage.

A twelve transponder European DBS satellite covering France, Germany, Italy, and Sweden is to be placed at 19 degrees West.

Europesat is to be a series of four high-power second-generation DBS satellites from Eutelsat. There will be three working Europesats plus a back-up in orbit, all at 19 degrees West.

Austria, France, Germany, Italy, the Netherlands and Switzerland are six of the eight countries with DBS locations allocated at 19 degrees West. They have joined Portugal, Sweden, and the former Yugoslavia in planning Europesat.

The first Europesat is to be launched in 1996, with new launches every six months. One-ft. dish antennas will provide reception on the ground.

The trio of Europesats will transmit a total of 36 channels, in PAL, Secam, MAC, or HD-MAC.

Since the nine participating countries are asking for a total of up to 54 channels, more Europesats may be built and slotted at 29 degrees East.

American DBS. They were slow to start, but direct broadcast satellite operations are coming to North America.

PrimeStar is a service inaugurated in 1990 by General Electric and nine cable operators, delivering programming with medium transmitter power in the Ku-band. PrimeStar uses Satcom-K1 at 85 degrees West. Its eleven channels include so-called super stations, Japan's NHK and three pay-per-view channels.

SkyPix wanted digital signal compression to cram 80 program channels into 10 Ku-band transponders on SBS-6 at 99 degrees West. However, SkyPix seems to be out of business.

Sky Cable was a 100-channel service backed by NBC, satellite-maker Hughes, cable operator Cablevision, and British Sky Broadcasting owner News Corporation, but Sky Cable was disbanded.

DirecTV is the best bet for a successful second-generation European-style DBS over North America. Hughes Communications and United States Satellite Broadcasting build and operate the DirecTV satellites.

Hughes is a unit of GM (General Motors) Hughes Electronics. It is the world's largest manufacturer and private operator of commercial communications satellites.

A DirecTV satellite, located at 101 degrees West, carries up to 150 channels, including regular cable channels, pay-per-view networks and high definition television (HDTV) services. The number of channels from one DirecTV satellite can be increased to 200 through the use of digital signal compression techniques.

An 18-in. antenna receives DirecTV's news, weather, sports, movies, cartoons, talk shows and other kinds of television programs.

U.S. cable systems say they would like to offer as many as 500 wired TV channels in the future. DBS companies hope their satellites will be the wireless cable systems of the future, replacing terrestrial wired cable systems, offering hundreds of channels appealing to various narrow interests.

HDTV, or high-definition television, with its digital-television pictures and new screen shape, is the wave of the future in electronics engineering. Satellite broadcasters hope to have a global HDTV standard by the year 2000.

Japan's NHK already transmits an HDTV system called Hi-Vision. The system will not become a standard since it is not digital. When the dust settles after the engineering competition of the 1990s, analog and mixed analog-digital HDTV systems will not become global standards.

The European Commission is trying to foist on European broadcasters an HDTV system called HD-MAC, but that won't become a standard as it is half digital and half analog. European broadcasters, like others around the globe, are looking for an all-digital standard.

The Federal Communications Commission (FCC) finalized an all-digital HDTV system for the United States which is likely to become the global HDTV standard.

Non-Video Signals

Broadcast satellites carry more than news and television entertainment programs. There are hundreds of audio, radio and data networks, stock market and commodity exchange news reports, transmissions by international press agencies, even telephone channels.

All of these are audio signals transmitted through the satellites as *subcarriers*.

Audio subcarriers are extra signals transmitted

alongside a video signal. The video information sent through a satellite transponder occupies a frequency range of zero to 4.2 MHz. Subcarriers are added at frequencies higher than video, anywhere between 5 and 8 MHz. Often, they are at 6.2 or 6.8 MHz.

North American satellites distribute more than 70 of these audio signals. Here is a sample:

Spacenet-2 satellite has Hispanic Radio Broadcasting on channel 9 at 7.75 and 7.93 MHz and, on channel 7, Becker Satellite Network at 7.50 MHz. Becker includes programs from Radio New York International, Satellite Information Service, Johnny Lightning Show, World Jazz Federation, DBI Satellite Radio Talk, and others.

Spacenet-3, channel 2, has Nebraska Public Radio at 5.76 and 5.94 MHz. Its channel 15 has KLON-FM broadcasting jazz and NPR at 5.58 and 5.76 MHz. Its channel 21 has Let's Talk Radio at 6.2 MHz, a daily program about radio listening and satellite television viewing from 6 p.m. to 1 a.m. Eastern time.

Galaxy-3, channel 7, has Southern Gospel Music at 5.36 MHz and Contemporary Hispanic Network at 5.94 MHz. Galaxy-3, channel 8 has Minnesota Public Radio at 8.235 MHz, Business Radio Network at 8.055 MHz, and In Touch for the visually impaired at 7.875 MHz. The Caribbean Satellite Network uplinks news, music, situation comedies and documentaries on Caribbean life from its Miami studios to Galaxy-3.

Galaxy-3, channel 11 has Classical Collections at 6.30 and 6.48 MHz, Soft Rock at 5.22 and 5.40 MHz, America's Country Favorites at 5.04 and 7.74 MHz, New Age of Jazz at 7.38 and 7.56 MHz, Light and Lively Rock at 5.94 and 6.12 MHz, Soft Sounds at 5.58 and 5.76 MHz, and Golden Oldies at 8.10 and 8.28 MHz.

Galaxy-3, channel 24, has two interesting C-Span audio channels. The cable-TV industry's public-service station C-Span regularly telecasts the U.S. Senate and House of Representatives. C-Span also carries two interesting audio subcarrier channels. One rebroadcasts BBC World Service 24 hours a day at 5.40 MHz. The other, at 5.22 MHz, rebroadcasts a variety of international broadcasters, including Radio Japan, Germany's Deutsche Welle, Radio Sweden and Canadian Broadcasting Corporation.

Galaxy-5, channel 22, has CNN Headline News for radio stations at 6.30 MHz. The same satellite, channel 17, 5.80 MHz, has the Project Saturn Global educational radio network from Los Angeles, with a library of 2000 programs for all ages, on subjects from fairy tales to computer technology, and hourly educational news.

Galaxy-6, channel 17, at 7.48 MHz, has KGAY, a radio network for gays and lesbians.

Satcom-1R, channel 15, has Radio Japan in Japanese at 7.38 MHz, while Satcom-C5, channel 24, has KSKA-FM Anchorage/NPR at 7.30 and 7.56 MHz.

Anik-E1, a Canadian satellite, has on channel 17 Radio France International at 5.41 and 6.12 MHz.

Anik-E2, another Canadian satellite, has on channel 16 CBC Radio at 5.41 MHz.

European satellites, including Astra, Eutelsat, Intelsat, Kopernikus, TDF, Telecom and Tele-X, use audio subcarriers to deliver hundreds of radio channels to listeners on the ground.

Astra satellites transmit many radio stations including BBC, CNN Radio, Country Music Radio, Deutsche Welle, Deutschlandfunk, Euronet, Hit Radio, Holland-FM, NDR 2, NDR 4, Power FM, Quality Europe FM, Radio Asia, Radio Eviva, Radio Sweden, Radio-Ropa, RMF, RTL Berlin, RTL-4 Radio, Sky Radio, Star*Sat Radio, Sudwestfunk 3, London Indian community commercial station Sunrise Radio, Super Gold and Swiss Radio International, United Christian Broadcasters, and ASDA storecast FM.

Other satellites relay audio-subcarrier programs:

Eutelsat-2-F1 relays BBC World Service, Voice of America (VOA), Deutsche Welle, Sky Radio and other audio-subcarrier programs.

Eutelsat-2-F2 broadcasts Turkish programs and Radio Liberty.

Eutelsat-2-F3 has audio programs from Portugal, Croatia, Egypt and Tunisia.

Eutelsat-2-F4 relays Turkish and Serbian audio programs.

Intelsat-512 has Norwegian audio channels.

Intelsat-601 carries BBC and CNN Radio.

Kopernikus DFS-3 carries German-language audio-subcarrier programs.

TDF-1 and **TDF-2** broadcast French audio programs.

Tele-X carries Radio Sweden and Radio Z. Svensk Radioutveckling (SRU) a service of major Swedish newspapers, relays private commercial radio stations across Sweden via Tele-X.

Telecom-2B relays French audio programs.

Asian and Pacific satellites also relay radio and audio programs using subcarriers.

One Japanese company relays an astounding 500 audio channels. Almost anything imaginable can be heard, even sounds of bells ringing, cows, dogs, horses, roosters, steam trains and waterfalls. The company may increase to 2,000 programs!

Digital Radio By Satellite

After Japan pioneered DBS telecasts with its BS-2A satellite in 1984, it launched another satellite, BS-3A, in 1990 to pioneer the world's first nationwide digital radio system.

Along with broadcasting DBS television programs, BS-3A at 110 degrees East uses pulse-code modulation (PCM) to broadcast radio-style audio programs with sound quality like compact discs directly to antennas on homes in Japan.

Digital satellite radio (DSR) is a German technique providing audio quality matching the sound from a compact disc (CD).

Sixteen German digital radio channels have been transmitting through Germany's satellite Kopernikus DFS-3 at 23.5 degrees East.

BBC World Service, Voice of America, Deutsche Welle, Radio France International and other international broadcasters have been transmitting DSR via eight audio channels in transponders 26 and 34A on the European satellite Eutelsat-2-F3 at 16 degrees East.

There are digital satellite radio (DSR) services in the United States:

Digital Cable Radio (DCR) was launched in 1990 as the first digital cable audio service in the United States. DCR transmits 56 CD-quality radio channels to American and Mexican cable systems and individual subscribers via transponder 9 on the satellite Satcom-C3. Individuals receive music, news, information, talk and foreign language programs. Business users receive various background music formats. DCR plans to upgrade to 250 channels and expand into the rest of the Americas, Europe and Asia.

Digital Music Express (DMX) distributes thirty blends of music, with no commercials or disc jockeys, to U.S. cable-television subscribers. The home cable decoder displays song, artist, album, composer and position on the pop-chart.

The music includes big band, blues, chamber music, classic jazz, classic rock, contemporary, country and western, golden oldies, heavy metal, hot American hits, hot European hits, jazz, opera, reggae and world beat.

DMX sends twenty music mixes from its Atlanta studio via satellite to Australia where it is microwaved to customers who have small roof-mounted antennas.

DMX also sends music through Intelsat-601 to European cable operators and to a broadcaster who uplinks it for subscribers of the satellite Astra-1C.

DAB. Eventually, DSR technology will have to do battle in the marketplace with a different technology known as digital audio broadcasting (DAB), also known as BSS-Sound.

A 1992 World Administrative Radio Conference (WARC) approved frequencies for satellite DAB.

DAB is supposed to replace old-fashioned FM radio stations on the ground sometime after DAB broadcasts start in 1995.

The high-quality digital audio broadcasts would be received directly from satellites with small portable receivers. Of course, replacing FM radio stations could take many years.

WARC countries generally considered a set of frequencies around 1500 MHz (1.5 GHz) to be best for DAB.

That frequency was said to offer best reception by portable receivers using small whip antennas, without interference from buildings or terrain. Satellites for that frequency would be cheaper to build, too.

Unfortunately, politics reared its ugly head at WARC. Many countries could not agree to 1500 MHz because they had existing users there, including their military forces.

WARC was deadlocked until a last-moment compromise provided three DAB allocations—one at 1500 MHz, another for the United States at 2300 MHz, and a third at 2600 MHz.

If satellites and receivers are built and sold quickly, DAB could start on schedule in some parts of the world.

SCPC. Another means of sending audio programs through a satellite is known as "single channel per carrier," or simply SCPC.

In SCPC, a satellite transponder is divided into individual channels with no video. SCPC transmissions can be either FM and single sideband (SSB). FM is favored for high-quality audio broadcasts and network relays. Business information and news services use SSB.

Tuning in SCPC. Receivers can be purchased. Universal Electronics of Reynoldsburg, Ohio, designed the first household SCPC decoder to be attached to an ordinary satellite receiver.

SCPC also can be monitored readily with a common household police scanner, or a VHF-UHF communications receiver, attached to an ordinary satellite-television receiver.

Older TVRO receivers have intermediate or downblock frequencies at 70 or 134 MHz. A scanner or VHF-UHF communications receiver can be used to monitor transponder frequencies 18 MHz on either side of a channel, although the audio quality may not be as good as sound from a real SCPC receiver.

A splitter can be installed in the cable from the LNB to a late-model TVRO satellite receiver. The extra output can be fed through a DC block to an FM scanner which covers the LNB's 950-1450 MHz range.

Miscellaneous data services using single-sideband can be monitored by attaching an SSB receiver to a TVRO receiver's 0-4.2 MHz output.

Many satellites carry telephone services. A transponder can hold 9,000 telephone channels. Telephone services can be found on such satellites as Satcom-5 and Galaxy-2 by tuning the SSB receiver between 3.720 and 4.180 MHz.

In addition, careful tuning will turn up radioteletype (RTTY) signals which can be interpreted with a radioteletype reader attached to the SSB communications receiver.

What's on SCPC? While SCPC isn't used much in Europe, it's popular for North America. Here are examples:

National Public Radio is a non-commercial radio network in the United States. NPR uses sixteen audio channels on Galaxy-6 transponders 2, 3, and 4.

American Public Radio (APR), a non-commercial rival of NPR, also uses Galaxy-6. BBC World Service is repeated eight hours a day on one of APR's SCPC channels.

NBC Desktop News is a news-on-demand data relay scheme delivering requested information via satellite. The system, promoted by NBC, IBM and NuMedia, lets personal-computer users scan headlines and retrieve text, graphics, video and sound about interesting stories.

International Broadcasters

International broadcasting has been an important medium of global communication for most of the 20th century. Worldwide broadcasts have been possible because shortwave radio signals "propagate" themselves over long distances.

For many decades, shortwave radio broadcasts have let listeners eavesdrop on the everyday workings of international commerce and politics. Many countries have broadcast information about themselves and native entertainment programs to the world in English and many other languages.

Shortwave listeners (SWLs) have been able to find out directly from original sources what the rest of the world thinks is important and what positions are being taken on various issues.

Recently, many nations have been moving their international radio broadcasts to satellites. The boldest also are moving into global telecasting.

One by one, the old-time, super-power shortwave stations are going dark. SWLs are beginning to fear an end to their hobby of listening to voices from other nations. However, they need not worry. International radio broadcasting is continuing to happen...by satellite.

Of course, shortwave transmitters and receivers use low frequencies from 3-30 MHz in the radio spectrum, while satellites broadcast at radically-higher frequencies around 4,000 and 11,000 MHz.

BBC, Radio France, Germany's Deutsche Welle and other long-time broadcasters transmit programs internationally via satellite 24 hours a day. Swiss Radio International and Radio Sweden program satellite channels in various languages. Radio Netherlands uses Panamsat's PAS-1 to broadcast digital audio to the Caribbean and Latin America. Sometimes these international broadcasts are picked up and repeated by local cable systems.

In the future, international broadcasters may share satellite channels. One channel would carry several English broadcasters, a different channel would broadcast French audio programs, another channel would relay German broadcasts, and so on. Local cable system operators would be able to mix and match international radio programs.

International DAB will be a path into other countries for cross-border broadcasters when small, inexpensive, portable, digital audio broadcast (DAB) satellite receivers have been distributed throughout the world. Nations will transmit through their satellites to audiences across international boundaries.

A Washington, D.C., organization known as Worldspace is starting a service called Afrispace to broadcast nine digital radio channels to the Middle East and Africa. Afrispace's satellite, Afristar-1, would be located at 12 degrees West. Portable $100 receivers would pick up its signals.

Radio Netherlands contracted with Afrispace for one of the channels, to broadcast digital signals 24 hours a day to Africa and the Middle East.

International Radio Satellite Corporation (RadioSat) is another U.S. company creating direct broadcast satellites for lease to international broadcasters. RadioSat plans to launch three high-power satellites, starting in 1995, with 200 channels each for broadcasters such as the Voice of America, BBC World Service and Radio Moscow.

Global television. Major international television broadcasters are linking the entire globe with large-scale satellite networks.

CNN, America's Cable News Network, broadcasts to North America via the high-power Galaxy-7 satellite and to Europe via the popular Astra satellite. The news network broadcasts directly into Asian and Middle Eastern homes via Japan's Superbird, Indonesia's Palapa, and the Arab League's Arabsat-1C. CNN relays news to the Mediterranean and southern Africa via Intelsat-601.

BBC World Service Television is a 24-hour news channel which grew out of international fascination with the Persian Gulf War in 1991, the so-called "first live TV war." Great Britain started the global service that same year by replacing an existing BBC-TV service to Europe relayed by Intelsat-601 (then known as Intelsat-6-F4). World Service Television expanded into Asian via ASIAsat in 1991. Service to Africa via Intelsat-601 began in 1992. It is relayed to North America on Intelsat-601 at 27.5 degrees West, C-band transponder 4. World Service Television will enter Japan and the Pacific, and Latin America, and may start an international entertainment channel.

CBC, the Canadian Broadcasting Corporation, and BBC World Service Television began exchanging news video daily in 1992. CBC broadcasts BBC news, weather reports and business data across Canada in English and French. CBC Newsworld, a 24-hour cable news channel, is distributed via the Canadian satellite Anik-E2, transponder 16.

Deutsche Welle, Germany's international radio broadcaster, has become a global satellite broadcaster, sharing satellite transponders around the world with the United States Information Agency (USIA) Worldnet. It broadcasts to Europe on Eutelsat-2-F1. It uses Intelsat-K to reach North America where it is repeated on Satcom-C4, transponder 5, and Spacenet-2, transponder 3. Deutsche Welle is relayed to Africa by Intelsat-601 at 27.5 degrees West and to Asia, the Pacific and Africa by Intelsat-505 at 66 degrees East. Deutsche Welle transmits to the Far East and the Pacific on Intelsat-508 at 180 degrees East.

TV5, a global French broadcaster, transmits to North America on Canada's Anik-E1, transponder 17, to Europe on Eutelsat-1-F1 and to Latin America on Panamsat-1. It broadcasts TV5-Afrique to the African continent via a Russian satellite Ghorizont-12 at 40 degrees East. Programs come from France's TF-1, France-2, France-3, Radio Canada in Quebec, Belgium's RTBF and Switzerland's SSR, and others.

Canal France International is relayed from France to the Middle East via Arabsat-1C and to Africa via Intelsat-505 and Intelsat-601.

MBC, the Middle East Broadcasting Centre, broadcasts to the Middle East and North Africa through Arabsats, to Europe via Eutelsats, and to North America.

Satellite Business Uses

Communications satellites provide myriad services to business in the 1990s. Some examples:

Pulsat is General Motors Corp.'s satellite communications system feeding data to dish antennas atop 10,000 automobile dealerships.

Pulsat lets GM headquarters communicate electronically with dealers, interconnect car sales operations and transmit electronic shop manuals, car and truck diagnostics, warranty and service data, and electronic vehicle invoicing.

The satellite network relays data back and forth between dealers and GM and broadcasts television programs instructing dealers about subjects from product information to sales and service skills.

Pulsat probably is the largest internal corporate satellite network in the world. Hughes Network Systems, a unit of Hughes Aircraft which is part of General Motors, provides the satellites. GM subsidiaries Electronic Data Systems (EDS) and GM Hughes Electronics built the network with a "Personal Earth Station" with radios, computers and six-ft. dish antenna at each dealership.

Chrysler serves another 5,000 automobile dealerships with its Chrysler Dealer Network.

ASTN is the independent Automotive Satellite Television Network operated by Westcott Communications Inc. of Texas and received by the GM and Chrysler dealer satellite networks.

Major retailers such as K-Mart, JC Penney, Silo, Sears, H-E-Butt Grocery Co., Pay 'N' Pak Stores and Mervyn's Department Stores improve store operations and customer service with satellite communications networks.

Mervyn's is based in Hayward, California, and has some 250 stores in fifteen states specializing in national brands and private-label casual and sports clothing at moderate prices.

Satellite ground stations in each Mervyn's store relay communications through GTE's Spacenet satellites. Mervyn's network carries inventory, personnel and other vital information from outlying locations to the Hayward hub. Point-of-sale information, including credit-card and check-verification information, are transmitted over the satellite network.

As Mervyn's expands, the satellite helps make store openings possible by providing a way to control costs, improve data flow and allow new stores to be added quickly to the network.

The store terminals form a V-SAT (very-small aperture terminal) network around GTE Spacenet's interactive Skystar system. It includes one-way television broadcasts from headquarters to stores.

GTE Spacenet, headquartered at McLean, Virgina, relays satellite communications for news organizations, businesses, educational institutions, and government agencies around the globe.

Montgomery Ward & Co., has a four-ft. satellite dish antenna on the roof and a V-SAT inside each of its 350 stores across the U.S.

A satellite 22,300 miles above Earth replaces long-distance telephone lines on the ground, relaying data from individual stores to a ground station outside Ward's Chicago headquarters.

The network, built by Hughes, also communicates price changes, check and credit card authorizations, inventory management, sales reports and in-store music.

Bankers use satellites to create nationwide automated teller machine (ATM) networks. The satellites eliminate computer "middle men" linking regional ATM systems. Banks also use satellites to relay financial information.

Huntington Bank of Columbus, Ohio, uses NASA's Advanced Communications Technology Satellite to move data to a check-processing center at Parma, Ohio. ACTS' narrow radio beam lets the bank use a portable, back-up ground station during breakdowns or disasters.

A financial company in Albany, New York, has its computer in Alaska while its offices are in Maine, Oregon and New York. It keeps in touch with all points by satellite. The firm uses fifteen-ft. satellite dish antennas at its data processing centers in Anchorage and Fairbanks to signal its offices in Augusta, Maine, Portland, Oregon, and Albany, New York. Data coded into radio transmissions is sent up to a satellite in Earth orbit. The satellite repeats them down to the offices on the ground.

Stamp collectors buy stamps at retail from stamp dealers who use space satellites to connect with other dealers across the United States.

The American Stamp Dealers Association (ASDA) lists merchandise for resale on a satellite bulletin-board (bbs) and sells goods to members at wholesale. Retailers selling baseball cards and coins also use the BBS.

Dealers advertised had been contacting each other through slower printed magazine, until a Lexington, Massachusetts, company figured out how to do it faster.

To use the satellite system, a dealer with a want ad telephones an operator who types the ad into a computer and transmits it to a communications satellite over North America. The satellite retransmits the information to FM radio stations across the country.

The stations can transmit invisible signals at the same time as they broadcast popular music for the general public. The hidden messages are decoded by special receivers in retail stores and fed to computers which display the want ads. If an ad catches a dealer's eye, he telephones the sender to close the deal.

Charities frequently take the high road today in search of treasure from big givers across North America. Rather than waste time going door to door, they cluster some potential big-bucks donors in motel ballrooms around the country and beam them a pitch via communications satellite.

CPAs must meet annual continuing education requirements to retain their certification. Local seminars via satellite make obtaining those credits easy. One professional class was telecast from St. Louis by satellite to 9,000 CPAs at 1,500 offices. The interactive broadcast allowed CPAs to ask questions of speakers. The American Institute of CPAs enrolls 300,000 certified public accounts.

Mobile Communications

You can drive the highways and byways of the world and stay in touch via satellite. Connecting mobile ground stations by satellite is one of the great growth industries of the 1990s, maybe even bigger than cellular telephones, digital paging, and personal computers.

American manufacturers of mobile satellite systems (MSS) are investing billions of dollars in industry infrastructure—satellites, ground stations and terminals. Two million MSS ground terminals may be moving around the U.S. by 1995.

A great variety of businesses can use mobile satellites to control field operations through constant location and status information on trucks, airplanes, oil rigs, pipelines, trains and ships.

Two-way mobile voice and data communication via satellite already is here. Mobile satellites are being used by trucking companies, banks, insurance companies, brokerage firms, shipping, airlines, manufacturers and oil companies.

Trucks have small on-board satellite dishes which transmit location, cargo temperature and other information automatically to satellites for relay to a transportation company. Headquarters can track all freight shipments instantaneous.

Drivers have keyboards to report traffic delays and send emergency messages to headquarters without stopping to make a phone call.

United Van Lines Safeguard Fleet and Snyder National Trucking Corp. use mobile satellites.

TMC Transportation Inc., a long-haul truck operator in Des Moines, Iowa, installed mobile-satellite ground terminals in its fleet of trucks.

American Mobile Satellite Corp.'s Skycell network gives TMC seamless two-way voice and fax dispatch communication and position reports nationwide, 24 hours a day in any weather.

TMC truckers use laptop computers for satellite terminals. They include Skycell communications transceivers and Navstar global-positioning system (GPS) satellite receivers. Messages can be sent between truck and dispatcher's office, between trucks, and to other locations worldwide.

GPS lets dispatchers know exactly where trucks are. Truck terminals can report positions to 40 destinations. Eight timers, which automatically report positions to different destinations on different schedules, can be programmed by the central office over the satellite link.

AMSC covers the continental U.S., Alaska, Hawaii, Puerto Rico, the Virgin Islands and 200 miles of coastal waters. AMSC is owned by Hughes Communications Inc., McCaw Cellular Communications Inc., Mtel Corp., and Singapore Telecom and General Dynamics.

Stock brokers like mobile satellites which give them "walkie-talkie" access via satellite to current stock market information.

Insurance adjusters can process claims in the field quickly via mobile satellite, a boon in hard times following major natural disasters such as hurricanes and earthquakes.

Emergency crews use mobile satellites to communicate in the aftermath of calamities.

Development of small and inexpensive satellite terminals makes use of satellites practical.

Companies selling mobile satellite systems include American Mobile Satellite Corp., C. Itoh, Eutelsat, Geostar, Hughes Network Systems, Ingenico, Inmarsat, INMOS International Plc, JRC, Locstar, Magnavox, Man Group, Motorola, Qualcomm, Sony, Telesat Mobile Inc., Thrane & Thrane, and Toshiba.

Inmarsat, the International Maritime Satellite Organization, organized in 1979 to provide satellite communications for ships at sea, serves 59 nations from its headquarters in London.

Inmarsat provides mobile telephone, telex, fax (facsimile) and data relays via several stationary satellites to some 9,000 ships and shore stations. It has expanded its services in several countries to include land-mobile and aviation communications.

The growing membership of Inmarsat includes Algeria, Argentina, Australia, Bahrain, Belgium, Brazil, Bulgaria, Canada, Chile, the People's Republic of China, Denmark, Egypt, Finland, France, Gabon, Germany, Great Britain, Greece, India, Indonesia, Iran, Iraq, Italy, Japan, Korea, Kuwait, Liberia, Malaysia, the Netherlands, New Zealand, Norway, Oman, Pakistan, the Philippines, Poland, Portugal, Russia and the CIS, Saudi Arabia, Singapore, Spain, Sri Lanka, Sweden, Tunisia, United Arab Emirates and the United States.

AMSC, mentioned earlier, is the American Mobile Satellite Corporation of Washington, D.C. AMSC is licensed by the U.S. Federal Communications Commission (FCC) to provide U.S. mobile communications via satellite.

Telesat Mobile Inc. (TMI) of Canada is licensed by the Canadian government to provide Canadian mobile communications via satellite.

British Telecom of London is a mobile communications satellite firm with connections in Australia, Hong Kong and Japan.

NASA researchers at the Jet Propulsion Lab at Pasadena, California, built a two-way-radio satellite antenna for cars from inexpensive off-the-shelf parts. The antenna would lock onto a stationary communications satellite, 22,300 miles out in space. The researchers strapped thc antenna to their station wagon and communicated up and down 300 miles of California roadways.

Inmarsat, which relays messages between ships at sea and home ports from above the western Pacific Ocean, was used for the antenna test. The antenna steered mechanically, keeping itself pointed toward Inmarsat by signal seeking and inertial reference.

In signal seeking, if the signal from the satellite is strong enough, the antenna simply locks onto that. On the other hand, the antenna's computer knows where it is on the ground and where the satellite is in the sky at all times. So, if you drive behind a hill or through heavy foliage which blocks the satellite signal, the antenna refers to its computer memory and points itself toward where it knows the satellite should be in the sky.

NASA thought the test was the first in which a steerable antenna on a passenger vehicle tracked a satellite in orbit, but amateur radio operators recalled doing the same trick a decade earlier with an OSCAR hamsat.

MSAT-X is NASA's name for its mobile satellite experiments. The agency hopes to make it possible for everybody to be in touch with somebody from anywhere anytime.

A fleet of mobile satellites would extend nationwide mobile phone service, complementing cellular telephones in cities. It also could provide data transmission to remote locales. Users might include forestry personnel and plane and ship passengers, as well as husbands and wives discussing the kids.

NASA expects private industry to take over the idea in the 1990s, building ground control stations and launching satellites.

Iridium is a cellular telephone system in the sky—a wireless global communications network for handheld telephone calls beamed through a constellation of 66 satellites to be launched to low-earth-orbit (LEO) after 1996.

People all over the world will be able to enjoy seamless voice and data communications via portable and mobile telephones and computers.

Motorola Inc. named its satellite network Iridium after the precious metal resembling platinum which has an atomic number of 77. Originally, there were to be 77 satellites, but improved antennas reduced the number needed.

The 66 satellites will work together as a digital switched communications network in space, handling both voice and data.

Iridium will boost the concept of a "global village" by offering real worldwide personal communications. People will communicate by phone from anywhere on land, at sea or in the air via portable radiotelephones. A caller using the

Iridium network will not need to know the location of a person being called. He merely will dial a person's number to be connected instantly.

Motorola Inc., of Schaumburg, Illinois, is known for two-way radios, cellular telephones, pagers, semiconductors and aerospace, automotive, defense and industrial electronics.

LEO satellites are not the same as stationary satellites commonly used for international communications. Iridium's low altitude will facilitate easy radio links with portable radiotelephones on Earth, needing only small antennas, not large satellite dishes.

Low satellites look down on small areas of Earth's surface at any one time. That permits the network to re-use radio frequencies, like cellular telephone systems.

There won't be a problem when one low-orbit satellite disappears over the horizon because the large number of satellites in space have an inter-satellite switching system, reminiscent of cellular telephone switching systems.

Iridium will use cellular principles, but it is expected to complement land-based cellular telephone systems. In high-population areas, terrestrial cellular systems probably will be the most efficient means of telephoning. Iridium will be efficient in remote, sparsely populated areas.

Iridium will be a alternative for terrestrial mobile telephone and telephone services in low-population areas. In underdeveloped areas of the world, Iridium could be a foundation for a local telephone system.

For airplanes and ships, Iridium will relay voice and data links as well as global positioning information, eliminating expensive, sophisticated on-board communications gear.

Iridium will not be dependent on land links in any one locality so it is likely to play a role in disaster-recovery efforts following hurricanes, earthquakes and other calamities.

Iridium satellites will be relatively small when compared with older commercial satellites. An Iridium will be about three feet in diameter and six feet tall, weigh about 700 lbs. and work five to six years in space.

Iridium satellites are called "smart" because of their ability to switch and route calls in space.

The numerous satellites will enable continuous line-of-sight coverage of any point on Earth's surface, as well as up to an altitude of 100 miles.

Individual satellites would be replaced at the end of their five-year working lives. An emergency replacement satellite could be launched on 36-hours notice.

Iridium telephones will be lightweight and portable, similar to Motorola's existing cellular telephones (Dyna-Tac). However, the phones will display latitude, longitude, altitude and Universal time (UTC, GMT). Base station and automobile and truck mobile telephones also will be sold.

Frequencies will be 1500-1600 MHz (1.5-1.6 GHz). In addition to voice, the digital system will relay 2400-baud data. Baud is short for bits-per-second (bps).

Uplink and downlink bandwidths will be 29 MHz and that can be expanded, if needed. Gateways and crosslinks will operate at 20 GHz.

Voices will be changed to digital signals for transmission through a satellite, then converted back to analog audio by the users handheld telephone and by the gateway making the public telephone connection.

Iridium and terrestrial cellular phones will be linked for seamless global communications.

Gateway ground stations will link Iridium satellites to public telephone networks. There will be twenty gateways around the globe, tracking each user's location, connecting users to local telephone networks and storing customer billing information.

Iridium will be able to support millions of users, with a capacity more than ten times greater than stationary satellites.

Iridium will be the largest and most sophisticated fleet of LEO comsats among several different satellite constellations which have been proposed by various companies.

Motorola has partners in the Iridium project in more than 20 countries, including Brazil, China, Hong Kong, Indonesia, Japan, Malaysia, Russia, Singapore, South Korea, Thailand and the U.S.

The international partners provide entry into markets in their regions, frequency allocations within their countries, and connections to their national telecommunications systems.

Iridium launches can be handled by a variety of rockets, including America's Delta and Atlas, Europe's Ariane and Russia's Proton. Three or more satellites could be launched in one flight. The U.S. Pegasus airplane-launched space booster could launch single satellites.

Lockheed Missiles and Space Co., Sunnyvale, California, is coordinating launch of the 66 satellites. The first five satellites in the Iridium mobile communications network are planned for launch in 1996.

Lockheed has experience in space, including one of America's most successful boosters, the Agena, which has flown more than 350 times. Lockheed built the U.S. Navy's solid-fuel Fleet Ballistic Missile carried by Polaris submarines in the 1950s and Poseidon, Trident-I and Trident-II.

Lockheed is designing a new rocket to ferry payloads weighing from 2,300 to 8,000 lbs. to Earth orbit. The new booster, called Lockheed Launch Vehicle (LLV), would be launched from Cape Canaveral, Florida, and Vandenberg Air Force Base in California.

Motorola said in 1993 it would launch some Iridium satellites on China's Long March rockets.

Lockheed is considering buying three launches to high-inclination orbits on Russian rockets—possibly powerful Proton boosters—from the Plesetsk launch site in Russia. Each flight could carry seven of the small satellites. Altogether, Proton could loft 21 of the 66 Iridium satellites.

Krunichev Enterprises sells commercial satellite launches for the Proton factory near Moscow. Export of American satellites to Russia is prohibited under U.S. technology-transfer laws, but exceptions have been made for China.

The Russian government would have to upgrade Plesetsk for commercial launches. Of course, that investment would create work for Russians idled by collapse of their defense industry.

Commercial launches also could be made by old USSR SS-19 intercontinental ballistic missiles (ICBMs) converted into space rockets.

Iridium orbits would be circular, polar, 475-mi.-high paths. Every point on Earth's surface would be in line of sight of one or more of the satellites continuously.

Viewing the constellation of satellites in space would reveal seven planes of eleven satellites each, all traveling in the same direction.

The seven planes would rotate around the globe together, towards the North Pole on one side of the Earth, crossing over the Pole, and traveling down to the South Pole on the other side of the Earth.

The eleven satellites in a plane would be spaced equally around their orbit. The satellites in planes 1, 3, 5 and 7 would be in phase with one another, while those in planes 2, 4, and 6 would be in phase with each other and halfway out of phase with 1, 3, 5 and 7.

To prevent collisions at the North and South Poles, there would be a "minimum miss distance." Each of the seven planes would be separated by 27 degrees. The seam between planes 1 and 7—plane 1 satellites going up on one side of the Earth and plane 7 satellites coming down in the adjacent plane—would be separated by 17 degrees.

"Telephone cells" would be the footprints of narrow radio beams on Earth. Each Iridium satellite will have a set of antennas projecting a pattern of 37 cells onto Earth's surface.

Each cell will originate as a very narrow radio beam from a high-gain antenna pointed at Earth's surface. Different cells will reuse frequencies for different customers on the same channel.

A satellite's 37 cell antennas will form a hexagonal pattern with one center cell surrounded by three rings of smaller cells. The three rings will consist of 6, 12, and 18 cells respectively. The cell antennas will be separated by a main antenna on each satellite.

Each of the 37 cells will be of the same shape and size. Each cell will be 400 miles in diameter. The aggregation of all cells on all Iridium satellites will cover Earth's surface.

To conserve energy, cells will be turned on only as needed to cover individual points on Earth.

Iridium cells will differ somewhat from terrestrial cellular phone cells. In a terrestrial system, a non-moving set of cells serves mobile users. Iridium is just the opposite. The users move at a slow pace, relative to the satellite which flies at 17,000 mph, so the users appear static while the cells move.

That makes it easier for a satellite to pass a mobile user from cell to cell. Because the satellite is traveling at high speed, most handoffs are in one direction as a caller travels from cell to cell. A handoff usually is to one of two cells, not to one of six adjacent cells as in a terrestrial system.

Crosslinks are connections between satellites. An Iridium satellite will be able to crosslink to the nearest satellite 2,500 miles ahead in its orbital plane and the nearest satellite 2,500 miles behind in its orbital plane.

The crosslinks will be by radio on frequencies near 20,000 MHz (20 GHz).

An Iridium satellite also could crosslink with as many as six of its relatives in other orbital planes. Such interplane crosslinks would vary in angle and distance up to 2,900 miles.

Signal would be a similar network of 48 communications satellites to compete with Motorola's Iridium network from the Russian space agency, NPO Energia.

The Russians found a recently-designed military communications satellite system left over from the Cold War and converted it for telephone service.

NPO Energia would use Cosmos rockets and Cyclone rockets to launch 48 lightweight, low-orbit Signals satellites to create a regional network over Europe and Russia by 1995. Global expansion might come later.

Telephone handsets would be built by the

Russian radio equipment research institute Radiopriborostroenia. NPO Energia formed the consortium KOSS to fund the commercial venture.

Odyssey is another worldwide comsat network relaying telephone calls, data and paging messages to mobile subscribers.

Odyssey is a twelve-satellite project of TRW, a Redondo Beach, California, conglomerate which builds high-tech equipment for space, automotive, defense and information markets. The Odyssey network would have the capacity to assist three million calls around the globe.

Satellite Builders

Several large North American and European electronics manufacturers build communications and broadcast satellites, including America's General Electric, Hughes, Martin Marietta, GTE and Loral companies, Canada's Spar, and Europe's Matra Espace, Aerospatiale, RCA, British Aerospace and Alenia. Here's a run-down on one of the manufacturers:

Hughes Aircraft Company is the world's largest manufacturer and private operator of commercial communications satellites.

Hughes' high-power communications satellite model known as HS-601 relays television and radio programs and general telephone, telex, fax and data communications. HS-601s have a cube-shaped body and two horizontal, three-panel solar array wings. Examples of HS-601s in space are Galaxy-4 and Galaxy-7.

HS-601s are built by Hughes Space and Communications Group (SCG) at El Segundo, California, and operated by Hughes Communications Inc. (HCI). Both are subsidiaries of Hughes Aircraft Company which is a unit of GM Hughes Electronics. GM is the automobile manufacturer General Motors.

Satellite communications networks are built for commercial customers by Hughes Network Systems, Inc., Washington, D.C., another subsidiary of Hughes Aircraft Company.

Hughes launched its first satellite in 1963, before it was owned by GM, and has built more than 40 percent of the satellites operating in space today. For instance, in 1993-94, Hughes built and launched communications satellites serving Asia, Australia, Brazil, Canada, Europe, Indonesia, Mexico and the United States.

Hughes builds satellites for its own use as well as for other satellite-operating companies and governments. For instance, both of Mexico's Morelos communications satellites were built by Hughes. They were launched in 1985 by U.S. space shuttles. With the first of those due to expire in 1994, the Mexican government's Secretariat of Communications and Transportation contracted with Hughes to build a pair of second-generation satellites to be known as Solidaridad.

Hughes' competitors bidding on the Mexican jobs were General Electric Co. of the U.S. and Matra Espace of France.

Solidaridad will have three times the capacity of Morelos. Solidaridads will relay television, telephone, data, facsimile (fax) and business network services in C-band and Ku-band as well as mobile communications in the L-band between 1,000 and 2,000 MHz (1-2 GHz).

Solidaridads will cover Mexico and the U.S. south. Their signals may reach as far north and east as Detroit, Chicago and New York. They definitely will reach east into the Caribbean and south to Columbia, Ecuador and Peru. In addition, they are likely to be seen even farther south at Buenos Aires, Montevideo and Santiago.

Hughes most popular communications satellite over the years has been the smaller, spin-stabilized model HS-376. A total of 37 HS-376s have been built over the years.

Hughes says its big HS-601 model is the most successful body-stabilized communications satellite in history. Since its introduction in 1987, HS-601s have been ordered by many companies and governments, including Aussat Pty. Ltd. of Australia, the European Satellite Society (SES) in Luxembourg, Telesat Mobile Inc. of Canada, American Mobile Satellite Corp., the U.S. Navy, and Hughes Communications Inc.

In one year alone, 1990, twelve companies and governments signed up with Hughes for HS-601s. Galaxy-7, Galaxy-4 and Astra-1C were the first HS-601s built at El Segundo and launched by Arianespace.

The Solidaridads were the 21st and 22nd HS-601s ordered from Hughes.

HS-601 satellites are sent to stationary orbits 22,300 miles above the equator where they extend electricity-generating solar-power wings and point antennas toward Earth.

Galaxy-7, an HS-601 model, was launched for Hughes in October 1992 on one of the European Space Agency's Ariane rockets from Kourou, French Guiana, to a stationary orbit over the equator at 91 degrees West longitude.

Astra-1C, also an HS-601, was launched by Ariane seven months later in May 1993 for Luxembourg's Societe Europeenne des Satellites (SES, the European Satellite Society) to stationary orbit at 19.2 degrees East.

Galaxy-4, another HS-601 weighing 6,574-lbs, was launched just six weeks after Astra, in June 1993, on another Ariane from Kourou to a Clarke Belt position at 99 degrees West longitude.

Galaxy-4 and Galaxy-7 relay communications across North America. Astra-1C relays SES's direct-to-home TV and radio broadcasts to Europe.

Ariane rockets are built and their launches are sold by the European Space Agency's commercial space-transportation company, Arianespace. The rockets blast off from ESA's Guiana Space Center on the northeast South American coast.

Armchair Astronauts

NASA makes it easy to be an armchair astronaut by televising complete launch-to-landing coverage of all American space shuttle flights.

NASA Select Television is transmitted in the clear for all to see via General Electric's stationary satellite Satcom-F2R at 72 degrees West Longitude. The transponder is 13; frequency is 3960 MHz; audio is at 6.2 and 6.8 MHz.

Edited replays of each flight day often are relayed by Telstar-301 at 96 degrees West Longitude, transponder 9, channel 18, frequency 4060 MHz.

Even when shuttles are not flying, NASA Select Television broadcasts a variety of space and astronomy programs on Satcom-F2R, transponder 13, seven days a week.

The NASA Select Television schedule is available on COMSTORE, the shuttle-mission TV schedule computer bulletin board (bbs) at (713) 483-5817. The schedule also can be found in the NASA area of the large commercial database service, CompuServe (Go SPACE).

Amateur Radio Satellites

Private groups of amateur radio operators around the globe have built and sent dozens of amateur radio communications and science satellites to orbit since OSCAR-1 was launched on December 12, 1961.

The high-tech spacecraft have been financed through donations of time, hardware and cash from hams in the United States, Canada, Great Britain, Australia, Russia, Germany, France, Italy, Japan, Brazil, Argentina, Belgium, South Korea, Finland, Israel, Mexico, South Africa and other nations.

OSCAR. A California group of amateur radio operators, calling itself Project OSCAR for Orbital Satellite Carrying Amateur Radio, built the first amateur radio satellite in 1961. Since then, most hamsats have been called OSCAR.

Amateur radio OSCARs are not the same as the U.S. Navy series of Oscar navigation satellites.

Radiosputnik, or RS for short, has been the name of most USSR and Russian amateur radio satellites. Three USSR hamsats were called Iskra, Russian for "spark."

British amateur radio satellites, built at the University of Surrey, are known as UoSAT. Japanese hamsats have been called Fuji which is Japanese for "wisteria."

Flourishing. The number of amateur radio satellites has been mushrooming. Only four were orbited in all of the 1960s. Six went to space in the 1970s. Seventeen amateur radio and amateur-related satellites were launched in the 1980s. In the early years of the 1990s, 21 amateur radio and amateur-related satellites were launched with another 18 planned for the later years of the 1990s.

Record launch years were 1981 and 1990, with eight hamsats each. Close behind was 1991 with four hamsats and eight amateur-related satellites launched. Most hamsats remain in orbit today and many still are in use.

Launches. Hamsats often receive free rides to space as ballast on U.S., Russian, European and Japanese government rockets which happen to be carrying other commercial or government satellites to orbit. However, with available space over-booked, paid tickets sometimes are required today.

Orbits. Most hamsats have been what radio amateurs call Phase-1 and Phase-2. They fly in north-south polar orbits or low east-west equatorial orbits from 200–1,000 miles altitude. That's just a bit higher than a space shuttle usually flies.

A low-flying hamsat circles the globe, coming within range of a ham station on the ground every hour or so. It stays overhead only 15 to 30 minutes. Polar-orbit satellites come within range of a ground station about the same time every day.

A few amateur satellites are in long elliptical orbits which keep them in view of ground stations for hours at a time. Radio amateurs refer to such hamsats as Phase-3. They range out 20,000-30,000 miles from Earth, but loop back around the planet, within 1,500-2,500 miles, every ten to twelve hours. Their long elliptical tracks are known as Molniya orbits after a class of Russian communications satellites which follow similar paths through space.

A Phase-4 hamsat would be an OSCAR in stationary orbit. There probably won't be one until after the year 2000.

Sky-high repeaters. For the most part, hamsats are communications repeaters in the sky. Their transponders relay voice, Morse code and digital-computer signals. Most amateur satellites carry gear for digital computer-to-computer communication and store-and-forward message bulletin-board systems (bbs). Sometimes they have transmitters for radio propagation tests, ionospheric research, radioteletype and meteor sounding; receivers for radioastronomy, radiolocation and other original science research; and television cameras for photos of Earth.

Hamsats are open for use by all appropriately-licensed amateur radio operators around the world. The satellites serve the public, as well, by training satellite trackers, relaying medical data, teaching school science groups and providing emergency communications for disaster relief.

Beacons. Early amateur satellites carried only one-way radio beacons which sent down telemetry information about conditions of satellite equipment and the space environment to anybody interested in receiving the data. Hamsats of the 1990s still transmit beacons, along with two-way communications transponders.

A hamsat monitors its solar cells and battery. Its telemetry beacon reports the amount of current being generated by the solar cells, the voltage available from the battery, the temperature of the battery, transmitter power, temperatures of other parts of the satellite, and other useful information. Such telemetry data is easily read by amateurs.

Ragchewing. Along with building their spacecraft, hamsat users enjoy an old-fashioned "rag chew" via OSCAR. Contests and achievement awards add spice to their time on the air.

Citations include Worked All States for contacting a station in each of the United States via satellite, Satellite DXCC for contacting hams in 100 countries via satellite, and a series of technical achievement awards.

Radio amateurs love to track down hidden transmitters in "fox hunts." In a SatFox Test, amateurs at home simulate a hidden transmitter hunt using a hamsat to find a hidden "fox."

ZRO Technical Achievement Award is earned for superior station performance in a sensitivity test of receiving weak satellite signals.

Weekly amateur radio meetings on the air, known as "nets," spread the latest news of the amateur space program.

AMSAT. The Radio Amateur Satellite Corporation (AMSAT) was founded in 1969 as a group of ham operators around the globe who want to communicate by satellite. AMSAT groups have constructed and operated numerous OSCARs.

Information is available from AMSAT, P.O. Box 27, Washington, D.C. 20044.

ARRL. Information also is available from the American Radio Relay League (ARRL), the national amateur radio fraternity at 225 Main Street, Newington, Connecticut 06111.

TAPR. Tucson Amateur Packet Radio (TAPR) was founded by members of an Arizona chapter of the IEEE Computer Society to develop amateur packet radio, including hamsat systems.

Information is available from TAPR at P.O. Box 12925, Tucson, Arizona 85732-2925.

Moonbounce

Earth's natural satellite, the Moon, has been used for communications as a passive radio signal reflector many times in recent decades.

However, before 1946, no one knew if radio waves could pass through Earth's ionosphere. Wondering if missiles could be detected above the ionosphere, the Pentagon ordered Project Diana, named after a mythical Moon goddess.

Meanwhile, fifteen-year-old John H. DeWitt Jr. of Nashville had become an amateur radio operator in 1921. He built his town's first broadcasting station in 1922. By 1940, he was using a radiotelescope to tune in natural radio noise generated by the Milky Way galaxy.

DeWitt joined the U.S. Army Electronic Branch in 1942. Two years later, he became director of the Evans Signal Laboratory at Belmar, New Jersey, and went to work on Project Diana.

The project's radar gear was quite different from common radio equipment of that time. It had a big antenna composed of 64 small antennas which could be pointed over the Atlantic Ocean toward the rising Moon. DeWitt found success on January 10, 1946, when the 112 MHz radar signal from the Signal Corps 3,000-watt transmitter reached the Moon and he was able to detect its faint reflection coming back to Earth.

The technique is known today as Earth-Moon-Earth (EME) or Moonbounce. Signals of very high power are transmitted from antennas pointed at the Moon. The Moon acts like a giant radio mirror in the sky, bouncing the radio waves back to Earth to be received by listeners.

The first ham radio signals to echo from the Moon were transmitted in 1953. Today, radio amateurs regularly beam signals to the Moon with their reflections painting a large area of our planet.

Hamsat Chronology

Here is a summary in chronological order of all amateur radio satellites, as well as all non-amateur-but-amateur-related satellites:

1961: OSCAR-1, The First Hamsat

Just four years after the USSR launched the Space Age with its Sputnik artificial satellite, a

10-lb. Orbital Satellite Carrying Amateur Radio, or OSCAR for short, was launched December 12, 1961, as ballast on a Thor-Agena rocket which carried the U.S. military satellite Discoverer-36.

The rocket left OSCAR in an elliptical orbit ranging from 152 to 295 miles above Earth's surface, just above our planet's atmosphere.

That first tiny 11-lb. hamsat measured 9 in. by 12 in. by 6 in. tall. OSCAR did not offer two-way communications. Its radio only transmitted HI in International Morse code with 140 milliwatts of power on a frequency of 144.983 MHz—fourteen times the power of the 10-milliwatt radio in Explorer-1, America's first satellite.

There was a bit of scientific value in OSCAR's BEEP-BEEP-BEEP-BEEP BEEP-BEEP greeting. The speed of the message was controlled by the temperature inside the satellite.

OSCAR's battery wasn't rechargeable and had only enough strength to power the transmitter for 22 days. During that time, hundreds of amateurs in 28 nations around the globe picked up OSCAR's call from space and mailed in reception reports.

The satellite's low altitude let it stay in orbit above Earth only 50 days. OSCAR slipped down into the atmosphere and burned January 31, 1962.

1962: OSCAR-2

The second hamsat was very similar to the first. OSCAR-2 was launched to a 240-mile-high orbit June 2, 1962, just six months after OSCAR-1.

OSCAR-2's transmitter power was lowered to 100 milliwatts to make its battery last longer. However, it operated only 19 days and fell into the atmosphere June 21, 1962.

Another OSCAR, built about the same time with a 250-milliwatt transmitter, wasn't launched.

1965: OSCAR-3

The third hamsat was the first-ever active telecommunications satellite with free access to all. OSCAR-3 was launched to a 590-mile-high orbit on March 9, 1965. It was the first hamsat to have a two-way signal-repeating transponder making it a radio-relay station in the sky.

OSCAR-3's transponder received signals and used a 1-watt transmitter to repeat those signals. The transponder worked for 18 days in space.

The satellite also had two radio beacons. One sent a continuous signal for tracking and propagation studies. The other sent telemetry data about temperatures and battery voltages.

More than 1,000 hams from 22 nations chatted via OSCAR-3 during its 18 days of operation. The first trans-Atlantic ham satellite link was made. The hamsat also carried the first direct contacts with hams in Eastern European countries such as Bulgaria and the nation known at that time as Czechoslovakia. Other long-distance (DX) contacts linked New Jersey with Spain, Massachusetts with Germany and New York with Alaska.

A few solar cells were attached to the satellite to recharge the battery powering the beacons, thus OSCAR-3 became the first amateur spacecraft to use solar power. However, the satellite was not fully solar powered. The solar cells allowed the beacons to continue transmitting months longer than the transponder. Without electricity, its radios are dead, but the satellite remains in orbit.

1965: OSCAR-4

To mark the fourth anniversary of OSCAR-1, the fourth hamsat was launched to orbit December 21, 1965. Unfortunately, OSCAR-4 was the first amateur satellite to have a partial launch failure.

The 30-lb. satellite was blasted into space aboard a Titan 3-C rocket, but the rocket's upper stage failed and the satellite did not make it to its intended 21,000-mi.-high circular orbit. Instead, the satellite ended up in a highly-elliptical orbit ranging from a low point of 122 miles altitude out to 20,875 miles.

OSCAR-4 was the first satellite to be powered fully by solar-cells generating electricity. It also was the first hamsat to use two bands, receiving signals on 144 MHz and transmiting three watts of power on 432 MHz. The first U.S.-to-USSR satellite contact was made through OSCAR-4.

No telemetry beacon was included, so hams were unable to know why OSCAR-4's radio failed after a few weeks. The battery may have overheated or radiation may have knocked out the solar cells.

The radio operated only 85 days. Then, in May 1966, the U.S. worldwide satellite tracking network lost track of Oscar-4, but the small spacecraft was found again April 15, 1972. The satellite stayed in space almost eleven years, falling into the atmosphere April 12, 1976.

Project Oscar was the U.S. West Coast group which designed, constructed and launched the first four OSCARs. In 1969, the Radio Amateur Satellite Corp. (AMSAT) was formed by an East Coast group of amateurs to build and fly hamsats.

1970: OSCAR-5

OSCAR-5, also known as Australis-OSCAR-5, was designed and built by students in the Astronautical Society and Radio Club at the University of Melbourne, Australia.

AMSAT managed launch of the satellite January 23, 1970, from Vandenberg Air Force Base, California, to a 925-mile-high polar orbit aboard a Delta rocket ferrying an American weather satellite to space.

OSCAR-5 had telemetry beacons transmitting seven kinds of data about the satellite at 29 and 144 MHz. It was the first amateur satellite to be controlled from the ground. It contained a command receiver which allowed ground stations to turn on and off its 29 MHz beacon transmitter.

OSCAR-5 had no solar cells and no two-way transponder, but it did have a magnetic attitude-stabilizing system. Telemetry operated only 52 days, however OSCAR-5 remains in space.

1972: OSCAR-6

OSCAR-5 was the last of the first generation of hamsats. OSCAR-6 started a second generation of amateur radio satellites known as Phase-2.

Parts of OSCAR-6 had been built in the U.S., West Germany and Australia and critical parts had redundant back-ups. It was launched to a 900-mile-high orbit alongside a government weather satellite on October 15, 1972.

The satellite's two-way communications transponder received signals from the ground on 146 MHz and repeated them at 29 MHz with a transmitter power of one watt. Low-power ground stations with simple antennas were successful in using the satellite.

OSCAR-6 had a sophisticated telemetry beacon which reported information about many parts of the spacecraft, including voltages, currents and temperatures. Where OSCAR-5 had seven kinds of data reported in its telemetry beacon, OSCAR-6 had 24. OSCAR-6 had a magnetic attitude-stabilizing system.

OSCAR-6 also had an elaborate ground-control system to turn off parts of the satellite selectively. The hamsat could react to 35 different commands from ground stations in Australia, Canada, Great Britain, Hungary, Morocco, New Zealand, West Germany and the United States.

Codestore was a digital store-and-forward message system built into the hamsat. Ground controllers in Canada sent messages to the satellite which were stored and repeated later to ground control stations in Australia.

Static in the satellite affected its computer which read the noise as a command to shut down. To overcome the problem, controllers sent a continuous stream of ON commands to the satellite to keep it turned on. Where OSCAR-5 had been commanded twice a week, OSCAR-6 received 80,000 a day. The trick worked. The solar-charged batteries allowed the radio to work 4.5 years in orbit. OSCAR-6 remains in space today.

1974: OSCAR-7

Amateur radio had two working satellites in orbit for the first time after OSCAR-7 was launched November 15, 1974. It was a second Phase-2 satellite, similar to OSCAR-6, but with improvements. For instance, OSCAR-7 had two transponders. One received at 146 MHz and repeated what it heard at 29 MHz while the other listened on 432 MHz and relayed the signals on 146 MHz. The latter had an eight-watt transmitter and was built by radio amateurs in West Germany.

In the first satellite-to-satellite link-up in history, a ham transmitted to OSCAR-7 which relayed the signal to OSCAR-6 which repeated it to a different station on the ground.

Australians built a telemetry encoder for the satellite and Canadians built a 435 MHz beacon.

Other beacons were at 146 MHz and 2304 MHz. The 2304 MHz beacon, with a transmitter power of 100 milliwatts, was built by the San Bernardino Microwave Society of California. Unfortunately, the FCC denied the hams permission to turn on their 2304 MHz beacon so it never was tested in space.

OSCAR-7's radio system worked 6.5 years. The satellite remains in space.

1978: OSCAR-8

The third Phase-2 hamsat was launched March 5, 1978, on a Delta rocket from Vandenberg Air Force Base, California, to a circular 570-mile-high polar orbit. OSCAR-8 spun around the globe every 103 minutes.

It had two transponders, including one designed by Japanese radio amateurs. It listened at 146 MHz and repeated what it heard through a transmitter on 435 MHz. The rest of the hardware was built by American, Canadian and West German amateurs.

Phase-2 fun continued for 5.3 years while OSCAR-8's radio worked. Unfortunately, its batteries died in the middle of 1983. OSCAR-8 remains in orbit today.

1978: Radiosputniks 1 and 2

The USSR's Sputnik had launched the Space Age in 1957, but Russian hams had been only spectators as Western radio amateurs launched seven OSCARs between 1961 and 1978.

In the mid-1970s, Soviet amateurs found themselves among engineers visiting the United States in preparation for the 1975 Apollo-Soyuz joint flight. While visiting NASA's Goddard Space Flight Center, Russian hams discussed OSCARs with AMSAT members.

Word of a USSR amateur satellite program leaked from behind the Iron Curtain in a 1975 article in the Russian electronics magazine, *Radio*. Transponders were said to be under construction in Moscow and Kiev.

In 1977, the USSR government notified the

International Frequency Registration Board that a series of hamsats would be launched. Radio magazine Radiosputnik (RS) hamsats.

A Russian F-2 rocket blasted off October 26, 1978, from the Northern Cosmodrome at Plesetsk carrying a government satellite and the first two Soviet hamsats—RS-1 and RS-2—to elliptical orbits 1,000 miles above Earth.

Each had a 145-to-29 MHz transponder. The satellites, sometimes also referred to as Radio-1 and Radio-2, circled the globe every 120 minutes.

The satellites transmitted a telemetry beacon in Morse Code, reading out temperature and voltage data. The hamsats had solar cells as well as a Codestore message store-and-forward mailbox. Ground control stations were at Moscow, Novosibirsk and Arseneyev near Vladivostok.

The satellites had very sensitive receivers and overload fuses to flip off whenever a ham on the ground would use excessive transmitter power. The circuit breaker could be reset from the ground when over the USSR. Western hams, transmitting thousands of watts of power, kept tripping the fuses and turning the Radiosputniks off. The Russian ground controllers kept resetting the circuit breakers, but most operation ended up being over the Soviet Union since Western hams kept shutting off the transponders when the satellites were over North America and Western Europe.

The Radiosputniks each weighed 88 lbs. and were cylinders 17 in. in diameter and 15 in. long. The batteries in RS-1 lasted only a few months. However, RS-2 was heard until 1981. Today, they are dead in orbit at altitudes around 1,050 miles.

Even in today's openness, we don't know for sure there was not a failed Radiosputnik prior to RS-1. We do know many more Radiosputniks were to come as well as three USSR hamsats known as Iskra or "spark" in Russian.

1980: Phase-3A

At the end of the 1970s, as more and more amateur radio operators were sharing the fun of talking via satellite, AMSAT began work on a new generation of larger Phase-3 satellites.

The third-generation would be more complex spacecraft, using higher radio frequencies, and flying Molniya orbits. Phase-3 satellites would be over ground stations for hours at a time.

The first Phase-3 spacecraft, nine years in planning and four years in construction, was built, integrated, and tested at Goddard. Hams in Canada, Hungary, Japan, West Germany and the United States built parts for the spacecraft.

The satellite was to be launched on the second flight of Europe's new Ariane rocket from a site outside Kourou, French Guiana, on the northeast coast of South America. In a promising development, the first Ariane rocket made a successful flight from Kourou in December 1979.

Unfortunately, the second Ariane and Phase-3A were destroyed May 23, 1980, in a European Space Agency launch failure during liftoff from Kourou.

With the launch window open that day, breakdowns and rain in French Guiana forced three hours of holds. Finally, a go order was signaled and the Ariane fired. Three minutes into the flight, as the Ariane was lumbering upward from the South American coastline, the Ariane's first stage failed. The Phase-3A hamsat fell to the Atlantic Ocean and sank hundreds of feet to the bottom.

Within weeks, AMSAT began work on Phase-3B which would go to space in 1983.

1981: Iskra 1

Most USSR amateur radio satellites were called Radiosputnik, but three had a different name, Iskra or "spark" in Russian. Students and radio amateurs at Moscow's Ordzhjonikidze Aviation Institute built the 62-lb. satellites, each powered by solar cells. Each had a transponder, telemetry beacon, ground-command radio, Codestore message bulletin board, and computer with memory.

The satellite transponders received at 21 MHz and transmitted at 28 MHz. Their telemetry beacons were near 29 MHz.

Controlled by ground stations at Moscow and Kaluga, Iskras were intended for communication among Eastern Bloc hams in Bulgaria, Cuba, Czechoslovakia, East Germany, Hungary, Laos, Mongolia, Poland, Romania, USSR and Vietnam.

Iskra-1 was launched July 10, 1981, on an A-1 rocket from the Northern Cosmodrome at Plesetsk to a 400-mi.-high polar orbit. After 13 weeks, it burned in the atmosphere October 7, 1981.

1981: UoSAT-OSCAR-9

Great Britain was an early leader in spaceflight and satellite technology, launching its Black Knight satellite from Australia in 1971. Students at the University of Surrey were eager to get hands on experience. Those who were radio amateurs designed and built a Phase-2 hamsat known as UoSAT, short for University of Surrey Satellite.

The 115-lb. science and education satellite was blasted to a 340-mi.-high polar orbit October 6, 1981, on a U.S. Delta rocket from Vandenberg Air Force Base, California.

In space, the hamsat was called UoSAT-OSCAR-9, or simply UO-9. Sometimes it was referred to as UoSAT-1.

UO-9 transmitted data and had a television

camera which sent pictures to Earth with 2 km or 1.24 mi. resolution. The satellite had one of the earliest two-dimensional charge-coupled device (CCD) arrays forming the first low-cost CCD television camera in orbit. The resulting images transmitted from space were spectacular, considering the freshness of its technology. UO-9 was not a stabilized Earth-pointing satellite so the areas covered by its photos were random.

The hamsat had a magnetometer and detectors for radiation and particles. Two particle counters detected solar activity and auroras as they interfered with radio signals.

UO-9 also had a synthesized radio voice with 150-word vocabulary to read out spacecraft condition reports. Students around the world learned about space science by tuning receivers to OSCAR-9's so-called Digitalker.

The hamsat's radio beacons transmitted at 145 and 435 MHz. For propagation studies, there were additional beacons at shortwave frequencies near 7, 14, 21 and 28 MHz and microwave frequencies near 2 and 10 GHz.

UO-9 had Codestore for messages and a voice synthesizer for educational demonstrations. A control computer in the satellite could be reprogrammed from the ground. The satellite did not have a transponder for general chatting.

In 1982, a software error mistakenly turned on both the 145 and 435 MHz beacons at the same time, preventing the hamsat's receiver from hearing signals from controllers. Surrey hams called on radio amateurs at Stanford University, California, to override the jamming. Stanford hams used a 150-ft. dish antenna to transmit power equal to 15 million watts toward the satellite. It worked. After six months of running out of control in space, OSCAR-9 heard and understood what it was to do. Later, after eight years in orbit more than 300 miles above Earth, OSCAR-9 burned in the atmosphere October 13, 1989.

1981: Radiosputniks 3-8

As the 1980s dawned, the amateur radio club at the University of Moscow was busy building a covey of new amateur radio satellites.

Like RS-1 and RS-2, the six new Radiosputniks each weighed 88 lbs. and were cylinders 17 in. in diameter and 15 in. long.

The Soviet government launched the sextuplet December 17, 1981, on one C-1 rocket from the Northern Cosmodrome at Plesetsk to 1,000 miles altitude. At that time, it was the largest clutch of amateur radio satellites ever orbited at one time.

Designated Radiosputnik-3 (RS-3) through Radiosputnik-8 (RS-8), they were in orbits similar to those used by RS-1 and RS-2. They circled the globe every 119 minutes. The satellites sometimes also were referred to as Radio-3 through Radio-8.

The six satellites had transponders receiving at 145 MHz and transmitting at 29 MHz. They also had store-and-forward mailboxes, solar cells and a Morse Code temperature and voltage data beacon.

Some carried an "autotransponder" electronic robot operator. Hams on the ground could call a satellite and the robot would respond with a greeting and signal report.

The usefulness of each RS satellite ended as its battery failed. RS-5 and RS-7 were able to stay on the air until 1988. Today, all six are dead in orbit at altitudes around 1,000 miles.

1982: Iskra 2

The USSR's Salyut-7 space station was launched to Earth orbit April 19, 1982, with the second 62-lb. Iskra bundled up inside. Cosmonauts Anatoli Berezovoi and Valentin Lebedev blasted off from Baikonur Cosmodrome May 13 in a Soyuz transport. They docked at Salyut-7 two days later on May 15 and opened up the new station.

The cosmonauts unwrapped Iskra-2 and pushed it out an airlock on May 17 at an altitude of 210 miles. Moscow TV showed live coverage of that "hand launch" allowing the Ordzhjonikidze Aviation Institute students to see their satellite go into its own orbit.

Iskra-2's telemetry was at 29 MHz. Since it started life in such a low orbit, the satellite was able to remain in space only about seven weeks before burning in the atmosphere July 9, 1982.

1982: Iskra 3

On November 18, Berezovoi and Lebedev, hand launched Iskra-3 from an airlock at an altitude of 220 miles. Even though Iskra-3 was much like Iskra-2, it suffered from internal overheating and didn't work as well. Iskra-3's telemetry beacon also was at 29 MHz. The third Iskra remained in space only four weeks before descending into the atmosphere and burning December 16, 1982.

1983: AMSAT-OSCAR-10

Shortly after its Phase-3A satellite sank in the Atlantic in 1980, AMSAT started work on Phase-3B. The 200-lb. clone was built mostly by German hams and launched on an Ariane June 16, 1983. It was named AMSAT-OSCAR-10.

Seconds after dropping off AO-10 in orbit, the Ariane bumped it, damaging an antenna on the hamsat and spinning the satellite wildly away. AMSAT had to wait for the satellite to stabilize in space before firing an internal thruster to change the orbit July 11.

The thruster didn't shut off as ordered and blasted 50 percent longer than planned. That threw the satellite into an exaggerated orbit taking it nearly twice as far away from Earth as planned.

Another kick-motor firing was attempted July 26, but helium had leaked from the satellite after the Ariane bump and fuel valves didn't operate. AO-10 ended in an uncontrollable orbit ranging from 2,390 miles to 22,126 miles.

The damaged antenna wouldn't work right. The orbit, farther from Earth than planned, exposed the satellite to more radiation damage. The incorrect attitude kept solar panels from orienting toward the Sun so batteries couldn't charge properly.

AO-10's transponders worked, but the broken antenna and low inclination made it less useful. Its signals were weak. Access time was limited. Even so, hundreds of radio amateurs used AO-10.

In 1986, intense subatomic particles trapped in Earth's magnetic field bombarded AO-10's computer memory chips, leaving false information behind. The memory began to turn up mysterious data bits and the satellite became harder to control.

AO-10's transponders switched off from time to time as voltage dropped when sunlight was low. The satellite required solar illumination 90 percent of the time, but sometimes received only 50. Then AO-10 would turn itself off and a command station would be required to transmit a reset order.

After the AO-10 launch, AMSAT started building a third Phase-3 satellite to be launched in 1988. AO-10 continues to transmit signals, but satellites launched later attracted most of its users.

1984: UoSAT-OSCAR-11

The second science and education satellite built by students at England's University of Surrey was UoSAT-B, launched March 2, 1984, from California to a 430-mi.-high polar orbit. The 132-lb. hamsat was renamed UoSAT-OSCAR-11 (UO-11). It also has been called UoSAT-2.

UO-11's beacons transmit at 145, 435 and 2401 MHz. It handles messages while photographing aurora over the Poles with a sensitive camera which stores the images in memory.

Digital telemetry beacons relay news bulletins from AMSAT and UoSAT, which is headquartered at the Spacecraft Engineering Research Unit, Electrical Engineering, University of Surrey, Guildford, Surrey, GU2 5XH, England.

1985: NUsat

Northern Utah Satellite (NUsat) was not an amateur-radio satellite, but an amateur-built satellite used as a radar calibration target. It was an important step into space for a college which later became a university and built other satellites.

North Pole Ski Trek

UO-11's Digitalker held listeners around the globe spellbound in 1988 as it guided skiers across the North Pole wilderness from Russia to Canada.

Four Canadians and nine Russians moved north onto smooth land ice beyond Cape Arctic at the top of the USSR's Severnaya Zemlya Islands March 3 for a three-month 1,075-mile trek across the frozen Arctic Ocean to Cape Columbia at the northern tip of Ellesmere Island in the Canadian Arctic.

It was the first trek across the Arctic without vehicles or dogs, and the first to be guided across the ice pack by a talking satellite. Their 100-lb. backpacks held bacon, dried fish and other high-protein, high-fat foods to maintain energy and body heat. The expedition observed glaciers and weather, measured geomagnetism and tested physiology and biochemistry limits of endurance and isolation. Obstacles included storms, open water, thin ice, pressure ridges and temperatures to minus-50 degrees Celsius.

As the skiers crossed the top of the world, they were in touch with Russian and Canadian ham radio stations at Resolute Bay on Cornwallis Island in Canada's Northwest Territories; Severnaya Zemlya; Ottawa; Toronto; Moscow; an ice island base; and Russia's North Pole Station 28.

They trekked a dozen miles over ten hours each day, then set up a twelve-man tent, ate, talked on their radio and slept. Mornings, they ate a quick breakfast, took down their tent, and turned on their location transmitter. They varied their diet of dried-meat pemmican with fruit, eggs and steak.

The party's position was computed thousands of miles away at a ground station and sent up from Surrey to the hamsat UO-11.

The skiers carried miniature receivers to hear the hamsat's computerized voice. The satellite's digital voice, known as Digitalker, announced the team's latitude and longitude in plain English around the clock. The satellite was in a polar orbit, so the trekkers could hear it every 100 minutes.

The team took 55 days to reach the North Pole where temperatures were 22º below zero. They arrived April 24 to be greeted by press and politicians flown in from Canada and the USSR. Then the team made landfall June 3 at Ward Hunt Island just north of Cape Columbia.

In 1990, UO-11 again supported an expedition, this one pulling sleds to the North Pole.

Students at a Center for AeroSpace Technology (CAST) at Weber State College, Ogden, Utah, designed NUsat. The 26-sided polyhedron satellite was built by Morton Thiokol, Inc., Brigham City, Utah, for WSC in coordination with the Federal Aviation Administration (FAA).

NUsat was ferried to an altitude of 219 miles in

shuttle Challenger and dropped overboard April 29, 1985—the first payload ejected from a NASA GAScan to an orbit apart from a shuttle's orbit.

A GAScan is a Get Away Special Canister, a container for small payloads. With the 115-lb. NUsat in it, the GAScan weighed 433 lbs.

NUsat was used to calibrate air traffic control radar systems by measuring antenna patterns for ground-based radars operated in the United States and in member countries of the International Civil Aviation Organization. After 595 days in orbit, NUsat burned December 15, 1986. CAST later built another satellite, WO-18, for launch in 1990.

1986: Fuji-OSCAR-12

Japanese radio amateurs built their first Japan Amateur Satellite (JAS-1a) and sent it to a 932-mi.-high orbit August 12, 1986, from Japan's Tanegashima Space Center. AMSAT labeled it OSCAR-12. Japanese hams called it Fuji, or wisteria. The Japanese name their satellites after flowers. It came to be known as Fuji-OSCAR-12.

FO-12 had a transponder which received at 145 and retransmitted at 435 MHz. Primarily a pacsat, Fuji's transponder could be used either as a message bulletin-board or as a voice repeater. Fuji's telemetry beacon sent data in 20-words-per minute Morse.

The mailbox in the sky received typewritten messages from individual ham stations and stored them in a 1.5 megabyte RAM memory. This electronic message center permitted amateurs on one side of the world to place messages on the satellite's bulletin board to be removed by others when the satellite was on the far side of the globe.

Users were disappointed when FO-12's solar generator was unable to produce sufficient electricity for Fuji's battery. Japanese controllers were forced to turn the satellite off November 5, 1989. FO-12 remains in orbit and a replacement hamsat, JAS-1b, was launched in 1990.

Radiosputnik-9

Soviet hams wanted to launch a new hamsat in the mid-1980s, but launch of Radiosputnik-9 was delayed repeatedly. Finally, the flight was canceled and the number RS-9 retired permanently.

1987: Radiosputniks 10 and 11

Soviet hams delighted the amateur satellite world June 23, 1987, with the launch of a combo package of hamsats, RS-10 and RS-11, aboard one large government spacecraft.

Radiosputnik-10 and Radiosputnik-11 went to a 621-mi.-high circular orbit as part of the Russian navigation satellite Cosmos 1861 which circles the globe every 105 minutes.

Part of a Russian system similar to the U.S. Navstar series of global positioning satellites (GPS), Cosmos 1861 helps Russian fishing fleets locate themselves on the world's oceans.

RS-10 and RS-11 were unique hamsats, at the time, in sharing space aboard the navsat. Hamsats usually get free piggy-back rides to space on government rockets, but in space they are separate payloads dropped off in at least slightly different orbits. By comparison, Cosmos 1861, RS-10 and RS-11 are just one big package with the ham-radio sections sharing electrical power generated by the Cosmos 1861 solar wings.

RS-10 and RS-11 telemetry beacons transmit near 29 and 145 MHz. The satellites have identical shortwave and vhf transponders, but frequencies are different. Hams on the ground send signals to RS-10 and RS-11 on frequencies near 21 and 145 MHz. Downlink signals from the satellites are at 29 and 145 MHz. Cosmos 1861 also has a transmitter, at a frequency near 150 MHz.

A robot radio operator, known as an autotransponder, is built into RS-10 and RS-11. To contact one of the robots, a ham on the ground transmits Morse code on 21 or 145 MHz. The satellite's computer returns a greeting, signal report and contact number on its telemetry beacon frequency, as if it were a human operator.

1988: AMSAT-OSCAR-13

With its second Phase-3 satellite safely in orbit in 1983, AMSAT started work on a third which was launched on an Ariane rocket June 15, 1988. Phase-3C was renamed AMSAT-OSCAR-13.

In its Molniya-style elliptical orbit, AO-13 swings from as close as 1,500 miles out to 22,000 miles from Earth.

AO-13 was the most complex amateur satellite to date. It offered four transponders for packet, facsimile (fax), slow-scan television (sstv), voice (ssb), radioteletype (rtty) and Morse code (cw).

Transponders receive at 435 and 1269 MHz and retransmit at 145, 435 and 2400 MHz. The satellite's computer follows a calendar in switching among transponders modes. Telemetry beacons transmit data via packet, cw and rtty.

AO-13 provides near-hemispheric coverage linking continents for up to eight hours at a time. During weekly on-the-air meetings, commentators talk about satellite topics via AO-13.

The hamsat delivered essential communications in 1992 after Hurricane Iniki leveled parts of Hawaii. After normal public circuits broke down, AO-13 and other hamsats moved health and welfare messages from devastated areas to anxious relatives and friends.

Solar power makes sunlight important. Even a partial eclipse of the Sun by the Moon can affect a satellite. During a 1992 eclipse, sunlight reaching AO-13 was cut 40 percent. Twilight was reported in the satellite's housekeeping telemetry as a drop in solar panel temperature. The low illumination necessitated turning off a transponder for a time to allow full recharging of the satellite's battery.

Mailboxes In Orbit

Packet radio is a relatively-new two-way communications technique in which radio replaces telephone lines and computers talk to computers.

On one end of the digital circuit, an operator types a message addressed to someone else into a computer and then orders the computer to radio the message to Pacsat. The Earth-orbiting satellite receives the message and holds it in memory.

When an operator at a distant location wants to read his electronic mail sometime later, he tells his computer to send up a go-ahead order. The satellite, determining that the intended recipient is on hand on the ground, takes the information from its memory bank and sends it down to Earth.

The receiving operator, after reading his mail, can compose a reply if he wishes and send it up to the satellite. The satellite will hold the mail electronically in its memory until the addressee sends up a radio signal. Then the satellite transmits the reply to the ground.

The radio signals are visualized as packets because information being transmitted is separated into groups of letters, numbers and spaces. A packet sometimes contains 128 characters of information. Other quantities are possible. Each set of characters is sent to or from a satellite by separate radio transmission. A complete message may contain many packets totaling thousands of characters. A computer in the satellite and computers at the ground stations check the messages to make sure they are error-free. Then they sort the packets so they are read in proper order.

Most new hamsats are outfitted as "pacsats" to receive and relay amateur packet signals. Pacsats usually have store-and-forward bulletin-board systems (bbs) to hold messages—creating electronic mailboxes in the sky. Typed messages sent to an orbiting mailbox usually are typewritten, but hamsats can have voice mailboxes.

Amateur radio satellites are experimental, of course, so glitches do turn up. For instance, ground controllers had trouble commanding AO-13 while the satellite was looking down on North America. Uplink signals were being corrupted by a government radar. Commands had to be forwarded to Australia for transmission to the satellite.

All hamsats have been successful experiments in the end, because amateurs have learned from each and applied their findings to later models.

AO-13's attitude can be changed on command by a technique known as magnetorquing.

Everything up must come down, of course. All satellites orbiting Earth are pulled slowly by gravity down toward the atmosphere where they will burn. Some fall more slowly than others.

AO-13 is no different from any other satellite. Its orbit is decaying. AO-13's demise is predicted for sometime around late 1996. A replacement, Phase-3D, is being built for launch by 1996.

Microsats

The 1986 shuttle Challenger explosion caused a temporary shortage of spaceflight opportunities for satellite makers around the world. Competition from commercial satellite owners for space aboard rockets meant amateur radio satellites, which had received many free rides over the years, would have to pay for some future launches.

Like civilian and military satellites, OSCARs had been getting heavier and larger. Smaller satellites would fit in places on rockets reserved for lead ballast and would need only modest launch services. The answer for amateur radio would be a radical design departure—microsatellites.

AMSAT's new standard spacecraft would be very small in size and weight, making possible cheap launches. Each microsat was to be so compact and lightweight it could be launched to orbit by even the smallest space booster.

The microsats could fit where larger satellites couldn't, making more launch opportunities available. They were even smaller than payloads designed for NASA shuttle GAScans.

Amateur satellite designers couldn't command vast budgets, supercomputers and thousands of technicians like major aerospace firms. Instead, the volunteer radio amateur community went back to its experimenter roots and showed that back-of-the-envelope designs could be turned into state-of-the-art spacecraft.

AMSAT received an opportunity in 1989 to show off its new technology when the European Space Agency needed to test a new payload carrier.

ESA had a new piece of hardware called ASAP—Ariane Structure for Auxiliary Payloads—which was a large flat ring to hold small satellites at equal distances around its level surface. The idea was to give small secondary payloads inexpensive rides to space alongside major satellites.

To test the new structure, AMSAT and UoSAT built six small amateur radio satellites for a free ride to space on an ASAP aboard an Ariane flight.

The six microsats launched January 22, 1990, to 500-mi.-high polar orbits were:

★UoSAT-OSCAR-14 (UO-14) and UoSAT-OSCAR-15 (UO-15) from Surrey,

★AMSAT-OSCAR-16 (AO-16) from North American AMSAT,

★DOVE-OSCAR-17 (DO-17), also called Peacetalker, from Brazilian radio amateurs,

★WEBERsat-OSCAR-18 (WO-18) from Weber State University, and

★LUsat-OSCAR-19 (LO-19), from Argentine radio amateurs.

Microsats shared a standard framework design, but each was outfitted with electronics suited to its particular mission. Compared with the massive proportions and tonnage weights of civilian and military communications satellites, the pee-wee microsats were nine-inch cubes under 25 lbs. each.

AO-16, DO-17, WO-18 and LO-19 were designed at Boulder, Colorado, and other cities in the U.S., Argentina, Brazil and Canada, by AMSAT-North America, AMSAT-Argentina, BRAMSAT (AMSAT-Brazil) and the Center for Aerospace Technology (CAST) at Weber State University, Ogden, Utah. AMSAT volunteers at the Microsat Lab in Boulder assembled their own AO-16 as well as DO-17 for BRAMSAT, WO-18 for Weber, and LO-19 for AMSAT-Argentina.

The hamsats were shipped to the European Space Agency launch site in French Guiana where they were attached to the ASAP.

The six ended up in nearly-perfect Sun-synchronous orbits passing over local areas at about the same time each morning and evening.

The six were the largest number of Western hamsats sent to space at one time. It was the biggest single proliferation since 1981 when USSR hams had sent up six in one flight. Here's the story on each of the microsats:

1990: UoSAT-OSCARs 14 and 15

By the end of the 1980s, University of Surrey space enthusiasts had a complex new satellite, UoSAT-C, ready to go to space. In fact, it was scheduled for launch in 1988 on a U.S. rocket, but the flight was postponed. The UoSAT team wasn't able to locate a ride to space for the heavy UoSAT-C, but they did obtain a launch for a pair of lighter satellites on the Ariane ASAP test.

Due to ASAP weight limits, the functions of UoSAT-C had to be split between two lighter replacement satellites—UoSAT-D and UoSAT-E.

Fortunately, UoSAT-C had been loaded with modules which could be pulled apart quickly to take advantage of the short-notice Ariane opportunity. Many mechanical and electrical parts of UoSAT-C were taken apart and reassembled. The UoSAT-C framework was shelved.

UoSAT-D and UoSAT-E were matching frameworks outfitted with identical housekeeping computer systems, but otherwise housing different electronic payloads. They would be low-orbit pacsats with message handling. They would study space radiation and its effects on semiconductors, develop a low-cost computerized spacecraft attitude control for precise Earth pointing, and photograph Earth with a low-cost charge-coupled device (CCD) television camera.

UoSAT-D and UoSAT-E were attached to the ASAP and launched January 22, 1990, alongside the four AMSAT microsats.

UoSAT-D was renamed UoSAT-OSCAR-14 (UO-14). It also has been called UoSAT-3.

UoSAT-E was renamed UoSAT-OSCAR-15 (UO-15). It also has been called UoSAT-4.

Surrey hams successfully commanded UO-14 and UO-15 on during their first day in space. Later that day, each spacecraft's computer software was sent up by radio to the satellites. Nominal telemetry was received from both hamsats.

Unfortunately, news turned bad 25 hours later when no UO-15 signals were received at Surrey.

Operators repeatedly transmitted commands to activate its redundant systems, with no luck. They tried for months to hear something, but no signals have been received since then from UO-15.

Stanford University had helped Surrey back in 1982, transmitting a strong signal to UO-9 to overcome blockage of that satellite's command receiver. Stanford hams tried again in 1990 to come to the rescue, using the same 150-ft. antenna with sophisticated digital signal processing equipment to look for an extremely weak signal from UO-15 oscillators. The big dish was able to hear UO-14 oscillators, but nothing from UO-15.

Attempts to restart UO-15 were abandoned. U.S. government radar continues to track the satellite, orbiting a mile or so higher than UO-14, amidst the pack of microsats launched January 22, 1990. UO-15 is simply dead in orbit.

The loss of UO-15 was mitigated by the good news from UO-14 which worked well. The tragedy spurred UoSAT to build a new small satellite, UoSAT-F, which was launched in 1991.

Meanwhile, UO-14 was working well in orbit handling lots of electronic-mail messages.

UO-14 has three computers for housekeeping and packet radio, including a 16-bit microprocessor with 4.5 megabytes of RAM. Four megabytes are used for bbs message storage.

The satellite's digital transponder receives at 145 MHz and transmits at 435 MHz. Ten watts of transmitter power make the satellite usable by small portable ground stations. Telemetry data packets are beaconed near 435 MHz.

UO-14 is an Earth-pointing satellite with a gravity-gradient boom and computer-controlled magnetorquing, an ideal system for small satellites in low-Earth orbit because it has no continuously-moving parts and expends no fuel.

To calculate its attitude, UO-14 carries a flux-gate magnetometer measuring Earth's geomagnetic field in the satellite's three axes.

Electricity is generated by gallium-arsenide solar arrays feeding nickel-cadmium rechargeable batteries. The satellite rotates slowly, distributing the Sun's light and heat evenly across the satellite.

Pacsats regularly relay messages around the globe to terrestrial packet networks in many regions. A gateway is a satellite ground station acting as a bridge between a pacsat and a terrestrial network. Automated gateways upload and download traffic without human operators.

In 1991, radio amateurs in Alaska and California created a gateway offering same-day delivery of dozens of messages to Alaska from the Lower 48 states. That success brought gateways to every continent. Today, scores of gateways cover North and South America, the Caribbean, Europe, the Middle East, South Africa, Oceania, Asia and even Baffin Island above the Arctic Circle. They deliver messages and responses within 24 hours. Such fast action is important in health-and-welfare traffic during emergencies and natural disasters.

In 1992, the first medical image transmitted via hamsat showed a fractured hip repaired with a compression hip screw. The hip had been pictured by a portable fluoroscopy X-ray displaying real-time images on a TV monitor for a physician during surgery. The image was stored on computer disk and transmitted to UO-14. The satellite kept it in memory for a few days, then sent it back to Earth, proving hamsats can help remote clinics get assessments from specialists.

Some of the construction costs of UO-14 had been paid by an American organization known as Volunteers In Technical Assistance. The non-profit organization had a large library of data on farming, windmills, stoves, ovens and other useful non-military subjects which it wanted to send to development workers in remote areas via UO-14's store-and-forward mailbox.

For a time, radio amateurs and VITA shared UO-14's transmitter, computer and 400-message memory. To accommodate both amateur and non-amateur users, the satellite switched back and forth between amateur and non-amateur frequencies.

When UO-14 transmitted on its non-amateur frequency, it was sending technical information to areas of the developing world poorly served by existing data communications. Such traffic was considered inappropriate for amateur channels. When the transmitter switches occurred, amateurs lost reception from their satellite for periods from a quarter-second to five seconds.

UO-14's uplink became congested with users. When 200 stations began using the satellite regularly, the satellite's 400 message limit was reached frequently. Amateur stations were limiting access for non-amateur VITA stations. After the British hamsat UO-22 was launched in 1991, Surrey decided to change UO-14's mission.

Amateur radio service was dropped. UO-14 stopped transmitting on its amateur frequency. Ham operations were moved to the new UO-22.

Today, UO-14's electronic mailbox links VITA with inexpensive, portable ground stations built around a computer, radio, battery and antenna. A ground station fits in a suitcase and works where no power lines exist. The pacsat is low enough for small ground stations to use simple whip antennas, made from coat hangers if necessary, to hear the satellite.

UO-14's bbs memory holds four million characters of information. When the satellite is overhead, a ground station can send up a message at 500 characters per second. During the few minutes the satellite is overhead, a ground station might send up 200,000 characters of information—equivalent to ten magazine articles.

Even if mail were picked up immediately by a recipient the next time the satellite passed over his head, it might be stored in the satellite from a few seconds up to twelve hours. Mail can stay in the satellite for days, of course, awaiting radio commands from an addressee.

A busy operator on the ground can put his satellite station on autopilot. He programs his computer to determine when the satellite will be overhead. Most pacsats pass over the North and South Poles every hour and a half and over any one point on the surface four times a day. One is overhead for only a dozen minutes or so.

At the appointed hour, the ground-station computer would fire up its radio and send up a signal asking the satellite if any messages were on hand. If messages for that ground station were stored in the satellite, the satellite's computer would order them sent down.

The computer on the ground then could store

them for future reading and turn off its radio as the satellite passed out of sight over the horizon.

A pacsat like UO-14 and others makes it cheap and easy to send messages, data and images in or out of developing regions. Scores of portable ground stations already are linking underdeveloped countries to medical, weather, agriculture and engineering databanks.

Volunteers in Africa, Asia and South America use portable ground stations to ask for technical assistance. Travelers in the most rugged terrain receive data. Relief workers communicate directly with emergency teams at natural disaster sites.

1990: AMSAT-OSCAR-16

Among the microsatellites launched January 22, 1990, was a small spacecraft known before launch as PACSAT-NA and after launch as AO-16 for AMSAT-OSCAR-16.

AMSAT ground controllers had no trouble commanding AO-16 on the air shortly after launch. The packet telemetry beacon was strong enough to be received easily with handheld radios using small, flexible, rubber-covered antennas. The message bbs was turned on in March 1990.

AO-16 downlink frequencies are near 437 and 2401 MHz. Uplink frequencies are near 145 MHz. Four packet stations on the ground can use the AO-16 bbs at one time.

The medical image of a fractured hip transmitted via UO-14 in 1992 also was relayed by AO-16. The Alaska-to-California message-traffic gateway created in 1991 used AO-16.

1990: DOVE-OSCAR-17

DOVE Peacetalker may be best known among the microsats launched in January 1990. It certainly has had the largest listening audience.

After launch to a low polar orbit alongside the other small satellites, DOVE Peacetalker was renamed DOVE-OSCAR-17. DOVE stands for Digital Orbiting Voice Encoder.

DO-17 was designed to educate the public about space by providing an easily-received satellite signal for demonstrations to children. DOVE was the first hamsat to transmit spoken messages promoting peace among nations.

The microsat has a digital recording system hooked to its receiver and a voice-synthesizer attached to its transmitter. School children around the globe are encouraged to write and speak messages which are transmitted to DOVE by Brazilian hams. DO-17 records the voices and then broadcasts the messages on 145.825 MHz to be heard by anyone with an inexpensive vhf fm receiver or vhf scanner radio of the kind used to monitor police and fire calls.

Cosmic Particles

While Earth's atmosphere and magnetic field shield the planet surface from most radiation from the Sun and other cosmic sources, satellites are not protected and receive high levels of radiation.

Cosmic particles, trapped in the geomagnetic field, oscillate between the magnetic poles in the Van Allen radiation belts which surround Earth, increasing the amount of radiation pelting satellites. In the very harsh space environment, radiation damage is the primary threat to electronic systems once a satellite has survived launch.

Satellites rely heavily on microprocessors and memories. Sometimes a satellite's computer will crash mysteriously or stored data will turn up corrupted. Such incidents usually result when cosmic particles penetrate memories and change their contents. Sometimes radiation causes them to fail altogether.

The British hamsats UO-9, UO-11 and UO-14 allowed radio amateurs to study the effects of radiation on satellite parts. UO-9 carried a Geiger counter to detect radiation. UO-11 monitored radiation effects with a particle wave experiment to measure electron flux at eight energy levels.

UO-14 had a detector to spot cosmic particles as they passed through a diode array, depositing a charge. The charges were measured to reveal the energy of the particles and the angle at which they entered the satellite.

UO-14 also had a total-dose monitor to assess shielding offered by the spacecraft structure. Radiation-detecting field-effect transistors spotted at seven places in the satellite are monitored to measure radiation doses absorbed.

In 1991, the satellite noticed a decrease in cosmic ray flux, then observed a large solar flare.

The satellite offers students easy access to space research data. Its voice is programmed with various languages so students around the world can understand and learn as DOVE reads out data from the satellite's sensors in synthesized speech.

No tracking is required to receive DO-17. Its signals have been received using a standard pull-up whip antenna attached to a receiver or a flexible antenna on a portable receiver. A high outdoor antenna is even better than an indoor antenna.

Listening from 8 am to 1 pm and 7 pm to 12 midnight local time may reveal the microsat flying overhead several times. When its 145.825 radio is on, it transmits for 2.5 minutes followed by an off period of 30 seconds during which it stands by for commands from the ground.

DOVE has beacons, but no active transponders. Besides 145 MHz, there is an S-band beacon relaying data in packet radio near 2401 MHz.

Boost From Space

The small school was called Weber State College and it was located in the northern Utah city of Ogden. Thus, the name Northern Utah Satellite, or NUsat, was applied to a satellite designed by Weber students and launched by shuttle in 1985.

Later a new kind of microsatellite was under construction at the growing institution of higher education. Known first as NUsat-2, it was to become WEBERsat-OSCAR-18.

Publicity surrounding design, construction and launch of the new microsats in 1990 showered on Weber State College. In fact, faculty and students had provided such significant time and materials to the project, and to the then-proposed Phase-4 hamsat, the entire state took notice.

Utah's Board of Regents in January 1990 overruled heated objections by the University of Utah and Utah State University and voted to change Weber State College to Weber State University. Regents said the successful WEBERsat launch that month was significant in influencing their decision.

WEBERsat's renown brought many elementary and high school students from the Salt Lake City-Ogden area to the WSU command station. WSU students enthusiastically explained WO-18 and its on-board science experiments, hoping visitors would start thinking about space and ham radio.

AMSAT has an amateur satellite projects agreement with WSU. Information is available from Center for Aero Space Technology (CAST), Weber Sate University, Ogden, Utah 84408-1805.

Many high schools collected telemetry data for science experiments. One California physics class monitored the rate at which DO-17 was spinning. It was supposed to be about three revolutions per minute (rpm). Analyzing telemetry sampled at intervals, the students understood the relationship between sample rate and spin rate. If the sample rate weren't fast enough, incorrect conclusions would have been drawn about rate and direction. Students concluded DOVE's spin rate had slowed.

A teacher's guide to using DO-17 with classroom exercises and experiments is available to schools from AMSAT Science Education Advisor, 421 N. Military, Dearborn, Michigan 48124 USA.

1990: WEBERsat-OSCAR-18

Say "cheese" when you look up; you may be looking into the business end of a hamsat camera. WEBERsat-OSCAR-18, designed at Utah's Weber State University, has been snapping photos of Earth since shortly after it was lobbed to a 500-mi.-high polar orbit alongside the other AMSAT microsats on January 22, 1990.

WO-18 has a charge-coupled device (CCD) television camera which stores images in memory and compresses them into packet radio bursts transmitted to Earth.

One picture fills about 200k of computer memory, but the hamsat can send it down to the WSU ground station in as little as seven seconds.

WO-18 has recorded such exotic locales as Ethiopia, the Bay of Bengal, Brazil, Africa's Lake Victoria, clouds over the Indian Ocean, Chile, the coastline of British Columbia near Vancouver Island, Australia, India, the coast of Peru, the U.S. Great Lakes, even the Moon, Sun and stars.

Wispy clouds in WO-18 pictures mark the possibility of satellite meteorology with very inexpensive imaging equipment.

WO-18 has a message bbs, a particle impact detector, a spectrometer, a magnetometer for navigation, an amateur television (atv) uplink receiver, and two downlink beacon transmitters.

1990: LUsat-OSCAR-19

AMSAT-Argentina had been contemplating a satellite of its own in 1988 when news of the six-payload Ariane ASAP launch became known. The Argentines immediately joined North American AMSAT in the microsat project.

The letters LU are an amateur radio callsign prefix for Argentina so the satellite was dubbed LUsat. It was licensed in Argentina and paid for by Argentines, but constructed in Utah at CAST.

In orbit, it was renamed LUsat-OSCAR-19. Commanded in space from Argentina, LO-19 is a pacsat with digital message bbs available for non-profit use by hams around the globe.

The first message sent to the bbs was from Carlos Saul Menem, president of the Argentine Republic and a ham operator himself.

One special telemetry beacon aboard LO-19 transmits data about the hamsat in easy-to-read 12 words-per-minute Morse code at 437.125 MHz. LO-19 broadcasts news bulletins continuously on Mondays on another frequency near 437 MHz.

LO-19 and other hamsats relayed essential public-service communications after Hurricane Iniki leveled parts of Hawaii in 1992. Health and welfare messages from devastated areas were forwarded to anxious relatives around the globe.

1990: Fuji-OSCAR-20

Japanese hams had a replacement satellite ready three months after their first had to be turned off in 1989 for lack of electricity.

Their first hamsat had been called Japan Amateur Satellite (JAS-1a) and Fuji-OSCAR-12. Fuji is Japanese for wisteria. The replacement was labeled JAS-1b and called Fuji-OSCAR-20.

Running With The Pack

The microsats launched in 1990—UO-14, UO-15, AO-16, DO-17, WO-18 and LO-19—were dropped off in similar orbits by their Ariane rocket.

After the rocket reached space, springs popped the hamsats away from its nose. The springs had varying degrees of springiness and the satellites had different weights, so each hamsat left the rocket at a slightly different velocity.

For a few days, all six remained close together in orbit. During that time, they passed over one place on the ground about the same time each day. Since then, they have been spreading apart. Although their altitudes are close, they always are separated by a few miles. The chance of two bumping in the future is almost nil.

All six will not bunch together again during their time in space, although groups of two or three may.

LO-19 is lowest. That makes its trip around the globe slightly shorter, so it has been able to move ahead of the pack. UO-14, on the other hand, is highest which makes it seem slower as it takes longest to circle the globe.

In 1991, LO-19 lapped UO-14, then the silent UO-15, and then AO-16. LO-19 did it again in 1993.

Even though UO-15 has not been heard since hours after launch, it still is running with the pack.

JAS-1b was the seventh hamsat hurled to space in 1990. Just 16 days after the European Space Agency launched the microsats, an H-1 rocket provided by Japan's National Space Development Agency ferried the 110-lb. JAS-1b to a 750-mi.-high orbit February 7. The H-1 also carried two government satellites, MOS-1b and Debut. It was the first time Japan had launched more than two satellites at one time.

Eclipses kept FO-12 from producing enough electricity, but FO-20 was in a more favorable orbit without as many periods of solar eclipse.

FO-20 uses torque generated by interaction of two permanent magnets with Earth's magnetic field to maintain its attitude.

FO-20 receives and retransmits voice and Morse code, permitting hams to chat over wide areas of the globe. It also provides an electronic mailbox, storing messages in 1.5 megabyte random-access memory (RAM) and delivering them on request to other stations at later times. Amateurs on one side of the world place messages on the satellite's electronic bulletin board to be read by others when the hamsat is on the far side of the globe.

The communications uplink is near 145 MHz. The downlink is near 435 MHz. Telemetry is in Morse code and packet radio near 435 MHz.

1990: Badr-1

Radio amateurs have pushed the state of the art in electronics in nations around the globe. In Pakistan, for instance, a number of engineers at the government's Space and Upper Atmosphere Research Commission (SUPARCO) are hams.

SUPARCO is at the University of the Punjab at Lahore, a prominent border city in eastern Pakistan not far from Delhi, India, and at the Arabian Sea port of Karachi in southern Pakistan. SUPARCO has fired small rockets on sub-orbital science flights from launch pads at its Maini Beach flight-range, 36 miles west of Karachi.

Several SUPARCO personnel completed masters degrees in engineering at England's University of Surrey—the institution which built and operates UO-9, UO-11 and UO-22 hamsats. While at Surrey, they worked on UoSAT projects. When the engineering students returned to Karachi and Lahore, they built ground stations and took part in digital communications experiments with the British hamsats UO-9 and UO-11.

SUPARCO hams also used knowledge gained at the university to build their own satellite. With support from the Pakistan Amateur Radio Society, they started building a small hamsat in the last half of 1986. They called it Badr, after the Urdu language word for "new moon."

The first satellite, Badr-1 or Badr-A, was to have been ferried to space in a U.S. shuttle, but that plan changed after the 1986 Challenger explosion delayed American flights.

Four pre-launch ground tests were successful. In 1989, Pakistan registered the planned satellite with the International Frequency Registration Bureau. Then the spacecraft was shipped to China's Xichang Launch Center in 1990.

China launched the 150-lb. Badr-1 on July 16, 1990, to a 375-mi.-high circular orbit on a Long March rocket. It was one of the eight hamsats sent aloft in 1990. The tiny Badr-1 circled the globe every 96 minutes, passing over Pakistan for 15 minutes three to four times a day.

The Pakistani satellite, shaped as a polyhedron with 26 surfaces or facets, was about 20 inches in diameter. It resembled the U.S. NUsat launched from an American shuttle in 1985, but Badr-A housed digital communications gear modeled after the radio system aboard the British satellite UO-11 launched in 1984.

Badr-1 offered one radio channel for digital store-and-forward communications. Uplink was near 435 MHz. Downlink was near 145 MHz. The telemetry beacon was near 145 MHz.

Badr-1's orbit was so low it could not sustain

itself in space more than 146 days. It fell into Earth's atmosphere and burned December 9, 1990.

SUPARCO hams built a second satellite, Badr-2, to be launched in 1994-95. Badr-2 will be more sophisticated than Badr-1, with a CCD camera for pictures of Earth and a system allowing ground stations to change the hamsat's direction in space.

1991: AMSAT-OSCAR-21/RS-14

It could have been just another amateur radio package riding piggyback on a big government spacecraft when it was launched in 1991. But, when ground controllers converted AMSAT-OSCAR-21 into a voice repeater in the sky in 1992, it immediately became one of the most popular hamsats orbiting the globe.

The Russian satellite had been called Radio M-1 as it rode to space from Russia's Northern Cosmodrome at Plesetsk on January 29, 1991.

Like those popular Russian dolls inside dolls, Radio M-1 was inside a large government spacecraft called INFORMATOR-1, which housed equipment from the Ministry of Geology and Science (GEOS). And then, inside Radio M-1, was a German digital transponder called RUDAK-2.

The Russian space industry had been in the habit of labeling improved spacecraft with the letter M for modified. For instance, when the Progress space freighter was redesigned, it was called Progress-M. Similarly, after the first hamsats had been called Radio-1 and Radio-2, the fourteenth was modified and called Radio M-1.

RUDAK is a German-language acronym for Regenerating Transponder for Digital Amateur Communications. RUDAK-2 was designed and built by amateur radio members of AMSAT-DL, the German affiliate of AMSAT. DL is Germany's amateur callsign prefix. RUDAK-1 had flown to space aboard AO-13, but didn't work.

The Russian amateur radio satellite club Orbita and the Adventure Club of Moscow built Radio M-1 as a joint project with German hams at Marburg, Munich and Hannover. The collaboration led to dual names for the new amateur radio satellite once it arrived in orbit: AMSAT-OSCAR-21 (AO-21) and Radiosputnik-14 (RS-14).

AO-21/RS-14 is in a 600-mile-high circular orbit. Its spacecraft-mate, the geological-survey satellite GEOS, should not be confused with the American GOES weather satellites. The combo is orbiting Earth every 104 minutes.

Amateur radio operators around the world use AO-21/RS-14 to communicate via frequency modulation (fm) voice, single-sideband (ssb) voice, Morse code (cw) telegraphy, and packet radio. RUDAK-2 is the packet-radio section of AO-21. A portion of computer memory is set aside as a mailbox where hams leave messages for others.

In 1992, ground controllers switched RUDAK to an fm repeater mode. It immediately became very popular with hams around the globe because it was very easy to use common ham gear with simple antennas to access the repeater in the sky. The uplink frequency was near 435 MHz. The downlink was at 145 MHz.

Non-human voices were heard from AO-21 in 1991 as ground controllers tested its talking-satellite experiment. The 145 MHz telemetry beacon was transmitting, "I am completely operational and all my circuits are functioning properly." The multilingual hamsat spoke Russian in 1992 when the speech synthesizer was greeting, "Hello to those on the ground and the cosmonauts in the space station Mir."

A Russian-Danish expedition, organized by the Adventure Club of Moscow, one of the original sponsors of AO-21, discovered the burial place of Danish explorer Vitus Bering (1682-1741) and his eight-man team at Commander Bay in 1991.

In 1728, the Danish navigator had been the first to traverse what we now know as the Bering Strait. He died in 1741 on Bering Island in the Commander Islands east of Kamchatka Peninsula in the Bering Sea.

A 1992 ceremony to re-bury Vitus Bering was held on his island. In honor of the event, AO-21's digital voice broadcast, "Greetings to the Russian-Denmark Expedition, which discovered the burial place of Vitus Bering and his team at the Commander Bay near Kamchatka."

Also in 1992, a female voice was heard on the AO-21 downlink speaking in the Slovenian language, the native tongue of Matjaz Vidmar, who was listening to a portable ham transceiver in his hospital bed. He was hospitalized following an automobile accident. Vidmar was known for his transponder designs, including S-Band transmitters for the hamsats AO-16, DO-17 and Phase-3D.

Celebrating the annual Army-Navy football game in 1992, Naval Academy midshipmen ran from Annapolis, Maryland, to the stadium at Philadelphia, Pennsylvania. They carried an unusual football helmet outfitted with a global-positioning satellite (GPS) receiver and a packet-radio satellite terminal which reported their positions along the route every two minutes. From 600 miles overhead, the hamsat AO-21 helped out along the way by relaying position reports from runners and chase vehicles.

1991: Radiosputniks 12 and 13

RS-10 and RS-11 worked very well after their

1987 launch and were popular with amateur satellite enthusiasts. Sticking with a proven design, Russian hams delivered another popular package February 5, 1991, in the launch of RS-12 and RS-13, just a week after AO-21/RS-14.

It was another flashy three-in-one space shot—the hamsats Radiosputnik-12 and Radiosputnik-13, and the government navigation satellite Cosmos 2123, all combined in one large spacecraft orbiting 600 miles above Earth.

Hamsats often ride piggy-back to space on government rockets, but there they separate into different orbits. By comparison, Cosmos 2123, RS-12 and RS-13 are one package with ham radio gear taking electricity from Cosmos' solar-panel wings. Cosmos 2123 helps Russian fishing fleets locate themselves on the world's oceans.

As with all hamsats since 1961, the Radiosputnik series is open for use by all amateurs around the globe. RS-12 and RS-13 telemetry beacons are near 29 and 145 MHz. They have identical transponders, but frequencies differ. Ground stations transmit on frequencies near 21 and 145 MHz. Downlink signals are near 29 and 145 MHz. Cosmos 2123 transmits at 150 MHz.

An innovation of earlier Radiosputniks, the popular autotransponder "robot" operator, was carried forward in RS-12 and RS-13. Hams also have fun sending slow-scan television (sstv) pictures to each other via RS-12 and RS-13.

1991: DARPA Microsats

One of the exciting spaceflight developments of the 1990s has been the winged Pegasus booster carrying small payloads from an airplane to low Earth orbit. The three-stage solid-fuel rocket is carried aloft to 40,000 feet under the wing of NASA's Boeing B-52 bomber and dropped. From there, it blasts upward to space.

Pegasus' second blast off led to interesting radio tests for U.S. amateur operators who were asked to help the military check out seven Lightsats.

In the July 17, 1991, flight, the winged rocket carried to orbit the seven DARPA communications microsatellites referred to as Lightsats. DARPA is the Defense Advanced Research Projects Agency.

The microsats were not amateur radio satellites, but related to amateur radio through the Military Affiliate Radio System (MARS). Radio amateurs operate MARS stations which are known for handling messages to and from service personnel and families. The military microsats had voice and packet radio transponders similar to amateur radio satellites launched earlier.

The rocket's first and second stages didn't separate smoothly and, later, the nosecone covering the payload of seven DARPA satellites didn't jettison as quickly as planned. As a result, the seven satellites were dropped off in lower-than-expected polar orbits at altitudes around 150 miles above Earth. The satellites were in such low orbits, they could stay in space only about six months, falling into the atmosphere and burning around January 25, 1992. During that time, however, radio amateurs in MARS managed some informative communications tests.

The microsat transponders were compatible with military transceivers. During demonstrations in 1991 and 1992, the spacecraft operated on military frequencies between 225-400 MHz.

Midshipmen and staff of the U.S. Naval Academy at Annapolis used training vessels and a 36-ft. dish antenna to test Navy and Marine Corps tactical communication via the microsats.

Voice contact was established via microsat with the Aegis-class destroyer USS Arleigh Burke at the U.S. Navy base, Norfolk, Virginia. A mobile operator made it a three-way conversation when he transmitted from a vehicle in northern Virginia.

Packet radio data beacons relayed training vessel position and status reports through the microsats to Annapolis. An amateur radio operator in Massachusetts received the beacon from a training vessel in a shipping channel near the Philadelphia airport. A Naval Space Command station on the roof of a building in Dahlgren, Virginia, contacted a training vessel under way in the Chesapeake Bay.

Three four-person and five-person tactical units of midshipmen waited in a helicopter landing zone at Quantico Marine Corps Base, Virginia, for a Microsat to pass overhead. The teams contacted each other and an Annapolis command station via the satellite, then moved in three directions 300 yards into woods around the helicopter pad. For 52 minutes, signals to and from the satellite were clear, even as the teams exercised in the woods.

While the military microsats were in low orbits 150 miles above Earth, U.S. Navy Fltsatcom communications satellites were a great deal higher in stationary orbits 23,300 miles above Earth. Even so, Naval Academy personnel at Annapolis were able to use a microsat to link stations in northern Virginia. The microsats also were used by MARS hams at Fort Gordon, Georgia.

The success of amateurs in acquiring microsat signals prompted Naval Academy hams to organize a network of monitoring stations. As the microsatellites moved north to south over the eastern U.S., they were tracked by many hams.

1991: UoSAT-OSCAR-22

For the first time ever, on July 17, 1991,

thirteen OSCARs were active in orbit at one time. They were AO-10, UO-11, AO-13, UO-14, AO-16, DO-17, WO-18, LU-19, FO-20, AO-21/RS-14, the RS-10/RS-11 combo, the RS-12/RS-13 combo and the brand-new UO-22.

Hams at the University of Surrey had designed and built yet another small satellite, UoSAT-F, which was launched that July 17 on a European Ariane rocket. In polar orbit, 480 miles above Earth, UoSAT-F was renamed UoSAT-OSCAR-22. It also was known as UoSAT-5.

Some of the costs of UO-22 had been borne by the organization SatelLife, formed by International Physicians for the Prevention of Nuclear War, an organization which had received a Nobel Peace Prize in 1985.

SatelLife used the pacsat to start HealthNet, an international not-for-profit E-mail network for health professionals. Early users included five African medical schools which linked up with HealthNet to receive fresh medical literature and exchange electronic mail by satellite.

HealthNet message packets were transmitted on non-amateur frequencies near 428 MHz, not far from UO-22's amateur downlinks at 435 MHz. When off-duty from HealthNet, UO-22 would switch to amateur radio frequencies.

In congestion very similar to the UO-14 overcrowding problem, hams using UO-22 were limiting access for non-amateur SatelLife stations even as the non-amateur transmissions interrupted amateur activity. Surrey had obligations to SatelLife and VITA, so it moved to resolve the congestion in 1992.

Amateur radio service was dropped from UO-14 altogether in favor of non-amateur SatelLife and VITA and all ham activity was moved to UO-22. Today, UO-22 works for radio amateurs while UO-14 works for SatelLife and VITA.

UO-22 is a pacsat with bbs, but the satellite's most remarkable feature may be its charge-coupled device (CCD) television camera with 110-degree wide-angle lens showing a field of view nearly the same size as the satellite's footprint.

UO-22 snaps three or four shots a day, each of a ground area 994 by 1118 miles. Notable pictures have included Italy, showing the familiar boot outlined by the Mediterranean, Adriatic, and Tyrrhenian Seas, and Yugoslavia and Greece; an Antarctic iceberg; Bulgaria and Romania; Denver; Cuba and Haiti; Denmark and the Netherlands; haze over Djibouti, Somalia and Yemen; French Guyana; eastern South Africa; Egypt and Sinai, the Nile Valley and the Upper Nile; the Gulf of Mexico; Equatorial Africa; Kuwait and Persian Gulf smoke plumes; Limerick, Ireland; the Balkans; North Africa; the Great Lakes; northern Australia; Florida and the Mississippi Delta; Spain, Portugal and Mahgreb; California; the Red Sea; Korea; and the Straits of Hormuz.

The 110-lb. hamsat also carries radiation dose experiments, horizon sensors, and magnetometers, one inside and the other on a small boom protruding above the spacecraft. UO-22 has a 15-ft. gravity-gradient boom with a five-lb. weight on the end which provides restoring torque to keep the camera lens and radio antenna pointed to Earth.

Her Majesty Queen Elizabeth II was touring the UoSAT control room in 1992 when UO-22 flew overhead and transmitted a synthesized-voice greeting. Then another hamsat, UO-14, delivered a message to Her Majesty from President Fredrick Chiluba of the Commonwealth nation of Zambia. The Queen left a reply message which was returned to Zambia by UO-14.

1991: SARA

One of the most exciting developments in recent years for radio hobbyists has been SARA, a first-of-its-kind amateur astronomy satellite made in France and launched to Earth orbit July 17, 1991, on the same European Ariane rocket that ferried UO-22 to space.

SARA is not an amateur radio communications satellite, but an orbiting amateur radiotelescope listening for high-frequency (hf) radio signals from the planet Jupiter. SARA stands for Satellite for Amateur Radio Astronomy.

SARA revolves around Earth every 100 minutes in a 500-mi.-high polar orbit.

The spacecraft was built by ESIEESPACE, a club of amateur astronomers at France's Ecole Superieure d'Ingenieurs en Electrotechnique et Electronique (ESIEE). Previously the club had built and launched payloads for suborbital rocket and balloon flights. SARA's simple gear was conservatively designed around off-the-shelf consumer equipment, rather than expensive space-qualified or military-specification components.

Jupiter is the fifth planet out from the Sun and may itself have been a prototype Sun which wasn't big enough to become a star.

Five times farther from the Sun than Earth is, Jupiter orbits the Sun every 11.9 years. It is largest of the nine planets with 318 times the mass of Earth. Jupiter is more than two-thirds of the total mass of all planets in our Solar System.

Although a thousand times smaller than the Sun, Jupiter is a big planet—as big as 1,317 Earths. Still, if it had been several times more massive, it might have become a star billions of years ago as the pressure and temperature at its core triggered nuclear fusion.

Today, Jupiter is a big gas bag, with a relatively low density, which somehow sends out naturally-generated radio noise in the high-frequency part of the electromagnetic spectrum.

Jupiter's radio emissions are powerful enough to wipe out other natural signals arriving at Earth. The flux from Jupiter is much stronger than the galactic background noise. Solar eruptions can be distinguished from Jovian signals by signal strength, length of time and correlation with signals received at other wavelengths.

There are three bands of powerful radio transmissions coming from the giant planet. They are 1-1000 KHz low frequency (lf), 2-40 MHz high frequency (hf) or shortwave, and 75-15000 MHz.

The low-frequency energy seems to come from Jupiter's famous Red Spot and the plasma link between Jupiter and its volcanic moon Io.

Shortwave radio signals sweeping Earth are most interesting to SARA users. They may be generated by the interaction between Jupiter and Io. Volcanoes on Io may spew out sulfur, oxygen and other atoms and molecules. That material then may be captured in Jupiter's magnetic field and ionized. The ionization would be so great, the natural frequency of the electrons gyrating in the magnetic donut, or torus, can reach 40 MHz. It is strong around 20 MHz.

The natural radio signals arrive at Earth as a beam of energy, appearing four minutes earlier each day, taking two hours to swing across Earth. Earth's atmosphere blocks the signals, making them hard to study from the ground.

If Jupiter is above the horizon and the ionosphere is thin, such as at night, the signal can make it down through Earth's ionosphere to the ground where it sometimes can be received with an indoor horizontal loop antenna around a ceiling.

Far better is a receiver in a satellite orbiting above Earth's atmosphere where it clearly can hear radio waves from Jupiter. Satellites have measured Jovian emissions before, but not for long periods. Before SARA, no Earth satellite had listened to Jupiter, during a period of peak solar activity, in the high-frequency range between 2 and 15 MHz. Voyager 1 had tuned in briefly in 1979 as it flew past Jupiter, but its receiver was noisy and it could hear only the strongest signals from Jupiter.

Electromagnetic energy from Jupiter actually is very strong. SARA needs only three pairs of 15-ft. antennas, made of 100-mm-wide steel tapes held perpendicular to each other, to snare the radio waves. The perpendicular arrangement allows radio waves to be captured and signal strength to be measured, no matter which direction the satellite is turned nor the polarity of the signal.

One of the antenna pairs doubles as a telemetry antenna, transmitting data down to Earth in the two-meter amateur radio band at 145 MHz.

SARA's receiver listens to eight 100-KHz channels spread across the spectrum between 2-15 MHz. The receiver switches among the eight channels and three antenna pairs many times during a 150-second period, averaging the signal strength of the eight channels. Averaging smoothes out any storm peaks on Jupiter.

The receiver is sensitive enough to detect galactic background noise which has a constant level of strength. Galactic noise gives the amateur radio astronomers on the ground a point of reference when Jupiter and the Sun are silent.

SARA is an 18-in. cube housing electronics. Antennas are on the outside. Solar cells on each side cover 60 percent of the satellite's exterior. The power supply inside is a storage battery charged by sunlight shining on the photovoltaic cells.

SARA is controlled by a built-in microcomputer which chooses receive frequencies and antennas, digitizes analog data and stores it, interprets commands received from the ground, and prepares the telemetry. The satellite continuously sends telemetry data to the ground at 145 MHz.

SARA offers an opportunity for teachers to introduce astronomy, spaceflight topics, orbital dynamics, even the history of science. Information is available from BELAMSAT, Thier des Critchions 2, B-4032 Chenee, Belgium.

1992: KITsat-OSCAR-23

As it had done in 1990, amateur radio led a nation into space again in 1992. Back in 1990, it had been Pakistan's first-ever satellite. This time, it was South Korea's first-ever satellite.

The microsat was built at Great Britain's University of Surrey for the Korean Advanced Institute of Science and Technology (KAIST) by ten Korean students working under the guidance of Surrey engineers. The 110-lb. South Korean hamsat was called KITsat-A.

A European Ariane rocket put KITsat-A into orbit on August 10, 1992. AMSAT called the satellite KITsat-OSCAR-23 (KO-23).

Its builders named the hamsat Uribyol, Korean for "Our Star." Uribyol No. 1 made South Korea the 22nd country with a satellite in orbit since 1957 when the USSR launched the first Sputnik.

KO-23 consolidated technical advances from Surrey's UO-14 and UO-22, cloning much of its electronics from UO-14, UO-15 and UO-22, shoehorning them into the same tiny UoSAT style of spacecraft "bus." It's a 14-in. cube with an 18-ft. gravity-gradient boom made of measuring-tape steel preformed so it extends from the spacecraft to form a tubular shape. A five-lb. weight is on the end. The boom provides restoring torque which keeps the TV camera lens pointed towards Earth.

KO-23's inclination of 66 degrees makes it available to users farther north and south than most amateur radio satellites. It circles Earth every 112 minutes. KO-23 circles Earth a dozen times a day, passing over the Korean peninsula seven times a day. The small satellite is photographing Earth, detecting cosmic particles and measuring cosmic rays, and providing an amateur radio electronic-mail system in orbit. It also makes digital voice-broadcast tests.

Spacecraft in orbit are showered by radiation from beyond Earth, which can damage integrated circuit (IC) chips and scramble data stored in solid-state memories. AO-10 and FO-12 were crippled by radiation damage. Recent hamsats have used even-more-delicate semiconductors in critical systems. UO-9, UO-11, UO-14 and UO-22 began amateur studies of radiation, measuring effects on electronics. However, those measurements were in relatively-benign low-altitude, high-inclination orbits. On the other hand, KO-23 is in a high-altitude, low-inclination orbit, where much worse radiation is found.

KO-23 has a cosmic ray experiment looking for high-energy cosmic rays and measuring the total radiation dose. Effects on computers, memories, power systems and solar panels are monitored.

KO-23 is a pacsat with two user uplink frequencies in the 145 MHz band, one 435 MHz transmitter, and a command uplink. It was the second amateur radio satellite to offer high-speed 9600-baud transfers. For message storage, the pacsat has 13 megabytes of CMOS random-access memory (RAM).

KO-23 has speech synthesis, store-and-forward speech relay, and high-speed modulation. It allows users to leave voice mail. KO-23 was reported playing military-style music in November 1992.

KITsat has an upgraded version of the UO-22 TV camera capable of shooting photographs with either four-kilometer or 400-meter resolution.

Like UO-22, one of KITsat's two CCD cameras provides a wide field of view with four-kilometer ground resolution. However, it covers a larger area of Earth than UO-22's camera. KITsat's second camera has a telephoto lens giving 400 meters ground resolution. The wide-angle camera is used to spot areas to be photographed in more detail. Then, detailed images are made with the narrow-field camera. Pictures snapped by KO-23 are stored as data in memory, to be downloaded by ground stations via packet radio.

KO-23 has recorded remarkable images around the globe, including Antarctica and the tip of Patagonia as a low Sun angle highlighted splendid cloud formations; wide-angle shots of Korea and Japan; telephoto shots of Kitakyushu, Hiroshima and the land bridge between the Japanese islands of Kyushu and Honshu; and the coastline of Burma, Laos, Thailand, and Bangladesh.

1993: ARSENE

France is one of the oldest spacefaring nations, having launched its Asterix from North Africa in 1965. Even so, the first French amateur radio satellite wasn't to be launched until 1993.

When the European Space Agency launched its first Ariane rocket in 1979, French amateur radio operators began to plan for a hamsat of their own.

In 1980, French hams founded RACE, the Radio Amateur Club de l'Espace, to build a satellite. They called their projected hamsat ARSENE, an acronym for Ariane Radio Amateur Satellite pour l'ENseignement de l'Espace.

The structure was designed by engineering students and electronic modules were built by amateurs from ATEPRA, Association Technique pour l'Experimentation du Packet Radio Amateur.

It took time to complete the hamsat and book a flight, but the pacsat finally was able to find a ride to space on an Ariane rocket. ARSENE was launched in 1993.

ARSENE is to be in an elliptical, equatorial orbit ranging from a high point (apogee) of 23,000 miles to a low point (perigee) of 12,500 miles. That's unusual because most hamsats have been launched to north-south polar orbits in which they pass over Earths poles. ARSENE's orbit, on the other hand, girdles the planet.

The high-altitude equatorial orbit provides stations on the ground, up to 40 degrees latitude, 12 hours a day access. It takes ARSENE 17 hours 30 minutes to orbit Earth. One-third of Earth is in view of the satellite at one time. ARSENE drifts slowly from west to east.

ARSENE is a three-axis stabilized hamsat with Sun and Earth sensors. It can control its own attitude, using gas jets about twice a year under ground command. At 350 lbs., it is the heaviest amateur radio satellite launched to date.

ARSENE carries three packet transponders, but no message bbs. A user contacts another station directly through the satellite.

1993: UNAMsat-1

Hamsats are a source of national pride. Pakistan's first satellite was Badr-1 in 1990. South Korea's first was KITsat-A in 1992. Amateur radio again captured the heart of a nation in 1993-94 with Mexico's first home-brewed satellite.

Mexico already owned two large American-made commercial satellites, called Morelos, in the Clarke Belt and two more American-made satellites, known as Solidaridad, were under construction for launch to stationary orbit. The nation even had an astronaut aboard an orbiting U.S. space shuttle in 1985. But, UNAMsat was the first satellite designed and built by Mexicans.

UNAMsat is named after the Spanish initials of the National Autonomous University of Mexico where hams and students assembled it.

UNAMsat was to be launched by the Russian Space Agency and Russian Academy of Sciences alongside a weather satellite on a decommissioned SS-18 intercontinental ballistic missile.

Orbiting at an altitude of 621 miles, UNAMsat was to pass over Mexico six times a day. While overhead for 22-24 minutes, it would transmit information to scientists at the university.

UNAMsat is a nine-in. cube patterned after a standard AMSAT microsat. As with earlier microsats, the framework houses five modules. AMSAT refers to the fifth module as "This Space For Rent" (TSFR). Four of UNAMsat's modules are updated copies of earlier microsat hardware and software. The TSFR module holds an innovative meteor sounder experiment.

UNAMsat is an amateur radio pacsat, but its first priority is detecting meteorites and measuring how frequently they fall into Earth's atmosphere, as well as supporting research into seismic activities, tide movements and volcanic activity in Mexico. When UNAMsat is not sounding meteors, it switches to digital mailbox work.

Meteors are streaks of light from burning small bits of rock or interplanetary debris, often the size of a grain of sand. They arrive regularly in Earth's atmosphere. Friction heats the particle plunging through Earth's atmosphere and makes it glow in the air, causing the streak of light also known as a shooting star or fireball. A meteorite is a chunk large enough to survive the fall to Earth's surface.

The non-amateur meteor sounder is a 60-watt transmitter pulsing every few seconds on a frequency near 40 MHz. The pulses strike meteors dropping through Earth's atmosphere. Echoes from the meteors are detected by a receiver which measures their Doppler shift. From that, scientists can calculate meteor speed and distribution.

UNAM scientists are searching for high-velocity meteors originating outside our Solar System. UNAMsat might shed light on whether comets originate in our Solar System. If Doppler shift shows speed greater than the escape velocity from our Solar System, that might suggest an origin beyond the Solar System.

Amateur radio operators on the ground also can use the hamsat as a meteor radar.

1993: ITAMsat-1

Paying for a ride on an Ariane rocket ASAP platform is a way for satellite builders to get their spacecraft to orbit in the 1990s. Four hamsats and two other microsats were launched on an ASAP in one Ariane flight in 1993. The completely-amateur satellites were ITAMsat-1 from Italy and KITsat-B from South Korea. Two partly-amateur satellites were the Portuguese university satellite POsat and Italy's EYEsat, a brother to ITAMsat.

The other two satellites on the ASAP were the medical-mail pacsat Healthsat and Stella, a German geodetic satellite similar to the American Laser Geodynamics Satellite (LAGEOS) which measured movement of Earth's crust by laser beam.

ITAMsat was the first satellite ever built by Italian amateur radio operators. North American AMSAT provided design experience to its affiliate in Italy during ITAMsat construction at Milan. Mechanically, ITAMsat was to be the same as earlier AMSAT microsats. All of it, except the solar panels, was built in Italy.

Like most amateur radio satellites launched in the 1990s, ITAMsat is a digital pacsat. It has a 1200-baud and 9600-baud transponder with uplinks at 145 MHz and downlinks at 435 MHz.

1993: EYEsat

On the ASAP with ITAMsat-1 was its part-commercial, part-amateur brother EYEsat.

The amateur payload is called Amrad. Uplinks are at 145 and downlinks at 436 MHz. EYEsat's primary mission is non-amateur.

1993: KITsat-B

South Korea's second hamsat, KITsat-B, was aboard the same ASAP as ITAMsat.

KITsat-A had been launched in 1992 to become KO-23. Its builders called it Uribyol, Korean for "Our Star." The 110-lb. KITsat-B was built in Korea, by the engineers trained at Surrey, to become Uribyol No. 2 as well as KITsat-OSCAR.

KITsat-B cloned KO-23 and added a GPS navsat receiver for position-location experiments. The small amateur radio satellite photographs Earth, detects cosmic particles, measures cosmic rays, stores and delivers E-mail messages and voice mail, and makes digital voice broadcasts.

KITsat-B has an upgraded version of the UO-22 imager shooting photographs with both four-kilometer and 400-meter resolution. There are two CCD cameras, one with wide-angle lens and one with telephoto lens. The wide-angle is used to spot areas to be photographed. Then, detailed images are made with the narrow-field camera.

KITsats were built at Great Britain's University of Surrey for the Korean Advanced Institute of Science and Technology by Korean students guided by Surrey engineers. KAIST and the Korean Amateur Radio League designed a dedicated hand-held amateur radio receiver for cheap-and-easy reception of KITsat synthesized-voice broadcasts.

1993: PoSAT-1

Portugal's 110-lb. UoSAT-size satellite is part commercial and part amateur. Known as PoSAT-A or PoSAT-1, it was launched on the same ASAP platform as ITAMsat, EYEsat and KITsat-B.

PoSAT carries two cameras seeing 600-ft. objects, cosmic ray instruments, GPS navigation receiver, CCD star sensor and pacsat mailbox.

PoSAT-1 was built by staff at the University of Surrey and a team of four Portugese industry engineers. Its primary mission is non-amateur. Stations equipped for UO-22 can download Earth photos and GPS data from PoSAT-1.

PoSAT-B will be completed and launched later.

1993: HealthSat-A

HealthSat-A is not an amateur radio satellite, but an amateur-related pacsat built by radio amateurs for a private corporation, SatelLife, and ferried to orbit alongside ITAMsat and KITsat-B.

Ham operators who are doctors and who have OSCAR experience use HealthSat-A to send and receive fresh medical literature and electronic mail via satellite to remote areas of the world. SatelLife operators are experienced with pacsats, having used UO-14 and UO-22 to test methods for Healthsat and to start HealthNet, an international non-profit satellite E-mail network for health workers.

1993: Radiosputnik-15

The low-flying Radiosputniks from RS-10 to RS-14 were popular with amateur radio satellite enthusiasts around the world. Sticking with a good idea, Russian hams offered the amateur satellite community a clone, RS-15.

The 155-lb. Russian hamsat was assembled by the same team which had constructed all previous Radiosputniks—NPO of Applied Mechanics in Krasnoyarsk. The hamsat's radio, labeled BRTK-11, was designed in the Laboratory of Space Technology at the Tsiolkovsky Museum of Cosmonautics at Kaluga.

The BRTK-11 electronics and communications "radiotechnical complex" includes transponder, bbs and two telemetry beacons.

RS-15 was to be launched in 1993 to a circular, 1430-mi.-high orbit. Launch and operation of the hamsat was coordinated by the Laboratory of Aero-Cosmic Technology of the Russian Defense and Technical Sports Organization.

Information is available from ROSTO, ul. Zemlynoi Wal 46/48, Moscow 103 064, Russia.

Konstantin Tsiolkovsky

RS-15's radio was designed in the Laboratory of Space Technology at the Tsiolkovsky Museum of Cosmonautics at Kaluga.

Theorist Konstantin Eduardovich Tsiolkovsky was born in 1857, the son of a Polish forester who had moved to Russia. He taught himself mathematics and became a high school teacher in the small town of Kaluga, 90 miles south of Moscow. Tsiolkovsky's astronautics writings brought him international recognition. He never built rockets, but encouraged young engineers who did, including Sergei Korolev who became chief designer of USSR spacecraft in the 1950s.

After Tsiolkovsky died in 1935, the Tsiolkovsky Museum of Cosmonautics was built at Kaluga. A crater on the back of the Moon was named for him.

1994: SEDsat-1

As smaller satellites do bigger jobs, amateurs become more creative in planning their spacecraft. For instance, SEDsat-1 is to be an unusual science and communications hamsat flying at the end of a long tether during one of NASA's Small Expendable Deployer System (SEDS) missions.

SEDsat stands for Students for the Exploration and Development of Space Satellite. SEDsat-1 was built by the University of Alabama–Huntsville (UAH) chapter of Students for the Exploration and Development of Space.

SEDS-1 was a 57-pound box the size of a portable TV set unreeled on twelve miles of cord in 1993. The tether was about the same thickness and texture as dental floss. It created the longest spacecraft ever flown by the space agency.

The tether was spooled out to its full length for ninety minutes, then cut loose. The box and its cord reentered Earth's atmosphere and burned.

SEDsat-1 will be a secondary payload on a rocket ferrying a large commercial satellite to orbit. After dropping off its main payload, the rocket's upper stage will move into an elliptical orbit at an altitude just above 400 miles.

There, as the rocket upper stage revolves around Earth, SEDS will unreel a 26-mi. tether with SEDsat-1 as a 50-lb. weight at the end.

From its end-of-the-string viewpoint, SEDsat-1 will send down television images of the cable. Then, the hitting of rocket brakes will snap the tether and flip SEDsat 30 miles higher where it will fly in its own orbit for five years.

The hamsat electronics are similar to earlier microsats—television camera, transponder, speech synthesizer to read news bulletins, packet-radio mailbox, telemetry beacon. In addition, SEDsat has an accelerometer to record debris hits on the spacecraft and send down data on frequency, magnitude and direction of impacts.

Adding a different kind of educational facet to its mission, SEDsat will house a potent parallel-processing mini-supercomputer. Science and engineering students around the globe will run their own programs through the SEDsat processor.

Spacecraft engineers would like to use tethers in the future to maneuver payloads in space from one orbit to another without using fuel; to generate electricity as a conductive tether sweeps through Earth's magnetic field; to drag recoverable probes through regions of Earth's upper atmosphere now inaccessible to direct study; and to create artificial gravity by spinning a spacecraft on each end.

1994: SUNsat

Another hamsat will give a boost to national pride, this time in South Africa.

Ham operators in the Southern African Amateur Radio Satellite Association (AMSAT-SA) helped build SUNsat for the University of Stellenbosch near Cape Town. SUNsat is an acronym for Southern Universities Satellite. In orbit the hamsat will be called SUNsat-OSCAR.

The university recently built a new technology park for its Bureau of Systems Engineering which is in charge of the satellite project. AMSAT-South Africa, the university and the South African electronics firms Altech, Grinaker Electronics and Plessey, worked together in building SUNsat.

Europe's spysat Helios is to be launched on an Ariane rocket. SUNsat will ride along to be dropped off in its own polar orbit, similar to the orbits of the earlier microsats UO-14 to UO-19.

SUNsat will be a combo spacecraft, reminiscent of some recent Russian hamsats. AMSAT-South Africa will operate the amateur radio section of the spacecraft's payload. SUNsat will be a pacsat with bbs. It also will offer position locating and photography of Earth.

The university will operate student science experiments carried in SUNsat and the electronics firms will operate other non-amateur commercial sections of the satellite.

1995: Guerwin-1 TECHsat

Israel's first hamsat, known as Guerwin-1 or TECHsat, will be launched from Russia's Baikonur Cosmodrome in Kazakhstan aboard a Russian rocket in March 1995.

Technion is the Israel Polytechnical Institute. Asher Space Research Institute is on the Technion campus in the northern city of Haifa. Engineering students there designed the new Guerwin-1 amateur radio communications and photography satellite.

The students started the hamsat project in 1990. Technical-design assistance was received from twelve companies including Israel Aircraft Industries, Digitron, Motorola, Tadiran and others.

Guerwin-1 was to have been launched on one of Europe's Ariane rockets in 1993, but was switched in 1993 to the 1995 Russian launch. It will be carried as a secondary payload alongside Russian and German satellites.

Russia and Israel had no diplomatic ties for twenty years after the 1967 Arab-Israeli War, then links were rebuilt at the end of the 1980s. More than 450,000 Russian Jews have moved to Israel since 1989. Russia's Space Research Institute will test the hamsat before launching it.

The 130-lb. microsat will have a bbs able to receive 1200- and 9600-baud signals at the same time. Downlinks will be at 29 and 435 MHz. Uplinks will be at 145, 1260 and 2400 MHz.

Guerwin-1 will have a TV camera to snap photos of Earth and let amateur meteorologists observe cloud formations. The hamsat will measure radiation in space and have a satellite receiver for GPS location experiments.

Israel launched two of its own small test satellites, in 1988 and 1990, but they since have burned in the atmosphere.

1995 CEsar-1

Amateur radio operators in the Radio Club Federation of Santiago, Chile, plan to build a microsat to be called CEsar-1 and launched in 1995. CEsar-1 will be similiar to AO-16, LO-19, WO-18, and DOVE, and in a 500-mi.-high polar orbit. CEsar-1 will carry scientific experiments constructed by students at three local universities with help from the Chilean Air Force.

1995: NANOsat

If 25-lb. amateur radio satellites are microsats, what should we call even smaller hamsats? Nanosats, after the metric prefix for one-billionth.

British radio amateurs reportedly have designed a five-lb. NANOsat to be ferried in a Progress space freighter to Russia's orbiting Mir space station and tossed by hand from a station airlock to its own low orbit. The transmitter would broadcast synthesized speech for two months.

1996: AMSAT-Phase-3D

AMSAT's third-generation satellites are larger, more complex spacecraft, using more modes and higher radio frequencies, and flying longer elliptical paths known as Molniya orbits after a class of Russian communications satellites.

The first, Phase-3A, was destroyed in a launch explosion in 1980. Phase-3B was launched in 1983 and is in orbit today as AO-10. Phase-3C, was launched in 1988 and is in space as AO-13.

After its launch in 1996, Phase-3D will marshal an extensive array of transponders and directional high-gain antennas.

At 950 lbs. and nine-feet diameter, Phase-3D will be larger than AO-13. It has a 250-watt transmitter and microwave transponders at 5.6, 10 and 24 GHz. Finnish hams built the 10 GHz gear.

AMSAT-Japan contributed a television camera called SCOPE with wide-angle and telephoto lenses—24 and 90 degrees—for color photos of Earth plus a lens to record stars and planets.

Phase-3D is a challenging project for AMSAT, costing $5 million to construct and launch. AMSAT-Germany designed it with government funds for research in satellite communications. American hams donated $1.5 million. Framework fabrication work was done on numerical-controlled machines at Weber State University, Ogden, Utah.

Phase-3D will be launched on a European rocket to an orbit ranging from 2,500 miles to 30,000 miles. It will cicrle Earth every 16 hours, staying over ground stations in North America, Europe, and Asia for hours at a time.

1997: HUTsat

Sponsored by the AMSAT group in Finland, HUTsat will be launched on a European rocket in 1997-98. The digital-signal-processing pacsat will have a bbs, transponder, telemetry beacons and frequencies near 435, 1260, and 2400 MHz.

2000: AMSAT-Phase-4

The dozens of Phase-1, Phase-2, and Phase-3 OSCARs and amateur radio-related satellites sent to space so far have not been in stationary orbits. A Phase-4 hamsat, on the othr hand, would be a large complex spacecraft in stationary orbit 23,300 miles above Earth. From there, it would seem to remain permanently over ground stations for use at any time of day or night.

Amateur radio satellites so far haven't been able to travel as fast as the Earth turns so they appear to pass overhead. They sail around the globe on lower paths which bring them over ham stations on the ground at regular intervals. They are not always in view of any one ground station.

At the end of the 1980s, AMSAT was planning a new spacecraft to stare down on the Northern Hemisphere continuously from stationary orbit. Senior engineering students at the Center for Aerospace Technology (CAST) at Weber State University, Ogden, Utah, volunteered 10,000 hours of labor to build a mock-up of Phase-4A.

The full-scale Phase-4A model displayed in 1989 was awesome. The 12-sided, 8-ft.-diameter, 30-in.-high spacecraft was the largest satellite AMSAT ever had designed. In fact, it was so large a student was able to crawl inside.

Unfortunately, by 1992, other demands on manpower and financial reserves forced AMSAT to put Phase-4A blueprints on hold, awaiting plans for the next century.

Search And Rescue Satellites

A vital spin-off from comsat technology has been the globe-circling complex of life-saving satellites known as SARSAT/COSPAS.

SARSAT is shorthand for Search and Rescue Satellite-Aided Tracking System. COSPAS is a Russian acronym for Space System for Search of Vehicles in Distress.

The satellites are weather, navigation, research and communications spacecraft with extra radio gear stowed aboard to receive and relay distress messages from emergency locator beacons carried by ships and planes. SARSAT ground stations span the globe.

Canada, France, Russia and the United States founded the network based on American and Soviet spacecraft in 1982. American SARSATs are aboard the U.S. National Oceanic and Atmospheric Administration's NOAA and GOES weather satellites. COSPAS satellites are navigational satellites for Russia's merchant marine. Usually, there are three Russian and three American SARSATs working in polar orbits.

Today, a dozen countries monitor the clutch of SARSATs to trace ships lost at seas and downed aircraft. Authorities credit SARSATs with having saved 2,400 persons around the world from shipwrecks, plane crashes and even a dog sled race which went awry.

Pioneers. A private amateur radio satellite built by ham operators in the 1970s led to today's international SARSAT/COSPAS system.

The Radio Amateur Satellite Corporation (AMSAT) is a global network of hams who build and operate communications and science satellites. AMSAT proved the SARSAT/COSPAS concept could work when its orbiting hamsat OSCAR-7 received low-power radio signals from the ground and repeated them to Goddard Space Flight Center in Maryland in December 1975.

OSCAR-7 showed that a weak signal from the ground could be tracked by satellite to within two to four miles of an emergency site.

Officially, SARSAT/COSPAS started in 1982. Since then, it is said 1,200 lives of air and sea accident victims—mostly Americans—have been saved by rescue teams on the ground alerted by SARSATs. Today, there are more than half a dozen government SARSAT/COSPAS satellites listening in space.

ELT. An emergency locator transmitter, or ELT, is activated upon impact to send out a warbling alert tone from a downed aircraft or automatically activated by a ship sinking at sea.

ELTs transmit on the international civilian aviation distress frequency of 121.5 MHz and on the international military distress frequency of 243 MHz. The newest ELTs also transmit on a new frequency of 406 MHz. Most aircraft are required to monitor emergency frequencies.

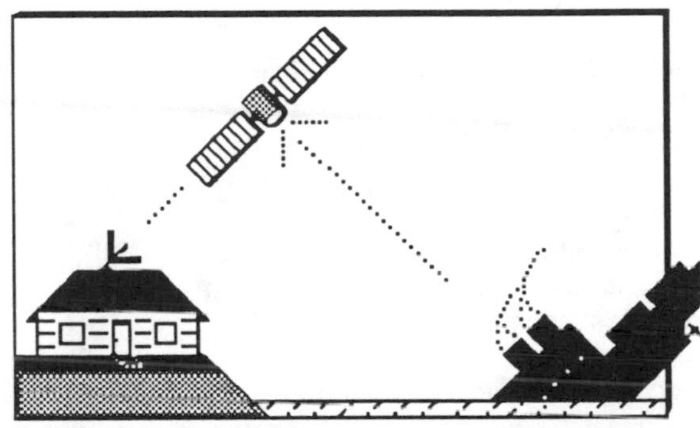

SARSATs receive and relay emergency beacon signals from ships and planes.

EPIRB. An EPIRB, or Emergency Position Indicating Radio Beacon, is a small transmitter used by boaters who are out of two-way radio range to summon aid.

A class-A EPIRB automatically floats free and turns on. A class-B EPIRB is activated manually.

The beacons transmit from 48 to 72 hours on 121.5 and 243.0 MHz. Since they send out only weak line-of-sight signals, they usually are found by nearby airplanes which can see sufficient horizon to cover a sizable area of land or sea.

The latest EPIRB is referred to as "406" because it transmits on 406 MHz. The 406-EPIRB transmitter is more powerful than older models. Its signal is coded with an identification number, letting a ground station know who is in trouble and where to call to verify the emergency.

A 406 also transmits on 121.5 MHz so planes and ships can home in as they get close. Unfortunately, 406's are expensive compared with older models. Old styles sell at prices around $200 while new types cost nearly $2,000.

PLB. A personal locator beacon, or PLB, is a small ELT worn by a camper, hunter and other person in a remote area.

SARSATs. While carrying out their assigned primary tasks, the SARSATs also continuously monitor those international radio distress frequencies for ELT signals.

When a SARSAT hears an ELT, it retransmits the signal to one of seven ground receiving stations, mostly in the Northern Hemisphere. The ground station crew analyzes the Doppler shift of the ELT signal to "fix" or pinpoint the emergency location to within two to four miles.

If an emergency signal is being transmitted by a 406-EPIRB, the SARSAT itself can calculate the location to within one mile, making it easier to find the lost soul and cutting rescue costs.

Meanwhile, boaters in trouble in the Southern Hemisphere sometimes are out of luck. Their signals may be heard and repeated by a satellite, but no ground station is beneath the satellite to receive the alert.

The satellites. A variety of satellites carry SARSAT/COSPAS equipment. For instance, America's NOAA-9, NOAA-10 and NOAA-11 weather satellites.

Launched in 1988, polar-orbiting NOAA-11 doubles as a SARSAT. It listens for distress signals from ships and planes as it circles Earth every 102 minutes at an altitude of 540 miles. As Earth rotates beneath NOAA-11, the satellite scans the entire globe every 12 hours. If it hears an emergency signal, NOAA-11 relays the distress call to a rescue service on the ground.

America's GOES weather satellites also house SARSAT receivers. They are stationary satellites, sending down cloud pictures and atmospheric temperatures from 22,300-mi.-high orbits.

SARSATs in low polar orbits have receivers tuned to the worldwide aircraft emergency beacon frequency of 121.5 MHz, while GOES satellites listen on a new SARSAT frequency of 406 MHz. The higher altitude of GOES satellites lets them hear distress calls earlier to signal the first alert if an emergency is occurring. After a GOES alert, satellites in lower polar orbits pinpoint the location of an emergency beacon signal.

The U.S. National Oceanic and Atmospheric Administration controls GOES and NOAA weather satellites from Camp Springs, Maryland.

Russian. After the COSPAS/SARSAT system started in 1982, the Soviet Union launched three navigation satellites carrying search-and-rescue receivers. They were Cosmos-1383 in 1982, Cosmos-1447 in 1983 and Cosmos-1574 in 1984.

Among several more COSPAS satellites was the civilian navigation satellite Nadezhda, or Hope, launched in 1989 to an altitude of 625 miles.

Another was Russia's Cosmos 1861 launched 600 miles above Earth in 1987. That satellite also housed the amateur RS-10 and RS-11 payloads.

Sailors. Two Maryland men sailing a 35-ft. boat from Massachusetts to Bermuda owe their lives to a SARSAT which relayed their distress signals in a storm.

The boat was tacking 140 miles southeast of Nantucket in 1987 when heavy wind and waves broke the mast. When the small craft took on more water than bilge pumps could handle, Chris Burtis of Annapolis and Troy Wilson of Bowie switched on their distress transmitter.

A SARSAT passing overhead picked up their signal and relayed it to Scott Air Force Base in Illinois. The Air Force used the distress signal to determine the location of the sailboat, then sent the information to a Coast Guard air station on Cape Cod. The Coast Guard rescued the sailors.

Ontario. The weather satellite NOAA-10 was unlimbering during its very first day of operation in space in 1986 when it saved the lives of Rory Johnston and three other Canadians after their small airplane crashed in a remote area of Ontario.

The Cessna lost power during takeoff in poor weather, forcing an emergency landing on a lake. The plane sank, nose down, in eight feet of water.

The ELT aboard Johnston's plane activated automatically upon impact.

With face cuts, bruised shoulder and dislocated wrist, Johnston swam 200 yards to shore where he found a canoe. He paddled back out to his sunken craft and brought his injured passengers ashore.

NOAA-10 had just been activated by ground controllers during its 76th trip around the globe. While passing over Canada during its 90th revolution, the satellite's receiver heard Johnston's emergency beacon. NOAA-10 reported its discovery to Canadian rescue forces at Trenton.

At the same time, a Russian COSPAS satellite also reported hearing Johnston's beacon beaming up from Ontario. And the pilot of a private plane heard and reported Johnston's emergency beacon to authorities. NOAA-10 reported the signal was still there as it completed its 91st circuit.

The combination of reports from American and Russian SARSATs and the private pilot alerted rescue teams in Edmonton, Alberta. A four-engine C-130 Hercules with paramedics aboard took to the air in search of Johnston. Poor weather prevented the C-130 crew from spotting Johnston's plane that night. When fog lifted the next morning, the C-130 crew found the Cessna exactly where the SARSAT/COSPAS satellites had said it was.

Two medical technicians parachuted in to

provide first aid. Johnston and his three passengers were flown to an airstrip at Sachigo Lake, transferred to another plane and taken to Winnipeg, Manitoba, where they were hospitalized.

Pacific. The 459-ft. freighter Independencia out of Port Washington, New York, was 143 miles northeast of Kanton Island, halfway between Hawaii and Australia, enroute from Ecuador to China, when it sank in the Pacific Ocean early one December morning in 1991.

As its crew of 19, including seventeen Pakistanis, one Kenyan and one Turk, abandoned ship between 3:20 and 5:15 a.m., a search and rescue team radioed the captain, asking him to turn on an EPIRB in his life raft.

The beacon transmitted a signal to outer space where it was heard by the umbrella of polar-orbiting SARSATs and tracked by Russian and American ground stations.

The entire Independencia crew of 19 was picked up from their 20-man life raft that afternoon by U.S. Coast Guard motor vessel Columbus Virginia and delivered to Honolulu harbor.

Alaska. Alaska's remote North Slope proved to be a good place to pack a PLB. The borough covers 92,000 square miles, an area the size of Utah, and has only eight towns and villages. There are few roads and most travelers use amphibious vehicles in summer and snowmobiles in winter.

It was fortunate for 41-year-old native Alaskan hunter John Brower that the borough's Search and Rescue Department had a PLB network in October 1992. He was snowmobiling in the wilderness four days out of Barrow on a Saturday afternoon when severe abdominal cramps developed.

Brower triggered his personal locator beacon. At 8:45 p.m., a Russian COSPAS picked up the distress signal and relayed it to a local-user terminal in Fairbanks which sent it through the Air Force Rescue Coordination Center in Anchorage to Alaskan officials and the North Slope borough Search and Rescue Department.

A Jet Ranger search helicopter was alerted at 9:52 P.M. and off the ground under a 200-foot ceiling at 10:42 p.m. The PLB coordinates relayed by satellite guided the pilots to within two-tenths of a mile of Brower's location. Just 2 hours 48 minutes after the PLB first was received on the ground, the rescuers were over him at 11:33 p.m.

Despite blowing snow and icing landing conditions, the pilots set down in the pitch black and reached Brower who was bent over with pain. Despite the very rapid response, officials reported Brower's first words were, "What took you so long?" The rescue team delivered Brower to a Barrow hospital at 12:24 am Sunday morning. Doctors found an intestinal blockage.

Ethiopia. In 1989, U.S. Rep. Mickey Leland and thirteen others, making their way from the Ethiopian capital of Addis Ababa to the village of Fugnido, were lost when their plane neither landed at Fugnido nor returned to Addis Ababa.

A NOAA weather satellite, circling the globe in a 527-mi.-high polar orbit, monitored distress signals from the missing Ethiopian government Twin Otter turboprop aircraft. The satellite relayed the ELT signals to a new tracking station under construction at Bangelore, India.

Even though the station in India was in a poor position for tracking an ELT in Ethiopia, its technicians stopped construction and activated their receivers. First, the Ethiopian ELT signal was fixed at a point 100 miles south-southeast of Addis Ababa. The next day, the SARSAT received a signal from 100 miles north-northeast of the city.

False alarms. Sometimes, ELTs and EPIRBs are turned on accidentally. In fact, they suffer a 95 percent false-alarm rate. Only three percent of EPIRB signals come from true emergencies. And older EPIRBs identify neither themselves nor the source of an emergency call.

Because, 406-EPIRBs send out a signal coded with an identification number, ground stations know who is in trouble and where to call to verify an emergency. False alarms are cut to five percent.

Use of an aircraft ELT is legal only when a plane has crashed. Shipboard ELTs and boaters' EPIRBs are legal only for Mayday emergencies in which a vessel is in danger of sinking or there is a medical emergency on board.

Weather Satellites

Weather satellites are vital tools for forecasters, showing storm systems, weather fronts, jet streams, upper-level troughs and ridges, fog, sea ice, snow cover, upper-level wind direction and speeds and cloud formations.

Coastal and island weather stations use satellite data to track tropical storms, hurricanes and typhoons. Visible-light and infrared satellite photos show remote desert, ocean and polar areas where weather would be unknown. The satellites even record ocean surface temperatures for fishing and shipping.

Television news programs often show pictures of cloud patterns snapped from overhead only an hour or so before air time. A weatherman points out differences between places with clear skies and swirling storm clouds. A series of time lapse photos made over a 24-hour period show movement of a storm.

As familiar as they are today, such weather satellite photos were not possible before 1960. That's when Harry Wexler, director of research for the U.S. government Weather Bureau, turned his dream into reality. Wexler had been a leading proponent of weather satellites. He had imagined that cameras in space satellites, orbiting above the clouds, would see thick bands of clouds swarming along weather fronts, popcorn fair weather clouds, and enormous, revolving hurricanes over the oceans.

Explorer. The earliest, experimental TV pictures of Earth's cloud cover were sent down by an American satellite called Explorer 6 after it was launched in August 1959.

TIROS. Eight months later, the prototype weather satellite Television and Infra-Red Observation Satellite (TIROS) was launched to a 600-mi.-high orbit in April 1, 1960.

Wexler was proven correct within hours of the TIROS-1 launch when government scientists showed U.S. President Dwight D. Eisenhower photos of clouds from space. The first weather satellite picture was of clouds over the Gulf of St. Lawrence. The 270-lb. TIROS-1 worked only 78 days, but sent down 22,952 photos.

TIROS became a series of ten experimental weather satellites, launched by NASA between 1960 and 1965 to test television cameras as well as Sun-angle and horizon scanners for meteorology. TIROS-2 was launched in November 1960. TIROS-3 in July 1961. The last in the series was TIROS-10 launched in 1965. The satellites orbited above the equator, carrying two TV cameras each, one with wide-angle lens and one with narrow-angle lens for more detail. Each had an horizon seeker, infrared sensors, tape recorder for storing images, and three radios for transmitting pictures and data to Earth. Solar cells converted sunlight to electricity stored in nickel-cadmium (NiCad) batteries.

TIROS Satellites

Each TIROS satellite is listed, including its working life in days and the number of pictures acquired.

TIROS	Launch	Days	Pictures
TIROS 1	1960 Apr 1	89	22,952
TIROS 2	1960 Nov 23	376	36,156
TIROS 3	1961 Jul 12	230	35,033
TIROS 4	1962 Feb 8	161	32,593
TIROS 5	1962 Jun 19	321	58,226
TIROS 6	1962 Sep 18	389	68,557
TIROS 7	1963 Jun 19	1809	125,331
TIROS 8	1963 Dec 21	1287	102,463
TIROS 9	1965 Jan 22	1238	88,892
TIROS 10	1965 Jul 2	730	78,874
Advanced	**Experimental**	**TIROS**	
TIROS M		1970 Jan 23	
TIROS N		1978 Oct 13	

Nimbus. Nimbus is a meteorologist's name for rain clouds. It also was America's second generation of weather satellites. Seven experimental Nimbus satellites were sent to polar orbit between 1964-1978 to take pictures at night as well as during daylight hours.

Nimbus-1, launched in 1964, was the first polar-orbiting U.S. weather satellite. The Nimbus series had better still-photo cameras, television cameras for mapping clouds, and infrared radiometers for night photography. Nimbus satellites flew at 600-mile-high altitudes.

It automatically sent down high-resolution TV pictures of cloud cover in visible and infrared light immediately after they were taken. Anyone beneath a Nimbus satellite with a proper radio and fax machine could receive weather photos.

High-resolution infrared radiometers, seeing differences between cloud and surface temperatures, helped meteorologists map night-time clouds.

Spectrometers and radiometers measured atmospheric temperature, ozone, water-vapor and solar radiation.

The 10-ft.-tall, butterfly-shaped Nimbus satellites were built by General Electric. Their weights ranged from 830 lbs. to 2,176 lbs. A five-ft. aluminum ring, attached to an upper compartment by a magnesium truss, housed sensors and electronic components and three TV cameras which delivered finer detail than TIROS cameras. An attitude-control system in the upper compartment pointed the cameras toward Earth. Outside were two solar panels sopping up sunlight to produce 550 watts of electrical power for Nimbus.

Some 1,440 pictures of a cross-section of Earth's surface 1,500 miles east-west and 500 miles north-south were taped each day for playback over ground tracking stations.

Nimbus Satellites

Each Nimbus satellite is described.

Nimbus	Launch	Results
Nimbus 1	1964 Aug 28	Transmitted 27,000 cloud-cover photos, until shut off Sept. 23, 1964.
Nimbus 2	1966 May 15	Transmitted infrared and TV cloud-cover photos.
Nimbus B	1968 May 18	Rocket guidance failed, destroyed enroute to orbit.
Nimbus 3	1969 Apr 14	Sent infrared, ultraviolet, television, geodetic data; first U.S. satellite to measure atmosphere day and night temperatures globally at various altitudes.
Nimbus 4	1970 Apr 8	Like Nimbus 3 with new & improved instruments.
Nimbus 5	1972 Dec 11	Like Nimbus 4 with new & improved instruments.
Nimbus 6	1975 Jun 12	Like Nimbus 5 with new & improved instruments.
Nimbus 7	1978 Oct 24	First satellite built to monitor Earth's atmosphere for natural and artificial pollutants.

ESSA Satellites

Nine ESSA satellites were launched.

ESSA	Launch
ESSA 1	1966 Feb 3
ESSA 2	1966 Feb 28
ESSA 3	1966 Oct 2
ESSA 4	1967 Jan 26
ESSA 5	1967 Apr 20
ESSA 6	1967 Nov 10
ESSA 7	1968 Aug 16
ESSA 8	1968 Dec 15
ESSA 9	1969 Feb 26

ESSA. In 1965, the U.S. Weather Bureau became part of a new Environmental Science Services Administration (ESSA) within the Department of Commerce and the next series of nine weather satellites launched from 1966-1969 came to be known as ESSA. The nine satellites formed the so-called TIROS Operational System (TOS).

The ESSA satellites, flying at altitudes around 900 miles above Earth, had the advanced cameras tested early in Nimbus satellites. ESSA cloud-cover photos were recorded in space and transmitted to Earth automatically. Anyone with the simple required equipment could receive them. Other nations installed gear and, today, more than 120 countries get their weather pictures from various U.S. satellites.

Since 1966, the entire Earth has been photographed at least once a day.

ITOS. The ESSA series of weather satellites was followed by the ITOS series, patterned after TIROS. ITOS stood for Improved TIROS Operational Satellite. The first to orbit was TIROS-M, launched January 23, 1970. In orbit, it became known as ITOS-1.

NOAA. The polar-orbiting satellites which followed ITOS-1 into orbit from 1970-1976 were called NOAA, for ESSA's successor—the National Oceanic and Atmospheric Administration.

America's first polar-orbiting NOAA satellite, NOAA-1, was launched December 11, 1970. It had a camera for continuous day and night pictures.

NOAA-1 to NOAA-5 were like ESSA satellites. They were launched into 900-mile orbits to see the entire Earth each day. Night cloud photos were made with infrared sensors. Radiometers measured vertical temperature profiles of the atmosphere. Each satellite covered the globe every 12 hours and could relay photos and radiometer data automatically or store them for playback later.

Third Generation. The third generation of U.S. polar-orbiting weather satellites began operation with the launch of TIROS-N in 1978. TIROS-N was a research prototype for the follow-on operational series of satellites which were called NOAA-A to NOAA-N before launch and numbered NOAA-6 and above after launch.

Beginning with NOAA-6 in 1979, the NOAA name and numbering sequence was continued. NOAA-6 to NOAA-13 are modernized TIROS satellites, part of a cooperative program between Canada, France, Great Britain, NASA and NOAA.

NOAA satellites are equipped with sophisticated cameras and radiometers as well as infrared, stratospheric, and microwave sounders. In addition, they provide readings of sea-surface temperatures, identify snow cover and ice at sea, and measure particle densities in the upper atmosphere in order to predict the onset of solar disturbances.

★NOAA-A was launched in 1979, renamed NOAA-6, and deactivated in 1987.

★NOAA-B was launched in 1980, but failed to reach its proper orbit.

★NOAA-C was launched in 1981, renamed NOAA-7, and deactivated in 1986 after electrical power failed.

★NOAA-D was dropped out of sequence in favor of NOAA-E, which was a longer spacecraft and could hold more equipment, including a search and rescue radio receiver. NOAA-E was launched in 1983, and renamed NOAA-8 in space. Its clock and electrical power system broke down and it stopped working in 1985.

★NOAA-F was launched in 1984, and renamed NOAA-9 in space. Today, it is on standby in orbit with some data still being processed.

★NOAA-G was launched in 1986, and designated NOAA-10 in space. Today, it works, except for its Earth Radiation Budget Experiment (ERBE) scanner, which sticks or hangs up sometimes.

★NOAA-H was launched in 1988, and renamed NOAA-11 in space. Its instruments continue to work well, however its attitude control system has lost two of four gyros and was being controlled by special software.

★NOAA-D was launched to a 522-mi.-high polar orbit in 1991 to become NOAA-12. Like others since NOAA-6, NOAA-12 was a TIROS-N class of satellite built by General Electric to provide day and night environmental data.

★NOAA-I was launched August 9, 1993, to be NOAA-13. It is a TIROS-N class of satellite with five primary instruments. Unfortunately, NOAA-13 failed in orbit August 21, 1993.

NOAA Satellites

National Oceanic and Atmospheric Administration polar-orbiting NOAA meteorology satellites.

NOAA	Launch
NOAA 1	1970 Dec 11
NOAA 2	1972 Oct 15
NOAA 3	1973 Nov 6
NOAA 4	1974 Nov 15
NOAA 5	1976 Jul 29
TIROS N	1978 Oct 13
NOAA 6	1979 Jun 27
NOAA 7	1981 Jun 23
NOAA 8	1983 Mar 28
NOAA 9	1984 Dec 12
NOAA 10	1986 Sep 17
NOAA 11	1988 Sep 24
NOAA 12	1991 May 24
NOAA 13	1993 Aug 9
NOAA J	1994
NOAA K	1996
NOAA L	1997
NOAA M	1999
NOAA N	2000

Stationary satellites. At 22,300 miles above Earth, a stationary weather satellite seems to hang stationary above Earth's surface. It is at a much higher altitude than the polar-orbiting TIROS and Nimbus.

Meteorologists noticed when the U.S. launched the world's first stationary communications satellite in 1963. By 1970, weather forecasters wanted cloud pcitures shot more frequently than once or twice a day. They needed a satellite parked overhead, sending down pictures continuously from stationary orbit.

SMS. NASA built a pair of experimental Synchronous Meteorological Satellites (SMS) to test weather satellites systems in stationary or geosynchronous orbits.

An SMS was an 11-ft.-tall, 6-ft.-diameter, aluminum cylinder weighing 535 lbs. A spin-scan radiometer in the satellite stared down continuously, day and night, to record visible light pictures with one-half-mile resolution and infrared images with five-mile resolution.

SMS-1 was launched in 1974 to a spot 22,591 miles above the equator at 45 degrees west longitude. The satellite later was moved to 75 degrees west longitude. SMS-2 was launched in 1975 to a point above the equator at 135 degrees west longitude.

The two observed the Western Hemisphere continuously, transmitting cloud pictures every 30 minutes. SMS was a forerunner of NOAA's

Geostationary Operational Environmental Satellite (GOES) series of stationary weather satellites. GOES-1 was launched eight months after SMS-2.

SMS Satellites

NASA's Synchronous Meteorological Satellites tested weather satellites in stationary orbits.

SMS	Launch
SMS-1	1974 May 17
SMS-2	1975 Feb 6

GOES. When forecasters demanded a satellite to take pictures continuously, the result was Geostationary Operational Environmental Satellite (GOES) built for NOAA and launched by NASA.

The first series was launched in the 1970s to stationary orbit 22,300 miles above Earth: GOES-1 in 1975, GOES-2 in 1977, GOES-3 in 1978.

A second-generation series of four improved satellites was launched in the 1980s: GOES-4 in 1980, GOES-5 in 1981, GOES-6 in 1983, and GOES-7 in 1987.

The future, third-generation GOES-I, GOES-J, GOES-K, GOES-L and GOES-M were to be lobbed to space in 1994, 1995, 1999, 2000, 2005 and renamed GOES-8 to GOES-12 in orbit.

Built by Hughes, GOES satellites are 12-ft.-long, 7-ft.-diameter, 1,382-lb. cylinders containing spin-scan radiometers and vertical atmospheric sounders (VAS) producing simultaneous three-dimensional photos in visible and infrared light.

The solar-powered satellites have 16-inch telescopes looking down at visible and infrared light, day and night, to observe clouds, cloud heights, winds and vertical temperature profiles.

While NOAA satellites are lower and see more detail over a smaller area, GOES satellites see an entire hemisphere. GOES can photograph one third of the entire planet every 30 minutes, while NOAA satellites in lower polar orbits pass over a hurricane, for example, only once a day.

NOAA's National Environmental Satellite Data and Information Service (NESDIS) operates GOES satellites. Ground stations are NOAA's Satellite Operations and Control Center (SOCC) at Suitland, Maryland, and the Command and Data Acquisition Station (CDA) at Wallops Island, Virginia. Raw weather data is transmitted from a GOES satellite to Wallops, which processes the information and retransmits it through the spacecraft to various data users, including the National Weather Service.

GOES transmissions are received by manned and unmanned stations on the ground. Forecasters interpret the data to predict weather, ocean currents and river levels. Cloud pictures seen on television by Americans usually are from GOES satellites.

GOES shortage. There are supposed to be two GOES satellites watching North America at all times, transmitting images every half hour from offshore over the Atlantic and Pacific oceans.

GOES-5, launched in 1981, monitored the eastern U.S., while GOES-6, launched in 1983, observed the west. When GOES-5 failed in 1984, NOAA moved GOES-6 to 108 degrees West longitude to monitor the entire country. During hurricane seasons, GOES-6 was moved to 98 degrees West to increase Caribbean coverage.

Unfortunately, a replacement for GOES-5 was destroyed when its rocket failed during launch from Cape Canaveral in 1986, forcing NOAA to use the last GOES it had on the shelf. It was launched in 1987 to become GOES-7.

Unhappily, GOES-6 failed in 1989. NOAA forecasters were forced to rely on the single remaining satellite, GOES-7.

Ground controllers moved GOES-7 over the center of the U.S. in winters to watch Pacific storms affecting Alaska, Hawaii and the West Coast, then back over the East Coast in summers for hurricane season.

It takes about four months for the satellite to drift west from the Atlantic coast to a mid-continent position. No fuel is used; the satellite just coasts across the continent.

However, fuel is burned in thruster rockets fired to keep the satellite at an assigned position in orbit. GOES-7 was running low on fuel in 1994. NASA said conservative use might leave enough fuel to keep GOES-7 in place until 1995. After it runs out of fuel, it will drift away from its position over the equator.

GOES Satellites

Geostationary Operational Environmental Satellite built for NOAA and launched by NASA.

GOES	Launch
GOES 1	1975 Oct 16
GOES 2	1977 Jun 16
GOES 3	1978 Jun 16
GOES 4	1980 Sep 9
GOES 5	1981 May 22
GOES 6	1983 Apr 28
GOES 7	1987 Feb 26
GOES I	1994
GOES J	1995
GOES K	1999
GOES L	2000
GOES M	2005

Emergency. The U.S. declared a weather satellite emergency in 1991. Lease of a foreign weather satellite was ordered and the European Space Agency moved its Meteosat-3 weather satellite west over the Atlantic Ocean to cover the U.S. East Coast. That relieved GOES-7 to move west for coverage of the rest of the U.S.

If GOES-7 were to run out of fuel before the next GOES is launched, the U.S. would have to rely on pictures from the low-flying NOAA satellites, as well as U.S. military meteorological satellites and European and Japanese satellites.

Proper global stationary weather satellite coverage includes a Japanese Sunflower satellite over the western Pacific, Russian Meteor satellite over the Indian Ocean, European Meteosat over the eastern Atlantic, and the American GOES over the western Atlantic and eastern Pacific.

GOES-NEXT. The second GOES series is being replaced by an improved flotilla of third-generation satellites known as GOES-NEXT.

The first GOES-NEXT was to have been launched in 1989, but remained on the ground into 1994 after mirror flaws, wiring problems and infrared detector defects were found.

Five GOES-NEXT satellites are being built with mirrors to reflect light to several sensors. Routine tests before the planned 1989 launch found that temperature extremes—such as exposure to raw sunlight in space—could warp the surface of a mirror. Engineers coated the mirrors to reduce susceptibility to temperature extremes.

GOES-I, -J, -K, -L and -M were to be lofted to orbit between 1994 and 2005 and renamed GOES-8 to GOES-12 in orbit.

Today's NOAA, Nimbus and GOES satellites are essential to the U.S. National Operational Meteorological System. About one-third of the WEFAX (weather facsimile) pictures intended for the U.S. come from one of the polar-orbiting NOAA satellites; the rest from GOES.

Weather satellite photos have been useful in public emergencies and disasters. For example, California emergency workers used GOES-6 photos to locate Big Sur wildfires in 1985.

ERBS. The Earth Radiation Budget Satellite (ERBS) was ferried by space shuttle to orbit 375 miles above Earth in 1984 to collect measurements of radiation emitted by our planet. For meteorologists, it also monitored the vertical distribution of areosols in the stratosphere as well as ozone and carbon dioxide.

Military satellites. Many U.S. military weather satellites have been developed secretly and launched over the years since 1960. Two Defense Meteorological Satellites (DMS) usually can be found circling the globe every twelve hours in 500-mi.-high, north-south polar orbits.

After launch from Vandenberg Air Force Base, California, DMS satellites track storms and scan 1,600-mi.-wide swaths of Earth's surface, feeding cloud cover and temperature information to the military services. NOAA sometimes uses DMS information for civilian weather forecasts.

The Weather Bureau

U.S. government meteorology started as the General Weather Service in 1870 when Congress directed the Army to forecast storms over the Atlantic and Pacific coasts and the Great Lakes.

Farmers needed forecasts of weather and long-term climate trends so forecasting was moved to a new civilian Weather Bureau in the Department of Agriculture in 1891.

When aviators asked for frequent observations and short-term forecasts, the Weather Bureau was transferred to the Department of Commerce. In 1965, the Bureau became part of a new Environmental Science Services Administration (ESSA), with climatology separated into a new Environmental Data Service (EDS).

In 1970, the Bureau was renamed the National Weather Service (NWS), a part of the National Oceanic and Atmospheric Administration (NOAA) headquartered in Rockville, Maryland. NOAA was set up within the U.S. Department of Commerce during a 1970 federal government reorganization.

The National Weather Service operates the National Meteorological Center at Camp Springs, Maryland, National Hurricane Center at Coral Gables, Florida, and National Severe Storms Forecast Center at Kansas City, Missouri. NWS has hundreds of meteorological, oceanographic and hydrological stations which release millions of forecasts about weather in the United States and its possessions to the public each year.

NOAA also conducts geodetic surveys and environmental-information services for the public. It operates the National Ocean Survey and National Marine Fisheries Service.

Technology. In addition to satellite data and pictures, meteorologists use data transmitted 24 hours a day from more than a thousand manned and unmanned weather stations to gauge ozone levels, measure water vapor, survey pollution levels, plot storms, follow jet streams, examine fronts, watch fog, estimate snow and ice cover, monitor river levels and detect forest fires.

Weather satellites record atmosphere and ocean temperatures at various altitudes and depths, gauge rainfall for forecasting of droughts and harvests, survey chlorophyll content for crop health studies, spot forest fires, map ocean currents, see volcanic eruptions, and chart ice in shipping lanes.

As they monitor the total global environment, some also use search-and-rescue radios to listen for distress signals from ships at sea and downed aircraft (described in chapter three).

Radiometer. NASA recently devised a satellite thermometer 100 times more sensitive than earlier equipment. The new infrared radiometer measures ocean surface temperatures from space.

In a major breaktrough, it can tell the difference between the surface of the ocean and the atmosphere immediately above it. Previously, it had been difficult to draw accurate climate maps because sensors couldn't find a distinction between surface and atmosphere. The radiometer gauges sea temperatures by receiving and analyzing infrared sunlight reflected naturally from the ocean surface.

The new infrared radiometer is passive—it only receives light in the infrared part of the energy spectrum. Weather radar, on the other hand, is an active device because it both transmits and receives radio signals to monitor weather patterns.

Meteorologists need the better temperature and wind maps from the radiometer to draw more-detailed pictures showing the interaction of worldwide weather patterns. They hope to learn how storms in the tropics spread to other latitudes. They want to understand powerful weather such as El Nino, the unusually-warm water in the eastern Pacific Ocean which can damage fishing industries.

Tornadoes. A different kind of radiometer can sound an alarm when severe hailstorms and tornadoes develop. Scientists discovered cloud tops rise rapidly just before a severe storm. A cloud-top radiometer sees the movement which signals formation of a hailstorm or tornado.

The 22-lb. radiometer, which looks something like an office wastebasket atop a stationary weather satellite, is outfitted with a small camera peering down through six spectral filters to monitor discrete depths of the atmosphere. The technique is known as sounding.

A stepper motor turns a wheel which moves each filter in front of the camera lens for a time. Controllers use a remote-control radio to point a mirror in the radiometer to focus the camera on clouds of interest. The cloud-top radiometer gives 25 minutes warning when a tornado threatens.

Satellites Of Other Nations

While American engineers were developing satellite meteorology technology, other nations were sending spacecraft to orbit to report on weather conditions from above the atmosphere.

Russia. The Union of Soviet Socialist Republics (USSR) launched its first weather satellite, Cosmos-4, in 1962. It also was the Soviet Union's first photo-reconnaissance spysat.

Four years later, the USSR tested its first polar-orbiting weather satellite system in Cosmos-122 launched in 1966.

Cosmos-144, launched to polar orbit 500 miles above Earth in 1967, started the Cosmos-Meteor series of weather satellites. Cosmos-144 had two daylight optical cameras, an infrared camera, and four instruments to measure radiation emitted by Earth.

Other Cosmos-Meteor weather satellites were Cosmos-156 and Cosmos-184 launched in 1967, and Cosmos-206 and Cosmos-226 in 1968.

One Cosmos could monitor an eighth of Earth's surface in one circuit around the globe. Data was stored in computer memory aboard the satellite and dumped to a ground station when over the Soviet Union.

Meteor. Today's Russian weather satellites continue use of the name Meteor. Many Meteors have been launched to polar orbit in three series known as Meteor-1, Meteor-2 and Meteor-3.

The first generation, Meteor-1, started with a satellite launched in 1969. It carried the same kind of instruments as the earlier Cosmos-Meteor series and added an automatic picture transmitter compatible with a system in American satellites.

Data was sent to three Hydrometeorological Service stations, at Moscow, Khabarovsk, and Novosibirsk in Siberia.

Altogether, 31 Meteor-1 weather satellites were launched between 1969 and 1981.

The second-generation Soviet weather satellites were called Meteor-2 and started with a satellite launched to a 500-mi.-high polar orbit in 1975. Meteor-2 used the same framework as Meteor-1, but housed improved electronics.

A Meteor-2 could show cloud distribution and snow cover in visible-light and infrared photos. It also could report atmospheric temperatures, cloud heights, ocean temperatures and Earth radiation.

Meteor-2 could dump accumulated weather data directly from its computer memory to the national Hydrometeorological Service ground stations on a radio frequency of 460-470 MHz, while sending pictures to local stations at 137-138 MHz. The twentieth Meteor-2 was launched in 1990.

The satellites were improved again, in a third-generation called Meteor-3, which started with a satellite launched in 1985. Meteor-3 weather satellites photograph Earth in visible, infrared and ultraviolet light. The fourth and fifth Meteor-3 weather satellites were launched in 1991.

There Are Problems...

Individuals who enjoy eavesdropping on foreign satellite operations turned up some exciting information in 1987. The USSR sent its fifteenth Meteor-2 weather satellite to a 600-mile-high orbit from the Northern Cosmodrome near Plesetsk on January 5, despite subzero temperatures which disrupted Europe and Asia that day.

Radio hobbyists in England reported signals from the new Meteor were interfering with transmissions from an older Meteor weather satellite in orbit and a secret ocean surveillance radar satellite known as Cosmos-1766. All three were transmitting on the same frequency.

On January 14, the USSR changed the new Meteor's frequency, but radio monitors still heard the old Meteor stuck on the old frequency and continuing to interfere with the military satellite.

France. In 1971, the U.S. launched France's weather-related satellite, EOLE-FR-2, the first application satellite from the French space agency, Centre National D'Etudes Spatiales (CNES).

At an altitude of 400-500 miles, the satellite received atmospheric pressures and temperatures from 500 stratospheric balloons, cruising at 39,000 ft. The data then was relayed to meteorologists waiting on Earth.

Europe. In 1972, the European Space Research Organization (ESRO) ordered work on two prototype weather satellites for stationary orbit over Europe. Later, ESRO became the European Space Agency (ESA).

Europeans called their weather satellites Meteosat. The first launch was on an American rocket in 1977. Meteosat-1's radiometer worked two years in stationary orbit, but stopped in 1979.

ESA launched Meteosat-2 on an Ariane rocket in 1981. Meteosat-3, produced from an engineering model built before Meteosat-1 and Meteosat-2, was launched on an Ariane in 1988.

MOP. Meteosat-4 was an improved version launched on an Ariane in 1989. Referred to as MOP-1, for Meteosat Operational Program, it could relay weather charts and written reports to weather bureaus in sixteen European nations.

Meteosat-5 was MOP-2 launched on an Ariane in 1991. Meteosat-6 was MOP-3 launched on an Ariane in 1993.

The last first-generation satellite, Meteosat-7, is scheduled to ride an Ariane to space in 1995-1996.

Meteosats are are 705-lb. cylinders, seven feet in diameter and ten feet tall, in stationary orbit some 22,300 miles above Earth.

A Meteosat actually is an orbiting combo of weather observer and radio relay in the sky. Its main payload is a sharp-eyed radiometer which shoots infrared, water-vapor and visible-light pictures of the Earth through a 15.6-in. telescope.

A Meteosat takes pictures every half hour and transmits them to Darmstadt for processing. The processed images then are radioed back up to the satellite for relay down to the 16 countries using the weather data.

Visible-light photos show objects down to 1.5 miles across. In infrared photos, objects as small as three miles across can be distinguished.

A Meteosat also has WEFAX to send photos to Earth and a system to collect data from automatic weather stations on land and sea.

Meteorologists use radiometer data received from Meteosats to calculate the direction of cloud movements, cloud-top altitudes and temperatures, sea surface temperatures, and the humidty of the upper troposphere.

Eumetsat. After launch, ESA turns Meteosats over to the European Meteorological Satellite Organization (Eumetsat) founded in 1983.

Meteosats are controlled from ESA's Operations Centre at Darmstadt, near Frankfurt, Germany.

Weather data is fed to national weather bureaus in Belgium, Denmark, Finland, France, Germany, Great Britain, Greece, Ireland, Italy, the Netherlands, Norway, Portugal, Spain, Sweden, Switzerland and Turkey.

The first-generation of MOPs is scheduled to end with the launch of Meteosat-7 in 1995-1996. It will be the last MOP and will be known as Meteosat Transition Program (MTP), bridging the gap between first-generation and second-generation Meteosats. The first second-generation Meteosat is scheduled for launch in the year 2000.

Ariane rockets are launched from the European Space Agency's Guiana Space Center at Kourou, French Guiana, on the northeastern Atlantic coast of South America.

Japan. Japan's weather satellites are known generally as Geostationary Meteoroligical Satellite (GMS). They are operated by Japan's National Space Development Agency (NASDA) and Japan's Meteorological Agency.

In orbit, Japanese satellites are named after flowers. The GMS weather satellites are called Himawari or Sunflower. They monitor weather across Japan and the eastern Pacific from stationary orbit 22,300 miles above the equator near Papua-

New Guinea at 140 degrees East longitude.

Sunflower No. 1 was blasted to orbit in 1977 on a U.S. rocket from Cape Canaveral.

The Japanese launched Sunflower No. 2 (Himawari-2 or GMS-2) in 1981 on their own two-stage N-2 rocket from Tanegashima Island.

Sunflower No. 3 (Himawari-3 or GMS-3) was launched in 1984 on an N-2 from Tanegashima.

A three-stage H-1 rocket carried Sunflower No. 4 (Himawari-4 or GMS-4) from Tanegashima Island to stationary orbit in 1989.

Tanegashima Island Space Center is 616 miles southwest of Tokyo at the southern tip of Japan off Kagoshima Island on the southern tip of Kyushu, Japan's southernmost main island.

India. In the 1970s, India's Space Research Organization (ISRO) built a pair of satellites to observe resources from space. They were called Satellite for Earth Observation (SEO) and renamed Bhaskara in orbit. They sent data to meteorologists about the amount of water vapor in the atmosphere over oceans surrounding the Indian sub-continent.

Bhaskara-1 and Bhaskara-2 were launched in 1979 and 1981 by the USSR to orbits about 325-350 miles above Earth.

India National Satellite (INSAT), launched in 1982, was a new series of dual-purpose satellites to collect weather data as well as relay television and long-distance telephone service across India.

They were built in the U.S. and boosted to stationary orbits by U.S. and European rockets.

INSATs were designed to collect weather data from 110 automatic ground stations strewn across the South Asian nation and to record weather photos in visible light and infrared light every half hour. The information is downloaded to the Meteorological Data Utilization Center at New Delhi. INSATs also relay television programs into rural communities, and telephone calls and data communications across the large country.

INSAT-1A was launched on an American rocket from Cape Canaveral in 1982. Its attitude control broke down and the satellite couldn't be used.

INSAT-1B rode a U.S. shuttle to space in 1983. It was maneuvered over the equator to 74 degrees East longitude.

INSAT-1C was to have been ferried to orbit by a U.S. shuttle, but was switched to an unmanned European rocket after the 1986 Challenger disaster. An Ariane booster lifted it to space in 1988.

The satellite was expected to work ten years, but part of its electrical power system broke down a week after arriving in orbit. That left INSAT-1C crippled with only half of its electronics connected to an operating power supply.

Receiving Satellite Pictures

Weather facsimile, or WEFAX, is the method of transmitting photos and weather maps via radio (wireless) and telephone (wired). In weather-satellite lingo, WEFAX is known as APT.

NOAA's National Environmental Satellite, Data and Information Service receives WEFAX photos from weather satellites.

NESDIS draws maps over the pictures and transmits the resulting weather maps by radio, telephone and satellite.

Individuals find it easy to receive pictures from low-flying, polar-orbiting satellites, such as NOAA, Meteor or Feng Yun. Their radio signals are transmitted from only 500-750 miles overhead.

Before launch to orbit, each satellite is assigned a transmitting frequency. For instance, NOAA and Meteor transmit near 137.50 MHz.

Satellites can be received with modest equipment and a simple antenna. Sometimes, even a scanner or monitor radio of the type used to receive local police and fire calls can receive signals from low-orbit weather satellites.

To receive weather photos faxed from low-flying, polar-orbiting satellites, a user needs a receiver capable of tuning to the satellite's transmitting frequency, an appropriate antenna, possibly a pre-amplifier to boost weak signals, an interface device between receiver and display, and a small computer with monitor or printer, or a special fax machine to print weather maps.

Since stationary-orbit weather satellites are much higher at 22,300 miles above Earth, their signals become progressively weaker as they travel the much longer distance to an antenna on Earth's surface.

Intercepting signals from a GOES weather satellite requires a large outside antenna and a more sensitive receiver capable of tuning to frequencies around 1690 MHz.

A source of information on home reception of weather satellite signals is the Dallas Remote Imaging Group with 4,000 members worldwide. Many DRIG hobbyists have automated tracking stations and upload their satellite pictures to DRIG's computer bulletin board system (bbs).

In operation since 1984, the bbs offers help with satellite pictures, frequencies, gear, tracking data, remote-sensing imagery, digital image and signal processing, and American, Chinese and Russian space programs.

Using a personal computer and modem to connect with the DRIG bbs at (214) 394-7438 brings a caller into contact with gigabytes of data and thousands of images from weather satellites, ham radio satellites, NASA's Voyager deep-space probes and many other Chinese, Russian and U.S. civilian and military satellites. The office telephone is (214) 394-7325. The fax line is (214) 492-7747.

A new resource-observing spacecraft, the India Remote Sensing (IRS) satellite, was launched by the USSR in 1988.

A fourth satellite in the INSAT series, INSAT-1D, was launched in 1990 from Cape Canaveral. The fifth and sixth were INSAT-2A and INSAT-2B launched in 1992 and 1993 on European rockets.

China. The People's Republic of China launched its first two weather satellites on Long March rockets to 500-mi.-high polar orbits in 1988 and 1990. Called Feng Yun or Wind and Cloud, they were similar to U.S. NOAA satellites.

They had infrared and visible-light sensors and sent data on clouds, ocean surface temperatures, marine water color, Earth's surface, vegetation growth, and ice and snow cover to ground stations.

Feng Yun No. 1 (Wind and Cloud No. 1) and Feng Yun No. 2 (Wind and Cloud No. 2) were fired to space from China's Taiyuan Space Center in Shanxi province in north-central China, 60 miles southwest of Beijing. The satellites transmit data to Urumqi in the autonomous region of Xinjiang in northwest China.

China first launched a weather satellite to stationary orbit in 1988.

Atmospheric Satellites

Since the world's first satellite, Sputnik-1, was launched by the Soviet Union in 1957, dozens of spacecraft have been lofted to study the atmosphere high above Earth. In fact, Sputnik itself immediately increased our knowledge of the density of the atmosphere.

Vanguard. The American satellite Vanguard-2, launched in 1959, carried four photocells and two optical telescopes for rudimentary observations of the clouds surrounding Earth.

More balloons. After the 100-ft. Echo balloon was successful as a communications satellite in 1960, four smaller 12-ft. balloons were launched by the U.S. to examine the composition of the upper thermosphere and lower exosphere.

All were launched on small Scout rockets from Wallops Island, Virginia, and Vandenberg Air Force Base, California, to be inflated in space. They found the density of air varied with latitude, season and local solar time.

The 15-lb. aluminum-coated Mylar spheres were the "polka-dot" balloon Explorer-9 launched from Wallops in 1961, Explorer-19 in 1963, Explorer-24 in 1964 and Explorer-39 in 1968.

More Explorers. Five NASA satellites launched between 1963 and 1975 read temperature, pressure, density and composition of Earth's atmosphere. They were Explorer-17 launched in 1963, Explorer-32 in 1966, Explorer-51 in 1973, Explorer-54 in 1975 and Explorer-55 in 1975.

San Marco. In 1964, Italy became the third nation to orbit a satellite with launch of the first of four San Marco atmospheric-science satellites.

San Marco-1 was launched on a Scout rocket from NASA's site at Wallops Island, Virginia.

An Italian team launched San Marco-2 in 1967 on a Scout rocket from a platform in the Indian Ocean off Kenya, East Africa. Known as the San Marco Equatorial Range, the launch site was a pair of platforms, similar to rigs used in offshore oil drilling. They were anchored three miles offshore in Kenya's Ngwana Bay, 90 miles north of Kenya's port city, Mombasa.

As mentioned, the launch pad is operated by the University of Rome. The site was established in 1964 and named for the patron saint of navigators. Launches from San Marco became so popular with the Italian public that Pope Paul VI blessed them.

The San Marco launches were the first firings of NASA rockets by foreign technicians. Since then, Italy has been the most-active foreign user of NASA's Scout space booster.

San Marco-3 was launched from the San Marco Range in 1971. San Marco-4 was launched from the San Marco Range in 1974.

San Marco satellites were designed by Centro Ricerche Aerospaziali (CRA), the Aerospace Research Center at the University of Rome. Each had a dynamometer called Bilancia Broglio, a balance named after the professor who designed it. It measured variations in atmospheric density and allowed scientists to calculate average temperature and molecular weight of air. NASA atmospheric instruments also were carried.

Ariel. Great Britain's Ariel-2 satellite was launched on a Scout from Wallops in 1964. Ariel-3 was launched from Vandenberg in 1967. The satellites also were known as UK-2 and UK-3.

They explored the upper atmosphere, measured a vertical column of ozone and oxygen in the atmosphere and checked on the incoming flow of micrometeorites. The American-built Ariels also listened for radio noise in the galaxy, measured the intensity of very-low frequency radiation, measured the temperature and density of electrons in the ionosphere, and watched for the flow of radio waves from natural phenomena such as storms.

Cosmos. The USSR sent up Cosmos-261 in 1968 to study the density of the atmosphere and aurora around the poles.

Aeros. West Germany offered a pair of atmospheric research satellites in 1972 and 1974. Aeros-1 and Aeros-2 measured the density and temperature of electrons, ions and neutral particles in the upper atmosphere. They also looked for ultraviolet light and X-rays from the Sun.

Both launched on Scout rockets from Vandenberg Air Force Base in California.

ISS. Japan launched two of its own atmospheric-science satellites, ISS-1 and ISS-2, in 1976 and 1978. ISS stood for Ionosphere Sounding Satellite. The satellites observed the ionosphere and radio interference. Also known as Ume, they were launched on N-1 rockets from Tanegashima Space Center.

SAGE. A Scout rocket in 1979 lobbed America's SAGE-AEM-2 satellite from Wallops Island to orbit. SAGE stood for Stratospheric Aerosol and Gas Experiment while AEM stood for Applications Explorer Mission.

The satellite's radiometer created a vertical map of carbon dioxide, ozone and aerosols, as well as molecular extinction, in the stratosphere.

China. The satellite China-9 was launched on a Long March rocket alongside China-10 and China-11 in 1981. China-9 was a balloon linked to a metal sphere by a thin wire to measure the density of the atmosphere. China-10 and China-11 measured infrared and ultraviolet light and X-rays from the Sun.

SME. The Solar Mesosphere Explorer (SME) looked into reactions between sunlight, ozone and other chemicals in the atmosphere. It was launched from Vandenberg in 1981.

San Marco D/L. An Italian team launched a new San Marco satellite, known as D/L, in 1986. It went to space on a Scout rocket from the San Marco Range in the Indian Ocean.

San Marco D/L was designed at the Centro Ricerche Aerospaziali (CRA), the Aerospace Research Center at the University of Rome. It had a Bilancia Broglio dynamometer and instruments from NASA, the universities of Texas and Michigan, and West Germany to measure how the atmosphere is affected by the solar wind.

The mission. Astronaut Sally K. Ride was the first American woman to go to space. As she was leaving NASA in 1987, Ride called for a "Mission to Planet Earth" to preserve our planet's environment. Mission to Planet Earth would be a coordinated research effort to study Earth's environment as a complete global system.

EOS. NASA then proposed Earth Observing System (EOS) as the centerpiece of the Mission to Planet Earth. From high above Earth, EOS would observe land, sea and atmosphere so scientists could predict changes in the environment.

NASA originally proposed six very large "platforms" in space, crammed with sensors. The satellites would have collected data over fifteen years, allowing scientists to study Earth as a total system—atmosphere, oceans, biological cycle, chemistry, and the "greenhouse" effect.

When Congress strangled NASA's budget in 1991, the space agency was forced to downsize its EOS plans. Now, nine small satellites will be orbited and they will focus only on global climate change. The first is to be launched in 1998. Some other satellite projects will be linked to the Mission to Planet Earth and EOS names.

There will be three series of small EOS spacecraft known as AM, PM and CHEM. Each series will include three satellites.

UARS. An early satellite linked to the Mission to Planet Earth and EOS names was the Upper Atmosphere Research Satellite (UARS) carried 372 miles above Earth by shuttle in 1991.

Ten sensors in the satellite explored winds, chemical composition and energy of the stratosphere, revealing how natural processes and man-made pollutants destroy the ozone layer.

Chlorine monoxide is one of the chief chemical culprits in the depletion of Earth's ozone layer and the main cause of the Antarctic ozone hole. The compound is left behind in the upper atmosphere when the Sun's ultraviolet radiation breaks down man-made chlorofluorocarbons. The more chlorine monoxide, the less ozone.

UARS completed the first global picture of chlorine monoxide in the lower stratosphere.

One of the science instruments aboard UARS was a Microwave Limb Sounder (MLS) which observed the chlorine monoxide as well as the global aftermath of the Mount Pinatubo eruption in 1991. The volcano in the Philippines ejected a voluminous fountain of sulfur dioxide into the stratosphere. Ozone was destroyed by the plume of aerosol sulfur dioxide particles.

In addition to chlorine monoxide, sulfur dioxide and ozone, UARS measured nitric oxide, water vapor, temperatures in the stratosphere, energetic particles which cause aurora, and the spectrum of ultraviolet light from the Sun.

Volcanoes. The American weather satellite Nimbus-7, launched in 1978, was the first spacecraft built to monitor Earth's atmosphere for natural and artificial pollutants.

A Total Ozone Mapping Spectrometer (TOMS) built into Nimbus-7 has been very successful in measuring the tragic environmental aftermath of volcano eruptions around the globe.

The toxic gas sulfur dioxide is a major cause of air pollution. In the atmosphere, it reacts with water to form a sulfuric acid aerosol. Volcanic aerosols can have measurable effects on climates.

TOMS monitors sulfur dioxide emissions and measures ozone levels. Scientists use its maps to track volcanic-eruption plumes and measure their sulfur dioxide output.

Researchers used to look at satellite cloud pictures to guess which shapes might be plumes of sulphur dioxide from volcanoes. In the mid 1980s, they discovered the Nimbus-7 ozone-mapping spectrometer was seeing the plumes and highlighting them on maps. Since then, Nimbus-7 has checked on two dozen erupting volcanoes.

Pinatubo. Nimbus-7's biggest success was monitoring gas flowing from the Mt. Pinatubo volcano in the Philippines after a massive eruption on June 16, 1991.

Nimbus-7's unblinking eye watched from above as a 5,000-mile-long cloud of sulfur dioxide seeped across the tropical Northern Hemisphere from the Mount Pinatubo plume.

In fact, Mt. Pinatubo may have been the largest volcano eruption of the 20th century. It belched fifteen million tons of sulfur dioxide into the stratosphere, double the large quantities which fanned out from the eruption of the El Chichon volcano in southern Mexico in 1982.

Water mixed with the Mt. Pinatubo ash and the jetstream carried the aerosol cloud around the globe, scattering sunlight and making the sky redder and brighter. Rosy sunsets were seen around the world. Sadly, the cloud encircled the globe for several years and may have reflected sunlight back into space, changing Earth's heat balance.

Ozone. Ozone is a molecule of three atoms of oxygen. A thin layer of ozone molecules in Earth's upper atmosphere absorbs harmful ultraviolet radiation from the Sun. Unfortunately, bathing atoms of chlorine and other chemicals in sunlight can strip an oxygen atom from an ozone molecule, leaving behind an oxygen molecule which does not absorb ultraviolet.

Chlorine from man-made chlorofluorocarbon compounds rises to the upper atmosphere where it may be the main cause of ozone holes. The Antarctic ozone hole is a large area where ozone is missing from the upper atmosphere in late summer and early fall.

In 1992, just fifteen months after Mt. Pinatubo loaded the upper atmosphere with sulfuric acid droplets, Nimbus-7 observed the largest Antarctic ozone hole ever seen. TOMS found the area of depleted ozone to cover more than 8.9 million square miles. That was fifteen percent larger than the depleted area observed just one year earlier.

Nimbus-7 had reported the stratospheric ozone depletion over Antarctica in 1990 as low as levels in 1987 and 1989, which were the smallest quantities ever observed.

ERBE. Did Mt. Pinatubo actually cool Earth? NASA's Earth Radiation Budget Experiment satellite found the eruption did cool Earth's surface temporarily by about one degree Fahrenheit.

As the Pinatubo dust cloud blanketed the globe, ERBE measured an increase in solar radiation reflected from Earth. Cooling was sixty percent greater than any climate disturbance since systematic global observations began in 1976. The surface will cool after the dust cloud finally settles back to the surface by 1994-95.

TIROS-N And The Greenhouse

The TIROS-N series of weather satellites has been flying nearly-circular Sun-synchronous orbits 530 miles above Earth since 1978, sending down visible-light and infrared-light images with half-mile resolution. A TIROS-N measures the temperature profile of the atmosphere from the surface to 20 miles altitude with an accuracy of 2 to 4 degrees Fahrenheit. More than 1,000 ground stations in 80 nations receive pictures.

Twelve years after its launch, TRIOS-N cast doubt in 1990 on the existence of a so-called greenhouse effect around planet Earth. Scientists at NASA's Marshall Space Flight Center and University of Alabama, Huntsville, said analysis of 12 years of data showed no conclusive evidence of global warming from a greenhouse effect.

Gases and ash spewing from volcanos penetrate up to thirty miles into the stratosphere. The debris spreads a dust veil over the globe, then slowly filters down to the surface over several years.

The dust veil cools Earth's surface by reflecting some of the Sun's rays back to space. At the same time, volcanic dust can start a greenhouse warming effect by trapping infrared rays emitted by Earth itself. Cooling and heating of the planet is determined by the balance between absorbed solar radiation and emitted infrared radiation.

Solar radiation not reflected by clouds, atmosphere or surface, is absorbed and causes the Earth to warm. The emitted infrared radiation cools the planet by carrying heat away from the planet.

The ERBE satellite, launched in 1984, is a forerunner to NASA's Mission to Planet Earth. It measures changes in emitted infrared and reflected solar radiation.

Meteor TOMS. As another element of the Mission To Planet Earth, NASA placed a second

TOMS aboard a Russian Meteor-3 weather satellite to monitor more sulfur dioxide and ozone.

The Meteor-3 was launched in 1991 on one of Russia's Tsiklon (Cyclone) rockets from the Plesetsk Cosmodrome in the northern Arkhangelsk region of Russia, 500 miles northeast of Moscow.

The satellite's orbit 980 miles above Earth let scientists observe aerosols and ozone at twilight.

Meteor-3/TOMS was the first major space research project between the United States and Russia since the 1975 Apollo-Soyuz flight. TOMS was the first U.S. instrument carried aboard a Russian satellite. Technicians from NASA's Goddard Space Flight Center were in the Moscow Flight Control Center during the launch.

The State Committee for Hydrometeorology at Dolgoprudny near Moscow provided the rocket, launch, operations and TOMS housekeeping data.

The project was implemented by scientists from the Hydrometeorology Committee's Central Aerological Observatory, and the All-Union Scientific Research Institute of Electromechanics.

Data from TOMS is downlinked to receiving stations at Obninsk, Russia, and NASA's Wallops Flight Facility at Wallops Island, Virginia.

SPOT. SPOT-3 went to polar orbit in 1993. It carried an ozone sensor known as Polar Ozone and Aerosol Measurement (POAM-2).

POAM-2's nine optical sensors collect data on ozone, water vapor, nitrogen dioxide in the polar atmosphere and stratospheric clouds.

As POAM-2 observes a sunrise and sunset on each 101-minute trip around the Earth, it measures changes in the intensity of sunlight passing through layers of the atmosphere, allowing researchers to analyze the chemical composition and properties in layers from top to bottom.

POAM-2, made at Waltham, Massachusetts, was sponsored by the Office of Naval Research and the Air Force Space Test Program. It supports University of Wyoming studies of the Antarctic ozone hole.

TOMS-EP. Another American Total Ozone Mapping Spectrometer (TOMS) will be on board the Total Ozone Mapping Spectrometer-Earth Probe (TOMS-EP) 600 miles above Earth's poles. TOMS-EP will be built in one of TRW's Eagle-class lightweight satellite frameworks.

ADEOS. And another TOMS will fly on the Japanese space agency's Advanced Earth Observation Satellite (ADEOS). The Earth-science satellite also will carry a NASA Scatterometer.

China. Even China is looking into ozone problems. In 1990, the giant Asian nation fired a sounding rocket 20 miles above Earth's surface to measure ozone depletion. Built by the Chinese Academy of Sciences, Institute of the Atmosphere, the probe had temperature, pressure and ozone sensors. Additional probes were launched later.

CRRES. NASA and the U.S. Air Force launched a satellite in 1990 to shed some light on Earth's trapped radiation belts and find how they effect electronic equipment.

Known as the Combined Release and Radiation Effects Satellite, CRRES illuminated Earth's invisible magnetic field in sensational displays resembling Northern Lights.

CRRES was loaded with 24 canisters of lithium, barium, calcium and strontium chemicals which were ejected at various altitudes over a year.

One or two canisters were ejected at a time. Nine were released into the upper ionosphere 250 miles above Earth, with the rest at lower altitudes.

Ultraviolet light arriving from the Sun ionized the released chemicals, changing negatively-charged electrons in each atom to give the gas an electrical charge. The resulting plasma cloud traveled along Earth's magnetic field lines, reminiscent of the way iron filings line up around a bar magnet to make its field visible.

As they were ionized, the chemical vapors painted the sky with glowing, 60-mi.-wide, aurora-like clouds which were visible to the naked eye as they flowed along the lines of Earth's magnetic field. Scientists observed the temporarily-visible magnetic field lines through telescopes on the ground and from specially-outfitted airplanes.

Instruments in the satellite also probed the inner and outer Van Allen radiation belts, intense zones which girdle Earth and endanger spacecraft electronics. Altogether, CRRES had five major experiments using 55 instruments.

Almaz. Russia's Almaz satellite is a late-model, high-tech civilian spacecraft carrying an all-weather radar to probe Earth's atmosphere and oceans. Almaz surveys Earth's surface in the interests of ecology, geology, oceanology, forestry, agriculture and land management. Areas struck by floods, forest fires and volcano eruptions can be observed by Almaz.

TIMED. NASA atmospheric researchers have dubbed Earth's upper atmosphere at altitudes between 40 and 110 miles the "ignorosphere." The region is difficult to study because it is too high for research balloons, which only float up to altitudes of 25 miles, and dense enough to cause a low-flying satellite's orbit to decay quickly.

NASA would like to launch two Thermosphere-Ionosphere-Mesosphere Energetics and Dynamics (TIMED) satellites to study the "ignorosphere."

Most atmospheric research has explored the lowest atmosphere layers—the troposphere where life exists and the stratosphere where planes fly and large weather patterns like the jet stream occur.

Above those are the mesosphere and lower thermosphere—the skin between outer space and the life-sustaining lower layers. They absorb most of the ultraviolet radiation and block high energy atomic particles from the Sun, and intercept particles accelerated within the magnetosphere which surrounds our planet like a giant cocoon.

Absorption of energy from particles can be seen in the Northern and Southern Lights, or aurora, which shimmers in Earth's night skies.

Satellite Scoreboard

A total of 32 nations and international organizations have sent satellites to orbit over the years. The table below shows total objects ever in orbit, including useless debris and valuable payloads. The 7,387 objects in orbit include 2,126 payloads and 5,261 chunks of debris. Since 1957, all nations have sent 22,718 payloads and pieces of debris to space. The total number of objects sent to space increases monthly. A total of 15,331 objects no longer are in orbit. That includes 2,311 payloads and 13,020 pieces of space junk. Most of the payloads and debris no longer in orbit burned as they descended. Debris in this table is large enough to be tracked by ground radar. See page 190 for abbeviations and acronym explanations. The master lists of all satellites in orbit and no longer in orbit start on pages 191 and 253.

PAYLOADS AND DEBRIS IN ORBIT
as of July 1, 1993

	Objects In Orbit			No Longer In Orbit		
Country	**Payload**	**Debris**	**Total**	**Payload**	**Debris**	**Total**
Arab Satellite Communications Organization (ASCO)	0	0	0	0	0	0
Argentina	1	0	1	0	0	0
ASIAsat Corp.	0	0	0	0	0	0
Australia	6	1	7	1	0	1
Brazil	4	0	4	0	0	0
Canada	16	0	16	1	0	1
Czechoslovakia	1	0	1	1	0	1
ESA (European Space Agency)	23	134	157	3	446	449
ESRO (European Space Research Organization)	0	0	0	7	3	10
France	21	16	37	7	59	66
France/FRG	2	0	2	0	0	0
FRG (Federal Republic of Germany)	12	2	14	4	5	9
India	8	2	10	6	8	14
Indonesia	6	0	6	1	1	2
Int'l Maritime Satellite Organization (IMSO)	3	0	3	0	0	0
Int'l Telecommunications Sat Org (ITSO, Intelsat)	43	0	43	1	0	1
Israel	0	0	0	2	2	4
Italy	3	0	3	5	0	5
Japan	49	50	99	9	71	80
Korea, South	1	0	1	0	0	0
Luxembourg	3	2	5	0	0	0
Mexico	2	0	2	0	0	0
NATO (North Atlantic Treaty Organization)	7	2	9	0	0	0
The Netherlands	0	0	0	1	3	4
Pakistan	0	0	0	1	0	1
PRC (People's Republic of China)	10	79	89	23	71	94
Saudi Arabia	3	0	3	0	0	0
Spain	2	2	4	0	0	0
Sweden	3	0	3	0	0	0
UK (United Kingdom of Great Britain & No. Ireland)	16	2	18	8	3	11
USA (United States of America)	612	2628	3240	639	2850	3489
USSR (Union of Soviet Socialist Republics)	1269	2341	3610	1591	9498	11089
Totals	2126	5261	7387	2311	13020	15331
Grand Total						22718

Earth-Observing Satellites

One of the great benefits of the Space Age has been our newfound ability to step back and look at Earth as a whole. After hundreds of generations stuck to the surface of our planet, mankind now has eyes in the sky.

Satellites circle hundreds of miles above the surface of the planet, observing land and sea, allowing us to map even the most remote areas, explore for natural resources and expose environmental conditions.

First there were the weather satellites of the 1960s. They made it obvious much was to be gained from having other kinds of eyes in the sky. Their rudimentary sensors forecast the complex remote-sensing, mapping and oceanography satellites which sweep the entire planet today.

Remote-sensing satellites function much like weather satellites. Weather satellites use remote-sensing techniques to observe the atmosphere above Earth's surface. Similarly, surface-observing satellites use remote-sensing techniques to observe land masses and oceans.

Like weather satellites, observation satellites fly a hundred times higher than airplanes. From their orbits hundreds of miles above Earth, they take pictures in visible, infrared, and ultraviolet light reflected from the planet's surface, and in microwave energy and X-rays radiating from the planet. Radar beams measure the planet and make images of the surface.

Some of the jobs handled by remote-sensing, mapping and oceanography satellites include:

★monitoring changes in Earth's environments,

★recording Earth's geological formations and unveiling concealed features of the land masses,

★recording detailed photographic images of the surface and drawing more accurate maps of Earth,

★measuring oceanographic features of the seas,

★searching for useful resources on the surface and finding minerals hidden within the planet, and

★recording the innumerable kinds of vegetation blanketing the globe.

In the photos sent down by the satellites, researchers can distinguish between soil and water, towns and pastures, ice and water in an ocean, wheat and corn in fields, and even distressed and healthy crops.

The many remote-sensing and Earth-observing satellites orbiting our planet include American weather satellites and LANDSATs, France's SPOT satellites, Japan's Maritime Observation Satellites (MOS) and Earth Resources Satellite (JERS-1), India's Remote Sensing (IRS) satellite, the European Remote Sensing (ERS-1) satellite and several of Russia's various Cosmos satellites.

LANDSAT Satellites

LANDSAT	Launch
LANDSAT-1	1972 Jul 23
LANDSAT-2	1975 Jan 22
LANDSAT-3	1978 Mar 5
LANDSAT-4	1982 Jul 16
LANDSAT-5	1984 Mar 1
LANDSAT-7	1995

LANDSAT-1. Satellites for sensing Earth remotely from polar orbits were invented in the United States in the 1960s.

NASA used the framework from a Nimbus weather satellite to build the first Earth Resources Technology Satellite (ERTS-1) in the early 1970s. The satellite's name was changed to LANDSAT-1 and it was launched to polar orbit in 1972.

LANDSAT-1 circled Earth every 103 minutes at an altitude of about 570 miles. It flew over the same point on Earth every 18 days, shooting photos and scanning a strip of the planet's surface fourteen times each day, recording forest, ocean, desert, cropland and urban environments.

The satellite's mirror oscillated thirteen times a second to scan a strip of Earth's surface 115 miles wide for green, red, infrared and near-infrared light. The resolution of its scanner images was 260 feet.

Three TV cameras took photographs in the same light with similar resolution. Two video recorders stored images while the satellite was out of range of ground control. The black-and-white pictures were converted to "false" colors on the ground.

LANDSAT-1 also could collect data from ground stations around the globe for relay to a central collection station.

Despite the breakdown of its two TV recorders in 1972 and 1974, LANDSAT-1 transmitted more than 300,000 pictures of Earth before it was turned off in 1978.

LANDSAT-2. A second LANDSAT was launched from California to a similar orbit in

1975. It was like LANDSAT-1, but with an added computer to store commands from the ground to be carried out when out of range of ground control.

LANDSAT-2's recorders also broke down, in 1977 and 1981. The satellite was turned off in 1979, turned back on in 1980, and turned off for good in 1982.

LANDSAT-3. A third satellite in the series, launched from California in 1978, was much the same as the first two. One of its recorders failed in 1979 and the TV cameras broke down in 1981. LANDSAT-3 was turned off in 1983.

Hills And Valleys In The Ocean

Without wind or waves, Earth's oceans would seem calm and flat. In reality, oceans have hills and valleys—like the hills and valleys on land.

An ocean's surface is like a blanket dropped over rugged terrain. At the bottom of the ocean, gravity pulls water down into valleys and pushes it up around mountains. The result, gentle hills and valleys on the ocean surface.

Ocean surfaces are so uneven, for example, looking north from Puerto Rico toward Bermuda, the ocean surface drops 60 feet over 30 miles.

NASA used radar-altimeter data from SEASAT-1 and the third Geodynamic Experimental Ocean Satellite (GEOS-3) to draw colored maps depicting hills and valleys of the world's ocean surfaces.

The radar altimeters in the satellites were so accurate, scientists could detect small surface features only three to six feet in height.

The maps showed surface wave crests and valleys, as well as the shape of the underlying ocean floor, helping scientists study earthquakes and volcanoes and understand how currents flow and oceans circulate.

HCMM. Also in 1978, a small observation satellite was launched 400 miles above Earth to measure variations in Earth's temperature. Known as Heat Capacity Mapping Mission (HCMM) and Applications Explorer Mission (AEM-1), it took daytime pictures in visible and near-infrared light and nighttime pictures in infrared.

HCCM sampled minimum and maximum temperatures every eleven hours over mid-latitude sites in the United States, Europe and Australia. The recorded surface temperature variations allowed researchers on the ground to distinguish between kinds of rocks and mineral deposits, and monitor snow-melt runoff and the condition of vegetation. HCCM worked into 1980.

SEASAT. America launched its first oceanography satellite in 1978. From 500 miles above Earth, Sea Satellite (SEASAT-1) monitored ocean surface temperatures, surface wind speeds, water-vapor content in the air, rate of rainfall, tides, storm surges, ice conditions, wave heights, currents, ocean topography and coastal storms. SEASAT's radar measured wave heights from 3 to 65 feet with an accuracy of 4 inches.

SEASAT-1 had an all-weather synthetic aperture radar (SAR) to photograph waves through clouds day or night, a radiometer to measure water surface temperature, a scatterometer to measure wind speed and direction, a radiometer for pictures in visible and infrared light, and a radar altimeter. The satellite circled Earth fourteen times a day and scanned 95 percent of the world's oceans every 36 hours. Unfortunately, its power failed after only 106 days in space.

Meanwhile, land vegetation was recorded by the NOAA-6 satellite after it was orbited in 1979. The Sahara Desert, tropical rainforests of South America and the spring greening of North American fields and forests were major images captured by NOAA-6.

LANDSAT-4. NASA moved up to an improved, second-generation observation satellite with the launch of LANDSAT-4 from California in 1982. From 425 miles high, the new satellite observed the entire globe every 16 days.

LANDSAT-4's new standard framework style also was used for other satellites, such as the Solar Max astronomy satellite. The generic frame held four modules: power, propulsion, attitude control and communications with data processing.

LANDSAT-4 had the same kind of scanner as earlier LANDSATs, plus a new thematic mapper.

It did not have recorders to break down. Rather, LANDSAT-4 sent data to the ground via NASA's orbiting network of TDRS data-relay satellites.

The thematic mapper stopped working in 1983, just seven months after launch. Later, an electrical power failure cut the satellite's energy in half.

LANDSAT-5. A twin to LANDSAT-4 was launched to a similar orbit in 1984.

In 1986, LANDSAT-5 photos were the first proof that disaster had struck the USSR's Chernobyl nuclear plant. In 1988, the U.S. Forest Service turned to LANDSAT-5 images in its battle against devastating national-park blazes. In 1989, LANDSAT-5's mid-infrared sensors viewed the extensive Valdez oil spill in Alaska.

LANDSAT-5 photos have every-day uses. Farmers use them to predict agricultural yields. Oil companies use them to find petroleum resources. Fast-food franchises look through them for restaurant sites. Insurance companies use them to adjust disaster claims. Governments use them to observe deforestation and urban growth.

The satellites have been used to assess forests and rangelands, explore for oil and minerals, check vegetation and soil conditions, measure worldwide grain production, trace ocean currents, conduct geologic mapping, assess environmental impact of development, plan urban and rural growth, inventory major crops, monitor volcanoes, size-up reservoirs and lakes for dam-safety studies and spot ponds and rivers for water-resource planning.

Remote-sensing satellites also are used for land use studies, pipeline monitoring, hazardous waste detection, urban infrastructure mapping, fishery development, wetlands monitoring and real-time disaster assessment.

Consumers Power Company used LANDSAT pictures to look at land use and soil composition near its six hydroelectric plants along Michigan's Au Sable River, to improve flood control for an area twice the size of Rhode Island.

EOSat. The LANDSATs launched between 1972 and 1984 were operated by the Commerce Dept., until 1985 when President Ronald Reagan handed operations over to the private Earth Observation Satellite Co. (EOSat), a joint venture of Hughes Aircraft Co. and General Electric Co.

EOSat was to receive $15 million a year to maintain LANDSAT operations for the Commerce Dept.'s National Oceanic and Atmospheric Administration (NOAA). The U.S. government also had planned to subsidize the privatization with $250 million to keep EOSat operating and send two new observation satellites to orbit, but the company got only one installment of $90 million in 1986. Meanwhile, the governments of France, Japan, India, the USSR and other nations continued to subsidize LANDSAT's competition.

LANDSAT-4 and -5 still were providing some useful data, but they were mostly broken down. It had been hard to budget for them since it was not known if they would continue to operate.

In voting funds for NOAA, Congress had provided only enough money to operate the LANDSAT system through March 31, 1989.

The government began preparing to get out of the space-photo business. With no money to continue LANDSAT operations after March, NOAA's National Environmental Satellite, Data and Information Service was forced to order EOSat to begin shutting down the satellites.

However, the order was rescinded after scientists at NASA's Jet Propulsion Laboratory discovered active earthquake faults in California's Mojave Desert in LANDSAT photos.

Disappearing Rain Forests

Tropical rain forests cover less than seven percent of Earth's surface, but are home to more than half of all plant and animal species.

Biodiversity, the range of life, has become a major environmental concern. Many tropical plants and animals produce chemicals useful in medicine and other industries.

Trees hold more carbon than crops. Converting forests for pasture and cropland releases carbon dioxide into the atmosphere, contributing to the "greenhouse effect" in which carbon dioxide and other gases in the atmosphere trap heat radiating from Earth, as glass traps heat in a greenhouse.

More than 64,000 square miles of tropical forest is destroyed around the world each year, leading to extinction of large numbers of plants and trees.

In South America, for instance, there has been major deforestation of Brazil's Amazon River basin, the largest continuous tropical forest region in the world. The basin covers two million square miles in eight states, including 1.6 million forested acres, 330,000 tropical-savanna acres and 35,000 acres of water. More than 6,000 square miles of the basin are deforested each year.

Checking the extent of deforestation in 1993, NASA examined 200 photos from LANDSAT-4 and LANDSAT-5, covering the entire forested portion of Brazil's Amazon Basin. The images confirmed the bad news—agricultural expansion has been deforesting the area rapidly for two decades.

"Edge effects" of wind, weather, foraging livestock, non-forest animals and humans are threatening the habitat of plants and animals living on the fringes of remaining forests fragments.

NASA helps Belize, Costa Rica, El Salvador, Guatemala, Honduras, Nicaragua and Panama protect the ecology of their rain forests by training technicians to use satellite data to measure Central American vegetation and forest cover and monitor endangered plant and animal habitats.

For instance, Guatemala uses satellites to supervise 14,000 square miles of forest and savannah in a northeastern reserve.

The faults in the central and eastern parts of the desert appeared in optical and infrared photos. Some faults were active; others inactive. All were part of a complex region between Death Valley and the San Andreas region. Scientists said they could not have made the discovery without LANDSAT.

The news startled the public and businesses which had been relying on American satellite data. They were surprised to learn that all LANDSAT services, including access to more than two million pictures of the planet in the archive, were to end. More than 100 members of Congress sent a letter to President George Bush asking him to keep LANDSATs-4 and -5 running through 1989.

The last two LANDSATs escaped shutdown that year, but money was not available to build a replacement. In 1992, President George Bush again asked Congress for funds to build a new LANDSAT, to be launched in the mid-1990s.

Atlantic Spring Bloom

A satellite peering down at the ocean since 1978 has charted all of the plankton near the surface of the North Atlantic.

Nimbus-7 stared at sunlight reflecting from a 500-mile-wide swath of ocean every time it flew overhead, noting the color of the surface of the sea. Researchers on the ground combined many Nimbus-7 photos to make an ocean-wide color picture of how the microscopic plant life phytoplankton looks to the satellite.

The most striking feature is Spring Bloom, a big concentration of plankton extending all the way across the North Atlantic into the North Sea.

Plants in the ocean take carbon dioxide from the atmosphere and water, and change it to plant tissue in a process called photosynthesis. Chlorophyll is a pigment which allows plants like phytoplankton to complete photosynthesis.

Nimbus-7 saw five colors of light until its ocean scanner quit working in 1986. Its pictures also revealed land and ocean plants on the edges of continents bordering the North Atlantic. Researchers put the satellite data through mathematical equations to compute how much chlorophyll was in the water.

Satellite Photo Business

The selling of satellite photos to governments, educators, news media, geologists, foresters, oil men, farmers, miners, geographers, developers and others interested in surface features and resources has become big business.

As soon as it went to work in space in 1986, France's SPOT satellite offered strong competition to LANDSAT. SPOT photos showed more detail than LANDSAT photos, but photos from Russian observation satellites turned out to be even sharper and more detailed.

LANDSAT photos offer 90-ft resolution. Japan's JERS-1 show objects as small as 60 feet. SPOT has 30-ft. resolution. Russian satellites beat all, making photos with 18-ft. resolution. A camera in space with 30-ft. resolution shows an area as small as half a tennis court.

News pix. Global media became interested in reproducing pictures from satellites after space photos revealed the melted Chernobyl nuclear reactor and a chemical weapons plant in Libya.

When LANDSAT-5 unveiled the Chernobyl nuclear power plant meltdown in 1986 in the former USSR, television networks, newspapers and magazines clamored for LANDSAT and SPOT pictures of the plant and a Soviet naval base.

Two years later, the U.S. Forest Service looked at LANDSAT photos as they battled forest fires in Yellowstone National Park.

LANDSAT's infrared camera photographed the Valdez oil spill in Alaska in 1989.

Not to be outdone by the American photo-sat, SPOT pictures exposed airfields and a chemical weapons plant in Libya.

Space-photo coverage has increased as reporters cover disasters, devastation, wars, anything on a scale large enough to be seen from hundreds of miles overhead. Floods, hurricanes, tornadoes, cyclones, typhoons, snow and rain storms, earthquake damage, nuclear disasters, military movements, nuclear missile installations, anything imaginable is photographed from space for reproduction in newspapers and TV newscasts.

Legality. Because the U.S. Defense Department feared civilian satellite photos could disclose military secrets, federal rules once prevented American companies from using private satellites to photograph objects on Earth.

The Pentagon barred private companies from operating satellites able to produce surface pictures showing objects smaller than about 33 feet across.

Of course, by 1988 the Soviet Union and French companies were operating picture-taking satellites able to resolve objects as small as six feet and selling satellite photographs at very competitive prices. Former President Ronald Reagan changed the rules in 1988 to make American companies more competitive in the international commercial space market.

Government rules today allow anyone looking for magnified pictures of things on Earth to buy private snapshots made by civilian space satellites.

TOPEX/Poseidon. Successes with the SEASAT and GEOS satellites led NASA's Jet Propulsion Laboratory and the French space agency, CNES, to design a new oceanography satellite known as TOPEX/Poseidon, which was launched to an altitude of 800 miles in 1992.

Poseidon was France's name for its radar altimeter measuring ocean surface topography.

TOPEX (Ocean Topography Experiment) was NASA's moniker for its microwave radiometer measuring atmospheric water vapor and a radar altimeter designed at Johns Hopkins Applied Physics Laboratory to measure mean sea level to an accuracy of a few centimeters.

The oceanography spacecraft was the second satellite in NASA's Mission to Planet Earth. It was designed to make a topographic map of 90 percent of the world's oceans, and determine their circulation patterns, variations, marine conditions, wave heights, wind velocities and ice distribution.

Scientists at NASA's Goddard Space Flight Center designed the maritime observation satellite, which then was built by Fairchild Space, an American subsidiary of France's Matra SA. CNES launched TOPEX/Poseidon on an Ariane rocket in 1992. It was the first NASA satellite ever carried to space by a European rocket.

NASA and CNES provided sensors to map surface features, sea levels, seasonal warming and cooling, and winds on global oceans, and study the climate-control role played by ocean circulation.

TOPEX/Poseidon has a pair of radar altimeters, a microwave radiometer, and two radiolocation systems. The American altimeter operates at C-band and Ku-band frequencies, while the French altimeter operates at Ku-band. The altimeters share thc same radar antenna, so they take turns measuring sea-wave and ocean surface heights.

The French altimeter recorded 24-foot seas in the Pacific Ocean around Hurricane Iniki in 1992.

Sea level changes reflect seasonal warming and cooling and winds. Ocean-surface currents transport heat around the globe. The satellite's altimeters are so accurate, scientists can see most of Earth's ocean-surface currents.

They expect to determine the general circulation pattern of the oceans, see surface features, and understand how oceans affect climates and influence worldwide weather patterns.

TOPEX/Poseidon measurements are compared with data from ships of 44 nations to see whether pollution is changing Earth's environment.

The satellite carries a global positioning system (GPS) satellite receiver to locate itself in space, along with a laser reflective array (LRA) and Doppler orbitography and radio positioning integrated from space (DORIS).

Ocean Color. Nimbus-7 color images led NASA to plan an Ocean Color Data Mission (OCDM) satellite to measure additional changes in ocean color where phytoplankton concentrate.

Studying such concentrations helps scientists understand the global carbon cycle. Understanding global storage and release of carbon will assist meteorologists in understanding climates.

The ocean color data will be used by Goddard Space Flight Center's Laboratory for Hydrospheric Processes and sold to commercial fishermen who find phytoplankton draws large schools of fish.

Sea Levels Rise And Fall

Scientists watching data flowing to Earth from a maritime observation satellite saw the level of sea waters in the Gulf Stream off the U.S. East Coast, and in the Kuroshio region east of Japan, drop by twelve inches from October 1992 to March 1993.

The changes in sea levels, observed by the French-American TOPEX-Poseidon oceanography satellite, probably resulted from winter cooling of the ocean by the cold continental air mass blown off the North American and Asian continents.

While the Gulf Stream and Japan are in the Northern Hemisphere, summertime warming of the atmosphere in the Southern Hemisphere brought a corresponding sea level rise at similar latitudes.

Change is larger in the Northern Hemisphere than in the Southern because the Northern Hemisphere's larger land mass creates colder continental air mass which cool ocean waters off the east coasts of North America and Asia.

Seasonal changes in trade winds cause the sea level to drop at the equator in both the Pacific and Atlantic oceans.

In the Indian Ocean, reversing seasonal monsoon winds causes a fall in sea level in the eastern and southern regions and a rise in sea level in the northwestern region.

A temperature increase or decrease of 1.8 degrees Fahrenheit in the average temperature of a water column 165 feet deep causes a sea level to rise or fall by 0.4 inches.

GFO. The U.S. Navy has a new oceanography satellite known as GEOSAT-Follow-On (GFO).

Its radar altimeter will measure sea surface heights, sea floor topography, water density and wind speed for the Navy's Space and Naval Warfare Systems Command, after GFO is ferried to space by an air-launched Pegasus rocket in 1995.

The relatively-tiny 660-lb. spacecraft will transmit data directly to 65 shipboard and shore stations. They will correlate the information with data from global-positioning satellites (GPS) to route ships, support amphibious operations and conduct anti-submarine warfare.

GFO was designed to work eight years in space. Two more may be built later for Pegasus launches.

Radar mapper. Engineers at NASA's Jet Propulsion Laboratory developed a radar system for more accurate mapping of Earth's topography.

Known as TOPSAR, for topographic synthetic-aperture radar, the interferometric radar is three times more accurate than earlier topographic mappers. JPL designed it in cooperation with the Italian Consortium for Research and Development of Advanced Remote Sensing Systems.

What Is El Nino?

El Nino is a global weather disturbance which regularly drenches the U.S. West Coast, chills the East Coast and warms other cool areas around the world as it strengthens and weakens.

An El Nino disruption starts when weakening trade winds allow a large pool of warm water in the western Pacific to slip eastward to South America.

The warmer waters arrive off Peru and Ecuador at Christmas time, thus the name El Nino which is a Spanish name for the Christ child.

Orbiting 800 miles above the world's oceans, the TOPEX-Poseidon oceanography satellite can detect the eastward expansion of warm Pacific water because heat makes water expand, raising sea level, which is measured by the satellite.

The waves occur regularly. For instance, one El Nino wave started in the western Pacific in Spring 1991. Its pulse of warm water arrived at South America in January 1992, before TOPEX-Poseidon was launched to orbit.

A year later, TOPEX-Poseidon was on duty in space to measure a giant wave of warm water, six inches above normal sea level, moving eastward at six mph in December 1992. It warmed the coasts of Ecuador and Peru in Spring 1993.

The pool of warm Pacific water meant El Nino was growing stronger again, promising rainstorms to drench drought-plagued California, a cool spring for the U.S. East Coast, and warm, dry summer weather in central South America, South Africa, Indonesia and even Europe where a French space scientist decided it would make for better wine.

Sea water in the El Nino pool increased more than nine degrees Fahrenheit above normal, killing some fish needing colder water.

The rise in sea level seen by TOPEX-Poseidon off the coast of South America in the eastern tropical Pacific Ocean was the remnant of the Kelvin wave pulses that began in December 1992.

The warm El Nino waters surging eastward along the Equator in the central Pacific Ocean also are referred to as a Kelvin Wave pulse.

Thermometer. NASA devised a satellite thermometer 100 times more sensitive than earlier equipment. The infrared radiometer measures ocean surface temperatures from space, distinguishing between the ocean surface and the atmosphere immediately above it.

The infrared radiometer is passive—it only receives light in the infrared part of the energy spectrum. Weather radar, on the other hand, is an active device because it both transmits and receives radio signals to monitor weather patterns.

It had been difficult to draw climate maps because sensors couldn't tell the difference between surface and atmosphere. The radiometer gauges sea temperatures by analyzing infrared light in sunlight reflected naturally from the ocean surface.

Meteorologists needed better temperature and wind maps to draw more-detailed pictures showing the interaction of worldwide weather patterns. For example, they hope to learn how storms in the tropics spread to other latitudes. Eventually, they want to understand powerful weather such as El Nino, the unusually-warm water in the eastern Pacific Ocean which can damage fishing industries.

EOS. As astronaut Sally K. Ride, the first American woman in space, was leaving NASA in 1987, she suggested a Mission to Planet Earth to preserve our planet through coordinated research into Earth's global environment.

Following her proposal, NASA planned a new series of satellites known as the Earth Observing System (EOS) to be the nucleus of the Mission to Planet Earth. From high above Earth, EOS satellites would monitor land, sea and atmosphere for changes in the environment.

Recently, the plan has included nine small EOS satellites to be launched from 1998 to focus only on global climate change. Some other satellite projects are linked to the Mission to Planet Earth and EOS names.

Satellites Of Other Nations

While American engineers were developing photo-observation satellite technology, others were sending spacecraft to orbit to search for resources and monitor land and sea surface conditions.

Russia. The USSR was able to launch observation satellites from 1962, when it sent up a combined weather and spy satellite.

By 1966, Russia was testing a polar-orbiting weather satellite in Cosmos-122. The Meteor series of weather satellites started with Cosmos-144 in 1967. It had two daylight optical cameras, an infrared camera, and four instruments to measure radiation emitted by Earth. Other Cosmos-Meteor weather satellites were Cosmos-156 and Cosmos-184 launched in 1967, and Cosmos-206 and Cosmos-226 in 1968.

Meteor. Later, many Russian weather satellites known as Meteor would be flown to polar orbit. The first generation, Meteor-1, began with a 1969 launch.

The second-generation, Meteor-2, started with a satellite launched in 1975. It showed snow cover and cloud distribution in visible-light and infrared photos. It reported atmospheric temperatures, cloud heights, ocean temperatures and Earth radiation.

A third-generation, Meteor-3, started with a

launch in 1985. Meteor-3 satellites photograph Earth in visible, infrared and ultraviolet light.

Altogether, Russia sent 31 Meteor-1 weather satellites to orbit. The 30th Meteor-1, lobbed to space in 1980, carried an experimental remote-sensing earth-resources package.

Meteor-Priroda. The next year, the Soviet Union's first satellite exclusively for observing Earth's resources, Meteor-Priroda-1, was launched to a 400-mi.-high polar orbit in 1981. Priroda is Nature in Russian.

Meteor-Priroda-1 carried cameras and scanners to observe the entire Earth every seven days.

Oceanography. The Soviet Union's first satellite exclusively for oceanography was Cosmos-1076 launched to a 400-mi.-high polar orbit in 1979. It was followed by three more in the series: Cosmos-1151 in 1980, Cosmos-1500 in 1983 and Cosmos-1602 in 1984.

Each had radar, radiometer and scanner for visible and infrared light and microwave data.

Cosmos-1766. The Russian oceanography satellite Cosmos-1766 found the drifting and lost Druzhnaya Antarctic ice station in 1986.

The station crew had set up shop on ice at Filchner Glacier, but were lost when the ice broke off and drifted away. A side-looking radar on the oceanography satellite spotted the drifting ice floe and revealed a sea path to save the crew.

Cosmos-1870. The largest unmanned civilian survey satellite ever ferried to space up to its time was Cosmos-1870. The size of a school bus, it was tossed 150 miles into space by one of the USSR's large Proton boosters in 1987.

The Russians had orbited large military radar stations before, but Cosmos-1870 was the first purely for non-military research.

Because it was so large, the satellite was seen in the sky easily with the naked eye, just before dawn and just after dusk.

The twenty-ton orbiting radar station mapped oceans and looked for natural resources on land as it revolved around the globe every hour and a half.

The satellite transmitted information on ground conditions regardless of weather or time of day, and could study charged particles in space. Its data went to the USSR Academy of Sciences.

Cosmos-1990. Cosmos 1990, launched in January 1989, photographed the areas of the Soviet Union, including Armenia, which had been devastated by a December 1988 earthquake.

Almaz. A huge new spacecraft, known as Almaz-1, was sent aloft on another mighty Proton booster in 1991 to continue the study of land and ocean surfaces started in 1987 by Cosmos-1870.

Orbiting 175 miles above Earth, Almaz-1 became the largest operational, unmanned, civilian, resources satellite launched to date.

Almaz is Russian for diamond, or diamond in the rough. Financially-strapped Russian officials wanted Alamz to bring in hard currency, while showing a military industry could convert to a civilian business. Almaz imagery was expected to be a major competitor to LANDSAT and SPOT.

Almaz-1 was Russia's first large commercial venture with Space Commerce Corp. of Houston which markets Russian launches to U.S. and South American satellite owners.

For 18 months in orbit, Almaz-1 carried an all-weather radar to probe Earth's oceans and atmosphere, survey surface ecology, geology, oceanology, forestry, agriculture and land management, and monitor areas struck by floods, forest fires and volcano eruptions.

Resurs-F. Russia has launched a series of satellites referred to as Resurs-F in recent years. They fly in orbits between 100 and 200 miles to photograph natural resources and ecological problems. The satellites stay in space only a few weeks, long enough to expose their photographic film and drop it for recovery on Earth.

India. Meanwhile, in the 1970s, India's Space Research Organization (ISRO) had built a pair of satellites to observe resources on the sub-continent and surrounding oceans.

They were called Satellite for Earth Observation (SEO), but renamed Bhaskara in orbit. Bhaskara-1 and Bhaskara-2 were launched by the USSR from its Kapustin Yar Cosmodrome to 330-mi.-high orbits in 1979 and 1981.

Each collected data on forests, geology and hydrology, using two cameras making pictures in visible light and near-infrared light. A microwave radiometer observed ocean surface conditions. The Bhaskara satellites recorded the amount of water vapor in the atmosphere over oceans surrounding the Indian sub-continent.

Bhaskara-1's television camera worked eleven months and the satellite was shut down in 1981.

INSAT. India National Satellite (INSAT) was a new series of dual-purpose spacecraft launched in 1982 to collect weather data and relay television and long-distance telephone service across India.

They were built in the U.S. and boosted to stationary orbits by U.S. and European rockets.

INSATs collect data from 110 automatic weather stations and record cloud photos in visible light and infrared light every half hour.

INSAT-1A was launched on an American rocket in 1982. INSAT-1B rode a U.S. shuttle to space in

1983. INSAT-1C was launched on a European rocket in 1988. INSAT-1D, was launched in 1990 from Cape Canaveral. The fifth and sixth satellites in the series, INSAT-2A and INSAT-2B, were launched in 1992 and 1993 on European rockets.

Exploring Asian Salt Domes

Russian satellite photos have unveiled a vast complex of salt domes in the Caspian Depression located on remote western lands of the central Asian nation of Kazakhstan.

Salt domes are underground formations of crystalline salt which also store potassium, iodine, magnesium and rare elements. The chemicals can be recovered from salt domes by the cheapest form of mining, quarrying.

One salt dome, the massive Chelkar deposit, covers 1,250 square miles and is nearly five miles deep. It alone could provide raw material for several chemical plants for half a century, according to the Institute of Geological Sciences at the Kazakhstan capital of Alma-Ata.

Explorers know the domes point to deposits of oil and gas. Before the satellite study, oil and gas fields had been developed on a commercial basis in the Caspian Depression, but only the Indorski salt dome had been mined for salt and borates.

The orbiting resource satellites mapped 1,200 salt domes, some with their tops cropped right out onto the surface. Unfortunately, Kazakhstan has been short on funds to tap its many salt domes.

IRS. Meanwhile, ISRO built two new India Remote Sensing (IRS) satellites in the 1980s. The first was IRS-1a launched in 1988 for India by the Soviet Union on a Vostok rocket from the Baikonur Cosmodrome to a 500-mi.-high orbit.

A second satellite, IRS-1b, was lofted in 1991 for India by the Russians to a similar orbit.

Each IRS had a bank of three cameras scanning India, Asia and Africa for information to be used in meteorology, mineral exploration, town planning, pest control, and agriculture, including forecasting of tea and coffee crop yields.

Ground stations equipped to receive data from SPOTs and LANDSATs can receive IRS-1b data.

Russia was to launch IRS-1c and IRS-1d in 1994 and 1996. Later, Russia was to launch one of India's Gramsat communications satellite.

India's rural satellite network, called Gramsat or Village Satellite, disseminates information on health, environment, family planning and literacy.

India and Russia had been cooperating in space projects since 1975 when the USSR launched India's satellite Ariabata.

An Indian citizen, Rakesh Sharma, flew in the Russia's Soyuz T-11 capsule to the space station Salyut-7 in 1984. He used yoga to combat the effects of weightlessness during eight days at the station. Sharma's flight was the first time eleven people had been in orbit at once—six cosmonauts at Salyut-7 and five astronauts in a U.S. shuttle.

India's National Remote Sensing Agency receives IRS data at Shadnagar near Hyderabad, capital of the southern state of Andhra Pradesh. NRSA also receives Europe's ERS-1 satellite at Shadnagar. Unlike IRS which gathers only surface data, ERS's microwaves penetrate up to five feet into soil and 20 feet into desert sands.

France. France launched its first photography satellite, SPOT-1, in 1986 on an Ariane rocket to a 500-mi.-high orbit. The telescope attached to a camera in the satellite could resolve objects as small as 30 feet across.

SPOT-2 was launched to orbit in 1990. The visible and infrared light telescopes in the SPOT satellites were designed to map the globe, study land use, investigate renewable resources on forests and farm lands, and explore for minerals and oil.

SPOT was an acronym for Satellite Probatoire de l'Observation de la Terre. SPOTs are owned by an organization of private and government interests in France, Belgium and Sweden. They are operated by the Centre National d'Etudes Spatiales (CNES), the French national space agency.

SPOT-3 went to a polar orbit in 1993. Along with the usual photo-mapping equipment, SPOT-3 has an ozone sensor known as Polar Ozone and Aerosol Measurement (POAM-2).

POAM-2's nine optical sensors collect data on ozone, water vapor, nitrogen dioxide in the polar atmosphere and stratospheric clouds.

As POAM-2 observes a sunrise and sunset on each 101-minute trip around the Earth, it measures changes in the intensity of sunlight passing through layers of the atmosphere. Researchers will be able to analyze the chemical composition and properties in layers from top to bottom.

The orbiting POAM-2 experiment, made at Waltham, Massachusetts, was sponsored by the Office of Naval Research and the Air Force Space Test Program and will aid studies of the Antarctic ozone hole by the University of Wyoming.

Japan. Japan's first remote-sensing satellite was the Maritime Observation Satellite (MOS-1a).

Built by Japan's National Space Development Agency (NASDA), it was blasted 560 miles above Earth in 1987 aboard a Mitsubishi N-2 rocket from Tanegashima Island Space Center.

MOS-1a focused on the world's oceans, noting color and temperature, and providing ocean data to sixty Japanese organizations and sixteen other countries. Of course, MOS-1a also looked at land surfaces when over them, checking on agriculture, fishing, forestry and environmental problems.

MOS-1a had three passive sensors to collect sunlight reflected from the surface of Earth and electromagnetic energy radiated from the planet's surface. One sensor sorted out the colors in visible and infrared light reflected from the surface of the oceans and from land surfaces.

Another sensor looked at wide swaths of Earth in visible and infrared light to measure water surface temperature and the amount of water vapor in the atmosphere. A third sensor scooped up microwave radiation from the surface to measure quantities of water vapor, snow and ice.

MOS-1a's data was valuable to agriculture, forestry, fishery, and environmental preservation. The satellite circled the globe every 103 minutes, completing 14 trips a day and covering the entire Earth every 17 days.

MOS-1b, a twin to MOS-1a, was launched in 1990 with two other satellites on one H-1 rocket.

It was the first time NASDA had lofted more than two satellites to space on a single rocket. The others were the amateur radio satellite JAS-1b, (Fuji-OSCAR-20) and a satellite called Debut preparing for Japan's role in the U.S.-international space station Freedom.

MOS satellites are controlled by NASDA's Tsukuba Space Center. Signals also are received by tracking stations at Katsuura, Masuda and Okinawa. NASDA's Earth Observation Center, established in 1978, is at Hatoyama-machi, Saitama prefecture, 30 miles northwest of downtown Tokyo.

JERS-1. NASDA launched its Japanese Earth Resources Satellite (JERS-1) in 1992 on an H-1 rocket from the Tanegashima Space Center to orbit at 350 miles altitude.

NASDA built the satellite framework and Japan's Ministry of International Trade and Industry (MITI) built the sensing instruments.

The observation satellite's optical sensors, visible-light and near-infrared radiometer and all-weather synthetic-aperture radar (SAR) scan the entire Earth every 44 days.

JERS-1 can make stereoscopic photos of objects as small as sixty feet during both day and night and under any weather conditions.

Scientists use the satellite imagery to analyze data in 250 fields of research, including tropical forests, movement of glaciers, and conditions before and after cases of environmental damage. The satellite monitors the global environment, natural resources, agricultural crops, forests and fish populations. It makes coastal surveys and helps with fire prevention.

ADEOS. NASA's first Total Ozone Mapping Spectrometer (TOMS) was highly successful in the Nimbus-7 weather satellite. Another TOMS will fly on the Japanese space agency's Advanced Earth Observation Satellite (ADEOS). The satellite also has a NASA Scatterometer. ADEOS is Japan's most ambitious Earth-resources program to date.

TRMM. NASA is planning a new Tropical Rainfall Measurement Mission (TRMM) satellite to measure rainfall over South America and Africa.

Several measuring instruments will be carried aboard the spacecraft, built by Hughes for NASA's Goddard Space Flight Center to be launched by Japan on an H-2 rocket in 1997.

TRMM will take the measure of tropical rainfall between 35 degrees North and South latitude, from a 220-mi.-high orbit.

One instrument will be a microwave imager which can see through clouds to measure rainfall rate and distribution, cloud and soil moisture levels, land and sea surface temperatures, and sea surface wind speeds.

The microwave imager will be similar to the special-sensor microwave imager used by the U.S. Air Force in its Defense Meteorological Satellites.

Those DMS satellites provide weather data to all American military services. For example, they were used to forecast weather for Operation Desert Storm in the Persian Gulf War.

Europe. The European Space Agency (ESA) launched its first remote-sensing satellite on an Ariane rocket in 1991.

Known as European Remote Sensing Satellite, or ERS-1, it was built in the satellite framework design used earlier in France's SPOT observation satellites. ERS-1 was the biggest, heaviest and most complex European spacecraft to that time.

The amateur radio satellite UoSAT-F (UO-22) also was launched on the same rocket.

ERS-1 monitors land, sea and coastlines from Earth orbit with an all-weather synthetic-aperture radar (SAR) which uses microwave radio signals for high-resolution photos of Earth's surface and for measuring wind-blown waves on the oceans.

The satellite's scatterometer measures speed and direction of surface winds. Its radar altimeter measures altitude, wind speeds, wave heights, currents, and ice topography. ERS-1 covers the entire Earth every three days.

ERS-1 may make earthquake prediction easier. Orbiting 488 miles out in space, the satellite detected movement of less than one centimeter (0.39 inches) by objects on Earth's surface.

The 1992 test, by scientists at the Polytechnic Institute of Milan and the University of Bonn, made the satellite a powerful tool for measuring motions in Earth's crust, enhancing the study of very small motions in the planet's tectonic plates.

Greenland's Ice Sheet

The European Remote Sensing satellite, ERS-1, carries a radar altimeter which can map precisely the topography of glacier ice sheets.

As global temperatures increase, they melt some polar ice and raise sea levels. At the same time, rising temperatures stimulate precipitation, increasing the size of ice sheets. Scientists need more data about the elevation of ice surfaces to know for sure whether ice sheets in Greenland and Antarctica are growing or shrinking.

They suspect melting of the Greenland glacier raises global sea levels. A nine-inch change in the average height of the central Greenland ice sheet is said to result in a 0.12-inch change in the sea level of the world's oceans.

But, sea levels are difficult to measure directly. Measuring polar-glacier ice sheets is an indirect way to calculate sea-level changes.

In recent years, to measure the surface elevation of polar glaciers, scientists from NASA's Wallops Island Flight Facility in Virginia have been reading the instruments aboard Europe's ERS-1 satellite when it is over Greenland.

Using data from ERS-1 and from several global positioning satellites (GPS), the researchers have been able to compute both buildup and melting of ice on the surface of the Greenland glacier.

For additional data, they flew an aircraft along a path beneath ERS-1 and sent a ground crew from Ohio State University beneath the satellite track to measure surface elevation. Repeated surveys over years show glaciers expanding and shrinking.

Canada. Some locales on Earth are shrouded by cloud cover nearly continuously, hindering visible-light photography. Microwave radar can record surface features regardless of cloud cover and at night. It measures altitude to provide three-dimensional images. Microwave radar avoids shadowing. Ice reflects radar differently than water, so a microwave radar system can track ice flows.

When Canada's RADARsat is launched to a 621-mi.-high Earth orbit on a U.S. Delta rocket by 1995, it will carry an all-weather synthetic-aperture radar (SAR) to monitor ice hindering navigation in the Northwest Passage and to protect oil platforms. RADARsat will perform the first year-round space monitoring of Arctic Sea routes.

The Canadian satellite will use its microwave radar to take detailed snapshots of Earth's surface in daylight, darkness and through clouds. The SAR will produce 30-ft.-resolution images.

The SAR's sensitivity to soil moisture and vegetation will supply crop assessments and information about forest operations. The satellite also will be on the lookout for drought, floods, forest fires and other natural disasters.

NASA and the U.S. National Oceanic and Atmospheric Administration (NOAA) are taking part in the Canadian Space Agency (CSA) project.

NASA is providing the rocket in return for data from RADARsat, while NOAA is providing one of its advanced very-high-resolution radiometers to measure temperatures and cloud distribution for various U.S. agencies. RADARsat also will have optical sensors from Germany.

RADARsat will draw the first comprehensive map of the Antarctic continental ice sheet. NASA plans to install a ground station at McMurdo Station in Antarctica to download data from the satellite. That data will be transmitted to the University of Alaska at Fairbanks for image processing and distribution to scientists.

Although designed for a five-year working life, RADARsat will be serviced by American space shuttles. New instruments may be installed by visiting Canadian astronauts.

China. The People's Republic of China (PRC) launched its first space satellite in 1970. By 1994, the Asian nation had launched three dozen remote sensing, communications and weather satellites for both civilian and military use.

In 1975, what China called a "homing satellite" made it the third nation capable of retrieving a satellite from orbit. Over the years, the PRC sent a dozen to orbit with packages to be retrieved. Those were dropped and recovered successfully.

China was able to launch three satellites to orbit on one rocket in 1981.

China launched its first weather satellites to 500-mi.-high polar orbits in 1988 and 1990. They were called Feng Yun, or Wind and Cloud, and were similar to U.S. NOAA weather satellites.

The satellites had infrared and visible-light sensors and sent down data on clouds, ocean surface temperatures, marine water color, Earth's surface, vegetation growth, and ice and snow cover to ground stations.

China first launched a weather satellite to stationary orbit in 1988.

The PRC used remote-sensing technology at the end of the 1980s to monitor floods in the lower and middle reaches of the Yangtze and

Yellow rivers. The nation also used satellites to map its forest resources and see society's impact on the ecology of northwestern, north-central and northeastern regions of China.

Among remote-sensing devices in its satellites are imaging spectrometers, infrared multispectral scanners, microwave radiometers and synthetic-aperture side-looking radars.

China-Brazil. Brazil has plans for a remote-sensing satellite—known as the China-Brazil Earth Resources Satellite (CBERS)—on the drawing board to be launched by China in 1995.

Asia. Nations in the Asian Pacific region need satellites to manage natural resources, according to the Asian Development Bank and the Economic and Social Commission for Asia and the Pacific.

They said remote sensing is vital where rapid deforestation is changing the tropical atmosphere. For instance, satellite surveys of water resources are needed where aquaculture and alternate land use are erasing Thailand's mangrove swamps. Satellite monitoring of cyclones is a necessity in storm-whipped Bangladesh, where howlers have resulted in many lives lost. And satellite observation of the red tide infestation plaguing the Philippines would help fisheries.

Of course, China already is using remote sensing to explore for oil, gas, coal, gold and other metal and mineral resources. Mongolia is using satellite data to improve mining, geology, erosion hazards, hydrogeology and land cover. Even the Philippines has used remote-sensing data, to map places hit by mud flows from the Mt. Pinatubo volcano eruption and pinpoint the best areas for rehabilitation and community resettlement.

India has offered to provide free remote-sensing training for technicians from developing nations.

Thailand upgraded its regional ground station to acquire data from the European Resource Satellite (ERS-1). That data is available to other countries.

The Asian Development Bank urged Japan, the U.S. and the industrialized world to train Asian engineers in radar technology.

A Lost Arabian Society

And Other Satellite Success Stories

Observation satellites staring down from space penetrated 600-ft. mountains of windswept sand to make a startling find on the fringe of the Arabian Desert. The faint shadow of a lost civilization turned up like a ghost in computer-enhanced radar images of ancient ground under the Rub al Khali desert in the sultanate of Oman. A timeworn network of roads under the dunes seems to point to the burial place of the legendary society of Ad.

Referred to in the Koran, the tales of The Arabian Nights and the Holy Bible, Ad probably was the bustling hub of the world's frankincense trade 5,000 years ago. Biblical archaeologists suggest wise men traded there for frankincense they bore as gifts for the infant Jesus.

Frankincense is an aromatic resin from the sap of Middle Eastern and East African trees—an incense used ages ago by monarchs and common folk alike, in cremations, religious rituals, ceremonies and imperial processions.

Shifting Sands

The world's largest desert fluctuated in size during the 1980s, according to a NASA study of observation-satellite data.

The Sahara Desert was observed in red light and infrared light reflected from the desert surface up to four orbiting American weather satellites, NOAA-6, -7, -9 and -10.

Desert fluctuations depended on the amount and distribution of the rainfall in the area. Rainfall controls the amount of vegetation seen from space. Scientists suggest changes in global desert area may be tied to global climate changes.

The Atlas Mountains and Mediterranean Sea make up a nearly immovable northern boundary, but the Sahara's southern boundary moved south 80 miles between 1980 and 1990.

After moving to the south between 1981 and 1984, the Sahara retreated northward 88 miles from 1985 to 1986. However, it migrated 34 miles south in 1987. The southern boundary retreated 62 miles to the north in 1988, then expanded 46 miles to the south in 1989 and 1990.

The grinding sand of the Rub al Khali had defeated a water-short British explorer's search for Ad's ancient trade routes in the 1930s. Modern archaeologists still are unable to search the entire perilous desert. Instead, they work in shirtsleeves in laboratories, feeding data from satellite radar to computers searching for long-lost clues.

In observation satellite photos made in the 1980s, archaeologists at NASA's Jet Propulsion Laboratory in California were able to see a 100-yard-wide hoof-trodden path hidden under tons of sand in giant dunes. Backed by a wealthy businessman, a scouting expedition of NASA, British and private explorers, in 1990, tracked the trail they had concluded was formed by frankincense traders riding camels. Following their satellite map, the party looked for geological evidence of a trail through the now-barren land to the once-thriving city of Ubar.

The explorers stumbled upon Ad artifacts—900

pottery shards and flint pieces—on the trade route in the sprawling Rub al Khali desert in 1990.

The adventurer T.E. Lawrence once described Ubar as the "Atlantis of the sands." Frankincense was an important commodity in the ancient world before the rise of Christianity and Ubar may have been the main shipping center of Ad. Worldwide shipments of frankincense to markets as far away as China and Rome could have started at Ubar.

Ad society lasted from 3000 B.C. to the 1st century A.D. In the end, it was victimized by politics, economics and climate after a drop in demand for the frankincense fragrance as Christianity preached burying bodies instead of burning them. The abandoned villages of Ad eventually were inundated by tides of shifting sands, and eventually dunes reaching heights of 200 to 600 feet.

Satellite Archaeology

Archaeologists want to know what happened to the ancient Mayans in Central America and how they managed their tropical forest economy.

At NASA's Stennis Space Center in Mississippi, an archaeologist uses remote-sensing satellites in his search for knowledge of primitive times.

Data soaked up by satellites remotely-sensing the rain forests helps local Central American governments protect ancient cultural sites of the Mayan civilization. After all, the ancient Mayan sites are key to the region's tourism.

As rain forests are felled, archaeological sites are exposed to looters. Governments want to preserve the ancient sites, while fostering settlement and tourism.

Rift Valley fever. Africa's Rift Valley stretches from Mozambique to the Red Sea.

In 1977, mosquitoes carried a virus from the Rift Valley on the eastern side of Africa across the continent to Egypt. Some 200,000 humans caught the fever while slaughtering sheep. More than 2,000 people and most of the sheep in Egypt were killed by the deadly fever.

A U.S. Army research lab at Maryland's Fort Detrick found a way to track the path of the disease by satellite. Researchers used space photos of land conditions across Africa to map vegetation density. That revealed clues to mosquito movements and let scientists predict the spread of the fever virus.

The 1977 outbreak was stemmed by animal vaccine from South Africa and human vaccine from the U.S. Army's Fort Detrick lab.

Orange trees. Florida regularly conducts a census of trees in thousands of acres of citrus groves. Counts which once took workers seven months can be done in a few days by observation satellites fitted with infrared cameras feeding data to computers on the ground.

Infrared satellite images also reveal the health of citrus trees, the presence of disease, the deterioration of crops, as well as poor soil and stressed regions that may not produce an abundant crop. The state uses the census to tax citrus groves. Unhealthy trees are not taxed.

Infrared satellite photography also has been used to count and analyze timber in midwestern states.

Wine louse. NASA used satellites in 1993 to help California wineries battle an aphid-like insect that had bugged the industry for centuries.

About 65 percent of vineyards in Napa and Sonoma counties were planted with a grape rootstock vulnerable to a new strain of phylloxera, an insect that kills plants by sucking juice from grapevine roots.

Infrared satellite photos of stressed plants were used to map root louse damage to wine grapes—a serious threat to California's huge wine industry. A century earlier, a strain of phylloxera had nearly destroyed the vineyards of California and France.

The satellites' visible and near-infrared cameras saw nutrient deficiencies within plants at an early stage—damage usually not visible until two or three years after the insect has been feeding on the plant. Then, the plant declines rapidly and cannot bring its fruit to maturity for harvesting.

Temperature is an indicator of plant health. Stressed plants are warmer because they cannot efficiently pass water through their membranes.

Thermal scans by observation satellites recorded subtle differences in grapevine temperatures. Grapevine leaves were short on nutrients and preparing to turn yellow.

The photos allowed researchers to see plant damage and map the spread of the insect. Vineyard managers could replant with resistant roots—the only way to rid the vineyards of the pest.

NASA, known to be an agency in love with acronyms, didn't fall short on the phylloxera project, labeling it GRAPES (Grapevine Remote sensing Analysis of Phylloxera Early Stress).

Pollution hunter. Garbage washing ashore along New Jersey's coast in 1987 was so bad federal officials looked to the skies for help.

Believing that ships illegally dump trash in the oceans, New Jersey's U.S. Sen. Bill Bradley called for a satellite to snap photos of the ocean surface regularly so federal inspectors could see polluters dumping junk. Bradley said he was "revolted and angered" by the floating filth tormenting his state.

Bradley and U.S. Deputy Commerce Secretary Clarence Brown handed out observation satellite photos detailing the New Jersey coastline. The 90-ft.-resolution photos—smaller than a football field, about the size of a baseball infield—showed a ship snaking a trail of trash along the New York-New Jersey coast. Bradley said a watchdog satellite would deter polluters who face jail for dumping.

Marijuana. Not everyone is happy with remote-sensing satellite applications. Mexicans were ticked off in 1990 when American law enforcement officers used observation-satellite photos to track down vast marijuana plantations below the Rio Grande. The problem: Mexico said it was not consulted in advance about the use of a satellite in the fight against drug trafficking.

Kiwis. New Zealanders didn't seem too happy, either, when France mounted a satellite "beacon" on the roof of a local post office.

Distressed by French nuclear bomb tests in the South Pacific, local laws forbade cooperation with nuclear military powers. New Zealanders were especially upset when a 1985 bombing by French agents sunk the Greenpeace ship Rainbow Warrior in Auckland harbor.

Meanwhile, France was overhead, with two SPOT observation satellites, the French-U.S. TOPEX/Poseidon oceanography satellite, and the European Earth Resources satellite ERS-1.

In support of SPOT, TOPEX/Poseidon and ERS-1, the French National Geodetic Institute in January 1988 went to New Zealand national postal authorities—an agency known then as New Zealand Post—and rented commercial space on the roof of a local post office on the remote Chatham Islands 500 miles east of New Zealand.

The Institute mounted an antenna, referred to in news reports in New Zealand as a beacon code-named DORIS, on that post office roof.

Apparently, the New Zealanders didn't know SPOT, TOPEX/Poseidon and ERS-1, all were civilian satellites, and neither military nor nuclear.

They go about peacefully, collecting data for civilians on ecology, ocean levels, plant coverage and other conditions on Earth's surface.

TOPEX/Poseidon, for example, does find its own position in Earth orbit with a radiolocation system known by the acronym DORIS. That's short for Doppler Orbitography and Radio positioning Integrated from Space.

Meanwhile, the French antenna didn't get much of a workout in the Chatham Islands. ERS-1 wasn't launched until 1991 and TOPEX/Poseidon didn't get off the ground until 1992. The antenna was tested briefly and then shut down.

Spying On California

The Metropolitan Water District of Southern California used a Landsat infrared-photography satellite scooting through space 450 miles overhead to ferret out water wasters.

The 5,200-sq.-mi. water district encompassed drought-stricken Los Angeles, Orange, Riverside, San Bernardino, Ventura and San Diego counties.

The district's spy-eye-in-the-sky photographs showed where residents were overwatering their crops, lawns and vegetation.

Landsat recorded false-color images of infrared light reflecting from plant leaves. The greener the vegetation, the redder it looked in satellite photos. The wetter the leaves, the brighter the red.

The water district didn't target individual homes. Rather, the infrared photographs were used to uncover over-watering in areas larger than a city block—irrigated croplands, parks, golf courses.

That didn't stop New Zealand Prime Minister Geoffrey Palmer and other senior politicians from losing their tempers. Palmer's government may or may not have known about DORIS before, but it "officially" learned about the beacon in January 1989 when local anti-nuke peace activist Owen Wilkes stumped the countryside, charging the beacon violated New Zealand's anti-nuclear law.

Government officials came forward to say New Zealand Post somehow had failed to let them know about the rooftop rental and installation.

"France is a nuclear weapons state," Palmer declared as his cabinet voted to have the antenna removed from the post office roof.

Agreeing with Wilkes that DORIS could be used by remote control to guide French ballistic missiles armed with nuclear warheads, Palmer's government sent DORIS packing.

Navigation Satellites

From the time of the earliest navigators, travelers have followed guide stars across land and sea. In particular, mysteries of the deep required ancient mariners to look to the skies as they steered themselves across dangerous oceans.

In the latter half of the 20th century, man-made guide stars with names like Transit and NAVSTAR have replaced nature's beacons in the sky for maritime navigation.

Today's man-made guide stars are navigation satellites orbiting a few hundred to a few thousand miles above Earth. They send down radio signals which modern travelers use to locate themselves with extreme precision on sea, land or in the air.

The Doppler Effect

When listening to a radio signal coming down from an orbiting satellite, a phenomenon known as Doppler Shift will make the signal's frequency seem higher when the satellite is coming toward you and lower when it's going away.

Navigation satellites allow a ship, plane, train, truck, car or even a hiker to use the Doppler effect to pinpoint a location on Earth' surface.

Since a satellite's orbit is known, an unknown position on Earth can be determined accurately by measuring the increase or decrease in the radio frequency emitted by a satellite as it orbits Earth. The change in frequency is the Doppler effect.

The earliest navigation satellites were developed by military forces to guide submarines launching ballistic missiles against distant targets. Each submarine had to know its location precisely so it could aim its missiles. Of course, surface navies also wanted to know their exact locations.

The United States launched the first navigation satellites, known as Transit, in 1959 and 1960. The USSR followed suit with its Tsikada navsats starting in 1978.

The U.S. recognized the private sector's need for navigation satellites in 1967, while the USSR started civilian systems in 1978. In the post-Cold War years of the 1990s, civilian uses for navsats moved to the forefront.

Transit. The first-ever navigation satellite was Transit-1A, launched from Cape Canaveral September 17, 1959. It failed to reach orbit. Transit-1B was launched April 13, 1960.

Johns Hopkins University had designed those first Transits for the U.S. Navy. The satellites had two double-frequency transmitters putting out radio signals at 162 and 216 MHz and 54 and 324 MHz. Later Transits had an electronic clock and memory to transmit position data every twelve hours.

To use the satellite navigation system, a surface ship or submarine would receive signals from a Transit and measure the Doppler-effect change in frequency to determine the satellite's position. Then the ship could calculate its own position.

Nuclear power. Transit-4A and -4B were the first satellites to carry nuclear-power electricity generators. The radioisotope generators were called SNAP for Supplementary Nuclear Power.

Operational. The Transit satellite system was considered experimental until the launch of Transit-5A1 December 19, 1962, from Vandenberg Air Force Base, California, to a 400-mi.-high polar orbit. With that launch, the system was declared operational. Unfortunately, Transit-5A1 had a faulty receiver and failed.

The Navy's operational all-weather navigation system also had 60,000 civilian users. It was used in commercial shipping, charting offshore oil and mineral deposits, and land surveying programs.

Solar power. The launch of Transit-5BN3 failed in 1964, not an unusual occurrence in the early days of space flight, but this satellite carried one of the plutonium-238 nuclear power supplies.

A kilogram of radioactive fuel dispersed into the atmosphere, igniting an international political flap which led the follow-on satellite, Transit-5C, to carry panels of solar cells for electrical power.

Oscar. The Navy also referred to some Transits as Oscar and Series O. They should not be confused with the OSCAR series of amateur radio communications satellites.

Each Oscar sprouted a 70-ft. stabilizing boom to counteract a satellite rocking motion caused by the Earth's magnetic field.

SOOS. When the Navy was able to launch two of its lightweight Oscars to 600-mi.-high polar orbits on one Scout rocket, the flight would be nicknamed SOOS for Stacked Oscars On Scout.

For instance, SOOS-2 was launched in 1987. SOOS-3 and SOOS-4 were launched in 1988.

At the time, Scout was America's smallest expendable launch vehicle (ELV) unmanned space rocket capable of placing a payload in orbit.

Civilian navsats. From 1967, civilian ships were allowed to use the Transit system which then acquired yet another name, Naval Navigation Satellite, or simply Navsat.

When signals were received from three Transit satellites at one time, a ship could calculate its position within 1,640 feet. The time between bearings was 90 to 110 minutes.

An improved system included six satellites in polar orbit at all times. They circled Earth at altitudes of about 680 miles, allowing fixes 24 times a day. Corrected time and position data were sent up to each satellite every twelve hours from a ground control station. Each satellite then broadcast its position and a time signal every two minutes at 150 and 400 MHz.

Receiving one frequency could tell a moving ship its position within 575 feet. Receiving two frequencies allowed ship-board operators to correct for atmospheric radio-wave bending, revealing a position within 330 feet.

At a receiving site on land, the accuracy was 82 feet. Averaging two dozen consecutive satellite passes could improve accuracy to 4.9 feet.

LOFTI. Meanwhile, the U.S. Navy was carrying out other navsat research projects.

LOFTI, an acronym for Low-Frequency Trans-Ionospheric Satellite, was the name of a pair of aluminum beach balls tossed to space in 1961 to see how well low-frequency radio signals penetrated the ionosphere. LOFTI-1 was launched alongside Transit-3B. LOFTI-2A was launched in 1963.

TRAAC. The experimental Transit Research And Attitude Control (TRAAC) satellite relied on gravity for attitude control after it was launched alongside Transit-4B in 1961. A 105-ft. boom extended from the satellite.

SECOR. The first U.S. Army navigation and position-location satellites were called Sequential Collation of Range (SECOR). Thirteen were launched from 1964 to 1969.

Each had two solid-state transponders. Signals were sent to the satellite from four positions and repeated by the transponders. Three of the positions were known and the fourth position was determined by the satellite.

SECOR-1 was launched to a 500-mi.-high orbit in a triple shot from Cape Canaveral, alongside GGSE-1 and Solrad-7A satellites.

TOPO. The Army Corps of Engineers topographic laboratory launched a SECOR satellite from Vandenberg Air Force Base in California to a 600-mi.-high orbit in 1970, but called it TOPO. The satellite tested triangulation for navigation.

Timation. The Navy's improved Transit was called Time Navigation, or Timation for short. The first improved Transit satellite, Timation-1, was launched from Vandenberg in 1967.

Timation satellites traveled higher than Transit satellites, at altitudes near 600 miles above Earth.

They carried precise quartz clocks for more accurate measurements in a technique known as "three-dimensional navigation."

California Light Show

The SOOS-3 launch on April 26, 1988, became a sky spectacular when its Scout rocket put on a high-altitude light show visible 100 miles from the launch pad at Vandenberg Air Force Base.

The launch left a colorful exhaust contrail visible as far south as Malibu. The contrail was a twilight phenomenon of spent fuel trailing behind the booster. As it traveled through the upper atmosphere, the fuel combined with water droplets which crystallized. The crystals were struck by high-altitude sunlight, causing aurora borealis visible for 15 to 20 minutes. The blue and green contrail over Orange County was blown by winds to resemble a giant pretzel.

The Western Space and Missile Center launch complex at Vandenberg is 125 miles northwest of Los Angeles on the Pacific Ocean coast.

NTS. The third Timation further improved on Transit. It was labeled Navigation Technology Satellite (NTS-1) and included results of U.S. Air Force research known as System-621B.

NTS-1 was launched in 1977 from Vandenberg to a much higher orbit, 8,000 miles above Earth.

It carried two quartz clocks plus two experimental rubidium-vapor atomic clocks. Radio frequencies were changed to 335 and 1,580 MHz.

NTS-2 carried an experimental cesium atomic clock when launched in 1977.

TIP. Alongside the Timation project, the Navy also tested a navigation satellite development known as Transit Improvement Program (TIP).

The TIP series had a system called Disturbance Compensation System (DISCOS), allowing a satellite to maintain its position carefully in orbit by correcting for disturbances to that orbit caused by atmospheric resistance and the solar wind.

TRIAD. The first spacecraft in the TIP series was called TRIAD-1, blasted from Vandenberg to a 500-mi.-high polar orbit in 1972. TRIAD-1 was nuclear powered, carrying a radioisotope thermal generator (RTG).

From the second satellite in the series, TIPs were constructed with solar cells for electricity.

TIP electronics and computers were improved further and ground control of clock adjustments was added.

NOVA. The TIP satellite design was standardized in 1981 and the name of the solar-powered satellites was changed to NOVA. Built by RCA, they had a 25-ft. boom for stabilization and received clock corrections from the ground.

NOVA-1 was fired from Vandenberg to a 700-mi.-high polar orbit in 1981.

Transat. A Transit satellite, fitted with a Satrack receiver and called Transat, was launched from Vandenberg to a 600-mi.-high orbit in 1977.

NAVSTAR GPS Satellites

The latest development in space navigation is the NAVSTAR global positioning system (GPS).

NAVSTAR is short for Navigation System using Timing And Ranging. The fleet of satellites provides latitude, longitude, altitude, direction of travel, travel velocity and correct time of day to anyone anywhere, day or night, in any weather.

Design of the extensive 24-satellite system was authorized in 1973 and the first satellite was launched in 1978.

Just as the Transit satellite navigation system gave more-precise location fixes than the Loran non-satellite navigation system, NAVSTAR gives more-precise location fixes more frequently than the Transit navigation-satellite system.

The U.S. Air Force created the constellation of NAVSTAR satellites circling Earth twice a day. At least four of the space beacons are in view from any spot on Earth at any time.

New deal. Where the older generations of American navigation satellites had flown only 600 miles above Earth, the NAVSTARs would be sent much higher to 12,625 miles above Earth.

Transits had transmitted on two frequencies, at 150 and 400 MHz, and NAVSTARs would transmit on two. But NAVSTAR's frequencies would be much higher, with 1575.42 MHz for civilians and 1227.60 MHz for military services.

Where the older system used six satellites in orbit, NAVSTAR would require 24 satellites flying in six rings of four satellites. Among the 24, three would be working spares stored in orbit.

In the old system, the time between bearings was 90-110 minutes. In the NAVSTAR scheme, any point on Earth would be in view of at least five satellites at all times so bearings could be received instantaneously whenever desired.

Ships, planes, trains, trucks, cars and even persons on foot could know their positions in latitude, longitude, and altitude within 58 feet, along with their velocity within 0.45 mph, and correct time to within one-millionth of a second.

System. The NAVSTAR system consists of a constellation of orbiting satellites, a ground control system, and thousands of GPS receivers.

A receiver on the ground has to hear from only three NAVSTAR satellites to find its own latitude and longitude location. Hearing from a fourth satellite lets the receiver calculate its own altitude.

Military. NAVSTAR is a Defense Department project. The NAVSTAR Control Center is at the U.S. Air Force space command, Falcon Air Force Base, Colorado Springs, Colorado.

Air Force receivers around the globe monitor all NAVSTARs for the Colorado Springs master control station which updates each satellite's navigation and health messages for all users.

Each satellite also monitors its own navigation data errors, signal availability, anti-spoof failures and clock failures. If something fails, users are notified within six seconds. Some failures are reported only to master control and take from 15 minutes to several hours to rectify.

Gulf War. U.S. and Allied military forces in Operation Desert Storm during the Persian Gulf War made extensive use of NAVSTARs. On land, sea and in the air, they received global, three-dimensional, position and speed information.

Forces could observe voice and radar radio silence during a rendezvous. Aircraft could link up for mid-air refueling without communications.

F-16 and B-52 aircraft used GPS receivers to deliver bombs and guided munitions on specified military targets. Navy cruise missiles used GPS position data to attack heavily defended high-priority targets. The Army had GPS receivers in Apache helicopters and M60 tanks.

Army and Marine troops followed their hand-held GPS receivers across the featureless Saudi desert. GPS was used to site radar, while forward observers used GPS to direct artillery fire on Iraqi positions. The Navy used GPS to map mine fields and to direct the rendezvous of supply ships.

Somalia. In Operation Restore Hope in Somalia, GPS receivers were aboard Air Force and civilian cargo aircraft for approaching and landing at makeshift airfields without the electronic aids available at most larger airports.

Civilian uses. NAVSTAR receivers guide planes, ships, trucks, buses, spacecraft and even private automobiles. Myriad applications include mapping, aerial refueling rendezvous, geodetic surveys and oceanography, law enforcement and fire fighting, air traffic control, space operations, exploring off-shore for oil, and search and rescue.

Nearly all land surveying today uses GPS receivers and NAVSTAR is being used to measure Earth movement along geological faults.

Receivers are so small and inexpensive that hikers are packing handheld personal navigators weighing under a pound into the wilderness.

Recreational use of GPS by boat owners is expanding rapidly. Receivers in private planes are permitting touch downs in the dark.

Dispatchers track GPS-equipped police, fire and rescue vehicles continuously on computer maps.

Refrigerated railcars and uranium shipments are tracked across the nation. Truck, train and ship terminal bottlenecks are relieved as dispatchers know ahead of time who is coming and when.

Dallas monitors its bus fleet with GPS, while a Louisiana, dispatcher uses a GPS vehicle locator to follow Lafayette's ambulance fleet through 22 southern Louisiana parishes and Baton Rouge on one electronic map.

The Coast Guard uses real-time differential GPS to navigate harbors. Using NAVSTAR for harbor and ship navigation reduces oil tanker accidents.

Robots. Engineers are predicting all ships, planes, trucks, buses and automobiles will have GPS receivers by the year 2000. Boats on inland waterways will avoid collisions with a warning device built around NAVSTAR. Downed aircraft and boats lost at sea will be located quickly as they transmit GPS position data to search parties.

Robot crop dusting aircraft will follow location information from NAVSTAR. Road grading will be done by robot equipment operators using NAVSTAR positions to move about.

Eventually, robots will do land surveys. Map making and wild-animal tracking will be completed by machines following NAVSTAR data. Ocean buoys will monitor themselves, while air traffic control will become automatic.

Highway maps. Ohio State University invented a time-saving, money-saving use for GPS satellites in 1991, with a machine to check on highway deterioration for the Federal Highway Administration, the transportation departments of 38 states and the Canadian province of Alberta.

The University's Center for Mapping mounted TV cameras and GPS receivers on a van to cruise secondary roads, cross bridges, and pass equipment sites, grass-mowing areas and railway crossings.

As the high-tech truck rolls along highways, its cameras scan the terrain for roadways needing repair, hazardous conditions and locations of fatal accidents, while the receivers record latitude and longitude. Results are stored as digital maps in a Geographic Information System computer.

A Heart Of Atoms

The heart of NAVSTAR is its accurate clock. Each satellite has an atomic oscillator with rubidium cell and cesium beam. The accuracy of that clock is within one second in 300,000 years!

In space, the beacons burble cheerily, 24 hours a day, sending never-ending messages telling the time. NAVSTAR Control Center at Colorado Springs, Colorado, checks the satellite clocks regularly to make sure they are beeping in sync.

A receiver on Earth checks its own clock, too, comparing it with time signals from whichever satellites happen to be overhead at the moment.

After checking four satellites, the computer in a receiver computes how far away the satellites are, then displays its own location on a screen.

To do that, a receiver calculates how long it has taken a signal to arrive from a satellite. Knowing the speed at which radio waves travel, distance to the satellite can be computed.

Signals from three satellites let a receiver calculate its distance, or range, from three points in the sky. The receiver's computer then uses trigonometry to find its own position from the known points in the sky.

Signals received four satellites let the receiver calculate its altitude.

A person using a GPS receiver can pinpoint his location and altitude to within either 58 or 328 feet anywhere in any weather. A receiver can compute velocity to better than one foot per second.

The GPS van was tested on roads in Louisiana, Ohio, Florida, Colorado, Virginia, and the West Coast. Unreported road deterioration costs the United States billions each year in wasted fuel and vehicle repairs. As a bonus, states can use the digital data to draw new road maps for motorists.

The Center uses GPS to manage land, ocean resources and erosion, and help relieve disasters.

Blocks. From 1973, the Air Force Space and Missile Systems Center at Los Angeles Air Force Base and Rockwell International Corp. designed and tested eleven, developmental, solar-powered NAVSTAR satellites.

First-generation NAVSTARs were block 1, while the second-generation were block 2.

Launch of eleven block-1 NAVSTARs from Vandenberg began with NAVSTAR-1 on February 22, 1978. The seventh attempt failed.

In 1983, the Pentagon ordered 28 improved second-generation NAVSTARs from Rockwell.

Block-2 satellites would include 21 working in orbit, 3 working spares in orbit, and 4 replenishment spares on the ground to be launched as needed to replace aging satellites.

The first block-2 NAVSTAR went to orbit February 14, 1989, on a Delta rocket from Cape Canaveral Air Force Station. Use of a Delta rocket resulted when loss of shuttle Challenger in 1986 grounded the American fleet nearly three years. With no way to send payloads to orbit, the Pentagon rushed to buy 21 Deltas from McDonnell Douglas Space Systems of California.

Block-1 NAVSTAR GPS launches	
1	1978 Feb 22
2	1978 May 13
3	1978 Oct 7
4	1978 Dec 11
5	1980 Feb 9
6	1980 Apr 26
7	failed to achieve orbit
8	1983 Jul 14
9	1984 Jun 13
10	1984 Sep 8
11	1985 Oct 9
Block-2 NAVSTAR GPS launches	
1	1989 Feb 14
2	1989 Jun 10
3	1989 Aug 18
4	1989 Oct 21
5	1989 Dec 11
6	1990 Jan 24
7	1990 Mar 26
8	1990 Aug 2
9	1990 Oct 1
Block-2a NAVSTAR GPS launches	
10	1990 Nov 26
11	1991 Jul 4
12	1992 Feb 23
13	1992 Apr 10
14	1992 Jul 7
15	1992 Sep 9
16	1992 Nov 22
17	1992 Dec 18
18	1993 Feb 3
19	1993 Mar 30
20	1993 May 13
21	1993 Jun 26
22	1993 Aug 30
23	1993 Oct 28±
24	1994 Mar 2±
25	when needed
26	when needed
27	when needed
28	when needed

when needed: replacement satellites to be launched when needed as older satellites fail

After nine block-2 satellites were fired to orbit, refinements were made and the Air Force started launching block-2a satellites to space in 1990.

Six block-2a NAVSTARs were launched in 1992 and again in 1993. In 35 years of American space satellite history, that was the most of any one kind of satellite sent to space in multiple launches in one year.

Identity. Since all NAVSTARs transmit on the same frequency, they are distinguished by their pseudo random noise (PRN) number.

Contents. Each satellite transmits a continuous stream of data, telling where it is at a given moment and what time it is at that moment.

Navigation data includes satellite ephemeris and atmospheric propagation corrections. System information includes clock bias.

A NAVSTAR also relays information about nearby satellites. In effect, it is transmitting a periodically-updated celestial almanac and a receiver on the ground is holding that data in memory.

NAVSTAR satellites also carry sensors for detecting nuclear explosions.

Shorting civilians. The Pentagon has reduced dramatically the accuracy of NAVSTAR GPS data available to civilians.

NAVSTAR satellites offer Precise Positioning Service (PPS) for military units and Standard Positioning Service (SPS) for civilians.

PPS gives horizontal positions accurate to within 58 feet and vertical positions to within 91 feet. SPS gives horizontal positions accurate to within 328 feet and vertical to within 512 feet.

Signals are sent from NAVSTAR satellites in codes and must be decoded by a GPS receiver. The codes are precise (P) and coarse/acquisition (C/A).

P-coded signals are available only for use by military forces, while the C/A code has been publicized so it can be used by civilians.

The C/A signal allows calculation of position to within 328 feet. The P signal offers far-more-accurate computation of position to within 58 feet.

Each GPS satellite broadcasts spread-spectrum signals on two L-band frequencies, known as channels L1 and L2. Channel L1 carries both P and C/A signals at 1575.42 MHz. Channel L2 transmits only P-coded signals at 1227.60 MHz.

Using a receiver. Despite the military heritage of the NAVSTAR system, handheld and mobile GPS receivers giving locations in latitude and longitude are sold to civilians by stores and catalog merchants around the world.

When a user turns his receiver on, it starts searching the airwaves for signals from the closest satellite. Finding a satellite, the receiver accepts almanac data and records it in memory.

Next the receiver listens to three nearby satellites for their positions and time signals. A fourth satellite is received for a reference time.

The receiver checks its clock to see when it received a time signal. It compares its time with the time reported by a satellite to see how long it took the radio signal to get to it from the satellite.

Using the known speed of radio signals, the receiver calculates the distance to the satellite. From the distance to three satellites, the receiver can triangulate or fix its position accurately.

2-D. Boats on water and other slow-moving vehicles not requiring altitude fixes can triangulate with only three satellites. With three satellites, the fix is two-dimensional.

3-D. Altitude or height above Earth's surface is the missing third dimension. For altitude, data must be received from a fourth satellite.

Satnav. NAVSTAR enhanced the services provided by previous satellite systems, bringing greater accuracy at lower cost.

Of course, Transit "Satnav" receivers still are used to obtain data from those older satellites still in orbit. Satnav receivers still rely on a satellite passing overhead every 90 minutes for a new position fix, while NAVSTAR provides instantaneous data in three dimensions.

Speed. A GPS satellite sends down two kinds of data—time-location and celestial almanac.

GPS receivers are separated into single-channel, 2-channel, 5-channel, up to 20-channel types, depending on amount of data received at one time.

A single-channel receiver takes in either time-location data or celestial-almanac data from one satellite at a time, leading to slower computations.

A two-channel receiver takes in time-location data and celestial-almanac data at the same time from one satellite. That lets it calculate faster.

A single-channel receiver takes ten to fifteen seconds to fix its position, while a two-channel receiver updates its position twice in one second.

Professional receivers used by surveyors and others are five-channel receivers. Four of the channels listen for time-and-place data from four satellites, while a fifth channel listens to celestial almanac data from one satellite. Position fixes from a five-channel receiver are extraordinarily accurate and updated many times each second.

Did The Landmarks Move?

Of course the American landmarks didn't actually move themselves. Navigation satellites moved them. Well...cartographers actually moved them on maps after NAVSTAR global positioning satellites pinpointed their true positions.

The biggest leap was the flagpole in front of the judiciary building in Honolulu, Hawaii, which hopped 1,480.8 feet to the southeast.

The Bismarck, North Dakota, water tank was relocated 101 feet west. The Empire State Building in New York City was moved 120.5 feet northwest.

The District of Columbia's Washington Monument turned up 94.8 feet northeast, while the head of the statue on the dome of the U.S. Capitol was moved 94.8 feet to the northeast.

Nearby, in Annapolis, the Maryland state capitol dome moved 98.5 feet northeast. In Richmond, Virginia, the war memorial moved 99.8 feet northeast. The border between the United States and Canada moved as much as 66 feet on maps.

Latitude and longitude form an imaginary grid of lines covering the world. The basic reference points had been measured last in 1927 when surveyors used portable towers, plumb lines and sighting telescopes.

In a 12-year project through the 1980s into the 1990s, some 250,000 basic reference points were recalculated for engineers, military, surveyors, highway builders, utilities, regional planners, and geologists studying the movements of continents.

NAVSTAR satellites radioed data to 5,000 receivers which calculated latitude and longitude. Those 5,000 readings were used to recalculate the other 245,000 basic reference points with 100 times the accuracy of the 1927 findings. The new margin of error was only one inch in six miles.

Homeowner property lines weren't changed, of course, but state and national boundaries did move a few feet.

Russian Navsats

While American engineers were developing navigation satellite technology, the USSR also was orbiting spacecraft to aid shipping navigation.

Cosmos. The generic name for most Russian satellites over the years has been Cosmos. The USSR launched navigation satellites as early as 1967 when Cosmos-192 was sent aloft.

It was followed by dozens of navsats in four technology generations comparable to the U.S. series of Transit navigation satellites. Most were launched to Earth orbit from the USSR's Northern Cosmodrome at Plesetsk.

The first-generation satellite system had three satellites in three orbital planes separated by 120 degrees. They were considered experimental until 1971 when the system became operational with the launch of the satellites Cosmos-385, Cosmos-422 and Cosmos-465.

The second generation started with Cosmos-514 in 1972. It had three satellites on orbital planes 60 degrees apart. The third generation, which arrived with Cosmos-700 in 1974, had six satellites on four orbital planes separated by 30 degrees.

A Satellite Map In Your Car

You don't notice the receiving antenna molded into your automobile roof, but it notices signals from a flotilla of NAVSTAR satellites orbiting 12,600 miles overhead. The antenna inhales the signals and feeds them to a GPS receiver lost in the jumble of computers and other electronics hidden behind your automobile's dashboard.

As you steer your car along a thoroughfare, a television picture tube sunk into the front of your dashboard displays a colorful map of city streets surrounded by country roads. A small black automobile icon rolls along the map, laying down a trail of dashed lines from where you've been.

When you feel it's safe, you switch the TV map to a heads-up display projected on the windshield.

The computer understands spoken words, so you need only say street names out loud. The gear responds by showing how to get from here to there.

You'll even know when you're up on Pike's Peak or down in Death Valley. With information from four NAVSTARs, your receiver displays altitude, as well as longitude and latitude. It also displays an incredibly accurate time of day.

Manufacturers have demonstrated automotive NAVSTAR systems. The price of a car receiver eventually will drop below $500, so all cars may be equipped with GPS receivers by the year 2000.

The fourth generation, starting with Cosmos-1000 in 1978, had four satellites on four planes separated by 45 degrees. Each Cosmos was about 620 miles above Earth, circled the globe every 105 minutes and transmitted a unique identification number on the frequencies of 150 and 400 MHz. Ships could locate themselves within 300 feet.

Tsikada. Soviet military domination of the USSR's satellite navigation system ended with the launch of Cosmos-1000 in 1978 when the system was renamed Tsikada and opened to civilians.

On Earth, tsikada (cicada in English) is a cricket. A male cricket makes a sharp chirping sound. The satellite name was inspired by the sound from a radio receiving a Tsikada satellite.

The Tsikada satellites Cosmos-1861, launched in 1987, and Cosmos-2123, in 1991, also carried amateur radio payloads RS-10/11 and RS-12/13.

After the international COSPAS-SARSAT system started in 1982, three Russian navigation satellites were equipped with search-and-rescue receivers to monitor distress frequencies for sinking ships and downed aircraft. They were Cosmos-1383 launched in 1982, Cosmos-1447 in 1983, and Cosmos-1574 in 1984.

GLONASS. The USSR matched America's NAVSTARs with a new generation of high-flying navigation satellites in 1982 when Cosmos-1413, Cosmos-1414 and Cosmos-1415 were launched on one rocket to start the Global Orbiting Navigation Satellite System (GLONASS).

GLONASS satellites work much the same as NAVSTARs, including flying 12,000 miles above Earth and transmitting on two frequencies in the 1200-1600 MHz range. Commercial GPS receivers sold today often are capable of receiving both NAVSTAR and GLONASS data.

Aircraft. Like the NAVSTAR global system, GLONASS satellites provide precise positions to aircraft in all weather anywhere in the world, including over oceans. Data sent to flight crews is more accurate than inertial navigation systems. Satellite receivers require no input from pilots, so navigation is less susceptible to human error.

In 1991, shortly after launching the GLONASS satellites Cosmos-2139, -2140 and -2141 on one rocket, Russia proposed that GLONASS be added to the cockpits of international civil airlines.

Later that year, Northwest Airlines became the first American airline to navigate by GLONASS when one of its Boeing-747 cargo planes, equipped with GLONASS and NAVSTAR receivers, made a demonstration flight from Anchorage, Alaska, through Russian Far Eastern airspace, to Tokyo.

The test flight indicated worldwide navigation by satellite could lead to shorter, more economical flights between the United States and Asia. Direct routes over Russian airspace provided shorter flying times to cities in Asia. Flight data showed the GLONASS and NAVSTAR systems worked together to enhance safety and accuracy.

In 1993, Continental Airlines became the first commercial air carrier to use GPS regularly for aircraft navigation and landing, with GPS receivers in McDonnell Douglas MD-80, Aerospatiale Aeritalia ATR-42, deHavilland Dash-7 aircraft.

Japan. The ETS-5 communications satellite was blasted to space in 1987 atop an H-1 rocket from Tanegashima Space Center on a small island at the southern tip of Japan. The satellite not only relays communications, but also works as a position guide for small ships and aircraft.

Military Satellites

Lurking just out of sight, far above the clouds, a massive armada of military satellites patrols the near reaches of space, peering at the world below with keen night eyes, cocking a watchful ear, threatening interlopers with laser beams and brick-bats, and babbling never-ending reports on nearly everything to be seen and heard on land, sea and in the air, from Earth's teeming cities to the most far-flung ocean wildernesses.

Military satellites are vital for communications, photo reconnaissance, electronic surveillance, radar imaging, navigation, early warning, meteorology, nuclear inspection, scientific and technical research and even as orbiting weapons.

Among the thousands of military payloads orbiting 100 to 22,300 miles above Earth are reconnaissance satellites spying on land, ocean and air; electronic eavesdroppers tuning in on two-way radio transmissions; inspection satellites watching for nuclear explosions and sounding early warnings of nuclear missile attacks.

Also, navigation satellites leading troops into battle and missiles to their targets; weather satellites keeping tabs on the atmosphere; communications satellites providing rapid transit for command and control messages and battlefield data; and even laser beams, anti-satellite weapons and orbiting bombs.

Satellites relay communications, help pinpoint navigation, report environmental conditions, scan land and sea for enemy deployments, monitor arms-control agreements and give early warning in the event of nuclear attack.

Military satellites in orbit spy on facilities and forces, scan radio waves as electronic ferrets, watch for nuclear explosions and missile launches, map Earth's surface and relay communications.

Some satellites are in low orbits, 100 to 300 miles above Earth. Others are at intermediate altitudes from 500 to 1,000 miles. Some orbit the equator, while others fly over the poles. Some pace the whirling Earth from 22,300 miles altitude, making them seem stationary in the sky.

Military technology. The Pentagon seldom discusses America's top-secret espionage satellites and other military spacecraft, but most of the space technology used by civilians today started life in military hardware.

The first communications satellite was a U.S. military payload launched in 1958. The U.S. launched the first navigation satellite in 1959 to guide submarines. The USSR's first weather satellite, launched in 1962, was a photo spysat.

Nearly 5,000 payloads have been launched to orbit so far in the Space Age, not counting debris and space junk. Most years have seen more than 100 military satellites launched by a variety of nations around the globe, while only a dozen or two civilian satellites may have been launched.

The need for military satellites has been reduced somewhat by the end of the Cold War.

Reconnaissance Satellites

Just as civilians can spy on the weather from space or ferret out natural resources by satellite, so military units can spy on each other.

Photography satellites are Peeping Toms flying hundreds of miles overhead. These photo-intelligence (PHOTOINT) satellites shoot clear pictures and record infrared images of installations on the ground from space.

Spysats also gather electronic intelligence (ELINT), signals intelligence (SIGINT) and radar intelligence (RADINT).

Radio-intercepting satellites are big electronic ears in the sky eavesdropping on conversations hundreds of miles below. Radar imaging satellites sweep Earth with their powerful beams, looking for targets on land and sea.

Specialized reconnaissance satellites watch over naval fleets on the world's oceans, provide early warning of impending missile attack, carry out nuclear inspections to confirm compliance with treaties, and, of course, report global weather.

Photo Spysats

The complete story of photo-observation satellites is in chapter five.

The U.S. experimented in 1959 with its first photo-reconnaissance satellite, Discoverer. It flew an elliptical path from 100-650 miles above Earth.

After its U-2 spy plane was shot down over the Soviet Union in 1960, the United States decided to get serious about satellite surveillance.

Sentry-Samos. An American spysat called Sentry-Samos was launched to a 300-mi.-high orbit in 1961.

Close Look. Another kind of U.S. reconnaissance satellite was Close Look, launched in 1962 to an elliptical orbit ranging from 150 to 400 miles altitude.

Key Hole. The worst-kept secret, but best eye in the sky anywhere, has been the U.S. government's recent series of photo-reconnaissance spysats humorously labeled Key Hole.

Recent American photo-reconnaissance satellite models have been identified as KH-9 and KH-11 and KH-12 The latter is the biggest, heaviest, most versatile of the Key Hole spysats.

KH-9. A KH-9 photo-reconnaissance satellite actually ejects camera-made photos in a re-entry capsule which drops more than 100 miles from space and floats by parachute to an ocean surface where it is picked up by ship.

The first KH-9 was sent to an elliptical orbit 100-150 mile high in 1966. The improved KH-9 model known as Big Bird first was launched in 1971 to orbit at 115-185 miles altitude.

KH-11. On the other hand, a KH-11 has a television camera peering down. The first KH-11, launched in 1976, was the first espionage satellite to convert live television pictures to digital signals and beam them to Earth in real time.

The first KH-11 was sent to an elliptical orbit ranging from 150 to 330 miles high in 1976.

A new model referred to as an Advanced KH-11 was ferried to orbit in shuttle Discovery in 1989. A KH-11 probably works three years in orbit.

Hysteria. The intelligence community was in a panic from 1985 to 1988 as the United States found itself in the ticklish position of being down to its last working photo-reconnaissance satellite.

The cargo of a Titan-34D rocket in 1985 probably was a KH-9 Big Bird. Unfortunately, the Titan exploded, destroying its payload.

Meanwhile, a worn out KH-11 had to be turned off in orbit that same year.

The United States wasn't completely blind, but was down to its last working spy eye in the sky. One KH-11 still was working in orbit to watch USSR military events, but that satellite was near the end of its three-year life.

In 1986, another Titan-34D rocket exploded, probably destroying a KH-11 meant to replace the one deactivated in 1985.

Things began to look better in October 1987 and November 1988 when one of two successful Titan-34D launches carried a KH-11. That Key Hole satellite would have been the last KH-11, an engineering test model dusted off and refurbished for flight. On the other hand, that last KH-11 might have been aboard a Titan 34-D when that rocket's third stage failed in September 1988, stranding its spysat payload in a useless orbit.

KH-12. A new generation of KH-12 spysats, capable of taking infrared pictures at night, began relieving the older KH-11s at the end of the 1980s.

The first KH-12 probably was ferried to space in shuttle Columbia in August 1989.

A KH-12 has more capability than a KH-11, with a TV camera plus radar to take pictures through clouds and bad weather. However, that makes a KH-12 very heavy. Only a shuttle or Titan-4 unmanned rocket can carry it to space.

The relationship between KH-12 and the secret new radarsat Lacrosse (page 107) isn't clear.

A KH-12 is said to be able to snoop on most of the world, making detailed snapshots with digital cameras and transmitting digital images to Earth where they are enhanced by Central Intelligence Agency and National Security Agency computers.

The best military photo-reconnaissance satellite photos made in visible light today may show objects as small six inches across.

A KH-12 reportedly carries considerable fuel so it can move around in orbit to focus on specific reconnaissance targets on the ground. Russia has similar spysats maneuvering over U.S. and allied installations on the ground.

Spying On A Spysat

After Challenger exploded in 1986, American space shuttles were grounded until 1989. The delay created such a backlog of spysats awaiting launch that five of the next twelve shuttle flights ferried secret reconnaissance satellites to space.

Amateur astronomers tipped off the world in 1989 when one of those spysats seemed to be tumbling uncontrollably in space.

The secret satellite was ferried to space in Columbia in August 1989 and dumped overboard by the five-man military crew.

Eight days later, the satellite boosted itself to a 286-mi.-high orbit in which it would fly over 80 percent of the Soviet Union.

However, shortly after arriving on station, the imaging satellite seemed to start an unbridled romp across space. Astronomers in seven countries were taken aback by the odd behavior and rushed to make time exposures of the satellite.

Dozens of photos confirmed the spysat was flashing brightly, then dimming, then brightening again, reflecting sunlight toward Earth once each second, like something tumbling out of control.

The amateur astronomers, watching through optical telescopes on the ground, said tumbling had slowed by October 1989 when the spysat maneuvered in its orbit. As expected, neither NASA nor the Air Force offered explanations.

Little is known about a second, smaller military payload released from Columbia during that same mysterious August flight.

KH-12 burns. Shuttle Atlantis hardly made it home from orbit in March 1990 when a spysat which it had left in space exploded.

Atlantis had been launched February 28 in a dazzling night blastoff, probably with a KH-12, known as AFP-731, in its cargo hold.

The orbiter flew the highest angle to the equator ever traveled by a shuttle to that time. That allowed the astronaut crew to eject the spysat into an orbit covering most of the Soviet Union, including areas of the far north where ships, submarines and missile bases were concentrated.

In fact, the spysat was supposed to fly over every point on the globe between 62 degrees north and south latitude, including most population and industrial centers in Europe and Asia.

By March 1, Atlantis astronauts had pushed the satellite overboard into its own 127-mi.-high orbit. The shuttle landed March 4.

The satellite must have malfunctioned almost immediately. By March 7, the USSR's Novosti News Agency was reporting the big satellite was disintegrating in space with four large remnants expected to plunge into the atmosphere.

The first of the four chunks of the 17-ton satellite burned in the sky March 19 over the Pacific Ocean 900 miles north of the Midway Islands. The second chunk disintegrated March 20. The two remaining pieces disintegrated in the atmosphere without reaching ground by mid-May.

It wasn't clear whether the satellite, worth at least $500 million, came apart unexpectedly or was exploded deliberately to keep secret gear from falling intact into Russian hands. The U.S. government only said "hardware elements are expected to re-enter Earth's atmosphere."

Russia. The Union of Soviet Socialist Republics (USSR) launched its first photo-reconnaissance satellite, Cosmos-4, in 1962. It also was the Soviet Union's first weather satellite. Cosmos-4 flew in a 175-200-mi.-high orbit.

Cosmos-22. Russia's second generation of spysats started with the launch of Cosmos-22 in 1963 to an elliptical orbit ranging from 120 to 240 miles above Earth.

Cosmos-208. Russia inaugurated a third-generation of reconnaissance satellites with Cosmos-208, launched in 1968 to a 130 to 170-mi.-high orbit.

Cosmos-758. The fourth generation of spysats began with Cosmos-758 in 1975, in an elliptical orbit from 110 to 200 miles above Earth.

Cosmos-1813. Cosmos-1813 was a spysat in a 250-mile-high orbit from which it had a good view of North America and Western Europe.

SPOTting A Chemical Plant

When several Western nations pointed fingers at Libya for producing chemical weapons, France's SPOT-1 and SPOT-2 civilian photo-observation satellites were drafted into military service.

Libya denied its "pharmaceutical factory," built by West German businessmen at Rabta, 50 miles south of Tripoli, was used to manufacture chemical weapons. After a fire at the plant, Libyan leader Colonel Moammar Ghadhafi suspected German spies of sabotaging the plant.

While the United States said the plant was ravaged by the fire, SPOT's 33-ft.-resolution photos from 520 miles above Libya showed only scorch marks on a few small outbuildings.

The Bonn government denied involvement in the fire as 2,000 Libyans demonstrated outside the German embassy in Tripoli.

Cosmos-1813 exploded in 1987 after two weeks in orbit, snapping photos of various Western military installations. Here's what happened.

Cosmos-1813 was supposed to drop film from its camera to a recovery team on the ground, but it wouldn't respond to dump commands. To prevent danger from an uncontrolled impact on Earth of the recovery capsule which weighed several tons, and to prevent recovery by other nations, controllers in the USSR exploded Cosmos-1813 in space.

The explosion created a voluminous cloud of space debris, blowing more than 100 spacecraft fragments out to 350 miles.

Cosmos-1871. Another Russian spysat, this time the ten-ton Cosmos-1871, failed to reach orbit in 1987 and fell harmlessly into the South Pacific 1,500 miles east of New Zealand near Antarctica.

While the USSR said Cosmos-1871 carried "scientific apparatus" for "investigation of the cosmos" and radio systems for adjusting its orbit and transmitting data, it probably was a military reconnaissance satellite planned for a polar orbit similar to other spysats.

Cosmos-1871 would have been the heaviest satellite ever placed in polar orbit, but it reached only 90 miles altitude on an SL-16 rocket.

Cosmos-1871 was the first time the USSR announced beforehand that one of its satellites was dropping. The last time before 1987 the Soviets had made any announcement about a premature return was in 1983 when Cosmos-1402 burned up over the Indian Ocean.

Cosmos-2030. Russia accelerated launches of military imaging spacecraft and civilian Earth-observation satellites in 1989.

Friendly Snoops

Not all spysats peer into the business of enemies; some check up on friends. For instance, the United States was said to be using a reconnaissance satellite in 1992 to monitor Jewish settlement construction in Israel's occupied West Bank and Gaza Strip.

Israel was the largest recipient of U.S. aide, receiving $3 billion in military and economic assistance a year. It had promised not to increase settlements on land it occupied in the 1967 Middle East war because the United States considered the settlements obstacles to peace.

The space photos followed frustrated requests for information on construction. The pictures showed construction more than doubling in 1991.

For example, the spycraft Cosmos-2030 was launched in July 1989 on an SL-4 Soyuz rocket from the Northern Cosmodrome at Plesetsk.

Two months later, ground commanders lost control of the 7.5-ton Cosmos-2030 and exploded it in orbit just ten days before U.S. space shuttle Columbia carried what may have been a new American KH-12 spysat into space in August.

Even as they were investigating the failure of Cosmos-2030, Soviet engineers were testing a different kind of spysat launched July 18.

With the launch of Columbia's payload, both the U.S. and the USSR had new strategic spacecraft undergoing space tests at the same time.

Cosmos-2077. Yet another example of a photo-reconnaissance satellite was Cosmos-2077.

The fourth-generation spysat was launched from the Northern Cosmodrome at Plesetsk in 1990 for a two-month tour of duty in space.

China-3. The People's Republic of China (PRC) launched its first spysat, China-3, in 1975 to an orbit ranging from 115 to 285 miles altitude.

Europe. Helios is an espionage spacecraft built by France, Italy and Spain under a General Directorate for Armament (DGA) to send images from polar orbit to receiving stations in each of the three nations.

The Netherlands and other nations plan to join France, Italy and Spain in building an even bigger spysat for launch around the year 2000.

The Dutch say photo-reconnaissance satellites are necessary for a strong Europe. They would like France to make Europe less dependent on America for intelligence-gathering from space. France is a member of the Atlantic Alliance, but does not take part in NATO's integrated military command.

Israel. Israel wants a spy eye in the sky to keep tabs on Arab enemies across the Middle East.

Israeli Aeronautic Industries fired the Hebrew nation's first experimental satellites, Horizon-1 (Ofek-1) and Horizon-2 (Ofek-2), to space in 1988 and 1990 on three-stage Comet (Shavit) rockets from a military launch pad in the Negev Desert.

Like first satellites launched by other nations, the first Horizons were small radio-beeping space beacons which stayed in orbit only a few months.

Israel planned to launch Horizon-3 (Ofek-3) in 1993-94. The larger, more-complex experimental intelligence-gathering spacecraft would test optical and infrared sensors, radio-intercepting receivers and communications systems needed for a future operational spysat. Israel may have received financial help from South Africa for Ofek-3.

Eavesdropping Satellites

Radio ears in the sky are electronic intelligence (ELINT) satellites. They are orbiting receivers used by ground stations to tune in radio and telephone conversations elsewhere on the ground. They also receive radar signals, missile telemetry, and other kinds of radio energy.

Elint-Ferret. The first U.S. electronic intelligence gathering satellite was Elint-Ferret, launched in 1962 to an orbit ranging from 200 to 400 miles altitude.

Subsequent American electronic eavesdropping satellites have carried names such as Vortex, Chalet and Jumpseat.

One American electronic-interception ELINT payload in a 22,300-mi.-high stationary orbit uses an antenna as large as a baseball diamond to eavesdrop on missile tests, radar, radio, telephone and electronic diplomatic and military communications.

Cosmos-389. The USSR's first electronic-intelligence spysat was Cosmos-389, launched 400 miles above Earth in 1970.

UAE. Intelligence agencies reportedly were unhappy in 1993 when the United States was said to be considering selling a signals intelligence (SIGINT) satellite to the United Arab Emirates.

The UAE, a moderate pro-Western nation in the volatile Middle East, wanted a spysat to guard against attacks from Iran. However, critics feared the sale might lead to the intelligence being used against the United States. Some also wondered if the UAE might use the satellite to snoop into its neighbors' economic affairs.

If approved, the U.S. would launch the satellite and retain control over what it hears and how data is used. Some SIGINT data from surveillance aircraft and satellites may already be supplied to the UAE by the National Security Agency.

Radar-Imaging Satellites

Additional stories about radar satellites are in chapters four and five.

Radar satellites sweep Earth with powerful beams of microwave energy and record reflections bouncing back from the surface. Measuring the time between when a pulse of energy leaves and when its reflection returns, translates into the distance to a point on the ground where the pulse has been reflected. The distances to many points on the surface below can be processed by a computer into photo-like images and maps of land and sea surface features.

Radar, an acronym for radio detecting and ranging, penetrates through clouds, day and night.

The level of detail distinguished by an imaging radar depends on the size of its antenna. The bigger the antenna, the smaller the size of objects noticeable on the ground.

On the other hand, smaller radar antennas can be used effectively sometimes because satellites move very rapidly through space. In fact, any satellite moves a considerable distance along its path between the time a radar pulse is emitted and when a reflection is received. Thus, a small antenna may have an "effective" diameter of several thousand feet. A radar relying on such an "effective" antenna is said to have a "synthetic aperture." The system is known as a synthetic-aperture radar (SAR).

SARs have been used in science satellites such as the American SEASAT oceanography satellite, the European Remote Sensing (ERS) satellite, the Japanese Earth Resources Satellite (JERS), and even the Magellan probe orbiting Venus.

Magellan's antenna is only twelve feet across, but it can see objects on the planet surface as small 350 feet across. By comparison, when unfolded in space, the antenna on the Lacrosse high-resolution military radarsat may be as wide as 150 feet allowing it to distinguish objects as small as six inches across.

Cosmos. The Russians have orbited many secret military radar satellites over the years. At least one was as large as the school-bus-sized Cosmos-1870 civilian radarsat in 1987.

As its orbit carries a military radar satellite around the globe every hour and a half, the spacecraft's radar beam penetrates clouds to sweep land and sea, day and night. Images are transmitted without delay in "real time" to ground control stations in Russia.

Lacrosse. A very similar American satellite, Lacrosse, probably was ferried to space eighteen months later in a secret 1988 flight of space shuttle Atlantis. Lacrosse was said to be in an orbit which passed over nearly all of Russia.

Lacrosse reportedly is the biggest U.S. military satellite in orbit. Its all-weather radar eye scans the planet surface, drawing digital images even through clouds. That powerful imaging radar is capable of taking photo-like pictures of surface features and ground targets, whether day or night, cloudy or not. Lacrosse probably can distinguish objects as small as two feet across.

The 18-ton Lacrosse was 150 feet long and probably cost at least $500 million. It was crammed with advanced technology. For instance, it reportedly housed a 10,000-watt imaging radar to penetrate clouds obscuring land masses and oceans.

The surveillance satellite also was said to have television cameras to snap photos of land and sea in visible and infrared light through clouds. It transmitted the photos and radar images to Earth as digital radio signals.

Lacrosse may have been designed to be refueled every few years by shuttle astronauts, or retrieved and returned to Earth for improvements.

Top secret. Before their secret Lacrosse flight, Atlantis astronauts were not allowed to discuss the mission. However, they did don Lone Ranger masks to poke fun at the secrecy.

Commander Robert "Hoot" Gibson told reporters he would answer questions only with, "Yes. No. I don't know. I can't tell you. I can tell you, but I'd have to kill you."

Although the public saw Atlantis liftoff on a northward track from Kennedy Space Center, there was a news blackout while Atlantis was in space. Still-up-there announcements were made every 24 hours. The landing at Edwards Air Force Base, California, was announced only hours in advance.

Cold War ends. One assignment for the big satellite at the time Lacrosse was launched was to find targets deep within Warsaw Pact countries for America's intercontinental ballistic missiles and the then-new stealth bomber.

Critics described such husky satellites as fat targets for future space warfare. With the end of the Cold War, the Warsaw Pact was disbanded and Russia and the U.S. became friends, although American officials said a threat still existed.

Canada. Some locales on Earth are shrouded by cloud cover nearly continuously, hindering visible-light photography. Microwave radar can record surface features regardless of cloud cover and at night. It measures altitude to provide three-dimensional images. Microwave radar avoids shadowing. Ice reflects radar differently than water, so a microwave radar system can track ice flows.

When Canada's RADARsat is launched to a 621-mi.-high Earth orbit on a U.S. Delta rocket by 1995, it will carry an all-weather synthetic-aperture radar (SAR) to monitor ice hindering navigation in the Northwest Passage and to protect oil platforms. RADARsat will perform the first year-round space monitoring of Arctic Sea routes.

The Canadian satellite will use its microwave radar to take detailed snapshots of Earth's surface in daylight, darkness and through clouds. The SAR will produce 30-ft.-resolution images.

The SAR's sensitivity to soil moisture and vegetation will supply crop assessments and information about forest operations. The satellite also will be on the lookout for drought, floods, forest fires and other natural disasters.

NASA and the U.S. National Oceanic and Atmospheric Administration (NOAA) are taking part in the Canadian Space Agency (CSA) project.

NASA is providing the rocket in return for data from RADARsat, while NOAA is providing one of its advanced very-high-resolution radiometers to measure temperatures and cloud distribution for various U.S. agencies. RADARsat also will have optical sensors from Germany.

RADARsat will draw the first comprehensive map of the Antarctic continental ice sheet. NASA plans to install a ground station at McMurdo Station in Antarctica to download data from the satellite. That data will be transmitted to the University of Alaska at Fairbanks for image processing and distribution to scientists.

Although designed for a five-year working life, RADARsat will be serviced by American space shuttles. New instruments may be installed by visiting Canadian astronauts.

Ocean Observation Sats

The complete story of oceanography satellites is in chapter five.

The United States and Russia watch over each other's naval fleets with radar satellites designed especially for clandestine ocean reconnaissance.

RORSats. Russia tracks movements of U.S. warships with a network of ocean surveillance satellites referred to as RORSat, which is short for Radar Ocean Reconnaissance Satellite.

The first RORSat was Cosmos-198 which was launched to a 160-mi.-high orbit in 1967.

Russia has launched several dozen RORSats over the years. Each satellite fires small thruster rockets twice a week to correct its orbit.

Radioactivity. Nuclear reactors are compact sources of large quantities of electrical power used in some science and military satellites. RORSats use nuclear reactors to operate the powerful radar sets they need to detect warships.

Satellite nuclear accidents have caused worldwide concern since a U.S. Transit-5BN-3 navigation satellite failed to achieve orbit in 1964. Its nuclear power pack disintegrated in the atmosphere, spewing plutonium-238 around the globe.

Similarly, Russian RORSats have nuclear reactors to generate electrical power for their radars. At the end of each satellite's working life in space, something has to be done with its radioactive core.

★It can be left inside the satellite with the intention it will burn up with the spacecraft body as it falls into the atmosphere.

★It can be jettisoned from the satellite body to make the radioactive material less protected in the hope it would be more likely to burn in the atmosphere.

★It can be ejected from the falling satellite and boosted up to a higher orbit from which it would take hundreds of years to descend.

Unfortunately, some of the dozens of nuclear-powered RORSats, which fell into the atmosphere over the years, made it all the way down to Earth.

Cosmos-954. When Cosmos-954 re-entered the atmosphere in January 1978, it did not eject the 110 lbs. of uranium fuel from its power generator. That spewed flakes of radioactive debris across 20,000 square miles of remote northwestern Canada. The USSR repaid Canada half of $3.5 million spent to clean up after Cosmos-954.

Russian nuclear engineers then redesigned the RORSat to allow the reactor to jettison its fuel rods so they would not have the protection of the satellite body as it re-entered the atmosphere. They hoped that would make the radioactive waste more likely to burn up in the atmosphere.

Cosmos-1402. One of the new-design RORSats, Cosmos-1402, dropped from orbit over the Indian Ocean in January 1983. No radioactive debris from it was reported on Earth and the reactor was said to have splashed into the Indian Ocean.

More changes. Publicity about dangerous nuclear satellites continued, leading Soviet satellite designers to change their ocean reconnaissance satellites again.

Some RORSats would follow an automatic sequence in which the radioactive fuel would be vaporized, then the reactor core would be separated and boosted to a 500-mi.-high graveyard orbit where hundreds of years would pass before it would fall into the atmosphere.

From the high altitude, the isotopes eventually would disintegrate to dust suspended in the upper atmosphere. Some, no larger than snowflakes, might settle slowly to the ground, offering little danger, according to nuclear engineers.

Cosmos-1818. And some ocean spysats would have nuclear power replaced altogether by solar panels to generate electricity. The solar-powered Cosmos-1818 was launched in 1987.

Cosmos-1900. A nuclear-powered RORSat launched in 1987 was labeled Cosmos-1900. When Western satellite watchers noticed in April 1988 that thruster rockets on Cosmos-1900 were not firing, Soviet controllers acknowledged they had lost radio contact.

Cosmos-1900 weighed 2,200 lbs. and was 20 ft. long and 7 ft. wide. Its reactor weighed 220 lbs., including 110 lbs. of uranium U-235 fuel. Using the powerplant to generate electricity had reduced most of the fuel to radioactive isotopes.

American alarm. A spent Soviet rocket, not Cosmos-1900, crashed in northeastern Canada August 21, 1988, lighting up the sky from New Mexico to Michigan. Police switchboards were flooded with calls from alarmed observers who thought Cosmos-1900 was falling on them.

It turned out the falling rocket body, a Soviet workhorse model SL-12, had been tracked by officials at the U.S. Space Surveillance Center at Cheyenne Mountain in Colorado. The SL-12 had been launched by the Soviets August 18 to carry a Ghorizont communications satellite to orbit.

The rocket came down just where the U.S. Space Surveillance Center had predicted it would. No injuries or damage were reported after the rocket plummeted to Earth at 1 a.m.

It's not unusual for a rocket body to leave orbit. There's nothing to keep it there. An average of one space vehicle reenters the Earth's atmosphere each day. U.S. officials don't try to recover parts of such rocket bodies because 95 percent of them burn up upon reentry.

End of the game. The worldwide alert for radioactive debris raining on Earth was called off October 1, 1988, after Cosmos-1900 properly separated nuclear power pack from instruments and shot the reactor to a graveyard orbit.

The satellite was designed to boost its powerplant to a 500-mi. orbit to delay re-entry until radioactive elements decayed to safe levels.

Cosmos-1900 was the third nuclear-powered satellite to fall to Earth in a decade, but the reactor from Cosmos-1900 did rocket up to an orbit ranging from 433 to 476 miles where it and some debris still circle Earth every 99.3 minutes.

One large, non-nuclear piece of Cosmos-1900 instruments splashed into the Atlantic Ocean off Africa's northwest coast. Another non-radioactive piece, orbiting for a time at 91 miles, burned up as it slid down into dense layers of the atmosphere.

French Fear Carnival Satellite

The possibility Cosmos-1900 could shower radioactive debris on them frightened Europeans. In fact, Paris sounded an alert when observers thought the radioactive satellite had fallen near there September 30, 1988.

Even though French space specialists had given Cosmos-1900 only one chance in a thousand of falling on France, a motorist near Peronne, 85 miles north of Paris, saw something smoking and shiny on the Paris-Lille superhighway about 9 p.m. Thinking it might be satellite debris, he phoned gendarmes who raced to the scene, diverted traffic and set up a security perimeter.

But, there was no need for alarm. It turned out the smoking sphere that closed the superhighway for two hours was a six-ft. aluminum ball which had fallen from a carnival truck, igniting its electrical circuits. The highway reopened at 11 p.m.

Space Ban. A Russian space official and an American scientist in 1988 called for a ban on nuclear reactors in Earth orbit.

Such a ban would have ended Russia's use of nuclear-powered RORSats, but also America's use of its SP-100 orbiting nuclear reactors to power Star Wars weapons. A ban would not apply to nuclear reactors aboard interplanetary spacecraft, such as an unmanned probe of Pluto or a manned mission to Mars.

SSU. The United States launched an ocean-surveillance satellite, known as SSU, to an 865-mi.-high orbit in 1975.

NOSS. America then launched another ocean-surveillance satellite, called NOSS, to a 700-mi.-high orbit in 1976.

White Cloud. U.S. Navy ocean spysats, known as White Cloud, eavesdrop on radar and radio signals from the Russian fleet.

A White Cloud developed from an SSU was launched in 1976 to a 700-mi.-high orbit.

Several small White Cloud satellites can be sent to space in a single rocket flight. For instance, a Titan-2 carried four in one 1988 flight.

SEASAT. America launched its first oceanography satellite, Sea Satellite (SEASAT-1), in 1978. From 500 miles above Earth, SEASAT's all-weather synthetic aperture radar (SAR) measured wave heights from 3 to 65 feet with an accuracy of 4 inches through clouds day or night.

TOPEX/Poseidon. SEASAT success led NASA and the French space agency, CNES, to design a new oceanography satellite known as TOPEX/Poseidon, which was launched to an altitude of 800 miles in 1992.

GFO. The U.S. Navy has a new oceanography satellite known as GEOSAT-Follow-On (GFO).

Its radar altimeter will measure sea surface heights, sea floor topography, water density and wind speed for the Navy's Space and Naval Warfare Systems Command, after GFO is ferried to space by an air-launched Pegasus rocket in 1995.

The relatively-tiny 660-lb. spacecraft will transmit data directly to 65 shipboard and shore stations. They will correlate the information with data from global-positioning satellites (GPS) to route ships, support amphibious operations and conduct anti-submarine warfare.

GFO was designed to work eight years in space. Two more may be built later for Pegasus launches.

Early Warning Satellites

Some reconnaissance satellites have one specific duty—detecting a nuclear missile attack and sounding an early warning.

Midas. The United States launched its first Midas early-warning satellite to a 300-mi.-high orbit in 1960.

MEWS. America launched a different kind of early-warning satellite, BMEWS, to a distant orbit ranging from 19,000 to 25,000 miles above Earth in 1968. In 1970, the U.S. launched the early-warning satellite IMEWS to a slightly lower orbit ranging from 16,000 to 23,000 miles above Earth.

DSP. The American early-warning spacecraft of today are the Defense Support Program (DSP) satellites providing early warning of strategic missile attack against the U.S. and its allies.

Each DSP satellite looks down with infrared sensors to see hot plumes of exhaust gases from ballistic missiles moving across Earth's surface, to provide 30 minutes warning of nuclear attack.

TRW has built DSP satellites for the Air Force at Redondo Beach, California, since the early 1970s. Sensor payloads are built by Aerojet Electronic Systems Division of Azusa, California.

The satellites have been upgraded five times over the years as they have been required to watch more and smaller targets. The first DSP weighed 2,000 lbs., had a 400-watt electrical powerplant, and was designed to work only 15 months in space. That compares with three DSP satellites being built for use later in the 1990s. Each will weigh 5,250 pounds, use 1,275 watts of electrical power and work about five years in orbit.

Iraq. In 1989, a U.S. nuclear-attack warning satellite spotted the fiery rocket exhaust when Iraq launched its first satellite to Earth orbit.

Iraq's first was not a separate satellite, but the third stage of a 48-ton rocket launched from the Al-Anbar Space Research Center 50 miles west of Baghdad to sweep around the globe for six revolutions before falling out of orbit.

The 75-ft. rocket was similar to a U.S. Scout rocket used to send small satellites to low orbits. The Iraqi booster may have been a modified version of Argentina's Condor ballistic missile.

The launch was the first time Iraq exposed its space research. North American Aerospace Defense Command tracked the third stage around the globe.

Used as a ballistic missile, such a rocket could carry a nuclear warhead 1,240 miles. Iraq already had a 600-mi. ballistic missile built around the USSR Scud. Scuds were launched against Iranian cities in the eight-year Persian Gulf war in the 1980s and Israel and Saudi Arabia in the 1991 Persian Gulf War. The Gulf War stopped Iraq's space research program for years.

Cosmos-520. The USSR launched its first early-warning satellite in 1972. Cosmos-520 was sent to an elliptical, 141- to 416-mi.-high orbit.

Nuclear Inspection Sats

A special kind of U.S. military reconnaissance satellite monitors compliance with arms treaties by countries around the world, including Russia and other new nations from the former Soviet Union. The satellites are designed to spot the bright flash of light from a nuclear explosion.

VELA. America started watching for nuclear explosions with its first VELA satellite in 1963. It was placed in an extremely high orbit ranging from 63,000 to 70,000 miles away from Earth.

NAVSTAR. NAVSTAR global positioning system (GPS) satellites also carry extra sensors, call Ionds, for detecting nuclear explosions.

Meteorology Satellites

The complete story of weather satellites is in chapter four.

Weather forecasting is essential in battle. Many military weather satellites have been developed secretly and launched over the years since 1960.

The U.S., Russia and other nations have weather satellites to find holes in cloud cover for reconnaissance work, help low-level bombers reach their targets and detect communications-disrupting solar flares. Today, the U.S. Air Force operates the Defense Meteorological Support Program (DMS or DMSP) satellite system.

The first DMSP was launched in 1966 to an orbit ranging from 440 to 555 miles above Earth.

The first improved Block-5D DMS satellite was launched in 1976 to a 500 mile high orbit.

Two DMS satellites usually are in polar orbits at a time, circling the globe in a north-south flight path every 12 hours, scanning Earth's surface for storms and feeding cloud cover and temperature information to the military.

They have a microwave imager which can see through clouds, measuring energy radiating from Earth to record land and sea temperatures, sea surface wind speeds, cloud and soil moisture levels, and rain rate and distribution.

NOAA, the National Oceanic and Atmospheric Administration, sometimes uses DMS information for civilian weather forecasts.

New weather satellites are launched regularly. Since the end of the Cold War, the Air Force has used refurbished Atlas intercontinental ballistic missiles (ICBMs) to ferry DMS satellites from Vandenberg Air Force Base to polar orbit.

Meteor. Russian civilian weather satellites usually double as military weather satellites.

Communications Satellites

The complete story of communications satellites is in chapter two.

Military research has driven space technology forward since the end of World War II. In fact, the first communications satellite was a simple U.S. military payload launched in 1958.

Communications for command and control were seen as important to military units, so satellite development moved ahead rapidly in the 1960s. The Pentagon blasted eight comsats to a high orbit 21,000 miles above Earth in one 1966 launch.

In 1969, the Pentagon launched TACsat, the first U.S. military communications satellite in stationary orbit. A British military stationary-orbit satellite called Skynet was launched the same year.

A satellite in a stationary orbit takes 24 hours to circle the globe once, which makes it appear to hang stationary in the sky as the Earth turns.

In 1970, a military communications satellite, NATO-A, was launched to stationary orbit for the North Atlantic Treaty Organization. Meanwhile, the USSR fired a military comsat, Cosmos-336, to a 900-mi.-high orbit in 1970.

The U.S. launched a Satellite Data System (SDS) spacecraft to a Molniya orbit in 1971. Its long path through space took SDS from only 250 miles above Earth to 21,000 miles from Earth.

Also in 1971, the U.S. launched a stationary-orbit military communications satellite, DSCS-1.

The U.S. Navy launched a new series of Fltsatcom stationary-orbit satellites in 1978.

The French space agency, CNES, modified a European Communications Satellite (ECS) into Telecom-1 and launched it to a stationary orbit in 1984. It relayed military and civilian phone calls, television, telegraph and data signals.

Satellite Buoys

U.S. Navy submarines under the Arctic ice use yard-long radio buoys to communicate via satellite with the continental United States.

When a sub under the polar cap wants to talk back home, it loads a message into a buoy and launches it. The buoy floats up to the bottom of the ice. Its flotation device anchors the buoy to the ice while a penetrator forces a radio antenna up through the ice.

A buoy is three inches in diameter and 39.5 inches long. The penetrator and radio antenna contain tiny microelectronic circuits, including a microprocessor with commands stored in memory and an activity-controlling timer.

Once above the surface, the antenna transmits its stored message to a satellite which relays the signal to a ground station in the continental U.S.

After the manufacturer, TRW, showed it was possible to penetrate upward through Arctic ice in 1985 and launched a penetrator from a submarine in 1988, the U.S. Space and Naval Warfare Systems Command bought thousands of buoys.

The 1988 test was the first time a submarine submerged under the ice pack had communicated with the continental U.S. via satellite.

The Navy already had purchased thousands of submarine communications buoys from Hazeltine Electro-Acoustic Systems Laboratory for use in satellite communications on the open ocean.

Much of the military traffic relayed today by communications spacecraft like Fltsatcom, UHF, DSCS, Milstar, SDS, Skynet, NATO and Cosmos is scrambled, or encoded, but occasional transmissions are in the clear.

Fltsatcom. The U.S. Navy's squadron of nine Fleet Satellite Communications (Fltsatcom) satellites are stationary in the Clarke Belt at 72, 75, 77, 100, 105, 145, and 177 degrees West.

The first Fltsatcom was lofted to orbit in 1978. The Navy, Air Force and Coast Guard rely on Fltsatcoms to reach forces scattered across the globe. Air Force channels are called AFSATCOM.

UHF. Today, the Fltsatcom flotilla is near the end of its useful life in space, so the Navy is replacing it with a new series of up to nine Ultra High Frequency (UHF) satellites during the 1990s.

Like Fltsatcom, the UHF satellites are to be in stationary orbits 22,300 miles above the Equator.

The first launch, UHF-F1, also referred to as AC-74/UHF, blasted off March 25, 1993.

UHF satellites are Hughes HS-601 satellites. They work with terminals already in service in shore stations and aboard ships and submarines in the Atlantic and Pacific fleets.

Those terminals, which offer protection against

enemy jamming, also will operate extremely-high-frequency (EHF) communications satellites such as the Milstar network.

DSCS. The U.S. Air Force operates a series of Defense Satellite Communications System (DSCS) spacecraft relaying classified conversations among military personnel.

Pronounced DISK-us, the satellites are in stationary orbits 22,300 miles above the Equator. DSCS-1 was launched in 1971.

A DSCS-3 joined eight older satellites in 1993 in the globe-spanning network of satellites routing military voice and data signals around the world.

Four late-model DSCS-3 satellites already were in orbit, along with four less-refined DSCS-2 satellites. DSCS-3 has protection against radiation and enemy jamming.

Altogether, the Air Force plans to launch fourteen DSCS-3 satellites in the 1990s. GE Astro Space builds the solar-powered, 3,000-lb., DSCS-3 satellites to work up to ten years in space.

Australian Parakeets

Australian military forces drive Parakeets. No, not birds, but six-wheeled Land Rovers sporting racks of 20-watt Ku-band satellite radios.

An Australian Department of Defence satellite dish antenna with eight-ft. petals pointing at a communications satellite sprouts from the top of each of the eighteen Land Rovers.

Milstar. The Air Force is launching a second-generation series of large U.S. Defense Department communications satellites known as Milstar.

Milstars are telephone switchboards in space to route all military message traffic and conversations around the world. For the first time, all services will be in contact with each other.

A Milstar satellite houses five communications systems, operating at extremely high frequency (EHF), ultra high frequency (UHF) and super high frequency (SHF), feeding more than a dozen antennas. The huge satellite has more than 100 electronic units, including four computers.

TRW began work on the first three Milstar-1 satellites in 1984. A fourth satellite, Milstar-2 with added equipment for tactical communications, was added later.

The super-secret satellites are being launched on Titan-4 rockets to stationary orbits 22,300 miles above the Equator. Ten slots have been set aside for Milstars in the Clarke Belt.

Backbone. The secure, survivable, jam-proof Milstar network is the command backbone for all American armed forces. The satellites allow national command authorities to communicate directly with worldwide tactical forces.

Milstar ground terminals are used by the Joint Chiefs of Staff, commanders-in-chief of unified and specified commands and major forces. Later Milstars have been upgraded to provide tactical communications between commanders in a theater of operations. Milstar also relays communications for certain other U.S. government agencies.

The Army uses Milstar terminals known as SMART-T, for Secure Mobile Anti-Jam Reliable Tactical Terminal. The Air Force accesses Milstars through Low Cost Terminals (LCT).

SDS. America's super-secret KH-11 photo-reconnaissance satellites report the results of their espionage work to commanders on the ground through a network of more than a dozen military communications satellites in polar orbit.

The Satellite Data System (SDS) spacecraft relay data from KH-11s to ground-control stations.

The first SDS was launched into a Molniya-style elliptical orbit in 1971. Its long path through space took it from a low 250 miles above Earth out to a high point 21,000 miles from Earth.

A 1988 launch may have orbited the last pair of SDS satellites. If so, NASA's Tracking and Data Relay Satellites (TDRS) may have taken over relaying data from the KH-11 satellites.

Lightsats. An important development of the 1990s has been Pegasus, a space rocket which carries small payloads from an airplane to low Earth orbit. The booster is carried to 40,000 feet under the wing of a Boeing B-52 bomber and dropped to blast on up to space.

Pegasus ferried seven military communications satellites to space in one 1991 flight. The seven miniature communications satellites, referred to as Lightsats or Microsats, were provided by DARPA, the Defense Advanced Research Projects Agency.

The Microsats had voice and data transponders compatible with military transceivers.

Midshipmen and staff of the U.S. Naval Academy at Annapolis, Maryland, used training vessels and a 36-ft. dish antenna to test Navy and Marine Corps tactical communication via the Microsats. Voice contact was established with Aegis-class destroyer USS Arleigh Burke at Norfolk, Virginia. A Naval Space Command station on the roof of a Virginia building contacted a training vessel under way in the Chesapeake Bay. Vessel positions and status reports were relayed by Microsat to Annapolis.

Three four-person and five-person tactical units of midshipmen waited in a helicopter landing zone at Quantico Marine Corps Base, Virginia, for a Microsat to pass overhead. Then the teams would contact each other and Annapolis command via the satellite and move in three directions 300 yards into woods around the helicopter pad. Signals to and from the tiny satellites were clear, even as the teams exercised in the woods.

The Microsats also were tested at Fort Gordon, Georgia. Amateur radio operators in the Military Affiliate Radio System (MARS) helped with tests.

NATO. NATO, the North Atlantic Treaty Organization, needs comsats to relay diplomatic and military conversations among Atlantic Alliance leaders in sixteen nations stretching across the globe from the United States to Turkey.

The treaty organization's first communications satellite, NATO-A, was launched to stationary orbit 22,300 miles above the Equator in 1970.

Today, NATO routes communications through two satellites in stationary orbit. NATO-4A and NATO-4B allow government officials, diplomats, military officers and troops in the field to stay in touch with NATO headquarters in Brussels. The satellites also are linked to American military satellites, ground stations, naval vessels and ground forces.

Great Britain's ministry of defense operates the satellite system for NATO. NATO-4A and -4B are similar to Great Britain's Skynet-4 military communications satellites. The boxy, 3,831-lb., solar-powered NATO satellites were built by British Aerospace and Marconi Space Systems and launched in the 1990s on U.S. Delta rockets.

Great Britain. The British have their own system of satellites in stationary orbit to channel military communications to submarines, surface ships, troops and ground stations around the globe.

The first of many was Skynet-1A, launched to stationary orbit in 1969.

The most recent launches have been Skynet-4A on a U.S. Titan-3 rocket, and Skynet-4B and Skynet-4C on European Ariane rockets.

Each solar-powered Skynet-4 has three radio transponders. The Skynet-4 satellites were built by British Aerospace for the UK Ministry of Defence. The first was lobbed to orbit in 1988.

Christmas calls. British Army troops in Bosnia during the 1992 Christmas season were able to benefit from free satellite telephone service by British Telecom. Ten phones in the town of Vitez were connected to an eight-ft. dish antenna.

Ten-minute phone calls were relayed through a Eutelsat-2 satellite to British Telecom's London Teleport where they were routed to families in Great Britain and Germany.

Polar BEAR In Space

The U.S. Air Force sent a polar bear to orbit in November 1986 aboard a Scout rocket. It wasn't a real bear, but rather a communications satellite known as Polar BEAR.

It was a 20-year-old Navy satellite which had been mothballed at the Smithsonian Institution. When the Navy built it in the 1960s as a backup spare, it was known as Oscar-17. Navy Oscars aren't the same as amateur radio OSCARs.

When the 270-lb. Oscar-17 went unused, it was placed on display in the Smithsonian's Air and Space Museum in 1971.

When the Air Force came up short in the 1980s, the government traded the Smithsonian an extra Transit-5A navigation satellite for the old Oscar-17 radio satellite.

Technicians spruced up the old bird, finding electronic equipment in good working order. Only one solar cell out of thousands on the satellite's aging solar power panels had to be replaced.

The satellite was renamed Polar BEAR, for Polar Beacon Experiment and Auroral Research, and launched into a 625-mile-high polar orbit from California aboard a 75-foot-tall Scout rocket.

The satellite was sent to space to photograph the Northern Lights while studying electrical particles and the Earth's magnetic field over the planet's poles.

All satellites depend heavily on radio to receive commands from Earth and to send data down from space. But the Northern Lights or aurora borealis create static which interferes with communication when satellites are near the North Pole. Polar BEAR took a close look at the problem.

Russia. The USSR first launched satellites for communications in the 1950s. Many Russian satellites over the years have been referred to by the generic name Cosmos.

The USSR launched its first active real-time communications satellite, Molniya-1A, in 1965.

Military and civilian communications were relayed by Molniya satellites which worked in elliptical orbits that included coverage of high latitudes beyond the reach of common stationary satellites which are too low on the horizon. Molniya satellites also carried TV cameras.

The USSR's Ghorizont, launched in 1978 and sometimes referred to as Statsionar, was the first stationary television satellite to serve Europe. It, too, relayed some military communications.

The Russian version of NASA's TDRS communication satellite is labeled Satellite Data Relay Network. SDRN satellites carry military and manned space voice communications, as well as public broadcasts.

Telephone Signals

A military communications satellite system left over from the Cold War is under conversion by the Russians for civilian telephone service. A Russian manufacturer plans to launch 48 lightweight Signals satellites to low orbits to create a regional telephone network over Europe and Russia.

Cosmos-1817 may have been a military comsat on its way to orbit 22,300 miles out in space when the fourth stage of its Proton rocket failed to fire in 1987. Cosmos-1817's secret mission ended in a useless orbit 150 miles above Earth.

Navigation Satellites

The complete story of navigation satellites is in chapter six.

Satellites like the U.S. Navy's old Transit system and the contemporary NAVSTAR global positioning system (GPS) have replaced stars as beacons for maritime navigation.

Navigation satellites travel in orbits from a few hundred to a few thousand miles above Earth, sending down radio signals which travelers use to locate themselves anywhere on the globe.

The earliest navigation satellites were developed by military forces to guide submarines launching ballistic missiles against distant targets. Each submarine had to know its location precisely so it could aim its missiles. Of course, surface navies also wanted to know their exact locations.

Transit. The United States launched the first navigation satellite, Transit-1A, in 1959. It failed to reach orbit. Transit-1B was launched in 1960.

To use the new navigation system, a surface ship or submarine would receive signals from a satellite and measure the change in its radio frequency to determine the satellite's position. The ship hen could calculate its own position.

Tsikada. The USSR followed suit with Tsikada navsats starting in 1978.

NAVSTAR. The latest development in space navigation is the NAVSTAR global positioning system (GPS). NAVSTAR is short for Navigation System using Timing And Ranging.

The satellites provide latitude, longitude, altitude, direction of travel, travel velocity and correct time of day, day or night, in any weather.

NAVSTAR gives more-precise location fixes more frequently than the Transit navigation-satellite system. Design of the extensive 24-satellite system was authorized in 1973 and the first satellite was launched in 1978.

The U.S. Air Force lofted the constellation of NAVSTAR satellites circling Earth twice a day. At least four of the space beacons are in view from any spot on Earth at any time.

NAVSTARs work 12,625 miles above Earth. The GPS system requires 24 satellites, including, three working spares stored in orbit.

Ships, planes, trains, trucks, cars and even persons on foot could know their positions in latitude, longitude, and altitude within 58 feet, along with their velocity within 0.45 mph, and correct time to within one-millionth of a second.

A receiver on the ground has to hear from only three NAVSTAR satellites to find its own latitude and longitude location. Hearing from a fourth satellite lets the receiver calculate its own altitude.

Generations. First-generation NAVSTARs were referred to as block 1, while the second-generation is known as block 2.

The Air Force Space and Missile Systems Center at Los Angeles Air Force Base and Rockwell International Corp. designed NAVSTAR satellites. Launch of eleven block-1 satellites from Vandenberg Air Force Base, California, began with NAVSTAR-1 on February 22, 1978.

In 1983, the Pentagon ordered 28 improved second-generation NAVSTARs from Rockwell. Block-2 satellites would include 21 working in orbit, 3 working spares in orbit, and 4 replenishment spares on the ground to be launched as needed to replace aging satellites.

The first block-2 NAVSTAR went to orbit February 14, 1989. After nine block-2 satellites were orbited, refinements were made and the Air Force started launching block-2a satellites in 1990.

Military. NAVSTAR is a Defense Department project. The NAVSTAR Control Center is at the U.S. Air Force space command, Falcon Air Force Base, Colorado Springs, Colorado.

Air Force receivers around the globe monitor all NAVSTARs for the Colorado Springs master control station which updates each satellite's navigation and health messages for all users.

GPS provides the pinpoint accuracy needed by the giant MX nuclear ICBM (intercontinental ballistic missile). NAVSTARs even allow troops to synchronize watches to billionths of a second, while moving about the land, sea and air.

Handheld navigation radios labeled Precision Lightweight GPS Receiver (PLGR) are popular with U.S. ground troops in the 1990s.

Civilian use. The Pentagon reduced the accuracy of NAVSTAR GPS data available to civilians. NAVSTAR satellites offer Precise Positioning Service (PPS) for military units and Standard Positioning Service (SPS) for civilians.

PPS gives horizontal positions accurate to

within 58 feet and vertical positions to within 91 feet. SPS gives horizontal positions accurate to within 328 feet and vertical to within 512 feet.

Signals are sent from NAVSTAR satellites in codes and must be decoded by a GPS receiver. The codes are precise (P) and coarse/acquisition (C/A). P-coded signals are available only for use by military forces, while the C/A code has been publicized so it can be used by civilians. The C/A signal allows calculation of position to within 328 feet. The P signal offers far-more-accurate computation of position to within 58 feet.

TAOS. Some satellites in low Earth orbits use NAVSTAR to navigate through space. For instance, the U.S. Air Force satellite called Technology for Autonomous Operation Survivability (TAOS) carries a GPS receiver, as do various Defense Advanced Research Projects Agency satellites (DARPASAT).

Radcal. Air Force radar calibration satellites, known as Radcal, have made ground radar stations around the globe more accurate in tracking military missiles, civilian rockets, military and civilian satellites, space shuttles, space stations, airplanes, and space junk.

After two older radar-calibrating satellites ended their tours of duty, a replacement Radcal was sent to orbit in 1993 for the Space and Missile Systems Center at Los Angeles Air Force Base in 1993. It was blasted from Vandenberg Air Force Base, California, to a 518-mi.-high polar orbit.

The small 192-lb. satellite broadcast radio signals from space for a year so its position could be tracked by seventy military, Energy Department and NASA radars around the world.

The satellite's exact location in space was determined precisely with data from NAVSTAR global positioning system (GPS) satellites. NAVSTAR measurements were compared to each ground radar's data to improve accuracy.

Research Satellites

Science and technology research spacecraft are covered in chapter eight.

Some satellites test new technology which may have future military applications. Other spacecraft carry out basic science research experiments.

Secor. The United States launched an early military research spacecraft, Secor, to a 575-mi.-high orbit in 1964.

OV-1. The next year, the U.S. launched an OV-1 military satellite to a highly elliptical orbit ranging from 250 to 2,150 miles above Earth.

OV-2. Also in 1965, a similar research spacecraft, OV-2, was launched to a low orbit ranging from 440 to 500 miles altitude.

OV-3. In 1966, the U.S. launched a research satellite, OV-3, to a highly-elliptical orbit ranging from 220 to 3600 miles above Earth.

OV-4. That same year, the U.S. launched the research satellite OV-4 to a 200-mi-high orbit.

OV-5/ERS. In 1967, America's OV-5, also known as ERS, was launched to a very high, very elliptical orbit from 5,000-70,000 miles altitude.

ASTEX. The research satellite ASTEX was launched in 1971 to a 500-mi.-high orbit.

Argos. The U.S. Air Force was to fire a Delta-2 rocket in 1995 to low orbit carrying a scientific research satellite known as Argos P91-1.

It was to be the first launch by the Air Force of a Delta-2 from Vandenberg Air Force Base.

The satellite would carry out nine experiments and demonstrate technology for various agencies.

STEP-Eagle. A pair of U.S. Air Force satellites, in a project known as Space Technology Experiments Platform (STEP), will be built in TRW's Eagle-class lightweight spacecraft frameworks.

Eagles are bantam-weight satellites which can house a range of payloads and be launched on a variety of small rockets. STEP will carry an assortment of electronic instruments and military experiments in low Earth orbit.

Tracking. The U.S. Air Force has a dozen Satellite Control Network (AFSCN) ground stations across the world to track and control, and download telemetry data from, orbiting Department of Defense satellites.

The Automated Remote Tracking Stations (ARTS) are in California, Colorado, Florida, Greenland, Guam, Hawaii, New Hampshire, Diego Garcia and the Seychelles in the Indian Ocean, a British-operated station in England, the satellite operations complex at Falcon Air Force Base, Colorado, and the Consolidated Space Test Center at Onizuka Air Force Base, California.

They are connected with the Air Force networks which control the NAVSTAR global positioning system (GPS) satellites and the Defense Meteorological System satellites. The stations also track ballistic missile test launches.

Manned Satellites

The stories of space stations and space shuttles are in chapter nine.

The U.S. Defense Department has cloaked details of its experiments, but space shuttles have ferried spysats to orbit and astronauts have been used for land reconnaissance, naval surveillance, and other military operations. They have conducted tests coordinated with ground troops, ships at sea and missile launches.

Mir. Russian cosmonauts orbiting in space station Mir 200 miles overhead also can carry out the same kinds of reconnaissance, surveillance and espionage, only over longer periods of time.

Weapons In Orbit

While almost none of the hundreds of military spacecraft sent to space by the United States and Russia are armed, the potential exists for using satellites as armament.

Satellites can be outfitted to attack other satellites in orbit as well as ballistic missiles traveling from one point on Earth to another. Satellites even can carry bombs to drop on targets on the surface of Earth.

FOBS. The Russians actually designed and tested an orbiting weapon called Fractional Orbit Bombardment System (FOBS). The idea was to send a warhead to orbit from where it could be dropped later onto any area on Earth.

The sole FOBS launch was a Cosmos satellite sent to an elliptical orbit ranging from 100 to 650 miles above Earth in 1966.

Satellite Killers

Reconnaissance satellites have great strategic value in modern warfare. Much of an enemy's defensive capability can be seen from space and his offensive plans can be judged from information ferreted out by the various kinds of satellites reconnoitering up there today.

During full-scale warfare, an opponent must blind and deafen his enemy's intelligence-gathering satellites, if not destroy them entirely.

Satellites can be overpowered by strong beams of microwave radio signals. They can be blinded by laser beams. They can be destroyed by hits from small solid objects. Some weapons are fired from the ground and some wait in orbit for commands to strike other targets in space.

A laser beam on the ground can be pointed at a satellite to blind its optical sensors. A rocket fired from the ground can smash a fast-moving satellite. An orbiting satellite can aim a laser across space or fire a solid projectile to shatter a satellite.

Cosmos-249. The Soviet Union orbited a satellite with the generic label Cosmos-249 in 1968. Ranging in altitude from 85 to 160 miles above Earth, Cosmos-249 was an attempt to create a satellite killer in orbit.

ASAT. The United States designed an anti-satellite rocket weapon fired from the ground. Known as ASAT, it was tested over the Pacific Ocean in 1984.

SOARS. Since spacecraft can be attacked directly, each needs its own early warning of impending attack.

The U.S. Air Force is building a device into all new military satellites so a satellite about to be attacked would be alerted to defend itself, while notifying ground controllers through a satellite threat and warning system.

The device is called Satellite Onboard Attack Reporting System (SOARS). It detects and reports attacks by threats nearby in space as well as from anti-satellite systems on the ground.

SOARS has a microwave radio-frequency sensor and a pair of mid- and high-irradiance laser sensors to detect threats. Its impact sensor feels damage from an object hitting the satellite.

SOARS is wired into the host satellite's data, command, telemetry and power systems. It can relay an alarm through the satellite's radio, but it also has an emergency S-band transmitter of its own to alert ground controllers.

Laser weapons. Laser beams like those in Walt Disney World shows were fired at satellites in 1990 for research into the use of battle mirrors in space to destroy incoming enemy missiles.

The satellites, known by the acronyms LACE and RME, recalled the fictional laser battle mirror in Tom Clancy's *The Cardinal of the Kremlin.*

LACE and RME were part of a test by the U.S. missile defense research group once known as Strategic Defense Initiative Organization (SDIO) and nicknamed Star Wars. In 1993, that group was renamed Ballistic Missile Defense Organization.

LACE. LACE was a satellite target in space for lasers fired from the Air Force Maui Optical Station on Mount Haleakala, Maui, Hawaii.

LACE was an acronym for Low-power Atmospheric Compensation Experiment. The Pentagon wanted to learn how to counteract the atmospheric pollution weakening of laser beams by up to 1,000 times. Before launch to its 334-mi.-high orbit, researchers mounted a reflector on a boom sticking out from LACE.

LACE had two cameras to track rocket exhaust plumes from space to pinpoint a missile's location while it is climbing into space.

Weakening. A laser light show reveals just how much the atmosphere affects laser beams, which can be seen only when reflecting from dust, water vapor and pollutants suspended in the air.

Military researchers used LACE's reflector to determine how a laser beam could be directed by a fast-steering mirror. A low-power laser on Mount Haleakala illuminated the orbiting reflector.

The returning laser beam was fed into a computer which figured out how much the atmosphere had distorted the light. Researchers then used a "rubber mirror" to distort the outgoing laser beam by the exact opposite amount.

RME. RME's mirror tested the concept of bouncing laser beams from the ground to attack enemy missiles and other targets.

RME was an acronym for Relay Mirror Experiment. The satellite was in a 288-mile-high orbit, from where it reflected a 10-watt laser beam from one ground station to another.

The primary ground site was the Optical Station on Mount Haleakala. A second ground station was a scoring control site.

As RME passed overhead, the optical station pointed a low-power laser at the spacecraft. RME's mirror reflected the beam back down to a target board located at the scoring control site. Twelve fast-steering mirrors included one 24-inch mirror.

Fast-steering mirrors also can used for space telescopes, tactical sensors, medical laser imaging, laser weapons systems and communications.

Alpha Laser. At San Juan Capistrano, California, an experimental laser beam weapon known as Alpha is the most powerful military laser in the American inventory.

It produces 2.2 million watts of energy which military planners hope will be enough to destroy targets in space. Alpha, a predecessor to an Earth orbiting laser beam satellite, was to be launched into space as an experiment on a 1994 flight.

BEAR. A neutral particle beam weapon fires an energized stream of hydrogen atoms which have no net electrical charge. The first firing of such a gun in space was during a 1989 test known as BEAR, for Beam Experiment Aboard Rocket.

The BEAR neutral particle beam accelerator was lifted into space from the White Sands Missile Range on a 15-minute suborbital flight aboard a small Aires sounding rocket. In space, the accelerator was turned on for four minutes to fire a series of five-second pulses, while engineers noted how its beam of particles moved.

It was a step forward in the Pentagon's search for futuristic weapons which can beam powerful atomic particles across space to shoot down nuclear missiles fired at the U.S. and its allies.

Unlike industrial lasers which burn through metal slowly, a neutral particle beam travels at half the speed of light and penetrates deep into a target where it releases its destructive energy all at once as intense heat. The experiment was conducted by the national laboratory at Los Alamos, New Mexico.

Smithereens. To find out how to smash an orbiting satellite, the unused satellite Oscar-22 was blasted to fragments and dust in 1992 as part of a U.S. Defense Nuclear Agency experiment.

To simulate debris colliding with a satellite in orbit, Oscar-22 was blasted with aluminum projectiles in a test chamber on the ground.

Oscar-22 was one of the U.S. Navy's Transit navigation satellites, a spare designed by the Applied Physics Laboratory of Johns Hopkins University and never launched to orbit. It was not OSCAR-22, the amateur radio satellite.

Star Wars

A ballistic missile and its warheads are a rocket and payloads launched from one place on Earth's surface on a sub-orbital path through space to attack another point on Earth.

Anti-missile and anti-warhead weapons can be fired from the ground or they can be orbited to wait in space for an attack.

Missiles and warheads can be destroyed in space just like satellites can be killed. Warhead target sensors can be overpowered and blinded by laser and microwave beams. Missiles and warheads can be destroyed by hits from solid objects.

SDI. President Ronald Reagan proposed in 1984 to build a shield in space to protect America from nuclear attack by intercontinental ballistic missile (ICBM).

The Strategic Defense Initiative Organization (SDIO), sometimes called Star Wars, was formed.

Reagan said atomic weapons would be rendered impotent by Star Wars, ending the danger of nuclear war. The orbiting shield would have been composed of several exotic future-technology space weapons conceived by American SDI researchers.

★Spy satellites linked to ground radar stations would have spotted incoming enemy missiles and warheads and sounded an alarm;

★Powerful laser beams from the ground would have destroyed ballistic missiles and warheads;

★Dozens of tiny, high-speed, killer rockets would have been fired from the ground;

★Killer satellites would have used laser beams and radar to guide rockets from Earth to smash into ballistic missiles or explode near them;

★So-called "dumb rocks" would have destroyed warheads by simply crashing into them;

★Exploding canisters would have destroyed warheads by exploding near them.

To see if defense hardware would work in orbit, SDIO blasted several rockets to space between 1986 and 1993, carrying Star Wars experiments.

For instance, two satellites in a 1986 flight tailed each other around the globe, charted the launch of another rocket on the ground, and finally demolished each other in a premeditated collision.

A cluster of small satellites whisked to orbit in a 1988 test simulated fifteen Russian nuclear missiles being tracked through space.

Delta Star. A three-ton missile-hunter with seven television cameras, laser radar and an infrared imager was shot into the sky aboard a Delta rocket from Cape Canaveral in 1989, on a six-month search for missile plumes on the ground.

Delta Star was a research satellite in the U.S. Strategic Defense Initiative to place a missile defense shield across space above America.

The secret spacecraft also had a detector to see lasers firing from the ground in an experiment to let future satellites take evasive action.

Delta Star peered down at ballistic missile and space-rocket launches to help engineers find a means of spotting enemy rockets in flight. Researchers were interested to learn if Delta Star's sensors could see rocket exhaust above the North Pole. In a war, most USSR ICBMs would have had to pass over the North Pole on their way to North America.

Delta Star had 48 small jets to aim the satellite at hot plumes of exhaust as missiles and rockets started on their way to space. Several test missiles and rockets were launched from Wallops Island, Virginia; Poker Flats, Alaska; White Sands, New Mexico; Barking Sands, Hawaii; and Cape Canaveral, Florida. Delta Star sensors recorded information on the heat and light in exhaust flames from various kinds of rockets against backgrounds such as ocean, land, space, horizon, atmospheric effects and aurora borealis northern lights over the North Pole.

The Russians inadvertently cooperated by continuing their usual space-rocket launch schedule while under observation by Delta Star.

Teal Ruby. The U.S. developed satellites to monitor flights of ballistic missiles and cruise missiles from space. Teal Ruby satellites watch from 460 miles above Earth for infrared light from the heat escaping from rocket bodies and exhaust.

Superfluid helium was used to cool the infrared sensors, similar to the equipment in the scientific Infra-Red Astronomy Satellite (IRAS).

Carrying a limited supply of superfluid helium reduced the working life of a satellite to one year.

Red Tigress. Not all tests require orbiting satellites. In 1993, the Air Force launched two sub-orbital sounding rockets from Cape Canaveral to see if sensors could tell the difference between incoming missile warheads and decoys.

The rockets carried Red Tigress-2 payloads to altitudes just over 200 miles. Thirteen experiments were ejected as twenty sensors on ships, aircraft, satellites and the ground monitored the 15-minute flight. The experiment packages separated into 60 objects. The rocket splashed down in the Atlantic Ocean 465 miles off the Florida coast.

End of Star Wars. The Pentagon wanted to launch hundreds of satellites to form the Star Wars shield by the end of the 1990s, but the multibillion-dollar scheme met years of resistance in a budget-minded Congress.

The administration of President George Bush limited Star Wars in 1989 when then-Vice President J. Danforth Quayle described President Reagan's plan as political jargon and unrealistic.

Recalling Reagan had said an impenetrable SDI shield would be completely leak proof, Quayle said such political semantics exaggerated capabilities.

The Bush administration tried to lower expectations about Star Wars' ability to deflect a first strike. Quayle said Star Wars could cope only with a limited Soviet attack or a strike of less-sophisticated rockets launched by a small country.

The administration began referring to Star Wars by the name of one of its weapons, Brilliant Pebbles. The Vice President said Brilliant Pebbles could not turn back an "all out, bolt out of the blue attack" from the USSR.

Brilliant Pebbles. Swarms of tiny rockets, known as Brilliant Pebbles, would track incoming ballistic missiles and shoot them down by crashing into them. The Pentagon wanted to place thousands the tiny of Brilliant Pebbles satellites in orbit. Each would have been a three-ft. weapon with a detector to lock onto an incoming missile. The Pebble would swing around to aim at an enemy warhead and launch a tiny interceptor rocket on a collision course.

BMDO. SDIO lost most of its political backing when the Cold War ended after 1990. The Red Tigress-2 suborbital test in 1993, described above, was SDIO's last act. The group was assigned a more limited mission in 1993 and renamed Ballistic Missile Defense Organization.

GPALS. The grand SDI plan was scaled back to cut costs. In the latest plan, the United States will rely only on missiles fired from the ground to knock out nuclear warheads in space.

The ground interceptors, known as Global Protection Against Limited Strikes (GPALS), will try to stop a few missiles or an accidental launch.

Brilliant Eyes. The end result of years of Star Wars satellite research will be Brilliant Eyes.

The GPALS system will have Brilliant Eyes missile-detecting satellites in space and 750-1,000 speedy little rockets on the ground to rush into space and hit an approaching warhead.

The first two Brilliant Eyes missile-tracking satellites will be launched for the U.S. Air Force before 1998. They will spot and track any ballistic missile launched against the United States and its allies, and alert the Air Force Space Command.

Brilliant Eyes satellites will observe the world below in visible and infrared light, watching for signs of launches of both theater and international ballistic missiles. If a launch is detected, a satellite would notify ground controllers who would fire Eris interceptors from the ground to space.

Brilliant Eyes satellites will communicate with each other, as well as with ground stations.

Brilliant Eyes also will monitor satellites in low and high Earth orbits for future targets. Later, GPALS may have up to a thousand killer satellites waiting in space to fire tiny interceptors at ballistic missiles rising from enemy launch pads.

Brilliant Eyes will be the primary sensors in the National Missile Defense (NMD). The first two are to be designed and flown by Rockwell International Corp. and TRW Inc. After a testing period, dozens of Brilliant Eyes satellites will be purchased from one of the companies.

Eris. Killer missiles may be able to find and destroy warheads in space. A warhead reportedly was overtaken by an Eris interceptor from Meck Island in the Kwajalein test range in the Pacific Ocean. A Minuteman-1 ICBM had been launched from California. At 21.5 minutes into its flight, an Eris was fired toward space. In 7.5 minutes, the killer missile reportedly found the dummy warhead amid decoys 160 miles above Earth. Eris extended a screen to increase the size of its killing area and slammed into the missile at 19,686 feet per second. News reports in 1993 questioned a 1984 intercept, saying it was rigged to fool the USSR.

The Persian Gulf War

Iraq invaded Kuwait on August 2, 1990, and declared the tiny nation its 19th province, igniting a brief, but high-tech, war which lasted only eight months until April 6, 1991.

Military satellites were crucial to U.S. allies in winning the Gulf War. For the first time, satellites were woven into the scheme of warfare.

Photosats. At least five American and three Russian spysats were detailed to fly over the warfront, snapping photos, drawing radar images and tuning in to eavesdrop on radio, telephone and data communications across the Middle East.

With each reconnaissance satellite passing overhead every two days, so many images of Iraqi forces flooded the U.S. Central Intelligence Agency (CIA) National Photo Interpretation Center, it was said analysts barely could keep up.

They worked around the clock to keep the pipeline filled with critical intelligence data for Operation Desert Storm commanders and the multi-national forces in Saudi Arabia.

Radarsat. The Lacrosse radarsat worked overtime to measure the effectiveness of bombing raids and identify new targets.

The radarsat maneuvered over much of the Persian Gulf region, passing over Iraq every few days. Its radar beam actually penetrated several feet into the sand so Desert Storm planners could identify camouflaged trenches and hidden features.

Early warning. Defense Support Program (DSP) early-warning satellites sounded the alarm as Scud missiles flew out of Iraq toward Israel and Saudi Arabia. U.S. Space Command then alerted missile batteries to fire Patriots at the Scuds.

Comsats. Fundamental changes in warfare resulted from lessons learned during Operation Desert Storm. Military officers depended on the flow of battlefield information, and satellites provided crucial links which could make or break units in conflict.

The Persian Gulf War also was the "first live TV war" broadcast around the world by many satellites which relayed television news broadcasts as well as military communications.

Live TV pictures from Baghdad, Iraq, and from the United Nations coalition warfront skipped around the globe on many satellites.

Mir station. Demonstrating the desirability of a high perch, Russian cosmonauts orbiting in space station Mir 200 miles overhead shot a three-minute film of thick black smoke billowing into the sky above Kuwait and an oil slick spreading across the Gulf. It was broadcast on Moscow TV.

Navsats. U.S. forces and allies in the Persian Gulf region found NAVSTAR global positioning system (GPS) satellites exceptionally useful.

Desert Storm troops maneuvered in forbidding territory among seemingly-endless miles of sand dunes. Maps were poor with landmarks few and far between. But, on land, as well as at sea and in the air, allied forces were able to receive instantaneous three-dimensional position and speed information from the NAVSTAR constellation of satellites.

They could observe radar and voice radio silence during a rendezvous. Aircraft could link up for mid-air refueling without communications.

F-16 and B-52 aircraft used GPS receivers to deliver bombs and guided munitions on specified military targets. Navy cruise missiles used GPS position data to attack heavily defended high-priority targets. The Army had GPS receivers in Apache helicopters and M60 tanks.

Army and Marine troops followed hand-held GPS receivers across the featureless Saudi desert. GPS was used to site radar. Forward observers used GPS to direct artillery fire on Iraqi positions. The Navy used GPS to map mine fields and to direct the rendezvous of supply ships.

After Desert Storm, commanders were said to wonder how they had operated in earlier conflicts without a NAVSTAR global positioning system.

Somalia. In Operation Restore Hope in Somalia in 1992-93, GPS receivers were aboard U.S. Air Force cargo aircraft for approaching and landing at makeshift airfields without the electronic aids available at most larger airports.

Science & Technology Satellites

Space satellites are forward movers. Many sciences rely on them today to move forward the frontier of knowledge, move forward our understanding of the environment around us, move forward the state of the art in various technologies, move forward our comprehension of the Universe.

Satellites orbiting Earth can study biology in space or track wild animals through forests and oceans on Earth, look outward to the Sun, planets and stars or focus downward on the movement of Earth's crusts, even make exotic materials and test new technologies for tomorrow.

Biological Satellites

Plants and animals have been rocketed to the region of Outer Space near Earth for decades in an ongoing quest for knowledge of how the space environment affects living things.

The menagerie sent to orbit over the decades has included monkeys, chimpanzees, dogs, guinea pigs, mice, rats, fish and fish eggs, turtles, newts, flies, drosophila, beetles, worms, tomatoes, orchids, planaria, paramecia and other animals, plants, seeds and single-celled organisms.

Albert. The first primates to fly near space were the monkeys Albert-1 and Albert-2. They reached 70 miles altitude and died in the nosecones of captured V-2 rockets during American flight tests at White Sands, New Mexico, in 1946-1952.

Aerobee. A monkey and eleven mice were shot to edge of space on a U.S. Aerobee rocket in 1951, but died when their parachute failed to open. Later that year, in the first successful space flight by living creatures, a different monkey and eleven mice were blasted to the edge of space on an Aerobee and were recovered alive.

Laika. The first higher life form in orbit was Laika, a dog on a life-support system in the Soviet Union's Sputnik-2, launched in 1957.

Sam. A monkey named Sam made a suborbital flight in 1959 in America's Mercury flight known as Little Joe 3.

Miss Sam. The next year, another monkey, Miss Sam, made a suborbital flight in Mercury Little Joe 4.

Belka and Strelka. The first successful recovery of a satellite from space was Sputnik-5 in 1960. It was the second unmanned test of a Vostok capsule. The satellite carried the dogs Belka and Strelka, and forty mice. A tape of a Russian choral group played over the radio during descent, while TV pictures showed a human dummy in a spacesuit. The dogs were ejected after re-entry and parachuted to Earth.

A Higher Life Form In Orbit

The first higher life form actually in orbit in space was Laika, a dog on a life-support system in the Soviet Union's second satellite, Sputnik-2, launched in 1957. Laika was in a capsule attached to the converted intercontinental ballistic missile (ICBM) in which she was launched.

The dog captured hearts around the world as her life slipped away a few days into her journey. Sputnik-2 burned in the atmosphere in 1958.

Six other dogs were launched to space by the end of the Sputnik satellite series in 1961 as the Russians prepared to send men to orbit.

Altogether, the seven dogs sent to orbit in the Sputnik series of tests were Laika (Barker in Russian), Belka (Squirrel), Strelka (Little Arrow), Pchelka (Little Bee), Mushka (Little Fly), Chernushka (Blackie) and Zvezdochka (Little Star). Laika, Pchelka, and Mushka died in flight.

Five years later, the dogs Veterok and Ugolek orbited 23 days in a Voskhod capsule.

Pchelka and Mushka. The dogs Pchelka and Mushka died in their burning Vostok capsule when it re-entered at too steep an angle in 1960. Sputnik-6 was the third unmanned Vostok test.

Ham. The United States sent the chimp Ham on a suborbital flight in 1961 in the Mercury capsule MR-2.

Chernushka. A fourth unmanned Vostok test, Sputnik-9 in 1961, carried the dog Chernushka and a human dummy in a spacesuit. When it descended after one orbit, the dog ejected and parachuted to Earth.

Zvezdochka. The fifth and final unmanned Vostok test, Sputnik-10 in 1961, carried the dog Zvezdochka and a human dummy in a spacesuit. When it descended after one orbit, the dog ejected and parachuted to Earth.

Discoverer. The U.S. launched the biosat Discoverer-29 in 1961.

Enos. The first primate in orbit was a chimp. Enos flew two orbits of Earth in 1961 in America's Mercury capsule MA-5.

Cosmos-4. The USSR launched two biosats in 1962, Cosmos-4 and Cosmos-7.

Cosmos-92. Three years later, the Soviet Union again launched two biosats, Cosmos-92 and Cosmos-94, in 1965.

Cosmos-110. The next year, Russia launched the biosat Cosmos-110, in 1966.

Veterok and Ugolek. In 1966, the dogs Veterok and Ugolek were launched in Voskhod-3 to be observed in orbit for 23 days via TV and biomedical telemetry.

Biosatellite. The United States launched a series of three biosats between 1966 and 1969. Biosatellite-1 was launched in 1966, followed by Biosatellite-2 in 1967 and Biosatellite-3 in 1969.

Moon turtles. The USSR tested equipment in its Zond-5 Mooncraft in 1968, preparing for a manned lunar landing. Turtles, worms, flies, plants and seeds were inside the unmanned capsule, which circled the Moon and returned to Earth. It was the first-ever flight to the Moon and back.

Biosat-3. Just three weeks before the first men were to land on the Moon in 1969, the United States launched Biosatellite-3 to orbit Earth.

Biosat's monkey passenger was supposed to stay in space for a month, but had to be brought down after only nine days when it fell ill from loss of body fluids. The monkey died shortly after landing. Later, both the U.S. and the USSR successfully launched and retrieved biological satellites carrying monkeys and many other kinds of animals and plants.

Bullfrogs. To study weightlessness, the United States launched two live bullfrogs in a biosatellite known as the Orbiting Frog Otolith (OFO-1). The satellite was blasted to orbit from Wallops Island, Virginia, in 1970.

Salyut. The world's first orbiting space station was Salyut-1 lofted by the USSR in 1971. It was dedicated to science, as opposed to military service, and carried a hydroponic garden and various biological specimens for study in the low gravity of near-Earth orbit.

Biocosmos-1. Success with its research on living organisms in space encouraged the USSR to continue with regular biosatellite launches. The biosat Cosmos-605 launched in 1973 also was known as Biocosmos-1.

Biocosmos-2. The Soviet Union launched the biosatellite Cosmos-690 in 1974. It also was known as Biocosmos-2.

Cooperation starts. Despite the Cold War, the United States and Russia cooperated in launching biosats studying the weightlessness of outer space. A 1971 agreement between NASA and the USSR Academy of Sciences set up a biology and medicine working group. In 1974, the USSR asked the U.S. to join in actual space flights.

The Soviets flew five biosats in the decade between 1975 and 1985. They were:

★Biocosmos-3, Cosmos-782, in 1975;
★Biocosmos-4, Cosmos-936, in 1977;
★Biocosmos-5, Cosmos-1129, in 1979;
★Biocosmos-6, Cosmos-1514, in 1983;
★Biocosmos-7, Cosmos-1667, in 1985.

Under the 1971 cooperation agreement, the U.S. contributed experiments to each flight.

Biocosmos-6 brought the first use of monkeys in the Russian space program. One of the monkeys, Bion, died shortly after landing. The flight also carried pregnant rats to test the effect of microgravity on embryos.

Biocosmos-7 carried two female macaque monkeys, named Versiyy and Gordyy, plus ten rats, ten newts, 1,500 flies and Iris flowers.

Cooperation ends. In 1980, U.S. President Ronald Reagan called the USSR an "evil empire" and stopped American participation in the joint working group.

Space tomatoes. The U.S. space shuttle Challenger carried 12.5 million tomato seeds to orbit in 1984 in a science-experiment carrier known as the Long Duration Exposure Facility.

Shuttle Columbia retrieved the LDEF satellite from space in 1990 and brought it down to Earth. NASA mailed the tomato seeds to schools.

Squirrel monkeys. Twenty-four rats and two squirrel monkeys went to space for seven days in 1985 in the European Space Agency's Spacelab module tucked into the cargo compartment of the U.S. space shuttle Challenger.

The flight became memorable when seals on the cages broke and food and animal droppings floated from Spacelab into Challenger's cockpit.

After the flight, researchers discovered the rats had suffered a fifty percent reduction in growth hormones. They lost muscle and bone strength.

Cooperation restarts. Nearly seven years went by before a new agreement on exploration and use of space for peaceful purposes was signed in 1987 by U.S. Secretary of State George Shultz and Soviet Foreign Minister Eduard Shevardnadze.

A new working group met in the USSR in 1987 and the U.S. was re-invited to contribute experiments to Russian biosats. The flights restarted that year with the launch of Cosmos-1887.

Biocosmos-8. The first American project with the Moscow Institute for Biomedical Problems under the 1987 treaty was Biocosmos-8, also known as Cosmos-1887, for research into space sickness, a problem for humans in orbit.

The 1987 flight from Plesetsk came at the 30th anniversary of the world's first satellite Sputnik-1.

Some sixty scientists from American colleges and NASA centers were involved in 26 life-science experiments aboard the unmanned, international-cooperation flight. The experimenters investigated the effects of space on bones, muscles, nerves, heart, liver, glands and blood. The United States also placed eight radiation detectors inside and outside the biosat to measure harmful dosages.

France, Hungary, East Germany, Poland, Rumania and the European Space Agency also had experiments aboard the Cosmos-1887 capsule.

Biocosmos-8 housed two rhesus monkeys, ten white Czechoslovakian rats, fish and fish eggs, newts, fruit flies, plants and seeds, and organisms as small as paramecia.

Rhesus monkeys. The rhesus monkeys were named Yerosha and Dryoma. Cosmos-1887 was the third Russian biosat to fly in Earth orbit with monkeys isolated in sealed chambers.

Fifteen sensors were implanted in their heads to detect changes in electrical impulses as the monkeys adapted to zero gravity. A telemetry radio transmitted data on the animals to Earth.

Yerosha, whose name signified a frisky or mischievous or trouble-making animal in Russian folk tales, was smaller and livelier than Dryoma, whose name meant drowsy or dreamer.

Scientists found out just how excitable Yerosha was when his pulse shot up to 200 beats per minute during blast off. Dryoma, on the other hand, reacted quietly to launch and weightlessness.

Bored in space. Yerosha was a lively and intelligent passenger, but he and Dryoma had been strapped in little chairs in a hermetically-sealed capsule for a two-week flight. After five dull days of orbiting, Yerosha decided to have some fun.

The monkey grabbed the attention of ground controllers on the fifth day when he appeared on TV screens without his metal name tag.

The tag had been pinned on his hat, which now was askew. Yerosha had managed to free his left front paw from a restraining cuff to play with the hat—and everything else he could reach.

His handlers thought about cutting the flight short to see what Yerosha had done with his name tag. They set up a mock capsule at Moscow, with a similar monkey and the same temptations, to see what it might do with a free hand. When nothing much happened, they decided Yerosha couldn't get into trouble and let the flight go on as planned.

Juice. Yerosha and Dryoma were trained to follow commands sent by radio and displayed as light signals. They were fed when they responded.

Yerosha apparently was so busy playing he forgot to do some activities to get all his food. Television pictures showed Yerosha flexing his free paw and sucking one of his food tubes.

In case a tube was blocked, ground control increased his supply of concentrated juices.

Russian Biocosmos Satellites

Russia launched the first in a long series of cooperative international biological satellite flights in 1973. American experiments were carried in the unmanned, recoverable biosats starting in 1975.

Year	Biocosmos	Cosmos
1973	Biocosmos-1	Cosmos-605
1974	Biocosmos-2	Cosmos-690
1975	Biocosmos-3	Cosmos-782
1977	Biocosmos-4	Cosmos-936
1979	Biocosmos-5	Cosmos-1129
1983	Biocosmos-6	Cosmos-1514
1985	Biocosmos-7	Cosmos-1667
1987	Biocosmos-8	Cosmos-1887
1989	Biocosmos-9	Cosmos-2044
1992	Biocosmos-10	Cosmos-2229

Russian media reported Yerosha "studies with interest the design of his space suit, but evidently treats with caution the helmet from which wires stretch to the information systems."

Telemetry indicated Yerosha's health was good, and, in the end, other than his bit of fun, the orbiting menagerie flew smoothly until landing.

Overshot. Cosmos-1887 returned to Earth after 13 days, on schedule, but in the wrong place. A faulty braking rocket dropped the capsule 1,000 miles off course to the surprise of residents near Mirnyy, a city in the diamond-rich Yakutia area of central Siberia 2,000 miles east of Plesetsk.

Helicopters from Mirnyy airport found Cosmos-1887 in a forest. The monkeys and other creatures inside the capsule were okay, despite a frosty early-morning landing in a pine-tree taiga 25 miles from Mirnyy. Biologists erected a tent over the capsule to raise the temperature, protecting the animals against minus 5 degrees Fahrenheit.

Having fun in orbit apparently hadn't prepared Yerosha to handle the stress of the harrowing re-entry flight as well as Dryoma. At the landing site, recovery workers had to use intensive methods to resuscitate Yerosha.

He and the rest of the menagerie was flown by airliner 2,500 miles to Moscow where the Soviet Institute of Biomedical Problems said Dryoma felt fine, but Yerosha seemed weak.

Eventually, Yerosha and Dryoma survived the harrowing re-entry and recovered fully. They were not sacrificed for science after the space trip. In fact, Yerosha was such an interesting space flier, researchers speculated that he might be assigned another space flight.

Hormones. An international brigade of biologists had welcomed the returning biosat. The scientists included eight Americans who went to the USSR for the landing of Cosmos-1887.

Tissue from five Biocosmos-8 rats was flown to the United States for examination. Their pituitary cells showed effects on growth hormones, while spleen and bone marrow cells revealed effects of low gravity on the immune system.

Diminished growth hormones had been found in American astronauts returning from short space flights. In adult humans, hormones affect growth and repair of tissue, muscle and bone, as well as the immune system and red blood cell production.

Understanding hormones is critical to planners of Mars flights and Moon settlements.

Astronauts experience weakness after space flights. Cosmonauts in orbit for extended periods have had trouble walking after landing.

Orchid

A blooming orchid, grown on Earth from seeds first planted in the orbiting space station Mir, was carried back to Mir in 1988 by cosmonauts in Russia's Soyuz TM-5 capsule.

China. The People's Republic of China used a Long March rocket to launch a Chinese Academy of Sciences biosatellite carrying animals and plants to space on October 5, 1990. Known as PRC-33, the biosat was the thirtieth satellite successfully launched by China.

It was blasted to orbit, from the Jiuquan Space Center in the northern Gobi Desert not far from Mongolia, to assess the response of its live animal and plant passengers to weightlessness. The passengers were returned to Earth after eight days.

Biocosmos-9. Russia invited the U.S. to place biomedical experiments aboard the 1989 flight of Cosmos-2044. The plants and animals from Russia, Europe and the United States included Rhesus monkeys, male Wistar rats, fish and fish eggs, newts, snails, beetles, ants, worms, drosophila fruit flies, plants, seeds, single-cell organisms and planaria.

The "orbiting zoo," as it was dubbed by the popular press, was studied carefully for effects of weightlessness and radiation on living organisms during space flight. Researchers looked for bone changes, muscle alterations, circadian rhythms, thermal regulation, nervous system changes and effects of radiation and gravity.

Monkeys. Biocosmos-7 and -8 had carried monkeys to space and Biocosmos-9 did too. The monkeys aboard Biocosmos-9 were named Zhakonya and Zabiyaka.

Biocosmos-9 was the ninth launch since the international program started in 1973 and the seventh in which NASA had participated.

NASA sponsored 85 researchers from 19 states and three foreign countries working on 29 Biocosmos-9 projects. Also sending experiments were scientists from Canada, Czechoslovakia, East Germany, the European Space Agency, France, Hungary, Poland and Romania. Altogether, eighty experiments were carried out by scientists from twenty countries.

Canadian Space Agency scientists sent along rats to study the effects of weightlessness on bone structure. The experiments exposed decalcification and weakening of bones in low gravity. Humans in space experience irreversible loss of bone mass and bone strength. A second Canadian experiment measured radiation in space.

Touch down. The menagerie was brought safely down to Earth after 14 days in space. The Cosmos-2044 capsule touched down softly south of the city of Kustanay, 1,000 miles southeast of Moscow in the USSR's Kazakh Republic. The monkeys were flown to Moscow for medical checkups, while thousands of other biology samples were sent to labs for analysis.

Some 3,000 samples were sent to labs across the United States. American space biology and medicine benefited from the Russian biosat flight because a two-week stay in space for rats and monkeys was about twice the time NASA could provide in a shuttle Spacelab flight.

Biocosmos flights also helped the U.S. develop its own space biology experiments, technology, and hardware for shuttle flights.

Tree frogs. The first Japanese to travel in space, Toyohiro Akiyama, took along green tree frogs from Japan when he visited Russia's orbiting space station in Mir in 1990.

SLS. The U.S. invited Russian biologists to analyze specimens from shuttle Columbia's 1991 Spacelab Life Sciences (SLS-1) flight and to send experiments along on Columbia SLS-2 in 1993. Russians also were invited to send experiments on SLS-3 in 1995, SLS-4 in 1997, SLS-5 in 1999.

Hornets. Endeavour was a real zoo in 1992 as the astronauts took a few friends along for a ride in space. The shuttle's laboratory-in-space featured oriental hornets, South African clawed frogs, Japanese carp, fruit flies and even chicken eggs.

Activity buzzed around the nest of hundreds of oriental hornets as the astronauts watched for signs of how weightlessness effected combs built by the large social wasps.

Meanwhile, the four female frogs laid 600 eggs in a "frog box" in the cargo bay. The eggs were fertilized and hatched after three days in orbit.

It was the first time eggs had been produced, fertilized and developed in microgravity. The astronauts took note of how the lack of gravity in space effected growth of the tadpoles.

Two silver-and-orange carp swam in a special tank to help scientists understand more about why astronauts get sick in space. Researchers looked for effects of weightlessness on an organ in the inner ear which helps a carp maintain its balance.

Hundreds of fruit flies were in an incubator aboard Endeavour to find out how cosmic radiation would effect them on a long trip. Their species, Drosophila melanogaster, was selected because radiation causes two or more hairs to grow on each cell on its wings. Usually, only one hair grows on each cell on a fruit fly's wing.

Thirty chicken eggs also were aboard in an experiment to see why bones and muscles weaken during long space flights.

Biocosmos-10. Scientists from the United States, Europe and Russia sent experiments to space in Russia's Cosmos-2229 in 1992.

Biocosmos-10 was launched from Plesetsk December 29 for biological, medical and cosmic radiation studies in orbit for eleven days. It was the eighth time NASA had participated in a Russian biosatellite space flight.

Nine Russian-American research teams were joined by scientists from Austria, Canada, China, France, Germany, Ukraine and the European Space Agency (ESA). They conducted 26 experiments on Rhesus monkeys, fish, beetles and other insects, reptiles, plants and seeds, micro-organisms and cell cultures. Equipment aboard the unmanned biosat monitored the cargo of flora and fauna.

Eight experiments were devised by Russia, Ukraine and five ESA nations. Three of those were in a "Biobox" programmable incubator for research into the biological effects of weightlessness.

Monkeys. The Rhesus monkeys were monitored for physiological, neurological and behavioral changes in space. Physiological measurements in space included body temperature, electrical activity of the heart and electrical currents generated in active muscles.

NASA scientists looked for changes in their bones, metabolism, eye movements and inner-ear gravity receptors. They analyzed urine and plasma samples for signs of bone, muscle and connective tissue breakdown. They also watched for changes in muscles, bone density and the endocrine system.

The scientists studied the immune system, temperature regulation and circadian rhythms, which are behavioral and physiological rhythms, such as metabolism and sleep rhythms, that follow approximate 24-hour cycles.

One monkey had minor problems with the food supply system after a few days in orbit, but both monkeys were okay after their flight.

Due to overheating, the satellite had to land two days earlier than planned in January 1993, but the payload was not affected greatly by the problem.

Russia's Institute of Biomedical Problems at Moscow runs the Biocosmos program, while NASA's Ames Research Center, Mountain View, California, manages American participation. Russia provides the satellite, rocket, flight hardware and biological specimens.

Wildlife-Tracking Satellites

A satellite looking down on a wide area of the globe from its high vantage point is a useful tool for biologists, naturalists and conservationists working with animals, birds and fish in their natural environments.

Scientists have attached tiny radio transmitters to wild animals for years, tracking the signals with radio receivers nearby on the ground. Now they can track wildlife ranging over much wider areas with receivers high above the ground.

A receiver in an orbiting space satellite can hear a transmitter attached to an animal. Even if the animal is out of sight over the horizon from a ground tracker, a satellite in space can hear the transmitter and repeat its signal down to trackers on the ground.

Whales. Tag, Notch and Baby were young male pilot whales nursed back to health after they beached themselves on Cape Cod in December 1986 and were found in shock and very ill.

They were named at Boston's New England Aquarium. After treatment, radio transmitters capable of reaching a satellite in a low orbit were attached to their dorsal fins and the whales were returned to the ocean in June 1987, the first time scientists had released more than one at a time.

Baby lost his radio as he entered the water and the battery in Notch's transmitter failed after a month, but scientists were able to track Tag into October 1987 when his radio batteries ran down.

Relay sat. It was the longest tracking ever of a whale or dolphin, up to that time. Signals from Tag's transmitter were received by a satellite orbiting overhead which relayed them to scientists at a research station on land.

Tag carried the transmitter on 200,000 dives in 95 days, staying underwater for an average of 34 seconds and as long as 7.75 minutes.

He swam north to Maine and south to Delaware. The last signals received indicated Tag was 20 miles off Cape Cod, the closest he had been to shore since being released. Aquarium scientists decided the whales, heading south, were not in danger of beaching.

Pod. Tag, Notch and Baby had to join a pod of pilot whales for survival and Tag was seen with a pod in August 1987. Researchers decided Notch and Baby had remained with Tag.

More than three months of satellite tracking that summer showed the giant mammals could be returned safely to the wild sea, but did not reveal why whales beach themselves. The tracking did add to our knowledge of whale diving and migration.

Four out of the previous five winters, whales had stranded themselves on Cape Cod beaches. More than seventy had run aground on the Cape and died in 1986. Some were pushed out to sea by concerned humans.

The strandings occurred among whales headed south to warmer waters. Scientists speculated the enclosed bay and unusual weather, or maybe the whale's herd instinct, caused mass beachings.

Researchers now know they can rehabilitate and return a whale to the ocean. They planned to send out the next whale with a longer-life battery.

Swans. The Wild Bird Society of Japan wanted to know exactly where swans go when they migrate to the Arctic each summer, so they used a satellite to track signals from tiny 1.4-oz. radio transmitters attached to the 12-lb. birds.

Japan's telephone company, Nippon Telegraph and Telephone Corp., made special miniature transmitters for the birders since larger transmitters used to track migrations of dolphins and seals were too heavy for swans to carry in flight.

The first radios were attached in Spring 1990 to four swans on Hokkaido, Japan's northernmost island. Signals beaming upward from the flying birds were received by Argos, an orbiting U.S.-French environmental satellite.

Information relayed by the satellite to scientists on the ground in Japan revealed the swans' flight path to their Arctic summer home.

Sea turtles. Japan's ministry of fisheries wondered where sea turtles go when they migrate, so they hung tracking radios on 100 members of the threatened species and tuned in by satellite.

Japan is the largest breeding ground for sea turtles in the Northern Hemisphere. About 6,000 lay their eggs along Japanese seashores each year.

The transmitters, weighing only 250 grams each, were attached during the breeding season in Spring 1993 to be observed for five years.

Storks. The German space agency Deutsche Agentur Fur Raumfahrt Angelegenheit (DARA) uses orbiting space satellites to track storks migrating across Europe and deer and other wildlife migrating along the Austrian/Czech border.

A Russian rocket in 1993 ferried to space an environmental research satellite known as Resurs-O (Resource-O). Among the electronics inside the satellite was Germany's SAFIR science payload, a forerunner of environmental satellites from the Department of Research and Technology (BMFT).

SAFIR contained an American five-channel GPS navigation receiver for wildlife tracking. The satellite also is used to relay data to a German research station in Antarctica.

Fishermen. Not just wild animals, but people can be tracked by satellite, also.

When the United States become concerned about illegal catches of salmon and steelhead trout by vessels harvesting squid with driftnets, Japan, South Korea and Taiwan required radio beacons aboard fishing boats so their driftnet fleets could be monitored by satellite.

Beacon signals from the boats are transmitted to an orbiting satellite which relays them to the U.S. National Oceanic and Atmospheric Administration.

Since 1990, NOAA and U.S. Coast Guard surveillance aircraft and surface patrols have used the location information to track and apprehend vessels illegally fishing unauthorized areas of the northern Pacific Ocean.

For instance, the Coast Guard used the satellite system in 1990 to nab the Daian Maru No. 65, one of six Japanese boats fishing in the northern Pacific 30 miles outside the squid boundary.

They uncovered 12.5 tons of illegal salmon in a freezer in the hold of Daian Maru No. 65.

The Coast Guard detained the boat, accused the fishermen of violating the International North Pacific Fisheries Convention, the U.S.–Japanese driftnet agreement and the Japanese domestic squid fishing regulations, then handed them over to a Japanese ocean patrol boat.

Europe. Euteltracs is a Europe-wide satellite tracking and position reporting system monitoring the movement of fishing boats within European Community (EC) waters. In 1993 tests for the

European Commission, a patrol boat sailed along a Euteltracs path from Newfoundland to Rostock, Germany, arriving with a navigation accuracy of 500 feet.

Rental cars. Satellites even can track rental cars, railroad cars, trucks, and container and cargo ships. Parked vehicles can be depicted on remote maps within fifty feet or a map can show the direction of a moving vehicle.

The same satellites can monitor the movements of individual people for roadside emergencies. They can monitor utility lines, storage tanks and pipelines in remote areas.

The Satellite Tracking Radiolocation and Communications system (SAT/TRAC) by Unisys Corp. and Energetics Satellite Corp. offers such global monitoring of equipment for companies controlling far-flung inventories.

Geodetic Satellites

Geodetic satellites are tools for measuring the movement of continents. Their results are used to predict when the Earth will quake and when volcanoes will erupt. Many geodetic satellites have been launched since the first in 1962.

ANNA. The first was an American military project known as ANNA. The name was derived from the initials of the Army, Navy, NASA and Air Force. The Defense Department and NASA cooperated on the early satellite project.

ANNA-1A was launched from the Cape in 1962, but its rocket failed. Five months later, ANNA-1B was lobbed successfully to orbit 700 miles above Earth's surface.

The small spacecraft's three instruments set the pace for future geodetic satellites:

★a flashing light which lit up at prescribed times for photography from down on Earth;

★a Sequential Collocation Of Range (Secor) radio transmitter;

★a Doppler radio system with transmitter frequency controlled precisely by a highly-stable quartz crystal oscillator.

The flashing light and Secor transmitter allowed measures with accuracy within 33 to 100 feet. The Doppler measurements gave measurements within 50 to 165 feet.

Cosmos-26. Two years later, the Russians launched the geodetic satellite Cosmos-26 from Kapustin Yar to a 200-mi.-high orbit, in 1974.

Seven months after that, the geodetic satellite Cosmos-49 went to a 300-mi.-high orbit, in 1974.

Unlike the American ANNA satellite, Cosmos-26 and Cosmos-49 used magnetometers to survey 75 percent of Earth's surface for magnetism. The result was establishment of what scientists refer to as the coefficient of Gauss and a table of values for terrestrial magnetism.

Beacon Explorer. A pair of ionospheric-research satellites launched from the United States in 1964-65 also were intended for geodetic studies.

Beacon Explorer-A, also known as S-66, was lost during a 1964 launch attempt.

Beacon Explorer-B, also known as Explorer-22, was launched successfully in 1964 from Cape Canaveral. The satellite's body was plastered with 360 silicon mirrors to reflect a laser light beam fired at it from Earth. The distance, or range, to the satellite could be measured precisely from the reflected laser light beam.

Beacon Explorer-C, also referred to sometimes as Explorer-27, was launched in 1965 from Vandenberg Air Force Base, California. It, too, had silicon laser reflectors, but also a transmitter with an ultra-stable oscillator for measuring the Doppler effect. That allowed it to detect irregularities in its own orbit and measure Earth's gravity field.

GEOS Satellites

NASA's Geodetic Earth Orbiting Satellites

GEOS	Launch	Explorer
GEOS-1	1965 Nov 6	Explorer 29
GEOS-2	1968 Jan 11	Explorer 36
GEOS-3	1975 Apr 9	--

GEOS. NASA's first purely-geodetic satellite was the Geodetic Earth Orbiting Satellite (GEOS), also known as Explorer-29, launched in 1965.

It had four packages of instruments for geodetic research and communications:

★a flashing light to be photographed at set intervals from Earth;

★a total of 440 quartz prisms to reflect laser light beamed at it from Earth;

★a radio transmitter for measuring the change in frequency known as the Doppler effect;

★a two-way radio transponder with a data transmission antenna.

The GEOS series of geodetic satellites should not be confused with GOES weather satellites.

Pageos. Balloons were popular in the 1960s as cheap-n-easy satellites. NASA sent up the 100-ft. balloons Echo-1 and Echo-2 as communications satellites and four 12-ft. science-satellite balloons to measure the composition of the atmosphere. Benzoic acid inflated the balloons in orbit.

The 100-ft. balloons were so large the sunlight they reflected could be seen by naked eye on Earth.

NASA sent up yet another of the popular Mylar

balloons in 1966. This 100-ft. aluminum-coated sphere was the Passive Geodetic Earth Orbiting Satellite (PAGEOS). Sunlight reflecting from its shiny surface made it seem as bright as the North Star. Scientists set up observation posts around the globe to photograph it for measurements of Earth's surface.

Diapason. The French space agency, Centre Nationale des Etudes Spatiales (CNES), launched three small geodetic satellites in 1966 and 1967 on Diamant rockets from a secret military base at Hammaguir, Algeria.

The first was Diapason D-1A launched in 1966. The second was Diademe D-1C launched in 1967 and the third was Diademe D-1D launched in 1967.

Each had a radio with a highly-stable oscillator for Doppler frequency-change measurements. In addition, the Diademe satellites sported reflectors as targets for laser beams on the ground.

GEOS-2. Another purely-geodetic spacecraft was the second Geodetic Earth Orbiting Satellite (GEOS-2), also known as Explorer-36, launched by NASA in 1968.

It carried the same instruments as GEOS-1: flashing light for photos; quartz prisms to reflect lasers; transmitter measuring the Doppler effect; and transponder with data transmission antenna. In addition, GEOS-2 had a C-band radar transponder to test geodetic measurements by satellite radar.

The GEOS series of geodetic satellites should not be confused with GOES weather satellites.

Peole. France returned to space in 1970 with a new geodetic satellite. Peole was launched on a Diamant-B rocket from French Guiana to test the transmitter for the future Eole weather satellite.

Peole was studded with 44 mirrors as targets for laser beams from Earth, part of a program by sixteen nations to measure the globe.

China. The People's Republic of China blasted its second satellite, Mao-2, to space from the Shuang Cheng Tse launch site in 1971. The tiny three-ft. sphere was a solar-powered science satellite to measure the magnetic field near Earth's surface and detect cosmic radiation.

GEOS-3. NASA launched a combined geodetic and oceanographic satellite in 1975 from Vandenberg. This time the name GEOS stood for Geodynamics Experimental Ocean Satellite.

GEOS-3 was not part of the Explorer series, as GEOS-1 and -2 had been. Instead, it was part of NASA's Earth and Ocean Physics Applications Program (EOPAP). GEOS satellites should not be confused with GOES weather satellites.

Although GEOS-3's exterior was similar to the earlier GEOS satellites, it housed a different set of five instruments: mirrors to reflect laser beams, a transmitter measuring Doppler effect, an S-band transponder, two C-band transponders, and a radar altimeter for oceanography.

GEOS-3 data was blended with data from the SEASAT oceanography spacecraft for more precise models of the Earth.

Starlette. France continued its measurement of Earth with the launch of a ten-inch sphere named Starlette in 1975. The miniature satellite, swathed with 60 reflectors to bounce laser beams back down to Earth, was lobbed 500 miles into space on a Diamant-B-P4 rocket from French Guiana. Starlette had a heart of uranium-235 to raise its weight to 110 lbs.

Pollux and Castor. France launched a pair of test satellites on one rocket in 1975. Pollux D-5A and Castor D-5B were blasted to orbit on a Diamant-B-P4 from French Guiana.

Pollux tested a trim-propulsion system while Castor tested microaccelerometers. Castor had 26 mirrors fixed to its exterior to reflect laser beams fired from Earth for geodetic measurements.

Lageos. The Laser Geodynamics Satellite (Lageos) was a NASA satellite dedicated to measuring movements of Earth's crust.

The two-ft. aluminum sphere, launched in 1976, was a passive-reflector satellite with 426 silicon mirrors attached to its surface to bounce back laser beams fired from Earth.

Lageos had a mass of brass at its heart to increase its weight to near 900 lbs. Measurements of crust movements with Lageos were as close as 0.4 to 0.8 inches.

Messages To The Future

Lageos was launched in 1976. At its altitude of 3,600 miles, it will be good for geodetic research work for fifty years, or until the year 2026.

Lageos won't fall from orbit in 2026, however. In fact, satellite trackers say it may be able to stay up there for eight million years. That would be until the year 8,001,976.

In case it lasts that long, NASA mounted two steel plates inside Lageos. The plates have three "Messages To The Future" written by Cornell University professor Carl Sagan. The texts describe Earth's history.

Cosmos. The USSR launched a series of four Cosmos satellites from Plesetsk in 1978-83 for geodetic research. Cosmos-1067 was launched in 1978. Cosmos-1312, in 1981. Cosmos-1410, in 1982. Cosmos-1510, in 1983. The satellites had laser reflectors and other geodetic systems.

Magsat. The Magnetic Field Satellite (Magsat) was NASA's first satellite dedicated to measuring Earth's magnetic field and magnetic anomalies of the planet's crust. It was launched to a polar orbit from Vandenberg in 1979.

Magsat had two magnetometers which measured magnetic anomalies to within 217 miles.

Intercosmos-17. The USSR built a series of satellites, called Intercosmos, for cooperative projects with its Eastern Bloc allies.

Hungary, Romania and Czechoslovakia cooperated with the USSR in a satellite to measure the energy of charged particles near Earth. Intercosmos-17 was launched in 1977. It carried a laser reflector for geodetic research.

Intercosmos-20. Two in the Intercosmos series, Intercosmos-20 and Intercosmos-21 were built to study land and ocean surfaces and launched from Plesetsk in 1979 and 1981.

Intercosmos-21. Sent to orbit in 1981, Intercosmos-21 was identical to Intercosmos-20.

Intercosmos-22. A satellite in the series with a different purpose was Intercosmos-22, also known as Bulgaria-1300. It was designed to report on Earth's ionosphere and magnetosphere.

The ionosphere is a layer in Earth's upper atmosphere where atoms have been ionized by solar radiation. Ionization is the electrical charging of an atom by removing an electron.

The magnetosphere is a region surrounding a planet where the planet's magnetic field is stronger than the interplanetary magnetic field.

Intercosmos-22 was lined with mirrors to reflect laser beams to Earth for geodetic measurements accurate within 16 to 32 feet.

Geosat. Johns Hopkins University built a geodetic satellite for the U.S. Naval Research Laboratory and the Naval Electronics System Command. Geosat was launched from Vandenberg in 1985. It had a radar altimeter to map the shape of Earth's land and sea surfaces.

Ajisai. Japan traditionally names its satellites for flowers after launch. The Experimental Geodetic Payload (EGP) was renamed Ajisai in orbit. An ajisai is an hydrangea, a flower which Japanese builders said the satellite resembled.

Ajisai, pronounced "ah-high-sy," was a totally passive mirrorball like a big rotating disco flasher in the sky. It had no electronics aboard.

The glowing seven-foot sphere was launched in 1986 for Japan's Hydrographic Department on an H-1 rocket from Tanegashima launch site on an island off the south coast of Kyushu.

Ajisai's job in space was to be a laser target for precise calculation of the locations of geodetic stations on Earth's surface. The laser ranging and experimental sky position satellite was clothed with hundreds of mirror reflectors.

Ajisai also reflected sunlight so brightly it was visible to a naked eye on the ground, like a bright star taking eighteen minutes to cross the sky.

The satellite, spinning at forty revolutions per minute, flashed two sunlight reflections each second. Its apparent brightness of first magnitude to fourth magnitude made it look as bright as a second magnitude star. Humans see fifth or sixth magnitude stars with the naked eye, so most observers were able to see Ajisai overhead easily.

How Laser Ranging Works

Ajisai was a target for precise calculation of locations on Earth's surface.

Some 85 percent of the exterior surface of the satellite was covered by 318 convex mirrors which reflected sunlight. The periphery of the satellite was lined with 1,436 cubes of polished fused silica glass for additional reflection.

The cubes were in 120 clusters of 12 cubes each, except one cluster which had only eight cubes to meet construction requirements. The cubes were corner reflectors used as targets for laser beams. Laser tracking was so precise the data had to be corrected for the 42-in. distance from corner reflectors to the center of the satellite.

Geodetic stations in Maryland, California, Australia, England and Switzerland fired laser beams at Ajisai. Extremely brief laser pulses, lasting from 200 picoseconds to one nanosecond, from two to 19 inches in physical length, were transmitted toward the satellite. The laser light bounced off Ajisai's reflectors back to Earth.

The time it took a laser beam to go up and back told scientists how far away Ajisai was.

Photography of its sunlight flashes showed its position against the background field of stars.

Lageos-2. Italy and America cooperated on a new geodetic satellite ferried to space by shuttle Columbia in 1992. Lageos-2 was a passive satellite much like the first Lageos launched by NASA in 1976.

The two-ft. sphere was covered by 426 silicon mirrors to reflect laser beams from geodetic stations around the globe. Lageos-2 was used to measure movement of tectonic plates, movement of the poles, and variations in ocean tides.

Stella. Germany launched a small geodetic satellite named Stella in 1993 alongside six other satellites on one European Ariane rocket. It was similar to the two earlier Lageos satellites.

Space Technology Satellites

Dozens of unmanned satellites have been sent to orbit to test spacecraft designs and new satellite technology. Some also carried science instruments as secondary payloads.

Explorer-13. In the early days of space flight, just about any satellite could be said to be testing spacecraft hardware. However, some satellites were designed specifically for research into space technology. For example, in 1961 the United States launched the space-technology satellite Explorer-13.

TRS. The next year, America launched TRS, another space-technology satellite.

Explorer-16. Also in 1962, the U.S. launched the Explorer-16 technology satellite.

SERT-1. The United States launched the Space Electric Rocket Test (SERT-1) on a sub-orbital flight in 1964. SERT-2 would be launched in 1970 to a 600-mi.-high orbit.

Asterix. Asterix (A-1) was the small satellite launched in 1965 from North Africa which made France the third nation with a rocket powerful enough to boost a satellite to Earth orbit.

France's network of receiving stations tracked the Asterix radio beacon for two days until its chemical battery ran down. *(Details, chapter one.)*

Wresat. Wresat was Australia's first satellite, a technology satellite launched in 1967.

TTS. The United States launched the space-technology satellite TTS in 1967.

SERT-2. The United States launched the Space Electric Rocket Test (SERT-2) satellite to a 600-mi.-high orbit in 1970. Among the satellite's instruments was a Reflector Erosion Experiment (REX)—two polished metal discs with temperature sensors that measured erosion and damage to the mirrored surfaces from the continual impact of tiny micrometeorites.

Erosion data was used by engineers at NASA's Lewis Research Center at Cleveland, Ohio, to compare the benefits of solar cells against the benefits of a solar dynamic power source.

Solar cells capture the Sun's energy and convert it to electricity. A solar dynamic power source has mirrors to heat fluids driving turbo generators.

SERT-2 had been turned off in space for a decade when it was revived in 1989 to test power sources for the U.S.-international space station Freedom, which is to orbit Earth by the year 2000.

Ohsumi. Japan launched its first satellite, Ohsumi-1, in 1970, making it the fourth nation with a space rocket.

Mao-1. Two months after the launch of Ohsumi, the People's Republic of China launched its first satellite. The 1970 satellite was named Mao-1 and has been called China-1 and PRC-1.

Tansei-1. The next year, Japan launched another space-technology satellite, Tansei-1.

Black Knight. Also in 1971, Great Britain launched its first satellite, Black Knight-1, from Australia. The space-technology satellite also was referred to as X-4 Prospero.

MAS STRET. Two tiny French space technology satellites were launched by the Soviet Union in 1972 and 1975. The Russians called them MAS while the French referred to the satellites as STRET. Each rode to orbit as ballast on rockets carrying large Molniya communications satellites to space. STRET-1 tested solar cells for satellites and should not be confused with the earlier American SERT satellite.

Explorer-46. That same year, the U.S. also launched the technology satellite Explorer-46.

Tansei-2. Japan launched Tansei-2 in 1974.

Miranda. Also in 1974, Great Britain launched another technology satellite X-3 Miranda.

MAS-2 STRET-2. The second of two tiny French space technology satellites was launched by the Soviet Union in 1975. The Russians called it MAS-2 while the French called it STRET-2.

Riding along as ballast on a rocket carrying a Molniya communications satellite, STRET-2 tested thermal insulation materials for future satellites. It should not be confused with the American satellite SERT-2.

Kiku. Japan launched Engineering Technology Satellite (ETS-1) in 1975. It was called Kiku.

Intercosmos-15. The USSR built a series of Intercosmos satellites for cooperative projects with Socialist allies in the Eastern Bloc. One of those satellites, launched in 1976, was a space-technology satellite labeled Intercosmos-15.

Tansei-3. Japan launched Tansei-3 in 1977.

Scatha. The United States launched the space-technology satellite Scatha in 1979.

Tansei-4. Japan launched Tansei-4 in 1980.

Rohini. India became the seventh nation with a space rocket when it launched Rohini Satellite (RS-D1) in 1980.

CAT. The first Ariane, launched in 1979 by the European Space Agency from French Guiana, carried CAT, ESA's first satellite.

Ofek. Israel lofted its first satellite, Horizon (Ofek or Ofeq), in 1988, then Horizon-2 in 1990. They had radio beacons. *(Details, chapter one.)*

Iraq. Iraq launched a three-stage rocket in 1989 on a six-orbit flight. It did not release a separate satellite, but the third stage did sweep around the globe for six times before falling out of orbit.

Official Space Art

America's first and only official space artwork was carried to orbit by shuttle Discovery in 1989. It also was to be the last for some time.

The seven-pound cube-shaped sculpture called Boundless Aperture, created by Boston artist Lowry Burgess, was approved for flight by NASA's Non-Scientific Payloads Committee in 1984.

Back then, the agency was accepting civilian requests to orbit non-science satellites. But, after the 1986 Challenger disaster killed civilian Christa McAuliffe and six other astronauts, the Non-Scientific Payloads Committee was disbanded.

Although the space agency cut off civilian access to space, an exception was granted for Boundless Aperture.

Burgess, a professor at Massachusetts College of Art, conceived Boundless Aperture in the 1960s as part of what he called The Quiet Axis.

By the time Boundless Aperture went to space, Burgess already had buried a sculpture in the floor of the Pacific Ocean off Easter Island and lugged another into central Afghanistan's Hindu Kush mountains.

Space Physics Satellites

Satellites are used to increase our understanding of the space environment near Earth, including the vital impact of radiation from the Sun on Earth's upper atmosphere. In fact, spacecraft have been used to carry out physics experiments in space near Earth from the time of the earliest satellites.

Some space-physics satellites perform dual duties by observing ultraviolet, infrared and X-ray radiation from sources beyond the Sun for astronomers, even while performing near-Earth astrophysics research experiments.

Satellites dedicated to astronomy are covered below, in the following section of this chapter.

1957-58 Space Physics

Sputnik. Earth's first man-made satellite was Sputnik-1, launched in October 1957 by Russia. The aluminum ball housed a thermometer and a battery-powered radio transmitter. The thermometer changed the radio frequency. *(Details, chapter one.)*

Explorer. Three months later, in January 1958, the U.S. Army launched America's first satellite, Explorer-1. It was designed by James Van Allen at the Jet Propulsion Lab in California and carried two micrometeorite detectors and a Geiger counter in search of charged particles.

Explorer-1 discovered intriguing belts of intense radiation circling planet Earth. Those belts were named after Van Allen. Discovery of the Van Allen belts spurred further work on artificial Earth satellites. *(Details in chapter one.)*

The United States has been launching Explorer satellites every since Explorer-1. More than 75 U.S. and international payloads have been sent to orbit in the Explorer program over the decades.

Vanguard. Two months later, in March, the U.S. Navy launched Vanguard-1. It carried a Geiger counter, a magnetometer and a micrometeorite counter. Vanguard revealed the shape of the Earth is reminiscent of a pear.

Sputnik-3. In May, the USSR launched a large geophysical observatory. Sputnik-3 assayed the upper atmosphere, measured the concentration of charged particles, took magnetic and electrostatic readings, and watched the flow of micrometeorites.

Explorer-3. Eleven days later in May, the U.S. Army launched Explorer-3. Its instruments measuring temperature, cosmic rays and meteorites were mostly the same as those in Explorer-1 with a data tape recorder added.

Explorer-4. In July, the U.S. Army launched Explorer-4 with improved versions of the instruments in Explorer-1 and Explorer-3, plus detectors looking for X-rays and cosmic rays.

Van Allen analyzed the data from Explorer-1, Explorer-3 and Explorer-4 to describe further the belts of radiation trapped in Earth's magnetic field.

1959-60 Space Physics

Explorer-6. A trio of physics satellites, Explorer-6, -10 and -45, were sent to orbit between 1959-1971 to report on Earth's magnetosphere.

Magnetosphere is the upper region surrounding a planet where the planet's magnetic field is stronger than the interplanetary magnetic field.

Launched in 1959, Explorer-6 studied electrical fields around Earth and discovered three radiation levels. The earliest television pictures of Earth's cloud cover were sent down by Explorer-6.

Explorer-8. A group of five space-physics satellites, Explorer-8, -20 and -22, -27 and -31, were sent to orbit between 1960-1965 to report on Earth's ionosphere. Explorer-8 was lofted in 1960.

The ionosphere is a layer in Earth's upper atmosphere where atoms have been ionized by solar radiation. Ionization is the electrical charging of an atom by removing an electron.

1961 Space Physics

Explorer-10. Lofted in 1961, Explorer-10 read the interplanetary magnetic field near Earth. It was in a trio of satellites orbited between 1959-1971 to report on Earth's magnetosphere.

Explorer-12. Four satellites, Explorer-12, -14, -15 and -26, were sent to orbit between 1961-1964 to find out how the radiation belts around Earth trap, hold and lose their charged particles. Explorer-12 was launched in 1961.

1962 Space Physics

Ariel-1. Ariel-1 was Great Britain's first space satellite. It was launched in 1962 by the United States for Great Britain.

Ariel-1 should not be confused with the Black Knight satellite lofted in 1971 which was the first satellite Great Britain was able to launch for itself.

Ariel-1 also was known as United Kingdom-1 (UK-1). It carried experiments to measure ions, electrons, solar radiation and cosmic rays.

Alouette. Canada's Alouette-1 was launched in 1962 to test the effect of the ionosphere on communications.

Explorer-14. Explorer-14 was among four satellites sent to orbit between 1961-1964 to examine how Earth's radiation belts trap, hold and lose their charged particles. Explorer-14 was launched in 1962.

Explorer-15. Explorer-15 was launched in 1962 to study Earth's radiation belts.

Cosmos-1. Over the years, Russia launched many satellites to study the interaction between the Earth and the Sun. The first in the series was Cosmos-1 launched from Kapustin Yar in 1962. Some of the others were Cosmos-2, -3, -26, -49, -97, -321, -381, -477, -481, -1463 and -1508.

Cosmos-2. Russia launched many satellites to study the interaction between the Earth and the Sun. The second in the series was Cosmos-2 launched from Kapustin Yar in 1962.

Cosmos-3. Launched in 1962, Cosmos-3 was third in a series of Russian satellites fired to orbit to examine the interaction between the Earth and the Sun.

1963 Space Physics

Explorer-18 IMP. A group of ten Explorer satellites launched by the U.S. between 1963-1973 were Interplanetary Monitoring Platforms (IMP).

They studied interplanetary radiation and magnetic fields between the Earth and the Moon, adding to knowledge of the energetic relationship between Earth, Moon and Sun.

Explorer-18, launched in 1963, first in the IMP series, confirmed the existence of a shock wave created when the solar wind bumps into Earth's magnetosphere. Explorer-18 also discovered the existence of Earth's magnetic tail. The satellite's orbit took it 123,000 miles from Earth.

1964 Space Physics

Elektron. The Soviet Union launched four Elektron satellites in 1964 to assess the outer and inner Van Allen radiation belts surrounding Earth.

Elektron-1 and Elektron-2 were launched on one rocket in January, while Elektron-3 and Elektron-4 were launched on a rocket in July.

Elektron-1 and Elektron-3 studied the inside belts out to 4.300 miles, while Elektron-2 and Elektron-4 studied the outside belts out to 42,000 miles and the space between the belts.

Explorer-20. In the group of five Explorer satellites sent to orbit in the first half of the 1960s to examine Earth's ionosphere, Explorer-20 was launched in 1964.

Explorer-21 IMP. The U.S. Explorer-21, an Interplanetary Monitoring Platform (IMP), was sent to space in 1964 to study radiation and magnetic fields between Earth, Moon and Sun. Explorer-21 didn't go as far as planned, but its orbit still took it 60,000 miles from Earth.

Explorer-22. Also launched in 1964 was Explorer-22 in the group of five Explorers orbited in the 1960s to test Earth's ionosphere. Explorer-22 also was called Beacon Explorer-B.

Explorer-25 Injun-4. A trio of Injun satellites launched by the U.S. between 1964-1974 studied Earth's magnetosphere.

They measured the flow of radiation into Earth's atmosphere and the distribution of charged-particle energy. Explorer-25, also known as Injun-4, was launched in 1964.

Explorer-26. Explorer-26 was among four launched between 1961-1964 to examine how Earth's radiation belts hold and lose charged particles. Explorer-26 was launched in 1964.

Cosmos-26. Launched in 1964, Cosmos-26 was one of many Russian satellites sent to orbit to study the interaction between Sun and Earth.

Cosmos-49. Launched in 1964, Cosmos-49 was one of many Russian satellites sent to orbit to study the interaction between Sun and Earth.

OGO-1. NASA launched a set of six Orbiting Geophysical Observatories (OGO) between 1964 and 1969. They studied Earth's atmosphere, magnetosphere and the space between the Earth and the Moon.

The first was OGO-1 lofted in 1964. Its orbit took it 71,000 miles from Earth.

OGO-1 had twenty experiments, including twelve studying particles, two looking at the magnetic field, and one each examining interplanetary dust, electric fields, Lyman-Alpha emissions, the Gegenschein effect, atmospheric mass, and radio astronomy.

1965 Space Physics

Explorer-27. Launched the next year, in 1965, in the group of five Explorers orbited in the 1960s to test Earth's ionosphere, was Explorer-27 which also was known as Beacon Explorer-C.

Explorer-28 IMP. Sent to space by the United States in 1965, Explorer-28 was one of ten satellites launched between 1963-1973 as Interplanetary Monitoring Platforms (IMP). Its orbit took it 165,000 miles from Earth.

Explorer-31. Explorer-31 also was launched in 1965. It was one of five satellites lofted in the 1960s to examine Earth's ionosphere. It studied the density and temperature of ions and electrons, the composition of ions, and corpuscular radiation.

Explorer-31 also was referred to as International Satellite for Ionospheric Studies (ISIS-X)/ADME. Canada joined the U.S. in the ISIS project.

Cosmos-97. Launched in 1965, Cosmos-97 was one of many Russian satellites sent to orbit to study the interaction between Sun and Earth.

OGO-2. OGO-2, launched in 1965, carried twenty instruments to test the ionosphere, the atmosphere, the magnetic field, the radiation belts, cosmic rays, micrometeorites, and solar energy blasts. It was second of six Orbiting Geophysical Observatories (OGO) launched by NASA between 1964 and 1969 to study Earth's atmosphere, magnetosphere and space between Earth and Moon.

FR-1. The French space agency, Centre Nationale des Etudes Spatiales (CNES), had a small satellite launched from Vandenberg in 1965 to check on anomalies in the ionosphere and magnetosphere. FR-1 experimented with very-low-frequency (vlf) radio waves.

1966 Space Physics

Explorer-33 IMP. Sent to space in 1966, Explorer-33 was one of ten satellites launched between 1963-1973 as Interplanetary Monitoring Platforms (IMP). They studied interplanetary radiation and magnetic fields between the Earth and the Moon, adding to knowledge of the energetic relationship between Earth, Moon and Sun.

Explorer-33 examined Earth's magnetic tail and revealed the Moon's magnetic field. Its orbit took it 307,000 miles from Earth.

OGO-3. Launched in 1966, OGO-3 was in an orbit which took it 76,000 miles from Earth. The satellite carried the same twelve experiments as OGO-1, including instruments to study particles, interplanetary dust, magnetic and electric fields, Lyman-Alpha emissions, atmospheric mass, the Gegenschein effect and radio astronomy. OGO-3 was the third Orbiting Geophysical Observatory (OGO) launched by NASA between 1964 and 1969 to study Earth's atmosphere, magnetosphere and space between the Earth and the Moon.

1967 Space Physics

Explorer-34 IMP. Sent to space in 1967, Explorer-34 was one of ten satellites launched between 1963-1973 as Interplanetary Monitoring Platforms (IMP) to study radiation and magnetic fields around Earth, Moon and Sun.

Explorer-34 discovered that Saturn emitted radio waves just like Jupiter and the Earth. Its orbit took it 133,000 miles from Earth.

Explorer-35 IMP. Explorer-35, launched in 1967, was one of ten Interplanetary Monitoring Platforms launched between 1963-1973. It was sent to orbit the Moon where it found solar wind directly behind the Moon. It measured the electrical conductivity and internal temperature of the Moon.

OGO-4. Launched in 1967, OGO-4 had the same twenty experiments as OGO-2, including instruments to examine the ionosphere, the atmosphere, the magnetic field, radiation belts, cosmic rays, solar radiation and micrometeorites.

OGO-4 was fourth of six Orbiting Geophysical Observatories (OGO) launched by NASA between 1964 and 1969 to study Earth's atmosphere, magnetosphere and space between Earth and Moon.

1968 Space Physics

Explorer-40 Injun-5. Explorer-40, also known as Injun-5, was launched in 1968 to measure radiation flowing into Earth's atmosphere and the distribution of charged-particle energy. It was one of three American satellites studying the magnetosphere between 1964-1974.

OGO-5. OGO-5 was launched in 1968 to an orbit which took it 92,000 miles from Earth. It carried 25 experiments of which seventeen studied particles, two measured the magnetic field, and one each studied radioastronomy, ultraviolet light, Lyman-Alpha emissions, X-rays from the Sun, plasma waves, and electric fields. OGO-5 was the fifth Orbiting Geophysical Observatories launched between 1964-1969 to study Earth's atmosphere, magnetosphere and space between Earth and Moon.

HEOS-1. The European Space Agency (ESA) built two nearly identical Highly Eccentric Orbit Satellites (HEOS). Each was packed with seven experiments to find interactions between solar particles and Earth's magnetosphere.

HEOS-1 was launched in 1968 from Cape Canaveral to an orbit ranging from a perigee of 260 miles to an apogee of 139,000 miles. HEOS-A2 was launched four years later in 1972.

Radioastronomy

The first person to tune in radio waves coming from somewhere in the sky, beyond Earth, was Karl Jansky who worked at Bell Telephone Laboratories in New Jersey in 1932. He was the first radioastronomer.

If Jansky was the father of radio astronomy, amateur radio operator Grote Reber of Wheaton, Illinois, was the father of the radiotelescope.

Reber built the first radiotelescope in 1937 in his backyard. He used the 31-ft. dish antenna to uncover the first discrete radio sources beyond Earth and mapped the spread of natural radio signals across the Milky Way. Reber worked alone in the field until World War II.

ESRO-2 IRIS. The European Space Research Organization (ESRO) was the predecessor of the European Space Agency (ESA).

ESRO-2 IRIS was the first space satellite of the European Space Research Organization. Launched in 1968, it was one of four ESRO satellites between 1968-1972 for space-physics research.

ESRO-2 IRIS had seven experiments studying radiation and cosmic particles.

ESRO-1 Aurorae. The second satellite sent to orbit by the European Space Research Organization was ESRO-1 Aurorae. Launched in 1968, it was one of four ESRO satellites between 1968-1972 for space-physics research.

ESRO-1 Aurorae carried eight instruments to examine the ionosphere and aurora over the poles.

1969 Space Physics

Explorer-41 IMP. Launched in 1969, Explorer-41 was another of the Interplanetary Monitoring Platforms. Its orbit took it 110,000 miles from Earth where it measured interplanetary radiation and magnetic fields.

OGO-6. Launched in 1969, OGO-6 had 26 experiments to study particles, the magnetic field, radioastronomy, ultraviolet light, Lyman-Alpha emissions, X-rays from the Sun, plasma waves, and electric fields. OGO-6 was the last of six Orbiting Geophysical Observatories (OGO) to be launched by NASA between 1964 and 1969 to study Earth's atmosphere, magnetosphere and space between Earth and Moon.

ESRO-1 Boreas. ESRO-1 Boreas was sent to orbit in 1969 by the European Space Research Organization. It was one of four ESRO satellites between 1968-1972 for space-physics research.

ESRO-1 Boreas carried eight instruments to examine the ionosphere and aurora over the poles.

ISIS-1. Canada's research into the effect of the ionosphere on communications, started in 1962 with Alouette satellites, was continued with new ISIS satellites. ISIS-1 was launched in 1969. ISIS-2 was launched two years later in 1971.

Intercosmos-1. The USSR built a series of satellites, called Intercosmos, for cooperative projects with its Socialist allies in the Eastern Bloc. The first Intercosmos satellite was launched by the Soviet Union at Kapustin Yar in 1969.

Intercosmos-1 was a space-physics research craft built in collaboration with Czechoslovakia and East Germany. Its instruments received X-rays and ultraviolet light from the Sun and watched how they interacted with Earth's upper atmosphere.

Intercosmos-2. The second Intercosmos satellite also was launched by the Soviet Union at Kapustin Yar in 1969. It was a collaboration with Bulgaria, East Germany and Czechoslovakia to count electrons and positive ions in the ionosphere and read the temperature of electrons close to the satellite. It measured the concentration of electrons between the satellite and Earth.

1970 Space Physics

Cosmos-321. Cosmos-321, sent to orbit in 1970, was one of many Russian satellites to study the interaction between the Sun and the Earth.

Cosmos-381. The Cosmos-381 satellite also studied Sun and Earth interactions after it was launched in 1970.

Intercosmos-3. The Soviet Union sent its third Intercosmos satellite to space from Kapustin Yar in 1970. It was built in collaboration with Czechoslovakia to assess interactions between the Sun and the Van Allen radiation belts.

Intercosmos-4. The fourth Intercosmos satellite also was launched from Kapustin Yar in 1970. The project with East Germany and Czechoslovakia searched for ultraviolet light and X-rays and watched their effect on the ionosphere.

DIAL. West Germany and France collaborated on a physics-research satellite launched from French Guiana in 1970. Dial/Wika was outfitted with equipment to measure particles belting Earth.

Four instruments measured the intensity of Lyman-Alpha radiation, the density of electrons, the energy in protons, electrons and alpha particles, and variations in Earth's magnetic field.

1971 Space Physics

Explorer-43 IMP. Explorer-43's orbit in 1971 took it 127,000 miles from Earth. The Interplanetary Monitoring Platform studied interplanetary radiation and magnetic fields between Sun, Earth and Moon.

Explorer-45. Orbited in 1971, Explorer-45 measured electrical activity and magnetic storms and looked for a relationship between aurora, magnetic storms and the acceleration of charged particles in the magnetosphere. It was in a trio of satellites launched between 1959-1971 to report on Earth's magnetosphere.

ISIS-2. Canadian research into the effect of the ionosphere on communications continued with its ISIS-2 launched in 1971. ISIS-1 had been launched in 1969.

Azur. The small satellite Azur-1 was launched from West Germany from Vandenberg in 1971.

It studied the magnetosphere and the Sun-Earth relationship, while measuring Earth's magnetic field. Experiments looked into the distribution over time of protons, electrons and alpha particles flowing in the Van Allen radiation belts as well as the flow of electrons in aurora over the poles.

Intercosmos-5. The USSR launched its fifth Intercosmos satellite from Kapustin Yar in 1971. Built in collaboration with Czechoslovakia, Intercosmos-5 studied the flow of charged particles and influence of the Van Allen radiation belts on the ionosphere.

Ariel-4. A radioastronomy satellite was fired to orbit for Great Britain from Vandenberg in 1971. Ariel-4, also known as UK-4, measured the intensity of radio emissions from galaxies, noise at low and extremely-low radio frequencies, and temperature and density of electrons.

Aureole-1. Three Russian physics research satellites built and launched by the Soviet Union in cooperation with France were called Aureole or Orcol. Aureole-1, launched from Plesetsk in 1971, carried a French experiment to measure protons and electrons in the solar wind.

Shinsei. Japan's Institute of Space and Astronautical Science (ISAS) built a satellite to measure cosmic rays and the ionosphere, the temperature of electrons and the density and temperature of ions, and high-frequency radio emissions from the Sun. Shinsei was launched from Kagoshima Space Center in 1971.

1972 Space Physics

HEOS-A2. The European Space Agency (ESA) built two nearly identical Highly Eccentric Orbit Satellites (HEOS). Each was packed with seven experiments to find interactions between solar particles and Earth's magnetosphere.

The second satellite, HEOS-A2, was launched in 1972 from Vandenberg to an orbit ranging from a perigee of 220 miles to an apogee of 148,000 miles. HEOS-1 had been launched in 1968.

Cosmos-477. One of several satellites launched by the USSR to study the interaction between the Sun and the Earth was Cosmos-477, sent to Earth orbit in 1972.

Cosmos-481. Another satellite launched by the USSR in 1972 to study the interaction between the Sun and the Earth was Cosmos-481.

Explorer-47 IMP. Sent to space in 1972, Explorer-47 was one of ten satellites launched between 1963-1973 as Interplanetary Monitoring Platforms. It studied Earth's magnetic tail and the middle field of radiation between Earth and Moon. Its orbit took it 147,000 miles from Earth.

ESRO-4. ESRO-4 was sent to orbit in 1972 by the European Space Research Organization. One of four ESRO physics satellites between 1968-1972, ESRO-4 carried six experiments to examine the ionosphere and solar particles.

Denpa. Less than a year after the launch of its Shinsei space-physics research satellite, Japan's Institute of Space and Astronautical Science (ISAS) launched another. Denpa was used to study plasma waves, flowing electrons, the magnetic field around Earth, and the propagation of radio waves in the magnetosphere. Denpa was launched from Kagoshima Space Center in 1972.

Intercosmos-6. Sixth in the series of Intercosmos satellites was launched from Kapustin Yar in 1972. Intercosmos-6 was built in collaboration with Czechoslovakia, Hungary, Poland, Romania and Mongolia to monitor micrometeorites, cosmic rays and the spectrum of electromagnetic energy.

Intercosmos-7. The USSR also launched Intercosmos-7 in 1972 from Kapustin Yar.

Intercosmos-7 was a space-physics research craft built in collaboration with Czechoslovakia and East Germany to measure X-rays and ultraviolet light and their effects on the upper atmosphere.

Intercosmos-8. The USSR also launched Intercosmos-8 in 1972 from Plesetsk.

Intercosmos-8 was a space-physics research craft built in collaboration with Czechoslovakia, East Germany and Bulgaria to measure electrons and positive ions in the ionosphere and the temperature of electrons.

1973 Space Physics

Explorer-50 IMP. Explorer-50 was sent to orbit in 1973. The Interplanetary Monitoring Platform studied the Earth's magnetic tail and the middle field of radiation between Earth and Moon. Its orbit took it 180,000 miles from Earth.

Aureole-2. Three Russian physics research satellites built and launched by the Soviet Union in cooperation with France were called Aureole or Oreol. Aureole-2, launched from Plesetsk in 1973, carried a package of French instruments to measure the solar wind, the magnetosphere, the ionosphere and the aurora over the poles.

Intercosmos-9. The Soviet Union's Intercosmos-9, launched in 1973 from Kapustin Yar, was a space-physics research craft built in collaboration with Czechoslovakia and Poland to measure X-rays and ultraviolet light and their effects on the upper atmosphere.

Intercosmos-10. The USSR also launched Intercosmos-10 in 1973 from Plesetsk. It was a space-physics research craft in collaboration with Czechoslovakia and East Germany to study the relationship of ionosphere and magnetosphere.

1974 Space Physics

Explorer-52 Hawkeye. Explorer-52, also referred to as Hawkeye, was launched in 1974 to observe the interaction between the solar wind and the magnetic field in the polar regions. Its orbit took it 79,000 miles from Earth as it mapped points and neutral lines during maximum and minimum solar activity periods.

Intercosmos-11. Intercosmos-11, launched in 1974 from Plesetsk, was built in collaboration with Czechoslovakia and East Germany to survey X-rays and ultraviolet light from the Sun and their effects on the upper atmosphere.

Intercosmos-12. The USSR also launched Intercosmos-12 in 1974 from Plesetsk.

Intercosmos-12 was a space-physics research craft built in collaboration with Czechoslovakia, East Germany, Hungary and Romania to gauge the atmosphere and the flow of micrometeorites.

IntaSat. The Spanish company Inta built Spain's IntaSat, which was launched to orbit from Vandenberg in 1974 to measure the density of the ionosphere and test radio propagation. It received signals from high-altitude balloons and relayed them to ground stations.

1975-76 Space Physics

Intercosmos-13. Intercosmos-13 was launched in 1975 from Plesetsk. It was built in collaboration with Czechoslovakia to look for dynamic processes in the magnetosphere and the ionosphere above the poles.

Intercosmos-14. The USSR also launched Intercosmos-14 in 1975, in collaboration with Czechoslovakia, Bulgaria and Hungary, to map the ionosphere, tune electromagnetic oscillations in the magnetosphere, and check intensity in the flow of micrometeorites.

Intercosmos-16. The 16th Intercosmos satellite was launched in 1976 from Kapustin Yar. Built in collaboration with Czechoslovakia, East Germany and Sweden, it looked for effects of the Sun's ultraviolet light and X-rays on Earth's upper atmosphere.

1977 Space Physics

Intercosmos-17. The USSR built a series of Intercosmos satellites for cooperative projects with its Eastern Bloc allies. Intercosmos-17 was launched in 1977 from Kapustin Yar.

Intercosmos-17 was a space-physics research craft built in collaboration with Czechoslovakia, Hungary and Romania to measure the energy of charged particles close to Earth.

ISEE. NASA and the European Space Agency built International Sun Earth Explorer (ISEE) satellites for an International Magnetosphere Study (IMS) in 1977-1978. NASA built ISEE-1 and ISEE-3. ESA built ISEE-2.

ISEE-1 and ISEE-2 were launched on one rocket from Cape Canaveral in 1977. There orbits took them more than 85,000 miles from Earth.

ISEE satellites measured the fall off in density of the magnetosphere in the region of space known as the plasma-pause. They also surveyed Earth's magneto-sheath, the magneto-pause where the solar wind strikes Earth's magnetic field and the shock wave there, and Earth's magnetic tail. ISEE-1 and ISEE-2 had thirteen and eight experiments.

GEOS-1. The Europe Space Agency built two magnetosphere satellites with identical sets of seven instruments to measure electromagnetic effects and the flow of electrons and protons. They checked on energetic particles at the 22,300-mi. altitude used by stationary satellites. GEOS-1 was launched from Cape Canaveral in 1977, but did not make to stationary orbit. It was left in an elliptical orbit ranging out to 24,000 miles.

ESA's GEOS satellites launched in 1977-1978 should not be confused with NASA's Geodetic Earth Orbiting Satellites (GEOS) from 1965-1975.

1978 Space Physics

Intercosmos-18. The Soviet Union built the Intercosmos-18 satellite with Czechoslovakia, East Germany, Poland, Hungary and Romania. It was launched in 1978 from Kapustin Yar to survey Earth's magnetosphere and the ionosphere.

ISEE-3. NASA's third International Sun Earth Explorer (ISEE-3), launched in 1978, had thirteen experiments similar to ISEE-1. ISEE-3 worked in a "halo" orbit around a point in space 716,000 miles from Earth, between Earth and Sun.

A Radar Trap In Space

Outbursts on the Sun trigger large magnetic storms on Earth. Observations by the ISEE-3 satellite from 1978-82 revealed just how important mass ejections from the Sun's corona are.

Storms in Earth's magnetic field make compass needles swing wildly, interrupt communication from satellites in space and disrupt electrical power grids on Earth. If only Earth could receive early warning of incoming solar mass outbursts before they reach Earth's magnetosphere.

A mass ejected from the Sun's corona is an immense blob blasted into space when the Sun's magnetic field swells like an inflating balloon and pops a billion tons of hot gas into the void.

When mass ejections from the corona first were observed in the 1970s, it seemed there were more ejections than there were geomagnetic storms.

But, when they looked again at data from the ISEE-3 satellite, plasma physics researchers at the Los Alamos National Laboratory found only one in six ejections caused a disturbance.

The key turned out to be speed. A slow-moving blob didn't spread much geomagnetic disturbance across Earth, while a fast mover was a shock wave rushing across interplanetary space, reminiscent of a sonic boom from a jet plane flying above Earth.

Traveling at two million mph, a fast-moving ejection overtakes the steady flow of electrons and atomic nuclei in the ambient solar wind. The leading edge of the ejection snowplows magnetic field lines and charged particles.

As the disturbance passes Earth, the planet's magnetic field is compressed and distorted by the shock wave. Charged particles dumped on Earth's atmosphere for many hours cause a geomagnetic storm with brilliant auroras. Strong currents of electricity in the upper atmosphere circle the globe above the Equator at high latitudes, actually weakening the magnetic field at lower latitudes.

Could Earth's residents receive early warning of geomagnetic storms? Three satellites might do it.

Tripping along our planet's path, ninety degrees ahead and behind Earth, two would act like a small-town radar trap, spotting ejections heading toward Earth, metering the mass's speed and sounding sirens on Earth days in advance of impact. The third satellite, parked between Sun and Earth, would sound off sixty minutes before an ejection and shock wave arrived at Earth.

Of course, the satellites wouldn't use radar guns like your local speed-trap operator, but would use particle-counting electromagnetic sniffers.

GEOS-2. European Space Agency built two magnetosphere satellites with an identical set of seven instruments to measure electromagnetic effects and the flow of electrons, protons, and energetic particles at the 22,300-mi. altitude used by stationary satellites. GEOS-2 was launched to stationary orbit from Cape Canaveral in 1978.

Kyokko. Japan launched its own satellite in 1978 from Kagoshima Space Center to measure the density, temperature and energy of electrons in space near Earth.

Kyokko observed the density of plasma and the aurora in ultraviolet light. Japan's Institute of Space and Astronautical Science (ISAS) built Kyokko, which also was known as EXOS-A.

Jikiken. Japan launched another satellite in 1978 from Kagoshima Space Center to survey the magnetosphere for plasma waves, electromagnetic fields, and the density of charged particles and electrons. Jikiken's orbit carried it 19,000 miles from Earth. ISAS built Jikiken, which also was known as EXOS-B.

1979 Space Physics

Intercosmos-19. The USSR built a series of Intercosmos satellites for cooperative projects with its Eastern Bloc allies. Intercosmos-19 was launched in 1979 from Kapustin Yar.

Intercosmos-19 was a space-physics research craft built in collaboration with Czechoslovakia, Bulgaria, Poland and Hungary to survey the magnetosphere and the ionosphere.

1981 Space Physics

Aureole-3. Three Russian physics research satellites built and launched by the Soviet Union in cooperation with France were called Aureole or Oreol. Aureole-3, launched from Plesetsk in 1981, carried a package of French instruments to measure the solar wind, the magnetosphere, the ionosphere and the aurora over the poles.

China-10 & -11. The People's Republic of China launched three satellites on one rocket in 1981. Two of the three were China-10 and China-11, also referred to as PRC-10 and PRC-11.

China-10 and -11 examined Earth's upper atmosphere, magnetic field, infrared radiation from Earth, ultraviolet light emitted by Earth and Sun, and X-rays from the Sun.

Dynamics Explorer. Two satellites sent to space on one rocket from Vandenberg in 1981 were labeled Dynamics Explorer-1 (DE-1) and Dynamics Explorer-2 (DE-2).

They watched how particles radiating from the Sun interacted with Earth's upper atmosphere and magnetic field. DE-1 ranged out to 15,000 miles from Earth.

Capturing All The Aurora

Dynamics Explorer-1 was the first satellite to acquire global images of Earth's aurora.

In its highly-elliptical orbit, ranging from 300 miles above Earth out to 14,500 miles, DE-1 was used by astronomers to study the energy, electric currents and mass coupling Earth's upper atmosphere, ionosphere and magnetosphere.

DE-1 was supposed to work only three years, but lasted nearly ten. NASA ended its use of DE-1 in 1991 after the satellite had refused to accept commands since 1990. Meanwhile, DE-2 had descended in 1983.

1983 Space Physics

Cosmos-1463. One of several Russian satellites launched to study the interaction between the Sun and the Earth was Cosmos-1463, sent to Earth orbit in 1983.

Cosmos-1508. Another of several Russian satellites launched to study the interaction between the Sun and the Earth, Cosmos-1508, was sent to Earth orbit in 1983.

HiLat. The United States modified a Transit navigation satellite to study the ionosphere at high latitudes. HiLat was launched in 1983.

1984 Space Physics

AMPTE. A group of British, German and American satellites named Active Magnetosphere Particle Tracer Explorers (AMPTE) were launched in 1984 on the same rocket from Cape Canaveral.

Great Britain's satellite in the trio was United Kingdom Subsatellite (UKS). West Germany's was Ion Release Module (IRM). The U.S. satellite was Charge Composition Explorer (CCE).

The orbits of IRM and UKS took them more than 70,000 miles from Earth. CCE's orbit carried it 31,000 miles from Earth.

IRM released clouds of lithium and barium to see how they would be affected by the solar wind. UKS observed the artificial clouds from close up, while CCE studied them from lower in the magnetosphere.

EXOS-C. Japan's Institute of Space and Astronautical Science (ISAS) built a third EXOS satellite and launched it in 1984 from Kagoshima to map the mesosphere and stratosphere and to analyze the reactions of plasma and charged particles in the upper atmosphere. EXOS-C also studied anomalies in the ionosphere above Earth's geomagnetic anomaly in Brazil. The satellite was part of the Middle Atmospheric Program (MAP). EXOS-A and EXOS-B had been launched in 1978.

1985 Space Physics

Prognoz-10. The Soviet Union sent ten Prognoz spacecraft to Earth orbit between 1972-1983 for space-physics research and astronomy.

Prognoz-1 to Prognoz-9 were solar astronomy satellites launched between 1972-1983. They are described below with other astronomy satellites.

The tenth Prognoz was the same style of spacecraft as Prognoz-1 to Prognoz-9. Launched by the Soviet Union in 1985 in collaboration with Czechoslovakia, Prognoz-10's orbit carried it 124,000 miles from Earth where it observed the shock wave as the solar wind struck Earth's magnetosphere.

1986-88 Space Physics

Viking. Sweden built a space-physics research satellite which was launched in 1986 on a European rocket from French Guiana.

From altitudes ranging from 2,500 to 10,000 miles above Earth, Viking watched hot and cold plasma along the lines of Earth's magnetic field at northern latitudes. It watched the acceleration of charged particles forming the aurora over the poles and natural radio waves. It saw ionospheric plasma injected into the magnetosphere.

Viking housed three instruments from Sweden and one each from Canada and the United States.

Foton. Russia launched Foton-1 in 1988, Foton-2 in 1989, Foton-3 in 1990 and Foton-4 in 1991.

1989 Space Physics

EXOS-D. Japan's Institute of Space and Astronautical Science (ISAS) built a fourth EXOS satellite and launched it in 1989 from Kagoshima Space Center to study space near Earth. EXOS-A and EXOS-B had been launched in 1978 and EXOS-C in 1984

1990 Space Physics

Hiten. Japan became the third nation to send a spacecraft to the Moon when its Muses-A blasted off from the Kagoshima Space Center in 1990.

After launch, the robot science explorer was renamed Hiten, which means space flyer.

Hiten carried a micrometeorite counter from West Germany's Munich Technical University to record weight, speed and direction of dust particles striking the craft.

Hiten also carried a basketball-sized satellite, Hagoromo, which was dropped off in lunar orbit as Hiten swung by the Moon.

As Hiten looped around Moon and Earth eight times, it scooped up data on space dust and other phenomena. By its fifth Moon fly-by, Hiten was traveling more than 600,000 miles from Earth.

Hiten's flight was in preparation for Geotail, an international Sun and Earth space-physics research flight launched to a long elliptical orbit in 1992.

Hiten was sponsored by Japan's government-funded Institute of Space and Astronautical Science (ISAS). Affiliated with the Education Ministry, ISAS focuses on science research in space.

Debut. Japan also launched in 1990 a small satellite in preparation for space station Freedom. Debut tested mechanical movements in space by unfolding and folding a boom and umbrella.

Pegsat. A small satellite carried to space in 1990 by America's then-new Pegasus air-launched rocket was called Pegsat. It emptied two canisters of barium over northern Canada to illuminate Earth's magnetic field in a display resembling the Northern Lights. Similar canisters were released from ten sounding rockets launched in 1991 from Kwajalein Atoll in the Republic of the Marshall Islands in the South Pacific and from Puerto Rico.

CRRES. Combined Release and Radiation Effects Satellite (CRRES) was a joint NASA and U.S. Air Force project launched in 1990 to map the Van Allen radiation belts surrounding Earth. CRRES also studied the effects of space radiation on microelectronic parts and how the ionosphere and magnetosphere interact.

CRRES was loaded with 24 canisters of lithium, barium, calcium and strontium chemicals to be ejected at various altitudes over the course of a year. Nine canisters were released into the upper ionosphere 250 miles above Earth and the rest at lower altitudes. The chemicals illuminated Earth's usually-invisible magnetic field in sensational displays resembling Northern Lights.

As they were ionized by ultraviolet light from the Sun, the chemicals painted the sky with glowing 60-mi. clouds. Ultraviolet light from the Sun ionized the chemicals, changing negatively-charged electrons in each atom to give the gas an electrical charge. The resulting plasma cloud traveled along Earth's magnetic field lines—like iron filings line up around a bar magnet. That made its field visible. The misty aurora could be seen flowing along lines of Earth's magnetic field.

The entire continent saw the barium clouds, which looked like Northern Lights or aurora borealis. Scientists watched the temporarily-visible magnetic field lines through telescopes on the ground and from specially-outfitted airplanes.

Researchers saw interactions between the clouds and the normal invisible charged particles in the ionosphere and magnetosphere.

CRRES traveled through the inner and outer Van Allen radiation belts, intense radiation zones which can harm spacecraft electronics. CRRES had five major experiments using 55 instruments. Microelectronic parts were exposed to radiation to see if they would fail. Instruments probed the Van Allen radiation belts which girdle Earth.

1991 Space Physics

Intercosmos-25. The Russian cooperative project Intercosmos-25 was built in collaboration with Czechoslovakia and others and launched from Russia in 1991. Also known as Aktivny-1K, it surveyed the magnetosphere surrounding the globe. A small Magion research satellite was released.

MAGION-3. The Czechoslovakian satellite Magion-3 was released from Intercosmos-25 to study the magnetosphere.

REX. The United States launched a Radiation Experiment (REX) satellite in 1991 to look into how irregularities in electron density affect radio signals passing through the ionosphere.

1992 Space Physics

SAMPEX. Small Explorers are inexpensive 400-lb. satellites which NASA launches to low orbits on Pegasus and Scout rockets. The first Small Explorer was SAMPEX (Solar, Anomalous and Magnetospheric Particle Explorer) launched on a Scout rocket in 1992. It carried a telescope to study energy particles from the Sun, cosmic rays from across the Milky Way galaxy and electrons from the magnetosphere. Designed at University of Maryland, SAMPEX involved ten German and American scientists. Small Explorers are operated by NASA's Goddard Space Flight Center.

FREJA. The Swedish Space Corporation built a space-physics satellite for research into aurora. FREJA was launched in 1992 from China's Jiuquan Satellite Launch Center. FREJA is controlled from Esrange, Kiruna, Sweden.

International Solar Physics

NASA, Russia's Space Research Institute, the European Space Agency (ESA), and the Japanese Institute of Space and Astronautical Sciences (ISAS) are coordinating the use of 35 satellites in 25 studies of Earth's magnetism in the 1990s.

The International Solar Terrestrial Physics Project (ISTP) is a seventeen-nation cooperative effort to launch at least eight new unmanned space-physics satellites to observe interactions between the Earth and the Sun from orbits around Earth.

Scientists hope all of those satellites exploring Earth's geomagnetic tail and plasma physics will increase understanding of the cocoon of magnetism around Earth. They are looking for impacts of the solar wind on Earth's magnetic tail, interplanetary shock waves, pulses in the interplanetary magnetic field, tail instabilities, storage and violent release of energy in magnetic storms, and transfer of energy between Earth and distant areas of the tail.

Electrical and magnetic interactions in the magnetosphere, the solar wind and interplanetary shock waves have caused storms in the magnetosphere which disturb communications and electrical power systems on Earth and electronic equipment in satellites in space.

ISTP. The ISTP spacecraft include Japan's Geotail in 1992, Russia's two Interbol satellites in 1993, NASA's Wind and Polar satellites in 1994, the European Space Agency's Cluster and SOHO satellites in 1995, Japan's Akebono, and NASA's Interplanetary Monitoring Platform.

Geotail. Solar wind particles swooshing past Earth form a magnetic tail on the leeward side of the planet, hence the satellite name Geotail. The United States launched Geotail in 1992 for Japan's Institute of Space and Aeronautical Science.

The satellite's seven instruments observe solar wind particles interacting with Earth's magnetic field over four years.

Geotail passed within 6,000 miles of the Moon in 1992. That fly-by boosted the satellite into a larger orbit which pushed it far out along Earth's magnetic tail. Geotail ranged from 32,000 to 871,000 miles from Earth.

In 1995 the orbit will slacken so the satellite can investigate the magnetic tail near Earth.

Geotail is controlled at Japan's Sagamihara Spacecraft Operations Center. Data transmitted to Earth from Geotail is received at NASA Deep Space Network ground stations and forwarded to the Goddard Space Flight Center in Maryland. From there, the data is distributed to researchers in Japan, the United States and elsewhere. The International Solar Terrestrial Physics (ISTP) program is managed by Goddard.

DUE. Below the nosecone covering Geotail, the American rocket housed a science instrument known as Diffuse Ultraviolet Experiment.

DUE was to measure the diffuse glow from the interstellar medium. That data would lead to better understanding of interstellar matter. DUE received power from the rocket and was turned on only after Geotail was dropped off in its own orbit.

Interbol. Two ISTP satellites are Russian-French joint projects known as Interbol. Two large Progress-style capsules will discharge two small Magion satellites to observe the magnetosphere, following 1991 work by Aktivny-1K and Magion.

Polar. Another of the eight ISTP satellites is NASA's Polar Auroral Plasma Physics (Polar) satellite observing the plasma physics of aurora above Earth's poles.

Wind. Another NASA ISTP satellite is Wind, which gauges the solar wind splashing onto the magnetosphere.

SOHO. The Solar and Heliospheric Observatory (SOHO) is an ISTP satellite to be launched in 1995 from the U.S. to investigate how the Sun's corona forms. The corona is a faint haze surrounding the Sun during a total eclipse.

SOHO also will study the Sun's interior structure from its highly elliptical Earth orbit. SOHO's on-board rocket motor will drive it one million miles from Earth during its two-year tour of duty. The SOHO project is managed by NASA's Lewis Research Center, Cleveland, Ohio.

ISEE-3 becomes ICE

One satellite actually was changed in space from near-Earth physics researcher to astronomer.

Among the dozens of Explorer spacecraft launched over the decades by the United States was the International Comet Explorer (ICE).

It wasn't launched as the International Comet Explorer, but as the third International Sun-Earth Explorer—a solar observatory launched by NASA in 1978 to a halo orbit at the Earth-Sun libration point where the pull of gravity from Earth and Sun are equal. Hanging about a million miles from Earth toward the Sun, ISEE-3 warned of solar flares before magnetic disturbances reached Earth.

That is, until the Comet Giacobini-Zinner was found rushing to pass Earth in September 1985. NASA quickly renamed ISEE-3 as ICE and pointed it toward the in-rushing comet.

Giacobini-Zinner wasn't to be ICE's last encounter with a comet. Just six months later in March 1986, the satellite was turned again. This time, ICE was pointed at the famous Comet Halley making its 76-year trek through the Solar System.

ICE observed Comet Halley from a distance of 17 million miles, sending home a wealth of information about that sooty, potato-shaped rock as big as a mountain flying around the Sun.

Astronomy Satellites

Man has studied the heavens since ancient times but the view has always been made fuzzy by the ocean of air between Earth's hard surface and the near-vacuum of space. Only just now are we able to put entire astronomical observatories in space where the seeing is not blurred by an atmosphere.

Many satellites orbiting Earth look out, toward the Sun, planets, stars, galaxies, pulsars, black holes, quasars and other far-away objects at the very edge of the Universe.

Their television cameras and imaging detectors stare through powerful telescopes, scanning their horizons for visible, infrared and ultraviolet light. Their receivers measure radio waves, gamma rays, X-rays, microwaves and exotic energies arriving from deep space.

All of these astronomy satellites are electronic robots, working alone for years at a time in space, transmitting data about what they see to eager astronomers on Earth.

Vanguard-3. The earliest satellite to carry out astronomical research was America's tiny Vanguard-3 launched in 1959. The Naval Research Lab packed its twenty-in. magnesium sphere with an ionization camera to measure X-rays, a micrometeorite detector and a magnetometer.

Explorer-7. Eleven Explorer satellites were launched by the United States between 1959-1975 for astronomy work in Earth orbit. The first was Explorer-7, launched on a Juno-2 rocket from Cape Canaveral in 1959. It observed X-rays from the Sun and how they influenced Earth's ionosphere.

1961 Astronomy

Explorer-11. Explorer-11, launched on a Juno-2 rocket from Cape Canaveral in 1961, was the first satellite to search for high-energy gamma rays arriving at Earth from deep space. Explorer-11 data led astronomers to abandon their steady-state theory of the Universe which had proposed a continuous generation of matter and anti-matter.

Explorer-11 was among eleven U.S. astronomy Explorers launched to orbit between 1959-1975.

1962 Astronomy

Cosmos-7. The USSR sent a great many astronomy satellites to Earth orbit over the decades. The first was Cosmos-7 lofted in 1962.

Cosmos-8. The second astronomy satellite from the USSR was Cosmos-8 in 1962.

OSO-1. Between 1962-75, the United States sent eight Orbiting Solar Observatory (OSO) satellites to Earth orbit to study the Sun. They received X-rays, gamma rays and ultraviolet and infrared light from the Sun and other objects deep in space. OSO-1 was launched in 1962.

1964 Astronomy

Cosmos-51. Launched in 1964, Cosmos-51 was one of many astronomy satellites launched by the USSR over the decades.

Explorer-23. Launched from Vandenberg in 1964, Explorer-23 measured the flow of micrometeorites. It was in the group of eleven U.S. astronomy Explorer satellites launched to Earth orbit between 1959-1975.

X-Ray Astronomy

By studying X-rays arriving at Earth from deep space, astronomers better understand how stars, galaxies and the Universe itself formed.

X-rays must be seen in space because they don't penetrate Earth's atmosphere.

Gamma rays and ultraviolet light from deep space also do not penetrate the atmosphere.

Until recently, knowledge of the Universe was limited to objects seen in visible light, infrared light and radio wave portions of the electromagnetic spectrum which reach Earth's surface.

X-ray astronomy shows a Universe quite different from what astronomers see at optical wavelengths.

1965 Astronomy

Explorer-30 SolRad-1. Launched by the U.S. from Vandenberg in 1965 to measure solar X-rays during the International Year of the Quiet Sun (IQSY), Explorer-30 was in the group of eleven U.S. astronomy Explorer satellites launched to Earth orbit between 1959-1975. Explorer-30 also was known as Solar Radiation satellite, or SolRad-1. More SolRads were launched in 1968 and 1971.

OSO-2. The U.S. sent eight Orbiting Solar Observatories (OSO) to Earth orbit between 1962-75 to study the Sun. They recorded X-rays, gamma rays and ultraviolet and infrared light. OSO-2 was launched in 1965.

Pegasus. NASA converted three second stages from Saturn-1 rockets into satellites which were launched in February, May and July 1965 to study micrometeorites.

The satellites were named Pegasus. Within a minute after it arrived in space, two fifty-ft. wings would pop out from the sides of a spent rocket stage—a giant micrometeorite catcher in space.

The wings held 208 panels of very thin sheets of aluminum, copper, Mylar and polyurethane. When a micrometeorite plunged through one of the panels, it would cause an electrical short circuit.

The frequency of micrometeorite hits could be measured from the number of punctures over time.

The size of a micrometeorite could be measured from the number of panels encompassed by a hit.

It even was possible to detect the direction of fall and strength of micrometeorite impact.

Solar cells to power the instruments also were built into the wings.

Proton-1. Four massive Soviet astronomy satellites were launched between 1965-1968. They were called Proton and carried instruments to detect high-energy cosmic rays, gamma rays and electrons arriving at Earth from across the galaxy.

The USSR's powerful "D" space rocket, known today as Proton, received its name after it carried the heavy satellites to orbit. The first Proton satellite was launched in 1965.

Proton-2. The USSR's second heavy Proton astronomy satellite also was launched in 1965. It also found cosmic rays, gamma rays and electrons.

1966 Astronomy

Proton-3. The third of four heavy Proton astronomy satellites was launched from the USSR in 1966. It detected high-energy cosmic rays, gamma rays and electrons from across the galaxy.

OAO. To peer even deeper into space, NASA built four Orbiting Astronomical Observatories (OAO). They were launched between 1966-1972, but only two worked. OAO satellites observed ultraviolet light from hot young stars, new sources of X-rays and interstellar gas.

OAO-1 was launched in 1966, but did not work correctly in orbit. OAO-4 also did not work correctly. OAO-2 and OAO-3 were launched in 1968 and 1972.

1967 Astronomy

Cosmos-163. The USSR launched two astronomy satellites in 1967. The first that year was Cosmos-163.

Cosmos-166. The second 1967 astronomy satellite from the Soviet Union was Cosmos-166.

OSO-3. America sent eight Orbiting Solar Observatories (OSO) to Earth orbit between 1962-75 to study the Sun. They received X-rays, gamma rays and ultraviolet and infrared light. OSO-3 was launched in 1967.

OSO-4. The United States also sent OSO-4 to orbit in 1967.

1968 Astronomy

Explorer-37 SolRad-2. Launched from Vandenberg in 1968 to measure solar X-rays, Explorer-37 was in the group of eleven U.S. astronomy Explorer satellites launched to Earth orbit between 1959-1975. Explorer-37 also was known as Solar Radiation satellite, or SolRad-2. The first SolRad had been launched in 1965 and a third SolRad was launched in 1971.

Explorer-38 RAE-1. Also launched from Vandenberg in 1968 was Explorer-38, known as the Radio Astronomy Explorer (RAE-1). It received low-frequency natural radio signals from the Sun, the planets and sources deep in space.

It was among eleven astronomy U.S. Explorers launched to Earth orbit between 1959-1975. The second RAE was launched in 1973.

Proton-4. The fourth of four heavy Proton astronomy satellites was launched by the Soviet Union in 1968. It recorded high-energy cosmic rays, gamma rays and electrons arriving from across the galaxy. The first Proton satellite had been launched in 1965.

OAO-2. NASA launched OAO-2 in 1968, one of four Orbiting Astronomical Observatories to peer deep into space between 1966-1972.

OAO-2 had four University of Wisconsin photometers and four one-ft.-diameter telescopes from the Smithsonian Astrophysical Observatory.

OAO satellites observed ultraviolet light from hot young stars, new sources of X-rays and interstellar gas.

Cosmos-208. Continuing its string of astronomy launches from 1967, the Soviet Union launched five astronomy satellites in 1968. The first that year was Cosmos-208.

Cosmos-215. The second 1968 astronomy satellite from the Soviet Union was Cosmos-215.

Cosmos-230. The third astronomy satellite launched by the USSR in 1968 was Cosmos-230.

Cosmos-251. The fourth 1968 astronomy satellite from the Soviet Union was Cosmos-251.

Cosmos-262. The fifth astronomy satellite launched from Russia in 1968 was Cosmos-262.

1969 Astronomy

Cosmos-264. Launched in 1969, Cosmos-264 was one of many astronomy satellites launched by the USSR over the decades.

OSO-5. The U.S. sent eight Orbiting Solar Observatories (OSO) to Earth orbit between 1962-75 to study the Sun. They recorded X-rays, gamma rays and ultraviolet and infrared light. OSO-5 was launched in 1969.

OSO-6. OSO-6 also was launched in 1969.

1970 Astronomy

Explorer-42 SAS-1 Uhuru. To look deeper into space, beyond our Solar System, NASA launched three Small Astronomical Satellites (SAS) between 1970-75.

The first was a satellite with three names launched from Italy's San Marco Range in 1970. It was known as Explorer-42, SAS-1 and Uhuru. Uhuru means freedom in Swahili.

Uhuru was among eleven astronomy Explorers launched to Earth orbit between 1959-1975.

After receiving X-rays from 200 sources in the sky, Uhuru drew the first X-ray map of the sky.

Uhuru detected a black hole in the constellation Cygnus in 1971. The satellite received X-rays and radio waves from the direction of a faint blue star which had a tiny, unseen, dark companion, more massive than a neutron star—the black hole.

Marjorie Townsend

The first woman ever to launch a spacecraft sent the U.S. Explorer-42 satellite (SAS-1) to Earth orbit from Italy's San Marco Platform in the Indian Ocean off Kenya on December 12, 1970.

She was Marjorie Townsend and she called the satellite Uhuru, a Swahili word for "freedom," in honor of Kenya's independence day December 12.

1971 Astronomy

Explorer-44 SolRad-3. Launched from Vandenberg in 1971 to measure X-rays and ultraviolet light from the Sun, Explorer-44 was among eleven U.S. astronomy Explorers launched to Earth orbit between 1959-1975.

Explorer-44 also was known as Solar Radiation satellite, or SolRad-3. SolRad-1 and SolRad-2 had been launched in 1965 and 1968.

OSO-7. OSO-7 was launched in 1971. It was among eight Orbiting Solar Observatories to orbit Earth for the U.S. between 1962-75, measuring X-rays, gamma rays and ultraviolet and infrared light.

Cosmos-461. The Soviet Union launched the astronomy satellite Cosmos-461 in 1971.

Tournesol. The French space agency, Centre Nationale des Etudes Spatiales (CNES), launched two small astronomy satellites on Diamant rockets from French Guiana in 1971 and 1975. Tournesol D2-A was sent to orbit in 1971.

Tournesol D2-A had five instruments to measure distribution of stellar hydrogen in Earth's atmosphere and observe the Sun's ultraviolet light.

Salyut-1. The Soviet Union's first Salyut space station carried astronomy telescopes to orbit in April 1971. They were a spectrogram telescope Orion-1 and a gamma ray telescope Anna-3.

1972 Astronomy

Explorer-48 SAS-2. NASA launched three Small Astronomical Satellites (SAS) to look beyond our Solar System between 1970-75. The second SAS was Explorer-48 launched from Italy's San Marco Range in 1972 to map gamma ray sources in the sky.

The first SAS had been launched in 1970 and the third would be launched in 1975. They were among eleven astronomy Explorers launched by the United States between 1959-1975.

Cosmos-484. The Soviet Union fired two Cosmos astronomy satellites to orbit in 1972. The first was Cosmos-484.

Cosmos-490. Another astronomy satellite launched by the USSR in 1972 was Cosmos-490.

OAO-3 Copernicus. OAO-3 astronomy satellite was launched by the U.S. in 1972, one of four Orbiting Astronomical Observatories to peer deep into space between 1966-1972.

OAO-3 was called Copernicus after Polish astronomy Nicolaus Copernicus (1473-1543).

OAO satellites observed ultraviolet light from hot young stars, new sources of X-rays and interstellar gas. OAO-3 had a ten-ft. ultraviolet telescope with a three-ft. mirror from Princeton University—the Princeton Experiment Package. Copernicus carried three small telescopes and an X-ray collimator from University College, London.

TD-1A. ESRO, the European Space Research Organization, was the predecessor of ESA, the European Space Agency. ESRO launched four space-physics satellites between 1968-1972. Last of the series was an astronomy satellite, TD-1A, launched from Vandenberg in 1972 with seven ultraviolet, gamma rays and X-rays instruments.

Prognoz-1. The USSR launched a series of nine Prognoz satellites to study the Sun between 1972-1983. The first Prognoz was lofted in 1972.

Prognoz-2. Also in 1972, the Soviet Union launched Prognoz-2 to study the Sun.

1973 Astronomy

Explorer-49 RAE-2. Explorer-49, known as Radio Astronomy Explorer (RAE-2), was launched by the U.S. in 1973 to receive low-frequency radio waves from the Sun, planets and sources deep in space. It was the second RAE among eleven Explorers launched between 1959-1975. The first RAE was launched in 1968.

Cosmos-561. The astronomy satellite Cosmos-561 was launched to Earth orbit in 1973.

Prognoz-3. USSR launched nine satellites to study the Sun between 1972-1983. The third Prognoz was launched in 1973.

Skylab-4. The last crew aboard the American space station Skylab in 1973 photographed Comet Kohoutek. They also photographed a solar flare for the first time from space.

Soyuz-13. By coincidence, in space at the same time as Skylab-4 were Soviet cosmonauts in the manned transport Soyuz-13.

Soyuz-13 was not a trip to a space station, but an independent astronomy flight. The cosmonauts used the Orion-2 ultraviolet camera to photograph the Sun and stars obscured by Earth's atmosphere. They made 10,000 spectrograms of 3,000 stars. They also looked for solar X-rays.

The Soyuz-13 cosmonauts were the first Soviets to spend December 25 in space. They did not report any unusual stars on Christmas Eve.

1974 Astronomy

ANS. The Astronomical Netherlands Satellite (ANS) was launched for Holland by the United States in 1974. The Scout rocket failed on its way to space from Vandenberg, leaving ANS-1 in an unexpected elliptical orbit.

The satellite carried a Cassegrain telescope to observe ultraviolet light from hot young stars. A Cassegrain telescope is a compact kind of reflector originated in the 17th century. ANS-1 also had two instruments detecting hard and soft X-rays.

Ariel-5. The fifth British satellite was called Ariel-5 and sometimes referred to as UK-5. It was launched by the United States for Great Britain in 1974 from Italy's San Marco Range in the Indian Ocean off Kenya. The small satellite housed six X-ray astronomy instruments, including two from the United States.

Salyut-4. The Soviet Union's fourth Salyut space station carried astronomy telescopes to Earth orbit in December 1974. They included infrared, X-ray and solar telescopes to observe the Solar System, Milky Way and other galaxies.

1975 Astronomy

Explorer-53 SAS-3. NASA launched three Small Astronomical Satellites (SAS) to look beyond our Solar System between 1970-75.

The third SAS was Explorer-53 launched from Italy's San Marco Range in 1975. It mapped X-ray sources in the sky.

The first and second SAS satellites had been launched in 1970 and 1972. The trio were among eleven astronomy Explorer satellites launched by the United States between 1959-1975.

OSO-8. The U.S. sent the eighth Orbiting Solar Observatory to Earth orbit in 1975 to study the Sun. It recorded X-rays, gamma rays and ultraviolet and infrared light.

Aura. French space agency CNES launched two small astronomy satellites on Diamant rockets from French Guiana in 1971 and 1975. Aura D2-B was launched in 1975 with five instruments to observe ultraviolet light from the Sun and measure stellar hydrogen in Earth's atmosphere.

Prognoz-4. The USSR sent nine satellites to Earth orbit to study the Sun between 1972-1983. The fourth Prognoz was launched in 1975.

Aryabhata. The Russians launched India's first satellite in 1975 from Kapustin Yar. The Aryabhata satellite, built by the Indian Space Research Organization (ISRO), conducted experiments in X-ray astronomy, aeronomics and solar physics.

COS-B. The U.S. launched the European Space Agency (ESA) gamma-ray astronomy satellite COS-B in 1975 from Vandenberg. The satellite carried one gamma-ray telescope.

SRATS Taiyo. The Japanese launched their own Solar Radiation And Thermospheric Structure satellite (SRATS) from the Kagoshima Space Center in 1975. In space, SRATS was renamed Taiyo, which means Sun.

Taiyo caught X-rays and extreme-ultraviolet light from the Sun. It also watched particles in the solar wind interact with Earth's magnetosphere, thermosphere and mesosphere. It measured density and temperature of electrons and ions.

The satellite was designed at the Institute of Space and Astronautical Science (ISAS) of the University of Tokyo.

1976 Astronomy

Prognoz-5. The Soviet Union launched nine Sun-observing satellites between 1972-1983. The fifth Prognoz was launched in 1976.

1977 Astronomy

Prognoz-6. USSR launched nine satellites to study the Sun between 1972-1983. The sixth Prognoz was launched in 1977.

HEAO-1. NASA launched three High Energy Astrophysical Observatories (HEAO) between 1977-1979 to record X-rays, gamma rays and cosmic rays from the Sun, stars and deep-space objects beyond the Milky Way galaxy.

HEAO-1 was sent to Earth orbit in 1977. Its X-ray telescope mapped 1,500 X-ray sources and a universal background gas enveloping galaxies. Its gamma-ray telescope caught the strange galaxy Centaurus-A emitting gamma rays.

Signe-3. Russia launched France's Signe-3, known in the USSR as Sneg-3, from Kapustin Yar in 1977 to observe X-rays and gamma rays.

1978 Astronomy

HEAO-2 Einstein. NASA's High Energy Astrophysical Observatories observed gamma rays, X-rays and cosmic rays between 1977-1979.

HEAO-2 was an X-ray telescope sent to orbit

300 miles above Earth in 1978. It was known as the Einstein Observatory in honor of the German-born U.S. physicist Albert Einstein (1879-1955).

The Einstein Observatory received X-rays from quasars billions of lightyears away. In fact, it was able to see X-rays coming from sources in deep space 1,000 times farther than previously seen. It also received X-rays from the planet Jupiter.

A lightyear is the distance light travels at 186,000 miles per second through space in one year, about 5.9 trillion miles.

Along with its X-ray telescope, HEAO-2 carried a television camera.

Prognoz-7. The USSR sent nine Sun-observing satellites to Earth orbit between 1972-1983. The seventh Prognoz was launched in 1978.

IUE. America's first satellite was Explorer-1 in 1958. NASA has continued launching satellites in the Explorer series for decades. More than 75 U.S. and international payloads have been sent to orbit in the program.

A notable example of an Explorer-series satellite is the International Ultraviolet Explorer (IUE) which went to space from Cape Canaveral in 1978, and sent down sufficient information from space for astronomers to write 1,400 articles for science journals.

IUE has a Ritchey-Chrétien telescope with an 18-in. aperture for ultraviolet images. Controlled by ground stations in Maryland and Spain, the telescope can be used in real time, as if it were on the ground. Tens of thousands of images have been made through the telescope, including ultraviolet light photos of planets, comets, interstellar dust and gas, supernovas, stars, galaxies and quasars.

IUE has been one of Earth's most productive telescopes, despite the fact it originally was expected to last only three years. Astronomers have used IUE to find halos of hot gas around our Milky Way galaxy, snap the first pictures of Comet Halley from space, peer at volcanoes on Jupiter's moon Io, and catch on film the intense blast of ultraviolet radiation from the exploding star, Supernova 1987a, 163,000 lightyears distant in the Large Magellanic Cloud galaxy.

IUE was a joint project of NASA, the European Space Agency and Great Britain's Scientific and Engineering Research Council (SERC).

The telescope is controlled 16 hours a day from Goddard Space Flight Center, Greenbelt, Maryland, near Washington, D.C. For the other eight hours each day, it is controlled from Villafranca Ground Station near Madrid, Spain.

Over the years, 800 astronomers at Goddard and 750 at Villafranca have used IUE. There are many larger telescopes on the ground, but being in space above Earth's hazy atmosphere improves images.

1979 Astronomy

HEAO-3. Three High Energy Astrophysical Observatories observed gamma rays, X-rays and cosmic rays for NASA between 1977-1979.

HEAO-3 was sent to orbit Earth in 1979 to observe gamma rays and cosmic rays.

Ariel-6. Great Britain's sixth satellite was the astronomy spacecraft Ariel-6, or UK-6, launched by the United States from Wallops Island, Virginia, in 1979. It carried two instruments to receive X-rays and one to collect cosmic rays.

CORSA-B Hakucho. Japan's Institute of Space and Astronautical Science (ISAS) launched its Cosmic Radiation Satellite (CORSA-B) in 1979 from Kagoshima Space Center to study X-rays arriving at Earth from deep space. In orbit, it was renamed Hakucho. Like most of Japan's space-physics and astronomy satellites, Hakucho was built by the electronics firm NEC.

Salyut-6. When the Soviet Union launched its sixth space station to Earth orbit in 1977, it was a new, second-generation manned spacecraft.

A second-generation Salyut had two spacecraft docking ports—one forward, one aft. With two docks, Salyut could hold a Soyuz manned transport and a Progress freighter in each, allowing more convenient resupply by freighter and short-term visits by cosmonauts.

Among the tasks carried out by cosmonaut crews aboard Salyut-6 was radioastronomy. A radiotelescope, known as KRT-10, was used to map our Milky Way galaxy in 1979. Its 33-ft. antenna was fished through the rear docking port and strung up outside the space station.

1980 Astronomy

Solar Max. Solar Max was launched from Cape Canaveral Valentine's Day 1980 on what then was to have been a two-year space flight to study the Sun's atmosphere, solar flares, and solar energy. Instead, it worked in space for ten years.

The Solar Maximum Mission (SMM or Solar Max) satellite was launched to monitor the Sun during peaks of solar activity. The satellite had gamma ray, ultraviolet and X-ray spectrographs, a chronograph and a radiometer.

The X-ray imaging spectrometer was an international project involving The Netherlands, Great Britain and the U.S.

Solar Max snapped pictures of 12,500 solar flares. It exposed 250,000 photos of the Sun's corona and recorded more than 100,000 images in ultraviolet light. Astronomers using Solar Max discovered ten sun-grazing comets. It was first to detect gamma rays from Supernova 1987a.

Solar flares are bursts of energy from sunspot areas in the Sun's corona, or outer atmosphere. They pump large amounts of electrically-charged particles and radiation into space. Lasting minutes to hours, solar flares release the energy of 10 billion one-megaton nuclear bombs, emissions which can disrupt communications on Earth.

From 1986-1989, scientists used Solar Max to look down to study the concentration of ozone in the Earth's mesosphere at altitudes of 34-47 miles with its ultraviolet spectrometer and polarimeter.

Repair in orbit. When it was sent to space, Solar Max sent down data for nine months—until a fuse blew and its attitude controls failed, leaving it unable to point correctly. Fortunately, Solar Max had been the first satellite designed to be retrieved by space shuttle.

NASA launched a space repair mission in 1984, sending shuttle Challenger with five astronauts to a 300-mi.-high orbit to grab the broken satellite and fix it. The crew included mission specialists James D. "Ox" Van Hoften and George D. "Pinky" Nelson. During their third day in space, Challenger was flown to within 200 feet of Solar Max.

Nelson and Van Hoften put on their spacesuits and went outside the crew cabin into the wide-open cargo bay. Nelson donned a manned maneuvering unit (MMU) and flew out to the satellite.

He tried to grasp Solar Max with a capture tool known as a Trunnion Pin Acquisition Device (TPAD). Three attempts to clamp TPAD onto the satellite failed.

Van Hoften then tried to grab the satellite using the shuttle's 60-ft. robot arm—the remote manipulating system (RMS)—but Solar Max started tumbling. The effort was called off.

While the astronauts slept that night, ground controllers were able to regain control of the satellite by radioing commands which ordered its magnetic torque bars to stop the tumbling. Solar Max went into a slow, regular spin.

Got it. The next day, Nelson and Van Hoften went back into the cargo bay to try again. This time they succeeded on the first try, using the RMS to place Solar Max on a cradle in the payload bay where they replaced the satellite's attitude control mechanism and the main electronics system of its chronograph instrument.

The refurbished Solar Max was pushed back overboard into its own orbit the next day, ending one of the most unusual rescue and repair missions in the history of the U.S. space program.

Return to duty. Solar Max resumed monitoring the Sun. Just a month after the overhaul in orbit, Solar Max recorded one of the largest flares ever seen erupting on the Sun. In fact, Solar Max recorded science data about the Sun for nearly ten years, covering almost one complete solar cycle. Highlights of the science harvest:

★Solar Max discovered the variability of astronomy's Solar Constant.

★Solar Max recorded more than 12,500 solar flares, measuring magnetic fields above the visible surface of the Sun for the first time.

★Solar Max confirmed the existence of neutrons in solar flares.

★The gamma ray spectrometer aboard Solar Max was first to detect gamma rays from a distant star exploding in a neighboring galaxy.

★Solar Max discovered ten comets skimming past or crashing into the Sun.

★The satellite's ultraviolet spectrometer uncovered increases in the level of high-altitude ozone above Earth north of the equator and decreases south of the equator.

Gamma Rays

Natural gamma rays are photons found in cosmic rays, probably generated by deep-space objects such as supernovas, quasars and neutron stars. Photons, produced in nuclear fission, are far more powerful than X-rays or light. Astronomers are not sure of the processes. Solar flares generate gamma rays, but Earth's atmosphere absorbs the radiation before it reaches the ground.

Nuclear blindness. The USSR had built nuclear reactors into several dozen satellites since the 1960s. The reactors generated electricity for satellite electronics, often radar sweeping the oceans for U.S. warships.

To save weight, satellites were designed to shield only their own equipment. Most radiation from reactors is gamma rays.

Solar Max was blinded eight times a day by gamma rays from those reactors in space. During those orbits, data could not be gathered.

The satellite's digital data storage memory would become saturated by positron-generated gamma rays eight times a day. That would blind Solar Max for the rest of each orbit until the data could be transmitted to the ground.

The energy flow from a nearby satellite nuclear reactor would be 50 times greater than flux from celestial sources. Knowing that, scientists were able to detect a satellite nuclear reactor in orbit up to 1,500 miles altitude. The fact that radiation from nuclear reactors in USSR spy satellites sometimes blinded Solar Max was kept secret for years by the U.S. government.

Positrons. Another problem for Solar Max was electrons and positrons, which are positively-charged electrons. The charged particles often were trapped for several minutes by Earth's magnetic field. Some positrons, colliding with ordinary matter in the satellite, were annihilated, yielding even more gamma rays to bother Solar Max.

Falling. NASA knew Solar Max was falling in 1989. The satellite was down to an altitude of 206 miles October 31, to 177 miles by November 17 and to 166 miles by November 22.

Goddard controllers started to put Solar Max to sleep November 17, shutting off electricity to three science instruments needing a stable orbit to work. Technicians radioed a command November 22 to jettison the high-gain antenna. Other instruments continued to record data to be sent down as Solar Max passed over ground stations.

Battery-charging solar panels were jettisoned November 24, ending science operations.

The 5,000-lb. satellite ended ten years of duty December 2, 1989, as it burned in the atmosphere over the Indian Ocean near Sri Lanka. There were no reports of debris falling on land.

Prognoz-8. USSR launched nine satellites to study the Sun between 1972-1983. The eighth Prognoz was launched in 1980.

1981 Astronomy

Astro-A Hinotori. Japan's Institute of Space and Astronautical Science (ISAS) launched its Astro-A astronomy satellite in 1981 from Kagoshima Space Center. The satellite observed X-rays emitted from 500 flares on the Sun. In orbit, Astro-A was renamed Hinotori.

1982 Astronomy

Salyut-7. The USSR's seventh Salyut space station carried astronomy gear to orbit in April 1982, including an improved X-ray telescope.

1983 Astronomy

IRAS. The Infra-Red Astronomical Satellite, one of the most sophisticated astronomy satellites ever orbited, was a cooperative project between the Netherlands, the United States and Great Britain.

IRAS was built by a Dutch manufacturer with help from three American firms. Like the International Ultraviolet Explorer (IUE) launched five years earlier in 1978, IRAS carried a Ritchey-Chrétien telescope with a 22-in. aperture.

The IRAS telescope was composed of 62 infrared detectors chilled to –270 degrees Celsius by 500 quarts of liquid helium. The satellite also had a spectrometer and two other detectors seeing different wavelengths of infrared light.

NASA launched IRAS from Vandenberg in January 1983 to an orbit 550 miles above Earth from where its telescope could see infrared light coming from more than a quarter-million deep-space objects, including 15,000 galaxies. In fact, during a year of work in space, IRAS created the most comprehensive deep-sky map to its time.

A group of British and Canadian astronomers used IRAS to chart 2,000 galaxies—some were 450 million light years away. It was the most comprehensive map to that time of deep space close to the Solar System.

IRAS discovered a ring of dust around the Solar System, clouds of graphite dust in interstellar space, possible planets orbiting the star Vega, five comets in our Solar System, and unidentified cold objects strewn across the Universe.

Results from IRAS challenged standard theories of how galaxies formed in space. No model of galaxy formation validates the existence of large clusters of galaxies and the vast voids between the clusters. On the IRAS map, covering so much of the Universe, there simply were too many galaxy clusters, and the clusters were just too large, to fit theories to date.

One popular theory had proposed something called "cold dark matter" (CDM). Supposedly, CDM explains why much less matter is actually seen in the Universe than the amount required to keep the Universe from collapsing. The CDM theory assumes the tug of gravity can be detected from some sort of unknown, invisible matter.

Proponents say such cold, dark matter would not be like anything found on Earth or in the Sun. Unfortunately for the proponents, their CDM model didn't explain the very large galaxy clusters.

IRAS made sky maps of infrared light coming from deep space for 11 months until its supply of cooling helium ran out.

Astro-B Tenma. Japan's Institute of Space and Astronautical Science (ISAS) launched its Astro-B astronomy satellite in February 1983 from Kagoshima Space Center. The satellite observed X-rays from stars, galaxies and nebulas. In orbit, Astro-B was renamed Tenma.

EXOSat. The European X-ray Observatory Satellite (EXOSat) was launched in May 1983 for the European Space Agency from Vandenberg.

EXOSat was in a highly elliptical orbit, from 200 to 120,000 miles from Earth, where it observed 2,000 X-ray sources, including stars, planets, quasars, supernova remnants, neutron stars, black holes and galaxy groups. EXOSat was controlled by the European Space Operations Center (ESOC), Darmstadt, Germany.

Prognoz-9. The USSR sent a series of nine Prognoz satellites to orbit to study the Sun from 1972-1983. Prognoz-9 was launched in July 1983.

Astron. The Russians modified their Venera spacecraft, used often to explore Venus, into the Earth-orbiting astronomy satellite Astron.

The first Astron, fired to space in March 1983, was equipped with a French ultraviolet telescope to observe galaxies. It also had an X-ray spectrometer and telescope. Astron flew a highly-elliptical orbit which carried it outside the interfering bands of Van Allen radiation belts surrounding Earth.

The Roentgen-style X-ray telescope aboard Astron was referred to as SKR-02 by the Soviets.

The ultraviolet telescope was a Russian-French project known as Spika. It was a double-reflecting Ritchey-Chrétien telescope designed at the Crimean Astrophysical Observatory. Spika's collecting surface was larger than the ultraviolet sensor in 1972's American OAO-3 Copernicus satellite. Spika had a 24-ft. focal length.

Astron was focused on the constellation Taurus which displays two concentrations of stars and the crab-like nebula which contains a pulsar.

Radioastron. Astron became the standard spacecraft for general-purpose Russian astronomy satellites carrying large telescopes and heavy science projects. For example, Radioastron was an observation satellite with a 33-ft. antenna to gather radio waves from deep space for five years. The radioastronomy satellite searched the radio portion of the electromagnetic spectrum.

The European Space Agency, the United States and seventeen other nations are participating in the Radioastron project. American astronomers helped plan the launch. The U.S. loaned recorders to capture data. NASA will help track the satellite.

Astronomers analyze data from the satellite for galaxies, black holes, neutron stars and quasars.

RASCAS. RASCAS is an even bigger radiotelescope satellite with 100-ft.-diameter antenna. The Radioastron uses the technique of interferometry in conjunction with telescopes on the ground in the Crimea to study natural radio signals received from deep space.

Granat. Another satellite built on the larger general-purpose Astron spacecraft model was Granat, a French-Russian telescope to study X-ray stars in deep space, following up on work done by Gamma-1 and the Roentgen satellite.

Golden Era In Space

U.S. Vice President Dan Quayle, then-chairman of the National Space Council, noted in 1990 that mankind was entering a golden era in space exploration.

U.S. help with Russia's Radioastron astronomy satellite was a small, significant step in evolving cooperation in space, Quayle said. He recalled that renowned Soviet physicist Andrei Sakharov, before he died, had asked for American involvement in the Radioastron project.

1987 Astronomy

Kvant-1. The Russians housed some heavier telescopes in a modified shell of the spacecraft used otherwise for Soyuz manned transports and Progress unmanned space freighters.

For example, the small Kvant module attached to the orbiting Mir station in 1987 was built in a Progress-style capsule. It is 19 feet long and 13.6 feet in diameter. Kvant is Russian for Quantum.

Kvant-1 is an observatory created by Russian, British, Dutch and West German astronomers. It has an X-ray telescope with electronic equipment from the European Space Agency.

It also has an ultraviolet telescope developed by Switzerland and the USSR. The Soviet Academy of Sciences, the European Space Agency, and West Germany each supplied special spectrometers. The detector in the hard-X-ray spectrometer is effectively six times larger than the earlier U.S. HEAO-3 satellite.

Great Britain and the Netherlands supplied a telescope for electronic images. Kvant-1 is used to gather data about remote reaches of the Universe. Such an observatory above Earth's clouds and dusty atmosphere sees the Sun, planets, Milky Way galaxy and deep space much more clearly.

Aelita. Another astronomy satellite built in a Progress-style spacecraft was Aelita. The Russian-French satellite was designed to study very-short-wavelength signals arriving from deep space.

Astro-C. Japan's Institute of Space and Astronautical Science (ISAS) launched Astro-C in 1987 from Kagoshima. It used instruments from the U.S. Los Alamos National Laboratories and the United Kingdom's University of Leicester and Rutherford Appleton Laboratory to observe X-rays from stars, galaxies and nebulas. Astro-C also carried a gamma ray burst detector.

1988 Astronomy

Mir. A block of X-ray detectors forming the lens of a Dutch telescope deteriorated in the harsh space environment on the outside wall of Kvant-1 in 1988. The Mir cosmonauts had to spacewalk twice that year to repair the antenna. During the first spacewalk, a wrench broke and new tools had to be sent up from the ground for a later spacewalk.

Kvant-1 was the small astronomy module attached to Mir space station in 1987. The larger Kvant-2 would be added to Mir in 1989.

1989 Astronomy

Kvant-2. The Soviet Union added another astronomy module to its orbiting space station Mir in 1989. Kvant-2, built in a spacecraft shell of the original Mir style, relieved overcrowding by doubling the size of the space station.

Kvant-1 had been added to the station in 1987. Kvant is Russian for Quantum.

Kvant-2 astronomy gear included three West European telescopes and one Russian instrument.

COBE. Another Explorer satellite was sent to space in 1989, this time to find out why the Universe has a lumpy consistency.

America's ultra-sensitive Cosmic Background Explorer (COBE) satellite was launched to a 570-mi.-high orbit in November from Vandenberg Air Force Base, California. It flew a two-year mission over Earth's poles and along the terminator line between night and day.

The deep space observatory was to see back to a possible beginning of the Universe, to study the Big Bang primeval explosion some think sent the Universe expanding outward 15 billion years ago.

COBE looked across the Universe, searching for explanations of why the far flung star-clouds we call galaxies are distributed so unevenly across the cosmos. Three receivers recorded weak radiation supposedly left over from the Big Bang explosion.

Two of COBE's three instruments looking at the spectrum of electromagnetic energy arriving at Earth from deep space saw infrared light. One saw microwave energy and COBE completed a scan of the entire sky for microwave radiation arriving at Earth from across the Universe.

Big Bang. Many scientists say the Universe exploded into existence 15 billion years ago in a hot, dense fireball which instantly started expanding in all directions. That's the so-called Big Bang, which should make the Universe smooth in its expansion, with an even distribution of matter everywhere. The problem is, the Universe looks uneven or lumpy, with great clusters of galaxies dotted across giant dark voids.

COBE sent down the best pictures ever seen of the center of our Milky Way galaxy. NASA combined near-infrared images from the satellite to make individual pictures, clear of the dust that usually blocks the view of the heart of our galaxy, exposing millions of stars in this galaxy.

COBE photographs of the Milky Way galaxy revealed the dust from which planets and galaxies are formed. They showed radiation from cold interstellar dust. The images highlighted differences in the distribution of stars, and the dust and gas clouds between stars, across the galaxy.

Cold tank. COBE was 18 feet long and eight feet in diameter on Earth. In space, the observatory expanded to 27.5 feet diameter. The spacecraft and its instruments—FIRAS, DMR and DIRBE—were built by NASA's Goddard Space Flight Center.

COBE carried those instruments in a large vacuum-insulated tank at the very cold temperature of minus-457 degrees Fahrenheit. Similar to one orbited in the Infrared Astronomical Satellite (IRAS) in 1983, the tank kept cold by evaporation of liquid helium.

Ground controllers use NASA's TDRS tracking and data relay satellite to communicate with COBE. Observed data was fed into tape recorders in the satellite, then radioed to the ground, analyzed at Goddard, and used to draw maps of the sky.

Hipparcos. Scientists weren't enthusiastic in 1966 when Pierre Lacroute of France's Strasbourg Observatory suggested a satellite could measure with a then-unbelievable precision.

He figured a precision satellite in orbit would be a great tool for drawing up a precise catalog of 120,000 stars visible from above Earth.

Lacroute thought a precision of one-hundredth of an arc second was possible. His fellow astronomers doubted it could be accomplished as they seldom had been able to achieve a precision better than one-tenth of an arc second. In fact, data from Southern and Northern Hemisphere observatories differed by arc seconds on any star.

It took the European Space Agency time to get the project off the ground, but Lacroute won the argument and Hipparcos was approved by ESA in 1980. Hipparcos was designed to a cost of $360 million and construction began in 1984.

The satellite seemed expensive to some since it had only one small reflecting telescope. Of course, it was very sophisticated. In the end, Hipparcos could measure positions of 120,000 stars for precise new star maps. It ended up with a precision ten times better than even Lacroute had imagined.

Twenty years after its proposal, Hipparcos assembly was completed in 1986 at a technical center in Noordwijk, The Netherlands. Hipparcos rode to space in 1989 on an Ariane rocket.

The Rocky Road To Space

A strange thing happened to the star-mapper on its way to space. Hipparcos almost didn't make it through a meteorite bombardment in Italy on its way to Europe's launch pad in the South America.

The rocky road to space for Hipparcos started in 1966 when Hipparcos was suggested by Pierre Lacroute of France's Strasbourg Observatory.

When the European Space Agency finished building Hipparcos in 1986 in the Netherlands, the shiny new satellite was packed into a nitrogen-filled container and sent to the Aeritalia company in Turin, Italy, to wait for a ride to space.

The satellite was to have been launched in July 1988, but Ariane troubles delayed ESA launches. Rather than ship the satellite to French Guiana, ESA kept Hipparcos in the Turin warehouse. That's where a really odd coincidence appeared.

On May 18, 1988, a two-lb. meteorite smacked into the warehouse parking lot beside a tiny Fiat automobile parked there, missing Hipparcos by less than 1,000 feet. Excited astronomers rushed out to the space rock, grabbed the biggest remaining fragment, a one-lb. chunk they named Aeritalia-A, and hauled it into their lab for study.

Unscathed, Hipparcos was shipped to Toulouse, France, for passage to Kourou and launch to space on Ariane flight V-33.

Hipparcos stands for High Precision Parallax Collecting Satellite. It is named for Hipparchus, a Greek astronomer who lived about 130 B.C.

Hipparcos was launched to an elliptical transfer orbit. It ranged from 125 to 22,300 miles above Earth. Some 37 hours after launch, Hipparcos was to supposed to kick itself into a circular orbit.

The small solid-fuel-rocket "kick motor" which was supposed to push Hipparcos to a stationary orbit over the equator above Africa failed, stranding the star-mapper in an elliptical orbit. European Space Agency engineers were aghast.

For the next week, commands were radioed to the satellite to induce its kick motor to ignite each time the satellite sailed through the high point of its orbit. Each time the command failed.

Frustrated engineers used tiny thrusters designed for minor orbit corrections to shift the satellite a bit higher, to a safer orbit from 315-22,300 miles altitude, from which it wouldn't fall for a time.

Hipparcos was built to spin so its telescope could scan the entire sky. Fortunately, ESA engineers were able to calculate how the satellite could complete most of its assignment from its low elliptical orbit. Hipparcos still would sweep 80 percent of the sky.

Hipparcos finally was able to get to work, cataloging the exact positions of 80 percent of the planned 120,000 stars. Data from Hipparcos was received by a ground station at Odenwald, Germany. ESA added ground stations at Perth, Australia, and Kourou, French Guiana, to compensate for the irregular orbit and help Hipparcos calculate data.

The satellite took in so much information it had to send its data continuously and very rapidly to Earth. It transmitted 2,000 measurements per second, eventually totaling more than 200 billion star positions collected at ESA's European Space Operations Centre at Darmstadt, near Frankfurt.

The data was dispatched across Europe to two separate groups of scientists who double-checked each other's mapping. The data was processed by two different computers, which will compose the final star catalog, showing distances to the stars and their directions of movement. It is expected to be available by the end of the 1990s.

Distances between stars in binary pairs can be estimated from the data. The mass of a star can be deduced from the data. That information helps astronomers calculate other physical properties of stars. Previously, useful star movement data had been available for only 1,500 stars.

Second mapper. Along with the main telescope, there was a less-precise star mapper on board to position the satellite. It always is directed toward a specific area of the sky so the satellite can keep track of its own location in space.

As a fringe benefit, that tracker sees and maps an additional 400,000 stars with a precision of about one-30th of an arc second, about the same precision achieved today by telescopes on the ground. It also measures the intensity of light coming from the stars in two ranges of the electromagnetic spectrum.

Astrometry

Astronomers find the distance from Earth to a star by measuring the star's position from each side of Earth's orbit. Finding the distance to a star is an astrometric measurement.

The diameter of Earth's orbit around the Sun is 184 million miles. During the time it takes Earth to get from one side of its orbit to the other, a star appears to move slightly—like an object held at arm's length seems to jump back and forth as you look with one eye and then the other.

This apparent motion of a star is small, but sufficient for astronomers to sketch a triangle with a 184 million mile base.

Trigonometry then is used to calculate the length of the other two sides of the triangle.

Hipparcos was the first European astrometry satellite. From its position above Earth's murky atmosphere, Hipparcos improved on earlier astrometric measurements as it swept the whole sky every ten hours. Hipparcos measurements will improve man's yardstick for measuring the size of the Universe, and deepen understanding of the evolution of stars, galaxies and the Universe.

1990 Astronomy

RoSat. German scientist Wilhelm Conrad Roentgen discovered X-rays in 1895. Today, a Roentgen is the international unit of measurement for X-rays and gamma rays.

Roentgen Satellite (RoSat) was a German spacecraft containing an X-ray telescope. It was to have gone to Earth orbit in a U.S. shuttle, but the 1986 Challenger disaster knocked it out of the schedule. Instead, the satellite was boosted 360 miles into space in 1990 on an unmanned rocket from Cape Canaveral.

RoSat was similar to NASA's Einstein Observatory (HEAO-2) launched in 1978, but with more sensitive detectors. In fact, RoSat was one thousand times more sensitive than any previous X-ray telescope. Its telescope equipped the satellite to carry out the first full-sky survey of deep-space objects emitting X-rays.

Sky survey. The robot observatory was the most ambitious X-ray project up to that time.

It was the largest imaging X-ray telescope ever built to map sources of X-rays arriving at Earth from deep space.

While astronomers previously had logged 5,000 objects emitting X-rays, they expected to find 100,000 more X-ray objects through RoSat.

In addition, RoSat focused on 1,400 targets of special interest.

RoSat's telescope focal length was 7.78 feet with a maximum aperture of 835 millimeters. While the satellite kept its electricity-generating solar panels pointing at the Sun, the telescope swiveled to survey the entire sky every 180 days.

RoSat also carried a British telescope, known as the Wide Field Camera, to observe extreme-ultraviolet light sources from deep space.

RoSat is controlled by the Oberpfaffenhofen ground control center near Munich, Germany.

Neutron star. NASA's High Resolution Imager aboard RoSat sent pictures to Earth of Cygnus X-2, a neutron star orbiting a normal stellar companion 3,000 lightyears from Earth.

A second photograph showed a supernova remnant, Cassiopea-A, in the constellation Cassiopea. It may have resulted from a supernova explosion in the Milky Way galaxy 320 years ago.

The X-rays in the images came from a hot plasma cloud of several million degrees produced when the blast wave from the supernova struck the surrounding interstellar material. The supernova remnant is 9,000-10,000 lightyears from Earth.

Galaxy clusters usually give off strong X-rays. A third X-ray photograph showed a cluster of galaxies known as Abell 2256.

The X-rays in the image came from hot gas swept out of colliding galaxies and accumulating between the galaxies. Such gasses are thought to be tens of million degrees in temperature.

X-rays are very high natural energy emissions—greater than one million degrees. Rich sources of X-rays are supernova remnants, galaxy clusters, quasars and binary star systems containing neutron stars or black holes.

X-rays let astronomers better understand how stars and galaxies are formed

Astro-1. Ultraviolet and X-ray telescopes flew to orbit in the U.S. space shuttle Columbia in December 1990, capturing extraordinary images and spectra of a variety of deep space objects, despite two crippling computer failures.

Astro-1 was an array of telescopes to study the Universe in ultraviolet light. A broad-band X-ray telescope known as BBXRT also was aboard Spacelab in the shuttle's cargo bay.

Astro-1 photographed stars and other deep-space objects impossible to see with visible-light telescopes on the ground. Astronomers returned to Earth with a shuttle load of extraordinary photos, including ultraviolet images of a spiral galaxy, a globular cluster and a supernova remnant.

The three Astro-1 ultraviolet instruments were:

★Hopkins Ultraviolet Telescope (HUT), which searched for faint astronomical objects such as quasars, active galactic nuclei and supernova remnants. It looked at the outer planets of the Solar System for auroras and the interactions of each planet's magnetosphere with the solar wind.

★Wisconsin Ultraviolet Photo-polarimeter Experiment (WUPPE) which studied polarization of ultraviolet light from deep-space objects such as quasars, hot stars and the nuclei of galaxies.

★Ultraviolet Imaging Telescope (UIT), which exposed 70mm film images of faint objects as it searched out the star content and history of star formation in galaxies, the nature of spiral structure and non-thermal sources in galaxies.

BBXRT. The broad-band X-Ray telescope (BBXRT) aboard Columbia actually was two imaging telescopes with cryogenically-cooled lithium-drifted silicon detectors.

High Energy Astronomy

Stars and galaxies shine at many different wavelengths, depending on chemical makeup and nuclear reactions.

The higher the energy, the shorter the wavelength of light and electromagnetic energy given off. X-rays fall between ultraviolet light and gamma rays on the energy spectrum.

Visible light from deep space reaches Earth's surface, but high-energy radiation such as X-rays, gamma rays and ultraviolet light mostly do not pass through the atmosphere so are invisible to observers on the ground.

X-ray satellites can look for high-energy X-rays coming from stars, black holes and those hot clouds of gas and dust which are remnants of supernovas, or exploding stars.

X-ray radiating bodies in deep space, existing at temperatures of one million to 100 million degrees, are many billions of miles away across the Universe. Such objects release enormous amounts of energy, sometimes revealing evidence of black holes and other strange objects unseen in visible light. Black holes are collapsed burned-out cinders of massive stars with gravity so intense visible light cannot escape.

Gas and dust being sucked into a black hole is accelerated by the extraordinary gravity of the collapsed star, heating the particles to extremes and sending off showers of X-rays which can be detected by satellites in Earth orbit.

By studying X-rays arriving at Earth from deep space, astronomers better understand how stars, galaxies and the Universe itself formed.

BBXRT viewed active galaxies, clusters of galaxies, supernova remnants and stars. BBXRT measured the amount of energy of each X-ray detected. It studied Supernova 1987a, the brightest exploding star seen by modern Earth astronomers.

It was the first U.S. shuttle flight in five years dedicated entirely to science. The four astronomer-astronauts aboard Columbia for the eight-day flight 200 miles above Earth studied quasars, binary stars, pulsars, black holes, galaxies and high-energy stars, including Supernova 1987a, the nearby supergiant star Betelgeuse, radio-quiet quasar Q1821, spiral-poor galaxy cluster Abell 2256, NGC 1633, NGC 1399 in the constellation Fornax, and Q1821+64.

Despite early control problems in orbit, the telescopes in the cargo bay ended up examining 135 deep space targets during 394 observations.

Gamma. Gamma was a joint Russian-French-Polish astrophysics satellite launched in 1990. It carried a heavy telescope to study gamma rays and X-rays arriving at Earth from deep space.

1991 Astronomy

Solar-A Yohkoh. The Sun may look like a hot old yellow ball sitting out in space doing nothing, but things aren't what they seem.

Our temperamental star goes through periods of intense activity followed by quiet periods. The ebb and flow over eleven years is the Sunspot Cycle.

During periods of peak activity, the Sun fires off magnificent high-energy solar flares which affect the transmission of radio and television signals on Earth. To look at these flares from a better vantage point, Japanese and American scientists placed an eye in the sky called Solar-A.

Solar-A was Japan's fourteenth science satellite. It was the only satellite dedicated to solar flares during very-high activity on the Sun in 1991-92.

Japan's Solar-A satellite was launched to a low orbit 344 miles above Earth in August 1991 from Kagoshima Space Center. In space, following Japanese custom, it was renamed Yohkoh, which means Sun ray, Sun beam or Sun light.

Yohkoh stares straight at the Sun, looking for solar flares on the face of our star. The satellite sees X-rays and gamma rays from the Sun.

Yohkoh is a project of Japan's Institute of Space and Astronautical Science (ISAS). The Japanese built the satellite's flight electronics, computer and power system. NASA scientists supplied the optics and detector for a soft X-ray telescope aboard Yohkoh. Other instruments are a Japanese hard X-Ray telescope, a wide-band spectrometer from Japan and Great Britain's Bragg crystal spectrometer.

Soft X-rays. The Sun's magnetic field ensnares hot corona gases in loops called magnetic bottles. Loop brightness reveals its temperature and gas density. NASA's soft X-ray telescope photographs intricate filaments of the corona extending far above the Sun's surface. Soft X-rays are weaker X-rays containing less energy. Two other instruments observe hard-X-ray flares.

To accomplish its objectives, the Japanese needed to design a satellite with three-axis stabilization. Solar-A is the first ISAS satellite with three-axis stabilization precise enough to point a high-resolution telescope.

During each 98-minute revolution around Earth, Yohkoh stores eighty megabytes of data in a magnetic-bubble recorder, then "dumped" by radio to NASA's Deep Space Network ground stations.

SARA. Not all astronomy satellites are billion-dollar government projects. A small craft named SARA was built by a club of amateur astronomers at a school in France and launched to orbit in 1991.

SARA is an acronym for Satellite for Amateur Radio Astronomy. The club was ESIEESPACE at France's Ecole Superieure d'Ingenieurs en Electrotechnique et Electronique (ESIEE).

For six years before SARA, the club had built payloads for suborbital rocket and balloon flights. The club's first-of-its-kind amateur radioastronomy satellite was launched to Earth orbit in July 1991 by a European Space Agency Ariane rocket.

SARA should not be confused with amateur radio satellites for communications among hams on the ground. Rather, SARA is a radioastronomy satellite listening for natural high-frequency (hf) radio signals from Jupiter.

SARA was built with off-the-shelf consumer equipment, rather than space-qualified or military-spec components. It will work years in orbit.

Jupiter is the fifth planet from the Sun and may itself have been a prototype Sun which wasn't big enough to become a star.

It is more than five times farther from the Sun than the Earth, at a distance of 483.6 million miles. It orbits the Sun every 11.9 years.

Jupiter is the largest of nine planets circling the Sun. In fact, Jupiter alone is more than two-thirds of the total mass of all planets in our Solar System. It's 318 times the mass of Earth. Jupiter is a big gas bag, with a relatively low density.

Although a thousand times smaller than the Sun, Jupiter is a big planet—as big as 1,317 Earths. Still, Jupiter wasn't big enough. If the planet had been several times more massive, it might have become a star as the pressure and temperature at its core triggered nuclear fusion.

Astronomers have known that Jupiter somehow sends out naturally-generated radio noise in the hf part of the electromagnetic spectrum. Unhappily, Earth's atmosphere, and sometimes our planet itself, block the signals from Jupiter, making them hard to study from the ground.

However, a receiver in a satellite orbiting above Earth's atmosphere can hear radio waves. Other satellites have measured Jovian emissions, but not for long periods. For instance, Voyager-1 tuned in briefly as it flew by Jupiter in 1979, but its receiver was noisy and Voyager could hear only the strongest signals coming from Jupiter.

Jupiter has at least sixteen moons. The radio signals sweeping Earth are generated by the interaction between Jupiter and its moon Io. The signal arrives at Earth as a beam of energy, appearing four minutes earlier each day, taking two hours to swing across Earth. Before SARA, no satellite had listened to Jupiter's hf signal during a time of peak solar activity.

Electromagnetic energy from Jupiter is very strong. SARA needs only three fifteen-ft. antennas to snare the waves. The antennas were made of 100-mm-wide steel tapes rolled up for launch, then unrolled in orbit. Their arrangement allows waves to be measured, no matter which direction the satellite is turned. One of the antennas doubles as a telemetry antenna, transmitting data to Earth.

SARA's receiver listens to eight blocks of hf frequencies between 2-15 MHz. It switches among the eight and the three antennas many times during a 150-second period. Averaging the signal strength smoothes out any storm peaks on Jupiter.

SARA's receiver is sensitive enough to detect galactic background noise which has a constant level of strength. Galactic noise gives the amateur astronomers on the ground a point of reference when Jupiter and the Sun are silent.

The computerized, solar-powered satellite circles Earth every 100 minutes in a polar orbit 478 miles above Earth, continuously sending data by teletype to the ground with a one-watt amplitude-modulated (AM) transmitter on 145.955 MHz.

1992 Astronomy

EUVE. NASA's four Extreme Ultraviolet Explorer (EUVE) astronomical telescopes were sent to orbit in June 1992 from Cape Canaveral.

Astronomers wanted to explore area of the electromagnetic energy spectrum, between X-rays and ultraviolet light, known as extreme-ultraviolet. They mounted UV telescopes in an Explorer module, a piece of standardized satellite hardware.

EUVE surveyed the entire sky, then mapped the brightest individual sources of extreme-ultraviolet light in deep space. A definitive sky map of the sky in the extreme-ultraviolet spectrum and a catalog of extreme-ultraviolet sources was drawn.

1993 Astronomy

Spartan. U.S. shuttle Discovery in 1993 carried the third Spartan satellite to orbit for astronomy work with two telescopes to study the outer atmosphere of the Sun and the solar wind.

The telescopes had been flown to space three times before, but only on sounding rockets from which they could gather fifteen minutes worth of data. Inside the retrievable Spartan, the telescopes spent fifty hours observing the Sun.

Payloads launched on sounding rockets do not go into orbit. They fly hundreds of miles into space only to fall back to Earth in fifteen minutes.

Spartan's instruments measured the temperature, and density of invisible protons, electrons and ions surging through the Sun's corona.

The solar wind is an invisible stream of charged particles of matter which start in the Sun's ultra-hot corona and blow through the Solar System at more than a million mph.

A gust of solar wind from an intense storm on the Sun can trigger major events on Earth. Solar storms have forced electrical power black outs, worldwide communications flickers and Northern Light displays all the way down to Florida.

Ellen Ochoa, NASA's first female Hispanic astronaut, operated the Discovery's robot arm to deploy and retrieve Spartan from the orbiter's cargo bay in 1993. While away from Discovery, Spartan recorded its telescope findings on tape recorders.

The first Spartan flew in Discovery in 1985. The second Spartan and its X-ray telescope to study Comet Halley were destroyed in 1986 in the Challenger explosion. Spartan will fly several more shuttle missions in the 1990s.

Astro-D. Japan's Institute of Space and Astronautical Science (ISAS) launched its Astro-D satellite for X-ray astronomy in 1993 from Kagoshima Space Center.

ALEXIS. The tiny U.S. astronomy satellite ALEXIS (Array of Low Energy X-ray Imaging Sensors) was launched to Earth orbit on a small Pegasus rocket April 25, 1993, to survey the entire sky for X-rays for a year.

The miniature 240-lb. satellite was 25 inches in diameter and 41 inches long. Unfortunately, its solar panel and communications antenna may have been torn off during launch. Scientists at Los Alamos National Laboratory were unable to contact ALEXIS, except for a brief burst of radio signal June 2.

Great Observatories

NASA's four orbiting Great Observatories in space are the Hubble Space Telescope (HST), the Compton Gamma Ray Observatory (GRO), the Advanced X-Ray Astrophysics Facility (AXAF) and the Space Infrared Telescope Facility (SIRTF).

Hubble and GRO are in orbit and NASA plans to launch AXAF about 1999, but SIRTF is off the schedule due to budget problems.

The high-tech satellites with strange-sounding acronyms, above the blurring atmosphere of Earth, open previously-unseen views of the Universe. Hubble works in visible and ultraviolet light. GRO measures gamma rays. AXAF will receive X-rays. SIRTF would record infrared light images.

All of the Great Observatories would be visited from time to time by maintenance crews in U.S. space shuttles from Earth and later from the U.S.-international space station Freedom.

Hubble. Hubble Space Telescope was the first of the Great Observatories. HST was the largest telescope ever sent to space when it was dropped overboard from U.S. shuttle Discovery in 1990 at an altitude of 380 miles above Earth.

Called the most powerful optical instrument ever built for use in space, the telescope was a joint project of the European Space Agency and NASA. Astronomers predicted the telescope would show astronomers far-flung objects fifteen billion lightyears away from our planet.

Looking at visible light and ultraviolet light, Hubble was designed to expand the observable volume of the Universe by several hundred times.

The powerful instrument was expected to relay the best pictures yet of the Universe. It would return pictures of the Universe much more distinct than those seen through telescopes on the ground.

Hubble has a 94.5-in. mirror, five sensitive cameras and other instruments to revolutionize optical astronomy from above Earth's obscuring atmosphere. NASA said Hubble was powerful enough to see the edge of the Universe.

The wide-field planetary camera alone was said to be so sensitive that, on Earth, it could see a baseball 200 miles away. A dime could be seen 20 miles away. In space, it would detect objects 100 times fainter than those visible from Earth's surface, with 10 times the sharpness or resolution.

Mirror. Within weeks after launch, Hubble was sending down photos of deep-space objects in detail never seen before. Unfortunately, even though the detail was better than Earth telescopes could achieve at that time, it was not as exquisite as had been predicted. The problem was faulty grinding of the telescope's main mirror.

Hubble mirrors were ground early in the 1980s by Perkin-Elmer Corp. of Connecticut. The company used a measuring instrument known as a reflective null corrector to position and check its grinder, which shaped the concave Hubble mirror to reflect visible and ultraviolet light.

An investigating panel found a tiny 1.3-mm spacing error in a Perkin-Elmer grinder. The faulty measuring device contained 1.3 mm-thick spacing washers, the same size as the error. Considered huge for high-tech optics, the 1.3-mm error was likened to a room built 3 feet too long.

Old records indicated a test in the early 1980s had found the error after the mirror was manufactured, but Perkin-Elmer engineers placed complete faith in their null corrector and disregarded the test results. NASA says it was not aware of the test results.

Perkin-Elmer Corp. was bought in 1989 by

Hughes and now is known as Hughes Danbury Optical Systems Inc. In 1991, NASA gave Hughes Danbury Optical Systems Inc. a contract to grind mirrors for another Great Observatory, the Advanced X-Ray Astrophysics Facility (AXAF) to be launched in 1998.

How fuzzy? Like a slightly-out-of-focus film projector. Like looking through someone else's glasses. That's how astronomers described the fuzzy pictures Hubble was capable of shooting.

Two mirrors in the satellite collect and reflect star light. The surfaces of both mirrors are perfectly smooth and curved to cause light to be magnified.

But, one was built with the tiny error. Its surface curve does not align perfectly with the curve of the other mirror so light reflected to the camera is scattered slightly, making starlight look fuzzy with no crisp, sharp edges.

Focus is off by only two microns—about four one-hundredths the diameter of a human hair—but enough to blur Hubble's vision.

Most affected by the spherical aberration were the Wide Field/Planetary Camera, the European Space Agency's Faint Object Camera and the High-Speed Photometer. The Goddard High Resolution Spectrograph and the Faint Object Spectrograph don't require finely focused light to complete their observations successfully.

Glasses. NASA came up in 1991 with a plan to fix the blurry vision—give the telescope eye glasses. Never short of pithy acronyms, the agency calls the eye glasses COSTAR for Corrective Optics Space Telescope Axial Replacement. If Hubble is the defective star on the public stage, COSTAR will be its...co-star.

COSTAR will have ten small mirrors, and mechanisms to position and support the mirrors, to correct Hubble's spherical aberration.

NASA had planned to send astronauts up in a space shuttle in 1994 for routine maintenance on the big satellite anyway. They will take COSTAR along and install it on Hubble.

Each of the ten mirrors is about the size of a quarter. COSTAR is to correct the aberration from the primary mirror, sending its light on to the three instruments that suffer from the aberration.

Astronomers have their fingers crossed that COSTAR will restore the major instruments—the Goddard High Resolution Spectrograph, the Faint Object Spectrograph and the Faint Object Camera.

Fix-it flight. The bad mirror was not the end of NASA's problems with Hubble. In 1991, gyroscopes which stabilize the satellite failed and a faulty power supply cut off a key spectrograph.

Anatomy Of A Delay

NASA started work on Hubble in 1977, wanting to launch the telescope in 1985. But, that date had slipped to Fall 1986 when launch was postponed by the January 1986 explosion of shuttle Challenger.

NASA changed Hubble software, upgraded the safe mode, and inserted new solar arrays and higher-power nickel-hydrogen batteries.

HST was to have cost $435 million, but delays ballooned that to $1.5 billion. It was years behind schedule in a California warehouse generating storage bills of $8 million a month.

Challenger set shuttle flights back three years. NASA had to fly the huge telescope to space in a shuttle, but the space agency had trouble scheduling a launch.

There was a shortage of booster fuel and Pentagon demands for shuttle rides pushed the flight date into a time when the Sun would be blossoming with sunspots. An eleven-year peak in the Sunspot Cycle was approaching.

Exploding gases on the star heat Earth's upper atmosphere, expanding it deeper into space and causing extra drag on satellites. That threatened to pull the 12-ton, 43-ft. telescope down to a low orbit from which it might fall to Earth like a meteor.

Sunspots gave NASA an exasperating choice: launch Hubble so high in the atmosphere a shuttle might not be able to reach it again for repair or else launch it so low shuttles would have to fly up every six months to boost it higher at $300 million per nudge. Experts said the agency would delay the launch to prevent frequent service flights.

Hubble finally was shipped in October 1989, in a special container meant for large military spy satellites aboard an Air Force C-5A cargo jet from Sunnyvale, California, to a cleanroom at Kennedy Space Center, Florida. The satellite was set up in NASA's large Vertical Processing Facility hangar where it was prepared for 15 years in space.

Hubble's wide-field planetary camera, as big as a baby grand piano, was not shipped to Florida. One of two computers to control the instrument had broken, so the camera had to be repaired at NASA's Jet Propulsion Laboratory at Pasadena. The camera was shipped to Florida in December.

At last, after the many delays, Hubble was loaded aboard shuttle Discovery and ferried April 24, 1990, to 380 miles above Earth. That was 70 miles higher than any previous shuttle had flown.

Hubble began its solo voyage in space when astronauts used the shuttle's fifty-ft. mechanical arm to lift the satellite out of the cargo bay and drop it overboard into its own orbit. Hubble's solar arrays and dish antenna were extended. Two astronauts almost had to spacewalk when a solar panel didn't unfold until the third try. Their work done, the astronauts flew the empty Discovery home April 29.

The spacewalking astronauts will not only fix the spherical aberration in 1994, but also fix the spectrograph power supply, replace gyroscopes and tighten loose solar panels.

The astronauts visiting Hubble will make repairs and replace batteries and fuel. Hubble's orbit will have deteriorated by then, so the shuttle will be used to boost the telescope back up to its original altitude.

With such regular maintenance, Hubble is expected to be used for at least 20 years. NASA's Space Telescope Science Institute, Baltimore, Maryland, coordinates research with the telescope.

Cassegrain. HST is a Cassegrain telescope. Starlight enters its tube and bounces off a 94.5-inch primary mirror, back up sixteen feet to a 12-inch secondary mirror, from which it reflects back down through a hole in the primary mirror to be captured and recorded.

The electronic wide-field planetary camera is one of five major science instruments on board the satellite. There is another camera, two light-splitting spectrographs and a photometer to study the intensity of starlight.

First light. Hubble sent its first two pictures to Earth May 20, 1990. The so-called "first light" recorded by the telescope had traversed 1,260 lightyears of deep space from an open cluster of stars known to astronomers as Theta Carina.

Hubble was 381 miles above Jayapura, New Guinea, when the shutter of its electronic wide-field planetary camera snapped the first rough-focus, black-and-white picture and stored the 30-second exposure in a recorder.

Two hours later, that image was transmitted in four parts to one of NASA's Tracking and Data Relay (TDRS) satellites, then to receivers on the ground at White Sands, New Mexico, then back up to another communications satellite which relayed them down to astronomers waiting at Goddard Space Flight Center, Greenbelt, Maryland.

A Goddard computer removed background noise and assembled the photo. In that first "raw" Hubble photo, stars in the cluster looked like vague points of light against a murky field.

Technicians immediately set about to improve the contrast, making the background blacker and sharpening the points of light for easier viewing.

The first picture clearly revealed two stars where astronomers on the ground previously had seen only one elongated object.

Interestingly, most light from the stars in the first photo was in the center of the image. Astronomers said that's where they would expect it to be from a mirror slightly out of focus.

Even with its deficiencies, Hubble probably will transform astronomy, shooting 30,000 photos during 15 years traveling around Earth.

The pictures will show more details of planets, stars and galaxies than ever before. Telescopes on the ground are hampered by particles in the air which absorb and distort starlight. The space telescope is above the atmosphere in the vacuum of space where the cameras, spectrographs and photometer will see much more clearly.

Remarkable. HST already has found clues to basic questions about the birth, age, size, shape, and even the ultimate fate of the Universe.

Despite the focusing problem, Hubble's photos of bright targets still have been better than the best produced by ground telescopes on a clear night.

In 1990, Hubble sent down photos showing how massive stars form in galaxies other than our Milky Way. Pictures uncovered new information about our neighboring galaxy, the Large Magellanic Cloud. One showed 30 Doradus, the most prolific stellar nursery in that galaxy. It showed hot stars in Cluster R136 generating more blue and ultraviolet light than cooler stars.

Hubble photographed Saturn's Great White Spot, a huge storm on that planet more than 50,000 miles in diameter and covering most of one hemisphere of the planet.

Astronomers used Hubble to photograph Saturn and its rings, a ring around Supernova 1987a, the nucleus of Seyfert galaxy M77, the globular cluster M14, quasar light bent by a gravitational lens, Orion Nebula, star cluster R136 in the Large Magellanic Cloud, a possible black hole in the crowded heart of galaxy NGC 7457, Pluto and its moon Charon, and hundreds of other targets.

GRO. The second of NASA's four Great Observatories to go to space was the huge Gamma Ray Observatory (GRO) satellite launched in April 1991. Its name was changed later to Compton Gamma Ray Observatory (GRO or CGRO).

The 17-ton satellite, carried to a 280-mi.-high Earth orbit in shuttle Atlantis, was the heaviest NASA satellite ever launched by the space agency.

GRO's searches for highly-energetic gamma rays blasted out by the most violent processes in the Universe. The invisible rays are fierce radiation created when the nuclei of atoms collide and matter is annihilated in the presence of antimatter.

GRO records high-energy gamma rays flowing from quasars, supernovas, pulsars, neutron stars, and black holes, the most violent natural processes known in the Universe.

Gear. GRO's great bulk was dictated by size and weight of instruments needed to trap gamma rays, which cannot be viewed through our planet's atmosphere. GRO has four devices to investigate the invisible energy rays which may hold clues to how the Universe was formed, including an Oriented Scintillation Spectrometer Experiment and an Energetic Gamma Ray Experiment.

Every 14 days, GRO's view is shifted a bit so astronomers can sweep the entire Universe. The satellite is able to keep itself at a 280-mi. altitude by correcting its own orbit to compensate for drag.

The satellite has arrays of electricity-generating solar cells and a high-gain radio antenna.

Science. GRO science operations began May 16, 1991, when the observatory turned to point toward a pulsar in the Crab Nebula and swept the sky for gamma rays.

Controllers at Goddard Space Flight Center then pointed the satellite at the Sun in June so GRO could peer at two X-class solar flares, the most powerful outbursts of intense solar radiation with 430,000-mi.-high geysers of hot gas.

GRO photographed the binary star system Cygnus X-3 some 30,000 lightyears from Earth in the disk of the Milky Way after astronomers found it spewing radio emissions. GRO photographed Vela Pulsar, Hercules X-1 and Nova Muscae.

Gamma rays are the strongest part of the energy spectrum. GRO receives a much wider range of wavelengths than earlier gamma ray observatories.

By the end of the year, GRO was detecting gamma ray bursts with more sensitivity than any previous receiver and in more detail than previously possible. It was observing a gamma ray burst almost every day, a rate of 250 per year.

Astronomers hope GRO will return clues about the question of whether most gamma rays arriving at Earth from across the Universe come from quasars and pulsars, or are there other objects in our Universe sending out gamma rays.

GRO gathers gamma rays generated in the far-flung reaches of the Universe as much as 15 billion years ago. In fact, GRO found what was described as the most-distant and most-luminous gamma-ray source ever seen—the variable quasar 3C279 in the constellation Virgo.

Seven billion lightyears from Earth, the Quasar 3C279 was blasting out a prodigious amount of gamma rays. The luminosity, or total energy, emitted by the quasar was ten million times the total from the Milky Way galaxy. The quasar was said to be variable because its intensity changes.

AXAF. The third Great Observatory will start revealing clues to hidden mysteries in some of the most violent processes in the Universe in 1999.

Earlier X-ray satellites in Earth orbit, such as the Uhuru Observatory in 1970 and the Einstein Observatory in 1978, tantalized astronomers with a taste of what may lurk across the vast reaches of the Universe, hidden from view in the X-ray spectrum. In fact, Einstein Observatory images were so valuable that astronomers immediately called for a long-lived orbiting X-ray observatory. That's when NASA started drawing up plans for an Advanced X-Ray Astrophysics Facility.

The 15-ton, 14-ft.-diameter, 45-ft.-long satellite should operate at least 15 years in a circular orbit 320 miles above Earth. It has to be above Earth's atmosphere because X-rays can't pass through the planet's sea of air to the ground.

Gear. AXAF's high-resolution mirror will feed four instruments: a charge-coupled-device (CCD) X-ray camera from Pennsylvania State University; a Bragg Crystal Spectrometer from the Massachusetts Institute of Technology (MIT); a high resolution camera from the Smithsonian Astrophysics Observatory; and an X-ray spectrometer from Goddard Space Flight Center.

The CCD imaging spectrometer will be up to 1,000 times more sensitive than current X-ray telescopes. Its X-ray images will be ten times as sharp. Astronomers will measure energy levels of X-rays from extremely faint celestial bodies.

Mirror. The company which made the Hubble Space Telescope imperfection built AXAF mirrors. Hughes Danbury Optical Systems Inc., formerly Perkin-Elmer, said in 1990 it could provide good mirrors.

The Hubble problem had nothing to do with AXAF, of course. A 1.3-mm spacing error was found in a measuring instrument used to grind the Hubble mirror in the early 1980s. Hughes Danbury used a different polishing method for AXAF mirrors ten years later.

AXAF is expected to deliver X-ray images with a resolution of 0.23 arc second, 10 times better than any previous X-ray telescope.

Dewar. AXAF will have a thermally-insulated container, known as a helium dewar, to cool its X-ray spectrometer. The dewar would operate five years—the longest time a cryogen cooler would have operated in space—allowing AXAF to be 100 times more sensitive than previous X-ray telescopes.

AXAF data is expected to add much knowledge in the fields of plasma physics, atomic and nuclear physics, general relativity, and cosmology.

The Smithsonian Astrophysical Observatory, Cambridge, Massachusetts, is the official AXAF Science Center to oversee the observation program and distribute AXAF data.

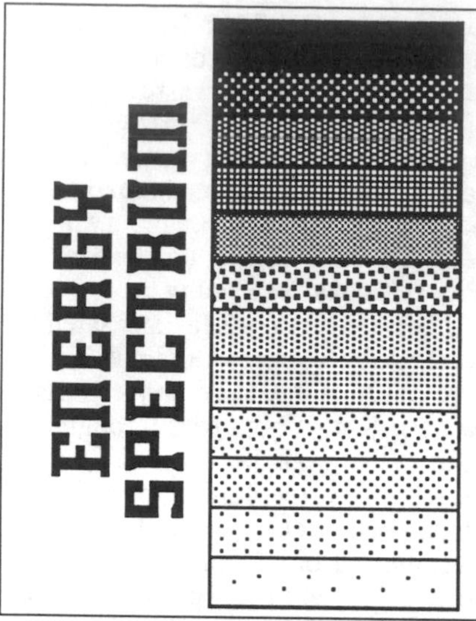

The Energy Spectrum
The spectrum of electromagnetic energy includes radio and light waves and extends from very low to extremely high frequencies. Only in this century, have we been able to "see" radio waves. Starting from the long-wave end of the spectrum, our ability to "see" across the spectrum has grown. AM broadcast stations transmit medium waves. FM broadcast stations use VHF. TV broadcast stations use VHF and UHF. Familiar microwave ovens and police speed-trap radars use the microwave portion of the spectrum. Visible light is flanked by infrared and ultraviolet we can't see; however, we can capture infrared, ultraviolet and X-rays on film. Today we have receivers small enough to send to space.

SIRTF. The fourth Great Observatory would be the Space Infrared Telescope Facility, but SIRTF is off the schedule due to budget problems. That disappoints astronomers who say the satellite would be a thousand times more sensitive than earlier infrared spacecraft and might be the one which could see planets in other solar systems.

SIRTF would be the most sophisticated infrared receiver ever flown, an evolutionary leap forward from the renowned helium-cooled systems such as IRAS in 1983 and COBE in 1989. SIRTF's telescope would watch the invisible infrared part of the electromagnetic spectrum, most of which does not penetrate Earth's atmosphere.

Future Astronomy

Small Explorers. Small Explorers are inexpensive lightweight satellites launched to low orbits on Scout, Pegasus and other small rockets.

The first Small Explorer was SAMPEX in 1992. Small Explorers are operated by NASA's Goddard Space Flight Center, Greenbelt, Maryland.

NASA has scheduled dozens more Small Explorers for launch by the year 2002:

FAST. The Small Explorer known as Fast Auroral Snapshot Explorer (FAST) will fly in 1994 on a Pegasus rocket. Scientists hope it will tell them more about how the aurora, known as Northern Lights and Southern Lights, occur in Earth's atmosphere.

HETE. A Small Explorer known as High Energy Transient Experiment (HETE) will be launched by Pegasus in 1994 to study the locations of gamma ray and X-ray sources.

SAC-B. On the same Pegasus rocket in 1994 will be Argentina's Small Explorer known as Satellite de Aplicaciones Cientificas-B (SAC-B).

The Argentine satellite will have a spectrometer to study hard X-rays from solar flares and cosmic transient X-ray emissions.

SWAS. Submillimeter Wave Astronomy Satellite (SWAS), to be launched in 1995 on a Pegasus rocket, will have a telescope to observe molecular clouds. U.S. and German scientists hope to discover how they collapse to stars and planets.

XTE. NASA will send an X-ray Timing Explorer (XTE) satellite to low Earth orbit in 1995. XTE will study a variety of X-ray sources in deep space, including white dwarfs and black holes.

TOMS. Two Small Explorers known as Total Ozone Mapping Spectrometer (TOMS) will be launched on Pegasus rockets in 1993 and 1997 to study ozone in Earth's stratosphere. Not astronomy satellites, they will provide daily maps of ozone around the globe, helping predict depletion. NASA labeled TOMS a high-priority Earth-observation mission, critical for monitoring long-term depletion of ozone from the stratosphere.

SAX. An X-ray astronomy satellite built for the Italian space agency—Agenzia Spaciale Italiana or ASI—and the Netherlands Agency for Space Programs is to be launched in 1994 from Cape Canaveral. Manufactured by Aeritalia, the satellite would complete 2,000-3,000 X-ray observations during its lifetime.

Automated Platforms

The automated platform is a new kind of space-research satellite used in the space shuttle era.

The reusable platform is a generic payload

carrier which can be loaded with a variety of payloads and ferried to orbit by shuttle.

The structure is dropped overboard into its own orbit 200 miles above Earth to be retrieved and returned to Earth by the same shuttle or by a flight months or years later.

Back on Earth, the miscellaneous payloads are removed from the platform, which then is packed with new payloads and shuttled to space again.

Automated platforms usually are laden with science experiments and technology research packages. The platform provides communications and power and controls its own attitude in space.

SPAS. The first platform was built by West Germany and flown to space by American shuttle Challenger in 1983. The Shuttle Pallet Satellite (SPAS) was a battery-powered, tubing framework surrounding eight remote sensing, environmental pollution and microgravity experiments. SPAS-01 flew free from shuttle Challenger for about ten hours in 1983. SPAS has been reflown.

LDEF. The Long Duration Exposure Facility (LDEF) was more like a huge trunk full of experiments left in the attic for years.

The entirely-passive container was a twelve-sided aluminum cylinder, thirty feet long, fourteen feet in diameter—big as a school bus.

Experiments were tucked away in lots of little cubby holes. For its first flight in 1984, LDEF was loaded with 57 science experiments and ferried to orbit by shuttle Challenger.

The platform was supposed to stay only a year in space, but retrieval was held up by scheduling problems, then postponed after shuttle Challenger exploded during launch in 1986. LDEF finally was retrieved in 1990 by shuttle Columbia.

Among the 57 experiments were 12.5 million tomato seeds which NASA mailed to schools so 3.5 million students could study the after-effects of the harsh space environment on growth.

Eureca. The European Space Agency built a large, instrumented, maneuverable platform known as the European Reusable Carrier (Eureca).

Eureca has two large wings of solar cells providing electrical power for experiments. The vehicle can maneuver itself in space.

Eureca was packed with fifteen materials and life-sciences experiments and ferried to Earth orbit for the first time in 1992 by shuttle Atlantis. It was retrieved in 1993 by shuttle Endeavour. Astronauts had to spacewalk to secure Eureca in Endeavour's cargo hold.

Mysteries Of The Deep (Space)

Enormous pulses of radio energy and light from the hearts of galaxies seem to many astronomers to be prima facie evidence for the existence of the vast gravity pits from which even light cannot escape—black holes. Are they correct?

Astrophysicists are puzzling over that and a panoply of other questions about quasars, pulsars, neutron stars, galaxy clusters and other deep space mysteries:

★Is the dark matter spread through the Universe actually dark stars, asteroids, planets, black holes, exotic unknown particles or what? An X-ray satellite like AXAF in orbit above Earth would expand human knowledge of dark matter.

★Would matter falling onto a black hole release strong X-rays? AXAF will show details of the immediate vicinity of any suspected black hole.

★Are all neutron stars pulsars? X-ray images of pulsating stars may show spinning neutron stars.

★Are quasars the most powerful energy sources anywhere in the Universe? AXAF will bring home new clues.

★What is the exact age of the Universe? From AXAF data, astronomers will compute a more accurate age.

★Are there differences among the many clusters of galaxy? AXAF will weigh invisible matter in clusters to highlight differences.

★Is the Universe open or a closed loop? Will it expand forever or will it retract, collapsing in on itself, triggering another Big Bang, recycling matter into a new expansion? AXAF will shed new light on this ultimate mystery.

Manned Satellites

A space station or a space shuttle is a satellite like any other, except it has people inside. Many unmanned satellites have been as large as a space station, have flown at the same altitudes and, in many ways, have been as complex.

The significant difference is the life-support system built into the station or shuttle to keep its human occupants alive and safe from the rigorous space environment.

More than 300 men and women have lived and worked aboard manned satellites in Earth orbit.

Like unmanned satellites, manned spacecraft have been used for science and technology research, observations of land, sea and air, astronomy, military reconnaissance and communications.

When in space, U.S. and Russian shuttles are manned satellites of Earth. They usually fly in orbits 200 or so miles above Earth. On rare occasion, they fly up to 300 or 400 miles altitude.

Russia's space station Mir is another manned satellite which maintains its own orbit just above 200 miles altitude. The U.S.-international space station Freedom will be a similar manned satellite of Earth around the year 2000.

Cosmonauts are men and women who fly in the spacecraft of Russia and the former Soviet Union.

Astronauts are those who fly in U.S. spacecraft. A few persons from various nations are both cosmonauts and astronauts. No other nations have originated manned flights, although Europe, Japan and China are preparing to do so.

People In Space

After World War II, manned space flight was studied in the U.S., but not planned until the USSR launched unmanned satellites Sputnik-1 and -2 in 1957. When the Russians indicated they were rushing toward manned flight, Congress started planning to send Americans to orbit.

The U.S. National Advisory Committee for Aeronautics (NACA) suggested in 1958 a wingless satellite could carry a person on a ballistic path to re-enter the atmosphere without subjecting the astronaut to damaging acceleration or temperatures.

The National Aeronautics and Space Act of 1958 created the National Aeronautics and Space Administration (NASA) from NACA.

Manned satellites. There was a vigorous race to send the first man to space, where he would orbit Earth in a satellite known as a capsule. The Soviet Union won in 1961 as cosmonaut Yuri Gagarin rode his Vostok-1 capsule around the world to land. He was the first man in space.

Vostok, meaning East in Russian, was the USSR's first man-in-space program. It lasted from 1961-63. The first woman to fly in space and the first woman to orbit Earth was Valentina Tereshkova, in her capsule Vostok-6 in 1963.

Moon Race. Meanwhile, U.S. President John F. Kennedy, in 1961, set a goal of landing a man on the Moon within the 1960s.

Three technology levels had to be developed—the single-seat Mercury capsule, the twin-seat Gemini capsule, and the triple-seat Apollo capsule.

NASA included parts of the NACA wingless-satellite proposal into Project Mercury. It and Gemini would orbit at low altitudes around Earth.

Apollo would fly to Earth orbit, travel away from Earth to the Moon, orbit the Moon, set down astronauts in a lander, blast them from the Moon back up to lunar orbit, and return to Earth.

Mercury. America's first men in space went there in Mercury capsules. Astronauts made two suborbital and four orbital flights in tiny 6-ft. by 9-ft. capsules they called "garbage cans."

Voskhod. The USSR's second man-in-space program was called Voskhod, meaning Sunrise in Russian. It featured three-man capsules.

USSR Premier Nikita Khrushchev, wanting to upstage America's two-man Gemini flights scheduled for 1965, had Soviet engineers remove life-support and safety equipment and bolt three seats into what otherwise would have been a two-man Vostok capsule. Voskhod flew in 1964.

There wasn't room for three cosmonauts in spacesuits so they wore shirt-sleeve-style work coveralls. Cosmonauts landing in Vostok had ejected and parachuted to Earth, but Voskhod didn't have room for ejection equipment and individual parachutes, so one large parachute was designed to float the entire capsule to a hard landing. The cosmonauts even dieted to reduce launch weight.

Gemini. Project Gemini was America's second man-in-space program and a bridge between Mercury and Apollo. Launches from 1964-66 featured two-person capsules boosted to low orbit.

Gemini was named for the third constellation of the zodiac which is said to have twin stars, Castor and Pollux. Gemini is Latin for twins.

Gemini tested men and hardware in flights of up

to two weeks in Earth orbit, to develop and practice rendezvous and docking with spacecraft in orbit, to move around space, to polish re-entry techniques, and to touch down on dry land. However, ocean splashdowns were continued and landing on solid ground was dropped in 1964.

Mercury had shown astronauts could live in orbit up to 34 hours. Now NASA needed to know if a sophisticated life-support system would work and astronauts could endure the time in weightless freefall needed for a round trip to the Moon.

Gemini used NASA's global tracking and communications network built for Mercury. There were 10 manned Gemini flights.

Apollo. Between 1968-1972, U.S. astronauts rode the first manned interplanetary transportation system, Apollo, to the Moon. Twelve men landed there, explored the lunar landscape and picked up samples at six sites on the near side of the Moon.

Soyuz. Soyuz is Russia's third man-in-space program, after the Vostok and Voskhod. Soyuz are two-seat and three-seat capsules which transport men and women to low Earth orbits. From 1967 to the present, independent flights have been made, as well as trips to Russia's space stations.

Space Stations

Europeans envisioned manned satellites in the 1920s as permanent science and technology bases orbiting Earth, the Moon or other planets.

They fancied large colonies built around factories, science labs, and observatories to look down on Earth and out to the Sun, planets and stars. Seventy years later, we have space stations which do those kinds of work, but there are no large human colonies in space. In fact, fewer than a dozen persons inhabit a station at one time.

Salyut. While the U.S. won the race to the Moon in 1969, the Soviet Union won the competition for first space station with the launch of its Salyut-1 in 1971.

The USSR space program advanced quickly with the launch of Salyut-2 in 1973, Salyut-3 in 1974, Salyut-4 in 1974 and Salyut-5 in 1976.

Skylab. Meanwhile, NASA enjoyed large budgets during the 1960s space race and wanted to continue into the 1970s. The space agency planned to expand Project Apollo into an Apollo Applications Program (AAP).

AAP would have been a space station in Earth orbit as a pit stop for astronauts on their way to the Moon and Mars. But, after six spectacular manned Apollo landings on the Moon, much of NASA's money dried up.

Many politicians said space spectaculars were unnecessary since the U.S. had won the race.

Forced to choose, NASA canceled three Moon landing flights and built one space station. The name AAP was changed to Skylab.

Workshop. Calling it an orbital workshop, NASA outfitted the space station on the ground and launched it to Earth orbit in 1973.

At 77.5 tons and 118 feet long, Skylab was one of the largest satellites ever sent to Earth orbit.

Astronauts were shipped to the station, three at a time, in left-over Apollo Moon capsules. Three groups totaled 172 days at Skylab in 1973-74.

NASA drew up plans for a larger space station, but sharply-reduced budgets again forced tough decisions. Skylab went unused after just three visits. Plans for a second station were dumped.

NASA preferred a reusable space transportation system to carry men and equipment to orbit, so the limited money available was switched to designing and building space shuttles. Skylab was allowed to fall from orbit in 1979.

Improvements. Meanwhile, the USSR's first-generation space station technology, displayed in Salyut-1 through Salyut-5, was superseded by an improved, second-generation station, Salyut-6, in 1977. Another second-generation station, Salyut-7, was sent to Earth orbit in 1982.

Mir. Mir is a Russian word for peace. It also is Russia's name for a third-generation space station.

Mir was launched in 1986, just 23 days after the fatal explosion of U.S. shuttle Challenger.

When it was launched, the station was as big as a house trailer in orbit—42,000 lbs., 43 feet long. Since then, add-on modules have quadrupled Mir.

Today, Mir is in permanent, full-time use. As it circles the globe every 92 minutes, its two big cruise engines and 32 small attitude control thrusters maneuver the station and keep it 215 or more miles above Earth.

For observers on Earth, Mir is the brightest man-made object in the sky. Its glint often is seen as sunlight reflects from its vast surface.

Living quarters. The living space inside Mir can be imagined as something like a two-room house. It has a large galley, several recreation facilities, big bathroom with shower, and private compartments for crew members. Cosmonauts adjust the inside temperature for shirt-sleeve work from 64 to 82 degrees Fahrenheit.

The private crew compartments, more palatable galley and nicer bathroom and shower make Mir seem more like a home than the earlier Salyuts.

Unmanned Progress freighters arrive every three or four months with food, water, fuel, supplies and mail for the station.

Space Station Launches

Nine stations have been placed in Earth orbit by Russia and the United States. This list shows launch year, country of origin, station name and the year it descended to burn in the atmosphere. The tenth station is expected to be Mir-2 in 1997. The eleventh would be Freedom by the year 2000.

Year	Country	Station	Down
1971	USSR	Salyut 1	1971
1973	USSR	Salyut 2	1973
1973	USA	Skylab	1979
1974	USSR	Salyut 3	1975
1974	USSR	Salyut 4	1977
1976	USSR	Salyut 5	1977
1977	USSR	Salyut 6	1982
1982	USSR	Salyut 7	1991
1986	USSR	Mir	--
1997	USSR	Mir-2	--
2000	USA	Freedom	--

Mir's orbit is low. It constantly is dragged down toward the atmosphere. While a Progress supply ship is docked at Mir, its engines are used to push the station back up to 215 miles altitude.

Months in space. Station activities over the years have given Russia the lead in man-hours in orbit and space biomedicine.

The usual crew is two cosmonauts, but eight can work in Mir. Several have made repeated visits to the station. Some have spent half a year to a year in space. Two cosmonauts, Vladimir Titov and Musa Manarov, hold the human single-trip-in-space endurance record of 366 days. Manarov holds the one-person time in space endurance record with 541 days total in two trips to Mir station. Titov and Manarov were launched to space in the transport Soyuz TM-4 on December 21, 1987. They returned to Earth December 21, 1988.

Recreation. During work hours, Mir crews do science and technology research, commercial development work and materials manufacturing, and military projects.

A long stay in space has its boring moments. Personal recreation time is important.

Cosmonauts have taken various games and musical instruments, as well as music tapes and movie videos, to the station for entertainment during rest periods.

They have passed the time playing video games and reading. The cosmonauts built their own ham shack—an amateur radio station—aboard Mir. They use it frequently to chat with hams on the ground around the world.

Mir-2. Russia is completing construction of a new and improved Mir station for launch in 1997. Mir-2 will be a fourth-generation space station.

The launch would be three years before the U.S. opens its second-generation station, Freedom, in space around the year 2000. Japan, Europe and the United States are expected to cooperate on Mir-2. Freedom and Mir-2 might be joined in space.

Freedom. In 1984, U.S. President Ronald Reagan announced an American project to build a permanently-manned space station in Earth orbit within 10 years. He called the station Freedom.

Canada, Japan and the European Space Agency joined the project. Station configurations and construction dates changed many times over subsequent years, but NASA is moving forward on constructing the station in orbit by the year 2000.

Europe. The European Space Agency has been firing large Ariane rockets to space since 1979. European astronauts have flown to Russia's Mir station and aboard U.S. shuttles.

Columbus is the name of Europe's manned, pressurized, science module to be attached permanently to Freedom station. ESA also may build the small manned space shuttle Hermes to carry men and supplies to Columbus.

Another Columbus module could be modified into an independent European space station after the year 2000. Hermes could fly cargo and astronauts to an independent Columbus.

Japan. Japan has been firing rockets to space since 1970. A Japanese astronaut flew in an American space shuttle and a Japanese cosmonaut visited Russia's Mir station.

Japan is building JEM, the Japanese Experimental Module, to be attached permanently to the U.S.-international space station Freedom.

JEM will be three enclosed modules and an exposed facility attached to the station truss.

One module would be a pressurized laboratory, two would be logistics modules, and the exposed facility would allow experiments to be open to the space environment.

Astronauts from Japan, Europe, Canada and the United States would process materials and do life sciences research in the pressurized laboratory, ferrying materials between the station and Earth in one of the logistics modules. The other module would store specimens, gases and consumables.

Japan has built Moon satellites and Venus probes. A space shuttle, Hope, has been planned. It's easy to imagine an independent station growing out of the JEM blueprints after 2001.

China. The People's Republic of China has been firing satellites to orbit since 1970, bringing some safely back to Earth, and preparing since the late 1970s to launch astronauts.

In 1980, a Chinese magazine showed astronauts training in a mock spaceship. Spacesuits and food have been in the works. The newspaper *People's Daily* said in 1986 the first manned flight was "not far off." Life-support systems and crew cabin were reported ready and a crew was being selected. China may be designing a four-place ship, twice the size of a U.S. Gemini capsule from the 1960s.

At a 1989 Beijing meeting of space authorities from Pacific nations, the Chinese announced an independent space station for completion by 2000.

Space Shuttles

The U.S. Air Force first conceived a winged spacecraft in 1962. The idea was to bolt an unmanned X-15 experimental rocket plane to the top of a B-52 bomber and fly it as high as the plane could go. Near space, the B-52 pilot would launch the unmanned rocket plane. The X-15 would continue on to space, drop off a satellite in orbit and fly back down to Earth.

While the X-15 didn't fly to space, didn't drop a satellite, and the Air Force shelved the plan in 1965, the idea of a reusable space shuttle hung on.

Forerunners. A number of experimental prototypes were flown before the American shuttle fleet was built, including the X-15 which was tested from 1959-1968. The 50-ft. rocket aircraft was launched 199 times from under the wing of a converted B-52. It flew at speeds up to Mach 6.7, reaching an altitude of 67 miles in 1963. Its fastest speed was 4,534 mph in 1967. Among its pilots were Neil Armstrong who became the first man on the Moon and Joe Engle who flew a space shuttle.

Other space shuttle forerunners included the X-20, flown from 1960-1963. It was a design for a small military suborbital and orbital shuttlecraft, called Dynasoar, to be launched by Titan rocket. It was expensive, never flown, and canceled in 1963.

Asset was the name for half a dozen launches in 1963-1965 of small Dynasoar models testing aerodynamics and thermal protection. M2 was a lifting body testing shuttle handling at trans-sonic speeds from 1966-1967. It flew in 1966, crashed in 1967, and was rebuilt for more tests. Prime was the name for three launches of maneuverable lifting bodies in 1966-1967 to near-orbital speeds. HL-10 was a manned lifting body testing shuttle handling at trans-sonic speeds in the 1960s.

X-24a was a manned lifting body testing shuttle handling at trans-sonic speeds in the 1970s. X-24b flew manned lifting body tests of handling at trans-sonic speeds. It flew 36 times in 1973-1975 with some landings simulating shuttle landings on concrete runways at Edwards Air Force Base.

Shuttle. NASA wanted a reusable space transportation system to carry men and equipment to orbit, so the space agency concentrated on designing and building space shuttles.

The shuttle was to be a winged spaceplane with wheeled landing gear and large cargo-carrying capacity. It would be launched to Earth orbit by rocket, then leave orbit to glide to a runway landing. The Space Transportation System (STS) would include the winged orbiter with a pair of solid-fuel booster rockets and a large external liquid-fuel tank strapped to it.

On the way to space, main engines in the orbiter would burn liquid fuel from the tank and be supplemented by the solid-fuel boosters.

Eventually, six shuttles were built, starting with Enterprise in 1974 and then Challenger, Columbia, Discovery, Atlantis and Endeavour. Dozens of week-long spaceflights have been made since 1981, but NASA never again enjoyed the powerful budgets it commanded in the 1960s.

Enterprise. Space shuttle Enterprise was the first orbiter, but it never orbited, never shuttled, never went to space.

Enterprise was designated OV-101 when work was started on it in June 1974. After 100,000 Star Trek TV fans wrote in, former U.S. President Gerald Ford named the orbiter Enterprise in 1976. NASA didn't like the name. The space agency wanted to call the shuttle Constitution.

Enterprise had no engines. The 130-ton orbiter only flew glide tests, bolted to the back of a Boeing 747 jet. The airplane-shuttle combo did not go to space. Instead, the plane dropped Enterprise to test its gliding ability.

The combo took off briefly in February 1977 to see if they could fly together. Then, for the first free flight in August 1977, astronauts Fred Haise Jr. and Charles Gordon Fullerton were at the controls when the 747 dropped Enterprise from an altitude of 4.5 miles over California. They glided 5.5 minutes to land at Edwards Air Force Base.

Five test flights were made over two years. Haise and Fullerton piloted flights one, three and five. Astronauts Joe Engle and Richard Truly piloted flights two and four.

Engle and Truly later flew shuttle Columbia on the second actual shuttle spaceflight. Truly went on to head the space agency as NASA Administrator from 1989 to 1992.

Challenger. After the glide tests, Enterprise was supposed to go back to the workshop for engines, but it became cheaper to convert a structure-test orbiter into a ready-to-fly orbiter, so the plan to fly Enterprise was abandoned.

The structure-test orbiter, numbered OV-099, became Challenger, the shuttle which exploded in

1986 after nine successful spaceflights.

Columbia. Another orbiter, number OV-101, became Columbia. In 1981, it became the first shuttle to fly in space. Challenger, built before Columbia, didn't fly in space until 1983.

Discovery. Vehicle number OV-102 was named Discovery and made its first spaceflight in 1984. That was the year President Reagan announced plans to build Freedom station in Earth orbit within 10 years.

Atlantis. Vehicle number OV-103 became Atlantis and went to space in 1984. NASA named its orbiters for historic sea vessels used in research and exploration.

Soyuz. During the race to the Moon in the 1960s, the Russians developed capsules to carry cosmonauts to Earth orbit and on to the Moon.

Over the decades, scores of cosmonauts have been launched in small Soyuz capsules in dozens of flights to Russia's Salyut and Mir space stations. Working at their stations gave the Russians the lead in man-hours in orbit. The cosmonauts have returned to Earth in the capsules which parachute to land.

Supplies have been sent to the stations in unmanned Progress cargo-freighter capsules which then are jettisoned to burn in the atmosphere.

Buran. As Americans pursued reusable space shuttles in the 1970s, the USSR designed its own shuttle. The first was named Buran—Russian for snowstorm or blizzard. The first unmanned flight test of Buran No. 1 was successful in 1988.

Tight budgets delayed a second flight—Buran No. 2—to 1995, or even later, when it may make an unmanned flight to Mir.

Space station visits by shuttles carrying men and supplies may increase, but it probably will remain less expensive to ferry men and supplies in Soyuz and Progress capsules.

Challenger. In nine distinguished flights, Challenger compiled an impressive record:

★first American women in space, Sally Ride;
★first black man in space, Guion S. Bluford Jr.;
★first spacewalk from a space shuttle;
★first satellite repair in orbit, on Solar Max;
★first American female spacewalk by astronaut Kathryn Sullivan;
★first untethered spacewalk using a manned maneuvering unit (MMU);
★and even the first Coke and Pepsi in orbit.

Challenger's tenth flight, on January 28, 1986, set a tragic record: first Americans to die during a spaceflight and the first persons anywhere to die enroute to space. Seven astronauts were killed.

After the Challenger disaster in 1986, an outpouring of national emotion supported the allocation of federal funds to build a replacement orbiter. Even then, some scientists and politicians continued to argue against manned spaceflights, forcing NASA to choose between shuttles and space stations vs. unmanned interplanetary probes and orbiting astronomy observatories.

The Challenger Seven

The seven astronauts aboard Challenger flight STS-51L (STS-25) on January 28, 1986, were:

★Francis R. "Dick" Scobee
★Michael J. Smith
★Judith A. Resnik
★Ellison S. Onizuka
★Ronald E. McNair
★Gregory B. Jarvis
★Sharon Christa McAuliffe

McAuliffe was a Concord, New Hampshire, high school social studies teacher. She and the other six were killed when a solid-fuel booster rocket leak led to an explosion during lift off from a Cape Canaveral launch pad.

People around the world watched as mission control reported at one minute thirteen seconds into flight, "Obviously a major malfunction. We have no downlink. The vehicle has exploded."

OV-105. Work was started immediately on a replacement orbiter, OV-105. NASA needed a name for the new shuttle after the name Challenger was retired in honor of the dead astronauts.

The agency wanted to continue the tradition of naming orbiters for the ships of famous explorers.

Directed by Congress to do so, NASA let American students choose the name. Elementary and secondary students in public and private schools in the United States and territories, Department of Defense overseas dependents schools and Bureau of Indian Affairs schools entered the competition. More than 6,100 teams, including 71,650 students, researched name choices. Each state, territory and agency announced a winner. NASA decided the final winner and announced the name in 1989.

Endeavour. Sailing ships of Capt. James Cook, the 18th-century British explorer, turned out to be most popular among American children.

Thirty-one of 111 state winners wanted the new shuttle to be named Endeavour, although some used the American spelling Endeavor. OV-105 was named Endeavour in 1989 and flew in 1992.

Extraterrestrial Satellites

Everything in our Solar System is a satellite of the Sun, of course. Whether the planet Mercury, Venus, Earth, Mars, Jupiter, Saturn, Uranus, Neptune or Pluto, or an asteroid or comet, everything is clasped tightly by the Sun's gravity.

Many large bodies have their own natural satellites. For instance, Earth has its Moon. Mars has Phobos and Deimos. Jupiter has at least 16 moons. Saturn has at least 22. Uranus has at least 15. Neptune has at least 8. Even tiny distant Pluto has its companion Charon.

Man learned how to simulate nature in 1957 when the artificial moon Sputnik was sent to orbit Earth. It was locked in step with Earth just as the planet is locked in step with the Sun.

Curiosity lead man to try to break the lock, to send a satellite away from Earth's grip, to look at other bodies. First we visited Earth's natural Moon. Even then we were marching around the Sun in lockstep with Earth. We felt the strong urge to get beyond Earth's grip, to see close up Mars, Venus, Comet Halley and the rest of our Solar System's mysteries.

To do that, we had to invent rockets strong enough to pull free from the powerful gravity which keeps us planted firmly on Earth. When we accomplished that, those rockets took our satellites, now referred to as interplanetary probes, to comets, asteroids and most of the nine planets.

Deep Space Probes

Spacecraft which leave Earth behind and fly on to the planets become satellites of the Sun. The power of their launching rocket boosts them to such a high speed, they are able to break away from the pull of Earth's gravity.

Having rocketed away from Earth, an interplanetary probe falls into orbit around another body. If the destination is a planet, the spacecraft may become a satellite of that planet.

For instance, the interplanetary probe Magellan flew away from Earth in 1989 and into orbit around Venus in 1990.

A planet probe which misses its target falls into its own long orbit around the Sun. An example is the USSR's probe Mars-4. It was launched in 1973, but is orbiting the Sun today because a braking rocket failed and the spacecraft overshot the Red Planet in 1974.

The pull of the Sun's gravity is much stronger than the tug of any one planet. The attraction of the Sun is so strong, in fact, few spacecraft have received sufficient boost to leave the Solar System. One which did was Pioneer-10, launched in 1972. The first spacecraft to cross the Asteroid Belt, it flew by Jupiter in 1973. Ten years later, it crossed an invisible boundary in 1983, departing the Solar System and entering interstellar space.

Venus. Unmanned interplanetary spacecraft from Earth have explored the Sun, the Moon, the Asteroid Belt and seven of the eight planets in our Solar System since 1959. The very first explorations were fly-bys.

Not long after their 1957 launch of Sputnik, space engineers in the Soviet Union fired the first rocket toward another celestial body. Venera-1, in 1961, was the first spacecraft sent beyond Earth.

The Magellan radar-mapping probe fired toward Venus from a U.S. shuttle in 1989 was the sixth American spacecraft and the 30th from Earth built to explore Venus. One of the six American flights failed, while nine of 24 Russian missions failed.

Mars. Space engineers in the 1960s were able to send probes on interplanetary science missions.

Over the decades, a total of 19 unmanned Mars explorers have been fired into interplanetary space from the U.S. and the USSR to look at the Red Planet and its moons Phobos and Deimos.

The USSR's Mars-1, launched in 1962, was the first attempt to probe Mars. Unfortunately, contact was lost with the spacecraft only 60 million miles along its route.

America's Mariner-4 launched in 1964 was the first successful probe to reach Mars, sending back 22 photos as it flew by in 1965.

Mariner-9 arrived in a 12-hour orbit around the Red Planet in 1971, the first man-made satellite to orbit a planet other than Earth.

Mariner-9 brought the first close-ups of the moons Phobos and Deimos. It had two television cameras which sent back 7,329 photos.

Later, Viking-1 and Viking-2 carried the American flag across millions of miles of interplanetary space to photograph Mars, Phobos and Deimos. Both landed on Mars in 1976.

Phobos-2 carried the USSR flag 111 million miles to Mars orbit in 1989. It detected water vapor in the Martian atmosphere.

Pioneer. The Pioneer series of spacecraft were American interplanetary probes to the Moon, Sun, Jupiter and Venus. A total of 14 launches were attempted between 1958-78.

Luna. Thirty Luna and Zond spacecraft from the USSR explored the Moon between 1959-1976.

Ranger. Ranger was a series of probes in the early 1960s, the first NASA space program to investigate another Solar System body—the Moon. At first, Rangers were supposed to bounce onto the lunar surface with a seismometer, but the first six failed.

After that, they were redesigned for photography. The first five were launched in 1961 and 1962. The sixth, in 1964. The last three Rangers, redesigned, were successful.

They snapped pictures as a Ranger approached its bulls-eye on the Moon. They made the first close-ups of the Moon surface, showing boulders and three-foot craters.

Lunar Orbiter. Lunar Orbiter was a U.S. series of five satellites launched in 1966 and 1967 to orbit and map the Moon. Lunar Orbiters-1, -2 and -3 sought Apollo landing sites, while Lunar Orbiters -4 and -5 did global mapping of the lunar landscape. They discovered excess concentrations of mass under the maria, known as mascons, and photographed 99 percent of the lunar surface.

Surveyor. Surveyor was a series of seven U.S. unmanned TV and trenching probes designed to land softly on the Moon, part of America's search for Apollo moonship landing sites.

They were launched between 1966-1968. Five soft landings were successful.

Surveyors had a moveable TV camera powered by solar cells and a trench digger along with the capability to retrieve and analyze a soil sample.

Surveyor-1 made the first soft landing on the Moon in 1966 for man's first close-up look at the surface. The three-legged spacecraft flew 240,000 miles to land within nine miles of target.

Over eight months it transmitted 11,150 pictures, including panoramas and close-ups. Surveyors found lunar maria to be basaltic and the lunar highlands rich in aluminum and calcium.

Within two years, four other Surveyors landed to study Moon chemistry and physics. Information retrieved by the five Surveyors was used in the manned Apollo flights from 1969-72.

Mariner. Mariner was a series of U.S. interplanetary science probes to the Solar System's inner planets in the 1960s and early 1970s. Later, Mariner-11 and Mariner-12 became Voyager-1 and Voyager-2 which went on grand tours of the Solar System's outer planets.

Voyager. The Voyager twins were to fly out through the Asteroid Belt to explore Jupiter and Saturn. The probes crossed the Asteroid Belt in 1978 and 1979. Voyager-1 swept past Jupiter in 1979, using the giant planet's gravity for a slingshot boost toward Saturn. Voyager-2 sped by Jupiter in 1979.

Voyager-1 flew past Saturn in 1980, using the planet's gravity to fling it up out of the Solar System plane to encounter Saturn's moon Titan. Voyager-2 went by Saturn in 1981.

Voyager-1 pulled above the swirling disk of planets surrounding our Sun, rushing for the edge of the Solar System and beyond on an endless space odyssey. Today, it is several billion miles from Earth and billions of miles above the plane of the Solar System.

The Jupiter and Saturn results were so electrifying, Voyager-2's assignment was extended to Uranus and Neptune—the so-called Grand Tour.

The small spacecraft called on those big outer planets for repeated gravity assists as it cruised on to encounter Uranus in 1986 and Neptune in 1989.

Today, Voyager 2 also is cruising out of the Solar System in interstellar space.

Spritzing Spacecraft

Small as coffee cans—that's how tiny interplanetary probes could be if NASA decided to spritz them across the Solar System.

Instead of launching one large expensive probe each decade, the space agency could launch ten to fifty midget spacecraft each year.

A two-lb. Coffee Can probe would be crammed with microelectronics, ferried to Earth orbit and blasted away to cruise the Solar System, sniffing around the Sun, planets, comets and asteroids.

It would be an inexpensive spacecraft and a cheap launch. One of the tiny science craft couldn't tote many instruments on its trek through the Solar System, but it could have a miniature camera, radio gear and power supply. It could carry a gamma-ray detector and seismometer to measure quake activity on planets, asteroids and moons. It even could be padded to land on a hard surface.

Midget spacecraft would ease the pain of probe failure since loss of one small instrument wouldn't be as drastic as the failure of a major all-the-eggs-in-one-basket spacecraft. Each Coffee Can would cost only a tiny fraction of the $300 to $800 million NASA pays for a jumbo probe.

In the late '50s and early '60s, at the dawn of the Space Age, satellites and probes under 50 lbs. were the norm, compared with today's spacecraft weighing hundreds or thousands of pounds.

Interplanetary Probes

This a selection of the best-known interplanetary probes. Most were at least somewhat successful. Others, not listed, were not successful. The chronological list shows year of launch or key event, name of the probe, the country which originated the flight and the current location or status of the interplanetary probe.

Year	Name	Source	Current Status
1959	Luna-1	USSR	orbiting the Sun
1959	Pioneer-4	US	orbiting the Sun
1959	Luna-2	USSR	smashed into the surface of the Moon
1959	Luna-3	USSR	was in Earth-Moon orbit, now decayed
1960	Pioneer-5	US	orbiting the Sun
1960	Luna-4	USSR	failed to orbit the Moon and was lost in space
1961	Venera-1	USSR	Venus probe now orbiting the Sun
1962	Ranger-3	US	missed the Moon and now is orbiting the Sun
1962	Ranger-4	US	smashed into the surface of the Moon
1962	Mariner-2	US	Venus flyby now orbiting the Sun
1962	Ranger-5	US	lunar flyby now orbiting the Sun
1962	Mars-1	USSR	contact lost on way to Mars
1963	Luna-4	USSR	in Earth-Moon orbit
1964	Ranger-6	US	smashed into the surface of the Moon
1964	Zond-1	USSR	Venus probe now orbiting the Sun
1964	Ranger-7	US	sent pictures of its impact on Moon
1964	Mariner-3	US	Mars-flyby attempt, now orbiting the Sun
1964	Mariner-4	US	flew by Mars, took the first Mars-photos, now orbiting the Sun
1964	Zond-2	USSR	contact lost enroute to Mars
1965	Ranger-8	US	sent pictures of its impact on Moon
1965	Ranger-9	US	sent pictures of its impact on Moon
1965	Luna-5	USSR	soft-lander failed, smashed into the Moon
1965	Luna-6	USSR	soft-lander missed the Moon and now is orbiting the Sun
1965	Zond-3	USSR	lunar flyby, photographed Moon, now orbiting the Sun
1965	Luna-7	USSR	soft-lander failed and smashed into the Moon
1965	Venera-2	USSR	Venus probe now orbiting the Sun
1965	Venera-3	USSR	Venus probe crashed on Venus
1965	Luna-8	USSR	lunar soft-lander failed and smashed into the Moon
1965	Pioneer-6	US	solar probe orbiting the Sun
1966	Luna-9	USSR	landed on the Moon
1966	Luna-10	USSR	in lunar orbit
1966	Surveyor-1	US	landed on the Moon
1966	Lunar Orbiter-1	US	orbited Moon, Far Side photos, crashed on Moon on command
1966	Pioneer-7	US	solar probe orbiting the Sun
1966	Luna-11	USSR	in lunar orbit
1966	Surveyor-2	US	lunar soft-lander failed and crashed on the Moon
1966	Luna-12	USSR	in lunar orbit
1966	Lunar Orbiter-2	US	orbited the Moon, photographed its Far Side and potential Apollo-landing sites, then crashed into the Moon on command
1966	Luna-13	USSR	landed on the Moon
1967	Lunar Orbiter-3	US	orbited the Moon, photographed its far side and an Apollo-12 landing site, then crashed into the Moon on command
1967	Surveyor-3	US	landed on the Moon and brought soil back to Earth
1967	Lunar Orbiter-4	US	Moon polar-orbit covers Earth side, crashed on command
1967	Venera-4	USSR	probably impacted on Venus
1967	Mariner-5	US	Venus flyby now orbiting the Sun
1967	Surveyor-4	US	lunar soft-lander failed and crashed on the Moon
1967	Lunar Orbiter-5	US	Moon polar-orbited for high-res images, crashed on command
1967	Surveyor-5	US	landed on the Moon
1967	Surveyor-6	US	landed on the Moon, then took off from the Moon
1967	Pioneer-8	US	solar probe orbiting the Sun
1968	Surveyor-7	US	landed on the Moon
1968	Luna-14	USSR	probe in lunar-solar orbit
1968	Zond-5	USSR	lunar fly-around and return to Earth

Year	Name	Source	Current Status
1968	Pioneer-9	US	solar probe orbiting the Sun
1968	Zond-6	USSR	lunar fly-around and return to Earth
1968	Apollo-8	US	first manned lunar fly-around and return to Earth
1969	Venera-5	USSR	probably impacted on Venus
1969	Venera-6	USSR	probably impacted on Venus
1969	Mariner-6	US	Mars-flyby, transmitted 75 photos, now orbiting the Sun
1969	Mariner-7	US	Mars-flyby, transmitted 126 photos, now orbiting the Sun
1969	Apollo-10	US	second manned lunar fly-around and return to Earth
1969	Luna-15	USSR	lunar orbiter finally landed on the Moon
1969	Apollo-11	US	first manned lunar landing, returned men and samples to Earth
1969	Zond-7	USSR	lunar fly-around and return to Earth
1969	Apollo-12	US	second manned lunar landing, returned men, samples to Earth
1970	Apollo-13	US	manned lunar fly-by and emergency-rescue return to Earth
1970	Luna-16	USSR	landed on Moon, picked up soil samples and returned to Earth
1970	Luna-17	USSR	landed on Moon, automated rover maneuvered around surface
1971	Apollo-14	US	third manned lunar landing, returned men and samples to Earth
1971	Mars-2	USSR	reached Mars orbit, lander crashed, but no data received
1971	Mars-3	USSR	reached Mars orbit, landed on Mars, failed after 20 seconds
1971	Mariner-8	US	was to orbit Mars, but failed to leave Earth orbit
1971	Mariner-9	US	orbiting Mars, mapped surface, 7329 photos transmitted
1971	Apollo-15	US	fourth manned lunar landing, returned men, samples to Earth
1971	subsatellite	US	subsatellite launched into lunar orbit from Apollo-15
1971	Luna-18	USSR	lunar orbiter finally impacted on lunar surface
1971	Luna-19	USSR	in lunar orbit
1971	Mariner-9	US	entered orbit around Mars
1971	Cosmos-419	USSR	was to orbit Mars and drop lander, but failed to leave Earth orbit
1972	Luna-20	USSR	landed on Moon, picked up soil samples and returned to Earth
1972	Pioneer-10	US	flew by Jupiter, then left the Solar System
1972	Venera-8	USSR	landed on Venus
1972	Apollo-16	US	fifth manned lunar landing, returned men and samples to Earth
1972	subsatellite	US	launched to lunar orbit from Apollo-16, later crashed on Moon
1972	Explorer-47	US	probe of space between Earth and Moon, in Earth-Moon orbit
1972	Apollo-17	US	sixth, last manned lunar landing, returned men and samples
1973	Luna-21	USSR	landed on Moon, automated rover maneuvered around surface
1973	Pioneer-11	US	flew by Jupiter, then flew by Saturn
1973	Explorer-49	US	solar physics probe in lunar orbit
1973	Mars-4	USSR	failed to orbit Mars, now orbiting the Sun
1973	Mars-5	USSR	orbiting Mars
1973	Mars-6	USSR	Mars lander failed and now is orbiting the Sun
1973	Mars-7	USSR	Mars flyby and its lander failed so now also orbiting the Sun
1973	Mariner-10	US	flew by Venus and Mercury and now is orbiting the Sun
1973	Pioneer-10	US	flew by Jupiter
1974	Luna-22	USSR	lunar probe now orbiting the Sun
1974	Luna-23	USSR	lunar probe crashed on the Moon
1974	Pioneer-11	US	flew by Jupiter
1974	Helios	US/FRG	solar probe orbiting the Sun
1975	Venera-9	USSR	orbited Venus, sent down lander
1975	Venera-10	USSR	orbited Venus, sent down lander
1975	Viking-1	US	orbited Mars and sent down lander
1975	Viking-2	US	orbited Mars and sent down lander
1976	Helios	US/FRG	solar probe orbiting the Sun
1976	Viking-1 Orbiter	US	in Mars orbit until year 2025
1976	Viking-1 Lander	US	landed on Mars
1976	Viking-2 Orbiter	US	in Mars orbit until year 2025
1976	Viking-2 Lander	US	landed on Mars
1976	Luna-24	USSR	landed on Moon, picked up soil samples and returned to Earth
1977	Voyager-2	US	flew by Jupiter and Saturn
1977	Voyager-1	US	flew by Jupiter and Saturn
1978	Pioneer-12	US	orbiting Venus as Pioneer-Venus Orbiter
1978	Pioneer-13	US	Pioneer-Venus Probes, entered atmosphere, down on surface

Year	Name	Source	Current Status
1978	ISEE-3/ICE	US	solar probe later to study Giacobini-Zinner and Comet Halley
1978	Venera-11	USSR	landed on Venus, sent back photos
1978	Venera-12	USSR	landed on Venus, sent back photos
1979	Voyager-1	US	flew by Jupiter
1979	Voyager-2	US	flew by Jupiter
1979	Pioneer-11	US	date of Saturn fly-by, escape trajectory
1980	Voyager-1	US	flew by Saturn
1981	Voyager-2	US	flew by Saturn
1982	Venera-13	USSR	landed on Venus, sent back photos
1982	Venera-14	USSR	landed on Venus, sent back photos
1983	Pioneer-10	US	first to leave the Solar System
1983	Venera-15	USSR	arrived at Venus, orbited Venus, used radar to map the surface
1983	Venera-16	USSR	arrived at Venus, orbited Venus, used radar to map the surface
1984	Vega-1	USSR	flew by Venus, then on to probe Comet Halley, now orbiting Sun
1984	Vega-2	USSR	flew by Venus, then on to probe Comet Halley, now orbiting Sun
1985	ISEE-3	US	flew by Comet Giacobini-Zinner
1986	Voyager-2	US	flew by Uranus
1986	Sakigake	Japan	flew by Comet Halley
1986	Vega-1	USSR	flew by Comet Halley
1986	Suisei	Japan	flew by Comet Halley
1986	Vega-2	USSR	flew by Comet Halley
1986	Giotto	ESA	flew by Comet Halley
1988	Phobos-1	USSR	lost through command error enroute to Mars
1988	Phobos-2	USSR	orbiting Mars, contact lost, didn't send down Phobos lander
1989	Magellan	US	radar mapper sent from U.S. space shuttle toward Venus
1989	Voyager-2	US	flew by Neptune
1990	Magellan	US	orbited Venus to map the planet by radar
1989	Voyager	US	flew by within 3,000 miles of Neptune
1989	Galileo	US	launched from U.S. space shuttle toward Jupiter
1990	Hiten	Japan	unmanned Moon probe in Earth-Moon orbit, dropped small Hagoromo satellite into lunar orbit
1990	Pioneer-11	US	left the Solar System
1990	Galileo	US	flew by Venus
1990	Voyager	US	takes first picture of planets from outside Solar System
1990	Pioneer-10	US	passed 50 AU from the Sun
1990	Ulysses	US-ESA	formerly International Solar Polar Mission (ISPM), solar probe launched from U.S. shuttle toward the Sun
1991	Galileo	US	flew by Earth and asteroid Gaspra enroute to Jupiter
1992	Ulysses	US-ESA	flew by Jupiter enroute to the Sun
1992	Giotto	ESA	flew by Comet Grigg-Skjellrup
1992	Galileo	US	flew by Earth enroute to Jupiter
1993	Mars Observer	US	radio signal lost just as probe was about to enter Mars orbit
1994	Ulysses	US-ESA	passes over Sun's South Pole
1995	Galileo	US	fires probe into Jupiter's atmosphere, enters Jupiter orbit
1995	Ulysses	US-ESA	passes over Sun's North Pole
1996	Planet-B	Japan	Venus probe launched
1996	Moon Lander	Japan	Moon lander launched
2000	Cassini	US-ESA	Saturn fly-by drops probe onto moon Titan
2000	Lunar Observer	US	lunar orbiter launched
2002	Solar Probe	US	launched to fly very near the Sun

Glossary

AAA

A. amateur radio satellite mode, transponder frequency designator, uplink 145 downlink 29 MHz.
abort. terminate the launch or flight of a space rocket.
A/D. Analog to Digital.
ADCS. attitude determination and control system.
ADEOS. Advanced Earth Observation Satellite, Japan.
Adventure Club. Moscow group participated with AMSAT-U-Orbita in Radiosputnik 14.
AE. antenna electronics.
AEM. Applications Explorer Mission satellite.
Aerobee. small space rocket.
AFB. Air Force Base.
AFI. automatic fault indication.
AFS. Air Force Station.
afsk. audio frequency shift keying, keyboard-to-keyboard communications; representing digital information by modulating a radio carrier with two or more audio frequencies.

Agena-64-31C. a space satellite.
AHRS. attitude and heading reference system.
AIMS. airspace traffic control radar beacon system IFF.
A/J. anti-jamming.
aka. also known as.
Al-Anbar. Iraq's space launch site.
Albert. monkey, first primate to fly near space.
ALEXIS. Array of Low Energy X-ray Imaging Sensors satellite.
ALINS. AMSAT Launch Information Net/Service.
Almaz. large Russian satellite to observe land and ocean surfaces.
Alouette. Canadian experimental communications satellite.
altitude. satellite orbital apogee or maximum altitude in miles.
am. amplitude modulation; also a.m.; radio modulation type; voice communication; commonly used in medium-wave and shortwave radio broadcasting and for vhf and uhf aircraft communications.
amateur radio. ham radio, a hobby in which persons are licensed to transmit and receive messages and conduct conversations among themselves on radio frequencies specified by a government for enjoyment, experimentation, public service and promotion of the radio art.
amateur radio operator. licensed user of frequencies allocated to radio amateurs; a ham.
amateur related. non-amateur satellite used by amateurs or in some way related to amateur radio.
amateur satellite. dozens of ham radio satellites have been launched to Earth orbit since 1961. Along with voice and Morse code communications equipment, many amateur radio satellites have contained microcomputers and packet mailboxes to store-and-forward messages for relay over long distances. Some have contained TV cameras to photograph the Earth and stars and transmit the pictures to the hams on the ground.
AMPTE. Active Magnetosphere Particle Tracer Explorers; British IKS, German IRM, American CCE satellites.
AMSAT. Radio Amateur Satellite Corporation, P.O. Box 27, Washington, D.C. 20044; founded in 1969; worldwide organization of amateur radio operators and others interested in building and using satellites for space communications and science research. AMSAT has local affiliates around the globe.
AMSC. American Mobile Satellite Corp.
angstrom. a unit of length = 1.0e-08cm.
Angular momentum. A quantity of motion caused by rotation.
Anik. Canadian geosynchronous satellites.
ANNA. American military project name derived from the initials of Army, Navy, NASA and Air Force.
ANS. Astronomical Netherlands Satellite, Holland.
AO. AMSAT-OSCAR; amateur radio satellite.
AOC. Auxiliary Output Chip.
AOS. acquisition of signal.
aphelion. farthest point in an orbit from the Sun.
apogee. high point in orbit, farthest from center; orbit location farthest from the body being orbited. For instance, the point where a satellite orbiting Earth is farthest from Earth.
APT. the satellite version of weather facsimile sending photos and weather maps by radio.
Arab League. parent of Arab Satellite Communication Organisation which builds Arabsats.
Arabsat. communications satellites from the Arab Satellite Communication Organisation, an Arab League group based in Riyadh, Saudi Arabia.
Ariane. European Space Agency satellite-launching rocket; launched the European Space Agency's first space satellite, CAT, from Kourou, French Guiana, and dozens since.
Ariel. British satellite.
ARRL. American Radio Relay League, primary national amateur radio fraternity in the U.S.
ARSENE. French acronym for Ariane Radio Amateur Satellite pour l'ENseignement de l'Espace.
artificial satellite. a man-made object or machine sent to orbit Earth or another body in Outer Space for purposes useful to mankind.
Aryabhata. India's first satellite.
A-S. anti-spoofing.
ASAP. Ariane structure for auxiliary payloads.

ASAT. United States anti-satellite rocket weapon fired from the ground.
ASC. a geosynchronous satellite.
ASCO. Arab Satellite Communication Organisation, an Arab League group building Arabsats.
ASIC. application specific integrated circuit.
Asterix. France's first space satellite; also a red-whiskered Celtic barbarian in French comics.
asteroid. a rock of debris orbiting the Sun; a minor planet or planetoid; thousands of asteroids, mostly in an orbit, known as the Asteroid Belt, are between the orbits of Mars and Jupiter.
Astra. satellite of Luxembourg's Societe Europeenne des Satellites, SES, the European Satellite Society.
Astro. ultraviolet light telescopes in space shuttle.
astronaut. man or woman who has flown, or is preparing to fly, in a spacecraft.

astronomical unit. AU, a unit of measurement. The average distance between Earth and Sun is about 92.96 million miles. Astronomers refer to that distance as one astronomical unit (AU). The outer edge of the Solar System is estimated to be 75 to 100 times the distance from the Earth to the Sun. That is 75 to 100 AU. The mean Sun-Earth distance; 1.496E+13cm or 214.94 solar radii.
ATE. automatic test equipment.
ATEPRA. Association Technique pour l'Experimentation du Packet Radio Amateur, France.
Atlantis. fourth U.S. space shuttle to fly to orbit.
Atlas. unmanned U.S. space rocket used for decades to lift medium to heavy-weight payloads to Earth orbit.
ATS. Applied Technology Satellite; series of American stationary communications satellites.
atv. amateur television, visual mode of communications. fstv and sstv.
AU. astronomical unit; the mean Sun-Earth distance; 1.496E+13cm or 214.94 solar radii.
Aureole. Oreol; Russian-French science satellite.
Aurora. a geosynchronous satellite.
aurora. faint visual phenomenon associated with geomagnetic activity, occurs mainly in Earth's high-latitude night sky; typical auroras are 100 to 250 km above ground; energy from the Sun strikes particles in Earth's upper atmosphere, causing a glow over polar territory.
Aussat. Australian comunications satellite.

Australis-OSCAR 5. OSCAR 5, Astronautical Society & Radio Club, Univ. of Melbourne, Australia, 1970.
autotransponder. electronic robot amateur radio operator built into Radiosputniks.
autumnal equinox. see equinox.
AX.25. amateur-X.25 packet digital protocol.
AXAF. Advanced X-Ray Astrophysics Facility, one of NASA's Great Observatories in space.
azimuth. direction to the horizon under an object in space; in satellite tracking, the direction in which to look, measured in degrees from north. 0 or 360 degrees is north. 90 is east. 180 is south. 270 is west.

BBB

B. amateur radio satellite mode, transponder frequency designator, uplink 70 cm. downlink 145 MHz.
Badr. "new moon" in the Urdu language; the Badr-A was the first Pakistani amateur radio satellite, launched by China in 1990.
Baikonur Cosmodrome. vast Russian southern spaceport, located outside of Russia in Kazakhstan.
band. a contiguous group of frequencies within the spectrum.
BBC. British Broadcasting Corporation.
bbs. store-and-forward electronic-mail message bulletin-board system.
BBXRT. broad-band X-ray telescope in space shuttle.
BCD. binary code decimal.
BCR. battery charge regulator.
beacon. one-way hamsat transmitter sending down telemetry information about conditions of satellite equipment and the space environment; locating radio signal, with or without information.
BEAR. Beam Experiment Aboard Rocket.
Belka. dog in Sputnik-5, Squirrel in Russian, first successful recovery of a satellite from space.
Bhaskara. Satellite for Earth Observation, SEO, India.
bi-stable. satellite can find gravity-gradient lock either pointing toward Earth or pointing into space.
Big Bird. American military spy satellite.
BIH. Bureau International de L'Heure.
BIPM. International Bureau of Weights and Measures.
BIT. built-in-test.
Black Arrow. rocket which launched Great Britain's first space satellite, Black Knight, from Woomera, Australia.
Black Knight-1. Great Britain's first space satellite.
BMDO. Ballistic Missile Defense Organization.
boosters. helper rockets strapped to a main rocket engine.
Boundless Aperture. space artwork.
bps. bit per second.
bpsk. binary phase shift keying.
Brilliant Pebbles. tiny rockets to track ballistic missiles in space and crash into them.
BRTK-11. Radiosputnik-15 hamsat radio.
BSB. Japanese geosynchronous satellite.
BSS-Sound. digital audio broadcasting, DAB.
bulletin-board system. bbs; electronic store-and-forward message system.
Buran. first orbiter of the Russian space shuttle fleet. Buran is Russian for snowstorm or blizzard.
burn. firing a rocket engine.

CCC

C-band. 4,000-8,000 MHz or 4-8 GHz, radio-frequency band in the electromagnetic spectrum.
C/A-code. coarse/acquisition-code.

C/No. carrier to noise ratio.
CADC. central air data computer.
CAP. Civil Air Patrol, volunteers who search for downed aircraft.
Cape Canaveral. vast American space launch site, in Florida, adjacent to Kennedy Space Center and Patrick Air Force Base.
capsule. pressurized compartment or spacecraft, atop a rocket, in which astronauts travel to space or in which material is dropped from space to Earth. For instance, Mercury, Gemini, Apollo, and Soyuz capsules.
CAST. Center for Aero Space Technology, Weber State University, Ogden, Utah.
CAT. the European Space Agency's first space satellite.
Cayenne. French Guiana city.
CBC. Canadian Broadcasting Corporation.
CBERS. China-Brazil Earth Resources Satellite, a remote-sensing satellite.
CCD. charge-coupled device, television camera technology used in satellites.
CCE. Charge Composition Explorer; the U.S. AMPTE satellite.
CDM. cold dark matter across the Universe.
CDMA. code division multiplex access.
CDU. control display unit.
celestial sphere, equator, poles. the celestial sphere is the imaginary inverted bowl of stars surrounding Earth. The celestial equator is a projection of Earth's equator onto the celestial sphere, equidistant from the celestial poles, dividing the celestial sphere into two hemispheres. The celestial poles are points on the celestial sphere above Earth's North Pole and South Pole. The celestial sphere seems to rotate around the celestial poles. The meridian is an imaginary great circle line crossing the celestial sphere, passing through both Poles and the Zenith. The prime meridian is the meridian passing through the vernal equinox. Zenith is the point in the sky directly above the observer. Nadir is the point on the celestial sphere opposite Zenith. Precession is the shift of celestial poles and equinox due to the gravitational tug of the Sun and the Moon on Earth's equatorial bulge. Right ascension is the eastward angle between an object and the vernal equinox, expressed in hours-minutes-seconds. Solstice is the sky positions where the Sun is at maximum angle or declination from the celestial equator. Declination is the angle between a celestial object and the celestial equator. A node is a point in orbit where an ascending or descending object crosses the ecliptic.
Centaur. a space satellite.
Centre Spatial Guianase. CSG; European Space Agency launch site in jungle near Kourou.
CEP. circular error probable.
CEsar. Amateur radio communications satellite from the Radio Club Federation of Santiago, Chile.
CGRO. Compton Gamma Ray Observatory, one of NASA's Great Observatories in space.
Challenger. second U.S. space shuttle. Challenger was destroyed in a 1986 launch explosion.
Chernushka. Blackie in Russian, Sputnik dog.
chromosphere. outer atmosphere layer of the Sun above the photosphere.
cislunar space. region between the Earth and the Moon.
Clarke Belt. stationary-orbit region of space at 22,300 miles altitude where satellites are synchronous, geostationary or geosynchronous.
CMOS. Complementary Metal Oxide Semiconductor.
CNES. Centre National d'Etudes Spatiales, the French space agency.
CNN. Cable News Network.
COBE. Cosmic Background Explorer satellite.
Codestore. satellite digital store-and-forward message system.
coherent cw. synchronous binary amplitude modulation.
Columbia. first U.S. space shuttle to fly to orbit.
Columbus. Europe's manned module in development for space station Freedom.
Comet. Shavit in Hebrew; the rocket which launched Israel's first space satellite, Horizon, from the Negev Desert.
comet. clumps of gas, dust, dirt, rock, ice; giant snowballs of dirty slush in long elliptical orbits around the Sun; more than 1,000 have been seen as orbital paths bring them near the Sun; best known is Comet Halley.
comm, comms. communication, communications.
communications relay. A communications satellite is a two-way relay device, or repeater, relaying voice, television or digital signals.
communications satellite. unmanned satellite in orbit to relay radio, television, data and other communications signals; the first, called SCORE, ECHO, TELSTAR, RELAY, and SYNCOM, were launched between 1958 and 1963.
communications station. manned or unmanned space station in orbit to relay radio, television, data and other communications signals.
Compton. Compton Gamma Ray Observatory, GRO, one of NASA's Great Observatories in space.
CompuServe. electronic data service circulating orbital data for popular satellites.
comsat, commsat. communications satellite.
COMSAT. trade name for one communications satellite operating company.
Comstar. a stationary communications satellite.
Coordinated Universal Time. UTC; by international agreement, the local time at the prime meridian which passes through Greenwich, England; thus, Greenwich Mean Time (GMT); also z or zulu time. Mean solar time at the zero-degree meridian at Greenwich, England.
CORSA. Cosmic Radiation Satellite, Japan's Hakucho.
COS-B. European gamma-ray astronomy satellite.
cosmic rays. energetic particles, or ions, traveling through space after being emitted by stars.

cosmonaut. man or woman who has flown, or is preparing to fly, Russian spacecraft.
Cosmos. generic name given to most of Russia's unmanned satellites.
Cosmos-1861. host for Radiosputniks 10 and 11.
Cosmos-2123. host for Radiosputniks 12 and 13.
COSPAR. Committee On Space and Atmospheric Research; an international numbering system for satellites, composed of year of launch plus sequence of launch within year plus a letter designating the object launched.
COSPAS. Russian acronym for Space System for Search of Vehicles in Distress, as in SARSAT/COSPAS.
countdown. high to low counting of final seconds leading to a rocket launch. Ten, nine, eight...
country. nation of origin.
Courier-1B. U.S. communications satellite.
CPE. cosmic particle experiment aboard hamsat.
CPU. microcomputer central processing unit.
CREDO. cosmic ray effects and docimetry experiment.
crew commander. U.S. shuttle astronaut who has overall responsibility for mission success and flight safety.
CRPA. controlled radiation pattern antenna.
CRRES. Combined Release and Radiation Effects Satellite.
CSA. Canadian Space Agency.
CSG. Centre Spatial Guianasc; Europcan Space Agency's launch site carved from jungle near Kourou.
CSOC. consolidated space operations center.
CTS. Communications Technology Satellite, a geosynchronous satellite.
cutoff. stopping combustion to end thrust by shutting off the flow of fuel in a rocket engine.
cw. continuous wave, radio communications using International Morse code, a transmitter's carrier is turned on and off in a pattern.
Cyrillic. Greek-based alphabet used in Russian and Slavic languages.

DDD

D-level. depot level.
D-region. a daytime layer of Earth's ionosphere approximately 50 to 90 km in altitude.
DAB. digital audio broadcasting, BSS-Sound.
DAC. digital to analog converter.
DAIRY. UO-14 computer operating system.
DARA. Deutsche Agentur Fur Raumfahrt Angelegenheit, German space agency.
DARPA. Defense Advanced Research Projects Agency, Arlington, Virginia; Microsats 1-7, 1991.
data. digital computer information, a mode of communications.
dB. decibel.
DCE. digital communications experiment.
decay. loss of altitude by a satellite, resulting from a reduction in kinetic energy caused by atmospheric friction.
declination. angle between satellite and celestial equator.
deep space. realm beyond Earth and Moon.
Delta. a U.S. space satellite launching rocket used for decades to lift medium-weight payloads to Earth orbit.
demo. demonstration.
DGPS. Differential GPS; more accurate than GPS.
Diamant. rocket which launched France's first space satellite, Asterix, from Algeria.
differential GPS. DGPS.
digi. digipeater.
DIGIMOON. proposed packet-radio bbs mailbox and tv camera on Moon.
digipeater. digital repeater, relays packet-radio.
digital. data transmission.
Digitalker. UoSAT synthesized-audio voice transmitted from space to Earth.
Discoverer-36. U.S. military payload on Thor-Agena rocket carrying OSCAR-1 to orbit in 1961.
Discovery. the third U.S. space shuttle to fly to orbit.
DLM. data loader module.
DLR. data loader receptacle.
DLS. data loader system.
DMA. Defense Mapping Agency.
DMS, DMSP. Defense Meteorological Support Program satellite system.
docking. mechanical linking of two spacecraft.
DOD. Department of Defense.
DOP. dilution of precision.
Doppler shift. when listening to a signal from an orbiting satellite, makes the signal's frequency seem higher when the satellite is coming toward you and lower when going away.
DORIS. Doppler Orbitography and Radio positioning Integrated from Space.
DOVE. Digital Orbiting Voice Encoder, Brazilian amateur radio satellite.
down. date about when a satellite descended into the atmosphere and burned up.
downlink. transmission downward from a satellite, shuttle or space station; transmission from space to Earth.
downlink ssb. ssb voice communications.
drag. resistance of air in the upper atmosphere to the motion of a satellite.
DRIG. Dallas Remote Imaging Group.
dRMS. distance root mean square.
DRS. dead reckoning system.
DSCS. Defense Satellite Communications System stationary satellite, relaying classified conversations among military personnel; DSCS often is pronounced DISK-us.
DSN. Deep Space Network, NASA.
DSP. Defense Support Program early-warning satellites.
DT&E. development test and evaluation.
DUE. Diffuse Ultraviolet Experiment.
DX. distance.

EEE

E-region. a daytime layer of Earth's ionosphere approximately between the altitudes of 85-140 km.
Earth. the fifth largest planet of the Solar System and third from the Sun.
Earth Imaging System. EIS; satellite television camera for space photography.
Earth satellite, artificial. object sent to orbit Earth for one or more purposes useful to mankind.
ECEF. Earth-centered-Earth-fixed.
ECHO. A 100-ft. aluminum-coated balloon used as a passive communiations satellite to bounce radio and television signals from New Jersey to France.
ECP. engineering change proposal.
EDM. electronic distance measurement.
Edwards. Edwards Air Force Base, California.
EFIS. electronic flight instrument system.
EGP. Experimental Geodetic Payload, Japan's Ajisai satellite.
ehf. extremely high frequency, 25,000-300,000 MHz or 25-300 GHz, radio-frequency band.
eirp. effective radiated power.
EIS. Earth Imaging System; satellite television camera for space photography.
electromagnetic. spectrum of wavelengths; gamma rays, X-rays, ultraviolet rays, visible light, infrared, microwaves, radio.
electrons. parts of atoms; the solar wind is a stream of negatively-charged electrons and positively-charged protons moving away from the Sun.
elevation. angle of an object in space above the horizon; in satellite tracking, the angle above the horizon. 0 degrees is the horizon, 90 degrees is directly overhead.
elf. extremely low frequency; portion of radio frequency spectrum from 30 to 3000 hertz.
ELINT. electronic intelligence satellite.
elliptical. oval or egg-shaped orbit.

El Nino. global weather disturbance which starts when weakening trade winds allow a large pool of warm water in the western Pacific to slip eastward to South America.
ELT. emergency locator transmitter.
EM. electro magnetic.
EMCON. emission control.
EME. Earth-Moon-Earth or moonbounce communications path using the Moon as a passive radio reflector.
Endeavour. the fifth U.S. shuttle to fly to orbit; replaced shuttle Challenger.
Energia. Russia's most-powerful space-launch rocket; used to launch shuttle Buran to Earth orbit.
engine. rocket, also known as motor.
Enos. a chimp, first primate in orbit, Project Mercury.
Enterprise. 1970s prototype U.S. shuttle; it was not flown to space.
EOS. Earth Observing System satellites, nucleus of NASA's Mission to Planet Earth.
EOSat. Earth Observation Satellite Co., joint venture of Hughes Aircraft Co. and General Electric Co. to maintain LANDSAT operations for the Commerce Dept.'s National Oceanic and Atmospheric Administration.
ephemeris. daily position tables for planets, comets and other celestial bodies.
EPIRB. Emergency Position Indicating Radio Beacon.
equatorial orbit. satellites flying east-west, circling the globe above the equator.
equinox. at equinox, the Sun crosses the equator and day and night are equal in length. Autumnal and vernal equinoxes are the points where the ecliptic crosses the celestial equator. Autumnal equinox occurs as the Sun's apparent north-to-south path crosses the celestial equator. Vernal equinox as the Sun's apparent south-to-north path crosses the celestial equator. Precession is the shift of the equinox due to the gravitational tug of the Sun and the Moon on Earth's equatorial bulge.
erp. effective radiated power.
ERS. European Remote Sensing satellite.
ERTS. Earth Resources Technology Satellite, NASA, LANDSAT.
ESA. European Space Agency; countries working together in space exploration through satellites.
escape velocity. speed required for a rocket to leave a planet or celestial body without falling into orbit around it.
ESGN. electrically suspended gyro navigator.
ESIEE. France's Ecole Superieure d'Ingenieurs en Electrotechnique et Electronique.
ESIEESPACE. amateur astronomy club at ESIEE.
ESOC. European Space Operations Center, Darmstadt, Germany.
ether. imaginary substance said to fill outer space.
ETR. Eastern Test Range, Florida space launch site.
ETS. Engineering Technology Satellite, Japan's Kiku.
European Space Agency. ESA; countries working together in space exploration through satellites; launch site, Kourou, French Guiana, South America.
EUV. extreme ultraviolet; portion of electromagnetic spectrum from 100 to 1000 angstroms.
EUVE. Extreme Ultraviolet Explorer; satellite with four astronomical telescopes.
EVA. extra-vehicular activity by an astronaut or cosmonaut, a spacewalk outside a shuttle or space station.
EXOS. Japanese space-physics satellites, Kyokko, Jikiken.
EXOSat. European X-ray Observatory Satellite.
Explorer-1. the first American space satellite.
Explorer-6. sent down the first television pictures of Earth's cloud.
external tank. a large liquid-fuel tank attached to a U.S. shuttle orbiter at launch.
extreme ultraviolet. EUV; portion of electromagnetic spectrum from 100 to 1000 angstroms.

extremely low frequency. ELF; portion of radio frequency spectrum from 30 to 3000 hertz.
EYEsat. Italian microsatellite.

FFF

FAA. Federal Aviation Administration.
facsimile. fax, the method used to send photographs and weather satellite maps by radio or telephone.
Far Side. back side of the Moon.
fax. facsimile, communications carrying images, a picture mode of radio communication, on hf.
FCC. Federal Communications Commission regulates radio in the United States and issues licenses.
Fellow Traveler of the Earth. Iskustvennyi Sputnik Zemli in Russian; Sputnik-1.
Feng Yun. Chinese weather satellite.
fireball. bright meteor or shooting star seen in Earth's sky.
Firewheel. main payload destroyed when Ariane launch failed May 23, 1980.
FLTSATCOM. U.S. Navy geosynchronous communications satellite.
fly-by. exploring a planet, moon, asteroid, comet, Sun or other body by a passing, remote-controlled, unmanned reconnaissance spacecraft.
flying saucer. imaginary flying machine said to be from a planet other than Earth; UFO.
fm. frequency modulation, wbfm, nbfm, voice communication mode; commonly used in vhf and uhf radios.
fMIN. lowest radio wave frequency reflected from the ionosphere.
FMS. foreign military sales.
FO-12. Fuji-OSCAR 12, first Japanese hamsat, twelfth OSCAR, launched in 1986.
FO-20. Fuji-OSCAR 20, second Japanese hamsat, twentieth OSCAR, launched in 1990.
FOBS. Fractional Orbit Bombardment System.
foEs. maximum ordinary mode radio wave frequency reflected from sporadic E region of ionosphere.
foF2. maximum ordinary mode radio wave frequency reflected from the F2 region of the ionosphere.
FOM. figure of merit.
Freedom. second-generation U.S.-international space station in development. U.S. President Ronald Reagan proposed in 1984 building permanently-manned station Freedom within 10 years, to include modules built by Europe, Japan, Canada and others. Some politicians and scientists objected to the expense. NASA's Johnson Space Center leads the project. European and Japanese agencies are developing modules to be attached. Freedom may be in operation by the year 2000.
F-region. upper layer of the ionosphere 120 to 1500 km in altitude; subdivided into F1 and F2; F2 is most dense and peaks between 200 and 600 km altitude; F1 is a smaller peak in electron density which forms at lower altitudes in daytime.
French Guiana. Kourou; European Space Agency launch site on the northeast coast of South America.
freq. frequency.
frequency. measured in cycles per second and labeled in hertz; a "spot on the radio dial."
FRPA. fixed radiation pattern antenna.
FRPA-GP. FRPA ground plane.
fsk. frequency shift keying, radioteletype, rtty, teleprinter, keyboard-to-keyboard communication. Representing digital information by shifting the frequency of a radio carrier in amounts representing ones and zeros.
fstv. fast-scan television, a visual mode of communication; faster than sstv.
fuel. liquid or solid propellant used in an internal reaction to provide thrust for a space rocket.
Fuji. Japanese for "wisteria."
Fuji-OSCAR 12. FO-12, first Japanese hamsat, twelfth OSCAR, launched in 1986.
Fuji-OSCAR 20. FO-20, second Japanese hamsat, twentieth OSCAR, launched in 1990.
fuzzy logic. controversial model of human logic outside of traditional logical systems which are intolerant of imprecision and partial truth.

GGG

GaAs. gallium arsenide.
Gagarin, Yuri. first man in space, first man in Earth orbit, in Vostok-1 from the USSR.
Galaxy. American geosynchronous broadcast communications satellite.
GAScan. Get-Away Special canister carried in the cargo hold of a U.S. space shuttle.
gateway. ground station linking communications satellites to other communications networks.
GDOP. geometric dilution of precision.
gegenschein. faint glow in the sky opposite the Sun, caused by light reflecting off cosmic dust.
geomagnetic. effect of Earth's magnetic field.
geomagnetic field. magnetic field in and around Earth; intensity at Earth's surface about 0.32 gauss at the equator and 0.62 gauss at North Pole.
geomagnetic storm. worldwide disturbance of Earth's magnetic field.
geomagnetic tail. Earth's geomagnetic field blown away from the Sun side of the planet by the solar wind.
GEOS. Geodetic Earth Orbiting Satellite; also Ministry of Geology and Science payload in Russian government satellite INFORMATOR-1.
geostationary. geosynchronous orbit; the path of a satellite circling Earth at an altitude of 22,300 miles at a velocity of 6,850 mph; where satellites revolve around the planet at the same speed at which Earth rotates, thus seeming to hang stationary over one spot on Earth's surface. Geostationary transfer orbit, GTO, is a much lower orbit, around 100 miles above Earth's surface, from where a satellite may be boosted on out to geosynchronous orbit.
geosynchronous. see geostationary.

GFO. GEOSAT-Follow-On, U.S. Navy oceanography satellite.
GHz. gigahertz, one billion hertz or one thousand megahertz or one million kilohertz.
gimbal. swiveling of a rocket exhaust nozzle to control direction of flight.
Glenn, John. first American in Earth orbit, in the Mercury MA-6 capsule Friendship 7.
GLONASS. Global Orbiting Navigation Satellite System; Russian satellites very similar to American NAVSTAR GPS.
GMT. Greenwich Mean Time.
Goddard, Robert. American whose 20th century experimental work confirmed a satellite might be launched by means of a rocket.
Goddard SFC. NASA's Goddard Space Flight Center, Greenbelt, Maryland.
GOES. Geostationary Operational Environmental Satellites; stationary orbit U.S. weather satellites.
GPALS. Global Protection Against Limited Strikes, ground interceptor rockets.
GPS. Global Positioning System navigation satellite.
Gramsat. India's rural satellite network, Village Satellite.
GRAPES. Grapevine Remote-sensing Analysis of Phylloxera Early Stress, NASA, California wine.
gravitation. force of attraction between bodies.
gravitation theory. English mathematician Isaac Newton, developing his theory of gravitation, mentioned in 1687 the theoretical possibility of establishing an artificial satellite of Earth.
gravity-gradient boom. rod with end weight to keep an orbiting satellite pointed at Earth.
Greenwich Mean Time. GMT; by international agreement, the local time at the prime meridian which passes through Greenwich, England; Coordinated Universal Time(UTC); also z or zulu time.
GRO. Compton Gamma Ray Observatory, one of NASA's Great Observatories in space.
ground control. a communications station on Earth where human operators direct the work of satellites orbiting hundreds or thousands of miles above our planet.
ground test. testing a rocket, tied to the ground, without a launch.
GStar. a geosynchronous communications satellite.
GTO. geostationary transfer orbit.
Guerwin. Israeli microsatellite.

HHH

Ham. American chimp on suborbital Mercury flight.
ham. amateur radio operator.
hamsat. amateur radio satellite for communications and science research.
hamshack. amateur radio operating station.
HCMM. Heat Capacity Mapping Mission satellite
HDOP. horizontal dilution of precision.
HealthNet. international non-profit electronic-mailbox microsat network for health professionals.
HealthSat. pacsat built by amateurs for private corporation SatelLife; doctors send medical literature and electronic mail to remote areas of the world.
HEAO. High Energy Astrophysical Observatorics.
HEOS. Highly Eccentric Orbit Satellites, Europe.
Hermes. manned space shuttle, Europe.
hertz. cycle per second, a measurement of radio frequency. Amount of movement of an electromagnetic wave.
hf. high frequency, radio frequency spectrum 3-30 MHz or 3,000-30,000 KHz, shortwave band.
HI. Morse-code greeting transmitted to Earth by OSCAR-1 and most hamsats since.
high frequency. hf; portion of radio frequency spectrum from 3-30 MHz.
highlands, lunar. lunar highlands, soil on the Moon rich in aluminum and calcium.
high latitudes. areas more than 60 degrees north or south of equator; zones of geomagnetic activity, "high latitudes" refers to 50-80 degrees geomagnetic.
Hipparchus. Greek astronomer about 130 B.C.
Hipparcos. High Precision Parallax Collecting Satellite, Europe.
Hispasat. Spain's geosynchronous communications satellite.
homing satellite. Chinese satellite which can be retrieved from orbit.
Hope. Nadezhda; Russian navigation satellite; also Japan's space shuttle under development.
Horizon-1. Ofeq in Hebrew; Israel's first satellite.
horse latitudes. areas of high pressure over oceans between trade winds and westerlies.
HOW. hand-over word.
HQ. headquarters.
HSI. horizontal situation indicator.
HST. Hubble Space Telescope, one of NASA's Great Observatories in space.
ht. handie-talkie, walkie-talkie, hand-held battery-powered portable two-way radio.
Hubble. HST, a NASA Great Observatory in space.
HUT. Hopkins Ultraviolet Telescope, Astro-1.
HUTsat. Finland hamsat, a pacsat, AMSAT-OH.
HV. host vehicle.
Hz. hertz; measurement of radio frequency; mount of movement of an electromagnetic wave.

III

I-level. intermediate level.
IARU. International Amateur Radio Union.
IC. integrated circuit.
ICBM. intercontinental ballistic missile.
ICD. interface control document.
ICE. International Comet Explorer, formerly ISEE-3.
ICS. initial control system.
IF. intermediate frequency.
IFF. identification friend or foe.
IFRB. International Frequency Registration Board.
IGY. International Geophysical Year.
IHU. integrated housekeeping unit, computer in a satellite.

ILS. instrument landing system.
IMP. Interplanetary Monitoring Platform.
IMS. International Magnetosphere Study.
inclination. angle between the orbital plane of a satellite and the plane of the ecliptic.
INFORMATOR-1. Russian spacecraft housing Ministry of Geology and Science payload and RS-14.
infrared telescope. equipment on Earth and in satellites sensing light at infrared wavelengths.
Injun. U.S. magnetosphere satellites.
INS. inertial navigation system.
INSAT. India National Satellite.
Intelsat. a series of geosynchronous communications satellites.
Intercosmos. Russian space satellites.
International Geophysical Year. IGY.
Internet. electronic service circulating orbital data on popular satellites.
interplanetary. within the Solar System but outside the atmosphere of any planet or the Sun. Between the planets. For example, a probe which flies away from Earth, beyond the Moon, to or near planets and other bodies in our Solar System is an interplanetary spacecraft.
interplanetary travel. hypothetical manned flights to other planets in the Solar System.
ION. Institute of Navigation.
ion. a charged nuclear particle having either an electron missing or added.
ion drive. an electric-drive space rocket which expels ions to create thrust.
ionization. electrically charging an atom by removing an electron.
ionosphere. a layer in Earth's upper atmosphere where atoms have been ionized by solar radiation; region of Earth's upper atmosphere containing a small percentage of free electrons and ions produced by photo-ionization of the atmosphere by solar ultraviolet radiation at very short wavelengths <1000 angstroms; significantly influences radio wave propagation of frequencies below 30 MHz.
ionospheric storm. disturbance in F-region of ionosphere, in connection with geomagnetic activity.
IOT&E. initial operational test and evaluation.
IP. Instrumentation Port.
IQSY. International Year of the Quiet Sun.
IRAS. Infra-Red Astronomy Satellite.
IRM. Ion Release Module; West Germany's AMPTE.
IRS. India Remote Sensing satellite.
ISAS. Institute of Space and Astronautical Science, Japan.
ISEE. International Sun Earth Explorer satellites.
Isis. Canadian communications satellite.
Iskra. Russian for spark; Iskra-1, -2, and -3 were USSR hamsats launched in 1981 and 1982.
Iskustvennyi Sputnik Zemli. Russian for Fellow Traveler of the Earth; Sputnik-1.
ISRO. India Space Research Organization.
ISTP. International Solar Terrestrial Physics project.
ITAMsat. Italian amateur radio pacsat.
ITS. intermediate level test set.
ITU. International Telecommunications Union.
IUE. International Ultraviolet Explorer satellite.

JJJ

J. hamsat mode, transponder frequency designator, uplink 145 MHz downlink 70 cm.
JA. hamsat mode, transponder frequency designator, uplink 145 MHz downlink 70 cm., analog, same as mode J; also the international amateur radio callsign prefix for Japan.
JAS. Japan Amateur Satellite, FO-12 and FO-20.
JD. hamsat mode, transponder frequency designator, uplink 145 MHz downlink 70 cm., digital mode J.
JEM. Japanese Experimental Module; laboratory to be attached to U.S.-international space station Freedom.
JERS. Japanese Earth Resources Satellite.
JL. hamsat mode, frequency designator, uplink 145 MHz & 23 cm. downlink 70 cm., modes J & L combined.
Johnson. NASA's Johnson Space Flight Center, Houston, Texas.
Jovian. of Jupiter.
JPL. NASA's Jet Propulsion Laboratory, Pasadena, California.
JPO. Joint Program Office.
J/S. jamming to signal ratio.
JSC. NASA's Johnson Space Flight Center, Houston, Texas.
JTIDS. joint tactical information distribution system.
Jupiter. fifth planet from the Sun.
Jupiter-C. American rocket which launched the first U.S. space satellite, Explorer-1, from Florida; a modified Redstone ballistic missile.

KKK

K. hamsat mode, transponder frequency designator, uplink 21 MHz downlink 29 MHz.
K-band. 18,000-27,000 MHz, 18-27 GHz, radio-frequency band of the electromagnetic spectrum.
K-index. a 3-hour local index of geomagnetic activity; range 0 to 9.
KA. hamsat mode, frequency designator, uplink 145 and 21 MHz downlink 29 MHz, modes K and A combined.
Ka-band. 27,000-40,000 MHz, 27-40 GHz, radio-frequency band of the electromagnetic spectrum.
Kagoshima. Japanese space launch site.
KAIST. Korean Advanced Institute Of Technology.
Kapustin Yar. Russia's central cosmodrome.
Kelvin. a unit of absolute temperature.
Kennedy. NASA's space center at Cape Canaveral adjacent to Patrick Air Force Base.
Kepler, Johannes. first worked out orbit of Mars and proved Copernican theory of the Solar System.
Keplerian elements. a set of numbers describing the path and motion of a satellite in orbit, used for

tracking the satellite.
Keps. short for Keplerian elements.
Key Hole. U.S. photo-reconnaissance satellite.
KHz. kilohertz, frequency in kilohertz, one thousand hertz.
kick motor. small rocket attached to a spacecraft to perfect its orbit.
kilohertz. one thousand Hertz; KHz; a measurement of radio frequency; amount of movement of an electromagnetic wave.
KISS. Keep It Simple Stupid.
KITsat. South Korean hamsat, KITsat-OSCAR.
knot. speed of one nautical mile per second.
KO-23. KITsat-OSCAR 23, first South Korean hamsat, twenty-third OSCAR, launched in 1992.
Kourou. French Guiana; European Space Agency launch site on the northeast coast of South America.
Kp-index. 3-hour planetary geomagnetic index generated in Gottingen, Germany, based on K-index at 12 to 13 stations around globe.
KSC. NASA's Kennedy Space Center, Florida.
KT. hamsat mode, transponder freq. designator, uplink 21 MHz downlink 29 & 145 MHz, modes K & T combined.
Ku-band. 12,000-18,000 MHz, 12-18 GHz, radio-frequency band of the electromagnetic spectrum.
Kvant. Russian for Quantum; space station Mir astronomy module.

LLL

L. hamsat mode, transponder frequency designator, uplink 23 c. downlink 70 cm.
L-band. 1,000-2,000 MHz, 1-2 GHz, radio-frequency band of the electromagnetic spectrum.
L1. GPS primary frequency, 1575.42 MHz.
L2. GPS secondary frequency, 1227.60 MHz.
LACE. acronym for Low-power Atmospheric Compensation Experiment.
Lageos. Laser Geodynamics Satellite.
Lagrangian point. the place in space where gravity from two large bodies is neutralized.
Laika. Barker in Russian, Sputnik dog, first higher life form in orbit, died in flight.
Lambda 4S-5. rocket which launched Japan's first space satellite, Ohsumi.
LAN. local area network. A set of communication channels connecting computers for distribution of information, over the air in packet radio or via wires between computers.
LANDSAT. U.S. photo-observation satellite, Earth Resources Technology Satellite, ERTS.
latitude. north-south position in degrees from zero to ninety above or below the equator.
latitudes, high. areas more than 60 degrees north or south of equator.
latitudes, horse. areas of high pressure over oceans between trade winds and westerlies.
latitudes, low. areas less than 30 degrees north or south of equator.
latitudes, middle. zones of geomagnetic activity; refers to 20 to 50 degrees geomagnetic.
launch. lift off toward space of a rocket. The start of space flight with the take off of a rocket.
launch pad. platform, tower or tube from which a rocket blasts off for space.
launch window. time interval during which a spacecraft must be launched to achieve a desired orbit or flight path away from Earth.
LDEF. Long Duration Exposure Facility.
Leasat. a communications satellite.
LEP. Linear Error Probable.
LES. a geosynchronous communications satellite.
LF. low frequency, 30-300 KHz radio frequency band.
life. satellite's operational life in orbit.
Lightsats. DARPA microsats launched in 1991.
liquid fuel. frozen gases used as fuel in space rockets.
liquid rocket. space rockets using frozen-gases for liquid fuels.
LO. local oscillator.
LOFTI. U.S. experimental Low-Frequency Trans-Ionospheric Satellite.
longitude. east-west position in degrees from zero to 180 along equator east or west of prime meridian.
Long March-1. rocket which launched China first space satellite, Mao-1, from Inner Mongolia.
LOS. loss of signal.
low frequency. LF; portion of radio frequency spectrum from 30 to 300 KHz.
low latitudes. areas less than 30 degrees north or south of equator.
low orbit. under 1,000 miles altitude.
LRA. laser reflective array.
LRIP. low rate initial production.
LRU. line replaceable unit.
lsb. lower-sideband, single-sideband voice communications.
Luna. spacecraft from the USSR explored the Moon.
lunar. referring to the Moon.
Lunar Orbiter. U.S. satellites which orbited and mapped the Moon; discovered excess concentrations of mass under the maria, known as mascons.
lunar highlands. soil on the Moon rich in aluminum and calcium.
lunar maria. lowland basaltic soil on the Moon.
LW. long wave, 30-300 KHz, radio frequency band.

MMM

M 3000. optimum HF radio wave with 3000 km range; reflects only once from the ionosphere; single hop transmission.
Magellan. interplanetary probe orbiting Venus, using radar to map the surface.
magnetic field. the Sun generates a powerful magnetic field which dominates the Solar System environment; planets have magnetic fields.
magnetic storm. heavy flow of energetic particles from the Sun which disrupts Earth's magnetic field.
magnetometer. astronomer's device to measure strength of a magnetic field.

magnetopause. boundary layer between solar wind and magnetosphere; the Earth-space boundary where the solar wind strikes Earth's magnetic field.
magnetosphere. the upper region surrounding a planet where the planet's magnetic field is stronger than the interplanetary magnetic field.
Magsat. Magnetic Field Satellite.
mailbox. store-and-forward packet-radio bulletin-board system in a pacsat.
major geomagnetic storm. magnetic storm with Ap index greater than 49 and less than 100.
man-made satellite. an object sent to orbit Earth for one or more purposes useful to mankind.
Manarov, Musa. cosmonaut with the most time accumulated in space.
Mao-1. China's first space satellite.
Marecs. a synchronous communications satellite.
maria. lunar maria, lowland basaltic soil on the Moon.
Mariner. series of U.S. interplanetary science probes to the Solar System's inner planets.
Mariner-4. the first successful probe to reach Mars; United States.
Mariner-9. the first man-made satellite to orbit a planet other than Earth; United States; also the first close-ups of the Martian moons Phobos and Deimos.
Mariner-11. better known as Voyager-1; fly by Jupiter and Saturn.
Mariner-12. better known as Voyager-2 on grand tour of four of the Solar System's outer planets.
Marisat. a synchronous communications satellite.
MARS. Military Affiliate Radio System, radio amateurs volunteering service to Defense Department.
Mars. seventh largest planet of the Solar System and fourth from the Sun; referred to as the Red Planet because it looks orange-red through binoculars or telescope. It's faint, but when closest to Earth can be as bright as Jupiter.
Mars-1. the first attempt to probe Mars; USSR.
Marshall. NASA's Marshall Space Flight Center, Huntsville, Alabama.
mascon. excess concentration of mass under the maria on the Moon.
mB. millibar.
MCS. master control station.
MCT. mean corrective maintenance time.
mcw. modulated cw, Morse code can be heard on an a.m.-only radio.
mean anomaly. MA; descibes a satellite's position in its orbit; an orbit is broken down into 256 equally spaced segments, from an angular perspective; mean anomoly 000 is the point in the orbit which is closest to the Earth, or perigee; mean anomoly 128 is the point in the orbit which is farthest away from the Earth, or apogee.
medium frequency. mf; radio frequency spectrum from 300-3000 KHz or 0.3 to 3 MHz.
megahertz. one million hertz; 1 MHz; measurement of radio frequency. Amount of movement of an electromagnetic wave.
meridian. an imaginary great circle line crossing the celestial sphere, passing through both Poles and the Zenith; prime meridian is the meridian passing through the vernal equinox.
meteor sounder. radio transmitter in UNAMsat to ping meteors passing through Earth's atmosphere.
Meteor. Russian weather satellite.
meteor. light streaks caused by meteoroids burning in Earth's atmosphere.
meteorite. meteoroid large enough to survive the fall to Earth's surface.
meteoroid. interplanetary debris, often small as a grain of sand or smaller, arriving in Earth's atmosphere from space; may be spewed out by a comet or be a piece broken off from an asteroid.
Meteosat. European synchronous weather satellite.
MeV. megaelectronvolt; million electron volt; unit of energy, total energy carried by particle or photon.
mf. medium frequency; radio frequency spectrum from 300-3000 KHz or 0.3 to 3 MHz.
MHz. megahertz, one million hertz; frequency measurement in megahertz, one million hertz or one thousand kilohertz. A measurement of radio frequency. Amount of movement of an electromagnetic wave.
microgravity. space science term for the nearly zero gravity of Earth orbit.
micrometeoroid. very fine grain of dust arriving in Earth's atmosphere from space.
microsat. microsatellite.
Microsats 1-7. DARPA communications sats.
microwave burst. radio wave signal from optical and X-ray flares on the Sun.
middle latitudes. zones of geomagnetic activity; refers to 20 to 50 degrees geomagnetic.
midshipmen. students at the U.S. Naval Academy, Annapolis, Maryland.
military satellites. spy, communications, weather and experimental satellites.
millimeter band. 40,000-300,000 MHz, 40-300 GHz, frequency band of the electromagnetic spectrum.
Milstar. U.S. Defense Department communications satellite.
minor geomagnetic storm. magnetic storm with Ap index greater than 29 and less than 50.
Mir. Russian for peace, the manned, third-generation, Russian space station orbiting Earth.
Miss Sam. monkey, suborbital Mercury flight.
mission commander. U.S. shuttle astronaut with overall responsibility for mission success and flight safety.
mission specialist. astronaut trained for general work aboard a U.S. space shuttle flight.
MITI. Ministry of International Trade and Industry, Japan.
MLV. medium launch vehicle.
MmaxCT. maximum corrective maintenance time.
MMU. manned maneuvering unit.
mode A. hamsat transponder frequency designator, uplink 145 MHz downlink 29 MHz.
mode B. hamsat transponder frequency designator, uplink 70 cm. downlink 145 MHz.

mode J. hamsat transponder frequency designator, uplink 145 MHz downlink 70 cm.
mode JA. hamsat transponder frequency designator, uplink 145 MHz downlink 70 cm., analog, same as mode J.
mode JD. hamsat transponder frequency designator, uplink 145 MHz downlink 70 cm., digital mode J.
mode JL. hamsat frequency designator, uplink 145 MHz and 23 cm. downlink 70 cm., modes J and L combined.
mode K. hamsat transponder frequency designator, uplink 21 MHz downlink 29 MHz.
mode KA. hamsat frequency designator, uplink 145 and 21 MHz downlink 29 MHz, modes K and A combined.
mode L. hamsat transponder frequency designator, uplink 23 c. downlink 70 cm.
mode S. hamsat transponder frequency designator, uplink 70 cm. downlink 13 cm.
mode T. hamsat transponder frequency designator, uplink 21 MHz downlink 145 MHz.
modem. modulator-demodulator; converts analog signals to digital in a binary data-stream series of ones and zeros for a tnc or computer.
modulation. application of intelligence to radio signals by varying the waveform's frequency or amplitude, as in frequency modulation or amplitude modulation.
Molniya. a class of Russian communications satellites in long elliptical orbits.
Moon. a natural satellite; a body in orbit around a planet. Earth has one, Mars two. There are more than 54 moons in the Solar System. Artificial moons are man-made satellites.
moonbounce. Earth-Moon-Earth communications path using the Moon as a passive radio reflector.
Moonray. proposed amateur radio lunar communications project.
Morelos. Mexican geosynchronous communications satellite.
Morse. International Morse code; also cw or continuous wave; radio communication in which a transmitter's carrier is turned on and off in a pattern.
MOS. Maritime Observation Satellite, Japan.
motor. rocket engine.
MOU. memorandum of understanding.
M/S. meters per second.
MSL. mean sea level.
MTBF. mean time between failure.
MTBM. mean time between maintenance.
muf. maximum usable frequency, shortwave.
Mushka. Little Fly in Russian, Sputnik dog, died in flight.
mux. multiplexed, a mode of communications.
mw. medium wave, 300-3,000 KHz or .3-3 MHz, radio frequency band.

NNN

NA. North America.
N/A. not applicable.
NACA. National Advisory Committee for Aeronautics, replaced in 1958 by the National Aeronautics and Space Administration.
Nadezhda. Hope; Russian navigation satellite.
nadir. point on the celestial sphere opposite zenith; zenith is the point in the sky above the observer.
NANOsat. very small British hamsat.
NASA. National Aeronautics and Space Administration, the U.S. space agency.
NASDA. National Space Development Agency, Japan.
NATO. North Atlantic Treaty Organization; NATO's geosynchronous communications satellites.
nautical mile. one nautical mile is 1.1516 statute mile.
NAV-msg. navigation message.
navigation satellite. The first, TRANSIT, was launched in 1960.
nbfm. narrow band frequency modulation, a mode of communications.
NDEA. National Defense Education Act; a 1958 crash program to finance science education in the U.S.
Negev Desert. site of Israel's space launches.
net. network; several individual stations meeting on a frequency at the same time.
NET/ROM. a "smart" digipeater node for networking packet stations.
network. interconnecting computers for distribution of information.
Newton, Isaac. English mathematician, theory of gravitation, mentioned in 1687 the theoretical possibility of establishing an artificial satellite of Earth.
NiCd. ni-cad; nickel-cadmium rechargeable battery.
NiMH. sealed nickel-metal hydride rechargeable battery.
NOAA. National Oceanic and Atmospheric Administration; NOAA also is the name of U.S. weather satellites in polar orbits.
node. the point in a satellite's orbit around Earth where it crosses the equator; a point in orbit where an ascending or descending object crosses the ecliptic; also a programmable digipeater in a packet network.
non-amateur. refers to a satellite not owned by amateurs but in some way related to amateur radio.
NORAD. North American Air Defense Command, now the U.S. Space Command; assigns numbers in sequential order to all objects it tracks in orbit.
Northern Cosmodrome. Russia's northern spaceport at Plesetsk.
NOSC. Naval Ocean Systems Center.
NOVA. standardized TIP navigation satellite.
NRL. Naval Research Labratory.
NS. nanosecond.
NSA. National Security Agency.
NTDS. Navy tactical data system.
NTS. Navigation Technology Satellite.
NUsat 1. Northern Utah Satellite.
nutation. a nodding motion of a body such as the Moon.

OOO

O-level. Organization Level.
OAO. Orbiting Astronomical Observatory satellite.
OBC. on-board computer in hamsat.
OBS. omni bearing select.
occultation. one celestial object covering another in the view from Earth.
OCS. operational control system.
Ofek-1. Ofek, Horizon in Hebrew; Israel's first space satellite.
OFO. Orbiting Frog Otolith, biosatellite carrying two live bullfrogs.
OGO. Orbiting Geophysical Observatories, NASA.
Ohsumi. Japan's first space satellite.
Okean. Russian oceanography satellite.
Old Number Seven. Soviet rocket which launched Sputnik-1.
Olympus. European geosynchronous communications satellite.
Oort Cloud. unseen but suspected ring of comets up to 100,000 times farther from the Sun than Earth.
opposition. when a superior planet, or satellite, is opposite the Sun from Earth.
ORBCOMM-X. non-amateur American microsatellite launched in 1991.
orbit. the closed elliptical path traveled through space by an object around another object, following Newton's laws of gravitation and motion. For instance, satellites orbit planets. In a satellite's orbit around Earth, perigee is closest to the planet and apogee is farthest away. The sidereal period is the time taken by a body in completing an orbit around another object. The synodic period is the time between oppositions of a satellite. A revolution is completed when a satellite in orbit passes over the longitude of its launch site. Period is the time interval of one orbit. The path traveled through space by an object, often a circle or ellipse. The closed path of one body around another. Planets, comets, asteroids and satellites orbit the Sun. The Sun and its Solar System orbit the center of the Milky Way galaxy. Perihelion is the point of a planet's orbit closest to the Sun. The farthest point from the Sun is aphelion. The sidereal period is the time taken by a body in completing an orbit around another object. The synodic period is the time between oppositions of a planet or satellite; the high-altitude path around the globe; achieved when an object is given a horizontal velocity of approximately 17,500 mph at sea level. At that velocity, Earth's surface curves away from the horizontal as fast as gravity pulls the object downward. As the altitude of a satellite increases, its velocity decreases and its period increases. A satellite 170 miles above Earth's surface would have a period of 1.5 hours and a velocity of 17,300 mph, but the same object 22,300 miles above Earth would have a period of 24 hours (the same as the Earth's rotation) and a velocity of only 6,850 mph. A satellite in that higher orbit is called a synchronous satellite. If such a satellite orbits in the equatorial plane, it is said to be geostationary because it seems to remain at the same point above the Earth's surface.
orbital velocity. the speed required for a rocket to go into orbit around a planet or celestial body. To reach Earth orbit from the surface, a velocity of five miles per second is required.
Ordzhjonikidze. aviation institute at Moscow.
Oreol. Aureole; Russian-French science satellite.
OSCAR. Orbital Satellite Carrying Amateur Radio.
Oscar. U.S. Navy series of navigation satellites.
OSO. Orbiting Solar Observatory satellite.
OTHT. over-the-horizon targeting.
Outer Space. region beyond Earth's atmosphere.
outer-space raspberry. the beep-beep radio signal from Sputnik-1 as described by Claire Booth Luce, member of Congress.

PPP

P-3-I. pre-planned product improvement.
P-Code. precise code.
packet. a short burst of information extracted from a longer stream of digital data.
packet radio. digital communication technique for relaying computer messages; short bursts or packets of information are extracted from longer data streams and transmitted one at a time from one station to another; has built-in error detection to provide error-free communications; a received packet is checked for errors and is displayed only if correct.
pacsat. packet satellite; relaying packet radio, usually including a store-and-forward message bbs.
PAGEOS. Passive Geodetic Earth Orbiting Satellite.
Palapa. Indonesia's geosynchronous communications satellite.
PAS. PanAmSat, stationary comsat.
Patrick. Patrick Air Force Base, adjacent to NASA's Kennedy Space Center, Florida.
payload. anything carried by a spacecraft beyond what is needed for operation.
payload commander. a U.S. space shuttle astronaut who is the lead mission specialist planning and coordinating shuttle cargo. Payload commanders are forerunners of space station mission commanders.
payload specialist. an astronaut trained to work with a payload aboard a U.S. space shuttle flight. NASA refers to payload specialists as members of a space shuttle flight crew, not as astronauts.
pbbs. packet bulletin board system.
PC. personal computer.
PC-1. Sometimes used to refer to hamsat Radiosputnik-1.
PCE. packet communications experiment.
Pchelka. Little Bee in Russian, Sputnik dog, died in flight.
PDOP. position dilution of precision.
Peacetalker. DOVE, DO-17, Brazilian amateur radio satellite.
Pegasus. airplane-launched rocket to lift lightweight satellites to Earth orbit.
perigee. low point in orbit, closest approach, the

orbit location closest to the body being orbited, the point in orbit where a satellite is closest to Earth.
perihelion. point of planet's orbit closest to Sun.
period. the time a satellite takes to circle the Earth; the time interval of one orbit.
Phase 1. the first five amateur radio satellites.
Phase 2. second-generation hamsats flying low north-south polar orbits or east-west equatorial orbits from 200–1,000 miles altitude; OSCAR 6 was first.
Phase 3. amateur satellites in long elliptical orbits which keep them in view of ground stations for hours at a time, ranging out 20,000–30,000 miles from Earth, but looping back around the planet, within 1,500–2,500 miles, every ten to twelve hours; known as Molniya orbits after a class of Russian communications satellites.
Phase 4. stationary-orbit OSCAR.
Phobos. two Russian Mars probes; Phobos-2 found water vapor in the Martian atmosphere.
PHOTOINT. photo-intelligence satellite.
photometer. a device to measure light.
photon. unit of light.
photovoltaic cells. solar cells generating electricity from sunlight for satellites in space.
pilot. astronaut especially trained to operate the flight controls of a U.S. space shuttle.
Pioneer. series of American interplanetary probes to the Moon, Sun, Jupiter and Venus.
pkt. packet.
planet. a non-luminous body in orbit around a star such as our Sun. The known planets are nine massive bodies in our Solar System revolving around the Sun and reflecting its light: Mercury, Venus, Earth, Mars, Jupiter, Saturn, Uranus, Neptune, Pluto. Astronomers say there may be a tenth plus planets orbiting other stars. A minor planet is an asteroid. There are thousands of minor planets, mostly in an orbit known as the Asteroid Belt between the orbits of Mars and Jupiter.
plasma. ionized gas; gas containing ions and electrons; stream of ionized gas flowing from Sun.
Plesetsk. Russia's Northern Cosmodrome, a busy spaceport inside Russia.
PLL. phase-locked loop.
PLSS. precision location strike system.
pm. phase modulation, a mode of communications.
POAM. Polar Ozone and Aerosol Measurement sensor aboard SPOT-3.
POLAR. Polar Auroral Plasma Physics satellite.
Polar BEAR. Polar Beacon Experiment and Auroral Research.
polarimeter. device to measure the polarization of a ray of light.
polar-orbit. satellites traveling a north-south path across the poles.
power satellite. giant satellite with several square miles of photovoltaic cells to collect sunlight and beam the electrical power down to Earth.
PPM. parts per million; also pulse per minute.
PPS. precise positioning service; also pulse per second.
PPS-SM. PPS security module.
PRC. People's Republic of China space satellite.
precession. the shift of celestial poles and equinox due to the gravitational tug of the Sun and the Moon on Earth's equatorial bulge; the turning of Earth on its axis like a spinning top, caused by the pull of Moon's gravity on Earth's equatorial bulge.
Preliminary Satellite. or simply PS, as Sputnik-1 was called before launch by its Korolev design team.
prime meridian. the meridian passing through the vernal equinox, on celestial sphere.
PRN. pseudo random noise number, a NAVSTAR satellite identity number as determined by a receiver; since all GPS satellites transmit on the same frequency, each is distinguished by its PRN code number. Some receivers list an SV (satellite vehicle) number, but actually mean PRN number.
probe. unmanned interplanetary spacecraft which flies away from Earth, beyond the Moon, to or near planets and other bodies in our Solar System.
Prognoz. Russian space-physics and astronomy satellites.
Progress. a Russian unmanned cargo freighter serving the orbiting Mir space station.
Project Moonray. proposed amateur radio lunar communications project.
Project OSCAR. In 1961, California group of radio amateurs built the first amateur radio satellite, OSCAR-1.
propagation. dissemination of radio signal from point to point across Earth.
propellant. liquid or solid fuel used in an internal reaction to provide thrust for a space rocket.
protocol. standard for communication between computers.
Proton. USSR/Russia's series of heavy-lifting space rockets.
protons. parts of atoms. Solar wind is a stream of positively-charged protons and negatively-charged electrons moving away from the Sun.
PS. Preliminary Satellite, as Sputnik-1 was called before launch by its Korolev design team; the technicians also would honor Korolev by reversing the letters to SP from his first two initials for Sergei Pavlovich.
psk. phase shift keying.
PTTI. precise time and time interval.
pulse. a modulation type, also transmissions by Omega or Loran-C navigational signals.
PVT. position velocity and time.
PWE. particle wave experiment.
pyroheliometer. device to measure energy radiating from the Sun.

QQQ

Q-signals. shorthand telegraph messages used in Morse code communication via satellites.
Quantum. Kvant in Russian; space station Mir astronomy module.

RRR

RACE. Radio Amateur Club de l'Espace, France.
radar. acronym for radio detecting and ranging.
radar astronomy. radar from Earth and orbiting satellites map Venus, Mercury, Solar System bodies.
RADARsat. Canadian observation satellite.
radfet. radiation-detecting field-effect transistor.
RADINT. radar intelligence satellite.
Radio. Russian electronics magazine which reported first USSR hamsat project in 1975.
Radio-1. refers to Radiosputnik-1 hamsat.
Radio M-1. Radiosputnik-14 hamsat in Russian government spacecraft INFORMATOR-1.
radio amateur. ham operator.
radio location. using radio signals for navigation, or determining position or location.
radio positioning. using radio signals for navigation, or determining position or location.
radioastronomy. study of sources of radio energy from deep space using reflecting radiotelescopes having one or more radio antennas to receive and measure radio waves; astronomers using radiotelescopes have mapped galaxies and discovered quasars, pulsars, and a large number of complex organic molecules in interstellar space.
radiogram. formal radio message traffic.
radioisotope thermoelectric generator. RTG; long-life electricity generator powered by the decay of radioactive plutonium 238 dioxide, used in Earth satellites and interplanetary probes.
Radiosputnik. RS; a USSR/Russian hamsat.
radiotelescope. reflecting telescope with one or more radio antennas to receive and measure natural radio waves from deep space, electromagnetic energy coming from stars, planets, the dust and gas between stars in our own galaxy, and galaxies and other sources beyond our Milky Way galaxy.

radioteletype. rtty, radioteletype, keyboard-computer terminal-teleprinter communications.
RAE. Radio Astronomy Explorer.
raised cosine modulation. pacsat transmitter shaping binary waveforms like halves of sine waves instead of square waves; reduce bandwidth and improve downlink signal-to-noise ratio.
RAM. random access memory; also reliability and maintainability.
Ranger. series of U.S. Moon probes, first NASA space program to investigate another Solar System body.
rbbs. radio bulletin board system; a digital communications network; see packet radio above.
rcpsk. raised cosine phase shift keying, a mode of communications.
rcv, rcvr. receive, receiver.
receiver. a device to take in a radio signal on a frequency.
reconnaissance satellites. military spy satellites, equipped with elaborate photographic and electronic equipment, missions are secret, America's SAMOS was the first.
Redstone. class of rocket used in five launches in the Project Mercury series.
RELAY. an early communications satellite.
relay. most communications satellites are relays repeating voice, TV and digital signals.
remote sensing. obtaining data about Earth's biosphere by a non-contact method, such as remote-sensing satellites; observing land masses and oceans.
repeater. a communications relay device; a radio signal bridge between two stations; most communications satellites are repeaters relaying voice, television and digital signals.
retrograde motion. passage of a planet or satellite through a star field in a backward or westerly direction, opposite the easterly motion of an observer on Earth.
retrograde orbit. the orbit of a satellite in a backward or westerly direction, opposite the easterly motion of an observer on Earth.
revolution. completed when a satellite in orbit passes over the longitude of its launch site.
REX. Radiation Experiment satellite.
RF. radio frequency.
RGB. red-green-blue, computer monitor or television.
right ascension. the eastward angle between an object and the vernal equinox, expressed in hours-minutes-seconds.
RME. acronym for Relay Mirror Experiment.
RMS. root mean square.
RNAV. area navigation.
robot. autotransponder, electronic robot amateur radio operator built into Radiosputniks.
rocket. transportation device containing at least one rocket engine, fuel and payload.
rocket engine. propulsion motor driving a rocket.
Rohini-1. India's first space satellite.
Roman. alphabet basis for English and some other European alphabets.
RORSat. Radar Ocean Reconnaissance Satellite.
RoSat. Roentgen Satellite; science spacecraft.
ROSTO. Russian Defense and Technical Sports Organization.
rpm. revolutions per minute.
rptr. repeater.
RS. Radiosputnik, USSR/Russian hamsats.
RS-232. computer interconnection standard for serial peripherals; in packet, RS-232 is a common interface between tnc and computer; RS-323-C.
RSS. root sum square.
RT. remote terminal.
RTCA. Radio Technical Commission for Aeronautics.
RTCM. Ratio Technical Commission for Maritime Services.
RTG. radioisotope thermoelectric generator; long-life electricity generator powered by the decay of radioactive plutonium 238 dioxide, used in interplanetary probes.
rtty. radioteletype, keyboard-computer terminal-teleprinter communications.

RUDAK. German acronym for Regenerating Transponder for Digital Amateur Communications; digital transponder aboard RS-14/AO-21.

SSS

S. amateur radio satellite mode, transponder frequency designator, uplink 70 cm. downlink 13 cm.
S80/T. vhf-mobile-communications microsatellite.
S-band. 2000–4000 MHz (2-4 GHz) microwave radio band of the spectrum.
S/A. selective availability.
Salyut. series of seven first-generation and second-generation USSR space stations, launched to Earth orbit in the 1970s and 1980s; Salyut is Russian for Salute, a name given to the first USSR space station in 1971 honoring the tenth anniversary of the flight of the first man in space, Yuri Gagarin.
Sam. monkey, suborbital Mercury flight.
SAMOS. the first spy satellite, U.S.
SAMPEX. Solar, Anomalous and Magnetospheric Particle Explorer.
SAMSO. Space and Missile Systems Organization.
San Marco Equatorial Range. Italy's space launch pad on a platform in the Indian Ocean off Kenya, East Africa.
San Marco-1. Italy's first satellite, launched from Kenya; first firing of a NASA rocket by foreign technicians.
SAR. search and rescue; also synthetic aperture radar: a satellite radar-imaging system in which the antenna size is, in effect, increased by the motion through space of the satellite.
SARA. French acronym for Satellite for Amateur Radio Astronomy; a radioastronomy satellite surveying Jupiter's electromagnetic energy.
SAREX. in amateur radio: Shuttle Amateur Radio Experiment; in CAP: Search And Rescue Exercise.
SARSAT. Search and Rescue Satellite-Aided Tracking System, SARSAT/COSPAS.
SAS. Small Astronomical Satellite.
sat, sats. satellite, satellites.
Satcom. synchronous communications satellite.
SatelLife. organization by International Physicians for the Prevention of Nuclear War.
satellite. an small body orbiting a larger body in Outer Space, such as a moon orbiting a planet; natural satellites are moons; an artificial satellite is a man-made object or machine sent to orbit Earth or another body in Outer Space.
satellite, artificial. an object sent to orbit Earth for one or more purposes useful to mankind.
satellite dish. round concave antenna to transmit to, or receive signals from, a satellite.
Satellite Launch Vehicle. rocket which launched India's first space satellite, Rohini, from Sriharikota Island.
satellite, man-made. an object sent to orbit Earth for one or more purposes useful to mankind.
SBB. smart buffer box.
SBS. a geosynchronous communications satellite.
SBTS. a geosynchronous satellite.
SC. special committee.
SCORE. an early communications satellite.
Scout. class of unmanned U.S. space launch rockets used for decades to lift lightweight payloads to Earth orbit.
SDIO. Strategic Defense Initiative Organization.
SDRN. Satellite Data Relay Network; Russian communications satellite system like NASA's TDRS.
SDS. Satellite Data System, U.S. military communications satellites.
SEASAT. U.S. oceanography satellite.
SECOR. Sequential Collation of Range satellite.
SEDS. Small Expendable Deployer System, NASA machine to unreel a tether in space; also Students for the Exploration and Development of Space.
SEDsat. Students for the Exploration and Development of Space Satellite, built by students at University of Alabama—Huntsville.
seismometer. device to measure impacts and quakes, such as Moonquakes.
sensor. device to receive energy.
SEO. Satellite for Earth Observation, Bhaskara, India.
SEP. spherical error probable.
separation. release and falling away of a space rocket's boosters after fuel for the boosters has been exhausted.
SERC. Scientific and Engineering Research Council, Great Britain.
Series O. Navy Oscar navigation satellites.
SERT. Space Electric Rocket Test.
SES. Societe Europeenne des Satellites, European Satellite Society, Luxembourg; Astra satellite.
SETI. search for extraterrestrial intelligence.
SEU. single event upsets, energetic cosmic particles penetrating and corrupting semiconductor memories.
severe geomagnetic storm. magnetic storm with Ap index of 100 or more.
shake test. simulating effects of rocket-launch vibrations on a spacecraft.
Shavit. Hebrew for comet; the rocket which launched Israel's first space satellite, Horizon, from the Negev Desert.
Shepard, Alan. first American in space, in the Mercury MR-3 capsule Freedom-7; not in orbit.
SHF. super high frequency, 3,000-25,000 MHz, 3-25 GHz, radio-frequency band.
shooting star. bright meteor or fireball seen in Earth's sky.
shortwave. high frequency (hf), 3-30 MHz, radio-frequency band in the electromagnetic spectrum.
shortwave fade. swf; ionospheric solar flare effect; sudden ionospheric disturbance (SID); vlf to hf radio signals absorbed for minutes to hours.
shuttle. partly-reusable manned spacecraft which can land on a runway.
Shuttle-C. proposed unmanned cargo version of a U.S. space shuttle.
SI. International System of Units.
SID. sudden ionospheric disturbance; HF

propagation anomaly due to ionospheric change from solar flares, proton event, geomagnetic storm.
sidereal period. time required for a body to complete an orbit around another object.
siderite. a nickel-iron meteorite.
SIGINT. signals intelligence satellite.
SIL. System Integration Laboratory.
SINS. shipborne INS.
SIRTF. Space Infrared Telescope Facility, one of NASA's Great Observatories in space.
sitor. keyboard-teleprinter communication.
Skylab. America's one and only space station; in orbit 1973-1979.
SMM. Solar Maximum Mission, Solar Max satellite.
smoothed sunspot number. average of 13 monthly numbers, centered on a month of interest.
SO. SUNsat-OSCAR.
SOARS. Satellite Onboard Attack Reporting System.
SOHO. Solar and Heliospheric Observatory, an ISTP satellite.
solar. referring to the Sun.
Solar-A. Japanese solar flare satellite Yohkoh.
solar cell. wafer of material, usually silicon, which converts sunlight into direct-current electricity; photovoltaic cell; electricity-generating panel charging batteries in a satellite.
solar constant. the quantity of energy from the Sun radiating onto the top of Earth's atmosphere—1.94 calories per minute per square centimeter.
solar cycle. 11-year quasi-periodic variation in frequency or number of solar events.
solar flare. a sudden brightening of a spot on the surface of the Sun from which a high-speed jet of hot plasma gas erupts.
solar heat collector. electricity generator for a space station; a dish-shaped collector heats a fluid to boiling; the resulting steam spins blades of a turbine to generate electricity.
Solar Max. Solar Maximum Mission, SMM satellite.
solar maximum. month of solar cycle when 12-month mean of monthly average sunspot numbers reaches maximum.
solar minimum. month of solar cycle when 12-month mean of monthly average sunspot numbers reaches minimum.
solar sail. spacecraft propelled by the solar wind radiation pressing against a large sail.
Solar System. the Solar System is the Sun, the surrounding bubble of energy it generates, and the nine known major planets, asteroids, comets and other matter. A solar system is a group of planets orbiting a star. The Solar System is a collection of bodies held near each other in space by the gravitational attraction of a star we call the Sun. The Solar System bodies, including major planets, their moons, minor planets, comets, asteroids, and debris, all orbit the Sun. Billions of years ago the major planets formed from a disk of matter which surrounded the Sun. Now all of the bodies of the Solar System swim in a thin soup of rocky dust particles. The rocks and dust may be left over from the original accretion formation process, or ejected by comets as they pass through the inner Solar System, or they may be debris from collisions between minor planets. The nine major planets in order from the Sun are Mercury, Venus, Earth, Mars, Jupiter, Saturn, Uranus, Neptune and Pluto. Some astronomers speculate there may be a tenth planet in our system. The Solar System also has moons, asteroids, minor planets, planetoids, comets and other objects.
solar wind. supersonic stream of energized particles, negatively-charged electrons and positively-charged protons, a stream of ionized hydrogen and helium, flowing away from the Sun through interplanetary space; outward flux of solar particles and magnetic fields from the Sun; velocities near 350 kilometers-per-second.
solid fuel. a chemical rocket fuel with a rubbery consistency, not gas or liquid.
solid rocket. a space rocket using a rubbery, so-called solid, fuel.
SolRad. Solar Radiation satellite.
solstice. the sky positions where the Sun is at maximum angle or declination from the celestial equator.
SOOS. Stacked Oscars On Scout; two navigation satellites launched on one rocket.
Soyuz. long series of USSR and Russian cosmonaut pace transports. Soyuz capsules ferry cosmonauts to and from space, especially to and from the orbiting Mir space station. Soyuz is Russian for Union, symbolizing the spacecraft's rendezvous and docking capability.
SP. same as PS or Preliminary Satellite, as Sputnik-1 was called before launch by its Korolev design team; the technicians would honor Korolev by reversing the letters from PS to SP from his first two initials for Sergei Pavlovich.
space. region beyond Earth's atmosphere.
Space Command, U.S. formerly NORAD; tracks thousands of satellites in Earth orbit; assigns numbers in sequential order to all objects it tracks in orbit.
spacecraft. vehicle traveling in Outer Space.
space flight. flight into or through Outer Space.
space flight participant. an astronaut traveling in Outer Space aboard a U.S. space shuttle flight.
space junk. dead satellites and spent rockets, and thousands of other pieces of debris orbiting near Earth.
spaceman. astronaut or cosmonaut.
Spacenet. a stationary communications satellite.
space science. study of aerospace science, aerospace engineering, rocketry, and astronomy.
spaceship. manned spacecraft in Outer Space.
space shuttle. partly-reusable, manned spacecraft combining rocket and glider aircraft, ferrying passengers and equipment to orbit in Outer Space and back to land on a runway.
space station. orbiting satellite base in space

from which Outer Space can be explored; structure with quarters for living and working in orbit around a planet or moon.
spacesuit. pressurized suit allowing the wearer to survive in Outer Space.
spacewalk. extra-vehicular activity, EVA, an Outer Space trip by an astronaut or cosmonaut wearing a space suit and moving around in Outer Space outside a shuttle or space station.
sparkles. random bit errors seen in photos made by satellites.
specific impulse. the ratio of pounds of thrust to pounds of fuel in a rocket.
spectrum. electromagnetic spectrum; wavelength range; gamma rays, X-rays, ultraviolet, visible light, infrared, microwaves, radio.
sporadic-E. phenomenon in ionosphere E region affects HF radio day or night; varies with latitude.
SPOT. French acronym for Systeme Probatoire d'Observation de la Terre; an observation satellite.
SPS. standard positioning service.
Sputnik. first artificial satellite, launched to Earth orbit by the USSR on October 4, 1957; initiated era of space exploration known as Space Age.
spy satellites. spysats, reconnaissance satellites equipped with photographic and electronic equipment for secret missions; SAMOS was the first, in 1961.
SRATS. Solar Radiation And Thermospheric Structure satellite, Japan's Taiyo.
SRB. solid rocket booster.
Sriharikota Island. India's space launch site.
SRU. shop replaceable unit.
SS. spread spectrum, a frequency-hopping mode of communications.
ssb. single sideband voice communication, less bandwidth than am or fm, commonly used for hf communication.
sstv. slow-scan television, visual communication, slower than fast-scan television.
stars. fireballs of bubbling-hot gas with surface temperatures ranging from 3,000 to 100,000 degrees Fahrenheit—hot enough to melt almost anything on Earth; surface temperature of the Sun is about 11,000 degrees Fahrenheit; a star's color tells how hot it is; a cool star, at about 3,600 degrees, is red; a hot star, at about 90,000 degrees, is blue; the Sun is yellow and provides light and warmth to the Earth; the Sun isn't a very large star, being only 865,400 miles in diameter; it is, however, 100 times the size of Earth and the heart of a Solar System with nine orbiting planets, comets and asteroids; the Sun is a yellow dwarf star, mostly hydrogen and helium, with about one percent of its mass in heavier elements.
stationary. orbit of satellite which seems always overhead at the same place in the sky; it actually moves around the globe at the same speed as Earth is turning at altitudes near 22,300 miles making it seem to hang over the same spot on the ground; circles Earth at a velocity of 6,850 mph.
stationary satellite. in an equatorial orbit 22,300 miles above Earth where it seems to stand still.
statute mile. one statute is 0.8684 nautical mile.
STDCDU. standard CDU.
stn. station.
store-and-forward. electronic message bulletin-board mailbox system.
Strelka. dog in Sputnik-5, Little Arrow in Russian, first successful recovery of a satellite from space.
STS. Space Transportation System; U.S. shuttles.
submillimeter. above 300,000 MHz or 300 GHz, radio-frequency band of the electromagnetic spectrum.
Sun. the star we call the Sun is one of at least a hundred billion stars orbiting the center of our Milky Way galaxy—a blazing pinwheel of stars 100,000 lightyears in diameter; the Sun is the only star known to be surrounded by a solar system, although astronomers suggest other stars with planets likely will be found nearby.
SUNsat. acronym for Southern Universities Satellite; South African hamsat, pacsat with television; SUNsat-OSCAR, SO.
sunspot. area seen as dark spot on photosphere of Sun; concentration of magnetic flux in bipolar clusters or groups; appear dark because they are cooler than surrounding photosphere; large storms in the Sun's atmosphere are called sunspots. The storms generate massive explosions from the surface, firing high-speed atomic particles outward. The spots range from specks to as much as 90,000 miles across. They usually are noticed by observers on Earth when they grow as large in diameter as Earth. The largest sunspots can be seen with the naked eye. However, never look directly at the Sun with the naked eye, nor with binoculars, telescope or camera. The eye can be damaged seriously, even though no pain is felt. Each sunspot has a dark core, known as umbra, and an outer gray band known as penumbra. The cool spots are about 7,000 degrees Fahrenheit, while surrounding areas are about 11,000 degrees. The spots change in size and position every day as the rotation of the Sun makes them seem to travel across the surface in about 14 days. The number of sunspots at one time varies with a maximum about every 11 years. Increased spots bring increased auroras on Earth as well as disruptions of radio, television, telephone and electrical systems.
sunspot number. daily index of sunspot activity.
SUPARCO. Space and Upper Atmosphere Research Commission, Pakistan.
Surrey. University of Surrey, Great Britain; UoSAT.
Surveyor. U.S. unmanned television and trenching probes which landed softly on the Moon in search of Apollo moonship landing sites; the three-legged Surveyors found lunar maria to be basaltic and the lunar highlands rich in aluminum and calcium.
Surveyor-1. first soft landing on the Moon, man's first close-up look at the surface.
SVN. Satellite Vehicle NAVSTAR number.
sw. shortwave, high frequency, hf, 3-30 MHz, 3-30 MHz, a radio-frequency band.
swf. short wave fade; ionospheric solar flare effect;

sudden ionospheric disturbances (SID); radio signals VLF to HF are absorbed for minutes to hours.
swl. shortwave listener, a radio hobbyist.
synchronous satellite. a satellite circling Earth at 22,300 miles altitude at 6,850 mph.
SYNCOM. an early communications satellite.
synodic period. the time between oppositions of a planet or satellite.
synthetic aperture radar. SAR; satellite radar-imaging system in which antenna size is effectively increased by its motion through space.

TTT

T. hamsat mode, transponder frequency designator, uplink 21 MHz downlink 145 MHz.
TACAN. tactical air navigation.
TAI. international atomic time.
TAPR. Tucson Amateur Packet Radio Association.
TAS. three-axis accelerometer system aboard SEDsat records debris impacts frequency, magnitude and direction.
tba. to be announced.
tbd. to be determined.
tca. time of closest approach.
TCP/IP. Transmission Control Protocol/Internet Protocol.
TDF. French stationary communications satellite.
TDOP. time dilution of precision.
TDPE. three Transputer parallel microcomputers simultaneously processing parts of a task to improve speed.
TDRS. Tracking and Data Relay Satellite; NASA's stationary communications satellites.
TECHsat. Israeli microsatellite Guerwin.
telemetry. control data, science data, an information beacon transmitted from rocket or spacecraft to Earth.
TELSTAR. an early communications satellite; today, geosynchronous communications satellites.
terminal node controller. tnc; packet radio tnc is a modem which divides a message into digital packets, keys a transmitter and sends the packets. Receiving packets, the tnc decodes, checks for errors, recombines the packets and displays the message.
terminator. line dividing dark and light hemispheres of a planet or satellite. The line between day and night seen from a spacecraft orbiting Earth.
tether. cable unreeled in orbit by SEDS satellite.
TFOM. time figure of merit.
thermal protection system. heat shield on a rocket or manned spacecraft.
Thor-Agena. rocket which carried OSCAR-1 to orbit in 1961.
thrust. propulsion force generated by a rocket during firing. A rocket's thrust is used to carry a satellite to orbit.
Timation. Time Navigation satellites.
Time Navigation. Timation satellites.
TIP. Transit Improvement Program.
TIROS-1. Television Infra-Red Orbital Satellite, the first weather satellite.
Titan. class of unmanned U.S. space launch rockets used to lift heavy payloads to Earth orbit.
Titov, Vladimir. cosmonauts Vladimir Titov and Musa Manarov had longest one-trip stay in space.
tlm. telemetry, remote data content of communications.
tnc. terminal node controller in packet radio; a modem which divides a message into digital packets, keys a transmitter and sends the packets; receiving packets, the tnc decodes, checks for errors, recombines the packets and displays the message.
TNT. Turner Network Television.
TOMS. Total Ozone Mapping Spectrometer, NASA.
TOPEX. Ocean Topography Experiment.
TOPEX/Poseidon. an oceanography satellite.
TOPSAR. topographic synthetic-aperture radar, an interferometry radar.
TPAD. Trunnion Pin Acquisition Device; Solar Max.
TRAAC. experimental Transit Research And Attitude Control satellite.
trade winds. easterly winds prevailing at low latitudes and blowing toward equator.
transceiver. a device which takes in a radio signal on a frequency and sends a radio signal on a frequency.
TRANSIT. the first navigation satellite; U.S.
transit. passage of object across an observer's meridian or an object across face of another object.
transmitter. a device to send a radio signal on a specific frequency.
transponder. transceiver to receive a radio signal on a frequency and retransmit it on a different frequency; a two-way relay device in a satellite to relay voice, Morse code, and digital computer-to-computer communications.
Transputer. parallel processing microcomputers.
Transtage. a space satellite.
TRIAD. first spacecraft in the TIP series.
TRMM. Tropical Rainfall Measurement Mission (TRMM) satellite to measure rainfall over South America and Africa.
tropopause. part of the atmosphere between troposphere and stratosphere.
TSFR. This Space For Rent.
Tsikada. Russian navigation satellite system, similar to American Transit system.
Tsiolkovsky. Konstantin Eduardovich Tsiolkovsky, Russian astronautics theorist born in 1857, whose 20th century theoretical work predicted a satellite might be launched by a rocket.
TTFF. time to first fix.
TUBsat. German microsatellite.
TV. television, video pictures, a visual mode of communication; ATV is amateur tv.
TVSat. German stationary broadcast satellite.

UUU

UAH. University of Alabama at Huntsville.
UARS. Upper Atmosphere Reasearch Satellite.
UE. user equipment.

UERE. user equivalent range error.
UFO. unidentified flying object; a hypothetical spacecraft from another planet; flying saucer.
Ugolek. dog in USSR Voskhod capsule.
UHF. Ultra High Frequency satellites, U.S. Navy communications satellites.
uhf. ultra high frequency, 300-3,000 MHz, 0.3-1 GHz, radio-frequency band.
Uhuru. freedom in Swahili; first Small Astronomical Satellite.
UIT. Ultraviolet Imaging Telescope, part of Astro-1.
UK. short for UKOGBANI, United Kingdom of Great Britain and Northern Ireland.
UKS. United Kingdom Subsatellite; Britain's AMPTE satellite.

ulf. ultra low frequency, below 30 KHz, 30 Hz to 300 Hz, radio band.
ultra high frequency. uhf; 300-3,000 MHz, 0.3-1 GHz, frequency band; line-of-sight communication.
ultraviolet telescope. equipment on Earth and aboard satellites sensing ultraviolet wavelengths.
umbilical. a bundle of wires and pipes carrying electricity and fluids between a tower on a launch pad and an upright rocket before launch. An umbilical also is a life support line between an astronaut outside a shuttle or space station on a spacewalk and the spacecraft.
UNAM. Autonomous University of Mexico.
UNAMsat. Spanish acronym from initials of National Autonomous University of Mexico.
universal time. UTC, coordinated universal time; by international agreement, UTC is the local time at the prime meridian, which passes through Greenwich, England; therefore, it once was known as Greenwich Mean Time (GMT); also z or zulu time; mean solar time at the zero-degree meridian at Greenwich, England.
University of Alabama. University of Alabama—Huntsville, UAH; SEDsat.
University of Surrey. home of UoSAT, builder of British amateur radio satellites with AMSAT-UK.
UO. UoSAT-OSCAR.
uplink. Earth-to-space radio signal; transmission upward to a satellite, shuttle or space station.
Uribyol. Our Star in Korean, name in orbit for South Korea's hamsats KITsat-OSCAR.
USA. United States of America.
usb. upper-sideband, single-sideband voice communications.
USNO. US Naval Observatory.
U.S. Space Command. formerly NORAD; tracks thousands of satellites in Earth orbit; assigns numbers in sequential order to all objects it tracks in orbit.
UT. universal time.
UTC. coordinated universal time; by international agreement, the local time at the prime meridian, which passes through Greenwich, England; therefore, it is known as Greenwich Mean Time (GMT); also z or zulu time.

VVV

V-2. German rocket carrying bomb in World War II.
Van Allen Belts. zones of electrified particles surrounding Earth; radiation trapped at high altitudes by the planet's magnetic field; discovered by Explorer-1 satellite.
Van Allen, James. designed America's first satellite, Explorer-1, at the Jet Propulsion Laboratory in California; belts of radiation surrounding Earth found by the first Explorers were named after him.
Vandenberg. Vandenberg Air Force Base, California; Western Space and Missile Center launch complex; launches satellites to polar orbit.
Vanguard-1. The U.S. Navy's first space satellite.
VDOP. vertical dilution of precision.
Venera-1. first spacecraft sent beyond Earth; USSR.
Venus. sixth largest and brightest planet of the Solar System and second from the Sun.
vernal equinox. see equinox.
very high frequency. vhf; frequency spectrum from 30 to 300 MHz, line-of-sight communications.
very low frequency. vlf; radio, 3-30 KHz.
Veterok. dog in USSR Voskhod capsule.
vhf. very high frequency, 30-300 MHz, frequency, spectrum, line-of-sight communication.
VHSIC. very high speed integrated circuit.
Viking. two American Mars landers, also photographed the Martian moons Phobos and Deimos.
Village Satellite. India's rural satellite network, Gramsat.
visible-light telescope. TV camera on Earth, aboard satellite or in interplanetary probe to sense light at wavelengths visible to the human eye.
VITA. Volunteers In Technical Assistance.
vlf. very low frequency, 3-30 KHz radio frequency band; audio frequencies.
VLSIC. very large scale integrated circuit.
voice mailboxes. satellite store-and-forward message bulletin-board (bbs) in which digitally-encoded voices are stored for later retrieval.
VOR. vhf omnidirectional range.
Voyager. twin interplanetary probes which flew through the Asteroid Belt to explore Jupiter and Saturn. Voyager-2 also flew by Uranus and Neptune—the so-called Grand Tour; the Voyager twins originally were known as Mariner-11 and Mariner-12.

WWW

WARC. World Administrative Radio Conference.
WBFM. wide band frequency modulation; wide-band fm, a mode of communications.
weather. the state of the atmosphere with respect to wind, temperature, moisture and other conditions.
weather fax. WEFAX, method to send photos and weather maps by radio; satellite version, APT.
weather satellite. an unmanned spacecraft photographing clouds and weather patterns from above Earth and sending down meteorological data.
Weber State. Ogden, Utah, university.

WEBERsat-OSCAR 18. WO-18, eighteenth OSCAR, launched in 1990.
WEFAX. weather fax or facsimile, transmitting cloud photos and weather maps via radio.
Westar. stationary communications satellite.
westerlies. winds from subtropical areas in west toward middle latitudes.
WGS-84. World Geodetic System—1984.
wisteria. "fuji" in Japanese, name for hamsats.
WO. WEBERsat-OSCAR.
WOD. whole orbit data.
wormhole. communications gateway; interconnect point between two computer networks such as wired and packet radio networks.
wpm. words per minute.
WSMC. U.S. Western Space and Missile Center launch complex at Vandenberg Air Force Base, California.
WUPPE. Wisconsin Ultraviolet Photo-polarimeter Experiment, part of Astro-1.
wx sat. weather satellite.

XXX

X-band. 8,000-12,000 MHz, 8-12 GHz, radio-frequency band of the electromagnetic spectrum.
xmit. transmit.

YYY

Yohkoh. Japanese solar flare satellite Solar-A.
YPG. Yuma Proving Ground.

ZZZ

z. zulu time; Greenwich Mean Time (GMT); Coordinated Universal Time (UTC).
zenith. point in the sky directly above observer; nadir is on celestial sphere opposite zenith.
zero-G. space science term for the nearly-zero gravity of Earth orbit.
zodiac. band of 12 constellations on the celestial sphere with the ecliptic as its middle line, including the paths of all the major planets except Pluto. The zodiac is a band of the celestial sphere, 8 degrees wide on each side of the ecliptic. It is divided into 12 equal zones of 30 degrees each, and used in astrology.
zodiacal light. a subtle cone of light reaching up from the horizon, caused by the scattering of sunlight by interplanetary particles. A faint glow in the East just before dawn and in the West just after sunset.
Zond. spacecraft from the USSR explored the Moon.
zulu time. z; Greenwich Mean Time (GMT); Coordinated Universal Time (UTC).
Zvezdochka. Little Star in Russian, Sputnik dog.

Master List Of All Satellites

Starting on the next page is a complete list of all satellites ever sent to orbit by any nation or international group. It includes working satellites, dead satellites, spent rockets, fragments of exploded satellites, astronaut trash and other space junk big enough to be tracked in orbit.

The master list is divided into two sections. The first section, starting on page 191, reports data on all satellites and debris in orbit as of July 1, 1993. The second section, starting on page 253, shows all satellites and debris which once were in orbit, but since have suffered "orbital decay" and have fallen from orbit. A nation-by-nation Satellite Scoreboard is on page 82.

Satellites In Orbit column headings

International Designation. An international numbering system, composed of the year the satellite was launched plus the launch sequence within the year plus a letter designating the object launched, is assigned by the Committee On Space and Atmospheric Research (COSPAR). For example, Mir space station was designated 1986-017A.

Name. The name commonly used to refer to a satellite. For example, SARA is the name for a French radioastronomy satellite designated 1991-050E.

Catalog Number. Sequential catalog numbers are assigned by U.S. Space Command, formerly North American Air Defense Command (NORAD), to all objects which it tracks in orbit. For example, Hubble Space Telescope is catalog number 20580.

Source. The nation or international group which supplied the satellite.

Launch Date. The date a satellite was launched.

Period In Minutes. The time it takes a satellite to complete one orbit or revolution around Earth. For example, the satellite OSCAR-14 takes 100.7 minutes to travel once around Earth.

Inclination. The angle between the orbital plane of a satellite and the equator.

Apogee In Kilometers. The point in its orbit where a satellite is farthest from Earth.

Perigee In Kilometers. The point in its orbit where a satellite is closest to Earth.

Notes. Special information about particular satellites.

Satellites No Longer In Orbit column headings

International Designation. same as in the Satellites In Orbit section

Name. same as in the Satellites In Orbit section

Catalog Number. same as in the Satellites In Orbit section

Source. same as in the Satellites In Orbit section

Launch Date. same as in the Satellites In Orbit section

Decay Date. The date a satellite fell from orbit.

Notes. same as in the Satellites In Orbit section

Countries and international groups under SOURCE include Arab Satellite Communications Organization (ASCO), Argentina (ARGNT), ASIAsat Corporation, Australia (AUST), Brazil, Canada (CANA), Czechoslovakia (CZCH), European Space Agency (ESA), European Space Research Organization (ESRO), France (FRAN), Federal Republic of Germany (FRG), India, Indonesia (INDON), International Maritime Satellite Organization (IMSO, Inmarsat), International Telecommunications Satellite Organization (ITSO, Intelsat), Israel (ISRAL), Italy, Japan, Korea, South, Luxembourg (LUXBRG), Mexico, North Atlantic Treaty Organization (NATO), The Netherlands (NETH), Pakistan, People's Republic of China (PRC), Saudi Arabia (SA), Spain, Sweden, United Kingdom of Great Britain and Northern Ireland (UK), United States of America (US), and Union of Soviet Socialist Republics (USSR) which today indicates the nation of Russia.

Data has been supplied by the National Aeronautics and Space Administration (NASA), including Goddard Space Flight Center, Johnson Space Flight Center, Jet Propulsion Laboratory and other NASA centers, the U.S. Space Command/NORAD, Smithsonian Astrophysical Observatory, European Space Agency, Arianespace, the governments of spacefaring nations around the world, satellite owners, commercial launch companies and others.

Data is presented for the personal information of the reader and may not be sufficiently precise for some science applications. Gravity continuously drags satellites down toward Earth. Orbital parameters more than a few weeks old may no longer have the precision required for some applications. Means of obtaining current Keplerian data is explained in chapter 1 page 9.

Satellites In Orbit

International Designation	Satellite Name	Catalog Number	Source	Launch Date	Period Mins	Incl-ination	Apogee KM	Perigee KM
1958 LAUNCHES								
BETA 1		16	US	17 MAR	137.7	34.3	4256	653
BETA 2	VANGUARD 1	5	US	17 MAR	133.2	34.3	3870	652
BETA 3		1576	US	17 MAR	126.7	34.2	3321	635
1959 LAUNCHES								
ALPHA 1	VANGUARD 2	11	US	17 FEB	122.8	32.9	3054	557
ALPHA 2		12	US	17 FEB	127.1	32.9	3436	559
ALPHA 4		14934	US	17 FEB	111.7	32.9	2079	533
ETA 1	VANGUARD 3	20	US	18 SEP	126.4	33.4	3417	512
IOTA 1	EXPLORER 7	22	US	13 OCT	98.6	50.3	861	524
MU 1	LUNA 1	112	USSR	02 JAN	HELIOCENTRIC ORBIT			
NU 1	PIONEER 4	113	US	03 MAR	HELIOCENTRIC ORBIT			
1960 LAUNCHES								
ALPHA 1	PIONEER 5	27	US	11 MAR	HELIOCENTRIC ORBIT			
BETA 2	TIROS 1	29	US	01 APR	98.3	48.4	695	658
BETA 4		115	US	01 APR	98.4	48.2	720	646
ETA 1	TRANSIT 2A	45	US	22 JUN	100.8	66.7	990	604
ETA 2	GREB	46	US	22 JUN	100.2	66.7	943	590
ETA 3		47	US	22 JUN	100.4	66.7	952	600
ETA 4		840	US	22 JUN	97.9	66.7	775	544
ETA 5		841	US	22 JUN	97.8	66.7	762	541
IOTA 2		50	US	12 AUG	118.1	47.2	1684	1502
IOTA 3		51	US	12 AUG	118.2	47.2	1687	1516
IOTA 4		52	US	12 AUG	NO CURRENT ELEMENTS			
IOTA 5		53	US	12 AUG	118.4	47.3	1686	1529
NU 1	COURIER 1B	58	US	04 OCT	107.1	28.3	1214	967
NU 2		59	US	04 OCT	106.6	28.2	1208	926
PI 1	TIROS 2	63	US	23 NOV	96.3	48.5	614	549
PI 5		5922	US	23 NOV	105.2	47.0	1035	974
XI 1	EXPLORER 8	60	US	03 NOV	102.5	49.9	1361	395
1961 LAUNCHES								
A DELTA 1	MIDAS 4	192	US	21 OCT	165.9	95.8	3762	3483
A DELTA 3		194	US	21 OCT	165.5	95.8	3856	3356
A DELTA 4		195	US	21 OCT	166.3	95.8	3863	3415
A DELTA 5		2009	US	21 OCT	165.7	95.8	3733	3493
A DELTA 6		2371	US	21 OCT	165.2	95.9	4713	2480
A ETA 1	TRANSIT 4B	202	US	15 NOV	105.7	32.4	1104	953
A ETA 2	TRAAC	205	US	15 NOV	105.8	32.4	1107	956
A ETA 3		204	US	15 NOV	105.6	32.4	1097	950
A ETA 4		10796	US	15 NOV	105.8	32.4	1106	955
DELTA 2		82	US	16 FEB	117.8	38.9	2530	637
DELTA 3		85	US	16 FEB	108.6	38.8	1736	582
DELTA 6		3927	US	16 FEB	109.7	38.9	1840	587
DELTA 7		4026	US	16 FEB	110.2	38.9	1888	582
GAMMA 1	VENERA 1	80	USSR	12 FEB	HELIOCENTRIC ORBIT			
NU 1	EXPLORER 11	107	US	27 APR	104.5	28.8	1465	479
NU 2		3739	US	27 APR	90.6	28.8	334	273
OMICRON 1	TRANSIT 4A	116	US	29 JUN	103.6	66.8	984	868
OMICRON 3 TO 297			US	29 JUN	SEE NOTE 1			
OMICRON 2	INJUN-SR-3	117	US	29 JUN	103.6	66.8	990	871
RHO 1	TIROS 3	162	US	12 JUL	100.0	47.9	791	723
RHO 2		165	US	12 JUL	98.1	47.9	688	647
RHO 3		166	US	12 JUL	90.3	47.9	294	282
RHO 4		167	US	12 JUL	101.5	47.9	900	759
SIGMA 1	MIDAS 3	163	US	12 JUL	161.4	91.2	3539	3344
SIGMA 3		188	US	12 JUL	161.1	91.2	3543	3312
SIGMA 4		196	US	12 JUL	161.8	91.2	3559	3357

International Designation	Satellite Name	Catalog Number	Source	Launch Date	Period Mins	Incl-ination	Apogee KM	Perigee KM
1962 LAUNCHES								
A ALPHA 1	TIROS 5	309	US	19 JUN	99.4	58.1	889	573
A ALPHA 3		312	US	19 JUN	100.1	58.3	948	574
A ALPHA 4		313	US	19 JUN	90.1	58.0	291	269
A EPSILON 1	TELSTAR 1	340	US	10 JUL	157.8	44.8	5642	947
A EPSILON 2		341	US	10 JUL	157.6	44.8	5626	945
A OMICRON 1		369	US	23 AUG	98.1	98.5	754	579
A OMICRON 4		388	US	23 AUG	94.9	98.6	550	480
A PSI 1	TIROS 6	397	US	18 SEP	97.6	58.3	652	635
A PSI 3		399	US	18 SEP	97.4	58.4	653	613
A PSI 5		19436	US	18 SEP	90.9	58.3	326	313
A RHO 1	MARINER 2	374	US	27 AUG	HELIOCENTRIC ORBIT			
A RHO 2		375	US	27 AUG	HELIOCENTRIC ORBIT			
ALPHA 1	RANGER 3	221	US	26 JAN	HETIOCENTRIC ORBIT			
ALPHA 2		222	US	26 JAN	HELIOCENTRIC ORBIT			
B ALPHA 1	ALOUETTE 1	424	CANADA	29 SEP	105.2	80.5	1022	987
B ALPHA 2		426	US	29 SEP	105.2	80.5	1019	991
B ALPHA 3		510	US	29 SEP	105.2	80.5	1013	989
B ALPHA 4		511	US	29 SEP	105.3	80.4	1031	982
B CHI 1	EXPLORER 16	506	US	16 DEC	104.1	52.0	1159	745
B ETA 1	RANGER 5	439	US	18 OCT	HELIOCENTRIC ORBIT			
B ETA 2		440	US	18 OCT	HELIOCENTRIC ORBIT			
B MU 1	ANNA 1B	446	US	31 OCT	107.9	50.1	1181	1075
B MU 2		447	US	31 OCT	107.6	50.1	1164	1065
B NU 3		450	USSR	01 NOV	HELIOCENTRIC ORBIT			
B UPSILON 1	RELAY 1	503	US	13 DEC	185.1	47.5	7436	1323
B UPSILON 2		515	US	13 DEC	184.8	47.5	7418	1322
BETA 1	TIROS 4	226	US	08 FEB	99.9	48.3	812	694
BETA 2		227	US	08 FEB	100.6	48.2	888	683
BETA 3		228	US	08 FEB	97.8	48 4	674	636
BETA 4		229	US	08 FEB	97.4	48.3	665	604
KAPPA 1		271	US	09 APR	152.9	86.7	3406	2782
KAPPA 3		273	US	09 APR	152.5	86.7	3362	2794
KAPPA 4		274	US	09 APR	153.3	86.7	3448	2767
KAPPA 7		18603	US	26 OCT	153.0	86.7	3420	2775
KAPPA 8		19981	US	26 OCT	152.7	86.7	3388	2784
MU 2		282	US	23 APR	HELIOCENTRIC ORBIT			
1963 LAUNCHES								
1963-004A	SYNCOM 1	553	US	14 FEB	NO CURRENT ELEMENTS			
1963-008B	LUNA 4	566	USSR	02 APR	BARYOCENTRIC ORBIT			
1963-013A	TELSTAR 2	573	US	07 MAY	225.3	42.8	10807	967
1963-013B		575	US	07 MAY	225.0	42.7	10788	964
1963-014A		574	US	09 MAY	166.4	87.3	3678	3605
1963-014D TO 014FH			US	09 MAY	SEE NOTE 2			
1963-014B	ERS 5	579	US	09 MAY	165.0	87.2	4901	2273
1963-014C	ERS 6	608	US	09 MAY	166.4	87.3	3714	3569
1963-014EN		19814	US	09 MAY	165.8	87.1	4571	2665
1963-022B		603	US	16 JUN	95.9	89.9	568	557
1963-024A	TIROS 7	604	US	19 JUN	92.7	58.2	415	398
1963-025B		614	US	27 JUN	114.6	82.1	2553	324
1963-030A	ERS 10	622	US	18 JUL	167.8	88.4	3724	3673
1963-030B	ERS 9	635	US	18 JUL	167.8	88.4	3730	3667
1963-030C		630	US	18 JUL	167.4	88.4	3751	3616
1963-030E		631	US	18 JUL	168.2	88.4	3804	3627
1963-030F		3121	US	18 JUL	167.8	88.5	3731	3666
1963-030G		3132	US	18 JUL	167.8	88.4	3765	3634
1963-030H		20153	US	18 JUL	162.1	88.7	5768	1168
1963-031A	SYNCOM 2	634	US	26 JUL	NO CURRENT ELEMENTS			
1963-038A		669	US	28 SEP	107.0	90.0	1104	1067
1963-038B		670	US	28 SEP	107.1	90.0	1126	1062
1963-038C	SN 39	671	US	28 SEP	107.1	90.0	1124	1061
1963-038D		672	US	28 SEP	106.2	90.0	1076	1021
1963-038E		745	US	28 SEP	106.5	89.9	1083	1049
1963-038F		2097	US	28 SEP	106.3	89.9	1083	1022
1963-038G		3166	US	28 SEP	107.1	90.0	1125	1061
1963-038J		12943	US	28 SEP	104.6	89.9	1075	876
1963-038K		20470	US	28 SEP	105.9	89.9	1050	1024
1963-039A		674	US	17 OCT	NO CURRENT ELEMENTS			
1963-039C		692	US	17 OCT	NO CURRENT ELEMENTS			
1963-047A	CENTAUR 2	694	US	27 NOV	104.6	30.4	1485	468

International Designation	Satellite Name	Catalog Number	Source	Launch Date	Period Mins	Incl-ination	Apogee KM	Perigee KM
1963-047D		698	US	27 NOV	106.2	29.9	1494	607
1963-047F		700	US	27 NOV	108.0	30.5	1690	577
1963-047G		701	US	27 NOV	105.8	30.0	1464	599
1963-047H		739	US	27 NOV	104.8	30.4	1487	486
1963-047K		2886	US	27 NOV	108.6	29.9	1653	667
1963-047L		3741	US	27 NOV	104.9	29.9	1350	625
1963-047Q		14528	US	27 NOV	106.0	29.6	1426	655
1963-047T		19106	US	27 NOV	104.1	30.5	1266	638
1963-049A		703	US	05 DEC	106.7	90.1	1083	1060
1963-049B		704	US	05 DEC	106.9	90.1	1111	1057
1963-049C		705	US	05 DEC	106.9	90.1	1109	1056
1963-049D		706	US	05 DEC	106.5	90.1	1087	1041
1963-049E		715	US	05 DEC	105.8	90.1	1048	1013
1963-049F		753	US	05 DEC	106.6	90.1	1095	1041
1963-049G		2432	US	05 DEC	106.9	90.1	1109	1054
1963-049H		2620	US	05 DEC	106.2	90.1	1062	1042
1963-053B		721	US	19 DEC	115.2	78.6	2329	597
1963-053C		722	US	19 DEC	110.2	78.6	1835	634
1963-053E		724	US	19 DEC	108.4	78.6	1701	600
1963-053G		726	US	19 DEC	105.8	78.6	1472	590
1963-053H		732	US	19 DEC	109.8	78.6	1809	620
1963-053J		3750	US	19 DEC	107.9	78.6	1634	621
1963-053K		17665	US	19 DEC	110.7	78.7	1876	640
1963-054A	TIROS 8	716	US	21 DEC	98.5	58.5	711	663
1963-054C		720	US	21 DEC	100.1	58.5	854	671
1963-054E		19396	US	21 DEC	98.1	58.5	698	634
1964 LAUNCHES								
1964-001A		727	US	11 JAN	103.2	69.9	924	899
1964-001B	GRAVITY GRADIENT 1	728	US	11 JAN	103.2	69.9	919	896
1964-001C	SECOR (EGRS) 1	729	US	11 JAN	103.3	69.9	925	900
1964-001D	SOLRAD 7A	730	US	11 JAN	103.2	69.9	923	899
1964-001E	GREB	731	US	11 JAN	103.2	69.9	923	899
1964-002A		733	US	19 JAN	100.7	99.0	819	768
1964-002B		734	US	19 JAN	100.9	99.1	811	790
1964-002C		735	US	19 JAN	100.9	99.1	815	791
1964-003A	RELAY 2	737	US	21 JAN	194.7	46.4	7535	1966
1964-003B		738	US	21 JAN	194.8	46.4	7541	1965
1964-004B		741	US	25 JAN	108.8	81.5	1300	1039
1964-004C		742	US	25 JAN	108.6	81.5	1294	1032
1964-004D		743	US	25 JAN	108.6	81.5	1294	1029
1964-006A	ELEKTRON 1	746	USSR	30 JAN	162.8	60.8	6590	407
1964-006C TO 006AE			USSR	30 JAN	SEE NOTE 4			
1964-006B	ELEKTRON 2	748	USSR	30 JAN	1356.4	61.1	61357	7065
1964-006N		18589	USSR	30 JAN	149.9	58.4	4246	1688
1964-016D	ZOND 1	785	USSR	02 APR	HELIOCENTRIC ORBIT			
1964-026A		801	US	04 JUN	102.2	90.5	901	824
1964-026B		805	US	04 JUN	102.1	89.9	887	827
1964-026C		806	US	04 JUN	98.9	90.8	746	668
1964-026D		809	US	04 JUN	102.5	90.5	914	838
1964-026E		2986	US	04 JUN	102.6	90.5	926	837
1964-031A		812	US	18 JUN	101.2	99.8	821	812
1964-031B		813	US	18 JUN	101.3	99.8	822	814
1964-031C		815	US	18 JUN	101.1	99.8	818	799
1964-038A	ELEKTRON 3	829	USSR	10 JUL	161.2	60.8	6465	399
1964-038C		831	USSR	10 JUL	138.2	60.8	4551	397
1964-040A		836	US	17 JUL	NO CURRENT ELEMENTS			
1964-040B		837	US	17 JUL	NO CURRENT ELEMENTS			
1964-041B		843	US	28 JUL	BARYOCENTRIC ORBIT			
1964-047A	SYNCOM 3	858	US	19 AUG	NO CURRENT ELEMENTS			
1964-047B		862	US	19 AUG	NO CURRENT ELEMENTS			
1964-049D	COSMOS 41	869	USSR	22 AUG	714.7	69.3	38823	1378
1964-049E		898	USSR	22 AUG	717.6	69.4	38964	1383
1964-049F		13091	USSR	22 AUG	715.9	69.3	39135	1127
1964-051A	EXPLORER 20	870	US	25 AUG	103.6	79.9	1001	855
1964-051B		871	US	25 AUG	103.2	79.9	977	843
1964-053A	COSMOS 44	876	USSR	28 AUG	98.7	65.1	803	588
1964-053B		877	USSR	28 AUG	99.0	65.1	758	659
1964-053C		21126	USSR	28 AUG	98.9	65.1	754	656
1964-054A	OGO 1	879	US	05 SEP	NO CURRENT ELEMENTS			
1964-063A	NNSS 30010	893	US	06 OCT	106.2	90.1	1067	1030
1964-063B		897	US	06 OCT	106.4	90.1	1073	1046

International Designation	Satellite Name	Catalog Number	Source	Launch Date	Period Mins	Incl-ination	Apogee KM	Perigee KM
1964-063C		900	US	06 OCT	105.5	90.1	1034	999
1964-063D		901	US	06 OCT	106.4	90.1	1069	1045
1964-063E		902	US	06 OCT	106.4	90.1	1075	1047
1964-063F		903	US	06 OCT	105.4	90.1	1027	995
1964-063G		18496	US	06 OCT	104.2	90.1	997	915
1964-064A	EXPLORER 22	899	US	10 OCT	104.3	79.7	1054	872
1964-064B		907	US	10 OCT	104.4	79.7	1057	876
1964-064C		976	US	10 OCT	103.0	79.3	997	806
1964-064D		977	US	10 OCT	104.8	80.0	1084	889
1964-073A	MARINER 3	923	US	05 NOV	HELIOCENTRIC ORBIT			
1964-076B	EXPLORER 25	932	US	21 NOV	114.6	81.3	2354	522
1964-076C		933	US	21 NOV	113.9	81.3	2283	524
1964-077A	MARINER 4	938	US	28 NOV	HELIOCENTRIC ORBIT			
1964-077B		942	US	28 NOV	HELIOCENTRIC ORBIT			
1964-078C	ZOND 2	945	USSR	30 NOV	HELIOCENTRIC ORBIT			
1964-083A	NNSS 30020	953	US	13 DEC	106.0	89.8	1063	1016
1964-083B		956	US	13 DEC	105.7	89.8	1053	1000
1964-083C		959	US	13 DEC	105.9	89.8	1065	1007
1964-083D		965	US	13 DEC	106.1	89.8	1079	1016
1964-083F		967	US	13 DEC	105.7	89.8	1053	1000
1964-083G		1099	US	13 DEC	105.9	89.8	1063	1008
1964-083J		1608	US	13 DEC	105.1	89.7	1024	969
1964-086A	EXPLORER 26	963	US	21 DEC	209.2	19.8	10305	284

1965 LAUNCHES

International Designation	Satellite Name	Catalog Number	Source	Launch Date	Period Mins	Incl-ination	Apogee KM	Perigee KM
1965-004A	TIROS 9	978	US	22 JAN	118.9	96.4	2564	702
1965-004B		979	US	22 JAN	118.7	96.4	2546	700
1965-004C		1312	US	22 JAN	117.5	96.4	2466	669
1965-004D		1313	US	22 JAN	120.0	96.4	2634	730
1965-008A		1001	US	11 FEB	145.4	32.1	2797	2766
1965-008B		1000	US	11 FEB	145.7	32.1	2802	2784
1965-008C		1002	US	11 FEB	145.8	32.1	2810	2783
1965-010B		1087	US	17 FEB	BARYOCENTRIC ORBIT			
1965-016A	GREB	1271	US	09 MAR	103.2	70.1	927	893
1965-016B	GRAVITY GRADIENT 2	1244	US	09 MAR	103.2	70.1	929	895
1965-016C	GRAVITY GRADIENT 3	1292	US	09 MAR	103.0	70.1	918	884
1965-016D	SOLRAD 7B	1291	US	09 MAR	103.3	70.1	931	897
1965-016E	SECOR (EGRS)3	1208	US	09 MAR	103.2	70.1	929	895
1965-016F	OSCAR 3	1293	US	09 MAR	102.8	70.1	904	873
1965-016H	SURCAL	1272	US	09 MAR	103.3	70.1	933	897
1965-016J		1245	US	09 MAR	103.2	70.1	925	891
1965-016K		12099	US	09 MAR	103.0	70.1	914	881
1965-020AC		1370	USSR	15 MAR	102.1	56.1	1190	524
1965-020AH		1392	USSR	15 MAR	104.4	55.9	1405	522
1965-020BB		1477	USSR	15 MAR	111.8	55.5	1798	823
1965-020BC		1478	USSR	15 MAR	109.6	56.1	1790	623
1965-020BD		1479	USSR	15 MAR	114.8	56.0	2087	805
1965-020BE		1480	USSR	15 MAR	114.6	56.1	2122	747
1965-020BV		1495	USSR	15 MAR	103.0	55.6	1195	608
1965-020CV		1549	USSR	15 MAR	114.4	56.2	2096	762
1965-020E		1335	USSR	15 MAR	106.1	56.1	1503	585
1965-020ED		1634	USSR	15 MAR	115.8	56.2	2172	806
1965-020EH		2334	USSR	15 MAR	110.7	55.7	1729	784
1965-020EM		2934	USSR	15 MAR	115.4	55.7	1760	1184
1965-020EN		3038	USSR	15 MAR	107.8	56.3	1661	589
1965-020ER		3708	USSR	15 MAR	102.6	56.4	1159	603
1965-020ES		3743	USSR	15 MAR	118.1	56.7	1803	1388
1965-020ET		3745	USSR	15 MAR	115.3	56.0	1586	1348
1965-020EU		3749	USSR	15 MAR	107.2	56.1	1581	613
1965-020EV		3931	USSR	15 MAR	116.6	56.1	1693	1362
1965-020EY		3965	USSR	15 MAR	117.7	56.3	1790	1368
1965-020FD		6252	USSR	15 MAR	117.1	56.0	1697	1404
1965-020FF		13517	USSR	15 MAR	109.2	55.6	1660	722
1965-020S		1347	USSR	15 MAR	101.6	56.0	1139	528
1965-023B		1298	US	21 MAR	HELIOCENTRIC ORBIT			
1965-027A		1314	US	03 APR	111.4	90.3	1316	1268
1965-027B TO 027BD			US	03 APR	SEE NOTE 5			
1965-027B	SECOR (EGRS)4	1315	US	03 APR	111.4	90.3	1314	1264
1965-028A	EARLY BIRD	1317	ITSO	06 APR	1437.3	14.6	35842	35775
1965-028B		1318	US	06 APR	NO CURRENT ELEMENTS			
1965-032A	EXPLORER 27	1328	US	29 APR	107.7	41.2	1312	929
1965-032B		1358	US	29 APR	107.7	41.2	1313	931

International Designation	Satellite Name	Catalog Number	Source	Launch Date	Period Mins	Incl-ination	Apogee KM	Perigee KM
1965-032D		2011	US	29 APR	108.3	41.2	1267	1027
1965-034A		1359	US	06 MAY	157.1	32.1	3745	2785
1965-034B		1360	US	06 MAY	309.9	32.2	14788	2792
1965-034C		1361	US	06 MAY	145.6	32.1	2801	2780
1965-034D		2529	US	06 MAY	309.9	32.1	14794	2786
1965-038A		1377	US	20 MAY	97.1	98.1	737	504
1965-038B		1378	US	20 MAY	93.8	97.9	505	418
1965-044A	LUNA 6	1393	USSR	08 JUN	HELIOCENTRIC ORBIT			
1965-048A	NNSS 30040	1420	US	24 JUN	106.6	90.1	1126	1014
1965-048B		1428	US	24 JUN	106.4	90.1	1105	1018
1965-048C		1425	US	24 JUN	106.7	90.1	1130	1021
1965-048D		1435	US	24 JUN	105.8	90.1	1082	979
1965-048E		2701	US	24 JUN	106.0	90.1	1084	995
1965-048F		3592	US	24 JUN	106.0	90.1	1085	1000
1965-048G		21945	US	24 JUN	105.6	90.1	1084	955
1965-051A	TIROS 10	1430	US	02 JUL	100.1	98.8	807	722
1965-051B		1433	US	02 JUL	99.5	98.7	772	699
1965-051C		1440	US	02 JUL	93.7	98.6	483	432
1965-051D		1529	US	02 JUL	101.4	99.0	854	799
1965-056A	ZOND 3	1454	USSR	18 JUL	HELIOCENTRIC ORBIT			
1965-058A		1458	US	20 JUL	NO CURRENT ELEMENTS			
1965-058B		1459	US	20 JUL	NO CURRENT ELEMENTS			
1965-063A	SECOR (EGRS) 5	1506	US	10 AUG	122.2	69.2	2419	1134
1965-063B		1502	US	10 AUG	122.2	69.2	2418	1136
1965-064A	CENTAUR 6	1503	US	11 AUG	BARYOCENTRIC ORBIT			
1965-065A	NNSS 30050	1504	US	13 AUG	107.7	90.0	1167	1070
1965-065B		1508	US	13 AUG	107.5	89.8	1141	1081
1965-065C		1510	US	13 AUG	105.5	90.0	1055	981
1965-065D		1511	US	13 AUG	107.9	90.0	1184	1078
1965-065E		1512	US	13 AUG	108.0	90.0	1186	1079
1965-065F		1514	US	13 AUG	107.9	90.0	1183	1076
1965-065G		1515	US	13 AUG	107.2	90.0	1145	1048
1965-065H		1520	US	13 AUG	107.9	90.1	1180	1075
1965-065J		1521	US	13 AUG	108.0	89.9	1187	1079
1965-065K		1577	US	13 AUG	107.9	89.9	1179	1076
1965-065L		1522	US	13 AUG	108.0	90.0	1187	1078
1965-065P		3810	US	13 AUG	107.2	90.0	1145	1047
1965-065Q		5265	US	13 AUG	107.8	89.8	1153	1093
1965-070A	COSMOS 80	1570	USSR	03 SEP	115.0	56.1	1537	1368
1965-070B	COSMOS 81	1571	USSR	03 SEP	115.3	56.1	1543	1395
1965-070C	COSMOS 82	1572	USSR	03 SEP	115.7	56.1	1553	1417
1965-070D	COSMOS 83	1573	USSR	03 SEP	116.0	56.1	1561	1443
1965-070E	COSMOS 84	1574	USSR	03 SEP	116.4	56.1	1569	1469
1965-070F		1575	USSR	03 SEP	114.6	56.1	1517	1354
1965-070G		3045	USSR	03 SEP	115.9	55.5	1726	1261
1965-072A		1580	US	10 SEP	101.2	98.5	995	635
1965-072D		1583	US	10 SEP	100.1	98.5	911	616
1965-072E		1931	US	10 SEP	101.7	99.0	1048	624
1965-072F		1932	US	10 SEP	97.9	98.2	735	578
1965-073A	COSMOS 86	1584	USSR	18 SEP	115.0	56.1	1623	1290
1965-073B	COSMOS 87	1585	USSR	18 SEP	115.4	56.1	1636	1315
1965-073C	COSMOS 88	1586	USSR	18 SEP	115.8	56.1	1648	1337
1965-073D	COSMOS 89	1587	USSR	18 SEP	116.2	56.0	1658	1364
1965-073E	COSMOS 90	1588	USSR	18 SEP	116.6	56.1	1667	1392
1965-073F		1589	USSR	18 SEP	116.8	56.0	1677	1395
1965-073G		1590	USSR	18 SEP	115.9	56.2	1614	1375
1965-073H		1591	USSR	18 SEP	116.2	56.1	1643	1378
1965-073J		1617	USSR	18 SEP	117.0	56.1	1737	1355
1965-073K		1618	USSR	18 SEP	117.3	56.3	1743	1377
1965-073L		2647	USSR	18 SEP	116.0	56.1	1645	1351
1965-078A		1613	US	05 OCT	117.7	144.2	2751	407
1965-078B		1616	US	05 OCT	116.3	144.2	2619	408
1965-082B TO 082UQ			US	15 OCT	SEE NOTE 7			
1965-089A	EXPLORER 29	1726	US	06 NOV	120.3	59.4	2274	1113
1965-089B		1729	US	06 NOV	120.3	59.4	2272	1112
1965-089C		2700	US	06 NOV	119.1	59.6	2215	1068
1965-089D		2888	US	06 NOV	121.3	59.2	2320	1156
1965-091A	VENERA 2	1730	USSR	12 NOV	HELIOCENTRIC ORBIT			
1965-092D		1736	USSR	16 NOV	HELIOCENTRIC ORBIT			
1965-093A	EXPLORER 30	1738	US	19 NOV	100.2	59.7	875	664
1965-093B		1739	US	19 NOV	99.8	59.7	825	673
1965-093C		2013	US	19 NOV	97.8	59.7	695	609
1965-093D		2088	US	19 NOV	100.0	59.7	827	687

International Designation	Satellite Name	Catalog Number	Source	Launch Date	Period Mins	Incl-ination	Apogee KM	Perigee KM
1965-096A	A-1	1778	FRANCE	26 NOV	107.6	34.3	1699	527
1965-096B		1805	FRANCE	26 NOV	106.1	34.3	1570	522
1965-096D		1996	FRANCE	26 NOV	101.1	34.2	1121	499
1965-098A	ALOUETTE 2	1804	CANADA	29 NOV	118.3	79.8	2708	501
1965-098B	EXPLORER 31	1806	US	29 NOV	120.0	79.8	2859	501
1965-098C		1807	US	29 NOV	118.8	79.8	2756	502
1965-098D		1808	US	29 NOV	105.3	79.8	1540	474
1965-098E		1944	US	29 NOV	104.0	79.7	1423	468
1965-098F		1948	US	29 NOV	113.2	79.9	2251	493
1965-098G		1951	US	29 NOV	113.2	79.7	2258	492
1965-098H		2092	US	29 NOV	118.5	79.9	2727	502
1965-098J		2153	US	29 NOV	118.2	79.7	2698	502
1965-101A	FR-1	1814	FRANCE	06 DEC	98.8	75.9	708	696
1965-101B		1815	US	06 DEC	98.5	75.9	695	683
1965-105A	PIONEER 6	1841	US	16 DEC	HELIOCENTRIC ORBIT			
1965-106A	COSMOS 100	1843	USSR	17 DEC	95.1	65.0	578	465
1965-106B		1844	USSR	17 DEC	94.1	65.0	490	461
1965-109A	NNSS 30060	1864	US	22 DEC	104.6	89.1	1060	891
1965-109B		1865	US	22 DEC	104.7	89.1	1066	895
1965-109C		2086	US	22 DEC	100.3	89.1	791	754
1965-109D		2226	US	22 DEC	106.8	89.1	1269	887
1965-109E		2353	US	22 DEC	104.9	89.4	1105	873
1965-112Q		1937	USSR	28 DEC	94.1	55.9	488	458

1966 LAUNCHES

International Designation	Satellite Name	Catalog Number	Source	Launch Date	Period Mins	Incl-ination	Apogee KM	Perigee KM
1966-005A	NNSS 30070	1952	US	28 JAN	105.5	89.9	1183	850
1966-005B		1953	US	28 JAN	105.6	89.9	1193	852
1966-005C		2140	US	28 JAN	107.2	90.1	1342	847
1966-005D		2141	US	28 JAN	103.3	89.8	1016	813
1966-005E		2889	US	28 JAN	109.4	89.5	1326	1068
1966-005F		2989	US	28 JAN	103.5	89.9	1013	835
1966-005J		11991	US	28 JAN	105.0	89.9	1146	842
1966-006D		2001	USSR	31 JAN	BARYOCENTRIC ORBIT			
1966-008A	ESSA 1	1982	US	03 FEB	99.7	97.8	806	684
1966-008B		1983	US	03 FEB	99.3	97.8	785	661
1966-008C		2085	US	03 FEB	96.6	97.6	608	578
1966-008D		2118	US	03 FEB	100.3	98.0	881	667
1966-008E		2154	US	03 FEB	99.1	97.8	761	670
1966-013A	D-1A	2016	FRANCE	17 FEB	115.9	34.1	2488	503
1966-013B		2017	FRANCE	17 FEB	114.6	34.1	2367	502
1966-013G		2161	FRANCE	17 FEB	107.7	34.1	1743	498
1966-016A	ESSA 2	2091	US	28 FEB	113.4	101.0	1412	1352
1966-016B		2096	US	28 FEB	113.4	101.1	1412	1350
1966-016C		2223	US	28 FEB	111.8	101.0	1381	1238
1966-016D		2224	US	28 FEB	115.0	101.1	1562	1346
1966-016E		6214	US	28 FEB	114.2	101.7	1509	1328
1966-024A	NNSS 30080	2119	US	26 MAR	104.9	89.7	1099	879
1966-024B		2120	US	26 MAR	105.0	89.8	1106	883
1966-025A	OV1-4	2121	US	30 MAR	104 0	144.5	1008	884
1966-025B	OV1-5	2122	US	30 MAR	105 6	144.6	1056	985
1966-025C		2123	US	30 MAR	105.6	144.6	1056	986
1966-025D		2124	US	30 MAR	104.0	144.5	1005	885
1966-025E		3611	US	30 MAR	102.1	144.6	906	810
1966-025G		5361	US	30 MAR	103.6	144.6	966	890
1966-025H		5599	US	30 MAR	102.2	144.6	904	823
1966-026A		2125	US	31 MAR	99.4	98.3	857	604
1966-026B		2129	US	31 MAR	97.1	98.1	690	546
1966-026D		2177	US	31 MAR	100.0	99.0	924	592
1966-027A	LUNA 10	2126	USSR	31 MAR	SELENOCENTRIC ORBIT			
1966-027D		2130	USSR	31 MAR	HELIOCENTRIC ORBIT			
1966-027E		2131	USSR	31 MAR	BARYOCENTRIC ORBIT			
1966-027F		2132	USSR	31 MAR	BARYOCENTRIC ORBIT			
1966-031A	OAO 1	2142	US	08 APR	100.6	35.0	793	783
1966-031B		2144	US	08 APR	100.2	35.0	776	763
1966-034A	OV3-1	2150	US	22 APR	132.5	82.4	4119	341
1966-034B		2167	US	22 APR	110.6	82.4	2186	321
1966-040A	NIMBUS 2	2173	US	15 MAY	108 0	100.6	1174	1091
1966-040B		2174	US	15 MAY	107 8	100.5	1166	1080
1966-041A	NNSS 30090	2176	US	19 MAY	102.8	90.1	951	835
1966-041B		2180	US	19 MAY	103.0	90.1	958	841
1966-041C		2225	US	19 MAY	98 6	90.0	716	669
1966-041D		2644	US	19 MAY	105.0	90.1	1157	828

International Designation	Satellite Name	Catalog Number	Source	Launch Date	Period Mins	Incl- ination	Apogee KM	Perigee KM
1966-041E		3591	US	19 MAY	102.9	90.1	951	836
1966-041F		4555	US	19 MAY	101.6	90.0	882	786
1966-045B		2187	US	30 MAY	BARYOCENTRIC ORBIT			
1966-049A	OGO 3	2195	US	07 JUN	NO CURRENT ELEMENTS			
1966-052A		2201	US	10 JUN	142.9	40.8	4703	645
1966-052B		2206	US	10 JUN	142.5	40.9	4673	643
1966-052C		2498	US	10 JUN	138.4	40.6	4381	585
1966-052D		2516	US	10 JUN	144.5	41.0	4782	703
1966-053A		2207	US	16 JUN	NO CURRENT ELEMENTS			
1966-053B		2215	US	16 JUN	1334.5	11.7	33889	33657
1966-053C		2216	US	16 JUN	NO CURRENT ELEMENTS			
1966-053D		2217	US	16 JUN	NO CURRENT ELEMENTS			
1966-053E		2218	US	16 JUN	NO CURRENT ELEMENTS			
1966-053F		2219	US	16 JUN	NO CURRENT ELEMENTS			
1966-053G		2220	US	16 JUN	NO CURRENT ELEMENTS			
1966-053H		2221	US	16 JUN	NO CURRENT ELEMENTS			
1966-053J		2222	US	16 JUN	1349.4	12.2	34738	33404
1966-056A	PAGEOS 1	2253	US	24 JUN	177.0	84.3	5599	2533
1966-056AH		9468	US	24 JUN	180.3	85.6	4513	3871
1966-056B		2255	US	24 JUN	181.1	87.0	4281	4171
1966-056C		2256	US	24 JUN	181.3	86.9	4275	4191
1966-056D		2511	US	24 JUN	181.5	87.0	4257	4220
1966-056G		8066	US	24 JUN	160.7	81.9	6372	450
1966-056H		8074	US	24 JUN	173.1	87.2	6028	1793
1966-058A	EXPLORER 33	2258	US	01 JUL	NO CURRENT ELEMENTS			
1966-058C		2260	US	01 JUL	NO CURRENT ELEMENTS			
1966-063B		2327	US	14 JUL	103.9	144.2	959	928
1966-063C		2328	US	14 JUL	105.2	144.2	1012	998
1966-063D		2329	US	14 JUL	104.5	144.3	971	967
1966-063E		2337	US	14 JUL	105.2	144.2	1006	997
1966-070A	OV3-3	2389	US	04 AUG	121.9	81.4	3182	348
1966-070D		2800	US	04 AUG	126.3	81.5	3507	411
1966-073B		2395	US	10 AUG	BARYOCENTRIC ORBIT			
1966-075A	PIONEER 7	2398	US	17 AUG	HELIOCENTRIC ORBIT			
1966-075C		2402	US	17 AUG	HELIOCENTRIC ORBIT			
1966-076A	NNSS 30100	2401	US	18 AUG	106.5	88.9	1089	1037
1966-076B		2413	US	18 AUG	106.6	88.9	1093	1042
1966-076C		2580	US	18 AUG	104.8	89.1	1059	911
1966-076D		2702	US	18 AUG	107.9	88.6	1198	1063
1966-077A		2403	US	19 AUG	167.4	89.7	3708	3658
1966-077B	SECOR (EGRS) 7	2411	US	19 AUG	167.5	89.7	3699	3672
1966-077C	ERS 15	2412	US	19 AUG	167.6	89.7	3700	3680
1966-078A	LUNA 11	2406	USSR	24 AUG	SELENOCENTRIC ORBIT			
1966-082A		2418	US	16 SEP	100.2	98.3	858	675
1966-082B		2422	US	16 SEP	100.1	98.3	850	672
1966-084B		2426	US	20 SEP	BARYOCENTRIC ORBIT			
1966-087A	ESSA 3	2435	US	02 OCT	114.5	100.9	1483	1384
1966-087B		2436	US	02 OCT	114.5	100.9	1482	1381
1966-087C		2518	US	02 OCT	115.8	100.8	1557	1429
1966-087D		2775	US	02 OCT	113.2	100.9	1470	1277
1966-087E		6213	US	02 OCT	112.5	101.9	1362	1323
1966-087F		8791	US	02 OCT	114.3	101.7	1535	1307
1966-089A		2481	US	05 OCT	167.5	90.0	3721	3656
1966-089B	SECOR (EGRS) 8	2520	US	05 OCT	167.6	90.0	3707	3675
1966-094A	LUNA 12	2508	USSR	22 OCT	SELENOCENTRIC ORBIT			
1966-095B		2513	US	25 OCT	BARYOCENTRIC ORBIT			
1966-096A	INTELSAT 2 F-1	2514	ITSO	26 OCT	717.7	16.9	37229	3123
1966-096C		11792	US	26 OCT	455.6	18.1	26021	447
1966-110A	ATS 1	2608	US	07 DEC	1436.0	14.3	35817	35750
1966-111A	OV1-9	2610	US	11 DEC	140.0	99.1	4629	473
1966-111B	OV1-10	2611	US	11 DEC	96.2	93.4	611	542
1966-111C		2621	US	11 DEC	97.8	93.4	706	601
1966-111D		2622	US	11 DEC	139.2	99.1	4562	474

1967 LAUNCHES

International Designation	Satellite Name	Catalog Number	Source	Launch Date	Period Mins	Incl- ination	Apogee KM	Perigee KM
1967-001A	INTELSAT 2 F-2	2639	ITSO	11 JAN	NO CURRENT ELEMENTS			
1967-001B TO 001AU			US	11 JAN	SEE NOTE 59			
1967-003A		2645	US	18 JAN	NO CURRENT ELEMENTS			
1967-003B		2649	US	18 JAN	NO CURRENT ELEMENTS			
1967-003C		2650	US	18 JAN	NO CURRENT ELEMENTS			
1967-003D		2651	US	18 JAN	NO CURRENT ELEMENTS			
1967-003E		2652	US	18 JAN	NO CURRENT ELEMENTS			

International Designation	Satellite Name	Catalog Number	Source	Launch Date	Period Mins	Incl- ination	Apogee KM	Perigee KM
1967-003F		2653	US	18 JAN	1336.4	9.0	34029	33594
1967-003G		2654	US	18 JAN	NO CURRENT ELEMENTS			
1967-003H		2655	US	18 JAN	NO CURRENT ELEMENTS			
1967-003J		2660	US	18 JAN	NO CURRENT ELEMENTS			
1967-006A	ESSA 4	2657	US	26 JAN	113.4	102.0	1437	1323
1967-006B		2661	US	26 JAN	113.5	102.1	1439	1338
1967-006C		2706	US	26 JAN	114.2	102.1	1446	1390
1967-006D		2707	US	26 JAN	112.5	101.8	1457	1228
1967-006E		5971	US	26 JAN	113.1	101.9	1453	1280
1967-010A		2669	US	08 FEB	101.1	99.1	846	771
1967-010B		2741	US	08 FEB	101.0	99.1	847	767
1967-011A	DIADEME 1	2674	FRANCE	08 FEB	101.2	40.0	1086	548
1967-011B		2671	FRANCE	08 FEB	102.3	40.0	1177	558
1967-014A	DIADEME 2	2680	FRANCE	15 FEB	108.5	39.5	1736	582
1967-014B		2682	FRANCE	15 FEB	109.0	39.5	1782	582
1967-014C		2684	FRANCE	15 FEB	106.2	40.0	1537	564
1967-014F		2685	FRANCE	15 FEB	105.3	38.9	1459	560
1967-014J		14505	FRANCE	15 FEB	104.5	38.8	1372	571
1967-014M		18911	FRANCE	15 FEB	108.4	38.9	1740	563
1967-014N		18928	FRANCE	15 FEB	93.9	39.4	533	392
1967-026A	INTELSAT 2 F-3	2717	ITSO	23 MAR	NO CURRENT ELEMENTS			
1967-027Z		18270	US	03 APR	111.3	90.3	1309	1263
1967-034A	NNSS 30120	2754	US	14 APR	106.2	90.1	1066	1035
1967-034B		2755	US	14 APR	106.4	90.1	1071	1044
1967-034C		2777	US	14 APR	103.2	90.3	1013	809
1967-034D		2778	US	14 APR	108.2	90.1	1237	1045
1967-034E		4843	US	14 APR	106.6	90.4	1091	1047
1967-034H		22172	US	14 APR	106.1	90.1	1076	1013
1967-035B		2764	US	17 APR	BARYOCENTRIC ORBIT			
1967-036A	ESSA 5	2757	US	20 APR	113.5	102.0	1419	1352
1967-036B		2758	US	20 APR	113.5	101.9	1417	1354
1967-036C		2976	US	20 APR	112.3	102.1	1408	1256
1967-036D		2977	US	20 APR	114.5	101.4	1481	1388
1967-040A		2765	US	28 APR	NO CURRENT ELEMENTS			
1967-040B		2766	US	28 APR	NO CURRENT ELEMENTS			
1967-040C	ERS 18	2767	US	28 APR	NO CURRENT ELEMENTS			
1967-040D	ERS 20	2768	US	28 APR	NO CURRENT ELEMENTS			
1967-040E	ERS 27	2769	US	28 APR	NO CURRENT ELEMENTS			
1967-040F		2770	US	28 APR	NO CURRENT ELEMENTS			
1967-045A	COSMOS 158	2801	USSR	15 MAY	100.3	74.0	811	729
1967-045B		2802	USSR	15 MAY	100.0	74.0	811	707
1967-048A	NNSS 30130	2807	US	18 MAY	106.7	89.6	1088	1060
1967-048B		2811	US	18 MAY	106.8	89.6	1090	1063
1967-053A		2826	US	31 MAY	101.5	69.9	831	823
1967-053B		2825	US	31 MAY	103.2	70.0	914	900
1967-053C	GRAVITY GRADIENT 4	2828	US	31 MAY	103.1	70.0	914	899
1967-053D	GRAVITY GRADIENT 5	2834	US	31 MAY	103.2	70.0	917	903
1967-053E		2847	US	31 MAY	102.9	70.0	903	891
1967-053F		2872	US	31 MAY	103.1	70.0	910	897
1967-053G		2873	US	31 MAY	103.1	70.0	913	899
1967-053H		2874	US	31 MAY	103.2	70.0	916	902
1967-053J		2909	US	31 MAY	101.2	70.0	818	808
1967-053K		19245	US	31 MAY	102.6	70.0	888	872
1967-060A	MARINER 5	2845	US	14 JUN	HELIOCENTRIC ORBIT			
1967-060B		2846	US	14 JUN	HELIOCENTRIC ORBIT			
1967-065A	SECOR (EGRS) 9	2861	US	29 JUN	172.1	90.1	3947	3791
1967-065B	AURORA 1	2876	US	29 JUN	172.1	90.1	3948	3791
1967-065C		2877	US	29 JUN	172.1	90.1	3951	3788
1967-066A	TITAN 3 C-14	2862	US	01 JUL	NO CURRENT ELEMENTS			
1967-066B		2863	US	01 JUL	NO CURRENT ELEMENTS			
1967-066C		2864	US	01 JUL	1311.7	11.1	33555	33075
1967-066D		2865	US	01 JUL	1313.6	11.1	33578	33129
1967-066E		2866	US	01 JUL	1316.1	11.1	33609	33197
1967-066F	DODGE	2867	US	01 JUL	1319.1	11.1	33670	33258
1967-066G		2868	US	01 JUL	1319.1	11.1	33677	33251
1967-068B		2883	US	14 JUL	BARYOCENTRIC ORBIT			
1967-070A	EXPLORER 35	2884	US	19 JUL	SELENOCENTRIC ORBIT			
1967-075B		2908	US	01 AUG	BARYOCENTRIC ORBIT			
1967-080A		2920	US	23 AUG	101.9	99.0	874	818
1967-080B		2940	US	23 AUG	101.8	98.9	871	815
1967-084B		2938	US	08 SEP	BARYOCENTRIC ORBIT			
1967-092A	NNSS 30140	2965	US	25 SEP	106.5	89.3	1100	1028
1967-092B		2967	US	25 SEP	106.5	89.3	1100	1032

International Designation	Satellite Name	Catalog Number	Source	Launch Date	Period Mins	Incl-ination	Apogee KM	Perigee KM
1967-092C		2994	US	25 SEP	103.7	89.4	1005	858
1967-092D		3122	US	25 SEP	108.8	89.1	1318	1027
1967-094A	INTELSAT 2 F-4	2969	ITSO	28 SEP	1435.4	14.3	35821	35724
1967-094C		2971	US	28 SEP	NO CURRENT ELEMENTS			
1967-096A		2980	US	11 OCT	99.2	99.2	797	639
1967-096B		2985	US	11 OCT	99.0	99.2	785	634
1967-104B		3019	USSR	27 OCT	95.5	64.1	649	438
1967-111A	ATS 3	3029	US	05 NOV	1436.1	14.2	35844	35730
1967-112B		3034	US	07 NOV	BARYOCENTRIC ORBIT			
1967-114A	ESSA 6	3035	US	10 NOV	114.8	102.2	1482	1407
1967-114B		3036	US	10 NOV	114.8	102.2	1482	1408
1967-114C		3051	US	10 NOV	114.1	101.5	1481	1343
1967-114D		3123	US	10 NOV	115.4	102.6	1493	1449
1967-114E		5443	US	10 NOV	114.6	101.7	1483	1386
1967-116A	COSMOS 192	3047	USSR	23 NOV	99.2	74.0	725	717
1967-116B		3048	USSR	23 NOV	99.1	74.0	719	708
1967-123A	PIONEER 8	3066	US	13 DEC	HELIOCENTRIC ORBIT			
1967-127A	COSMOS 198	3081	USSR	27 DEC	103.4	65.1	947	887

1968 LAUNCHES

International Designation	Satellite Name	Catalog Number	Source	Launch Date	Period Mins	Incl-ination	Apogee KM	Perigee KM
1968-001B		3092	US	07 JAN	BARYOCENTRIC ORBIT			
1968-002A	EXPLORER 36	3093	US	11 JAN	112.2	105.8	1572	1079
1968-002B		3094	US	11 JAN	112.1	105.8	1562	1079
1968-002C		3126	US	11 JAN	112.3	106.1	1579	1083
1968-002D		3127	US	11 JAN	112.1	105.3	1570	1073
1968-011A	COSMOS 203	3129	USSR	20 FEB	109.2	74.0	1199	1180
1968-011B		3131	USSR	20 FEB	109.2	74.0	1202	1180
1968-012A	NNSS 30180	3133	US	02 MAR	106.7	90.0	1130	1013
1968-012B		3137	US	02 MAR	106.7	90.0	1133	1017
1968-012C		3213	US	02 MAR	104.6	90.0	1078	874
1968-012D		3214	US	02 MAR	108.6	90.1	1304	1015
1968-013A	ZOND 4	3134	USSR	02 MAR	HELIOCENTRIC ORBIT			
1968-014A	OGO 5	3138	US	04 MAR	NO CURRENT ELEMENTS			
1968-014B		3145	US	04 MAR	NO CURRENT ELEMENTS			
1968-023A	COSMOS 209	3158	USSR	22 MAR	103.0	65.3	932	872
1968-026A	OV1-13	3173	US	06 APR	198.8	99.9	9238	568
1968-026B	OV1-14	3174	US	06 APR	207.1	100.0	9882	552
1968-026C		3177	US	06 APR	207.0	100.0	9876	547
1968-026D		3212	US	06 APR	198.3	99.9	9194	579
1968-027A	LUNA 14	3178	USSR	07 APR	SELENOCENTRIC ORBIT			
1968-040A	COSMOS 220	3229	USSR	07 MAY	98.1	74.0	700	639
1968-040B		3230	USSR	07 MAY	97.8	74.0	683	622
1968-042A		3266	US	23 MAY	101.8	98.9	883	806
1968-042B		3271	US	23 MAY	101.8	98.8	880	803
1968-050A		3284	US	13 JUN	1335.2	11.8	33858	33715
1968-050B		3285	US	13 JUN	NO CURRENT ELEMENTS			
1968-050C		3286	US	13 JUN	NO CURRENT ELEMENTS			
1968-050D		3287	US	13 JUN	NO CURRENT ELEMENTS			
1968-050E		3288	US	13 JUN	NO CURRENT ELEMENTS			
1968-050F		3289	US	13 JUN	NO CURRENT ELEMENTS			
1968-050G		3290	US	13 JUN	NO CURRENT ELEMENTS			
1968-050H		3291	US	13 JUN	NO CURRENT ELEMENTS			
1968-050J		3292	US	13 JUN	1363.7	12.5	35027	33687
1968-055A	EXPLORER 38	3307	US	04 JUL	224.2	120.8	5869	5825
1968-055B		3315	US	04 JUL	155.7	120.7	5726	689
1968-055C		3848	US	04 JUL	224.1	120.9	5864	5823
1968-055D		4841	US	04 JUL	155.3	120.8	5748	633
1968-063A		3334	US	06 AUG	NO ELEMENTS AVAILABLE			
1968-066B	EXPLORER 40	3338	US	08 AUG	117.9	80.7	2494	677
1968-066C		3341	US	08 AUG	117.7	80.7	2480	678
1968-066D		3342	US	08 AUG	107.0	80.6	1544	630
1968-066E		3343	US	08 AUG	102.6	80.6	1188	578
1968-066F		3390	US	08 AUG	108.1	80.6	1636	641
1968-066G		3391	US	08 AUG	106.9	80.7	1546	617
1968-066H		3392	US	08 AUG	110.9	80.7	1866	665
1968-066J		3393	US	08 AUG	108.7	80.6	1684	645
1968-069A	ESSA 7	3345	US	16 AUG	114.9	101.4	1471	1428
1968-069B		3346	US	16 AUG	114.8	101.4	1463	1426
1968-069C		3416	US	16 AUG	113.6	101.9	1485	1299
1968-069D		3417	US	16 AUG	116.1	102.2	1557	1455
1968-069E		3974	US	16 AUG	114.9	102.1	1477	1421
1968-069F		3975	US	16 AUG	114.8	101.5	1482	1414

International Designation	Satellite Name	Catalog Number	Source	Launch Date	Period Mins	Incl-ination	Apogee KM	Perigee KM
1968-069G		4499	US	16 AUG	115.1	101.4	1480	1435
1968-081A	OV2-5	3428	US	26 SEP	1417.9	12.3	35779	35081
1968-081C	ERS 21	3430	US	26 SEP	NO CURRENT ELEMENTS			
1968-081D	LES 6	3431	US	26 SEP	1435.7	12.6	35807	35751
1968-081E		3432	US	26 SEP	1418.5	12.3	35835	35048
1968-091A	COSMOS 249	3504	USSR	20 OCT	111.5	62.3	2094	493
1968-0918 TO 091DQ			USSR	20 OCT	SEE NOTE 8			
1968-092A		3510	US	23 OCT	101.0	98.6	829	784
1968-092B		3522	US	23 OCT	100.9	98.8	823	779
1968-097A	COSMOS 252	3530	USSR	01 NOV	112.0	62.3	2110	529
1968-097B TO 097EU			USSR	01 NOV	SEE NOTE 10			
1968-100A	PIONEER 9	3533	US	08 NOV	HELIOCENTRIC ORBIT			
1968-106A	COSMOS 256	3576	USSR	30 NOV	109.3	74.0	1221	1169
1968-106B		3577	USSR	30 NOV	109.2	74.0	1216	1161
1968-110A	OAO-A2	3597	US	07 DEC	99.9	35.0	759	750
1968-110B		3598	US	07 DEC	99.6	35.0	777	698
1968-112B		3605	US	12 DEC	114.3	80.4	1466	1378
1968-112C		3617	US	12 DEC	114.0	80.2	1444	1372
1968-112D		3618	US	12 DEC	114.7	80.5	1506	1373
1968-112E		3840	US	12 DEC	114.4	80.6	1453	1403
1968-114A	ESSA 8	3615	US	15 DEC	114.6	101.8	1461	1411
1968-114B		3616	US	15 DEC	115.0	101.8	1467	1446
1968-114C		3811	US	15 DEC	112.8	101.9	1462	1248
1968-114D		3812	US	15 DEC	116.3	102.4	1571	1458
1968-116A	INTELSAT 3 F-2	3623	ITSO	19 DEC	1475.2	14.7	37122	35974
1968-118B		3627	US	21 DEC	HELIOCENTRIC ORBIT			

1969 LAUNCHES

International Designation	Satellite Name	Catalog Number	Source	Launch Date	Period Mins	Incl-ination	Apogee KM	Perigee KM
1969-009A	ISIS 1	3669	CANADA	30 JAN	127.7	88.4	3471	574
1969-009B		3670	US	30 JAN	126.6	88.4	3378	573
1969-010B		3673	US	05 FEB	114.0	80.4	1430	1390
1969-010C		3841	US	05 FEB	113.7	80.2	1419	1370
1969-011A	INTELSAT 3 F-3	3674	ITSO	06 FEB	NO CURRENT ELEMENTS			
1969-011B		5977	US	06 FEB	463.6	29.5	26634	296
1969-013A		3691	US	09 FEB	NO CURRENT ELEMENTS			
1969-013B		3692	US	09 FEB	NO CURRENT ELEMENTS			
1969-014A	MARINER 6	3759	US	25 FEB	HELIOCENTRIC ORBIT			
1969-014B		3760	US	25 FEB	HELIOCENTRIC ORBIT			
1969-016A	ESSA 9	3764	US	26 FEB	115.2	101.4	1503	1422
1969-016B		3767	US	26 FEB	115.1	101.4	1497	1417
1969-018B		3770	US	03 MAR	HELIOCENTRIC ORBIT			
1969-024A	COSMOS 272	3818	USSR	17 MAR	109.2	74.0	1206	1176
1969-024B		3819	USSR	17 MAR	109.1	74.0	1194	1177
1969-024C		6289	USSR	17 MAR	108.8	74.0	1180	1162
1969-025C	OV1-19	3825	US	18 MAR	151.5	104.7	5590	478
1969-025E		3827	US	18 MAR	150.3	104.7	5488	487
1969-029A	METEOR	3835	USSR	26 MAR	96.0	81.2	579	552
1969-030A	MARINER 7	3837	US	27 MAR	HELIOCENTRIC ORBIT			
1969-030B		3845	US	27 MAR	HELIOCENTRIC ORBIT			
1969-036A		3889	US	13 APR	NO ELEMENTS AVAILABLE			
1969-037A	NIMBUS 3	3890	US	14 APR	107.2	100.0	1128	1069
1969-037B	SECOR (EGRS) 13	3891	US	14 APR	107.2	100.0	1127	1067
1969-037C		3892	US	14 APR	107.3	100.0	1132	1072
1969-043B		3943	US	18 MAY	HELIOCENTRIC ORBIT			
1969-043C	LM/DESCENT	3948	US	18 MAY	SELENOCENTRIC ORBIT			
1969-043D	LM/ASCENT	3949	US	18 MAY	HELIOCENTRIC ORBIT			
1969-045A	INTELSAT 3 F-4	3947	ITSO	22 MAY	NO CURRENT ELEMENTS			
1969-046A	OV5-5/ERS-29	3950	US	23 MAY	NO CURRENT ELEMENTS			
1969-046B	OV5-6	3951	US	23 MAY	NO CURRENT ELEMENTS			
1969-046C	OV5-9	3952	US	23 MAY	NO CURRENT ELEMENTS			
1969-046D		3954	US	23 MAY	6700.9	61.6	145637	77081
1969-046E		3955	US	23 MAY	6700.7	61.0	150633	72080
1969-046F		3956	US	23 MAY	NO CURRENT ELEMENTS			
1969-053B		3993	US	21 JUN	NO CURRENT ELEMENTS			
1969-059B		4040	US	16 JUL	HELIOCENTRIC ORBIT			
1969-059C	LUNAR MODULE	4041	US	16 JUL	SELENOCENTRIC ORBIT			
1969-062A		4047	US	23 JUL	100.9	98.8	834	768
1969-062B		4048	US	23 JUL	100.8	98.7	829	765
1969-064C		4053	US	26 JUL	122.6	30.3	3334	262
1969-069A	ATS 5	4068	US	12 AUG	1447.5	13.9	36031	35986
1969-069B		4069	US	12 AUG	703.3	16.8	37238	2398
1969-069C		5991	US	12 AUG	682.2	17.2	36497	2086

International Designation	Satellite Name	Catalog Number	Source	Launch Date	Period Mins	Incl-ination	Apogee KM	Perigee KM
1969-069D		21052	US	12 AUG	1466.7	14.1	36913	35852
1969-070A	COSMOS 292	4070	USSR	13 AUG	99.3	74.0	733	720
1969-070B		4071	USSR	13 AUG	99.0	74.0	720	697
1969-070C		4084	USSR	13 AUG	99.7	74.1	760	727
1969-070D		18912	USSR	13 AUG	98.2	74.0	705	644
1969-082L TO 082LF			US	30 SEP	SEE NOTE 11			
1969-082B		4256	US	30 SEP	103.1	70.0	921	891
1969-082C		4257	US	30 SEP	103.2	70.0	928	895
1969-082D		4259	US	30 SEP	103.3	70.0	930	897
1969-082E		4237	US	30 SEP	103.3	70.0	929	896
1969-082F		4247	US	30 SEP	103.3	70.0	928	896
1969-082G		4295	US	30 SEP	103.3	70.0	929	896
1969-082H		4168	US	30 SEP	103.2	70.0	928	896
1969-082J		4166	US	30 SEP	101.0	70.0	812	798
1969-082K		4132	US	30 SEP	102.1	70.0	868	849
1969-084A	METEOR	4119	USSR	06 OCT	95.4	81.2	553	519
1969-084B		4120	USSR	06 OCT	94.0	81.2	508	433
1969-091A	COSMOS 304	4138	USSR	21 OCT	99.6	74.0	749	731
1969-091B		4139	USSR	21 OCT	98.9	74.0	710	703
1969-097A	GRS-A/AZUR	4221	FRG	08 NOV	110.8	102.8	2155	371
1969-097B		4222	US	08 NOV	100.8	102.8	1243	349
1969-099B		4226	US	14 NOV	NO CURRENT ELEMENTS			
1969-101A	SKYNET A	4250	UK	22 NOV	1436.1	13.4	35895	35676
1969-101B		4251	US	22 NOV	NO CURRENT ELEMENTS			
1969-103A	COSMOS 312	4254	USSR	24 NOV	108.5	74.0	1173	1138
1969-103B		4255	USSR	24 NOV	108.3	74.0	1156	1138

1970 LAUNCHES

International Designation	Satellite Name	Catalog Number	Source	Launch Date	Period Mins	Incl-ination	Apogee KM	Perigee KM
1970-003A	INTELSAT 3 F-6	4297	ITSO	15 JAN	NO CURRENT ELEMENTS			
1970-003B		4298	US	15 JAN	521.7	27.7	29856	320
1970-008A	ITOS 1	4320	US	23 JAN	115.0	101.3	1477	1431
1970-008B	OSCAR 5	4321	AUSTRL	23 JAN	115.0	101.3	1475	1431
1970-008C		4322	US	23 JAN	115.0	101.4	1477	1432
1970-009A	SERT 2	4327	US	04 FEB	106.0	99.2	1044	1038
1970-011A	OHSUMI	4330	JAPAN	11 FEB	114.6	31.0	2548	325
1970-012A		4331	US	11 FEB	100.8	98.9	840	750
1970-012B		4332	US	11 FEB	100.8	98.9	844	751
1970-021A	NATO 1	4353	NATO	20 MAR	1436.2	12.9	35798	35778
1970-021B		4354	US	20 MAR	520.9	25.2	29795	338
1970-021C		5975	US	20 MAR	538.2	25.7	30749	327
1970-025A	NIMBUS 4	4362	US	08 APR	107.1	99.9	1096	1086
1970-025C TO 025QM			US	08 APR	SEE NOTE 12			
1970-025B	TOPO 1	4363	US	08 APR	106.9	99.8	1084	1082
1970-027A		4366	US	08 APR	6691.0	61.2	121226	101260
1970-027B		4368	US	08 APR	6707.9	57.3	119313	103569
1970-028A	COSMOS 332	4369	USSR	11 APR	99.4	74.C	736	727
1970-028B		4370	USSR	11 APR	99.1	74.0	727	703
1970-028C		14814	USSR	11 APR	98.4	74.0	688	674
1970-032A	INTELSAT 3 F-7	4376	ITSO	23 APR	NO CURRENT ELEMENTS			
1970-032B		4377	US	23 APR	NO CURRENT ELEMENTS			
1970-034A	MAO 1	4382	PRC	24 APR	111.6	68.4	2169	432
1970-034B		4392	PRC	24 APR	100.4	68.4	1146	411
1970-036A	COSMOS 336	4383	USSR	25 APR	115.4	74.0	1483	1462
1970-036B	COSMOS 337	4384	USSR	25 APR	116.2	74.0	1550	1465
1970-036C	COSMOS 338	4385	USSR	25 APR	115.8	74.0	1516	1465
1970-036D	COSMOS 339	4386	USSR	25 APR	115.0	74.0	1467	1442
1970-036E	COSMOS 340	4387	USSR	25 APR	114.6	74.0	1468	1405
1970-036F	COSMOS 341	4388	USSR	25 APR	113.9	74.0	1467	1339
1970-036G	COSMOS 342	4389	USSR	25 APR	113.5	74.0	1466	1308
1970-036H	COSMOS 343	4390	USSR	25 APR	114.2	74.0	1466	1373
1970-036J		4391	USSR	25 APR	116.6	74.0	1585	1466
1970-037A	METEOR	4393	USSR	28 APR	95.9	81.2	582	538
1970-037B		4394	USSR	28 APR	96.6	81.2	665	524
1970-046A		4418	US	19 JUN	NO ELEMENTS AVAILABLE			
1970-046B		4511	US	19 JUN	NO ELEMENTS AVAILABLE			
1970-047A	METEOR	4419	USSR	23 JUN	101.8	81.2	873	815
1970-047B		4420	USSR	23 JUN	102.0	81.2	919	790
1970-055A	INTELSAT 3 F-8	4478	ITSO	23 JUL	1408.2	13.9	36634	33842
1970-055B		4486	US	23 JUL	NO CURRENT ELEMENTS			
1970-062A	SKYNET B	4493	UK	19 AUG	NO CURRENT ELEMENTS			
1970-067A	NNSS 30190	4507	US	27 AUG	106.7	90.0	1203	944
1970-067B		4515	US	27 AUG	106.8	90.0	1207	946

International Designation	Satellite Name	Catalog Number	Source	Launch Date	Period Mins	Incl-ination	Apogee KM	Perigee KM
1970-067C		5036	US	27 AUG	102.8	90.1	908	870
1970-067D		5447	US	27 AUG	109.1	90.0	1429	942
1970-069A		4510	US	01 SEP	NO ELEMENTS AVAILABLE			
1970-070A		4512	US	03 SEP	100.6	98.9	837	739
1970-070B		4513	US	03 SEP	100.7	99.0	843	741
1970-079A	COSMOS 367	4564	USSR	03 OCT	104.5	65.3	1005	933
1970-083A	COSMOS 371	4578	USSR	12 OCT	99.3	74.0	728	723
1970-083B		4579	USSR	12 OCT	99.0	74.0	721	703
1970-085A	METEOR	4583	USSR	15 OCT	93.8	81.2	462	461
1970-085B		4584	USSR	15 OCT	94.6	81.2	546	454
1970-086A	COSMOS 372	4588	USSR	16 OCT	100.4	74.1	787	767
1970-086B		4589	USSR	16 OCT	100.2	74.1	781	749
1970-086C		5357	USSR	16 OCT	98.2	74.0	680	669
1970-086D		5358	USSR	16 OCT	99.1	74.0	720	711
1970-089A	COSMOS 374	4594	USSR	23 OCT	106.9	63.0	1649	514
1970-089B TO 089DG			USSR	23 OCT	SEE NOTE 13			
1970-091A	COSMOS 375	4598	USSR	30 OCT	111.3	62.8	1999	574
1970-091B TO 091AX			USSR	30 OCT	SEE NOTE 15			
1970-093A		4630	US	06 NOV	1197.9	16.3	36121	25847
1970-093B		4632	US	06 NOV	1197.7	16.3	36150	25812
1970-102A	COSMOS 381	4783	USSR	02 DEC	104.8	74.0	1005	959
1970-102B		4784	USSR	02 DEC	104.6	74.0	997	957
1970-102D		5225	USSR	02 DEC	104.0	74.0	960	934
1970-102E		8764	USSR	02 DEC	104.2	74.0	971	939
1970-102F		9794	USSR	02 DEC	98.6	74.0	691	688
1970-103A	COSMOS 382	4786	USSR	02 DEC	171.0	55.9	5264	2389
1970-103B		4789	USSR	02 DEC	158.8	51.6	5086	1585
1970-103C		4790	USSR	02 DEC	159.1	51.6	5089	1608
1970-103G		12854	USSR	02 DEC	144.1	50.3	3951	1505
1970-106A	NOAA 1	4793	US	11 DEC	114.8	101.3	1470	1421
1970-106B		4794	US	11 DEC	114.9	101.3	1477	1420
1970-106C		8828	US	11 DEC	116.4	102.4	1537	1496
1970-108A	COSMOS 385	4799	USSR	12 DEC	104.6	74.0	978	972
1970-108B		4800	USSR	12 DEC	104.5	74.0	975	962
1970-109B		4802	FRANCE	12 DEC	96.2	15.0	600	549
1970-113A	COSMOS 389	4813	USSR	18 DEC	95.9	81.2	574	545
1970-113B		4814	USSR	18 DEC	96.5	81.2	633	544

1971 LAUNCHES

International Designation	Satellite Name	Catalog Number	Source	Launch Date	Period Mins	Incl-ination	Apogee KM	Perigee KM
1971-003A	METEOR	4849	USSR	20 JAN	95.6	81.2	554	540
1971-003B		4850	USSR	20 JAN	95.3	81.2	582	488
1971-003C		18277	USSR	20 JAN	93.6	81.2	479	426
1971-006A	INTELSAT 4 F-2	4881	ITSO	26 JAN	1456.9	12.6	36246	36140
1971-006B		4882	US	26 JAN	653.3	27.5	36480	644
1971-009A	NATO 2	4902	NATO	03 FEB	1436.1	13.7	35830	35744
1971-009B		4903	US	03 FEB	NO CURRENT ELEMENTS			
1971-009D		5986	US	03 FEB	NO CURRENT ELEMENTS			
1971-010A	COSMOS 394	4922	USSR	09 FEB	95.4	65.8	560	517
1971-011A	TANSEI 1	4952	JAPAN	16 FEB	106.1	29.7	1106	987
1971-011B		5126	JAPAN	16 FEB	104.8	29.7	994	974
1971-012A		4953	US	17 FEB	100.3	98.7	800	741
1971-012B		4954	US	17 FEB	100.4	98.7	802	747
1971-015A	COSMOS 397	4964	USSR	25 FEB	113.2	65.7	2167	575
1971-015B TO 015DV			USSR	25 FEB	SEE NOTE 16			
1971-016A	COSMOS 398	4966	USSR	26 FEB	112.7	51.5	2510	190
1971-020A	COSMOS 400	5050	USSR	18 MAR	104.9	65.8	1000	981
1971-020B		5051	USSR	18 MAR	104.7	65.8	1023	938
1971-020C		5052	USSR	18 MAR	104.9	65.8	996	981
1971-021A		5053	US	21 MAR	NO ELEMENTS AVAILABLE			
1971-021B		5054	US	21 MAR	NO ELEMENTS AVAILABLE			
1971-024A	ISIS 2	5104	CANADA	01 APR	113.5	88.2	1421	1355
1971-024B		5106	US	01 APR	113.5	88.2	1418	1351
1971-024C		5360	US	01 APR	113.5	88.3	1420	1357
1971-025A	COSMOS 402	5105	USSR	01 APR	104.9	65.0	1030	946
1971-028A	COSMOS 405	5117	USSR	07 APR	96.7	81.2	604	597
1971-028B		5118	USSR	07 APR	96.9	81.2	659	564
1971-028D		5724	USSR	07 APR	95.9	81.2	564	557
1971-031B		5143	USSR	17 APR	94.6	81.2	531	462
1971-035A	COSMOS 407	5174	USSR	23 APR	100.6	74.0	800	773
1971-035B		5175	USSR	23 APR	100.4	74.0	799	753
1971-035C		5300	USSR	23 APR	99.4	74.0	736	720
1971-035D		5301	USSR	23 APR	99.8	74.0	761	737

International Designation	Satellite Name	Catalog Number	Source	Launch Date	Period Mins	Incl-ination	Apogee KM	Perigee KM
1971-038A	COSMOS 409	5180	USSR	28 APR	109.2	74.0	1208	1175
1971-038B		5181	USSR	28 APR	109.0	74.0	1222	1140
1971-039A		5204	US	05 MAY	NO ELEMENTS AVAILABLE			
1971-039B		5205	US	05 MAY	NO ELEMENTS AVAILABLE			
1971-041A	COSMOS 411	5210	USSR	07 MAY	113.8	74.0	1488	1313
1971-041B	COSMOS 412	5211	USSR	07 MAY	116.1	74.0	1532	1478
1971-041C	COSMOS 413	5212	USSR	07 MAY	115.7	74.0	1505	1471
1971-041D	COSMOS 414	5213	USSR	07 MAY	115.1	74.0	1491	1425
1971-041E	COSMOS 415	5214	USSR	07 MAY	115.4	74.0	1498	1448
1971-041F	COSMOS 416	5215	USSR	07 MAY	114.4	74.0	1490	1368
1971-041G	COSMOS 417	5216	USSR	07 MAY	114.1	74.0	1490	1340
1971-041H	COSMOS 418	5217	USSR	07 MAY	114.7	74.0	1490	1397
1971-041J		5218	USSR	07 MAY	116.8	74.0	1591	1484
1971-045A	MARS 2	5234	USSR	19 MAY	MARS ORBIT			
1971-046A	COSMOS 422	5238	USSR	22 MAY	104.9	74.0	1003	980
1971-046B		5239	USSR	22 MAY	104.8	74.0	994	977
1971-049A	MARS 3	5252	USSR	28 MAY	MARS ORBIT			
1971-051A	MARINER 9	5261	US	30 MAY	MARS ORBIT			
1971-051B		5267	US	30 MAY	HELIOCENTRIC ORBIT			
1971-052A	COSMOS 426	5281	USSR	04 JUN	99.8	74.0	1143	355
1971-052B		5282	USSR	04 JUN	100.8	74.0	1237	358
1971-059B		5328	USSR	16 JUL	94.8	81.2	545	474
1971-063D	APOLLO 15 SUBSATELLITE	5377	US	26 JUL	SELENOCENTRIC ORBIT			
1971-067B	OV1-21	5397	US	07 AUG	101.7	87.6	898	776
1971-067E		5398	US	07 AUG	101.0	87.6	859	754
1971-067J		5405	US	07 AUG	96.3	87.6	602	564
1971-067K		5395	US	07 AUG	100.9	87.6	852	751
1971-067L		5399	US	07 AUG	96.8	87.6	624	582
1971-067M		5400	US	07 AUG	96.5	87.6	607	571
1971-067N		5384	US	07 AUG	101.4	87.6	885	760
1971-069C		5426	USSR	12 AUG	99.5	49.6	817	655
1971-071A	EOLE 1	5435	FRANCE	16 AUG	99.7	50.2	837	652
1971-071B		5438	US	16 AUG	99.6	50.2	832	648
1971-071C		5440	US	16 AUG	96.5	50.7	637	545
1971-073B		5449	USSR	02 SEP	SELENOCENTRIC ORBIT			
1971-080A	SHINSEI	5485	JAPAN	28 SEP	113.1	32.1	1865	873
1971-080B		5498	JAPAN	28 SEP	111.9	32.0	1756	870
1971-082A	LUNA 19	5488	USSR	28 SEP	SELENOCENTRIC ORBIT			
1971-082C		5490	USSR	28 SEP	SELENOCENTRIC ORBIT			
1971-086A	COSMOS 444	5547	USSR	13 OCT	114.1	74.0	1505	1319
1971-086B	COSMOS 445	5548	USSR	13 OCT	114.4	74.0	1509	1348
1971-086C	COSMOS 446	5549	USSR	13 OCT	114.8	74.0	1511	1377
1971-086D	COSMOS 447	5550	USSR	13 OCT	115.1	74.0	1512	1408
1971-086E	COSMOS 448	5551	USSR	13 OCT	115.5	74.0	1514	1438
1971-086F	COSMOS 449	5552	USSR	13 OCT	116.2	74.0	1540	1480
1971-086G	COSMOS 450	5553	USSR	13 OCT	115.8	74.0	1527	1459
1971-086H	COSMOS 451	5554	USSR	13 OCT	116.6	74.0	1570	1488
1971-086J		5555	USSR	13 OCT	117.3	74.0	1621	1500
1971-087A		5557	US	14 OCT	101.1	99.2	851	772
1971-087B		5556	US	14 OCT	101.3	99.2	868	774
1971-089A		5560	US	17 OCT	99.8	92.7	764	736
1971-093A	PROSPERO	5580	UK	28 OCT	104.5	82.0	1407	531
1971-093B		5581	UK	28 OCT	104.6	82.0	1416	531
1971-095A		5587	US	03 NOV	1436.0	13.2	35816	35751
1971-095B		5588	US	03 NOV	1436.2	13.1	35793	35784
1971-095C		5589	US	03 NOV	1481.7	13.9	37369	35980
1971-099A	COSMOS 457	5614	USSR	20 NOV	109.4	74.0	1215	1181
1971-099B		5615	USSR	20 NOV	109.3	74.0	1208	1176
1971-110A		5678	US	14 DEC	NO ELEMENTS AVAILABLE			
1971-110B		5679	US	14 DEC	NO ELEMENTS AVAILABLE			
1971-110C		5680	US	14 DEC	NO ELEMENTS AVAILABLE			
1971-110D		5681	US	14 DEC	NO ELEMENTS AVAILABLE			
1971-110E		5682	US	14 DEC	NO ELEMENTS AVAILABLE			
1971-111A	COSMOS 465	5683	USSR	15 DEC	104.8	74.0	1004	963
1971-111B		5685	USSR	15 DEC	104.6	74.0	994	958
1971-114A	COSMOS 468	5705	USSR	17 DEC	100.4	74.0	789	769
1971-114B		5707	USSR	17 DEC	100.3	74.0	790	755
1971-114C		5778	USSR	17 DEC	99.7	74.0	753	732
1971-114D		5858	USSR	17 DEC	99.6	74.0	745	728
1971-116A	INTELSAT 4 F-3	5709	ITSO	20 DEC	1445.5	10.3	36013	35928
1971-117A	COSMOS 469	5721	USSR	25 DEC	104.6	64.5	993	962
1971-119A	OREOL 1	5729	USSR	27 DEC	109.0	74.0	1967	389
1971-119B		5730	USSR	27 DEC	108.2	73.9	1901	387

International Designation	Satellite Name	Catalog Number	Source	Launch Date	Period Mins	Incl-ination	Apogee KM	Perigee KM
1971-120A	METEOR	5731	USSR	29 DEC	102.5	81.3	912	837
1971-120B		5732	USSR	29 DEC	102.0	81.3	873	837
1971-120C		8826	USSR	29 DEC	100.8	81.2	810	783
1971-120D		8827	USSR	29 DEC	101.9	81.3	858	838
1971-120F		15344	USSR	29 DEC	97.4	81.2	645	623

1972 LAUNCHES

International Designation	Satellite Name	Catalog Number	Source	Launch Date	Period Mins	Incl-ination	Apogee KM	Perigee KM
1972-003A	INTELSAT 4 F-4	5775	ITSO	23 JAN	1442.4	9.7	35921	35896
1972-003B		5816	US	23 JAN	652.9	28.4	36498	606
1972-007B		5836	USSR	14 FEB	SELENOCENTRIC ORBIT			
1972-009A	COSMOS 475	5846	USSR	25 FEB	104.6	74.0	993	961
1972-009B		5847	USSR	25 FEB	104.4	74.0	989	945
1972-010A		5851	US	01 MAR	NO ELEMENTS AVAILABLE			
1972-010B		5854	US	01 MAR	NO ELEMENTS AVAILABLE			
1972-011B		5853	USSR	01 MAR	93.1	81.2	446	409
1972-012A	PIONEER 10	5860	US	03 MAR	HELIOCENTRIC ORBIT			
1972-012B		5861	US	03 MAR	HELIOCENTRIC ORBIT			
1972-018A		5903	US	24 MAR	101.3	98.9	858	782
1972-018B		5904	US	24 MAR	101.3	98.9	854	783
1972-019A	COSMOS 480	5905	USSR	25 MAR	109.1	83.0	1197	1168
1972-019B		5907	USSR	25 MAR	108.9	83.0	1190	1161
1972-022A	METEOR	5917	USSR	30 MAR	102.3	81.2	877	854
1972-022B		5918	USSR	30 MAR	102.5	81.2	917	832
1972-023E		6073	USSR	31 MAR	157.1	52.2	6322	211
1972-029A	PROGNOZ	5941	USSR	14 APR	NO CURRENT ELEMENTS			
1972-031C	LUNAR MODULE	6005	US	16 APR	SELENOCENTRIC ORBIT			
1972-035A	COSMOS 489	6019	USSR	06 MAY	104.7	74.0	996	960
1972-035B		6020	USSR	06 MAY	104.5	74.0	985	954
1972-041A	INTELSAT 4 F-5	6052	ITSO	13 JUN	1438.6	10.7	35858	35811
1972-041B		6058	US	13 JUN	650.1	26.9	36411	549
1972-043A	COSMOS 494	6059	USSR	23 JUN	100.4	74.1	786	772
1972-043B		6061	USSR	23 JUN	100.2	74.1	783	751
1972-043C		6063	USSR	23 JUN	99.3	74.1	733	721
1972-043D		6065	USSR	23 JUN	99.7	74.1	754	729
1972-049A	METEOR	6079	USSR	30 JUN	102.7	81.2	893	876
1972-049B		6080	USSR	30 JUN	102.8	81.2	927	856
1972-049C		20348	USSR	30 JUN	102.8	81.2	926	855
1972-057A	COSMOS 504	6117	USSR	20 JUL	113.9	74.0	1493	1319
1972-057B	COSMOS 505	6118	USSR	20 JUL	114.3	74.0	1494	1349
1972-057C	COSMOS 506	6119	USSR	20 JUL	114.6	74.0	1494	1379
1972-057D	COSMOS 507	6120	USSR	20 JUL	114.9	74.0	1494	1409
1972-057E	COSMOS 508	6121	USSR	20 JUL	115.3	74.0	1493	1441
1972-057F	COSMOS 509	6122	USSR	20 JUL	115.6	74.0	1496	1471
1972-057G	COSMOS 510	6123	USSR	20 JUL	116.0	74.0	1507	1493
1972-057H	COSMOS 511	6124	USSR	20 JUL	116.4	74.0	1542	1494
1972-057J		6125	USSR	20 JUL	117.0	74.0	1600	1489
1972-058A	LANDSAT 1	6126	US	23 JUL	103.0	99.3	908	896
1972-058B TO 058JL			US	23 JUL	SEE NOTE 17			
1972-062A	COSMOS 514	6148	USSR	16 AUG	104.2	83.0	965	949
1972-062B		6149	USSR	16 AUG	104.1	83.0	961	946
1972-062C		6277	USSR	16 AUG	104.1	82.9	956	945
1972-062D		7560	USSR	16 AUG	102.7	83.0	938	835
1972-065A	COPERNICUS	6153	US	21 AUG	99.2	35.0	725	713
1972-065B		6155	US	21 AUG	98.7	35.0	730	664
1972-066A	COSMOS 516	6154	USSR	21 AUG	104.5	64.8	1021	922
1972-069A	TRIAD 01-1X	6173	US	02 SEP	99.9	90.0	796	707
1972-069B		6180	US	02 SEP	99.4	90.0	767	690
1972-069C		6250	US	02 SEP	97.8	89.7	686	618
1972-072A	COSMOS 520	6192	USSR	19 SEP	715.3	68.4	36294	3936
1972-072E		6302	USSR	19 SEP	706.7	68.2	35955	3850
1972-073A	EXPLORER 47	6197	US	23 SEP	NO CURRENT ELEMENTS			
1972-074A	COSMOS 521	6206	USSR	29 SEP	104.9	65.8	1005	974
1972-074B		6207	USSR	29 SEP	104.7	65.8	999	962
1972-074C		6210	USSR	29 SEP	104.9	65.8	1004	972
1972-076A		6212	US	02 OCT	97.4	98.6	640	630
1972-076B		6217	US	02 OCT	98.7	98.7	706	688
1972-076C		6218	US	02 OCT	99.1	98.5	724	707
1972-076D		6221	US	02 OCT	96.8	98.6	611	601
1972-079C		6822	US	10 OCT	114.7	95.7	1463	1416
1972-079D		6823	US	10 OCT	114.7	95.8	1483	1403
1972-079E		6824	US	10 OCT	114.6	95.5	1443	1430
1972-082A	NOAA 2	6235	US	15 OCT	114.9	102.0	1453	1446

International Designation	Satellite Name	Catalog Number	Source	Launch Date	Period Mins	Incl-ination	Apogee KM	Perigee KM
1972-082B	AMSAT-OSCAR 6	6236	US	15 OCT	114.9	102.0	1452	1446
1972-082C		6237	US	15 OCT	109.2	102.8	1464	914
1972-085A	METEOR	6256	USSR	26 OCT	102.3	81.2	879	852
1972-085B		6257	USSR	26 OCT	102.4	81.3	915	830
1972-087A	COSMOS 528	6262	USSR	01 NOV	114.1	74.0	1465	1363
1972-087B	COSMOS 529	6264	USSR	01 NOV	114.5	74.0	1465	1400
1972-087C	COSMOS 530	6265	USSR	01 NOV	113.7	74.0	1466	1330
1972-087D	COSMOS 531	6266	USSR	01 NOV	114.7	74.0	1466	1419
1972-087E	COSMOS 532	6267	USSR	01 NOV	113.4	74.0	1465	1298
1972-087F	COSMOS 533	6268	USSR	01 NOV	113.6	74.0	1466	1314
1972-087G	COSMOS 534	6269	USSR	01 NOV	113.9	74.0	1466	1346
1972-087H	COSMOS 535	6270	USSR	01 NOV	114.3	74.0	1466	1381
1972-087J		6271	USSR	01 NOV	116.6	74.0	1591	1464
1972-089A		6275	US	09 NOV	101.2	98.7	841	786
1972-089B		6276	US	09 NOV	101.4	98.7	854	798
1972-090A	ANIK A1	6278	CANADA	10 NOV	1457.1	10.8	36258	36136
1972-097A	NIMBUS 5	6305	US	11 DEC	107.1	99.8	1099	1086
1972-097B		6306	US	11 DEC	111.7	99.8	1514	1098
1972-101A		6317	US	20 DEC	NO CURRENT ELEMENTS			
1972-101B		6318	US	20 DEC	NO CURRENT ELEMENTS			
1972-102A	COSMOS 539	6319	USSR	21 DEC	112.9	74.0	1378	1339
1972-102B		6320	USSR	21 DEC	112.7	74.0	1372	1332
1972-104A	COSMOS 540	6323	USSR	25 DEC	100.4	74.1	790	763
1972-104B		6324	USSR	25 DEC	100.0	74.1	767	752
1972-104C		6391	USSR	25 DEC	98.7	74.1	703	691
1972-104D		6396	USSR	25 DEC	98.6	74.0	702	682

1973 LAUNCHES

International Designation	Satellite Name	Catalog Number	Source	Launch Date	Period Mins	Incl-ination	Apogee KM	Perigee KM
1973-005A	COSMOS 546	6350	USSR	26 JAN	95.7	50.7	565	538
1973-009A	PROGNOZ 3	6364	USSR	15 FEB	NO CURRENT ELEMENTS			
1973-013A		6380	US	06 MAR	NO ELEMENTS AVAILABLE			
1973-015A	METEOR	6392	USSR	20 MAR	102.4	81.2	878	861
1973-015B		6393	USSR	20 MAR	102.5	81.3	921	833
1973-019A	PIONEER 11	6421	US	06 APR	HELIOCENTRIC ORBIT			
1973-019B		6425	US	06 APR	HELIOCENTRIC ORBIT			
1973-023A	ANIK A2	6437	CANADA	20 APR	1443.0	9.4	35970	35873
1973-034A	METEOR	6659	USSR	29 MAY	102.2	81.2	880	843
1973-034B		6660	USSR	29 MAY	102.5	81.2	909	840
1973-037A	COSMOS 564	6675	USSR	08 JUN	114.6	74.0	1477	1393
1973-037B	COSMOS 565	6676	USSR	08 JUN	115.3	74.0	1488	1446
1973-037C	COSMOS 566	6677	USSR	08 JUN	115.0	74.0	1480	1431
1973-037D	COSMOS 567	6678	USSR	08 JUN	114.8	74.0	1480	1411
1973-037E	COSMOS 568	6679	USSR	08 JUN	114.4	74.0	1478	1373
1973-037F	COSMOS 569	6680	USSR	08 JUN	114.1	74.0	1478	1354
1973-037G	COSMOS 570	6681	USSR	08 JUN	113.9	74.0	1478	1335
1973-037H	COSMOS 571	6682	USSR	08 JUN	113.7	74.0	1477	1316
1973-037J		6683	USSR	08 JUN	116.8	74.0	1593	1482
1973-039A	EXPLORER 49	6686	US	10 JUN	SELENOCENTRIC ORBIT			
1973-039D		6689	US	10 JUN	NO CURRENT ELEMENTS			
1973-039F		6725	US	10 JUN	SELENOCENTRIC ORBIT			
1973-039G		6726	US	10 JUN	SELENOCENTRIC ORBIT			
1973-040A		6691	US	12 JUN	NO ELEMENTS AVAILABLE			
1973-040B		11940	US	12 JUN	NO ELEMENTS AVATLABLE			
1973-042A	COSMOS 574	6707	USSR	20 JUN	104.9	83.0	1008	975
1973-042B		6708	USSR	20 JUN	104.8	82.9	995	976
1973-047A	MARS 4	6742	USSR	21 JUL	HELIOCENTRIC ORBIT			
1973-049A	MARS 5	6754	USSR	25 JUL	MARS ORBIT			
1973-052A	MARS 6	6768	USSR	05 AUG	MARS ORBIT			
1973-053A	MARS 7	6776	USSR	09 AUG	MARS ORBIT			
1973-053D	CAPSULE	7224	USSR	09 AUG	HELIOCENTRIC ORBIT			
1973-054A		6787	US	17 AUG	100.9	98.8	820	780
1973-054B		6788	US	17 AUG	101.1	98.9	831	789
1973-056A		6791	US	21 AUG	NO ELEMENTS AVAILABLE			
1973-056B		6792	US	21 AUG	NO ELEMENTS AVAILABLE			
1973-058A	INTELSAT 4 F-7	6796	ITSO	23 AUG	1452.4	9.7	36138	36072
1973-058B		6797	US	23 AUG	651.9	27.7	36482	571
1973-064A	COSMOS 585	6825	USSR	08 SEP	113.5	74.0	1401	1373
1973-064B		6826	USSR	08 SEP	113.4	74.0	1402	1358
1973-065A	COSMOS 586	6828	USSR	14 SEP	104.7	82.9	1001	959
1973-065B		6829	USSR	14 SEP	104.6	82.9	992	956
1973-069A	COSMOS 588	6845	USSR	02 OCT	115.3	74.0	1490	1446
1973-069B	COSMOS 589	6846	USSR	02 OCT	114.9	74.0	1487	1411

International Designation	Satellite Name	Catalog Number	Source	Launch Date	Period Mins	Incl-ination	Apogee KM	Perigee KM
1973-069C	COSMOS 590	6847	USSR	02 OCT	115.1	74.0	1485	1431
1973-069D	COSMOS 591	6848	USSR	02 OCT	114.1	74.0	1484	1344
1973-069E	COSMOS 592	6849	USSR	02 OCT	113.9	74.0	1481	1329
1973-069F	COSMOS 593	6850	USSR	02 OCT	114.3	74.0	1483	1361
1973-069G	COSMOS 594	6851	USSR	02 OCT	114.5	74.0	1484	1378
1973-069H	COSMOS 595	6852	USSR	02 OCT	114.7	74.0	1484	1396
1973-069J		6853	USSR	02 OCT	117.1	74.0	1620	1482
1973-078A	EXPLORER 50	6893	US	26 OCT	NO CURRENT ELEMENTS			
1973-078C		6895	US	26 OCT	96.5	28.8	861	324
1973-078D		6896	US	26 OCT	NO CURRENT ELEMENTS			
1973-081A	NNSS 30200	6909	US	30 OCT	105.2	89.8	1123	885
1973-081B		6910	US	30 OCT	105.3	89.8	1127	885
1973-081C		15764	US	30 OCT	105.8	90.5	1169	889
1973-084A	COSMOS 606	6916	USSR	02 NOV	718.8	69.1	37061	3345
1973-084D		6939	USSR	02 NOV	706.6	67.5	37080	2719
1973-085A	MARINER 10	6919	US	03 NOV	HELIOCENTRIC ORBIT			
1973-086A	NOAA 3	6920	US	06 NOV	116.1	102.2	1508	1499
1973-086B TO 086HF			US	06 NOV	SEE NOTE 18			
1973-088D		6938	US	10 NOV	114.5	96.9	1453	1413
1973-088E		7559	US	10 NOV	114.6	96.8	1476	1400
1973-098A	COSMOS 614	6965	USSR	04 DEC	100.2	74.1	787	751
1973-098B		6966	USSR	04 DEC	100.1	74.1	778	744
1973-098C		6967	USSR	04 DEC	98.3	74.1	690	669
1973-098D		9569	USSR	04 DEC	99.4	74.1	743	717
1973-100A		6973	US	13 DEC	1474.6	13.3	36666	36408
1973-100B		6974	US	13 DEC	1436.2	12.9	35793	35783
1973-100D		6976	US	13 DEC	1515.0	13.7	38520	36116
1973-104A	COSMOS 617	6985	USSR	19 DEC	113.9	74.0	1482	1331
1973-104B	COSMOS 618	6986	USSR	19 DEC	115.2	74.0	1484	1442
1973-104C	COSMOS 619	6987	USSR	19 DEC	115.0	74.0	1484	1422
1973-104D	COSMOS 620	6988	USSR	19 DEC	115.4	74.0	1491	1456
1973-104E	COSMOS 621	6989	USSR	19 DEC	114.7	74.0	1482	1404
1973-104F	COSMOS 622	6990	USSR	19 DEC	114.3	74.0	1483	1367
1973-104G	COSMOS 623	6991	USSR	19 DEC	114.5	74.0	1483	1384
1973-104H	COSMOS 624	6992	USSR	19 DEC	114.1	74.0	1483	1348
1973-104J		6993	USSR	19 DEC	117.0	74.0	1620	1473
1973-107A	OREOL 2	7003	USSR	26 DEC	103.5	74.0	1457	387
1973-107B		7004	USSR	26 DEC	102.8	74.0	1402	378
1973-108A	COSMOS 626	7005	USSR	27 DEC	103.9	65.4	983	906
1973-109A	COSMOS 627	7008	USSR	29 DEC	104.9	83.0	1013	964
1973-109B		7009	USSR	29 DEC	104.6	83.0	990	959

1974 LAUNCHES

International Designation	Satellite Name	Catalog Number	Source	Launch Date	Period Mins	Incl-ination	Apogee KM	Perigee KM
1974-001A	COSMOS 628	7094	USSR	17 JAN	104.7	83.0	1008	950
1974-001B		7095	USSR	17 JAN	104.5	83.0	998	942
1974-011A	METEOR	7209	USSR	05 MAR	101.9	81.2	877	821
1974-011B		7210	USSR	05 MAR	102.0	81.2	910	792
1974-013A	UK-X4	7213	UK	09 MAR	100.3	97.9	867	677
1974-013B		7228	US	09 MAR	100.4	97.9	863	688
1974-015A		7218	US	16 MAR	100.9	99.1	845	758
1974-015B		7219	US	16 MAR	101.2	99.1	863	766
1974-017A	COSMOS 637	7229	USSR	26 MAR	1428.9	13.0	35805	35485
1974-017F		11567	USSR	26 MAR	1425.7	12.9	35762	35404
1974-020B		7244	US	10 APR	NO ELEMENTS AVAILABLE			
1974-022A	WESTAR 1	7250	US	13 APR	1441.6	9.1	35907	35879
1974-024A	COSMOS 641	7265	USSR	23 APR	114.5	74.0	1479	1385
1974-024B	COSMOS 642	7266	USSR	23 APR	113.7	74.0	1478	1316
1974-024C	COSMOS 643	7267	USSR	23 APR	114.1	74.0	1479	1351
1974-024D	COSMOS 644	7268	USSR	23 APR	113.9	74.0	1479	1333
1974-024E	COSMOS 645	7269	USSR	23 APR	114.3	74.0	1479	1367
1974-024F	COSMOS 646	7270	USSR	23 APR	114.7	74.0	1483	1401
1974-024G	COSMOS 647	7271	USSR	23 APR	114.9	74.0	1482	1419
1974-024H	COSMOS 648	7272	USSR	23 APR	115.1	74.0	1487	1435
1974-024J		7273	USSR	23 APR	117.0	74.0	1604	1486
1974-025A	METEOR	7274	USSR	24 APR	102.3	81.2	881	852
1974-025B		7275	USSR	24 APR	102.4	81.2	913	831
1974-026A	MOLNIYA 2-9	7276	USSR	26 APR	640.6	62.3	35596	878
1974-026E		7373	USSR	26 APR	699.2	62.3	38646	789
1974-028A	COSMOS 650	7281	USSR	29 APR	113.4	74.0	1398	1366
1974-028B		7284	USSR	29 APR	113.2	74.0	1387	1361
1974-029A	COSMOS 651	7291	USSR	15 MAY	103.4	65.0	940	897
1974-032A	COSMOS 654	7297	USSR	17 MAY	104.4	64.9	1023	908

International Designation	Satellite Name	Catalog Number	Source	Launch Date	Period Mins	Incl-ination	Apogee KM	Perigee KM
1974-033A	SMS 1	7298	US	17 MAY	1460.3	14.9	36301	36215
1974-037A	LUNA 22	7315	USSR	29 MAY	SELENOCENTRIC ORBIT			
1974-039A	ATS 6	7318	US	30 MAY	1412.1	12.5	35440	35190
1974-039C		7324	US	30 MAY	1430.6	12.7	35790	35569
1974-044A	COSMOS 660	7337	USSR	18 JUN	104.4	83.0	1549	383
1974-044B		7338	USSR	18 JUN	101.6	83.0	1290	382
1974-048A	COSMOS 663	7349	USSR	27 JUN	104.7	82.9	997	961
1974-048B		7350	USSR	27 JUN	104.5	82.9	985	959
1974-050C		7354	USSR	29 JUN	682.6	62.5	38557	46
1974-052A	METEOR	7363	USSR	09 JUL	102.9	81.2	907	883
1974-052B		7364	USSR	09 JUL	102.5	81.2	907	843
1974-054A		7369	US	14 JUL	468.7	125.1	13772	13448
1974-054C		8599	US	14 JUL	NO CURRENT ELEMENTS			
1974-056A	MOLNIYA 2-10	7376	USSR	23 JUL	717.2	62.0	39835	491
1974-056D		7382	USSR	23 JUL	731.9	61.8	40367	681
1974-060A	MOLNIYA 1-S	7392	USSR	29 JUL	1437.1	13.3	35858	35753
1974-060F		20836	USSR	29 JUL	1436.4	13.3	35869	35714
1974-063A		7411	US	09 AUG	101.1	98.8	844	780
1974-063B		7412	US	09 AUG	101.3	98.8	854	786
1974-066B		7418	USSR	16 AUG	93.8	81.2	477	445
1974-066C		8424	USSR	16 AUG	91.2	81.2	337	325
1974-069A	COSMOS 675	7424	USSR	29 AUG	113.6	74.1	1421	1361
1974-069B		7426	USSR	29 AUG	113.5	74.1	1420	1351
1974-071A	COSMOS 676	7433	USSR	11 SEP	100.7	74.0	799	779
1974-071B		7434	USSR	11 SEP	100.5	74.0	796	764
1974-071C		8756	USSR	11 SEP	99.5	74.1	743	731
1974-071D		8829	USSR	11 SEP	100.2	74.1	779	752
1974-072A	COSMOS 677	7435	USSR	19 SEP	114.4	74.0	1464	1394
1974-072B	COSMOS 678	7436	USSR	19 SEP	115.9	74.0	1530	1464
1974-072C	COSMOS 679	7437	USSR	19 SEP	115.7	74.0	1507	1464
1974-072D	COSMOS 680	7438	USSR	19 SEP	115.5	74.0	1489	1464
1974-072E	COSMOS 681	7439	USSR	19 SEP	115.3	74.0	1470	1463
1974-072F	COSMOS 682	7440	USSR	19 SEP	115.0	74.0	1464	1450
1974-072G	COSMOS 683	7441	USSR	19 SEP	114.8	74.0	1463	1433
1974-072H	COSMOS 684	7442	USSR	19 SEP	114.6	74.0	1463	1414
1974-072J		7443	USSR	19 SEP	117.7	74.0	1682	1473
1974-075A	WESTAR 2	7466	US	10 OCT	1442.2	8.9	35928	35883
1974-075C		7468	US	10 OCT	156.8	24.3	6320	190
1974-079A	COSMOS 689	7476	USSR	18 OCT	104.9	82.9	1014	968
1974-079B		7477	USSR	18 OCT	104.8	82.9	1010	959
1974-083A	METEOR	7490	USSR	28 OCT	102.2	81.2	888	835
1974-083B		7493	USSR	28 OCT	102.3	81.2	901	835
1974-083C		15521	USSR	28 OCT	102.3	81.2	900	833
1974-089A	NOAA 4	7529	US	15 NOV	114.9	101.9	1457	1442
1974-089D TO 089FF			US	15 NOV	SEE NOTE 19			
1974-089B	AMSAT-OSCAR 7	7530	US	15 NOV	114.8	101.9	1457	1437
1974-089C	INTASAT	7531	SPAIN	15 NOV	114.8	101.9	1457	1439
1974-093A	INTELSAT 4 F-8	7544	ITSO	21 NOV	1443.0	8.1	35949	35894
1974-093B		7545	US	21 NOV	652.4	25.5	36543	537
1974-094A	SKYNET 2B	7547	UK	23 NOV	1436.9	11.6	35828	35775
1974-097A	HELIOS 1	7567	FRG	10 DEC	HELIOCENTRIC ORBIT			
1974-097B		7568	US	10 DEC	NO CURRENT ELEMENTS			
1974-097C		7569	US	10 DEC	HELIOCENTRIC ORBIT			
1974-097D		7570	FRG	10 DEC	HELIOCENTRIC ORBIT			
1974-099A	METEOR	7574	USSR	17 DEC	102.1	81.2	870	844
1974-099B		7575	USSR	17 DEC	102.1	81.2	895	819
1974-101A	SYMPHONIE-A	7578	FR/FRG	19 DEC	1440.6	11.9	35896	35853
1974-101G		9330	US	19 DEC	652.6	13.0	36668	419
1974-105A	COSMOS 700	7593	USSR	26 DEC	104.6	83.0	992	958
1974-105B		7594	USSR	26 DEC	104.5	82.9	980	958
1975 LAUNCHES								
1975-004A	LANDSAT 2	7615	US	22 JAN	103.1	98.8	911	899
1975-004B TO 004HR			US	22 JAN	SEE NOTE 20			
1975-007A	COSMOS 706	7625	USSR	30 JAN	717.6	67.7	34774	5571
1975-007D		7629	USSR	30 JAN	716.9	67.6	35573	4739
1975-010A	STARLETTE	7646	FRANCE	06 FEB	104.2	49.8	1108	805
1975-010B		7647	FRANCE	06 FEB	104.3	49.8	1126	801
1975-010C		7654	FRANCE	06 FEB	103.6	49.9	1064	795
1975-010D		7655	FRANCE	06 FEB	103.7	49.8	1071	794
1975-010E		7659	FRANCE	06 FEB	103.8	49.8	1083	793
1975-011A	SMS 2	7648	US	06 FEB	1446.4	10.9	36169	35808

International Designation	Satellite Name	Catalog Number	Source	Launch Date	Period Mins	Incl-ination	Apogee KM	Perigee KM
1975-011F		20835	US	06 FEB	1460.7	13.0	36676	35856
1975-012A	COSMOS 708	7663	USSR	12 FEB	113.5	69.2	1408	1367
1975-012B		7665	USSR	12 FEB	113.3	69.2	1397	1362
1975-016A	COSMOS 711	7678	USSR	28 FEB	115.4	74.0	1490	1459
1975-016B	COSMOS 712	7679	USSR	28 FEB	114.9	74.0	1487	1409
1975-016C	COSMOS 713	7680	USSR	28 FEB	114.6	74.0	1485	1393
1975-016D	COSMOS 714	7681	USSR	28 FEB	115.2	74.0	1488	1443
1975-016E	COSMOS 715	7682	USSR	28 FEB	115.7	74.0	1502	1467
1975-016F	COSMOS 716	7683	USSR	28 FEB	115.9	74.0	1512	1477
1975-016G	COSMOS 717	7684	USSR	28 FEB	116.1	74.0	1533	1477
1975-016H	COSMOS 718	7685	USSR	28 FEB	115.0	74.0	1487	1427
1975-016J		7686	USSR	28 FEB	117.9	74.0	1718	1457
1975-017A		7687	US	10 MAR	NO ELEMENTS AVAILABLE			
1975-017B		7688	US	10 MAR	NO ELEMENTS AVAILABLE			
1975-023A	METEOR	7714	USSR	01 APR	102.3	81.2	883	853
1975-023B		7715	USSR	01 APR	102.4	81.2	909	833
1975-024A	COSMOS 723	7718	USSR	02 APR	103.6	64.7	953	907
1975-025A	COSMOS 724	7727	USSR	07 APR	102.9	65.6	940	856
1975-027A	GEOS 3	7734	US	09 APR	101.6	115.0	851	815
1975-027B		7735	US	09 APR	101.3	115.0	855	780
1975-027C		10728	US	09 APR	101.4	115.2	878	772
1975-027E		10730	US	09 APR	103 5	115.0	1002	846
1975-028A	COSMOS 726	7736	USSR	11 APR	104 5	83.0	988	951
1975-028B		7737	USSR	11 APR	104.4	83.0	977	950
1975-029D		7741	USSR	14 APR	726.6	62.3	40651	139
1975-034A	COSMOS 729	7768	USSR	22 APR	104.8	83.0	1004	970
1975-034B		7769	USSR	22 APR	104.7	83.0	995	969
1975-036A	MOLNIYA 1-29	7780	USSR	29 APR	717.6	61.8	39464	880
1975-036D		7800	USSR	29 APR	732.4	62.1	40493	582
1975-038A	ANIK A3	7790	CANADA	07 MAY	1439.5	8.2	35872	35833
1975-038D		7794	US	07 MAY	386.0	24.6	22091	275
1975-042A	INTELSAT 4 F-1	7815	ITSO	22 MAY	1450.8	8.1	36133	36015
1975-042B		7902	US	22 MAY	652.9	26.0	36531	574
1975-043A		7816	US	24 MAY	NO ELEMENTS AVAILABLE			
1975-043B		7817	US	24 MAY	NO ELEMENTS AVAILABLE			
1975-045A	COSMOS 732	7820	USSR	28 MAY	114.6	74.0	1467	1402
1975-045B	COSMOS 733	7822	USSR	28 MAY	116.2	74.0	1552	1467
1975-045C	COSMOS 734	7823	USSR	28 MAY	115.0	74.0	1469	1441
1975-045D	COSMOS 735	7824	USSR	28 MAY	115.2	74.0	1472	1458
1975-045E	COSMOS 736	7825	USSR	28 MAY	115.5	74.0	1484	1467
1975-045F	COSMOS 737	7826	USSR	28 MAY	115.9	74.0	1526	1468
1975-045G	COSMOS 738	7827	USSR	28 MAY	115.7	74.0	1506	1467
1975-045H	COSMOS 739	7828	USSR	28 MAY	114.8	74.0	1468	1422
1975-045J		7831	USSR	28 MAY	117.9	74.0	1692	1484
1975-049B	SRET 2	7910	FRANCE	05 JUN	NO CURRENT ELEMENTS			
1975-050A	VENERA 9	7915	USSR	08 JUN	VENUS ORBIT			
1975-051C	SSU 1	7937	US	08 JUN	113.5	95.1	1393	1382
1975-051D		7938	US	08 JUN	113.2	95.0	1401	1343
1975-051E		7939	US	08 JUN	113.9	95.2	1424	1381
1975-052A	NIMBUS 6	7924	US	12 JUN	107.4	99.8	1111	1098
1975-052B TO 052JX			US	12 JUN	SEE NOTE 56			
1975-054A	VENERA 10	7947	USSR	14 JUN	VENUS ORBIT			
1975-055A		7963	US	18 JUN	NO ELEMENTS AVAILABLE			
1975-055B		7964	US	18 JUN	NO ELEMENTS AVAILABLE			
1975-056B		7969	USSR	20 JUN	94.2	81.2	493	461
1975-063A	MOLNIYA 2-13	8015	USSR	08 JUL	716.4	62.0	39450	835
1975-063D		8018	USSR	08 JUL	732.6	62.1	40517	564
1975-064A	METEOR 2	8026	USSR	11 JUL	102.2	81.3	878	844
1975-064B		8027	USSR	11 JUL	102.3	81.3	908	831
1975-064C		8039	USSR	11 JUL	102.2	81.3	874	853
1975-064D		8110	USSR	11 JUL	102.1	81.3	881	833
1975-072A	COS-B	8062	ESA	09 AUG	NO CURRENT ELEMENTS			
1975-072B		8063	US	09 AUG	120.1	89.2	3051	315
1975-074A	COSMOS 755	8072	USSR	14 AUG	104.8	82.9	1004	965
1975-074B		8073	USSR	14 AUG	104.7	82.9	993	963
1975-075A	VIKING ORBITER 1	8108	US	20 AUG	MARS ORBIT			
1975-075B		8111	US	20 AUG	HELIOCENTRIC ORBIT			
1975-076B		8128	USSR	22 AUG	94.8	81.2	535	485
1975-077A	SYMPHONIE-B	8132	FR/FRG	27 AUG	1440.4	12.1	35880	35861
1975-077B		8133	US	27 AUG	102.6	25.3	1369	398
1975-077C		8134	US	27 AUG	637.7	13.7	35918	409
1975-081A	MOLNIYA 2-14	8195	USSR	09 SEP	717.6	61.7	39417	927
1975-081D		8418	USSR	09 SEP	732.5	62.0	40464	613

International Designation	Satellite Name	Catalog Number	Source	Launch Date	Period Mins	Incl-ination	Apogee KM	Perigee KM
1975-082A	KIKU	8197	JAPAN	09 SEP	106.0	47.0	1103	975
1975-082B		8352	JAPAN	09 SEP	105.9	47.0	1100	973
1975-083A	VIKING ORBITER 2	8199	US	09 SEP	MARS ORBIT			
1975-083B		8272	US	09 SEP	HELIOCENTRIC ORBIT			
1975-086A	COSMOS 761	8285	USSR	17 SEP	114.6	74.0	1480	1397
1975-086B	COSMOS 762	8286	USSR	17 SEP	115.1	74.0	1482	1436
1975-086C	COSMOS 763	8287	USSR	17 SEP	115.8	74.0	1508	1471
1975-086D	COSMOS 764	8288	USSR	17 SEP	116.0	74.0	1524	1476
1975-086E	COSMOS 765	8289	USSR	17 SEP	116.3	74.0	1548	1476
1975-086F	COSMOS 766	8290	USSR	17 SEP	114.9	74.0	1483	1415
1975-086G	COSMOS 767	8291	USSR	17 SEP	115.3	74.0	1484	1454
1975-086H	COSMOS 768	8292	USSR	17 SEP	115.5	74.0	1489	1469
1975-086J		8295	USSR	17 SEP	117.8	74.0	1682	1479
1975-087A	METEOR	8293	USSR	18 SEP	102.0	81.3	911	798
1975-087B		8294	USSR	18 SEP	102.2	81.3	917	812
1975-089A	COSMOS 770	8325	USSR	24 SEP	109.1	83.0	1206	1161
1975-089B		8326	USSR	24 SEP	108.9	83.0	1196	1158
1975-091A	INTELSAT 4A F-1	8330	ITSO	26 SEP	1441.0	8.1	35914	35852
1975-091B		8331	US	26 SEP	652.1	21.4	36588	472
1975-094A	COSMOS 773	8343	USSR	30 SEP	100.5	74.1	789	773
1975-094B		8344	USSR	30 SEP	100.3	74.1	788	753
1975-094C		8346	USSR	30 SEP	98.2	74.0	683	665
1975-094D		14865	USSR	30 SEP	99.9	74.0	755	749
1975-097A	COSMOS 775	8357	USSR	08 OCT	1436.6	13.0	35852	35740
1975-097F		11676	USSR	08 OCT	1438.5	13.0	35935	35732
1975-100A	GOES 1	8366	US	16 OCT	1435.7	11.7	35801	35756
1975-100C		8368	US	16 OCT	139.3	23.4	4794	250
1975-100F		20962	US	16 OCT	1412.7	12.4	36488	34167
1975-103A	COSMOS 778	8419	USSR	04 NOV	104.8	83.0	999	966
1975-103B		8421	USSR	04 NOV	104.6	83.0	994	958
1975-105A	MOLNIYA 3-3	8425	USSR	14 NOV	717.4	61.8	39483	851
1975-105D		8462	USSR	14 NOV	733.3	62.0	40515	603
1975-112A	COSMOS 783	8458	USSR	28 NOV	100.6	74.1	797	779
1975-112B		8459	USSR	28 NOV	100.4	74.1	790	767
1975-112C		8757	USSR	28 NOV	99.3	74.0	727	723
1975-112D		14801	USSR	28 NOV	100.2	74.1	774	763
1975-112E		18500	USSR	28 NOV	100.4	74.1	789	765
1975-116A	COSMOS 785	8473	USSR	12 DEC	104.2	65.1	1019	892
1975-117A	RCA SATCOM I	8476	US	13 DEC	1445.8	8.2	36080	35873
1975-118A		8482	US	14 DEC	NO ELEMENTS AVAILABLE			
1975-118C		8516	US	14 DEC	NO ELEMENTS AVAILABLE			
1975-118D		8517	US	14 DEC	NO CURRENT ELEMENTS			
1975-121A	MOLNIYA 2-15	8492	USSR	17 DEC	416.9	62.8	24113	103
1975-122A	PROGNOZ 4	8510	USSR	22 DEC	NO CURRENT ELEMENTS			
1975-123A	RADUGA 1	8513	USSR	22 DEC	1436.4	12.7	35818	35768
1975-123D		8546	USSR	22 DEC	380.2	46.0	21616	395
1975-123E		8547	USSR	22 DEC	289.2	46.7	16021	196
1975-123F		11568	USSR	22 DEC	1433.1	12.7	35797	35658
1975-124A	METEOR	8519	USSR	25 DEC	102.1	81.2	875	843
1975-124B		8520	USSR	25 DEC	102.2	81.3	894	835

1976 LAUNCHES

International Designation	Satellite Name	Catalog Number	Source	Launch Date	Period Mins	Incl-ination	Apogee KM	Perigee KM
1976-003A	HELIOS 2	8582	FRG	15 JAN	HELIOCENTRIC ORBIT			
1976-003B		8583	US	15 JAN	HELIOCENTRIC ORBIT			
1976-003C		8584	US	15 JAN	HELIOCENTRIC ORBIT			
1976-004A	CTS	8585	CANADA	17 JAN	1437.1	12.2	35887	35726
1976-005A	COSMOS 789	8591	USSR	20 JAN	104.9	83.0	1011	965
1976-005B		8597	USSR	20 JAN	104.7	83.0	1000	963
1976-006A	MOLNIYA 1-32	8601	USSR	22 JAN	720.3	62.4	39794	683
1976-006D		8701	USSR	22 JAN	695.2	62.2	38417	815
1976-008A	COSMOS 791	8607	USSR	28 JAN	114.7	74.1	1483	1400
1976-008B	COSMOS 792	8608	USSR	28 JAN	115.1	74.0	1488	1433
1976-008C	COSMOS 793	8609	USSR	28 JAN	114.9	74.0	1487	1416
1976-008D	COSMOS 794	8610	USSR	28 JAN	115.3	74.0	1492	1449
1976-008E	COSMOS 795	8611	USSR	28 JAN	115.6	74.1	1496	1464
1976-008F	COSMOS 796	8612	USSR	28 JAN	115.8	74.0	1513	1470
1976-008G	COSMOS 797	8613	USSR	28 JAN	116.0	74.1	1527	1477
1976-008H	COSMOS 798	8614	USSR	28 JAN	116.3	74.1	1552	1476
1976-008J		8615	USSR	28 JAN	117.9	74.0	1693	1481
1976-010A	INTELSAT 4A F-2	8620	ITSO	29 JAN	1444.5	8.3	35968	35933
1976-010B		8621	US	29 JAN	653.7	22.0	36530	615
1976-011A	COSMOS 800	8645	USSR	03 FEB	104.9	83.0	1008	975

International Designation	Satellite Name	Catalog Number	Source	Launch Date	Period Mins	Incl-ination	Apogee KM	Perigee KM
1976-011B		8646	USSR	03 FEB	104.8	83.0	990	980
1976-014A	COSMOS 803	8688	USSR	12 FEB	95.3	65.8	561	501
1976-017A	MARISAT 1	8697	US	19 FEB	1436.1	10.4	35797	35777
1976-017C		8702	US	19 FEB	151.4	24.4	5810	251
1976-019A	UME	8709	JAPAN	29 FEB	105.0	69.7	1003	989
1976-019B		8710	JAPAN	29 FEB	105.1	69.7	1007	991
1976-022A	COSMOS 807	8744	USSR	12 MAR	104.8	82.9	1581	383
1976-022B		8745	USSR	12 MAR	101.2	82.9	1268	367
1976-023A	LES 8	8746	US	15 MAR	1436.3	17.8	35833	35747
1976-023B	LES 9	8747	US	15 MAR	1436.1	17.8	35885	35688
1976-023C	SOLRAD 11A	8748	US	15 MAR	NO CURRENT ELEMENTS			
1976-023D	SOLRAD 11B	8749	US	15 MAR	NO CURRENT ELEMENTS			
1976-023F		8751	US	15 MAR	1465.6	18.2	36999	35723
1976-023G		8752	US	15 MAR	NO CURRENT ELEMENTS			
1976-023H		8753	US	15 MAR	NO CURRENT ELEMENTS			
19Z6-023J		8832	US	15 MAR	NO CURRENT ELEMENTS			
1976-023K		13753	US	15 MAR	1420.9	10.7	35493	35483
1976-024A	COSMOS 808	8754	USSR	16 MAR	92.4	81.2	394	389
1976-024B		8755	USSR	16 MAR	93.8	81.3	475	440
1976-029A	RCA SATCOM II	8774	US	26 MAR	1460.1	7.8	36501	36010
1976-032A	METEOR	8799	USSR	07 APR	102.0	81.3	880	830
1976-032B		8800	USSR	07 APR	102.2	81.2	931	790
1976-035A	NATO III-A	8808	NATO	22 APR	1442.3	10.1	36008	35806
1976-038A		8818	US	30 APR	NO ELEMENTS AVAILABLE			
1976-038B		8819	US	30 APR	NO ELEMENTS AVAILABLE			
1976-038C	SSU-1	8835	US	30 APR	NO ELEMENTS AVAILABLE			
1976-038D	SSU-2	8836	US	30 APR	NO ELEMENTS AVAILABLE			
1976-038E		8839	US	30 APR	NO ELEMENTS AVAILABLE			
1976-038F		8842	US	30 APR	NO ELEMENTS AVAILABLE			
1976-038G		8843	US	30 APR	NO ELEMENTS AVAILABLE			
1976-038H		8859	US	30 APR	NO ELEMENTS AVAILABLE			
1976-038J	SSU-3	8884	US	30 APR	NO ELEMENTS AVAILABLE			
1976-038K		9796	US	30 APR	NO ELEMENTS AVAILABLE			
1976-038L		9996	US	30 APR	NO ELEMENTS AVAILABLE			
1976-039A	LAGEOS	8820	US	04 MAY	225.4	109.9	5945	5838
1976-039C		8822	US	04 MAY	225.4	109.9	5941	5837
1976-039D		14514	US	04 MAY	99.0	109.9	1160	258
1976-041A	MOLNIYA 3-5	8833	USSR	12 MAY	664.3	62.0	37590	93
1976-041D		8844	USSR	12 MAY	710.5	61.9	39885	111
1976-042A	COMSTAR 1	8838	US	13 MAY	1442.6	8.0	35921	35905
1976-042B		8840	US	13 MAY	648.2	21.3	36244	620
1976-043A	METEOR	8845	USSR	15 MAY	102.0	81.3	885	825
1976-043B		8846	USSR	15 MAY	102.3	81.2	901	828
1976-047A	P 76-5	8860	US	22 MAY	105.4	99.6	1044	981
1976-047B		8861	US	22 MAY	105.5	99.5	1046	984
1976-047C		8867	US	22 MAY	106.3	99.3	1110	997
1976-047D		8868	US	22 MAY	104.5	100.0	1012	934
1976-050A		8871	US	02 JUN	NO ELEMENTS AVAILABLE			
1976-050B		8872	US	02 JUN	NO ELEMENTS AVAILABLE			
1976-051A	COSMOS 823	8873	USSR	02 JUN	104.9	83.0	1004	970
1976-051B		8874	USSR	02 JUN	104.7	83.0	1000	964
1976-053A	MARISAT 2	8882	US	10 JUN	1436.1	9.5	35813	35760
1976-053F		8910	US	10 JUN	468.1	25.2	26922	263
1976-054A	COSMOS 825	8889	USSR	15 JUN	114.6	74.0	1485	1392
1976-054B	COSMOS 826	8890	USSR	15 JUN	116.2	74.0	1542	1480
1976-054C	COSMOS 827	8891	USSR	15 JUN	114.9	74.0	1487	1410
1976-054D	COSMOS 828	8892	USSR	15 JUN	115.1	74.0	1486	1431
1976-054E	COSMOS 829	8893	USSR	15 JUN	115.3	74.0	1488	1448
1976-054F	COSMOS 830	8894	USSR	15 JUN	115.5	74.0	1491	1466
1976-054G	COSMOS 831	8895	USSR	15 JUN	115.7	74.0	1505	1472
1976-054H	COSMOS 832	8896	USSR	15 JUN	116.0	74.0	1518	1480
1976-054J		8897	USSR	15 JUN	117.9	74.0	1686	1485
1976-059A		8916	US	26 JUN	NO ELEMENTS AVAILABLE			
1976-059C		8918	US	26 JUN	NO ELEMENTS AVAILABLE			
1976-059D		8919	US	26 JUN	NO CURRENT ELEMENTS			
1976-061A	COSMOS 836	8923	USSR	29 JUN	100.6	74.1	801	773
1976-061B		8924	USSR	29 JUN	100.4	74.1	790	766
1976-061C		9572	USSR	29 JUN	99.1	74.1	720	712
1976-061D		14815	USSR	29 JUN	99.3	74.1	737	715
1976-065C		9008	US	08 JUL	NO ELEMENTS AVAILABLE			
1976-066A	PALAPA 1	9009	INDNSA	08 JUL	1439.1	8.0	35867	35821
1976-066C		9017	US	08 JUL	310.2	24.5	17355	243
1976-067A	COSMOS 839	9011	USSR	08 JUL	115.6	65.9	2065	903

International Designation	Satellite Name	Catalog Number	Source	Launch Date	Period Mins	Incl-ination	Apogee KM	Perigee KM
1976-067B TO 067BZ			USSR	08 JUL	SEE NOTE 21			
1976-069A	COSMOS 841	9022	USSR	15 JUL	100.4	74.0	789	769
1976-069B		9023	USSR	15 JUL	100.3	74.0	784	756
1976-069C		9704	USSR	15 JUL	99.3	74.1	731	720
1976-069D		13499	USSR	15 JUL	100.3	74.1	789	758
1976-070A	COSMOS 842	9025	USSR	21 JUL	104.8	83.0	1005	963
1976-070B		9044	USSR	21 JUL	104.6	83.0	988	965
1976-073A	COMSTAR 2	9047	US	22 JUL	1436.2	7.9	35791	35784
1976-073B		9329	US	22 JUL	645.7	21.4	36204	530
1976-077A	NOAA 5	9057	US	29 JUL	116.2	102.1	1518	1505
1976-077B TO 077FR			US	29 JUL	SEE NOTE 22			
1976-078A	COSMOS 846	9061	USSR	29 JUL	104.6	82.9	1007	945
1976-078B		9062	USSR	29 JUL	104.5	82.9	991	947
1976-080A		9270	US	06 AUG	NO ELEMENTS AVAILABLE			
1976-080B		9271	US	06 AUG	NO ELEMENTS AVAILABLE			
1976-091A	DMSP-F1	9415	US	11 SEP	NO ELEMENTS AVAILABLE			
1976-091B		9419	US	11 SEP	NO ELEMENTS AVAILABLE			
1976-091C		9420	US	11 SEP	NO ELEMENTS AVAILABLE			
1976-091F		9484	US	11 SEP	NO ELEMENTS AVAILABLE			
1976-091G		9518	US	11 SEP	NO ELEMENTS AVAILABLE			
1976-092A	RADUGA 2	9416	USSR	11 SEP	1436.2	12.5	35889	35686
1976-092F		17872	USSR	11 SEP	1435.4	12.4	35831	35713
1976-098A	COSMOS 858	9443	USSR	29 SEP	100.6	74.0	794	774
1976-098B		9444	USSR	29 SEP	100.4	74.0	786	765
1976-098C		14816	USSR	29 SEP	100.3	74.0	784	756
1976-098D		14817	USSR	29 SEP	99.3	74.1	738	712
1976-098E		18504	USSR	29 SEP	99.7	74.1	757	726
1976-101A	MARISAT 3	9478	US	14 OCT	1436.0	10.9	35791	35779
1976-102A	METEOR	9481	USSR	15 OCT	102.2	81.3	887	836
1976-102B		9482	USSR	15 OCT	102.3	81.3	912	825
1976-103A	COSMOS 860	9486	USSR	17 OCT	104.3	64.7	1016	902
1976-103F		19297	USSR	17 OCT	99.7	64.9	838	652
1976-104A	COSMOS 861	9494	USSR	21 OCT	104.2	64.9	1006	910
1976-105A	COSMOS 862	9495	USSR	22 OCT	717.8	64.8	39450	907
1976-105D		9506	USSR	22 OCT	711.8	63.3	39387	670
1976-105E		9888	USSR	22 OCT	NO CURRENT ELEMENTS			
1976-105F		9889	USSR	22 OCT	718.2	64.3	39595	780
1976-105G		9890	USSR	22 OCT	NO CURRENT ELEMENTS			
1976-105H		9891	USSR	22 OCT	864.1	64.7	46003	1344
1976-105J		9892	USSR	22 OCT	717.9	64.6	39509	849
1976-105K		9893	USSR	22 OCT	NO CURRENT ELEMENTS			
1976-105L		9894	USSR	22 OCT	702.5	63.9	39040	558
1976-105M		9895	USSR	22 OCT	718.7	65.9	38931	1468
1976-105N		9896	USSR	22 OCT	725.4	65.0	39655	1074
1976-105P		9902	USSR	22 OCT	727.1	64.6	39702	1111
1976-107A	EKRAN	9503	USSR	26 OCT	1435.4	12.4	36060	35487
1976-107F		11569	USSR	26 OCT	1419.3	12.3	35506	35408
1976-108A	COSMOS 864	9509	USSR	29 OCT	104.7	82.9	1003	957
1976-108B		9510	USSR	29 OCT	104.6	82.9	993	953
1976-112A	PROGNOZ 5	9557	USSR	25 NOV	NO CURRENT ELEMENTS			
1976-118A	COSMOS 871	9588	USSR	07 DEC	114.6	74.0	1462	1415
1976-118B	COSMOS 872	9589	USSR	07 DEC	114.4	74.0	1461	1397
1976-118C	COSMOS 873	9590	USSR	07 DEC	115.5	74.0	1493	1462
1976-118D	COSMOS 874	9591	USSR	07 DEC	115.7	74.0	1514	1462
1976-118E	COSMOS 875	9592	USSR	07 DEC	114.8	74.0	1463	1433
1976-118F	COSMOS 876	9593	USSR	07 DEC	116.0	74.0	1537	1461
1976-118G	COSMOS 877	9594	USSR	07 DEC	115.0	74.0	1463	1451
1976-118H	COSMOS 878	9595	USSR	07 DEC	115.3	74.0	1472	1462
1976-118J		9598	USSR	07 DEC	117.6	74.0	1682	1463
1976-120AT		11216	USSR	09 DEC	95.0	65.9	544	487
1976-120AU		11217	USSR	09 DEC	89.7	65.8	264	253
1976-120AY		11221	USSR	09 DEC	94.5	65.8	519	468
1976-122A	COSMOS 883	9610	USSR	15 DEC	104.7	83.Q	1004	952
1976-122B		9613	USSR	15 DEC	104.6	83.0	996	950
1976-126A	COSMOS 886	9634	USSR	27 DEC	114.7	65.8	2299	585
1976-126B TO 126CG			USSR	27 DEC	SEE NOTE 25			
1976-128A	COSMOS 887	9637	USSR	28 DEC	104.6	82.9	1010	944
1976-128B		9638	USSR	28 DEC	104.5	82.9	994	947

1977 LAUNCHES

International Designation	Satellite Name	Catalog Number	Source	Launch Date	Period Mins	Incl-ination	Apogee KM	Perigee KM
1977-002A	METEOR 2-2	9661	USSR	06 JAN	102.7	81.3	892	876

International Designation	Satellite Name	Catalog Number	Source	Launch Date	Period Mins	Incl-ination	Apogee KM	Perigee KM
1977-002B		9662	USSR	06 JAN	102.8	81.3	929	853
1977-002C		9663	USSR	06 JAN	102.7	81.3	890	880
1977-002D		9664	USSR	06 JAN	102.7	81.3	893	880
1977-004A	COSMOS 890	9737	USSR	20 JAN	105.0	83.0	1013	974
1977-004B		9738	USSR	20 JAN	104.8	83.0	997	974
1977-005A	NATO III-B	9785	NATO	28 JAN	1436.2	9.9	35789	35788
1977-005B		9786	US	28 JAN	103.7	28.0	1252	618
1977-005D		9809	US	28 JAN	NO CURRENT ELEMENTS			
1977-005E		9810	US	28 JAN	NO CURRENT ELEMENTS			
1977-005F		9811	US	28 JAN	NO CURRENT ELEMENTS			
1977-007A		9803	US	06 FEB	NO ELEMENTS AVAILABLE			
1977-007C		9855	US	06 FEB	NO ELEMENTS AVAILABLE			
1977-007D		9856	US	06 FEB	NO CURRENT ELEMENTS			
1977-010A	MOLNIYA 2-17	9829	USSR	11 FEB	717.5	62.4	39935	405
1977-010E		9850	USSR	11 FEB	730.1	63.1	40497	464
1977-012A	TANSEI 3	9841	JAPAN	19 FEB	134.1	65.8	3799	801
1977-012C		9843	JAPAN	19 FEB	134.1	65.7	3795	800
1977-012E		9981	JAPAN	19 FEB	133.3	65.2	3732	796
1977-012F		9982	JAPAN	19 FEB	133.5	65.9	3768	780
1977-012G		9983	JAPAN	19 FEB	134.1	65.6	3793	808
1977-012H		12857	JAPAN	19 FEB	133.9	66.3	3776	811
1977-012J		13133	JAPAN	19 FEB	133.0	65.8	3704	799
1977-012K		14512	JAPAN	19 FEB	133.8	65.7	3771	799
1977-012L		19314	JAPAN	19 FEB	133.3	65.4	3885	643
1977-013A	COSMOS 894	9846	USSR	21 FEB	104.8	82.9	1008	962
1977-013B		9848	USSR	21 FEB	104.7	82.9	992	967
1977-014A	KIKU 2	9852	JAPAN	23 FEB	1439.7	11.1	35862	35852
1977-015B		9854	USSR	26 FEB	94.1	81.2	501	451
1977-018A	PALAPA 2	9862	INDNSA	10 MAR	1439.5	6.9	35873	35831
1977-021A	MOLNIYA 1-36	9880	USSR	24 MAR	717.7	62.2	39921	428
1977-021D		9927	USSR	24 MAR	732.5	63.0	40698	381
1977-024A	METEOR	9903	USSR	05 APR	102.3	81.3	889	843
1977-024B		9904	USSR	05 APR	102.4	81.3	910	831
1977-024C		9907	USSR	05 APR	102.5	82.9	894	857
1977-027A	COSMOS 903	9911	USSR	11 APR	717.4	67.5	37668	2669
1977-027D		9921	USSR	11 APR	723.9	67.9	37947	2709
1977-027E		10946	USSR	11 APR	NO CURRENT ELEMENTS			
1977-029A	ESA-GEOS	9931	ESA	20 APR	734.1	26.6	38283	2874
1977-032A	MOLNIYA 3-7	9941	USSR	28 APR	717.5	62.2	39925	414
1977-034A		10000	US	12 MAY	1489.6	11.6	36905	36750
1977-034B		10001	US	12 MAY	1509.1	11.2	37359	37047
1977-034C		10002	US	12 MAY	1506.9	11.9	38452	35870
1977-036A	COSMOS 909	10010	USSR	19 MAY	117.0	65.9	2100	992
1977-036B		10011	USSR	19 MAY	116.9	65.9	2089	990
1977-036C		10013	USSR	19 MAY	117.0	65.9	2098	993
1977-038A		10016	US	23 MAY	NO ELEMENTS AVAILABLE			
1977-038B		10017	US	23 MAY	NO ELEMENTS AVAILABLE			
1977-038C		15422	US	23 MAY	NO ELEMENTS AVAILABLE			
1977-039A	COSMOS 911	10019	USSR	25 MAY	104.7	82.9	997	962
1977-039B		10020	USSR	25 MAY	104.5	82.9	994	948
1977-041A	INTELSAT 4A F-4	10024	ITSO	26 MAY	1448.1	7.0	36075	35966
1977-041B		10025	US	26 MAY	647.9	21.1	36276	571
1977-044A	DMSP-F2	10033	US	05 JUN	NO ELEMENTS AVAILABLE			
1977-044B		10034	US	05 JUN	NO ELEMENTS AVAILABLE			
1977-044C		10037	US	05 JUN	NO ELEMENTS AVAILABLE			
1977-044D		10085	US	05 JUN	NO ELEMENTS AVAILABLE			
1977-047A	COSMOS 917	10059	USSR	16 JUN	717.4	67.6	35536	4797
1977-047D		10089	USSR	16 JUN	722.5	67.4	36579	4007
1977-048A	GOES 2	10061	US	16 JUN	1435.8	10.2	35797	35762
1977-048B		10062	US	16 JUN	108.3	28.4	1721	574
1977-048F		10409	US	16 JUN	NO CURRENT ELEMENTS			
1977-048G		20799	US	16 JUN	1431.8	11.5	36590	34813
1977-053A		10091	US	23 JUN	718.1	64.1	20280	20089
1977-053B		10960	US	23 JUN	314.4	64.3	17038	835
1977-054D		10155	USSR	24 JUN	691.3	62.9	38800	238
1977-055A	COSMOS 921	10095	USSR	24 JUN	97.1	75.8	652	588
1977-055B		10096	USSR	24 JUN	97.2	75.8	657	589
1977-057A	METEOR	10113	USSR	29 JUN	91.9	97.4	373	362
1977-057B		10114	USSR	29 JUN	96.3	97.9	591	569
1977-059A	COSMOS 923	10120	USSR	01 JUL	100.7	74.0	800	781
1977-059B		10121	USSR	01 JUL	100.5	74.1	796	766
1977-059C		14802	USSR	01 JUL	100.0	74.1	766	746
1977-059D		14818	USSR	01 JUL	99.6	74.1	746	729

International Designation	Satellite Name	Catalog Number	Source	Launch Date	Period Mins	Incl-ination	Apogee KM	Perigee KM
1977-061B		10135	USSR	07 JUL	94.3	81.2	502	464
1977-062A	COSMOS 926	10137	USSR	08 JUL	104.9	82.9	1016	965
1977-062B		10138	USSR	08 JUL	104.8	82.9	1002	969
1977-064A	COSMOS 928	10141	USSR	13 JUL	104.6	83.0	1003	947
1977-064B		10142	USSR	13 JUL	104.5	83.0	1000	938
1977-065A	HIMAWARI	10143	JAPAN	14 JUL	1451.0	10.4	36152	36001
1977-065B TO 65GC			US	14 JUL	SEE NOTE 26			
1977-068A	COSMOS 931	10150	USSR	20 JUL	718.7	66.5	36016	4382
1977-068D		10167	USSR	20 JUL	720.8	67.0	35475	5027
1977-068E		12906	USSR	20 JUL	717.6	67.7	34579	5765
1977-068F		12996	USSR	20 JUL	704.4	61.8	38095	1596
1977-068G		14000	USSR	20 JUL	718.2	65.1	36876	3500
1977-068J		19881	USSR	20 JUL	666.3	59.9	37436	346
1977-071A	RADUGA 3	10159	USSR	23 JUL	1435.0	12.1	35809	35719
1977-071F		11570	USSR	23 JUL	1473.4	12.5	36551	36474
1977-076A	VOYAGER 2	10271	US	20 AUG	HELIOCENTRIC ORBIT			
1977-076B		10272	US	20 AUG	HELIOCENTRIC ORBIT			
1977-076C		10273	US	20 AUG	HELIOCENTRIC ORBIT			
1977-079A	COSMOS 939	10282	USSR	24 AUG	114.8	74.0	1461	1429
1977-079B	COSMOS 940	10286	USSR	24 AUG	114.4	74.0	1460	1391
1977-079C	COSMOS 941	10287	USSR	24 AUG	114.6	74.0	1460	1410
1977-079D	COSMOS 942	10288	USSR	24 AUG	115.9	74.0	1531	1459
1977-079E	COSMOS 943	10289	USSR	24 AUG	115.0	74.0	1460	1447
1977-079F	COSMOS 944	10290	USSR	24 AUG	115.2	74.0	1468	1459
1977-079G	COSMOS 945	10291	USSR	24 AUG	115.4	74.0	1488	1460
1977-079H	COSMOS 946	10292	USSR	24 AUG	115.6	74.0	1508	1459
1977-079J		10293	USSR	24 AUG	117.5	74.0	1675	1460
1977-080A	SIRIO	10294	ITALY	25 AUG	1438.7	8.3	35925	35750
1977-080B		10295	US	25 AUG	115.5	27.1	2082	874
1977-082A	MOLNIYA 1-38	10315	USSR	30 AUG	681.6	62.6	38268	284
1977-082E		10369	USSR	30 AUG	634.4	63.9	35919	238
1977-084A	VOYAGER 1	10321	US	05 SEP	HELIOCENTRIC ORBIT			
1977-084B		10322	US	05 SEP	HELIOCENTRIC ORBIT			
1977-084C		10323	US	05 SEP	HELIOCENTRIC ORBIT			
1977-087A	COSMOS 951	10352	USSR	13 SEP	104.8	83.0	1009	959
1977-087B		10355	USSR	13 SEP	104.7	83.0	1004	954
1977-088A	COSMOS 952	10358	USSR	16 SEP	104.1	64.9	996	905
1977-091A	COSMOS 955	10362	USSR	20 SEP	94.7	81.2	512	499
1977-091B		10363	USSR	20 SEP	95.0	81.2	542	495
1977-092A	EKRAN	10365	USSR	20 SEP	1435.4	12.0	35897	35646
1977-092G		11571	USSR	20 SEP	1421.8	11.8	35550	35461
1977-093A	PROGNOZ 6	10370	USSR	22 SEP	NO CURRENT ELEMENTS			
1977-102D		10425	US	22 OCT	NO CURRENT ELEMENTS			
1977-105A	MOLNIYA 3-8	10455	USSR	28 OCT	717.4	63.1	39895	438
1977-105E		10485	USSR	28 OCT	731.5	63.4	40240	791
1977-106A	NNSS 30110	10457	US	28 OCT	106.8	89.7	1096	1060
1977-106B		10462	US	28 OCT	106.8	89.7	1097	1062
1977-106C		12858	US	28 OCT	106.9	89.5	1095	1066
1977-107A	COSMOS 962	10459	USSR	28 OCT	104.7	82.9	1003	960
1977-107B		10461	USSR	28 OCT	104.6	82.9	999	951
1977-108A	METEOSAT 1	10489	ESA	23 NOV	1435.9	11.3	35815	35748
1977-108B		10490	US	23 NOV	115.1	28.3	2431	491
1977-109A	COSMOS 963	10491	USSR	24 NOV	109.2	82.9	1205	1174
1977-105B		10492	USSR	24 NOV	109.1	82.9	1200	1169
1977-112A		10502	US	08 DEC	NO ELEMENTS AVAILABLE			
1977-112B		10504	US	08 DEC	NO ELEMENTS AVAILABLE			
1977-112C		10528	US	08 DEC	NO ELEMENTS AVAILABLE			
1977-112D		10529	US	08 DEC	NO ELEMENTS AVAILABLE			
1977-112E		10544	US	08 DEC	NO ELEMENTS AVAILABLE			
1977-112F		10594	US	08 DEC	NO ELEMENTS AVAILABLE			
1977-112G		10595	US	08 DEC	NO ELEMENTS AVAILABLE			
1977-112H		12859	US	08 DEC	NO ELEMENTS AVAILABLE			
1977-114A		10508	US	11 DEC	NO ELEMENTS AVAILABLE			
1977-114B		10509	US	11 DEC	NO ELEMENTS AVAILABLE			
1977-116A	COSMOS 967	10512	USSR	13 DEC	104.7	65.8	992	969
1977-116B		10513	USSR	13 DEC	104.5	65.8	987	953
1977-116C		10518	USSR	13 DEC	104.6	65.8	987	966
1977-116D		10526	USSR	13 DEC	104.7	65.8	1009	953
1977-117A	METEOR 2-3	10514	USSR	14 DEC	102.2	81.2	874	847
1977-117B		10515	USSR	14 DEC	102.3	81.2	898	832
1977-117C		14950	USSR	14 DEC	102.3	81.2	899	831
1977-118A	SAKURA	10516	JAPAN	15 DEC	1455.8	9.8	36182	36162
1977-118B		10517	US	15 DEC	109.2	28.7	1895	483

International Designation	Satellite Name	Catalog Number	Source	Launch Date	Period Mins	Incl-ination	Apogee KM	Perigee KM
1977-118C		10519	US	15 DEC	109.7	29.1	1882	542
1977-119A	COSMOS 968	10520	USSR	16 DEC	100.4	74.0	791	765
1977-119B		10521	USSR	16 DEC	100.2	74.0	783	750
1977-119C		10524	USSR	16 DEC	99.8	74.0	762	739
1977-119D		10525	USSR	16 DEC	99.9	74.0	766	741
1977-119E		18512	USSR	16 DEC	99.7	74.0	748	736
1977-121A	COSMOS 970	10531	USSR	21 DEC	105.9	65.9	1145	926
1977-121B TO 121BY			USSR	21 DEC	SEE NOTE 27			
1977-122A	COSMOS 971	10536	USSR	23 DEC	104.9	82.9	1002	972
1977-122B		10537	USSR	23 DEC	104.7	82.9	995	964
1977-123A	COSMOS 972	10539	USSR	27 DEC	103.7	75.8	1156	710
1977-123B		10541	USSR	27 DEC	103.7	75.8	1153	712
1978 LAUNCHES								
1978-002A	INTELSAT 4A F-3	10557	ITSO	07 JAN	1441.4	6.5	35901	35877
1978-002B		10722	US	17 JAN	650.1	21.1	36346	617
1918-004A	COSMOS 975	10561	USSR	10 JAN	95.1	81.2	530	517
1978-004B		10582	USSR	10 JAN	95.8	81.2	583	527
1978-005A	COSMOS 976	10581	USSR	10 JAN	115.0	74.0	1461	1453
1978-005B	COSMOS 977	10584	USSR	10 JAN	114.4	74.0	1461	1398
1978-005C	COSMOS 978	10585	USSR	10 JAN	114.6	74.0	1461	1417
1978-005D	COSMOS 979	10586	USSR	10 JAN	114.8	74.0	1460	1436
1978-005E	COSMOS 980	10587	USSR	10 JAN	115.3	74.0	1473	1461
1978-005F	COSMOS 981	10588	USSR	10 JAN	115.5	74.0	1493	1462
1978-005G	COSMOS 982	10589	USSR	10 JAN	115.7	74.0	1513	1462
1978-005H	COSMOS 983	10590	USSR	10 JAN	116.0	74.0	1535	1462
1978-005J		10591	USSR	10 JAN	117.7	74.0	1691	1461
1978-007A	COSMOS 985	10599	USSR	17 JAN	104.6	82.9	1014	936
1978-007B		10600	USSR	17 JAN	104.5	82.9	1006	934
1978-012A	IUE	10637	US	26 JAN	1435.6	33.8	41343	30210
1978-012C		10723	US	26 JAN	536.7	29.2	30780	217
1978-014A	KYOKKO	10664	JAPAN	04 FEB	134.0	65.4	3942	650
1978-014C		12329	JAPAN	04 FEB	133.7	65.3	3927	640
1978-014D		12330	JAPAN	04 FEB	133.9	65.4	3941	644
1978-014E		12331	JAPAN	04 FEB	132.4	64.8	3807	647
1978-014F		12406	JAPAN	04 FEB	133.0	65.9	3855	650
1978-016A	FLTSATCOM 1	10669	US	09 FEB	1436.1	10.5	35798	35776
1978-016C		12908	US	09 FEB	193.1	26.4	9129	247
1978-018A	UME 2	10674	JAPAN	16 FEB	107.2	69.4	1215	974
1978-018B		10675	JAPAN	16 FEB	107.1	69.4	1210	974
1978-018C		13132	JAPAN	16 FEB	107.9	69.2	1289	968
1978-019A	COSMOS 990	10676	USSR	17 FEB	100.4	74.0	791	765
1978-019B		10677	USSR	17 FEB	100.2	74.0	780	756
1978-019C		14803	USSR	17 FEB	99.2	74.0	723	714
1978-019D		13500	USSR	17 FEB	99.9	74.1	763	743
1978-019E		18501	USSR	17 FEB	100.0	74.1	767	745
1978-020A		10684	US	22 FEB	727.0	64.3	20587	20221
1978-020B		10801	US	22 FEB	268.4	64.1	13930	876
1978-021A		10688	US	25 FEB	NO ELEMENTS AVAILABLE			
1978-021B		10689	US	25 FEB	NO ELEMENTS AVAILABLE			
1978-022A	COSMOS 991	10692	USSR	28 FEB	104.6	83.0	1003	950
1978-022B		10693	USSR	28 FEB	104.5	83.0	988	958
1978-026A	LANDSAT 3	10702	US	05 MAR	103.1	98.8	916	894
1978-026C TO 026HT			US	05 MAR	SEE NOTE 28			
1978-026B	AMSAT-OSCAR-8	10703	US	05 MAR	103.0	99.2	904	893
1978-028A	COSMOS 994	10731	USSR	15 MAR	104.9	82.9	1005	969
1978-028B		10732	USSR	15 MAR	104.7	82.9	994	965
1978-029B		10734	US	16 MAR	NO ELEMENTS AVAILABLE			
1978-031A	COSMOS 996	10744	USSR	28 MAR	104.6	82.9	1004	948
1978-031B		10745	USSR	28 MAR	104.5	82.9	995	943
1978-034A	COSMOS 1000	10776	USSR	31 MAR	104.7	82.9	1006	954
1978-034B		10777	USSR	31 MAR	104.6	82.9	993	953
1978-035A	INTELSAT 4A F-6	10778	ITSO	31 MAR	1435.6	6.5	35801	35753
1978-035B		10779	US	31 MAR	647.8	21.7	36209	635
1978-038A		10787	US	07 APR	NO ELEMENTS AVAILABLE			
1978-038B		10788	US	07 APR	NO ELEMENTS AVAILABLE			
1978-039A	YURI	10792	JAPAN	07 APR	1435.2	11.0	35796	35740
1978-039B		10793	US	07 APR	110.9	28.2	1962	574
1978-039C		10794	US	07 APR	167.9	26.9	7187	219
1978-042A		10820	US	01 MAY	100.7	98.7	800	786
1978-044A	OTS-2	10855	ESA	11 MAY	1452.6	8.5	36124	36092
1978-044B		10856	US	11 MAY	139.9	27.9	3526	1573

International Designation	Satellite Name	Catalog Number	Source	Launch Date	Period Mins	Incl-ination	Apogee KM	Perigee KM
1978-044C		10857	US	11 MAY	NO CURRENT ELEMENTS			
1978-045A	COSMOS 1005	10860	USSR	12 MAY	94.6	81.2	502	494
1978-045B		10861	USSR	12 MAY	95.9	81.2	591	536
1978-047A		10893	US	13 MAY	714.2	63.8	20628	19547
1978-047B		10894	US	13 MAY	286.7	64.4	15039	1007
1978-051A	PIONEER VENUS ORBITER	10911	US	20 MAY	VENUS IMPACT			
1978-051B		10912	US	20 MAY	HELIOCENTRIC ORBIT			
1978-053A	COSMOS 1011	10917	USSR	23 MAY	104.7	82.9	1006	954
1978-053B		10918	USSR	23 MAY	104.6	82.9	996	951
1978-055A	MOLNIYA 1-40	10925	USSR	02 JUN	718.0	63.1	39787	576
1978-055E		10949	USSR	02 JUN	732.4	63.3	40046	1029
1978-056A	COSMOS 1013	10930	USSR	07 JUN	116.3	74.0	1552	1476
1978-056B	COSMOS 1014	10931	USSR	07 JUN	116.0	74.0	1529	1476
1978-056C	COSMOS 1015	10932	USSR	07 JUN	115.8	74.0	1515	1471
1978-056D	COSMOS 1016	10933	USSR	07 JUN	115.6	74.0	1496	1469
1978-056E	COSMOS 1017	10934	USSR	07 JUN	115.4	74.0	1489	1456
1978-056F	COSMOS 1018	10935	USSR	07 JUN	115.2	74.0	1485	1441
1978-056G	COSMOS 1019	10936	USSR	07 JUN	115.0	74.0	1485	1422
1978-056H	COSMOS 1020	10937	USSR	07 JUN	114.8	74.0	1482	1405
1978-056J		10938	USSR	07 JUN	117.8	74.0	1690	1477
1978-058A		10941	US	10 JUN	NO ELEMENTS AVAILABLE			
1978-058B		10942	US	10 JUN	NO ELEMENTS AVAILABLE			
1978-062A	GOES 3	10953	US	16 JUN	1436.0	9.1	35808	35761
1978-062B		10954	US	16 JUN	107.3	28.4	1651	555
1978-062D		20801	US	16 JUN	1451.1	11.3	39990	32168
1978-063A	COSMOS 1023	10961	USSR	21 JUN	100.4	74.1	787	766
1978-063B		10962	USSR	21 JUN	100.2	74.1	785	748
1978-063C		13497	USSR	21 JUN	100.2	74.1	781	755
1978-063D		14804	USSR	21 JUN	98.4	74.0	691	675
1978-064A	SEASAT 1	10967	US	27 JUN	100.1	108.0	765	761
1978-066A	COSMOS 1024	10970	USSR	28 JUN	718.0	67.4	34844	5521
1978-066D		10998	USSR	28 JUN	720.1	67.4	35183	5287
1978-067A	COSMOS 1025	10973	USSR	28 JUN	95.9	82.5	572	552
1978-067B		10974	USSR	28 JUN	97.2	82.5	635	609
1978-068A	COMSTAR 3	10975	US	29 JUN	1451.8	6.3	36181	36004
1978-068B		10976	US	29 JUN	648.7	21.9	36253	637
1978-071A	ESA GEOS 2	10981	ESA	14 JUL	1449.1	11.1	36056	36023
1978-071C		10983	US	14 JUL	409.5	25.7	23536	241
1978-073A	RADUGA 4	10987	USSR	18 JUL	1435.5	11.6	35816	35735
1978-073D		11074	USSR	18 JUL	565.5	46.3	31893	657
1978-073E		11941	USSR	18 JUL	1475.9	12.0	36629	36494
1978-074A	COSMOS 1027	10991	USSR	27 JUL	104.6	82.9	996	956
1978-074B		10992	USSR	27 JUL	104.5	82.9	986	958
1978-075A		10993	US	05 AUG	NO ELEMENTS AVAILABLE			
1978-075B		10994	US	05 AUG	NO ELEMENTS AVAILABLE			
1978-078C		11003	US	08 AUG	HELIOCENTRIC ORBIT			
1978-079A	ICE	11004	US	12 AUG	HELIOCENTRIC ORBIT			
1978-079C		11006	US	12 AUG	118.5	28.2	2640	587
1978-079D		13413	US	12 AUG	NO ELEMENTS AVAILABLE			
1978-080A	MOLNIYA 1-42	11007	USSR	22 AUG	709.9	62.5	39703	262
1978-080D		11075	USSR	22 AUG	731.0	63.2	40724	279
1978-083A	COSMOS 1030	11015	USSR	06 SEP	718.7	66.4	35945	4454
1978-083D		11076	USSR	06 SEP	723.2	67.4	35473	5149
1978-083E		12907	USSR	06 SEP	711.4	64.0	36813	3227
1978-083F		12919	USSR	06 SEP	719.5	64.0	37421	3020
1978-083G		13959	USSR	06 SEP	721.7	63.7	37601	2948
1978-084A	VENERA 11	11020	USSR	09 SEP	HELIOCENTRIC ORBIT			
1978-086A	VENERA 12	11025	USSR	14 SEP	HELIOCENTRIC ORBIT			
1978-087A	IJIKI'KEN	11027	JAPAN	16 SEP	376.2	31.3	21525	242
1978-087B		11028	JAPAN	16 SEP	318.8	31.2	17910	251
1978-091A	COSMOS 1034	11042	USSR	04 OCT	114.9	74.0	1478	1420
1978-091B	COSMOS 1035	11044	USSR	04 OCT	114.6	74.0	1476	1401
1978-091C	COSMOS 1036	11045	USSR	04 OCT	115 1	74.0	1479	1439
1978-091D	COSMOS 1037	11046	USSR	04 OCT	115 3	74.0	1480	1458
1978-091E	COSMOS 1038	11047	USSR	04 OCT	115.5	74.0	1484	1475
1978-091F	COSMOS 1039	11048	USSR	04 OCT	116.3	74.0	1550	1476
1978-091G	COSMOS 1040	11049	USSR	04 OCT	116 0	74.0	1525	1477
1978-091H	COSMOS 1041	11050	USSR	04 OCT	115 8	74.0	1506	1475
1978-091J		11051	USSR	04 OCT	117.9	74.0	1696	1480
1978-093A		11054	US	07 OCT	744.2	63.6	20941	20705
1978-094A	COSMOS 1043	11055	USSR	10 OCT	93.6	81.2	455	448
1978-094B		11056	USSR	10 OCT	95.0	81.2	544	491
1978-095A	MOLNIYA 3-10	11057	USSR	13 OCT	717.8	63.0	39741	612

International Designation	Satellite Name	Catalog Number	Source	Launch Date	Period Mins	Incl-ination	Apogee KM	Perigee KM
1978-095E		11079	USSR	13 OCT	734.3	63.0	40280	884
1978-096A	TIROS-N	11060	US	13 OCT	101.7	98.7	845	829
1978-096B		11061	US	13 OCT	100.1	98 9	763	758
1978-096C		11062	US	13 OCT	100.0	98 9	762	757
1978-098A	NIMBUS 7	11080	US	24 OCT	104.0	99.1	955	940
1978-098B	CAMEO	11081	US	24 OCT	104.0	99.6	966	924
1978-100A	COSMOS 1045	11084	USSR	26 OCT	120.3	82.5	1702	1683
1978-100D TO 100AZ			USSR	26 OCT	SEE NOTE 29			
1978-100B	RADIO 1	11085	USSR	26 OCT	120.3	82.5	1704	1683
1978-100C	RADIO 2	11086	USSR	26 OCT	120.3	82.5	1702	1683
1978-105A	COSMOS 1048	11111	USSR	16 NOV	100.5	74.0	797	769
1978-105B		11112	USSR	16 NOV	100.4	74.0	804	750
1978-105C		11113	USSR	16 NOV	99.8	74.0	757	741
1978-105D		11114	USSR	16 NOV	99.6	74.0	745	731
1978-106A	NATO III-C	11115	NATO	19 NOV	1462.2	6.9	36307	36283
1978-109A	COSMOS 1051	11128	USSR	05 DEC	114.6	74.0	1484	1392
1978-109B	COSMOS 1052	11129	USSR	05 DEC	114.8	74.0	1487	1407
1978-109C	COSMOS 1053	11130	USSR	05 DEC	115.0	74.0	1486	1427
1978-109D	COSMOS 1054	11131	USSR	05 DEC	115.2	74.0	1487	1444
1978-109E	COSMOS 1055	11132	USSR	05 DEC	115.4	74.0	1488	1162
1978-109F	COSMOS 1056	11133	USSR	05 DEC	115.7	74.0	1501	1470
1978-109G	COSMOS 1057	11134	USSR	05 DEC	115.9	74.0	1513	1478
1978-109H	COSMOS 1058	11135	USSR	05 DEC	116.1	74.0	1536	1477
1978-109J		11136	USSR	05 DEC	118.1	74.0	1698	1489
1978-112A		11141	US	11 DEC	746.5	64.3	21012	20749
1978-112B		11142	US	11 DEC	269.5	63.8	14253	628
1978-113A		11144	US	14 DEC	1435.5	9.3	35792	35755
1978-113B		11145	US	14 DEC	1451.8	9.0	36104	36082
1978-113D		11147	US	14 DEC	1533.4	10.9	38847	36494
1978-116A	ANIK B1	11153	CANADA	16 DEC	1442.7	5.8	35943	35887
1978-117A	COSMOS 1063	11155	USSR	19 DEC	95.2	81.2	529	522
1978-117B		11156	USSR	19 DEC	95.3	81.3	555	507
1978-118A	GORIZONT 1	11158	USSR	19 DEC	1436.0	21.7	49357	22210
1978-118C		11926	USSR	19 DEC	1417.4	21.4	48725	22114
1978-121A	COSMOS 1066	11165	USSR	23 DEC	102.0	81.2	891	818
1978-121B		11166	USSR	23 DEC	101.9	81.2	898	799
1978-121C		19643	USSR	23 DEC	101.9	81.2	895	798
1978-122A	COSMOS 1067	11168	USSR	26 DEC	109.0	83.0	1208	1155
1978-122B		11170	USSR	26 DEC	108.9	83.0	1194	1157

1979 LAUNCHES

International Designation	Satellite Name	Catalog Number	Source	Launch Date	Period Mins	Incl-ination	Apogee KM	Perigee KM
1979-003A	COSMOS 1072	11238	USSR	16 JAN	104.8	82.9	1011	957
1979-003B		11239	USSR	16 JAN	104.7	82.9	1009	949
1979-004A	MOLNIYA 3-11	11240	USSR	18 JAN	717.6	63.5	39784	561
1979-004D		11553	USSR	18 JAN	733.0	63.9	40190	911
1979-005A	METEOR 1-29	11251	USSR	25 JAN	96.2	97.7	599	548
1979-005B		11252	USSR	25 JAN	94.7	97.4	510	497
1979-007A	SCATHA	11256	US	30 JAN	1418.4	9.4	42737	28140
1979-009A	AYAME 1	11261	JAPAN	06 FEB	1312.8	2.0	37404	29269
1979-011A	COSMOS 1076	11266	USSR	12 FEB	95.2	82.5	536	521
1979-011B		11267	USSR	12 FEB	97.1	82.5	631	607
1979-012A	COSMOS 1077	11268	USSR	13 FEB	94.5	81.2	500	491
1979-012B		11269	USSR	13 FEB	95.0	81.2	551	490
1979-015A	EKRAN 3	11273	USSR	21 FEB	1436.4	11.3	35951	35632
1979-015D		13900	USSR	21 FEB	1421.0	11.1	35536	35446
1979-017AM		16084	US	24 FEB	93.7	97.9	467	442
1979-017AN		16085	US	24 FEB	94.2	97.8	486	474
1979-017BX		16308	US	24 FEB	93.1	97.8	432	419
1979-017CA		16311	US	24 FEB	93.7	97.8	466	446
1979-017FE		16480	US	24 FEB	93.7	97.8	462	449
1979-017FF		16484	US	24 FEB	92.1	97.6	385	370
1979-017FG		16485	US	24 FEB	90.4	97.4	300	285
1979-017GJ		16551	US	24 FEB	93.0	97.8	430	410
1979-017GX		16564	US	24 FEB	94.2	97.8	482	474
1979-017JF		16878	US	24 FEB	93.9	97.9	474	458
1979-017JH		17094	US	24 FEB	93.9	97.8	484	449
1979-020A	INTERCOSMOS 19	11285	USSR	27 FEB	96.5	74.0	723	455
1979-020B		11286	USSR	27 FEB	96.7	74.0	742	454
1979-021A	METEOR 2-4	11288	USSR	01 MAR	102.0	81.2	872	836
1979-021B		11289	USSR	01 MAR	102.0	81.2	912	799
1979-021C		11290	USSR	01 MAR	102.1	81.2	882	833
1979-021D		14632	USSR	01 MAR	102.8	81.3	931	853

International Designation	Satellite Name	Catalog Number	Source	Launch Date	Period Mins	Incl-ination	Apogee KM	Perigee KM
1979-024A	COSMOS 1081	11296	USSR	15 MAR	114.5	74.0	1464	1401
1979-024B	COSMOS 1082	11297	USSR	15 MAR	114.7	74.0	1403	1421
1979-024C	COSMOS 1083	11298	USSR	15 MAR	114.9	74.0	1463	1441
1979-024D	COSMOS 1084	11299	USSR	15 MAR	115.1	74.0	1462	1459
1979-024E	COSMOS 1085	11300	USSR	15 MAR	115.6	74.0	1502	1463
1979-024F	COSMOS 1086	11301	USSR	15 MAR	115.4	74.0	1480	1463
1979-024G	COSMOS 1087	11302	USSR	15 MAR	115.8	74.0	1522	1463
1979-024H	COSMOS 1088	11303	USSR	15 MAR	116.1	74.0	1544	1464
1979-024J		11304	USSR	15 MAR	117.6	74.0	1688	1457
1979-025B		11306	US	16 MAR	NO ELEMENTS AVAILABLE			
1979-026A	COSMOS 1089	11308	USSR	21 MAR	104.7	83.0	997	964
1979-026B		11309	USSR	21 MAR	104.6	83.0	988	961
1979-028A	COSMOS 1091	11320	USSR	07 APR	104.8	82.9	1006	959
1979-028B		11321	USSR	07 APR	104.6	82.9	991	962
1979-030A	COSMOS 1092	11326	USSR	11 APR	104.7	82.9	1002	957
1979-030B		11327	USSR	11 APR	104.6	82.9	997	954
1979-031A	MOLNIYA 1-43	11328	USSR	12 APR	100.4	63.7	1460	99
1979-031D		11551	USSR	12 APR	620.8	64.1	35321	132
1979-032A	COSMOS 1093	11331	USSR	14 APR	94.4	81.2	497	484
1979-032B		11332	USSR	14 APR	95.7	81.2	585	518
1979-035A	RADUGA 5	11343	USSR	25 APR	1435.8	11.2	35788	35772
1979-035E		17873	USSR	25 APR	1437.3	11.2	35934	35685
1979-038A	FLTSATCOM 2	11353	US	04 MAY	1461.3	9.2	36334	36222
1979-046A	COSMOS 1104	11378	USSR	31 MAY	104.7	82.9	1004	953
1979-046B		11379	USSR	31 MAY	104.6	82.9	990	957
1979-050A		11389	US	06 JUN	NO ELEMENTS AVAILABLE			
1979-050B		11403	US	06 JUN	NO ELEMENTS AVAILABLE			
1979-050C		11408	US	06 JUN	NO ELEMENTS AVAILABLE			
1979-050D		11410	US	06 JUN	NO ELEMENTS AVAILABLE			
1979-050G		11534	US	06 JUN	NO ELEMENTS AVAILABLE			
1979-053A		11397	US	10 JUN	NO ELEMENTS AVAILABLE			
1979-053C		11436	US	10 JUN	NO ELEMENTS AVAILABEE			
1979-053D		20364	US	10 JUN	NO ELEMENTS AVAILABLE			
1979-057A	NOAA 6	11416	US	27 JUN	100.7	98.6	801	786
1979-057B		11419	US	27 JUN	98.9	98.3	707	702
1979-057C		11634	US	27 JUN	98.9	98.3	710	705
1979-058A	COSMOS 1109	11417	USSR	27 JUN	718.3	63.3	39424	957
1979-058D		11555	USSR	27 JUN	721.6	67.9	38581	1960
1979-058E		12833	USSR	27 JUN	715.2	67.2	38326	1900
1979-058F		12834	USSR	27 JUN	719.3	67.7	34707	5722
1979-058G		12909	USSR	27 JUN	719.6	68.2	39114	1329
1979-058H		12995	USSR	27 JUN	698.9	66.3	38251	1169
1979-058J		13960	USSR	27 JUN	720.1	67.5	38503	1966
1979-060A	COSMOS 1110	11425	USSR	28 JUN	100.6	74.0	797	776
1979-060B		11427	USSR	28 JUN	100.4	74.0	792	762
1979-060C		14866	USSR	28 JUN	99.4	74.0	731	726
1979-060D		15784	USSR	28 JUN	99.8	74.0	753	747
1979-062A	GORIZONT 2	11440	USSR	05 JUL	1436.4	10.8	35809	35776
1979-062D		14005	USSR	05 JUL	1474.4	11.1	36553	36511
1979-067B		11458	USSR	20 JUL	94.7	81.2	533	479
1979-070A	MOLNIYA 1-44	11474	USSR	31 JUL	717.9	63.7	39380	978
1979-070D		11556	USSR	31 JUL	733.1	64.0	39596	1511
1979-072A	WESTAR 3	11484	US	10 AUG	1441.0	4.6	35889	35874
1979-072C		13940	US	11 JAN	207.5	24.3	10281	179
1979-077A	COSMOS 1124	11509	USSR	28 AUG	716.9	67.4	35129	5182
1979-077D		11550	USSR	28 AUG	723.8	67.4	35576	5076
1979-077E		12814	USSR	28 AUG	720.2	68.3	38472	2001
1979-077F		12815	USSR	28 AUG	715.5	67.1	35239	5004
1979-077G		12816	USSR	28 AUG	686.5	63.6	36895	1904
1979-077H		12817	USSR	28 AUG	720.7	68.4	37999	2497
1979-078A	COSMOS 1125	11510	USSR	28 AUG	100.6	74.0	796	778
1979-078B		11511	USSR	28 AUG	100.4	74.0	790	767
1979-078C		14805	USSR	28 AUG	99.4	74.1	734	725
1979-078D		14806	USSR	28 AUG	100.4	74.0	780	772
1979-078E		18650	USSR	28 AUG	99.3	74.1	730	718
1979-084A	COSMOS 1130	11538	USSR	25 SEP	114.6	74.0	1478	1395
1979-084B	COSMOS 1131	11539	USSR	25 SEP	114.8	74.0	1480	1408
1979-084C	COSMOS 1132	11540	USSR	25 SEP	114.9	74.0	1480	1423
1979-084D	COSMOS 1133	11541	USSR	25 SEP	115.1	74.0	1480	1437
1979-084E	COSMOS 1134	11542	USSR	25 SEP	115.3	74.0	1481	1452
1979-084F	COSMOS 1135	11543	USSR	25 SEP	115.4	74.0	1490	1460
1979-084G	COSMOS 1136	11544	USSR	25 SEP	115.6	74.0	1495	1470
1979-084H	COSMOS 1137	11545	USSR	25 SEP	115.8	74.0	1512	1470

International Designation	Satellite Name	Catalog Number	Source	Launch Date	Period Mins	Incl-ination	Apogee KM	Perigee KM
1979-084J		11546	USSR	25 SEP	117.8	74.0	1682	1479
1979-086A		11558	US	01 OCT	NO ELEMENTS AVAILABLE			
1979-086C		11560	US	01 OCT	NO ELEMENTS AVAILABLE			
1979-087A	EKRAN 4	11561	USSR	03 OCT	1436.3	10.9	35853	35727
1979-087C		17939	USSR	03 OCT	1433.4	10.9	35880	35587
1979-089A	COSMOS 1140	11573	USSR	11 OCT	100.4	74.1	787	765
1979-089B		11574	USSR	11 OCT	100.2	74.1	779	755
1979-089C		14345	USSR	11 OCT	99.9	74.0	764	742
1979-089D		14807	USSR	11 OCT	99.3	74.1	732	718
1979-089E		19048	USSR	11 OCT	100.0	74.0	773	743
1979-090A	COSMOS 1141	11585	USSR	16 OCT	104.6	82.9	996	952
1979-090B		11586	USSR	16 OCT	104.4	82.9	988	947
1979-090C		11587	USSR	16 OCT	102.4	82.9	893	850
1979-091A	MOLNIYA 1-45	11589	USSR	20 OCT	717.5	61.8	39804	535
1979-091D		11602	USSR	20 OCT	731.7	61.9	40380	657
1979-093A	COSMOS 1143	11600	USSR	26 OCT	95.3	81.2	535	533
1979-093B		11601	USSR	26 OCT	95.8	81.2	584	528
1979-095A	METEOR 2-5	11605	USSR	31 OCT	102.4	81.2	879	861
1979-095B		11608	USSR	31 OCT	102.4	81.2	914	833
1979-098A		11621	US	21 NOV	1435.9	8.6	35787	35778
1979-098B		11622	US	21 NOV	1436.1	8.6	35799	35774
1979-098C		11623	US	21 NOV	1510.8	10.0	38531	35940
1979-099A	COSMOS 1145	11629	USSR	27 NOV	94.6	81.2	503	496
1979-099B		11630	USSR	27 NOV	95.5	81.2	574	513
1979-101A	RCA SATCOM III	11635	US	07 DEC	788.9	8.2	35423	8385
1979-105A	GORIZONT 3	11648	USSR	28 DEC	1435.4	10.5	35801	35744
1979-105E		11684	USSR	28 DEC	1459.2	10.6	36316	36159

1980 LAUNCHES

International Designation	Satellite Name	Catalog Number	Source	Launch Date	Period Mins	Incl-ination	Apogee KM	Perigee KM
1980-003A	COSMOS 1150	11667	USSR	14 JAN	104.8	83.0	1011	962
1980-003B		11668	USSR	14 JAN	104.7	82.9	996	962
1980-004A	FLTSATCOM 3	11669	US	18 JAN	1436.7	8.4	35885	35710
1980-005A	COSMOS 1151	11671	USSR	23 JAN	96.3	82.5	591	570
1980-005B		11672	USSR	23 JAN	97.2	82.5	635	609
1980-007A	COSMOS 1153	11680	USSR	25 JAN	104.8	82.9	1014	958
1980-007B		11681	USSR	25 JAN	104.7	82.9	1008	952
1980-008A	COSMOS 1154	11682	USSR	30 JAN	95.6	81.2	552	542
1980-008B		11683	USSR	30 JAN	96.0	81.2	599	533
1980-011A		11690	US	09 FEB	718.0	64.7	20548	19817
1980-011B		11705	US	09 FEB	289.4	63.7	15610	619
1980-012A	COSMOS 1156	11691	USSR	11 FEB	114.5	74.0	1472	1396
1980-012B	COSMOS 1157	11692	USSR	11 FEB	114.7	74.0	1475	1412
1980-012C	COSMOS 1158	11693	USSR	11 FEB	115.0	74.0	1475	1431
1980-012D	COSMOS 1159	11694	USSR	11 FEB	115.2	74.0	1477	1447
1980-012E	COSMOS 1160	11695	USSR	11 FEB	115.4	74.0	1481	1463
1980-012F	COSMOS 1161	11696	USSR	11 FEB	115.6	74.0	1500	1466
1980-012G	COSMOS 1162	11697	USSR	11 FEB	115.8	74.0	1516	1470
1980-012H	COSMOS 1163	11698	USSR	11 FEB	116.1	74.0	1541	1469
1980-012J		11699	USSR	11 FEB	117.8	74.0	1691	1467
1980-016A	RADUGA 6	11708	USSR	20 FEB	1435.3	10.7	35778	35763
1980-016D		11728	USSR	20 FEB	1475.1	11.0	36617	36476
1980-018A	AYAME 2	11715	JAPAN	22 FEB	1386.6	1.4	36839	32785
1980-018C		11718	JAPAN	22 FEB	317.7	24.5	17816	271
1980-019A		11720	US	03 MAR	NO ELEMENTS AVAILABLE			
1980-019B		11721	US	03 MAR	NO ELEMENTS AVAILABLE			
1980-019C		11731	US	03 MAR	NO ELEMENTS AVAILABLE			
1980-019D		11732	US	03 MAR	NO ELEMENTS AVAILABLE			
1980-019E		11733	US	03 MAR	NO ELEMENTS AVAILABLE			
1980-019F		11734	US	03 MAR	NO ELEMENTS AVAILABLE			
1980-019G		11745	US	03 MAR	NO ELEMENTS AVAILABLE			
1980-019H		11746	US	03 MAR	NO ELEMENTS AVAILABLE			
1980-022A	COSMOS 1168	11735	USSR	17 MAR	104.7	82.9	1007	956
1980-022B		11736	USSR	17 MAR	104.6	82.9	1000	952
1980-022C		12404	USSR	17 MAR	103.0	82.9	923	881
1980-026A	COSMOS 1171	11750	USSR	03 APR	104.8	65.8	1005	965
1980-026B		11751	USSR	03 APR	104.6	65.8	992	960
1980-026C		11752	USSR	03 APR	104.8	65.8	1002	963
1980-028A	COSMOS 1172	11758	USSR	12 APR	719.1	65.4	38669	1748
1980-028E		11762	USSR	12 APR	722.3	65.8	39174	1401
1980-030A	COSMOS 1174	11765	USSR	18 APR	103.0	66.1	1426	376
1980-030AE		12360	USSR	18 APR	94.6	66.0	620	376
1980-030AM		13929	USSR	18 APR	93.5	65.7	501	386

International Designation	Satellite Name	Catalog Number	Source	Launch Date	Period Mins	Incl-ination	Apogee KM	Perigee KM
1980-030AQ		13932	USSR	18 APR	95.8	65.6	660	450
1980-030AX		15781	USSR	18 APR	103.8	65.8	1280	598
1980-030AY		18644	USSR	18 APR	101.5	67.0	850	805
1980-030J		11777	USSR	18 APR	99.7	66.0	1041	446
1980-030K		11778	USSR	18 APR	105.9	66.5	1450	625
1980-030N		11781	USSR	18 APR	100.4	66.2	1153	397
1980-030R		12343	USSR	18 APR	104.8	66.4	1338	632
1980-030V		12347	USSR	18 APR	96.0	66.0	730	403
1980-030Y		12354	USSR	18 APR	102.6	66.1	1357	408
1980-032A		11783	US	26 APR	707.8	63.0	20454	19404
1980-032B		11791	US	26 APR	201.3	63.1	9805	194
1980-032C		21944	US	26 APR	229.7	62.7	11652	440
1980-034A	COSMOS 1176	11788	USSR	29 APR	103.4	64.8	953	882
1980-034D		11971	USSR	29 APR	103.1	64.8	940	867
1980-039A	COSMOS 1181	11803	USSR	20 MAY	104.8	82.9	1002	967
1980-039B		11804	USSR	20 MAY	104.7	82.9	994	962
1980-044A	COSMOS 1184	11821	USSR	04 JUN	95.4	81.2	541	531
1980-044B		11822	USSR	04 JUN	96.3	81.3	607	550
1980-049A	GORIZONT 4	11841	USSR	14 JUN	1460.1	10.2	36267	36242
1980-049F		11862	USSR	14 JUN	1470.4	10.4	36578	36332
1980-050A	COSMOS 1188	11844	USSR	14 JUN	717.4	67.5	38047	2286
1980-050B		11847	USSR	14 JUN	722.9	67.5	38259	2347
1980-051B		11849	USSR	18 JUN	96.1	97.6	589	552
1980-052C		11852	US	18 JUN	NO ELEMENTS AVAILABLE			
1980-056A	COSMOS 1190	11869	USSR	01 JUL	100.6	74.0	792	778
1980-056B		11870	USSR	01 JUL	100.4	74.1	789	767
1980-056C		14808	USSR	01 JUL	100.7	74.0	806	778
1980-056D		14809	USSR	01 JUL	100.5	74.0	792	773
1980-057A	COSMOS 1191	11871	USSR	02 JUL	716.6	67.6	34679	5618
1980-057D		11888	USSR	02 JUL	722.0	67.4	35424	5135
1980-057E		13999	USSR	02 JUL	708.6	65.7	37658	2245
1980-058A	COSMOS 1192	11875	USSR	09 JUL	114.5	74.0	1472	1394
1980-058B	COSMOS 1193	11876	USSR	09 JUL	114.7	74.0	1472	1412
1980-058C	COSMOS 1194	11877	USSR	09 JUL	114.9	74.0	1471	1430
1980-058D	COSMOS 1195	11878	USSR	09 JUL	115.1	74.0	1472	1448
1980-058E	COSMOS 1196	11879	USSR	09 JUL	115.3	74.0	1473	1465
1980-058F	COSMOS 1197	11880	USSR	09 JUL	115.5	74.0	1490	1468
1980-058G	COSMOS 1198	11881	USSR	09 JUL	115.7	74.0	1506	1471
1980-058H	COSMOS 1199	11882	USSR	09 JUL	116.0	74.0	1528	1471
1980-058J		11883	USSR	09 JUL	117.6	74.0	1680	1467
1980-060A	EKRAN 5	11890	USSR	14 JUL	1436.1	0.0	35834	35737
1980-060F		14193	USSR	14 JUL	1417.3	10.3	35492	35341
1980-063A	MOLNIYA 3-13	11896	USSR	18 JUL	717.8	63.1	38696	1658
1980-063D		11909	USSR	18 JUL	732.5	63.3	39328	1751
1980-069A	COSMOS 1206	11932	USSR	15 AUG	95.3	81.2	538	529
1980-069B		11933	USSR	15 AUG	96.0	81.2	601	535
1980-073A	METEOR 2-6	11962	USSR	09 SEP	102.1	81.2	887	832
1980-073B		11963	USSR	09 SEP	102.2	81.2	908	816
1980-074A	GOES 4	11964	US	09 SEP	1451.3	8.6	36713	35453
1980-074C		11970	US	09 SEP	1767.3	0.1	49745	34341
1980-081A	RADUGA 7	12003	USSR	05 OCT	1435.5	10.2	35794	35754
1980-081F		12447	USSR	05 OCT	1440.5	10.3	35887	35859
1980-085A	COSMOS 1217	12032	USSR	24 OCT	716.7	67.2	37991	2309
1980-085D		12035	USSR	24 OCT	722.0	67.5	38602	1959
1980-087A	FLTSATCOM 4	12046	US	31 OCT	1436.1	8.5	35798	35775
1980-087B		12069	US	31 OCT	178.6	26.2	7997	260
1980-089A	COSMOS 1220	12054	USSR	04 NOV	97.7	65.0	755	538
1980-091A	SBS 1	12065	US	15 NOV	1442.5	5.3	35946	35878
1980-092A	MOLNIYA 1-48	12066	USSR	16 NOV	713.8	62.6	39279	880
1980-092D		12070	USSR	16 NOV	733.5	62.6	40288	840
1980-093A	COSMOS 1222	12071	USSR	21 NOV	95.9	81.2	565	560
1980-093B		12072	USSR	21 NOV	96.0	81.2	606	531
1980-095A	COSMOS 1223	12078	USSR	27 NOV	718.7	68.4	35242	5155
1980-095E		12086	USSR	27 NOV	723.3	68.0	35970	4656
1980-097A	COSMOS 1225	12087	USSR	05 DEC	104.8	82.9	1023	942
1980-097B		12088	USSR	05 DEC	104.6	82.9	1010	939
1980-098A	INTELSAT 5 F-2	12089	ITSO	06 DEC	1436.2	3.8	35806	35769
1980-098B		12445	US	06 DEC	227.7	23.6	11569	380
1980-099A	COSMOS 1226	12091	USSR	10 DEC	104.8	82.9	1007	958
1980-099B		12092	USSR	10 DEC	104.6	82.9	997	954
1980-100A		12093	US	13 DEC	NO ELEMENTS AVAILABLE			
1980-100B		12094	US	13 DEC	NO ELEMENTS AVAILABLE			
1980-102A	COSMOS 1228	12107	USSR	23 DEC	114.4	74.0	1463	1390

International Designation	Satellite Name	Catalog Number	Source	Launch Date	Period Mins	Incl- ination	Apogee KM	Perigee KM
1980-102B	COSMOS 1229	12108	USSR	23 DEC	114.6	74.0	1463	1411
1980-102C	COSMOS 1230	12109	USSR	23 DEC	114.4	74.0	1463	1396
1980-102D	COSMOS 1231	12110	USSR	23 DEC	114.5	74.0	1463	1403
1980-102E	COSMOS 1232	12111	USSR	23 DEC	114.6	74.0	1463	1410
1980-102F	COSMOS 1233	12112	USSR	23 DEC	114.7	74.0	1463	1416
1980-102G	COSMOS 1234	12113	USSR	23 DEC	114.6	74.0	1463	1407
1980-102H	COSMOS 1235	12114	USSR	23 DEC	114.6	74.0	1463	1410
1980-102J		12115	USSR	23 DEC	114.9	74.0	1466	1436
1980-103A	PROGNOZ 8	12116	USSR	25 DEC	5687.8	65.8	197364	978
1980-104A	EKRAN 6	12120	USSR	26 DEC	1436.4	10.1	35801	35783
1980-104E		12471	USSR	26 DEC	1420.9	10.0	35630	35347

1981 LAUNCHES

International Designation	Satellite Name	Catalog Number	Source	Launch Date	Period Mins	Incl- ination	Apogee KM	Perigee KM
1981-002A	MOLNIYA 3-14	12133	USSR	09 JAN	717.8	63.8	39407	946
1981-002B		12134	USSR	09 JAN	732.2	64.2	39617	1445
1981-003A	COSMOS 1238	12138	USSR	16 JAN	106.3	83.0	1716	395
1981-003B		12139	USSR	16 JAN	104.8	83.0	1582	392
1981-006A	COSMOS 1241	12149	USSR	21 JAN	104.9	65.8	1005	975
1981-006B		12150	USSR	21 JAN	104.6	65.8	1022	930
1981-006C		12151	USSR	21 JAN	104.8	65.8	1003	971
1981-008A	COSMOS 1242	12154	USSR	27 JAN	96.3	81.2	587	571
1981-008B		12155	USSR	27 JAN	96.4	81.2	629	543
1981-009A	MOLNIYA 1-49	12156	USSR	30 JAN	717.5	63.8	38287	2053
1981-009D		12159	USSR	30 JAN	731.6	64.1	38632	2402
1981-012A	KIKU 3	12295	JAPAN	11 FEB	378.5	28.4	21651	255
1981-012C		12787	JAPAN	11 FEB	513.0	28.4	29433	265
1981-013A	COSMOS 1244	12297	USSR	12 FEB	104.7	83.C	1004	958
1981-013B		12298	USSR	12 FEB	104.6	83.0	998	953
1981-016A	COSMOS 1247	12303	USSR	19 FEB	711.1	67.3	35256	4768
1981-016E		12311	USSR	19 FEB	703.5	67.2	34924	4723
1981-016F		12984	USSR	19 FEB	710.5	67.3	35159	4833
1981-016G		12985	USSR	19 FEB	710.1	65.4	37292	2682
1981-016H		12992	USSR	19 FEB	706.6	65.8	38696	1106
1981-018A	COMSTAR 4	12309	US	21 FEB	1436.2	6.4	35791	35785
1981-018B		12363	US	21 FEB	649.7	20.5	36332	606
1981-021A	COSMOS 1249	12319	USSR	05 MAR	103.9	65.0	990	890
1981-021C		12551	USSR	05 MAR	103.5	65.0	968	883
1981-022A	COSMOS 1250	12320	USSR	06 MAR	114.4	74.0	1469	1387
1981-022B	COSMOS 1251	12321	USSR	06 MAR	114.6	74.0	1470	1401
1981-022C	COSMOS 1252	12322	USSR	06 MAR	114.7	74.0	1469	1416
1981-022D	COSMOS 1253	12323	USSR	06 MAR	115.6	74.0	1495	1465
1981-022E	COSMOS 1254	12324	USSR	06 MAR	114.9	74.0	1469	1430
1981-022F	COSMOS 1255	12325	USSR	06 MAR	115.0	74.0	1469	1443
1981-022G	COSMOS 1256	12326	USSR	06 MAR	115.2	74.0	1474	1454
1981-022H	COSMOS 1257	12327	USSR	06 MAR	115.4	74.0	1477	1466
1981-022J		12328	USSR	06 MAR	117.6	74.0	1694	1454
1981-025A		12339	US	16 MAR	NO ELEMENTS AVAILABLE			
1981-025C		12371	US	16 MAR	NO ELEMENTS AVAILABLE			
1981-027A	RADUGA 8	12351	USSR	18 MAR	1434.3	10.1	36101	35401
1981-027F		14194	USSR	18 MAR	1474.5	10.4	36614	36454
1981-028BE		13682	USSR	20 MAR	95.4	65.0	557	521
1981-031A	COSMOS 1261	12376	USSR	31 MAR	716.9	67.6	35652	4658
1981-031D		12384	USSR	31 MAR	707.5	67.5	34956	4888
1981-031E		12892	USSR	31 MAR	719.4	68.1	35352	5083
1981-031F		12893	USSR	31 MAR	716.1	64.2	37401	2868
1981-031G		12894	USSR	31 MAR	718.4	65.2	37293	3094
1981-033A	COSMOS 1263	12388	USSR	09 APR	106.0	83.0	1695	388
1981-033B		12389	USSR	09 APR	103.8	83.0	1502	371
1981-036E		12427	USSR	16 APR	102.3	99.0	985	751
1981-037A	COSMOS 1266	12409	USSR	21 APR	103.6	64.8	948	905
1981-037D		12435	USSR	21 APR	103.3	64.8	927	906
1981-038A		12418	US	24 APR	NO ELEMENTS AVAILABLE			
1981-038B		12446	US	24 APR	NO ELEMENTS AVAILABLE			
1981-041A	COSMOS 1269	12442	USSR	07 MAY	100.7	74.1	798	781
1981-041B		12443	USSR	07 MAY	100.5	74.1	786	782
1981-041C		13498	USSR	07 MAY	100.1	74.0	772	754
1981-041D		14346	USSR	07 MAY	99.6	74.0	750	733
1981-043A	METEOR 2-7	12456	USSR	14 MAY	102.2	81.3	888	835
1981-043B		12457	USSR	14 MAY	102.4	81.3	915	824
1981-043C		15769	USSR	14 MAY	102.4	81.3	916	825
1981-044A	NNSS 30480	12458	US	15 MAY	108.9	90.1	1184	1162
1981-046A	COSMOS 1271	12464	USSR	19 MAY	96.2	81.2	584	569

International Designation	Satellite Name	Catalog Number	Source	Launch Date	Period Mins	Incl-ination	Apogee KM	Perigee KM
1981-046B		12465	USSR	19 MAY	96.7	81.2	634	562
1981-049A	GOES 5	12472	US	22 MAY	1436.6	5.7	35808	35785
1981-050A	INTELSAT 5 F-1	12474	ITSO	23 MAY	1438.2	4.4	35856	35799
1981-0S0B		12497	US	23 MAY	217.9	24.0	10918	311
1981-053A	COSMOS 1275	12504	USSR	04 JUN	104.7	83.0	1005	953
1981-053B TO 053MT			USSR	04 JUN	SEE NOTE 34			
1981-054A	MOLNIYA 3-16	12512	USSR	09 JUN	717.6	63.8	39808	536
1981-054E		12519	USSR	09 JUN	733.5	64.1	40219	909
1981-057A	METEOSAT 2	12544	ESA	19 JUN	1458.7	5.6	36330	36124
1981-057B	APPLE	12545	INDIA	19 JUN	1439.5	9.2	35957	35748
1981-057C		12546	ESA	19 JUN	515.7	10.5	29596	253
1981-057F		20837	ESA	19 JUN	1449.1	9.2	36363	35718
1981-058A	COSMOS 1278	12547	USSR	19 JUN	717.4	67.2	37304	3033
1981-058D		12561	USSR	19 JUN	724.0	67.5	38295	2365
1981-058E		17256	USSR	19 JUN	716.9	67.5	37257	3054
1981-059A	NOAA 7	12553	US	23 JUN	101.7	98.9	847	829
1981-059B		12559	US	23 JUN	100.9	98.9	804	796
1981-059C		12560	US	23 JUN	100.9	98.9	804	797
1981-061A	ERRAN 7	12564	USSR	25 JUN	1435.4	9.8	35787	35756
1981-061F		12851	USSR	25 JUN	1425.6	9.7	35587	35575
1981-065A	METEOR 1-31	12585	USSR	10 JUL	96.6	97.9	615	574
1981-065B		12586	USSR	10 JUL	96.8	97.9	618	589
1981-069A	RADUGA 9	12618	USSR	30 JUL	1437.1	9.7	35815	35797
1981-069F		12850	USSR	30 JUL	1473.9	9.9	36626	36421
1981-070A	DE 1	12624	US	03 AUG	409.7	88.8	23286	505
1981-070E		12679	US	03 AUG	410.8	88.9	23364	491
1981-070J		14620	US	03 AUG	393.9	88.8	22294	546
1981-070K		14621	US	03 AUG	396.8	88.9	22475	545
1981-070L		19478	US	03 AUG	402.9	88.8	22862	522
1981-071A	COSMOS 1285	12627	USSR	04 AUG	727.0	67.5	36156	4651
1981-071D		12680	USSR	04 AUG	722.7	67.6	35768	4829
1981-071E		12993	USSR	04 AUG	727.7	67.6	36087	4756
1981-071F		13961	USSR	04 AUG	726.8	68.1	36964	3834
1981-073A	FLTSATCOM 5	12635	US	06 AUG	1460.4	8.1	36311	36209
1981-074A	COSMOS 1287	12636	USSR	06 AUG	115.7	74.0	1510	1462
1981-074B	COSMOS 1288	12637	USSR	06 AUG	115.5	74.0	1490	1462
1981-074C	COSMOS 1289	12638	USSR	06 AUG	114.7	74.0	1463	1423
1981-074D	COSMOS 1290	12639	USSR	06 AUG	114.9	74.0	1463	1439
1981-074E	COSMOS 1291	12640	USSR	06 AUG	115.1	74.0	1463	1455
1981-074F	COSMOS 1292	12641	USSR	06 AUG	115.3	74.0	1474	1462
1981-074G	COSMOS 1293	12642	USSR	06 AUG	114.6	74.0	1464	1406
1981-074H	COSMOS 1294	12643	USSR	06 AUG	114.4	74.0	1463	1390
1981-074J		12644	USSR	06 AUG	117.4	74.0	1668	1462
1981-075A	INTERCOSMOS	12645	USSR	07 AUG	101.6	81.2	881	789
1981-075B		12646	USSR	07 AUG	101.8	81.2	890	792
1981-076A	GMS 2	12677	JAPAN	10 AUG	1446.5	8.3	36037	35943
1981-077A	COSMOS 1295	12681	USSR	12 AUG	104.6	82.9	1008	944
1981-077B		12682	USSR	12 AUG	104.5	82.9	995	944
1981-081A	COSMOS 1299	12783	USSR	24 AUG	103.9	65.1	978	910
1981-082A	COSMOS 1300	12785	USSR	24 AUG	96.7	82.5	613	589
1981-082B		12786	USSR	24 AUG	97.3	82.5	643	614
1981-084A	COSMOS 1302	12791	USSR	28 AUG	100.5	74.0	797	771
1981-084B		12792	USSR	28 AUG	100.4	74.0	787	766
1981-084C		12793	USSR	28 AUG	100.0	74.0	760	755
1981-084D		14810	USSR	28 AUG	100.6	74.0	801	769
1981-087A	COSMOS 1304	12803	USSR	04 SEP	103.8	82.9	972	903
1981-087B		12804	USSR	04 SEP	103.7	82.9	965	900
1981-088A	COSMOS 1305	12818	USSR	11 SEP	263.7	63.5	13245	1240
1981-088F		12827	USSR	11 SEP	262.4	63.5	13205	1189
1981-088G		14131	USSR	11 SEP	247.4	63.3	12494	855
1981-088H		18598	USSR	11 SEP	251.1	63.7	12764	845
1981-091A	COSMOS 1308	12835	USSR	18 SEP	104.7	82.9	1000	959
1981-091B		12836	USSR	18 SEP	104.6	82.9	993	959
1981-094A	OREOL 3	12848	USSR	21 SEP	106.0	82.5	1687	394
1981-094B		12849	USSR	21 SEP	108.1	82.5	1877	400
1981-096A	SBS 2	12855	US	24 SEP	1436.2	4.4	35797	35778
1981-098A	COSMOS 1312	12879	USSR	30 SEP	115.9	82.6	1499	1489
1981-098B		12880	USSR	30 SEP	115.8	82.6	1498	1486
1981-100C		12889	US	06 OCT	118.7	99.9	2695	550
1981-102A	RADUGA 10	12897	USSR	09 OCT	1436.2	9.4	35806	35769
1981-102F		14195	USSR	09 OCT	1437.0	9.5	35860	35749
1981-103A	COSMOS 1315	12903	USSR	13 OCT	96.4	81.2	603	572
1981-103B		12904	USSR	13 OCT	96.9	81.2	637	580

International Designation	Satellite Name	Catalog Number	Source	Launch Date	Period Mins	Incl-ination	Apogee KM	Perigee KM
1981-105A	MOLNIYA 3-17	12915	USSR	17 OCT	713.8	63.0	39371	784
1981-105E		12920	USSR	17 OCT	733.2	63.4	40308	806
1981-106A	VENERA 13	12927	USSR	30 OCT	HELIOCENTRIC ORBIT			
1981-107A		12930	US	31 OCT	NO ELEMENTS AVAILABLE			
1981-107C		12932	US	31 OCT	NO ELEMENTS AVAILABLE			
1981-108A	COSMOS 1317	12933	USSR	31 OCT	718.8	68.3	35290	5115
1981-108D		12940	USSR	31 OCT	723.3	68.2	36070	4555
1981-108E		14734	USSR	31 OCT	713.6	65.3	36709	3438
1981-108F		14735	USSR	31 OCT	714.7	65.1	36478	3725
1981-108G		14736	USSR	31 OCT	719.9	68.4	35351	5110
1981-110A	VENERA 14	12938	USSR	04 NOV	HELIOCENTRIC ORBIT			
1981-113A	MOLNIYA 1-51	12959	USSR	17 NOV	713.7	63.7	39943	211
1981-113D		12986	USSR	17 NOV	698.7	63.9	39025	382
1981-114A	RCA SATCOM IIIR	12967	US	20 NOV	1438.6	1.8	35846	35826
1981-116A	COSMOS 1320	12975	USSR	28 NOV	117.2	74.0	1633	1478
1981-116B	COSMOS 1321	12976	USSR	28 NOV	117.2	74.0	1629	1479
1981-116C	COSMOS 1322	12977	USSR	28 NOV	117.2	74.0	1627	1479
1981-116D	COSMOS 1323	12978	USSR	28 NOV	117.1	74.0	1622	1479
1981-116E	COSMOS 1324	12979	USSR	28 NOV	117.1	74.0	1618	1479
1981-116F	COSMOS 1325	12980	USSR	28 NOV	117.0	74.0	1615	1479
1981-116G	COSMOS 1326	12981	USSR	28 NOV	117.0	74.0	1609	1478
1981-116H	COSMOS 1327	12982	USSR	28 NOV	116.9	74.0	1601	1479
1981-116J		12983	USSR	28 NOV	117.5	74.0	1661	1478
1981-117A	COSMOS 1328	12987	USSR	03 DEC	96.9	82.5	622	597
1981-117B		12988	USSR	03 DEC	97.3	82.5	645	614
1981-119A	INTELSAT 5 F-3	12994	ITSO	15 DEC	1436.1	3.4	35801	35770
1981-119B		13007	US	15 DEC	220.6	23.6	11150	278
1981-120A	RADIO 3	12997	USSR	17 DEC	118.4	83.0	1655	1562
1981-120B	RADIO 8	12998	USSR	17 DEC	119.6	83.0	1679	1649
1981-120C	RADIO 5	12999	USSR	17 DEC	119.4	83.0	1668	1642
1981-120D	RADIO 4	13000	USSR	17 DEC	119.3	83.0	1662	1633
1981-120E	RADIO 7	13001	USSR	17 DEC	119.1	83.0	1658	1620
1981-120F	RADIO 6	13002	USSR	17 DEC	118.6	83 0	1656	1579
1981-120G		13003	USSR	17 DEC	120.8	83 0	1782	1650
1981-122A	MARECS A	13010	ESA	20 DEC	1436.0	5.6	35795	35775
1981-122B	CAT 4	13011	ESA	20 DEC	543.4	10.5	31094	267
1981-123A	MOLNIYA 1-52	13012	USSR	23 DEC	717.8	63.8	38085	2272
1981-123D		13016	USSR	23 DEC	695.2	64.1	37030	2206

1982 LAUNCHES

International Designation	Satellite Name	Catalog Number	Source	Launch Date	Period Mins	Incl-ination	Apogee KM	Perigee KM
1982-001A	COSMOS 1331	13027	USSR	07 JAN	100.4	74.0	794	758
1982-001B		13028	USSR	07 JAN	100.3	74.0	789	759
1982-001C		13029	USSR	07 JAN	100.0	74.0	768	747
1982-001D		13030	USSR	07 JAN	99.5	74.0	750	715
1982-003A	COSMOS 1333	13033	USSR	14 JAN	104.9	82.9	1011	963
1982-003B		13034	USSR	14 JAN	104.7	82.9	1004	957
1982-004A	RCA SATCOM IV	13035	US	16 JAN	1446.0	1.1	35988	35970
1982-006C		13103	US	21 JAN	NO ELEMENTS AVAILABLE			
1982-006D		13104	US	21 JAN	NO ELEMENTS AVAILABLE			
1982-006E		13105	US	21 JAN	NO ELEMENTS AVAILABLE			
1982-006F		13152	US	21 JAN	NO ELEMENTS AVAILABLE			
1982-009A	EKRAN 8	13056	USSR	05 FEB	1440.8	9.3	35996	35761
1982-009F		14117	USSR	05 FEB	1426.0	9.2	35752	35424
1982-012A	COSMOS 1339	13065	USSR	17 FEB	104.7	82.9	1012	946
1982-012B		13066	USSR	17 FEB	104.6	82.9	1006	940
1982-013A	COSMOS 1340	13067	USSR	19 FEB	96.7	81.2	606	591
1982-013B		13068	USSR	19 FEB	96.7	81.2	625	580
1982-014A	WESTAR 4	13069	US	26 FEB	1443.4	1.1	35934	35923
1982-015A	MOLNIYA 1-53	13070	USSR	26 FEB	717.8	63.1	39145	1212
1982-015D		13075	USSR	26 FEB	730.9	63.6	39816	1180
1982-016A	COSMOS 1341	13080	USSR	03 MAR	717.9	67.5	35686	4673
1982-016D		13090	USSR	03 MAR	708.9	67.3	35777	4140
1982-017A	INTELSAT 5 F-4	13083	ITSO	05 MAR	1435.3	3.4	35791	35751
1982-019A		13086	US	06 MAR	NO ELEMENTS AVAILABLE			
1982-019B		13089	US	06 MAR	NO ELEMENTS AVAILABLE			
1982-020A	GORIZONT 5	13092	USSR	15 MAR	1461.5	9.2	36431	36133
1982-020F		13899	USSR	15 MAR	1460.0	9.3	36375	36129
1982-024A	COSMOS 1344	13110	USSR	24 MAR	104.8	82.9	1005	964
1982-024B		13111	USSR	24 MAR	104.7	82.9	1009	948
1982-025A	METEOR 2	13113	USSR	25 MAR	104.0	82.5	957	933
1982-025B		13114	USSR	25 MAR	104.0	82.5	957	934
1982-027A	COSMOS 1346	13120	USSR	31 MAR	96.5	81.2	604	574

International Designation	Satellite Name	Catalog Number	Source	Launch Date	Period Mins	Incl-ination	Apogee KM	Perigee KM
1982-027B		13121	USSR	31 MAR	96.8	81.2	635	576
1982-029A	COSMOS 1348	13124	USSR	07 APR	718.4	68.2	35489	4897
1982-029D		13169	USSR	07 APR	705.4	68.1	35339	4402
1982-030A	COSMOS 1349	13127	USSR	08 APR	104.8	82.9	1007	963
1982-030B		13128	USSR	08 APR	104.7	82.9	1000	957
1982-031A	INSAT-1A	13129	INDIA	10 APR	1434.2	0.1	35936	35562
1982-037A	COSMOS 1354	13148	USSR	28 APR	100.7	74.0	798	783
1982-037B		13149	USSR	28 APR	100.5	74.0	793	772
1982-037C		14344	USSR	03 AUG	100.6	74.1	801	772
1982-037D		14811	USSR	28 APR	100.7	74.0	811	774
1982-039A	COSMOS 1356	13153	USSR	05 MAY	96.7	81.2	615	588
1982-039B		13154	USSR	05 MAY	97.2	81.2	661	583
1982-040A	COSMOS 1357	13160	USSR	06 MAY	114.6	74.0	1477	1397
1982-040B	COSMOS 1358	13161	USSR	06 MAY	114.8	74.0	1479	1413
1982-040C	COSMOS 1359	13162	USSR	06 MAY	115.0	74.0	1478	1430
1982-040D	COSMOS 1360	13163	USSR	06 MAY	115.2	74.0	1479	1444
1982-040E	COSMOS 1361	13164	USSR	06 MAY	115.3	74.0	1481	1459
1982-040F	COSMOS 1362	13165	USSR	06 MAY	115.5	74.0	1493	1465
1982-040G	COSMOS 1363	13166	USSR	06 MAY	115.7	74.0	1503	1471
1982-040H	COSMOS 1364	13167	USSR	06 MAY	115.9	74.0	1522	1471
1982-040J		13168	USSR	06 MAY	117.7	74.1	1685	1470
1982-041C		13172	US	11 MAY	NO ELEMENTS AVAILABLE			
1982-043A	COSMOS 1365	13175	USSR	14 MAY	103.6	65.1	976	884
1982-043D		13594	USSR	14 MAY	103.3	65.1	964	869
1982-044A	COSMOS 1366	13177	USSR	17 MAY	1435.9	8.8	35802	35762
1982-044F		14114	USSR	17 MAY	1437.8	8.8	35894	35746
1982-045A	COSMOS 1367	13205	USSR	20 MAY	716.9	67.7	35930	4379
1982-045D		13215	USSR	20 MAY	704.0	67.2	36106	3568
1982-051A	COSMOS 1371	13241	USSR	01 JUN	100.7	74.0	801	782
1982-051B		13242	USSR	01 JUN	100.5	74.0	799	763
1982-051C		14398	USSR	01 JUN	100.3	74.0	777	772
1982-051D		18502	USSR	01 JUN	100.4	74.1	776	775
1982-051E		18509	USSR	01 JUN	100.3	74.0	784	765
1982-051F		18510	USSR	01 JUN	100.4	74.0	790	763
1982-051G		19102	USSR	01 JUN	100.3	74.1	784	761
1982-052A	COSMOS 1372	13243	USSR	01 JUN	103.9	64.9	962	923
1982-052D		13416	USSR	01 JUN	103.6	64.9	931	924
1982-055A	COSMOS 1375	13259	USSR	06 JUN	105.0	65.8	1014	975
1982-055B TO 055BM			USSR	06 JUN	SEE NOTE 37			
1982-058A	WESTAR 5	13269	US	09 JUN	1451.4	0.8	36149	36023
1982-059A	COSMOS 1378	13271	USSR	10 JUN	96.9	82.5	620	596
1982-059B		13272	USSR	10 JUN	97.3	82.5	644	615
1982-064A	COSMOS 1382	13295	USSR	25 JUN	718.0	68.0	35333	5030
1982-064D		13298	USSR	25 JUN	708.4	67.7	35189	4701
1982-066A	COSMOS 1383	13301	USSR	29 JUN	105.2	82.9	1024	983
1982-066B		13302	USSR	29 JUN	105.1	82.9	1027	968
1982-069A	COSMOS 1386	13353	USSR	07 JUL	104.6	83.0	1005	947
1982-069B		13354	USSR	07 JUL	104.5	83.0	1007	930
1982-072A	LANDSAT 4	13367	US	16 JUL	98.8	98.3	705	693
1982-073A	COSMOS 1388	13375	USSR	21 JUL	114.5	74.0	1471	1391
1982-073B	COSMOS 1389	13376	USSR	21 JUL	114.7	74.0	1472	1408
1982-073C	COSMOS 1390	13377	USSR	21 JUL	114.9	74.0	1472	1425
1982-073D	COSMOS 1391	13378	USSR	21 JUL	115.0	74.0	1473	1441
1982-073E	COSMOS 1392	13379	USSR	21 JUL	115.2	74.0	1472	1458
1982-073F	COSMOS 1393	13380	USSR	21 JUL	115.4	74.0	1481	1467
1982-073G	COSMOS 1394	13381	USSR	21 JUL	115.6	74.0	1494	1472
1982-073H	COSMOS 1395	13382	USSR	21 JUL	115.8	74.0	1514	1471
1982-073J		13386	USSR	21 JUL	117.9	74.0	1710	1462
1982-074D		13390	USSR	21 JUL	688.3	63.3	38695	192
1982-079A	COSMOS 1400	13402	USSR	05 AUG	96.4	81.2	592	578
1982-079B		13403	USSR	05 AUG	96.9	81.2	645	573
1982-082A	ANIK D-1	13431	CANADA	26 AUG	1438.5	1.5	35851	35814
1982-083A	MOLNIYA 3-19	13432	USSR	27 AUG	717.1	64.0	38327	1993
1982-083E		13446	USSR	27 AUG	733.1	64.3	38944	2161
1982-087A	ETS 3	13492	JAPAN	03 SEP	107.2	44.6	1226	966
1982-087B		13493	JAPAN	03 SEP	105.1	44.6	1007	992
1982-087C		13510	JAPAN	03 SEP	107.0	44.6	1220	950
1982-087D		14569	JAPAN	03 SEP	106.3	44.9	1147	962
1982-092A	COSMOS 1408	13552	USSR	16 SEP	96.8	82.6	617	591
1982-092B		13553	USSR	16 SEP	97.3	82.6	646	615
1982-093A	EKRAN 9	13554	USSR	16 SEP	1437.1	8.8	35903	35708
1982-093F		14115	USSR	16 SEP	1422.1	8.7	35546	35479
1982-095A	COSMOS 1409	13585	USSR	22 SEP	718.6	66.5	36694	3702

International Designation	Satellite Name	Catalog Number	Source	Launch Date	Period Mins	Incl-ination	Apogee KM	Perigee KM
1982-095D		13591	USSR	22 SEP	707.2	66.0	37288	2540
1982-096A	COSMOS 1410	13589	USSR	24 SEP	115.9	82.6	1499	1489
1982-096B		13590	USSR	24 SEP	115.8	82.6	1498	1487
1982-097A	INTELSAT 5 F 5	13595	ITSO	28 SEP	1436.1	2.9	35819	35754
1982-099A	COSMOS 1412	13600	USSR	02 OCT	103.9	64.8	980	904
1982-099E		13653	USSR	02 OCT	103.6	64.8	957	897
1982-100A	COSMOS 1413	13603	USSR	12 OCT	673.3	64.7	19075	19062
1982-100D	COSMOS 1414	13606	USSR	12 OCT	675.7	64.7	19210	19048
1982-100E	COSMOS 1415	13607	USSR	12 OCT	673.5	64.7	190,4	19072
1982-100F		13608	USSR	12 OCT	286.3	52.2	15724	298
1982-100G		13609	USSR	12 OCT	306.0	52.1	16989	336
1982-100H		13610	USSR	12 OCT	672.9	64.7	19072	19043
1982-102A	COSMOS 1417	13617	USSR	19 OCT	104.7	83.0	1006	954
1982-102B		13618	USSR	19 OCT	104.6	83.0	998	951
1982-103A	GORIZONT 6	13624	USSR	20 OCT	1436.0	8.4	35797	35769
1982-103E		13630	USSR	20 OCT	1435.0	8.3	35824	35707
1982-105A	RCA SATCOM-V	13631	US	28 OCT	1436.2	1.7	35795	35779
1982-106A		13636	US	30 OCT	1436.1	5.7	35947	35626
1982-106B		13637	US	30 OCT	1436.1	2.5	35792	35779
1982-106D		13643	US	30 OCT	1449.0	7.0	36203	35875
1982-109A	COSMOS 1420	13648	USSR	11 NOV	100.6	74.0	800	769
1982-109B		13649	USSR	11 NOV	100.4	74.0	793	763
1982-109D		15528	USSR	11 NOV	100.2	74.0	780	760
1982-110B	SBS 3	13651	US	11 NOV	1436.2	1.2	35799	35776
1982-110C	ANIK C-3	13652	CANADA	12 NOV	1436.1	1.3	35796	35778
1982-110D		13658	US	11 NOV	629.4	23.2	35509	389
1982-110E		13666	US	11 NOV	628.4	23.0	35459	388
1982-113A	RADUGA 11	13669	USSR	26 NOV	1473.8	7.9	36694	36348
1982-113F		13954	USSR	26 NOV	1475.9	8.0	36659	36463
1982-116A	METEOR 2-9	13718	USSR	14 DEC	101.8	81.2	882	803
1982-116B		13719	USSR	14 DEC	101.8	81.3	896	796
1982-116C		13720	USSR	14 DEC	101.8	81.2	882	802
1982-116D		17755	USSR	14 DEC	101.8	81.3	897	793
1982-118A		13736	US	21 DEC	101.0	98.6	810	797
1982-118C		13738	US	21 DEC	97.9	98.5	658	656

1983 LAUNCHES

International Designation	Satellite Name	Catalog Number	Source	Launch Date	Period Mins	Incl-ination	Apogee KM	Perigee KM
1983-001A	COSMOS 1428	13757	USSR	12 JAN	104.6	82.9	999	949
1983-001B		13758	USSR	12 JAN	104.5	82.9	989	948
1983-001C		14568	USSR	12 JAN	103.3	82.9	935	896
1983-002A	COSMOS 1429	13761	USSR	19 JAN	115.8	74.0	1517	1464
1983-002B	COSMOS 1430	13762	USSR	19 JAN	115.6	74.0	1497	1464
1983-002C	COSMOS 1431	13763	USSR	19 JAN	115.4	74.0	1481	1463
1983-002D	COSMOS 1432	13764	USSR	19 JAN	115.2	74.0	1465	1461
1983-002E	COSMOS 1433	13765	USSR	19 JAN	115.0	74.0	1465	1444
1983-002F	COSMOS 1434	13766	USSR	19 JAN	114.8	74.0	1464	1429
1983-002G	COSMOS 1435	13767	USSR	19 JAN	114.6	74.0	1466	1412
1983-002H	COSMOS 1436	13768	USSR	19 JAN	114.5	74.0	1465	1396
1983-002J		13769	USSR	19 JAN	117.9	74.0	1692	1477
1983-003A	COSMOS 1437	13770	USSR	20 JAN	96.6	81.2	608	582
1983-003B		13771	USSR	20 JAN	96.8	81.2	638	574
1983-004A	IRAS	13777	US	26 JAN	102.9	99.0	903	884
1983-004B		13778	US	26 JAN	102.3	100.0	882	851
1983-004C		13783	US	26 JAN	102.8	99.0	902	878
1983-006A	CS-2A	13782	JAPAN	04 FEB	1448.7	5.0	36075	35990
1983-006B		13786	JAPAN	04 FEB	141.8	28.5	5034	224
1983-008A		13791	US	09 FEB	NO ELEMENTS AVAILABLE			
1983-008B		13792	US	09 FEB	NO ELEMENTS AVAILABLE			
1983-008C		13834	US	09 FEB	NO ELEMENTS AVAILABLE			
1983-008D		13835	US	09 FEB	NO ELEMENTS AVAILABLE			
1983-008E		13844	US	09 FEB	NO ELEMENTS AVAILABLE			
1983-008F		13845	US	09 FEB	NO ELEMENTS AVAILABLE			
1983-008G		13849	US	09 FEB	NO ELEMENTS AVAILABLE			
1983-008H		13874	US	09 FEB	NO ELEMENTS AVAILABLE			
1983-010A	COSMOS 1441	13818	USSR	16 FEB	96.4	81.1	589	580
1983-010B		13819	USSR	16 FEB	96.7	81.1	634	562
1983-015A	MOLNIYA 3-20	13875	USSR	11 MAR	717.3	64.0	38573	1757
1983-015E		13882	USSR	11 MAR	732.0	64.3	39348	1703
1983-016A	EKRAN 10	13878	USSR	12 MAR	1515.4	9.3	37474	37174
1983-016F		14086	USSR	12 MAR	1424.5	8.8	35630	35486
1983-019A	MOLNIYA 1-56	13890	USSR	16 MAR	720.5	64.0	39253	1236
1983-019D		13897	USSR	16 MAR	732.7	64.0	39887	1200

International Designation	Satellite Name	Catalog Number	Source	Launch Date	Period Mins	Incl-ination	Apogee KM	Perigee KM
1983-020A	ASTRON	13901	USSR	23 MAR	5929.9	33.7	178073	26218
1983-020D		20413	USSR	23 MAR	5835.4	33.9	177564	24415
1983-021A	COSMOS 1447	13916	USSR	24 MAR	104.7	82.9	1010	952
1983-021B		13917	USSR	24 MAR	104.6	82.9	997	954
1983-022A	NOAA 8	13923	US	28 MAR	101.0	98.5	817	793
1983-023A	COSMOS 1448	13949	USSR	30 MAR	104.7	83.0	1001	955
1983-023B		13950	USSR	30 MAR	104.6	83.0	1004	946
1983-025A	MOLNIYA 1-57	13964	USSR	02 APR	717.4	63.9	39016	1320
1983-025D		13967	USSR	02 APR	699.3	64.3	37961	1476
1983-026B	TDRS 1	13969	US	04 APR	1436.1	6.6	35797	35777
1983-026C		13970	US	04 APR	1089.7	4.5	35353	22044
1983-026D		13971	US	04 APR	529.6	25.9	30335	276
1983-028A	RADUGA 12	13974	USSR	08 APR	1436.5	7.6	35795	35792
1983-028F		13983	USSR	08 APR	1439.3	7.6	35960	35738
1983-030A	RCA SATCOM VI	13984	US	11 APR	1442.0	0.1	35956	35847
1983-030B		13985	US	11 APR	112.9	25.4	2428	295
1983-031A	COSMOS 1452	13991	USSR	12 APR	100.6	74.0	800	774
1983-031B		13992	USSR	12 APR	100.5	74.1	785	775
1983-031D		14812	USSR	12 APR	100.7	74.1	808	777
1983-037A	COSMOS 1455	14032	USSR	23 APR	96.8	82.5	617	596
1983-037B		14033	USSR	23 APR	97.4	82.5	645	619
1983-038A	COSMOS 1456	14034	USSR	25 APR	719.0	66.7	37959	2457
1983-038E		14041	USSR	25 APR	707.2	66.6	37636	2195
1983-038H		14297	USSR	25 APR	719.0	66.9	36977	3436
1983-038J		14301	USSR	25 APR	789.5	67.0	43591	246
1983-038K		14306	USSR	25 APR	720.6	64.3	39697	795
1983-041A	GOES 6	14050	US	28 APR	1435.4	4.5	35785	35758
1983-041B		14051	US	28 APR	115.4	25.3	2542	406
1983-041C		14069	US	28 APR	1707.4	10.9	49205	32679
1983-042A	COSMOS 1459	14057	USSR	06 MAY	104.6	83.0	1012	940
1983-042B		14059	USSR	06 MAY	104.5	83.0	1002	938
1983-044A	COSMOS 1461	14064	USSR	07 MAY	98.6	65.0	828	552
1983-044AA		15707	USSR	07 MAY	96.3	65.1	662	503
1983-044EL		18480	USSR	07 MAY	98.0	65.0	782	539
1983-047A	INTELSAT 5 F-6	14077	ITSO	19 MAY	1436.2	1.9	35797	35779
1983-048A	COSMOS 1464	14084	USSR	24 MAY	104.8	82.9	1004	961
1983-048B		14085	USSR	24 MAY	104.6	82.9	997	958
1983-051B		14096	US	26 MAY	119.1	72.3	2519	758
1983-053A	VENERA 15	14104	USSR	02 JUN	VENUS ORBIT			
1983-054A	VENERA 16	14107	USSR	07 JUN	VENUS ORBIT			
1983-056A		14112	US	09 JUN	NO ELEMENTS AVAILABLE			
1983-056B		14113	US	09 JUN	NO ELEMENTS AVAILABLE			
1983-056C		14143	US	09 JUN	NO ELEMENTS AVAILABLE			
1983-056D		14144	US	09 JUN	NO ELEMENTS AVAILABLE			
1983-056E		14145	US	09 JUN	NO ELEMENTS AVAILABLE			
1983-056F		14146	US	09 JUN	NO ELEMENTS AVAILABLE			
1983-056G		14180	US	09 JUN	NO ELEMENTS AVAILABLE			
1983-056H		14181	US	09 JUN	NO ELEMENTS AVAILABLE			
1983-058A	ECS 1	14128	ESA	16 JUN	1436.1	3.0	35802	35770
1983-058B	OSCAR 10	14129	FRG	16 JUN	699.5	27.0	35427	4022
1983-058C		14130	ESA	16 JUN	331.8	8.4	18715	281
1983-058F		17331	ESA	16 JUN	116.4	7.6	2730	307
1983-059B	ANIK C2	14133	CANADA	18 JUN	1436.1	1.2	35793	35780
1983-059C	PALAPA B1	14134	INDNSA	18 JUN	1436.1	2.4	35790	35784
1983-059D		14135	US	18 JUN	602.6	23.4	34151	357
1983-059E		14136	US	18 JUN	620.5	25.6	35136	305
1983-060C		14139	US	20 JUN	NO ELEMENTS AVAILABLE			
1983-061A	COSMOS 1470	14147	USSR	22 JUN	97.0	82.5	629	596
1983-061B		14148	USSR	22 JUN	97.4	82.5	650	616
1983-063A		14154	US	27 JUN	100.6	82.0	819	754
1983-063B		14155	US	27 JUN	100.5	82.0	813	750
1983-063C		14222	US	27 JUN	99.7	82.4	760	723
1983-063D		14223	US	27 JUN	100.8	81.7	842	751
1983-065A	GALAXY 1	14158	US	28 JUN	1436.1	0.0	35791	35782
1983-065C		14168	US	28 JUN	272.6	23 2	14897	201
1983-066A	GORIZONT 7	14160	USSR	30 JUN	1464.3	7.3	36386	36286
1983-066E		14167	USSR	30 JUN	147.2	46.5	5514	200
1983-066F		15141	USSR	30 JUN	1475.2	7.4	36598	36497
1983-067A	PROGNOZ 9	14163	USSR	01 JUL	NO CURRENT ELEMENTS			
1983-069A	COSMOS 1473	14171	USSR	06 JUL	114.4	74.0	1461	1392
1983-069B	COSMOS 1474	14172	USSR	06 JUL	114.6	74.0	1462	1408
1983-069C	COSMOS 1475	14173	USSR	06 JUL	114.7	74.0	1461	1426
1983-069D	COSMOS 1476	14174	USSR	06 JUL	114.9	74.0	1461	1443

International Designation	Satellite Name	Catalog Number	Source	Launch Date	Period Mins	Incl-ination	Apogee KM	Perigee KM
1983-069E	COSMOS 1477	14175	USSR	06 JUL	115.1	74.0	1463	1459
1983-069F	COSMOS 1478	14176	USSR	06 JUL	115.3	74.0	1479	1461
1983-069G	COSMOS 1479	14177	USSR	06 JUL	115.5	74.0	1497	1461
1983-069H	COSMOS 1480	14178	USSR	06 JUL	115.8	74.0	1517	1460
1983-069J		14179	USSR	06 JUL	117.4	74.0	1668	1461
1983-070A	COSMOS 1481	14182	USSR	08 JUL	707.3	67.2	36368	3469
1983-070D		14191	USSR	08 JUL	707.9	67.3	36191	3676
1983-070E		14192	USSR	08 JUL	708.9	67.3	36463	3449
1983-070F		20412	USSR	08 JUL	705.8	67.5	36835	2925
1983-072A		14189	US	14 JUL	718.0	63.9	20547	19816
1983-072B		14190	US	14 JUL	371.8	64.1	20219	1278
1983-073A	MOLNIYA 1-58	14199	USSR	19 JUL	502.7	63.8	29006	123
1983-075A	COSMOS 1484	14207	USSR	24 JUL	96.1	97.5	595	548
1983-075B		14208	USSR	24 JUL	96.8	97.6	629	581
1983-075C		14209	USSR	24 JUL	96.5	97.6	622	560
1983-075D		14229	USSR	24 JUL	97.1	97.8	646	593
1983-075E		14631	USSR	24 JUL	96.4	97.6	604	566
1983-075F		14928	USSR	24 JUL	96.8	97.6	629	582
1983-077A	TELSTAR 3A	14234	US	28 JUL	1436.2	0.1	35796	35780
1983-077C		14236	US	28 JUL	212.9	22.7	10637	224
1983-078A		14237	US	31 JUL	NO ELEMENTS AVAILABLE			
1983-078B		14238	US	31 JUL	NO ELEMENTS AVAILABLE			
1983-079A	COSMOS 1486	14240	USSR	03 AUG	100.6	74.1	794	775
1983-079B		14241	USSR	03 AUG	100.5	74.1	794	766
1983-079D		14813	USSR	03 AUG	100.8	74.0	809	778
1983-079E		15756	USSR	03 AUG	99.8	74.1	759	741
1983-081A	CS-2B	14248	JAPAN	05 AUG	1457.4	4.3	36212	36191
1983-084A	COSMOS 1490	14258	USSR	10 AUG	675.7	64.8	19167	19091
1983-084B	COSMOS 1491	14259	USSR	10 AUG	668.4	64.7	19066	18823
1983-084C	COSMOS 1492	14260	USSR	10 AUG	676.8	64.8	19159	19154
1983-084F		14264	USSR	10 AUG	676.3	64.7	19168	19118
1983-084G		14277	USSR	10 AUG	325.0	51.9	18238	322
1983-084H		14278	USSR	10 AUG	323.1	52.0	18143	296
1983-088A	RADUGA 13	14307	USSR	25 AUG	1466.9	7.3	36431	36341
1983-088F		14333	USSR	25 AUG	1475.2	7.4	36619	36477
1983-089B	INSAT 1B	14318	INDIA	31 AUG	1436.2	3.0	35811	35765
1983-089C		14524	US	31 AUG	554.7	24.6	31704	266
1983-090A	MOLNIYA 3-21	14313	USSR	30 AUG	717.0	64.4	38246	2067
1983-090D		14319	USSR	30 AUG	731.2	64.3	38972	2043
1983-094A	RCA SATCOM VII	14328	US	08 SEP	1436.2	0.0	35803	35772
1983-094B		14329	US	08 SEP	108.7	25.5	2049	279
1983-098A	GALAXY 2	14365	US	22 SEP	1436.2	0.0	35792	35783
1983-099A	COSMOS 1500	14372	USSR	28 SEP	96.9	82.5	625	595
1983-099B		14373	USSR	28 SEP	97.4	82.5	649	615
1983-100A	EKRAN 11	14377	USSR	30 SEP	1435.3	8.0	35782	35760
1983-100F		14394	USSR	30 SEP	1425.0	7.9	35653	35485
1983-103A	COSMOS 1503	14401	USSR	12 OCT	100.7	74.0	800	779
1983-103B		14402	USSR	12 OCT	100.5	74.1	802	760
1983-105A	INTELSAT 5 F-7	14421	ITSO	19 OCT	1436.1	2.2	35806	35765
1983-108A	COSMOS 1506	14450	USSR	26 OCT	104.6	82.9	1008	946
1983-108B		14451	USSR	26 OCT	104.5	82.9	996	947
1983-109A	METEOR 2-10	14452	USSR	28 OCT	101.1	81.2	876	743
1983-109B		14453	USSR	28 OCT	101.2	81.2	889	736
1983-109C		14454	USSR	28 OCT	101.1	81.2	879	739
1983-111A	COSMOS 1508	14483	USSR	11 NOV	107.0	82.9	1784	391
1983-111B		14484	USSR	11 NOV	104.8	82.9	1597	368
1983-113A		14506	US	18 NOV	101.1	98.4	819	801
1983-113E		14610	US	18 NOV	98.1	98.5	671	663
1983-114A	MOLNIYA 1-59	14516	USSR	23 NOV	717.5	64.2	38375	1963
1983-114D		14520	USSR	23 NOV	699.2	64.3	37485	1947
1983-115A	COSMOS 1510	14521	USSR	24 NOV	116.0	73.6	1522	1477
1983-115B		14522	USSR	24 NOV	115.9	73.6	1518	1477
1983-118A	GORIZONT 8	14532	USSR	30 NOV	1465.4	7.0	36473	36243
1983-118F		14548	USSR	30 NOV	1436.3	6.8	35985	35596
1983-120A	COSMOS 1513	14546	USSR	08 DEC	104.8	82.9	1013	955
1983-120B		14547	USSR	08 DEC	104.6	82.9	1010	938
1983-122A	COSMOS 1515	14551	USSR	15 DEC	96.9	82.5	621	598
1983-122B		14552	USSR	15 DEC	97.4	82.5	646	619
1983-123A	MOLNIYA 3-22	14570	USSR	21 DEC	713.9	64.4	39873	290
1983-123D		14582	USSR	21 DEC	732.4	64.7	40513	561
1983-126A	COSMOS 1518	14587	USSR	28 DEC	714.0	67.0	37506	2662
1983-126D		14596	USSR	28 DEC	705.4	66.9	37265	2476
1983-127A	COSMOS 1519	14590	USSR	29 DEC	675.7	66.5	19187	19071

International Designation	Satellite Name	Catalog Number	Source	Launch Date	Period Mins	Incl-ination	Apogee KM	Perigee KM
1983-127B	COSMOS 1520	14591	USSR	29 DEC	675.7	66.5	19148	19110
1983-127C	COSMOS 1521	14592	USSR	29 DEC	673.4	66.5	19148	18993
1983-127F		14595	USSR	29 DEC	673.1	66.4	19151	18974
1983-127G		14607	USSR	29 DEC	326.4	52.1	18224	425
1983-127H		14608	USSR	29 DEC	330.6	51.6	18587	333
1983-127J		21752	USSR	29 DEC	230.9	53.1	11628	548
1983-127K		21753	USSR	29 DEC	250.5	52.1	13139	428
1983-127M		21935	USSR	29 DEC	317.1	52.2	17476	577
1984 LAUNCHES								
1984-001A	COSMOS 1522	14611	USSR	05 JAN	115.4	74.0	1490	1460
1984-001B	COSMOS 1522	14612	USSR	05 JAN	114.4	74.0	1458	1395
1984-001C	COSMOS 1524	14613	USSR	05 JAN	114.6	74.0	1459	1410
1984-001D	COSMOS 1525	14614	USSR	05 JAN	114.7	74.0	1458	1426
1984-001E	COSMOS 1526	14615	USSR	05 JAN	114.9	74.0	1460	1440
1984-001F	COSMOS 1527	14616	USSR	05 JAN	114.7	74.0	1457	1427
1984-001G	COSMOS 1528	14617	USSR	05 JAN	115.3	74.0	1474	1459
1984-001H	COSMOS 1529	14618	USSR	05 JAN	115.6	74.0	1509	1460
1984-001J		14619	USSR	05 JAN	117.5	74.0	1671	1467
1984-003A	COSMOS 1531	14624	USSR	11 JAN	104.9	82.9	1006	977
1984-003B		14625	USSR	11 JAN	104.8	82.9	1002	967
1984-005A	BS-2A	14659	JAPAN	23 JAN	1453.8	4.1	36181	36085
1984-008A	PRC 14	14670	PRC	29 JAN	162.0	36.1	6462	470
1984-009A		14675	US	31 JAN	NO ELEMENTS AVAILABLE			
1984-009C		14677	US	31 JAN	NO ELEMENTS AVAILABLE			
1984-010A	COSMOS 1535	14679	USSR	02 FEB	104.7	83.0	1012	951
1984-010B		14680	USSR	02 FEB	104.6	83.0	1003	948
1984-011E		14693	US	06 FEB	95.6	28.2	825	265
1984-011F		14694	US	03 FEB	97.9	27.7	1018	299
1984-012A		14690	US	05 FEB	NO ELEMENTS AVAILABLE			
1984-012B		14691	US	05 FEB	NO ELEMENTS AVAILABLE			
1984-012C		14728	US	05 FEB	NO ELEMENTS AVAILABLE			
1984-012D		14729	US	05 FEB	NO ELEMENTS AVAILABLE			
1984-012F		14795	US	05 FEB	NO ELEMENTS AVAILABLE			
1984-012J		15347	US	05 FEB	NO ELEMENTS AVAILABLE			
1984-012K		15348	US	05 FEB	NO ELEMENTS AVAILABLE			
1984-012L		15349	US	05 FEB	NO ELEMENTS AVAILABLE			
1984-013A	COSMOS 1536	14699	USSR	08 FEB	97.1	82.5	633	603
1984-013B		14700	USSR	08 FEB	97.4	82.5	648	617
1984-016A	RADUGA 14	14725	USSR	15 FEB	1435.8	6.8	35797	35763
1984-016F		17874	USSR	15 FEB	1436.9	6.8	35946	35658
1984-019A	COSMOS 1538	14759	USSR	21 FEB	100.6	74.0	800	769
1984-019B		14760	USSR	21 FEB	100.5	74.1	802	760
1984-019C		15785	USSR	21 FEB	100.1	74.0	770	752
1984-019D		18519	USSR	21 FEB	100.1	74.1	771	751
1984-021A	LANDSAT 5	14780	US	01 MAR	98.8	98.2	703	695
1984-021B	UOSAT 2	14781	UK	01 MAR	98.0	97.8	670	653
1984-022A	COSMOS 1540	14783	USSR	02 MAR	1436.1	7.5	35817	35756
1984-022F		14948	USSR	02 MAR	1442.1	7.6	36007	35799
1984-023A	INTELSAT 5 F-8	14786	ITSO	05 MAR	1436.1	1.6	35802	35771
1984-024A	COSMOS 1541	14790	USSR	06 MAR	718.9	66.6	36330	4078
1984-024D		14796	USSR	06 MAR	709.8	66.6	36029	3929
1984-027A	COSMOS 1544	14819	USSR	15 MAR	96.8	82.5	618	595
1984-027B		14820	USSR	15 MAR	97.4	82.5	645	619
1984-028A	EKRAN 12	14821	USSR	16 MAR	1499.1	8.3	37059	36962
1984-028D		14828	USSR	16 MAR	624.7	46.6	35418	238
1984-028F		15139	USSR	16 MAR	1419.8	7.9	35548	35386
1984-029A	MOLNIYA 1-60	14825	USSR	16 MAR	717.5	64.5	39750	589
1984-029D		14830	USSR	16 MAR	731.0	64.7	40313	689
1984-031A	COSMOS 1546	14867	USSR	29 MAR	1436.2	6.7	35881	35696
1984-031D		14887	USSR	29 MAR	566.9	45.3	32280	345
1984-031F		14951	USSR	29 MAR	1448.4	6.7	36093	35961
1984-033A	COSMOS 1547	14884	USSR	04 APR	717.6	67.5	36431	3914
1984-033D		14894	USSR	04 APR	706.5	67.3	36220	3577
1984-035A	PRC 15	14899	PRC	08 APR	1435.9	5.3	35847	35718
1984-035B		14900	PRC	08 APR	624.2	31.0	35202	430
1984-037A		14930	US	14 APR	NO ELEMENTS AVAILABLE			
1984-037B		14931	US	14 APR	NO ELEMENTS AVAILABLE			
1984-041A	GORIZONT 9	14940	USSR	22 APR	1436.7	6.5	35808	35788
1984-041D		14943	USSR	22 APR	1460.1	6.6	36312	36196
1984-043A	COSMOS 1550	14965	USSR	11 MAY	104.9	83.0	1007	970
1984-043B		14966	USSR	11 MAY	104.8	83.0	995	972

International Designation	Satellite Name	Catalog Number	Source	Launch Date	Period Mins	Incl-ination	Apogee KM	Perigee KM
1984-046A	COSMOS 1553	14973	USSR	17 MAY	104.7	82.9	1004	957
1984-046B		14974	USSR	17 MAY	104.6	82.9	1008	940
1984-047A	COSMOS 1554	14977	USSR	19 MAY	675.7	66.8	19181	19077
1984-047B	COSMOS 1555	14978	USSR	19 MAY	675.7	66.5	19164	19094
1984-047C	COSMOS 1556	14979	USSR	19 MAY	676.3	66.5	19167	19121
1984-047F		14984	USSR	19 MAY	675.5	66.5	19165	19083
1984-047G		15053	USSR	19 MAY	332.2	52.1	18610	409
1984-047H		15054	USSR	19 MAY	312.1	52.0	17408	319
1984-049A	SPACENET 1	14985	US	23 MAY	1436.1	0.0	35792	35781
1984-052A	COSMOS 1559	14998	USSR	28 MAY	115.7	74.0	1508	1467
1984-052B	COSMOS 1560	14999	USSR	28 MAY	115.5	74.0	1490	1468
1984-052C	COSMOS 1561	15000	USSR	28 MAY	115.4	74.0	1483	1459
1984-052D	COSMOS 1562	15001	USSR	28 MAY	115.2	74.0	1474	1452
1984-052E	COSMOS 1563	15002	USSR	28 MAY	115.0	74.0	1473	1436
1984-052F	COSMOS 1564	15003	USSR	28 MAY	114.8	74.0	1473	1422
1984-052G	COSMOS 1565	15004	USSR	28 MAY	114.7	74.0	1473	1407
1984-052H	COSMOS 1566	15005	USSR	28 MAY	114.5	74.0	1472	1392
1984-052J		15006	USSR	28 MAY	117.6	74.0	1677	1471
1984-055A	COSMOS 1569	15027	USSR	06 JUN	718.6	65.9	37176	3219
1984-055D		15030	USSR	06 JUN	706.9	66.2	37095	2721
1984-056A	COSMOS 1570	15031	USSR	08 JUN	100.7	74.1	800	782
1984-056B		15032	USSR	08 JUN	100.6	74.1	797	772
1984-056C		15033	USSR	08 JUN	100.7	74.1	803	780
1984-056D		15757	USSR	08 JUN	95.7	74.0	553	552
1984-059A		15039	US	13 JUN	718.0	63.7	20302	20060
1984-059B		15040	US	13 JUN	362.3	62.0	20690	223
1984-062A	COSMOS 1574	15055	USSR	21 JUN	104.8	83.0	1003	964
1984-062B		15056	USSR	21 JUN	104.6	83.0	995	960
1984-063A	RADUGA 15	15057	USSR	22 JUN	1435.7	6.5	35813	35742
1984-063E		15076	USSR	22 JUN	349.6	46.9	19910	210
1984-063F		15693	USSR	22 JUN	1394.2	6.3	35035	34888
1984-065C		15071	US	25 JUN	NO ELEMENTS AVAILABLE			
1984-067A	COSMOS 1577	15077	USSR	27 JUN	104.7	83.0	1006	954
1984-067B		15078	USSR	27 JUN	104.6	83.0	992	956
1984-069A	COSMOS 1579	15085	USSR	29 JUN	103.9	65.1	988	896
1984-069D		15330	USSR	29 JUN	103.6	65.1	963	893
1984-069E		19453	USSR	29 JUN	102.6	65.8	936	828
1984-071A	COSMOS 1581	15095	USSR	03 JUL	720.9	67.9	36305	4201
1984-071D		15098	USSR	03 JUL	705.6	67.6	35830	3923
1984-072A	METEOR 2-11	15099	USSR	05 JUL	104.0	82.5	955	937
1984-072B		15100	USSR	05 JUL	104.0	82.5	954	938
1984-078A	GORIZONT 10	15144	USSR	01 AUG	1437.3	6.2	35817	35803
1984-078F		15181	USSR	01 AUG	1437.3	6.2	35917	35703
1984-079A	COSMOS 1586	15147	USSR	02 AUG	717.4	65.7	36763	3572
1984-079D		15156	USSR	02 AUG	705.8	65.8	36419	3341
1984-080A	GMS 3	15152	JAPAN	02 AUG	1436.2	4.1	35792	35786
1984-080C		15157	JAPAN	02 AUG	192.5	28.8	9150	183
1984-080E		22266	JAPAN	02 AUG	1443.6	5.8	36512	35354
1984-081A	ECS 2	15158	ESA	04 AUG	1436.1	2.2	35799	35775
1984-081B	TELECOM 1A	15159	FRANCE	04 AUG	1463.4	2.1	36462	36177
1984-081D		15166	ESA	04 AUG	597.6	6.5	33651	597
1984-081E		20674	ESA	04 AUG	600.3	6.6	33702	688
1984-084A	COSMOS 1589	15171	USSR	08 AUG	115.9	82.6	1500	1489
1984-084B		15172	USSR	08 AUG	115.8	82.6	1499	1487
1984-085A	MOLNIYA 1-61	15182	USSR	10 AUG	715.8	64.3	38482	1775
1984-085D		15188	USSR	10 AUG	730.9	64.4	39217	1784
1984-088A	CCE	15199	US	16 AUG	939.4	4.8	49758	1032
1984-088B	IRM	15200	FRG	16 AUG	2653.4	27.0	113818	402
1984-088C	UKS	15201	UK	16 AUG	2659.6	26.9	113417	1002
1984-088D		15202	US	16 AUG	133.9	28.9	4031	549
1984-088E		15205	US	16 AUG	132.9	28.7	3947	552
1984-088F		15206	US	16 AUG	919.2	28.5	49354	520
1984-088G		19008	US	16 AUG	131.7	28.7	3842	551
1984-088H		19599	US	16 AUG	133.0	28.7	3951	552
1984-089A	MOLNIYA 1-62	15214	USSR	24 AUG	735.2	64.0	40261	948
1984-089D		15223	USSR	24 AUG	738.9	64.1	40007	1385
1984-090A	EKRAN 13	15219	USSR	24 AUG	1499.7	7.4	37094	36949
1984-090F		17875	USSR	24 AUG	1422.1	7.0	35587	35437
1984-091A		15226	US	28 AUG	NO ELEMENTS AVAILABLE			
1984-091B		15227	US	28 AUG	NO ELEMENTS AVAILABLE			
1984-093B	SBS 4	15235	US	31 AUG	1436.2	0.0	35795	35780
1984-093C	SYNCOM IV-2	15236	US	31 AUG	1436.0	4.1	35787	35779
1984-093D	TELSTAR 3C	15237	US	01 SEP	1436.2	0.0	35793	35783

International Designation	Satellite Name	Catalog Number	Source	Launch Date	Period Mins	Incl-ination	Apogee KM	Perigee KM
1984-093E		15244	US	31 AUG	259.0	27.2	13826	335
1984-093F		15245	US	31 AUG	595.7	22.6	33791	357
1984-093G		15246	US	01 SEP	640.4	25.3	36105	361
1984-09SA	COSMOS 1593	15259	USSR	04 SEP	675.7	64.8	19175	19083
1984-095B		15260	USSR	04 SEP	677.2	64.8	19186	19146
1984-095C		15261	USSR	04 SEP	675.7	64.7	19172	19086
1984-095F		15264	USSR	04 SEP	675.9	64.7	19179	19089
1984-095G		15265	USSR	04 SEP	327.4	52.2	18383	328
1984-095H		15266	USSR	04 SEP	330.7	52.0	18587	340
1984-096A	COSMOS 1596	15267	USSR	07 SEP	717.0	67.9	36284	4032
1984-096D		15270	USSR	07 SEP	703.2	67.6	35832	3799
1984-097A		15271	US	08 SEP	NO ELEMENTS AVAILABLE			
1984-097B		15272	US	08 SEP	369.2	63.7	20340	999
1984-100A	COSMOS 1598	15292	USSR	13 SEP	104.9	82.9	1012	964
1984-100B		15293	USSR	13 SEP	104.7	82.9	999	966
1984-101A	GALAXY 3	15308	US	21 SEP	1436.2	0.1	35793	35782
1984-104L		15982	USSR	27 SEP	90.5	65.8	305	289
1984-105A	COSMOS 1602	15331	USSR	28 SEP	97.0	82.5	629	600
1984-105B		15332	USSR	28 SEP	97.4	82.5	649	617
1984-106A	COSMOS 1603	15333	USSR	28 SEP	101.9	71.0	857	837
1984-106C		15335	USSR	28 SEP	101.4	66.5	840	811
1984-106F		15338	USSR	28 SEP	101.7	66.6	846	833
1984-107A	COSMOS 1604	15350	USSR	04 OCT	717.6	67.8	36283	4061
1984-107D		15355	USSR	04 OCT	708.1	67.7	35982	3891
1984-108B	ERBS	15354	US	05 OCT	96.4	57.0	590	578
1984-109A	COSMOS 1605	15359	USSR	11 OCT	104.7	82.9	1014	947
1984-109B		15360	USSR	11 OCT	104.6	82.9	1007	945
1984-110A		15362	US	12 OCT	108.9	89.9	1199	1149
1984-111A	COSMOS 1606	15369	USSR	18 OCT	96.9	82.5	623	595
1984-111B		15370	USSR	18 OCT	97.3	82.5	646	616
1984-112A	COSMOS 1607	15378	USSR	31 OCT	104.1	65.0	974	927
1984-112C		15503	USSR	31 OCT	103.8	65.0	946	926
1984-113B	ANIK D2	15383	CANADA	09 NOV	1436.2	0.0	35796	35780
1984-113C	SYNCOM IV-1	15384	US	10 NOV	1466.8	2.8	36427	36341
1984-113D		15387	US	09 NOV	616.6	26.0	34916	320
1984-113E		15390	US	10 NOV	259.3	27.1	13860	324
1984-113F		18904	US	10 NOV	241.0	26.7	12578	325
1984-114A	SPACENET 2	15385	US	10 NOV	1436.2	0.0	35794	35783
1984-114B	MARECS B2	15386	ESA	10 NOV	1436.2	3.5	35803	35773
1984-114C		15388	ESA	10 NOV	600.9	7.1	34049	372
1984-115A	NATO III-D	15391	NATO	14 NOV	1436.2	1.4	35796	35780
1984-115B		15392	US	14 NOV	115.8	21.5	2310	677
1984-115C		15402	US	14 NOV	635.1	22.8	35815	377
1984-118A	COSMOS 1610	15398	USSR	15 NOV	104.8	82.9	1009	962
1984-118B		15399	USSR	15 NOV	104.7	82.9	1002	954
1984-122A		15423	US	04 DEC	NO ELEMENTS AVAILABLE			
1984-123A	NOAA 9	15427	US	12 DEC	101.8	99.1	854	834
1984-123B		15440	US	12 DEC	98.1	99.0	669	666
1984-124A	MOLNIYA 1-63	15429	USSR	14 DEC	718.1	63.9	38913	1457
1984-124H		15439	USSR	14 DEC	733.4	64.3	39587	1532
1984-125A	VEGA 1	15432	USSR	15 DEC	HELIOCENTRIC ORBIT			
1984-125D		15447	USSR	15 DEC	HELIOCENTRIC ORBIT			
1984-128A	VEGA 2	15449	USSR	21 DEC	HELIOCENTRIC ORBIT			
1984-128B		15450	USSR	21 DEC	HELIOCENTRIC ORBIT			
1984-129A		15453	US	22 DEC	NO ELEMENTS AVAILABLE			
1984-129B		15454	US	22 DEC	NO ELEMENTS AVAILABLE			
1985 LAUNCHES								
1985-001A	MS-T5	15464	JAPAN	07 JAN	HELIOCENTRIC ORBIT			
1985-001B		15465	JAPAN	07 JAN	HELIOCENTRIC ORBIT			
1985-003A	COSMOS 1617	15469	USSR	15 JAN	114.0	82.6	1411	1410
1985-003B	COSMOS 1618	15470	USSR	15 JAN	114.0	82.6	1410	1405
1985-003C	COSMOS 1619	15471	USSR	15 JAN	113.7	82.6	1410	1381
1985-003D	COSMOS 1620	15472	USSR	15 JAN	113.8	82.6	1410	1388
1985-003E	COSMOS 1621	15473	USSR	15 JAN	113.8	82.6	1410	1393
1985-003F	COSMOS 1622	15474	USSR	15 JAN	113.9	82.6	1410	1398
1985-003G		15475	USSR	15 JAN	114.7	82.6	1469	1411
1985-004A	MOLNIYA 3-23	15476	USSR	16 JAN	717.9	64.8	39279	1079
1985-004D		15481	USSR	16 JAN	731.7	65.0	39716	1323
1985-006A	COSMOS 1624	15482	USSR	17 JAN	100.6	74.0	798	777
1985-006B		15483	USSR	17 JAN	100.5	74.0	795	765
1985-006C		15490	USSR	17 JAN	100.3	74.0	772	768

International Designation	Satellite Name	Catalog Number	Source	Launch Date	Period Mins	Incl-ination	Apogee KM	Perigee KM
1985-006D		15491	USSR	17 JAN	100.7	74.0	805	777
1985-007A	GORIZONT 11	15484	USSR	18 JAN	1436.0	5.8	35796	35772
1985-007D		15487	USSR	18 JAN	1397.8	5.5	35101	34963
1985-007F		15489	USSR	18 JAN	380.1	46.9	21813	196
1985-009A	COSMOS 1626	15494	USSR	24 JAN	96.8	82.5	621	591
1985-009B		15495	USSR	24 JAN	97.3	82.5	646	615
1985-010B		15543	US	24 JAN	NO ELEMENTS AVAILABLE			
1985-010C		15544	US	24 JAN	NO ELEMENTS AVAILABLE			
1985-010D		15545	US	24 JAN	NO ELEMENTS AVAILABLE			
1985-011A	COSMOS 1627	15505	USSR	01 FEB	104.8	82.9	1014	952
1985-011B		15506	USSR	01 FEB	104.7	82.9	1005	951
1985-013A	METEOR 2-12	15516	USSR	06 FEB	103.9	82.5	957	930
1985-013B		15517	USSR	06 FEB	103.9	82.5	956	932
1985-014A		15546	US	08 FEB	NO ELEMENTS AVAILABLE			
1985-014B		15547	US	08 FEB	NO ELEMENTS AVAILABLE			
1985-015A	ARABSAT 1	15560	SA	08 FEB	1435.1	1.7	35819	35713
1985-015B	SBTS 1	15561	BRAZIL	08 FEB	1436.2	0.0	35791	35786
1985-015C		15562	ESA	08 FEB	577.0	6.6	32846	314
1985-016A	COSMOS 1629	15574	USSR	21 FEB	1436.0	5.9	35807	35764
1985-016F		15581	USSR	21 FEB	1448.6	6.0	36139	35921
1985-020A	COSMOS 1633	15592	USSR	05 MAR	96.8	82.5	613	599
1985-020B		15593	USSR	05 MAR	97.3	82.5	638	622
1985-021A	GEOSAT	15595	US	13 MAR	100.4	108.1	779	776
1985-021B		15596	US	13 MAR	100.3	108.1	795	745
1985-021D		15614	US	13 MAR	99.3	108.2	741	712
1985-021E		15615	US	13 MAR	100.4	107.8	816	736
1985-021F		15616	US	13 MAR	100.3	107.5	845	701
1985-022A	COSMOS 1634	15597	USSR	14 MAR	104.7	82.9	1007	955
1985-022B		15598	USSR	14 MAR	L04.6	82.9	991	960
1985-023A	COSMOS 1635	15617	USSR	21 MAR	115.8	74.1	1510	1471
1985-023B	COSMOS 1636	15618	USSR	21 MAR	115.6	74.1	1492	1472
1985-023C	COSMOS 1637	15619	USSR	21 MAR	115.4	74.1	1487	1462
1985-023D	COSMOS 1638	15620	USSR	21 MAR	115.2	74.0	1478	1454
1985-023E	COSMOS 1639	15621	USSR	21 MAR	115 1	74.1	1478	1439
1985-023F	COSMOS 1640	15622	USSR	21 MAR	114 9	74.1	1477	1424
1985-023G	COSMOS 1641	15623	USSR	21 MAR	114.8	74.1	1477	1410
1985-023H	COSMOS 1642	15624	USSR	21 MAR	114.6	74.1	1475	1397
1985-023J		15625	USSR	21 MAR	118.0	74.1	1709	1473
1985-024A	EKRAN 14	15626	USSR	22 MAR	1519.1	7.0	37464	37327
1985-024D		15630	USSR	22 MAR	1422.6	6.6	35588	35455
1985-025A	INTELSAT VF10	15629	ITSO	22 MAR	1436.1	0.5	35804	35770
1985-025B		15631	US	22 MAR	322.7	23.0	18225	184
1985-028B	ANIK C1	15642	CANADA	13 APR	1436.1	0.0	35796	35778
1985-028C	SYNCOM IV-3	15643	US	12 APR	1436.2	3.3	35803	35772
1985-028D		15644	US	13 APR	591.3	22.9	33551	366
1985-028E		16229	US	12 APR	274.4	26.9	14885	330
1985-033A	PROGNOZ 10	15661	USSR	26 APR	5783.7	76.8	194734	5975
1985-033D		15664	USSR	26 APR	5784.8	65.0	200315	420
1985-035A	GSTAR 1	15677	US	08 MAY	1436.2	0.1	35792	35782
1985-035B	TELECOM 1B	15678	FRANCE	08 MAY	1435.6	4.4	35787	35767
1985-035C		15679	ESA	08 MAY	469.4	6.9	27022	234
1985-035D		15680	ESA	08 HAY	304.8	6.7	16426	822
1985-037A	COSMOS 1650	15697	USSR	17 MAY	675.7	64.9	19177	19081
1985-037B	COSMOS 1651	15698	USSR	17 MAY	675.6	64.8	19146	19109
1985-037C	COSMOS 1652	15699	USSR	17 MAY	675.8	64.8	19154	19110
1985-037F		15702	USSR	17 MAY	675.0	64.8	19166	19057
1985-037G		15714	USSR	17 MAY	333.7	51.9	18723	393
1985-037H		15715	USSR	17 MAY	330.5	52.0	18534	377
1985-040A	MOLNIYA 3-24	15738	USSR	29 MAY	717.5	64.2	39202	1135
1985-040D		15741	USSR	29 MAY	732.3	64.4	39664	1402
1985-041A	COSMOS 1655	15751	USSR	30 MAY	105.0	82.9	1012	973
1985-041B		15752	USSR	30 MAY	104.9	82.9	1006	970
1985-042A	COSMOS 1656	15755	USSR	30 MAY	101.5	71.1	854	800
1985-042D		15772	USSR	30 MAY	101.4	71.1	853	798
1985-042E		15773	USSR	30 MAY	101.1	66.6	831	789
1985-042F		15774	USSR	30 MAY	101.2	66.6	830	802
1985-042G		18764	USSR	30 MAY	100.1	66.6	832	693
1985-042H		18765	USSR	30 MAY	101.2	66.6	815	811
1985-042J		18766	USSR	30 MAY	102.5	66.6	937	820
1985-042K		18767	USSR	30 MAY	104.3	66.6	1088	831
1985-042L		18819	USSR	30 MAY	101.8	66.6	889	796
1985-045A	COSMOS 1658	15808	USSR	11 JUN	718.6	65.4	36916	3478
1985-045D		15811	USSR	11 JUN	709.3	65.8	36754	3182

International Designation	Satellite Name	Catalog Number	Source	Launch Date	Period Mins	Incl-ination	Apogee KM	Perigee KM
1985-047A	COSMOS 1660	15821	USSR	14 JUN	116.0	73.6	1523	1479
1985-047B		15822	USSR	14 JUN	116.0	73.6	1519	1479
1985-048B	MORELOS A	15824	MEXICO	17 JUN	1436.1	0.0	35793	35781
1985-048C	ARABSAT 1B	15825	SA	18 JUN	1434.4	1.0	35891	35614
1985-048D	TELSTAR 3D	15826	US	19 JUN	1436.1	0.0	35789	35783
1985-048F		15832	US	17 JUN	624.9	25.8	35289	376
1985-048G		15836	US	18 JUN	618.3	27.0	34923	402
1985-048H		15837	US	18 JUN	651.4	25.9	36603	425
1985-049A	COSMOS 1661	15827	USSR	18 JUN	716.9	67.4	36635	3677
1985-049D		15830	USSR	18 JUN	724.8	67.8	37168	3534
1985-055A	INTELSAT VA F11	15873	ITSO	30 JUN	1436.1	0.1	35804	35769
1985-055B		15874	US	30 JUN	544.3	23.3	31101	308
1985-056A	GIOTTO	15875	ESA	02 JUL	HELIOCENTRIC ORBIT			
1985-056B		15876	ESA	02 JUL	498.2	8.1	28592	288
1985-056C		17255	ESA	02 JUL	598.8	8.5	33993	314
1985-056D		17325	ESA	02 JUL	548.2	7.2	31297	319
1985-056E		17332	ESA	02 JUL	437.4	7.9	25086	334
1985-058A	COSMOS 1666	15889	USSR	08 JUL	96.9	82.5	625	595
1985-058B		15890	USSR	08 JUL	97.4	82.5	650	616
1985-058C		19241	USSR	08 JUL	96.6	82.5	609	580
1985-061A	MOLNIYA 3-25	15909	USSR	17 JUL	716.9	64.8	38847	1465
1985-061D		15916	USSR	17 JUL	737.9	64.5	39951	1389
1985-064A	COSMOS 1670	15930	USSR	01 AUG	104.1	64.9	1003	897
1985-066A	NNSS 30300	15935	US	03 AUG	107.9	89.9	1255	999
1985-066B	NNSS 30240	15936	US	03 AUG	107.9	89.9	1256	999
1985-066C		15938	US	03 AUG	107.9	89.9	1256	1000
1985-066D		15950	US	03 AUG	106.6	89.9	1181	958
1985-066E		15951	US	03 AUG	106.8	89 9	1190	965
1985-066F		16020	US	03 AUG	107.5	90 2	1219	1004
1985-066G		17164	US	03 AUG	108.2	89.3	1296	986
1985-066H		21878	US	03 AUG	107 8	89.9	1252	997
1985-069A	COSMOS 1674	15944	USSR	08 AUG	96 9	82.5	623	595
1985-069B		15945	USSR	08 AUG	97 3	82 5	647	615
1985-070A	RADUGA 16	15946	USSR	08 AUG	1435 5	5 4	35784	35763
1985-070F		15963	USSR	08 AUG	1472 3	5.6	36550	36435
1985-071A	COSMOS 1675	15952	USSR	12 AUG	718 8	67.4	36985	3420
1985-071D		15955	USSR	12 AUG	708.2	67.3	36585	3298
1985-073A	PLANET A	15967	JAPAN	18 AUG	HELIOCENTRIC ORBIT			
1985-073C		15969	JAPAN	18 AUG	HELIOCENTRIC ORBIT			
1985-074A	MOLNIYA 1-64	15977	USSR	22 AUG	717.8	64.8	38436	1918
1985-074D		15983	USSR	22 AUG	732.4	65.0	39110	1960
1985-075A	COSMOS 1677	15986	USSR	23 AUG	103.9	64.7	980	900
1985-076B	AUSSAT 1	15993	AUSTRL	27 AUG	1436 1	O.0	35798	35777
1985-076C	ASC 1	15994	US	27 AUG	1436 1	0 0	35794	35778
1985-076D	SYNCOM IV-4	15995	US	29 AUG	1438.1	3 2	35843	35809
1985-076E		15996	US	27 AUG	629.7	26.1	35531	382
1985-076F		16001	US	29 AUG	277 7	27 4	15061	380
1985-076G		16007	US	29 AUG	627 8	26 7	35445	374
1985-077K		18608	USSR	29 AUG	100.2	74.1	779	757
1985-079A	COSMOS 1680	16011	USSR	04 SEP	100.6	74.1	799	774
1985-079B		16012	USSR	04 SEP	100 5	74.1	792	768
1985-079C		17754	USSR	04 SEP	100 6	74 0	804	773
1985-084A	COSMOS 1684	16064	USSR	24 SEP	717.4	65 5	36430	3905
1985-084D		16070	USSR	24 SEP	706.0	66.0	36163	3606
1985-087A	INTELSAT VA F-12	16101	ITSO	29 SEP	1436 1	0.1	35801	35772
1985-087B		16102	US	29 SEP	500 3	23 1	28721	275
1985-088A	COSMOS 1687	16103	USSR	30 SEP	717.4	67 3	36504	3833
1985-088D		16106	USSR	30 SEP	703.6	67.3	36150	3502
1985-090A	COSMOS 1689	16110	USSR	03 OCT	95.2	97.6	551	502
1985-090B		16111	USSR	03 OCT	96.4	97.6	621	550
1985-091A	MOLNIYA 3-26	16112	USSR	03 OCT	718.3	64.6	38322	2059
1985-091D		16125	USSR	03 OCT	734.0	64.8	38919	2232
1985-092B		16116	US	03 OCT	NO ELEMENTS AVAILABLE			
1985-092C		16117	US	03 OCT	NO ELEMENTS AVAILABLE			
1985-092D		16118	US	03 OCT	NO ELEMENTS AVAILABLE			
1985-092E		16119	US	03 OCT	NO ELEMENTS AVAILABLE			
1985-093A		16129	US	09 OCT	718.0	64.4	20537	19826
1985-093B		16137	US	09 OCT	368.2	64.1	20139	1138
1985-094A	COSMOS 1690	16138	USSR	09 OCT	113.7	82.6	1414	1378
1985-094B	COSMOS 1691	16139	USSR	09 OCT	114.0	82.6	1413	1409
1985-094C	COSMOS 1692	16140	USSR	09 OCT	113.8	82.6	1414	1386
1985-094D	COSMOS 1693	16141	USSR	09 OCT	113.8	82.6	1414	1390
1985-094E	COSMOS 1694	16142	USSR	09 OCT	113.9	82.6	1414	1395

International Designation	Satellite Name	Catalog Number	Source	Launch Date	Period Mins	Incl-ination	Apogee KM	Perigee KM
1985-094F	COSMOS 1695	16143	USSR	09 OCT	114.0	82.6	1414	1402
1985-094G		16144	USSR	09 OCT	114.7	82.6	1469	1412
1985-094K		16266	USSR	09 OCT	114.0	82.6	1427	1390
1985-094L		16267	USSR	09 OCT	112.8	82.6	1397	1310
1985-094M		16268	USSR	09 OCT	114.9	82.7	1513	1389
1985-094N		16269	USSR	09 OCT	114.1	82.6	1423	1402
1985-094P		16270	USSR	09 OCT	113.7	82.7	1598	1193
1985-094Q		16271	USSR	09 OCT	114.0	82.6	1414	1402
1985-094R		16272	USSR	09 OCT	113.4	82.6	1417	1347
1985-094S		17168	USSR	09 OCT	113.0	82.6	1421	1304
1985-094T		18282	USSR	09 OCT	113.7	82.6	1409	1385
1985-094U		18777	USSR	09 OCT	114.0	82.6	1412	1410
1985-094V		19111	USSR	09 OCT	113.8	82.6	1410	1387
1985-097A	COSMOS 1697	16181	USSR	22 OCT	101.9	71.0	854	842
1985-097B		16182	USSR	22 OCT	101.7	71.0	845	834
1985-097C		16389	USSR	22 OCT	104.8	71.0	1125	840
1985-097D		16390	USSR	22 OCT	105.1	71.0	1153	840
1985-097E		16391	USSR	22 OCT	104.7	71.0	1125	837
1985-097F		16392	USSR	22 OCT	104.9	71.0	1140	837
1985-098A	COSMOS 1698	16183	USSR	22 OCT	718.4	66.6	36550	3837
1985-098D		16186	USSR	22 OCT	707.9	66.7	36206	3658
1985-099A	MOLNIYA 1-65	16187	USSR	23 OCT	718.2	64.6	38720	1654
1985-099E		16197	USSR	23 OCT	698.0	64.6	37705	1668
1985-100A	METEOR 3	16191	USSR	24 OCT	109.3	82.5	1208	1178
1985-100B		16194	USSR	24 OCT	110.2	82.6	1245	1222
1985-102A	COSMOS 1700	16199	USSR	25 OCT	1436.4	5.1	35815	35767
1985-102D		16214	USSR	25 OCT	1431.1	5.1	35779	35596
1985-103A	MOLNIYA 1-66	16220	USSR	28 OCT	717.9	64.1	39007	1354
1985-103D		16223	USSR	28 OCT	701.1	64.5	38003	1525
1985-105A	COSMOS 1701	16235	USSR	09 NOV	718.2	67.2	37014	3359
1985-105D		16243	USSR	09 NOV	706.2	67.3	36687	3093
1985-107A	RADUGA 17	16250	USSR	15 NOV	148.0	5.2	35875	35773
1985-107F		16339	USSR	15 NOV	1477.0	5.3	36674	36492
1985-108A	COSMOS 1703	16262	USSR	22 NOV	96.9	82.5	625	596
1985-108B		16263	USSR	22 NOV	97.4	82.5	650	617
1985-109B	MORELOS B	16274	MEXICO	27 NOV	1436.1	0.0	35793	35780
1985-109C	AUSSAT 2	16275	AUSTRL	27 NOV	1436.2	0.0	35796	35779
1985-109D	SATCOM KU2	16276	US	28 NOV	1436.2	0.0	35797	35779
1985-109F		16293	US	27 NOV	637.9	25.8	35939	395
1985-109G		16294	US	27 NOV	633.9	26.2	35734	396
1985-109H		16295	US	28 NOV	617.2	26.7	34877	393
1985-110A	COSMOS 1704	16291	USSR	28 NOV	104.7	82.9	1004	960
1985-110B		16292	USSR	28 NOV	104.6	82.9	996	953
1985-113A	COSMOS 1707	16326	USSR	12 DEC	96.9	82.5	626	597
1985-113B		16327	USSR	12 DEC	97.4	82.5	648	618
1985-116A	COSMOS 1709	16368	USSR	19 DEC	104.8	82.9	1009	957
1985-116B		16369	USSR	19 DEC	104.6	82.9	1002	950
1985-117A	MOLNIYA 3-27	16393	USSR	24 DEC	712.8	64.2	38637	1470
1985-117F		16402	USSR	24 DEC	732.7	64.0	39488	1597
1985-118A	COSMOS 1710	16396	USSR	24 DEC	675.7	66.2	19171	19088
1985-118B	COSMOS 1711	16397	USSR	24 DEC	675.7	66.3	19147	19111
1985-118C	COSMOS 1712	16398	USSR	24 DEC	676.3	66.2	19158	19129
1985-118F		16404	USSR	24 DEC	675.5	66.2	19151	19097
1985-118R		16445	USSR	24 DEC	340.3	65.0	18978	557
1985-118L		16446	USSR	24 DEC	339.9	65.0	18997	512
1985-118M		21960	USSR	24 DEC	332.5	64.4	18249	791
1985-119A	METEOR 2-13	16408	USSR	26 DEC	103.9	82.5	955	933
1985-119B		16409	USSR	26 DEC	104.0	82.5	955	935

1986 LAUNCHES

International Designation	Satellite Name	Catalog Number	Source	Launch Date	Period Mins	Incl-ination	Apogee KM	Perigee KM
1986-002A	COSMOS 1716	16449	USSR	09 JAN	115.5	74.0	1490	1461
1986-002B	COSMOS 1717	16450	USSR	09 JAN	115.8	74.0	1511	1473
1986-002C	COSMOS 1718	16451	USSR	09 JAN	115.6	74.0	1493	1473
1986-002D	COSMOS 1719	16452	USSR	09 JAN	115.3	74.0	1483	1452
1986-002E	COSMOS 1720	16453	USSR	09 JAN	115.1	74.0	1482	1437
1986-002F	COSMOS 1721	16454	USSR	09 JAN	114.9	74.0	1482	1424
1986-002G	COSMOS 1722	16455	USSR	09 JAN	114.8	74.0	1482	1409
1986-002H	COSMOS 1723	16456	USSR	09 JAN	114.6	74.0	1480	1397
1986-002J		16457	USSR	09 JAN	117.9	74.0	1694	1479
1986-003B	SATCOM KU1	16482	US	12 JAN	1436.2	0.0	35796	35780
1986-003C		16483	US	12 JAN	615.4	27.0	34786	388
1986-005A	COSMOS 1725	16493	USSR	17 JAN	104.8	82.9	1000	965

International Designation	Satellite Name	Catalog Number	Source	Launch Date	Period Mins	Incl-ination	Apogee KM	Perigee KM
1986-005B		16494	USSR	17 JAN	104.6	82.9	992	960
1986-006A	COSMOS 1726	16495	USSR	17 JAN	96.8	82.5	618	594
1986-006B		16496	USSR	17 JAN	97.3	82.5	645	617
1986-007A	RADUGA 18	16497	USSR	17 JAN	1457.4	5.1	36506	35897
1986-007E		16501	USSR	17 JAN	647.7	47.0	36585	252
1986-007F		16870	USSR	17 JAN	1472.4	5.3	36636	36352
1986-008A	COSMOS 1727	16510	USSR	23 JAN	104.8	82.9	1013	955
1986-008B		16511	USSR	23 JAN	104.7	82.9	1003	955
1986-010A	PRC 18	16526	PRC	01 FEB	1436.5	4.0	35810	35778
1986-010B		16528	PRC	01 FEB	627.1	30.3	35387	393
1986-011A	COSMOS 1729	16527	USSR	01 FEB	718.3	64.8	36787	3590
1986-011F		16533	USSR	01 FEB	705.7	65.2	36319	3436
1986-014A		16591	US	09 FEB	NO ELEMENTS AVAILABLE			
1986-014B		16592	US	09 FEB	NO ELEMENTS AVAILABLE			
1986-014C		16622	US	09 FEB	NO ELEMENTS AVAILABLE			
1986-014D		16623	US	09 FEB	NO ELEMENTS AVAILABLE			
1986-014E		16624	US	09 FEB	NO ELEMENTS AVAILABLE			
1986-014F		16625	US	09 FEB	NO ELEMENTS AVAILABLE			
1986-014G		16630	US	09 FEB	NO ELEMENTS AVAILABLE			
1986-014H		16631	US	09 FEB	NO ELEMENTS AVAILABLE			
1986-015A	COSMOS 1732	16593	USSR	11 FEB	116.0	73.6	1523	1477
1986-015B		16594	USSR	11 FEB	115.9	73.6	1520	1476
1986-016A	BS-2B	16597	JAPAN	12 FEB	1450.3	2.1	36147	35980
1986-016C		16600	JAPAN	12 FEB	382.4	28.0	21904	243
1986-017A	MIR	16609	USSR	19 FEB	92.5	51.6	395	393
1986-017B TO 017GY			USSR	19 FEB	SEE NOTE 60			
1986-018A	COSMOS 1733	16611	USSR	19 FEB	96.9	82.5	620	596
1986-018B		16612	USSR	19 FEB	97.3	82.5	644	617
1986-019A	SPOT 1	16613	FRANCE	22 FEB	101.3	98.7	825	817
1986-019C TO 019VL			ESA	22 FEB	SEE NOTE 45			
1986-019B	VIKING	16614	SWEDEN	22 FEB	261.6	98.8	13521	818
1986-022C		16863	USSR	13 MAR	89.5	51.6	253	244
1986-024A	COSMOS 1736	16647	USSR	21 MAR	104.4	65.0	1007	923
1986-024E		16809	USSR	21 MAR	104.2	65.0	990	923
1986-026A	GSTAR 2	16649	US	28 MAR	1436.1	0.0	35793	35780
1986-026B	SBTS 2	16650	BRAZIL	28 MAR	1436.2	0.1	35789	35786
1986-026C		16657	ESA	28 MAR	651.1	6.5	36540	470
1986-026E		17253	ESA	28 MAR	537.3	8.6	30480	547
1986-026F		17254	ESA	28 MAR	532.9	8.4	30259	528
1986-027A	COSMOS 1738	16667	USSR	04 APR	1434.5	4.9	35821	35690
1986-027F		16676	USSR	04 APR	1474.1	5.0	36684	36369
1986-030A	COSMOS 1741	16681	USSR	18 APR	100.6	74.0	801	773
1986-030B		16682	USSR	18 APR	100.5	74.0	791	771
1986-030C		17842	USSR	18 APR	100.7	74 0	808	778
1986-030D		17843	USSR	18 APR	100.7	74 0	805	778
1986-030E		18274	USSR	18 APR	100.1	74.1	794	735
1986-030F		18526	USSR	18 APR	100.2	74.0	771	767
1986-030G		18681	USSR	18 APR	100.8	74.0	808	779
1986-030H		19235	USSR	18 APR	103.9	74.0	947	942
1986-031A	MOLNIYA 3-28	16683	USSR	18 APR	717.8	64.8	38539	1813
1986-031D		16686	USSR	18 APR	733.5	64.9	39117	2007
1986-034A	COSMOS 1743	16719	USSR	15 MAY	96.9	82.6	622	597
1986-034B		16720	USSR	15 MAY	97.4	82.6	647	618
1986-037A	COSMOS 1745	16727	USSR	23 MAY	104.8	83.0	1008	960
1986-037B		16728	USSR	23 MAY	104.6	83.0	999	954
1986-038A	EKRAN 15	16729	USSR	24 MAY	1491.6	5.9	36921	36809
1986-038D		16732	USSR	24 MAY	1420.6	5.6	35576	35387
1986-038E		16733	USSR	24 MAY	254.4	47.9	13280	559
1986-039A	METEOR 2-14	16735	USSR	27 MAY	104.0	82.5	955	935
1986-039B		16736	USSR	27 MAY	104.0	82.5	954	935
1986-042A	COSMOS 1748	16758	USSR	06 JUN	115.1	74.0	1467	1451
1986-042B	COSMOS 1749	16759	USSR	06 JUN	114.4	74.0	1466	1392
1986-042C	COSMOS 1750	16760	USSR	06 JUN	114.6	74.0	1467	1407
1986-042D	COSMOS 1751	16761	USSR	06 JUN	115.6	74.0	1504	1464
1986-042E	COSMOS 1752	16762	USSR	06 JUN	115.4	74.0	1485	1465
1986-042F	COSMOS 1753	16763	USSR	06 JUN	115.3	74.0	1475	1459
1986-042G	COSMOS 1754	16764	USSR	06 JUN	114.9	74.0	1467	1436
1986-042H	COSMOS 1755	16765	USSR	06 JUN	114.8	74.0	1467	1422
1986-042J		16766	USSR	06 JUN	117.7	74.0	1681	1470
1986-044A	GORIZONT 12	16769	USSR	10 JUN	1436.6	4.6	35798	35794
1986-044F		16797	USSR	10 JUN	1474.4	4.7	36579	36486
1986-046A	COSMOS 1758	16791	USSR	12 JUN	97.1	82.5	639	602
1986-046B		16792	USSR	12 JUN	97.4	82.5	654	613

International Designation	Satellite Name	Catalog Number	Source	Launch Date	Period Mins	Incl-ination	Apogee KM	Perigee KM
1986-047A	COSMOS 1759	16798	USSR	18 JUN	104.7	82.9	999	963
1986-047B		16799	USSR	18 JUN	104.6	82.9	1024	923
1986-049A	MOLNIYA 3-29	16802	USSR	19 JUN	717.0	64.9	38691	1622
1986-049D		16805	USSR	19 JUN	733.1	65.1	39424	1684
1986-050A	COSMOS 1761	16849	USSR	05 JUL	717.9	66.6	36692	3667
1986-050D		16854	USSR	05 JUL	710.1	66.7	36479	3493
1986-052A	COSMOS 1763	16860	USSR	16 JUL	100.3	74.0	794	749
1986-052B		16864	USSR	16 JUL	100.2	74.0	793	744
1986-052C		16865	USSR	16 JUL	99.4	74.0	747	710
1986-052D		16866	USSR	16 JUL	99.3	74.0	743	706
1986-052E		16867	USSR	16 JUL	99.6	74.0	757	721
1986-055A	COSMOS 1766	16881	USSR	28 JUL	97.1	82.5	631	603
1986-055B		16882	USSR	28 JUL	97.4	82.5	649	619
1986-057A	MOLNIYA 1-67	16885	USSR	30 JUL	717.3	64.7	38500	1828
1986-057D		16889	USSR	30 JUL	731.6	65.0	39230	1804
1986-061A	EGP	16908	JAPAN	12 AUG	115.7	50.0	1497	1479
1986-061B	JAS-1	16909	JAPAN	12 AUG	115.7	50.0	1497	1480
1986-061C		16910	JAPAN	12 AUG	116.9	50.0	1595	1484
1986-062A	COSMOS 1771	16917	USSR	20 AUG	104.2	65.0	999	910
1986-062C		17035	USSR	20 AUG	103.8	65.0	974	906
1986-065A	COSMOS 1774	16922	USSR	28 AUG	718.0	65.0	37165	3198
1986-065D		16925	USSR	28 AUG	707.0	65.4	36603	3216
1986-068A	MOLNIYA 1-68	16934	USSR	05 SEP	717.8	64.6	38173	2182
1986-068D		16939	USSR	05 SEP	731.2	64.8	38631	2384
1986-070A	COSMOS 1777	16952	USSR	10 SEP	100.6	74.0	803	769
1986-070B		16953	USSR	10 SEP	100.4	74.0	784	768
1986-071A	COSMOS 1778	16961	USSR	16 SEP	675.7	64.9	19135	19123
1986-071B	COSMOS 1779	16962	USSR	16 SEP	675.7	64.8	19138	19120
1986-071C	COSMOS 1780	16963	USSR	16 SEP	675.7	64.9	19188	19069
1986-071F		16968	USSR	16 SEP	675.2	64.9	19135	19097
1986-073A	NOAA 10	16969	US	17 SEP	101.0	98.5	816	796
1986-073B		16982	US	17 SEP	97.2	98.6	625	620
1986-074A	COSMOS 1782	16986	USSR	30 SEP	97.1	82.5	632	604
1986-074B		16987	USSR	30 SEP	97.4	82.5	649	618
1986-075A	COSMOS 1783	16993	USSR	03 OCT	358.0	63.6	19439	1209
1986-075D		16996	USSR	03 OCT	357.0	63.6	19372	1209
1986-078A	COSMOS 1785	17031	USSR	15 OCT	718.7	67.0	37307	3093
1986-078D		17037	USSR	15 OCT	707.7	67.6	36907	2946
1986-079A	MOLNIYA 3-30	17038	USSR	20 OCT	717.0	64.9	38365	1949
1986-079D		17041	USSR	20 OCT	699.0	64.9	37630	1793
1986-082A	RADUGA 19	17046	USSR	25 OCT	1436.0	4.3	35794	35774
1986-082D		17052	USSR	25 OCT	637.1	45.9	36043	252
1986-082E		17053	USSR	25 OCT	101.8	46.4	1551	132
1986-082F		17065	USSR	25 OCT	1475.5	4.5	36682	36424
1986-086A	COSMOS 1791	17066	USSR	13 NOV	104.7	82.9	1008	949
1986-086B		17067	USSR	13 NOV	104.5	82.9	999	947
1986-086C		18552	USSR	13 NOV	103.7	82.9	956	912
1986-088A	POLAR BEAR	17070	US	14 NOV	104.8	89.6	1014	954
1986-088B		17071	US	14 NOV	104.8	89.6	1012	955
1986-088C		18426	US	14 NOV	105.1	89.1	1051	944
1986-088D		18525	US	14 NOV	104.2	89.8	963	951
1986-089A	MOLNIYA 1-69	17078	USSR	15 NOV	717.2	63.8	38693	1632
1986-089D		17081	USSR	15 NOV	735.8	63.6	39180	2058
1986-090A	GORIZONT 13	17083	USSR	18 NOV	1488.8	4.2	36867	36755
1986-090D		17125	USSR	18 NOV	1436.4	4.1	35838	35747
1986-090F		17149	USSR	18 NOV	632.9	47.3	35812	268
1986-091A	COSMOS 1793	17134	USSR	20 NOV	717.5	67.1	37249	3091
1986-091D		17147	USSR	20 NOV	705.9	67.5	36865	2899
1986-092A	COSMOS 1794	17138	USSR	21 NOV	115.6	74.0	1497	1464
1986-092B	COSMOS 1795	17139	USSR	21 NOV	115.4	74.0	1479	1464
1986-092C	COSMOS 1796	17140	USSR	21 NOV	115.2	74.0	1476	1453
1986-092D	COSMOS 1797	17141	USSR	21 NOV	115.0	74.0	1470	1441
1986-092E	COSMOS 1798	17142	USSR	21 NOV	114.8	74.0	1470	1426
1986-092F	COSMOS 1799	17143	USSR	21 NOV	114.7	74.0	1470	1411
1986-092G	COSMOS 1800	17144	USSR	21 NOV	114.5	74.0	1471	1396
1986-092H	COSMOS 1801	17145	USSR	21 NOV	114.4	74.0	1470	1382
1986-092J		17146	USSR	21 NOV	117.6	74.0	1665	1482
1986-093A	COSMOS 1802	17159	USSR	24 NOV	104.9	82.9	1020	958
1986-093B		17160	USSR	24 NOV	104.8	82.9	1011	954
1986-094A	COSMOS 1803	17177	USSR	02 DEC	115.9	82.6	1500	1494
1986-094B		17178	USSR	02 DEC	115.9	82.6	1497	1493
1986-094C		20284	USSR	02 DEC	117.3	83.2	1738	1382
1986-096A		17181	US	05 DEC	1436.2	0.4	35849	35728

International Designation	Satellite Name	Catalog Number	Source	Launch Date	Period Mins	Incl-ination	Apogee KM	Perigee KM
1986-097A	COSMOS 1805	17191	USSR	10 DEC	96.9	82.5	622	601
1986-097B		17192	USSR	10 DEC	97.4	82.5	644	620
1986-098A	COSMOS 1806	17213	USSR	12 DEC	718.3	64.6	36767	3611
1986-098D		17216	USSR	12 DEC	705.9	65.1	36398	3366
1986-100A	COSMOS 1808	17239	USSR	17 DEC	105.0	82.9	1015	968
1986-100B		17240	USSR	17 DEC	104.8	82.9	1008	963
1986-100C		18545	USSR	17 DEC	104.1	82.9	971	933
1986-101A	COSMOS 7809	17241	USSR	18 DEC	104.1	82.5	960	939
1986-101B		17242	USSR	18 DEC	104.1	82.5	960	940
1986-101C		17268	USSR	18 DEC	103.7	82.6	953	910
1986-101D		17269	USSR	18 DEC	104.1	82.6	965	941
1986-101E		17270	USSR	18 DEC	104.0	82.4	949	941
1986-101F		17271	USSR	18 DEC	103.4	82.4	941	898
1986-101G		17272	USSR	18 DEC	103.3	82.5	922	905
1986-101H		17273	USSR	18 DEC	103.2	82.5	919	901
1986-101J		17274	USSR	18 DEC	104.1	82.5	976	929
1986-101K		17844	USSR	18 DEC	103.3	82.5	925	900
1986-101L		18680	USSR	18 DEC	103.3	82.5	925	904
1986-103A	MOLNIYA 1-70	17264	USSR	26 DEC	717.7	64.0	39433	916
1986-103D		17267	USSR	26 DEC	698.8	64.3	38428	986
1987 LAUNCHES								
1987-001A	METEOR 2-15	17290	USSR	05 JAN	104.0	82.5	954	937
1987-001B		17291	USSR	05 JAN	104.0	82.5	953	938
1987-003A	COSMOS 1812	17295	USSR	14 JAN	96.9	82.5	625	599
1987-003B		17296	USSR	14 JAN	97.4	82.5	648	618
1987-004GR		18273	USSR	15 JAN	90.4	72.8	321	265
1987-006A	COSMOS 1814	17303	USSR	21 JAN	100.5	74.1	800	762
1987-006B		17304	USSR	21 JAN	100.4	74.1	794	756
1987-006C		18257	USSR	21 JAN	100.1	74.0	769	758
1987-008A	MOLNIYA 3-31	17328	USSR	22 JAN	717.8	63.7	38948	1407
1987-008D		17333	USSR	22 JAN	730.8	63.7	39368	1623
1987-009A	COSMOS 1816	17359	USSR	29 JAN	104.8	82.9	1008	958
1987-009B		17360	USSR	29 JAN	104.6	82.9	998	953
1987-011A	COSMOS 1818	17369	USSR	01 FEB	100.7	65.0	797	781
1987-012B		17481	JAPAN	05 FEB	94.0	31.1	507	431
1987-015A		17506	US	12 FEB	NO ELEMENTS AVAILABLE			
1987-015B		17507	US	12 FEB	NO ELEMENTS AVAILABLE			
1987-017A	COSMOS 1821	17525	USSR	18 FEB	104.8	82.9	1013	957
1987-017B		17526	USSR	18 FEB	104.6	82.9	1006	945
1987-018A	MOS-1	17527	JAPAN	19 FEB	103.2	99.1	909	908
1987-018B		17528	JAPAN	19 FEB	99.8	97.4	870	625
1987-020A	COSMOS 1823	17535	USSR	20 FEB	116.0	73.6	1521	1478
1987-020B TO 020DQ			USSR	20 FEB	SEE NOTE 49			
1987-022A	GOES 7	17561	US	26 FEB	1436.2	0.4	35800	35775
1987-022B		17562	US	26 FEB	89.7	21.7	342	179
1987-022C		17563	US	26 FEB	611.8	17.7	34730	259
1987-024A	COSMOS 1825	17566	USSR	03 MAR	96.9	82.5	623	596
1987-024B		17567	USSR	03 MAR	97.4	82.5	646	617
1987-026A	COSMOS 1827	17582	USSR	13 MAR	113.8	82.6	1408	1393
1987-026B	COSMOS 1828	17583	USSR	13 MAR	113.7	82.6	1408	1382
1987-026C	COSMOS 1829	17584	USSR	13 MAR	114.0	82.6	1412	1408
1987-026D	COSMOS 1830	17585	USSR	13 MAR	113.9	82.6	1408	1405
1987-026E	COSMOS 1831	17586	USSR	13 MAR	113.8	82.6	1408	1389
1987-026F	COSMOS 1832	17587	USSR	13 MAR	113.9	82.6	1408	1399
1987-026G		17588	USSR	13 MAR	114.6	82.6	1468	1408
1987-027A	COSMOS 1833	17589	USSR	18 MAR	101.9	70.9	853	845
1987-027B		17590	USSR	18 MAR	101.7	71.0	842	833
1987-027C		18416	USSR	18 MAR	104.7	71.0	1121	839
1987-027D		18417	USSR	18 MAR	104.9	71.0	1144	838
1987-027E		18527	USSR	18 MAR	104.8	71.0	1133	839
1987-027F		18550	USSR	18 MAR	104.6	71.0	1111	837
1987-028A	RADUGA 20	17611	USSR	19 MAR	1500.6	4.4	37171	36908
1987-028D		17705	USSR	19 MAR	1442.0	4.4	36020	35785
1987-029A	PALAPA B-2P	17706	INDNSA	20 MAR	1436.2	0.0	35788	35788
1987-030A	KVANT 1	17845	USSR	31 MAR	92.5	51.6	395	394
1987-036K		21622	USSR	24 APR	143.2	64.8	4087	1292
1987-036L		21623	USSR	24 APR	212.5	62.8	10400	438
1987-036M		21657	USSR	24 APR	149.7	64.8	4762	1162
1987-036N		21725	USSR	24 APR	149.3	64.7	4680	1206
1987-038A	COSMOS 1842	17911	USSR	27 APR	97.0	82.5	630	603
1987-038B		17912	USSR	27 APR	97.4	82.5	650	618

International Designation	Satellite Name	Catalog Number	Source	Launch Date	Period Mins	Incl-ination	Apogee KM	Perigee KM
1987-040A	GORIZONT 14	17969	USSR	11 MAY	1474.6	6.0	36669	36405
1987-040D		17972	USSR	11 MAY	1397.9	5.8	35109	34960
1987-040E		18111	USSR	11 MAY	537.1	46.9	30892	128
1987-040F		18112	USSR	11 MAY	597.7	47.0	34107	144
1987-041A	COSMOS 1844	17973	USSR	13 MAY	101.9	70.9	853	842
1987-041B		17974	USSR	13 MAY	101.6	71.0	847	825
1987-041C		18410	USSR	13 MAY	105.0	71.0	1143	842
1987-041D		18411	USSR	13 MAY	104.8	71.0	1125	842
1987-041E		18412	USSR	13 MAY	104.8	71.0	1127	839
1987-041F		18476	USSR	13 MAY	105.0	71.0	1150	841
1987-043A		17997	US	15 MAY	NO ELEMENTS AVAILABLE			
1987-043B		17998	US	15 MAY	NO ELEMENTS AVAILABLE			
1987-043C		18007	US	15 MAY	NO ELEMENTS AVAILABLE			
1987-043D		18008	US	15 MAY	NO ELEMENTS AVAILABLE			
1987-043E		18009	US	15 MAY	NO ELEMENTS AVAILABLE			
1987-043F		18010	US	15 MAY	NO ELEMENTS AVAILABLE			
1987-043G		18024	US	15 MAY	NO ELEMENTS AVAILABLE			
1987-043H		18025	US	15 MAY	NO ELEMENTS AVAILABLE			
1987-048A	COSMOS 1849	18083	USSR	04 JUN	717.8	67.2	37666	2689
1987-048D		18086	USSR	04 JUN	706.2	67.3	37128	2653
1987-049A	COSMOS 1850	18095	USSR	09 JUN	100.6	74.0	798	775
1987-049B		18096	USSR	09 JUN	100.5	74.0	793	767
1987-050A	COSMOS 1851	18103	USSR	12 JUN	716.9	64.3	37157	3155
1987-050D		18106	USSR	12 JUN	707.3	64.3	36755	3083
1987-051A	COSMOS 1852	18113	USSR	16 JUN	115.6	74.0	1497	1471
1987-051B	COSMOS 1853	18114	USSR	16 JUN	115.4	74.0	1480	1471
1987-051C	COSMOS 1854	18115	USSR	16 JUN	115.3	74.0	1479	1456
1987-051D	COSMOS 1855	18116	USSR	16 JUN	115.1	74.0	1475	1444
1987-051E	COSMOS 1856	18117	USSR	16 JUN	114.9	74.0	1476	1428
1987-051F	COSMOS 1857	18118	USSR	16 JUN	114.8	74.0	1475	1414
1987-051G	COSMOS 1858	18119	USSR	16 JUN	114.6	74.0	1476	1399
1987-051H	COSMOS 1859	18120	USSR	16 JUN	114.4	74.0	1475	1384
1987-051J		18121	USSR	16 JUN	117.8	74.0	1686	1475
1987-052A	COSMOS 1860	18122	USSR	18 JUN	104.0	65.0	991	901
1987-052D		18241	USSR	18 JUL	103.7	65.0	969	894
1987-053A		18123	US	20 JUN	101.7	98.8	848	828
1987-053B		18127	US	20 JUN	99.8	98.8	759	739
1987-053C		18128	US	20 JUN	98.6	98.7	694	688
1987-053D		18154	US	20 JUN	98.2	98.7	677	668
1987-053E		18159	US	20 JUN	99.8	98.7	756	740
1987-054A	COSMOS 1861	18129	USSR	23 JUN	104.9	82.9	998	978
1987-054B		18130	USSR	23 JUN	104.6	82.9	990	964
1987-054C		18131	USSR	23 JUN	105.0	82.9	1018	968
1987-055A	COSMOS 1862	18152	USSR	01 JUL	97.1	82.5	636	605
1987-055B		18153	USSR	01 JUL	97.4	82.5	649	617
1987-057A	COSMOS 1864	18160	USSR	06 JUL	104.6	82.9	999	950
1987-057B		18161	USSR	06 JUL	104.7	82.9	1003	954
1987-060A	COSMOS 1867	18187	USSR	10 JUL	100.7	65.0	799	781
1987-062A	COSMOS 1869	18214	USSR	16 JUL	97.1	82.5	631	606
1987-062B		18215	USSR	16 JUL	97.4	82.5	649	619
1987-065C		19033	USSR	01 AUG	115.4	102.1	1499	1447
1987-068A	METEOR 2-16	18312	USSR	18 AUG	104.0	82.6	955	938
1987-068B		18313	USSR	18 AUG	104.0	82.6	954	938
1987-070A	ETS-V	18316	JAPAN	27 AUG	1436.2	1.4	35807	35768
1987-073A	EKRAN 16	18328	USSR	04 SEP	1492.5	4.5	36903	36864
1987-073D		18331	USSR	04 SEP	1420.4	4.3	35574	35384
1987-073E		18332	USSR	04 SEP	462.5	46.9	26531	334
1987-074A	COSMOS 1875	18334	USSR	07 SEP	113.7	82.6	1407	1383
1987-074B	COSMOS 1876	18335	USSR	07 SEP	114.0	82.6	1412	1407
1987-074C	COSMOS 1877	18336	USSR	07 SEP	113.9	82.6	1407	1406
1987-074D	COSMOS 1878	18337	USSR	07 SEP	113.9	82.6	1407	1399
1987-074E	COSMOS 1879	18338	USSR	07 SEP	113.8	82.6	1407	1394
1987-074F	COSMOS 1880	18339	USSR	07 SEP	113.8	82.6	1407	1390
1987-074G		18340	USSR	07 SEP	114.6	82.6	1470	1407
1987-078A	AUSSAT K3	18350	AUSTRL	16 SEP	1436.1	0.0	35796	35778
1987-078B	ECS 4	18351	ESA	16 SEP	1436.1	0.0	35811	35761
1987-078E		18571	ESA	16 SEP	535.2	7.3	30650	265
1987-079A	COSMOS 1883	18355	USSR	16 SEP	675.7	65.9	19147	19111
1987-079B	COSMOS 1884	18356	USSR	16 SEP	675.7	65.9	19163	19095
1987-079C	COSMOS 1885	18357	USSR	16 SEP	675.7	65.9	19158	19100
1987-079F		18360	USSR	16 SEP	674.7	65.9	19136	19072
1987-079G		18374	USSR	16 SEP	339.6	65.4	18832	661
1987-079H		18375	USSR	16 SEP	339.6	65.3	18787	703

International Designation	Satellite Name	Catalog Number	Source	Launch Date	Period Mins	Incl-ination	Apogee KM	Perigee KM
1987-080A		18361	US	16 SEP	107.1	90.4	1178	1011
1987-080B		18362	US	16 SEP	107.2	90.4	1180	1010
1987-080C		18363	US	16 SEP	107.2	90.4	1180	1012
1987-080E		18365	US	16 SEP	106.9	90.3	1164	1005
1987-080F		18530	US	16 SEP	106.3	90.4	1116	991
1987-080G		18561	US	16 SEP	107.0	90.4	1161	1009
1987-080H		18562	US	16 SEP	107.8	90.2	1262	986
1987-084A	COSMOS 1888	18384	USSR	01 OCT	1436.2	3.3	35817	35757
1987-084D		18387	USSR	01 OCT	1439.2	3.3	35940	35753
1987-087A	COSMOS 1891	18402	USSR	14 OCT	104.8	82.9	1023	949
1987-087B		18403	USSR	14 OCT	104.6	82.9	1019	934
1987-088A	COSMOS 1892	18421	USSR	20 OCT	96.9	82.5	623	599
1987-088B		18422	USSR	20 OCT	97.4	82.5	648	619
1987-090A		18441	US	26 OCT	NO ELEMENTS AVAILABLE			
1987-091A	COSMOS 1894	18443	USSR	28 OCT	1437.8	3.4	35825	35812
1987-091D		18446	USSR	28 OCT	1435.3	3.3	35886	35655
1987-091F		18448	USSR	28 OCT	599.4	46.8	34204	135
1987-095A	TVSAT 1	18570	FRG	21 NOV	1452.6	3.6	36170	36047
1987-096A	COSMOS 1897	18575	USSR	26 NOV	1436.9	3.2	35836	35767
1987-096D		18578	USSR	26 NOV	1431.9	3.2	35802	35604
1987-097A		18583	US	29 NOV	NO ELEMENTS AVAILABLE			
1987-097B		18584	US	29 NOV	NO ELEMENTS AVAILABLE			
1987-098A	COSMOS 1898	18585	USSR	01 DEC	100.6	74.0	800	770
1987-098B		18586	USSR	01 DEC	100.4	74.0	795	762
1987-098C		18697	USSR	01 DEC	100.3	74.0	777	767
1987-098D		18698	USSR	01 DEC	100.7	74.0	803	777
1987-100A	RADUGA 21	18631	USSR	10 DEC	1436.4	3.3	35796	35786
1987-100D		18634	USSR	10 DEC	1392.6	3.1	34995	34868
1987-100G		21620	USSR	10 DEC	193.4	46.6	9230	167
1987-101A	COSMOS 1900	18665	USSR	12 DEC	99.2	66.1	740	698
1987-105A	COSMOS 1903	18701	USSR	21 DEC	718.6	63.2	37990	2403
1987-105D		18704	USSR	21 DEC	705.1	64.4	37201	2527
1987-106A	COSMOS 1904	18709	USSR	23 DEC	104.8	82.9	1003	964
1987-106B		18710	USSR	23 DEC	104.7	82.9	997	959
1987-109A	EKRAN 17	18715	USSR	27 DEC	1436.5	3.0	35811	35777
1987-109D		18718	USSR	27 DEC	1428.1	2.9	35897	35361
1987-109E		18719	USSR	27 DEC	444.5	46.9	25553	276
1988 LAUNCHES								
1988-001A	COSMOS 1908	18748	USSR	06 JAN	97.0	82.5	630	599
1988-001B		18749	USSR	06 JAN	97.4	82.5	649	617
1988-002A	COSMOS 1909	18788	USSR	15 JAN	114.0	82.6	1410	1408
1988-002B	COSMOS 1910	18789	USSR	15 JAN	113.9	82.6	1408	1104
1988-002C	COSMOS 1911	18790	USSR	15 JAN	113.9	82.6	1408	1397
1988-002D	COSMOS 1912	18791	USSR	15 JAN	113.8	82.6	1408	1392
1988-002E	COSMOS 1913	18792	USSR	15 JAN	113.7	82.6	1408	1387
1988-002F	COSMOS 1914	18793	USSR	15 JAN	113.7	82.6	1408	1381
1988-002G		18794	USSR	15 JAN	114.6	82.6	1469	1408
1988-005A	METEOR 2-17	18820	USSR	30 JAN	103.9	82.5	955	933
1988-005B		18821	USSR	30 JAN	103.9	82.5	953	934
1988-006A		18822	US	03 FEB	101.2	98.5	818	807
1988-006D		18955	US	03 FEB	96.8	98.7	606	605
1988-006F		18984	US	03 FEB	98.9	98.6	708	701
1988-012A	CS-3A	18877	JAPAN	19 FEB	1436.2	0.0	35790	35787
1988-012C		18879	JAPAN	19 FEB	398.9	27.6	22916	227
1988-012D		20760	JAPAN	19 FEB	523.2	27.1	29682	575
1988-013A	COSMOS 1922	18881	USSR	26 FEB	718.8	64.7	37184	3222
1988-013C		18883	USSR	26 FEB	705.8	65.2	36757	3005
1988-014A	PRC 22	18922	PRC	07 MAR	1436.2	0.1	35802	35775
1988-016A	COSMOS 1924	18937	USSR	11 MAR	115.7	74.0	1513	1458
1988-016B	COSMOS 1925	18938	USSR	11 MAR	115.5	74.0	1495	1457
1988-016C	COSMOS 1926	18939	USSR	11 MAR	115.3	74.0	1477	1458
1988-016D	COSMOS 1927	18940	USSR	11 MAR	115.1	74.0	1465	1452
1988-016E	COSMOS 1928	18941	USSR	11 MAR	114.9	74.0	1459	1442
1988-016F	COSMOS 1929	18942	USSR	11 MAR	114.7	74.0	1459	1426
1988-016G	COSMOS 1930	18943	USSR	11 MAR	114.6	74.0	1459	1411
1988-016H	COSMOS 1931	18944	USSR	11 MAR	114.4	74.0	1459	1395
1988-016J		18945	USSR	11 MAR	117.6	74.0	1686	1460
1988-016K		19451	USSR	11 MAR	117.5	74.0	1682	1456
1988-017A	MOLNIYA 1-71	18946	USSR	11 MAR	717.8	63.5	38864	1489
1988-017D		18949	USSR	11 MAR	695.6	63.6	37868	1385
1988-018A	SPACENET 3R	18951	US	11 MAR	1436.2	0.0	35790	35785

International Designation	Satellite Name	Catalog Number	Source	Launch Date	Period Mins	Incl-ination	Apogee KM	Perigee KM
1988-018B	TELECOM 1C	18952	FRANCE	11 MAR	1436.1	0.1	36000	35573
1988-018C		18953	ESA	11 MAR	570.3	7.0	32539	266
1988-019A	COSMOS 1932	18957	USSR	14 MAR	104.4	65.0	992	936
1988-019D		19162	USSR	14 MAR	104.0	65.0	964	934
1988-020A	COSMOS 1933	18958	USSR	15 MAR	97.1	82.5	631	607
1988-020B		18959	USSR	15 MAR	97.4	82.5	644	620
1988-021A	IRS-1A	18960	INDIA	17 MAR	103.1	98.9	913	894
1988-021B		18961	USSR	17 MAR	102.8	98.8	932	847
1988-022A	MOLNIYA 1-72	18980	USSR	17 MAR	717.8	64.8	38341	2014
1988-022D		18983	USSR	17 MAR	731.7	64.9	38985	2054
1988-023A	COSMOS 1934	18985	USSR	22 MAR	104.6	83.0	1006	945
1988-023B		18986	USSR	22 MAR	104.5	83.0	993	946
1988-023C		21912	USSR	22 MAR	104.6	83.0	1003	943
1988-028A	GORIZONT 15	19017	USSR	31 MAR	1472.0	3.0	36633	36341
1988-028D		19020	USSR	31 MAR	1472.7	3.0	36593	36405
1988-028E		19036	USSR	31 MAR	640.4	46.4	36344	122
1988-028F		19037	USSR	31 MAR	621.6	46.4	35148	346
1988-029A	COSMOS 1937	19038	USSR	05 APR	100.5	74.0	798	761
1988-029B		19039	USSR	05 APR	100.3	74.0	796	751
1988-032A	COSMOS 1939	19045	USSR	20 APR	96.5	97.7	605	572
1988-032B		19046	USSR	20 APR	97.1	97.7	657	585
1988-033A		19070	US	26 APR	108.5	90.3	1302	1013
1988-033B		19071	US	26 APR	108.5	90.3	1300	1012
1988-033C		19072	US	26 APR	108.5	90.3	1303	1013
1988-033D		19077	US	26 APR	108.0	90.3	1272	999
1988-033E		19078	US	26 APR	107.6	90.6	1236	994
1988-033F		19140	US	26 APR	107.9	90.3	1258	997
1988-033G		19181	US	26 APR	109.1	90.1	1376	993
1988-034A	COSMOS 1940	19073	USSR	26 APR	1430.5	3.0	35779	35574
1988-034D		19076	USSR	26 APR	1438.7	2.9	35949	35727
1988-034E		19082	USSR	26 APR	639.3	48.6	36022	387
1988-034F		19083	USSR	26 APR	649.6	47.3	36714	223
1988-036A	EKRAN 18	19090	USSR	06 MAY	1513.5	3.9	37346	37229
1988-036E		19094	USSR	06 MAY	1424.1	3.7	35653	35447
1988-039A	COSMOS 1943	19119	USSR	15 MAY	101.8	71.0	854	835
1988-039B		19120	USSR	15 MAY	101.5	71.0	848	813
1988-039C		19125	USSR	15 MAY	104.6	71.0	1109	837
1988-039D		19126	USSR	15 MAY	104.7	71.0	1118	839
1988-039E		19127	USSR	15 MAY	105.1	71.0	1156	839
1988-039F		19128	USSR	15 MAY	105.0	71.0	1152	840
1988-040A	INTELSAT 5A F-13	19121	ITSO	17 MAY	1436.2	0.1	35798	35779
1988-040B		19122	ESA	17 MAY	634.0	7.3	35637	501
1988-043A	COSMOS 1946	19163	USSR	21 MAY	675.7	64.9	19146	19112
1988-043B	COSMOS 1947	19164	USSR	21 MAY	675.7	64.9	19139	19119
1988-043C	COSMOS 1948	19165	USSR	21 MAY	675.7	64.9	19139	19119
1988-043F		19168	USSR	21 MAY	674.5	65.0	19109	19087
1988-043G		19169	USSR	21 MAY	339.8	65.6	18789	715
1988-043H		19170	USSR	21 MAY	339.9	65.4	18758	751
1988-044A	MOLNIYA 3-32	19189	USSR	26 MAY	714.3	64.6	38267	1917
1988-044B		19190	USSR	26 MAY	733.0	64.8	38963	2137
1988-046A	COSMOS 1950	19195	USSR	30 MAY	116.0	73.6	1519	1482
1988-046B		19196	USSR	30 MAY	116.0	73.6	1514	1482
1988-050A	COSMOS 1953	19210	USSR	14 JUN	97.2	82.5	637	607
1988-050B		19211	USSR	14 JUN	97.4	82.5	649	618
1988-051A	METEOSAT	19215	ESA	15 JUN	1436.3	0.6	35796	35783
1988-051B	OSCAR 13	19216	US	15 JUN	686.6	57.5	38130	674
1988-051C	PAS-1	19217	US	15 JUN	1436.2	0.0	35799	35777
1988-051D		19218	ESA	15 JUN	350.7	10.2	20012	178
1988-051E		19219	ESA	15 JUN	598.6	10.1	34002	296
1988-051F		19220	ESA	15 JUN	435.0	9.9	24996	279
1988-051G		19857	ESA	15 JUN	631.8	7.0	35324	697
1988-051H		19951	ESA	15 JUN	632.3	8.0	35209	842
1988-052A		19223	US	16 JUN	108.9	90.0	1199	1149
1988-053A	COSMOS 1954	19256	USSR	21 JUN	100.6	74.0	797	772
1988-053B		19257	USSR	21 JUN	100.4	74.0	794	760
1988-053C		19260	USSR	21 JUN	100.4	74.1	779	771
1988-053D		19261	USSR	21 JUN	100.4	74.1	784	769
1988-056A	OKEAN 1	19274	USSR	05 JUL	97.1	82.5	634	607
1988-056B		19275	USSR	05 JUL	97.4	82.5	650	619
1988-058A	PHOBOS 1	19281	USSR	07 JUL	MARS ORBIT			
1988-058B		19282	USSR	07 JUL	HELIOCENTRIC ORBIT			
1988-059A	PHOBOS 2	19287	USSR	12 JUL	MARS ORBIT			
1988-059B		19288	USSR	12 JUL	HELIOCENTRIC ORBIT			

International Designation	Satellite Name	Catalog Number	Source	Launch Date	Period Mins	Incl-ination	Apogee KM	Perigee KM
1988-062A	COSMOS 1959	19324	USSR	18 JUL	104.6	83.0	1003	951
1988-062B		19325	USSR	18 JUL	104.5	83.0	995	950
1988-063A	INSAT 1C	19330	INDIA	21 JUL	1436.3	2.9	35823	35758
1988-063B	ECS 5	19331	ESA	21 JUL	1436.1	0.0	35835	35737
1988-063C		19332	ESA	21 JUL	449.7	7.5	25838	293
1988-063E		20127	ESA	21 JUL	630.3	7.3	35497	450
1988-063F		20488	ESA	21 JUL	307.9	7.4	17084	370
1988-063G		22101	INDIA	21 JUL	288.9	7.8	15801	393
1988-064A	METEOR 3-2	19336	USSR	26 JUL	109.3	82.5	1206	1180
1988-064B		19337	USSR	26 JUL	109.3	82.5	1204	1182
1988-065N		20380	USSR	28 JUL	91.0	65.8	323	319
1988-065ST		20378	USSR	28 JUL	90.8	65 8	318	308
1988-066A	COSMOS 1961	19344	USSR	01 AUG	1436.2	2.5	35801	35774
1988-066D		19347	USSR	01 AUG	1459.7	2.6	36392	36100
1988-066E		19348	USSR	01 AUG	421.5	46.8	24237	252
1988-069A	MOLNIYA 1-73	19377	USSR	12 AUG	717.7	64.9	38897	1454
1988-069D		19380	USSR	12 AUG	730.8	65.1	39604	1392
1988-071A	GORIZONT 16	19397	USSR	18 AUG	1440.3	2.5	35919	35819
1988-071D		19400	USSR	18 AUG	1432.0	2.5	35789	35624
1988-071E		19401	USSR	18 AUG	600.1	46.7	34244	131
1988-071F		19402	USSR	18 AUG	251.8	46.8	13480	178
1988-074A		19419	US	25 AUG	107.3	89.9	1175	1030
1988-074B		19420	US	25 AUG	107.3	89.9	1173	1031
1988-074C		19421	US	25 AUG	107.4	89.9	1176	1031
1988-074D		19515	US	25 AUG	107.2	89 9	1169	1021
1988-074E		19516	US	25 AUG	107.1	89 9	1162	1023
1988-074F		19559	US	25 AUG	107.2	89.4	1167	1026
1988-074G		19577	US	25 AUG	107.2	90.5	1167	1029
1988-076A	COSMOS 1966	19445	USSR	30 AUG	717.8	67.0	38225	2127
1988-076D		19448	USSR	30 AUG	705.5	67.2	37701	2047
1988-077A		19458	US	02 SEP	NO ELEMENTS AVAILABLE			
1988-077B		19459	US	02 SEP	NO ELEMENTS AVAILABLE			
1988-077C		19490	US	02 SEP	NO ELEMENTS AVAILABLE			
1988-078A		19460	US	05 SEP	NO ELEMENTS AVAILABLE			
1988-078B		19461	US	05 SEP	NO ELEMENTS AVAILABLE			
1988-080A	FENGYUN 1	19467	PRC	06 SEP	102.7	99.3	937	833
1988-080B		19468	PRC	06 SEP	102.7	99.3	895	873
1988-081A	GSTAR 3	19483	US	08 SEP	1436.2	4.7	35793	35782
1988-081B	SBS 5	19484	US	08 SEP	1436.1	0.0	35796	35777
1988-081C		19485	ESA	08 SEP	420.9	7.5	24181	268
1988-085A	COSMOS 1970	19501	USSR	16 SEP	675.7	65.7	19157	19101
1988-085B	COSMOS 1971	19502	USSR	16 SEP	675.7	65.7	19158	19100
1988-085C	COSMOS 1972	19503	USSR	16 SEP	675.7	65.7	19155	19103
1988-085E		19505	USSR	16 SEP	674.9	65.7	19115	19101
1988-085F		19535	USSR	16 SEP	339.2	65.5	18736	730
1988-085G		19537	USSR	16 SEP	339.2	65.3	18717	750
1988-085H		21751	USSR	16 SEP	214.2	64.4	10070	890
1988-086A	CS-3B	19508	JAPAN	16 SEP	1436.2	0.0	35791	35786
1988-086C		19558	JAPAN	16 SEP	629.3	27.9	35800	95
1988-089A	NOAA 11	19531	US	24 SEP	101.9	99.1	855	838
1988-089B		19532	US	24 SEP	98.1	98.9	671	665
1988-090A	MOLNIYA 3-33	19541	USSR	29 SEP	717.1	64.9	38797	1521
1988-090D		19544	USSR	29 SEP	698.1	64.8	37839	1541
1988-091B	TDRS 3	19548	US	29 SEP	1436.2	0.1	35804	35772
1988-091C		19549	US	29 SEP	603.2	26.2	34283	257
1988-091D		19550	US	29 SEP	1433.1	1.7	35799	35657
1988-092A	COSMOS 1974	19554	USSR	03 OCT	718.1	63.6	37698	2672
1988-092D		19557	USSR	03 OCT	705.5	63.7	37140	2604
1988-093A	COSMOS 1975	19573	USSR	11 OCT	97.1	82.5	636	606
1988-093B		19574	USSR	11 OCT	97.4	82.5	649	617
1988-093C		20471	USSR	11 OCT	96.1	82.5	585	561
1988-095A	RADUGA 22	19596	USSR	20 OCT	1436.2	2.3	35793	35785
1988-095D		19600	USSR	20 OCT	602.6	46.6	34363	144
1988-095E		19601	USSR	20 OCT	545.2	46.6	31321	138
1988-095F		19777	USSR	20 OCT	1470.3	2.4	36511	36396
1988-096A	COSMOS 1977	19608	USSR	25 OCT	718.9	63.9	37421	2987
1988-096D		19611	USSR	25 OCT	704.9	64.3	36873	2845
1988-098A	TDF-1	19621	FRANCE	28 OCT	1436.2	0.1	35799	35776
1988-098B		19622	ESA	28 OCT	567.2	4.2	32322	315
1988-098C		20132	ESA	26 OCT	347.2	3.9	19736	234
1988-099A		19625	US	06 NOV	NO ELEMENTS AVAILABLE			
1988-099B		19626	US	06 NOV	NO ELEMENTS AVAILABLE			
1988-102A	COSMOS 1980	19649	USSR	23 NOV	101.9	71.0	851	842

International Designation	Satellite Name	Catalog Number	Source	Launch Date	Period Mins	Incl-ination	Apogee KM	Perigee KM
1988-102B		19650	USSR	23 NOV	101.7	71.0	850	831
1988-102C		19656	USSR	23 NOV	105.1	71.0	1159	841
1988-102D		19657	USSR	23 NOV	105.1	71.0	1157	840
1988-102E		19658	USSR	23 NOV	104.9	71.0	1136	842
1988-102F		19659	USSR	23 NOV	104.7	71.0	1119	842
1988-102H		19813	USSR	23 NOV	105.1	71.0	1162	839
1988-102J		20301	USSR	23 NOV	101.9	71.0	860	833
1988-106B		19671	US	02 DEC	NO ELEMENTS AVAILABLE			
1988-108A	EKRAN 19	19683	USSR	08 DEC	1436.2	2.2	35801	35776
1988-108D		19686	USSR	08 DEC	1418.5	2.2	35509	35373
1988-109A	SKYNET 4B	19687	UK	11 DEC	1436.1	0.6	35855	35716
1988-109B	ASTRA 1A	19688	LUXBRG	11 DEC	1436.1	0.0	35821	35753
1988-109C		19689	ESA	11 DEC	638.6	7.3	35961	413
1988-109D		19690	ESA	11 DEC	314.5	7.1	17690	190
1988-111A	PRC 25	19710	PRC	22 DEC	1436.2	0.1	35790	35786
1988-112A	MOLNIYA 3-34	19713	USSR	22 DEC	717.7	63.5	39365	983
1988-112D		19716	USSR	22 DEC	696.1	63 5	38402	877
1988-113H		19764	USSR	23 DEC	94.1	73 5	477	470
1988-115A	MOLNIYA 1-74	19730	USSR	28 DEC	717.5	64.9	39365	975
1988-115D		19733	USSR	28 DEC	695.7	64.9	38235	1023

1989 LAUNCHES

International Designation	Satellite Name	Catalog Number	Source	Launch Date	Period Mins	Incl-ination	Apogee KM	Perigee KM
1989-001A	COSMOS 1987	19749	USSR	10 JAN	675.7	64.9	19143	19115
1989-001B	COSMOS 1988	19750	USSR	10 JAN	675.7	64.9	19145	19113
1989-001C	COSMOS 1989	19751	USSR	10 JAN	675.5	64.9	19151	19099
1989-001E		19753	USSR	10 JAN	675.5	64.9	19150	19098
1989-001F		19754	USSR	10 JAN	674.7	64.9	19105	19099
1989-001G		19755	USSR	10 JAN	339.6	65 4	18749	743
1989-001H		19856	USSR	10 JAN	339.6	65 5	18769	721
1989-004A	GORIZONT 17	19765	USSR	26 JAN	1436.0	2.1	35793	35775
1989-004E		19771	USSR	26 JAN	310.9	46.7	17451	198
1989-004F		19776	USSR	26 JAN	1469.5	2.1	36545	36329
1989-005A	COSMOS 1992	19769	USSR	26 JAN	100.5	74.1	799	764
1989-005B		19770	USSR	26 JAN	100.3	74.0	777	769
1989-005C		19831	USSR	26 JAN	100.3	74.1	787	757
1989-005D		19945	USSR	26 JAN	100.6	74.1	806	764
1989-006A	INTELSAT 5A F-15	19772	ITSO	27 JAN	1436.2	0.1	35802	35772
1989-006B		19773	ESA	27 JAN	636.9	8.1	35772	513
1989-009A	COSMOS 1994	19785	USSR	10 FEB	113.9	82.6	1414	1392
1989-009B	COSMOS 1995	19786	USSR	10 FEB	114.1	82.6	1414	1410
1989-009C	COSMOS 1996	19787	USSR	10 FEB	114.0	82 6	1414	1404
1989-009D	COSMOS 1997	19788	USSR	10 FEB	113.9	82 6	1414	1397
1989-009E	COSMOS 1998	19789	USSR	10 FEB	113.8	82.6	1414	1387
1989-009F	COSMOS 1999	19790	USSR	10 FEB	113.7	82.6	1414	1381
1989-009G		19791	USSR	10 FEB	114.7	82.6	1468	1414
1989-011A	COSMOS 2001	19796	USSR	14 FEB	717.8	66 1	38185	2167
1989-011D		19799	USSR	14 FEB	705.7	66 6	37688	2070
1989-013A		19802	US	14 FEB	718.0	55.1	20291	20071
1989-014A	MOLNIYA 1-75	19807	USSR	15 FEB	717.9	63.4	38207	2151
1989-014D		19810	USSR	15 FEB	694.4	63 4	37159	2036
1989-016A	EXOS-D	19822	JAPAN	21 FEB	188.8	75 1	8779	270
1989-016C		19824	JAPAN	21 FEB	175.0	75.1	7703	267
1989-016K		19952	JAPAN	21 FEB	144.2	75.6	5207	255
1989-016M		19963	JAPAN	21 FEB	167.3	75.1	7086	271
1989-016N		20021	JAPAN	21 FEB	154.6	74.4	6064	264
1989-016P		20034	JAPAN	21 FEB	89.8	74.8	364	164
1989-017A	COSMOS 2004	19826	USSR	22 FEB	104.9	83.0	1014	969
1989-017B		19827	USSR	22 FEB	104.8	83.0	1007	963
1989-018A	METEOR 2-18	19851	USSR	28 FEB	104.0	82.5	955	935
1989-018B		19852	USSR	28 FEB	104.0	82.5	959	935
1989-020A	JCSAT-1	19874	JAPAN	06 MAR	1436.2	0.0	35794	35782
1989-020B	MOP-1	19876	ESA	06 MAR	1436.2	0.2	35793	35782
1989-020E		20800	UK	06 MAR	1433.8	1.9	36271	35211
1989-021B	TDRS-D	19883	US	13 MAR	1436.2	0.1	35799	35776
1989-021C		19884	US	13 MAR	544.0	26.8	31152	238
1989-021D		19913	US	13 MAR	1431.1	5.2	35798	35580
1989-025A	COSMOS 2008	19902	USSR	24 MAR	114.4	74.0	1469	1391
1989-025B	COSMOS 2009	19903	USSR	24 MAR	114.6	74.0	1469	1407
1989-025C	COSMOS 2010	19904	USSR	24 MAR	114.8	74.0	1470	1421
1989-025D	COSMOS 2011	19905	USSR	24 MAR	115.0	74.0	1470	1436
1989-025E	COSMOS 2012	19906	USSR	24 MAR	115.1	74.0	1470	1453
1989-025F	COSMOS 2013	19907	USSR	24 MAR	115.3	74.0	1478	1462

International Designation	Satellite Name	Catalog Number	Source	Launch Date	Period Mins	Incl-ination	Apogee KM	Perigee KM
1989-025G	COSMOS 2014	19908	USSR	24 MAR	115.5	74.0	1488	1468
1989-025H	COSMOS 2015	19909	USSR	24 MAR	115.7	74.0	1507	1467
1989-025J		19910	USSR	24 MAR	117.7	74.0	1682	1472
1989-027A	TELE-X	19919	SWEDEN	02 APR	1436.1	0.0	35823	35750
1989-027B		19920	ESA	02 APR	313.3	4.0	17399	402
1989-028A	COSMOS 2016	19921	USSR	04 APR	104.7	83.0	1011	950
1989-028B		19922	USSR	04 APR	104.6	83.0	1001	947
1989-030A	RADUGA 23	19928	USSR	14 APR	1436.1	1.9	35789	35783
1989-030D		19931	USSR	14 APR	1470.6	2.0	36539	36377
1989-030F		19933	USSR	14 APR	597.5	46.8	34087	154
1989-033B	MAGELLAN	19969	US	04 MAY	VENUS ORBIT			
1989-033C		19970	US	04 MAY	426.0	27.9	24471	280
1989-033D		19971	US	04 MAY	VENUS ORBIT			
1989-035A		19976	US	10 MAY	NO ELEMENTS AVAILABLE			
1989-035B		19977	US	10 MAY	NO ELEMENTS AVAILABLE			
1989-035C		19983	US	10 MAY	NO ELEMENTS AVAILABLE			
1989-039A	COSMOS 2022	20024	USSR	31 MAY	675.7	65.5	19142	19116
1989-039B	COSMOS 2023	20025	USSR	31 MAY	675.7	65.5	19170	19088
1989-039C	COSMOS 2024	20026	USSR	31 MAY	675.4	65.5	19146	19096
1989-039E		20028	USSR	31 MAY	674.5	65.5	19142	19055
1989-039F		20044	USSR	31 MAY	675.4	65.5	19151	19090
1989-039G		20081	USSR	31 MAY	339.4	65.2	18841	636
1989-039H		20082	USSR	31 MAY	339.4	65.3	18816	662
1989-041A	SUPERBIRD A	20040	JAPAN	05 JUN	1443.8	1.9	35952	35922
1989-041B		20041	FRG	05 JUN	1436.1	0.0	35851	35720
1989-041C		20042	ESA	05 JUN	437.6	6.4	25173	254
1989-042A	COSMOS 2026	20045	USSR	07 JUN	104.6	82.9	1006	948
1989-042B		20046	USSR	07 JUN	104.5	82.9	997	946
1989-043A	MOLNIYA 3-35	20052	USSR	08 JUN	717.7	64.8	39325	1023
1989-043D		20055	USSR	08 JUN	733.3	65.0	40127	989
1989-044A		20061	US	10 JUN	718.0	54.9	20478	19886
1989-046A		20066	US	14 JUN	NO ELEMENTS AVAILABLE			
1989-046B		20067	US	14 JUN	NO ELEMENTS AVAILABLE			
1989-046C		20068	US	14 JUN	NO ELEMENTS AVAILABLE			
1989-046D		20069	US	14 JUN	NO ELEMENTS AVAILABLE			
1989-046E		20319	US	14 JUN	NO ELEMENTS AVAILABLE			
1989-048A	RADUGA 1-1	20083	USSR	21 JUN	1436.2	1.7	35798	35777
1989-048D		20086	USSR	21 JUN	1471.0	1.8	36572	36362
1989-048F		20094	USSR	21 JUN	453.8	46.5	26182	186
1989-050A	NADEZHDA	20103	USSR	04 JUL	104.8	83.0	1011	954
1989-050B		20104	USSR	04 JUL	104.6	83.0	1003	952
1989-052A	GORIZONT 18	20107	USSR	05 JUL	1436.2	1.6	35796	35780
1989-052D		20110	USSR	05 JUL	1397.3	1.5	35152	34894
1989-052F		20116	USSR	05 JUL	536.9	46.8	30749	257
1989-053A	OLYMPUS	20122	ESA	12 JUL	1436.2	0.7	35796	35779
1989-053B		20123	ESA	12 JUL	367.1	6.5	20971	237
1989-053C		20229	ESA	12 JUN	636.9	6.7	35862	425
1989-059A	COSMOS 2034	20149	USSR	25 JUL	104.8	82.9	1011	962
1989-059B		20150	USSR	25 JUL	104.7	82.9	1001	957
1989-061B		20167	US	08 AUG	NO ELEMENTS AVAILABLE			
1989-061C		20172	US	08 AUG	NO ELEMENTS AVAILABLE			
1989-061D		20344	US	08 AUG	NO ELEMENTS AVAILABLE			
1989-061E		22263	US	08 AUG	NO CURRENT ELEMENTS			
1989-061F		22264	US	08 AUG	NO CURRENT ELEMENTS			
1989-061G		22265	US	08 AUG	NO CURRENT ELEMENTS			
1989-061H		22267	US	08 AUG	NO CURRENT ELEMENTS			
1989-061J		22268	US	08 AUG	NO CURRENT ELEMENTS			
1989-062A	TV-SAT 2	20168	FRG	08 AUG	1436.2	0.0	35814	35761
1989-062B	HIPPARCOS	20169	ESA	08 AUG	638.7	7.2	35861	518
1989-062C		20170	ESA	08 AUG	621.7	7.7	35142	360
1989-064A		20185	US	18 AUG	717.9	54.9	20207	20154
1989-067A	BSB-R1	20193	UK	27 AUG	1436.2	0.0	35793	35783
1989-067C		20195	US	27 AUG	644.7	23.3	36412	272
1989-068A	COSMOS 2037	20196	USSR	28 AUG	116.0	73.6	1522	1482
1989-068B		20197	USSR	28 AUG	116.0	73.6	1520	1482
1989-069A		20202	US	04 SEP	NO ELEMENTS AVAILABLE			
1989-069B		20203	US	04 SEP	NO ELEMENTS AVAILABLE			
1989-069D		20205	US	04 SEP	NO ELEMENTS AVAILABLE			
1989-070A	GMS-4	20217	JAPAN	05 SEP	1436.2	0.3	35791	35784
1989-070B		20230	JAPAN	05 SEP	482.5	28.0	27819	179
1989-070C		20317	JAPAN	05 SEP	1458.1	0.9	37197	35234
1989-072A		20220	US	06 SEP	NO ELEMENTS AVAILABLE			
1989-072B		20221	US	06 SEP	NO ELEMENTS AVAILABLE			

International Designation	Satellite Name	Catalog Number	Source	Launch Date	Period Mins	Incl-ination	Apogee KM	Perigee KM
1989-074A	COSMOS 2038	20232	USSR	14 SEP	113.8	82.6	1408	1389
1989-074B	COSMOS 2039	20233	USSR	14 SEP	113.7	82.6	1408	1383
1989-074C	COSMOS 2040	20234	USSR	14 SEP	114.0	82.6	1413	1408
1989-074D	COSMOS 2041	20235	USSR	14 SEP	113.8	82.6	1408	1394
1989-074E	COSMOS 2042	20236	USSR	14 SEP	113.9	82.6	1408	1399
1989-074F	COSMOS 2043	20237	USSR	14 SEP	113.9	82.6	1408	1406
1989-074G		20238	USSR	14 SEP	114.7	82.6	1472	1407
1989-077A		20253	US	25 SEP	1436.1	2.9	35801	35774
1989-078A	MOLNIYA 1-76	20255	USSR	27 SEP	717.7	64.4	39751	598
1989-078D		20258	USSR	27 SEP	698.3	64.5	38761	627
1989-080A	INTER-COSMOS 24	20261	USSR	28 SEP	115.4	82.6	2453	497
1989-080B		20281	USSR	28 SEP	115.4	82.6	2446	496
1989-080C		20262	USSR	28 SEP	115.6	82.6	2472	496
1989-081A	GORIZANT 19	20263	USSR	28 SEP	1436.1	1.5	35795	35779
1989-081D		20266	USSR	28 SEP	1431.3	1.2	36032	35354
1989-084B	GALILEO	20298	US	18 OCT	HELIOCENTRIC ORBIT			
1989-084C		20299	US	18 OCT	446.1	34.4	25703	218
1989-084D		20300	US	18 OCT	HELIOCENTRIC ORBIT			
1989-085A		20302	US	21 OCT	718.0	53.9	20218	20144
1989-085B		20303	US	21 OCT	98.6	35.6	899	488
1989-086A	METEOR 3-3	20305	USSR	24 OCT	109.4	82.6	1208	1186
1989-086B		20306	USSR	24 OCT	109.4	82.6	1208	1185
1989-087A	INTELSAT 6A	20315	ITSO	27 OCT	1436.1	0.0	35809	35764
1989-087B		20316	ESA	27 OCT	589.6	7.4	33531	293
1989-089A	COBE	20322	US	18 NOV	102.6	99.0	885	873
1989-089B		20323	US	18 NOV	99.8	97.1	803	690
1989-089C		20324	US	18 NOV	102.3	99.0	879	860
1989-089D		20328	US	18 NOV	102.7	99.0	885	884
1989-090B		20355	US	23 NOV	NO ELEMENTS AVAILABLE			
1989-090C		20356	US	23 NOV	NO ELEMENTS AVAILABLE			
1989-090D		20357	US	23 NOV	NO ELEMENTS AVAILABLE			
1989-091A	COSMOS 2050	20330	USSR	23 NOV	717.8	62.7	38646	1710
1989-091D		20333	USSR	23 NOV	705.2	63.5	37973	1756
1989-093A	KVANT -2	20335	USSR	26 NOV	92.5	51.6	395	394
1989-094A	MOLNIYA 3-36	20338	USSR	28 NOV	717.6	64.4	39806	538
1989-094B		20339	USSR	28 NOV	732.1	64.5	40518	541
1989-096A	GRANAT	20352	USSR	01 DEC	5900.7	86.7	160909	42670
1989-096C		20354	USSR	01 DEC	5775.9	86.2	159600	40916
1989-097A		20361	US	11 DEC	718.0	55.2	20360	20003
1989-097B		20362	US	11 DEC	98.5	35.6	885	484
1989-098A	RADUGA 24	20367	USSR	15 DEC	1435.9	1.2	35793	35770
1989-098D		20370	USSR	15 DEC	1471.6	1.3	36559	36398
1989-100A	COSMOS 2053	20389	USSR	27 DEC	93.2	73.5	436	426
1989-100B		20390	USSR	27 DEC	94.5	73.5	501	482
1989-101A	COSMOS 2054	20391	USSR	27 DEC	1436.1	1.2	35792	35782
1989-101D		20394	USSR	27 DEC	1465.7	1.2	36411	36316
1989-101E		20399	USSR	27 DEC	477.7	46.8	27480	247
1989-101G		21648	USSR	27 DEC	NO CURRENT ELEMENTS			

1990 LAUNCHES

International Designation	Satellite Name	Catalog Number	Source	Launch Date	Period Mins	Incl-ination	Apogee KM	Perigee KM
1990-001A	SKYNET 4A	20401	UK	01 JAN	1436.2	1.8	35796	35780
1990-001B	JCSAT	20402	JAPAN	01 JAN	1436.2	0.0	35794	35781
1990-001D		20404	US	01 JAN	605.1	21.3	34347	290
1990-001F		20406	US	01 JAN	328.3	26.6	18477	295
1990-002B	LEASAT 5	20410	US	09 JAN	1436.2	2.7	35815	35759
1990-002C		20411	US	09 JAN	266.6	27.2	14346	337
1990-004A	COSMOS 2056	20432	USSR	18 JAN	100.6	74.0	804	769
1990-004B		20433	USSR	18 JAN	100.5	74.0	807	755
1990-004C		20434	USSR	18 JAN	100.8	74.0	809	779
1990-004D		20435	USSR	18 JAN	100.2	74.0	785	754
1990-005A	SPOT-2	20436	FRANCE	22 JAN	101.3	98.7	822	822
1990-005B	OSCAR 14	20437	UK	22 JAN	100.7	98.6	797	782
1990-005C	OSCAR 15	20438	UK	22 JAN	100.7	98.6	799	784
1990-005D	OSCAR 16	20439	US	22 JAN	100.7	98.6	797	781
1990-005E	OSCAR 17	20440	BRAZIL	22 JAN	100.6	98.6	797	781
1990-005F	OSCAR 18	20441	US	22 JAN	100.6	98.6	797	780
1990-005G	OSCAR 19	20442	ARGNT	22 JAN	100.6	98.6	797	780
1990-005H		20443	ESA	22 JAN	100.5	98.5	789	774
1990-006A	MOLNIYA 3	20444	USSR	23 JAN	717.7	64.8	39715	635
1990-006C		20446	USSR	23 JAN	696.7	64.7	38626	681
1990-007A	MUSES A	20448	JAPAN	24 JAN	NO ELEMENTS AVAILABLE			
1990-0078	HAGOROMO	20618	JAPAN	24 JAN	SELENOCENTRIC ORBIT			

International Designation	Satellite Name	Catalog Number	Source	Launch Date	Period Mins	Incl-ination	Apogee KM	Perigee KM
1990-007D		20451	JAPAN	24 JAN	NO ELEMENTS AVAILABLE			
1990-008A		20452	US	24 JAN	717.9	54.2	20313	20049
1990-008B		20453	US	24 JAN	101.6	35.6	1217	446
1990-008C		20450	US	24 JAN	176.9	37.6	7934	184
1990-010A	COSMOS 2058	20465	USSR	30 JAN	97.4	82.5	648	616
1990-010B		20466	USSR	30 JAN	97.5	82.5	655	621
1990-011A	PRC-26	20473	PRC	04 FEB	1436.2	0.0	35791	35785
1990-011B		20474	PRC	04 FEB	597.6	30.2	33937	306
1990-012C		20481	USSR	06 FEB	NO CURRENT ELEMENTS			
1990-013A	MOS 1B	20478	JAPAN	07 FEB	103.2	99.1	909	908
1990-013B	DEBUT	20479	JAPAN	07 FEB	112.2	99.1	1742	908
1990-013C	JAS 1-B	20480	JAPAN	07 FE8	112.2	99.1	1743	908
1990-013D		20491	JAPAN	07 FEB	110.5	99.1	1606	889
1990-015A		20496	US	14 FEB	94.2	43.1	492	469
1990-016A	RADUGA 25	20499	USSR	15 FEB	1436.1	1.1	35795	35776
1990-016D		20502	USSR	15 FEB	1439.8	1.1	36012	35706
1990-017A	NADEZHDA-2	20508	USSR	27 FEB	104.8	83.0	1016	952
1990-017B		20509	USSR	27 FEB	104.7	83.0	1010	949
1990-018A	OKEAN-2	20510	USSR	28 FEB	97.4	82.5	647	618
1990-018B		20511	USSR	28 FEB	97.6	82.5	656	628
1990-019B		20516	US	28 FEB	NO ELEMENTS AVAILABLE			
1990-019C		20517	US	28 FEB	NO ELEMENTS AVAILABLE			
1990-019D		20518	US	28 FEB	NO ELEMENTS AVAILABLE			
1990-019E		20519	US	28 FEB	NO ELEMENTS AVAILABLE			
1990-019F		20520	US	28 FEB	NO ELEMENTS AVAILABLE			
1990-019G		20521	US	28 FEB	NO ELEMENTS AVAILABLE			
1990-021A	INTELSAT-6	20523	ITSO	14 MAR	1436.2	0.1	35792	35784
1990-023A	COSMOS 2061	20527	USSR	20 MAR	104.9	82.9	1014	968
1990-023B		20528	USSR	20 MAR	104.8	82.9	1007	965
1990-025A		20533	US	26 MAR	717.9	55.2	20281	20080
1990-025C		20535	US	26 MAR	142.3	37.4	5121	179
1990-026A	COSMOS 2063	20536	USSR	27 MAR	717.5	63.8	39058	1281
1990-026D		20539	USSR	27 MAR	709.3	64.7	38600	1334
1990-028A	PEGSAT	20546	US	05 APR	94.1	94.1	539	410
1990-028B		20547	US	05 APR	95.9	94.1	643	481
1990-029A	COSMOS 2064	20549	USSR	06 APR	115.4	74.0	1488	1460
1990-029B	COSMOS 2065	20550	USSR	06 APR	115.2	74.0	1473	1459
1990-029C	COSMOS 2066	20551	USSR	06 APR	114.3	74.0	1461	1384
1990-029D	COSMOS 2067	20552	USSR	06 APR	114.4	74.0	1461	1398
1990-029E	COSMOS 2068	20553	USSR	06 APR	114.6	74.0	1460	1413
1990-029F	COSMOS 2069	20554	USSR	06 APR	114.8	74.0	1460	1427
1990-029G	COSMOS 2070	20555	USSR	06 APR	114.9	74.0	1461	1441
1990-029H	COSMOS 2071	20556	USSR	06 APR	115.1	74.0	1460	1457
1990-029J		20557	USSR	06 APR	117.7	74.0	1696	1460
1990-030A	ASIASAT 1	20558	UK	07 APR	1436.2	0.1	35789	35785
1990-030B		20559	PRC	07 APR	583.2	31.1	33229	258
1990-031A		20560	US	11 APR	NO ELEMENTS AVAILABLE			
1990-031B		20561	US	11 APR	NO ELEMENTS AVAILABLE			
1990-031C		20562	US	11 APR	NO ELEMENTS AVAILABLE			
1990-031D		20563	US	11 APR	NO ELEMENTS AVAILABLE			
1990-031E		20564	US	11 APR	NO ELEMENTS AVAILABLE			
1990-031F		20565	US	11 APR	NO ELEMENTS AVAILABLE			
1990-031G		20575	US	11 APR	NO ELEMENTS AVAILABLE			
1990-031H		20576	US	11 APR	NO ELEMENTS AVAILABLE			
1990-034A	PALAPA B2R	20570	INDO	13 APR	1436.2	0.0	35792	35785
1990-034B		20571	US	13 APR	103.8	22.7	1376	497
1990-034C		20572	US	13 APR	358.0	18.6	20454	189
1990-036A	COSMOS 2074	20577	USSR	20 APR	104.7	82.9	1003	961
1990-036B		20578	USSR	20 APR	104.6	82.9	992	962
1990-037B	HST	20580	US	24 APR	96.6	28.5	598	591
1990-039A	MOLNIYA 1-77	20583	USSR	26 APR	717.7	63.8	39606	743
1990-039D		20586	USSR	26 APR	733.1	63.8	40336	770
1990-040A	COSMOS 2076	20596	USSR	28 APR	718.7	63.0	39146	1253
1990-040D		20599	USSR	28 APR	707.7	63.7	38523	1330
1990-043A	SCOUT M-1	20607	US	09 MAY	98.3	89.9	755	601
1990-043B		20608	US	09 MAY	98.3	89.9	752	600
1990-043C		20609	US	09 MAY	97.9	89.9	731	588
1990-043D		20610	US	09 MAY	97.4	89.9	698	567
1990-043E		20611	US	09 MAY	97.2	89.9	688	562
1990-043F		20612	US	09 MAY	97.2	89.9	677	569
1990-043H		20614	US	09 MAY	97.1	89.9	673	567
1990-043K		20651	US	09 MAY	97.9	90.2	758	556
1990-043L		20759	US	09 MAY	96.2	89.7	603	545

International Designation	Satellite Name	Catalog Number	Source	Launch Date	Period Mins	Incl- ination	Apogee KM	Perigee KM
1990-045A	COSMOS 2079	20619	USSR	19 MAY	675.7	65.3	19195	19063
1990-045B	COSMOS 2080	20620	USSR	19 MAY	675.7	65.3	19144	19113
1990-045C	COSMOS 2081	20621	USSR	19 MAY	675.7	65.3	19168	19090
1990-045E		20623	USSR	19 MAY	674.7	65.3	19134	19074
1990-045F		20630	USSR	19 MAY	339.6	65.1	18891	602
1990-045G		20631	USSR	19 MAY	339.4	65.0	18877	604
1990-046A	COSMOS 2082	20624	USSR	22 MAY	101.9	71.0	853	841
1990-046B		20625	USSR	22 MAY	101.8	71.0	855	834
1990-046C		20626	USSR	22 MAY	105.1	71.0	1153	840
1990-046D		20627	USSR	22 MAY	105.2	71.0	1165	841
1990-046E		20628	USSR	22 MAY	105.1	71.0	1156	840
1990-046F		20629	USSR	22 MAY	104.9	71.0	1141	842
1990-048A	KRISTALL	20635	USSR	31 MAY	92.5	51.6	395	394
1990-049A	ROSAT	20638	FRG	01 JUN	95.6	53.0	557	542
1990-050A		20641	US	08 JUN	NO ELEMENTS AVAILABLE			
1990-050B		20682	US	08 JUN	NO ELEMENTS AVAILABLE			
1990-050C		20691	US	08 JUN	NO ELEMENTS AVAILABLE			
1990-050D		20692	US	08 JUN	NO ELEMENTS AVAILABLE			
1990-050E		20642	US	08 JUN	NO ELEMENTS AVAILABLE			
1990-050F		21916	US	08 JUN	NO ELEMENTS AVAILABLE			
1990-050G		21917	US	08 JUN	NO ELEMENTS AVAILABLE			
1990-050H		22171	US	08 JUN	NO ELEMENTS AVAILABLE			
1990-051A	INSAT-1D	20643	INDIA	12 JUN	1436.2	0.0	35798	35777
1990-052A	MOLNIYA 3-38	20646	USSR	13 JUN	717.7	63.1	39541	809
1990-052D		20649	USSR	13 JUN	734.1	63.1	40292	866
1990-054A	GORIZONT 20	20659	USSR	20 JUN	1436.2	0.8	35806	35769
1990-054D		20662	USSR	20 JUN	1433.0	0.7	35789	35664
1990-054E		20704	USSR	20 JUN	487.9	47.1	28002	298
1990-055A	COSMOS 2084	20663	USSR	21 JUN	97.8	62 8	767	535
1990-055D		20666	USSR	21 JUN	97.6	62 8	753	536
1990-056A	INTELSAT	20667	ITSO	23 JUN	1436.1	0.0	35931	35641
1990-056C		20669	US	23 JUN	664.8	24.5	37412	297
1990-057A	METEOR 2-19	20670	USSR	27 JUN	104.0	82.5	957	935
1990-057B		20671	USSR	27 JUN	104.0	82.5	956	935
1990-061A	COSMOS 2085	20693	USSR	18 JUL	1436.1	0.7	35791	35782
1990-061D		20696	USSR	18 JUL	1437.2	0.7	35949	35668
1990-061F		20698	USSR	18 JUL	516.7	47.0	29522	379
1990-063A	TDF-2	20705	FRANCE	24 JUL	1436.2	0.1	35799	35776
1990-063B	DFS-2	20706	FRG	24 JUL	1436.1	0.0	35827	35745
1990-063C		20717	ESA	24 JUL	633.7	3.8	35746	375
1990-063D		20718	ESA	24 JUL	579.3	4.2	32996	288
1990-064A	COSMOS 2087	20707	USSR	25 JUL	717.7	64.5	38488	1861
1990-064D		20710	USSR	25 JUL	703.9	64.7	37895	1774
1990-065A	CRRES	20712	US	25 JUL	614.4	18.0	34781	345
1990-065B TO 065S			US	25 JUL	SEE NOTE 57			
1990-066A	COSMOS 2088	20720	USSR	30 JUL	116.0	73.6	1521	1482
1990-066B		20721	USSR	30 JUL	116.0	73.6	1518	1481
1990-068A		20724	US	02 AUG	718.0	54.7	20457	19906
1990-070A	COSMOS 2090	20735	USSR	08 AUG	113.8	82.6	1410	1388
1990-070B	COSMOS 2091	20736	USSR	08 AUG	114.0	82.6	1412	1409
1990-070C	COSMOS 2092	20737	USSR	08 AUG	114.0	82.6	1411	1404
1990-070D	COSMOS 2093	20738	USSR	08 AUG	113.9	82.6	1411	1397
1990-070E	COSMOS 2094	20739	USSR	08 AUG	113.8	82.6	1410	1393
1990-070F	COSMOS 2095	20740	USSR	08 AUG	113.7	82.6	1410	1381
1990-070G		20741	USSR	08 AUG	114.6	82.6	1466	1410
1990-071A	MOLNIYA 1-78	20742	USSR	10 AUG	717.8	63.2	39118	1236
1990-071D		20745	USSR	10 AUG	732.7	63.2	39806	1279
1990-074A	BSB-R2	20762	UK	18 AUG	1436.1	0.0	35794	35779
1990-074B		20763	US	18 AUG	102.2	24.8	1253	476
1990-074C		20764	US	18 AUG	670.1	20.9	37480	493
1990-076A	COSMOS 2097	20767	USSR	28 AUG	717.6	64.7	39093	1252
1990-076D		20770	USSR	28 AUG	707.8	65.1	38621	1241
1990-077A	BS-3A	20771	JAPAN	28 AUG	1436.2	0.0	35800	35778
1990-078A	COSMOS 2098	20774	USSR	28 AUG	108.2	83.0	1896	391
1990-078B		20775	USSR	28 AUG	107.5	83.0	1843	375
1990-079A	SKYNET 4C	20776	UK	30 AUG	1436.1	2.6	35796	35776
1990-079B	EUTELSAT II F1	20777	ESA	30 AUG	1436.0	0.0	36167	35403
1990-079C		20778	ESA	30 AUG	330.8	7.5	18775	157
1990-081A	FENGYUN 1-2	20788	PRC	03 SEP	102.7	98.9	897	875
1990-081D TO 081CH			PRC	03 SEP	SEE NOTE 52			
1990-083A	COSMOS 2100	20804	USSR	14 SEP	104.8	82.9	1011	956
1990-083B		20805	USSR	14 SEP	104.7	82.9	1003	953
1990-084A	MOLNIYA 3-39	20813	USSR	20 SEP	717.8	63.0	39023	1330

International Designation	Satellite Name	Catalog Number	Source	Launch Date	Period Mins	Incl-ination	Apogee KM	Perigee KM
1990-084D		20816	USSR	20 SEP	731.6	63.1	39792	1242
1990-086A	METEOR 2-20	20826	USSR	28 SEP	104.0	82.5	958	938
1990-086B		20827	USSR	28 SEP	104.0	82.5	958	938
1990-088A		20830	US	01 OCT	718.0	55.2	20368	19995
1990-090B	ULYSSES	20842	US	06 OCT	HELIOCENTRIC ORBIT			
1990-090C		20843	US	06 OCT	551.9	28.4	31500	320
1990-090D		20844	US	06 OCT	HELIOCENTRIC ORBIT			
1990-090E		20845	US	06 OCT	HELIOCENTRIC ORBIT			
1990-091A	SBS-6	20872	US	12 OCT	1436.2	0.0	35795	35781
1990-091B	GALAXY VI	20873	US	12 OCT	1436.1	0.0	35796	35777
1990-091C		20874	ESA	12 OCT	594.8	7.4	33824	277
1990-093A	INMARSAT 2 F1	20918	UK	30 OCT	1436.1	1.8	35805	35768
1990-093B		20919	US	30 OCT	97.9	24.7	943	371
1990-094A	GORIZONT 21	20923	USSR	03 NOV	1436.1	0.5	35791	35779
1990-094D		20926	USSR	03 NOV	1427.8	0.4	35791	35456
1990-094E		20927	USSR	03 NOV	256.5	46.5	13835	150
1990-095A		20929	US	13 NOV	NO ELEMENTS AVAILABLE			
1990-095C		20931	US	13 NOV	NO ELEMENTS AVAILABLE			
1990-095D		20932	US	13 NOV	NO ELEMENTS AVAILABLE			
1990-097B		20963	US	15 NOV	NO ELEMENTS AVAILABLE			
1990-097C		20964	US	15 NOV	NO ELEMENTS AVAILABLE			
1990-097D		20965	US	15 NOV	NO ELEMENTS AVAILABLE			
1990-099A	COSMOS 2105	20941	USSR	20 NOV	717.9	65.1	39163	1198
1990-099D		20944	USSR	20 NOV	707.4	65.3	38577	1262
1990-100A	SATCOM I	20945	US	20 NOV	1436.1	0.0	35795	35778
1990-100B	GSTAR IV	20946	US	20 NOV	1436.1	0.0	35788	35787
1990-100C		20947	ESA	20 NOV	607.2	7.4	34430	319
1990-101A	MOLNIYA 1-79	20949	USSR	23 NOV	717.6	64.4	39584	759
1990-101D		20952	USSR	23 NOV	730.6	64.5	40218	766
1990-102A	GORIZONT 22	20953	USSR	23 NOV	1436.0	0.4	35790	35777
1990-102D		21046	USSR	23 NOV	1471.4	0.4	36556	36394
1990-103A		20959	US	26 NOV	718.0	55.0	20359	20005
1990-103B		20960	US	26 NOV	96.1	21.4	654	486
1990-104A	COSMOS 2106	20966	USSR	28 NOV	94.1	82.5	483	466
1990-104B		20967	USSR	28 NOV	94.7	82.5	511	492
1990-104G		21069	USSR	28 NOV	92.8	82.5	423	399
1990-105A		20978	US	01 DEC	100.5	98.8	838	725
1990-105AA		21124	US	01 DEC	94.6	98.8	533	462
1990-105AB		21125	US	01 DEC	95.7	98.9	614	491
1990-105AE		21690	US	01 DEC	99.2	98.8	761	675
1990-105B		20979	US	01 DEC	98.0	98.8	702	624
1990-105M		20998	US	01 DEC	96.4	98.9	675	496
1990-105S		21073	US	01 DEC	98.3	98.8	720	636
1990-105Z		21080	US	01 DEC	99.5	98.8	778	688
1990-110A	COSMOS 2109	21006	USSR	08 DEC	675.7	64.9	19284	18973
1990-110B	COSMOS 2110	21007	USSR	08 DEC	675.7	64.9	19227	19031
1990-110C	COSMOS 2111	21008	USSR	08 DEC	675.7	64.9	19151	19107
1990-110F		21011	USSR	08 DEC	675.2	64.9	19145	19087
1990-110G		21012	USSR	08 DEC	340.1	65.2	18891	629
1990-110H		21013	USSR	08 DEC	340.1	65.2	18914	607
1990-111A	COSMOS 2112	21014	USSP	10 DEC	100.6	74.0	808	764
1990-111B		21015	USSR	10 DEC	100.5	74.1	799	764
1990-111C		21255	USSR	10 DEC	100.7	74.0	802	775
1990-112A	RADUGA 26	21016	USSR	20 DEC	1436.2	0.4	35797	35779
1990-112D		21019	USSR	20 DEC	1439.4	0.3	35983	35720
1990-112F		21025	USSR	20 DEC	460.9	46.9	26529	245
1990-114A	COSMOS 2114	21028	USSR	22 DEC	114.0	82.6	1411	1408
1990-114B	COSMOS 2115	21029	USSR	22 DEC	113.9	82.6	1407	1405
1990-114C	COSMOS 2116	21030	USSR	22 DEC	113.9	82.6	1408	1399
1990-114D	COSMOS 2117	21031	USSR	22 DEC	113.8	82.6	1408	1393
1990-114E	COSMOS 2118	21032	USSR	22 DEC	113.7	82.6	1408	1389
1990-114F	COSMOS 2119	21033	USSR	22 DEC	113.7	82.6	1408	1382
1990-114G		21034	USSR	22 DEC	114.6	82.6	1470	1407
1990-116A	RADUGA 1-2	21038	USSR	27 DEC	1435.9	0.3	35791	35775
1990-116D		21041	USSR	27 DEC	1470.2	0.3	36598	36305
1990-116F		21045	USSR	27 DEC	321.5	46.6	18151	182
1990-116G		21961	USSR	27 DEC	351.1	46.6	20023	194

1991 LAUNCHES

International Designation	Satellite Name	Catalog Number	Source	Launch Date	Period Mins	Incl-ination	Apogee KM	Perigee KM
1991-001A	NATO IVA	21047	NATO	08 JAN	1436.2	3.1	35790	35785
1991-001B		21048	NATO	08 JAN	121.6	18.5	2721	787
1991-001C		21049	NATO	08 JAN	635.9	25.9	35452	782

International Designation	Satellite Name	Catalog Number	Source	Launch Date	Period Mins	Incl-ination	Apogee KM	Perigee KM
1991-003A	ITALSAT-1	21055	ITALY	15 JAN	1436.1	0.1	35973	35598
1991-003B	EUTELSAT	21056	ESA	15 JAN	1436.1	0.1	35813	35760
1991-003C		21057	ESA	15 JAN	586.0	7.1	33359	275
1991-003D		21058	ESA	15 JAN	464.2	6.9	26739	221
1991-006A	INFORMTR-1	21087	USSR	29 JAN	104.7	82 9	1007	953
1991-006B		21088	USSR	29 JAN	104.6	82.9	993	956
1991-007A	COSMOS 2123	21089	USSR	05 FEB	104.7	82.9	1003	961
1991-007B		21090	USSR	05 FEB	104.6	82.9	993	961
1991-007C		21091	USSR	05 FEB	104.6	82.9	995	953
1991-009A	COSMOS 2125	21100	USSR	12 FEB	115.2	74.0	1471	1454
1991-009J TO 009CJ			USSR	12 FEB	SEE NOTE 54			
1991-009B	COSMOS 2126	21101	USSR	12 FEB	115.5	74.0	1495	1464
1991-009C	COSMOS 2127	21102	USSR	12 FEB	115.3	74.0	1476	1464
1991-009D	COSMOS 2128	21103	USSR	12 FEB	115.0	74.0	1466	1443
1991-009E	COSMOS 2129	21104	USSR	12 FEB	114.8	74.0	1466	1428
1991-009F	COSMOS 2130	21105	USSR	12 FEB	114.5	74.0	1467	1399
1991-009G	COSMOS 2131	21106	USSR	12 FEB	114.4	74.0	1466	1385
1991-009H	COSMOS 2132	21107	USSR	12 FEB	114.7	74.0	1466	1413
1991-010A	COSMOS 2133	21111	USSR	14 FEB	1436.2	0.8	35798	35778
1991-010D		21114	USSR	14 FEB	408.2	46.8	23485	216
1991-010F		21129	USSR	14 FEB	1438.6	0.9	35909	35762
1991-012A	MOLNIYA 1-80	21118	USSR	15 FEB	717.8	63.1	39038	1318
1991-012D		21121	USSR	15 FEB	700.5	63.0	38196	1302
1991-012E		21122	USSR	15 FEB	588.3	47.2	33287	471
1991-013A	COSMOS 2135	21130	USSR	26 FEB	104.5	82.8	1017	920
1991-013B		21131	USSR	26 FEB	104.3	82.8	1009	918
1991-014A	RADUGA 27	21132	USSR	28 FEB	1436.4	0.5	35804	35780
1991-014D		21135	USSR	28 FEB	1392.3	0.5	35034	34813
1951-015A	ASTRA 1-B	21139	LUXEM	02 MAR	1436.1	0.1	35802	35772
1991-015B	MOP-2	21140	ESA	02 MAR	1436.3	0.1	35799	35782
1991-015C		21141	ESA	02 MAR	547.1	6.9	31350	212
1991-015D		21142	ESA	02 MAR	390.8	6.8	22394	260
1991-015E		21904	ESA	02 MAR	1437.9	0.7	36476	35166
1991-017A		21147	US	08 MAR	NO ELEMENTS AVAILABLE			
1991-017B		21148	US	08 MAR	NO ELEMENTS AVAILABLE			
1991-018A	INMARSAT-2	21149	UK	08 MAR	1436.2	2.2	35793	35782
1991-018B		21150	US	08 MAR	99.5	25.0	1066	407
1991-018C		21151	US	08 MAR	559.4	23.4	32003	216
1991-019A	NADEZHDA	21152	USSR	12 MAR	104.8	82.9	1014	954
1991-019B		21153	USSR	12 MAR	104.7	82.9	1005	953
1991-021A	COSMOS 2137	21190	USSR	19 MAR	92.5	65.8	418	375
1991-021B		21191	USSR	19 MAR	90.1	65.8	283	274
1991-022A	MOLNIYA 3-40	21196	USSR	22 MAR	717.6	63.1	39012	1331
1991-022D		21199	USSR	22 MAR	700.1	63.1	38133	1347
1991-025A	COSMOS 2139	21216	USSR	04 APR	675.7	65.0	19156	19102
1991-025B	COSMOS 2140	21217	USSR	04 APR	675.7	65.1	19162	19096
1991-025C	COSMOS 2141	21218	USSR	04 APR	675.7	65.0	19152	19106
1991-025E		21221	USSR	04 APR	675.5	65.0	19130	19117
1991-025F		21220	USSR	04 APR	339.3	64.9	18988	485
1991-025G		21226	USSR	04 APR	339.2	64.8	18991	477
1991-026A	ANIK E-2	21222	CANADA	05 APR	1436.1	0.0	35796	35778
1991-026B		21223	ESA	05 APR	635.1	3.7	35765	426
1991-027B	GRO	21225	US	05 APR	92.0	28.5	376	370
1991-028A	ASC 2 SPACENET 5	21227	US	13 APR	1436.2	0.0	35795	35780
1991-028B		21228	US	13 APR	115.5	24.0	2397	560
1991-028C		21229	US	13 APR	655.6	21.7	35828	1411
1991-029A	COSMOS 2142	21230	USSR	16 APR	104.9	83.0	1016	959
1991-029B		21231	USSR	16 APR	104.7	82.9	1006	953
1991-030A	METEOR 3-4	21232	USSR	24 APR	109.3	82.5	1208	1180
1991-030B		21233	USSR	24 APR	109.3	82.5	1210	1183
1991-030C		21234	USSR	24 APR	109.3	82.5	1209	1183
1991-031C		21262	US	28 APR	NO ELEMENTS AVAILABLE			
1991-032A	NOAA-12	21263	US	14 MAY	101.2	98.7	824	806
1991-032B		21267	US	14 MAY	100.6	98.7	789	782
1991-032C		21298	US	14 MAY	100.6	98.7	790	782
1991-033A	COSMOS 2143	21299	USSR	16 MAY	113.9	82.6	1414	1396
1991-033B	COSMOS 2144	21300	USSR	16 MAY	114.0	82.6	1414	1409
1991-033C	COSMOS 2145	21301	USSR	16 MAY	114.0	82.6	1414	1402
1991-033D	COSMOS 2146	21302	USSR	16 MAY	113.8	82.6	1413	1391
1991-033E	COSMOS 2147	21303	USSR	16 MAY	113.8	82.6	1413	1386
1991-033F	COSMOS 2148	21304	USSR	16 MAY	113.7	82.6	1414	1379
1991-033G		21305	USSR	16 MAY	114.7	82.6	1470	1413
1991-035C		21479	USSR	21 MAY	85.8	82.2	90	65

International Designation	Satellite Name	Catalog Number	Source	Launch Date	Period Mins	Incl-ination	Apogee KM	Perigee KM
1991-037A	AURORA-II	21392	US	29 MAY	1436.1	0.0	35792	35781
1991-037B		21393	US	29 MAY	112.6	25.0	2282	406
1991-037C		21394	US	29 MAY	648.8	23.6	35398	1494
1991-039A	OKEAN 3	21397	USSR	04 JUN	97.5	82.5	655	624
1991-039B		21398	USSR	04 JUN	97.6	82.5	659	627
1991-039C		21842	USSR	04 JUN	97.3	82.5	644	613
1991-041A	COSMOS 2150	21418	USSR	11 JUN	100.7	74.0	805	778
1991-041B		21419	USSR	11 JUN	100.6	74.0	800	774
1991-041C		21420	USSR	11 JUN	100.8	74.1	803	792
1991-041D		21711	USSR	11 JUN	100.6	74.0	799	773
1991-042A	COSMOS 2151	21422	USSR	13 JUN	97.5	82.5	653	628
1991-042B		21423	USSR	13 JUN	97.6	82.5	655	630
1991-043A	MOLNIYA 1-81	21426	USSR	18 JUN	717.8	63.3	39392	965
1991-043D		21429	USSR	18 JUN	732.2	63.3	40106	959
1991-045A	REX	21527	US	29 JUN	101.3	89 6	867	769
1991-045B		21528	US	29 JUN	101.1	89 6	853	766
1991-045C		21529	US	29 JUN	101.3	89.6	868	767
1991-045D		21532	US	29 JUN	101.2	89.5	870	761
1991-045E		21691	US	29 JUN	100.4	89 8	792	763
1991-045F		21712	US	29 JUN	101.9	89 3	955	741
1991-046A	GORIZONT 23	21533	USSR	02 JUL	1436.3	0.6	35850	35730
1991-046D		21536	USSR	02 JUL	1426.7	0.3	35753	35452
1991-046E		21538	USSR	02 JUL	471.1	46.9	27090	266
1991-047A		21552	US	04 JUL	718.0	55.5	20302	20061
1991-047D		21555	US	04 JUL	267.0	34.6	14532	180
1991-050A	ERS-1	21574	ESA	17 JUL	100.5	98.6	782	780
1991-050B	UOSAT-F	21575	UK	17 JUL	100.2	98.5	772	760
1991-050C	ORBCOMM-X	21576	US	17 JUL	100.2	98.5	771	765
1991-050D	TUBSAT	21577	FRG	17 JUL	100.2	98.5	772	762
1991-050E	SARA	21578	FRANCE	17 JUL	100.1	98.5	765	757
1991-050F		21610	ESA	17 JUL	100.3	98.4	776	771
1991-053A	MOLNIYA 1-82	21630	USSR	01 AUG	717.7	64.0	39777	571
1991-053D		21633	USSR	01 AUG	733.1	64.1	40551	558
1991-054B	TDRS-5	21639	US	02 AUG	1436.1	0.0	35793	35779
1991-054C		21640	US	02 AUG	619.4	26.3	35077	304
1991-054D		21641	US	02 AUG	1434.5	1.4	35896	35613
1991-054E		21642	US	02 AUG	618.8	27.0	35079	271
1991-055A	INTELSAT 6 F-5	21653	ITSO	14 AUG	1436.2	0.0	35791	35785
1991-055B		21654	ESA	14 AUG	597.3	7.3	33960	271
1991-056A	METEOR 3-5	21655	USSR	15 AUG	109.3	82.6	1205	1183
1991-056B		21656	USSR	15 AUG	109.3	82.6	1204	1183
1991-059A	COSMOS 2154	21666	USSR	22 AUG	104.8	82.9	1005	969
1991-059B		21667	USSR	22 AUG	104.7	82.9	1000	962
1991-060A	BS-3B	21668	JAPAN	25 AUG	1436.2	0.0	35813	35764
1991-061A		21688	INDIA	29 AUG	103.1	99.1	918	889
1991-061B		21689	INDIA	29 AUG	102.8	99.2	914	864
1991-062A	SOLAR-A	21694	JAPAN	30 AUG	97.5	31.3	761	519
1991-062B		21695	JAPAN	30 AUG	97.5	31.3	763	516
1991-062D		21697	JAPAN	30 AUG	94.7	31.3	551	451
1991-062F		21699	JAPAN	30 AUG	95.2	31.3	585	465
1991-062H		21802	JAPAN	30 AUG	97.5	31.5	717	557
1991-063B	UARS	21701	US	12 SEP	96.2	57.0	580	573
1991-064A	COSMOS 2155	21702	USSR	13 SEP	1437.7	0.2	35838	35795
1991-064B		21703	USSR	13 SEP	1441.7	0.2	35929	35861
1991-065A	MOLNIYA 3-41	21706	USSR	17 SEP	717.7	63.0	39255	1096
1991-065D		21709	USSR	17 SEP	733.3	63.0	39991	1124
1991-067A	ANIK E1	21726	CANADA	26 SEP	1436.1	0.1	35805	35768
1991-067B		21727	ESA	26 SEP	637.3	4.5	35871	435
1991-068A	COSMOS 2157	21728	USSR	28 SEP	114.0	82.6	1411	1404
1991-068B	COSMOS 2158	21729	USSR	28 SEP	113.9	82.6	1406	1402
1991-068C	COSMOS 2159	21730	USSR	28 SEP	113.7	82.6	1406	1386
1991-068D	COSMOS 2160	21731	USSR	28 SEP	113.8	82.6	1407	1396
1991-068E	COSMOS 2161	21732	USSR	28 SEP	113.8	82.6	1407	1392
1991-068F	COSMOS 2162	21733	USSR	28 SEP	114.0	82.6	1417	1405
1991-068G		21734	USSR	28 SEP	114.7	82.6	1477	1406
1991-074A	GORIZONT 24	21759	USSR	23 OCT	1436.2	0.5	35790	35788
1991-074D		21762	USSR	23 OCT	1444.3	0.5	36183	35711
1991-075A	INTELSAT F1 V1	21765	ESA	29 OCT	1436.2	0.0	35795	35780
1991-075B		21766	ESA	29 OCT	613.1	7.6	34750	304
1991-076A	USA 72	21775	US	08 NOV	NO ELEMENTS AVAILABLE			
1991-076B		21776	US	08 NOV	NO ELEMENTS AVAILABLE			
1991-076C		21799	US	08 NOV	NO ELEMENTS AVAILABLE			
1991-076D	USA 76	21808	US	08 NOV	NO ELEMENTS AVAILABLE			

International Designation	Satellite Name	Catalog Number	Source	Launch Date	Period Mins	Incl-ination	Apogee KM	Perigee KM
1991-076E	USA 77	21809	US	08 NOV	NO ELEMENTS AVAILABLE			
1991-076F		21956	US	08 NOV	NO CURRENT ELEMENTS			
1991-077A	COSMOS 2165	21779	USSR	12 NOV	113.8	82.6	1410	1391
1991-077B	COSMOS 2166	21780	USSR	12 NOV	113.9	82.6	1410	1403
1991-077C	COSMOS 2167	21781	USSR	12 NOV	113.9	82.6	1410	1397
1991-077D	COSMOS 2168	21782	USSR	12 NOV	113.8	82.6	1410	1387
1991-077E	COSMOS 2169	21783	USSR	12 NOV	113.7	82.6	1410	1380
1991-077F	COSMOS 2170	21784	USSR	12 NOV	114.0	82.6	1410	1410
1991-077G		21785	USSR	12 NOV	114.7	82.6	1471	1410
1991-079A	COSMOS 2172	21789	USSR	22 NOV	1436.2	0.5	35794	35781
1991-079D		21792	USSR	22 NOV	1460.2	0.5	36280	36232
1991-079F		21794	USSR	22 NOV	350.8	46.5	20036	158
1991-080B	USA 75	21805	US	25 NOV	NO ELEMENTS AVAILABLE			
1991-080C		21806	US	25 NOV	NO ELEMENTS AVAILABLE			
1991-080D		21807	US	25 NOV	NO ELEMENTS AVAILABLE			
1991-081A	COSMOS 2173	21796	USSR	26 NOV	104.7	83.0	1015	944
1991-081B		21797	USSR	26 NOV	104.6	83.0	1003	944
1991-082A	USA 73	21798	US	28 NOV	101.8	98.9	853	835
1991-082B		21800	US	28 NOV	101.5	98.9	832	822
1991-082C		21801	US	28 NOV	101.5	98.9	835	826
1991-082D		21825	US	28 NOV	101.6	98.9	846	819
1991-082E		21836	US	28 NOV	101.6	99.0	842	825
1991-083A	EUTELSAT II	21803	ESA	07 DEC	1435.7	0.2	36087	35469
1991-083B		21804	ESA	07 DEC	753.6	16.8	41188	921
1991-084A	TELECOM 2A	21813	FRANCE	16 DEC	1436.1	0.0	35799	35775
1991-084B	INMARSAT 2 F-3	21814	ITSO	16 DEC	1436.1	1.9	35798	35774
1991-084C		21815	ESA	16 DEC	643.4	3.8	36206	410
1991-084D		21818	ESA	16 DEC	621.8	4.0	35166	342
1991-086A	INTERCOSMOS 25	21819	USSR	18 DEC	121.4	82.6	3055	435
1991-086B		21820	USSR	18 DEC	121.5	82.6	3061	434
1991-086C		21826	USSR	18 DEC	120.4	82.6	2967	433
1991-086D		21827	USSR	18 DEC	120.6	82.5	2980	433
1991-086E	MAGION 3	21835	CZECH	18 DEC	121.4	82.6	3054	435
1991-086F		21905	USSR	18 DEC	121.3	82.6	3033	440
1991-087A	RADUGA 28	21821	USSR	19 DEC	1436.0	0.6	35790	35778
1991-087D		21824	USSR	19 DEC	1469.1	0.6	36495	36366
1991-087F		21829	USSR	19 DEC	453.2	46.5	26186	148
1991-088A	PRC 34	21833	PRC	28 DEC	632.8	31.5	33963	2112

1992 LAUNCHES

International Designation	Satellite Name	Catalog Number	Source	Launch Date	Period Mins	Incl-ination	Apogee KM	Perigee KM
1992-003A	COSMOS 2176	21847	USSR	24 JAN	717.7	64.0	39392	958
1992-003D		21850	USSR	24 JAN	706.1	64.1	38845	932
1992-005A	COSMOS 2177	21853	USSR	29 JAN	675.7	64.8	19148	19110
1992-005B	COSMOS 2178	21854	USSR	31 JAN	675.7	64.8	19170	19088
1992-005C	COSMOS 2179	21855	USSR	29 JAN	675.7	64.9	19148	19110
1992-005F		21858	USSR	29 JAN	675.3	64.8	19145	19094
1992-005G		21862	USSR	29 JAN	340.1	64.8	19113	410
1992-005H		21863	USSR	29 JAN	340.1	65.0	19120	401
1992-006A	USA 78	21873	US	10 FEB	NO ELEMENTS AVAILABLE			
1992-006B		21874	US	10 FEB	NO ELEMENTS AVAILABLE			
1992-006C		21877	US	10 FEB	NO ELEMENTS AVAILABLE			
1992-007A	JERS-1	21867	JAPAN	11 FEB	96.0	97.7	570	566
1992-007B		21868	JAPAN	11 FEB	93.7	97.7	500	414
1992-008A	COSMOS 2180	21875	USSR	17 FEB	104.8	82.9	1013	957
1992-008B		21876	USSR	17 FEB	104.7	82.9	1005	957
1992-009A	USA 79	21890	US	23 FEB	718.0	54.5	20336	20029
1992-009B		21891	US	23 FEB	98.3	20.0	723	631
1992-009C		21892	US	23 FEB	319.0	34.7	17975	201
1992-010A	SUPERBIRD B1	21893	JAPAN	26 FEB	1436.1	0.0	35807	35768
1992-010B	ARABSAT 1C	21894	SA	26 FEB	1436.1	0.0	35852	35722
1992-010C		21895	ESA	26 FEB	509.8	6.5	29337	187
1992-011A	MOLNIYA 1-83	21897	USSR	04 MAR	717.8	62.9	39513	841
1992-011D		21900	USSR	04 MAR	698.4	62.9	38560	832
1992-012A	COSMOS 2181	21902	USSR	09 MAR	104.9	82.9	1010	969
1992-012B		21903	USSR	09 MAR	104.8	82.9	1005	959
1992-013A	GALAXY 5	21906	US	14 MAR	1436.1	0.0	35791	35783
1992-013B		21907	US	14 MAR	638.5	19.4	35241	1125
1992-017A	GORIZONT 25	21922	USSR	02 APR	1436.2	0.8	35801	35774
1992-017D		21925	USSR	02 APR	1424.6	0.8	35646	35474
1992-019A	USA 80	21930	US	10 APR	718.0	55.3	20367	19997
1992-019B		21931	US	10 APR	97.2	21.2	712	535
1992-019C		21932	US	10 APR	323.3	34.5	18250	202

International Designation	Satellite Name	Catalog Number	Source	Launch Date	Period Mins	Incl-ination	Apogee KM	Perigee KM
1992-020A	COSMOS 2184	21937	USSR	15 APR	104.9	82.9	1011	963
1992-020B		21938	USSR	15 APR	104.7	82.9	1002	960
1992-021A	TELECOM 2B	21939	FRANCE	15 APR	1436.1	0.0	35804	35770
1992-021B	INMARSAT 2 F4	21940	IM	15 APR	1436.2	2.2	35804	35773
1992-021C		21941	ESA	15 APR	634.8	3.7	35921	255
1992-021D		21942	ESA	15 APR	622.9	3.7	35274	292
1992-023A	USA 81	21949	US	25 APR	NO ELEMENTS AVAILABLE			
1992-023B		21950	US	25 APR	NO ELEMENTS AVAILABLE			
1992-027A	PALAPA-B4	21964	INDO	14 MAY	1436.3	0.0	35795	35783
1992-027B		21965	US	14 MAY	119.3	19.7	2790	508
1992-027C		21966	US	14 MAY	701.4	23.0	36700	2845
1992-030A	COSMOS 2187	21976	USSR	03 JUN	114.6	74.0	1478	1399
1992-030B	COSMOS 2188	21977	USSR	03 JUN	114.5	74.0	1476	1386
1992-030C	COSMOS 2189	21978	USSR	03 JUN	114.8	74.0	1477	1414
1992-030D	COSMOS 2190	21979	USSR	03 JUN	115.0	74.0	1478	1428
1992-030E	COSMOS 2191	21980	USSR	03 JUN	115.7	74.0	1500	1470
1992-030F	COSMOS 2192	21981	USSR	03 JUN	115.5	74.0	1484	1469
1992-030G	COSMOS 2193	21982	USSR	03 JUN	115.1	74.0	1478	1444
1992-030H	COSMOS 2194	21983	USSR	03 JUN	115.3	74.0	1482	1455
1992-030J		21984	USSR	03 JUN	117.8	74.0	1680	1481
1992-031A	EUVE	21987	US	07 JUN	95.1	28.4	529	514
1992-032A	INTELSAT K	21989	ITSO	10 JUN	1436.2	0.0	35793	35782
1992-032B		21990	US	10 JUN	612.3	26.8	34810	206
1992-036A	COSMOS 2195	22006	USSR	01 JUL	104.7	82.9	1009	953
1992-036B		22007	USSR	01 JUL	104.6	82.9	999	947
1992-037A	USA 82	22009	US	02 JUL	NO ELEMENTS AVAILABLE			
1992-037B		22010	US	02 JUL	NO ELEMENTS AVAILABLE			
1992-037C		22011	US	02 JUL	NO ELEMENTS AVAILABLE			
1992-038A	SAMPEX	22012	US	03 JUL	96.6	81.7	679	509
1992-038B		22013	US	03 JUL	96.6	81.7	680	510
1992-039A	USA 83	22014	US	07 JUL	718.0	55.0	20396	19968
1992-039B		22015	US	07 JUL	97.7	20.7	731	567
1992-039C		22016	US	07 JUL	325.7	34.7	18417	190
1992-040A	COSMOS 2196	22017	USSR	08 JUL	718.0	63.6	39633	732
1992-040D		22020	USSR	08 JUL	705.7	63.4	39023	733
1992-041A	INSAT-2A	22027	INDIA	09 JUL	1436.1	0.1	35833	35741
1992-041B	EUTELSAT 2 F4	22028	FRANCE	09 JUL	1436.0	0.0	35816	35754
1992-041C		22032	ESA	09 JUL	631.6	7.5	35749	261
1992-041D		22033	ESA	09 JUL	602.2	7 3	34249	237
1992-042A	COSMOS 2197	22034	USSR	13 JUL	113.9	82 6	1413	1395
1992-042B	COSMOS 2198	22035	USSR	12 JUL	114.0	82.6	1413	1407
1992-042C	COSMOS 2199	22036	USSR	13 JUL	114.2	82.6	1423	1410
1992-042D	COSMOS 2200	22037	USSR	13 JUL	114.0	82.6	1413	1402
1992-042E	COSMOS 2201	22038	USSR	13 JUL	114.1	82.6	1418	1409
1992-042F	COSMOS 2202	22039	USSR	13 JUL	114.0	82.6	1415	1405
1992-042G		22040	USSR	13 JUL	114.7	82.6	1471	1408
1992-043A	GORIZONT 26	22041	USSR	14 JUL	1436.2	1.1	35808	35767
1992-043D		22044	USSR	14 JUL	1471.9	1.1	36600	36370
1992-043F		22048	USSR	14 JUL	643.2	46.7	36406	204
1992-044A	GEOTAIL	22049	JAPAN	24 JUL	4750.6	22.4	508542	41363
1992-044C		22051	US	24 JUL	3368.2	28.7	360227	184
1992-047A	COSMOS 2204	22056	USSR	30 JUL	675.7	64.9	19148	19110
1992-047B	COSMOS 2205	22057	USSR	30 JUL	675.7	64.9	19148	19110
1992-047C	COSMOS 2206	22058	USSR	30 JUL	675.7	64.9	19167	19091
1992-047E		22060	USSR	30 JUL	87.8	64.8	183	166
1992-047F		22061	USSR	30 JUL	675.0	64.9	19133	19088
1992-047G		22066	USSR	30 JUL	340.2	64.9	19116	409
1992-047H		22067	USSR	30 JUL	340.1	64.8	19116	408
1992-049B	EURECA-1	22065	ESA	31 JUL	94.6	28.5	503	499
1992-050A	MOLNIYA 1-84	22068	USSR	06 AUG	717.7	62.9	39520	830
1992-050D		22071	USSR	06 AUG	733.4	62.9	40293	826
1992-052A	TOPEX	22076	US	10 AUG	112.4	66.0	1342	1330
1992-052B	KITSAT A	22077	KOREA	10 AUG	111.9	66.1	1325	1304
1992-052C	SSO/T	22078	FRANCE	10 AUG	111.9	66.1	1324	1303
1992-052D		22079	FRANCE	10 AUG	112.7	66.0	1396	1302
1992-053A	COSMOS 2208	22080	USSR	12 AUG	100.8	74.0	803	785
1992-053B		22081	USSR	12 AUG	100.6	74.0	803	775
1992-054A	AUSSAT B1	22087	AUSTRL	13 AUG	1436.2	0.0	35792	35783
1992-054C		22089	AUSTRL	13 AUG	664.3	23.0	37265	417
1992-057A	SATCOM-C4	22096	US	31 AUG	1436.1	0.0	35791	35782
1992-057B		22097	US	31 AUG	132.0	25.2	2643	1772
1992-057C		22098	US	31 AUG	662.0	20.1	35757	1810
1992-058A	USA 84	22108	US	09 SEP	718.0	54.5	20457	19908

International Designation	Satellite Name	Catalog Number	Source	Launch Date	Period Mins	Incl-ination	Apogee KM	Perigee KM
1992-058B		22109	US	09 SEP	98.7	19.8	728	663
1992-058C		22110	US	09 SEP	342.8	34.7	19491	201
1992-059A	COSMOS 2209	22112	USSR	10 SEP	1436.1	1.1	35808	35765
1992-059D		22115	USSR	10 SEP	1442.8	1.1	35963	35872
1992-060A	HISPASAT 1A	22116	SPAIN	10 SEP	1436.2	0.0	35794	35783
1992-060B	SATCOM C3	22117	US	10 SEP	1436.1	0.1	35788	35785
1992-060C		22118	SPAIN	10 SEP	625.2	6.8	35472	213
1992-060D		22119	SPAIN	10 SEP	613.4	6.9	34841	229
1992-063A	MARS	22136	US	25 SEP	MARS ORBIT			
1992-063C		22138	US	25 SEP	MARS ORBIT			
1992-064A	FREJA	22161	SWEDEN	06 OCT	108.9	63.0	1759	596
1992-066A	DFS 3	22175	FRG	12 OCT	1436.6	0.0	35797	35793
1992-066B		22176	US	12 OCT	132.7	25.1	3075	1408
1992-066C		22177	US	12 OCT	658.5	19.7	35851	1536
1992-067A	MOLNIYA 3-42	22178	USSR	14 OCT	NO CURRENT ELEMENTS			
1992-067D		22181	USSR	14 OCT	733.6	62.8	40576	557
1992-068A	COSMOS 2211	22182	USSR	20 OCT	113.9	82.6	1410	1398
1992-068B	COSMOS 2212	22183	USSR	20 OCT	114.0	82.6	1410	1406
1992-068C	COSMOS 2213	22184	USSR	20 OCT	114.0	82.6	1411	1409
1992-068D	COSMOS 2214	22185	USSR	20 OCT	114.1	82.6	1420	1409
1992-068E	COSMOS 2215	22186	USSR	20 OCT	114.2	82.6	1426	1409
1992-068F	COSMOS 2216	22187	USSR	20 OCT	114.0	82.6	1413	1408
1992-068G		22188	USSR	20 OCT	114.7	82.6	1477	1405
1992-069A	COSMOS 2217	22189	USSR	21 OCT	717.7	63.0	39622	727
1992-069D		22192	USSR	21 OCT	709.9	63.2	39226	738
1992-070B	LAGEOS II	22195	ITALY	22 OCT	222.5	52.7	5950	5616
1992-070D		22196	US	22 OCT	152.8	41.2	5888	294
1992-070E		22197	US	22 OCT	222.4	52.7	5949	5615
1992-070F		22198	US	22 OCT	222.4	52.7	5950	5615
1992-072A	GALAXY VII	22205	US	28 OCT	1436.2	0.0	35798	35779
1992-072B		22206	FRANCE	28 OCT	457.8	7.1	26420	176
1992-073A	COSMOS 2218	22207	USSR	29 OCT	104.9	82.9	1013	963
1992-073B		22208	USSR	29 OCT	104.7	82.9	1005	956
1992-074A	EKRAN 20	22210	USSR	30 OCT	1436.0	1.3	35791	35780
1992-074D		22213	USSR	30 OCT	1423.1	2.3	35583	35479
1992-074E		22215	USSR	30 OCT	628.4	47.2	35583	266
1992-076A	COSMOS 2219	22219	USSR	17 NOV	101.9	71.0	854	845
1992-076B		22220	USSR	17 NOV	101.7	71.0	850	829
1992-076C		22221	USSR	17 NOV	104.7	71.0	1117	842
1992-076D		22222	USSR	17 NOV	105.2	71.0	1166	843
1992-076E		22223	USSR	17 NOV	104.8	71.0	1130	841
1992-076F		22224	USSR	17 NOV	104.9	71.0	1133	843
1992-078A	MSTI	22229	US	21 NOV	91.2	96.7	378	292
1992-079A	USA 85	22231	US	22 NOV	NO CURRENT ELEMENTS			
1992-079B		22232	US	22 NOV	97.3	21.2	724	532
1992-079C		22233	US	22 NOV	349.2	34.7	19919	180
1992-080A	COSMOS 2221	22236	USSR	24 NOV	97.7	82.5	662	631
1992-080B		22237	USSR	24 NOV	97.7	82.5	662	630
1992-081A	COSMOS 2222	22238	USSR	25 NOV	718.0	62.9	39705	661
1992-081D		22241	USSR	25 NOV	707.8	63.1	39205	654
1992-082A	GORIZONT 27	22245	USSR	27 NOV	1471.7	1.5	36509	36452
1992-082D		22248	USSR	27 NOV	1469.1	1.4	36483	36374
1992-082E		22249	USSR	27 NOV	645.9	47.2	36482	262
1992-082F		22250	USSR	27 NOV	647.6	47.2	36514	319
1992-083A	USA 86	22251	US	28 NOV	NO ELEMENTS AVAILABLE			
1992-083B		22252	US	28 NOV	NO ELEMENTS AVAILABLE			
1992-084A	SUPERBIRD A1	22253	JAPAN	01 DEC	1436.2	0.1	35800	35775
1992-084B		22254	FRG	01 DEC	631.7	6.9	35805	214
1992-085A	MOLNIYA 3-43	22255	USSR	02 DEC	717.7	62.9	39947	403
1992-085D		22258	USSR	02 DEC	697.8	62.9	38956	406
1992-087A	COSMOS 2223	22260	USSR	09 DEC	89.8	64.7	290	238
1992-088A	COSMOS 2224	22269	USSR	17 DEC	1448.3	2.3	36070	35977
1992-088D		22272	USSR	17 DEC	1439.4	2.3	35889	35813
1992-088E		22273	USSR	17 DEC	635.8	47.4	35881	346
1992-088F		22274	USSR	17 DEC	631.7	46.9	35801	216
1992-089A	USA 87	22275	US	18 DEC	718.0	54.7	20327	20036
1992-089B		22276	US	18 DEC	98.3	20.3	740	613
1992-089C		22277	US	18 DEC	339.1	34.9	19295	167
1992-090A	AUSSAT B2	22278	AUSTRL	21 DEC	96.5	28.1	972	210
1992-092A	COSMOS 2226	22282	USSR	22 DEC	116.0	73.6	1522	1477
1992-092B		22283	USSR	22 DEC	115.3	73.6	1515	1424
1992-092AP		22350	USSR	25 DEC	102.6	70.9	930	829
1992-092BX		22382	USSR	25 DEC	104.5	70.9	1105	834

International Designation	Satellite Name	Catalog Number	Source	Launch Date	Period Mins	Incl-ination	Apogee KM	Perigee KM
1992-093A	COSMOS 2227	22284	USSR	25 DEC	101.9	71.0	856	842
1992-093B TO 093HW			USSR	25 DEC	SEE NOTE 61			
1992-094A	COSMOS 2228	22286	USSR	25 DEC	97.7	82.5	665	630
1992-094B		22287	USSR	25 DEC	97.7	82.5	664	629
1992-094HD		22543	USSR	25 DEC	103.9	71.3	1055	831
1993 LAUNCHES								
1993-001A	COSMOS 2230	22307	USSR	01 JAN	104.8	82.9	1006	967
1993-001B		22308	USSR	01 JAN	104.7	82.9	1001	956
1993-002A	MOLNIYA 1-85	22309	USSR	13 JAN	717.7	63.0	39768	583
1993-002D		22312	USSR	13 JAN	732.0	63.0	40483	569
1993-003B	TDRS F6	22314	US	13 JAN	1436.2	0.1	35791	35785
1993-003C		22315	US	13 JAN	632.3	26.8	35756	293
1993-003D		22316	US	13 JAN	1438.0	1.9	36125	35520
1993-005A	SOYUZ TM-16	22319	USSR	24 JAN	92.4	51.6	394	393
1993-006A	COSMOS 2232	22321	USSR	26 JAN	717.9	62.8	39717	642
1993-006D		22324	USSR	26 JAN	706.8	62.8	39176	637
1993-007A	USA 88	22446	US	03 FEB	717.9	54.8	20352	20009
1993-007B		22447	US	03 FEB	97.7	20.9	728	566
1993-007C		22448	US	03 FEB	346.0	34.8	19714	180
1993-008A	COSMOS 2233	22487	USSR	09 FEB	104.7	82.9	1006	951
1993-008B		22488	USSR	09 FEB	104.5	82.9	997	948
1993-009A	OXP-1	22489	US	09 FEB	100.1	25.0	796	730
1993-009B	SCD 1	22490	BRAZIL	09 FEB	100.1	25.0	794	729
1993-009C		22491	US	09 FEB	99.8	25.0	789	712
1993-010A	COSMOS 2234	22512	USSR	17 FEB	675.7	64.8	19152	19106
1993-010B	COSMOS 2235	22513	USSR	17 FEB	675.7	64.8	19148	19110
1993-010C	COSMOS 2236	22514	USSR	17 FEB	675.7	64.8	19163	19095
1993-010F		22517	USSR	17 FEB	674.7	64.8	19137	19072
1993-010G		22524	USSR	17 FEB	340.2	64.9	19126	405
1993-010H		22528	USSR	17 FEB	340.2	64.9	19126	405
1993-011A	ASTRO D	22521	JAPAN	20 FEB	96.6	31.1	649	540
1993-011B		22522	JAPAN	22 FEB	96.1	31.1	616	530
1993-011C		22523	JAPAN	20 FEB	95.8	31.1	593	523
1993-011D		22534	JAPAN	20 FEB	96.3	30.9	614	551
1993-011E		22587	JAPAN	20 FEB	96.0	31.1	608	528
1993-013A	RADUGA 29	22557	USSR	25 MAR	1454.3	1.4	36273	36011
1993-013E		22569	USSR	25 MAR	647.8	47.4	36511	331
1993-013F		22570	USSR	25 MAR	646.7	47.3	36505	280
1993-014A	START-1	22561	USSR	25 MAR	101.4	75.8	967	680
1993-014B		22562	USSR	25 MAR	101.0	75.8	930	680
1993-014C		22567	USSR	25 MAR	101.5	75.8	978	681
1993-014D		22568	USSR	25 MAR	101.5	75.8	979	679
1993-015A	UHF F1	22563	US	25 MAR	622.2	26.9	35308	217
1993-015B		22564	US	25 MAR	192.4	27.3	9102	220
1993-016A	COSMOS 2237	22565	USSR	26 MAR	101.9	71.0	851	846
1993-016B		22566	USSR	26 MAR	101.8	71.0	852	835
1993-016C		22575	USSR	26 MAR	105.0	71.0	1149	842
1993-016D		22576	USSR	26 MAR	105.0	71.0	1143	843
1993-016E		22577	USSR	26 MAR	100.0	71.0	844	668
1993-016F		22578	USSR	26 MAR	99.9	71.1	844	661
1993-016G		22579	USSR	26 MAR	100.1	71.0	847	679
1993-016H		22580	USSR	26 MAR	100.3	71.2	847	696
1993-017A	USA 90	22581	US	30 MAR	723.7	54.9	20452	20193
1993-017C		22583	US	30 MAR	101.0	36.2	1303	308
1993-017D		22584	US	30 MAR	356.4	34.8	20356	190
1993-018A	COSMOS 2238	22585	USSR	30 MAR	92.7	65.0	415	402
1993-019A	PROGRESS M-17	22588	USSR	31 MAR	92.4	51.6	394	393
1993-020A	COSMOS 2239	22590	USSR	01 APR	104.7	82.9	997	961
1993-020B		22591	USSR	01 APR	104.6	82.9	988	960
1993-022A	COSMOS 2241	22594	USSR	06 APR	717.6	63.2	39619	725
1993-022D		22597	USSR	06 APR	703.1	63.3	38907	722
1993-024A	COSMOS 2242	22626	USSR	16 APR	97.7	82.5	665	629
1993-024B		22627	USSR	16 APR	97.7	82.5	665	627
1993-025A	MOLNIYA 3-44	22633	USSR	21 APR	717.9	62.8	39671	689
1993-025D		22636	USSR	21 APR	734.9	62.8	40482	714
1993-026A	ALEXIS	22638	US	25 APR	NO ELEMENTS AVAILABLE			
1993-026B		22639	US	25 APR	NO ELEMENTS AVAILABLE			
1993-029A	COSMOS 2244	22643	USSR	28 APR	92.7	65.0	416	401
1993-030A	COSMOS 2245	22646	USSR	11 MAY	113.9	82.6	1414	1395
1993-030B	COSMOS 2246	22647	USSR	11 MAY	113.9	82.6	1414	1398
1993-030C	COSMOS 2247	22648	USSR	11 MAY	114.0	82.6	1414	1401

International Designation	Satellite Name	Catalog Number	Source	Launch Date	Period Mins	Incl-ination	Apogee KM	Perigee KM
1993-030D	COSMOS 2248	22649	USSR	11 MAY	113.9	82.6	1414	1399
1993-030E	COSMOS 2249	22650	USSR	11 MAY	114.0	82.6	1414	1401
1993-030F	COSMOS 2250	22651	USSR	11 MAY	113.9	82.6	1414	1398
1993-030G		22652	USSR	11 MAY	114.0	82.6	1414	1402
1993-031A	ASTRA-1C	22653	LUX	12 MAY	1436.3	0.0	35930	35650
1993-031B	ARASENE	22654	FRAN	12 MAY	1012.7	1.1	36868	17186
1993-031C		22655	LUX	12 MAY	639.3	5.3	36125	283
1993-031D		22656	LUX	12 MAY	622.3	5.4	35305	227
1993-032A	USA 91	22657	US	13 MAY	717.9	55.0	20328	20032
1993-032B		22658	US	13 MAY	96.8	21.9	736	477
1993-032C		22659	US	13 MAY	351.2	34.9	20023	195
1993-034A	PROGRESS M-18	22666	USSR	22 MAY	92.4	51.6	391	389
1993-035A	MOLNIYA 1-86	22671	USSR	26 MAY	171.7	62.8	39866	483
1993-035D		22674	USSR	26 MAY	733.0	62.9	40621	480
1993-036A	COSMOS 2251	22675	USSR	16 JUN	100.7	74.0	803	778
1993-036B		22676	USSR	16 JUN	100.6	74.0	798	773
1993-037A	STS 57	22684	US	21 JUN	93.3	28.4	484	390
1993-038A	COSMOS 2252	22687	USSR	24 JUN	114.0	82.6	1413	1401
1993-038B	COSMOS 2253	22688	USSR	24 JUN	114.1	82.6	1422	1407
1993-038C	COSMOS 2254	22689	USSR	24 JUN	113.8	82.6	1411	1390
1993-038D	COSMOS 2255	22690	USSR	24 JUN	113.9	82.6	1412	1402
1993-038E	COSMOS 2256	22691	USSR	24 JUN	113.9	82.6	1411	1398
1993-038F	COSMOS 2257	22692	USSR	24 JUN	114.0	82.6	1417	1406
1993-038G		22693	USSR	24 JUN	114.7	82.6	1477	1406
1993-039A	GALAXY 4	22694	US	25 JUN	1437.1	0.1	35911	35700
1993-040A	RESURS F-18	22696	USSR	25 JUN	88.9	82.5	228	216
1993-041A	RADCAL	22698	US	25 JUN	101.3	89.6	885	755
1993-041B		22699	US	25 JUN	101.3	89.6	882	753
1993-042A	USA 92	22700	US	26 JUN	717.2	54.7	20240	20085
1993-042B		22701	US	26 JUN	94.4	25.7	785	194
1993-042C		22702	US	26 JUN	354.1	34.7	20217	185

Notes on Satellites In Orbit

1. 297 OBJECTS IDENTIFIED AS LAUNCHED WITH 1961 OMICRON 1 AND 2. SEE 1961 NO LONGER IN ORBIT.
2. 153 OBJECTS IDENTIFIED AS LAUNCHED WITH 1963 014A, B, C. SEE 1963 NO LONGER IN ORBIT SECTION.
4. 29 OBJECTS IDENTIFIED AS LAUNCHED WITH 1964 006A. SEE ALSO 1964 NO LONGER IN ORBIT SECTION.
5. 51 OBJECTS IDENTIFIED AS LAUNCHED WITH 1965 027A. SEE ALSO 1965 NO LONGER IN ORBIT SECTION.
7. 473 OBJECTS IDENTIFIED AS LAUNCHED WITH 1965 082A. SEE ALSO 1965 NO LONGER IN ORBIT SECTION.
8. 111 OBJECTS IDENTIFIED AS LAUNCHED WITH 1968 091A. SEE ALSO 1968 NO LONGER IN ORBIT SECTION.
10. 139 OBJECTS IDENTIFIED AS LAUNCHED WITH 1968 097A. SEE ALSO 1968 NO LONGER IN ORBIT SECTION.
11. 270 OBJECTS IDENTIFIED AS LAUNCHED WITH 1969 082A-082 K. SEE 1969 NO LONGER IN ORBIT SECTION.
12. 373 OBJECTS IDENTIFIED AS LAUNCHED WITH 1970 025A, B. SEE ALSO 1970 NO LONGER IN ORBIT SECTION.
13. 103 OBJECTS IDENTIFIED AS LAUNCHED WITH 1970 089A. SEE ALSO 1970 NO LONGER IN ORBIT SECTION.
15. 46 OBJECTS IDENTIFIED AS LAUNCHED WITH 1970 091A. SEE ALSO 1970 NO LONGER IN ORBIT SECTION.
16. 120 OBJECTS IDENTIFIED AS LAUNCHED WITH 1971 015A. SEE ALSO 1971 NO LONGER IN ORBIT SECTION.
17. 229 OBJECTS IDENTIFIED AS LAUNCHED WITH 1972 058A. SEE ALSO 1972 NO LONGER IN ORBIT SECTION.
18. 198 OBJECTS IDENTIFIED AS LAUNCHED WITH 1973 086A. SEE ALSO 1973 NO LONGER IN ORBIT SECTION.
19. 150 OBJECTS IDENTIFIED AS LAUNCHED WITH 1974 089A, B, C. SEE 1974 NO LONGER IN ORBIT SECTION.
20. 208 OBJECTS IDENTIFIED AS LAUNCHED WITH 1975 004A. SEE ALSO 1975 NO LONGER IN ORBIT SECTION.
21. 72 OBJECTS IDENTIFIED AS LAUNCHED WITH 1976 067A. SEE ALSO 1976 NO LONGER IN ORBIT SECTION.
22. 159 OBJECTS IDENTIFIED AS LAUNCHED WITH 1976 077A. SEE ALSO 1976 NO LONGER IN ORBIT SECTION.
25. 79 OBJECTS IDENTIFIED AS LAUNCHED WITH 1976 126A. SEE ALSO 1976 NO LONGER IN ORBIT SECTION.
26. 172 OBJECTS IDENTIFIED AS LAUNCHED WITH 1977 065A. SEE ALSO 1977 NO LONGER IN ORBIT SECTION.
27. 70 OBJECTS IDENTIFIED AS LAUNCHED WITH 1977 121A. SEE ALSO 1977 NO LONGER IN ORBIT SECTION.
28. 210 OBJECTS IDENTIFIED AS LAUNCHED WITH 1978 026A, B. SEE ALSO 1978 NO LONGER IN ORBIT SECTION.
29. 402 OBJECTS IDENTIFIED AS LAUNCHED WITH 1978 100A, B, C. SEE 1978 NO LONGER IN ORBIT SECTION.
34. 307 OBJECTS IDENTIFIED AS LAUNCHED WITH 1981 053A. SEE ALSO 1981 NO LONGER IN ORBIT SECTION.
37. 60 OBJECTS IDENTIFIED AS LAUNCHED WITH 1982 055A. SEE ALSO 1982 NO LONGER IN ORBIT SECTION.
45. 499 OBJECTS IDENTIFIED AS LAUNCHED WITH 1986 019A. SEE ALSO 1986 NO LONGER IN ORBIT SECTION.
49. 112 OBJECTS IDENTIFIED AS LAUNCHED WITH 1987 020A. SEE ALSO 1987 NO LONGER IN ORBIT SECTION.
52. 79 OBJECTS IDENTIFIED AS LAUNCHED WITH 1990 081A. SEE ALSO 1990 NO LONGER IN ORBIT SECTION.
54. 73 OBJECTS IDENTIFIED AS LAUNCHED WITH 1991 009A. SEE ALSO 1991 NO LONGER IN ORBIT SECTION.
56. 235 OBJECTS IDENTIFIED AS LAUNCHED WITH 1975 052A. SEE ALSO 1975 NO LONGER IN ORBIT SECTION.
57. 16 OBJECTS IDENTIFIED AS LAUNCHED WITH 1990 065A. SEE ALSO 1990 NO LONGER IN ORBIT SECTION.
59. 43 OBJECTS IDENTIFIED AS LAUNCHED WITH 1967 001A. SEE ALSO 1967 NO LONGER IN ORBIT SECTION.
60. 190 OBJECTS IDENTIFIED AS LAUNCHED WITH 1986 017A. SEE ALSO 1986 NO LONGER IN ORBIT SECTION.
61. 224 OBJECTS IDENTIFIED AS LAUNCHED WITH 1992 093A. SEE ALSO 1992 NO LONGER IN ORBIT SECTION.

Satellites No Longer In Orbit

International Designation	Satellite Name	Number	Source	Launch	Decay	Note
1957 LAUNCHES						
ALPHA 1		1	USSR	04 OCT	01 DEC 57	
ALPHA 2	SPUTNIK 1	2	USSR	04 OCT	01 JAN 58	
BETA 1	SPUTNIK 2	3	USSR	03 NOV	14 APR 58	
1958 LAUNCHES						
ALPHA 1	EXPLORER 1	4	US	01 FEB	31 MAR 70	
GAMMA 1	EXPLORER 3	6	US	26 MAR	28 JUN 5 8	
DELTA 1		7	USSR	15 MAY	03 DEC 58	
DELTA 2	SPUTNIK 3	8	USSR	15 MAY	06 APR 60	
EPSILON 1	EXPLORER 4	9	US	26 JUL	23 OCT 59	
ETA 1	PIONEER 1	110	US	11 OCT	12 OCT 58	
THETA 1	PIONEER 3	111	US	06 DEC	07 DEC 58	
ZETA 1	SCORE	10	US	18 DEC	21 JAN 59	
1959 LAUNCHES						
ALPHA 3		5807	US	17 FEB	22 NOV 82	
BETA 1	DISCOVERER 1	13	US	28 FEB	01 MAR 59	
GAMMA 1	DISCOVERER 2	14	US	13 APR	26 APR 59	
DELTA 1	EXPLORER 6	15	US	07 AUG	01 JUL 61	
DELTA 2		17	US	07 AUG	01 JUL 61	
EPSILON 1	DISCOVERER 5	18	US	13 AUG	28 SEP 59	
EPSILON 2	CAPSULE	26	US	13 AUG	11 FEB 61	
ZETA 1	DISCOVERER 6	19	US	19 AUG	20 OCT 59	
XI 1	LUNA 2	114	USSR	12 SEP	15 SEP 59	A
THETA 1	LUNA 3	21	USSR	04 OCT	29 APR 60	
IOTA 2		23	US	13 OCT	16 JUL 89	
KAPPA 1	DISCOVERER 7	24	US	07 NOV	26 NOV 59	
LAMBDA 1	DISCOVERER 8	25	US	20 NOV	08 MAR 60	
1960 LAUNCHES						
BETA 1		28	US	01 APR	02 JUL 91	
BETA 3		101	US	01 APR	17 APR 83	
GAMMA 1		30	US	13 APR	18 AUG 60	
GAMMA 2	TRANSIT 1B	31	US	13 APR	05 OCT 67	
GAMMA 3		33	US	13 APR	01 JUL 60	
GAMMA 4		99	US	13 APR	17 JUL 7 9	
DELTA 1	DISCOVERER 11	32	US	15 APR	26 APR 60	
EPSILON 1	SPUTNIK 4	34	USSR	15 MAY	05 SEP 62	
EPSILON 2		35	USSR	15 MAY	17 JUL 60	
EPSILON 3		36	USSR	15 MAY	15 OCT 65	
EPSILON 4		37	USSR	15 MAY	01 JUL 61	
EPSILON 5		38	USSR	15 MAY	01 OCT 60	
EPSILON 6		39	USSR	15 MAY	24 SEP 60	
EPSILON 7		40	USSR	15 MAY	24 SEP 60	
EPSILON 8		41	USSR	15 MAY	01 OCT 60	
EPSILON 9		42	USSR	15 MAY	01 OCT 60	
ZETA 1	MIDAS 2	43	US	24 MAY	07 FEB 74	
ZETA 2		44	US	24 MAY	05 DEC 60	
THETA 1	DISCOVERER 13	48	US	10 AUG	14 NOV 60	
THETA 1	CAPSULE	61002	US	10 AUG	11 AUG 60	B
IOTA 1	ECHO 1	49	US	12 AUG	24 MAY 68	
KAPPA 1	DISCOVERER 14	54	US	18 AUG	16 SEP 60	
KAPPA 1	CAPSULE	61003	US	18 AUG	19 AUG 60	B
LAMBDA 1	SPUTNIK 5	55	USSR	19 AUG	20 AUG 60	C
LAMBDA 2		56	USSR	19 AUG	23 SEP 60	
MU 1	DISCOVERER 15	57	US	13 SEP	18 OCT 60	
MU 1	CAPSULE	61004	US	13 SEP	15 SEP 60	D
XI 2		62	US	03 NOV	27 OCT 85	
XI 3		69	US	03 NOV	16 FEB 70	
XI 4		105	US	03 NOV	03 MAR 72	
OMICRON 1	DISCOVERER 17	61	US	12 NOV	29 DEC 60	
OMICRON 1	CAPSULE	61005	US	12 NOV	14 NOV 60	
PI 2		64	US	23 NOV	23 SEP 81	
PI 3		74	US	23 NOV	10 MAY 89	
PI 4		75	US	23 NOV	25 JAN 90	
RHO 1	SPUTNIK 6	65	USSR	01 DEC	02 DEC 60	
RHO 2		66	USSR	01 DEC	02 DEC 60	
SIGMA 1	DISCOVERER 18	67	US	07 DEC	02 APR 61	
SIGMA 1	CAPSULE	61006	US	07 DEC	10 DEC 60	B
TAU 1	DISCOVERER 19	68	US	20 DEC	23 JAN 61	
1961 LAUNCHES						
ALPHA 1	SAMOS 2	70	US	31 JAN	21 OCT 73	
ALPHA 2		79	US	31 JAN	09 OCT 70	
BETA 1	SPUTNIK 7	71	USSR	04 FEB	26 FEB 61	
BETA 2		72	USSR	04 FEB	13 FEB 61	
BETA 3		73	USSR	04 FEB	17 MAR 61	
GAMMA 2		76	USSR	12 FEB	18 FEB 61	
GAMMA 3	SPUTNIK 8	77	USSR	12 FEB	25 FEB 61	
GAMMA 4		78	USSR	12 FEB	18 FEB 61	
DELTA 1	EXPLORER 9	81	US	16 FEB	09 APR 64	
DELTA 4		86	US	16 FEB	01 JUL 61	
DELTA 5		3738	US	16 FEB	29 MAR 82	
DELTA 8		5237	US	16 FEB	20 JUL 74	
EPSILON 1	DISCOVERER 20	83	US	17 FEB	28 JUL 62	
EPSILON 2		88	US	17 FEB	02 APR 61	
EPSILON 3		89	US	17 FEB	20 APR 61	
EPSILON 4		90	US	17 FEB	31 OCT 61	
ZETA 1	DISCOVERER 21	84	US	18 FEB	20 APR 62	
ETA 1	TRANSIT 3B & LOFTI-1	87	US	22 FEB	30 MAR 61	
THETA 1	SPUTNIK 9	91	USSR	09 MAR	09 MAR 61	C
THETA 2		92	USSR	09 MAR	10 MAR 61	
THETA 3		93	USSR	09 MAR	10 MAR 61	
THETA 4		94	USSR	09 MAR	10 MAR 61	
IOTA 1	SPUTNIK 10	95	USSR	25 MAR	25 MAR 61	
IOTA 2		96	USSR	25 MAR	26 MAR 61	
IOTA 3		97	USSR	25 MAR	26 MAR 61	
KAPPA 1	EXPLORER 10	98	US	25 MAR	01 JUN 68	
LAMBDA 1	DISCOVERER 23	100	US	08 APR	16 APR 62	
LAMBDA 2	CAPSULE	102	US	08 APR	23 MAY 62	
LAMBDA 3		106	US	08 APR	10 SEP 61	
MU 1	VOSTOK 1	103	USSR	12 APR	12 APR 61	E
MU 2		104	USSR	12 APR	16 APR 61	
XI 1	DISCOVERER 25	108	US	16 JUN	12 JUL 61	
XI 2		109	US	16 JUN	19 JUN 61	
XI 1	CAPSULE	61007	US	16 JUN	18 JUN 61	B
OMICRON 101		328	US	29 JUN	03 MAY 70	
OMICRON 104		331	US	29 JUN	24 MAR 89	
OMICRON 105		332	US	29 JUN	20 SEP 70	
OMICRON 113		352	US	29 JUN	10 DEC 66	
OMICRON 117		356	US	29 JUN	02 FEB 81	
OMICRON 119		358	US	29 JUN	31 AUG 66	
OMICRON 120		359	US	29 JUN	31 AUG 65	
OMICRON 121		401	US	29 JUN	25 JAN 89	
OMICRON 131		411	US	29 JUN	20 JUN 78	
OMICRON 132		412	US	29 JUN	30 JUN 80	
OMICRON 133		413	US	29 JUN	02 DEC 89	
OMICRON 135		415	US	29 JUN	27 MAY 89	
OMICRON 138		418	US	29 JUN	05 NOV 91	
OMICRON 14		129	US	29 JUN	31 AUG 81	
OMICRON 141		421	US	29 JUN	02 OCT 84	
OMICRON 145		464	US	29 JUN	18 DEC 89	
OMICRON 147		466	US	29 JUN	26 NOV 88	
OMICRON 149		468	US	29 JUN	03 DEC 81	
OMICRON 154		473	US	29 JUN	02 MAR 72	
OMICRON 158		477	US	29 JUN	18 JAN 82	
OMICRON 164		531	US	29 JUN	15 APR 71	
OMICRON 167		539	US	29 JUN	24 FEB 91	
OMICRON 168		540	US	29 JUN	20 AUG 91	
OMICRON 170		542	US	29 JUN	06 JUL 91	
OMICRON 175		547	US	29 JUN	20 NOV 89	
OMICRON 178		550	US	29 JUN	03 FEB 91	
OMICRON 179		551	US	29 JUN	16 FEB 90	
OMICRON 181		556	US	29 JUN	28 MAY 9a	
OMICRON 182		557	US	29 JUN	14 FEB 67	
OMICRON 190		646	US	29 JUN	07 JUN 72	
OMICRON 194		650	US	29 JUN	27 MAY 82	
OMICRON 198		654	US	29 JUN	03 JUL 68	
OMICRON 206		662	US	29 JUN	27 DEC 91	
OMICRON 208		1501	US	29 JUN	23 JAN 79	
OMICRON 211		2010	US	29 JUN	06 MAY 80	
OMICRON 212		2087	US	29 JUN	13 MAR 79	
OMICRON 213		3037	US	29 JUN	01 FEB 68	
OMICRON 218		3740	US	29 JUN	25 DEC 74	
OMICRON 22		137	US	29 JUN	15 DEC 80	
OMICRON 224		3968	US	29 JUN	0 JUN 69	
OMICRON 225		3969	US	29 JUN	27 JUN 69	
OMICRON 230		4027	US	29 JUN	19 DEC 70	
OMICRON 234		5716	US	29 JUN	09 APR 82	
OMICRON 237		5747	US	29 JUN	03 JAN 77	
OMICRON 24		139	US	29 JUN	08 JUL 78	
OMICRON 242		5787	US	29 JUN	28 AUG 74	
OMICRON 247		5877	US	29 JUN	01 MAY 79	
OMICRON 248		5972	US	29 JUN	30 OCT 76	
OMICRON 249		5985	US	29 JUN	07 DEC 76	
OMICRON 25		140	US	29 JUN	30 SEP 62	
OMICRON 250		6179	US	29 JUN	20 NOV 76	
OMICRON 251		6746	US	29 JUN	21 DEC 74	
OMICRON 256		9969	US	29 JUN	05 APR 78	
OMICRON 263		13941	US	29 JUN	21 SEP 88	
OMICRON 265		14499	US	29 JUN	20 MAR91	
OMICRON 266		14500	US	29 JUN	06 MAY 89	
OMICRON 268		15339	US	29 JUN	07 DEC 87	
OMICRON 27		142	US	29 JUN	25 DEC 67	
OMICRON 270		15760	US	29 JUN	12 AUG 87	
OMICRON 272		17648	US	29 JUN	12 FEB 90	
OMICRON 277		17653	US	29 JUN	11 AUG 90	
OMICRON 278		17711	US	29 JUN	29 DEC 89	
OMICRON 279		17712	US	29 JUN	01 AUG 88	
OMICRON 28		143	US	29 JUN	16 JUN 62	

International Designation	Satellite Name	Number	Source	Launch	Decay	Note
OMICRON 281		17714	US	29 JUN	15 JAN 90	
OMICRON 282		18498	US	29 JUN	21 NOV 89	
OMICRON 286		18890	US	29 JUN	30 JUN 89	
OMICRON 288		19031	US	29 JUN	26 JUN 89	
OMICRON 293		19428	US	29 JUN	08 MAR 90	
OMICRON 294		19430	US	29 JUN	19 MAY 90	
OMICRON 31		146	US	29 JUN	07 JUL 69	
OMICRON 34		149	US	29 JUN	03 MAR 79	
OMICRON 36		151	US	29 JUN	02 DEC 67	
OMICRON 38		153	US	29 JUN	22 APR 81	
OMICRON 42		157	US	29 JUN	08 JAN 81	
OMICRON 45		161	US	29 JUN	29 AUG 81	
OMICRON 46		171	US	29 JUN	29 JAN 62	
OMICRON 49		176	US	29 JUN	12 FEB 82	
OMICRON 55		212	US	29 JUN	11 AUG 88	
OMICRON 56		216	US	29 JUN	13 NOV 89	
OMICRON 57		220	US	29 JUN	21 FEB 66	
OMICRON 62		231	US	29 JUN	13 FEB 65	
OMICRON 67		236	US	29 JUN	13 MAY 69	
OMICRON 68		237	US	29 JUN	12 APR 69	
OMICRON 69		238	US	29 JUN	07 DEC 69	
OMICRON 70		239	US	29 JUN	17 JUN 67	
OMICRON 75		250	US	29 JUN	13 SEP 81	
OMICRON 84		265	US	29 JUN	12 OCT 78	
OMICRON 88		300	US	29 JUN	15 DEC 79	
OMICRON 89		301	US	29 JUN	08 APR 81	
OMICRON 90		317	US	29 JUN	11 JAN 81	
OMICRON 92		319	US	29 JUN	23 JUL 81	
OMICRON 94		321	US	29 JUN	26 FEB 69	
OMICRON 95		322	US	29 JUN	06 JUN 69	
OMICRON 99		326	US	29 JUN	07 OCT 79	
PI 1	DISCOVERER 26	160	US	07 JUL	05 DEC 61	
PI 1	CAPSULE	61008	US	07 JUL	09 JUL 61	B
SIGMA 2		164	US	12 JUL	24 JUL 61	
SIGMA 5		6215	US	12 JUL	30 OCT 76	
TAU 1	VOSTOK 2	168	USSR	06 AUG	07 AUG 61	E
TAU 2		169	USSR	06 AUG	09 AUG 61	
UPSILON 1	EXPLORER 12	170	US	16 AUG	01 SEP 63	
PHI 1	RANGER 1	173	US	23 AUG	30 AUG 61	
PHI 2		174	US	23 AUG	03 SEP 61	
CHI 1	EXPLORER 13	180	US	25 AUG	28 AUG 61	
PSI 1	DISCOVERER 29	181	US	30 AUG	10 SEP 61	
PSI 1	CAPSULE	61009	US	30 AUG	04 SEP 61	
OMEGA 1	DISCOVERER 30	182	US	12 SEP	11 DEC 61	
OMEGA 2		185	US	12 SEP	18 SEP 61	
OMEGA 3		187	US	12 SEP	28 SEP 61	
OMEGA 1	CAPSULE	61010	US	12 SEP	15 SEP 61	B
A-ALPHA 1	MA 4	183	US	13 SEP	13 SEP 61	B
A-ALPHA 2		184	US	13 SEP	13 SEP 61	
A-BETA 1	DISCOVERER 31	186	US	17 SEP	26 OCT 61	
A-GAMMA 1	DISCOVERER 32	189	US	13 OCT	13 NOV 61	
A-GAMMA 2		190	US	13 OCT	25 OCT 61	B
A-GAMMA 3		191	US	13 OCT	16 OCT 61	
A-GAMMA 1	CAPSULE	61011	US	13 OCT	14 OCT 61	B
A-DELTA 2		193	US	21 OCT	05 DEC 61	
A-EPSILON 1	DISCOVERER 34	197	US	05 NOV	07 DEC 62	
A-EPSILON 2		198	US	05 NOV	30 NOV 61	
A-EPSILON 3		199	US	05 NOV	09 DEC 61	
A-EPSILON 4		200	US	05 NOV	10 DEC 61	
A-EPSILON 5		203	US	05 NOV	12 DEC 61	
A-ZETA 1	DISCOVERER 35	201	US	15 NOV	03 DEC 61	
A-ZETA 2		207	US	15 NOV	23 NOV 61	
A-ZETA 1	CAPSULE	61012	US	15 NOV	16 NOV 61	B
A-THETA 1	RANGER 2	206	US	18 NOV	20 NOV 61	
A-IOTA 1	MA 5	208	US	29 NOV	29 NOV 61	B
A-IOTA 2		209	US	29 NOV	30 NOV 61	
A-KAPPA 1	DISCOVERER 36	213	US	12 DEC	08 MAR 62	
A-KAPPA 2	OSCAR 1	214	US	12 DEC	31 JAN 62	
A-KAPPA 3		215	US	12 DEC	19 DEC 61	
A-KAPPA 1	CAPSULE	61013	US	12 DEC	16 DEC 61	B
A-LAMBDA 1		217	US	22 DEC	14 AUG 62	
A-LAMBDA 2		218	US	22 DEC	31 DEC 61	
A-LAMBDA 3		219	US	22 DEC	09 JAN 62	

1962 LAUNCHES

International Designation	Satellite Name	Number	Source	Launch	Decay	Note
GAMMA 1	FRIENDSHIP 7	240	US	20 FEB	20 FEB 62	F
GAMMA 2		241	US	20 FEB	21 FEB 62	
DELTA 1		242	US	21 FEB	04 MAR 62	
EPSILON 1	DISCOVERER 38	247	US	27 FEB	21 MAR 62	
EPSILON 2		248	US	27 FEB	03 MAR 62	
EPSILON 3		249	US	27 FEB	03 MAR 62	
EPSILON 4		251	US	27 FEB	07 MAR 62	
EPSILON 1	CAPSULE	61014	US	27 FEB	03 MAR 62	B
ETA 1		256	US	07 MAR	07 JUN 63	
ETA 2		258	US	07 MAR	31 MAR 62	
ETA 3		259	US	07 MAR	03 NOV 62	
ZETA 1	OSO 1	255	US	07 MAR	08 OCT 81	
ZETA 2		257	US	07 MAR	03 MAY 70	
THETA 1	COSMOS 1	266	USSR	16 MAR	25 MAY 62	
THETA 2		267	USSR	16 MAR	18 JUN 62	
IOTA 1	COSMOS 2	269	USSR	06 APR	20 AUG 63	
IOTA 2		270	USSR	06 APR	06 OCT 62	
KAPPA 2		272	US	09 APR	04 MAY 62	
KAPPA 5		4494	US	09 APR	22 MAY 77	
KAPPA 6		5703	US	09 APR	07 NOV 76	
LAMBDA 1		276	US	18 APR	28 MAY 62	
LAMBDA 2		277	US	18 APR	20 APR 62	
LAMBDA 3		278	US	18 APR	21 APR 62	
LAMBDA 4		279	US	18 APR	21 APR 62	
MU 1	RANGER 4	280	US	23 APR	26 APR 62	G
NU 1	COSMOS 3	281	USSR	24 APR	17 OCT 62	
NU 2		283	USSR	24 APR	05 AUG 62	
OMICRON 1	ARIEL 1	285	UK	26 APR	24 MAY 76	
OMICRON 2		288	US	26 APR	28 APR 71	
PI 1		286	US	26 APR	28 APR 62	
XI 1	COSMOS 4	287	USSR	26 APR	29 APR 62	
XI 2		284	USSR	26 APR	17 JUN 62	
XI 3		289	USSR	26 APR	03 MAY 62	
RHO 1		290	US	29 APR	26 MAY 62	
RHO 2		291	US	29 APR	01 MAY 62	
SIGMA 1		292	US	15 MAY	26 NOV 63	
SIGMA 2		293	US	15 MAY	03 JUL 62	
SIGMA 3		294	US	15 MAY	13 JUL 62	
TAU 1	AURORA 7	295	US	24 MAY	24 MAY 62	F
TAU 2		296	US	24 MAY	25 MAY 62	
UPSILON 1	COSMOS 5	297	USSR	28 MAY	02 MAY 63	
UPSILON 2		298	USSR	28 MAY	15 DEC 62	
PHI 1		302	US	30 MAY	11 JUN 62	
PHI 2		303	US	30 MAY	02 JUN 62	
CHI 1		304	US	02 JUN	28 JUN 62	
CHI 2	OSCAR 2	305	US	02 JUN	21 JUN 62	
CHI 3		306	US	02 JUN	06 JUN 62	
PSI 1		307	US	17 JUN	18 JUN 62	
OMEGA 1		308	US	18 JUN	30 OCT 63	
OMEGA 2		310	US	18 JUN	12 JUL 62	
OMEGA 3		314	US	18 JUN	14 JUL 62	
A ALPHA 2		311	US	19 JUN	28 NOV 90	
A ALPHA 5		6251	US	19 JUN	30 OCT 76	
A-BETA 1		315	US	23 JUN	07 JUL 62	
A-GAMMA 1		316	US	28 JUN	14 SEP 62	
A-DELTA 1	COSMOS 6	338	USSR	30 JUN	08 AUG 62	
A-DELTA 2		339	USSR	30 JUN	08 SEP 62	
A-ZETA 1		342	US	18 JUL	25 JUL 62	
A-ZETA 2		343	US	18 JUL	27 JUL 62	
A-ETA 1		344	US	21 JUL	14 AUG 62	
A-IOTA 1	COSMOS 7	346	USSR	28 JUL	01 AUG 62	
A-IOTA 2		347	USSR	28 JUL	21 AUG 62	
A-IOTA 3		348	USSR	28 JUL	31 JUL 62	
A-IOTA 4		349	USSR	28 JUL	30 JUL 62	
A-THETA 1		345	US	28 JUL	24 AUG 62	
A-KAPPA 1		360	US	02 AUG	26 AUG 62	
A-KAPPA 2		362	US	02 AUG	08 AUG 62	
A-LAMBDA 1		361	US	05 AUG	06 AUG 62	
A-MU 1	VOSTOK 3	363	USSR	11 AUG	15 AUG 62	E
A-MU 2		364	USSR	11 AUG	14 AUG 62	
A-NU 1	VOSTOK 4	365	USSR	12 AUG	15 AUG 62	E
A-NU 2		366	USSR	12 AUG	14 AUG 62	
A-XI 1	COSMOS 8	367	USSR	18 AUG	17 AUG 63	
A-XI 2		368	USSR	18 AUG	19 AUG 62	
A OMICRON 2		370	US	23 AUG	02 MAR 80	
A OMICRON 3		378	US	23 AUG	12 AUG 92	
A-PI 1		371	USSR	25 AUG	28 AUG 62	
A-PI 2		372	USSR	25 AUG	02 SEP 62	
A-PI 3		373	USSR	25 AUG	31 AUG 62	
A-PI 4		379	USSR	25 AUG	05 SEP 62	
A-PI 5		380	USSR	25 AUG	30 AUG 62	
A-PI 6		376	USSR	25 AUG	06 SEP 62	
A-PI 7		386	USSR	25 AUG	08 SEP 62	
A-PI 8		387	USSR	25 AUG	05 SEP 62	
A-SIGMA 1		377	US	29 AUG	10 SEP 62	
A-TAU 1		381	USSR	01 SEP	06 SEP 62	
A-TAU 2		382	USSR	01 SEP	03 SEP 62	
A-TAU 3		383	USSR	01 SEP	01 OCT 62	
A-TAU 4		384	USSR	01 SEP	21 SEP 62	
A-UPSILON 1		385	US	01 SEP	26 OCT 64	
A-PHI 1		389	USSR	12 SEP	14 SEP 62	
A-PHI 2		390	USSR	12 SEP	12 SEP 62	
A-PHI 3		391	USSR	12 SEP	17 SEP 62	
A-PHI 4		392	USSR	12 SEP	13 SEP 62	
A-PHI 5		393	USSR	12 SEP	13 SEP 62	
A-PHI 6		394	USSR	12 SEP	16 SEP 62	
A-PHI 7		395	USSR	12 SEP	15 SEP 62	
A-CHI 1	ERS 2	396	US	17 SEP	16 NOV 62	
A PSI 2		398	US	18 SEP	10 JAN 83	
A PSI 4		400	US	18 SEP	05 APR 89	
A-OMEGA 1		422	USSR	27 SEP	01 OCT 62	
A-OMEGA 2		423	USSR	27 SEP	22 DEC 62	
A-OMEGA 3		425	USSR	27 SEP	06 OCT 62	
A-OMEGA 4		428	USSR	27 SEP	08 OCT 62	
A-OMEGA 5		429	USSR	27 SEP	04 OCT 62	
A-OMEGA 6		430	USSR	27 SEP	08 OCT 62	
A-OMEGA 7		431	USSR	27 SEP	03 OCT 62	
A-OMEGA 8		435	USSR	27 SEP	06 OCT 62	
B-BETA 1		427	US	29 SEP	14 OCT 62	
B GAMMA 1	EXPLORER 14	432	US	02 OCT	01 JUL 66	
B GAMMA 2		61000	US	02 OCT	02 DEC 62	
B-DELTA 1	SIGMA 7	433	US	03 OCT	03 OCT 62	F
B-DELTA 2		434	US	03 OCT	04 OCT 62	
B-EPSILON 1		436	US	09 OCT	16 NOV 62	
B-ZETA 1		437	USSR	17 OCT	21 OCT 62	
B-ZETA 2		438	USSR	17 OCT	05 NOV 62	
B-THETA 1		441	USSR	20 OCT	18 MAY 64	
B-THETA 2		442	USSR	20 OCT	05 JUN 63	
B-IOTA 1	SPUTNIK 29	443	USSR	24 OCT	29 OCT 62	
B-IOTA 10		485	USSR	24 OCT	10 DEC 62	

International Designation	Satellite Name	Number	Source	Launch	Decay	Note
B-IOTA 11		486	USSR	24 OCT	30 NOV 62	
B-IOTA 12		487	USSR	24 OCT	01 DEC 62	
B-IOTA 13		491	USSR	24 OCT	25 DEC 62	
B-IOTA 14		492	USSR	24 OCT	15 DEC 62	
B-IOTA 15		493	USSR	24 OCT	21 DEC 62	
B-IOTA 16		494	USSR	24 OCT	17 DEC 62	
B-IOTA 17		495	USSR	24 OCT	26 DEC 62	
B-IOTA 18		496	USSR	24 OCT	23 DEC 62	
B-IOTA 19		497	USSR	24 OCT	26 DEC 62	
B-IOTA 2		456	USSR	24 OCT	02 DEC 62	H
B-IOTA 20		498	USSR	24 OCT	10 DEC 62	
B-IOTA 21		499	USSR	24 OCT	09 DEC 62	
B-IOTA 22		500	USSR	24 OCT	28 DEC 62	
B-IOTA 23		501	USSR	24 OCT	26 DEC 62	
B-IOTA 24		512	USSR	24 OCT	31 DEC 62	
B-IOTA 3		457	USSR	24 OCT	26 FEB 63	
B-IOTA 4		458	USSR	24 OCT	29 NOV 62	
B-IOTA 5		459	USSR	24 OCT	24 DEC 62	
B-IOTA 6		460	USSR	24 OCT	08 DEC 62	
B-IOTA 7		482	USSR	24 OCT	10 DEC 62	
B-IOTA 8		483	USSR	24 OCT	02 JAN 63	
B-IOTA 9		484	USSR	24 OCT	04 DEC 62	
B-KAPPA 1		444	US	26 OCT	05 OCT 67	
B LAMBDA 1	EXPLORER 15	445	US	27 OCT	19 DEC 78	
B LAMBDA 3		5992	US	27 OCT	01 DEC 78	
B LAMBDA 2		61001	US	27 OCT	27 DEC 62	
B-NU 1		448	USSR	01 NOV	03 NOV 62	
B-NU 2		449	USSR	01 NOV	03 NOV 62	
B-XI 1		451	USSR	04 NOV	05 NOV 62	
B-XI 2		452	USSR	04 NOV	08 NOV 62	
B-XI 3		454	USSR	04 NOV	19 JAN 63	
B-XI 4		488	USSR	04 NOV	25 DEC 62	
B-XI 5		489	USSR	04 NOV	27 DEC 62	
B-OMICRON 1		453	US	05 NOV	03 DEC 62	
B-PI 1		455	US	11 NOV	12 NOV 62	
B-RHO 1		481	US	24 NOV	13 DEC 62	
B-SIGMA 1		490	US	04 DEC	08 DEC 62	
B-TAU 1		502	US	13 DEC	09 FEB 67	
B-TAU 2	INJUN 3	504	US	13 DEC	25 AUG 68	
B-TAU 3		507	US	13 DEC	01 JUL 63	
B-TAU 4		508	US	13 DEC	18 JAN 66	
B-TAU 5		513	US	13 DEC	05 FEB 67	
B-TAU 6		520	US	13 DEC	24 MAR 68	
B-PHI 1		505	US	14 DEC	08 JAN 63	
B PSI 1	TRANSIT 5A	509	US	19 DEC	25 SEP 86	
B PSI 3		519	US	19 DEC	21 OCT 89	
B PSI 4		523	US	19 DEC	30 NOV 80	
B PSI 5		7258	US	19 DEC	19 JUL 77	
B-PSI 2		514	US	19 DEC	22 APR 69	
B-OMEGA 1		517	USSR	22 DEC	30 DEC 62	
B-OMEGA 2		518	USSR	22 DEC	22 JAN 63	

1963 LAUNCHES

International Designation	Satellite Name	Number	Source	Launch	Decay	Note
1963-001A	SPUTNIK	521	USSR	04 JAN	05 JAN 63	
1963-001B		522	USSR	04 JAN	11 JAN 63	
1963-001C		524	USSR	04 JAN	11 JAN 63	
1963-002A		525	US	07 JAN	24 JAN 63	
1963-002B		526	US	07 JAN	16 JAN 63	
1963-003A		527	US	16 JAN	09 JAN 69	
1963-003B		528	US	16 JAN	08 NOV 63	
1963-003C		529	US	16 JAN	31 DEC 63	
1963-004B		532	US	14 FEB	02 NOV 81	
1963-005A		533	US	19 FEB	26 DEC 79	
1963-005B		534	US	19 FEB	29 MAY 84	
1963-005C		535	US	19 FEB	29 JUN 69	
1963-005D		536	US	19 FEB	18 JUL 78	
1963-006A	COSMOS 13	554	USSR	21 MAR	29 MAR 63	
1963-006B		555	USSR	21 MAR	09 APR 63	
1963-007A		562	US	01 APR	26 APR 63	
1963-008A		563	USSR	02 APR	03 APR 63	
1963-009A	EXPLORER 17	564	US	03 APR	24 NOV 66	
1963-009B		565	US	03 APR	24 NOV 63	
1963-010A	COSMOS 14	567	USSR	13 APR	29 AUG 63	
1963-010B		568	USSR	13 APR	06 JUL 63	
1963-011A		569	USSR	22 APR	27 APR 63	
1963-011B		570	USSR	22 APR	01 MAY 63	
1963-012A		571	USSR	28 APR	08 MAY 63	
1963-012B		572	USSR	28 APR	20 MAY 63	
1963-014U		2373	US	09 MAY	23JUN 93	
1963-014AE		2497	US	09 MAY	30 MAR 81	
1963-014AF		2499	US	09 MAY	12 APR 77	
1963-014AG		2500	US	09 MAY	19 NOV 74	
1963-014AH		2522	US	09 MAY	18 JAN 78	
1963-014AL		2532	US	09 MAY	17 OCT 87	
1963-014AM		2533	US	09 MAY	03 JAN 77	
1963-014AN		2638	US	09 MAY	15 DEC 81	
1963-014AP		2793	US	09 MAY	25 FEB 71	
1963-014AW		3105	US	09 MAY	07 JUN 73	
1963-014AX		3159	US	09 MAY	18 FEB 68	
1963-014AY		3175	US	09 MAY	21 MAY 68	
1963-014BA		3236	US	09 MAY	10 MAY 79	
1963-014BF		3241	US	09 MAY	28 MAY 70	
1963-014BG		3242	US	09 MAY	17 OCT 87	
1963-014BH		3243	US	09 MAY	17 OCT 87	
1963-014BK		3245	US	09 MAY	30 OCT 76	
1963-014BL		3246	US	09 MAY	17 OCT 87	
1963-014BM		3247	US	09 MAY	04 APR 77	
1963-014BN		3248	US	09 MAY	17 OCT 87	
1963-014BP		3249	US	09 MAY	28 SEP 80	
1963-014BQ		3250	US	09 MAY	18 JAN 90	
1963-014BR		3251	US	09 MAY	17 OCT 87	
1963-014BS		3252	US	09 MAY	30 MAY 73	
1963-014BT		3253	US	09 MAY	17 OCT 87	
1963-014BU		3254	US	09 MAY	02 NOV 69	
1963-014BV		3255	US	09 MAY	30 OCT 76	
1963-014BW		3256	US	09 MAY	12 APR 77	
1963-014BY		3258	US	09 MAY	08 AUG 69	
1963-014BZ		3259	US	09 MAY	30 OCT 76	
1963-014CA		3260	US	09 MAY	17 JUN 69	
1963-014CB		3261	US	09 MAY	17 OCT 87	
1963-014CC		3262	US	09 MAY	04 JUL 69	
1963-014CD		3263	US	09 MAY	17 OCT 87	
1963-014CE		3264	US	09 MAY	17 OCT 87	
1963-014CG		3268	US	09 MAY	01 AUG 69	
1963-014CH		3324	US	09 MAY	14 DEC 90	
1963-014CJ		3329	US	09 MAY	17 OCT 87	
1963-014CK		3330	US	09 MAY	03 FEB 70	
1963-014CL		3756	US	09 MAY	17 OCT 87	
1963-014CM		5035	US	09 MAY	24 SEP 84	
1963-014CN		5182	US	09 MAY	17 OCT 87	
1963-014CP		5500	US	09 MAY	23 JAN 77	
1963-014CQ		5503	US	09 MAY	03 JAN 77	
1963-014CR		5511	US	09 MAY	23 JAN 75	
1963-014CS		5512	US	09 MAY	05 DEC 76	
1963-014CT		5513	US	09 MAY	19 FEB 75	
1963-014CU		5702	US	09 MAY	17 OCT 87	
1963-014CV		5916	US	09 MAY	26 APR 72	
1963-014CW		5953	US	09 MAY	21 SEP 85	
1963-014CX		5954	US	09 MAY	22 NOV 82	
1963-014CZ		5993	US	09 MAY	20 AUG 74	
1963-014DB		5995	US	09 MAY	17 OCT 87	
1963-014DC		5996	US	09 MAY	30 OCT 76	
1963-014DD		5997	US	09 MAY	21 NOV 73	
1963-014DE		9694	US	09 MAY	22 NOV 82	
1963-014DF		11581	US	09 MAY	17 OCT 87	
1963-014EF		19058	US	09 MAY	28 MAR 89	
1963-014ES		19880	US	09 MAY	16 DEC 91	
1963-015A	FAITH 7	576	US	15 MAY	16 MAY 63	F
1963-015B		577	US	15 MAY	16 MAY 63	
1963-016A		578	US	18 MAY	27 MAY 63	
1963-017A	COSMOS 17	580	USSR	22 MAY	02 JUN 65	
1963-017B		581	USSR	22 MAY	26 AUG 63	
1963-017C		582	USSR	22 MAY	12 MAY 65	
1963-017D		583	USSR	22 MAY	30 AUG 63	
1963-017E		584	USSR	22 MAY	25 JUL 63	
1963-017F		585	USSR	22 MAY	30 SEP 63	
1963-017G		588	USSR	22 MAY	02 APR 64	
1963-018A	COSMOS 18	586	USSR	24 MAY	02 JUN 63	
1963-018B		587	USSR	24 MAY	08 JUN 63	
1963-019A		590	US	12 JUN	11 JUL 63	
1963-020A	VOSTOK 5	591	USSR	14 JUN	19 JUN 63	E
1963-020B		592	USSR	14 JUN	16 JUN 63	
1963-021A		593	US	15 JUN	07 AUG 63	
1963-021B	LOFTI-2A	601	US	15 JUN	18 JUL 63	
1963-021C	SOLRAD 6	599	US	15 JUN	01 AUG 63	
1963-021D	RADOSE	600	US	15 JUN	30 JUL 63	
1963-021E		598	US	15 JUN	27 JUL 63	
1963-021F		597	US	15 JUN	05 JUL 63	
1963-021G		617	US	15 JUN	03 JUL 63	
1963-022A		594	US	16 JUN	03 AUG 90	
1963-022C		610	US	16 JUN	15 NOV 91	
1963-022D		611	US	16 JUN	01 MAR 71	
1963-023A	VOSTOK 6	595	USSR	16 JUN	19 JUN 63	E
1963-023B		596	USSR	16 JUN	18 JUN 63	
1963-024B		605	US	19 JUN	23 DEC 78	
1963-024C		606	US	19 JUN	09 JUN 88	
1963-024D		607	US	19 JUN	01 FEB 80	
1963-025A		609	US	27 JUN	26 JUL 63	
1963-026A	RESEARCH SATELLITE FOR GEOPHYSICS	612	US	28 JUN	14 DEC 83	
1963-027A		613	US	29 JUN	26 OCT 69	
1963-027B		615	US	29 JUN	15 NOV 64	
1963-027C		616	US	29 JUN	10 JUL 63	
1963-028A		618	US	12 JUL	18 JUL 63	
1963-028B		619	US	12 JUL	13 JUL 63	
1963-028C		620	US	12 JUL	16 JUL 63	
1963-029A		621	US	18 JUL	13 AUG 63	
1963-029B		623	US	18 JUL	29 JUL 63	
1963-030D		624	US	18 JUL	12 APR 71	
1963-031B		625	US	26 JUL	09 JUL 89	
1963-032A		626	US	31 JUL	11 AUG 63	
1963-032B		627	US	31 JUL	10 AUG 63	
1963-033A	COSMOS 19	632	USSR	06 AUG	30 MAR 64	
1963-033B		633	USSR	06 AUG	09 DEC 63	
1963-034A		636	US	24 AUG	12 SEP 63	
1963-034B		642	US	24 AUG	08 SEP 63	
1963-035A		637	US	29 AUG	07 NOV 63	
1963-035B		638	US	29 AUG	29 SEP 63	
1963-035C		639	US	29 AUG	02 SEP 63	
1963-035D		640	US	29 AUG	01 SEP 63	
1963-036A		641	US	06 SEP	13 SEP 63	
1963-036B		663	US	06 SEP	14 SEP 63	
1963-036C		664	US	06 SEP	10 SEP 63	
1963-036D		665	US	06 SEP	08 SEP 63	
1963-036E		666	US	06 SEP	11 SEP 63	

International Designation	Satellite Name	Number	Source	Launch	Decay	Note
1963-036F		667	US	06 SEP	10 SEP 63	
1963-037A		668	US	24 SEP	12 OCT 63	
1963-038H		7259	US	28 SEP	06 NOV 82	
1963-039B	ERS 12	675	US	17 OCT	01 JUL 65	
1963-040A	COSMOS 20	673	USSR	18 OCT	26 OCT 63	
1963-040B		676	USSR	18 OCT	31 OCT 63	
1963-041A		677	US	25 OCT	29 OCT 63	
1963-041B		678	US	25 OCT	29 OCT 63	
1963-041C		679	US	25 OCT	29 OCT 63	
1963-041D		680	US	25 OCT	29 OCT 63	
1963-042A		681	US	29 OCT	21 JAN 64	
1963-042B		682	US	29 OCT	23 MAY 65	
1963-042C		691	US	29 OCT	27 NOV 63	
1963-043A	POLYOT 1	683	USSR	01 NOV	16 OCT 82	
1963-043B		684	USSR	01 NOV	23 NOV 66	
1963-043C		685	USSR	01 NOV	29 OCT 65	
1963-043D		686	USSR	01 NOV	18 DEC 66	
1963-044A	COSMOS 21	687	USSR	11 NOV	14 NOV 63	
1963-044B		688	USSR	11 NOV	12 NOV 63	
1963-045A	COSMOS 22	689	USSR	16 NOV	22 NOV 63	
1963-045B		690	USSR	16 NOV	03 DEC 63	
1963-046A	EXPLORER 18	693	US	27 NOV	30 DEC 65	
1963-047B		696	US	27 NOV	20 MAR 92	
1963-047C		697	US	27 NOV	24 JUN 89	
1963-047E		699	US	27 NOV	04 MAR 85	
1963-047J		1994	US	27 NOV	30 JUL 89	
1963-047M		3922	US	27 NOV	10 APR 91	
1963-047N		6004	US	27 NOV	17 FEB 80	
1963-047P		6158	US	27 NOV	22 NOV 73	
1963-047R		14529	US	27 NOV	15 OCT 91	
1963-047S		14841	US	27 NOV	22 JAN 90	
1963-047U		19110	US	27 NOV	26 FEB 90	
1963-048A		695	US	27 NOV	15 DEC 63	
1963-049J		2930	US	05 DEC	14 OCT 79	
1963-049K		4586	US	05 DEC	01 JUN 79	
1963-049L		6182	US	05 DEC	09 AUG 80	
1963-049M		6283	US	05 DEC	20 JUN 79	
1963-050A	COSMOS 23	707	USSR	13 DEC	27 MAR 64	
1963-050B		708	USSR	13 DEC	06 MAR 64	
1963-050C		709	USSR	13 DEC	03 JAN 64	
1963-050D		710	USSR	13 DEC	01 JAN 64	
1963-051A		711	US	18 DEC	20 DEC 63	
1963-052A	COSMOS 24	712	USSR	19 DEC	28 DEC 63	
1963-052B		713	USSR	19 DEC	25 JAN 64	
1963-053A	EXPLORER 19	714	US	19 DEC	10 MAY 81	
1963-053D		723	US	19 DEC	12 NOV 82	
1963-053F		725	US	19 DEC	23 DEC 74	
1963-054B		717	US	21 DEC	05 OCT 90	
1963-054D		736	US	21 DEC	25 NOV 81	
1963-055A		718	US	21 DEC	08 JAN 64	
1963-055B		719	US	21 DEC	07 NOV 64	
1964 LAUNCHES						
1964-004A	ECHO 2	740	US	25 JAN	07 JUN 69	
1964-004E		749	US	25 JAN	23 FEB 66	
1964-005A	SATURN 5	744	US	29 JAN	30 APR 66	
1964-006AE		21621	USSR	30 JAN	06 AUG 91	
1964-006E		14427	USSR	30 JAN	03 JAN 90	
1964-006F		14428	USSR	30 JAN	02 FEB 89	
1964-006G		15786	USSR	30 JAN	03 JAN 90	
1964-006H		16544	USSR	30 JAN	17 APR 90	
1964-006J		16545	USSR	30 JAN	02 FEB 89	
1964-006K		16546	USSR	30 JAN	02 FEB 89	
1964-006L		16547	USSR	30 JAN	03 JAN 90	
1964-006M		16548	USSR	30 JAN	03 JAN 90	
1964-006P		18686	USSR	30 JAN	17 OCT 88	
1964-006Q		18966	USSR	30 JAN	03 JAN 90	
1964-006R		19010	USSR	30 JAN	03 JAN 90	
1964-006V		19992	USSR	30 JAN	15 MAR 90	
1964-007A	RANGER 6	747	US	30 JAN	02 FEB 64	
1964-008A		752	US	15 FEB	09 MAR 64	
1964-008B		755	US	15 FEB	21 FEB 64	
1964-009A		754	US	25 FEB	01 MAR 64	
1964-009B		756	US	25 FEB	26 FEB 64	
1964-010A	COSMOS 25	757	USSR	27 FEB	21 NOV 64	
1964-010B		758	USSR	27 FEB	18 JUN 64	
1964-010C		762	USSR	27 FEB	04 MAR 64	
1964-010D		763	USSR	27 FEB	23 APR 64	
1964-011A		759	US	28 FEB	19 FEB 69	
1964-011B		760	US	28 FEB	17 AUG 65	
1964-011C		761	US	28 FEB	21 SEP 65	
1964-011D		2663	US	28 FEB	01 MAR 67	
1964-012A		764	US	11 MAR	16 MAR 64	
1964-012B		765	US	11 MAR	12 MAY 64	
1964-013A	COSMOS 26	766	USSR	18 MAR	28 SEP 64	
1964-013B		767	USSR	18 MAR	17 MAY 64	
1964-013C		768	USSR	18 MAR	04 MAY 64	
1964-013D		769	USSR	18 MAR	25 NOV 64	
1964-014A	COSMOS 27	770	USSR	27 MAR	28 MAR 64	
1964-014B		772	USSR	27 MAR	29 MAR 64	
1964-014C		773	USSR	27 MAR	31 MAR 64	
1964-014D		774	USSR	27 MAR	04 APR 64	
1964-015A	ARIEL 2	771	UK	27 MAR	18 NOV 67	
1964-015B		775	US	27 MAR	13 APR 64	
1964-015C		847	US	27 MAR	06 MAY 68	
1964-016A	ZOND 1	776	USSR	02 APR	03 APR 64	
1964-016B		777	USSR	02 APR	02 APR 64	
1964-016C		778	USSR	02 APR	05 APR 64	
1964-017A	COSMOS 28	779	USSR	04 APR	12 APR 64	
1964-017B		780	USSR	04 APR	03 MAY 64	
1964-017C		781	USSR	04 APR	07 APR 64	
1964-018A	GEMINI 1	782	US	08 APR	12 APR 64	
1964-019A		783	USSR	12 APR	08 MAY 64	
1964-019B	POLYOT 2	784	USSR	12 APR	08 JUN 64	
1964-020A		786	US	23 APR	29 APR 64	
1964-020B		787	US	23 APR	26 APR 64	
1964-020C		788	US	23 APR	25 APR 64	
1964-020D		789	US	23 APR	25 APR 64	
1964-020E		790	US	23 APR	25 APR 64	
1964-021A	COSMOS 29	791	USSR	25 APR	02 MAY 64	
1964-021B		792	USSR	25 APR	11 MAY 64	
1964-021C		793	USSR	25 APR	27 APR 64	
1964-021D		794	USSR	25 APR	28 APR 64	
1964-021E		795	USSR	25 APR	27 APR 64	
1964-022A		796	US	27 APR	26 MAY 64	
1964-023A	COSMOS 30	797	USSR	18 MAY	26 MAY 64	
1964-023B		798	USSR	18 MAY	07 JUN 64	
1964-024A		799	US	19 MAY	22 MAY 64	
1964-025A	SATURN 6	800	US	28 MAY	01 JUN 64	
1964-026F		12142	US	04 JUN	25 OCT 81	
1964-026G		13360	US	04 JUN	11 JAN 88	
1964-027A		802	US	04 JUN	18 JUN 64	
1964-028A	COSMOS 31	803	USSR	06 JUN	20 OCT 64	
1964-028B		804	USSR	06 JUN	16 AUG 64	
1964-029A	COSMOS 32	807	USSR	10 JUN	18 JUN 64	
1964-029B		808	USSR	10 JUN	14 JUL 64	
1964-029C		810	USSR	10 JUN	12 JUN 64	
1964-030A		811	US	13 JUN	02 JUN 65	
1964-030B		820	US	13 JUN	07 AUG 64	
1964-032A		814	US	19 JUN	6 JUL 64	
1964-032B		821	US	19 JUN	02 JUL 64	
1964-033A	COSMOS 33	816	USSR	23 JUN	01 JUL 64	
1964-033B		817	USSR	23 JUN	10 JUL 64	
1964-033C		818	USSR	23 JUN	24 JUN 64	
1964-033D		819	USSR	23 JUN	24 JUN 64	
1964-034A	COSMOS 34	822	USSR	01 JUL	09 JUL 64	
1964-034B		823	USSR	01 JUL	15 JUL 64	
1964-035A		824	US	02 JUL	07 AUG 69	
1964-036A		825	US	06 JUL	08 JUL 64	
1964-036B		826	US	06 JUL	03 JAN 65	
1964-036C		827	US	06 JUL	19 JUL 64	
1964-037A		828	US	10 JUL	06 AUG 64	
1964-038B	ELEKTRON 4	830	USSR	10 JUL	12 OCT 83	
1964-038D		832	USSR	10 JUL	24 AUG 83	
1964-038E		5892	USSR	10 JUL	20 FEB 74	
1964-038F		5893	USSR	10 JUL	20 NOV 74	
1964-039A	COSMOS 35	833	USSR	15 JUL	23 JUL 64	
1964-039B		834	USSR	15 JUL	01 AUG 64	
1964-039C		835	USSR	15 JUL	31 JUL 64	
1964-039D		839	USSR	15 JUL	25 JUL 64	
1964-040C	ERS 13	838	US	17 JUL	01 JUL 66	
1964-041A	RANGER 7	842	US	28 JUL	31 JUL 64	G
1964-042A	COSMOS 36	844	USSR	30 JUL	28 FEB 65	
1964-042B		845	USSR	30 JUL	29 NOV 64	
1964-043A		846	US	05 AUG	01 SEP 64	
1964-044A	COSMOS 37	848	USSR	14 AUG	22 AUG 64	
1964-044B		849	USSR	14 AUG	03 SEP 64	
1964-045A		850	US	14 AUG	23 AUG 64	
1964-045B		851	US	14 AUG	08 MAR 79	
1964-045C		852	US	14 AUG	16 AUG 64	
1964-046A	COSMOS 38	853	USSR	18 AUG	08 NOV 64	
1964-046B	COSMOS 39	854	USSR	18 AUG	17 NOV 64	
1964-046C	COSMOS 40	855	USSR	18 AUG	18 NOV 64	
1964-046D		856	USSR	18 AUG	19 FEB 65	
1964-046E		857	USSR	18 AUG	10 SEP 64	
1964-046F		859	USSR	18 AUG	06 SEP 64	
1964-046G		860	USSR	18 AUG	06 SEP 64	
1964-048A		861	US	21 AUG	31 MAR 65	
1964-049A		863	USSR	22 AUG	15 SEP 64	
1964-049B		865	USSR	22 AUG	28 SEP 64	
1964-049C		868	USSR	22 AUG	30 AUG 64	
1964-050A	COSMOS 42	864	USSR	22 AUG	19 DEC 65	
1964-050B		866	USSR	22 AUG	02 AUG 65	
1964-050C	COSMOS 43	867	USSR	22 AUG	27 DEC 65	
1964-051C		873	US	25 AUG	18 APR 68	
1964-051D		874	US	25 AUG	17 JUN 68	
1964-051E		875	US	25 AUG	16 SEP 68	
1964-052A	NIMBUS 1	872	US	28 AUG	16 MAY 74	
1964-052B		878	US	28 AUG	13 AUG 74	
1964-055A	COSMOS 45	880	USSR	13 SEP	18 SEP 64	
1964-055B		881	USSR	13 SEP	27 SEP 64	
1964-056A		882	US	14 SEP	06 OCT 64	
1964-056B		888	US	14 SEP	22 SEP 64	
1964-057A	SATURN 7	883	US	18 SEP	22 SEP 64	
1964-058A		884	US	23 SEP	28 SEP 64	
1964-058B		887	US	23 SEP	26 SEP 64	
1964-059A	COSMOS 46	885	USSR	24 SEP	02 OCT 64	
1964-059B		886	USSR	24 SEP	07 OCT 64	
1964-060A	EXPLORER 21	889	US	04 OCT	30 JAN 66	
1964-061A		890	US	05 OCT	26 OCT 64	
1964-062A	COSMOS 47	891	USSR	06 OCT	07 OCT 64	
1964-062B		892	USSR	06 OCT	14 OCT 64	
1964-062C		894	USSR	06 OCT	09 OCT 64	
1964-062D		895	USSR	06 OCT	12 OCT 64	
1964-062E		896	USSR	06 OCT	09 OCT 64	

International Designation	Satellite Name	Number	Source	Launch	Decay	Note
1964-065A	VOSKHOD 1	904	USSR	12 OCT	13 OCT 64	
1964-065B		905	USSR	12 OCT	20 OCT 64	
1964-065C		906	USSR	12 OCT	13 OCT 64	
1964-066A	COSMOS 48	908	USSR	14 OCT	20 OCT 64	
1964-066B		909	USSR	14 OCT	28 OCT 64	
1964-066C		910	USSR	14 OCT	16 OCT 64	
1964-067A		911	US	17 OCT	04 NOV 64	
1964-068A		912	US	23 OCT	28 OCT 64	
1964-068B		914	US	23 OCT	23 FEB 65	
1964-068C		916	US	23 OCT	04 NOV 64	
1964-068D		918	US	23 OCT	29 OCT 64	
1964-069A	COSMOS 49	913	USSR	24 OCT	21 AUG 65	
1964-069B		917	USSR	24 OCT	17 NOV 64	
1964-069C		915	USSR	24 OCT	10 FEB 65	
1964-070A	COSMOS 50	919	USSR	28 OCT	05 NOV 64	
1964-070AA		1012	USSR	28 OCT	10 NOV 64	
1964-070AB		1013	USSR	28 OCT	09 NOV 64	
1964-070AC		1014	USSR	28 OCT	17 NOV 64	
1964-070AD		1015	USSR	28 OCT	12 NOV 64	
1964-070AE		1016	USSR	28 OCT	13 NOV 64	
1964-070AF		1017	USSR	28 OCT	11 NOV 64	
1964-070AG		1018	USSR	28 OCT	13 NOV 64	
1964-070AH		1019	USSR	28 OCT	11 NOV 64	
1964-070AJ		1020	USSR	28 OCT	14 NOV 64	
1964-070AK		1021	USSR	28 OCT	13 NOV 64	
1964-070AL		1022	USSR	28 OCT	12 NOV 64	
1964-070AM		1023	USSR	28 OCT	12 NOV 64	
1964-070AN		1024	USSR	28 OCT	14 NOV 64	
1964-070AP		1025	USSR	28 OCT	13 NOV 64	
1964-070AQ		1026	USSR	28 OCT	12 NOV 64	
1964-070AR		1027	USSR	28 OCT	12 NOV 64	
1964-070AS		1028	USSR	28 OCT	13 NOV 64	
1964-070AT		1029	USSR	28 OCT	12 NOV 64	
1964-070AU		1030	USSR	28 OCT	13 NOV 64	
1964-070AV		1031	USSR	28 OCT	13 NOV 64	
1964-070AW		1032	USSR	28 OCT	16 NOV 64	
1964-070AX		1033	USSR	28 OCT	10 NOV 64	
1964-070AY		1034	USSR	28 OCT	09 NOV 64	
1964-070AZ		1035	USSR	28 OCT	13 NOV 64	
1964-070B		920	USSR	28 OCT	02 NOV 64	J
1964-070BA		1036	USSR	28 OCT	12 NOV 64	
1964-070BB		1037	USSR	28 OCT	12 NOV 64	
1964-070BC		1038	USSR	28 OCT	15 NOV 64	
1964-070BD		1039	USSR	28 OCT	13 NOV 64	
1964-070BE		1040	USSR	28 OCT	13 NOV 64	
1964-070BF		1041	USSR	28 OCT	12 NOV 64	
1964-070BG		1042	USSR	28 OCT	13 NOV 64	
1964-070BH		1043	USSR	28 OCT	12 NOV 64	
1964-070BJ		1044	USSR	28 OCT	17 NOV 64	
1964-070BK		1045	USSR	28 OCT	17 NOV 64	
1964-070BL		1046	USSR	28 OCT	12 NOV 64	
1964-070BM		1047	USSR	28 OCT	11 NOV 64	
1964-070BN		1048	USSR	28 OCT	14 NOV 64	
1964-070BP		1049	USSR	28 OCT	15 NOV 64	
1964-070BQ		1050	USSR	28 OCT	14 NOV 64	
1964-070BR		1051	USSR	28 OCT	12 NOV 64	
1964-070BS		1052	USSR	28 OCT	13 NOV 64	
1964-070BT		1053	USSR	28 OCT	13 NOV 64	
1964-070BU		1054	USSR	28 OCT	13 NOV 64	
1964-070BV		1055	USSR	28 OCT	11 NOV 64	
1964-070BW		1056	USSR	28 OCT	13 NOV 64	
1964-070BX		1057	USSR	28 OCT	12 NOV 64	
1964-070BY		1058	USSR	28 OCT	13 NOV 64	
1964-070BZ		1059	USSR	28 OCT	11 NOV 64	
1964-070C		928	USSR	28 OCT	13 NOV 64	
1964-070CA		1060	USSR	28 OCT	10 NOV 64	
1964-070CB		1061	USSR	28 OCT	09 NOV 64	
1964-070CC		1062	USSR	28 OCT	16 NOV 64	
1964-070CD		1063	USSR	28 OCT	11 NOV 64	
1964-070CE		1064	USSR	28 OCT	15 NOV 64	
1964-070CF		1065	USSR	28 OCT	17 NOV 64	
1964-070CG		1066	USSR	28 OCT	11 NOV 64	
1964-070CH		1067	USSR	28 OCT	17 NOV 64	
1964-070CJ		1068	USSR	28 OCT	14 NOV 64	
1964-070CK		1069	USSR	28 OCT	16 NOV 64	
1964-070CL		1070	USSR	28 OCT	13 NOV 64	
1964-070CM		1071	USSR	28 OCT	12 NOV 64	
1964-070CN		1072	USSR	28 OCT	10 NOV 64	
1964-070CP		1073	USSR	28 OCT	12 NOV 64	
1964-070CQ		1074	USSR	28 OCT-	13 NOV 64	
1964-070CR		1075	USSR	28 OCT	13 NOV 64	
1964-070CS		1076	USSR	28 OCT	13 NOV 64	
1964-070CT		1077	USSR	28 OCT	14 NOV 64	
1964-070CU		1078	USSR	28 OCT	12 NOV 64	
1964-070CV		1079	USSR	28 OCT	12 NOV 64	
1964-070CW		1080	USSR	28 OCT	17 NOV 64	
1964-070CX		1081	USSR	28 OCT	11 NOV 64	
1964-070CY		1082	USSR	28 OCT	11 NOV 64	
1964-070CZ		1083	USSR	28 OCT	11 NOV 64	
1964-070D		929	USSR	28 OCT	15 NOV 6	
1964-070DA		1084	USSR	28 OCT	11 NOV 64	
1964-070E		989	USSR	28 OCT	13 NOV 64	
1964-070F		990	USSR	28 OCT	11 NOV 64	
1964-070G		991	USSR	28 OCT	08 NOV 64	
1964-070H		992	USSR	28 OCT	13 NOV 64	
1964-070J		993	USSR	28 OCT	12 NOV 64	
1964-070K		994	USSR	28 OCT	13 NOV 64	
1964-070L		995	USSR	28 OCT	11 NOV 64	
1964-070M		996	USSR	28 OCT	12 NOV 64	
1964-070N		997	USSR	28 OCT	14 NOV 64	
1964-070P		998	USSR	28 OCT	10 NOV 64	
1964-070Q		999	USSR	28 OCT	12 NOV 64	
1964-070R		1003	USSR	28 OCT	11 NOV 64	
1964-070S		1004	USSR	28 OCT	14 NOV 64	
1964-070T		1005	USSR	28 OCT	12 NOV 64	
1964-070U		1006	USSR	28 OCT	12 NOV 64	
1964-070V		1007	USSR	28 OCT	12 NOV 64	
1964-070W		1008	USSR	28 OCT	12 NOV 64	
1964-070X		1009	USSR	28 OCT	13 NOV 64	
1964-070Y		1010	USSR	28 OCT	13 NOV 64	
1964-070Z		1011	USSR	28 OCT	09 NOV 64	
1964-071A		921	US	02 NOV	28 NOV 64	
1964-072A		922	US	04 NOV	05 NOV 69	
1964-072B		925	US	04 NOV	02 OCT 67	
1964-072C		926	US	04 NOV	30 JUN 66	
1964-072D		927	US	04 NOV	30 JUN 66	
1964-074A	EXPLORER 23	924	US	06 NOV	29 JUN 83	
1964-074B		3747	US	06 NOV	13 JUN 73	
1964-074C		4019	US	06 NOV	15 JUL 69	
1964-074D		6731	US	06 NOV	02 JAN 74	
1964-074E		6732	US	06 NOV	29 MAY 74	
1964-075A		930	US	18 NOV	06 DEC 64	
1964-076A	EXPLORER 24	931	US	21 NOV	18 OCT 68	
1964-076D		934	US	21 NOV	02 OCT 82	
1964-076E		935	US	21 NOV	28 MAY 84	
1964-076F		936	US	21 NOV	03 APR 68	
1964-076G		937	US	21 NOV	19 APR 81	
1964-076H		939	US	21 NOV	30 APR 68	
1964-076J		941	US	21 NOV	22 FEB 79	
1964-076K		960	US	21 NOV	30 APR 89	
1964-076L		1411	US	21 NOV	05 MAR 68	
1964-076M		4079	US	21 NOV	21 OCT 71	
1964-076N		940	US	21 NOV	13 MAR 81	
1964-078A		943	USSR	30 NOV	01 DEC 64	
1964-078B		944	USSR	30 NOV	02 DEC 64	
1964-079A		946	US	04 DEC	05 DEC 64	
1964-080A	COSMOS 51	947	USSR	09 DEC	14 NOV 65	
1964-080B		948	USSR	09 DEC	11 MAY 65	
1964-080C		950	USSR	09 DEC	29 DEC 64	
1964-080D		952	USSR	09 DEC	01 JAN 65	
1964-080E		954	USSR	09 DEC	01 JAN 65	
1964-080F		955	USSR	09 DEC	02 JAN 65	
1964-081A		949	US	10 DEC	13 DEC 64	
1964-082A	CENTAUR 4	951	US	11 DEC	12 DEC 64	
1964-083E		966	US	13 DEC	10 JUN 82	
1964-083H		1528	US	13 DEC	07 JAN 80	
1964-083K		2798	US	13 DEC	24 SEP 80	
1964-083L		5444	US	13 DEC	18 JAN 79	
1964-083M		5540	US	13 DEC	20 FEB 79	
1964-084A	SAN MARCO 1	957	ITALY	15 DEC	13 SEP 65	
1964-084B		958	US	15 DEC	07 FEB 65	
1964-084C		962	US	15 DEC	21 JAN 65	
1964-085A		961	US	19 DEC	14 JAN 65	
1964-086B		4925	US	21 DEC	19 DEC 83	
1964-086C		4926	US	21 DEC	07 MAY 82	
1964-086D		5820	US	21 DEC	01 FEB 72	
1964-087A		964	US	21 DEC	11 JAN 65	

1965 LAUNCHES

International Designation	Satellite Name	Number	Source	Launch	Decay	Note
1965-001A	COSMOS 52	968	USSR	11 JAN	19 JAN 65	
1965-001B		969	USSR	11 JAN	29 JAN 65	
1965-001C		970	USSR	11 JAN	13 JAN 65	
1965-001D		971	USSR	11 JAN	14 JAN 65	
1965-002A		972	US	15 JAN	09 FEB 65	
1965-002B		982	US	15 JAN	30 JAN 65	
1965-003A		973	US	19 JAN	13 JUL 79	
1965-003B		974	US	19 JAN	17 JUN 65	
1965-003C		975	US	19 JAN	09 JUL 65	
1965-005A		980	US	23 JAN	29 JAN 65	
1965-005B		981	US	23 JAN	25 JAN 65	
1965-006A	COSMOS 53	983	USSR	30 JAN	12 AUG 66	
1965-006B		984	USSR	30 JAN	11 JAN 66	
1965-006C		985	USSR	30 JAN	05 MAR 65	
1965-006D		986	USSR	30 JAN	04 MAR 65	
1965-007A	OSO 2	987	US	03 FEB	09 AUG 89	
1965-007B		988	US	03 FEB	11 SEP 77	
1965-009A	PEGASUS 1	1085	US	16 FEB	17 SEP 78	
1965-009B		1088	US	16 FEB	10 JUL 85	
1965-010A	RANGER 8	1086	US	17 FEB	20 FEB 65	G
1965-011A	COSMOS 54	1089	USSR	21 FEB	15 SEP 68	
1965-011B	COSMOS 55	1090	USSR	21 FEB	02 FEB 68	
1965-011C	COSMOS 56	1091	USSR	21 FEB	02 NOV 67	
1965-011D		1092	USSR	21 FEB	23 DEC 69	
1965-011E		1094	USSR	21 FEB	08 APR 66	
1965-012A	COSMOS 57	1093	USSR	22 FEB	22 FEB 65	
1965-012AA		1122	USSR	22 FEB	01 MAR 65	
1965-012AB		1123	USSR	22 FEB	27 FEB 65	
1965-012AC		1124	USSR	22 FEB	02 MAR 65	
1965-012AD		1125	USSR	22 FEB	02 MAR 65	
1965-012AE		1126	USSR	22 FEB	01 MAR 65	
1965-012AF		1127	USSR	22 FEB	28 FEB 65	
1965-012AG		1128	USSR	22 FEB	05 MAR 65	
1965-012AH		1129	USSR	22 FEB	05 MAR 65	
1965-012AJ		1130	USSR	22 FEB	11 MAR 65	
1965-012AK		1131	USSR	22 FEB	07 MAR 65	

International Designation	Satellite Name	Number	Source	Launch	Decay	Note
1965-012AL		1132	USSR	22 FEB	07 MAR 65	
1965-012AM		1133	USSR	22 FEB	21 MAR 65	
1965-012AN		1134	USSR	22 FEB	19 MAR 65	
1965-012AP		1135	USSR	22 FEB	13 MAR 65	
1965-012AQ		1136	USSR	22 FEB	17 MAR 65	
1965-012AR		1137	USSR	22 FEB	09 MAR 65	
1965-012AS		1138	USSR	22 FEB	08 MAR 65	
1965-012AT		1139	USSR	22 FEB	10 MAR 65	
1965-012AU		1140	USSR	22 FEB	25 MAR 65	
1965-012AV		1141	USSR	22 FEB	10 MAR 65	
1965-012AW		1142	USSR	22 FEB	14 MAR 65	
1965-012AX		1143	USSR	22 FEB	20 MAR 65	
1965-012AY		1144	USSR	22 FEB	21 MAR 65	
1965-012AZ		1145	USSR	22 FEB	21 MAR 65	
1965-012B		1095	USSR	22 FEB	05 MAR 65	K
1965-012BA		1146	USSR	22 FEB	17 MAR 65	
1965-012BB		1147	USSR	22 FEB	16 MAR 65	
1965-012BC		1148	USSR	22 FEB	11 MAR 65	
1965-012BD		1149	USSR	22 FEB	19 MAR 65	
1965-012BE		1150	USSR	22 FEB	17 MAR 65	
1965-012BF		1151	USSR	22 FEB	10 MAR 65	
1965-012BG		1152	USSR	22 FEB	16 MAR 65	
1965-012BH		1153	USSR	22 FEB	07 MAR 65	
1965-012BJ		1154	USSR	22 FEB	07 MAR 65	
1965-012BK		1155	USSR	22 FEB	07 MAR 65	
1965-012BL		1156	USSR	22 FEB	18 MAR 65	
1965-012BM		1157	USSR	22 FEB	13 MAR 65	
1965-012BN		1158	USSR	22 FEB	04 MAR 65	
1965-012BP		1159	USSR	22 FEB	18 MAR 65	
1965-012BQ		1160	USSR	22 FEB	04 MAR 65	
1965-012BR		1161	USSR	22 FEB	05 MAR 65	
1965-012BS		1162	USSR	22 FEB	25 MAR 65	
1965-012BT		1163	USSR	22 FEB	05 MAR 65	
1965-012BU		1164	USSR	22 FEB	13 MAR 65	
1965-012BV		1165	USSR	22 FEB	09 MAR 65	
1965-012BW		1166	USSR	22 FEB	07 MAR 65	
1965-012BX		1167	USSR	22 FEB	20 MAR 65	
1965-012BY		1168	USSR	22 FEB	06 MAR 65	
1965-012BZ		1169	USSR	22 FEB	05 MAR 65	
1965-012C		1100	USSR	22 FEB	09 MAR 65	
1965-012CA		1170	USSR	22 FEB	04 MAR 65	
1965-012CB		1171	USSR	22 FEB	05 MAR 65	
1965-012CC		1172	USSR	22 FEB	05 MAR 65	
1965-012CD		1173	USSR	22 FEB	04 MAR 65	
1965-012CE		1174	USSR	22 FEB	22 MAR 65	
1965-012CF		1175	USSR	22 FEB	05 MAR 65	
1965-012CG		1176	USSR	22 FEB	05 MAR 65	
1965-012CH		1177	USSR	22 FEB	27 MAR 65	
1965-012CJ		1178	USSR	22 FEB	05 MAR 65	
1965-012CK		1179	USSR	22 FEB	16 MAR 65	
1965-012CL		1180	USSR	22 FEB	04 MAR 65	
1965-012CM		1181	USSR	22 FEB	06 MAR 65	
1965-012CN		1182	USSR	22 FEB	07 MAR 65	
1965-012CP		1183	USSR	22 FEB	08 MAR 65	
1965-012CQ		1184	USSR	22 FEB	07 MAR 65	
1965-012CR		1185	USSR	22 FEB	11 MAR 65	
1965-012CS		1186	USSR	22 FEB	01 MAR 65	
1965-012CT		1187	USSR	22 FEB	01 MAR 65	
1965-012CU		1188	USSR	22 FEB	01 MAR 65	
1965-012CV		1189	USSR	22 FEB	01 MAR 65	
1965-012CW		1190	USSR	22 FEB	01 MAR 65	
1965-012CX		1194	USSR	22 FEB	16 MAR 65	
1965-012CY		1195	USSR	22 FEB	09 MAR 65	
1965-012CZ		1196	USSR	22 FEB	09 MAR 65	
1965-012D		1101	USSR	22 FEB	25 FEB 65	
1965-012DA		1197	USSR	22 FEB	21 MAR 65	
1965-012DB		1198	USSR	22 FEB	04 MAR 65	
1965-012DC		1199	USSR	22 FEB	17 MAR 65	
1965-012DD		1200	USSR	22 FEB	11 MAR 65	
1965-012DE		1201	USSR	22 FEB	04 MAR 65	
1965-012DF		1202	USSR	22 FEB	09 MAR 65	
1965-012DG		1203	USSR	22 FEB	14 MAR 65	
1965-012DH		1204	USSR	22 FEB	15 MAR 65	
1965-012DJ		1205	USSR	22 FEB	16 MAR 65	
1965-012DK		1206	USSR	22 FEB	09 MAR 65	
1965-012DL		1209	USSR	22 FEB	11 MAR 65	
1965-012DM		1210	USSR	22 FEB	04 MAR 65	
1965-012DN		1211	USSR	22 FEB	16 MAR 65	
1965-012DP		1212	USSR	22 FEB	12 MAR 65	
1965-012DQ		1213	USSR	22 FEB	14 MAR 65	
1965-012DR		1214	USSR	22 FEB	20 MAR 65	
1965-012DS		1215	USSR	22 FEB	16 MAR 65	
1965-012DT		1216	USSR	22 FEB	17 MAR 65	
1965-012DU		1217	USSR	22 FEB	28 MAR 65	
1965-012DV		1218	USSR	22 FEB	20 MAR 65	
1965-012DW		1219	USSR	22 FEB	14 MAR 65	
1965-012DX		1220	USSR	22 FEB	12 MAR 65	
1965-012DY		1221	USSR	22 FEB	18 MAR 65	
1965-012DZ		1222	USSR	22 FEB	12 MAR 65	
1965-012E		1102	USSR	22 FEB	26 FEB 65	
1965-012EA		1223	USSR	22 FEB	31 MAR 65	
1965-012EB		1224	USSR	22 FEB	16 MAR 65	
1965-012EC		1225	USSR	22 FEB	14 MAR 65	
1965-012ED		1226	USSR	22 FEB	18 MAR 65	
1965-012EE		1227	USSR	22 FEB	11 MAR 65	
1965-012EF		1229	USSR	22 FEB	29 MAR 65	
1965-012F		1103	USSR	22 FEB	02 MAR 65	
1965-012G		1104	USSR	22 FEB	10 MAR 65	
1965-012H		1105	USSR	22 FEB	25 FEB 65	
1965-012J		1106	USSR	22 FEB	02 MAR 65	
1965-012K		1107	USSR	22 FEB	25 MAR 65	
1965-012L		1108	USSR	22 FEB	01 MAR 65	
1965-012M		1109	USSR	22 FEB	26 FEB 65	
1965-012N		1110	USSR	22 FEB	27 FEB 65	
1965-012P		1111	USSR	22 FEB	02 MAR 65	
1965-012Q		1112	USSR	22 FEB	01 MAR 65	
1965-012R		1113	USSR	22 FEB	25 FEB 65	
1965-012S		1114	USSR	22 FEB	03 MAR 65	
1965-012T		1115	USSR	22 FEB	27 FEB 65	
1965-012U		1116	USSR	22 FEB	04 MAR 65	
1965-012V		1117	USSR	22 FEB	02 MAR 65	
1965-012W		1118	USSR	22 FEB	01 MAR 65	
1965-012X		1119	USSR	22 FEB	01 MAR 65	
1965-012Y		1120	USSR	22 FEB	02 MAR 65	
1965-012Z		1121	USSR	22 FEB	01 MAR 65	
1965-012EG		1230	USSR	23 FEB	13 MAR 65	
1965-012EH		1231	USSR	23 FEB	17 MAR 65	
1965-012EJ		1232	USSR	23 FEB	24 MAR 65	
1965-012EK		1233	USSR	23 FEB	20 MAR 65	
1965-012EL		1234	USSR	23 FEB	20 MAR 65	
1965-012EM		1235	USSR	23 FEB	17 MAR 65	
1965-012EN		1236	USSR	23 FEB	17 MAR 65	
1965-012EP		1237	USSR	23 FEB	20 MAR 65	
1965-012EQ		1238	USSR	23 FEB	17 MAR 65	
1965-012ER		1239	USSR	23 FEB	29 MAR 65	
1965-012ES		1240	USSR	23 FEB	17 MAR 65	
1965-012ET		1241	USSR	23 FEB	17 MAR 65	
1965-012EU		1242	USSR	23 FEB	24 MAR 65	
1965-012EV		1243	USSR	23 FEB	17 MAR 65	
1965-012EW		1254	USSR	23 FEB	12 MAR 65	
1965-012EX		1255	USSR	23 FEB	20 MAR 65	
1965-012EY		1256	USSR	23 FEB	21 MAR 65	
1965-012EZ		1257	USSR	23 FEB	21 MAR 65	
1965-012FA		1258	USSR	23 FEB	17 MAR 65	
1965-012FB		1261	USSR	23 FEB	06 APR 65	
1965-012FC		1262	USSR	23 FEB	15 MAR 65	
1965-012FD		1263	USSR	23 FEB	21 MAR 65	
1965-012FE		1264	USSR	23 FEB	26 MAR 65	
1965-012FF		1265	USSR	23 FEB	15 MAR 65	
1965-012FG		1266	USSR	23 FEB	06 MAR 65	
1965-012FH		1275	USSR	23 FEB	21 MAR 65	
1965-012FJ		1276	USSR	23 FEB	20 MAR 65	
1965-012FK		1277	USSR	23 FEB	20 MAR 65	
1965-012FL		1278	USSR	23 FEB	22 MAR 65	
1965-012FM		1279	USSR	23 FEB	31 MAR 65	
1965-012FN		1280	USSR	23 FEB	24 MAR 65	
1965-012FP		1281	USSR	23 FEB	19 MAR 65	
1965-012FQ		1282	USSR	23 FEB	21 MAR 65	
1965-012FR		1283	USSR	23 FEB	23 MAR 65	
1965-012FS		1284	USSR	23 FEB	19 MAR 65	
1965-012FT		1295	USSR	23 FEB	20 MAR 65	
1965-012FU		1296	USSR	23 FEB	15 MAR 65	
1965-012FV		1297	USSR	23 FEB	22 MAR 65	
1965-012FW		1299	USSR	23 FEB	20 MAR 65	
1965-012FX		1300	USSR	23 FEB	23 MAR 65	
1965-012FY		1304	USSR	23 FEB	28 MAR 65	
1965-012FZ		1309	USSR	23 FEB	28 MAR 65	
1965-013A		1096	US	25 FEB	18 MAR 65	
1965-013B		1193	US	25 FEB	12 MAR 65	
1965-014A	COSMOS 58	1097	USSR	26 FEB	25 FEB 90	
1965-014B		1098	USSR	26 FEB	25 JUL 88	
1965-015A	COSMOS 59	1191	USSR	07 MAR	15 MAR 65	
1965-015B		1192	USSR	07 MAR	19 MAR 65	
1965-015C		1207	USSR	07 MAR	10 MAR 65	
1965-016G	SURCAL	1310	US	09 MAR	27 MAR 81	
1965-017A	NNSS 30030	1303	US	11 MAR	14 JUN 65	
1965-017B	SECOR (EGRS) 2	1250	US	11 MAR	26 FEB 68	
1965-017C		1228	US	11 MAR	04 SEP 67	
1965-017D		1248	US	11 MAR	29 JUL 67	
1965-017E		1251	US	11 MAR	01 AUG 65	
1965-017F		1249	US	11 MAR	10 SEP 65	
1965-017G		1319	US	11 MAR	28 JUN 65	
1965-017H		1323	US	11 MAR	23 DEC 65	
1965-018A	COSMOS 60	1246	USSR	12 MAR	17 MAR 65	
1965-018B		1252	USSR	12 MAR	17 MAR 65	
1965-018C		1253	USSR	12 MAR	15 MAR 65	
1965-019A		1247	US	12 MAR	17 MAR 65	
1965-019B		1259	US	12 MAR	16 MAR 65	
1965-019C		1260	US	12 MAR	21 MAR 65	
1965-020A	COSMOS 61	1267	USSR	15 MAR	15 JAN 68	
1965-020AA		1355	USSR	15 MAR	09 OCT 66	
1965-020AB		1356	USSR	15 MAR	14 AUG 65	
1965-020AD		1371	USSR	15 MAR	26 SEP 80	
1965-020AE		1372	USSR	15 MAR	24 JUN 89	
1965-020AF		1373	USSR	15 MAR	04 FEB 79	
1965-020AG		1375	USSR	15 MAR	14 OCT 65	
1965-020AJ		1397	USSR	15 MAR	06 APR 68	
1965-020AK		1398	USSR	15 MAR	25 JUN 70	
1965-020AL		1400	USSR	15 MAR	05 OCT 65	
1965-020AM		1401	USSR	15 MAR	20 MAR 66	
1965-020AN		1402	USSR	15 MAR	05 MAY 67	
1965-020AP		1403	USSR	15 MAR	03 DEC 65	
1965-020AQ		1409	USSR	15 MAR	19 MAY 66	
1965-020AR		1410	USSR	15 MAR	17 FEB 67	
1965-020AS		1416	USSR	15 MAR	05 MAY 67	
1965-020AT		1417	USSR	15 MAR	09 SEP 65	

International Designation	Satellite Name	Number	Source	Launch	Decay	Note
1965-020AU		1418	USSR	15 MAR	10 AUG 65	
1965-020AV		1419	USSR	15 MAR	01 SEP 66	
1965-020AW		1436	USSR	15 MAR	20 MAY 65	
1965-020AX		1437	USSR	15 MAR	28 OCT 65	
1965-020AY		1438	USSR	15 MAR	03 NOV 80	
1965-020AZ		1439	USSR	15 MAR	16 JAN 79	
1965-020B	COSMOS 62	1268	USSR	15 MAR	24 SEP 68	
1965-020BA		1476	USSR	15 MAR	21 SEP 67	
1965-020BF		1481	USSR	15 MAR	23 MAY 79	
1965-020BG		1482	USSR	15 MAR	09 JAN 66	
1965-020BH		1483	USSR	15 MAR	27 FEB 68	
1965-020BJ		1484	USSR	15 MAR	24 OCT 68	
1965-020BK		1485	USSR	15 MAR	15 APR 68	
1965-020BL		1486	USSR	15 MAR	10 JUN 67	
1965-020BM		1487	USSR	15 MAR	18 APR 67	
1965-020BN		1488	USSR	15 MAR	14 NOV 67	
1965-020BP		1489	USSR	15 MAR	26 OCT 66	
1965-020BQ		1490	USSR	15 MAR	03 JAN 66	
1965-020BR		1491	USSR	15 MAR	10 SEP 68	
1965-020BS		1492	USSR	15 MAR	23 NOV 65	
1965-020BT		1493	USSR	15 MAR	04 NOV 66	
1965-020BU		1494	USSR	15 MAR	04 SEP 65	
1965-020BW		1496	USSR	15 MAR	15 SEP 66	
1965-020BX		1497	USSR	15 MAR	28 MAY 67	
1965-020BY		1498	USSR	15 MAR	06 MAY 66	
1965-020BZ		1499	USSR	15 MAR	07 SEP 66	
1965-020C	COSMOS 63	1269	USSR	15 MAR	04 NOV 67	
1965-020CA		1530	USSR	15 MAR	04 JUL 69	
1965-020CB		1531	USSR	15 MAR	05 MAR 67	
1965-020CC		1532	USSR	15 MAR	11 JAN 66	
1965-020CD		1533	USSR	15 MAR	27 DEC 65	
1965-020CE		1534	USSR	15 MAR	05 MAY 71	
1965-020CF		1535	USSR	15 MAR	14 DEC 65	
1965-020CG		1536	USSR	15 MAR	26 APR 66	
1965-020CH		1537	USSR	15 MAR	17 AUG 66	
1965-020CJ		1538	USSR	15 MAR	03 DEC 74	
1965-020CK		1539	USSR	15 MAR	26 SEP 66	
1965-020CL		1540	USSR	15 MAR	27 FEB 68	
1965-020CM		1541	USSR	15 MAR	20 MAR 66	
1965-020CN		1542	USSR	15 MAR	31 JUL 66	
1965-020CP		1543	USSR	15 MAR	19 SEP 68	
1965-020CQ		1544	USSR	15 MAR	22 FEB 67	
1965-020CR		1545	USSR	15 MAR	24 JAN 66	
1965-020CS		1546	USSR	15 MAR	16 AUG 66	
1965-020CT		1547	USSR	15 MAR	24 JUL 66	
1965-020CU		1548	USSR	15 MAR	27 MAY 78	
1965-020CW		1550	USSR	15 MAR	08 JUL 70	
1965-020CX		1551	USSR	15 MAR	19 DEC 69	
1965-020CY		1552	USSR	15 MAR	20 MAR 67	
1965-020CZ		1553	USSR	15 MAR	23 JUL 66	
1965-020D		1270	USSR	15 MAR	05 AUG 66	L
1965-020DA		1554	USSR	15 MAR	06 JAN 67	
1965-020DB		1555	USSR	15 MAR	12 FEB 68	
1965-020DC		1556	USSR	15 MAR	02 MAY 67	
1965-020DD		1557	USSR	15 MAR	12 FEB 67	
1965-020DE		1558	USSR	15 MAR	12 JUN 68	
1965-020DF		1559	USSR	15 MAR	18 APR 67	
1965-020DG		1560	USSR	15 MAR	13 OCT 65	
1965-020DH		1561	USSR	15 MAR	12 APR 67	
1965-020DJ		1562	USSR	15 MAR	22 FEB 66	
1965-020DK		1563	USSR	15 MAR	28 OCT 69	
1965-020DL		1564	USSR	15 MAR	10 SEP 81	
1965-020DM		1565	USSR	15 MAR	02 FEB 67	
1965-020DN		1566	USSR	15 MAR	08 DEC 79	
1965-020DP		1567	USSR	15 MAR	12 SEP 74	
1965-020DQ		1568	USSR	15 MAR	01 MAY 66	
1965-020DR		1569	USSR	15 MAR	18 JUL 66	
1965-020DS		1592	USSR	15 MAR	27 OCT 70	
1965-020DT		1593	USSR	15 MAR	28 SEP 70	
1965-020DU		1594	USSR	15 MAR	19 DEC 66	
1965-020DV		1595	USSR	15 MAR	06 APR 72	
1965-020DW		1596	USSR	15 MAR	29 OCT 67	
1965-020DX		1597	USSR	15 MAR	28 DEC 88	
1965-020DY		1598	USSR	15 MAR	12 APR 76	
1965-020DZ		1599	USSR	15 MAR	16 JAN 71	
1965-020EA		1600	USSR	15 MAR	10 MAR 78	
1965-020EB		1601	USSR	15 MAR	17 JAN 67	
1965-020EC		1633	USSR	15 MAR	31 JAN 81	
1965-020EE		1635	USSR	15 MAR	20 OCT 67	
1965-020EF		2203	USSR	15 MAR	20 APR 67	
1965-020EG		2204	USSR	15 MAR	17 OCT 66	
1965-020EJ		2339	USSR	15 MAR	12 NOV 66	
1965-020EK		2387	USSR	15 MAR	19 OCT 66	
1965-020EL		2637	USSR	15 MAR	19 FEB 67	
1965-020EP		3163	USSR	15 MAR	16 JUL 68	
1965-020EQ		3707	USSR	15 MAR	13 OCT 79	
1965-020EW		3963	USSR	15 MAR	09 JUL 91	
1965-020EX		3964	USSR	15 MAR	29 OCT 80	
1965-020EZ		4016	USSR	15 MAR	13 APR 81	
1965-020F		1336	USSR	15 MAR	24 JUL 68	
1965-020FA		5733	USSR	15 MAR	11 MAR 72	
1965-020FB		5813	USSR	15 MAR	13 FEB 81	
1965-020FC		5983	USSR	15 MAR	09 AUG 72	
1965-020FE		6736	USSR	15 MAR	12 JUL 80	
1965-020G		1337	USSR	15 MAR	29 SEP 66	
1965-020H		1338	USSR	15 MAR	30 APR 67	
1965-020J		1339	USSR	15 MAR	14 OCT 69	
1965-020K		1340	USSR	15 MAR	30 MAY 68	

International Designation	Satellite Name	Number	Source	Launch	Decay	Note
1965-020L		1341	USSR	15 MAR	13 FEB 69	
1965-020M		1342	USSR	15 MAR	09 SEP 69	
1965-020N		1343	USSR	15 MAR	20 DEC 66	
1965-020P		1344	USSR	15 MAR	14 NOV 69	
1965-020Q		1345	USSR	15 MAR	26 NOV 88	
1965-020R		1346	USSR	15 MAR	18 MAY 75	
1965-020T		1348	USSR	15 MAR	15 NOV 81	
1965-020U		1349	USSR	15 MAR	09 APR 79	
1965-020V		1350	USSR	15 MAR	19 DEC 65	
1965-020W		1351	USSR	15 MAR	06 AUG 72	
1965-020X		1352	USSR	15 MAR	03 JUL 67	
1965-020Y		1353	USSR	15 MAR	30 JAN 66	
1965-020Z		1354	USSR	15 MAR	15 MAY 68	
1965-021A		1273	US	18 MAR	31 DEC 89	
1965-021B		1288	US	18 MAR	05 OCT 65	
1965-021C		1289	US	18 MAR	06 SEP 79	
1965-021D		1290	US	18 MAR	01 OCT 65	
1965-021E		1376	US	18 MAR	01 JAN 70	
1965-021F		1463	US	18 MAR	18 MAY 79	
1965-022A	VOSKHOD 2	1274	USSR	18 MAR	19 MAR 65	
1965-022B		1285	USSR	18 MAR	27 MAR 65	
1965-022C		1286	USSR	18 MAR	25 MAR 65	
1965-022D		1287	USSR	18 MAR	19 MAR 65	
1965-023A	RANGER 9	1294	US	21 MAR	24 MAR 65	G
1965-024A	GEMINI 3	1301	US	23 MAR	23 MAR 65	F
1965-024B		1302	US	23 MAR	24 MAR 65	
1965-025A	COSMOS 64	1305	USSR	25 MAR	02 APR 65	
1965-025B		1306	USSR	25 MAR	04 APR 65	
1965-025C		1308	USSR	25 MAR	26 APR 65	
1965-026A		1307	US	25 MAR	04 APR 65	
1965-026B		1311	US	25 MAR	01 APR 65	
1965-027AG		18670	US	03 APR	19 JAN 91	
1965-027X		18268	US	03 APR	02 FEB 89	
1965-029A	COSMOS 65	1320	USSR	17 APR	25 APR 65	
1965-029B		1321	USSR	17 APR	06 MAY 65	
1965-029C		1322	USSR	17 APR	20 APR 65	
1965-030A	MOLNIYA 1-1	1324	USSR	23 APR	16 AUG 79	
1965-030B		1325	USSR	23 APR	20 JUL 65	
1965-030C		1326	USSR	23 APR	02 JUL 65	
1965-030D		1967	USSR	23 APR	04 JUN 79	
1965-031A		1327	US	28 APR	03 MAY 65	
1965-031B		1329	US	28 APR	31 OCT 69	
1965-031C		1331	US	28 APR	30 APR 65	
1965-031D		1332	US	28 APR	30 APR 65	
1965-031E		1333	US	28 APR	08 MAY 65	
1965-031F		1334	US	28 APR	08 MAY 65	
1965-031G		1357	US	28 APR	18 MAR 66	
1965-032C		1995	US	29 APR	13 JAN 90	
1965-033A		1330	US	29 APR	26 MAY 65	
1965-033B		1367	US	29 APR	08 JUN 65	
1965-035A	COSMOS 66	1362	USSR	07 MAY	15 MAY 65	
1965-035B		1363	USSR	07 MAY	23 MAY 65	
1965-035C		1364	USSR	07 MAY	09 MAY 65	
1965-035D		1365	USSR	07 MAY	09 MAY 65	
1965-036A	LUNA 5	1366	USSR	09 MAY	12 MAY 65	A
1965-036B		1368	USSR	09 MAY	10 MAY 65	
1965-036C		1369	USSR	09 MAY	10 MAY 65	
1965-037A		1374	US	18 MAY	15 JUN 65	
1965-037B		1415	US	18 MAY	02 JUN 65	
1965-038C		1379	US	20 MAY	07 APR 83	
1965-038D		1380	US	20 MAY	02 MAY 66	
1965-038E		1461	US	20 MAY	20 MAR 89	
1965-038F		1462	US	20 MAY	23 APR 80	
1965-038G		1475	US	20 MAY	19 FEB 88	
1965-039A	PEGASUS 2	1381	US	25 MAY	03 NOV 79	
1965-039B		1385	US	25 MAY	08 JUL 89	
1965-040A	COSMOS 67	1382	USSR	25 MAY	02 JUN 65	
1965-040B		1383	USSR	25 MAY	04 JUN 65	
1965-040C		1384	USSR	25 MAY	31 MAY 65	
1965-040D		1387	USSR	25 MAY	31 MAY 65	
1965-041A		1386	US	27 MAY	01 JUN 65	
1965-042A	EXPLORER 28	1388	US	29 MAY	04 JUL 68	
1965-043A	GEMINI 4	1390	US	03 JUN	07 JUN 65	F
1965-043B		1391	US	03 JUN	05 JUN 65	
1965-044B		1394	USSR	08 JUN	12 JUN 65	
1965-044C		1395	USSR	08 JUN	10 JUN 65	
1965-045A		1396	US	09 JUN	22 JUN 65	
1965-045B		1414	US	09 JUN	20 JUN 65	
1965-046A	COSMOS 68	1404	USSR	15 JUN	23 JUN 65	
1965-046B		1405	USSR	15 JUN	07 JUL 65	
1965-046C		1406	USSR	15 JUN	20 JUN 65	
1965-046D		1407	USSR	15 JUN	19 JUN 65	
1965-046E		1408	USSR	15 JUN	19 JUN 65	
1965-047A		1412	US	18 JUN	29 JUN 65	
1965-047B		1413	US	18 JUN	21 JUN 65	
1965-048G		11952	US	24 JUN	24 FEB 89	
1965-048H		11953	US	24 JUN	07 MAR 89	
1965-048J		12516	US	24 JUN	04 DEC 81	
1965-048K		12582	US	24 JUN	09 MAR 82	
1965-048L		19062	US	24 JUN	11 MAY 91	
1965-048M		19838	US	24 JUN	31 OCT 89	
1965-049A	COSMOS 69	1421	USSR	25 JUN	03 JUL 65	
1965-049B		1423	USSR	25 JUN	06 JUL 65	
1965-049C		1429	USSR	25 JUN	30 JUN 65	
1965-050A		1422	US	25 JUN	22 AUG 68	
1965-050B		1424	US	25 JUN	30 JUN 65	
1965-050C		1426	US	25 JUN	27 JUN 65	
1965-050D		1427	US	25 JUN	26 NOV 65	

International Designation	Satellite Name	Number	Source	Launch	Decay	Note
1965-052A	COSMOS 70	1431	USSR	02 JUL	18 DEC 66	
1965-052B		1432	USSR	02 JUL	06 JUN 66	
1965-052C		1434	USSR	02 JUL	13 AUG 65	
1965-053A	COSMOS 71	1441	USSR	16 JUL	11 AUG 70	
1965-053B	COSMOS 72	1442	USSR	16 JUL	24 AUG 79	
1965-053C	COSMOS 73	1443	USSR	16 JUL	20 MAR 74	
1965-053D	COSMOS 74	1444	USSR	16 JUL	13 DEC 79	
1965-053E	COSMOS 75	1445	USSR	16 JUL	28 SEP 79	
1965-053F		1448	USSR	16 JUL	21 NOV 81	
1965-053G		1449	USSR	16 JUL	03 MAY 67	
1965-053H		1473	USSR	16 JUL	13 MAR 69	
1965-053J		2338	USSR	16 JUL	12 NOV 67	
1965-054A	PROTON 1	1466	USSR	16 JUL	11 OCT 65	
1965-054B		1451	USSR	16 JUL	18 SEP 65	
1965-054C		1450	USSR	16 JUL	24 JUL 65	
1965-054D		1446	USSR	16 JUL	31 JUL 65	
1965-055A		1447	US	17 JUL	18 DEC 68	
1965-055B		1452	US	17 JUL	05 JUL 66	
1965-055C		1455	US	17 JUL	24 JUL 66	
1965-055D		1744	US	17 JUL	16 APR 66	
1965-055E		1745	US	17 JUL	23 MAR 66	
1965-056B		1453	USSR	18 JUL	20 JUL 65	
1965-056C		1456	USSR	18 JUL	21 JUL 65	
1965-057A		1457	US	19 JUL	18 AUG 65	
1965-058C	ERS 17	1460	US	20 JUL	01 JUL 68	
1965-059A	COSMOS 76	1464	USSR	23 JUL	16 MAR 66	
1965-059B		1465	USSR	23 JUL	04 DEC 65	
1965-060A	PEGASUS 3	1467	US	30 JUL	04 AUG 69	
1965-060B		1468	US	30 JUL	22 NOV 75	
1965-060C		6841	US	30 JUL	02 OCT 73	
1965-061A	COSMOS 77	1469	USSR	03 AUG	11 AUG 65	
1965-061B		1470	USSR	03 AUG	08 AUG 65	
1965-062A		1471	US	03 AUG	07 AUG 65	
1965-062B		1472	US	03 AUG	17 JUN 68	
1965-062C		1474	US	03 AUG	06 AUG 65	
1965-065M		2335	US	13 AUG	13 APR 81	
1965-065N		3809	US	13 AUG	12 NOV 82	
1965-065R		5363	US	13 AUG	09 NOV 80	
1965-065S		6290	US	13 AUG	02 JUN 81	
1965-065T		11855	US	13 AUG	15 JUN 90	
1965-082NT		3425	US	13 AUG	17 OCT 81	
1965-066A	COSMOS 78	1505	USSR	14 AUG	22 AUG 65	
1965-066B		1507	USSR	14 AUG	03 SEP 65	
1965-066C		1509	USSR	14 AUG	17 AUG 65	
1965-067A		1513	US	17 AUG	11 OCT 65	
1965-068A	GEMINI 5	1516	US	21 AUG	29 AUG 65	F
1965-068B		1517	US	21 AUG	24 AUG 65	
1965-068C		1518	US	21 AUG	27 AUG 65	
1965-068D		1519	US	21 AUG	26 AUG 65	
1965-069A	COSMOS 79	1523	USSR	25 AUG	02 SEP 65	
1965-069B		1524	USSR	25 AUG	07 SEP 65	
1965-069C		1525	USSR	25 AUG	26 AUG 65	
1965-069D		1526	USSR	25 AUG	31 AUG 65	
1965-069E		1527	USSR	25 AUG	31 AUG 65	
1965-071A	COSMOS 85	1578	USSR	09 SEP	17 SEP 65	
1965-071B		1579	USSR	09 SEP	18 SEP 65	
1965-072B		1581	US	10 SEP	25 FEB 67	
1965-072C		1582	US	10 SEP	06 APR 67	
1965-074A		1602	US	22 SEP	11 OCT 65	
1965-075A	COSMOS 91	1603	USSR	23 SEP	01 OCT 65	
1965-075B		1604	USSR	23 SEP	04 OCT 65	
1965-075C		1605	USSR	23 SEP	29 SEP 65	
1965-075D		1606	USSR	23 SEP	29 SEP 65	
1965-075E		1607	USSR	23 SEP	29 SEP 65	
1965-076A		1609	US	30 SEP	05 OCT 65	
1965-076B		1614	US	30 SEP	03 OCT 65	
1965-077A	LUNA 7	1610	USSR	04 OCT	07 OCT 65	A
1965-077B		1611	USSR	04 OCT	04 OCT 65	
1965-077C		1612	USSR	04 OCT	05 OCT 65	
1965-079A		1615	US	05 OCT	29 OCT 65	
1965-080A	MOLNIYA 1-2	1621	USSR	13 OCT	17 MAR 67	
1965-080B		1619	USSR	13 OCT	08 NOV 65	
1965-080C		1622	USSR	13 OCT	22 OCT 65	
1965-080D		1623	USSR	13 OCT	16 NOV 65	
1965-080E		2735	USSR	13 OCT	18 JAN 67	
1965-081A	OGO 2	1620	US	14 OCT	17 SEP 81	
1965-081B		1625	US	14 OCT	27 NOV 87	
1965-082A	TITAN 3 C-4	1624	US	15 OCT	27 JUL 72	
1965-082AC		1665	US	15 OCT	28 JAN 81	
1965-082AD		1666	US	15 OCT	16 JUN 79	
1965-082AE		1667	US	15 OCT	02 APR 82	
1965-082AG		1669	US	15 OCT	15 AUG 71	
1965-082AH		1670	US	15 OCT	25 FEB 89	
1965-082AM		1674	US	15 OCT	18 MAR 69	
1965-082AN		1675	US	15 OCT	08 APR 79	
1965-082AR		1678	US	15 OCT	30 MAY 80	
1965-082AS		1679	US	15 OCT	24 MAY 80	
1965-082AU		1681	US	15 OCT	13 AUG 91	
1965-082AW		1683	US	15 OCT	11 JUL 66	
1965-082B		1640	US	15 OCT	25 AUG 82	
1965-082BB		1688	US	15 OCT	01 MAY 81	
1965-082BC		1689	US	15 OCT	29 DEC 78	
1965-082BD		1690	US	15 OCT	11 JUN 81	
1965-082BE		1691	US	15 OCT	12 MAR 71	
1965-082BF		1692	US	15 OCT	22 SEP 70	
1965-082BH		1694	US	15 OCT	26 OCT 70	
1965-082BJ		1695	US	15 OCT	28 MAR 89	
1965-082BK		1696	US	15 OCT	18 JAN 68	
1965-082BL		1697	US	15 OCT	08 JAN 78	
1965-082BM		1698	US	15 OCT	09 MAY 72	
1965-082BN		1699	US	15 OCT	21 NOV 66	
1965-082BP		1700	US	15 OCT	03 APR 70	
1965-082BQ		1709	US	15 OCT	07 APR 82	
1965-082BR		1710	US	15 OCT	01 MAR 73	
1965-082BS		1711	US	15 OCT	20 MAR 71	
1965-082BT		1712	US	15 OCT	16 SEP 81	
1965-082BU		1713	US	15 OCT	02 JAN 69	
1965-082BV		1714	US	15 OCT	09 AUG 70	
1965-082BW		1715	US	15 OCT	04 FEB 93	
1965-082BY		1717	US	15 OCT	02 FEB 69	
1965-082BZ		1718	US	15 OCT	07 JUN 81	
1965-082CA		1719	US	15 OCT	03 MAR 81	
1965-082CB		1720	US	15 OCT	04 MAR 67	
1965-082CE		1723	US	15 OCT	06 NOV 78	
1965-082CF		1724	US	15 OCT	26 APR 68	
1965-082CH		1746	US	15 OCT	03 MAY 78	
1965-082CJ		1747	US	15 OCT	19 AUG 80	
1965-082CK		1748	US	15 OCT	30 MAY 91	
1965-082CL		1749	US	15 OCT	09 DEC 79	
1965-082CM		1750	US	15 OCT	08 FEB 70	
1965-082CN		1751	US	15 OCT	21 OCT 79	
1965-082CP		1752	US	15 OCT	08 DEC 80	
1965-082CQ		1753	US	15 OCT	30 APR 90	
1965-082CR		1754	US	15 OCT	04 AUG 68	
1965-082CS		1755	US	15 OCT	30 APR 81	
1965-082CT		1756	US	15 OCT	13 SEP 87	
1965-082CU		1757	US	15 OCT	05 MAY 81	
1965-082CV		1758	US	15 OCT	13 JAN 73	
1965-082CW		1759	US	15 OCT	21 OCT 80	
1965-082CX		1760	US	15 OCT	31 MAR 71	
1965-082CZ		1762	US	15 OCT	22 JAN 68	
1965-082DA		1763	US	15 OCT	08 OCT 72	
1965-082DB		1764	US	15 OCT	28 SEP 77	
1965-082DC		1765	US	15 OCT	09 JUN 73	
1965-082DD		1766	US	15 OCT	03 APR 71	
1965-082DE		1767	US	15 OCT	19 FEB 71	
1965-082DF		1768	US	15 OCT	27 NOV 70	
1965-082DG		1769	US	15 OCT	07 NOV 81	
1965-082DH		1770	US	15 OCT	10 SEP 79	
1965-082DJ		1819	US	15 OCT	20 JUN 79	
1965-082DK		1820	US	15 OCT	30 MAR 81	
1965-082DL		1821	US	15 OCT	05 MAY 67	
1965-082DN		1823	US	15 OCT	21 APR 74	
1965-082DP		1824	US	15 OCT	11 JUL 67	
1965-082DQ		1825	US	15 OCT	03 JUL 89	
1965-082DR		1826	US	15 OCT	22 MAR 66	
1965-082DS		1827	US	15 OCT	30 DEC 80	
1965-082DT		1828	US	15 OCT	02 JAN 79	
1965-082DV		1830	US	15 OCT	18 JUN 91	
1965-082DW		1831	US	15 OCT	20 NOV 67	
1965-082DZ		1834	US	15 OCT	10 JUL 67	
1965-082E		1643	US	15 OCT	25 DEC 89	
1965-082EB		1836	US	15 OCT	17 APR 93	
1965-082EC		1837	US	15 OCT	01 MAY 68	
1965-082ED		1861	US	15 OCT	22 DEC 89	
1965-082EE		1862	US	15 OCT	11 SEP 68	
1965-082EF		1848	US	15 OCT	17 SEP 69	
1965-082EG		1849	US	15 OCT	19 JUL 67	
1965-082EH		1852	US	15 OCT	17 JAN 68	
1965-082EJ		1851	US	15 OCT	29 SEP 66	
1965-082EK		1850	US	15 OCT	22 APR 70	
1965-082EL		1853	US	15 OCT	29 NOV 66	
1965-082EM		1854	US	15 OCT	27 APR 66	
1965-082EN		1855	US	15 OCT	12 SEP 74	
1965-082EP		1856	US	15 OCT	06 FEB 67	
1965-082ER		1858	US	15 OCT	03 JAN 67	
1965-082ET		1860	US	15 OCT	19 APR 69	
1965-082EU		1879	US	15 OCT	22 MAR 66	
1965-082EV		1880	US	15 OCT	11 SEP 66	
1965-082EW		1881	US	15 OCT	07 JAN 89	
1965-082EX		1882	US	15 OCT	14 JUL 70	
1965-082EZ		1884	US	15 OCT	14 OCT 90	
1965-082F		1644	US	15 OCT	14 AUG 80	
1965-082FA		1885	US	15 OCT	30 JAN 70	
1965-082FB		1886	US	15 OCT	01 MAR 68	
1965-082FC		1887	US	15 OCT	23 MAY 79	
1965-082FD		1888	US	15 OCT	09 MAY 67	
1965-082FE		1889	US	15 OCT	09 JAN 68	
1965-082FF		1890	US	15 OCT	28 MAR 67	
1965-082FG		1891	US	15 OCT	11 MAR 68	
1965-082FH		1892	US	15 OCT	16 APR 67	
1965-082FJ		1893	US	15 OCT	02 NOV 67	
1965-082FK		1894	US	15 OCT	14 SEP 69	
1965-082FL		1895	US	15 OCT	08 SEP 66	
1965-082FM		1896	US	15 OCT	23 SEP 66	
1965-082FN		1897	US	15 OCT	07 FEB 67	
1965-082FP		1898	US	15 OCT	09 NOV 83	
1965-082FQ		1899	US	15 OCT	11 JUL 67	
1965-082FR		1900	US	15 OCT	14 NOV 66	
1965-082FS		1908	US	15 OCT	23 MAR 66	
1965-082FT		1909	US	15 OCT	19 JUL 66	
1965-082FU		1910	US	15 OCT	09 MAY 66	
1965-082FV		1911	US	15 OCT	21 JUN 66	
1965-082FW		1912	US	15 OCT	14 AUG 66	
1965-082FX		1913	US	15 OCT	26 SEP 79	
1965-082FY		1914	US	15 OCT	09 DEC 79	

International Designation	Satellite Name	Number	Source	Launch	Decay	Note
1965-082G		1645	US	15 OCT	04 MAY 79	
1965-082GA		1916	US	15 OCT	01 FEB 81	
1965-082GB		1917	US	15 OCT	07 APR 82	
1965-082GC		1918	US	15 OCT	10 NOV 78	
1965-082GD		1919	US	15 OCT	02 DEC 78	
1965-082GE		1920	US	15 OCT	08 FEB 80	
1965-082GF		1921	US	15 OCT	24 FEB 67	
1965-082GG		1922	US	15 OCT	05 MAR 79	
1965-082GH		1923	US	15 OCT	15 JUN 68	
1965-082GJ		1924	US	15 OCT	21 JUN 66	
1965-082GK		1925	US	15 OCT	15 OCT 67	
1965-082GL		1926	US	15 OCT	17 MAY 70	
1965-082GM		1927	US	15 OCT	03 AUG 73	
1965-082GN		1928	US	15 OCT	28 FEB 68	
1965-082GP		1929	US	15 OCT	11 APR 82	
1965-082GQ		1930	US	15 OCT	12 SEP 66	
1965-082GR		1956	US	15 OCT	12 AUG 68	
1965-082GS		1957	US	15 OCT	02 FEB 79	
1965-082GT		1958	US	15 OCT	23 MAY 79	
1965-082GV		1960	US	15 OCT	23 MAR 82	
1965-082HA		1965	US	15 OCT	06 MAR 70	
1965-082HB		1969	US	15 OCT	08 SEP 69	
1965-082HC		1970	US	15 OCT	11 OCT 69	
1965-082HD		1971	US	15 OCT	22 SEP 70	
1965-082HE		1972	US	15 OCT	02 FEB 67	
1965-082HF		1973	US	15 OCT	05 MAY 81	
1965-082HG		1974	US	15 OCT	05 MAR 81	
1965-082HH		1975	US	15 OCT	23 MAY 81	
1965-082HJ		1976	US	15 OCT	14 APR 82	
1965-082HK		1977	US	15 OCT	14 JUL 82	
1965-082HL		1978	US	15 OCT	09 MAR 80	
1965-082HM		1979	US	15 OCT	07 OCT 67	
1965-082HN		1980	US	15 OCT	25 MAY 69	
1965-082HQ		1984	US	15 OCT	29 MAR 74	
1965-082HR		1985	US	15 OCT	18 AUG 68	
1965-082HS		1986	US	15 OCT	08 JAN 80	
1965-082HT		1987	US	15 OCT	04 SEP 69	
1965-082HU		1988	US	15 OCT	14 NOV 70	
1965-082HV		1989	US	15 OCT	13 FEB 70	
1965-082HW		1990	US	15 OCT	09 JAN 69	
1965-082HX		1991	US	15 OCT	06 FEB 68	
1965-082HY		1992	US	15 OCT	12 NOV 80	
1965-082HZ		1993	US	15 OCT	30 MAR 80	
1965-082JA		1999	US	15 OCT	20 APR 67	
1965-082JB		2041	US	15 OCT	16 MAR 73	
1965-082JC		2042	US	15 OCT	21 NOV 79	
1965-082JD		2043	US	15 OCT	01 SEP 68	
1965-082JE		2044	US	15 OCT	20 NOV 80	
1965-082JF		2045	US	15 OCT	20 OCT 67	
1965-082JG		2046	US	15 OCT	15 NOV 67	
1965-082JH		2047	US	15 OCT	16 NOV 67	
1965-082JJ		2048	US	15 OCT	27 JAN 67	
1965-082JK		2049	US	15 OCT	14 JAN 70	
1965-082JM		2051	US	15 OCT	16 SEP 81	
1965-082JN		2052	US	15 OCT	26 JUN 72	
1965-082JP		2053	US	15 OCT	27 FEB 73	
1965-082JQ		2054	US	15 OCT	23 JAN 67	
1965-082JR		2055	US	15 OCT	28 JUN 66	
1965-082JS		2079	US	15 OCT	12 APR 83	
1965-082JT		2080	US	15 OCT	23 SEP 66	
1965-082JU		2081	US	15 OCT	08 NOV 66	
1965-082JV		2082	US	15 OCT	16 JUN 79	
1965-082JW		2083	US	15 OCT	20 JUN 89	
1965-082JX		2084	US	15 OCT	29 AUG 81	
1965-082JY		2100	US	15 OCT	13 DEC 68	
1965-082JZ		2101	US	15 OCT	19 DEC 80	
1965-082K		1648	US	15 OCT	02 MAR 68	
1965-082KA		2102	US	15 OCT	15 APR 69	
1965-082KB		2103	US	15 OCT	27 NOV 67	
1965-082KC		2158	US	15 OCT	27 OCT 66	
1965-082KD		2333	US	15 OCT	29 SEP 67	
1965-082KE		2340	US	15 OCT	11 JUL 73	
1965-082KF		2384	US	15 OCT	29 SEP 73	
1965-082KG		2385	US	15 OCT	16 APR 89	
1965-082KH		2390	US	15 OCT	27 OCT 66	
1965-082KJ		2493	US	15 OCT	16 FEB 73	
1965-082KK		2573	US	15 OCT	15 MAR 71	
1965-082KL		2574	US	15 OCT	03 DEC 68	
1965-082KM		2699	US	15 OCT	03 JUL 89	
1965-082KN		2786	US	15 OCT	12 FEB 79	
1965-082KP		2787	US	15 OCT	19 SEP 80	
1965-082KS		2790	US	15 OCT	13 AUG 82	
1965-082KT		2791	US	15 OCT	29 APR 81	
1965-082KU		2792	US	15 OCT	26 DEC 71	
1965-082KV		2794	US	15 OCT	03 FEB 81	
1965-082KW		2795	US	15 OCT	14 SEP 78	
1965-082KY		2887	US	15 OCT	21 MAR 80	
1965-082KZ		2898	US	15 OCT	27 JUN 71	
1965-082L		1649	US	15 OCT	16 OCT 79	
1965-082LA		2899	US	15 OCT	06 MAY 79	
1965-082LB		2929	US	15 OCT	23 APR 79	
1965-082LC		2932	US	15 OCT	26 APR 90	
1965-082LE		2991	US	15 OCT	20 FEB 79	
1965-082LF		2992	US	15 OCT	29 MAR 69	
1965-082LG		3039	US	15 OCT	05 FEB 79	
1965-082LH		3040	US	15 OCT	29 MAR 71	
1965-082LJ		3046	US	15 OCT	07 OCT 88	
1965-082LK		3049	US	15 OCT	06 JAN 68	
1965-082LL		3050	US	15 OCT	03 JUL 79	
1965-082LM		3064	US	15 OCT	25 DEC 79	
1965-082LN		3065	US	15 OCT	26 NOV 89	
1965-082LP		3149	US	15 OCT	03 DEC 70	
1965-082LQ		3325	US	15 OCT	17 MAR 69	
1965-082LR		3331	US	15 OCT	01 DEC 79	
1965-082LS		3355	US	15 OCT	03 NOV 92	
1965-082LT		3356	US	15 OCT	07 MAR 79	
1965-082LU		3357	US	15 OCT	09 APR 84	
1965-082LV		3358	US	15 OCT	27 MAR 82	
1965-082LW		3359	US	15 OCT	26 SEP 90	
1965-082LX		3360	US	15 OCT	11 JUN 72	
1965-082LZ		3362	US	15 OCT	03 DEC 81	
1965-082M		1650	US	15 OCT	18 OCT 79	
1965-082MA		3363	US	15 OCT	12 APR 82	
1965-082MB		3364	US	15 OCT	30 NOV 80	
1965-082MC		3365	US	15 OCT	09 MAR 80	
1965-082MD		3366	US	15 OCT	13 APR 81	
1965-082ME		3367	US	15 OCT	21 JAN 80	
1965-082MF		3368	US	15 OCT	01 MAR 70	
1965-082MG		3369	US	15 OCT	26 FEB 90	
1965-082MH		3370	US	15 OCT	30 MAR 80	
1965-082MJ		3371	US	15 OCT	10 AUG 82	
1965-082MK		3372	US	15 OCT	01 NOV 78	
1965-082MM		3374	US	15 OCT	10 APR 79	
1965-082MN		3376	US	15 OCT	30 MAY 80	
1965-082MP		3377	US	15 OCT	16 OCT 80	
1965-082MQ		3378	US	15 OCT	17 JAN 79	
1965-082MR		3379	US	15 OCT	30 JUL 79	
1965-082MS		3380	US	15 OCT	25 SEP 90	
1965-082MU		3382	US	15 OCT	06 SEP 78	
1965-082MV		3383	US	15 OCT	29 OCT 91	
1965-082MW		3384	US	15 OCT	18 JUN 81	
1965-082MX		3385	US	15 OCT	12 APR 69	
1965-082MZ		3387	US	15 OCT	11 MAR 90	
1965-082N		1651	US	15 OCT	27 JAN 91	
1965-082NA		3400	US	15 OCT	10 MAR 79	
1965-082NB		3401	US	15 OCT	01 MAR 72	
1965-082NC		3402	US	15 OCT	05 JUL 78	
1965-082ND		3403	US	15 OCT	06 NOV 80	
1965-082NE		3404	US	15 OCT	03 MAR 70	
1965-082NF		3405	US	15 OCT	21 MAY 69	
1965-082NG		3406	US	15 OCT	08 MAY 69	
1965-082NH		3407	US	15 OCT	17 APR 78	
1965-082NJ		3410	US	15 OCT	05 OCT 73	
1965-082NK		3411	US	15 OCT	06 OCT 80	
1965-082NL		3412	US	15 OCT	14 MAY 70	
1965-082NM		3413	US	15 OCT	08 MAR 69	
1965-082NN		3420	US	15 OCT	13 APR 81	
1965-082NP		3421	US	15 OCT	13 JAN 69	
1965-082NQ		3422	US	15 OCT	24 SEP 69	
1965-082NR		3423	US	15 OCT	13 JUL 80	
1965-082NS		3424	US	15 OCT	22 APR 79	
1965-082NU		3426	US	15 OCT	18 FEB 81	
1965-082NV		3427	US	15 OCT	28 MAY 80	
1965-082NW		3433	US	15 OCT	19 OCT 78	
1965-082NX		3434	US	15 OCT	13 APR 82	
1965-082NY		3435	US	15 OCT	20 DEC 79	
1965-082NZ		3436	US	15 OCT	08 MAR 69	
1965-082P		1652	US	15 OCT	15 MAR 79	
1965-082PA		3437	US	15 OCT	28 NOV 81	
1965-082PB		3438	US	15 OCT	13 OCT 71	
1965-082PC		3439	US	15 OCT	31 JAN 80	
1965-082PD		3440	US	15 OCT	13 APR 81	
1965-082PE		3441	US	15 OCT	04 DEC 88	
1965-082PF		3442	US	15 OCT	12 JUN 69	
1965-082PG		3443	US	15 OCT	15 APR 80	
1965-082PH		3444	US	15 OCT	01 MAY 81	
1965-082PJ		3445	US	15 OCT	25 JUL 80	
1965-082PK		3446	US	15 OCT	13 FEB 81	
1965-082PL		3447	US	15 OCT	01 JUL 79	
1965-082PM		3448	US	15 OCT	01 JAN 83	
1965-082PN		3453	US	15 OCT	29 APR 69	
1965-082PQ		3455	US	15 OCT	24 DEC 80	
1965-082PR		3456	US	15 OCT	18 FEB 82	
1965-082PS		3461	US	15 OCT	19 JUL 84	
1965-082PU		3463	US	15 OCT	21 MAY 82	
1965-082PV		3464	US	15 OCT	21 OCT 81	
1965-082PW		3465	US	15 OCT	19 SEP 72	
1965-082PX		3466	US	15 OCT	18 NOV 91	
1965-082PZ		3468	US	15 OCT	17 OCT 80	
1965-082QA		3475	US	15 OCT	03 MAY 73	
1965-082QB		3476	US	15 OCT	19 APR 72	
1965-082QC		3477	US	15 OCT	07 APR 71	
1965-082QD		3478	US	15 OCT	04 MAR 80	
1965-082QE		3480	US	15 OCT	14 MAR 69	
1965-082QF		3481	US	15 OCT	10 FEB 80	
1965-082QG		3482	US	15 OCT	05 MAR 75	
1965-082QH		3483	US	15 OCT	29 MAY 78	
1965-082QK		3489	US	15 OCT	03 JUN 70	
1965-082QL		3490	US	15 OCT	13 JAN 78	
1965-082QM		3491	US	15 OCT	08 JUN 80	
1965-082QP		3493	US	15 OCT	21 JAN 69	
1965-082QR		3495	US	15 OCT	21 SEP 69	
1965-082QS		3496	US	15 OCT	05 APR 79	
1965-082QT		3497	US	15 OCT	19 MAR 80	
1965-082QU		3498	US	15 OCT	27 OCT 91	
1965-082QV		3499	US	15 OCT	22 MAY 80	

International Designation	Satellite Name	Number	Source	Launch	Decay	Note
1965-082QW		3500	US	15 OCT	14 JUL 70	
1965-082QX		3501	US	15 OCT	23 SEP 71	
1965-082QY		3502	US	15 OCT	27 MAR 79	
1965-082QZ		3718	US	15 OCT	21 SEP 88	
1965-082R		1654	US	15 OCT	16 OCT 79	
1965-082RA		3719	US	15 OCT	26 MAR 71	
1965-082RB		3720	US	15 OCT	07 NOV 79	
1965-082RC		3721	US	15 OCT	27 JUL 72	
1965-082RD		3722	US	15 OCT	15 MAY 70	
1965-082RE		3723	US	15 OCT	14 JUL 79	
1965-082RF		3724	US	15 OCT	06 DEC 69	
1965-082RG		3725	US	15 OCT	11 NOV 69	
1965-082RH		3726	US	15 OCT	06 APR 69	
1965-082RJ		3727	US	15 OCT	22 JUN 69	
1965-082RK		3728	US	15 OCT	01 FEB 70	
1965-082RL		3729	US	15 OCT	26 JUN 69	
1965-082RM		3730	US	15 OCT	21 DEC 77	
1965-082RN		3731	US	15 OCT	16 AUG 71	
1965-082RP		3732	US	15 OCT	07 JUN 78	
1965-082RQ		3733	US	15 OCT	05 SEP 79	
1965-082RR		3734	US	15 OCT	02 DEC 69	
1965-082RS		3735	US	15 OCT	23 SEP 69	
1965-082RT		3736	US	15 OCT	05 APR 78	
1965-082RU		3746	US	15 OCT	29 NOV 80	
1965-082RV		3751	US	15 OCT	24 JAN 83	
1965-082RW		3838	US	15 OCT	30 NOV 80	
1965-082RX		3839	US	15 OCT	23 FEB 77	
1965-082RY		3925	US	15 OCT	30 SEP 80	
1965-082RZ		3926	US	15 OCT	26 FEB 79	
1965-082S		1655	US	15 OCT	19 APR 71	
1965-082SA		3928	US	15 OCT	17 APR 83	
1965-082SB		3929	US	15 OCT	27 DEC 70	
1965-082SC		3930	US	15 OCT	19 NOV 82	
1965-082SD		3936	US	15 OCT	22 APR 70	
1965-082SF		3980	US	15 OCT	24 APR 70	
1965-082SG		3981	US	15 OCT	19 FEB 70	
1965-082SH		3982	US	15 OCT	25 OCT 69	
1965-082SJ		3996	US	15 OCT	26 DEC 69	
1965-082SK		3998	US	15 OCT	06 FEB 71	
1965-082SL		4003	US	15 OCT	18 FEB 81	
1965-082SM		4004	US	15 OCT	05 JAN 70	
1965-082SN		4017	US	15 OCT	13 AUG 79	
1965-082SP		4020	US	15 OCT	16 JUN 79	
1965-082SQ		4023	US	15 OCT	02 FEB 70	
1965-082SR		4025	US	15 OCT	16 OCT 69	
1965-082SS		4032	US	15 OCT	30 DEC 69	
1965-082ST		4033	US	15 OCT	03 SEP 70	
1965-082SU		4021	US	15 OCT	06 FEB 81	
1965-082SW		5056	US	15 OCT	03 MAY 71	
1565-082SX		5057	US	15 OCT	26 MAY 78	
1965-082SY		5058	US	15 OCT	07 MAR 81	
1965-082SZ		5085	US	15 OCT	12 SEP 80	
1965-082T		1656	US	15 OCT	25 NOV 81	
1965-082TA		5312	US	15 OCT	02 APR 72	
1965-082TB		5344	US	15 OCT	25 MAR 73	
1965-082TC		5345	US	15 OCT	25 JUN 80	
1965-082TD		5372	US	15 OCT	22 NOV 82	
1965-082TE		5375	US	15 OCT	19 APR 78	
1965-082TF		5376	US	15 OCT	07 DEC 91	
1965-082TG		5471	US	15 OCT	31 AUG 79	
1965-082TH		5567	US	15 OCT	21 NOV 72	
1965-082TJ		5568	US	15 OCT	24 JAN 81	
1965-082TK		5569	US	15 OCT	22 JAN 80	
1965-082TL		5571	US	15 OCT	06 APR 81	
1965-082TM		5572	US	15 OCT	24 AUG 75	
1965-082TN		5711	US	15 OCT	21 JUN 80	
1965-082TP		5719	US	15 OCT	30 JUL 75	
1965-082TQ		5734	US	15 OCT	03 JUL 80	
1965-082TR		5744	US	15 OCT	21 NOV 78	
1965-082TS		5768	US	15 OCT	01 MAR 72	
1965-082TT		5811	US	15 OCT	18 JUL 79	
1965-082TU		5844	US	15 OCT	13 MAY 74	
1965-082TV		5845	US	15 OCT	17 SEP 82	
1965-082TW		5856	US	15 OCT	23 OCT 72	
1965-082TX		5857	US	15 OCT	03 AUG 79	
1965-082TY		5955	US	15 OCT	22 FEB 78	
1965-082TZ		5968	US	15 OCT	11 MAY 77	
1965-082U		1657	US	15 OCT	17 JUN 70	
1965-082UA		5970	US	15 OCT	26 JUL 73	
1965-082UB		5982	US	15 OCT	31 JAN 73	
1965-082UC		6041	US	15 OCT	04 JUL 72	
1965-082UD		6042	US	15 OCT	28 AUG 73	
1965-082UE		6141	US	15 OCT	23 FEB 73	
1965-082UF		6183	US	15 OCT	15 SEP 74	
1965-082UG		6253	US	15 OCT	15 APR 83	
1965-082UH		6783	US	15 OCT	27 JUL 80	
1965-082UJ		9840	US	15 OCT	21 MAY 78	
1965-082UK		14842	US	15 OCT	11 SEP 89	
1965-082UL		14843	US	15 OCT	27 APR 88	
1965-082UN		18874	US	15 OCT	28 FEB 90	
1965-082UP		19182	US	15 OCT	08 FEB 91	
1965-082V		1658	US	15 OCT	08 MAY 70	
1965-082W		1659	US	15 OCT	20 NOV 69	
1965-082X		1660	US	15 OCT	23 SEP 69	
1965-082Y		1661	US	15 OCT	15 MAY 71	
1965-082Z		1662	US	15 OCT	31 AUG 89	
1965-082UM		18258	US	16 OCT	04 FEB 89	
1965-083A	COSMOS 92	1626	USSR	16 OCT	24 OCT 65	
1965-083B		1627	USSR	16 OCT	29 OCT 65	
1965-083C		1628	USSR	16 OCT	25 OCT 65	
1965-084A	COSMOS 93	1629	USSR	19 OCT	03 JAN 66	
1965-084B		1630	USSR	19 OCT	17 NOV 65	
1965-084C		1631	USSR	19 OCT	25 OCT 65	
1965-084D		1632	USSR	19 OCT	26 OCT 65	
1965-084E		2098	USSR	19 OCT	06 JUL 68	
1965-085A	COSMOS 94	1636	USSR	28 OCT	05 NOV 65	
1965-085B		1638	USSR	28 OCT	03 NOV 65	
1965-086A		1637	US	28 OCT	17 NOV 65	
1965-086B		1639	US	28 OCT	01 NOV 65	
1965-087A	PROTON 2	1701	USSR	02 NOV	06 FEB 66	
1965-087B		1702	USSR	02 NOV	04 JAN 66	
1965-087C		1703	USSR	02 NOV	16 NOV 65	
1965-087D		1704	USSR	02 NOV	08 NOV 65	
1965-087E		1705	USSR	02 NOV	08 NOV 65	
1965-088A	COSMOS 95	1706	USSR	04 NOV	18 JAN 66	
1965-088B		1707	USSR	04 NOV	28 NOV 65	M
1965-088C		1708	USSR	04 NOV	11 NOV 65	
1965-088D		1740	USSR	04 NOV	06 JAN 66	
1965-088E		1741	USSR	04 NOV	13 DEC 65	
1965-088F		1784	USSR	04 NOV	11 NOV 65	
1965-088G		1785	USSR	04 NOV	16 NOV 65	
1965-088H		1786	USSR	04 NOV	09 NOV 65	
1965-088J		1787	USSR	04 NOV	09 NOV 65	
1965-088K		1788	USSR	04 NOV	19 NOV 65	
1965-088L		1789	USSR	04 NOV	28 NOV 65	
1965-088M		1790	USSR	04 NOV	12 NOV 65	
1965-088N		1791	USSR	04 NOV	07 NOV 65	
1965-088P		1792	USSR	04 NOV	16 NOV 65	
1965-088Q		1793	USSR	04 NOV	07 NOV 65	
1965-088R		1794	USSR	04 NOV	28 NOV 65	
1965-088S		1795	USSR	04 NOV	08 NOV 65	
1965-088T		1796	USSR	04 NOV	15 NOV 65	
1965-088U		1797	USSR	04 NOV	11 NOV 65	
1965-088V		1798	USSR	04 NOV	28 NOV 65	
1965-088W		1799	USSR	04 NOV	28 NOV 65	
1965-088X		1800	USSR	04 NOV	28 NOV 65	
1965-088Y		1801	USSR	04 NOV	26 NOV 65	
1965-088Z		1802	USSR	04 NOV	28 NOV 65	
1965-090A		1727	US	08 NOV	11 NOV 65	
1965-090B		1728	US	08 NOV	09 NOV 65	
1965-091B		1731	USSR	12 NOV	17 NOV 65	
1965-091C		1732	USSR	12 NOV	25 NOV 65	
1965-092A	VENERA 3	1733	USSR	16 NOV	01 MAR 66	N
1965-092B		1734	USSR	16 NOV	26 NOV 65	
1965-092C		1735	USSR	16 NOV	03 DEC 65	
1965-092E		1737	USSR	16 NOV	16 NOV 65	
1965-094A	COSMOS 96	1742	USSR	23 NOV	09 DEC 65	
1965-094B		1743	USSR	23 NOV	04 DEC 65	
1965-094C		1771	USSR	23 NOV	28 NOV 65	
1965-094D		1772	USSR	23 NOV	28 NOV 65	
1965-094E		1773	USSR	23 NOV	29 NOV 65	
1965-094F		1774	USSR	23 NOV	28 NOV 65	
1965-094G		1775	USSR	23 NOV	08 DEC 65	
1965-094H		1776	USSR	23 NOV	08 DEC 65	
1965-095A	COSMOS 97	1777	USSR	26 NOV	02 APR 67	
1965-095B		1779	USSR	26 NOV	21 FEB 67	
1965-095C		1782	USSR	26 NOV	21 JAN 66	
1965-095D		1783	USSR	26 NOV	21 JAN 66	
1965-096C		1938	FRAN	26 NOV	25 MAR 67	
1965-097A	COSMOS 98	1780	USSR	27 NOV	05 DEC 65	
1965-097B		1781	USSR	27 NOV	23 JAN 66	
1965-097C		1803	USSR	27 NOV	08 DEC 65	
1965-098K		20833	US	29 NOV	11 FEB 92	
1965-099A	LUNA 8	1810	USSR	03 DEC	06 DEC 65	A
1965-099B		1809	USSR	03 DEC	05 DEC 65	
1965-099C		1811	USSR	03 DEC	06 DEC 65	
1965-100A	GEMINI 7	1812	US	04 DEC	18 DEC 65	F
1965-100B		1813	US	04 DEC	07 DEC 65	
1965-100C		1845	US	04 DEC	27 JAN 66	
1965-101C		1934	FRAN	06 DEC	10 DEC 81	
1965-101D		1935	FRAN	06 DEC	10 NOV 80	
1965-102A		1816	US	09 DEC	26 DEC 65	
1965-102B		1838	US	09 DEC	15 DEC 65	
1965-103A	COSMOS 99	1817	USSR	10 DEC	18 DEC 65	
1965-103B		1818	USSR	10 DEC	28 DEC 65	
1965-104A	GEMINI 6	1839	US	15 DEC	16 DEC 65	F
1965-104B		1840	US	15 DEC	16 DEC 65	
1965-105B		1842	US	16 DEC	06 SEP 68	
1965-107A	COSMOS 101	1846	USSR	21 DEC	12 JUL 66	
1965-107B		1847	USSR	21 DEC	16 APR 66	
1965-108A	TITAN 3 C-8	1863	US	21 DEC	17 AUG 75	
1965-108B	LES 4	1870	US	21 DEC	01 AUG 77	
1965-108C	OSCAR 4	1902	US	21 DEC	12 APR 76	
1965-108D	LES 3	1941	US	21 DEC	06 APR 68	
1965-108E		3999	US	21 DEC	21 AUG 69	
1965-108F		4429	US	21 DEC	03 NOV 75	
1965-108G		4476	US	21 DEC	01 AUG 77	
1965-108H		5472	US	21 DEC	25 NOV 72	
1965-108J		5474	US	21 DEC	03 FEB 73	
1965-108K		5980	US	21 DEC	12 APR 76	
1965-110A		1866	US	24 DEC	20 JAN 66	
1965-110B		1901	US	24 DEC	06 JAN 66	
1965-111A	COSMOS 102	1867	USSR	27 DEC	13 JAN 66	
1965-112A	COSMOS 103	1868	USSR	28 DEC	02 JAN 90	
1965-112B		1869	USSR	28 DEC	15 JUN 82	P
1965-112C		1871	USSR	28 DEC	25 APR 67	

International Designation	Satellite Name	Number	Source	Launch	Decay	Note
1965-112D		1872	USSR	28 DEC	27 DEC 67	
1965-112E		1873	USSR	28 DEC	21 APR 67	
1965-112F		1874	USSR	28 DEC	01 MAY 68	
1965-112G		1875	USSR	28 DEC	17 FEB 68	
1965-112H		1876	USSR	28 DEC	12 MAR 68	
1965-112J		1877	USSR	28 DEC	22 MAR 68	
1965-112K		1878	USSR	28 DEC	02 JUL 70	
1965-112L		1906	USSR	28 DEC	19 OCT 68	
1965-112M		1907	USSR	28 DEC	16 OCT 68	
1965-112N		1933	USSR	28 DEC	05 JUL 67	
1965-112P		1936	USSR	28 DEC	07 APR 69	
1966 LAUNCHES						
1966-000A		2428	US	20 JUL	18 SEP 77	Q
1966-000B		2429	US	09 MAY	01 JAN 79	Q
1966-000C		2430	US	09 MAY	01 JAN 79	Q
1966-000D		2633	US	12 NOV	30 DEC 66	Q
1966-001A	COSMOS 104	1903	USSR	07 JAN	15 JAN 66	
1966-001B		1904	USSR	07 JAN	24 JAN 66	
1966-001C		1905	USSR	07 JAN	12 JAN 66	
1966-002A		1939	US	19 JAN	25 JAN 66	
1966-002B		1940	US	19 JAN	23 JAN 66	
1966-003A	COSMOS 105	1945	USSR	22 JAN	30 JAN 66	
1966-003B		1946	USSR	22 JAN	10 FEB 66	
1966-003C		1947	USSR	22 JAN	24 JAN 66	
1966-004A	COSMOS 106	1949	USSR	25 JAN	14 NOV 66	
1966-004B		1950	USSR	25 JAN	16 JUL 66	
1966-005G		4587	US	28 JAN	04 OCT 81	
1966-005H		11990	US	28 JAN	11 JAN 83	
1966-005K		14226	US	28 JAN	15 JUL 89	
1966-005L		15341	US	28 JAN	10 DEC 89	
1966-005M		19366	US	28 JAN	24 APR 90	
1966-006A	LUNA 9	1954	USSR	31 JAN	03 FEB 66	R
1966-006B		1955	USSR	31 JAN	02 FEB 66	
1966-006C		1966	USSR	31 JAN	31 JAN 66	
1966-007A		1968	US	02 FEB	27 FEB 66	
1966-009A		1997	US	09 FEB	26 SEP 69	
1966-009B		2003	US	09 FEB	18 NOV 66	
1966-009C		2004	US	09 FEB	18 NOV 66	
1966-010A	COSMOS 107	1998	USSR	10 FEB	18 FEB 66	
1966-010B		2000	USSR	10 FEB	28 FEB 66	
1966-010C		2005	USSR	10 FEB	12 FEB 66	
1966-010D		2006	USSR	10 FEB	12 FEB 66	
1966-011A	COSMOS 108	2002	USSR	11 FEB	21 NOV 66	
1966-011B		2007	USSR	11 FEB	01 JUN 66	
1966-011C		2008	USSR	11 FEB	27 MAR 66	
1966-012A		2012	US	15 FEB	22 FEB 66	
1966-012AA		2061	US	15 FEB	19 FEB 66	
1966-012AB		2062	US	15 FEB	22 FEB 66	
1966-012AC		2063	US	15 FEB	17 FEB 66	
1966-012AD		2064	US	15 FEB	22 FEB 66	
1966-012AE		2065	US	15 FEB	19 FEB 66	
1966-012AF		2066	US	15 FEB	19 FEB 66	
1966-012AG		2067	US	15 FEB	17 FEB 66	
1966-012AH		2068	US	15 FEB	17 FEB 66	
1966-012AJ		2069	US	15 FEB	20 FEB 66	
1966-012AK		2072	US	15 FEB	18 FEB 66	
1966-012AL		2073	US	15 FEB	17 FEB 66	
1966-012AM		2074	US	15 FEB	18 FEB 66	
1966-012AN		2075	US	15 FEB	17 FEB 66	
1966-012AP		2076	US	15 FEB	19 FEB 66	
1966-012AQ		2077	US	15 FEB	18 FEB 66	
1966-012AR		2078	US	15 FEB	19 FEB 66	
1966-012B		2014	US	15 FEB	16 FEB 66	
1966-012C		2015	US	15 FEB	22 FEB 66	
1966-012D		2025	US	15 FEB	18 FEB 66	S
1966-012E		2026	US	15 FEB	17 FEB 66	
1966-012F		2027	US	15 FEB	17 FEB 66	
1966-012G		2028	US	15 FEB	18 FEB 66	
1966-012H		2029	US	15 FEB	19 FEB 66	
1966-012J		2030	US	15 FEB	19 FEB 66	
1966-012K		2031	US	15 FEB	20 FEB 66	
1966-012L		2032	US	15 FEB	17 FEB 66	
1966-012M		2033	US	15 FEB	18 FEB 66	
1966-012N		2034	US	15 FEB	19 FEB 66	
1966-012P		2035	US	15 FEB	17 FEB 66	
1966-012Q		2036	US	15 FEB	18 FEB 66	
1966-012R		2037	US	15 FEB	20 FEB 66	
1966-012S		2038	US	15 FEB	19 FEB 66	
1966-012T		2039	US	15 FEB	16 FEB 66	
1966-012U		2040	US	15 FEB	18 FEB 66	
1966-012V		2056	US	15 FEB	20 FEB 66	
1966-012W		2057	US	15 FEB	21-FEB 66	
1966-012X		2058	US	15 FEB	18 FEB 66	
1966-012Y		2059	US	15 FEB	18 FEB 66	
1966-012Z		2060	US	15 FEB	18 FEB 66	
1966-013C		2018	FRAN	17 FEB	05 DEC 67	
1966-013D		2020	FRAN	17 FEB	11 SEP 67	
1966-013E		2021	FRAN	17 FEB	25 AUG 67	
1966-013F		2023	FRAN	17 FEB	16 JUL 91	
1966-013H		2993	FRAN	17 FEB	01 MAR 68	
1966-013J		3970	FRAN	17 FEB	28 MAY 81	
1966-013K		5420	FRAN	17 FEB	22 NOV 82	
1966-013L		5959	FRAN	17 FEB	21 NOV 78	
1966-013M		6784	FRAN	17 FEB	15 MAR 74	
1966-013N		14844	FRAN	17 FEB	08 NOV 88	
1966-013P		14845	FRAN	17 FEB	23 APR 89	
1966-013Q		19049	FRAN	17 FEB	22 JAN 89	
1966-014A	COSMOS 109	2019	USSR	19 FEB	27 FEB 66	
1966-014B		2022	USSR	19 FEB	27 FEB 66	
1966-014C		2024	USSR	19 FEB	21 FEB 66	
1966-015A	COSMOS 110	2070	USSR	22 FEB	16 MAR 66	C
1966-015B		2071	USSR	22 FEB	29 APR 66	
1966-015C		2089	USSR	22 FEB	05 MAR 66	
1966-015D		2090	USSR	22 FEB	05 MAR 66	
1966-017A	COSMOS 111	2093	USSR	01 MAR	03 MAR 66	
1966-017B		2094	USSR	01 MAR	01 MAR 66	
1966-017C		2095	USSR	01 MAR	02 MAR 66	
1966-018A		2099	US	09 MAR	29 MAR 66	
1966-018B		2113	US	09 MAR	22 MAR 66	
1966-019A	AGENA TARGET	2104	US	16 MAR	15 SEP 67	
1966-020A	GEMINI 8	2105	US	16 MAR	17 MAR 66	F
1966-020B		2106	US	16 MAR	17 MAR 66	
1966-020C		2115	US	16 MAR	18 APR 66	
1966-021A	COSMOS 112	2107	USSR	17 MAR	25 MAR 66	
1966-021B		2108	USSR	17 MAR	17 MAY 66	
1966-021C		2110	USSR	17 MAR	25 MAR 66	
1966-021D		2111	USSR	17 MAR	25 MAR 66	
1966-022A		2109	US	18 MAR	24 MAR 66	
1966-022B		2112	US	18 MAR	23 MAR 66	
1966-023A	COSMOS 113	2114	USSR	21 MAR	29 MAR 66	
1966-023B		2116	USSR	21 MAR	30 MAR 66	
1966-023C		2117	USSR	21 MAR	22 MAR 66	
1966-024C		2386	US	26 MAR	14 FEB 83	
1966-024D		3590	US	26 MAR	18 APR 78	
1966-024E		12856	US	26 MAR	20 JUL 88	
1966-025F		4007	US	30 MAR	12 APR 77	
1966-026C		2162	US	31 MAR	11 FEB 67	
1966-026E		2178	US	31 MAR	06 JAN 83	
1966-026F		2179	US	31 MAR	20 APR 93	
1966-027B		2127	USSR	31 MAR	04 APR 66	
1966-027C		2128	USSR	31 MAR	02 APR 66	
1966-028A	COSMOS 114	2133	USSR	06 APR	14 APR 66	
1966-028B		2134	USSR	06 APR	18 APR 66	
1966-028C		2135	USSR	06 APR	10 APR 66	
1966-028D		2137	USSR	06 APR	10 APR 66	
1966-028E		2138	USSR	06 APR	09 APR 66	
1966-029A		2136	US	07 APR	26 APR 66	
1966-030A	SURVEYOR MODULE	2139	US	08 APR	05 MAY 66	
1966-030B		2143	US	08 APR	17 APR 66	
1966-031C		2145	US	08 APR	31 OCT 81	
1966-032A		2146	US	19 APR	26 APR 66	
1966-032B		2148	US	19 APR	22 APR 66	
1966-033A	COSMOS 115	2147	USSR	20 APR	28 APR 66	
1966-033B		2149	USSR	20 APR	30 APR 66	
1966-034C		2208	US	22 APR	05 DEC 76	
1966-034D		2209	US	22 APR	04 DEC 89	
1966-034E		9998	US	22 APR	01 DEC 82	
1966-035A	MOLNIYA 1-3	2151	USSR	25 APR	11 JUN 73	
1966-035B		2155	USSR	25 APR	09 MAY 66	
1966-035C		2156	USSR	25 APR	08 MAY 66	
1966-035D		2157	USSR	25 APR	29 APR 66	
1966-035E		2160	USSR	25 APR	29 APR 66	
1966-035F		3222	USSR	25 APR	18 JUN 73	
1966-036A	COSMOS 116	2152	USSR	26 APR	03 DEC 66	
1966-036B		2159	USSR	26 APR	15 AUG 66	
1966-037A	COSMOS 117	2163	USSR	06 MAY	14 MAY 66	
1966-037B		2164	USSR	06 MAY	21 MAY 66	
1966-037C		2165	USSR	06 MAY	07 MAY 66	
1966-037D		2166	USSR	06 MAY	08 MAY 66	
1966-037E		2170	USSR	06 MAY	16 MAY 66	
1966-038A	COSMOS 118	2168	USSR	11 MAY	23 NOV 88	
1966-038B		2169	USSR	11 MAY	12 JUL 89	
1966-039A		2171	US	14 MAY	21 MAY 66	
1966-039B		2172	US	14 MAY	27 OCT 70	
1966-039C		2175	US	14 MAY	17 MAY 66	
1966-042A		2181	US	24 MAY	09 JUN 66	
1966-043A	COSMOS 119	2182	USSR	24 MAY	30 NOV 66	
1966-044A	EXPLORER 32	2183	US	25 MAY	22 FEB 85	
1966-044B		2184	US	25 MAY	19 JUN 69	
1966-044C		2336	US	25 MAY	31 AUG 69	
1966-045A	SURVEYOR 1	2185	US	30 MAY	02 JUN 66	T
1966-046A	ATDA	2186	US	01 JUN	11 JUN 66	
1966-046AA		2248	US	01 JUN	23 JUN 66	
1966-046AB		2249	US	01 JUN	04 JUL 66	
1966-046AC		2250	US	01 JUN	29 JUN 66	
1966-046AD		2251	US	01 JUN	25 JUN 66	
1966-046AE		2252	US	01 JUN	23 JUN 66	
1966-046AF		2264	US	01 JUN	04 JUL 66	
1966-046AG		2265	US	01 JUN	04 JUL 66	
1966-046AH		2266	US	01 JUN	04 JUL 66	
1966-046AJ		2267	US	01 JUN	04 JUL 66	
1966-046AK		2268	US	01 JUN	04 JUL 66	
1966-046AL		2269	US	01 JUN	04 JUL 66	
1966-046AM		2270	US	01 JUN	04 JUL 66	
1966-046AN		2271	US	01 JUN	04 JUL 66	
1966-046AP		2272	US	01 JUN	04 JUL 66	
1966-046AQ		2273	US	01 JUN	04 JUL 66	
1966-046AR		2274	US	01 JUN	04 JUL 66	
1966-046AS		2275	US	01 JUN	04 JUL 66	
1966-046AT		2276	US	01 JUN	04 JUL 66	
1966-046AU		2277	US	01 JUN	04 JUL 66	
1966-046AV		2278	US	01 JUN	04 JUL 66	
1966-046AW		2279	US	01 JUN	04 JUL 66	
1966-046AX		2280	US	01 JUN	04 JUL 66	

International Designation	Satellite Name	Number	Source	Launch	Decay	Note
1966-046AY		2281	US	01 JUN	04 JUL 66	
1966-046AZ		2282	US	01 JUN	04 JUL 66	
1966-046B		2188	US	01 JUN	22 JUN 66	U
1966-046BA		2283	US	01 JUN	04 JUL 66	
1966-046BB		2284	US	01 JUN	04 JUL 66	
1966-046BC		2285	US	01 JUN	04 JUL 66	
1966-046BD		2286	US	01 JUN	04 JUL 66	
1966-046BE		2287	US	01 JUN	04 JUL 66	
1966-046BF		2288	US	01 JUN	04 JUL 66	
1966-046C		2189	US	01 JUN	05 JUN 66	
1966-046D		2190	US	01 JUN	05 JUN 66	
1966-046E		2228	US	01 JUN	30 JUN 66	
1966-046F		2229	US	01 JUN	30 JUN 66	
1966-046G		2230	US	01 JUN	30 JUN 66	
1966-046H		2231	US	01 JUN	30 JUN 66	
1966-046J		2232	US	01 JUN	30 JUN 66	
1966-046K		2233	US	01 JUN	21 JUN 66	
1966-046L		2234	US	01 JUN	23 JUN 66	
1966-046M		2235	US	01 JUN	23 JUN 66	
1966-046N		2236	US	01 JUN	22 JUN 66	
1966-046P		2237	US	01 JUN	23 JUN 66	
1966-0460		2238	US	01 JUN	22 JUN 66	
1966-046R		2239	US	01 JUN	22 JUN 66	
1966-046S		2240	US	01 JUN	23 JUN 66	
1966-046T		2241	US	01 JUN	22 JUN 66	
1966-046U		2242	US	01 JUN	27 JUN 66	
1966-046V		2243	US	01 JUN	21 JUN 66	
1966-046W		2244	US	01 JUN	23 JUN 66	
1966-046X		2245	US	01 JUN	22 JUN 66	
1966-046Y		2246	US	01 JUN	21 JUN 66	
1966-046Z		2247	US	01 JUN	21 JUN 66	
1966-047A	GEMINI 9	2191	US	03 JUN	06 JUN 66	F
1966-047B		2193	US	03 JUN	04 JUN 66	
1966-048A		2192	US	03 JUN	09 JUN 66	
1966-048B		2194	US	03 JUN	09 JUN 66	
1966-050A	COSMOS 120	2196	USSR	08 JUN	16 JUN 66	
1966-050B		2197	USSR	08 JUN	12 JUN 66	
1966-050C		2198	USSR	08 JUN	11 JUN 66	
1966-050D		2199	USSR	08 JUN	09 JUN 66	
1966-051A		2200	US	09 JUN	03 DEC 66	
1966-051B	SECOR (EGRS) 6	2205	US	09 JUN	06 JUL 67	
1966-051C	ERS 16	2202	US	09 JUN	12 MAR 67	
1966-054A	COSMOS 121	2210	USSR	17 JUN	25 JUN 66	
1966-054B		2211	USSR	17 JUN	26 JUN 66	
1966-054C		2212	USSR	17 JUN	19 JUN 66	
1966-054D		2213	USSR	17 JUN	20 JUN 66	
1966-054E		2214	USSR	17 JUN	20 JUN 66	
1966-055A		2227	US	21 JUN	14 JUL 66	
1966-055B		2261	US	21 JUN	04 JUL 66	
1966-056AA		9461	US	24 JUN	16 DEC 77	
1966-056AB		9462	US	24 JUN	09 FEB 77	
1966-056AC		9463	US	24 JUN	13 APR 81	
1966-056AD		9464	US	24 JUN	22 AUG 80	
1966-056AE		9465	US	24 JUN	13 APR 81	
1966-056AF		9466	US	24 JUN	02 FEB 89	
1966-056AG		9467	US	24 JUN	18 APR 77	
1966-056AJ		9469	US	24 JUN	25 NOV 82	
1966-056AK		9739	US	24 JUN	20 JAN 77	
1966-056AL		9740	US	24 JUN	20 JAN 77	
1966-056AM		9741	US	24 JUN	20 JAN 77	
1966-056AN		9742	US	24 JUN	20 JAN 77	
1966-056AP		9743	US	24 JUN	20 JAN 77	
1966-056AQ		9744	US	24 JUN	20 JAN 77	
1966-056AR		9745	US	24 JUN	20 JAN 77	
1966-056AS		9746	US	24 JUN	20 JAN 77	
1966-056AT		9747	US	24 JUN	20 JAN 77	
1966-056AU		9748	US	24 JUN	20 JAN 77	
1966-056AV		9749	US	24 JUN	20 JAN 77	
1966-056AW		9750	US	24 JUN	20 JAN 77	
1966-056AX		9751	US	24 JUN	20 JAN 77	
1966-056AY		9752	US	24 JUN	20 JAN 77	
1966-056AZ		9753	US	24 JUN	20 JAN 77	
1966-056BA		9754	US	24 JUN	20 JAN 77	
1966-056BB		9755	US	24 JUN	20 JAN 77	
1966-056BC		9756	US	24 JUN	20 JAN 77	
1966-056BD		9757	US	24 JUN	20 JAN 77	
1966-056BE		9758	US	24 JUN	20 JAN 77	
1966-056BF		9759	US	24 JUN	20 JAN 77	
1966-056BG		9760	US	24 JUN	20 JAN 77	
1966-056BH		9761	US	24 JUN	20 JAN 77	
1966-056BJ		9762	US	24 JUN	20 JAN 77	
1966-056BK		9763	US	24 JUN	20 JAN 77	
1966-056BL		9764	US	24 JUN	20 JAN 77	
1966-056BM		9765	US	24 JUN	20 JAN 77	
1966-056BN		9766	US	24 JUN	20 JAN 77	
1966-056BP		9767	US	24 JUN	20 JAN 77	
1966-056BQ		9768	US	24 JUN	20 JAN 77	
1966-056BR		9769	US	24 JUN	20 JAN 77	
1966-056BS		9770	US	24 JUN	20 JAN 77	
1966-056BT		9771	US	24 JUN	20 JAN 77	
1966-056BU		9772	US	24 JUN	20 JAN 77	
1966-056BV		9773	US	24 JUN	20 JAN 77	
1966-056BW		9774	US	24 JUN	20 JAN 77	
1966-056BX		9775	US	24 JUN	20 JAN 77	
1966-056BY		9776	US	24 JUN	20 JAN 77	
1966-056BZ		9777	US	24 JUN	20 JAN 77	
1966-056CA		9778	US	24 JUN	20 JAN 77	
1966-056CB		9779	US	24 JUN	20 JAN 77	
1966-056CC		9780	US	24 JUN	20 JAN 77	
1966-056CD		9781	US	24 JUN	20 JAN 77	
1966-056CE		9782	US	24 JUN	20 JAN 77	
1966-056CF		9783	US	24 JUN	20 JAN 77	
1966-056CG		13795	US	24 JUN	25 APR 87	
1966-056CH		18641	US	24 JUN	20 DEC 89	
1966-056CJ		18672	US	24 JUN	13 FEB 90	
1966-056CK		18690	US	24 JUN	02 FEB 89	
1966-056E		8064	US	24 JUN	03 JAN 90	
1966-056F		8065	US	24 JUN	18 APR 78	V
1966-056J		8075	US	24 JUN	19 FEB 77	
1966-056K		8076	US	24 JUN	04 SEP 84	
1966-056L		8077	US	24 JUN	01 APR 77	
1966-056M		8078	US	24 JUN	19 FEB 77	
1966-056N		8079	US	24 JUN	12 JAN 86	
1966-056P		8080	US	24 JUN	21 JAN 77	
1966-056Q		8081	US	24 JUN	22 MAY 77	
1966-056R		9452	US	24 JUN	08 MAR 77	
1966-056S		9453	US	24 JUN	13 APR 81	
1966-056T		9454	US	24 JUN	08 FEB 77	
1966-056U		9455	US	24 JUN	03 MAR 77	
1966-056V		9456	US	24 JUN	14 FEB 77	
1966-056W		9457	US	24 JUN	20 AUG 77	
1966-056X		9458	US	24 JUN	29 JUN 77	
1966-056Y		9459	US	24 JUN	02 JUL 77	
1966-056Z		9460	US	24 JUN	09 JAN 90	
1966-057A	COSMOS 122	2254	USSR	25 JUN	14 NOV 89	
1966-057B		2257	USSR	25 JUN	20 JAN 84	
1966-058B		2259	US	01 JUL	17 NOV 66	
1966-058D		2262	US	01 JUL	07 SEP 66	
1966-058E		2263	US	01 JUL	29 AUG 66	
1966-059A	AS 203	2289	US	05 JUL	05 JUL 66	
1966-059AA		2319	US	05 JUL	20 JUL 66	
1966-059AB		2320	US	05 JUL	19 JUL 66	
1966-059AC		2321	US	05 JUL	18 JUL 66	
1966-059AD		2341	US	05 JUL	15 JUL 66	
1966-059AE		2342	US	05 JUL	20 JUL 66	
1966-059AF		2343	US	05 JUL	20 JUL 66	
1966-059AG		2344	US	05 JUL	20 JUL 66	
1966-059AH		2345	US	05 JUL	18 JUL 66	
1966-059AJ		2346	US	05 JUL	19 JUL 66	
1966-059AK		2347	US	05 JUL	19 JUL 66	
1966-059AL		2352	US	05 JUL	22 JUL 66	
1966-059B		2291	US	05 JUL	18 JUL 66	W
1966-059C		2297	US	05 JUL	12 JUL 66	
1966-059D		2298	US	05 JUL	12 JUL 66	
1966-059E		2299	US	05 JUL	11 JUL 66	
1966-059F		2300	US	05 JUL	12 JUL 66	
1966-059G		2301	US	05 JUL	11 JUL 66	
1966-059H		2302	US	05 JUL	10 JUL 66	
1966-059J		2303	US	05 JUL	12 JUL 66	
1966-059K		2304	US	05 JUL	11 JUL 66	
1966-059L		2305	US	05 JUL	11 JUL 66	
1966-059M		2306	US	05 JUL	11 JUL 66	
1966-059N		2307	US	05 JUL	19 JUL 66	
1966-059P		2308	US	05 JUL	14 JUL 66	
1966-059Q		2309	US	05 JUL	15 JUL 66	
1966-059R		2310	US	05 JUL	19 JUL 66	
1966-059S		2311	US	05 JUL	16 JUL 66	
1966-059T		2312	US	05 JUL	15 JUL 66	
1966-059U		2313	US	05 JUL	15 JUL 66	
1966-059V		2314	US	05 JUL	16 JUL 66	
1966-059W		2315	US	05 JUL	16 JUL 66	
1966-059X		2316	US	05 JUL	16 JUL 66	
1966-059Y		2317	US	05 JUL	17 JUL 66	
1966-059Z		2318	US	05 JUL	18 JUL 66	
1966-060A	PROTON 3	2290	USSR	06 JUL	16 SEP 66	
1966-060B		2293	USSR	06 JUL	21 AUG 66	
1966-060C		2294	USSR	06 JUL	19 JUL 66	
1966-060D		2292	USSR	06 JUL	11 JUL 66	
1966-061A	COSMOS 123	2295	USSR	08 JUL	10 DEC 66	
1966-061B		2296	USSR	08 JUL	02 OCT 66	
1966-062A		2322	US	12 JUL	20 JUL 66	
1966-062B		2323	US	12 JUL	14 JUL 66	
1966-063A	OV1-8	2324	US	1a JUL	04 JAN 78	
1966-064A	COSMOS 124	2325	USSR	14 JUL	22 JUL 66	
1966-064B		2326	USSR	14 JUL	19 JUL 66	
1966-064C		2330	USSR	14 JUL	17 JUL 66	
1966-064D		2331	USSR	14 JUL	16 JUL 66	
1966-064E		2332	USSR	14 JUL	19 JUL 66	
1966-065A	AGENA TARGET	2348	US	18 JUL	29 DEC 66	
1966-066A	GEMINI 10	2349	US	18 JUL	21 JUL 66	F
1966-066B		2350	US	18 JUL	19 JUL 66	
1966-066C		2354	US	18 JUL	14 AUG 66	
1966-066D		2355	US	18 JUL	18 SEP 66	
1966-066E		2356	US	18 JUL	18 APR 67	
1966-066F		2357	US	18 JUL	15 OCT 66	
1966-066G		2358	US	18 JUL	12 AUG 66	
1966-066H		2382	US	18 JUL	20 OCT 66	
1966-066J		2383	US	18 JUL	17 JAN 67	
1966-066K		2388	US	18 JUL	26 FEB 67	
1966-067A	COSMOS 125	2351	USSR	20 JUL	02 AUG 66	
1966-068A	COSMOS 126	2368	USSR	28 JUL	06 AUG 66	
1966-068B		2369	USSR	28 JUL	10 AUG 66	
1966-068C		2370	USSR	28 JUL	30 JUL 66	
1966-069A		2376	US	28 JUL	06 AUG 66	
1966-070B		2404	US	04 AUG	03 JAN 92	
1966-070C		2521	US	04 AUG	30 JUN 74	

International Designation	Satellite Name	Number	Source	Launch	Decay	Note
1966-071A	COSMOS 127	2391	USSR	08 AUG	16 AUG 66	
1966-071B		2392	USSR	08 AUG	12 AUG 66	
1966-072A		2393	US	09 AUG	11 SEP 66	
1966-073A	LUNAR ORBITER 1	2394	US	10 AUG	29 OCT 66	G
1966-074A		2396	US	16 AUG	24 AUG 66	
1966-074B		2397	US	16 AUG	05 MAR 70	
1966-074C		2400	US	16 AUG	18 AUG 66	
1966-075B		2399	US	17 AUG	07 AUG 67	
1966-075D		2405	US	17 AUG	26 FEB 67	
1966-075E		2572	US	17 AUG	09 MAR 67	
1966-076E		7558	US	18 AUG	02 SEP 81	
1966-076F		8000	US	18 AUG	11 SEP 81	
1966-076G		9401	US	18-AUG	25 MAR 82	
1966-078B		2407	USSR	24 AUG	25 AUG 66	
1966-078C		2408	USSR	24 AUG	26 AUG 66	
1966-079A	COSMOS 128	2409	USSR	27 AUG	04 SEP 66	
1966-079B		2410	USSR	27 AUG	07 SEP 66	
1966-080A	AGENA TARGET	2414	US	12 SEP	30 DEC 66	
1966-081A	GEMINI 11	2415	US	12 SEP	15 SEP 66	F
1966-081B		2416	US	12 SEP	13 SEP 66	
1966-081C		2417	US	12 SEP	08 OCT 66	
1966-081D		2421	US	12 SEP	20 SEP 66	
1966-083A		2419	US	16 SEP	23 SEP 66	
1966-083B		2420	US	16 SEP	09 MAY 68	
1966-083C		2423	US	16 SEP	18 SEP 66	
1966-083D		2424	US	16 SEP	30 OCT 66	
1966-088A		2437	USSR	17 SEP	11 NOV 66	
1966-088AA		2461	USSR	17 SEP	24 SEP 66	
1966-088AB		2462	USSR	17 SEP	27 SEP 66	
1966-088AC		2463	USSR	17 SEP	26 SEP 66	
1966-088AD		2464	USSR	17 SEP	27 SEP 66	
1966-088AE		2465	USSR	17 SEP	25 SEP 66	
1966-088AF		2466	USSR	17 SEP	28 SEP 66	
1966-088AG		2467	USSR	17 SEP	26 SEP 66	
1966-088AH		2468	USSR	17 SEP	22 SEP 66	
1966-088AJ		2469	USSR	17 SEP	24 SEP 66	
1966-088AK		2470	USSR	17 SEP	24 SEP 66	
1966-088AL		2471	USSR	17 SEP	25 SEP 66	
1966-088AM		2472	USSR	17 SEP	25 SEP 66	
1966-088AN		2473	USSR	17 SEP	27 SEP 66	
1966-088AP		2474	USSR	17 SEP	23 SEP 66	
1966-088AQ		2475	USSR	17 SEP	23 SEP 66	
1966-088AR		2476	USSR	17 SEP	27 SEP 66	
1966-088AS		2477	USSR	17 SEP	30 SEP 66	
1966-088AT		2478	USSR	17 SEP	27 SEP 66	
1966-088AU		2479	USSR	17 SEP	03 OCT 66	
1966-088AV		2480	USSR	17 SEP	25 SEP 66	
1966-088AW		2482	USSR	17 SEP	18 OCT 66	
1966-088AX		2483	USSR	17 SEP	08 OCT 66	
1966-088AY		2484	USSR	17 SEP	09 FEB 67	
1966-088AZ		2485	USSR	17 SEP	12 NOV 66	
1966-088B		2438	USSR	17 SEP	04 MAR 67	X
1966-088BA		2486	USSR	17 SEP	21 OCT 66	
1966-088BB		2487	USSR	17 SEP	18 OCT 66	
1966-088BC		2488	USSR	17 SEP	10 OCT 66	
1966-088BD		2506	USSR	17 SEP	21 OCT 66	
1966-088BE		2507	USSR	17 SEP	21 OCT 66	
1966-088C		2439	USSR	17 SEP	10 OCT 66	
1966-088D		2440	USSR	17 SEP	18 OCT 66	
1966-088E		2441	USSR	17 SEP	23 JUN 67	
1966-088F		2442	USSR	17 SEP	25 SEP 66	
1966-088G		2443	USSR	17 SEP	20 SEP 66	
1966-088H		2444	USSR	17 SEP	25 SEP 66	
1966-088J		2445	USSR	17 SEP	24 SEP 66	
1966-088K		2446	USSR	17 SEP	03 OCT 66	
1966-088L		2447	USSR	17 SEP	03 OCT 66	
1966-088M		2448	USSR	17 SEP	01 OCT 66	
1966-088N		2449	USSR	17 SEP	30 SEP 66	
1966-088P		2450	USSR	17 SEP	26 SEP 66	
1966-088Q		2451	USSR	17 SEP	02 OCT 66	
1966-088R		2452	USSR	17 SEP	17 SEP 66	
1966-088S		2453	USSR	17 SEP	04 OCT 66	
1966-088T		2454	USSR	17 SEP	02 OCT 66	
1966-088U		2455	USSR	17 SEP	04 OCT 66	
1966-088V		2456	USSR	17 SEP	28 SEP 66	
1966-088W		2457	USSR	17 SEP	26 SEP 66	
1966-088X		2458	USSR	17 SEP	30 SEP 66	
1966-088Y		2459	USSR	17 SEP	03 OCT 66	
1966-088Z		2460	USSR	17 SEP	22 SEP 66	
1966-084A	SURVEYOR 2	2425	US	20 SEP	23 SEP 66	G
1966-085A		2427	US	20 SEP	12 OCT 66	
1966-085B		2434	US	20 SEP	08 OCT 66	
1966-086A		2433	US	28 SEP	07 OCT 66	
1966-090A		2489	US	12 OCT	20 OCT 66	
1966-090B		2490	US	12 OCT	21 OCT 66	
1966-090C		2494	US	12 OCT	16 OCT 66	
1966-090D		2495	US	12 OCT	16 OCT 66	
1966-091A	COSMOS 129	2491	USSR	14 OCT	21 OCT 66	
1966-091B		2492	USSR	14 OCT	23 OCT 66	
1966-092A	MOLNIYA 1-4	2501	USSR	20 OCT	11 SEP 68	
1966-092B		2503	USSR	20 OCT	10 NOV 66	
1966-092C		2505	USSR	20 OCT	02 NOV 66	
1966-092D		3223	USSR	20 OCT	05 JUN 69	
1966-093A	COSMOS 130	2502	USSR	20 OCT	28 OCT 66	
1966-093B		2504	USSR	20 OCT	29 OCT 66	
1966-094B		2509	USSR	22 OCT	24 OCT 66	
1966-094C		2510	USSR	22 OCT	23 OCT 66	
1966-095A	AC 9	2512	US	26 OCT	06 NOV 66	
1966-096B		2515	US	26 OCT	09 FEB 67	
1966-097A	OV3-2	2517	US	28 OCT	29 SEP 71	
1966-097B		2519	US	28 OCT	28 JAN 70	
1966-097C		2613	US	28 OCT	20 SEP 67	
1966-097D		2614	US	28 OCT	20 AUG 76	
1966-098A		2523	US	02 NOV	10 NOV 66	
1966-098B		2525	US	02 NOV	16 NOV 66	
1966-099A		2524	US	02 NOV	09 JAN 67	
1966-099B		2526	US	02 NOV	05 JAN 67	
1966-099C		2527	US	02 NOV	31 DEC 66	
1966-099D		2528	US	02 NOV	11 JAN 67	
1966-101A		2536	USSR	02 NOV	17 NOV 66	
1966-101AA		2561	USSR	02 NOV	07 NOV 66	
1966-101AB		2562	USSR	02 NOV	08 NOV 66	
1966-101AC		2563	USSR	02 NOV	08 NOV 66	
1966-101AD		2564	USSR	02 NOV	09 NOV 66	
1966-101AE		2570	USSR	02 NOV	22 NOV 66	
1966-101AF		2571	USSR	02 NOV	29 NOV 66	
1966-101AG		2575	USSR	02 NOV	11 NOV 66	
1966-101AH		2576	USSR	02 NOV	13 NOV 66	
1966-101AJ		2581	USSR	02 NOV	23 NOV 66	
1966-101AK		2582	USSR	02 NOV	14 NOV 66	
1966-101AL		2583	USSR	02 NOV	14 NOV 66	
1966-101AM		2584	USSR	02 NOV	12 NOV 66	
1966-101AN		2585	USSR	02 NOV	14 NOV 66	
1966-101AP		2586	USSR	02 NOV	12 NOV 66	
1966-101AQ		2587	USSR	02 NOV	16 NOV 66	
1966-101AR		2598	USSR	02 NOV	17 NOV 66	
1966-101AS		2931	USSR	02 NOV	26 APR 76	
1966-101B		2538	USSR	02 NOV	22 NOV 66	Y
1966-101C		2539	USSR	02 NOV	15 NOV 66	
1966-101D		2540	USSR	02 NOV	13 NOV 66	
1966-101E		2541	USSR	02 NOV	29 NOV 66	
1966-101F		2542	USSR	02 NOV	21 NOV 66	
1966-101G		2543	USSR	02 NOV	06 MAY 67	
1966-101H		2544	USSR	02 NOV	29 NOV 66	
1966-101J		2545	USSR	02 NOV	17 NOV 66	
1966-101K		2546	USSR	02 NOV	08 NOV 66	
1966-101L		2547	USSR	02 NOV	12 NOV 66	
1966-101M		2548	USSR	02 NOV	29 NOV 66	
1966-101N		2549	USSR	02 NOV	08 NOV 66	
1966-101P		2550	USSR	02 NOV	24 NOV 66	
1966-101Q		2551	USSR	02 NOV	08 NOV 66	
1966-101R		2552	USSR	02 NOV	06 NOV 66	
1966-101S		2553	USSR	02 NOV	08 NOV 66	
1966-101T		2554	USSR	02 NOV	07 NOV 66	
1966-101U		2555	USSR	02 NOV	20 NOV 66	
1966-101V		2556	USSR	02 NOV	08 NOV 66	
1966-101W		2557	USSR	02 NOV	08 NOV 66	
1966-101X		2558	USSR	02 NOV	07 NOV 66	
1966-101Y		2559	USSR	02 NOV	09 NOV 66	
1966-101Z		2560	USSR	02 NOV	07 NOV 66	
1966-100A	LUNAR ORBITER 2	2534	US	06 NOV	11 OCT 67	G
1966-100B		2535	US	06 NOV	15 NOV 66	
1966-102A		2537	US	08 NOV	29 NOV 66	
1966-103A	AGENA TARGET	2565	US	11 NOV	23 DEC 66	
1966-104A	GEMINI 12	2566	US	11 NOV	15 NOV 66	F
1966-104B		2567	US	11 NOV	12 NOV 66	Z
1966-104C		2577	US	11 NOV	15 NOV 66	
1966-104D		2578	US	11 NOV	15 NOV 66	
1966-104E		2579	US	11 NOV	16 NOV 66	
1966-104F		2588	US	11 NOV	04 DEC 66	
1966-104G		2589	US	11 NOV	01 DEC 66	
1966-104H		2590	US	11 NOV	18 NOV 66	
1966-104J		2591	US	11 NOV	18 NOV 66	
1966-104K		2592	US	11 NOV	18 NOV 66	
1966-104L		2593	US	11 NOV	18 NOV 66	
1966-104M		2594	US	11 NOV	18 NOV 66	
1966-104N		2595	US	11 NOV	18 NOV 66	
1966-104P		2596	US	11 NOV	18 NOV 66	
1966-104Q		2597	US	11 NOV	18 NOV 66	
1966-105A	COSMOS 131	2568	USSR	12 NOV	20 NOV 66	
1966-105B		2569	USSR	12 NOV	22 NOV 66	
1966-106A	COSMOS 132	2599	USSR	19 NOV	27 NOV 66	
1966-106B		2600	USSR	19 NOV	27 NOV 66	
1966-107A	COSMOS 133	2601	USSR	28 NOV	30 NOV 66	
1966-107B		2602	USSR	28 NOV	29 NOV 66	
1966-108A	COSMOS 134	2603	USSR	03 DEC	11 DEC 66	
1966-108B		2604	USSR	03 DEC	10 DEC 66	
1966-108C		2605	USSR	03 DEC	04 DEC 66	
1966-109A		2606	US	05 DEC	14 DEC 66	
1966-109B		2607	US	05 DEC	08 DEC 66	
1966-110B		2609	US	07 DEC	15 OCT 74	
1966-112A	COSMOS 135	2612	USSR	12 DEC	12 APR 67	
1966-112B		2615	USSR	12 DEC	15 MAR 67	
1966-112C		2616	USSR	12 DEC	28 DEC 66	
1966-112D		2617	USSR	12 DEC	28 DEC 66	
1966-112E		2708	USSR	12 DEC	28 MAY 67	
1966-113A		2618	US	14 DEC	24 DEC 66	
1966-114A	BIOSATELLITE 1	2632	US	14 DEC	15 FEB 67	
1966-114B		2623	US	14 DEC	22 JAN 67	
1966-114C		2619	US	14 DEC	10 JAN 67	
1966-114D		2631	US	14 DEC	10 JAN 67	
1966-115A	COSMOS 136	2624	USSR	19 DEC	27 DEC 66	
1966-115B		2625	USSR	19 DEC	28 DEC 66	
1966-116A	LUNA 13	2626	USSR	21 DEC	24 DEC 66	R
1966-116B		2628	USSR	21 DEC	23 DEC 66	
1966-116C		2629	USSR	21 DEC	22 DEC 66	

International Designation	Satellite Name	Number	Source	Launch	Decay	Note
1966-117A	COSMOS 137	2627	USSR	21 DEC	23 NOV 67	
1966-117B		2630	USSR	21 DEC	25 SEP 67	
1966-118A		2634	US	29 DEC	05 APR 69	
1966-118B		2635	US	29 DEC	23 MAR 67	
1966-118C		2636	US	29 DEC	23 MAR 67	
1967 LAUNCHES						
1967-001B		2640	US	11 JAN	13 APR 67	
1967-001C		2641	US	11 JAN	24 FEB 67	
1967-001E		4901	US	11 JAN	09 FEB 71	
1967-001F		5122	US	11 JAN	30 APR 72	
1967-001G		5123	US	11 JAN	06 OCT 71	
1967-001H		5245	US	11 JAN	17 JUN 71	
1967-001J		5269	US	11 JAN	29 OCT 78	
1967-001K		5333	US	11 JAN	14 OCT 71	
1967-001L		5735	US	11 JAN	09 MAR 73	
1967-001M		5766	US	11 JAN	20 JAN 72	
1967-001N		5951	US	11 JAN	26 JUN 73	
1967-001P		5960	US	11 JAN	26 JUN 73	
1967-001Q		5961	US	11 JAN	26 JUN 73	
1967-001R		5969	US	11 JAN	14 SEP 75	
1967-001T		5988	US	11 JAN	29 JAN 93	
1967-001U		5989	US	11 JAN	07 NOV 76	
1967-001V		5990	US	11 JAN	14 OCT 92	
1967-001Z		13905	US	11 JAN	20 SEP 92	
1967-002A		2642	US	14 JAN	02 FEB 67	
1967-002B		2662	US	14 JAN	28 JAN 67	
1967-004A	COSMOS 138	2646	USSR	19 JAN	27 JAN 67	
1967-004B		2648	USSR	19 JAN	28 JAN 67	
1967-005A	COSMOS 139	2656	USSR	25 JAN	25 JAN 67	
1967-005B		2658	USSR	25 JAN	25 JAN 67	
1967-005C		2659	USSR	25 JAN	25 JAN 67	
1967-007A		2664	US	02 FEB	12 FEB 67	
1967-007B		2665	US	02 FEB	04 FEB 67	
1967-008A	LUNAR ORBITER 3	2666	US	05 FEB	09 OCT 67	
1967-009A	COSMOS 140	2667	USSR	07 FEB	09 FEB 67	
1967-009B		2668	USSR	07 FEB	08 FEB 67	
1967-011C		2673	FRAN	08 FEB	13 DEC 67	AB
1967-011D		2675	FRAN	08 FEB	03 JAN 68	
1967-011E		2676	FRAN	08 FEB	05 DEC 67	
1967-011F		2677	FRAN	08 FEB	12 FEB 68	
1967-011G		2688	FRAN	08 FEB	22 JUN 80	
1967-011H		2689	FRAN	08 FEB	20 JUN 90	
1967-011J		2690	FRAN	08 FEB	08 MAY 68	
1967-011K		2691	FRAN	08 FEB	29 OCT 67	
1967-011L		2692	FRAN	08 FEB	21 APR 89	
1967-011M		2900	FRAN	08 FEB	23 MAY 80	
1967-011N		2990	FRAN	08 FEB	07 FEB 80	
1967-011P		3742	FRAN	08 FEB	27 AUG 80	
1967-011Q		4009	FRAN	08 FEB	12 NOV 82	
1967-012A	COSMOS 141	2670	USSR	08 FEB	16 FEB-67	
1967-012B		2672	USSR	08 FEB	16 FEB 67	
1967-013A	COSMOS 142	2678	USSR	14 FEB	06 JUL 67	
1967-013B		2679	USSR	14 FEB	15 JUN 67	
1967-014D		2681	FRAN	15 FEB	16 SEP 68	
1967-014E		2683	FRAN	15 FEB	12 NOV 82	
1967-014G		3589	FRAN	15 FEB	20 NOV 79	
1967-014H		3935	FRAN	15 FEB	02 AUG 81	
1967-014K		14633	FRAN	15 FEB	30 AUG 91	
1967-014L		15531	FRAN	15 FEB	19 AUG 91	
1967-014P		18929	FRAN	15 FEB	10 NOV 91	
1967-015A		2686	US	22 FEB	11 MAR 67	
1967-016A		2687	US	24 FEB	06 MAR 67	
1967-017A	COSMOS 143	2693	USSR	27 FEB	07 MAR 67	
1967-017B		2694	USSR	27 FEB	09 MAR 67	
1967-018A	COSMOS 144	2695	USSR	28 FEB	14 SEP 82	
1967-018B		2696	USSR	28 FEB	01 SEP 89	
1967-018C		3095	USSR	28 FEB	08 APR 71	
1967-019A	COSMOS 145	2697	USSR	03 MAR	08 MAR 68	
1967-019B		2698	USSR	03 MAR	01 DEC 67	
1967-020A	OSO 3	2703	US	08 MAR	04 APR 82	
1967-020B		2704	US	08 MAR	19 APR 70	
1967-020C		2987	US	08 MAR	30 JAN 70	
1967-021A	COSMOS 146	2705	USSR	10 MAR	18 MAR 67	
1967-021B		2709	USSR	10 MAR	19 MAR 67	
1967-021C		2815	USSR	10 MAR	11 MAR 67	
1967-022A	COSMOS 147	2710	USSR	13 MAR	21 MAR 67	
1967-022B		2711	USSR	13 MAR	22 MAR 67	
1967-023A	COSMOS 148	2712	USSR	16 MAR	07 MAY 67	
1967-023B		2713	USSR	16 MAR	14 APR 67	
1967-024A	COSMOS 149	2714	USSR	21 MAR	07 APR 67	
1967-024B		2716	USSR	21 MAR	24 MAR 67	AC
1967-024C		2724	USSR	21 MAR	26 MAR 67	
1967-024D		2725	USSR	21 MAR	26 MAR 67	
1967-024E		2726	USSR	21 MAR	27 MAR 67	
1967-024F		2727	USSR	21 MAR	27 MAR 67	
1967-024G		2728	USSR	21 MAR	27 MAR 67	
1967-024H		2729	USSR	21 MAR	12 APR 67	
1967-024J		2730	USSR	21 MAR	03 APR 67	
1967-024K		2731	USSR	21 MAR	03 APR 67	
1967-024L		2732	USSR	21 MAR	09 APR 67	
1967-024M		2733	USSR	21 MAR	28 MAR 67	
1967-024N		2734	USSR	21 MAR	28 MAR 67	
1967-024P		2737	USSR	21 MAR	27 MAR 67	
1967-024Q		2738	USSR	21 MAR	27 MAR 67	
1967-024R		2739	USSR	21 MAR	27 MAR 67	
1967-024S		2748	USSR	21 MAR	31 MAR 67	
1967-024T		2749	USSR	21 MAR	04 APR 67	
1967-025A	COSMOS 150	2715	USSR	22 MAR	30 MAR 67	
1967-025B		2718	USSR	22 MAR	30 MAR 67	
1967-026B		2719	US	23 MAR	07 JUL 67	
1967-026C		11984	US	23 MAR	20 JUN 81	
1967-027A	COSMOS 151	2720	USSR	24 MAR	06 MAY 91	
1967-027B		2721	USSR	24 MAR	03 JUL 88	
1967-027C		2776	USSR	24 MAR	28 NOV 67	
1967-027D		2797	USSR	24 MAR	12 OCT 70	
1967-028A	COSMOS 152	2722	USSR	25 MAR	05 AUG 67	
1967-028B		2723	USSR	25 MAR	22 MAY 67	
1967-029A		2736	US	30 MAR	17 APR 67	
1967-030A	COSMOS 153	2740	USSR	04 APR	12 APR 67	
1967-030B		2742	USSR	04 APR	11 APR 67	
1967-031A	ATS 2	2743	US	06 APR	02 SEP 69	
1967-031B		2744	US	06 APR	29 JUN 68	
1967-032A	COSMOS 154	2745	USSR	08 APR	10 APR 67	
1967-032B		2747	USSR	08 APR	09 APR 67	
1967-032C		2746	USSR	08 APR	19 APR 67	
1967-033A	COSMOS 155	2750	USSR	12 APR	20 APR 67	
1967-033B		2751	USSR	12 APR	16 APR 67	
1967-033C		2752	USSR	12 APR	13 APR 67	
1967-033D		2753	USSR	12 APR	16 APR 67	
1967-034F		6718	US	14 APR	06 MAR 80	
1967-034G		7670	US	14 APR	02 MAY 81	
1967-035A	SURVEYOR 3	2756	US	17 APR	20 APR 67	T
1967-037A	SOYUZ 1	2759	USSR	23 APR	24 APR 67	AD
1967-037B		2760	USSR	23 APR	24 APR 67	
1967-038A	SAN MARCO 2	2761	ITALY	26 APR	14 OCT 67	
1967-038B		2771	US	26 APR	29 MAY 67	
1967-039A	COSMOS 156	2762	USSR	27 APR	23 OCT 89	
1967-039B		2763	USSR	27 APR	27 OCT 91	
1967-041A	LUNAR ORBITER 4	2772	US	04 MAY	06 OCT 67	G
1967-042A	ARIEL 3	2773	UK	05 MAY	14 DEC 70	
1967-042B		2774	US	05 MAY	05 DEC 70	
1967-042C		2859	US	05 MAY	05 MAY 68	
1967-042D		2860	US	05 MAY	09 MAR 68	
1967-043A		2779	US	09 MAY	13 JUL 67	
1967-043B		2780	US	09 MAY	14 MAR 93	
1967-043C		2824	US	09 MAY	06 JUN 67	
1967-044A	COSMOS 157	2781	USSR	12 MAY	20 MAY 67	
1967-044B		2782	USSR	12 MAY	24 MAY 67	
1967-044C		2783	USSR	12 MAY	13 MAY 67	
1967-044D		2784	USSR	12 MAY	12 MAY 67	
1967-044E		2785	USSR	12 MAY	14 MAY 67	
1967-045C		2823	USSR	15 MAY	21 DEC 91	
1967-045D		3737	USSR	15 MAY	16 NOV 89	
1967-046A	COSMOS 159	2805	USSR	17 MAY	11 NOV 77	
1967-046B		2804	USSR	17 MAY	29 MAY 67	
1967-046C		2803	USSR	17 MAY	02 JUN 67	
1967-046D		2808	USSR	17 MAY	22 MAY 67	
1967-046E		2809	USSR	17 MAY	22 MAY 67	
1967-046F		2924	USSR	17 MAY	14 APR 78	
1967-047A	COSMOS 160	2806	USSR	17 MAY	18 MAY 67	
1967-047B		2810	USSR	17 MAY	18 MAY 67	
1967-048C		17723	US	18 MAY	03 FEB 92	
1967-048D		19222	US	18 MAY	09 JAN 93	
1967-049A	COSMOS 161	2812	USSR	22 MAY	30 MAY 67	
1967-049B		2814	USSR	22 MAY	28 MAY 67	
1967-050A		2813	US	22 MAY	30 MAY 67	
1967-050B		2816	US	22 MAY	27 MAY 67	
1967-051A	EXPLORER 34	2817	US	24 MAY	03 MAY 69	
1967-052A	MOLNIYA 1-5	2822	USSR	24 MAY	26 NOV 71	
1967-052B		2818	USSR	24 MAY	05 JUN 67	
1967-052C		2819	USSR	24 MAY	15 JUN 67	
1967-052D		2820	USSR	24 MAY	28 MAY 67	
1967-052E		2821	USSR	24 MAY	28 MAY 67	
1967-052F		3224	USSR	24 MAY	16 DEC 71	
1967-052G		5439	USSR	24 MAY	01 SEP 71	
1967-054A	COSMOS 162	2827	USSR	01 JUN	09 JUN 67	
1967-054B		2829	USSR	01 JUN	04 JUN 67	
1967-054C		2830	USSR	01 JUN	02 JUN 67	
1967-055A		2831	US	04 JUN	12 JUN 67	
1967-055B		2835	US	04 JUN	06 JUN 67	
1967-056A	COSMOS 163	2832	USSR	05 JUN	11 OCT 67	
1967-056B		2833	USSR	05 JUN	12 SEP 67	
1967-057A	COSMOS 164	2836	USSR	08 JUN	14 JUN 67	
1967-057B		2837	USSR	08 JUN	17 JUN 67	
1967-057C		2838	USSR	08 JUN	09 JUN 67	
1967-058A	VENERA 4	2840	USSR	12 JUN	18 OCT 67	N
1967-058B		2841	USSR	12 JUN	13 JUN 67	
1967-058C		2839	USSR	12 JUN	12 JUN 67	
1967-058D		2844	USSR	12 JUN	13 JUN 67	
1967-059A	COSMOS 165	2842	USSR	12 JUN	15 JAN 68	
1967-059B		2843	USSR	12 JUN	01 NOV 67	
1967-061A	COSMOS 166	2848	USSR	16 JUN	25 OCT 67	
1967-061B		2849	USSR	16 JUN	11 OCT 67	
1967-062A		2850	US	16 JUN	20 JUL 67	
1967-062B		2851	US	16 JUN	22 OCT 68	
1967-063A	COSMOS 167	2852	USSR	17 JUN	25 JUN 67	
1967-063B		2853	USSR	17 JUN	21 JUN 67	
1967-063C		2856	USSR	17 JUN	26 JUN 67	
1967-063D		2854	USSR	17 JUN	18 JUN 67	
1967-063E		2855	USSR	17 JUN	19 JUN 67	
1967-063F		2857	USSR	17 JUN	28 JUN 67	
1967-064A		2858	US	20 JUN	30 JUN 67	
1967-067A	COSMOS 168	2869	USSR	04 JUL	12 JUL 67	
1967-067B		2870	USSR	04 JUL	08 JUL 67	

International Designation	Satellite Name	Number	Source	Launch	Decay	Note
1967-067C		2871	USSR	04 JUL	08 JUL 67	
1967-068A	SURVEYOR 4	2875	US	14 JUL	17 JUL 67	G
1967-069A	COSMOS 169	2878	USSR	17 JUL	17 JUL 67	
1967-069B		2879	USSR	17 JUL	18 JUL 67	
1967-069C		2880	USSR	17 JUL	18 JUL 67	
1967-069D		2881	USSR	17 JUL	18 JUL 67	
1967-069E		2882	USSR	17 JUL	18 JUL 67	
1967-070B		2885	US	19 JUL	31 AUG 67	
1967-071A		2890	US	25 JUL	05 JUN 69	
1967-071B		2891	US	25 JUL	21 AUG 67	
1967-071C		2892	US	25 JUL	16 OCT 67	
1967-071D		2978	US	25 JUL	17 OCT 67	
1967-072A	OV1-86	2893	US	27 JUL	22 FEB 72	
1967-072B		2894	US	27 JUL	15 NOV 70	
1967-072C		2897	US	27 JUL	02 MAY 77	
1967-072D	OV1-12	2901	US	27 JUL	22 JUL 80	
1967-072E		2918	US	27 JUL	02 NOV 67	
1967-073A	OGO 4	2895	US	28 JUL	16 AUG 72	
1967-073B		2896	US	28 JUL	15 MAY 74	
1967-074A	COSMOS 170	2902	USSR	31 JUL	31 JUL 67	
1967-074B		2903	USSR	31 JUL	01 AUG 67	
1967-074C		2904	USSR	31 JUL	01 AUG 67	
1967-074D		2905	USSR	31 JUL	02 AUG 67	
1967-074E		2906	USSR	31 JUL	01 AUG 67	
1967-075A	LUNAR ORBITER 5	2907	US	01 AUG	31 JAN 68	G
1967-076A		2910	US	07 AUG	01 SEP 67	
1967-076B		2923	US	07 AUG	25 AUG 67	
1967-077A	COSMOS 171	2911	USSR	08 AUG	08 AUG 67	
1967-077B		2912	USSR	08 AUG	09 AUG 67	
1967-077C		2913	USSR	08 AUG	09 AUG 67	
1967-077D		2915	USSR	08 AUG	08 AUG 67	
1967-077E		2916	USSR	08 AUG	09 AUG 67	
1967-078A	COSMOS 172	2914	USSR	09 AUG	17 AUG 67	
1967-078B		2917	USSR	09 AUG	13 AUG 67	
1967-079A		2919	US	16 AUG	29 AUG 67	
1967-081A	COSMOS 173	2921	USSR	24 AUG	17 DEC 67	
1967-081B		2922	USSR	24 AUG	30 OCT 67	
1967-082A	COSMOS 174	2925	USSR	31 AUG	30 DEC 68	
1967-082B		2926	USSR	31 AUG	15 SEP 67	
1967-082C		2927	USSR	31 AUG	19 SEP 67	
1967-082D		2928	USSR	31 AUG	05 SEP 67	
1967-083A	BIOSATELLITE 2	2935	US	07 SEP	04 OCT 67	
1967-083B		2936	US	07 SEP	19 JAN 68	
1967-083A	CAPSULE	61015	US	07 SEP	09 SEP 67	
1967-084A	SURVEYOR 5	2937	US	08 SEP	11 SEP 67	T
1967-085A	COSMOS 175	2939	USSR	11 SEP	19 SEP 67	
1967-085B		2941	USSR	11 SEP	23 SEP 67	
1967-085C		2943	USSR	11 SEP	13 SEP 67	
1967-086A	COSMOS 176	2942	USSR	12 SEP	03 MAR 68	
1967-086B		2945	USSR	12 SEP	12 DEC 67	
1967-086C		2944	USSR	12 SEP	04 OCT 67	
1967-086D		2956	USSR	12 SEP	21 SEP 67	
1967-086E		2957	USSR	12 SEP	06 APR 68	
1967-086F		2958	USSR	12 SEP	12 MAR 75	
1967-086G		2959	USSR	12 SEP	02 APR 68	
1967-086H		2960	USSR	12 SEP	03 OCT 67	
1967-086J		2961	USSR	12 SEP	06 OCT 67	
1967-086K		2979	USSR	12 SEP	13 JAN 68	
1967-086L		2983	USSR	12 SEP	04 NOV 67	
1967-086M		2984	USSR	12 SEP	31 OCT 67	
1967-087A		2946	US	15 SEP	04 OCT 67	
1967-088A	COSMOS 177	2947	USSR	16 SEP	24 SEP 67	
1967-088B		2948	USSR	16 SEP	19 SEP 67	
1967-089A	COSMOS 178	2951	USSR	19 SEP	19 SEP 67	
1967-089B		2953	USSR	19 SEP	20 SEP 67	
1967-089C		2954	USSR	19 SEP	19 SEP 67	
1967-089D		2955	USSR	19 SEP	19 SEP 67	
1967-090A		2952	US	19 SEP	30 SEP 67	
1967-091A	COSMOS 179	2962	USSR	22 SEP	22 SEP 67	
1967-091B		2963	USSR	22 SEP	23 SEP 67	
1967-091C		2964	USSR	22 SEP	22 SEP 67	
1967-092E		12554	US	25 SEP	20 MAR 89	
1967-092F		12555	US	25 SEP	02 FEB 89	
1967-092G		17176	US	25 SEP	03 OCT 91	
1967-092H		20009	US	25 SEP	12 SEP 91	
1967-092J		20652	US	25 SEP	26 SEP 91	
1967-093B		2968	USSR	25 SEP	05 OCT 67	
1967-093A	COSMOS 180	2966	USSR	26 SEP	04 OCT 67	
1967-094B		2970	US	28 SEP	26 DEC 67	
1967-094D		2972	US	28 SEP	27 OCT 67	
1967-095A	MOLNIYA 1-6	2973	USSR	03 OCT	04 MAR 69	
1967-095B		2974	USSR	03 OCT	17 OCT 67	
1967-095C		2975	USSR	03 OCT	21 OCT 67	
1967-095D		3225	USSR	03 OCT	10 FEB 69	
1967-097A	COSMOS 181	2981	USSR	11 OCT	19 OCT 67	
1967-097B		2982	USSR	11 OCT	19 OCT 67	
1967-098A	COSMOS 182	2995	USSR	16 OCT	24 OCT 67	
1967-098B		2996	USSR	16 OCT	24 OCT 67	
1967-098C		2999	USSR	16 OCT	24 OCT 67	
1967-099A	COSMOS 183	3001	USSR	18 OCT	18 OCT 67	
1967-099B		3002	USSR	18 OCT	19 OCT 67	
1967-099C		3003	USSR	18 OCT	18 OCT 67	
1967-100A	OSO 4	3000	US	18 OCT	15 JUN 82	
1967-100B		3004	US	18 OCT	02 MAY 80	
1967-101A	MOLNIYA 1-7	3008	USSR	22 OCT	31 DEC 69	
1967-101B		3005	USSR	2-2 OCT	07 NOV 67	
1967-101C		3006	USSR	22 OCT	03 NOV 67	
1967-101D		3007	USSR	22 OCT	28 OCT 67	

International Designation	Satellite Name	Number	Source	Launch	Decay	Note
1967-101E		3009	USSR	22 OCT	27 OCT 67	
1967-101F		3226	USSR	22 OCT	16 JAN 70	
1967-102A	COSMOS 184	3010	USSR	24 OCT	02 APR 89	
1967-102B		3011	USSR	24 OCT	19 DEC 91	
1967-103A		3012	US	25 OCT	05 NOV 67	
1967-104A	COSMOS 185	3013	USSR	27 OCT	14 JAN 69	
1967-105A	COSMOS 186	3014	USSR	27 OCT	31 OCT 67	
1967-105B		3015	USSR	27 OCT	29 OCT 67	
1967-106A	COSMOS 187	3016	USSR	28 OCT	28 OCT 67	
1967-106B		3017	USSR	28 OCT	28 OCT 67	
1967-106C		3018	USSR	28 OCT	28 OCT 67	
1967-107A	COSMOS 188	3020	USSR	30 OCT	02 NOV 67	
1967-107B		3022	USSR	30 OCT	02 NOV 67	
1967-108A	COSMOS 189	3021	USSR	30 OCT	08 JUN 78	
1967-108B		3023	USSR	30 OCT	07 MAR 80	
1967-108C		3028	USSR	30 OCT	18 JUN 68	
1967-108D		3041	USSR	30 OCT	23 DEC 67	
1967-108E		3042	USSR	30 OCT	24 JUN 68	
1967-108F		6362	USSR	30 OCT	27 FEB 74	
1967-108G		6696	USSR	30 OCT	27 FEB 74	
1967-108H		6697	USSR	30 OCT	29 MAR 77	
1967-108J		6698	USSR	30 OCT	13 MAY 74	
1967-108K		6699	USSR	30 OCT	29 MAR 74	
1967-108L		6701	USSR	30 OCT	10 FEB 74	
1967-109A		3024	US	02 NOV	02 DEC 67	
1967-109B		3025	US	02 NOV	28 MAR 69	
1967-110A	COSMOS 190	3026	USSR	03 NOV	11 NOV 67	
1967-110B		3027	USSR	03 NOV	09 NOV 67	
1967-111B		3030	US	05 NOV	08 AUG 68	
1967-112A	SURVEYOR 6	3031	US	07 NOV	10 NOV 67	T
1967-113A	APOLLO 4	3032	US	09 NOV	09 NOV 67	B
1967-113B		3033	US	09 NOV	09 NOV 67	
1967-115A	COSMOS 191	3043	USSR	21 NOV	02 MAR 68	
1967-115B		3044	USSR	21 NOV	14 JAN 68	
1967-117A	COSMOS 193	3052	USSR	25 NOV	03 DEC 67	
1967-117B		3053	USSR	25 NOV	01 DEC 67	
1967-118A	WRESAT	3054	AUST	29 NOV	10 JAN 68	
1967-119A	COSMOS 194	3055	USSR	03 DEC	11 DEC 67	
1967-119B		3056	USSR	03 DEC	09 DEC 67	
1967-120A	OV3-6	3057	US	05 DEC	09 MAR 69	
1967-120B		3059	US	05 DEC	08 APR 68	
1967-120C		3060	US	05 DEC	18 JUN 68	
1967-120D		3061	US	05 DEC	25 DEC 67	
1967-120E		3062	US	05 DEC	27 MAR 68	
1967-121A		3058	US	05 DEC	16 DEC 67	
1967-122A		3063	US	09 DEC	25 DEC 67	
1967-123B	TATS 1	3067	US	13 DEC	28 APR 68	
1967-123C		3068	US	13 DEC	01 APR 68	
1967-123D		3069	US	13 DEC	12 FEB 68	
1967-123E		3070	US	13 DEC	12 FEB 68	
1967-124A	COSMOS 195	3071	USSR	16 DEC	24 DEC 67	
1967-124B		3072	USSR	16 DEC	23 DEC 67	
1967-124C		3073	USSR	16 DEC	18 DEC 67	
1967-125A	COSMOS 196	3074	USSR	19 DEC	07 JUL 68	
1967-125B		3075	USSR	19 DEC	20 FEB 68	
1967-125C		3076	USSR	19 DEC	07 JAN 68	
1967-125D		3077	USSR	19 DEC	07 JAN 68	
1967-125E		3078	USSR	19 DEC	07 JAN 68	
1967-126A	COSMOS 197	3079	USSR	26 DEC	30 JAN 68	
1967-126B		3080	USSR	26 DEC	08 JAN 68	
1967-126C		3083	USSR	26 DEC	10 JAN 68	
1967-126D		3084	USSR	26 DEC	05 JAN 68	
1967-126E		3085	USSR	26 DEC	18 JAN 68	
1967-126F		3086	USSR	26 DEC	26 JAN 68	
1967-126G		3087	USSR	26 DEC	03 JAN 68	
1967-126H		3089	USSR	26 DEC	30 DEC 67	
1967-126J		3090	USSR	26 DEC	30 DEC 67	
1967-127B		3082	USSR	27 DEC	21 JAN 68	
1967-127C		3088	USSR	27 DEC	31 DEC 67	

1968 LAUNCHES

International Designation	Satellite Name	Number	Source	Launch	Decay	Note
1968-001A	SURVEYOR 7	3091	US	07 JAN	10 JAN 68	T
1968-003A	COSMOS 199	3099	USSR	16 JAN	01 FEB 68	
1968-003B		3096	USSR	16 JAN	25 JAN 68	
1968-003C		3115	USSR	16 JAN	30 JAN 68	
1968-003D		3117	USSR	16 JAN	03 FEB 68	
1968-004A		3097	US	17 JAN	07 JUL 70	
1968-004B		3101	US	17 JAN	29 MAR 68	
1968-004C		3116	US	17 JAN	22 MAR 68	
1968-005A		3098	US	18 JAN	04 FEB 68	
1968-006A	COSMOS 200	3100	USSR	19 JAN	24 FEB 73	
1968-006B		3102	USSR	19 JAN	04 FEB 72	
1968-006C		3103	USSR	19 JAN	30 SEP 80	
1968-006D		3104	USSR	19 JAN	06 APR 68	
1968-006E		3109	USSR	19 JAN	11 APR 68	
1968-006F		3110	USSR	19 JAN	16 JUN 68	
1968-006G		3111	USSR	19 JAN	15 APR 68	
1968-006H		3112	USSR	19 JAN	16 JUN 68	
1968-007A	LM-1/ASCENT	3106	US	22 JAN	24 JAN 68	
1968-007B	LM-1/DESCENT	3107	US	22 JAN	12 FEB 68	
1968-007C		3108	US	22 JAN	23 JAN 68	
1968-008A		3113	US	24 JAN	27 FEB 68	
1968-008B		3114	US	24 JAN	04 MAR 70	
1968-008C		3120	US	24 JAN	12 FEB 68	
1968-009A	COSMOS 201	-3118	USSR	06 FEB	14 FEB 68	
1968-009B		3119	USSR	06 FEB	13 FEB 68	
1968-010A	COSMOS 202	3128	USSR	20 FEB	24 MAR 68	

International Designation	Satellite Name	Number	Source	Launch	Decay	Note
1968-010B		3130	USSR	20 FEB	04 MAR 68	
1968-011C		3147	USSR	20 FEB	27 MAY 82	
1968-012E		18594	US	02 MAR	19 APR 91	
1968-013B		3135	USSR	02 MAR	07 MAR 68	
1968-013C		3136	USSR	02 MAR	07 MAR 68	
1968-013D		3144	USSR	02 MAR	05 MAR 68	
1968-015A	COSMOS 204	3139	USSR	05 MAR	02 MAR 69	
1968-015B		3143	USSR	05 MAR	13 OCT 68	
1968-016A	COSMOS 205	3140	USSR	05 MAR	13 MAR 68	
1968-016B		3142	USSR	05 MAR	09 MAR 68	
1968-017A	EXPLORER 37	3141	US	05 MAR	16 NOV 90	
1968-017B		3146	US	05 MAR	25 DEC 80	
1968-017C		3215	US	05 MAR	11 APR 69	
1968-017D		3328	US	05 MAR	08 SEP 91	
1968-017E		5743	US	05 MAR	25 OCT 77	
1968-018A		3148	US	13 MAR	24 MAR 68	
1968-019A	COSMOS 206	3150	USSR	14 MAR	22 APR 89	
1968-019B		3151	USSR	14 MAR	05 NOV 92	
1968-020A		3152	US	14 MAR	10 APR 68	
1968-020B		3153	US	14 MAR	03 JAN 70	
1968-021A	COSMOS 207	3154	USSR	16 MAR	24 MAR 68	
1968-021B		3155	USSR	16 MAR	22 MAR 68	
1968-022A	COSMOS 208	3156	USSR	21 MAR	02 APR 68	
1968-022B		3157	USSR	21 MAR	25 MAR 68	
1968-022C		3167	USSR	21 MAR	06 APR 68	
1968-023B		3160	USSR	22 MAR	29 MAR 68	
1968-023C		3161	USSR	22 MAR	25 MAR 68	
1968-023D		3162	USSR	22 MAR	10 APR 68	
1968-023E		3165	USSR	22 MAR	04 MAY 68	
1968-024A	COSMOS 210	3168	USSR	03 APR	11 APR 68	
1968-024B		3169	USSR	03 APR	12 APR 68	
1968-024C		3172	USSR	03 APR	05 APR 68	
1968-025A	APOLLO 6	3170	US	04 APR	04 APR 68	
1968-025B		3171	US	04 APR	25 APR 68	AE
1968-025C		3185	US	04 APR	28 APR 68	
1968-025D		3186	US	04 APR	05 MAY 68	
1968-025E		3187	US	04 APR	03 SEP 68	
1968-025F		3188	US	04 APR	29 APR 68	
1968-025G		3189	US	04 APR	20 APR 68	
1968-025H		3190	US	04 APR	17 APR 68	
1968-025J		3191	US	04 APR	17 APR 68	
1968-025K		3192	US	04 APR	17 APR 68	
1968-025L		3195	US	04 APR	01 MAY 68	
1968-025M		3196	US	04 APR	22 APR 68	
1968-025N		3197	US	04 APR	16 APR 68	
1968-025P		3198	US	04 APR	17 APR 68	
1968-025Q		3200	US	04 APR	17 APR 68	
1968-025R		3201	US	04 APR	16 APR 68	
1968-025S		3202	US	04 APR	17 APR 68	
1968-027B		3179	USSR	07 APR	09 APR 68	
1968-027C		3180	USSR	07 APR	09 APR 68	
1968-028A	COSMOS 211	3181	USSR	09 APR	10 NOV 68	
1968-028B		3182	USSR	09 APR	17 AUG 68	
1968-029A	COSMOS 212	3183	USSR	14 APR	19 APR 68	
1968-029B		3184	USSR	14 APR	16 APR 68	
1968-030A	COSMOS 213	3193	USSR	15 APR	20 APR 68	
1968-030B		3194	USSR	15 APR	19 APR 68	
1968-031A		3199	US	17 APR	29 APR 68	
1968-032A	COSMOS 214	3203	USSR	18 APR	26 APR 68	
1968-032B		3204	USSR	18 APR	29 APR 68	
1968-033A	COSMOS 215	3205	USSR	18 APR	30 JUN 68	
1968-033B		3206	USSR	18 APR	19 MAY 68	
1968-034A	COSMOS 216	3207	USSR	20 APR	28 APR 68	
1968-034B		3208	USSR	20 APR	23 APR 68	
1968-035A	MOLNIYA 1-8	3209	USSR	21 APR	29 JAN 74	
1968-035B		3211	USSR	21 APR	11 MAY 68	
1968-035C		3210	USSR	21 APR	29 MAY 68	
1968-035D		3514	USSR	21 APR	09 JUL 74	
1968-036A	COSMOS 217	3216	USSR	24 APR	26 APR 68	
1968-037A	COSMOS 218	3217	USSR	25 APR	25 APR 68	
1968-037B		3219	USSR	25 APR	25 APR 68	
1968-037C		3218	USSR	25 APR	25 APR 68	
1968-038A	COSMOS 219	3220	USSR	26 APR	02 MAR 69	
1968-038B		3221	USSR	26 APR	24 FEB 69	
1968-038C		3227	USSR	26 APR	03 JUN 68	
1968-039A		3228	US	01 MAY	15 MAY 68	
1968-039B		3232	US	01 MAY	20 MAY 68	
1968-040C		3231	USSR	07 MAY	09 SEP 90	
1968-041A	IRIS	3233	ESRO	17 MAY	08 MAY 71	
1968-041B		3234	US	17 MAY	31 MAR 70	
1968-041C		3265	US	17 MAY	04 NOV 68	
1968-043A	COSMOS 221	3269	USSR	24 MAY	31 AUG 69	
1968-043B		3270	USSR	24 MAY	13 APR 69	
1968-044A	COSMOS 222	3272	USSR	30 MAY	11 OCT 68	
1968-044B		3273	USSR	30 MAY	13 AUG 68	
1968-045A	COSMOS 223	3274	USSR	01 JUN	09 JUN 68	
1968-045B		3275	USSR	01 JUN	09 JUN 68	
1968-046A	COSMOS 224	3276	USSR	04 JUN	12 JUN 68	
1968-046B		3277	USSR	04 JUN	07 JUN 68	
1968-047A		3278	US	05 JUN	17 JUN 68	
1968-048A	COSMOS 225	3279	USSR	11 JUN	02 NOV 68	
1968-048B		3280	USSR	11 JUN	15 AUG 68	
1968-048C		3281	USSR	11 JUN	19 JUN 68	
1968-048D		3293	USSR	11 JUN	21 JUN 68	
1968-049A	COSMOS 226	3282	USSR	12 JUN	18 OCT 83	
1968-049B		3283	USSR	12 JUN	26 JAN 91	
1968-051A	COSMOS 227	3294	USSR	18 JUN	26 JUN 68	
1968-051B		3295	USSR	18 JUN	21 JUN 68	

International Designation	Satellite Name	Number	Source	Launch	Decay	Note
1968-052A		3296	US	20 JUN	16 JUL 68	
1968-052B		3297	US	20 JUN	11 JAN 70	
1968-053A	COSMOS 228	3298	USSR	21 JUN	03 JUL 68	
1968-053B		3300	USSR	21 JUN	24 JUN 68	
1968-053C		3299	USSR	21 JUN	23 JUN 68	
1968-053D		3301	USSR	21 JUN	23 JUN 68	
1968-053E		3302	USSR	21 JUN	23 JUN 68	
1968-053F		3303	USSR	21 JUN	23 JUN 68	
1968-053G		3306	USSR	21 JUN	07 JUL 68	
1968-054A	COSMOS 229	3304	USSR	26 JUN	04 JUL 68	
1968-054B		3305	USSR	26 JUN	04 JUL 68	
1968-056A	COSMOS 230	3308	USSR	05 JUL	02 NOV 68	
1968-056B		3309	USSR	05 JUL	16 OCT 68	
1968-057A	MOLNIYA 1-9	3310	USSR	06 JUL	15 MAY 71	
1968-057B		3311	USSR	06 JUL	09 AUG 68	
1968-057C		3312	USSR	06 JUL	22 AUG 68	
1968-057D		3313	USSR	06 JUL	08 JUL 68	
1968-057E		3314	USSR	06 JUL	18 JUL 68	
1968-057F		3515	USSR	06 JUL	02 SEP 70	
1968-058A	COSMOS 231	3316	USSR	10 JUL	18 JUL 68	
1968-058B		3317	USSR	10 JUL	20 JUL 68	
1968-059A	OV1-15	3318	US	11 JUL	06 NOV 68	
1968-059B	OV1-16	3319	US	11 JUL	19 AUG 68	
1968-059C		3320	US	11 JUL	16 AUG 68	
1968-059D		3321	US	11 JUL	16 JUL 68	
1968-060A	COSMOS 232	3322	USSR	16 JUL	24 JUL 68	
1968-060B		3323	USSR	16 JUL	26 JUL 68	
1968-061A	COSMOS 233	3326	USSR	18 JUL	07 FEB 69	
1968-061B		3327	USSR	18 JUL	01 JAN 69	
1968-062A	COSMOS 234	3332	USSR	30 JUL	05 AUG 68	
1968-062B		3333	USSR	30 JUL	07 AUG 68	
1968-064A		3335	US	06 AUG	16 AUG 68	
1968-065A		3336	US	07 AUG	27 AUG 68	
1968-066A	EXPLORER 39	3337	US	08 AUG	22 JUN 81	
1968-067A	COSMOS 235	3339	USSR	09 AUG	17 AUG 68	
1968-067B		3340	USSR	09 AUG	14 AUG 68	
1968-068A	ATS 4	3344	US	10 AUG	17 OCT 68	
1968-070A	COSMOS 236	3347	USSR	27 AUG	04 MAR 90	
1968-070B		3349	USSR	27 AUG	08 OCT 82	
1968-071A	COSMOS 237	3348	USSR	27 AUG	04 SEP 68	
1968-071B		3350	USSR	27 AUG	04 SEP 68	
1968-072A	COSMOS 238	3351	USSR	28 AUG	01 SEP 68	
1968-0728		3352	USSR	28 AUG	30 AUG 68	
1968-073A	COSMOS 239	3353	USSR	05 SEP	13 SEP 68	
1968-073B		3354	USSR	05 SEP	09 SEP 68	
1968-074A		3375	US	10 SEP	25 SEP 68	
1968-075A	COSMOS 240	3388	USSR	14 SEP	21 SEP 68	
1968-075B		3389	USSR	14 SEP	17 SEP 68	
1968-076A	ZOND 5	3394	USSR	14 SEP	21 SEP 68	C
1968-076B		3395	USSR	14 SEP	16 SEP 68	
1968-076C		3396	USSR	14 SEP	18 SEP 68	
1968-076D		3397	USSR	14 SEP	16 SEP 68	
1968-077A	COSMOS 241	3398	USSR	16 SEP	24 SEP 68	
1968-077B		3399	USSR	16 SEP	22 SEP 68	
1968-078A		3408	US	18 SEP	08 OCT 68	
1968-078B		3409	US	18 SEP	28 SEP 69	
1968-079A	COSMOS 242	3414	USSR	20 SEP	13 NOV 68	
1968-079B		3415	USSR	20 SEP	22 OCT 68	
1968-080A	COSMOS 243	3418	USSR	23 SEP	04 OCT 68	
1968-080B		3419	USSR	23 SEP	29 SEP 68	
1968-080C		3452	USSR	23 SEP	12 OCT 68	
1968-081B	ERS 28	3429	US	26 SEP	15 FEB 71	
1968-081F		5999	US	26 SEP	14 MAR 75	
1968-082A	COSMOS 244	3449	USSR	02 OCT	02 OCT 68	
1968-082B		3451	USSR	02 OCT	03 OCT 68	
1968-082C		3450	USSR	02 OCT	02 OCT 68	
1968-083A	COSMOS 245	3457	USSR	03 OCT	15 JAN 69	
1968-083B		3458	USSR	03 OCT	16 NOV 68	
1968-084A	ESRO 1	3459	ESRO	03 OCT	26 JUN 70	
1968-084B		3460	US	03 OCT	13 AUG 69	
1968-084C		3479	US	03 OCT	12 APR 69	
1968-084D		3916	US	03 OCT	09 MAY 69	
1968-085A	MOLNIYA 1-10	3469	USSR	05 OCT	16 JUL 76	
1968-085B		3470	USSR	05 OCT	30 OCT 68	
1968-085C		3471	USSR	05 OCT	04 NOV 68	
1968-085D		3513	USSR	05 OCT	07 MAY 75	
1968-086A		3472	US	05 OCT	26 MAR 71	
1968-087A	COSMOS 246	3473	USSR	07 OCT	12 OCT 68	
1968-087B		3474	USSR	07 OCT	08 OCT 68	
1968-088A	COSMOS 247	3484	USSR	11 OCT	19 OCT 68	
1968-088B		3485	USSR	11 OCT	18 OCT 68	
1968-089A	APOLLO 7	3486	US	11 OCT	22 OCT 68	F
1968-089B		3487	US	11 OCT	18 OCT 68	
1968-090A	COSMOS 248	3503	USSR	19 OCT	26 FEB 80	
1968-090B		3571	USSR	19 OCT	28 DEC 68	
1968-090C		3572	USSR	19 OCT	20 FEB 69	
1968-090D		3573	USSR	19 OCT	06 FEB 69	
1968-090E		3580	USSR	19 OCT	13 DEC 68	
1968-091AB		3609	USSR	20 OCT	26 DEC 82	
1968-091AC		3677	USSR	20 OCT	29 MAY 80	
1968-091AD		3678	USSR	20 OCT	07 JUN 81	
1968-091AF		3680	USSR	20 OCT	09 JUN 81	
1968-091AG		3681	USSR	20 OCT	15 JUN 92	
1968-091AH		3682	USSR	20 OCT	14 DEC 77	
1968-091AK		3693	USSR	20 OCT	05 DEC 76	
1968-091AM		3697	USSR	20 OCT	03 JAN 78	
1968-091AP		3702	USSR	20 OCT	05 DEC 76	
1968-091AQ		3706	USSR	20 OCT	05 JUL 70	

International Designation	Satellite Name	Number	Source	Launch	Decay	Note
1968-091AS		3754	USSR	20 OCT	21 MAY 75	
1968-091AW		3789	USSR	20 OCT	02 JUL 75	
1968-091AY		3791	USSR	20 OCT	29 JAN 90	
1968-091AZ		3792	USSR	20 OCT	13 APR 81	
1968-091B		3505	USSR	20 OCT	21 OCT 68	
1968-091BC		3795	USSR	20 OCT	21 MAY 80	
1968-091BD		3796	USSR	20 OCT	23 JAN 77	
1968-091BE		3797	USSR	20 OCT	10 DEC 88	
1968-091BF		3872	USSR	20 OCT	15 FEB 70	
1968-091BG		3873	USSR	20 OCT	03 NOV 89	
1968-091BJ		3875	USSR	20 OCT	27 MAY 81	
1968-091BK		3876	USSR	20 OCT	15 JUN 75	
1968-091BM		3909	USSR	20 OCT	16 MAR 75	
1968-091BN		3910	USSR	20 OCT	18 MAY 82	
1968-091BS		3934	USSR	20 OCT	24 AUG 77	
1968-091BT		3959	USSR	20 OCT	29 OCT 79	
1968-091BU		3960	USSR	20 OCT	26 NOV 78	
1968-091BV		3961	USSR	20 OCT	12 NOV 82	
1968-091BX		4002	USSR	20 OCT	03 AUG 70	
1968-091BZ		4500	USSR	20 OCT	12 DEC 70	
1968-091CC		5717	USSR	20 OCT	14 MAY 82	
1968-091CE		6045	USSR	20 OCT	06 MAR 81	
1968-091CH		6751	USSR	20 OCT	27 SEP 89	
1968-091CK		6764	USSR	20 OCT	27 OCT 79	
1968-091CP		14413	USSR	20 OCT	24 AUG 88	
1968-091CR		14415	USSR	20 OCT	12 APR 86	
1968-091CU		14429	USSR	20 OCT	12 DEC 89	
1968-091CW		15518	USSR	20 OCT	01 APR 88	
1968-091D		3507	USSR	20 OCT	09 AUG 89	
1968-091DB		15776	USSR	20 OCT	06 APR 86	
1968-091DC		17667	USSR	20 OCT	14 JUN 92	
1968-091DK		19476	USSR	20 OCT	01 OCT 88	
1968-091DM		19879	USSR	20 OCT	30 APR 89	
1968-091G		3518	USSR	20 OCT	13 NOV 79	
1968-091H		3519	USSR	20 OCT	06 MAR 82	
1968-091J		3520	USSR	20 OCT	29 MAY 81	
1968-091K		3521	USSR	20 OCT	06 APR 74	
1968-091N		3525	USSR	20 OCT	19 OCT 81	
1968-091P		3581	USSR	20 OCT	07 SEP 82	
1968-091Q		3582	USSR	20 OCT	28 MAR 80	
1968-091T		3585	USSR	20 OCT	09 FEB 90	
1968-091U		3586	USSR	20 OCT	31 MAR 70	
1968-091W		3588	USSR	20 OCT	05 MAY 69	
1968-091Z		3607	USSR	20 OCT	20 MAR 91	
1968-093A	SOYUZ 2	3511	USSR	25 OCT	28 OCT 68	
1968-093B		3512	USSR	25 OCT	27 OCT 68	
1968-091DP		20015	USSR	26 OCT	24 JUL 89	
1968-094A	SOYUZ 3	3516	USSR	26 OCT	30 OCT 68	
1968-094B		3517	USSR	26 OCT	28 OCT 68	
1968-095A	COSMOS 250	3526	USSR	30 OCT	15 FEB 78	
1968-095B		3527	USSR	30 OCT	08 NOV 78	
1968-095C		3659	USSR	30 OCT	08 APR 69	
1968-095D		3660	USSR	30 OCT	05 MAY 69	
1968-095E		3661	USSR	30 OCT	17 APR 69	
1968-09SF		3662	USSR	30 OCT	26 MAR 69	
1968-095G		8780	USSR	30 OCT	17 OCT 76	
1968-096A	COSMOS 251	3528	USSR	31 OCT	18 NOV 68	
1968-096B		3529	USSR	31 OCT	02 NOV 68	
1968-096C		3540	USSR	31 OCT	16 NOV 68	
1968-096D		3541	USSR	31 OCT	13 NOV 68	
1968-097AA		3610	USSR	01 NOV	24 AUG 91	
1968-097AB		3684	USSR	01 NOV	28 MAR 82	
1968-097AC		3685	USSR	01 NOV	02 OCT 91	
1968-097AF		3688	USSR	01 NOV	15 SEP 78	
1968-097AG		3689	USSR	01 NOV	13 JUL 74	
1968-097AJ		3694	USSR	01 NOV	28 OCT 80	
1968-097AK		3696	USSR	01 NOV	10 MAR 70	
1968-097AM		3704	USSR	01 NOV	09 SEP 74	
1968-097AN		3711	USSR	01 NOV	22 APR 69	
1968-097AP		3712	USSR	01 NOV	26 FEB 73	
1968-097AR		3714	USSR	01 NOV	20 DEC 80	
1968-097AS		3715	USSR	01 NOV	14 DEC 76	
1968-097AT		3716	USSR	01 NOV	12 SEP 81	
1968-097AY		3786	USSR	01 NOV	21 DEC 79	
1968-097AZ		3787	USSR	01 NOV	12 NOV 82	
1968-097BB		3813	USSR	01 NOV	03 NOV 81	
1968-097BD		3815	USSR	01 NOV	04 OCT 69	
1968-097BE		3816	USSR	01 NOV	25 JUN 82	
1968-097BF		3817	USSR	01 NOV	07 DEC 82	
1968-097BH		3896	USSR	01 NOV	12 NOV 82	
1968-097BJ		3897	USSR	01 NOV	02 JAN 82	
1968-097BP		3902	USSR	01 NOV	02 DEC 81	
1968-097BQ		3903	USSR	01 NOV	12 NOV 82	
1968-097BR		3904	USSR	01 NOV	24 OCT 79	
1968-097BT		3946	USSR	01 NOV	27 MAY 69	
1968-097BU		3953	USSR	01 NOV	13 APR 81	
1968-097BV		3979	USSR	01 NOV	02 JUL 71	
1968-097BW		3997	USSR	01 NOV	22 NOV 82	
1968-097BX		4010	USSR	01 NOV	10 SEP 69	
1968-097BY		4011	USSR	01 NOV	12 NOV 82	
1968-097BZ		4012	USSR	01 NOV	09 MAR 79	
1968-097C		3550	USSR	01 NOV	08 MAY 88	
1960-097CA		4013	USSR	01 NOV	16 OCT 80	
1968-097CB		4018	USSR	01 NOV	10 NOV 82	
1968-097CC		4024	USSR	01 NOV	06 SEP 91	
1968-097CD		4030	USSR	01 NOV	02 FEB 82	
1968-097CE		4031	USSR	01 NOV	30 JUL 82	
1968-097CG		5030	USSR	01 NOV	26 FEB 78	
1968-097CH		5223	USSR	01 NOV	02 MAY 83	
1968-097CK		5334	USSR	01 NOV	23 FEB 72	
1968-097CM		5418	USSR	01 NOV	19 NOV 82	
1968-097CN		5460	USSR	01 NOV	29 OCT 83	
1968-097CP		5514	USSR	01 NOV	05 JAN 81	
1968-097CQ		5515	USSR	01 NOV	06 FEB 78	
1968-097CR		5516	USSR	01 NOV	10 OCT 79	
1968-097CS		5517	USSR	01 NOV	08 OCT 82	
1968-097CT		5518	USSR	01 NOV	20 NOV 76	
1968-097CU		5592	USSR	01 NOV	22 JUL 79	
1968-097CV		5594	USSR	01 NOV	25 NOV 78	
1968-097CX		5631	USSR	01 NOV	15 OCT 81	
1968-097CY		5632	USSR	01 NOV	09 MAR 73	
1968-097CZ		5749	USSR	01 NOV	12 OCT 76	
1968-097DA		5788	USSR	01 NOV	05 SEP 81	
1968-097DB		5789	USSR	01 NOV	21 SEP 89	
1968-097DC		5790	USSR	01 NOV	18 AUG 81	
1968-097DE		5827	USSR	01 NOV	19 NOV 82	
1968-097DF		5828	USSR	01 NOV	19 NOV 82	
1968-097DG		5829	USSR	01 NOV	31 MAR 78	
1968-097DH		5830	USSR	01 NOV	21 FEB 79	
1968-097DJ		5870	USSR	01 NOV	20 JUN 81	
1968-097DL		5872	USSR	01 NOV	16 FEB 90	
1968-097DM		5873	USSR	01 NOV	16 JUN 79	
1968-097DN		5874	USSR	01 NOV	25 OCT 79	
1968-097DP		5875	USSR	01 NOV	09 MAY 81	
1968-097DQ		5889	USSR	01 NOV	19 NOV 82	
1968-097DR		5890	USSR	01 NOV	01 MAR 77	
1968-097DS		5891	USSR	01 NOV	31 JAN 77	
1968-097DT		5966	USSR	01 NOV	03 JAN 77	
1968-097DU		6175	USSR	01 NOV	27 SEP 75	
1968-097DX		10394	USSR	01 NOV	09 APR 91	
1968-097DZ		14418	USSR	01 NOV	30 SEP 88	
1968-097E		3552	USSR	01 NOV	06 JUL 89	
1968-097EA		14419	USSR	01 NOV	15 FEB 89	
1968-097EC		14840	USSR	01 NOV	26 JAN 88	
1968-097EH		17668	USSR	01 NOV	11 FEB 92	
1968-097EL		18488	USSR	01 NOV	19 JUL 90	
1968-097EM		18489	USSR	01 NOV	03 JAN 90	
1968-097EN		18490	USSR	01 NOV	25 JUL 90	
1968-097EP		18556	USSR	01 NOV	03 JAN 90	
1968-097ER		18618	USSR	01 NOV	02 JAN 89	
1968-097EV		21749	USSR	01 NOV	29 MAR 92	
1968-097J		3556	USSR	01 NOV	20 NOV 82	
1968-097T		3567	USSR	01 NOV	02 MAR 82	
1968-097U		3568	USSR	01 NOV	19 APR 79	
1968-097W		3570	USSR	01 NOV	10 APR 82	
1968-097X		3593	USSR	01 NOV	09 OCT 81	
1968-097Y		3600	USSR	01 NOV	03 NOV 81	
1968-097Z		3601	USSR	01 NOV	30 JUN 80	
1968-098R		3531	US	03 NOV	23 NOV 68	
1968-099A		3532	US	06 NOV	20 NOV 68	
1968-100B	TETR 2	3534	US	08 NOV	19 SEP 79	
1968-100C		3546	US	08 NOV	06 OCT 71	
1968-100D		3547	US	08 NOV	13 APR 78	
1968-100E		3548	US	08 NOV	06 JAN 72	
1968-101A	ZOND 6	3535	USSR	10 NOV	17 NOV 68	C
1968-101B		3536	USSR	10 NOV	12 NOV 68	
1968-101C		3537	USSR	10 NOV	13 NOV 68	
1968-101D		3538	USSR	10 NOV	13 NOV 68	
1968-101E		3539	USSR	10 NOV	13 NOV 68	
1968-102A	COSMOS 253	3542	USSR	13 NOV	18 NOV 68	
1968-102B		3543	USSR	13 NOV	20 NOV 68	
1968-103A	PROTON 4	3544	USSR	16 NOV	24 JUL 69	
1968-103B		3545	USSR	16 NOV	25 JAN 69	
1968-104A	COSMOS 254	3562	USSR	21 NOV	29 NOV 68	
1968-104B		3563	USSR	21 NOV	27 NOV 68	
1968-105A	COSMOS 255	3574	USSR	29 NOV	07 DEC 68	
1968-105B		3575	USSR	29 NOV	05 DEC 68	
1968-107A	COSMOS 257	3578	USSR	03 DEC	05 MAR 69	
1968-107B		3579	USSR	03 DEC	16 JAN 69	
1968-108A		3594	US	04 DEC	12 DEC 68	
1968-109A	HEOS-A1	3595	ESRO	05 DEC	28 OCT 75	
1968-109B		3596	US	05 DEC	27 DEC 68	
1968-111A	COSMOS 258	3602	USSR	10 DEC	18 DEC 68	
1968-111B		3603	USSR	10 DEC	17 DEC 68	
1968-112A		3604	US	12 DEC	28 DEC 68	
1968-113A	COSMOS 259	3612	USSR	14 DEC	05 MAY 69	
1968-113B		3613	USSR	14 DEC	10 MAY 69	
1968-113C		3614	USSR	14 DEC	13 MAY 69	
1968-115A	COSMOS 260	3619	USSR	16 DEC	09 JUL 73	
1968-115B		3620	USSR	16 DEC	06 FEB 69	
1968-115C		3621	USSR	16 DEC	18 JAN 69	
1968-115D		3622	USSR	16 DEC	21 SEP 73	
1968-116B		5978	US	19 DEC	26 JUN 73	
1968-117A	COSMOS 261	3624	USSR	19 DEC	12 FEB 69	
1968-117B		3625	USSR	19 DEC	07 JAN 69	AF
1968-117C		3628	USSR	19 DEC	06 JAN 69	
1968-117D		3631	USSR	19 DEC	01 FEB 69	
1968-117E		3632	USSR	19 DEC	22 JAN 69	
1968-117F		3633	USSR	19 DEC	17 FEB 69	
1968-117G		3634	USSR	19 DEC	18 AUG 69	
1968-117H		3635	USSR	19 DEC	15 MAR 69	
1968-117J		3636	USSR	19 DEC	01 FEB 69	
1968-117K		3637	USSR	19 DEC	22 JAN 69	
1968-117L		3638	USSR	19 DEC	13 FEB 69	
1968-117M		3639	USSR	19 DEC	02 MAY 69	
1968-117N		3640	USSR	19 DEC	10 FEB 69	

International Designation	Satellite Name	Number	Source	Launch	Decay	Note
1968-117P		3641	USSR	19 DEC	07 MAY 69	
1968-117Q		3644	USSR	19 DEC	06 MAR 69	
1968-117R		3645	USSR	19 DEC	08 MAR 69	
1968-117S		3647	USSR	19 DEC	19 MAR 69	
1968-117T		3698	USSR	19 DEC	20 MAR 69	
1968-117U		3700	USSR	19 DEC	14 MAR 69	
1968-117V		3703	USSR	19 DEC	17 FEB 69	
1968-117W		3705	USSR	19 DEC	15 FEB 69	
1968-117X		3709	USSR	19 DEC	21 FEB 69	
1968-117Y		3710	USSR	19 DEC	04 MAR 69	
1968-117Z		3768	USSR	19 DEC	04 MAR 69	
1968-118A	APOLLO 8	3626	US	21 DEC	27 DEC 68	AG
1968-119A	COSMOS 262	3629	USSR	26 DEC	18 JUL 69	
1968-119B		3630	USSR	26 DEC	30 APR 69	
1969 LAUNCHES						
1969-001A	VENERA 5	3542	USSR	05 JAN	16 MAY 69	N
1969-001B		3643	USSR	05 JAN	06 JAN 69	
1969-001C		3646	USSR	05 JAN	07 JAN 69	
1969-002A	VENERA 6	3648	USSR	10 JAN	17 MAY 69	N
1969-002B		3649	USSR	10 JAN	11 JAN 69	
1969-002C		3650	USSR	10 JAN	13 JAN 69	
1969-003A	COSMOS 263	3651	USSR	12 JAN	20 JAN 69	
1969-003B		3652	USSR	12 JAN	18 JAN 69	
1969-003C		3653	USSR	12 JAN	19 JAN 69	
1969-004A	SOYUZ 4	3654	USSR	14 JAN	17 JAN 69	E
1969-004B		3655	USSR	14 JAN	15 JAN 69	
1969-005A	SOYUZ 5	3656	USSR	15 JAN	18 JAN 69	E
1969-005B		3657	USSR	15 JAN	17 JAN 69	
1969-005C		3658	USSR	15 JAN	18 JAN 69	
1969-006A	OSO 5	3663	US	22 JAN	02 APR 84	
1969-006B		3664	US	22 JAN	08 DEC 79	
1969-007A		3665	US	22 JAN	03 FEB 69	
1969-007B		3666	US	22 JAN	26 JAN 69	
1969-008A	COSMOS 264	3667	USSR	23 JAN	05 FEB 69	
1969-008B		2668	USSR	23 JAN	30 JAN 69	
1969-008C		3671	USSR	23 JAN	13 FEB 69	
1969-010A		3672	US	05 FEB	24 FEB 69	
1969-012A	COSMOS 265	3675	USSR	07 FEB	01 MAY 69	
1969-012B		3676	USSR	07 FEB	17 MAY 69	
1969-015A	COSMOS 266	3761	USSR	25 FEB	05 MAR 69	
1969-015B		3762	USSR	25 FEB	04 MAR 69	
1969-015C		3763	USSR	25 FEB	26 FEB 69	
1969-017A	COSMOS 267	3765	USSR	26 FEB	06 MAR 69	
1969-017B		3766	USSR	26 FEB	05 MAR 69	
1969-018A	APOLLO 9	3769	US	03 MAR	13 MAR 69	F
1969-018C	LM/ASCENT	3771	US	03 MAR	23 OCT 81	
1969-018D	LM/DESCENT	3780	US	03 MAR	22 MAR 69	
1969-019A		3772	US	04 MAR	18 MAR 69	
1969-020A	COSMOS 268	3773	USSR	05 MAR	09 MAY 70	
1969-020B		3774	USSR	05 MAR	11 FEB 70	
1969-021A	COSMOS 269	3775	USSR	05 MAR	21 OCT 78	
1969-021B		3776	USSR	05 MAR	20 JUL 78	AH
1969-021C		3779	USSR	05 MAR	24 APR 69	
1969-021D		3781	USSR	05 MAR	07 MAY 69	
1969-021E		3782	USSR	05 MAR	02 APR 70	
1969-021F		3798	USSR	05 MAR	02 MAY 70	
1969-021G		3799	USSR	05 MAR	26 NOV 69	
1969-021H		3800	USSR	05 MAR	08 SEP 69	
1969-021J		3801	USSR	05 MAR	04 APR 70	
1969-021K		3802	USSR	05 MAR	02 MAY 70	
1969-021L		3803	USSR	05 MAR	11 JUN 69	
1969-021M		3804	USSR	05 MAR	19 MAR 69	
1969-021N		3805	USSR	05 MAR	04 SEP 69	
1969-021P		3806	USSR	05 MAR	29 JUN 69	
1969-021Q		3820	USSR	05 MAR	07 APR 69	
1969-021R		3821	USSR	05 MAR	17 MAY 69	
1969-021S		3822	USSR	05 MAR	09 APR 69	
1969-021T		3844	USSR	05 MAR	19 JUL 83	
1969-021U		3924	USSR	05 MAR	30 DEC 69	
1969-021V		3937	USSR	05 MAR	13 AUG 69	
1969-021W		3976	USSR	05 MAR	16 MAY 70	
1969-021X		3977	USSR	05 MAR	18 AUG 69	
1969-021Y		5425	USSR	05 MAR	20 DEC 72	
1969-022A	COSMOS 270	3777	USSR	06 MAR	14 MAR 69	
1969-022B		3778	USSR	06 MAR	12 MAR 69	
1969-023A	COSMOS 271	3807	USSR	15 MAR	23 MAR 69	
1969-023B		3808	USSR	15 MAR	22 MAR 69	
1969-025A	OV1-17	3823	US	18 MAR	05 MAR 70	
1969-025B	OV1-18	3824	US	18 MAR	27 AUG 72	
1969-025D	OV1-17A	3826	US	18 MAR	24 MAR 69	
1969-025F		3828	US	18 MAR	16 MAR 79	
1969-025G		4121	US	18 MAR	06 NOV 69	
1969-026A		3829	US	19 MAR	24 MAR 69	
1969-026B		3830	US	19 MAR	06 DEC 71	
1969-027A	COSMOS 273	3831	USSR	22 MAR	30 MAR 69	
1969-027B		3832	USSR	22 MAR	28 MAR 69	
1969-028A	COSMOS 274	3833	USSR	24 MAR	01 APR 69	
1969-028B		3834	USSR	24 MAR	29 MAR 69	
1969-029AA		3880	USSR	26 MAR	21 AUG 69	
1969-029AB		3881	USSR	26 MAR	06 JAN 72	
1969-029AC		3882	USSR	26 MAR	19 NOV 77	
1969-029AD		3905	USSR	26 MAR	12 SEP 69	
1969-029AF		3917	USSR	26 MAR	01 JUL 69	
1969-029AG		3918	USSR	26 MAR	10 MAY 69	
1969-029AH		3921	USSR	26 MAR	19 JUL 71	
1969-029AJ		3933	USSR	26 MAR	06 AUG 70	
1969-029AK		3966	USSR	26 MAR	10 SEP 69	
1969-029AL		3967	USSR	26 MAR	26 JUN 69	
1969-029AM		4289	USSR	26 MAR	26 DEC 70	
1969-029AN		9279	USSR	26 MAR	16 JUN 79	
1969-029AP		9692	USSR	26 MAR	18 JUL 78	
1969-029B		3836	USSR	26 MAR	10 JUN 89	
1969-029C		3849	USSR	26 MAR	05 NOV 80	AK
1969-029D		3850	USSR	26 MAR	08 SEP 69	
1969-029E		3851	USSR	26 MAR	12 NOV 86	
1969-029F		3852	USSR	26 MAR	13 JUL 70	
1969-029G		3853	USSR	26 MAR	07 OCT 70	
1969-029H		3858	USSR	26 MAR	18 MAY 70	
1969-029J		3859	USSR	26 MAR	28 NOV 79	
1969-029K		3860	USSR	26 MAR	26 DEC 71	
1969-029L		3861	USSR	26 MAR	04 JUN 74	
1969-029M		3862	USSR	26 MAR	13 DEC 70	
1969-029N		3863	USSR	26 MAR	18 JUN 72	
1969-029P		3864	USSR	26 MAR	23 FEB 70	
1969-029Q		3865	USSR	26 MAR	25 APR 73	
1969-029R		3866	USSR	26 MAR	01 JUN 69	
1969-029S		3867	USSR	26 MAR	13 FEB 71	
1969-029T		3868	USSR	26 MAR	04 FEB 70	
1969-029U		3869	USSR	26 MAR	01 JUN 71	
1969-029V		3870	USSR	26 MAR	26 AUG 70	
1969-029W		3871	USSR	26 MAR	03 JUN 82	
1969-029X		3877	USSR	26 MAR	23 NOV 79	
1969-029Y		3878	USSR	26 MAR	23 JUN 69	
1969-029Z		3879	USSR	26 MAR	14 OCT 87	
1969-031A	COSMOS 275	3846	USSR	28 MAR	07 FEB 70	
1969-031B		3847	USSR	28 MAR	16 SEP 69	
1969-029AE		3913	USSR	29 MAR	07 MAR 70	
1969-032A	COSMOS 276	3854	USSR	04 APR	11 APR 69	
1969-032B		3856	USSR	04 APR	14 APR 69	
1969-033A	COSMOS 277	3855	USSR	04 APR	06 JUL 69	
1969-033B		3857	USSR	04 APR	18 MAY 69	
1969-034A	COSMOS 278	3883	USSR	09 APR	17 APR 69	
1969-034B		3884	USSR	09 APR	16 APR 69	
1969-035A	MOLNIYA 1-11	3885	USSR	11 APR	17 APR 74	
1969-035B		3887	USSR	11 APR	01 MAY 69	
1969-035C		3886	USSR	11 APR	01 JUL 74	
1969-035D		3888	USSR	11 APR	05 MAY 69	
1969-038A	COSMOS 279	3893	USSR	15 APR	23 APR 69	
1969-038B		3894	USSR	15 APR	18 APR 69	
1969-039A		3895	US	15 APR	30 APR 69	
1969-040A	COSMOS 280	3906	USSR	23 APR	06 MAY 69	
1969-040B		3907	USSR	23 APR	26 APR 69	
1969-040C		3919	USSR	23 APR	08 MAY 69	
1969-040D		3920	USSR	23 APR	08 MAY 69	
1969-041A		3914	US	02 MAY	23 MAY 69	
1969-041B		3915	US	02 MAY	16 FEB 70	
1969-041C		3942	US	02 MAY	19 MAY 69	
1969-042A	COSMOS 281	3939	USSR	13 MAY	21 MAY 69	
1969-042B		3940	USSR	13 MAY	17 MAY 69	
1969-043A	APOLLO 10	3941	US	18 MAY	26 MAY 69	AG
1969-044A	COSMOS 282	3944	USSR	20 MAY	28 MAY 69	
1969-044B		3945	USSR	20 MAY	28 MAY 69	
1969-045B		5976	US	22 MAY	07 MAR 77	
1969-047A	COSMOS 283	3957	USSR	27 MAY	10 DEC 69	
1969-047B		3958	USSR	27 MAY	01 OCT 69	
1969-048A	COSMOS 284	3971	USSR	29 MAY	06 JUN 69	
1969-048B		3972	USSR	29 MAY	04 JUN 69	
1969-048C		3973	USSR	29 MAY	30 MAY 69	
1969-049A	COSMOS 285	3983	USSR	03 JUN	07 OCT 69	
1969-049B		3985	USSR	03 JUN	05 AUG 69	
1969-050A		3984	US	03 JUN	14 JUN 69	
1969-051A	OGO 6	3986	US	05 JUN	12 OCT 79	
1969-051B		3987	US	05 JUN	12 JUN 80	
1969-052A	COSMOS 286	3988	USSR	15 JUN	23 JUN 69	
1969-052B		3989	USSR	15 JUN	23 JUN 69	
1969-053A	EXPLORER 41	3990	US	21 JUN	23 DEC 72	
1969-054A	COSMOS 287	3991	USSR	24 JUN	02 JUL 69	
1969-054B		3992	USSR	24 JUN	27 JUN 69	
1969-055A	COSMOS 288	3994	USSR	27 JUN	05 JUL 69	
1969-055B		3995	USSR	27 JUN	02 JUL 69	
1969-056A	BIOSATELLITE 3	4000	US	29 JUN	20 JAN 70	B
1969-056B		4001	US	29 JUN	01 DEC 69	
1969-056C		4022	US	29 JUN	13 JUL 69	
1969-056A	CAPSULE	61016	US	29 JUN	07 JUL 69	
1969-057A	COSMOS 289	4034	USSR	10 JUL	15 JUL 69	
1969-057B		4035	USSR	10 JUL	16 JUL 69	
1969-058A	LUNA 15	4036	USSR	13 JUL	21 JUL 69	A
1969-058B		4037	USSR	13 JUL	16 JUL 69	
1969-058C		4038	USSR	13 JUL	16 JUL 69	
1969-059A	APOLLO 11	4039	US	16 JUL	24 JUL 69	AJ
1969-060A	COSMOS 290	4042	USSR	22 JUL	30 JUL 69	
1969-060B		4044	USSR	22 JUL	29 JUL 69	
1969-061A	MOLNIYA 1-12	4043	USSR	22 JUL	18 JUN 71	
1969-061B		4045	USSR	22 JUL	28 AUG 69	
1959-061C		4046	USSR	22 JUL	23 AUG 69	
1969-061D		4049	USSR	22 JUL	10 JUN 71	
1969-061E		5134	USSR	22 JUL	04 MAY 71	
1969-061F		5288	USSR	22 JUL	15 JUN 71	
1969-061G		5302	USSR	22 JUL	18 JUN 71	
1969-063A		4050	US	24 JUL	23 AUG 69	
1969-064A	INTELSAT 3 F-5	4051	ITSO	26 JUL	14 OCT 88	
1969-064AA		13914	US	26 JUL	23 OCT 91	
1969-064AB		13915	US	26 JUL	23 DEC 87	
1969-064AC		14846	US	26 JUL	29 JAN 86	

International Designation	Satellite Name	Number	Source	Launch	Decay	Note
1969-064B		4052	US	26 JUL	17 JUL 70	AL
1969-064D		4055	US	26 JUL	16 FEB 70	
1969-064E		4056	US	26 JUL	04 JUL 70	
1969-064F		4057	US	26 JUL	22 OCT 69	
1969-064G		4059	US	26 JUL	30 DEC 69	
1969-064H		4060	US	26 JUL	19 DEC 69	
1969-064J		4085	US	26 JUL	18 SEP 69	
1969-064K		4086	US	26 JUL	18 FEB 70	
1969-064L		4087	US	26 JUL	09 JAN 70	
1969-064M		4088	US	26 JUL	21 NOV 69	
1969-064N		4089	US	26 JUL	27 NOV 69	
1969-064P		4090	US	26 JUL	09 OCT 69	
1969-064Q		4091	US	26 JUL	08 NOV 69	
1969-064R		4099	US	26 JUL	02 JAN 70	
1969-064S		4100	US	26 JUL	19 FEB 70	
1969-064T		4101	US	26 JUL	23 SEP 69	
1969-064U		4108	US	26 JUL	30 JAN 70	
1969-064V		4109	US	26 JUL	04 SEP 70	
1969-064W		4110	US	26 JUL	24 OCT 69	
1969-064X		5473	US	26 JUL	05 JAN 75	
1969-064Z		13913	US	26 JUL	06 MAR 88	
1969-065A		4054	US	31 JUL	04 JAN 73	
1969-066A	COSMOS 291	4058	USSR	06 AUG	08 SEP 69	
1969-066B		4061	USSR	06 AUG	11 AUG 69	
1969-067A	ZOND 7	4062	USSR	07 AUG	14 AUG 69	
1969-067B		4063	USSR	07 AUG	10 AUG 69	
1969-067C		4064	USSR	07 AUG	12 AUG 69	
1969-067D		4067	USSR	07 AUG	13 AUG 69	
1969-068A	OSO 6	4065	US	09 AUG	07 MAR 81	
1969-068B	PAC 1	4066	US	09 AUG	28 APR 77	
1969-071A	COSMOS 293	4072	USSR	16 AUG	28 AUG 69	
1969-071B		4073	USSR	16 AUG	20 AUG 69	
1969-072A	COSMOS 294	4074	USSR	19 AUG	27 AUG 69	
1969-072B		4075	USSR	19 AUG	27 AUG 69	
1969-073A	COSMOS 295	4076	USSR	22 AUG	01 DEC 69	
1969-073B		4077	USSR	22 AUG	11 OCT 69	
1969-074A		4078	US	23 AUG	07 SEP 69	
1969-075A	COSMOS 296	4080	USSR	29 AUG	06 SEP 69	
1969-075B		4081	USSR	29 AUG	05 SEP 69	
1969-076A	COSMOS 297	4082	USSR	02 SEP	10 SEP 69	
1969-076B		4083	USSR	02 SEP	11 SEP 69	
1969-077A	COSMOS 298	4092	USSR	15 SEP	15 SEP 69	
1969-077B		4093	USSR	15 SEP	16 SEP 69	
1969-077C		4094	USSR	15 SEP	16 SEP 69	
1969-077D		4095	USSR	15 SEP	16 SEP 69	
1969-077E		4096	USSR	15 SEP	16 SEP 69	
1969-078A	COSMOS 299	4097	USSR	18 SEP	22 SEP 69	
1969-078B		4098	USSR	18 SEP	24 SEP 69	
1969-079A		4102	US	22 SEP	12 OCT 69	
1969-079B		4103	US	22 SEP	16 MAY 71	
1969-080A	COSMOS 300	4104	USSR	23 SEP	27 SEP 69	
1969-080B		4105	USSR	23 SEP	27 SEP 69	
1969-081A	COSMOS 301	4106	USSR	24 SEP	02 OCT 69	
1969-081B		4107	USSR	24 SEP	29 SEP 69	
1969-082A		4111	US	30 SEP	30 OCT 70	
1969-082AE		4162	US	30 SEP	14 APR 79	
1969-082AF		4163	US	30 SEP	09 JAN 81	
1969-082AH		4165	US	30 SEP	02 DEC 89	
1969-082AK		4167	US	30 SEP	04 SEP 89	
1969-082AN		4170	US	30 SEP	23 JAN 90	
1969-082AQ		4172	US	30 SEP	24 OCT 91	
1969-082AT		4175	US	30 SEP	14 DEC 92	
1969-082AU		4176	US	30 SEP	11 MAY 80	
1969-082AX		4179	US	30 SEP	13 OCT 82	
1969-082AZ		4181	US	30 SEP	10 JAN 89	
1969-082BB		4188	US	30 SEP	24 DEC 80	
1969-082BC		4189	US	30 SEP	15 DEC 80	
1969-082BF		4192	US	30 SEP	15 OCT 81	
1969-082BH		4194	US	30 SEP	12 OCT 82	
1969-082BK		4196	US	30 SEP	07 MAY 73	
1969-082BL		4197	US	30 SEP	02 APR 80	
1969-082BM		4198	US	30 SEP	08 APR 81	
1969-082BN		4199	US	30 SEP	31 AUG 81	
1969-082BR		4202	US	30 SEP	08 JUN 80	
1969-082BS		4203	US	30 SEP	08 JAN 82	
1969-082BT		4204	US	30 SEP	08 SEP 79	
1969-082CB		4212	US	30 SEP	06 JUL 91	
1969-082CC		4213	US	30 SEP	13 APR 81	
1969-082CK		4228	US	30 SEP	25 MAR 81	
1969-082CL		4229	US	30 SEP	10 DEC 79	
1969-082CM		4230	US	30 SEP	07 OCT 81	
1969-082CQ		4238	US	30 SEP	30 DEC 78	
1969-082CR		4239	US	30 SEP	19 NOV 79	
1969-082CS		4240	US	30 SEP	24 MAY 80	
1969-082CT		4241	US	30 SEP	09 OCT 80	
1969-082CU		4244	US	30 SEP	11 JUN 79	
1969-082CV		4245	US	30 SEP	25 JUL 86	
1969-082CX		4248	US	30 SEP	21 NOV 78	
1969-082DA		4113	US	30 SEP	21 OCT 69	
1969-082DB		4258	US	30 SEP	16 SEP 89	
1969-082DC		4115	US	30 SEP	21 OCT 69	
1969-082DD		4260	US	30 SEP	29 SEP 90	
1969-082DE		4268	US	30 SEP	19 JAN 90	
1969-082DF		4269	US	30 SEP	29 MAR 79	
1969-082DG		4270	US	30 SEP	16 JUN 80	
1969-082DH		4271	US	30 SEP	23 FEB 80	
1969-082DJ		4272	US	30 SEP	09 SEP 70	
1969-082DM		4278	US	30 SEP	13 AUG 88	
1969-082DN		4288	US	30 SEP	12 SEP 70	
1969-082DQ		4310	US	30 SEP	07 APR 82	
1969-082DS		4312	US	30 SEP	26 MAY 83	
1969-082DT		4317	US	30 SEP	12 NOV 82	
1969-082DV		4319	US	30 SEP	06 NOV 81	
1969-082DW		4323	US	30 SEP	03 FEB 71	
1969-082DX		4324	US	30 SEP	10 MAY 70	
1969-082EA		4335	US	30 SEP	07 MAY 73	
1969-082EB		4348	US	30 SEP	28 NOV 90	
1969-082EC		4359	US	30 SEP	01 OCT 80	
1969-082EE		4433	US	30 SEP	19 APR 79	
1969-082EF		4434	US	30 SEP	05 JAN 80	
1969-082EH		4436	US	30 SEP	20 NOV 90	
1969-082EJ		4437	US	30 SEP	22 JUN 81	
1969-082EL		4439	US	30 SEP	09 MAY 72	
1969-082EM		4440	US	30 SEP	13 OCT 88	
1969-082EN		4441	US	30 SEP	02 FEB 92	
1969-082EQ		4443	US	30 SEP	28 JUL 70	
1969-082ET		4448	US	30 SEP	16 SEP 80	
1969-082EV		4450	US	30 SEP	18 DEC 81	
1969-082EX		4452	US	30 SEP	25 APR 85	
1969-082EZ		4454	US	30 SEP	25 OCT 81	
1969-082FA		4457	US	30 SEP	19 OCT 70	
1969-082FC		4459	US	30 SEP	19 FEB 80	
1969-082FD		4460	US	30 SEP	08 MAR 84	
1969-082FE		4461	US	30 SEP	13 JUN 74	
1969-082FF		4462	US	30 SEP	25 MAR 82	
1969-082FG		4463	US	30 SEP	08 JUL 78	
1969-082FH		4164	US	30 SEP	02 APR 81	
1969-082FJ		4465	US	30 SEP	18 SEP 89	
1969-082FK		4466	US	30 SEP	13 JUL 90	
1969-082FL		4467	US	30 SEP	10 MAR 90	
1969-082FN		4469	US	30 SEP	30 NOV 80	
1969-082FP		4470	US	30 SEP	15 SEP 81	
1969-082FQ		4471	US	30 SEP	17 OCT 83	
1969-082FR		4472	US	30 SEP	21 JUL 80	
1969-082FS		4473	US	30 SEP	18 FEB 81	
1969-082FT		4474	US	30 SEP	14 NOV 79	
1969-082FV		4518	US	30 SEP	05 JUN 93	
1969-082FX		4520	US	30 SEP	20 DEC 77	
1969-082FY		4521	US	30 SEP	18 OCT 80	
1969-082FZ		4522	US	30 SEP	08 MAY 78	
1969-082GC		4531	US	30 SEP	31 JUL 90	
1969-082GD		4532	US	30 SEP	04 MAR 79	
1969-082GE		4533	US	30 SEP	01 MAY 81	
1969-082GF		4534	US	30 SEP	10 APR 83	
1969-082GG		4535	US	30 SEP	13 JUN 78	
1969-082GH		4540	US	30 SEP	03 FEB 72	
1969-082GJ		4541	US	30 SEP	20 DEC 79	
1969-082GK		4542	US	30 SEP	28 DEC 70	
1969-082GL		4543	US	30 SEP	03 AUG 90	
1969-082GM		4544	US	30 SEP	22 JUL 80	
1969-082GP		4548	US	30 SEP	14 FEB 82	
1969-082GQ		4549	US	30 SEP	12 DEC 79	
1969-082GR		4550	US	30 SEP	05 JAN 78	
1969-082GS		4551	US	30 SEP	26 OCT 81	
1969-082GU		4876	US	30 SEP	13 APR 81	
1969-082GW		4887	US	30 SEP	16 JUL 78	
1969-082GX		4888	US	30 SEP	21 JUN 83	
1969-082GY		4889	US	30 SEP	01 AUG 78	
1969-082HA		4891	US	30 SEP	01 JAN 72	
1969-082HB		4892	US	30 SEP	24 AUG 72	
1969-082HC		4893	US	30 SEP	07 MAY 73	
1969-082HF		4897	US	30 SEP	11 MAR 81	
1969-082HG		4906	US	30 SEP	17 SEP 81	
1969-082HH		4937	US	30 SEP	12 MAY 91	
1969-082HK		4939	US	30 SEP	27 OCT 80	
1969-082HM		5158	US	30 SEP	07 AUG 79	
1969-082HN		5159	US	30 SEP	30 DEC 72	
1969-082HP		5195	US	30 SEP	15 JUN 81	
1969-082HQ		5270	US	30 SEP	20 JAN 74	
1969-082HR		5271	US	30 SEP	07 APR 73	
1969-082HT		5339	US	30 SEP	10 JUN 82	
1969-082HU		5362	US	30 SEP	30 SEP 80	
1969-082HV		5429	US	30 SEP	16 JUN 79	
1969-082HX		5459	US	30 SEP	24 SEP 80	
1969-082JA		5521	US	30 SEP	27 JUL 79	
1969-082JB		5522	US	30 SEP	06 DEC 71	
1969-082JC		5593	US	30 SEP	02 MAY 73	
1969-082JD		5633	US	30 SEP	09 DEC 89	
1969-082JE		5634	US	30 SEP	13 APR 81	
1969-082JG		5636	US	30 SEP	13 MAY 78	
1969-082JJ		5750	US	30 SEP	23 JAN 77	
1969-082JK		5751	US	30 SEP	02 JAN 80	
1969-082JL		5752	US	30 SEP	12 OCT 76	
1969-082JM		5753	US	30 SEP	06 OCT 79	
1969-082JN		5754	US	30 SEP	19 JUL 83	
1969-082JQ		5756	US	30 SEP	26 OCT 74	
1969-082JR		5757	US	30 SEP	20 DEC 91	
1969-082JT		5791	US	30 SEP	03 JAN 77	
1969-082JU		5792	US	30 SEP	05 APR 72	
1969-082JV		5793	US	30 SEP	08 APR 72	
1969-082JX		5823	US	30 SEP	09 APR 72	
1969-082JY		5859	US	30 SEP	10 MAY 82	
1969-082JZ		5868	US	30 SEP	23 OCT 79	
1969 082KA		5869	US	30 SEP	22 FEB 82	
1969-082KB		5899	US	30 SEP	02 DEC 84	
1969-082KC		5900	US	30 SEP	19 MAR 73	

International Designation	Satellite Name	Number	Source	Launch	Decay	Note
1969-082KD		5964	US	30 SEP	13 AUG 81	
1969-082KE		5965	US	30 SEP	12 JUL 89	
1969-082KF		6044	US	30 SEP	03 DEC 81	
1969-082KG		6178	US	30 SEP	16 DEC 79	
1969-082KJ		7285	US	30 SEP	19 JUL 74	
1969-082KK		8606	US	30 SEP	28 JUN 78	
1969-082KL		9980	US	30 SEP	13 APR 81	
1969-082KM		15548	US	30 SEP	19 MAR 89	
1969-082KN		15549	US	30 SEP	31 AUG 89	
1969-082KP		15550	US	30 SEP	20 APR 89	
1969-082KT		15554	US	30 SEP	20 JAN 90	
1969-082KV		17659	US	30 SEP	26 OCT 91	
1969-082KX		17661	US	30 SEP	20 DEC 91	
1969-082KY		17724	US	30 SEP	03 JUL 89	
1969-082KZ		18256	US	30 SEP	07 SEP 89	
1969-082L		4142	US	30 SEP	03 NOV 80	
1969-082LF		19581	US	30 SEP	27 APR 91	
1969-082Q		4146	US	30 SEP	18 DEC 88	
1969-082R		4147	US	30 SEP	16 OCT 81	
1969-082S		4148	US	30 SEP	25 JAN 82	
1969-082U		4152	US	30 SEP	03 FEB 84	
1969-082V		4153	US	30 SEP	14 APR 71	
1969-083A	ESRO-1B	4114	ESRO	01 OCT	23 NOV 69	
1969-083B		4117	US	01 OCT	03 NOV 69	
1969-083C		4116	US	01 OCT	21 OCT 69	
1969-083D		4118	US	01 OCT	25 OCT 69	
1969-085A	SOYUZ 6	4122	USSR	11 OCT	16 OCT 69	
1969-085B		4123	USSR	11 OCT	12 OCT 69	
1969-086A	SOYUZ 7	4124	USSR	12 OCT	17 OCT 69	
1969-086B		4125	USSR	12 OCT	14 OCT 69	
1969-087A	SOYUZ 8	4126	USSR	13 OCT	18 OCT 69	
1969-087B		4127	USSR	13 OCT	15 OCT 69	
1969-088A	INTERCOSMOS 1	4128	USSR	14 OCT	02 JAN 70	
1969-088B		4129	USSR	14 OCT	16 DEC 69	
1969-089A	COSMOS 302	4130	USSR	17 OCT	25 OCT 69	
1969-089B		4131	USSR	17 OCT	24 OCT 69	
1969-090A	COSMOS 303	4136	USSR	18 OCT	23 JAN 70	
1969-090B		4137	USSR	18 OCT	07 DEC 69	
1969-092A	COSMOS 305	4150	USSR	22 OCT	24 OCT 69	
1969-092B		4151	USSR	22 OCT	24 OCT 69	
1969-093A	COSMOS 306	4182	USSR	24 OCT	05 NOV 69	
1969-093B		4183	USSR	24 OCT	29 OCT 69	
1969-094A	COSMOS 307	4184	USSR	24 OCT	30 DEC 70	
1969-094B		4185	USSR	24 OCT	20 JUL 70	
1969-095A		4186	US	24 OCT	08 NOV 69	
1969-096A	COSMOS 308	4219	USSR	04 NOV	04 JAN 70	
1969-096B		4220	USSR	04 NOV	06 DEC 69	
1969-097C		4242	FRG	08 NOV	07 OCT 81	
1969-097D		4243	FRG	08 NOV	24 AUG 81	
1969-097E		4261	FRG	08 NOV	11 JUL 80	
1969-097F		4265	FRG	08 NOV	13 APR 81	
1969-098A	COSMOS 309	4223	USSR	12 NOV	20 NOV 69	
1969-098B		4224	USSR	12 NOV	22 NOV 69	
1969-098C		4234	USSR	12 NOV	26 NOV 69	
1969-098D		4235	USSR	12 NOV	25 NOV 69	
1969-098E		4236	USSR	12 NOV	30 NOV 69	
1969-099A	APOLLO 12	4225	US	14 NOV	24 NOV 69	AJ
1969-099C	LUNAR MODULE	4246	US	14 NOV	20 NOV 69	AM
1969-100A	COSMOS 310	4232	USSR	15 NOV	23 NOV 69	
1969-100B		4233	USSR	15 NOV	23 NOV 69	
1969-102A	COSMOS 311	4252	USSR	24 NOV	10 MAR 70	
1969-102B		4253	USSR	24 NOV	19 JAN 70	
1969-104A	COSMOS 313	4262	USSR	03 DEC	15 DEC 69	
1969-104B		4263	USSR	03 DEC	06 DEC 69	
1969-105A		4264	US	04 DEC	10 JAN 70	
1969-106A	COSMOS 314	4266	USSR	11 DEC	22 MAR 70	
1969-106B		4267	USSR	11 DEC	27 JAN 70	
1969-107A	COSMOS 315	4273	USSR	20 DEC	25 MAR 79	
1969-107B		4274	USSR	20 DEC	02 APR 79	
1969-107C		4275	USSR	20 DEC	29 APR 70	
1969-107D		4279	USSR	20 DEC	10 OCT 71	
1969-107E		4287	USSR	20 DEC	13 JUN 70	
1969-107F		4294	USSR	20 DEC	14 MAY 70	
1969-108A	COSMOS 316	4282	USSR	23 DEC	28 AUG 70	
1969-108B		4283	USSR	23 DEC	01 JAN 70	
1969-108C		4284	USSR	23 DEC	28 JAN 70	
1969-109A	COSMOS 317	4280	USSR	23 DEC	05 JAN 70	
1969-109B		4281	USSR	23 DEC	28 DEC 69	
1969-109C		4290	USSR	23 DEC	08 JAN 70	
1969-109D		4291	USSR	23 DEC	09 JAN 70	
1969-110A	INTERCOSMOS 2	4285	USSR	25 DEC	07 JUN 70	
1969-110B		4286	USSR	25 DEC	21 MAR 70	

1970 LAUNCHES

International Designation	Satellite Name	Number	Source	Launch	Decay	Note
1970-001A	COSMOS 318	4292	USSR	09 JAN	21 JAN 70	
1970-001B		4293	USSR	09 JAN	15 JAN 70	
1970-002A		4296	US	14 JAN	01 FEB 70	
1970-004A	COSMOS 319	4299	USSR	15 JAN	01 JUL 70	
1970-004B		4300	USSR	15 JAN	01 MAY 70	
1970 005A	COSMOS 320	4301	USSR	16 JAN	10 FEB 70	
1970-005B		4302	USSR	16 JAN	28 JAN 70	
1970-005C		4303	USSR	16 JAN	24 JAN 70	
1970-005D		4304	USSR	16 JAN	11 FEB 70	
1970-005E		4305	USSR	16 JAN	03 MAR 70	
1970-005F		4306	USSR	16 JAN	24 JAN 70	
1970-005G		4307	USSR	16 JAN	29 JAN 70	
1970-006A	COSMOS 321	4308	USSR	20 JAN	23 MAR 70	
1970-006B		4309	USSR	20 JAN	07 MAR 70	
1970-006C		4313	USSR	20 JAN	25 JAN 70	
1970-006D		4314	USSR	20 JAN	29 JAN 70	
1970-007A	COSMOS 322	4315	USSR	21 JAN	29 JAN 70	
1970-007B		4316	USSR	21 JAN	25 JAN 70	
1970-010A	COSMOS 323	4328	USSR	10 FEB	18 FEB 70	
1970-010B		4329	USSR	10 FEB	15 FEB 70	
1970-013A	MOLNIYA 1-13	4336	USSR	19 FEB	29 SEP 75	
1970-013B		4333	USSR	19 FEB	09 MAR 70	
1970-013C		4334	USSR	19 FEB	11 MAR 70	
1970-013D		4337	USSR	19 FEB	09 JAN 76	
1970-014A	COSMOS 324	4338	USSR	27 FEB	23 MAY 70	
1970-014B		4339	USSR	27 FEB	11 APR 70	
1970-015A	COSMOS 325	4340	USSR	04 MAR	12 MAR 70	
1970-015B		4341	USSR	04 MAR	10 MAR 70	
1970-016A		4342	US	04 MAR	26 MAR 7Q	
1970-016B		4343	US	04 MAR	10 NOV 71	
1970-017A	DIAL-WIKA	4344	FRG	10 MAR	05 OCT 78	
1970-017B		4345	FRAN	10 MAR	09 SEP 74	
1970-018A	COSMOS 326	4346	USSR	13 MAR	21 MAR 70	
1970-018B		4347	USSR	13 MAR	23 MAR 70	
1970-019A	METEOR	4349	USSR	17 MAR	18 NOV 83	
1970-019B		4350	USSR	17 MAR	26 NOV 82	
1970-019C		5573	USSR	17 MAR	03 NOV 74	
1970-020A	COSMOS 327	4351	USSR	18 MAR	19 JAN 71	
1970-020B		4352	USSR	18 MAR	27 SEP 70	
1970-022A	COSMOS 328	4355	USSR	27 MAR	09 APR 70	
1970-022B		4356	USSR	27 MAR	01 APR 70	
1970-023A	COSMOS 329	4357	USSR	03 APR	15 APR 70	
1970-023B		4358	USSR	03 APR	05 APR 70	
1970-024A	COSMOS 330	4360	USSR	07 APR	12 JUN 79	
1970-024B		4361	USSR	07 APR	10 NOV 79	
1970-025AE		4646	US	08 APR	25 OCT 91	
1970-025AX		4674	US	08 APR	13 AUG 82	
1970-025BB		4678	US	08 APR	20 FEB 80	
1970-025BD		4680	US	08 APR	16 DEC 88	
1970-025BH		4684	US	08 APR	08 AUG 90	
1970-025BP		4701	US	08 APR	22 JUL 72	
1970-025BQ		4702	US	08 APR	07 JUL 72	
1970-025BR		4703	US	08 APR	01 DEC 78	
1970-025BS		4704	US	08 APR	27 JAN 92	
1970-025BV		4707	US	08 APR	14 OCT 83	
1970-025CF		4732	US	08 APR	06 FEB 72	
1970-025CP		4740	US	08 APR	10 AUG 72	
1970-025CY		4750	US	08 APR	23 AUG 90	
1970-025DG		4758	US	08 APR	18 MAR 82	
1970-025DQ		4770	US	08 APR	16 MAY 89	
1970-025DR		4771	US	08 APR	22 FEB 83	
1970-025DV		4775	US	08 APR	31 JUL 81	
1970-025E		4602	US	08 APR	14 OCT 82	
1970-025EN		4858	US	08 APR	23 MAY 79	
1970-025EQ		4860	US	08 APR	06 MAY 90	
1970-025EV		4865	US	08 APR	26 OCT 81	
1970-025FA		4870	US	08 APR	09 NOV 79	
1970-025FD		4898	US	08 APR	10 JUN 82	
1970-025FF		4908	US	08 APR	27 OCT 91	
1970-025FM		4915	US	08 APR	08 AUG 71	
1970-025FX		4931	US	08 APR	04 AUG 78	
1970-025FZ		4933	US	08 APR	08 JUN 79	
1970-025GC		4936	US	08 APR	22 JUL 80	
1970-025GH		4947	US	08 APR	18 JUL 81	
1970-025GK		4949	US	08 APR	05 MAR 72	
1970-025GS		4972	US	08 APR	29 MAY 71	
1970-025GU		4974	US	08 APR	02 JUL 71	
1970-025GV		4975	US	08 APR	02 JUN 78	
1970-025GY		4978	US	08 APR	13 APR 81	
1970-025H		4605	US	08 APR	11 AUG 86	
1970-025HE		4984	US	08 APR	11 APR 72	
1970-025HR		4995	US	08 APR	02 OCT 89	
1970-025HV		4999	US	08 APR	28 OCT 79	
1970-025HX		5023	US	08 APR	29 SEP 79	
1970-025J		4606	US	08 APR	14 AUG 79	
1970-025JB		5027	US	08 APR	17 JAN 91	
1970-025JC		5037	US	08 APR	18 JUL 80	
1970-025JD		5038	US	08 APR	18 JUL 71	
1970-025JK		5065	US	08 APR	30 JAN 84	
1970-025JV		5152	US	08 APR	13 MAR 91	
1970-025KD		5190	US	08 APR	10 NOV 79	
1970-025KF		5219	US	08 APR	28 OCT 88	
1970-025KK		5346	US	08 APR	14 OCT 71	
1970-025KM		5348	US	08 APR	04 MAR 92	
1970-025KW		5391	US	08 APR	16 FEB 82	
1970-025KZ		5421	US	08 APR	29 AUG 79	
1970-025L		4608	US	08 APR	26 JAN 92	
1970-025LB		5423	US	08 APR	05 DEC 71	
1970-025LC		5424	US	08 APR	19 MAR 82	
1970-025LD		5445	US	08 APR	28 OCT 88	
1970-025LF		5501	US	08 APR	26 DEC 78	
1970-025LH		5504	US	08 APR	23 JUN 91	
1970-025LJ		5505	US	08 APR	20 MAR 91	
1970-025LK		5506	US	08 APR	28 MAR 91	
1970-025LM		5508	US	08 APR	23 NOV 91	
1970-025LN		5509	US	08 APR	11 APR 72	
1970-025LP		5510	US	08 APR	18 MAR 91	
1970-025LQ		5535	US	08 APR	24 MAR 72	
1970-025LR		5600	US	08 APR	16 APR 90	
1970-025LS		5601	US	08 APR	12 NOV 88	
1970-025LT		5621	US	08 APR	21 MAR 72	

International Designation	Satellite Name	Number	Source	Launch	Decay	Note
1970-025LU		5622	US	08 APR	06 AUG 91	
1970-025LY		5693	US	08 APR	29 MAR 91	
1970-025M		4609	US	08 APR	15 MAY 91	
1970-025MA		5695	US	08 APR	13 DEC 84	
1970-025MC		5697	US	08 APR	01 APR 91	
1970-025ME		5699	US	08 APR	15 DEC 89	
1970-025MF		5700	US	08 APR	16 JUN 79	
1970-025MG		5701	US	08 APR	27 JAN 90	
1970-025MK		5825	US	08 APR	03 DEC 72	
1970-025MM		5843	US	08 APR	23 FEB 89	
1970-025MS		6015	US	08 APR	19 FEB 73	
1970-025MT		6051	US	08 APR	17 MAR 76	
1970-025MU		6084	US	08 APR	12 MAY 74	
1970-025MW		6140	US	08 APR	20 MAR 79	
1970-025NA		6161	US	08 APR	30 OCT 88	
1970-025NC		6184	US	08 APR	06 NOV 90	
1970-025ND		6185	US	08 APR	08 JUN 81	
1970-025NN		9784	US	08 APR	24 NOV 89	
1970-025NP		9997	US	08 APR	04 JAN 78	
1970-025NU		14468	US	08 APR	31 JAN 92	
1970-025NX		14479	US	08 APR	04 JAN 90	
1970-025NY		15520	US	08 APR	06 JUN 89	
1970-025NZ		15537	US	08 APR	20 DEC 88	
1970-025PB		17169	US	08 APR	28 FEB 92	
1970-025PG		17718	US	08 APR	22 DEC 89	
1970-025PK		17721	US	08 APR	17 AUG 89	
1970-025PM		18419	US	08 APR	20 JAN 92	
1970-025PR		18811	US	08 APR	14 AUG 89	
1970-026A	COSMOS 331	4364	USSR	08 APR	16 APR 70	
1970-026B		4365	USSR	08 APR	16 APR 70	
1970-029A	APOLLO 13	4371	US	11 APR	17 APR 70	AN
1970-029B		4372	US	11 APR	15 APR 70	G
1970-030A	COSMOS 333	4373	USSR	15 APR	28 APR 70	
1970-030B		4374	USSR	15 APR	18 APR 70	
1970-030C		4395	USSR	15 APR	02 MAY 70	
1970-030D		4396	USSR	15 APR	09 MAY 70	
1970-031A		4375	US	15 APR	06 MAY 70	
1970-033A	COSMOS 334	4378	USSR	23 APR	09 AUG 70	
1970-033B		4379	USSR	23 APR	16 MAY 70	
1970-033C		4397	USSR	23 APR	08 JUL 70	
1970-033D		4398	USSR	23 APR	25 MAY 70	
1970-033E		4399	USSR	23 APR	16 MAY 70	
1970-034C		4400	PRC	24 APR	23 MAR 83	
1970-035A	COSMOS 335	4380	USSR	24 APR	22 JUN 70	
1970-035B		4381	USSR	24 APR	17 MAY 70	
1970-037C		5055	USSR	28 APR	26 FEB 79	
1970-038A	COSMOS 344	4401	USSR	12 MAY	20 MAY 70	
1970-038B		4402	USSR	12 MAY	20 MAY 70	
1970-039A	COSMOS 345	4403	USSR	20 MAY	28 MAY 70	
1970-039B		4404	USSR	20 MAY	23 MAY 70	
1970-040A		4405	US	20 MAY	17 JUN 70	
1970-040B		4406	US	20 MAY	08 MAR 74	
1970-041A	SOYUZ 9	4407	USSR	01 JUN	19 JUN 70	
1970-041B		4408	USSR	01 JUN	03 JUN 70	
1970-042A	COSMOS 346	4409	USSR	10 JUN	17 JUN 70	
1970-042B		4410	USSR	10 JUN	13 JUN 70	
1970-043A	COSMOS 347	4411	USSR	12 JUN	07 NOV 71	
1970-043B		4412	USSR	12 JUN	14 FEB 71	
1970-044A	COSMOS 348	4413	USSR	13 JUN	25 JUL 70	
1970-044B		4414	USSR	13 JUN	09 JUL 70	
1970-044C		4415	USSR	13 JUN	17 JUN 70	
1970-045A	COSMOS 349	4416	USSR	17 JUN	25 JUN 70	
1970-045B		4417	USSR	17 JUN	22 JUN 70	
1970-048A		4422	US	25 JUN	06 JUL 70	
1970-049A	MOLNIYA 1-14	4430	USSR	26 JUN	16 FEB 76	
1970-049B		4424	USSR	26 JUN	25 JUL 70	
1970-049C		4423	USSR	26 JUN	17 JUL 70	
1970-049D		4431	USSR	26 JUN	02 MAR 76	
1970-050A	COSMOS 350	4425	USSR	26 JUN	08 JUL 70	
1970-050B		4426	USSR	26 JUN	29 JUN 70	
1970-051A	COSMOS 351	4427	USSR	27 JUN	13 OCT 70	
1970-051B		4428	USSR	27 JUN	18 AUG 70	
1970-052A	COSMOS 352	4446	USSR	07 JUL	15 JUL 70	
1970-052B		4447	USSR	07 JUL	12 JUL 70	
1970-053A	COSMOS 353	4455	USSR	09 JUL	21 JUL 70	
1970-053B		4456	USSR	09 JUL	16 JUL 70	
1970-054A		4477	US	23 JUL	19 AUG 70	
1970-056A	COSMOS 354	4481	USSR	28 JUL	28 JUL 70	
1970-056B		4480	USSR	28 JUL	29 JUL 70	
1970-056C		4479	USSR	28 JUL	29 JUL 70	
1970-057A	INTERCOSMOS 3	4482	USSR	07 AUG	06 DEC 70	
1970-057B		4483	USSR	07 AUG	17 NOV 70	
1970-058A	COSMOS 355	4484	USSR	07 AUG	15 AUG 70	
1970-058B		4485	USSR	07 AUG	14 AUG 70	
1970-059A	COSMOS 356	4487	USSR	10 AUG	02 OCT 70	
1970-059B		4488	USSR	10 AUG	01 OCT 70	
1970-060A	VENERA 7	4489	USSR	17 AUG	15 DEC 70	N
1970-060B		4490	USSR	17 AUG	18 AUG 70	
1970-060C		4491	USSR	17 AUG	18 AUG 70	
1970-061A		4492	US	18 AUG	03 SEP-70	
1970-063A	COSMOS 357	4495	USSR	19 AUG	24 NOV 70	
1970-063B		4496	USSR	19 AUG	15 OCT 70	
1970-064A	COSMOS 358	4497	USSR	20 AUG	26 JUN 90	
1970-064B		4498	USSR	20 AUG	07 JUL 79	
1970-064C		4514	USSR	20 AUG	29 OCT 70	
1970-065A	COSMOS 359	4501	USSR	22 AUG	06 NOV 70	
1970-065B		4502	USSR	22 AUG	29 AUG 70	
1970-065C		4504	USSR	22 AUG	11 SEP 70	
1970-065D		4505	USSR	22 AUG	06 OCT 71	
1970-065E		4506	USSR	22 AUG	09 SEP 70	
1970-066A	NNSS 30190	4503	US	26 AUG	26 MAR 75	
1970-067E		6372	US	27 AUG	13 APR 81	
1970-068A	COSMOS 360	4508	USSR	29 AUG	08 SEP 70	
1970-068B		4509	USSR	29 AUG	02 SEP 70	
1970-068C		4516	USSR	29 AUG	09 SEP 70	
1970-068D		4517	USSR	29 AUG	11 SEP 70	
1970-068E		4526	USSR	29 AUG	11 SEP 70	
1970-071A	COSMOS 361	4524	USSR	08 SEP	21 SEP 70	
1970-071B		4525	USSR	08 SEP	14 SEP 70	
1970-071C		4545	USSR	08 SEP	05 OCT 70	
1970-071D		4546	USSR	08 SEP	26 SEP 70	
1970-072A	LUNA 16	4527	USSR	12 SEP	24 SEP 70	AP
1970-072B		4528	USSR	12 SEP	15 SEP 70	
1970-072C		4529	USSR	12 SEP	15 SEP 70	
1970-073A	COSMOS 362	4536	USSR	16 SEP	13 OCT 71	
1970-073B		4537	USSR	16 SEP	28 MAR 71	
1970-074A	COSMOS 363	4538	USSR	17 SEP	29 SEP 70	
1970-074B		4539	USSR	17 SEP	22 SEP 70	
1970-075A	COSMOS 364	4553	USSR	22 SEP	02 OCT 70	
1970-075B		4554	USSR	22 SEP	28 SEP 70	
1970-075C		4563	USSR	22 SEP	09 OCT 70	
1970-076A	COSMOS 365	4556	USSR	25 SEP	25 SEP 70	
1970-076B		4557	USSR	25 SEP	26 SEP 70	
1970-076C		4558	USSR	25 SEP	25 SEP 70	
1970-077A	MOLNIYA 1-15	4569	USSR	29 SEP	20 MAR 76	
1970-077B		4559	USSR	29 SEP	23 OCT 70	
1970-077C		4560	USSR	29 SEP	16 OCT 70	
1970-077D		4570	USSR	29 SEP	31 MAY 72	
1970-077E		5950	USSR	29 SEP	30 APR 72	
1970-078A	COSMOS 366	4561	USSR	01 OCT	13 OCT 70	
1970-078B		4562	USSR	01 OCT	06 OCT 70	
1970-079B		4565	USSR	03 OCT	06 OCT 70	
1970-079C		4566	USSR	03 OCT	31 OCT 70	
1970-079D		4567	USSR	03 OCT	08 OCT 70	
1970-079E		4568	USSR	03 OCT	12 OCT 70	
1970-080A	COSMOS 368	4571	USSR	08 OCT	14 OCT 70	
1970-080B		4572	USSR	08 OCT	20 OCT 70	
1970-080C		4575	USSR	08 OCT	07 JAN 71	
1970-080D		4582	USSR	08 OCT	19 OCT 70	
1970-080E		4585	USSR	08 OCT	04 NOV 70	
1970-081A	COSMOS 369	4573	USSR	08 OCT	22 JAN 71	
1970-081B		4574	USSR	08 OCT	30 NOV 70	
1970-082A	COSMOS 370	4576	USSR	09 OCT	22 OCT 70	
1970-082B		4577	USSR	09 OCT	13 OCT 70	
1970-082C		4593	USSR	09 OCT	27 OCT 70	
1970-084A	INTERCOSMOS 4	4580	USSR	14 OCT	17 JAN 71	
1970-084B		4581	USSR	14 OCT	17 DEC 70	
1970-085C		6330	USSR	15 OCT	10 MAR 90	
1970-087A	COSMOS 373	4590	USSR	20 OCT	08 MAR 80	
1970-087B		4595	USSR	20 OCT	18 JAN 71	
1970-088A	ZOND 8	4591	USSR	20 OCT	27 OCT 70	
1970-088B		4592	USSR	20 OCT	26 OCT 70	
1970-089AA		5029	USSR	23 OCT	03 JUN 81	
1970-089AB		5031	USSR	23 OCT	30 MAR 90	
1970-089AC		5032	USSR	23 OCT	14 AUG 83	
1970-089AD		5033	USSR	23 OCT	01 MAR 90	
1970-089AF		5068	USSR	23 OCT	05 DEC 76	
1970-089AG		5069	USSR	23 OCT	22 FEB 72	
1970-089AJ		5071	USSR	23 OCT	02 OCT 78	
1970-089AK		5072	USSR	23 OCT	13 APR 81	
1970-089AL		5073	USSR	23 OCT	26 JUL 71	
1970-089AM		5074	USSR	23 OCT	13 APR 81	
1970-089AR		5078	USSR	23 OCT	22 APR 72	
1970-089AS		5079	USSR	23 OCT	26 MAR 79	
1970-089AT		5080	USSR	23 OCT	29 JUN 81	
1970-089AU		5081	USSR	23 OCT	29 OCT 81	
1970-089AV		5096	USSR	23 OCT	01 MAY 79	
1970-089AW		5132	USSR	23 OCT	10 NOV 82	
1970-089AX		5136	USSR	23 OCT	09 NOV 81	
1970-089AY		5138	USSR	23 OCT	24 MAR 72	
1970-089BA		5189	USSR	23 OCT	10 NOV 82	
1970-089BC		5199	USSR	23 OCT	05 DEC 76	
1970-089BD		5202	USSR	23 OCT	23 AUG 77	
1970-089BE		5224	USSR	23 OCT	22 DEC 78	
1970-089BG		5242	USSR	23 OCT	17 NOV 92	
1970-089BH		5276	USSR	23 OCT	16 JUN 79	
1970-089BJ		5278	USSR	23 OCT	06 MAR 83	
1970-089BK		5279	USSR	23 OCT	22 NOV 82	
1970-089BL		5293	USSR	23 OCT	05 DEC 76	
1970-089BM		5294	USSR	23 OCT	16 FEB 72	
1970-089BN		5295	USSR	23 OCT	16 DEC 91	
1970-089BP		5296	USSR	23 OCT	20 MAY 91	
1970-089BQ		5321	USSR	23 OCT	05 DEC 76	
1970-089BR		5335	USSR	23 OCT	20 NOV 76	
1970-089BT		5338	USSR	23 OCT	16 JUN 92	
1970-089BU		5411	USSR	23 OCT	02 JUN 82	
1970-089BV		5412	USSR	23 OCT	30 APR 80	
1970-089BW		5413	USSR	23 OCT	24 JUL 80	
1970-089BX		5431	USSR	23 OCT	26 JAN 72	
1970-089BZ		5433	USSR	23 OCT	01 JAN 82	
1970-089CA		5454	USSR	23 OCT	21 APR 80	
1970-089CB		5455	USSR	23 OCT	03 SEP 75	
1970-089CC		5456	USSR	23 OCT	03 SEP 77	
1970-089CE		5458	USSR	23 OCT	12 APR 76	
1970-089CF		5523	USSR	23 OCT	24 FEB 81	
1970-089CG		5524	USSR	23 OCT	07 APR 79	

International Designation	Satellite Name	Number	Source	Launch	Decay	Note
1970-089CM		5796	USSR	23 OCT	06 AUG 72	
1970-089CN		5797	USSR	23 OCT	05 MAY 74	
1970-089CP		5896	USSR	23 OCT	02 APR 73	
1970-089CQ		5897	USSR	23 OCT	22 JUL 81	
1970-089CR		5898	USSR	23 OCT	05 NOV 81	
1970-089CS		5963	USSR	23 OCT	23 JUL 79	
1970-089CU		6177	USSR	23 OCT	04 JAN 82	
1970-089CV		13936	USSR	23 OCT	20 OCT 83	
1970-089CX		14834	USSR	23 OCT	08 MAY 84	
1970-089CY		14835	USSR	23 OCT	21 APR 84	
1970-089CZ		14836	USSR	23 OCT	07 APR 89	
1970-089DA		15782	USSR	23 OCT	08 DEC 91	
1970-089DB		17669	USSR	23 OCT	06 JUL 91	
1970-089DE		19433	USSR	23 OCT	25 SEP 89	
1970-089DG		20049	USSR	23 OCT	15 JUN 89	
1970-089J		4666	USSR	23 OCT	25 SEP 90	
1970-089M		4669	USSR	23 OCT	08 AUG 80	
1970-089N		4670	USSR	23 OCT	03 APR 78	
1970-089P		4671	USSR	23 OCT	27 JAN 79	
1970-089U		4824	USSR	23 OCT	21 OCT 79	
1970-089Y		5022	USSR	23 OCT	24 JUN 72	
1970-090A		4596	US	23 OCT	11 NOV 70	
1970-091AA		4729	USSR	30 OCT	22 OCT 80	
1970-091AB		4730	USSR	30 OCT	26 NOV 78	
1970-091AC		4731	USSR	30 OCT	24 FEB 81	
1970-091AD		4826	USSR	30 OCT	05 AUG 81	
1970-091AF		4828	USSR	30 OCT	09 NOV 80	
1970-091AG		4896	USSR	30 OCT	16 SEP 77	
1970-091AL		5082	USSR	30 OCT	29 AUG 88	
1970-091AN		5084	USSR	30 OCT	22 NOV 82	
1970-091AP		5089	USSR	30 OCT	05 DEC 76	
1970-091AS		14527	USSR	30 OCT	29 AUG 90	
1970-091AT		14838	USSR	30 OCT	01 NOV 89	
1970-091AU		18798	USSR	30 OCT	23 JAN 89	
1970-091AW		19023	USSR	30 OCT	27 DEC 89	
1970-091F		4626	USSR	30 OCT	20 FEB 80	
1970-091G		4627	USSR	30 OCT	25 NOV 88	
1970-091K		4633	USSR	30 OCT	21 FEB 84	
1970-091P		4662	USSR	30 OCT	07 APR 82	
1970-091Q		4663	USSR	30 OCT	24 AUG 88	
1970-091W		4725	USSR	30 OCT	27 AUG 88	
1970-092A	COSMOS 376	4599	USSR	30 OCT	12 NOV 70	
1970-092B		4600	USSR	30 OCT	05 NOV 70	
1970-092C		4700	USSR	30 OCT	20 NOV 70	
1970-093C		4631	US	06 NOV	07 NOV 70	
1970-094A	OFO 1	4690	US	09 NOV	09 MAY 71	
1970-094B	RMS	4692	US	09 NOV	07 FEB 71	
1970-094C		4693	US	09 NOV	18 DEC 70	
1970-094D		4747	US	09 NOV	25 JAN 71	
1970-094E		4844	US	09 NOV	21 JAN 71	
1970-095A	LUNA 17	4691	USSR	10 NOV	17 NOV 70	R
1970-095B		4694	USSR	10 NOV	13 NOV 70	
1970-095C		4697	USSR	10 NOV	13 NOV 70	
1970-096A	COSMOS 377	4695	USSR	11 NOV	23 NOV 70	
1970-096B		4696	USSR	11 NOV	16 NOV 70	
1970-097A	COSMOS 378	4713	USSR	17 NOV	17 AUG 72	
1970-097B		4714	USSR	17 NOV	30 SEP 72	
1970-098A		4721	US	18 NOV	11 DEC 70	
1970-098B		4722	US	18 NOV	14 SEP 77	
1970-099A	COSMOS 379	4760	USSR	24 NOV	21 SEP 83	
1970-099B		4761	USSR	24 NOV	26 NOV 70	
1970-099C		4780	USSR	24 NOV	13 FEB 71	
1970-099D		4817	USSR	24 NOV	02 JAN 71	
1970-100A	COSMOS 380	4762	USSR	24 NOV	17 JUN 71	
1970-100B		4763	USSR	24 NOV	09 APR 71	
1970-101A	MOLNIYA 1-16	4779	USSR	27 NOV	25 NOV 75	
1970-101B		4781	USSR	27 NOV	11 DEC 70	
1970-101C		4782	USSR	27 NOV	17 DEC 70	
1970-101D		4785	USSR	27 NOV	02 JAN 76	
1970-102C		4840	USSR	02 DEC	23 APR 92	
1970-103D		5316	USSR	02 DEC	07 NOV 90	
1970-103E		5326	USSR	02 DEC	03 JAN 77	
1970-103F		6072	USSR	02 DEC	03 JAN 77	
1970-104A	COSMOS 383	4787	USSR	03 DEC	16 DEC 70	
1970-104B		4788	USSR	03 DEC	09 DEC 70	
1970-105A	COSMOS 384	4791	USSR	10 DEC	22 DEC 70	
1970-105B		4792	USSR	10 DEC	15 DEC 70	
1970-105C		4795	USSR	10 DEC	18 DEC 70	
1970-105D		4796	USSR	10 DEC	18 DEC 70	
1970-105E		4810	USSR	10 DEC	27 DEC 70	
1970-107A	EXPLORER 42	4797	US	12 DEC	05 APR 79	
1970-107B		4798	US	12 DEC	15 DEC 78	
1970-109A	PEOLE	4801	FRAN	12 DEC	16 JUN 80	
1970-109C		4803	FRAN	12 DEC	14 MAY 80	
1970-109D		4839	FRAN	12 DEC	20 APR 79	
1970-109E		5264	FRAN	12 DEC	06 JAN 80	
1970-109F		5952	FRAN	12 DEC	30 OCT 76	
1970-110A	COSMOS 386	4804	USSR	15 DEC	28 DEC 70	
1970-110B		4805	USSR	15 DEC	19 DEC 70	
1970-110C		4808	USSR	15 DEC	17 DEC 70	
1970-110D		4835	USSR	15 DEC	12 JAN 71	
1970-110E		4836	USSR	15 DEC	03 JAN 71	
1970-110F		4838	USSR	15 DEC	02 JAN 71	
1970-111A	COSMOS 387	4806	USSR	16 DEC	19 JAN 80	
1970-111B		4807	USSR	16 DEC	25 MAR 80	
1970-111C		4809	USSR	16 DEC	12 DEC 71	
1970-111D		4815	USSR	16 DEC	03 SEP 71	
1970-111E		4816	USSR	16 DEC	24 AUG 71	
1970-112A	COSMOS 388	4811	USSR	18 DEC	10 MAY 71	
1970-112B		4812	USSR	18 DEC	17 FEB 71	
1970-113C		4834	USSR	18 DEC	19 SEP 72	
1970-114A	MOLNIYA 1-17	4829	USSR	25 DEC	22 DEC 75	
1970-114B		4830	USSR	25 DEC	23 JAN 71	
1970-114C		4831	USSR	25 DEC	10 FEB 71	
1970-114D		4832	USSR	25 DEC	02 JAN 71	
1970-114E		4833	USSR	25 DEC	02 JAN 71	
1970-114F		4837	USSR	25 DEC	03 MAR 73	
1970-114G		6313	USSR	25 DEC	07 FEB 73	

1971 LAUNCHES

International Designation	Satellite Name	Number	Source	Launch	Decay	Note
1971-000A		4924	US	27 OCT	22 DEC 83	14
1971-000D		5309	US	26 JUL	26 JUN 73	AQ
1971-000E		5310	US	01 JAN	26 SEP 90	14
1971-001A	COSMOS 390	4845	USSR	12 JAN	25 JAN 71	
1971-001B		4846	USSR	12 JAN	17 JAN 71	
1971-001C		4878	USSR	12 JAN	01 FEB 71	
1971-001D		4879	USSR	12 JAN	27 JAN 71	
1971-001E		4883	USSR	12 JAN	27 JAN 71	
1971-001F		4886	USSR	12 JAN	27 JAN 71	
1971-002A	COSMOS 391	4847	USSR	14 JAN	21 FEB 72	
1971-002B		4848	USSR	14 JAN	21 AUG 71	
1971-004A	COSMOS 392	4872	USSR	21 JAN	02 FEB 71	
1971-004B		4873	USSR	21 JAN	26 JAN 71	
1971-005A		4874	US	21 JAN	09 FEB 71	
1971-005B		4875	US	21 JAN	22 JAN 71	
1971-005C		4909	US	21 JAN	09 FEB 71	
1971-007A	COSMOS 393	4884	USSR	26 JAN	16 JUN 71	
1971-007B		4885	USSR	26 JAN	31 MAR 71	
1971-008A	APOLLO 14	4900	US	31 JAN	09 FEB 71	AJ
1971-008B		4904	US	31 JAN	04 FEB 71	G
1971-008C	LUNAR MODULE	4905	US	31 JAN	07 FEB 71	AM
1971-009C		5974	US	03 FEB	10 AUG 75	
1971-015DH		18548	USSR	05 FEB	12 MAR 88	
1971-010B		4923	USSR	09 FEB	21 MAR 84	
1971-010C		4927	USSR	09 FEB	18 APR 81	
1971-011C		5419	JAPAN	16 FEB	26 JUL 80	
1971-012C		4957	US	17 FEB	17 OCT 89	
1971-012D		4958	US	17 FEB	20 SEP 89	
1971-012E		4963	US	17 FEB	07 JAN 90	
1971-013A	COSMOS 395	4955	USSR	17 FEB	06 APR 80	
1971-013B		4956	USSR	17 FEB	19 JUL 80	
1971-013C		5001	USSR	17 FEB	15 FEB 72	
1971-013D		5002	USSR	17 FEB	17 JAN 72	
1971-013E		5227	USSR	17 FEB	06 SEP 71	
1971-014A	COSMOS 396	4959	USSR	18 FEB	03 MAR 71	
1971-014B		4960	USSR	18 FEB	25 FEB 71	
1971-014C		4961	USSR	18 FEB	20 FEB 71	
1971-014D		4962	USSR	18 FEB	21 FEB 71	
1971-014E		5005	USSR	18 FEB	08 MAR 71	
1971-014F		5006	USSR	18 FEB	04 MAR 71	
1971-015AA		5101	USSR	25 FEB	24 AUG 89	
1971-015AB		5102	USSR	25 FEB	09 NOV 91	
1971-015AD		5130	USSR	25 FEB	29 OCT 80	
1971-015AN		5191	USSR	25 FEB	28 JAN 93	
1971-015AQ		5193	USSR	25 FEB	08 MAR 80	
1971-015AR		5194	USSR	25 FEB	25 MAY 80	
1971-015AS		5196	USSR	25 FEB	22 OCT 90	
1971-015AU		5200	USSR	25 FEB	13 FEB 91	
1971-015AV		5201	USSR	25 FEB	09 AUG 77	
1971-015AX		5241	USSR	25 FEB	12 JUL 83	
1971-015AY		5243	USSR	25 FEB	15 DEC 88	
1971-015B		4965	USSR	25 FEB	04 MAR 71	
1971-015BA		5262	USSR	25 FEB	04 OCT 74	
1971-015BC		5273	USSR	25 FEB	24 APR 91	
1971-015BD		5274	USSR	25 FEB	19 AUG 80	
1971-015BG		5323	USSR	25 FEB	13 APR 81	
1971-015BH		5324	USSR	25 FEB	19 NOV 80	
1971-015BJ		5325	USSR	25 FEB	25 FEB 91	
1971-015BR		5336	USSR	25 FEB	12 FEB 92	
1971-015BM		5415	USSR	25 FEB	09 JUL 78	
1971-015BP		5434	USSR	25 FEB	19 MAR 91	
1971-015BQ		5526	USSR	25 FEB	23 JAN 89	
1971-015BS		5528	USSR	25 FEB	21 FEB 78	
1971-015BU		5530	USSR	25 FEB	11 AUG 89	
1971-015BW		5532	USSR	25 FEB	16 OCT 81	
1971-015BY		5534	USSR	25 FEB	28 DEC 90	
1971-015C		5009	USSR	25 FEB	11 APR 93	
1971-015CC		5760	USSR	25 FEB	12 FEB 92	
1971-015CE		5762	USSR	25 FEB	05 OCT 81	
1971-015CG		6043	USSR	25 FEB	31 JUL 79	
1971-015CH		6299	USSR	25 FEB	01 OCT 90	
1971-015CJ		6749	USSR	25 FEB	11 JUN 74	
1971-015CM		10398	USSR	25 FEB	28 FEB 91	
1971-015CP		13944	USSR	25 FEB	24 MAR 89	
1971-015CT		15342	USSR	25 FEB	23 DEC 92	
1971-015CU		15343	USSR	25 FEB	17 SEP 89	
1971-015CW		15761	USSR	25 FEB	16 DEC 86	
1971-015CX		15777	USSR	25 FEB	31 JUL 91	
1971-015CY		15787	USSR	25 FEB	29 MAY 90	
1971-015CZ		17029	USSR	25 FEB	24 SEP 88	
1971-015DC		17664	USSR	25 FEB	05 JUN 89	
1971-015DE		18497	USSR	25 FEB	11 NOV 89	
1971-015DF		18521	USSR	25 FEB	06 NOV 88	
1971-015DG		18531	USSR	25 FEB	23 SEP 89	
1971-015DK		18613	USSR	25 FEB	02 SEP 89	

International Designation	Satellite Name	Number	Source	Launch	Decay	Note
1971-015DL		18692	USSR	25 FEB	23 APR 88	
1971-015DM		18976	USSR	25 FEB	03 JAN 90	
1971-015DP		19064	USSR	25 FEB	02 FEB 89	
1971-015DQ		19114	USSR	25 FEB	02 FEB 89	
1971-015DT		19829	USSR	25 FEB	27 FEB 91	
1971-015DV		20091	USSR	25 FEB	20 JUN 89	
1971-015R		5090	USSR	25 FEB	10 OCT 89	
1971-015T		5092	USSR	25 FEB	15 MAY 88	
1971-015V		5096	USSR	25 FEB	18 APR 92	
1971-015Y		5099	USSR	25 FEB	27 APR 90	
1971-015Z		5100	USSR	25 FEB	10 FEB 82	
1971-016B		4967	USSR	26 FEB	01 MAR 71	
1971-016C		5008	USSR	26 FEB	12 MAY 71	
1971-016D		5039	USSR	26 FEB	28 MAR 71	
1971-016E		5040	USSR	26 FEB	23 MAR 71	
1971-017A	COSMOS 399	5003	USSR	03 MAR	17 MAR 71	
1971-017B		5004	USSR	03 MAR	08 MAR 71	
1971-017C		5047	USSR	03 MAR	25 MAR 71	
1971-017D		5048	USSR	03 MAR	31 MAR 71	
1971-017E		5049	USSR	03 MAR	23 MAR 71	
1971-018A	MAO 2	5007	PRC	03 MAR	17 JUN 79	
1971-018B		5041	PRC	03 MAR	16 FEB 76	
1971-019A	EXPLORER 43	5043	US	13 MAR	02 OCT 74	
1971-019B		5044	US	13 MAR	22 MAY 71	
1971-019C		5045	US	13 MAR	15 MAR 75	
1971-019D		5046	US	13 MAR	19 APR 71	
1971-019E		5112	US	13 MAR	19 APR 71	
1971-022A		5059	US	24 MAR	12 APR 71	
1971-023A	COSMOS 401	5086	USSR	27 MAR	09 APR 71	
1971-023B		5087	USSR	27 MAR	02 APR 71	
1971-023C		5119	USSR	27 MAR	16 APR 71	
1971-023D		5120	USSR	27 MAR	12 APR 71	
1971-023E		5121	USSR	27 MAR	12 APR 71	
1971-025B		5107	USSR	01 APR	06 MAY 71	
1971-025C		5109	USSR	01 APR	06 APR 71	
1971-025D		5110	USSR	01 APR	05 APR 71	
1971-026A	COSMOS 403	5108	USSR	02 APR	14 APR 71	
1971-026B		5111	USSR	02 APR	05 APR 71	
1971-027A	COSMOS 404	5113	USSR	04 APR	04 APR 71	
1971-027B		5114	USSR	04 APR	12 APR 71	
1971-027C		5115	USSR	04 APR	06 APR 71	
1971-027D		5116	USSR	04 APR	07 APR 71	
1971-028C		5311	USSR	07 APR	12 MAY 74	
1971-029A	COSMOS 406	5124	USSR	14 APR	24 APR 71	
1971-029B		5125	USSR	14 APR	17 APR 71	
1971-029C		5127	USSR	14 APR	14 APR 71	
1971-029D		5169	USSR	14 APR	24 APR 71	
1971-029E		5170	USSR	14 APR	28 APR 71	
1971-029F		5179	USSR	14 APR	26 APR 71	
1971-030A	TOURNESOL	5128	FRAN	15 APR	28 JAN 80	
1971-030B		5135	FRAN	15 APR	09 JUL 79	
1971-030C		5139	FRAN	15 APR	15 JUL 72	
1971-030D		5140	FRAN	15 APR	15 JUL 77	
1971-030E		5141	FRAN	15 APR	15 APR 76	
1971-030F		5164	FRAN	15 APR	03 FEB 77	
1971-030G		5209	FRAN	15 APR	13 JUL 72	
1971-030H		5268	FRAN	15 APR	22 MAR 75	
1971-031A	METEOR	5142	USSR	17 APR	10 JAN 91	
1971-032A	SALYUT 1	5160	USSR	19 APR	11 OCT 71	
1971-032B		5161	USSR	19 APR	20 APR 71	
1971-032C		5162	USSR	19 APR	20 APR 71	
1971-032D		5163	USSR	19 APR	21 APR 71	
1971-032E		5166	USSR	19 APR	24 APR 71	
1971-032F		5167	USSR	19 APR	21 APR 71	
1971-032G		5168	USSR	19 APR	21 APR 71	
1971-033A		5171	US	22 APR	13 MAY 71	
1971-034A	SOYUZ 10	5172	USSR	22 APR	24 APR 71	E
1971-034B		5173	USSR	22 APR	25 APR 71	
1971-036A	SAN MARCO 3	5176	ITALY	24 APR	29 NOV 71	
1971-037A	COSMOS 408	5177	USSR	24 APR	29 DEC 71	
1971-037B		5178	USSR	24 APR	29 SEP 71	
1971-039C		5206	US	05 MAY	05 MAY 71	
1971-040A	COSMOS 410	5207	USSR	06 MAY	18 MAY 71	
1971-040B		5208	USSR	06 MAY	11 MAY 71	
1971-040C		5228	USSR	06 MAY	25 MAY 71	
1971-040D		5229	USSR	06 MAY	19 MAY 71	
1971-042A	COSMOS 419	5221	USSR	10 MAY	12 MAY 71	
1971-043A	COSMOS 420	5230	USSR	18 MAY	29 MAY 71	
1971-043B		5231	USSR	18 MAY	21 MAY 71	
1971-043C		5250	USSR	18 MAY	03 JUN 71	
1971-043D		5251	USSR	18 MAY	30 MAY 71	
1971-043E		5260	USSR	18 MAY	31 MAY 71	
1971-044A	COSMOS 421	5232	USSR	19 MAY	08 NOV 71	
1971-044B		5233	USSR	19 MAY	23 AUG 71	
1971-045B		5235	USSR	19 MAY	21 MAY 71	
1971-045C		5236	USSR	19 MAY	21 MAY 71	
1971-045D	CAPSULE	5739	USSR	19 MAY	27 NOV 71	AR
1971-047A	COSMOS 423	5246	USSR	27 MAY	26 NOV 71	
1971-047B		5247	USSR	27 MAY	29 AUG 71	
1971-048A	COSMOS 424	5248	USSR	28 MAY	10 JUN 71	
1971-048B		5249	USSR	28 MAY	03 JUN 71	
1971-048C		5286	USSR	28 MAY	15 JUN 71	
1971-048D		5287	USSR	28 MAY	11 JUN 71	
1971-048E		5292	USSR	28 MAY	11 JUN 71	
1971-049B		5255	USSR	28 MAY	31 MAY 71	
1971-049C		5256	USSR	28 MAY	31 MAY 71	
1971-049D		5263	USSR	28 MAY	01 JUN 71	
1971-049E	CAPSULE	5667	USSR	28 MAY	02 DEC 71	AR
1971-050A	COSMOS 425	5253	USSR	29 MAY	15 JAN 80	
1971-050B		5254	USSR	29 MAY	28 NOV 79	
1971-050C		5257	USSR	29 MAY	22 MAR 72	
1971-050D		5258	USSR	29 MAY	06 APR 72	
1971-050E		5259	USSR	29 MAY	09 JUN 72	
1971-050F		5299	USSR	29 MAY	14 MAY 72	
1971-052C		5722	USSR	04 JUN	24 MAR 73	
1971-052D		5723	USSR	04 JUN	21 SEP 74	
1971-052E		5782	USSR	04 JUN	22 JUN 73	
1971-052F		5956	USSR	04 JUN	08 DEC 77	
1971-053A	SOYUZ 11	5283	USSR	06 JUN	29 JUN 71	AD
1971-053B		5284	USSR	06 JUN	07 JUN 71	
1971-054A		5285	US	08 JUN	31 JAN 82	
1971-055A	COSMOS 427	5289	USSR	11 JUN	23 JUN 71	
1971-055B		5290	USSR	11 JUN	20 JUN 71	
1971-055C		5291	USSR	11 JUN	15 JUN 71	
1971-055D		5303	USSR	11 JUN	25 JUN 71	
1971-055E		5304	USSR	11 JUN	26 JUN 71	
1971-056A		5297	US	15 JUN	06 AUG 71	
1971-056B		5298	US	15 JUN	20 JUN 71	
1971-057A	COSMOS 428	5305	USSR	24 JUN	06 JUL 71	
1971-057B		5306	USSR	24 JUN	28 JUN 71	
1971-057C		5307	USSR	24 JUN	27 JUN 71	
1971-057D		5308	USSR	24 JUN	27 JUN 71	
1971-057E		5313	USSR	24 JUN	07 JUL 71	
1971-057F		5314	USSR	24 JUN	05 JUL 71	
1971-057G		5315	USSR	24 JUN	13 JUL 71	
1971-058A	EXPLORER 44	5317	US	08 JUL	15 DEC 79	
1971-058B		5318	US	08 JUL	28 JAN 76	
1971-058C		5319	US	08 JUL	22 SEP 72	
1971-058D		5320	US	08 JUL	07 NOV 75	
1971-059A	METEOR	5327	USSR	16 JUL	27 AUG 91	
1971-059C		5330	USSR	16 JUL	25 AUG 74	
1971-060A		5329	US	16 JUL	31 AUG 78	
1971-061A	COSMOS 429	5331	USSR	20 JUL	02 AUG 71	
1971-061B		5332	USSR	20 JUL	23 JUL 71	
1971-061C		5370	USSR	20 JUL	03 AUG 71	
1971-061D		5371	USSR	20 JUL	04 AUG 71	
1971-061E		5373	USSR	20 JUL	04 AUG 71	
1971-062A	COSMOS 430	5340	USSR	23 JUL	05 AUG 71	
1971-062B		5341	USSR	23 JUL	29 JUL 71	
1971-062C		5342	USSR	23 JUL	26 JUL 71	
1971-062D		5343	USSR	23 JUL	25 JUL 71	
1971-062E		5374	USSR	23 JUL	06 AUG 71	
1971-062F		5378	USSR	23 JUL	09 AUG 71	
1971-062G		5385	USSR	23 JUL	07 AUG 71	
1971-063A	APOLLO 15	5351	US	26 JUL	07 AUG 71	AJ
1971-063B		5352	US	26 JUL	30 JUL 71	G
1971-063C	LUNAR MODULE	5366	US	26 JUL	03 AUG 71	AM
1971-064A	MOLNIYA 1-18	5367	USSR	28 JUL	19 JUL 77	
1971-064B		5353	USSR	28 JUL	29 AUG 71	
1971-064C		5354	USSR	28 JUL	24 AUG 71	
1971-064D		5368	USSR	28 JUL	14 AUG 77	
1971-064E		5355	USSR	28 JUL	04 AUG 71	
1971-064F		5356	USSR	28 JUL	02 AUG 71	
1971-065A	COSMOS 431	5364	USSR	30 JUL	11 AUG 71	
1971-065B		5365	USSR	30 JUL	04 AUG 71	
1971-065C		5369	USSR	30 JUL	05 AUG 71	
1971-066A	COSMOS 432	5379	USSR	05 AUG	18 AUG 71	
1971-066B		5380	USSR	05 AUG	09 AUG 71	
1971-066C		5381	USSR	05 AUG	08 AUG 71	
1971-066D		5436	USSR	05 AUG	21 AUG 71	
1971-066E		5437	USSR	05 AUG	18 AUG 71	
1971-067A	OV1-20	5394	US	07 AUG	28 AUG 71	
1971-067C	OAR-901	5382	US	07 AUG	31 JAN 72	
1971-067D	OAR-907	5383	US	07 AUG	19 SEP 71	
1971-067F		5396	US	07 AUG	11 JUN 72	
1971-067G		5401	US	07 AUG	14 APR 79	
1971-067H		5406	US	07 AUG	02 NOV 79	
1971-067P		5410	US	07 AUG	01 SEP 81	
1971-068A	COSMOS 433	5402	USSR	08 AUG	09 AUG 71	
1971-068B		5403	USSR	08 AUG	10 AUG 71	
1971-068C		5404	USSR	08 AUG	10 AUG 71	
1971-069A	COSMOS 434	5407	USSR	12 AUG	22 AUG 81	
1971-069B		5408	USSR	12 AUG	18 AUG 71	
1971-069D		5427	USSR	12 AUG	19 SEP 71	
1971-069E		5428	USSR	12 AUG	28 OCT 71	
1971-070A		5409	US	12 AUG	03 SEP 71	
1971-070B		5452	US	12 AUG	04 SEP 71	
1971-072A	COSMOS 435	5441	USSR	27 AUG	28 JAN 72	
1971-072B		5442	USSR	27 AUG	20 NOV 71	
1971-073A	LUNA 18	5448	USSR	02 SEP	11 SEP 71	A
1971-073C		5450	USSR	02 SEP	07 SEP 71	
1971-073D		5451	USSR	02 SEP	07 SEP 71	
1971-073E		5453	USSR	02 SEP	03 SEP 71	
1971-074A	COSMOS 436	5461	USSR	07 SEP	04 JAN 80	
1971-074B		5462	USSR	07 SEP	13 MAR 80	AS
1971-074C		5463	USSR	07 SEP	12 MAY 72	
1971-074D		5464	USSR	07 SEP	22 JAN 72	
1971-074E		5465	USSR	07 SEP	09 MAY 72	
1971-074F		5470	USSR	07 SEP	11 MAY 72	
1971-074G		5583	USSR	07 SEP	30 JUL 72	
1971-074H		5584	USSR	07 SEP	19 APR 72	
1971-074J		5718	USSR	07 SEP	07 JUN 72	
1971-074K		5776	USSR	07 SEP	06 MAY 72	
1971-074L		5777	USSR	07 SEP	11 JUN 72	
1971-074M		5779	USSR	07 SEP	11 APR 72	
1971-074N		5780	USSR	07 SEP	05 JUL 72	

International Designation	Satellite Name	Number	Source	Launch	Decay	Note
1971-074P		5781	USSR	07 SEP	27 MAY 72	
1971-074Q		5783	USSR	07 SEP	12 JUL 72	
1971-075A	COSMOS 437	5466	USSR	10 SEP	29 MAR 80	
1971-075B		5467	USSR	10 SEP	06 JAN 80	
1971-075C		8779	USSR	10 SEP	23 FEB 77	
1971-076A		5468	US	10 SEP	05 OCT 71	
1971-076B		5469	US	10 SEP	03 FEB 76	
1971-076C		6181	US	10 SEP	29 MAR 73	
1971-077A	COSMOS 438	5475	USSR	14 SEP	27 SEP 71	
1971-077B		5476	USSR	14 SEP	22 SEP 71	
1971-077C		5477	USSR	14 SEP	16 SEP 71	
1971-077D		5482	USSR	14 SEP	28 SEP 71	
1971-077E		5483	USSR	14 SEP	27 SEP 71	
1971-077F		5484	USSR	14 SEP	30 SEP 71	
1971-078A	COSMOS 439	5478	USSR	21 SEP	02 OCT 71	
1971-078B		5479	USSR	21 SEP	27 SEP 71	
1971-079A	COSMOS 440	5480	USSR	24 SEP	29 OCT 72	
1971-079B		5481	USSR	24 SEP	10 MAY 72	
1971-081A	COSMOS 441	5486	USSR	28 SEP	10 OCT 71	
1971-081B		5487	USSR	28 SEP	03 OCT 71	
1971-081C		5495	USSR	28 SEP	06 OCT 71	
1971-081D		5496	USSR	28 SEP	28 SEP 71	
1971-081E		5541	USSR	28 SEP	13 OCT 71	
1971-081F		5542	USSR	28 SEP	11 OCT 71	
1971-081G		5543	USSR	28 SEP	12 OCT 71	
1971-082B		5489	USSR	28 SEP	01 OCT 71	
1971-082D		5494	USSR	28 SEP	01 OCT 71	
1971-083A	OSO 7	5491	US	29 SEP	09 JUL 74	
1971-083B	TETR 4	5492	US	29 SEP	19 SEP 78	
1971-083C		5499	US	29 SEP	24 JAN 74	
1971-083D		5610	US	29 SEP	20 DEC 71	
1971-084A	COSMOS 442	5493	USSR	29 SEP	12 OCT 71	
1971-084B		5497	USSR	29 SEP	05 OCT 71	
1971-084C		5544	USSR	29 SEP	14 OCT 71	
1971-084D		5545	USSR	29 SEP	17 OCT 71	
1971-084E		5546	USSR	29 SEP	19 OCT 71	
1971-085A	COSMOS 443	5536	USSR	07 OCT	19 OCT 71	
1971-085B		5537	USSR	07 OCT	13 OCT 71	
1971-085C		5538	USSR	07 OCT	14 OCT 71	
1971-085D		5539	USSR	07 OCT	10 OCT 71	
1971-085E		5561	USSR	07 OCT	21 OCT 71	
1971-085F		5562	USSR	07 OCT	30 OCT 71	
1971-088A	COSMOS 452	5558	USSR	14 OCT	27 OCT 71	
1971-088B		5559	USSR	14 OCT	18 OCT 71	
1971-088C		5577	USSR	14 OCT	28 OCT 71	
1971-088D		5578	USSR	14 OCT	31 OCT 71	
1971-088E		5579	USSR	14 OCT	30 OCT 71	
1971-088F		5590	USSR	14 OCT	16 NOV 71	
1971-090A	COSMOS 453	5563	USSR	19 OCT	19 MAR 72	
1971-090B		5564	USSR	19 OCT	01 JAN 72	
1971-090C		5596	USSR	19 OCT	24 DEC 71	
1971-091A		5565	US	21 OCT	21 JUL 72	
1971-091B		5566	US	21 OCT	18 NOV 71	
1971-091C		5570	US	21 OCT	09 FEB 72	
1971-091D		5576	US	21 OCT	23 JUL 83	
1971-091E		5602	US	21 OCT	01 JUL 74	
1971-092A		5575	US	23 OCT	17 NOV 71	
1971-092B		5606	US	23 OCT	17 NOV 71	
1971-093C		5582	UK	28 OCT	16 DEC 79	
1971-094A	COSMOS 454	5585	USSR	02 NOV	16 NOV 71	
1971-094B		5586	USSR	02 NOV	06 NOV 71	
1971-094C		5603	USSR	02 NOV	22 NOV 71	
1971-094D		5604	USSR	02 NOV	25 NOV 71	
1971-094E		5605	USSR	02 NOV	22 NOV 71	
1971-094F		5607	USSR	02 NOV	21 NOV 71	
1971-096A	EXPLCRER 45	5598	US	15 NOV	10 JAN 92	
1971-096B		5973	US	15 NOV	03 MAR 87	
1971-096C		21859	US	15 NOV	27 JUN 92	
1971-097A	COSMOS 455	5608	USSR	17 NOV	09 APR 72	
1971-097B		5609	USSR	17 NOV	08 FEB 72	
1971-098A	COSMOS 456	5611	USSR	19 NOV	02 DEC 71	
1971-098B		5612	USSR	19 NOV	25 NOV 71	
1971-098C		5613	USSR	19 NOV	20 NOV 71	
1971-098D		5616	USSR	19 NOV	20 NOV 71	
1971-098E		5639	USSR	19 NOV	10 DEC 71	
1971-098F		5640	USSR	19 NOV	06 DEC 71	
1971-098G		5653	USSR	19 NOV	05 DEC 71	
1971-098H		5657	USSR	19 NOV	06 DEC 71	
1971-100A	MOLNIYA 2-1	5620	USSR	24 NOV	10 MAY 76	
1971-100B		5617	USSR	24 NOV	19 DEC 71	
1971-100C		5618	USSR	24 NOV	30 DEC 71	
1971-100D		5619	USSR	24 NOV	29 NOV 71	
1971-100E		5689	USSR	24 NOV	05 JAN 74	
1971-101A	COSMOS 458	5623	USSR	29 NOV	20 APR 72	
1971-101B		5624	USSR	29 NOV	14 FEB 72	
1971-102A	COSMOS 459	5625	USSR	29 NOV	27 DEC 71	
1971-102B		5626	USSR	29 NOV	11 DEC 71	
1971-102C		5627	USSR	29 NOV	04 DEC 71	
1971-103A	COSMOS 460	5628	USSR	30 NOV	05 MAR 80	
1971-103B		5629	USSR	30 NOV	04 MAY 80	
1971-103C		5630	USSR	30 NOV	21 NOV 72	
1971-104A	INTERCOSMOS 5	5641	USSR	02 DEC	07 APR 72	
1971-104B		5642	USSR	02 DEC	02 MAR 72	
1971-104C		5645	USSR	02 DEC	13 DEC 71	
1971-105A	COSMOS 461	5643	USSR	02 DEC	21 FEB 79	
1971-105B		5644	USSR	02 DEC	29 APR 79	
1971-105C		5647	USSR	02 DEC	03 APR 72	
1971-105D		5648	USSR	02 DEC	03 APR 72	
1971-106A	COSMOS 462	5646	USSR	03 DEC	04 APR 75	
1971-106AA		5832	USSR	03 DEC	15 MAR 72	
1971-106AB		5833	USSR	03 DEC	25 FEB 72	
1971-106AC		5834	USSR	03 DEC	19 MAR 72	
1971-106AD		5866	USSR	03 DEC	11 JUL 72	
1971-106AE		5867	USSR	03 DEC	22 MAR 72	
1971-106B		5649	USSR	03 DEC	02 FEB 72	AT
1971-106C		5650	USSR	03 DEC	05 JAN 72	
1971-106D		5651	USSR	03 DEC	09 FEB 72	
1971-106E		5652	USSR	03 DEC	24 DEC 71	
1971-106F		5654	USSR	03 DEC	10 FEB 72	
1971-106G		5655	USSR	03 DEC	31 MAR 72	
1971-106H		5656	USSR	03 DEC	24 DEC 71	
1971-106J		5658	USSR	03 DEC	19 DEC 71	
1971-106K		5659	USSR	03 DEC	18 FEB 72	
1971-106L		5660	USSR	03 DEC	20 MAR 72	
1971-106M		5664	USSR	03 DEC	11 DEC 71	
1971-106N		5665	USSR	03 DEC	11 DEC 71	
1971-106P		5666	USSR	03 DEC	09 MAR 72	
1971-106Q		5668	USSR	03 DEC	26 JAN 72	
1971-106R		5672	USSR	03 DEC	13 DEC 71	
1971-106S		5673	USSR	03 DEC	18 MAR 72	
1971-106T		5798	USSR	03 DEC	02 MAR 72	
1971-106U		5799	USSR	03 DEC	18 MAR 72	
1971-106V		5800	USSR	03 DEC	08 MAR 72	
1971-106W		5801	USSR	03 DEC	06 FEB 73	
1971-106X		5802	USSR	03 DEC	29 OCT 72	
1971-106Y		5803	USSR	03 DEC	26 FEB 72	
1971-106Z		5831	USSR	03 DEC	16 MAR 72	
1971-107A	COSMOS 463	5661	USSR	06 DEC	11 DEC 71	
1971-107B		5662	USSR	06 DEC	11 DEC 71	
1971-107C		5663	USSR	06 DEC	09 DEC 71	
1971-107D		5669	USSR	06 DEC	15 DEC 71	
1971-108A	COSMOS 464	5670	USSR	10 DEC	16 DEC 71	
1971-108B		5671	USSR	10 DEC	24 DEC 71	
1971-108C		5674	USSR	10 DEC	27 DEC 71	
1971-108D		5684	USSR	10 DEC	17 DEC 71	
1971-108E		5686	USSR	10 DEC	19 DEC 71	
1971-109A	ARIEL 4	5675	UK	11 DEC	12 DEC 78	
1971-109B		5676	US	11 DEC	09 AUG 78	
1971-109C		5677	US	11 DEC	18 JAN 73	
1971-109D		5979	US	11 DEC	20 OCT 72	
1971-112A	COSMOS 466	5687	USSR	16 DEC	27 DEC 71	
1971-112B		5688	USSR	16 DEC	21 DEC 71	
1971-112C		5725	USSR	16 DEC	03 JAN 72	
1971-112D		5726	USSR	16 DEC	31 DEC 71	
1971-113A	COSMOS 467	5704	USSR	17 DEC	18 APR 72	
1971-113B		5706	USSR	17 DEC	17 FEB 72	
1971-115A	MOLNIYA 1-19	5712	USSR	19 DEC	13 APR 77	
1971-115B		5713	USSR	19 DEC	22 SEP 77	
1971-115C		5708	USSR	19 DEC	26 JAN 72	
1971-115D		5710	USSR	19 DEC	27 JAN 72	
1971-117B		5737	USSR	25 DEC	09 FEB 72	
1971-117C		5738	USSR	25 DEC	07 JAN 72	
1971-118A	COSMOS 470	5727	USSR	27 DEC	06 JAN 72	
1971-118B		5728	USSR	27 DEC	30 DEC 71	
1971-118C		5740	USSR	27 DEC	06 JAN 72	
1971-118D		5741	USSR	27 DEC	06 JAN 72	
1971-118E		5742	USSR	27 DEC	07 JAN 72	
1971-119C		5962	USSR	27 DEC	19 FEB 78	
1971-119D		6163	USSR	27 DEC	02 OCT 73	
1971-119E		6190	USSR	27 DEC	25 OCT 77	
1971-120E		9800	USSR	29 DEC	24 APR 84	

1972 LAUNCHES

International Designation	Satellite Name	Number	Source	Launch	Decay	Note
1972-001A	COSMOS 471	5764	USSR	12 JAN	25 JAN 72	
1972-001B		5765	USSR	12 JAN	20 JAN 72	
1972-001C		5767	USSR	12 JAN	15 JAN 72	
1972-001D		5773	USSR	12 JAN	24 JAN 72	
1972-001E		5774	USSR	12 JAN	24 JAN 72	
1972-001F		5806	USSR	12 JAN	26 JAN 72	
1972-001G		5812	USSR	12 JAN	26 JAN 72	
1972-002A		5769	US	20 JAN	29 FEB 72	
1972-002B		5770	US	20 JAN	23 JAN 72	
1972-002C		5771	US	20 JAN	23 JAN 72	
1972-002D		5772	US	20 JAN	17 APR 79	
1972-004A	COSMOS 472	5804	USSR	25 JAN	18 AUG 72	
1972-004B		5805	USSR	25 JAN	06 MAY 72	
1972-005A	HEOS-A2	5814	ESRO	31 JAN	02 AUG 74	
1972-005B		5815	US	31 JAN	27 SEP 78	
1972-005C		5817	US	31 JAN	01 SEP 74	
1972-005D		5818	US	31 JAN	08 JUN 75	
1972-005E		5819	US	31 JAN	02 SEP 75	
1972-006A	COSMOS 473	5821	USSR	03 FEB	15 FEB 72	
1972-006B		5822	USSR	03 FEB	11 FEB 72	
1972-007A	LUNA 20	5835	USSR	14 FEB	25 FEB 72	AP
1972-007C		5837	USSR	14 FEB	17 FEB 72	
1972-007D		5838	USSR	14 FEB	17 FEB 72	
1972-007E		5841	USSR	14 FEB	21 FEB 72	R
1972-008A	COSMOS 474	5839	USSR	16 FEB	29 FEB 72	
1972-008B		5840	USSR	16 FEB	23 FEB 72	
1972-008C		5848	USSR	16 FEB	29 FEB 72	
1972-008D		5849	USSR	16 FEB	28 FEB 72	
1972-008E		5850	USSR	16 FEB	01 MAR 72	
1972-010C		5855	US	01 MAR	03 MAR 72	
1972-011A	COSMOS 476	5852	USSR	01 MAR	25 OCT 91	

International Designation	Satellite Name	Number	Source	Launch	Decay	Note
1972-011C		20128	USSR	01 MAR	13 JAN 90	
1972-012C		5878	US	03 MAR	18 AUG 72	
1972-013A	COSMOS 477	5862	USSR	04 MAR	16 MAR 72	
1972-013B		5863	USSR	04 MAR	10 MAR 72	
1972-013C		5864	USSR	04 MAR	05 MAR 72	
1972-013D		5865	USSR	04 MAR	05 MAR 72	
1972-013E		5882	USSR	04 MAR	23 MAR 72	
1972-013F		5883	USSR	04 MAR	17 MAR 72	
1972-014A	TD-1A	5879	ESRO	12 MAR	09 JAN 80	
1972-014B		5880	US	12 MAR	09 OCT 80	
1972-014C		5881	ESRO	12 MAR	22 MAY 72	
1972-014D		5884	ESRO	12 MAR	15 NOV 75	
1972-014E		5932	ESRO	12 MAR	27 OCT 75	
1972-015A	COSMOS 478	5885	USSR	15 MAR	28 MAR 72	
1972-015B		5886	USSR	15 MAR	21 MAR 72	
1972-015C		5887	USSR	15 MAR	17 MAR 72	
1972-015D		5909	USSR	15 MAR	28 MAR 72	
1972-015E		5910	USSR	15 MAR	30 MAR 72	
1972-015F		5915	USSR	15 MAR	30 MAR 72	
1972-016A		5888	US	17 MAR	11 APR 72	
1972-017A	COSMOS 479	5894	USSR	22 MAR	13 APR 80	
1972-017B		5895	USSR	22 MAR	12 MAY 80	
1972-017C		5901	USSR	22 MAR	26 NOV 72	
1972-017D		5902	USSR	22 MAR	01 NOV 72	
1972-017E		5911	-USSR	22 MAR	05 NOV 72	
1972-020A	COSMOS 481	5906	USSR	25 MAR	02 SEP 72	
1972-020B		5908	USSR	25 MAR	11 JUN 72	
1972-021A	VENERA 8	5912	USSR	27 MAR	22 JUL 72	N
1972-021B		5913	USSR	27 MAR	29 MAR 72	
1972-021C		5914	USSR	27 MAR	29 MAR 72	
1972-023C		5921	USSR	27 MAR	02 APR 72	
1972-023A	COSMOS 482	5919	USSR	31 MAR	05 MAY 81	
1972-023B		5920	USSR	31 MAR	01 APR 72	
1972-023D		5923	USSR	31 MAR	20 FEB 83	
1972-024A	COSMOS 483	5924	USSR	03 APR	15 APR 72	
1972-024B		5925	USSR	03 APR	11 APR 72	
1972-024C		5926	USSR	03 APR	12 APR 72	
1972-024D		5943	USSR	03 APR	19 APR 72	
1972-024E		5944	USSR	03 APR	18 APR 72	
1972-025A	MOLNIYA 1-20	5927	USSR	04 APR	30 JAN 74	
1972-025B	SRET 1	5928	FRAN	04 APR	26 FEB 74	
1972-025C		5929	USSR	04 APR	28 APR 72	
1972-025D		5930	USSR	04 APR	08 MAY 72	
1972-025E		5931	USSR	04 APR	10 APR 72	
1972-025F		5935	USSR	04 APR	10 APR 72	
1972-025G		5940	USSR	04 APR	08 MAR 74	
1972-026A	COSMOS 484	5933	USSR	06 APR	18 APR 72	
1972-026B		5934	USSR	06 APR	08 APR 72	
1972-026C		6002	USSR	06 APR	18 APR 72	
1972-027A	INTERCOSMOS 6	5936	USSR	07 APR	11 APR 72	
1972-027B		5937	USSR	07 APR	10 APR 72	
1972-028A	COSMOS 485	5938	USSR	11 APR	30 AUG 72	
1972-028B		5939	USSR	11 APR	16 JUN 72	
1972-029B		5942	USSR	14 APR	11 MAY 72	
1972-029C		5946	USSR	14 APR	17 MAY 72	
1972-029D		5949	USSR	14 APR	19 APR 72	
1972-029E		6008	USSR	14 APR	01 JUN 72	
1972-030A	COSMOS 486	5945	USSR	14 APR	27 APR 72	
1972-030B		5947	USSR	14 APR	17 APR 72	
1972-030C		5948	USSR	14 APR	22 APR 72	
1972-030D		6010	USSR	14 APR	27 APR 72	
1972-030E		6011	USSR	14 APR	26 APR 72	
1972-030F		6012	USSR	14 APR	28 APR 72	
1972-031A	APOLLO 16	6000	US	16 APR	27 APR 72	AJ
1972-031B		6001	US	16 APR	19 APR 72	G
1972-031D	APOLLO 16 SUBSATELLITE	6009	US	16 APR	29 MAY 72	G
1972-032A		6003	US	19 APR	12 MAY 72	
1972-033A	COSMOS 487	6006	USSR	21 APR	24 SEP 72	
1972-033B		6007	USSR	21 APR	03 JUL 72	
1972-033C		6013	USSR	21 APR	29 APR 72	
1972-034A	COSMOS 488	6016	USSR	05 MAY	18 MAY 72	
1972-034B		6017	USSR	05 MAY	11 MAY 72	
1972-034C		6018	USSR	05 MAY	07 MAY 72	
1972-034D		6023	USSR	05 MAY	20 MAY 72	
1972-034E		6024	USSR	05 MAY	23 MAY 72	
1972-034F		6025	USSR	05 MAY	20 MAY 72	
1972-034G		6026	USSR	05 MAY	20 MAY 72	
1972-034H		6027	USSR	05 MAY	19 MAY 72	
1972-036A	COSMOS 490	6021	USSR	17 MAY	29 MAY 72	
1972-036B		6022	USSR	17 MAY	23 MAY 72	
1972-036C		6039	USSR	17 MAY	30 MAY 72	
1972-036D		6040	USSR	17 MAY	06 JUN 72	
1972-037A	MOLNIYA 2-2	6031	USSR	19 MAY	22 MAR 77	
1972-037B		6028	USSR	19 MAY	08 JUN 72	
1972-037C		6029	USSR	19 MAY	09 JUN 72	
1972-037D		6030	USSR	19 MAY	06 JUN 72	
1972-037E		6032	USSR	19 MAY	26 MAY 72	
1972-037F		6033	USSR	19 MAY	01 JUN 72	
1972-037G		6034	USSR	19 MAY	01 JUN 77	
1972-038A	COSMOS 491	6035	USSR	25 MAY	08 JUN 72	
1972-038B		6036	USSR	25 MAY	30 MAY 72	
1972-038C		6038	USSR	25 MAY	29 MAY 72	
1972-038D		6046	USSR	25 MAY	09 JUN 72	
1972-038E		6047	USSR	25 MAY	11 JUN 72	
1972-038F		6048	USSR	25 MAY	10 JUN 72	
1972-039A		6037	US	25 MAY	04 JUN 72	
1972-040A	COSMOS 492	6049	USSR	09 JUN	22 JUN 72	
1972-040B		6050	USSR	09 JUN	18 JUN 72	
1972-040C		6055	USSR	09 JUN	24 JUN 72	
1972-040D		6057	USSR	09 JUN	27 JUN 72	
1972-040E		6064	USSR	09 JUN	27 JUN 72	
1972-053E		6164	USSR	10 JUN	29 MAR 73	
1972-042A	COSMOS 493	6053	USSR	21 JUN	03 JUL 72	
1972-042B		6054	USSR	21 JUN	25 JUN 72	
1972-042C		6056	USSR	21 JUN	23 JUN 72	
1972-043E		6162	USSR	23 JUN	23 MAR 82	
1972-044A	COSMOS 495	6060	USSR	23 JUN	06 JUL 72	
1972-044B		6062	USSR	23 JUN	29 JUN 72	
1972-044C		6088	USSR	23 JUN	06 JUL 72	
1972-044D		6089	USSR	23 JUN	08 JUL 72	
1972-044E		6093	USSR	23 JUN	07 JUL 72	
1972-045A	COSMOS 496	6066	USSR	26 JUN	02 JUL 72	
1972-045B		6067	USSR	26 JUN	02 JUL 72	
1972-045C		6083	USSR	26 JUN	04 JUL 72	
1972-046A	PROGNOZ 2	6068	USSR	29 JUN	15 DEC 82	
1972-046B		6069	USSR	29 JUN	01 AUG 72	
1972-046C		6070	USSR	29 JUN	13 AUG 72	
1972-046D		6071	USSR	29 JUN	19 JUL 72	
1972-046E		6074	USSR	29 JUN	04 JUL 72	
1972-047A	INTERCOSMOS 7	6075	USSR	30 JUN	05 OCT 72	
1972-047B		6078	USSR	30 JUN	29 SEP 72	
1972-047C		6081	USSR	30 JUN	01 SEP 72	
1972-048A	COSMOS 497	6076	USSR	30 JUN	07 NOV 73	
1972-048B		6077	USSR	30 JUN	17 MAR 73	
1972-050A	COSMOS 498	6086	USSR	05 JUL	25 NOV 72	
1972-050B		6087	USSR	05 JUL	27 SEP 72	
1972-051A	COSMOS 499	6090	USSR	06 JUL	17 JUL 72	
1972-051B		6091	USSR	06 JUL	10 JUL 72	
1972-051C		6092	USSR	06 JUL	10 JUL 72	
1972-051D		6112	USSR	06 JUL	19 JUL 72	
1972-051E		6113	USSR	06 JUL	20 JUL 72	
1972-052A		6094	US	07 JUL	13 SEP 72	
1972-052B		6095	US	07 JUL	09 JUL 72	
1972-052C		6096	US	07 JUL	06 MAY 78	
1972-053A	COSMOS 500	6097	USSR	10 JUL	29 MAR 80	
1972-053B		6098	USSR	10 JUL	27 APR 80	
1972-053C		6101	USSR	10 JUL	25 MAR 73	
1972-053D		6102	USSR	10 JUL	15 APR 73	
1972-053F		6174	USSR	10 JUL	14 APR 73	
1972-054A	COSMOS 501	6099	USSR	12 JUL	09 MAY 74	
1972-054B		6100	USSR	12 JUL	10 JUN 73	
1972-054C		6103	USSR	12 JUL	05 SEP 72	
1972-054D		6104	USSR	12 JUL	03 AUG 72	
1972-054E		6107	USSR	12 JUL	07 AUG 72	
1972-054F		6108	USSR	12 JUL	06 AUG 72	
1972-054G		6109	USSR	12 JUL	25 JUL 72	
1972-054H		6110	USSR	12 JUL	25 JUL 72	
1972-054J		6133	USSR	12 JUL	18 SEP 72	
1972-055A	COSMOS 502	6105	USSR	13 JUL	25 JUL 72	
1972-055B		6106	USSR	13 JUL	18 JUL 72	
1972-055C		6111	USSR	13 JUL	15 JUL 72	
1972-055D		6128	USSR	13 JUL	25 JUL 72	
1972-055E		6129	USSR	13 JUL	26 JUL 72	
1972-056A	COSMOS 503	6114	USSR	19 JUL	01 AUG 72	
1972-056B		6115	USSR	19 JUL	27 JUL 72	
1972-056C		6116	USSR	19 JUL	23 JUL 72	
1972-056D		6132	USSR	19 JUL	01 AUG 72	
1972-056E		6134	USSR	19 JUL	03 AUG 72	
1972-058AA		7854	US	23 JUL	17 APR 89	
1972-058AC		7856	US	23 JUL	09 MAY 79	
1972-058AE		7858	US	23 JUL	28 DEC 81	
1972-058AF		7859	US	23 JUL	09 APR 80	
1972-058AG		7860	US	23 JUL	18 JUL 79	
1972-058AH		7861	US	23 JUL	30 JUL 79	
1972-058AJ		7862	US	23 JUL	11 JUL 80	
1972-058AK		7863	US	23 JUL	02 NOV 79	
1972-058AL		7864	US	23 JUL	14 AUG 81	
1972-058AN		7866	US	23 JUL	26 AUG 90	
1972-058AQ		7868	US	23 JUL	31 AUG 91	
1972-058AR		7869	US	23 JUL	05 JAN 90	
1972-058AT		7871	US	23 JUL	07 MAY 90	
1972-058AU		7872	US	23 JUL	02 MAR 82	
1972-058AV		7873	US	23 JUL	31 DEC 78	
1972-058AX		7875	US	23 JUL	25 MAR 89	
1972-058AY		7876	US	23 JUL	05 DEC 89	
1972-058AZ		7882	US	23 JUL	13 NOV 80	
1972-058B		6127	US	23 JUL	12 FEB 77	
1972-058BA		7883	US	23 JUL	04 NOV 81	
1972-058BB		7884	US	23 JUL	23 NOV 75	
1972-058BC		7886	US	23 JUL	03 MAY 80	
1972-058BD		7887	US	23 JUL	17 MAR 80	
1972-058BE		7888	US	23 JUL	14 DEC 89	
1972-058BF		7889	US	23 JUL	10 OCT 92	
1972-058BH		7891	US	23 JUL	25 SEP 81	
1972-058BL		7894	US	23 JUL	25 APR 80	
1972-058BM		7895	US	23 JUL	24 OCT 79	
1972-058BN		7896	US	23 JUL	27 NOV 79	
1972-058BP		7897	US	23 JUL	12 APR 80	
1972-058BQ		7898	US	23 JUL	09 JUL 81	
1972-058BR		7899	US	23 JUL	23 JUL 81	
1972-058BT		7908	US	23 JUL	16 JUN 79	
1972-058BU		7909	US	23 JUL	06 DEC 79	
1972-058BV		7914	US	23 JUL	28 JUN 81	
1972-058BW		7928	US	23 JUL	09 MAY 80	
1972-058BX		7929	US	23 JUL	16 JUL 80	

International Designation	Satellite Name	Number	Source	Launch	Decay	Note
1972-058BZ		7931	US	23 JUL	30 OCT 88	
1972-058CA		7932	US	23 JUL	22 MAY 82	
1972-058CC		7934	US	23 JUL	08 DEC 91	
1972-058CD		7935	US	23 JUL	17 APR 92	
1972-058CE		7936	US	23 JUL	07 SEP 80	
1972-058CH		7942	US	23 JUL	16 APR 93	
1972-058CL		7945	US	23 JUL	03 MAR 90	
1972-058CR		7961	US	23 JUL	12 JAN 80	
1972-058CS		7966	US	23 JUL	05 DEC 81	
1972-058CU		7973	US	23 JUL	14 NOV 78	
1972-058CV		7974	US	23 JUL	04 JUN 89	
1972-058CW		7975	US	23 JUL	31 MAR 80	
1972-058CY		7977	US	23 JUL	14 JAN 79	
1972-058CZ		7978	US	23 JUL	09 MAR 80	
1972-058D		7833	US	23 JUL	29 JUN 81	
1972-058DB		7980	US	23 JUL	18 FEB 79	
1972-058DC		8001	US	23 JUL	18 JUL 78	
1972-058DD		8002	US	23 JUL	27 AUG 80	
1972-058DE		8003	US	23 JUL	26 JAN 79	
1972-058DG		8005	US	23 JUL	10 SEP 91	
1972-058DH		8082	US	23 JUL	11 JAN 80	
1972-058DJ		8083	US	23 JUL	22 MAR 91	
1972-058DL		8085	US	23 JUL	23 DEC 78	
1972-058DM		8086	US	23 JUL	12 APR 80	
1972-058DN		8087	US	23 JUL	05 MAY 77	
1972-058DP		8088	US	23 JUL	25 MAY 80	
1972-058DQ		8089	US	23 JUL	25 NOV 81	
1972-058DR		8090	US	23 JUL	13 JAN 80	
1972-058DS		8091	US	23 JUL	07 FEB 79	
1972-058DU		8093	US	23 JUL	18 JUL 78	
1972-058DV		8094	US	23 JUL	30 APR 80	
1972-058DW		8095	US	23 JUL	09 JUN 83	
1972-058DX		8096	US	23 JUL	28 JUN 76	
1972-058DY		8097	US	23 JUL	15 NOV 75	
1972-058DZ		8098	US	23 JUL	02 FEB 80	
1972-058E		7834	US	23 JUL	09 MAY 81	
1972-058EA		8099	US	23 JUL	27 APR 91	
1972-058EB		8123	US	23 JUL	13 JAN 89	
1972-058EC		8124	US	23 JUL	09 NOV 78	
1972-058ED		8125	US	23 JUL	12 OCT 75	
1972-058EE		8126	US	23 JUL	10 JUN 82	
1972-058EF		8313	US	23 JUL	25 MAR 82	
1972-058EH		8315	US	23 JUL	06 NOV 79	
1972-058EJ		8316	US	23 JUL	12 MAY 81	
1972-058EX		8317	US	23 JUL	08 MAR 81	
1972-058EL		8318	US	23 JUL	10 JUL 89	
1972-058EN		8320	US	23 JUL	15 APR 82	
1972-058EP		8321	US	23 JUL	14 APR 79	
1972-058EQ		8374	US	23 JUL	16 JUN 79	
1972-058ER		8375	US	23 JUL	27 MAY 81	
1972-058ES		8376	US	23 JUL	05 MAR 83	
1972-058ET		8377	US	23 JUL	22 NOV 82	
1972-058EU		8378	US	23 JUL	28 MAR 78	
1972-058EV		8379	US	23 JUL	09 MAR 81	
1972-058EW		8380	US	23 JUL	30 MAR 79	
1972-058EX		B381	US	23 JUL	23 SEP 79	
1972-058EY		8382	US	23 JUL	13 OCT 80	
1972-058EZ		8383	US	23 JUL	09 OCT 79	
1972-058F		7835	US	23 JUL	23 FEB 77	
1972-058FA		8384	US	23 JUL	07 DEC 79	
1972-058FB		8385	US	23 JUL	13 NOV 78	
1972-058FD		8387	US	23 JUL	16 JUN 79	
1972-058FE		8388	US	23 JUL	25 JUN 81	
1972-058FF		8389	US	23 JUL	19 DEC 80	
1972-058FG		8390	US	23 JUL	07 MAR 91	
1972-058FH		8391	US	23 JUL	30 MAR 8	
1972-058FJ		8392	US	23 JUL	04 NOV 82	
1972-058FK		8393	US	23 JUL	16 MAR 77	
1972-058FL		8408	US	23 JUL	14 APR 79	
1972-058FN		8433	US	23 JUL	03 AUG 82	
1972-058FP		8434	US	23 JUL	31 JAN 76	
1972-058FR		8496	US	23 JUL	13 APR 81	
1972-058FS		8497	US	23 JUL	16 MAR 92	
1972-058FT		8498	US	23 JUL	03 NOV 80	
1972-058FU		8499	US	23 JUL	12 NOV 79	
1972-058FV		8500	US	23 JUL	28 JUN 79	
1972-058FW		8501	US	23 JUL	21 NOV 79	
1972-058FX		8502	US	23 JUL	26 SEP 78	
1972-058FY		8503	US	23 JUL	21 MAR 78	
1972-058FZ		8504	US	23 JUL	21 MAR 81	
1972-058GA		8505	US	23 JUL	27 JAN 83	
1972-058GB		8506	US	23 JUL	23 FEB 77	
1972-058GC		8507	US	23 JUL	29 JAN 80	
1972-058GD		8508	US	23 JUL	24 AUG 80	
1972-058GE		8509	US	23 JUL	05 AUG 80	
1972-058GF		8532	US	23 JUL	26 JUL 78	
1972-058GH		8534	US	23 JUL	16 NOV 80	
1972-058GJ		8535	US	23 JUL	16 FEB 79	
1972-058GK		8536	US	23 JUL	22 JAN 78	
1972-058GL		8537	US	23 JUL	31 MAR 88	
1972-058GP		8540	US	23 JUL	02 NOV 77	
1972-058GO		8541	US	23 JUL	07 JAN 79	
1972-058GS		8543	US	23 JUL	15 MAY 80	
1972-058GU		8545	US	23 JUL	29 JUL 88	
1972-058GV		8568	US	23 JUL	19 AUG 78	
1972-058GW		8569	US	23 JUL	15 MAR 86	
1972-058GX		8570	US	23 JUL	06 OCT 82	
1972-058GY		8571	US	23 JUL	05 MAR 81	
1972-058GZ		8572	US	23 JUL	04 JUL 80	
1972-058H		7837	US	23 JUL	25 SEP 81	
1972-058HA		8573	US	23 JUL	18 NOV 79	
1972-058HB		8574	US	23 JUL	16 JUN 79	
1972-058HC		8575	US	23 JUL	02 JAN 78	
1972-058HD		8576	US	23 JUL	06 FEB 80	
1972-058HF		8578	US	23 JUL	01 MAY 80	
1972-058HG		8579	US	23 JUL	24 JAN 81	
1972-058HH		8580	US	23 JUL	28 NOV 79	
1972-058HJ		8581	US	23 JUL	08 FEB 80	
1972-058HK		9669	US	23 JUL	24 FEB 79	
1972-058HL		9670	US	23 JUL	08 MAY 82	
1972-058HM		9671	US	23 JUL	14 MAR 80	
1972-058HN		9672	US	23 JUL	02 JUL 78	
1972-058HP		9673	US	23 JUL	16 JUN 79	
1972-058HR		9675	US	23 JUL	23 MAY 79	
1972-058HS		9676	US	23 JUL	28 JUL 81	
1972-058HU		9728	US	23 JUL	10 APR 79	
1972-058HV		9729	US	23 JUL	21 APR 77	
1972-058HW		9788	US	23 JUL	13 APR 81	
1972-058HX		9970	US	23 JUL	10 OCT 80	
1972-058HY		9971	US	23 JUL	21 NOV 89	
1972-058HZ		9972	US	23 JUL	29 JUN 79	
1972-058JA		13933	US	23 JUL	24 SEP 89	
1972-058JB		13934	US	23 JUL	05 AUG 86	
1972-058JC		13935	US	23 JUL	27 MAY 85	
1972-058JD		14476	US	23 JUL	23 APR 91	
1972-058JE		14874	US	23 JUL	05 OCT 89	
1972-058JF		14875	US	23 JUL	21 FEB 92	
1972-058JG		14876	US	23 JUL	18 DEC 86	
1972-058JH		14877	US	23 JUL	06 FEB 92	
1972-058JJ		14878	US	23 JUL	16 AUG 89	
1972-058JK		15539	US	23 JUL	22 JUL 91	
1972-058L		7840	US	23 JUL	16 MAR 91	
1972-058M		7841	US	23 JUL	28 MAY 90	
1972-058P		7843	US	23 JUL	10 NOV 88	
1972-058Q		7844	US	23 JUL	27 MAR 82	
1972-058R		7845	US	23 JUL	08 JAN 88	
1972-058U		7848	US	23 JUL	20 JUL 89	
1972-059A	COSMOS 512	6130	USSR	28 JUL	09 AUG 72	
1972-059B		6131	USSR	28 JUL	01 AUG 72	
1972-060A	COSMOS 513	6135	USSR	02 AUG	15 AUG 72	
1972-060B		6136	USSR	02 AUG	09 AUG 72	
1972-060C		6137	USSR	02 AUG	06 AUG 72	
1972-060D		6143	USSR	02 AUG	20 AUG 72	
1972-060E		6144	USSR	02 AUG	24 AUG 72	
1972-060F		6147	USSR	02 AUG	18 AUG 72	
1972-061A	EXPLORER 46	6142	US	13 AUG	02 NOV 79	
1972-061B		6145	US	13 AUG	05 MAR 79	
1972-061C		6146	US	13 AUG	25 MAY 78	
1972-063A	COSMOS 515	6150	USSR	18 AUG	31 AUG 72	
1972-063B		6151	USSR	18 AUG	23 AUG 72	
1972-063C		6165	USSR	18 AUG	03 SEP 72	
1972-063D		6166	USSR	18 AUG	04 SEP 72	
1972-063E		6171	USSR	18 AUG	02 SEP 72	
1972-064A	DENPA	6152	JAPAN	19 AUG	19 MAY 80	
1972-064B		6332	JAPAN	19 AUG	28 APR 81	
1972-065C		6156	US	21 AUG	07 JAN 82	
1972-066B		6196	USSR	21 AUG	25 SEP 72	
1972-066C		6199	USSR	21 AUG	20 OCT 72	
1972-067A	COSMOS 517	6168	USSR	30 AUG	11 SEP 72	
1972-067B		6169	USSR	30 AUG	03 SEP 72	
1972-067C		6170	USSR	30 AUG	08 SEP 72	
1972-068A		6172	US	01 SEP	30 SEP 72	
1972-070A	COSMOS 518	6186	USSR	15 SEP	24 SEP 72	
1972-070B		6187	USSR	15 SEP	22 SEP 72	
1972-070C		6198	USSR	15 SEP	27 SEP 72	
1972-071A	COSMOS 519	6188	USSR	16 SEP	26 SEP 72	
1972-071B		6189	USSR	16 SEP	24 SEP 72	
1972-071C		6191	USSR	16 SEP	19 SEP 72	
1972-071D		6200	USSR	16 SEP	14 OCT 72	
1972-071E		6201	USSR	16 SEP	30 SEP 72	
1972-072B		6193	USSR	19 SEP	20 OCT 72	
1972-072C		6194	USSR	19 SEP	05 NOV 72	
1972-072D		6195	USSR	19 SEP	30 SEP 72	
1972-073B		6202	US	23 SEP	26 SEP 72	
1972-073C		6203	US	23 SEP	01 FEB 76	
1972-073D		6204	US	23 SEP	01 FEB 76	
1972-073E		6205	US	23 SEP	26 SEP 72	
1972-075A	MOLNIYA 2-3	6208	USSR	30 SEP	12 JAN 78	
1972-075B		6209	USSR	30 SEP	26 OCT 72	
1972-075C		6211	USSR	30 SEP	01 NOV 72	
1972-075D		6303	USSR	30 SEP	21 FEB 78	
1972-076E		6224	US	02 OCT	10 DEC 80	
1972-077A	COSMOS 522	6219	USSR	04 OCT	17 OCT 72	
1972-077B		6220	USSR	04 OCT	14 OCT 72	
1972-077C		6238	USSR	04 OCT	24 OCT 72	
1972-077D		6247	USSR	04 OCT	20 OCT 72	
1972-078A	COSMOS 523	6222	USSR	05 OCT	07 MAR 73	
1972-078B		6223	USSR	05 OCT	07 DEC 72	
1972-078C		6225	USSR	05 OCT	17 NOV 72	
1972-078D		6226	USSR	05 OCT	08 NOV 72	
1972-078E		6239	USSR	05 OCT	17 NOV 72	
1972-078F		6240	USSR	05 OCT	22 NOV 72	
1972-078G		6241	USSR	05 OCT	12 NOV 72	
1972-078H		6242	USSR	05 OCT	25 OCT 72	
1972-078J		6243	USSR	05 OCT	19 OCT 72	
1972-078K		6244	USSR	05 OCT	12 NOV 72	

International Designation	Satellite Name	Number	Source	Launch	Decay	Note
1972-078L		6245	USSR	05 OCT	22 OCT 72	
1972-078M		6246	USSR	05 OCT	23 OCT 72	
1972-079A		6227	US	10 OCT	08 JAN 73	
1972-079B		6228	US	10 OCT	12 OCT 72	
1972-080A	COSMOS 524	6229	USSR	11 OCT	25 MAR 73	
1972-080B		6230	USSR	11 OCT	20 DEC 72	
1972-081A	MOLNIYA 1-21	6231	USSR	14 OCT	01 NOV 77	
1972-081B		6232	USSR	14 OCT	04 NOV 72	
1972-081C		6233	USSR	14 OCT	16 NOV 72	
1972-081D		6234	USSR	14 OCT	14 NOV 72	
1972-081E		6304	USSR	14 OCT	16 MAR 75	
1972-083A	COSMOS 525	6248	USSR	18 OCT	29 OCT 72	
1972-083B		6249	USSR	18 OCT	22 OCT 72	
1972-083C		6258	USSR	18 OCT	01 NOV 72	
1972-083D		6259	USSR	18 OCT	29 OCT 72	
1972-084A	COSMOS 526	6254	USSR	25 OCT	08 APR 73	
1972-084B		6255	USSR	25 OCT	08 'AN 73	
1972-086A	COSMOS 527	6260	USSR	31 OCT	13 NOV 72	
1972-086B		6261	USSR	31 OCT	08 NOV 72	
1972-086C		6263	USSR	31 OCT	03 NOV 72	
1972-086D		6280	USSR	31 OCT	15 NOV 72	
1972-086E		6281	USSR	31 OCT	18 NOV 72	
1972-086F		6284	USSR	31 OCT	18 NOV 72	
1972-088A	COSMOS 536	6272	USSR	03 NOV	20 JUL 80	
1972-088B		6273	USSR	03 NOV	19 JUL 80	
1972-088C		6274	USSR	03 NOV	19 DEC 73	
1972-088D		6700	USSR	03 NOV	28 FEB 83	
1972-090B		6279	US	10 NOV	07 MAY 73	
1972-091A	EXPLORER 48	6282	US	15 NOV	20 AUG 80	
1972-091B		6800	US	15 NOV	01 MAY 79	
1972-092A	ESRO IV	6285	ESRO	22 NOV	15 APR 74	
1972-092B		6286	US	22 NOV	09 JUL 73	
1972-093A	COSMOS 537	6287	USSR	25 NOV	07 DEC 72	
1972-093B		6288	USSR	25 NOV	02 DEC 72	
1972-094A	INTERCOSMOS 8	6291	USSR	30 NOV	02 MAR 73	
1972-094B		6292	USSR	30 NOV	12 JAN 73	
1972-094C		6293	USSR	30 NOV	02 JAN 73	
1972-095A	MOLNIYA 1-22	6294	USSR	02 DEC	11 FEB 76	
1972-095B		6295	USSR	02 DEC	10 JAN 73	
1972-095C		6296	USSR	02 DEC	02 FEB 73	
1972-095D		6297	USSR	02 DEC	12 DEC 72	
1972-095E		6298	USSR	02 DEC	11 DEC 72	
1972-095F		6322	USSR	02 DEC	25 FEB 75	
1972-096A	APOLLO 17	6300	US	07 DEC	19 DEC 72	AJ
1972-096B		6301	US	07 DEC	10 DEC 72	G
1972-096C	LUNAR MODULE	6307	US	07 DEC	15 DEC 72	AM
1972-098A	MOLNIYA 2-4	6308	USSR	12 DEC	22 JAN 75	
1972-098B		6309	USSR	12 DEC	13 JAN 73	
1972-098C		6310	USSR	12 DEC	19 JAN 73	
1972-098D		6341	USSR	12 DEC	07 OCT 74	
1972-099A	COSMOS 538	6311	USSR	14 DEC	27 DEC 72	
1972-099B		6312	USSR	14 DEC	22 DEC 72	
1972-099C		6314	USSR	14 DEC	18 DEC 72	
1972-099D		6325	USSR	14 DEC	04 JAN 73	
1972-099E		6331	USSR	14 DEC	04 JAN 73	
1972-100A	AEROS	6315	FRG	16 DEC	22 AUG 73	
1972-100B		6316	US	16 DEC	25 FEB 73	
1972-103A		6321	US	21 DEC	23 JAN 73	
1972-105A	COSMOS 541	6326	USSR	27 DEC	08 JAN 73	
1972-105B		6327	USSR	27 DEC	12 JAN 73	
1972-105C		6336	USSR	27 DEC	11 JAN 73	
1972-105D		6337	USSR	27 DEC	09 JAN 73	
1972-105E		6338	USSR	27 DEC	12 JAN 73	
1972-105F		6342	USSR	27 DEC	15 JAN 73	
1972-106A	COSMOS 542	6328	USSR	28 DEC	09 OCT 83	
1972-106B		6329	USSR	28 DEC	23 JUN 88	
1973 LAUNCHES						
1973-001A	LUNA 21	6333	USSR	08 JAN	15 JAN 73	R
1973-001B		6334	USSR	08 JAN	12 JAN 73	
1973-001C		6335	USSR	08 JAN	13 JAN 73	
1973-002A	COSMOS 543	6339	USSR	11 JAN	24 JAN 73	
1973-002B		6340	USSR	11 JAN	22 JAN 73	
1973-002C		6346	USSR	11 JAN	28 JAN 73	
1973-002D		6347	USSR	11 JAN	25 JAN 73	
1973-003A	COSMOS 544	6343	USSR	20 JAN	15 JUN 80	
1973-003B		6344	USSR	20 JAN	18 MAY 80	
1973-003C		6345	USSR	20 JAN	14 OCT 75	
1973-003D		6352	USSR	20 JAN	21 MAR 75	
1973-004A	COSMOS 545	6348	USSR	24 JAN	31 JUL 73	
1973-004B		6349	USSR	24 JAN	21 APR 73	
1973-005B		6351	USSR	26 JAN	02 MAR 88	
1973-006A	COSMOS 547	6353	USSR	01 FEB	13 FEB 73	
1973-006B		6354	USSR	01 FEB	09 FEB 73	
1973-006C		6355	USSR	01 FEB	02 FEB 73	
1973-007A	MOLNIYA 1-23	6356	USSR	03 FEB	23 OCT 77	
1973-007B		6357	USSR	03 FEB	13 MAR 73	
1973-007C		6358	USSR	03 FEB	18 MAR 73	
1973-007D		6363	USSR	03 FEB	10 MAR 73	
1973-007E		6368	USSR	03 FEB	04 AUG 78	
1973-008A	COSMOS 548	6359	USSR	08 FEB	21 FEB 73	
1973-008B		6360	USSR	08 FEB	16 FEB 73	
1973-008C		6361	USSR	08 FEB	09 FEB 73	
1973-008D		6367	USSR	08 FEB	25 FEB 73	
1973-008E		6369	USSR	08 FEB	23 FEB 73	
1973-008F		6370	USSR	08 FEB	26 FEB 73	
1973-008G		6371	USSR	08 FEB	23 FEB 73	
1973-009B		6365	USSR	15 FEB	22 MAR 73	
1973-009C		6366	USSR	15 FEB	23 MAR 73	
1973-010A	COSMOS 549	6373	USSR	28 FEB	29 JUN 80	
1973-010B		6374	USSR	28 FEB	17 SEP 80	
1973-010C		6375	USSR	28 FEB	24 JUN 74	
1973-011A	COSMOS 550	6376	USSR	01 MAR	11 MAR 73	
1973-011B		6377	USSR	01 MAR	09 MAR 73	
1973-011C		6384	USSR	01 MAR	18 MAR 73	
1973-011D		6385	USSR	01 MAR	12 MAR 73	
1973-011E		6386	USSR	01 MAR	12 MAR 73	
1973-011F		6387	USSR	01 MAR	12 MAR 73	
1973-012A	COSMOS 551	6378	USSR	06 MAR	20 MAR 73	
1973-012B		6379	USSR	06 MAR	13 MAR 73	
1973-012C		6381	USSR	06 MAR	07 MAR 73	
1973-012D		6389	USSR	06 MAR	22 MAR 73	
1973-012E		6390	USSR	06 MAR	19 MAR 73	
1973-013B		6388	US	06 MAR	04 JUN 74	
1973-014A		6382	US	09 MAR	19 MAY 73	
1973-014B		6383	US	09 MAR	11 MAR 73	
1973-016A	COSMOS 552	6394	USSR	22 MAR	03 APR 73	
1973-016B		6395	USSR	22 MAR	30 MAR 73	
1973-016C		6397	USSR	22 MAR	09 APR 73	
1973-017A	SALYUT 2	6398	USSR	03 APR	28 MAY 73	
1973-017AA		6430	USSR	03 APR	12 MAY 73	
1973-017AB		6431	USSR	03 APR	17 APR 73	
1973-017B		6399	USSR	03 APR	06 APR 73	AU
1973-017C		6400	USSR	03 APR	10 APR 73	
1973-017D		6401	USSR	03 APR	09 APR 73	
1973-017E		6402	USSR	03 APR	12 APR 73	
1973-017F		6403	USSR	03 APR	24 APR 73	
1973-017G		6404	USSR	03 APR	30 APR 73	
1973-017H		6405	USSR	03 APR	06 APR 73	
1973-017J		6406	USSR	03 APR	26 APR 73	
1973-017K		6407	USSR	03 APR	22 APR 73	
1973-017L		6408	USSR	03 APR	08 APR 73	
1973-017M		6409	USSR	03 APR	15 APR 73	
1973-017N		6410	USSR	03 APR	06 APR 73	
1973-017P		6411	USSR	03 APR	08 APR 73	
1973-017Q		6412	USSR	03 APR	08 APR 73	
1973-017R		6413	USSR	03 APR	08 APR 73	
1973-017S		6414	USSR	03 APR	06 APR 73	
1973-017T		6415	USSR	03 APR	12 APR 73	
1973-017U		6416	USSR	03 APR	08 APR 73	
1973-017V		6417	USSR	03 APR	09 APR 73	
1973-017W		6422	USSR	03 APR	09 APR 73	
1973-017X		6423	USSR	03 APR	10 APR 73	
1973-017Y		6424	USSR	03 APR	10 APR 73	
1973-017Z		6429	USSR	03 APR	13 MAY 73	
1973-018A	MOLNIYA 2-5	6418	USSR	05 APR	06 JAN 79	
1973-018B		6419	USSR	05 APR	24 APR 73	
1973-018C		6420	USSR	05 APR	30 APR 73	
1973-018D		6439	USSR	05 APR	05 DEC 79	
1973-019C		6426	US	06 APR	04 NOV 74	
1973-020A	COSMOS 553	6427	USSR	12 APR	11 NOV 73	
1973-020B		6428	USSR	12 APR	07 AUG 73	
1973-021A	COSMOS 554	6432	USSR	19 APR	27 MAY 73	
1973-021AA		6467	USSR	19 APR	28 MAY 73	
1973-021AB		6468	USSR	19 APR	06 JUL 73	
1973-021AC		6469	USSR	19 APR	29 JAN 74	
1973-021AD		6470	USSR	19 APR	03 JUL 73	
1973-021AE		6471	USSR	19 APR	29 NOV 73	
1973-021AF		6472	USSR	19 APR	22 MAY 73	
1973-021AG		6473	USSR	19 APR	30 MAY 73	
1973-021AH		6474	USSR	19 APR	17 MAY 73	
1973-021AJ		6475	USSR	19 APR	01 JUN 73	
1973-021AK		6476	USSR	19 APR	19 MAY 73	
1973-021AL		6477	USSR	19 APR	16 MAY 73	
1973-021AM		6478	USSR	19 APR	13 MAY 73	
1973-021AN		6479	USSR	19 APR	27 MAY 73	
1973-021AP		6480	USSR	19 APR	15 MAY 73	
1973-021AQ		6481	USSR	19 APR	18 MAY 73	
1973-021AR		6482	USSR	19 APR	19 MAY 73	
1973-021AS		6483	USSR	19 APR	15 MAY 73	
1973-021AT		6484	USSR	19 APR	08 JUN 73	
1973-021AU		6485	USSR	19 APR	19 MAY 73	
1973-021AV		6486	USSR	19 APR	19 MAY 73	
1973-021AW		6487	USSR	19 APR	18 MAY 73	
1973-021AX		6488	USSR	19 APR	21 MAY 73	
1973-021AY		6489	USSR	19 APR	13 JUN 73	
1973-021AZ		6490	USSR	19 APR	14 AUG 73	
1973-021B		6434	USSR	19 APR	26 APR 73	AV
1973-021BA		6491	USSR	19 APR	14 JUL 73	
1973-021BB		6492	USSR	19 APR	30 MAY 73	
1973-021BC		6493	USSR	19 APR	07 JUL 73	
1973-021BD		6494	USSR	19 APR	27 MAY 73	
1973-021BE		6495	USSR	19 APR	25 MAY 73	
1973-021BF		6496	USSR	19 APR	09 JUN 73	
1973-021BG		6497	USSR	19 APR	27 MAY 73	
1973-021BH		6500	USSR	19 APR	27 MAY 73	
1973-021BJ		6501	USSR	19 APR	25 JUN 73	
1973-021BK		6502	USSR	19 APR	02 JAN 74	
1973-021BL		6503	USSR	19 APR	01 JUL 73	
1973-021BM		6504	USSR	19 APR	18 MAY 73	
1973-021BN		6505	USSR	19 APR	21 MAY 73	
1973-021BP		6506	USSR	19 APR	17 MAY 73	
1973-021BQ		6507	USSR	19 APR	15 MAY 73	
1973-021BR		6508	USSR	19 APR	19 MAY 73	
1973-021BS		6509	USSR	19 APR	18 JUN 73	

International Designation	Satellite Name	Number	Source	Launch	Decay	Note
1973-021BT		6510	USSR	19 APR	06 JUN 73	
1973-021BU		6511	USSR	19 APR	12 AUG 73	
1973-021BV		6512	USSR	19 APR	26 MAY 73	
1973-021BW		6513	USSR	19 APR	26 MAY 73	
1973-021BX		6514	USSR	19 APR	18 MAY 73	
1973-021BY		6515	USSR	19 APR	28 MAY 73	
1973-021BZ		6516	USSR	19 APR	14 JUN 73	
1973-021C		6436	USSR	19 APR	25 APR 73	
1973-021CA		6517	USSR	19 APR	26 MAY 73	
1973-021CB		6518	USSR	19 APR	23 MAY 73	
1973-021CC		6519	USSR	19 APR	03 JUL 73	
1973-021CD		6520	USSR	19 APR	25 MAY 73	
1973-021CE		6521	USSR	19 APR	06 JUN 73	
1973-021CF		6522	USSR	19 APR	25 MAY 73	
1973-021CG		6523	USSR	19 APR	22 MAY 73	
1973-021CH		6524	USSR	19 APR	09 JUN 73	
1973-021CJ		6525	USSR	19 APR	16 SEP 73	
1973-021CK		6526	USSR	19 APR	26 MAY 73	
1973-021CL		6527	USSR	19 APR	26 MAY 73	
1973-021CM		6528	USSR	19 APR	26 MAY 73	
1973-021CN		6529	USSR	19 APR	25 MAY 73	
1973-021CP		6530	USSR	19 APR	02 JUN 73	
1973-021CQ		6531	USSR	19 APR	25 JUN 73	
1973-021CR		6532	USSR	19 APR	17 MAY 73	
1973-021CS		6533	USSR	19 APR	22 MAY 73	
1973-021CT		6534	USSR	19 APR	26 MAY 73	
1973-021CU		6535	USSR	19 APR	22 MAY 73	
1973-021CV		6536	USSR	19 APR	30 JUN 73	
1973-021CW		6537	USSR	19 APR	22 MAY 73	
1973-021CX		6538	USSR	19 APR	25 MAY 73	
1973-021CY		6539	USSR	19 APR	24 MAY 73	
1973-021CZ		6540	USSR	19 APR	31 DEC 73	
1973-021D		6442	USSR	19 APR	30 APR 73	
1973-021DA		6541	USSR	19 APR	27 MAY 73	
1973-021DB		6542	USSR	19 APR	07 MAY 73	
1973-021DC		6543	USSR	19 APR	07 MAY 73	
1973-021DD		6544	USSR	19 APR	07 MAY 73	
1973-021DE		6545	USSR	19 APR	07 MAY 73	
1973-021DF		6546	USSR	19 APR	07 MAY 73	
1973-021DG		6547	USSR	19 APR	07 MAY 73	
1973-021DH		6548	USSR	19 APR	07 MAY 73	
1973-021DJ		6549	USSR	19 APR	07 MAY 73	
1973-021DK		6550	USSR	19 APR	07 MAY 73	
1973-021DL		6551	USSR	19 APR	07 MAY 73	
1973-021DM		6552	USSR	19 APR	07 MAY 73	
1973-021DN		6553	USSR	19 APR	07 MAY 73	
1973-021DP		6554	USSR	19 APR	07 MAY 73	
1973-021DQ		6555	USSR	19 APR	07 MAY 73	
1973-021DR		6556	USSR	19 APR	07 MAY 73	
1973-021DS		6557	USSR	19 APR	07 MAY 73	
1973-021DT		6558	USSR	19 APR	07 MAY 73	
1973-021DU		6559	USSR	19 APR	07 MAY 73	
1973-021DV		6560	USSR	19 APR	07 MAY 73	
1973-021DW		6561	USSR	19 APR	07 MAY 73	
1973-021DX		6562	USSR	19 APR	07 MAY 73	
1973-021DY		6563	USSR	19 APR	07 MAY 73	
1973-021DZ		6564	USSR	19 APR	07 MAY 73	
1973-021E		6443	USSR	19 APR	01 MAY 73	
1973-021EA		6565	USSR	19 APR	07 MAY 73	
1973-021EB		6566	USSR	19 APR	07 MAY 73	
1973-021EC		6567	USSR	19 APR	07 MAY 73	
1973-021ED		6568	USSR	19 APR	07 MAY 73	
1973-021EE		6569	USSR	19 APR	07 MAY 73	
1973-021EF		6570	USSR	19 APR	07 MAY 73	
1973-021EG		6571	USSR	19 APR	07 MAY 73	
1973-021EH		6572	USSR	19 APR	07 MAY 73	
1973-021EJ		6573	USSR	19 APR	07 MAY 73	
1973-021EK		6574	USSR	19 APR	07 MAY 73	
1973-021EL		6575	USSR	19 APR	07 MAY 73	
1973-021EM		6576	USSR	19 APR	07 MAY 73	
1973-021EN		6577	USSR	19 APR	07 MAY 73	
1973-021EP		6578	USSR	19 APR	07 MAY 73	
1973-021EQ		6579	USSR	19 APR	07 MAY 73	
1973-021ER		6580	USSR	19 APR	07 MAY 73	
1973-021ES		6581	USSR	19 APR	07 MAY 73	
1973-021ET		6582	USSR	19 APR	07 MAY 73	
1973-021EU		6583	USSR	19 APR	07 MAY 73	
1973-021EV		6584	USSR	19 APR	07 MAY 73	
1973-021EW		6585	USSR	19 APR	07 MAY 73	
1973-021EX		6586	USSR	19 APR	07 MAY 73	
1973-021EY		6587	USSR	19 APR	07 MAY 73	
1973-021EZ		6588	USSR	19 APR	07 MAY 73	
1973-021F		6448	USSR	19 APR	24 JUN 73	
1973-021FA		6589	USSR	19 APR	07 MAY 73	
1973-021FB		6590	USSR	19 APR	07 MAY 73	
1973-021FC		6591	USSR	19 APR	07 MAY 73	
1973-021FD		6592	USSR	19 APR	07 MAY 73	
1973-021FE		6593	USSR	19 APR	07 MAY 73	
1973-021FF		6594	USSR	19 APR	07 MAY 73	
1973-021FG		6595	USSR	19 APR	07 MAY 73	
1973-021FH		6596	USSR	19 APR	07 MAY 73	
1973-021FJ		6597	USSR	19 APR	07 MAY 73	
1973-021FK		6598	USSR	19 APR	07 MAY 73	
1973-021FL		6599	USSR	19 APR	07 MAY 73	
1973-021FM		6600	USSR	19 APR	07 MAY 73	
1973-021FN		6601	USSR	19 APR	07 MAY 73	
1973-021FP		6602	USSR	19 APR	07 MAY 73	
1973-021FQ		6603	USSR	19 APR	07 MAY 73	
1973-021FR		6604	USSR	19 APR	07 MAY 73	
1973-021FS		6605	USSR	19 APR	07 MAY 73	
1973-021FT		6606	USSR	19 APR	07 MAY 73	
1973-021FU		6607	USSR	19 APR	07 MAY 73	
1973-021FV		6608	USSR	19 APR	07 MAY 73	
1973-021FW		6609	USSR	19 APR	07 MAY 73	
1973-021FX		6610	USSR	19 APR	07 MAY 73	
1973-021FY		6611	USSR	19 APR	07 MAY 73	
1973-021FZ		6612	USSR	19 APR	07 MAY 73	
1973-021G		6449	USSR	19 APR	21 MAY 73	
1973-021GA		6613	USSR	19 APR	07 MAY 73	
1973-021GB		6614	USSR	19 APR	07 MAY 73	
1973-021GC		6615	USSR	19 APR	07 MAY 73	
1973-021GD		6616	USSR	19 APR	07 MAY 73	
1973-021GE		6617	USSR	19 APR	07 MAY 73	
1973-021GF		6618	USSR	19 APR	07 MAY 73	
1973-021GG		6619	USSR	19 APR	07 MAY 73	
1973-021GH		6620	USSR	19 APR	07 MAY 73	
1973-021GJ		6621	USSR	19 APR	07 MAY 73	
1973-021GK		6622	USSR	19 APR	07 MAY 73	
1973-021GL		6623	USSR	19 APR	07 MAY 73	
1973-021GM		6624	USSR	19 APR	07 MAY 73	
1973-021GN		6625	USSR	19 APR	07 MAY 73	
1973-021GP		6626	USSR	19 APR	07 MAY 73	
1973-021GQ		6627	USSR	19 APR	07 MAY 73	
1973-021GR		6628	USSR	19 APR	07 MAY 73	
1973-021GS		6629	USSR	19 APR	07 MAY 73	
1973-021GT		6669	USSR	19 APR	23 SEP 73	
1973-021GU		6670	USSR	19 APR	26 JUN 73	
1973-021GV		6671	USSR	19 APR	23 AUG 73	
1973-021GW		6672	USSR	19 APR	25 JUN 73	
1973-021GX		6673	USSR	19 APR	19 JUN 73	
1973-021GY		6692	USSR	19 APR	15 JUN 73	
1973-021GZ		6693	USSR	19 APR	24 JUN 73	
1973-021H		6450	USSR	19 APR	16 MAY 73	
1973-021HA		6737	USSR	19 APR	02 SEP 73	
1973-021HB		6738	USSR	19 APR	14 OCT 73	
1973-021HC		6739	USSR	19 APR	02 SEP 73	
1973-021HD		6740	USSR	19 APR	20 SEP 73	
1973-021HE		6785	USSR	19 APR	18 NOV 73	
1973-021J		6451	USSR	19 APR	24 MAY 73	
1973-021K		6452	USSR	19 APR	16 MAY 73	
1973-021L		6453	USSR	19 APR	20 MAY 73	
1973-021M		6454	USSR	19 APR	27 MAY 73	
1973-021N		6455	USSR	19 APR	12 MAY 73	
1973-021P		6456	USSR	19 APR	17 MAY 73	
1973-021Q		6457	USSR	19 APR	16 MAY 73	
1973-021R		6458	USSR	19 APR	09 JUN 73	
1973-021S		6459	USSR	19 APR	16 MAY 73	
1973-021T		6460	USSR	19 APR	12 MAY 73	
1973-021U		6461	USSR	19 APR	20 MAY 73	
1973-021V		6462	USSR	19 APR	12 MAY 73	
1973-021W		6463	USSR	19 APR	21 MAY 73	
1973-021X		6464	USSR	19 APR	18 MAY 73	
1973-021Y		6465	USSR	19 APR	27 MAY 73	
1973-021Z		6466	USSR	19 APR	04 JUN 73	
1973-022A	INTERCOSMOS 9 COPERNICUS 500	6433	USSR	19 APR	15 OCT 73	
1973-022B		6435	USSR	19 APR	17 OCT 73	
1973-023B		6438	US	20 APR	18 OCT 75	
1973-024A	COSMOS 555	6440	USSR	25 APR	07 MAY 73	
1973-024B		6441	USSR	25 APR	28 APR 73	
1973-024C		6444	USSR	25 APR	06 MAY 73	
1973-024D		6445	USSR	25 APR	09 MAY 73	
1973-025A	COSMOS 556	6446	USSR	05 MAY	14 MAY 73	
1973-025B		6447	USSR	05 MAY	08 MAY 73	
1973-025C		6630	USSR	05 MAY	24 MAY 73	
1973-025D		6631	USSR	05 MAY	15 MAY 73	
1973-025E		6632	USSR	05 MAY	17 MAY 73	
1973-026A	COSMOS 557	6498	USSR	11 MAY	22 MAY 73	
1973-026B		6499	USSR	11 MAY	17 MAY 73	
1973-027A	SKYLAB 1	6633	US	14 MAY	11 JUL 79	
1973-027B		6634	US	14 MAY	11 JAN 75	AW
1973-027C		6635	US	14 MAY	10 SEP 73	
1973-027D		6636	US	14 MAY	25 JUN 73	
1973-027E		6637	US	14 MAY	05 AUG 75	
1973-027F		6638	US	14 MAY	03 APR 75	
1973-027G		6639	US	14 MAY	04 OCT 73	
1973-027H		6641	US	14 MAY	01 NOV 76	
1973-027J		6642	US	14 MAY	19 JAN 74	
1973-027K		6643	US	14 MAY	10 FEB 75	
1973-027L		6644	US	14 MAY	01 NOV 73	
1973-027M		6651	US	14 MAY	12 JUN 75	
1973-027N		6704	US	14 MAY	13 APR 74	
1973-027P		6705	US	14 MAY	24 SEP 73	
1973-027Q		6729	US	14 MAY	16 OCT 73	
1973-027R		6730	US	14 MAY	14 SEP 73	
1973-027S		6733	US	14 MAY	23 OCT 74	
1973-027T		6735	US	14 MAY	20 AUG 75	
1973-027U		6734	US	14 MAY	15 FEB 74	
1973-027V		6774	US	14 MAY	20 SEP 73	
1973-027W		6780	US	14 MAY	12 DEC 74	
1973-027X		6781	US	14 MAY	08 NOV 74	
1973-027Y		6782	US	14 MAY	17 JUL 74	
1973-027Z		6786	US	14 MAY	16 JAN 75	
1973-028A		6640	US	16 MAY	13 JUN 73	
1973-029A	COSMOS 558	6645	USSR	17 MAY	22 DEC 73	
1973-029B		6646	USSR	17 MAY	10 SEP 73	

International Designation	Satellite Name	Number	Source	Launch	Decay	Note
1973-030A	COSMOS 559	6647	USSR	18 MAY	23 MAY 73	
1973-030B		6648	USSR	18 MAY	28 MAY 73	
1973-030C		6649	USSR	18 MAY	29 MAY 73	
1973-030D		6650	USSR	18 MAY	12 JUN 73	
1973-031A	COSMOS 560	6652	USSR	23 MAY	05 JUN 73	
1973-031B		6653	USSR	23 MAY	01 JUN 73	
1973-031C		6654	USSR	23 MAY	01 JUN 73	
1973-031D		6663	USSR	23 MAY	12 JUN 73	
1973-031E		6664	USSR	23 MAY	06 JUN 73	
1973-032A	SKYLAB 2	6655	US	25 MAY	22 JUN 73	F
1973-032B		6656	US	25 MAY	25 MAY 73	
1973-033A	COSMOS 561	6657	USSR	25 MAY	06 JUN 73	
1973-033B		6658	USSR	25 MAY	02 JUN 73	
1973-033C		6661	USSR	25 MAY	07 JUN 73	
1973-033D		6662	USSR	25 MAY	20 JUN 73	
1973-035A	COSMOS 562	6665	USSR	05 JUN	07 JAN 74	
1973-035B		6666	USSR	05 JUN	29 SEP 73	
1973-036A	COSMOS 563	6667	USSR	06 JUN	18 JUN 73	
1973-036B		6668	USSR	06 JUN	13 JUN 73	
1973-036C		6674	USSR	06 JUN	15 JUN 73	
1973-036D		6703	USSR	06 JUN	25 JUN 73	
1973-036E		6706	USSR	06 JUN	23 JUN 73	
1973-038A	COSMOS 572	6684	USSR	10 JUN	23 JUN 73	
1973-038B		6685	USSR	10 JUN	16 JUN 73	
1973-038C		6711	USSR	10 JUN	26 JUN 73	
1973-038D		6712	USSR	10 JUN	23 JUN 73	
1973-039B		6687	US	10 JUN	01 JUL 78	
1973-039C		6688	US	10 JUN	29 JAN 74	
1973-039E		6690	US	10 JUN	19 SEP 73	
1973-039H		6773	US	10 JUN	27 SEP 73	
1973-041A	COSMOS 573	6694	USSR	15 JUN	17 JUN 73	
1973-041B		6695	USSR	15 JUN	21 JUN 73	
1973-041C		6702	USSR	15 JUN	21 JUN 73	
1973-043A	COSMOS 575	6709	USSR	21 JUN	03 JUL 73	
1973-043B		6710	USSR	21 JUN	27 JUN 73	
1973-044A	COSMOS 576	6713	USSR	27 JUN	09 JUL 73	
1973-044B		6714	USSR	27 JUN	07 JUL 73	
1973-044C		6715	USSR	27 JUN	30 JUN 73	
1973-044D		6716	USSR	27 JUN	30 JUN 73	
1973-044E		6717	USSR	27 JUN	29 JUN 73	
1973-044F		6719	USSR	27 JUN	11 JUL 73	
1973-044G		6720	USSR	27 JUN	11 JUL 73	
1973-044H		6721	USSR	27 JUN	14 JUL 73	
1973-045A	MOLNIYA 2-6	6722	USSR	11 JUL	05 AUG 78	
1973-045B		6723	USSR	11 JUL	22 AUG 73	
1973-045C		6724	USSR	11 JUL	03 AUG 73	
1973-045D		6741	USSR	11 JUL	19 SEP 78	
1973-046A		6727	US	13 JUL	12 OCT 73	
1973-046B		6728	US	13 JUL	15 JUL 73	
1973-047B		6743	USSR	21 JUL	23 JUL 73	
1973-047C		6744	USSR	21 JUL	26 JUL 73	
1973-048A	COSMOS 577	6745	USSR	25 JUL	07 AUG 73	
1973-048B		6753	USSR	25 JUL	01 AUG 73	
1973-048C		6770	USSR	25 JUL	07 AUG 73	
1973-048D		6772	USSR	25 JUL	12 AUG 73	
1973-048E		6775	USSR	25 JUL	09 AUG 73	
1973-049B		6755	USSR	25 JUL	27 JUL 73	
1973-049C		6756	USSR	25 JUL	27 JUL 73	
1973-050A	SKYLAB 3	6757	US	28 JUL	25 SEP 73	F
1973-050B		6758	US	28 JUL	28 JUL 73	
1973-051A	COSMOS 578	6759	USSR	01 AUG	13 AUG 73	
1973-051B		6760	USSR	01 AUG	09 AUG 73	
1973-051C		6761	USSR	01 AUG	05 AUG 73	
1973-051D		6762	USSR	01 AUG	02 AUG 73	
1973-051E		6763	USSR	01 AUG	02 AUG 73	
1973-052B		6769	USSR	05 AUG	07 AUG 73	
1973-052C		6771	USSR	05 AUG	07 AUG 73	
1973-052D	CAPSULE	7223	USSR	05 AUG	12 MAR 74	AR
1973-053B		6777	USSR	09 AUG	11 AUG 73	
1973-053C		6778	USSR	09 AUG	11 AUG 73	
1973-055A	COSMOS 579	6789	USSR	21 AUG	03 SEP 73	
1973-055B		6790	USSR	21 AUG	28 AUG 73	
1973-055C		6795	USSR	21 AUG	25 AUG 73	
1973-055D		6813	USSR	21 AUG	05 SEP 73	
1973-055E		6814	USSR	21 AUG	04 SEP 73	
1973-057A	COSMOS 580	6793	USSR	22 AUG	01 APR 74	
1973-057B		6794	USSR	22 AUG	30 NOV 73	
1973-059A	COSMOS 581	6798	USSR	24 AUG	06 SEP 73	
1973-059B		6799	USSR	24 AUG	31 AUG 73	
1973-059C		6801	USSR	24 AUG	25 AUG 73	
1973-059D		6816	USSR	24 AUG	06 SEP 73	
1973-059E		6817	USSR	24 AUG	09 SEP 73	
1973-060A	COSMOS 582	6802	USSR	28 AUG	05 SEP 80	
1973-060B		6803	USSR	28 AUG	01 JAN 81	
1973-060C		6804	USSR	28 AUG	20 JUN 82	
1973-060D		6998	USSR	28 AUG	05 MAR 75	
1973-060E		6999	USSR	28 AUG	07 NOV 74	
1973-061A	MOLNIYA 1-24	6805	USSR	30 AUG	05 DEC 79	
1973-061B		6806	USSR	30 AUG	01 OCT 73	
1973-061C		6807	USSR	30 AUG	08 OCT 73	
1973-061D		6808	USSR	30 AUG	06 SEP 73	
1973-061E		6812	USSR	30 AUG	07 SEP 73	
1973-061F		6815	USSR	30 AUG	05 DEC 79	
1973-062A	COSMOS 583	6809	USSR	30 AUG	12 SEP 73	
1973-062B		6810	USSR	30 AUG	06 SEP 73	
1973-062C		6811	USSR	30 AUG	03 SEP 73	
1973-063A	COSMOS 584	6818	USSR	06 SEP	20 SEP 73	
1973-063B		6819	USSR	06 SEP	17 SEP 73	
1973-063C		6820	USSR	06 SEP	09 SEP 73	
1973-063D		6821	USSR	06 SEP	07 SEP 73	
1973-063E		6830	USSR	06 SEP	28 SEP 73	
1973-063F		6831	USSR	06 SEP	25 SEP 73	
1973-063G		6835	USSR	06 SEP	25 SEP 73	
1973-064C		6827	USSR	08 SEP	12 DEC 73	
1973-066A	COSMOS 587	6832	USSR	21 SEP	04 OCT 73	
1973-066B		6833	USSR	21 SEP	28 SEP 73	
1973-066C		6834	USSR	21 SEP	23 SEP 73	
1973-066D		6854	USSR	21 SEP	08 OCT 73	
1973-066E		6855	USSR	21 SEP	04 OCT 73	
1973-067A	SOYUZ 12	6836	USSR	27 SEP	29 SEP 73	E
1973-067B		6838	USSR	27 SEP	29 SEP 73	
1973-067C		6839	USSR	27 SEP	21 JAN 74	
1973-067D		6840	USSR	27 SEP	03 NOV 73	
1973-068A		6837	US	27 SEP	29 OCT 73	
1973-070A	COSMOS 596	6856	USSR	03 OCT	09 OCT 73	
1973-070B		6857	USSR	03 OCT	09 OCT 73	
1973-070C		6861	USSR	03 OCT	17 OCT 73	
1973-071A	COSMOS 597	6858	USSR	06 OCT	12 OCT 73	
1973-071B		6859	USSR	06 OCT	14 OCT 73	
1973-071C		6860	USSR	06 OCT	08 OCT 73	
1973-071D		6865	USSR	06 OCT	16 OCT 73	
1973-071E		6866	USSR	06 OCT	18 OCT 73	
1973-072A	COSMOS 598	6862	USSR	10 OCT	16 OCT 73	
1973-072B		6863	USSR	10 OCT	21 OCT 73	
1973-072C		6864	USSR	10 OCT	11 OCT 73	
1973-072D		6871	USSR	10 OCT	20 OCT 73	
1973-072E		6872	USSR	10 OCT	18 OCT 73	
1973-073A	COSMOS 599	6867	USSR	15 OCT	28 OCT 73	
1973-073B		6868	USSR	15 OCT	21 OCT 73	
1973-073C		6869	USSR	15 OCT	24 OCT 73	
1973-073D		6870	USSR	15 OCT	24 OCT 73	
1973-074A	COSMOS 600	6873	USSR	16 OCT	23 OCT 73	
1973-074B		6874	USSR	16 OCT	28 OCT 73	
1973-074C		6887	USSR	16 OCT	24 OCT 73	
1973-074D		6888	USSR	16 OCT	28 OCT 73	
1973-074E		6899	USSR	16 OCT	29 OCT 73	
1973-075A	COSMOS 601	6875	USSR	16 OCT	15 AUG 74	
1973-075B		6876	USSR	16 OCT	06 NOV 73	AX
1973-075C		6880	USSR	16 OCT	28 MAR 74	
1973-075D		6881	USSR	16 OCT	04 DEC 73	
1973-075E		6882	USSR	16 OCT	07 FEB 74	
1973-075F		6883	USSR	16 OCT	04 DEC 73	
1973-075G		6884	USSR	16 OCT	21 FEB 74	
1973-075H		6889	USSR	16 OCT	04 FEB 74	
1973-075J		6890	USSR	16 OCT	04 DEC 73	
1973-075K		6891	USSR	16 OCT	01 MAR 74	
1973-075L		6892	USSR	16 OCT	04 DEC 73	
1973-075M		6897	USSR	16 OCT	18 JAN 74	
1973-075N		6922	USSR	16 OCT	19 NOV 73	
1973-075P		6923	USSR	16 OCT	04 DEC 73	
1973-076A	MOLNIYA 2-7	6877	USSR	19 OCT	08 JUL 83	
1973-076B		6878	USSR	19 OCT	03 DEC 73	
1973-076C		6879	USSR	19 OCT	18 DEC 73	
1973-076D		6898	USSR	19 OCT	22 JUN 83	
1973-077A	COSMOS 602	6885	USSR	20 OCT	29 OCT 73	
1973-077B		6886	USSR	20 OCT	31 OCT 73	
1973-077C		6905	USSR	20 OCT	31 OCT 73	
1973-077D		6906	USSR	20 OCT	01 NOV 73	
1973-078B		6894	US	26 OCT	29 OCT 73	
1973-078E		6902	US	26 OCT	27 OCT 73	
1973-079A	COSMOS 603	6900	USSR	27 OCT	09 NOV 73	
1973-079B		6901	USSR	27 OCT	09 NOV 73	
1973-079C		6903	USSR	27 OCT	29 OCT 73	
1973-079D		6904	USSR	27 OCT	29 OCT 73	
1973-079E		6924	USSR	27 OCT	10 NOV 73	
1973-079F		6925	USSR	27 OCT	13 NOV 73	
1973-080A	COSMOS 604	6907	USSR	29 OCT	19 JAN 92	
1973-080B		6908	USSR	29 OCT	30 SEP 92	
1973-082A	INTERCOSMOS 10	6911	USSR	30 OCT	01 JUL 77	
1973-082B		6912	USSR	30 OCT	08 OCT 77	
1973-082C		6915	USSR	30 OCT	07 DEC 73	
1973-083A	COSMOS 605	6913	USSR	31 OCT	22 NOV 73	
1973-083B		6914	USSR	31 OCT	28 NOV 73	
1973-083C		6947	USSR	31 OCT	16 JAN 74	
1973-083D		6948	USSR	31 OCT	05 DEC 73	
1973-083E		6949	USSR	31 OCT	09 JAN 74	
1973-083F		6956	USSR	31 OCT	30 DEC 73	
1973-084B		6917	USSR	02 NOV	06 JAN 74	
1973-084C		6918	USSR	02 NOV	29 JAN 74	
1973-086BE		7062	US	06 NOV	10 SEP 82	
1973-086BK		7067	US	06 NOV	26 SEP 82	
1973-086BN		7070	US	06 NOV	04 NOV 81	
1973-086BV		7077	US	06 NOV	02 APR 87	
1973-086CJ		7090	US	06 NOV	27 JAN 82	
1973-086CP		7125	US	06 NOV	26 JAN 91	
1973-086DJ		7144	US	06 NOV	15 FEB 83	
1973-086DX		7159	US	06 NOV	13 MAY 81	
1973-086FF		7197	US	06 NOV	09 JUN 80	
1973-086FH		7199	US	06 NOV	17 JUL 74	
1973-086FK		7201	US	06 NOV	17 NOV 79	
1973-086FN		7204	US	06 NOV	13 DEC 77	
1973-086FP		7205	US	06 NOV	14 DEC 76	
1973-086FQ		7206	US	06 NOV	06 OCT 74	
1973-086FR		7207	US	06 NOV	13 MAY 74	
1973-086FS		7208	US	06 NOV	22 DEC 78	
1973-086FY		9700	US	06 NOV	25 OCT 79	

International Designation	Satellite Name	Number	Source	Launch	Decay	Note
1973-087A	COSMOS 607	6926	USSR	10 NOV	22 NOV 73	
1973-087B		6927	USSR	10 NOV	23 NOV 73	
1973-087C		6930	USSR	10 NOV	15 NOV 73	
1973-087D		6945	USSR	10 NOV	28 NOV 73	
1973-087E		6946	USSR	10 NOV	23 NOV 73	
1973-088A		6928	US	10 NOV	13 MAR 74	
1973-088B		6929	US	10 NOV	13 NOV 73	
1973-088C		6931	US	10 NOV	26 DEC 78	
1973-089A	MOLNIYA 1-25	6932	USSR	14 NOV	26 MAY 79	
1973-089B		6933	USSR	14 NOV	02 JAN 74	
1973-089C		6934	USSR	14 NOV	26 NOV 73	
1973-089D		6935	USSR	14 NOV	26 JAN 74	
1973-089E		6940	USSR	14 NOV	20 AUG 79	
1973-090A	SKYLAB 4	6936	US	16 NOV	08 FEB 74	F
1973-090B		6937	US	16 NOV	16 NOV 73	
1973-091A	COSMOS 608	6941	USSR	20 NOV	10 JUL 74	
1973-091B		6942	USSR	20 NOV	23 MAR 74	
1973-092A	COSMOS 609	6943	USSR	21 NOV	04 DEC 73	
1973-092B		6944	USSR	21 NOV	04 DEC 73	
1973-092C		6963	USSR	21 NOV	06 DEC 73	
1973-092D		6964	USSR	21 NOV	05 DEC 73	
1973-093A	COSMOS 610	6950	USSR	27 NOV	15 SEP 80	
1973-093B		6951	USSR	27 NOV	06 NOV 80	
1973-093C		6962	USSR	27 NOV	29 JUN 75	
1973-094A	COSMOS 611	6952	USSR	28 NOV	19 JUN 74	
1973-094B		6955	USSR	28 NOV	26 MAR 74	
1973-095A	COSMOS 612	6953	USSR	28 NOV	11 DEC 73	
1973-095B		6954	USSR	28 NOV	12 DEC 73	
1973-095C		6968	USSR	28 NOV	16 DEC 73	
1973-095D		6969	USSR	28 NOV	12 DEC 73	
1973-095E		6970	USSR	28 NOV	11 DEC 73	
1973-096A	COSMOS 613	6957	USSR	30 NOV	29 JAN 74	
1973-096B		6960	USSR	30 NOV	04 DEC 73	
1973-096C		7103	USSR	30 NOV	09 MAR 74	
1973-097A	MOLNIYA 1-26	6958	USSR	30 NOV	09 JUN 85	
1973-097B		6959	USSR	30 NOV	30 DEC 73	
1973-097C		6961	USSR	30 NOV	31 DEC 73	
1973-097D		7178	USSR	30 NOV	17 APR 86	
1973-099A	COSMOS 615	6971	USSR	13 DEC	17 DEC 75	
1973-099B		6972	USSR	13 DEC	24 JAN 75	
1973-100C		6975	US	13 DEC	17 DEC 73	
1973-101A	EXPLORER 51	6977	US	16 DEC	12 DEC 78	
1973-101B		6978	US	16 DEC	14 AUG 74	
1973-102A	COSMOS 616	6979	USSR	17 DEC	28 DEC 73	
1973-102B		6980	USSR	17 DEC	31 DEC 73	
1973-102C		6981	USSR	17 DEC	14 JAN 74	
1973-102D		6994	USSR	17 DEC	02 JAN 74	
1973-102E		7006	USSR	17 DEC	02 JAN 74	
1973-102F		7007	USSR	17 DEC	02 JAN 74	
1973-103A	SOYUZ 13	6982	USSR	18 DEC	26 DEC 73	E
1973-103B		6983	USSR	18 DEC	22 DEC 73	
1973-103C		6984	USSR	18 DEC	14 JAN 74	
1973-105A	COSMOS 625	6995	USSR	21 DEC	03 JAN 74	
1973-105B		6996	USSR	21 DEC	01 JAN 74	
1973-105C		6997	USSR	21 DEC	22 DEC 73	
1973-105D		7010	USSR	21 DEC	02 JAN 74	
1973-105E		7011	USSR	21 DEC	06 JAN 74	
1973-106A	MOLNIYA 2-8	7000	USSR	25 DEC	24 NOV 84	
1973-106B		7001	USSR	25 DEC	24 JAN 74	
1973-106C		7002	USSR	25 DEC	17 JAN 74	
1973-106D		7372	USSR	25 DEC	24 MAY 85	
1973-108B		7113	USSR	27 DEC	17 FEB 74	
1973-108C		7114	USSR	27 DEC	23 FEB 74	
1973-108D		7115	USSR	27 DEC	22 MAR 74	
1973-108E		7116	USSR	27 DEC	16 FEB 74	
1974 LAUNCHES						
1974-002A	SKYNET 2A	7096	UK	19 JAN	25 JAN 74	
1974-002B		7097	US	19 JAN	20 JAN 74	
1974-002C		7098	US	19 JAN	23 JAN 74	
1974-002D		7099	US	19 JAN	23 JAN 74	
1974-003A	COSMOS 629	7100	USSR	24 JAN	05 FEB 74	
1974-003B		7101	USSR	24 JAN	01 FEB 74	
1974-003C		7102	USSR	24 JAN	26 JAN 74	
1974-003D		7108	USSR	24 JAN	09 FEB 74	
1974-004A	COSMOS 630	7104	USSR	30 JAN	13 FEB 74	
1974-004B		7105	USSR	30 JAN	14 FEB 74	
1974-004C		7106	USSR	30 JAN	06 FEB 74	
1974-004D		7107	USSR	30 JAN	05 FEB 74	
1974-004E		7119	USSR	30 JAN	14 FEB 74	
1974-004F		7120	USSR	30 JAN	16 FEB 74	
1974-005A	COSMOS 631	7109	USSR	06 FEB	03 OCT 80	
1974-005B		7110	USSR	06 FEB	07 DEC 80	
1974-005C		7111	USSR	06 FEB	02 JUL 78	
1974-005D		7112	USSR	06 FEB	11 OCT 75	
1974-005E		7257	USSR	06 FEB	11 JAN 78	
1974-006A	COSMOS 632	7117	USSR	12 FEB	26 FEB 74	
1974-006B		7118	USSR	12 FEB	17 FEB 74	
1974-006C		7171	USSR	12 FEB	27 FEB 74	
1974-006D		7172	USSR	12 FEB	02 MAR 74	
1974-006E		7189	USSR	12 FEB	10 MAR 74	
1974-007A		7121	US	13 FEB	17 MAR 74	
1974-008A	TANSEI 2	7122	JAPAN	16 FEB	22 JAN 83	
1974-008B		7123	JAPAN	16 FEB	21 JAN 83	
1974-009A	SAN MARCO 4	7154	ITALY	18 FEB	04 MAY 76	
1974-009B		7157	US	18 FEB	09 JUN 74	
1974-010A	COSMOS 633	7187	USSR	27 FEB	04 OCT 74	
1974-010B		7188	USSR	27 FEB	11 JUN 74	
1974-012A	COSMOS 634	7211	USSR	05 MAR	09 OCT 74	
1974-012B		7212	USSR	05 MAR	17 JUN 74	
1974-012C		7290	USSR	05 MAR	10 SEP 74	
1974-013C		7215	UK	09 MAR	05 APR 84	
1974-013D		7214	UK	09 MAR	04 MAY 83	
1974-014A	COSMOS 635	7216	USSR	14 MAR	26 MAR 74	
1974-014B		7217	USSR	14 MAR	25 MAR 74	
1974-014C		7220	USSR	14 MAR	23 MAR 74	
1974-014D		7221	USSR	14 MAR	20 MAR 74	
1974-014E		7222	USSR	14 MAR	07 APR 74	
1974-016A	COSMOS 636	7225	USSR	20 MAR	03 APR 74	
1974-016B		7226	USSR	20 MAR	24 MAR 74	
1974-016C		7227	USSR	20 MAR	25 MAR 74	
1974-016D		7232	USSR	20 MAR	04 APR 74	
1974-016E		7233	USSR	20 MAR	03 APR 74	
1974-016F		7239	USSR	20 MAR	05 APR 74	
1974-017B		7230	USSR	26 MAR	27 MAR 74	
1974-017C		7231	USSR	26 MAR	28 MAR 74	
1974-017D		7248	USSR	26 MAR	16 NOV 76	
1974-017E		7249	USSR	26 MAR	16 NOV 76	
1974-018A	COSMOS 638	7234	USSR	03 APR	13 APR 74	
1974-018B		7235	USSR	03 APR	09 APR 74	
1974-018C		7236	USSR	03 APR	06 APR 74	
1974-018D		7237	USSR	03 APR	05 APR 74	
1974-018E		7238	USSR	03 APR	06 APR 74	
1974-019A	COSMOS 639	7240	USSR	04 APR	15 APR 74	
1974-019B		7241	USSR	04 APR	07 APR 74	
1974-019C		7253	USSR	04 APR	16 APR 74	
1974-019D		7254	USSR	04 APR	17 APR 74	
1974-019E		7255	USSR	04 APR	16 APR 74	
1974-019F		7256	USSR	04 APR	16 APR 74	
1974-020A		7242	US	10 APR	28 JUL 74	
1974-020C		7247	US	10 APR	22 FEB 80	
1974-020D		7243	US	10 APR	12 APR 74	
1974-021A	COSMOS 640	7245	USSR	11 APR	23 APR 74	
1974-021B		7246	USSR	11 APR	14 APR 74	
1974-022B		7251	US	13 APR	25 MAY 74	
1974-022C		7252	US	13 APR	25 NOV 74	
1974-023A	MOLNIYA 1-27	7260	USSR	20 APR	17 NOV 83	
1974-023B		7261	USSR	20 APR	20 JUN 74	
1974-023C		7262	USSR	20 APR	04 JUL 74	
1974-023D		7263	USSR	20 APR	04 MAY 74	
1974-023E		7264	USSR	20 APR	21 DEC 83	
1974-026B		7277	USSR	26 APR	21 MAY 74	
1974-026C		7278	USSR	26 APR	17 MAY 74	
1974-026D		7279	USSR	26 APR	28 APR 74	
1974-027A	COSMOS 649	7280	USSR	29 APR	11 MAY 74	
1974-027B		7282	USSR	29 APR	03 MAY 74	
1974-027C		7283	USSR	29 APR	30 APR 74	
1974-027D		7286	USSR	29 APR	16 MAY 74	
1974-027E		7287	USSR	29 APR	11 MAY 74	
1974-027F		7288	USSR	29 APR	10 MAY 74	
1974-027G		7289	USSR	29 APR	13 MAY 74	
1974-029B		7387	USSR	15 MAY	30 JUL 74	
1974-029C		7388	USSR	15 MAY	05 SEP 74	
1974-030A	COSMOS 652	7292	USSR	15 MAY	23 MAY 74	
1974-030B		7294	USSR	15 MAY	19 MAY 74	
1974-030C		7308	USSR	15 MAY	27 MAY 74	
1974-030D		7309	USSR	15-MAY	25 MAY 74	
1974-030E		7311	USSR	15 MAY	05 JUN 74	
1974-030F		7310	USSR	15 MAY	05 JUN 74	
1974-030G		7312	USSR	15 MAY	25 MAY 74	
1974-031A	COSMOS 653	7293	USSR	15 MAY	27 MAY 74	
1974-031B		7295	USSR	15 MAY	20 MAY 74	
1974-031C		7296	USSR	15 MAY	28 MAY 74	
1974-032B		7395	USSR	17 MAY	04 AUG 74	
1974-032C		7396	USSR	17 MAY	30 AUG 74	
1974-032D		7397	USSR	17 MAY	07 SEP 74	
1974-032E		7398	USSR	17 MAY	30 AUG 74	
1974-033B		7300	US	17 MAY	20 MAY 74	
1974-033C		7301	US	17 MAY	21 MAY 74	
1974-033D		7303	US	17 MAY	12 NOV 74	
1974-033E		7304	US	17 MAY	18 MAY 74	
1974-034A	NTERCOSMOS 11	7299	USSR	17 MAY	06 SEP 79	
1974-034B		7302	USSR	17 MAY	07 JAN 80	
1974-034C		7305	USSR	17 MAY	29 DEC 74	
1974-035A	COSMOS 655	7306	USSR	21 MAY	19 NOV 80	
1974-035B		7307	USSR	21 MAY	05 OCT 80	
1974-035C		7515	USSR	21 MAY	10 JUL 81	
1974-035D		7516	USSR	21 MAY	22 APR 81	
1974-035E		7517	USSR	21 MAY	31 OCT 79	
1974-035F		7518	USSR	21 MAY	21 OCT 79	
1974-035G		7519	USSR	21 MAY	08 SEP 79	
1974-035H		8778	USSR	21 MAY	11 JAN 78	
1974-036A	COSMOS 656	7313	USSR	27 MAY	29 MAY 74	
1974-036B		7314	USSR	27 MAY	02 JUN 74	
1974-037B		7316	USSR	29 MAY	02 JUN 74	
1974-037C		7322	USSR	29 MAY	02 JUN 74	
1974-037D		7323	USSR	29 MAY	02 JUN 74	
1974-038A	COSMOS 657	7317	USSR	30 MAY	13 JUN 74	
1974-038B		7320	USSR	30 MAY	02 JUN 74	
1974-038C		7321	USSR	30 MAY	02 JUN 74	
1974-038D		7332	USSR	30 MAY	18 JUN 74	
1974-038E		7333	USSR	30 MAY	14 JUN 74	
1974-039B		7319	US	30 MAY	04 JUN 74	
1974-040A	EXPLORER 52	7325	US	03 JUN	30 APR 78	
1974-040B		7326	US	03 JUN	11 OCT 77	

International Designation	Satellite Name	Number	Source	Launch	Decay	Note
1974-040C		7327	US	03 JUN	10 NOV 74	
1974-040D		7331	US	03 JUN	21 NOV 74	
1974-041A	COSMOS 658	7328	USSR	06 JUN	18 JUN 74	
1974-041B		7329	USSR	06 JUN	12 JUN 74	
1974-042A		7330	US	06 JUN	23 JUL 74	
1974-043A	COSMOS 659	7334	USSR	13 JUN	26 JUN 74	
1974-043B		7335	USSR	13 JUN	18 JUN 74	
1974-043C		7336	USSR	13 JUN	16 JUN 74	
1974-043D		7346	USSR	13 JUN	27 JUN 74	
1974-045A	COSMOS 661	7339	USSR	21 JUN	27 AUG 80	
1974-045B		7340	USSR	21 JUN	16 OCT 80	
1974-045C		7341	USSR	21 JUN	02 FEB 78	
1974-046A	SALYUT 3	7342	USSR	24 JUN	24 JAN 75	
1974-046B		7343	USSR	24 JUN	03 JUL 74	
1974-046C		7344	USSR	24 JUN	27 JUN 74	
1974-046D		7345	USSR	24 JUN	28 JUN 74	
1974-047A	COSMOS 662	7347	USSR	26 JUN	28 AUG 76	
1974-047B		7348	USSR	26 JUN	26 OCT 75	
1974-049A	COSMOS 664	7351	USSR	29 JUN	11 JUL 74	
1974-049B		7355	USSR	29 JUN	09 JUL 74	
1974-049C		7356	USSR	29 JUN	08 JUL 74	
1974-049D		7357	USSR	29 JUN	01 JUL 74	
1974-049E		7358	USSR	29 JUN	08 JUL 74	
1974-049F		7359	USSR	29 JUN	08 JUL 74	
1974-049G		7365	USSR	29 JUN	13 JUL 74	
1974-049H		7366	USSR	29 JUN	13 JUL 74	
1974-050A	COSMOS 665	7352	USSR	29 JUN	06 JUL 90	
1974-050B		7353	USSR	29 JUN	01 SEP 74	
1974-050D		7360	USSR	29 JUN	13 AUG 74	
1974-051A	SOYUZ 14	7361	USSR	03 JUL	19 JUL 74	E
1974-051B		7362	USSR	03 JUL	05 JUL 74	
1974-051C		7374	USSR	03 JUL	17 JUL 74	
1974-053A	COSMOS 666	7367	USSR	12 JUL	25 JUL 74	
1974-053B		7368	USSR	12 JUL	18 JUL 74	
1974-053C		7379	USSR	12 JUL	25 JUL 74	
1974-053D		7381	USSR	12 JUL	30 JUL 74	
1974-054B		7370	US	14 JUL	26 NOV 80	
1974-055A	AEROS 2	7371	FRG	16 JUL	25 SEP 75	
1974-055B		7375	US	16 JUL	09 OCT 74	
1974-055C		7380	FRG	16 JUL	15 OCT 74	
1974-056B		7377	USSR	23 JUL	26 AUG 74	
1974-056C		7378	USSR	23 JUL	28 AUG 74	
1974-057A	COSMOS 667	7383	USSR	25 JUL	07 AUG 74	
1974-057B		7384	USSR	25 JUL	30 JUL 74	
1974-057C		7407	USSR	25 JUL	10 AUG 74	
1974-057D		7408	USSR	25 JUL	18 AUG 74	
1974-058A	COSMOS 668	7385	USSR	25 JUL	21 FEB 75	
1974-058B		7386	USSR	25 JUL	15 NOV 74	
1974-059A	COSMOS 669	7389	USSR	26 JUL	08 AUG 74	
1974-059B		7390	USSR	26 JUL	30 JUL 74	
1974-059C		7391	USSR	26 JUL	29 JUL 74	
1974-059D		7401	USSR	26 JUL	07 AUG 74	
1974-059E		7402	USSR	26 JUL	08 AUG 74	
1974-059F		7403	USSR	26 JUL	07 AUG 74	
1974-059G		7404	USSR	26 JUL	11 AUG 74	
1974-060B		7393	USSR	29 JUL	01 AUG 74	
1974-060C		7394	USSR	29 JUL	31 JUL 74	
1974-060D		7399	USSR	29 JUL	26 JAN 79	
1974-060E		7400	USSR	29 JUL	11 DEC 76	
1974-061A	COSMOS 670	7405	USSR	06 AUG	09 AUG 74	
1974-061B		7406	USSR	06 AUG	13 AUG 74	
1974-062A	COSMOS 671	7409	USSR	07 AUG	20 AUG 74	
1974-062B		7410	USSR	07 AUG	14 AUG 74	
1974-062C		7419	USSR	07 AUG	09 SEP 74	
1974-062D		7420	USSR	07 AUG	23 AUG 74	
1974-064A	COSMOS 672	7413	USSR	12 AUG	18 AUG 74	
1974-064B		7414	USSR	12 AUG	14 AUG 74	
1974-064C		7415	USSR	12 AUG	16 AUG 74	
1974-065A		7416	US	14 AUG	29 SEP 74	
1974-066A	COSMOS 673	7417	USSR	16 AUG	01 JUN 91	
1974-066D		18913	USSR	16 AUG	27 MAR 89	
1974-067A	SOYUZ 15	7421	USSR	26 AUG	28 AUG 74	E
1974-067B		7422	USSR	26 AUG	28 AUG 74	
1974-068A	COSMOS 674	7423	USSR	29 AUG	07 SEP 74	
1974-068B		7425	USSR	29 AUG	04 SEP 74	
1974-068C		7431	USSR	29 AUG	14 SEP 74	
1974-068D		7432	USSR	29 AUG	10 SEP 74	
1974-070A	ANS	7427	NETH	30 AUG	14 JUN 77	
1974-070B		7428	US	30 AUG	19 NOV 75	
1974-070C		7429	NETH	30 AUG	20 JUL 75	
1974-070D		7430	NETH	30 AUG	04 OCT 74	
1974-070E		7694	NETH	30 AUG	10 APR 75	
1974-073A	COSMOS 685	7445	USSR	20 SEP	02 OCT 74	
1974-073B		7444	USSR	20 SEP	26 SEP 74	
1974-073C		7446	USSR	20 SEP	25 SEP 74	
1974-074A	COSMOS 686	7447	USSR	26 SEP	01 MAY 75	
1974-074B		7448	USSR	26 SEP	29 OCT 74	AY
1974-074C		7449	USSR	26 SEP	19 MAR 75	
1974-074D		7450	USSR	26 SEP	01 JAN 76	
1974-074E		7451	USSR	26 SEP	13 NOV 74	
1974-074F		7452	USSR	26 SEP	18 MAR 75	
1974-074G		7453	USSR	26 SEP	05 JUL 75	
1974-074H		7454	USSR	26 SEP	03 NOV 74	
1974-074J		7455	USSR	26 SEP	03 NOV 74	
1974-074K		7456	USSR	26 SEP	03 NOV 74	
1974-074L		7457	USSR	26 SEP	17 NOV 74	
1974-074M		7458	USSR	26 SEP	26 NOV 74	
1974-074N		7459	USSR	26 SEP	05 MAY 75	
1974-074P		7460	USSR	26 SEP	20 OCT 74	
1974-074Q		7461	USSR	26 SEP	18 OCT 74	
1974-074R		7462	USSR	26 SEP	23 OCT 74	
1974-074S		7463	USSR	26 SEP	10 NOV 74	
1974-074T		7464	USSR	26 SEP	01 NOV 74	
1974-074U		7465	USSR	26 SEP	21 FEB 75	
1974-074V		7526	USSR	26 SEP	27 NOV 74	
1974-075B		7467	US	10 OCT	23 DEC 78	
1974-076A	COSMOS 687	7469	USSR	11 OCT	05 FEB 78	
1974-076B		7470	USSR	11 OCT	04 JAN 77	
1974-077A	ARIEL 5	7471	UK	15 OCT	14 MAR 80	
1974-077B		7472	US	15 OCT	13 MAY 79	
1974-078A	COSMOS 688	7473	USSR	18 OCT	30 OCT 74	
1974-078B		7474	USSR	18 OCT	24 OCT 74	
1974-078C		7475	USSR	18 OCT	23 OCT 74	
1974-078D		7496	USSR	18 OCT	31 OCT 74	
1974-080A	COSMOS 690	7478	USSR	22 OCT	12 NOV 74	
1974-080B		7479	USSR	22 OCT	10 NOV 74	
1974-080C		7521	USSR	22 OCT	21 NOV 74	
1974-080D		7522	USSR	22 OCT	03 DEC 74	
1974-080E		7523	USSR	22 OCT	14 DEC 74	
1974-080F		7524	USSR	22 OCT	16 NOV 74	
1974-080G		7527	USSR	22 OCT	23 NOV 74	
1974-080H		7528	USSR	22 OCT	24 NOV 74	
1974-081A	MOLNIYA 1-28	7480	USSR	24 OCT	29 DEC 85	
1974-081B		7481	USSR	24 OCT	31 DEC 74	
1974-081C		7482	USSR	24 OCT	03 JAN 75	
1974-081D		7485	USSR	24 OCT	27 OCT 86	
1974-081E		7486	USSR	24 OCT	04 NOV 74	
1974-082A	COSMOS 691	7483	USSR	25 OCT	06 NOV 74	
1974-082B		7484	USSR	25 OCT	29 OCT 74	
1974-082C		7511	USSR	25 OCT	10 NOV 74	
1974-082D		7512	USSR	25 OCT	06 NOV 74	
1974-084A	LUNA 23	7491	USSR	28 OCT	06 NOV 74	R
1974-084B		7492	USSR	28 OCT	01 NOV 74	
1974-084C		7494	USSR	28 OCT	01 NOV 74	
1974-085A		7495	US	29 OCT	19 MAR 75	
1974-085B		7498	US	29 OCT	23 JAN 80	
1974-085C		7499	US	29 OCT	26 MAY 75	
1974-085D		7497	US	29 OCT	31 OCT 74	
1974-086A	INTERCOSMOS 12	7500	USSR	31 OCT	11 JUL 75	
1974-086B		7501	USSR	31 OCT	23 JUL 75	
1974-086C		7507	USSR	31 OCT	11 NOV 74	
1974-086D		7508	USSR	31 OCT	11 NOV 74	
1974-087A	COSMOS 692	7502	USSR	01 NOV	13 NOV 74	
1974-087B		7503	USSR	01 NOV	06 NOV 74	
1974-087C		7504	USSR	01 NOV	05 NOV 74	
1974-087D		7505	USSR	01 NOV	05 NOV 74	
1974-087E		7506	USSR	01 NOV	05 NOV 74	
1974-087F		7520	USSR	01 NOV	23 NOV 74	
1974-087G		7525	USSR	01 NOV	18 NOV 74	
1974-088A	COSMOS 693	7509	USSR	04 NOV	16 NOV 74	
1974-088B		7510	USSR	04 NOV	09 NOV 74	
1974-088C		7513	USSR	04 NOV	06 NOV 74	
1974-088D		7514	USSR	04 NOV	16 NOV 74	
1974-088E		7534	USSR	04 NOV	17 NOV 74	
1974-089AM		8171	US	15 NOV	08 MAY 82	
1974-089AS		8176	US	15 NOV	17 OCT 89	
1974-089BC		8186	US	15 NOV	27 DEC 78	
1974-089BE		8297	US	15 NOV	10 JAN 91	
1974-089BM		8304	US	15 NOV	10 NOV 89	
1974-089BV		8312	US	15 NOV	03 JUN 82	
1974-089CU		8557	US	15 NOV	10 JUN 82	
1974-089CV		8558	US	15 NOV	28 JAN 78	
1974-089CW		8559	US	15 NOV	10 JUN 82	
1974-089DC		8565	US	15 NOV	16 MAY 88	
1974-089DP		8787	US	15 NOV	24 AUG 77	
1974-089DT		8909	US	15 NOV	15 NOV 81	
1974-089EL		13130	US	15 NOV	14 SEP 90	
1974-089EV		15522	US	15 NOV	04 SEP 88	
1974-089EY		18508	US	15 NOV	15 MAR 89	
1974-089FB		18901	US	15 NOV	11 JUL 89	
1974-089FF		20681	US	15 NOV	02AUG 90	
1974-089G		8138	US	15 NOV	14 NOV 92	
1974-090A	COSMOS 694	7533	USSR	16 NOV	29 NOV 74	
1974-090B		7535	USSR	16 NOV	25 NOV 74	
1974-090C		7536	USSR	16 NOV	17 NOV 74	
1974-090D		7537	USSR	16 NOV	18 NOV 74	
1974-090E		7555	USSR	16 NOV	03 DEC 74	
1974-090F		7556	USSR	16 NOV	30 NOV 74	
1974-090G		7557	USSR	16 NOV	03 DEC 74	
1974-091A	COSMOS 695	7538	USSR	20 NOV	15 JUL 75	
1974-091B		7539	USSR	20 NOV	15 MAR 75	
1974-092A	MOLNIYA 3-1	7540	USSR	21 NOV	15 MAY 86	
1974-092B		7541	USSR	21 NOV	27 JAN 75	
1974-092C		7542	USSR	21 NOV	11 JAN 75	
1974-092D		7543	USSR	21 NOV	10 JAN 75	
1974-092E		7546	USSR	21 NOV	09 MAR 86	
1974-094B		7548	US	23 NOV	07 JAN 75	
1974-094C		7549	US	23 NOV	26 DEC 74	
1974-094D		7550	US	23 NOV	08 DEC 89	
1974-094E		9530	US	23 NOV	04 NOV 76	
1974-095A	COSMOS 696	7551	USSR	27 NOV	09 DEC 74	
1974-095B		7552	USSR	27 NOV	05 DEC 74	
1974-095C		7553	USSR	27 NOV	29 NOV 74	
1974-095D		7554	USSR	27 NOV	03 DEC 74	
1974-096A	SOYUZ 16	7561	USSR	02 DEC	08 DEC 74	E
1974-096B		7562	USSR	02 DEC	07 DEC 74	

International Designation	Satellite Name	Number	Source	Launch	Decay	Note
1974-096C		7563	USSR	02 DEC	08 DEC 74	
1974-096D		7564	USSR	02 DEC	03 DEC 74	
1974-096E		7565	USSR	02 DEC	08 DEC 74	
1974-096F		7566	USSR	02 DEC	08 DEC 74	
1974-098A	COSMOS 697	7571	USSR	13 DEC	25 DEC 74	
1974-098B		7572	USSR	13 DEC	21 DEC 74	
1974-098C		7573	USSR	13 DEC	16 DEC 74	
1974-098D		7589	USSR	13 DEC	26 DEC 74	
1974-098E		7590	USSR	13 DEC	26 DEC 74	
1974-100A	COSMOS 698	7576	USSR	18 DEC	09 DEC 80	
1974-100B		7577	USSR	18 DEC	08 NOV 80	
1974-100C		7582	USSR	18 DEC	06 MAR 77	
1974-100D		7636	USSR	18 DEC	23 SEP 81	
1974-100E		7695	USSR	18 DEC	29 SEP 77	
1974-100F		7722	USSR	18 DEC	20 OCT 79	
1974-100G		7723	USSR	18 DEC	14 APR 80	
1974-100H		7724	USSR	18 DEC	02 MAR 78	
1974-100J		9793	USSR	18 DEC	01 AUG 77	
1974-101B		7579	US	19 DEC	19 APR 78	
1974-101C		7580	US	19 DEC	24 APR 77	
1974-101D		7581	US	19 DEC	02 JAN 77	
1974-101E		8739	US	19 DEC	08 SEP 77	
1974-101F		8740	US	19 DEC	25 DEC 77	
1974-102A	MOLNIYA 2-11	7583	USSR	21 DEC	07 JUL 88	
1974-102B		7584	USSR	21 DEC	12 MAR 75	
1974-102C		7585	USSR	21 DEC	08 MAR 75	
1974-102D		7586	USSR	21 DEC	05 JAN 89	
1974-103A	COSMOS 699	7587	USSR	24 DEC	16 OCT 77	
1974-103AA		8047	USSR	24 DEC	26 FEB 78	
1974-103AB		8048	USSR	24 DEC	22 MAR 76	
1974-103AC		8049	USSR	24 DEC	10 JUN 78	
1974-103AD		8050	USSR	24 DEC	12 NOV 75	
1974-103AE		8051	USSR	24 DEC	06 NOV 76	
1974-103AF		8052	USSR	24 DEC	04 AUG 75	
1974-103AG		8056	USSR	24 DEC	10 MAR 76	
1974-103AH		8057	USSR	24 DEC	21 NOV 78	
1974-103AJ		8100	USSR	24 DEC	20 SEP 75	
1974-103AK		8101	USSR	24 DEC	13 MAR 76	
1974-103AL		8102	USSR	24 DEC	12 MAR 78	
1974-103AM		8103	USSR	24 DEC	30 SEP 75	
1974-103AN		8104	USSR	24 DEC	29 JAN 76	
1974-103AP		8105	USSR	24 DEC	03 MAY 79	
1974-103AQ		8106	USSR	24 DEC	12 APR 76	
1974-103AR		8109	USSR	24 DEC	24 SEP 75	
1974-103AS		8112	USSR	24 DEC	04 MAR 78	
1974-103AT		8113	USSR	24 DEC	15 JUL 78	
1974-103AU		8114	USSR	24 DEC	09 MAY 76	
1974-103AV		8115	USSR	24 DEC	28 JAN 76	
1974-103AW		8116	USSR	24 DEC	23 SEP 77	
1974-103AX		8117	USSR	24 DEC	14 APR 76	
1974-103AY		8118	USSR	24 DEC	19 JUN 78	
1974-103AZ		8119	USSR	24 DEC	01 NOV 75	
1974-103B		7588	USSR	24 DEC	25 DEC 74	AZ
1974-103BA		8120	USSR	24 DEC	10 APR 78	
1974-103BB		8121	USSR	24 DEC	28 DEC 77	
1974-103BC		8122	USSR	24 DEC	28 JUN 78	
1974-103C		7748	USSR	24 DEC	08 JAN 77	
1974-103D		7749	USSR	24 DEC	12 APR 76	
1974-103E		7750	USSR	24 DEC	27 FEB 77	
1974-103F		7754	USSR	24 DEC	01 JUN 75	
1974-103G		7755	USSR	24 DEC	20 JAN 76	
1974-103H		7756	USSR	24 DEC	23 DEC 76	
1974-103J		7757	USSR	24 DEC	18 OCT 78	
1974-103K		7758	USSR	24 DEC	16 AUG 75	
1974-103L		7759	USSR	24 DEC	06 JUN 76	
1974-103M		7760	USSR	24 DEC	30 OCT 75	
1974-103N		7762	USSR	24 DEC	29 JUL 75	
1974-103P		7763	USSR	24 DEC	12 APR 76	
1974-103Q		7764	USSR	24 DEC	26 AUG 75	
1974-103R		7765	USSR	24 DEC	23 SEP 77	
1974-103S		7766	USSR	24 DEC	25 JUL 75	
1974-103T		7767	USSR	24 DEC	09 JUL 75	
1974-103U		7783	USSR	24 DEC	06 JUL 75	
1974-103V		7784	USSR	24 DEC	10 OCT 75	
1974-103W		7793	USSR	24 DEC	09 JUL 75	
1974-103X		7814	USSR	24 DEC	09 SEP 75	
1974-103Y		7999	USSR	24 DEC	06 DEC 78	
1974-103Z		8046	USSR	24 DEC	01 NOV 75	
1974-104A	SALYUT 4	7591	USSR	26 DEC	02 FEB 77	
1974-104B		7592	USSR	26 DEC	01 JAN 75	BA
1974-104C		7595	USSR	26 DEC	28 DEC 74	
1974-104D		7598	USSR	26 DEC	22 JAN 75	
1974-104E		7608	USSR	26 DEC	29 APR 75	
1974-104F		7628	USSR	26 DEC	08 APR 75	
1974-104G		7630	USSR	26 DEC	15 APR 75	
1974-104H		7634	USSR	26 DEC	15 APR 75	
1974-104J		7656	USSR	26 DEC	18 APR 75	
1974-104K		7657	USSR	26 DEC	26 APR 75	
1974-104L		7658	USSR	26 DEC	07 MAY 75	
1974-104M		7661	USSR	26 DEC	13 APR 75	
1974-104N		7662	USSR	26 DEC	10 SEP 75	
1974-104P		7704	USSR	26 DEC	16 JUN 75	
1974-104Q		7995	USSR	26 DEC	16 SEP 75	
1974-104R		7996	USSR	26 DEC	18 SEP 75	
1974-104S		7997	USSR	26 DEC	25 MAR 76	
1974-104T		7998	USSR	26 DEC	15 SEP 75	
1974-104U		8055	USSR	26 DEC	27 OCT 75	
1974-106A	COSMOS 701	7596	USSR	27 DEC	09 JAN 75	
1974-106B		7597	USSR	27 DEC	07 JAN 75	
1974-106C		7599	USSR	27 DEC	02 JAN 75	
1974-106D		7600	USSR	27 DEC	07 JAN 75	
1974-106E		7601	USSR	27 DEC	11 JAN 75	
1974-106F		7602	USSR	27 DEC	15 JAN 75	
1974-106G		7603	USSR	27 DEC	15 JAN 75	

1975 LAUNCHES

International Designation	Satellite Name	Number	Source	Launch	Decay	Note
1975-001A	SOYUZ 17	7604	USSR	10 JAN	09 FEB 75	E
1975-001B		7605	USSR	10 JAN	14 JAN 75	
1975-001C		7610	USSR	10 JAN	25 MAR 75	
1975-001D		7613	USSR	10 JAN	12 APR 75	
1975-001E		7614	USSR	10 JAN	08 MAY 75	
1975-001F		7622	USSR	10 JAN	24 OCT 75	
1975-001G		7633	USSR	10 JAN	15 APR 75	
1975-001H		7635	USSR	10 JAN	23 MAR 75	
1975-001J		7640	USSR	10 JAN	27 OCT 75	
1975-002A	COSMOS 702	7606	USSR	17 JAN	29 JAN 75	
1975-002B		7607	USSR	17 JAN	29 JAN 75	
1975-002C		7609	USSR	17 JAN	21 JAN 75	
1975-003A	COSMOS 703	7611	USSR	21 JAN	20 NOV 75	
1975-003B		7612	USSR	21 JAN	22 AUG 75	
1975-004AA		8944	US	22 JAN	18 FEB 80	
1975-004AB		8945	US	22 JAN	25 APR 80	
1975-004AC		8946	US	22 JAN	15 JUL 78	
1975-004AD		8947	US	22 JAN	08 APR 81	
1975-004AE		8948	US	22 JAN	10 AUG 81	
1975-004AF		8949	US	22 JAN	06 APR 79	
1975-004AG		8950	US	22 JAN	19 DEC 81	
1975-004AJ		8952	US	22 JAN	29 MAR 79	
1975-004AK		8953	US	22 JAN	07 JUN 89	
1975-004AM		8955	US	22 JAN	23 JUN 77	
1975-004AP		8957	US	22 JAN	04 NOV 79	
1975-004AS		8960	US	22 JAN	08 NAR 92	
1975-004AT		8961	US	22 JAN	24 APR 79	
1975-004AU		8962	US	22 JAN	29 DEC 79	
1975-004AX		8965	US	22 JAN	03 NOV 78	
1975-004AY		8966	US	22 JAN	09 NOV 78	
1975-004AZ		8967	US	22 JAN	16 OCT 89	
1975-004BD		8971	US	22 JAN	23 OCT 79	
1975-004BE		8972	US	22 JAN	05 NOV 79	
1975-004BF		8973	US	22 JAN	13 JAN 80	
1975-004BG		8974	US	22 JAN	31 JAN 77	
1975-004BH		8975	US	22 JAN	06 SEP 79	
1975-004BJ		8976	US	22 JAN	21 NOV 79	
1975-004BK		8977	US	22 JAN	28 JUL 81	
1975-004BQ		8982	US	22 JAN	23 JUN 80	
1975-004BR		8983	US	22 JAN	11 SEP 79	
1975-004BS		8984	US	22 JAN	23 NOV 79	
1975-004BT		8985	US	22 JAN	30 MAY 79	
1975-004BU		8986	US	22 JAN	17 OCT 79	
1975-004BV		8987	US	22 JAN	16 DEC 79	
1975-004BW		8988	US	22 JAN	25 APR 79	
1975-004BX		8989	US	22 JAN	16 AUG 78	
1975-004BY		8990	US	22 JAN	27 FEB 82	
1975-004BZ		8991	US	22 JAN	09 DEC 83	
1975-004C		8675	US	22 JAN	18 SEP 90	
1975-004CA		8992	US	22 JAN	23 APR 81	
1975-004CB		8993	US	22 JAN	01 DEC 86	
1975-004CC		8994	US	22 JAN	16 JUN 79	
1975-004CD		8995	US	22 JAN	11 OCT 86	
1975-004CE		8996	US	22 JAN	10 DEC 83	
1975-004CG		8998	US	22 JAN	24 MAR 79	
1975-004CH		8999	US	22 JAN	08 MAY 82	
1975-004CJ		9000	US	22 JAN	13 DEC 91	
1975-004CK		9001	US	22 JAN	20 FEB 80	
1975-004CL		9002	US	22 JAN	26 DEC 80	
1975-004CN		9004	US	22 JAN	03 JUL 77	
1975-004CP		9005	US	22 JAN	26 DEC 80	
1975-004CQ		9026	US	22 JAN	09 NOV 78	
1975-004CR		9027	US	22 JAN	04 FEB 78	
1975-004CS		9028	US	22 JAN	01 MAY 78	
1975-004CT		9029	US	22 JAN	22 OCT 80	
1975-004CU		9030	US	22 JAN	11 JAN 80	
1975-004CV		9031	US	22 JAN	14 DEC 79	
1975-004CW		9032	US	22 JAN	04 FEB 77	
1975-004CX		9033	US	22 JAN	21 SEP 79	
1975-004CY		9034	US	22 JAN	09 APR 81	
1975-004CZ		9035	US	22 JAN	12 OCT 78	
1975-004D		8676	US	22 JAN	05 NOV 79	
1975-004DA		9036	US	22 JAN	03 FEB 80	
1975-004DB		9037	US	22 JAN	03 OCT 79	
1975-004DC		9038	US	22 JAN	31 JAN 77	
1975-004DD		9039	US	22 JAN	06 OCT 78	
1975-004DF		9041	US	22 JAN	13 SEP 80	
1975-004DH		9282	US	22 JAN	10 DEC 79	
1975-004DJ		9283	US	22 JAN	02 DEC 79	
1975-004DK		9284	US	22 JAN	04 OCT 79	
1975-004DL		9285	US	22 JAN	31 JAN 77	
1975-004DM		9286	US	22 JAN	25 APR 77	
1975-004DN		9287	US	22 JAN	20 APR 79	
1975-004DP		9288	US	22 JAN	01 OCT 77	
1975-004DQ		9289	US	22 JAN	22 JAN 78	
1975-004DR		9290	US	22 JAN	27 MAY 77	
1975-004DS		9291	US	22 JAN	30 JUN 80	
1975-004DT		9292	US	22 JAN	24 MAY 84	
1975-004DV		9294	US	22 JAN	26 FEB 79	

International Designation	Satellite Name	Number	Source	Launch	Decay	Note
1975-004DW		9295	US	22 JAN	27 JAN 79	
1975-004DX		9296	US	22 JAN	18 MAY 81	
1975-004DY		9297	US	22 JAN	31 JAN 77	
1975-004DZ		9298	US	22 JAN	09 FEB 77	
1975-004E		8677	US	22 JAN	14 SEP 79	
1975-004EA		9299	US	22 JAN	30 JUN 78	
1975-004EB		9300	US	22 JAN	19 NOV 82	
1975-004EC		9301	US	22 JAN	04 JAN 79	
1975-004ED		9302	US	22 JAN	15 APR 79	
1975-004EE		9303	US	22 JAN	06 APR 80	
1975-004EF		9304	US	22 JAN	05 SEP 80	
1975-004EG		9305	US	22 JAN	24 JAN 89	
1975-004EH		9306	US	22 JAN	21 DEC 80	
1975-004EJ		9307	US	22 JAN	21 OCT 79	
1975-004EK		9308	US	22 JAN	09 MAY 81	
1975-004EL		9309	US	22 JAN	11 AUG 81	
1975-004EM		9310	US	22 JAN	07 JUL 77	
1975-004EN		9311	US	22 JAN	16 JUN 79	
1975-004EP		9312	US	22 JAN	08 MAY 78	
1975-004ER		9314	US	22 JAN	01 DEC 82	
1975-004ES		9315	US	22 JAN	13 FEB 79	
1975-004ET		9316	US	22 JAN	28 DEC 77	
1975-004EU		9317	US	22 JAN	21 APR 81	
1975-004EV		9318	US	22 JAN	04 JUN 79	
1975-004EW		9319	US	22 JAN	04 MAY 80	
1975-004EY		9321	US	22 JAN	05 FEB 80	
1975-004EZ		9322	US	22 JAN	12 APR 77	
1975-004FB		9324	US	22 JAN	08 SEP 79	
1975-004FC		9325	US	22 JAN	11 OCT 80	
1975-004FD		9326	US	22 JAN	17 APR 78	
1975-004FE		9327	US	22 JAN	23 MAR 80	
1975-004FF		9543	US	22 JAN	28 NOV 77	
1975-004FG		9544	US	22 JAN	12 MAY 79	
1975-004FH		9545	US	22 JAN	21 SEP 89	
1975-004FJ		9546	US	22 JAN	07 OCT 77	
1975-004FK		9547	US	22 JAN	19 JAN 80	
1975-004FL		9548	US	22 JAN	01 FEB 79	
1975-004FM		9549	US	22 JAN	08 JUL 78	
1975-004FN		9550	US	22 JAN	14 APR 79	
1975-004FQ		9571	US	22 JAN	31 DEC 78	
1975-004FR		9616	US	22 JAN	02 NOV 81	
1975-004FT		9618	US	22 JAN	31 OCT 80	
1975-004FW		9621	US	22 JAN	31 MAY 88	
1975-004FX		9622	US	22 JAN	18 JAN 79	
1975-004FY		9677	US	22 JAN	11 JUN 81	
1975-004FZ		9678	US	22 JAN	14 JUN 79	
1975-004GA		9679	US	22 JAN	20 SEP 90	
1975-004GB		9680	US	22 JAN	17 JUL 79	
1975-004GD		9682	US	22 JAN	16 JAN 92	
1975-004GE		9683	US	22 JAN	19 JUN 81	
1975-004GF		9684	US	22 JAN	20 FEB 79	
1975-004GG		9685	US	22 JAN	17 JUL 86	
1975-004GH		9686	US	22 JAN	16 JUN 79	
1975-004GJ		9687	US	22 JAN	30 NOV 79	
1975-004GK		9688	US	22 JAN	11 SEP 80	
1975-004GL		9689	US	22 JAN	12 SEP 90	
1975-004GM		9690	US	22 JAN	06 JUN 80	
1975-004GN		9691	US	22 JAN	21 JUN 78	
1975-004GP		9705	US	22 JAN	28 APR 78	
1975-004GO		9706	US	22 JAN	04 DEC 78	
1975-004GR		9707	US	22 JAN	16 NOV 79	
1975-004GS		9708	US	22 JAN	02 MAY 79	
1975-004GT		9709	US	22 JAN	12 MAR 78	
1975-004GW		9712	US	22 JAN	03 JAN 79	
1975-004GX		9713	US	22 JAN	18 JUL 78	
1975-004GY		9714	US	22 JAN	07 JAN 79	
1975-004GZ		9715	US	22 JAN	09 MAY 79	
1975-004H		8680	US	22 JAN	10 FEB 79	
1975-004HA		9716	US	22 JAN	25 NOV 77	
1975-004HB		9717	US	22 JAN	11 APR 79	
1975-004HC		9948	US	22 JAN	02 MAR 85	
1975-004HD		9949	US	22 JAN	26 OCT 81	
1975-004HE		9950	US	22 JAN	14 JUL 79	
1975-004HF		10397	US	22 JAN	06 JAN 81	
1975-004HG		14882	US	22 JAN	02 JUL 89	
1975-004HH		17729	US	22 JAN	04 MAY 92	
1975-004HJ		17730	US	22 JAN	09 MAY 89	
1975-004HM		19053	US	22 JAN	13 DEC 91	
1975-004HR		19431	US	22 JAN	14 DEC 90	
1975-004K		8683	US	22 JAN	05 APR 77	
1975-004L		8684	US	22 JAN	28 DEC 79	
1975-004N		8687	US	22 JAN	18 FEB 77	
1975-004P		8691	US	22 JAN	22 MAY 82	
1975-004Q		8692	US	22 JAN	28 AUG 89	
1975-004R		8693	US	22 JAN	29 AUG 81	
1975-004S		8936	US	22 JAN	20 NOV 78	
1975-004T		8937	US	22 JAN	09 DEC 79	
1975-004U		8938	US	22 JAN	18 APR 78	
1975-004V		8939	US	22 JAN	05 MAY 81	
1975-004X		8941	US	22 JAN	18 APR 78	
1975-004Y		8942	US	22 JAN	26 SEP 81	
1975-004Z		8943	US	22 JAN	17 JUL 82	
1975-005A	COSMOS 704	7617	USSR	23 JAN	06 FEB 75	
1975-005B		7618	USSR	23 JAN	03 FEB 75	
1975-005C		7619	USSR	23 JAN	26 JAN 75	
1975-005D		7620	USSR	23 JAN	26 JAN 75	
1975-005E		7621	USSR	23 JAN	26 JAN 75	
1975-005F		7631	USSR	23 JAN	07 FEB 75	
1975-005G		7632	USSR	23 JAN	06 FEB 75	
1975-005H		7644	USSR	23 JAN	08 FEB 75	
1975-005J		7649	USSR	23 JAN	09 FEB 75	
1975-006A	COSMOS 705	7623	USSR	28 JAN	18 NOV 75	
1975-006B		7624	USSR	28 JAN	25 JUN 75	
1975-007B		7626	USSR	30 JAN	04 APR 75	
1975-007C		7627	USSR	30 JAN	01 MAR 75	
1975-008A	COSMOS 707	7637	USSR	05 FEB	07 SEP 80	
1975-008B		7638	USSR	05 FEB	27 AUG 80	
1975-008C		7639	USSR	05 FEB	14 DEC 76	
1975-008D		7645	USSR	05 FEB	17 APR 77	
1975-008E		7777	USSR	05 FEB	29 APR 77	
1975-009A	MOLNIYA 2-12	7641	USSR	06 FEB	04 JUL 85	
1975-009B		7642	USSR	06 FEB	25 APR 75	
1975-009C		7643	USSR	06 FEB	23 APR 75	
1975-009D		7653	USSR	06 FEB	30 SEP 85	
1975-011B		7650	US	06 FEB	06 APR 82	
1975-011C		7651	US	06 FEB	16 FEB 75	
1975-011D		7652	US	06 FEB	16 FEB 75	
1975-011E		7660	US	06 FEB	21 FEB 75	
1975-013A	COSMOS 709	7664	USSR	12 FEB	25 FEB 75	
1975-013B		7666	USSR	12 FEB	17 FEB 75	
1975-013C		7667	USSR	12 FEB	15 FEB 75	
1975-013D		7668	USSR	12 FEB	15 FEB 75	
1975-013E		7669	USSR	12 FEB	15 FEB 75	
1975-013F		7672	USSR	12 FEB	01 MAR 75	
1975-013G		7673	USSR	12 FEB	26 FEB 75	
1975-014A	SRATS (TAIYO)	7671	JAPAN	24 FEB	29 JUN 80	
1975-014B		7674	JAPAN	24 FEB	17 MAY 80	
1975-015A	COSMOS 710	7675	USSR	26 FEB	12 MAR 75	
1975-015B		7676	USSR	26 FEB	03 MAR 75	
1975-015C		7677	USSR	26 FEB	03 MAR 75	
1975-015D		7689	USSR	26 FEB	12 MAR 75	
1975-015E		7690	USSR	26 FEB	16 MAR 75	
1975-018A	COSMOS 719	7691	USSR	12 MAR	25 MAR 75	
1975-018B		7692	USSR	12 MAR	16 MAR 75	
1975-018C		7693	USSR	12 MAR	15 MAR 75	
1975-018D		7701	USSR	12 MAR	26 MAR 75	
1975-018E		7702	USSR	12 MAR	30 MAR 75	
1975-018F		7703	USSR	12 MAR	28 MAR 75	
1975-019A	COSMOS 720	7696	USSR	21 MAR	01 APR 75	
1975-019B		7697	USSR	21 MAR	30 MAR 75	
1975-019C		7698	USSR	21 MAR	22 MAR 75	
1975-019D		7699	USSR	21 MAR	25 MAR 75	
1975-019E		7700	USSR	21 MAR	22 MAR 75	
1975-019F		7717	USSR	21 MAR	05 APR 75	
1975-019G		7719	USSR	21 MAR	04 APR 75	
1975-019H		7725	USSR	21 MAR	10 APR 75	
1975-019J		7726	USSR	21 MAR	09 APR 75	
1975-020A	COSMOS 721	7705	USSR	26 MAR	07 APR 75	
1975-020B		7706	USSR	26 MAR	29 MAR 75	
1975-020C		7707	USSR	26 MAR	29 MAR 75	
1975-020D		7708	USSR	26 MAR	31 MAR 75	
1975-020E		7720	USSR	26 MAR	07 APR 75	
1975-020F		7721	USSR	26 MAR	10 APR 75	
1975-021A	COSMOS 722	7709	USSR	27 MAR	09 APR 75	
1975-021B		7711	USSR	27 MAR	09 APR 75	
1975-021C		7712	USSR	27 MAR	29 MAR 75	
1975-021D		7716	USSR	27 MAR	02 APR 75	
1975-021E		7728	USSR	27 MAR	10 APR 75	
1975-021F		7729	USSR	27 MAR	12 APR 75	
1975-021G		7732	USSR	27 MAR	14 APR 75	
1975-021H		7733	USSR	27 MAR	10 APR 75	
1975-022A	INTERCOSMOS 13	7710	USSR	27 MAR	02 SEP 80	
1975-022B		7713	USSR	27 MAR	21 APR 80	
1975-024B		7795	USSR	02 APR	21 MAY 75	
1975-024C		7796	USSR	02 APR	19 MAY 75	
1975-024D		7797	USSR	02 APR	15 JUL 75	
1975-024E		7798	USSR	02 APR	24 MAY 75	
1975-024F		7799	USSR	02 APR	25 MAY 75	
1975-025B		7922	USSR	07 APR	07 AUG 75	
1975-025C		7923	USSR	07 APR	17 JUN 75	
1975-025D		18281	USSR	07 APR	05 OCT 88	
1975-026A	COSMOS 725	7730	USSR	08 APR	06 JAN 76	
1975-026B		7731	USSR	08 APR	09 SEP 75	
1975-027D		10729	US	09 APR	07 DEC 83	
1975-029A	MOLNIYA 3-2	7738	USSR	14 APR	29 NOV 88	
1975-029B		7739	USSR	14 APR	20 JUN 75	
1975-029C		7740	USSR	14 APR	04 JUN 75	
1975-030A	COSMOS 727	7742	USSR	16 APR	28 APR 75	
1975-030B		7743	USSR	16 APR	21 APR 75	
1975-030C		7744	USSR	16 APR	19 APR 75	
1975-030D		7774	USSR	16 APR	05 MAY 75	
1975-030E		7775	USSR	16 APR	01 MAY 75	
1975-031A	COSMOS 728	7745	USSR	18 APR	29 APR 75	
1975-031B		7746	USSR	18 APR	30 APR 75	
1975-031C		7751	USSR	18 APR	20 APR 75	
1975-031D		7761	USSR	18 APR	22 APR 75	
1975-031E		7776	USSR	18 APR	03 MAY 75	
1975-031F		7778	USSR	18 APR	28 APR 75	
1975-031G		7779	USSR	18 APR	16 MAY 75	
1975-032A		7747	US	18 APR	05 JUN 75	
1975-033A	ARIABAT	7752	INDIA	19 APR	11 FEB 92	
1975-033B		7753	USSR	19 APR	06 NOV 88	
1975-033C		8058	USSR	19 APR	12 MAR 78	
1975-033D		9839	USSR	19 APR	21 JUN 77	
1975-035A	COSMOS 730	7770	USSR	24 APR	06 MAY 75	
1975-035B		7771	USSR	24 APR	29 APR 75	

International Designation	Satellite Name	Number	Source	Launch	Decay	Note
1975-035C		7772	USSR	24 APR	01 MAY 75	
1975-035D		7773	USSR	24 APR	28 APR 75	
1975-035E		7785	USSR	24 APR	07 MAY 75	
1975-035F		7786	USSR	24 APR	06 MAY 75	
1975-035G		7787	USSR	24 APR	24 MAY 75	
1975-036B		7781	USSR	29 APR	29 MAY 75	
1975-036C		7782	USSR	29 APR	03 JUN 75	
1975-037A	EXPLORER 53	7788	US	07 MAY	09 APR 79	
1975-037B		7789	US	07 MAY	09 SEP 79	
1975-038B		7791	US	07 MAY	03 NOV 76	
1975-038C		7792	US	07 MAY	08 MAY 75	
1975-039A	POLLUX	7801	FRAN	17 MAY	05 AUG 75	
1975-039B	CASTOR	7802	FRAN	17 MAY	18 FEB 79	
1975-039C		7803	FRAN	17 MAY	07 AUG 76	
1975-039D		7804	FRAN	17 MAY	26 SEP 76	
1975-039E		7805	FRAN	17 MAY	18 JUN 76	
1975-039F		7806	FRAN	17 MAY	30 SEP 76	
1975-039G		8035	FRAN	17 MAY	29 JAN 78	
1975-040A		7807	US	20 MAY	26 MAY 75	
1975-040B		7808	US	20 MAY	26 MAY 75	E
1975-040C		7809	US	20 MAY	21 MAY 75	
1975-041A	COSMOS 731	7810	USSR	21 MAY	02 JUN 75	
1975-041B		7811	USSR	21 MAY	30 MAY 75	
1975-041C		7812	USSR	21 MAY	24 MAY 75	
1975-041D		7813	USSR	21 MAY	26 MAY 75	
1975-041E		7879	USSR	21 MAY	06 JUN 75	
1975-041F		7880	USSR	21 MAY	09 JUN 75	
1975-041G		7881	USSR	21 MAY	11 JUN 75	
1975-041H		7885	USSR	21 MAY	19 JUN 75	
1975-044A	SOYUZ 18	7818	USSR	24 MAY	26 JUL 75	E
1975-044B		7819	USSR	24 MAY	27 MAY 75	BB
1975-044C		7911	USSR	24 MAY	15 MAY 76	
1975-044D		7912	USSR	24 MAY	20 SEP 75	
1975-044E		7948	USSR	24 MAY	28 MAR 76	
1975-044F		7949	USSR	24 MAY	17 JUN 75	
1975-044G		7953	USSR	24 MAY	04 OCT 75	
1975-044H		7955	USSR	24 MAY	01 OCT 75	
1975-044J		7956	USSR	24 MAY	24 SEP 75	
1975-044X		7962	USSR	24 MAY	15 SEP 75	
1975-044L		7984	USSR	24 MAY	12 OCT 75	
1975-044M		8019	USSR	24 MAY	07 OCT 75	
1975-044N		8020	USSR	24 MAY	06 MAR 76	
1975-044P		8021	USSR	24 MAY	09 OCT 75	
1975-044Q		8022	USSR	24 MAY	09 OCT 75	
1975-044R		8023	USSR	24 MAY	29 MAR 76	
1975-044S		8024	USSR	24 MAY	12 FEB 76	
1975-044T		8025	USSR	24 MAY	05 OCT 75	
1975-044U		8038	USSR	24 MAY	20 OCT 75	
1975-046A	COSMOS 740	7821	USSR	28 MAY	10 JUN 75	
1975-046B		7829	USSR	28 MAY	03 JUN 75	
1975-046C		7830	USSR	28 MAY	04 JUN 75	
1975-046D		7920	USSR	28 MAY	18 JUN 75	
1975-046E		7921	USSR	28 MAY	18 JUN 75	
1975-047A	COSMOS 741	7877	USSR	30 MAY	11 JUN 75	
1975-047B		7878	USSR	30 MAY	04 JUN 75	
1975-048A	COSMOS 742	7900	USSR	03 JUN	15 JUN 75	
1975-048B		7901	USSR	03 JUN	11 JUN 75	
1975-048C		7913	USSR	03 JUN	12 JUN 75	
1975-048D		7951	USSR	03 JUN	16 JUN 75	
1975-048E		7952	USSR	03 JUN	15 JUN 75	
1975-049A	MOLNIYA 1-30	7903	USSR	05 JUN	12 AUG 87	
1975-049C		7904	USSR	05 JUN	16 JUL 75	
1975-049D		7905	USSR	05 JUN	09 JUL 75	
1975-049E		7906	USSR	05 JUN	16 AUG 75	
1975-049F		8548	USSR	05 JUN	13 OCT 90	
1975-050B		7916	USSR	08 JUN	09 JUN 75	
1975-050C		7917	USSR	08 JUN	09 JUN 75	
1975-050D	DESCENT CRAFT	8411	USSR	08 JUN	22 OCT 75	N
1975-051A		7918	US	08 JUN	05 NOV 75	
1975-051B		7919	US	08 JUN	11 JUN 75	
1975-052BA		21332	US	12 JUN	31 JUL 91	
1975-052C		7965	US	12 JUN	01 JUL 81	
1975-052CB		21357	US	12 JUN	18 AUG 92	
1975-052DM		21391	US	12 JUN	05 JUN 91	
1975-052DP		21431	US	12 JUN	11 DEC 91	
1975-052DQ		21432	US	12 JUN	03 MAR 92	
1975-052DU		21436	US	12 JUN	19 DEC 91	
1975-052EH		21449	US	12 JUN	28 AUG 91	
1975-052EK		21451	US	12 JUN	29 OCT 91	
1975-052EP		21455	US	12 JUN	20 AUG 91	
1975-052EQ		21456	US	12 JUN	30 SEP 91	
1975-052FC		21468	US	12 JUN	12 NOV 91	
1975-052FJ		21474	US	12 JUN	05 JUL 91	
1975-052FP		21494	US	12 JUN	05 AUG 91	
1975-052FQ		21495	US	12 JUN	17 AUG 91	
1975-052FX		21502	US	12 JUN	12 NOV 91	
1975-052FY		21503	US	12 JUN	04 AUG 92	
1975-052GA		21505	US	12 JUN	24 OCT 91	
1975-052GC		21507	US	12 JUN	10 AUG 91	
1975-052GD		21508	US	12 JUN	03 SEP 91	
1975-052GE		21509	US	12 JUN	29 AUG 91	
1975-052GF		21510	US	12 JUN	31 OCT 91	
1975-052GG		21511	US	12 JUN	10 JUL 91	
1975-052GJ		21513	US	12 JUN	04 JAN 92	
1975-052GN		21517	US	12 JUN	29 SEP 91	
1975-052GP		21518	US	12 JUN	29 OCT 91	
1975-052GQ		21519	US	12 JUN	13 NOV 91	
1975-052GR		21520	US	12 JUN	21 NOV 91	
1975-052GT		21522	US	12 JUN	10 NOV 91	
1975-052GV		21539	US	12 JUN	31 AUG 91	
1975-052HG		21562	US	12 JUN	07 AUG 91	
1975-052HH		21563	US	12 JUN	09 OCT 91	
1975-052HJ		21564	US	12 JUN	03 OCT 91	
1975-052HP		21569	US	12 JUN	16 NOV 91	
1975-052HR		21597	US	12 JUN	16 MAR 93	
1975-052HU		21600	US	12 JUN	10 NOV 91	
1975-052HX		21671	US	12 JUN	27 OCT 91	
1975-052HY		21672	US	12 JUN	24 DEC 91	
1975-052HZ		21673	US	12 JUN	11 NOV 91	
1975-052JB		21675	US	12 JUN	01 DEC 91	
1975-052JF		21679	US	12 JUN	19 OCT 91	
1975-052JH		21681	US	12 JUN	06 SEP 91	
1975-052JK		21683	US	12 JUN	17 OCT 92	
1975-052JL		21684	US	12 JUN	28 AUG 91	
1975-052JM		21685	US	12 JUN	01 NOV 91	
1975-052JP		21687	US	12 JUN	08 NOV 91	
1975-053A	COSMOS 743	7925	USSR	12 JUN	25 JUN 75	
1975-053B		7926	USSR	12 JUN	20 JUN 75	
1975-053C		7927	USSR	12 JUN	18 JUN 75	
1975-053D		7972	USSR	12 JUN	29 JUN 75	
1975-053E		7981	USSR	12 JUN	26 JUN 75	
1975-053F		7987	USSR	12 JUN	26 JUN 75	
1975-054B		7950	USSR	14 JUN	15 JUN 75	
1975-054C		7954	USSR	14 JUN	15 JUN 75	
1975-054D	DESCENT CRAFT	8423	USSR	14 JUN	25 OCT 75	N
1975-056A	COSMOS 744	7968	USSR	20 JUN	12 OCT 91	
1975-057A	OSO 8	7970	US	21 JUN	09 JUL 86	
1975-057B		7971	US	21 JUN	09 APR 81	
1975-058A	COSMOS 745	7982	USSR	24 JUN	12 MAR 76	
1975-058B		7983	USSR	24 JUN	14 NOV 75	
1975-059A	COSMOS 746	7985	USSR	25 JUN	08 JUL 75	
1975-059B		7986	USSR	25 JUN	01 JUL 75	
1975-059C		7988	USSR	25 JUN	26 JUN 75	
1975-059D		7989	USSR	25 JUN	26 JUN 75	
1975-059E		8011	USSR	25 JUN	12 JUL 75	
1975-059F		8012	USSR	25 JUN	09 JUL 75	
1975-060A	COSMOS 747	7990	USSR	27 JUN	09 JUL 75	
1975-060B		7991	USSR	27 JUN	03 JUL 75	
1975-060C		7992	USSR	27 JUN	28 JUN 75	
1975-060D		7993	USSR	27 JUN	02 JUL 75	
1975-060E		7994	USSR	27 JUN	29 JUN 75	
1975-060F		8013	USSR	27 JUN	17 JUL 75	
1975-060G		8014	USSR	27 JUN	09 JUL 75	
1975-061A	COSMOS 748	8006	USSR	03 JUL	16 JUL 75	
1975-061B		8007	USSR	03 JUL	09 JUL 75	
1975-061C		8008	USSR	03 JUL	15 JUL 75	
1975-061D		8028	USSR	03 JUL	15 JUL 75	
1975-061E		8029	USSR	03 JUL	16 JUL 75	
1975-061F		8034	USSR	03 JUL	17 JUL 75	
1975-062A	COSMOS 749	8009	USSR	04 JUL	26 SEP 80	
1975-062B		8010	USSR	04 JUL	26 DEC 80	
1975-062C		8107	USSR	04 JUL	27 AUG 81	
1975-063B		8016	USSR	08 JUL	09 AUG 75	
1975-063C		8017	USSR	08 JUL	09 AUG 75	
1975-065A	SOYUZ 19	8030	USSR	15 JUL	21 JUL 75	BC
1975-065B		8031	USSR	15 JUL	17 JUL 75	
1975-066A	APOLLO 18	8032	US	15 JUL	24 JUL 75	BC
1975-066B		8033	US	15 JUL	16 JUL 75	
1975-066C		8042	US	15 JUL	02 AUG 75	
1975-067A	COSMOS 750	8036	USSR	17 JUL	29 SEP 77	
1975-067B		8037	USSR	17 JUL	17 NOV 76	
1975-067C		8045	USSR	17 JUL	28 JUL 76	
1975-068A	COSMOS 751	8040	USSR	23 JUL	04 AUG 75	
1975-068B		8041	USSR	23 JUL	30 JUL 75	
1975-069A	COSMOS 752	8043	USSR	24 JUL	28 FEB 81	
1975-069B		8044	USSR	24 JUL	05 NOV 79	
1975-070A	MAO 3	8053	PRC	26 JUL	14 SEP 75	
1975-070B		8054	PRC	26 JUL	25 AUG 75	
1975-071A	COSMOS 753	8059	USSR	31 JUL	13 AUG 75	
1975-071B		8060	USSR	31 JUL	07 AUG 75	
1975-071C		8061	USSR	31 JUL	01 AUG 75	
1975-071D		8067	USSR	31 JUL	16 AUG 75	
1975-071E		8068	USSR	31 JUL	13 AUG 75	
1975-073A	COSMOS 754	8069	USSR	13 AUG	26 AUG 75	
1975-073B		8070	USSR	13 AUG	25 AUG 75	
1975-073C		8071	USSR	13 AUG	15 AUG 75	
1975-073D		8129	USSR	13 AUG	31 AUG 75	
1975-073E		8130	USSR	13 AUG	27 AUG 75	
1975-073F		8131	USSR	13 AUG	27 AUG 75	
1975-075C	VIKING LANDER 1	9024	US	20 AUG	20 JUL 76	BD
1975-076A	COSMOS 756	8127	USSR	22 AUG	05 NOV 92	
1975-077D		8135	US	27 AUG	23 SEP 76	
1975-078A	COSMOS 757	8147	USSR	27 AUG	09 SEP 75	
1975-078B		8148	USSR	27 AUG	03 SEP 75	
1975-078C		8149	USSR	27 AUG	02 SEP 75	
1975-078D		8150	USSR	27 AUG	02 SEP 75	
1975-078E		8193	USSR	27 AUG	08 SEP 75	
1975-078F		8194	USSR	27 AUG	11 SEP 75	
1975-079A	MOLNIYA 1-31	8187	USSR	02 SEP	19 NOV 85	
1975-079B		8188	USSR	02 SEP	10 NOV 75	
1975-079C		8189	USSR	02 SEP	01 NOV 75	
1975-079D		8190	USSR	02 SEP	12 OCT 75	
1975-079E		8274	USSR	02 SEP	08 JUL 86	
1975-080A	COSMOS 758	8191	USSR	05 SEP	25 SEP 75	
1975-080AA		8222	USSR	05 SEP	09 SEP 75	
1975-080AB		8223	USSR	05 SEP	09 SEP 75	

International Designation	Satellite Name	Number	Source	Launch	Decay	Note
1975-080AC		8224	USSR	05 SEP	09 SEP 75	
1975-080AD		8225	USSR	05 SEP	09 SEP 75	
1975-080AE		8226	USSR	05 SEP	09 SEP 75	
1975-080AF		8227	USSR	05 SEP	09 SEP 75	
1975-080AG		8228	USSR	05 SEP	09 SEP 75	
1975-080AH		8229	USSR	05 SEP	09 SEP 75	
1975-080AJ		8230	USSR	05 SEP	09 SEP 75	
1975-080AK		8231	USSR	05 SEP	09 SEP 75	
1975-080AL	AURA	8232	USSR	05 SEP	09 SEP 75	
1975-080AM		8233	USSR	05 SEP	09 SEP 75	
1975-080AN		8234	USSR	05 SEP	09 SEP 75	
1975-080AP		8235	USSR	05 SEP	09 SEP 75	
1975-080AQ		8236	USSR	05 SEP	09 SEP 75	
1975-080AR		8237	USSR	05 SEP	09 SEP 75	
1975-080AS		8238	USSR	05 SEP	09 SEP 75	
1975-080AT		8239	USSR	05 SEP	09 SEP 75	
1975-080AU		8240	USSR	05 SEP	09 SEP 75	
1975-080AV		8241	USSR	05 SEP	09 SEP 75	
1975-080AW		8242	USSR	05 SEP	09 SEP 75	
1975-080AX		8243	USSR	05 SEP	09 SEP 75	
1975-080AY		8244	USSR	05 SEP	09 SEP 75	
1975-080AZ		8245	USSR	05 SEP	09 SEP 75	
1975-080B		8192	USSR	05 SEP	10 SEP 75	BE
1975-080BA		8246	USSR	05 SEP	09 SEP 75	
1975-080BB		8247	USSR	05 SEP	09 SEP 75	
1975-080BC		8248	USSR	05 SEP	09 SEP 75	
1975-080BD		8249	USSR	05 SEP	09 SEP 75	
1975-080BE		8250	USSR	05 SEP	09 SEP 75	
1975-080BF		8251	USSR	05 SEP	09 SEP 75	
1975-080BG		8252	USSR	05 SEP	09 SEP 75	
1975-080BH		8253	USSR	05 SEP	09 SEP 75	
1975-080BJ		8254	USSR	05 SEP	09 SEP 75	
1975-080BK		8255	USSR	05 SEP	09 SEP 75	
1975-080BL		8256	USSR	05 SEP	09 SEP 75	
1975-080BM		8257	USSR	05 SEP	09 SEP 75	
1975-080BN		8258	USSR	05 SEP	09 SEP 75	
1975-080BP		8259	USSR	05 SEP	09 SEP 75	
1975-080BQ		8260	USSR	05 SEP	09 SEP 75	
1975-080BR		8261	USSR	05 SEP	09 SEP 75	
1975-080BS		8262	USSR	05 SEP	09 SEP 75	
1975-080BT		8263	USSR	05 SEP	09 SEP 75	
1975-080BU		8264	USSR	05 SEP	09 SEP 75	
1975-080BV		8265	USSR	05 SEP	09 SEP 75	
1975-080BW		8266	USSR	05 SEP	09 SEP 75	
1975-080BX		8267	USSR	05 SEP	09 SEP 75	
1975-080BY		8268	USSR	05 SEP	09 SEP 75	
1975-080BZ		8269	USSR	05 SEP	09 SEP 75	
1975-080C		8200	USSR	05 SEP	09 SEP 75	
1975-080CA		8270	USSR	05 SEP	09 SEP 75	
1975-080CB		8271	USSR	05 SEP	09 SEP 75	
1975-080CC		8273	USSR	05 SEP	17 SEP 75	
1975-080CD		8280	USSR	05 SEP	03 OCT 75	
1975-080CE		8283	USSR	05 SEP	26 SEP 75	
1975-080D		8201	USSR	05 SEP	09 SEP 75	
1975-080E		8202	USSR	05 SEP	09 SEP 75	
1975-080F		8203	USSR	05 SEP	09 SEP 75	
1975-080G		8204	USSR	05 SEP	09 SEP 75	
1975-080H		8205	USSR	05 SEP	09 SEP 75	
1975-080J		8206	USSR	05 SEP	09 SEP 75	
1975-080K		8207	USSR	05 SEP	09 SEP 75	
1975-080L		8208	USSR	05 SEP	09 SEP 75	
1975-080M		8209	USSR	05 SEP	09 SEP 75	
1975-080M		8210	USSR	05 SEP	09 SEP 75	
1975-080P		8211	USSR	05 SEP	09 SEP 75	
1975-080Q		8212	USSR	05 SEP	09 SEP 75	
1975-080R		8213	USSR	05 SEP	09 SEP 75	
1975-080S		8214	USSR	05 SEP	09 SEP 75	
1975-080T		8215	USSR	05 SEP	09 SEP 75	
1975-080U		8216	USSR	05 SEP	09 SEP 75	
1975-080V		8217	USSR	05 SEP	09 SEP 75	
1975-080W		8218	USSR	05 SEP	09 SEP 75	
1975-080X		8219	USSR	05 SEP	09 SEP 75	
1975-080Y		8220	USSR	05 SEP	09 SEP 75	
1975-080Z		8221	USSR	05 SEP	09 SEP 75	
1975-081B		8196	USSR	09 SEP	09 OCT 75	
1975-081C		8198	USSR	09 SEP	18 OCT 75	
1975-083C	VIKING LANDER 2	9408	US	09 SEP	03 SEP 76	BD
1975-084A	COSMOS 759	8275	USSR	12 SEP	23 SEP 75	
1975-084B		8276	USSR	12 SEP	25 SEP 75	
1975-084C		8277	USSR	12 SEP	14 SEP 75	
1975-084D		8278	USSR	12 SEP	17 SEP 75	
1975-084E		8279	USSR	12 SEP	16 SEP 75	
1975-084F		8324	USSR	12 SEP	26 SEP 75	
1975-084G		8329	USSR	12 SEP	26 SEP 75	
1975-085A	COSMOS 760	8281	USSR	16 SEP	30 SEP 75	
1975-085B		8282	USSR	16 SEP	21 SEP 75	
1975-085C		8284	USSR	16 SEP	18 SEP 75	
1975-085D		8334	USSR	16 SEP	05 OCT 75	
1975-085E		8335	USSR	16 SEP	30 SEP 75	
1975-085F		8347	USSR	16 SEP	01 OCT 75	
1975-088A	COSMOS 769	8322	USSR	23 SEP	05 OCT 75	
1975-088B		8323	USSR	23 SEP	02 OCT 75	
1975-088C		8349	USSR	23 SEP	13 OCT 75	
1975-090A	COSMOS 771	8327	USSR	25 SEP	08 OCT 75	
1975-090B		8328	USSR	25 SEP	29 SEP 75	
1975-090C		8355	USSR	25 SEP	12 OCT 75	
1975-090D		8356	USSR	25 SEP	08 OCT 75	
1975-092A	D2-B	8332	FRAN	27 SEP	30 SEP 82	
1975-092B		8333	FRAN	27 SEP	30 MAR 82	
1975-092C		8336	FRAN	27 SEP	17 OCT 78	
1975-092D		8337	FRAN	27 SEP	16 MAY 78	
1975-092E		8340	FRAN	27 SEP	27 OCT 78	
1975-092F		8341	FRAN	27 SEP	01 JAN 80	
1975-092G		8342	FRAN	27 SEP	20 NOV 79	
1975-093A	COSMOS 772	8338	USSR	29 SEP	02 OCT 75	
1975-093B		8339	USSR	29 SEP	06 OCT 75	
1975-095A	COSMOS 774	8345	USSR	01 OCT	15 OCT 75	
1975-095B		8348	USSR	01 OCT	10 OCT 75	
1975-095C		8350	USSR	01 OCT	08 OCT 75	
1975-095D		8351	USSR	01 OCT	08 OCT 75	
1975-095E		8362	USSR	01 OCT	14 OCT 75	
1975-095F		8363	USSR	01 OCT	16 OCT 75	
1975-095G		8365	USSR	01 OCT	14 OCT 75	
1975-096A	EXPLORER 54	8353	US	06 OCT	12 MAR 76	
1975-096B		8354	US	06 OCT	01 APR 76	
1975-097B		8358	USSR	08 OCT	10 OCT 75	
1975-097C		8359	USSR	08 OCT	09 OCT 75	
1975-097D		8414	USSR	08 OCT	01 MAR 78	
1975-097E		8415	USSR	08 OCT	17 OCT 78	
1975-098A		8360	US	09 OCT	30 NOV 75	
1975-099A	TIP 2	8361	US	12 OCT	26 MAY 91	
1975-099B		8364	US	12 OCT	28 MAR 78	
1975-099C		8409	US	12 OCT	12 NOV 78	
1975-099D		8410	US	12 OCT	31 MAY 76	
1975-100B		8367	US	16 OCT	05 JAN 76	
1975-100D		8371	US	16 OCT	11 NOV 75	
1975-100E		8372	US	16 OCT	01 NOV 75	
1975-101A	COSMOS 776	8369	USSR	17 OCT	29 OCT 75	
1975-101B		8370	USSR	17 OCT	23 OCT 75	
1975-101C		8373	USSR	17 OCT	08 NOV 75	
1975-101D		8412	USSR	17 OCT	01 NOV 75	
1975-101E		8413	USSR	17 OCT	31 OCT 75	
1975-102A	COSMOS 777	8416	USSR	29 OCT	03 JUN 76	
1975-102AA		8644	USSR	29 OCT	08 JAN 77	
1975-102AB		8647	USSR	29 OCT	05 JUN 78	
1975-102AC		8648	USSR	29 OCT	18 APR 79	
1975-102AD		8649	USSR	29 OCT	18 JUN 77	
1975-102AE		8650	USSR	29 OCT	27 MAR 76	
1975-102AF		8651	USSR	29 OCT	16 MAR 77	
1975-102AG		8652	USSR	29 OCT	20 NOV 76	
1975-102AH		8653	USSR	29 OCT	13 DEC 76	
1975-102AJ		8654	USSR	29 OCT	15 NOV 76	
1975-102AK		8655	USSR	29 OCT	26 JAN 77	
1975-102AL		8656	USSR	29 OCT	30 OCT 77	
1975-102AM		8657	USSR	29 OCT	13 DEC 76	
1975-102AN		8711	USSR	29 OCT	14 DEC 76	
1975-102AP		8712	USSR	29 OCT	01 MAY 77	
1975-102AQ		8713	USSR	29 OCT	01 APR 76	
1975-102AR		8714	USSR	29 OCT	20 NOV 76	
1975-102AS		8715	USSR	29 OCT	09 FEB 77	
1975-102AT		8716	USSR	29 OCT	04 DEC 76	
1975-102AU		8717	USSR	29 OCT	23 FEB 77	
1975-102AV		8718	USSR	29 OCT	13 DEC 76	
1975-102AW		8719	USSR	29 OCT	13 DEC 76	
1975-102AX		8720	USSR	29 OCT	25 FEB 77	
1975-102AY		8721	USSR	29 OCT	03 MAR 76	
1975-102AZ		8722	USSR	29 OCT	07 DEC 76	
1975-102B		8417	USSR	29 OCT	30 OCT 75	BF
1975-102BA		8723	USSR	29 OCT	14 APR 77	
1975-102BB		8724	USSR	29 OCT	15 NOV 76	
1975-102BC		8725	USSR	29 OCT	02 APR 76	
1975-102BD		8726	USSR	29 OCT	20 APR 76	
1975-102BE		8727	USSR	29 OCT	15 NOV 76	
1975-102BF		8728	USSR	29 OCT	29 MAR 76	
1975-102BG		8729	USSR	29 OCT	08 JAN 77	
1975-102BH		8730	USSR	29 OCT	13 DEC 76	
1975-102BJ		8731	USSR	29 OCT	05 NOV 77	
1975-102BK		8732	USSR	29 OCT	08 SEP 76	
1975-102BL		8733	USSR	29 OCT	08 JAN 77	
1975-102BM		8734	USSR	29 OCT	27 JAN 77	
1975-102BN		8735	USSR	29 OCT	25 APR 76	
1975-102BP		8736	USSR	29 OCT	09 JUN 77	
1975-102BQ		9563	USSR	29 OCT	27 NOV 76	
1975-102C		8622	USSR	29 OCT	04 DEC 76	
1975-102D		8623	USSR	29 OCT	13 DEC 76	
1975-102E		8624	USSR	29 OCT	15 NOV 76	
1975-102F		8625	USSR	29 OCT	15 MAR 76	
1975-102G		8626	USSR	29 OCT	08 JAN 77	
1975-102H		8627	USSR	29 OCT	23 FEB 77	
1975-102J		8628	USSR	29 OCT	15 NOV 76	
1975-102K		8629	USSR	29 OCT	15 NOV 76	
1975-102L		8630	USSR	29 OCT	19 AUG 76	
1975-102M		8631	USSR	29 OCT	21 AUG 77	
1975-102N		8632	USSR	29 OCT	20 JUL 77	
1975-102P		8633	USSR	29 OCT	09 DEC 76	
1975-102Q		8634	USSR	29 OCT	04 DEC 76	
1975-102R		8635	USSR	29 OCT	15 NOV 76	
1975-102S		8636	USSR	29 OCT	20 JAN 77	
1975-102T		8637	USSR	29 OCT	04 APR 77	
1975-102U		8638	USSR	29 OCT	14 MAR 76	
1975-102V		8639	USSR	29 OCT	04 DEC 76	
1975-102W		8640	USSR	29 OCT	15 NOV 76	
1975-102X		8641	USSR	29 OCT	30 JUN 77	
1975-102Y		8642	USSR	29 OCT	15 NOV 76	
1975-102Z		8643	USSR	29 OCT	04 DEC 76	
1975-104A	COSMOS 779	8420	USSR	04 NOV	18 NOV 75	

International Designation	Satellite Name	Number	Source	Launch	Decay	Note
1975-104B		8422	USSR	04 NOV	10 NOV 75	
1975-104C		8428	USSR	04 NOV	19 NOV 75	
1975-104D		8429	USSR	04 NOV	17 NOV 75	
1975-105B		8426	USSR	14 NOV	15 DEC 75	
1975-105C		8427	USSR	14 NOV	05 DEC 75	
1975-106A	SOYUZ 20	8430	USSR	17 NOV	16 FEB 76	C
1975-106B		8431	USSR	17 NOV	20 NOV 75	
1975-107A	EXPLORER 55	8440	US	20 NOV	10 JUN 81	
1975-107B		8441	US	20 NOV	19 MAR 76	
1975-108A	COSMOS 780	8442	USSR	21 NOV	03 DEC 75	
1975-108B		8443	USSR	21 NOV	27 NOV 75	
1975-108C		8446	USSR	21 NOV	22 NOV 75	
1975-108D		8460	USSR	21 NOV	11 DEC 75	
1975-108E		8461	USSR	21 NOV	02 DEC 75	
1975-109A	COSMOS 781	8444	USSR	21 NOV	26 NOV 80	
1975-109B		8445	USSR	21 NOV	20 DEC 80	
1975-109C		8447	USSR	21 NOV	28 JUN 78	
1975-109D		8448	USSR	21 NOV	25 NOV 77	
1975-109E		8449	USSR	21 NOV	09 OCT 77	
1975-109F		8776	USSR	21 NOV	18 NOV 77	
1975-109G		8777	USSR	21 NOV	18 JUL 78	
1975-109H		9693	USSR	21 NOV	11 SEP 77	
1975-109J		9994	USSR	21 NOV	11 FEB 78	
1975-109X		9995	USSR	21 NOV	12 JAN 78	
1975-109L		10433	USSR	21 NOV	19 MAY 78	
1975-110A	COSMOS 782	8450	USSR	25 NOV	15 DEC 75	
1975-110B		8451	USSR	25 NOV	23 DEC 75	
1975-110C		8486	USSR	25 NOV	31 DEC 75	
1975-110D		8487	USSR	25 NOV	10 FEB 76	
1975-110E		8518	USSR	25 NOV	26 JAN 76	
1975-111A	MAO 4	8452	PRC	26 NOV	29 DEC 75	
1975-111B		8453	PRC	26 NOV	22 DEC 75	
1975-111C		8454	PRC	26 NOV	29 NOV 75	
1975-111D		8455	PRC	26 NOV	28 NOV 75	
1975-111E		8456	PRC	26 NOV	28 NOV 75	
1975-111F		8457	PRC	26 NOV	28 NOV 75	
1975-113A	COSMOS 784	8463	USSR	03 DEC	15 DEC 75	
1975-113B		8464	USSR	03 DEC	07 DEC 75	
1975-113C		8465	USSR	03 DEC	04 DEC 75	
1975-113D		8466	USSR	03 DEC	08 DEC 75	
1975-113E		8481	USSR	03 DEC	15 DEC 75	
1975-113F		8484	USSR	03 DEC	20 DEC 75	
1975-113G		8485	USSR	03 DEC	21 DEC 75	
1975-114A		8467	US	04 DEC	01 APR 76	
1975-114B		8468	US	04 DEC	01 MAY 78	
1975-114C		8469	US	04 DEC	06 DEC 75	
1975-114D		8470	US	04 DEC	03 JUN 76	
1975-115A	INTERCOSMOS 14	8471	USSR	11 DEC	27 FEB 83	
1975-115B		8472	USSR	11 DEC	28 NOV 82	
1975-115C		8474	USSR	11 DEC	11 NOV 78	
1975-115D		8475	USSR	11 DEC	30 MAY 82	
1975-115E		8765	USSR	11 DEC	13 APR 81	
1975-115F		8766	USSR	11 DEC	29 MAY 78	
1975-116B		8477	USSR	12 DEC	14 DEC 75	
1975-116C		8480	USSR	12 DEC	05 FEB 76	
1975-117B		8478	US	13 DEC	07 AUG 76	
1975-117C		8479	US	13 DEC	10 NOV 82	
1975-118B		8483	US	14 DEC	19 DEC 75	
1975-119A	MAO 5	8488	PRC	16 DEC	27 JAN 76	
1975-119B		8491	PRC	16 DEC	10 JAN 76	
1975-120A	COSMOS 786	8489	USSR	16 DEC	29 DEC 75	
1975-120B		8490	USSR	16 DEC	21 DEC 75	
1975-120C		8523	USSR	16 DEC	03 JAN 76	
1975-120D		8524	USSR	16 DEC	29 DEC 75	
1975-120E		8528	USSR	16 DEC	30 DEC 75	
1975-121B		8493	USSR	17 DEC	14 JAN 76	
1975-121C		8494	USSR	17 DEC	11 JAN 76	
1975-121D		8529	USSR	17 DEC	13 OCT 86	
1975-122B		8511	USSR	22 DEC	22 FEB 76	
1975-122C		8512	USSR	22 DEC	09 FEB 76	
1975-123B		8514	USSR	22 DEC	22 DEC 75	
1975-123C		8515	USSR	22 DEC	25 DEC 75	
1975-125A	MOLNIYA 3-4	8521	USSR	27 DEC	12 AUG 86	
1975-125B		8522	USSR	27 DEC	23 JAN 76	
1975-125C		8525	USSR	27 DEC	19 JAN 76	
1975-125D		8526	USSR	27 DEC	28 DEC 75	
1975-125E		8527	USSR	27 DEC	08 JAN 76	
1975-125F		8600	USSR	27 DEC	30 JUL 86	

1976 LAUNCHES

International Designation	Satellite Name	Number	Source	Launch	Decay	Note
1976-001A	COSMOS 787	8530	USSR	06 JAN	12 DEC 80	
1976-001B		8531	USSR	06 JAN	15 MAR 81	
1976-001C		8549	USSR	06 JAN	20 OCT 77	
1976-001D		8550	USSR	06 JAN	24 MAR 77	
1976-001E		9731	USSR	06 JAN	15 JUL 78	
1976-001F		9790	USSR	06 JAN	05 AUG 77	
1976-001G		9791	USSR	06 JAN	01 DEC 77	
1976-001H		9792	USSR	06 JAN	12 DEC 77	
1976-001J		9837	USSR	06 JAN	27 AUG 77	
1976-001K		9838	USSR	06 JAN	30 JUL 77	
1976-002A	COSMOS 788	8551	USSR	07 JAN	20 JAN 76	
1976-002B		8552	USSR	07 JAN	14 JAN 76	
1976-002C		8553	USSR	07 JAN	12 JAN 76	
1976-002D		8588	USSR	07 JAN	27 JAN 76	
1976-002E		8589	USSR	07 JAN	23 JAN 76	
1976-002F		8590	USSR	07 JAN	20 JAN 76	
1976-004B		8586	US	17 JAN	14 FEB 76	
1976-004C		8587	US	17 JAN	14 APR 76	
1976-004D		8598	US	17 JAN	17 MAY 77	
1976-006B		8602	USSR	22 JAN	02 APR 76	
1976-006C		8603	USSR	22 JAN	19 APR 76	
1976-007A	COSMOS 790	8604	USSR	22 JAN	12 NOV 80	
1976-007B		8605	USSR	22 JAN	09 DEC 80	
1976-009A	COSMOS 799	8616	USSR	29 JAN	10 FEB 76	
1976-009B		8617	USSR	29 JAN	11 FEB 76	
1976-009C		8618	USSR	29 JAN	31 JAN 76	
1976-009D		8619	USSR	29 JAN	09 FEB 76	
1976-012A	COSMOS 801	8658	USSR	05 FEB	05 JAN 78	
1976-012B		8659	USSR	05 FEB	24 MAR 77	BG
1976-012C		8660	USSR	05 FEB	14 APR 77	
1976-012D		8661	USSR	05 FEB	28 JUL 76	
1976-012E		8662	USSR	05 FEB	11 OCT 76	
1976-012F		8663	USSR	05 FEB	05 APR 76	
1976-012G		8664	USSR	05 FEB	11 OCT 76	
1976-012H		8665	USSR	05 FEB	17 AUG 76	
1976-012J		8666	USSR	05 FEB	14 JUN 76	
1976-012K		8667	USSR	05 FEB	18 MAR 76	
1976-012L		8668	USSR	05 FEB	11 OCT 76	
1976-012M		8669	USSR	05 FEB	11 OCT 76	
1976-012N		8670	USSR	05 FEB	11 OCT 76	
1976-012P		8671	USSR	05 FEB	28 JUL 76	
1976-012Q		8672	USSR	05 FEB	11 OCT 76	
1976-012R		8673	USSR	05 FEB	25 SEP 76	
1976-012S		8674	USSR	05 FEB	21 JUN 76	
1976-013A	COSMOS 802	8681	USSR	11 FEB	25 FEB 76	
1976-013B		8686	USSR	11 FEB	17 FEB 76	
1976-013C		8706	USSR	11 FEB	25 FEB 76	
1976-013D		8707	USSR	11 FEB	29 FEB 76	
1976-013E		8708	USSR	11 FEB	26 FEB 76	
1976-014B		8689	USSR	12 FEB	09 NOV 83	
1976-014C		8690	USSR	12 FEB	09 DEC 81	
1976-015A	COSMOS 804	8694	USSR	16 FEB	16 FEB 76	
1976-015B		8695	USSR	16 FEB	26 FEB 76	
1976-016A		8696	US	19 FEB	19 FEB 76	
1976-017B		8698	US	19 FEB	16 MAR 76	
1976-017D		8704	US	19 FEB	02 MAR 76	
1976-017E		8705	US	19 FEB	29 FEB 76	
1976-018A	COSMOS 805	8699	USSR	20 FEB	11 MAR 76	
1976-018B		8700	USSR	20 FEB	27 FEB 76	
1976-018C		8703	USSR	20 FEB	22 FEB 76	
1976-020A	COSMOS 806	8737	USSR	10 MAR	23 MAR 76	
1976-020B		8738	USSR	10 MAR	15 MAR 76	
1976-020C		8768	USSR	10 MAR	30 MAR 76	
1976-020D		8769	USSR	10 MAR	26 MAR 76	
1976-020E		8771	USSR	10 MAR	25 MAR 76	
1976-021A	MOLNIYA 1-33	8741	USSR	11 MAR	10 OCT 90	
1976-021B		8742	USSR	11 MAR	13 APR 76	
1976-021C		8743	USSR	11 MAR	05 APR 76	
1976-021D		9411	USSR	11 MAR	01 JUL 91	
1976-023E		8750	US	15 MAR	15 MAR 76	
1976-025A	COSMOS 809	8758	USSR	18 MAR	30 MAR 76	
1976-025B		8759	USSR	18 MAR	26 MAR 76	
1976-025C		8760	USSR	18 MAR	19 MAR 76	
1976-025D		8761	USSR	18 MAR	19 MAR 76	
1976-026A	MOLNIYA 1-34	8762	USSR	19 MAR	14 MAY 85	
1976-026B		8763	USSR	19 MAR	27 APR 76	
1976-026C		8767	USSR	19 MAR	07 MAY 76	
1976-026D		8792	USSR	19 MAR	02 APR 78	
1976-027A		8770	US	22 MAR	18 MAY 76	
1976-028A	COSMOS 810	8772	USSR	26 MAR	08 APR 76	
1976-028B		8773	USSR	26 MAR	01 APR 76	
1976-028C		8796	USSR	26 MAR	07 APR 76	
1976-028D		8797	USSR	26 MAR	10 APR 76	
1976-028E		8798	USSR	26 MAR	07 APR 76	
1976-029B		8775	US	26 MAR	02 DEC 76	
1976-029C		8793	US	26 MAR	28 JUN 78	
1976-030A	COSMOS 811	8781	USSR	31 MAR	12 APR 76	
1976-030B		8782	USSR	31 MAR	10 APR 76	
1976-030C		8783	USSR	31 MAR	13 APR 76	
1976-030D		8803	USSR	31 MAR	14 APR 76	
1976-030E		8804	USSR	31 MAR	15 APR 76	
1976-030F		8805	USSR	31 MAR	15 APR 76	
1976-031A	COSMOS 812	8794	USSR	06 APR	30 OCT 80	
1976-031B		8795	USSR	06 APR	05 JAN 81	
1976-031C		9836	USSR	06 APR	18 JUL 78	
1976-031D		9990	USSR	06 APR	27 FEB 78	
1976-033A	COSMOS 813	8801	USSR	09 APR	21 APR 76	
1976-033B		8802	USSR	09 APR	14 APR 76	
1976-034A	COSMOS 814	8806	USSR	13 APR	13 APR 76	
1976-034B		8807	USSR	13 APR	16 APR 76	
1976-035B		8809	US	22 APR	30 MAY 76	
1976-035C		8810	US	22 APR	18 OCT 77	
1976-036A	COSMOS 815	8811	USSR	28 APR	11 MAY 76	
1976-036B		8813	USSR	28 APR	02 MAY 76	
1976-036C		8815	USSR	28 APR	01 MAY 76	
1976-036D		8816	USSR	28 APR	01 MAY 76	
1976-036E		8817	USSR	28 APR	03 MAY 76	
1976-036F		8830	USSR	28 APR	20 MAY 76	
1976-036G		8831	USSR	28 APR	13 MAY 76	
1976-036H		8841	USSR	28 APR	15 MAY 76	
1976-037A	COSMOS 816	8812	USSR	28 APR	24 NOV 79	
1976-037AA		9514	USSR	28 APR	17 OCT 77	
1976-037B		8814	USSR	28 APR	16 NOV 79	BH
1976-037C		8885	USSR	28 APR	12 JUL 77	
1976-037D		8886	USSR	28 APR	26 JUL 77	

International Designation	Satellite Name	Number	Source	Launch	Decay	Note
1976-037E		9396	USSR	28 APR	12 JUN 77	
1976-037F		9397	USSR	28 APR	17 SEP 77	
1976-037G		9428	USSR	28 APR	09 AUG 77	
1976-037H		9429	USSR	28 APR	10 AUG 77	
1976-037J		9430	USSR	28 APR	29 AUG 77	
1976-037K		9431	USSR	28 APR	24 AUG 77	
1976-037L		9432	USSR	28 APR	20 SEP 77	
1976-037M		9438	USSR	28 APR	19 AUG 77	
1976-037N		9475	USSR	28 APR	02 SEP 77	
1976-037P		9476	USSR	28 APR	16 SEP 77	
1976-037Q		9485	USSR	28 APR	28 OCT 77	
1976-037R		9487	USSR	28 APR	07 SEP 77	
1976-037S		9488	USSR	28 APR	27 SEP 77	
1976-037T		9489	USSR	28 APR	24 SEP 77	
1976-037U		9490	USSR	28 APR	15 SEP 77	
1976-037V		9507	USSR	28 APR	01 AUG 77	
1976-037W		9508	USSR	28 APR	17 SEP 77	
1976-037X		9511	USSR	28 APR	12 OCT 77	
1976-037Y		9512	USSR	28 APR	03 SEP 77	
1976-037Z		9513	USSR	28 APR	17 SEP 77	
1976-039B		8821	US	04 MAY	22 SEP 91	
1976-040A	COSMOS 817	8823	USSR	05 MAY	18 MAY 76	
1976-040B		8824	USSR	05 MAY	10 MAY 76	
1976-040C		8825	USSR	05 MAY	06 MAY 76	
1976-040D		8848	USSR	05 MAY	24 MAY 76	
1976-040E		8849	USSR	05 MAY	19 MAY 76	
1976-040F		8850	USSR	05 MAY	18 MAY 76	
1976-041B		8834	USSR	12 MAY	05 AUG 76	
1976-041C		8837	USSR	12 MAY	08 SEP 76	
1976-041E		8847	USSR	12 MAY	04 JUL 76	
1976-044A	COSMOS 818	8851	USSR	18 MAY	07 MAR 77	
1976-044B		8852	USSR	18 MAY	28 OCT 76	
1976-045A	COSMOS 819	8853	USSR	20 MAY	01 JUN 76	
1976-045B		8854	USSR	20 MAY	30 MAY 76	
1976-045C		8855	USSR	20 MAY	22 MAY 76	
1976-046A	COSMOS 820	8856	USSR	21 MAY	02 JUN 76	
1976-046B		8857	USSR	21 MAY	24 MAY 76	
1976-046C		8858	USSR	21 MAY	21 MAY 76	
1976-046D		8869	USSR	21 MAY	04 JUN 76	
1976-046E		8870	USSR	21 MAY	10 JUN 76	
1976-048A	COSMOS 821	8862	USSR	26 MAY	08 JUN 76	
1976-048B		8863	USSR	26 MAY	07 JUN 76	
1976-048C		8864	USSR	26 MAY	02 JUN 76	
1976-048D		8875	USSR	26 MAY	14 JUN 76	
1976-048E		8876	USSR	26 MAY	09 JUN 76	
1976-048F		8879	USSR	26 MAY	09 JUN 76	
1976-049A	COSMOS 822	8865	USSR	28 MAY	08 AUG 78	
1976-049B		8866	USSR	28 MAY	23 DEC 77	
1976-052A	COSMOS 824	8877	USSR	08 JUN	21 JUN 76	
1976-052B		8878	USSR	08 JUN	20 JUN 76	
1976-052C		8880	USSR	08 JUN	16 JUN 76	
1976-052D		8881	USSR	08 JUN	10 JUN 76	
1976-052E		8905	USSR	08 JUN	27 JUN 76	
1976-052F		8906	USSR	08 JUN	23 JUN 76	
1976-052G		8907	USSR	08 JUN	22 JUN 76	
1976-053B		8883	US	10 JUN	25 AUG 77	
1976-053C		8887	US	10 JUN	02 JUL 76	
1976-053D		8888	US	10 JUN	30 JUN 76	
1976-053E		8902	US	10 JUN	01 JUL 76	
1976-055A	COSMOS 833	8898	USSR	16 JUN	29 JUN 76	
1976-055B		8899	USSR	16 JUN	23 JUN 76	
1976-055C		8900	USSR	16 JUN	17 JUN 76	
1976-055D		8901	USSR	16 JUN	21 JUN 76	
1976-055E		8920	USSR	16 JUN	08 JUL 76	
1976-055F		8921	USSR	16 JUN	01 JUL 76	
1976-055G		8926	USSR	16 JUN	30 JUN 76	
1976-056A	INTERCOSMOS 15	8903	USSR	19 JUN	18 NOV 79	
1976-056B		8904	USSR	19 JUN	15 JAN 80	
1976-057A	SALYUT 5	8911	USSR	22 JUN	08 AUG 77	
1976-057B		8912	USSR	22 JUN	30 JUN 76	
1976-057C		8913	USSR	22 JUN	24 JUN 76	
1976-057D		9059	USSR	22 JUN	13 AUG 76	
1976-057E		9060	USSR	22 JUN	13 AUG 76	
1976-057F		9275	USSR	22 JUN	17 AUG 76	
1976-057G		9276	USSR	22 JUN	05 SEP 76	
1976-057H		9277	USSR	22 JUN	05 SEP 76	
1976-057J		9385	USSR	22 JUN	13 SEP 76	
1976-057K		9386	USSR	22 JUN	22 AUG 76	
1976-058A	COSMOS 834	8914	USSR	24 JUN	06 JUL 76	
1976-058B		8915	USSR	24 JUN	30 JUN 76	
1976-059B		8917	US	26 JUN	30 JUN 76	
1976-060A	COSMOS 835	8922	USSR	29 JUN	12 JUL 76	
1976-060B		8925	USSR	29 JUN	04 JUL 76	
1976-060C		9014	USSR	29 JUN	20 JUL 76	
1976-060D		9015	USSR	29 JUN	14 JUL 76	
1976-060E		9018	USSR	29 JUN	16 JUL 76	
1976-062A	COSMOS 837	8927	USSR	01 JUL	18 NOV 83	
1976-062B		8928	USSR	01 JUL	05 AUG 76	
1976-062C		8929	USSR	01 JUL	21 JUL 76	
1976-062D		8930	USSR	01 JUL	20 JUL 76	
1976-062E		8931	USSR	01 JUL	26 SEP 82	
1976-063A	COSMOS 838	8932	USSR	02 JUL	23 AUG 77	
1976-063AA		10099	USSR	02 JUL	05 SEP 77	
1976-063AB		10100	USSR	02 JUL	29 OCT 77	
1976-063AC		10101	USSR	02 JUL	20 SEP 77	
1976-063AD		10102	USSR	02 JUL	25 OCT 77	
1976-063AE		10103	USSR	02 JUL	03 JUL 78	
1976-063AF		10104	USSR	02 JUL	19 JUN 78	
1976-063AG		10105	USSR	02 JUL	17 AUG 77	
1976-063AH		10106	USSR	02 JUL	18 APR 78	
1976-063AJ		10107	USSR	02 JUL	03 SEP 77	
1976-063AK		10108	USSR	02 JUL	27 AUG 77	
1976-063AL		10109	USSR	02 JUL	22 SEP 77	
1976-063AM		10110	USSR	02 JUL	03 SEP 78	
1976-063AN		10122	USSR	02 JUL	09 NOV 77	
1976-063AP		10123	USSR	02 JUL	18 AUG 77	
1976-063AQ		10124	USSR	02 JUL	17 AUG 77	
1976-063AR		10125	USSR	02 JUL	28 JUL 77	
1976-063AS		10126	USSR	02 JUL	12 JAN 78	
1976-063B		8933	USSR	02 JUL	03 JUL 76	BJ
1976-063C		10047	USSR	02 JUL	13 SEP 77	
1976-063D		10048	USSR	02 JUL	14 AUG 77	
1976-063E		10049	USSR	02 JUL	10 APR 78	
1976-063F		10050	USSR	02 JUL	23 JUL 77	
1976-063G		10051	USSR	02 JUL	01 NOV 77	
1976-063H		10052	USSR	02 JUL	25 MAR 78	
1976-063J		10053	USSR	02 JUL	15 NOV 77	
1976-063K		10054	USSR	02 JUL	08 AUG 78	
1976-063L		10055	USSR	02 JUL	02 DEC 77	
1976-063M		10056	USSR	02 JUL	03 NOV 77	
1976-063N		10057	USSR	02 JUL	15 APR 78	
1976-063P		10058	USSR	02 JUL	20 SEP 77	
1976-063Q		10076	USSR	02 JUL	13 FEB 78	
1976-063R		10077	USSR	02 JUL	12 MAR 78	
1976-063S		10078	USSR	02 JUL	16 NOV 77	
1976-063T		10079	USSR	02 JUL	03 OCT 77	
1976-063U		10080	USSR	02 JUL	19 SEP 78	
1976-063V		10081	USSR	02 JUL	18 APR 78	
1976-063W		10082	USSR	02 JUL	23 SEP 77	
1976-063X		10083	USSR	02 JUL	27 JUL 77	
1976-063Y		10097	USSR	02 JUL	03 MAR 78	
1976-063Z		10098	USSR	02 JUL	23 JUL 77	
1976-064A	SOYUZ 21	8934	USSR	06 JUL	24 AUG 76	E
1976-064B		8935	USSR	06 JUL	09 JUL 76	
1976-065A		9006	US	08 JUL	13 DEC 76	
1976-065B		9007	US	08 JUL	24 APR 86	
1976-065D		9010	US	08 JUL	10 JUL 76	
1976-066B		9012	US	08 JUL	10 JUL 76	
1976-067AL		10467	USSR	08 JUL	03 JUN 82	
1976-067BH		15523	USSR	08 JUL	03 JAN 90	
1976-067BQ		18673	USSR	08 JUL	10 SEP 88	
1976-068A	COSMOS 840	9019	USSR	14 JUL	26 JUL 76	
1976-068B		9020	USSR	14 JUL	25 JUL 76	
1976-068C		9021	USSR	14 JUL	17 JUL 76	
1976-071A	COSMOS 843	9043	USSR	21 JUL	21 JUL 76	
1976-071B		9045	USSR	21 JUL	23 JUL 76	
1976-072A	COSMOS 844	9046	USSR	22 JUL	30 AUG 76	
1976-072AA		9086	USSR	22 JUL	02 AUG 76	
1976-072AB		9087	USSR	22 JUL	02 AUG 76	
1976-072AC		9088	USSR	22 JUL	02 AUG 76	
1976-072AD		9089	USSR	22 JUL	02 AUG 76	
1976-072AE		9090	USSR	22 JUL	02 AUG 76	
1976-072AF		9091	USSR	22 JUL	02 AUG 76	
1976-072AG		9092	USSR	22 JUL	02 AUG 76	
1976-072AH		9093	USSR	22 JUL	02 AUG 76	
1976-072AJ		9094	USSR	22 JUL	02 AUG 76	
1976-072AK		9095	USSR	22 JUL	02 AUG 76	
1976-072AL		9096	USSR	22 JUL	02 AUG 76	
1976-072AM		9097	USSR	22 JUL	02 AUG 76	
1976-072AN		9098	USSR	22 JUL	02 AUG 76	
1976-072AP		9099	USSR	22 JUL	02 AUG 76	
1976-072AQ		9100	USSR	22 JUL	02 AUG 76	
1976-072AR		9101	USSR	22 JUL	02 AUG 76	
1976-072AS		9102	USSR	22 JUL	02 AUG 76	
1976-072AT		9103	USSR	22 JUL	02 AUG 76	
1976-072AU		9104	USSR	22 JUL	02 AUG 76	
1976-072AV		9105	USSR	22 JUL	02 AUG 76	
1976-072AW		9106	USSR	22 JUL	02 AUG 76	
1976-072AX		9107	USSR	22 JUL	02 AUG 76	
1976-072AY		9108	USSR	22 JUL	02 AUG 76	
1976-072AZ		9109	USSR	22 JUL	02 AUG 76	
1976-072B		9048	USSR	22 JUL	30 JUL 76	BK
1976-072BA		9110	USSR	22 JUL	02 AUG 76	
1976-072BB		9111	USSR	22 JUL	02 AUG 76	
1976-072BC		9112	USSR	22 JUL	02 AUG 76	
1976-072BD		9113	USSR	22 JUL	02 AUG 76	
1976-072BE		9114	USSR	22 JUL	02 AUG 76	
1976-072BF		9115	USSR	22 JUL	02 AUG 76	
1976-072BG		9116	USSR	22 JUL	02 AUG 76	
1976-072BH		9117	USSR	22 JUL	02 AUG 76	
1976-072BJ		9118	USSR	22 JUL	02 AUG 76	
1976-072BK		9119	USSR	22 JUL	02 AUG 76	
1976-072BL		9120	USSR	22 JUL	02 AUG 76	
1976-072BM		9121	USSR	22 JUL	02 AUG 76	
1976-072BN		9122	USSR	22 JUL	02 AUG 76	
1976-072BP		9123	USSR	22 JUL	02 AUG 76	
1976-072BQ		9124	USSR	22 JUL	02 AUG 76	
1976-072BR		9125	USSR	22 JUL	02 AUG 76	
1976-072BS		9126	USSR	22 JUL	02 AUG 76	
1976-072BT		9127	USSR	22 JUL	02 AUG 76	
1976-072BU		9128	USSR	22 JUL	02 AUG 76	
1976-072BV		9129	USSR	22 JUL	02 AUG 76	
1976-072BW		9130	USSR	22 JUL	02 AUG 76	
1976-072BX		9131	USSR	22 JUL	02 AUG 76	
1976-072BY		9132	USSR	22 JUL	02 AUG 76	
1976-072BZ		9133	USSR	22 JUL	02 AUG 76	

International Designation	Satellite Name	Number	Source	Launch	Decay	Note
1976-072C		9064	USSR	22 JUL	02 AUG 76	
1976-072CA		9134	USSR	22 JUL	02 AUG 76	
1976-072CB		9135	USSR	22 JUL	02 AUG 76	
1976-072CC		9136	USSR	22 JUL	02 AUG 76	
1976-072CD		9137	USSR	22 JUL	02 AUG 76	
1976-072CE		9138	USSR	22 JUL	02 AUG 76	
1976-072CF		9139	USSR	22 JUL	02 AUG 76	
1976-072CG		9140	USSR	22 JUL	02 AUG 76	
1976-072CH		9141	USSR	22 JUL	02 AUG 76	
1976-072CJ		9142	USSR	22 JUL	02 AUG 76	
1976-072CK		9143	USSR	22 JUL	02 AUG 76	
1976-072CL		9144	USSR	22 JUL	02 AUG 76	
1976-072CM		9145	USSR	22 JUL	02 AUG 76	
1976-072CN		9146	USSR	22 JUL	02 AUG 76	
1976-072CP		9147	USSR	22 JUL	02 AUG 76	
1976-072CQ		9148	USSR	22 JUL	02 AUG 76	
1976-072CR		9149	USSR	22 JUL	02 AUG 76	
1976-072CS		9150	USSR	22 JUL	02 AUG 76	
1976-072CT		9151	USSR	22 JUL	02 AUG 76	
1976-072CU		9152	USSR	22 JUL	02 AUG 76	
1976-072CV		9153	USSR	22 JUL	02 AUG 76	
1976-072CW		9154	USSR	22 JUL	02 AUG 76	
1976-072CX		9155	USSR	22 JUL	02 AUG 76	
1976-072CY		9156	USSR	22 JUL	02 AUG 76	
1976-072CZ		9157	USSR	22 JUL	02 AUG 76	
1976-072D		9065	USSR	22 JUL	02 AUG 76	
1976-072DA		9158	USSR	22 JUL	02 AUG 76	
1976-072DB		9159	USSR	22 JUL	02 AUG 76	
1976-072DC		9160	USSR	22 JUL	02 AUG 76	
1976-072DD		9161	USSR	22 JUL	02 AUG 76	
1976-072DE		9162	USSR	22 JUL	02 AUG 76	
1976-072DF		9163	USSR	22 JUL	02 AUG 76	
1976-072DG		9164	USSR	22 JUL	02 AUG 76	
1976-072DH		9165	USSR	22 JUL	02 AUG 76	
1976-072DJ		9166	USSR	22 JUL	02 AUG 76	
1976-072DK		9167	USSR	22 JUL	02 AUG 76	
1976-072DL		9168	USSR	22 JUL	02 AUG 76	
1976-072DM		9169	USSR	22 JUL	02 AUG 76	
1976-072DN		9170	USSR	22 JUL	02 AUG 76	
1976-072DP		9171	USSR	22 JUL	02 AUG 76	
1976-072DQ		9172	USSR	22 JUL	02 AUG 76	
1976-072DR		9173	USSR	22 JUL	02 AUG 76	
1976-072DS		9174	USSR	22 JUL	02 AUG 76	
1976-072DT		9175	USSR	22 JUL	02 AUG 76	
1976-072DU		9176	USSR	22 JUL	02 AUG 76	
1976-072DV		9177	USSR	22 JUL	02 AUG 76	
1976-072DW		9178	USSR	22 JUL	02 AUG 76	
1976-072DX		9179	USSR	22 JUL	02 AUG 76	
1976-072DY		9180	USSR	22 JUL	02 AUG 76	
1976-072DZ		9181	USSR	22 JUL	02 AUG 76	
1976-072E		9066	USSR	22 JUL	02 AUG 76	
1976-072EA		9182	USSR	22 JUL	02 AUG 76	
1976-072EB		9183	USSR	22 JUL	02 AUG 76	
1976-072EC		9184	USSR	22 JUL	02 AUG 76	
1976-072ED		9185	USSR	22 JUL	02 AUG 76	
1976-072EE		9186	USSR	22 JUL	02 AUG 76	
1976-072EF		9187	USSR	22 JUL	02 AUG 76	
1976-072EG		9188	USSR	22 JUL	02 AUG 76	
1976-072EH		9189	USSR	22 JUL	02 AUG 76	
1976-072EJ		9190	USSR	22 JUL	02 AUG 76	
1976-072EK		9191	USSR	22 JUL	02 AUG 76	
1976-072EL		9192	USSR	22 JUL	02 AUG 76	
1976-072EM		9193	USSR	22 JUL	02 AUG 76	
1976-072EN		9194	USSR	22 JUL	02 AUG 76	
1976-072EP		9195	USSR	22 JUL	02 AUG 76	
1976-072EQ		9196	USSR	22 JUL	02 AUG 76	
1976-072ER		9197	USSR	22 JUL	02 AUG 76	
1976-072ES		9198	USSR	22 JUL	02 AUG 76	
1976-072ET		9199	USSR	22 JUL	02 AUG 76	
1976-072EU		9200	USSR	22 JUL	02 AUG 76	
1976-072EV		9201	USSR	22 JUL	02 AUG 76	
1976-072EW		9202	USSR	22 JUL	02 AUG 76	
1976-072EX		9203	USSR	22 JUL	02 AUG 76	
1976-072EY		9204	USSR	22 JUL	02 AUG 76	
1976-072EZ		9205	USSR	22 JUL	02 AUG 76	
1976-072F		9067	USSR	22 JUL	02 AUG 76	
1976-072FA		9206	USSR	22 JUL	02 AUG 76	
1976-072FB		9207	USSR	22 JUL	02 AUG 75	
1976-072FC		9208	USSR	22 JUL	02 AUG 76	
1976-072FD		9209	USSR	22 JUL	02 AUG 76	
1976-072FE		9210	USSR	22 JUL	02 AUG 76	
1976-072FF		9211	USSR	22 JUL	02 AUG 76	
1976-072FG		9212	USSR	22 JUL	02 AUG 76	
1976-072FH		9213	USSR	22 JUL	02 AUG 76	
1976-072FJ		9217	USSR	22 JUL	05 AUG 76	
1976-072FK		9218	USSR	22 JUL	05 AUG 76	
1976-072FL		9219	USSR	22 JUL	05 AUG 76	
1976-072FM		9220	USSR	22 JUL	05 AUG 76	
1976-072FN		9221	USSR	22 JUL	05 AUG 76	
1976-072FP		9222	USSR	22 JUL	05 AUG 76	
1976-072FQ		9223	USSR	22 JUL	05 AUG 76	
1976-072FR		9224	USSR	22 JUL	05 AUG 76	
1976-072FS		9225	USSR	22 JUL	05 AUG 76	
1976-072FT		9226	USSR	22 JUL	05 AUG 76	
1976-072FU		9227	USSR	22 JUL	05 AUG 76	
1976-072FV		9228	USSR	22 JUL	05 AUG 76	
1976-072FW		9229	USSR	22 JUL	05 AUG 76	
1976-072FX		9230	USSR	22 JUL	05 AUG 76	
1976-072FY		9231	USSR	22 JUL	05 AUG 76	
1976-072FZ		9232	USSR	22 JUL	05 AUG 76	
1976-072G		9068	USSR	22 JUL	02 AUG 76	
1976-072GA		9233	USSR	22 JUL	05 AUG 76	
1976-072GB		9234	USSR	22 JUL	05 AUG 76	
1976-072GC		9235	USSR	22 JUL	05 AUG 76	
1976-072GD		9236	USSR	22 JUL	05 AUG 76	
1976-072GE		9237	USSR	22 JUL	05 AUG 76	
1976-072GF		9238	USSR	22 JUL	05 AUG 76	
1976-072GG		9239	USSR	22 JUL	05 AUG 76	
1976-072GH		9240	USSR	22 JUL	05 AUG 76	
1976-072GJ		9241	USSR	22 JUL	05 AUG 76	
1976-072GK		9242	USSR	22 JUL	05 AUG 76	
1976-072GL		9243	USSR	22 JUL	05 AUG 76	
1976-072GM		9244	USSR	22 JUL	05 AUG 76	
1976-072GN		9245	USSR	22 JUL	05 AUG 76	
1976-072GP		9246	USSR	22 JUL	05 AUG 76	
1976-072GQ		9247	USSR	22 JUL	05 AUG 76	
1976-072GR		9248	USSR	22 JUL	05 AUG 76	
1976-072GS		9249	USSR	22 JUL	05 AUG 76	
1976-072GT		9250	USSR	22 JUL	05 AUG 76	
1976-072GU		9251	USSR	22 JUL	05 AUG 76	
1976-072GV		9252	USSR	22 JUL	05 AUG 76	
1976-072GW		9253	USSR	22 JUL	05 AUG 76	
1976-072GX		9254	USSR	22 JUL	05 AUG 76	
1976-072GY		9255	USSR	22 JUL	05 AUG 76	
1976-072GZ		9256	USSR	22 JUL	05 AUG 76	
1976-072H		9069	USSR	22 JUL	02 AUG 76	
1976-072HA		9257	USSR	22 JUL	05 AUG 76	
1976-072HB		9258	USSR	22 JUL	05 AUG 76	
1976-072HC		9259	USSR	22 JUL	05 AUG 76	
1976-072HD		9260	USSR	22 JUL	05 AUG 76	
1976-072HE		9261	USSR	22 JUL	05 AUG 76	
1976-072HF		9262	USSR	22 JUL	05 AUG 76	
1976-072HG		9263	USSR	22 JUL	05 AUG 76	
1976-072HH		9264	USSR	22 JUL	05 AUG 76	
1976-072HJ		9265	USSR	22 JUL	05 AUG 76	
1976-072HK		9266	USSR	22 JUL	05 AUG 76	
1976-072HL		9331	USSR	22 JUL	13 AUG 76	
1976-072HM		9332	USSR	22 JUL	13 AUG 76	
1976-072HN		9333	USSR	22 JUL	13 AUG 76	
1976-072HP		9334	USSR	22 JUL	13 AUG 76	
1976-072HQ		9335	USSR	22 JUL	13 AUG 76	
1976-072HR		9336	USSR	22 JUL	13 AUG 76	
1976-072HS		9337	USSR	22 JUL	13 AUG 76	
1976-072HT		9338	USSR	22 JUL	13 AUG 76	
1976-072HU		9339	USSR	22 JUL	13 AUG 76	
1976-072HV		9340	USSR	22 JUL	13 AUG 76	
1976-072HW		9341	USSR	22 JUL	13 AUG 76	
1976-072HX		9342	USSR	22 JUL	13 AUG 76	
1976-072HY		9343	USSR	22 JUL	13 AUG 76	
1976-072HZ		9344	USSR	22 JUL	13 AUG 76	
1976-072J		9070	USSR	22 JUL	02 AUG 76	
1976-072JA		9345	USSR	22 JUL	13 AUG 76	
1976-072JB		9346	USSR	22 JUL	13 AUG 76	
1976-072JC		9347	USSR	22 JUL	13 AUG 76	
1976-072JD		9348	USSR	22 JUL	13 AUG 76	
1976-072JE		9349	USSR	22 JUL	13 AUG 76	
1976-072JF		9350	USSR	22 JUL	13 AUG 76	
1976-072JG		9351	USSR	22 JUL	13 AUG 76	
1976-072JH		9352	USSR	22 JUL	13 AUG 76	
1976-072JJ		9353	USSR	22 JUL	13 AUG 76	
1976-072JK		9354	USSR	22 JUL	13 AUG 76	
1976-072JL		9355	USSR	22 JUL	13 AUG 76	
1976-072JM		9356	USSR	22 JUL	13 AUG 76	
1976-072JN		9357	USSR	22 JUL	13 AUG 76	
1976-072JP		9358	USSR	22 JUL	13 AUG 76	
1976-072JQ		9359	USSR	22 JUL	13 AUG 76	
1976-072JR		9360	USSR	22 JUL	13 AUG 76	
1976-072JS		9361	USSR	22 JUL	13 AUG 76	
1976-072JT		9362	USSR	22 JUL	13 AUG 76	
1976-072JU		9363	USSR	22 JUL	13 AUG 76	
1976-072JV		9364	USSR	22 JUL	13 AUG 76	
1976-072JW		9365	USSR	22 JUL	13 AUG 76	
1976-072JX		9366	USSR	22 JUL	13 AUG 76	
1976-072JY		9367	USSR	22 JUL	13 AUG 76	
1976-072JZ		9368	USSR	22 JUL	13 AUG 76	
1976-072K		9071	USSR	22 JUL	02 AUG 76	
1976-072KA		9369	USSR	22 JUL	13 AUG 76	
1976-072KB		9370	USSR	22 JUL	13 AUG 76	
1976-072KC		9371	USSR	22 JUL	13 AUG 76	
1976-072KD		9372	USSR	22 JUL	13 AUG 76	
1976-072KE		9373	USSR	22 JUL	13 AUG 76	
1976-072KF		9378	USSR	22 JUL	17 AUG 76	
1976-072KG		9379	USSR	22 JUL	17 AUG 76	
1976-072KH		9380	USSR	22 JUL	17 AUG 76	
1976-072KJ		9381	USSR	22 JUL	17 AUG 76	
1976-072L		9072	USSR	22 JUL	02 AUG 76	
1976-072M		9073	USSR	22 JUL	02 AUG 76	
1976-072N		9074	USSR	22 JUL	02 AUG 76	
1976-072P		9075	USSR	22 JUL	02 AUG 76	
1976-072Q		9076	USSR	22 JUL	02 AUG 76	
1976-072R		9077	USSR	22 JUL	02 AUG 76	
1976-072S		9078	USSR	22 JUL	02 AUG 76	
1976-072T		9079	USSR	22 JUL	02 AUG 76	
1976-072U		9080	USSR	22 JUL	02 AUG 76	
1976-072V		9081	USSR	22 JUL	02 AUG 76	
1976-072W		9082	USSR	22 JUL	02 AUG 76	

International Designation	Satellite Name	Number	Source	Launch	Decay	Note
1976-072X		9083	USSR	22 JUL	02 AUG 76	
1976-072Y		9084	USSR	22 JUL	02 AUG 76	
1976-072Z		9085	USSR	22 JUL	02 AUG 76	
1976-074A	MOLNIYA 1-35	9049	USSR	23 JUL	29 MAY 87	
1976-074B		9050	USSR	23 JUL	02 OCT 76	
1976-074C		9051	USSR	23 JUL	18 SEP 76	
1976-074D		9052	USSR	23 JUL	01 SEP 76	
1976-074E		9269	USSR	23 JUL	09 SEP 91	
1976-075A	COSMOS 845	9053	USSR	27 JUL	15 NOV 80	
1976-075B		9054	USSR	27 JUL	28 FEB 81	
1976-075C		9058	USSR	27 JUL	18 JUL 78	
1976-075D		9267	USSR	27 JUL	14 OCT 78	
1976-075E		9268	USSR	27 JUL	17 FEB 79	
1976-075F		9665	USSR	27 JUL	18 JUL 78	
1976-075G		9718	USSR	27 JUL	27 MAR 78	
1976-075H		9789	USSR	27 JUL	24 JUN 78	
1976-076A	INTERCOSMOS 16	9055	USSR	27 JUL	10 JUL 79	
1976-076B		9056	USSR	27 JUL	21 FEB 80	
1976-077BA		10753	US	29 JUL	02 JAN 82	
1976-077BX		10827	US	29 JUL	11 SEP 88	
1976-077DV		12541	US	29 JUL	10 MAY 92	
1976-077EW		16999	US	29 JUL	01 AUG 88	
1976-079A	COSMOS 847	9214	USSR	04 AUG	17 AUG 76	
1976-079B		9215	USSR	04 AUG	11 AUG 76	
1976-079C		9216	USSR	04 AUG	05 AUG 76	
1976-079D		9376	USSR	04 AUG	23 AUG 76	
1976-079E		9377	USSR	04 AUG	18 AUG 76	
1976-081A	LUNA 24	9272	USSR	09 AUG	18 AUG 76	R
1976-081B		9273	USSR	09 AUG	09 AUG 76	
1976-081C		9274	USSR	09 AUG	15 AUG 76	
1976-081D		9278	USSR	09 AUG	14 AUG 76	
1976-081E	LUNA 24 BL DESCENT CRAF	9384	USSR	09 AUG	22 AUG 76	
1976-082A	COSMOS 848	9280	USSR	12 AUG	25 AUG 76	
1976-082B		9281	USSR	12 AUG	24 AUG 76	
1976-082C		9328	USSR	12 AUG	17 AUG 76	
1976-082D		9374	USSR	12 AUG	01 SEP 76	
1976-082E		9375	USSR	12 AUG	24 AUG 76	
1976-083A	COSMOS 849	9382	USSR	18 AUG	24 AUG 78	
1976-083B		9383	USSR	18 AUG	02 NOV 77	
1976-083C		9407	USSR	18 AUG	16 FEB 77	
1976-084A	COSMOS 850	9387	USSR	26 AUG	16 MAY 77	
1976-084B		9388	USSR	26 AUG	14 JAN 77	
1976-085A	COSMOS 851	9389	USSR	27 AUG	05 AUG 89	
1976-085B		9390	USSR	27 AUG	22 MAR 92	
1976-086A	COSMOS 852	9391	USSR	28 AUG	10 SEP 76	
1976-086B		9392	USSR	28 AUG	02 SEP 76	
1976-086C		9393	USSR	28 AUG	29 AUG 76	
1976-086D		9412	USSR	28 AUG	16 SEP 76	
1976-086E		9413	USSR	28 AUG	11 SEP 76	
1976-086F		9414	USSR	28 AUG	11 SEP 76	
1976-087A	MAO 6	9394	PRC	30 AUG	25 NOV 78	
1976-087B		9395	PRC	30 AUG	04 FEB 78	
1976-088A	COSMOS 853	9398	USSR	01 SEP	31 DEC 76	
1976-088B		9399	USSR	01 SEP	10 OCT 76	
1976-088C		9400	USSR	01 SEP	20 MAY 77	
1976-088D		9402	USSR	01 SEP	09 DEC 76	
1976-089A	TIP 3	9403	US	01 SEP	30 MAY 81	
1976-089B		9404	US	01 SEP	29 JUN 78	
1976-089C		9409	US	01 SEP	01 MAR 78	
1976-089D		9410	US	01 SEP	17 MAY 78	
1976-090A	COSMOS 854	9405	USSR	03 SEP	16 SEP 76	
1976-090B		9406	USSR	03 SEP	07 SEP 76	
1976-090C		9422	USSR	03 SEP	19 SEP 76	
1976-090D		9423	USSR	03 SEP	17 SEP 76	
1976-091D		9474	US	11 SEP	20 SEP 80	
1976-091E		9483	US	11 SEP	24 AUG 83	
1976-092B		9417	USSR	11 SEP	14 SEP 76	
1976-092C		9418	USSR	11 SEP	12 SEP 76	
1976-092D		9440	USSR	11 SEP	19 JUL 77	
1976-092E		9441	USSR	11 SEP	20 MAY 77	
1976-093A	SOYUZ 22	9421	USSR	15 SEP	23 SEP 76	E
1976-093B		9424	USSR	15 SEP	20 SEP 76	
1976-093C		9425	USSR	15 SEP	16 SEP 76	
1976-094A		9426	US	15 SEP	05 NOV 76	
1976-094B		9427	US	15 SEP	17 SEP 76	
1976-095A	COSMOS 855	9433	USSR	21 SEP	03 OCT 76	
1976-095B		9434	USSR	21 SEP	02 OCT 76	
1976-095C		9445	USSR	21 SEP	06 OCT 76	
1976-095D		9446	USSR	21 SEP	08 OCT 76	
1976-096A	COSMOS 856	9435	USSR	22 SEP	05 OCT 76	
1976-096B		9436	USSR	22 SEP	30 SEP 76	
1976-096C		9437	USSR	22 SEP	23 SEP 76	
1976-096D		9447	USSR	22 SEP	08 OCT 76	
1976-096E		9448	USSR	22 SEP	16 OCT 76	
1976-097A	COSMOS 857	9439	USSR	24 SEP	07 OCT 76	
1976-097B		9442	USSR	24 SEP	29 SEP 76	
1976-097C		9449	USSR	24 SEP	15 OCT 76	
1976-097D		9450	USSR	24 SEP	08 OCT 76	
1976-097E		9451	USSR	24 SEP	09 OCT 76	
1976-097F		9470	USSR	24 SEP	11 OCT 76	
1976-099A	COSMOS 859	9471	USSR	10 OCT	21 OCT 76	
1976-099B		9472	USSR	10 OCT	15 OCT 76	
1976-099C		9473	USSR	10 OCT	11 OCT 76	
1976-099D		9491	USSR	10 OCT	24 OCT 76	
1976-099E		9492	USSR	10 OCT	25 OCT 76	
1976-100A	SOYUZ 23	9477	USSR	14 OCT	16 OCT 76	E
1976-100B		9480	USSR	14 OCT	16 OCT 76	
1976-101B		9479	US	14 OCT	15 NOV 76	
1976-101C		9493	US	14 OCT	01 NOV 76	
1976-101D		9498	US	14 OCT	04 NOV 76	
1976-101E		9542	US	14 OCT	12 MAY 77	
1976-103B		9531	USSR	17 OCT	29 DEC 76	
1976-103C		9537	USSR	17 OCT	14 NOV 76	
1976-103D		9538	USSR	17 OCT	15 NOV 76	
1976-103E		9539	USSR	17 OCT	12 NOV 76	
1976-104B		9633	USSR	21 OCT	21 OCT 76	
1976-104C		9631	USSR	21 OCT	04 FEB 77	
1976-104D		9632	USSR	21 OCT	25 DEC 76	
1976-105B		9496	USSR	22 OCT	27 DEC 76	
1976-105C		9497	USSR	22 OCT	10 DEC 76	
1976-106A	COSMOS 863	9499	USSR	25 OCT	05 NOV 76	
1976-106B		9500	USSR	25 OCT	01 NOV 76	
1976-106C		9501	USSR	25 OCT	29 OCT 76	
1976-106D		9502	USSR	25 OCT	26 OCT 76	
1976-106E		9521	USSR	25 OCT	11 NOV 76	
1976-106F		9522	USSR	25 OCT	07 NOV 76	
1976-106G		9524	USSR	25 OCT	07 NOV 76	
1976-106H		9527	USSR	25 OCT	09 NOV 76	
1976-106J		9528	USSR	25 OCT	10 NOV 76	
1976-106K		9529	USSR	25 OCT	07 NOV 76	
1976-107B		9504	USSR	26 OCT	29 OCT 76	
1976-107C		9505	USSR	26 OCT	27 OCT 76	
1976-107D		9540	USSR	26 OCT	26 JUN 77	
1976-107E		9541	USSR	26 OCT	20 MAY 77	
1976-109A	COSMOS 865	9515	USSR	01 NOV	13 NOV 76	
1976-109B		9516	USSR	01 NOV	14 NOV 76	
1976-109C		9517	USSR	01 NOV	05 NOV 76	
1976-109D		9519	USSR	01 NOV	07 NOV 76	
1976-109E		9520	USSR	01 NOV	22 NOV 76	
1976-109F		9523	USSR	01 NOV	10 NOV 76	
1976-109G		9525	USSR	01 NOV	10 FEB 77	
1976-109H		9526	USSR	01 NOV	05 FEB 77	
1976-110A	COSMOS 866	9532	USSB	11 NOV	23 NOV 76	
1976-110B		9533	USSR	11 NOV	14 NOV 76	
1976-110C		9534	USSR	11 NOV	16 NOV 76	
1976-110D		9535	USSR	11 NOV	14 NOV 76	
1976-110E		9536	USSR	11 NOV	13 NOV 76	
1976-110F		9551	USSR	11 NOV	24 NOV 76	
1976-110G		9553	USSR	11 NOV	23 NOV 76	
1976-111A	COSMOS 867	9552	USSR	23 NOV	06 DEC 76	
1976-111B		9554	USSR	23 NOV	01 FEB 77	
1976-111C		9555	USSR	23 NOV	16 DEC 76	
1976-111D		9556	USSR	23 NOV	08 DEC 76	
1976-111E		9566	USSR	23 NOV	11 DEC 76	
1976-111F		9582	USSR	23 NOV	07 JAN 78	
1976-111G		9583	USSR	23 NOV	26 MAY 77	
1976-111H		9584	USSR	23 NOV	20 MAR 77	
1976-111J		9585	USSR	23 NOV	07 JUL 77	
1976-111K		9586	USSR	23 NOV	05 MAR 77	
1976-112B		9558	USSR	25 NOV	23 JAN 77	
1976-112C		9559	USSR	25 NOV	05 JAN 77	
1976-112D		9560	USSR	25 NOV	24 JAN 77	
1976-113A	COSMOS 868	9561	USSR	26 NOV	08 JUL 78	
1976-113B		9562	USSR	26 NOV	27 NOV 76	
1976-114A	COSMOS 869	9564	USSR	29 NOV	17 DEC 76	
1976-114B		9565	USSR	29 NOV	06 DEC 76	
1976-115A	COSMOS 870	9573	USSR	02 DEC	20 DEC 80	
1976-115B		9576	USSR	02 DEC	28 FEB 81	
1976-115C		9577	USSR	02 DEC	10 FEB 78	
1976-115D		9991	USSR	02 DEC	30 APR 78	
1976-115E		9992	USSR	02 DEC	11 JUN 78	
1976-115F		9993	USSR	02 DEC	07 MAR 80	
1976-116A	MOLNIYA 2-16	9574	USSR	02 DEC	21 FEB 91	
1976-116B		9575	USSR	02 DEC	04 MAR 77	
1976-116C		9578	USSR	02 DEC	07 FEB 77	
1976-116D		9579	USSR	02 DEC	19 SEP 91	
1976-116E		9580	USSR	02 DEC	13 FEB 77	
1976-116F		9581	USSR	02 DEC	11 JAN 77	
1976-117A	MAO 7	9587	PRC	07 DEC	02 JAN 77	
1976-117B		9596	PRC	07 DEC	29 DEC 76	
1976-117C		9597	PRC	07 DEC	09 DEC 76	
1976-119A	COSMOS 879	9599	USSR	09 DEC	22 DEC 76	
1976-119B		9600	USSR	09 DEC	12 DEC 76	
1976-119C		9602	USSR	09 DEC	03 JAN 77	
1976-119D		9603	USSR	09 DEC	15 DEC 76	
1976-120A	COSMOS 880	9601	USSR	09 DEC	08 OCT 79	
1976-120AA		11199	USSR	09 DEC	18 FEB 79	
1976-120AB		11200	USSR	09 DEC	10 JUL 79	
1976-120AC		11201	USSR	09 DEC	09 DEC 79	
1976-120AD		11202	USSR	09 DEC	13 JUL 79	
1976-120AE		11203	USSR	09 DEC	10 SEP 79	
1976-120AF		11204	USSR	09 DEC	01 AUG 91	
1976-120AG		11205	USSR	09 DEC	11 JUL 90	
1976-120AH		11206	USSR	09 DEC	26 FEB 79	
1976-120AJ		11207	USSR	09 DEC	20 NOV 85	
1976-120AK		11208	USSR	09 DEC	11 NOV 79	
1976-120AL		11209	USSR	09 DEC	01 DEC 82	
1976-120AM		11210	USSR	09 DEC	30 JUL 83	
1976-120AN		11211	USSR	09 DEC	29 MAR 79	
1976-120AP		11212	USSR	09 DEC	15 JUL 84	
1976-120AQ		11213	USSR	09 DEC	23 NOV 90	
1976-120AR		11214	USSR	09 DEC	15 SEP 91	
1976-120AS		11215	USSR	09 DEC	24 MAY 79	
1976-120AV		11218	USSR	09 DEC	05 FEB 80	

International Designation	Satellite Name	Number	Source	Launch	Decay	Note
1976-120AW		11219	USSR	09 DEC	19 NOV 82	
1976-120AX		11220	USSR	09 DEC	19 MAY 83	
1976-120AZ		11222	USSR	09 DEC	16 JUN 79	
1976-120B		9604	USSR	09 DEC	21 DEC 84	
1976-120BA		11223	USSR	09 DEC	14 AUG 91	
1976-120BB		11224	USSR	09 DEC	03 JUN 79	
1976-120BC		11225	USSR	09 DEC	19 AUG 79	
1976-120C		9605	USSR	09 DEC	01 NOV 81	
1976-120D		11178	USSR	09 DEC	12 JUN 81	
1976-120E		11179	USSR	09 DEC	06 MAR 80	
1976-120F		11180	USSR	09 DEC	25 FEB 79	
1976-120G		11181	USSR	09 DEC	20 JUL 79	
1976-120H		11182	USSR	09 DEC	27 OCT 79	
1976-120J		11183	USSR	09 DEC	14 APR 79	
1976-120K		11184	USSR	09 DEC	06 JUL 79	
1976-120L		11185	USSR	09 DEC	05 APR 79	
1976-120M		11186	USSR	09 DEC	13 NOV 80	
1976-120N		11187	USSR	09 DEC	02 APR 79	
1976-120P		11188	USSR	09 DEC	05 MAR 79	
1976-120Q		11189	USSR	09 DEC	23 FEB 79	
1976-120R		11190	USSR	09 DEC	26 MAR 79	
1976-120S		11191	USSR	09 DEC	16 JUN 79	
1976-120T		11192	USSR	09 DEC	05 JUL 79	
1976-120U		11193	USSR	09 DEC	03 SEP 91	
1976-120V		11194	USSR	09 DEC	12 JUL 79	
1976-120W		11195	USSR	09 DEC	23 FEB 79	
1976-120X		11196	USSR	09 DEC	07 APR 79	
1976-120Y		11197	USSR	09 DEC	31 MAR 79	
1976-120Z		11198	USSR	09 DEC	15 MAR 79	
1976-121A	COSMOS 881	9606	USSR	15 DEC	15 DEC 76	
1976-121B	COSMOS 882	9607	USSR	15 DEC	15 DEC 76	
1976-121C		9608	USSR	15 DEC	20 DEC 76	
1976-121D		9609	USSR	15 DEC	23 DEC 76	
1976-121E		9611	USSR	15 DEC	26 DEC 76	
1976-121F		9612	USSR	15 DEC	21 DEC 76	
1976-123A	COSMOS 884	9614	USSR	17 DEC	29 DEC 76	
1976-123B		9623	USSR	17 DEC	23 DEC 76	
1976-123C		9624	USSR	17 DEC	31 JAN 77	
1976-123D		9639	USSR	17 DEC	31 DEC 76	
1976-123E		9640	USSR	17 DEC	01 JAN 77	
1976-123F		9646	USSR	17 DEC	03 JAN 77	
1976-124A	COSMOS 885	9615	USSR	17 DEC	14 OCT 79	
1976-124B		9625	USSR	17 DEC	05 SEP 79	BM
1976-124C		9626	USSR	17 DEC	29 OCT 77	
1976-124D		9897	USSR	17 DEC	22 NOV 77	
1976-124E		9901	USSR	17 DEC	11 NOV 77	
1976-124F		9914	USSR	17 DEC	09 DEC 77	
1976-124G		9915	USSR	17 DEC	29 NOV 77	
1976-124H		9916	USSR	17 DEC	16 NOV 77	
1976-124J		9917	USSR	17 DEC	05 NOV 77	
1976-124K		9918	USSR	17 DEC	20 NOV 77	
1976-124L		9919	USSR	17 DEC	10 NOV 77	
1976-124M		9920	USSR	17 DEC	22 DEC 77	
1976-124N		9922	USSR	17 DEC	16 NOV 77	
1976-124P		9923	USSR	17 DEC	09 DEC 77	
1976-124Q		9924	USSR	17 DEC	14 DEC 77	
1976-124R		9925	USSR	17 DEC	22 NOV 77	
1976-124S		9926	USSR	17 DEC	01 DEC 77	
1976-124T		9999	USSR	17 DEC	17 OCT 77	
1976-124U		10391	USSR	17 DEC	11 JAN 78	
1976-125A		9627	US	19 DEC	28 JAN 79	
1976-125B		9628	US	19 DEC	09 MAY 77	
1976-125C		9629	US	19 DEC	23 FEB 77	
1976-125D		9630	US	19 DEC	09 FEB 77	
1976-126AD		9817	USSR	27 DEC	12 JUL 82	
1976-126AM		9825	USSR	27 DEC	12 MAR 82	
1976-126AS		9952	USSR	27 DEC	11 OCT 80	
1976-126AZ		9959	USSR	27 DEC	20 JUN 82	
1976-126B		9643	USSR	27 DEC	30 DEC 76	
1976-126BH		9967	USSR	27 DEC	05 MAR 91	
1976-126BJ		10396	USSR	27 DEC	08 JAN 84	
1976-126BU		15762	USSR	27 DEC	03 JAN 90	
1976-126BX		18891	USSR	27 DEC	03 JAN 90	
1976-126C		9644	USSR	27 DEC	31 DEC 76	
1976-126CB		19221	USSR	27 DEC	03 JAN 90	
1976-126CE		19480	USSR	27 DEC	03 JAN 90	
1976-126D		9645	USSR	27 DEC	14 JAN 77	
1976-126P		9657	USSR	27 DEC	06 JAN 91	
1976-126Q		9719	USSR	27 DEC	11 JUL 78	
1976-126R		9720	USSR	27 DEC	09 OCT 81	
1976-126W		9725	USSR	27 DEC	21 DEC 92	
1976-127A	MOLNIYA 3-6	9635	USSR	28 DEC	06 FEB 90	
1976-127B		9636	USSR	28 DEC	16 MAR 77	
1976-127C		9641	USSR	28 DEC	15 FEB 77	
1976-127D		9642	USSR	28 DEC	23 FEB 77	
1976-127E		9647	USSR	28 DEC	30 SEP 87	
1977 LAUNCHES						
1977-001A	COSMOS 888	9658	USSR	06 JAN	19 JAN 77	
1977-001B		9659	USSR	06 JAN	11 JAN 77	
1977-001C		9660	USSR	06 JAN	11 JAN 77	
1977-001D		9732	USSR	06 JAN	23 JAN 77	
1977-001E		9733	USSR	06 JAN	21 JAN 77	
1977-001F		9734	USSR	06 JAN	20 JAN 77	
1977-003A	COSMOS 889	9735	USSR	20 JAN	01 FEB 77	
1977-003B		9736	USSR	20 JAN	01 FEB 77	
1977-005C		9787	US	28 JAN	06 FEB 81	
1977-006A	COSMOS 891	9801	USSR	02 FEB	04 FEB 81	
1977-006B		9802	USSR	02 FEB	18 DEC 79	
1977-007B		9808	US	06 FEB	06 FEB 77	
1977-008A	SOYUZ 24	9804	USSR	07 FEB	25 FEB 77	E
1977-008B		9805	USSR	07 FEB	12 FEB 77	
1977-008C		9806	USSR	07 FEB	12 FEB 77	
1977-009A	COSMOS 892	9812	USSR	09 FEB	22 FEB 77	
1977-009B		9813	USSR	09 FEB	15 FEB 77	
1977-009C		9814	USSR	09 FEB	10 FEB 77	
1977-009D		9845	USSR	09 FEB	26 FEB 77	
1977-009E		9847	USSR	09 FEB	24 FEB 77	
1977-009F		9849	USSR	09 FEB	23 FEB 77	
1977-009G		9851	USSR	09 FEB	23 FEB 77	
1977-010B		9830	USSR	11 FEB	13 MAR 77	
1977-010C		9831	USSR	11 FEB	02 MAR 77	
1977-010D		9832	USSR	11 FEB	27 FEB 77	
1977-011A	COSMOS 893	9833	USSR	15 FEB	06 OCT 84	
1977-011B		9834	USSR	15 FEB	05 NOV 77	
1977-011C		10486	USSR	15 FEB	18 MAR 78	
1977-012B		9842	JAPAN	19 FEB	21 MAR 79	
1977-012D		9844	JAPAN	19 FEB	15 MAY 78	
1977-014B		9859	JAPAN	23 FEB	14 NOV 91	
1977-015A	COSMOS 895	9853	USSR	26 FEB	22 MAR 92	
1977-016A	COSMOS 896	9857	USSR	03 MAR	16 MAR 77	
1977-016B		9858	USSR	03 MAR	04 MAR 77	
1977-016C		9865	USSR	03 MAR	25 MAR 77	
1977-016D		9869	USSR	03 MAR	17 MAR 77	
1977-016E		9870	USSR	03 MAR	17 MAR 77	
1977-016F		9873	USSR	03 MAR	24 MAR 77	
1977-016G		9874	USSR	03 MAR	23 MAR 77	
1977-016H		9875	USSR	03 MAR	21 MAR 77	
1977-017A	COSMOS 897	9860	USSR	10 MAR	23 MAR 77	
1977-017B		9861	USSR	10 MAR	15 MAR 77	
1977-017C		9876	USSR	10 MAR	31 MAR 77	
1977-017D		9877	USSR	10 MAR	27 MAR 77	
1977-017E		9878	USSR	10 MAR	26 MAR 77	
1977-017F		9879	USSR	10 MAR	20 MAY 77	
1977-017G		9885	USSR	10 MAR	27 MAR 77	
1977-018B		9864	US	10 MAR	25 APR 82	
1977-018C		9866	US	10 MAR	17 DEC 77	
1977-018D		9867	US	10 MAR	13 OCT 77	
1977-018E		9868	US	10 MAR	15 FEB 78	
1977-019A		9863	US	13 MAR	26 MAY 77	
1977-019B		10012	US	13 MAR	21 MAY 77	
1977-020A	COSMOS 898	9871	USSR	17 MAR	30 MAR 77	
1977-020B		9872	USSR	17 MAR	22 MAR 77	
1977-020C		9886	USSR	17 MAR	30 MAR 77	
1977-020D		9887	USSR	17 MAR	03 APR 77	
1977-021B		9881	USSR	24 MAR	19 APR 77	
1977-021C		9882	USSR	24 MAR	12 APR 77	
1977-022A	COSMOS 899	9883	USSR	24 MAR	19 OCT 80	
1977-022B		9884	USSR	24 MAR	25 DEC 80	
1977-022C		9900	USSR	24 MAR	11 APR 78	
1977-023A	COSMOS 900	9898	USSR	29 MAR	11 OCT 79	
1977-023B		9899	USSR	29 MAR	06 SEP 79	
1977-025A	COSMOS 901	9905	USSR	05 APR	28 JUN 78	
1977-025B		9906	USSR	05 APR	25 MAR 78	
1977-026A	COSMOS 902	9908	USSR	07 APR	20 APR 77	
1977-026B		9909	USSR	07 APR	09 APR 77	
1977-026C		9910	USSR	07 APR	10 APR 77	
1977-026D		9928	USSR	07 APR	20 APR 77	
1977-026E		9929	USSR	07 APR	23 APR 77	
1977-027B		9912	USSR	11 APR	06 JUN 77	
1977-027C		9913	USSR	11 APR	01 MAY 77	
1977-028A	COSMOS 904	9930	USSR	20 APR	04 MAY 77	
1977-028B		9935	USSR	20 APR	29 APR 77	
1977-028C		9936	USSR	20 APR	23 APR 77	
1977-029B		9932	US	20 APR	20 APR 77	
1977-029C		9933	US	20 APR	17 DEC 89	
1977-029D		9934	US	20 APR	20 APR 77	
1977-030A	COSMOS 905	9937	USSR	26 APR	26 MAY 77	
1977-030B		9939	USSR	26 APR	01 MAY 77	
1977-030C		9940	USSR	26 APR	14 MAY 77	
1977-030D		10023	USSR	26 APR	26 MAY 77	
1977-031A	COSMOS 906	9938	USSR	27 APR	23 MAR 80	
1977-032B		9942	USSR	28 APR	15 MAY 77	
1977-032C		9943	USSR	28 APR	15 MAY 77	
1977-033A	COSMOS 907	9944	USSR	05 MAY	16 MAY 77	
1977-033B		9945	USSR	05 MAY	13 MAY 77	
1977-033C		9946	USSR	05 MAY	06 MAY 77	
1977-033D		9947	USSR	05 MAY	06 MAY 77	
1977-033E		10004	USSR	05 MAY	17 MAY 77	
1977-033F		10005	USSR	05 MAY	20 MAY 77	
1977-033G		10006	USSR	05 MAY	24 MAY 77	
1977-034D		10003	US	12 MAY	14 MAY 77	
1977-035A	COSMOS 908	10007	USSR	17 MAY	31 MAY 77	
1977-035B		10008	USSR	17 MAY	21 MAY 77	
1977-035C		10009	USSR	17 MAY	22 MAY 77	
1977-035D		10027	USSR	17 MAY	03 JUN 77	
1977-037A	COSMOS 910	10014	USSR	23 MAY	23 MAY 77	
1977-037B		10015	USSR	23 MAY	27 MAY 77	
1977-037C		10018	USSR	23 MAY	24 MAY 77	
1977-040A	COSMOS 912	10021	USSR	26 MAY	08 JUN 77	
1977-040B		10022	USSR	26 MAY	31 MAY 77	
1977-040C		10026	USSR	26 MAY	27 MAY 77	
1977-040D		10035	USSR	26 MAY	12 JUN 77	
1977-040E		10036	USSR	26 MAY	18 JUN 77	
1977-042A	COSMOS 913	10028	USSR	30 MAY	29 DEC 79	

International Designation	Satellite Name	Number	Source	Launch	Decay	Note
1977-042B		10029	USSR	30 MAY	02 NOV 79	BN
1977-042C		10237	USSR	30 MAY	18 APR 78	
1977-042D		10238	USSR	30 MAY	18 FEB 78	
1977-042E		10329	USSR	30 MAY	26 FEB 78	
1977-042F		10330	USSR	30 MAY	23 FEB 78	
1977-042G		10331	USSR	30 MAY	23 FEB 78	
1977-042H		10332	USSR	30 MAY	18 APR 78	
1977-042J		10333	USSR	30 MAY	11 MAR 78	
1977-042K		10334	USSR	30 MAY	15 FEB 78	
1977-042L		10335	USSR	30 MAY	15 FEB 78	
1977-042M		10336	USSR	30 MAY	04 FEB 78	
1977-042N		10337	USSR	30 MAY	23 FEB 78	
1977-042P		10349	USSR	30 MAY	18 APR 78	
1977-042Q		10350	USSR	30 MAY	15 MAR 78	
1977-042R		10418	USSR	30 MAY	18 JUN 78	
1977-042S		10487	USSR	30 MAY	22 MAR 78	
1977-042T		10488	USSR	30 MAY	18 APR 78	
1977-042U		10494	USSR	30 MAY	18 MAR 78	
1977-042V		10495	USSR	30 MAY	18 MAR 78	
1977-042W		10496	USSR	30 MAY	18 APR 78	
1977-042X		10497	USSR	30 MAY	27 MAR 78	
1977-043A	COSMOS 914	10030	USSR	31 MAY	13 JUN 77	
1977-043B		10031	USSR	31 MAY	11 JUN 77	
1977-043C		10032	USSR	31 MAY	16 JUN 77	
1977-043D		10045	USSR	31 MAY	18 JUN 77	
1977-043E		10046	USSR	31 MAY	02 JUL 77	
1977-045A	COSMOS 915	10038	USSR	08 JUN	21 JUN 77	
1977-045B		10039	USSR	08 JUN	11 JUN 77	
1977-045C		10073	USSR	08 JUN	25 JUN 77	
1977-045D		10074	USSR	08 JUN	23 JUN 77	
1977-045E		10084	USSR	08 JUN	23 JUN 77	
1977-046A	COSMOS 916	10040	USSR	10 JUN	21 JUN 77	
1977-046B		10041	USSR	10 JUN	01 JUL 77	
1977-046C		10042	USSR	10 JUN	29 JUN 77	
1977-046D		10043	USSR	10 JUN	18 JUN 77	
1977-046E		10044	USSR	10 JUN	25 JUN 77	
1977-046F		10088	USSR	10 JUN	27 JUN 77	
1977-046G		10090	USSR	10 JUN	01 JUL 77	
1977-047B		10060	USSR	16 JUN	17 AUG 77	
1977-047C		10066	USSR	16 JUN	14 JUL 77	
1977-048C		10063	US	16 JUN	12 JUL 77	
1977-048D		10067	US	16 JUN	29 JUN 77	
1977-048E		10075	US	16 JUN	11 JUL 77	
1977-049A	SIGNE 3	10064	FRAN	17 JUN	20 JUN 79	
1977-049B		10069	USSR	17 JUN	05 FEB 80	
1977-050A	COSMOS 918	10065	USSR	17 JUN	18 JUN 77	
1977-050B		10068	USSR	17 JUN	18 JUN 77	
1977-051A	COSMOS 919	10070	USSR	18 JUN	28 AUG 78	
1977-051B		10071	USSR	18 JUN	21 APR 78	
1977-051C		10072	USSR	18 JUN	15 JUL 77	
1977-051D		10808	USSR	18 JUN	18 JUL 78	
1977-052A	COSMOS 920	10086	USSR	22 JUN	05 JUL 77	
1977-052B		10087	USSR	22 JUN	27 JUN 77	
1977-052C		10127	USSR	22 JUN	10 JUL 77	
1977-052D		10128	USSR	22 JUN	08 JUL 77	
1977-052E		10133	USSR	22 JUN	10 JUL 77	
1977-054A	MOLNIYA 1-37	10092	USSR	24 JUN	20 MAY 93	
1977-054B		10093	USSR	24 JUN	24 AUG 77	
1977-054C		10094	USSR	24 JUN	04 AUG 77	
1977-054E		21962	USSR	24 JUN	19 JUN 92	
1977-056A		10111	US	27 JUN	23 DEC 77	
1977-056B		10112	US	27 JUN	29 JUN 77	
1977-057C		10116	USSR	29 JUN	02 AUG 77	
1977-058A	COSMOS 922	10115	USSR	30 JUN	13 JUL 77	
1977-058B		10117	USSR	30 JUN	08 JUL 77	
1977-058C		10118	USSR	30 JUN	02 JUL 77	
1977-058D		10119	USSR	30 JUN	02 JUL 77	
1977-060A	COSMOS 924	10129	USSR	04 JUL	10 FEB 81	
1977-060B		10130	USSR	04 JUL	20 JAN 81	
1977-060C		10131	USSR	04 JUL	12 JAN 78	
1977-060D		10132	USSR	04 JUL	03 DEC 77	
1977-061A	COSMOS 925	10134	USSR	07 JUL	29 APR 93	
1977-063A	COSMOS 927	10139	USSR	12 JUL	25 JUL 77	
1977-063B		10140	USSR	12 JUL	18 JUL 77	
1977-063C		10162	USSR	12 JUL	27 JUL 77	
1977-063D		10163	USSR	12 JUL	26 JUL 77	
1977-065AA		10197	US	14 JUL	18 JUL 81	
1977-065AC		10199	US	14 JUL	30 NOV 80	
1977-065AD		10200	US	14 JUL	15 MAY 81	
1977-065AG		10203	US	14 JUL	01 JAN 86	
1977-065AJ		10205	US	14 JUL	09 AUG 81	
1977-065AN		10209	US	14 JUL	26 SEP 82	
1977-065AP		10210	US	14 JUL	17 NOV 81	
1977-065AS		10213	US	14 JUL	13 APR 81	
1977-065AU		10215	US	14 JUL	13 APR 81	
1977-065AV		10216	US	14 JUL	05 JUN 82	
1977-065AZ		10222	US	14 JUL	24 DEC 78	
1977-065BB		10224	US	14 JUL	20 JUN 81	
1977-065BC		10225	US	14 JUL	09 OCT 80	
1977-065BF		10228	US	14 JUL	11 MAR 81	
1977-065BJ		10231	US	14 JUL	20 JUL 82	
1977-065BK		10232	US	14 JUL	21 JAN 91	
1977-065BL		10233	US	14 JUL	28 AUG 90	
1977-065BP		10236	US	14 JUL	27 AUG 80	
1977-065BQ		10239	US	14 JUL	27 APR 84	
1977-065BT		10242	US	14 JUL	29 APR 81	
1977-065BV		10244	US	14 JUL	22 SEP 91	
1977-065BW		10245	US	14 JUL	04 NOV 77	
1977-065BY		10247	US	14 JUL	09 AUG 79	
1977-065BZ		10248	US	14 JUL	14 FEB 79	
1977-065C		10145	US	14 JUL	31 MAR 78	
1977-065CA		10249	US	14 JUL	24 OCT 89	
1977-065CE		10253	US	14 JUL	20 FEB 78	
1977-065CF		10254	US	14 JUL	04 JUN 81	
1977-065CG		10255	US	14 JUL	04 JAN 90	
1977-065CH		10256	US	14 JUL	05 FEB 82	
1977-065CK		10258	US	14 JUL	04 SEP 92	
1977-065CL		10259	US	14 JUL	07 DEC 82	
1977-065CM		10260	US	14 JUL	11 APR 83	
1977-065CQ		10263	US	14 JUL	01 MAY 81	
1977-065CR		10264	US	14 JUL	14 MAR 89	
1977-065CS		10265	US	14 JUL	22 JUN 80	
1977-065CT		10266	US	14 JUL	06 AUG 85	
1977-065CU		10267	US	14 JUL	21 JUN 82	
1977-065CV		10268	US	14 JUL	13 AUG 79	
1977-065CX		10270	US	14 JUL	11 SEP 89	
1977-065CY		10302	US	14 JUL	16 APR 89	
1977-065CZ		10303	US	14 JUL	15 JUN 90	
1977-065D		10156	US	14 JUL	06 NOV 81	
1977-065DA		10304	US	14 JUL	08 NOV 82	
1977-065DD		10307	US	14 JUL	06 JUN 81	
1977-065DE		10308	US	14 JUL	16 JUN 79	
1977-065DG		10310	US	14 JUL	11 JUL 80	
1977-065DH		10311	US	14 JUL	26 MAR 89	
1977-065DK		10313	US	14 JUL	31 MAY 93	
1977-065DP		10340	US	14 JUL	12 NOV 83	
1977-065DR		10342	US	14 JUL	26 JAN 83	
1977-065DS		10343	US	14 JUL	14 OCT 78	
1977-065DT		10344	US	14 JUL	02 JUL 79	
1977-065DW		10347	US	14 JUL	06 JUL 81	
1977-065DX		10348	US	14 JUL	07 AUG 92	
1977-065DZ		10390	US	14 JUL	06 DEC 81	
1977-065E		10176	US	14 JUL	22 MAY 79	
1977-065EA		10468	US	14 JUL	15 SEP 80	
1977-065EE		10472	US	14 JUL	16 MAR 81	
1977-065EF		10473	US	14 JUL	07 JAN 82	
1977-065EG		10474	US	14 JUL	12 JUN 78	
1977-065EH		10475	US	14 JUL	10 NOV 81	
1977-065EK		10477	US	14 JUL	10 DEC 81	
1977-065EN		10480	US	14 JUL	18 JUN 92	
1977-065EP		10481	US	14 JUL	03 FEB 87	
1977-065EQ		10482	US	14 JUL	10 NOV 82	
1977-065ER		10483	US	14 JUL	20 AUG 81	
1977-065ET		10624	US	14 JUL	27 JUN 80	
1977-065EU		10625	US	14 JUL	30 JUL 82	
1977-065EX		10628	US	14 JUL	21 MAR 90	
1977-065FA		10631	US	14 JUL	08 OCT 80	
1977-065FB		10632	US	14 JUL	03 FEB 89	
1977-065FC		10633	US	14 JUL	03 JAN 90	
1977-065FE		10635	US	14 JUL	30 SEP 80	
1977-065FF		10636	US	14 JUL	20 MAY 83	
1977-065FK		14494	US	14 JUL	09 JUL 88	
1977-065FL		14495	US	14 JUL	10 JUN 89	
1977-065FU		18643	US	14 JUL	27 NOV 88	
1977-065GC		19965	US	14 JUL	20 OCT 90	
1977-065H		10179	US	14 JUL	25 FEB 91	
1977-065J		10180	US	14 JUL	23 OCT 90	
1977-065L		10182	US	14 JUL	01 JAN 81	
1977-065P		10185	US	14 JUL	23 JUN 85	
1977-065Q		10186	US	14 JUL	25 JAN 92	
1977-065R		10187	US	14 JUL	27 OCT 78	
1977-065T		10189	US	14 JUL	16 FEB 79	
1977-065U		10190	US	14 JUL	24 APR 89	
1977-065V		10191	US	14 JUL	04 DEC 78	
1977-065W		10192	US	14 JUL	17 AUG 83	
1977-065X		10193	US	14 JUL	17 APR 79	
1977-065Y		10194	US	14 JUL	22 SEP 85	
1977-065Z		10196	US	14 JUL	10 NOV 82	
1977-066A	COSMOS 929	10146	USSR	17 JUL	02 FEB 78	
1977-066B		10147	USSR	17 JUL	29 JUL 77	
1977-066C		10148	USSR	17 JUL	24 JUL 77	
1977-067A	COSMOS 930	10149	USSR	19 JUL	12 MAY 80	
1977-068B		10151	USSR	20 JUL	22 SEP 77	
1977-068C		10152	USSR	20 JUL	17 AUG 77	
1977-068H		18528	USSR	20 JUL	02 JAN 89	
1977-069A	COSMOS 932	10153	USSR	20 JUL	02 AUG 77	
1977-069B		10154	USSR	20 JUL	24 JUL 77	
1977-069C		10170	USSR	20 JUL	03 AUG 77	
1977-069D		10171	USSR	20 JUL	03 AUG 77	
1977-070A	COSMOS 933	10157	USSR	22 JUL	01 NOV 78	
1977-070B		10158	USSR	22 JUL	07 JUN 78	
1977-071B		10160	USSR	23 JUL	26 JUL 77	
1977-071C		10161	USSR	23 JUL	26 JUL 77	
1977-071D		10276	USSR	23 JUL	05 JUN 78	
1977-071E		10277	USSR	23 JUL	25 MAR 78	
1977-072A	COSMOS 934	10164	USSR	27 JUL	09 AUG 77	
1977-072B		10165	USSR	27 JUL	05 AUG 77	
1977-072C		10166	USSR	27 JUL	02 AUG 77	
1977-072D		10174	USSR	27 JUL	17 AUG 77	
1977-072E		10175	USSR	27 JUL	10 AUG 77	
1977-072F		10195	USSR	27 JUL	12 AUG 77	
1977-073A	COSMOS 935	10168	USSR	29 JUL	11 AUG 77	
1977-073B		10169	USSR	29 JUL	03 AUG 77	
1977-097BT		11893	USSR	29 JUL	04 SEP 80	
1977-074A	COSMOS 936	10172	USSR	03 AUG	22 AUG 77	
1977-074B		10173	USSR	03 AUG	28 AUG 77	

International Designation	Satellite Name	Number	Source	Launch	Decay	Note
1977-074C		10274	USSR	03 AUG	05 SEP 77	
1977-074D		10275	USSR	03 AUG	08 OCT 77	
1977-075A	HEAO 1	10217	US	12 AUG	15 MAR 79	
1977-075B		10218	US	12 AUG	25 NOV 77	
1977-077A	COSMOS 937	10278	USSR	24 AUG	19 OCT 78	
1977-077B		10279	USSR	24 AUG	25 AUG 77	
1977-077C		10280	USSR	24 AUG	25 AUG 77	
1977-078A	COSMOS 938	10281	USSR	24 AUG	06 SEP 77	
1977-078B		10283	USSR	24 AUG	30 AUG 77	
1977-078C		10284	USSR	24 AUG	11 SEP 77	
1977-078D		10285	USSR	24 AUG	21 SEP 77	
1977-078E		10324	USSR	24 AUG	07 SEP 77	
1977-078F		10325	USSR	24 AUG	06 SEP 77	
1977-080C		10296	US	25 AUG	10 NOV 82	
1977-080D		10297	US	25 AUG	18 JAN 78	
1977-080E		10298	US	25 AUG	07 DEC 77	
1977-081A	COSMOS 947	10299	USSR	27 AUG	09 SEP 77	
1977-081B		10300	USSR	27 AUG	07 SEP 77	
1977-081C		10301	USSR	27 AUG	07 SEP 77	
1977-082B		10316	USSR	30 AUG	26 SEP 77	
1977-082C		10317	USSR	30 AUG	23 SEP 77	
1977-082D		10318	USSR	30 AUG	23 SEP 77	
1977-083A	COSMOS 948	10319	USSR	02 SEP	15 SEP 77	
1977-083B		10320	USSR	02 SEP	07 SEP 77	
1977-083C		10356	USSR	02 SEP	20 SEP 77	
1977-083D		10357	USSR	02 SEP	15 SEP 77	
1977-085A	COSMOS 949	10326	USSR	06 SEP	06 OCT 77	
1977-085B		10327	USSR	06 SEP	11 SEP 77	
1977-085C		10328	USSR	06 SEP	10 SEP 77	
1977-086A	COSMOS 950	10351	USSR	13 SEP	27 SEP 77	
1977-086B		10353	USSR	13 SEP	18 SEP 77	
1977-086C		10354	USSR	13 SEP	20 SEP 77	
1977-088B		10399	USSR	16 SEP	07 NOV 77	
1977-088C		10400	USSR	16 SEP	11 OCT 77	
1977-089A	COSMOS 953	10359	USSR	16 SEP	29 SEP 77	
1977-089B		10360	USSR	16 SEP	21 SEP 77	
1977-089C		10380	USSR	16 SEP	30 SEP 77	
1977-089D		10381	USSR	16 SEP	29 SEP 77	
1977-090A	COSMOS 954	10361	USSR	18 SEP	24 JAN 78	
1977-097AA		11092	USSR	19 SEP	12 DEC 78	
1977-097AB		11093	USSR	19 SEP	26 NOV 78	
1977-097AD		11095	USSR	19 SEP	15 DEC 78	
1977-097AF		11097	USSR	19 SEP	19 DEC 78	
1977-097AG		11100	USSR	19 SEP	16 FEB 79	
1977-091C		10364	USSR	20 SEP	03 JUN 82	
1977-092B		10366	USSR	20 SEP	21 SEP 77	
1977-092C		10367	USSR	20 SEP	21 SEP 77	
1977-092D		10368	USSR	20 SEP	25 SEP 77	
1977-092E		10408	USSR	20 SEP	30 APR 78	
1977-092F		10484	USSR	20 SEP	18 JUN 78	
1977-093B		10371	USSR	22 SEP	29 OCT 77	
1977-093C		10372	USSR	22 SEP	23 OCT 77	
1977-093D		10373	USSR	22 SEP	28 SEP 77	
1977-094A		10374	US	23 SEP	08 DEC 77	
1977-095A	COSMOS 956	10375	USSR	24 SEP	27 JUN 82	
1977-095B		10378	USSR	24 SEP	01 NOV 82	
1977-096A	INTERCOSMOS 17	10376	USSR	24 SEP	08 NOV 79	
1977-096B		10377	USSR	24 SEP	11 SEP 79	
1977-096C		10379	USSR	24 SEP	08 APR 78	
1977-097A	SALYUT 6	10382	USSR	29 SEP	29 JUL 82	
1977-097AC		11094	USSR	29 SEP	28 JAN 79	
1977-097AE		11096	USSR	29 SEP	30 JUL 79	
1977-097AH		11338	USSR	29 SEP	22 JUN 79	
1977-097AJ		11339	USSR	29 SEP	25 MAY 79	
1977-097AK		11340	USSR	29 SEP	03 MAY 79	
1977-097AL		11341	USSR	29 SEP	09 JUL 79	
1977-097AM		11342	USSR	29 SEP	25 MAY 79	
1977-097AN		11350	USSR	29 SEP	17 JUL 79	
1977-097AP		11431	USSR	29 SEP	04 SEP 79	
1977-097AQ		11432	USSR	29 SEP	05 OCT 79	
1977-097AR		11433	USSR	29 SEP	16 JUL 79	
1977-097AS		11434	USSR	29 SEP	27 NOV 79	
1977-097AT		11435	USSR	29 SEP	25 AUG 79	
1977-097AU		11467	USSR	29 SEP	23 OCT 79	
1977-097AV		11468	USSR	29 SEP	17 OCT 79	
1977-097AW		11469	USSR	29 SEP	17 OCT 79	
1977-097AX		11470	USSR	29 SEP	23 OCT 79	
1977-097AY		11471	USSR	29 SEP	15 OCT 79	
1977-097AZ		11472	USSR	29 SEP	16 MAR 80	
1977-097B		10383	USSR	29 SEP	05 OCT 77	BP
1977-097BA		11473	USSR	29 SEP	26 OCT 79	
1977-097BB		11480	USSR	29 SEP	15 OCT 79	
1977-097BC		11489	USSR	29 SEP	15 OCT 79	
1977-097BD		11493	USSR	29 SEP	26 AUG 79	
1977-097BE		11494	USSR	29 SEP	31 DEC 79	
1977-097BF		11495	USSR	29 SEP	11 OCT 79	
1977-097BG		11500	USSR	29 SEP	21 DEC 79	
1977-097BH		11501	USSR	29 SEP	28 DEC 79	
1977-097BJ		11502	USSR	29 SEP	11 SEP 79	
1977-097BK		11503	USSR	29 SEP	21 DEC 79	
1977-097BL		11504	USSR	29 SEP	22 MAR 80	
1977-097BM		11508	USSR	29 SEP	20 OCT 79	
1977-097BN		11523	USSR	29 SEP	19 JAN 80	
1977-097BP		11563	USSR	29 SEP	09 NOV 79	
1977-097BQ		11572	USSR	29 SEP	27 OCT 79	
1977-097BR		11740	USSR	29 SEP	28 MAY 80	
1977-097BS		11832	USSR	29 SEP	05 SEP 80	
1977-097BU		11930	USSR	29 SEP	15 OCT 80	
1977-097BV		11931	USSR	29 SEP	21 OCT 80	
1977-097BW		11934	USSR	29 SEP	08 SEP 80	
1977-097BX		11935	USSR	29 SEP	15 OCT 80	
1977-097BY		11936	USSR	29 SEP	14 SEP 80	
1977-097BZ		11937	USSR	29 SEP	16 OCT 80	
1977-097C		10384	USSR	29 SEP	03 OCT 77	
1977-097CA		11942	USSR	29 SEP	14 SEP 80	
1977-097CB		11947	USSR	29 SEP	16 OCT 80	
1977-097CC		11955	USSR	29 SEP	01 OCT 80	
1977-097CD		11956	USSR	29 SEP	27 OCT 80	
1977-097CE		11957	USSR	29 SEP	21 OCT 80	
1977-097CF		11958	USSR	29 SEP	24 SEP 80	
1977-097CG		11966	USSR	29 SEP	23 NOV 80	
1977-097CH		11967	USSR	29 SEP	06 OCT 80	
1977-097CJ		11972	USSR	29 SEP	03 NOV 80	
1977-097CK		11973	USSR	29 SEP	15 OCT 80	
1977-097CL		11974	USSR	29 SEP	13 NOV 80	
1977-097CM		11975	USSR	29 SEP	13 OCT 80	
1977-097CN		12090	USSR	29 SEP	26 JAN 81	
1977-097CP		12390	USSR	29 SEP	08 MAY 81	
1977-097CQ		12391	USSR	29 SEP	22 JUN 81	
1977-097CR		12392	USSR	29 SEP	29 APR 81	
1977-097CS		12393	USSR	29 SEP	25 APR 81	
1977-097CT		12394	USSR	29 SEP	05 MAY 81	
1977-097CU		12395	USSR	29 SEP	08 MAY 81	
1977-097CV		12398	USSR	29 SEP	27 JUN 81	
1977-097CW		12436	USSR	29 SEP	27 JUL 81	
1977-097CX		12437	USSR	29 SEP	06 JUL 81	
1977-097CY		12438	USSR	29 SEP	12 JUN 81	
1977-097CZ		12439	USSR	29 SEP	06 JUL 81	
1977-097D		10533	USSR	29 SEP	22 FEB 78	
1977-097DA		12440	USSR	29 SEP	11 JUN 81	
1977-097DB		12441	USSR	29 SEP	24 MAY 81	
1977-097DC		12444	USSR	29 SEP	10 JUL 81	
1977-097DD		12476	USSR	29 SEP	22 JUL 81	
1977-097DE		12477	USSR	29 SEP	14 SEP 81	
1977-097DF		12478	USSR	29 SEP	21 JUL 81	
1977-097DG		12479	USSR	29 SEP	19 JUL 81	
1977-097DH		12488	USSR	29 SEP	09 SEP 81	
1977-097DJ		12489	USSR	29 SEP	07 OCT 81	
1977-097DK		12493	USSR	29 SEP	09 SEP 81	
1977-097E		10545	USSR	29 SEP	24 APR 78	
1977-097F		10547	USSR	29 SEP	12 MAY 78	
1977-097G		10548	USSR	29 SEP	25 MAR 78	
1977-097H		10553	USSR	29 SEP	27 MAY 78	
1977-097J		10596	USSR	29 SEP	14 MAY 78	
1977-097K		10597	USSR	29 SEP	21 MAR 78	
1977-097L		10598	USSR	29 SEP	08 FEB 78	
1977-097M		10685	USSR	29 SEP	06 MAR 78	
1977-097N		10686	USSR	29 SEP	06 MAR 78	
1977-097P		10687	USSR	29 SEP	06 MAR 78	
1977-097Q		10738	USSR	29 SEP	26 APR 78	
1977-097R		10740	USSR	29 SEP	15 JUN 78	
1977-097S		10743	USSR	29 SEP	17 JUN 78	
1977-097T		10781	USSR	29 SEP	29 MAY 78	
1977-097U		11032	USSR	29 SEP	21 OCT 78	
1977-097V		11033	USSR	29 SEP	14 OCT 78	
1977-097W		11034	USSR	29 SEP	03 DEC 78	
1977-097X		11035	USSR	29 SEP	26 NOV 78	
1977-097Y		11036	USSR	29 SEP	06 MAY 79	
1977-097Z		11091	USSR	29 SEP	14 APR 79	
1977-098A	COSMOS 957	10385	USSR	30 SEP	13 OCT 77	
1977-098B		10386	USSR	30 SEP	05 OCT 77	
1977-098C		10387	USSR	30 SEP	13 OCT 77	
1977-098D		10388	USSR	30 SEP	13 OCT 77	
1977-098E		10406	USSR	30 SEP	15 OCT 77	
1977-098F		10407	USSR	30 SEP	13 OCT 77	
1977-099A	SOYUZ 25	10401	USSR	09 OCT	11 OCT 77	AD
1977-099B		10402	USSR	09 OCT	11 OCT 77	
1977-100A	COSMOS 958	10403	USSR	11 OCT	24 OCT 77	
1977-100B		10404	USSR	11 OCT	09 NOV 77	
1977-100C		10405	USSR	11 OCT	31 OCT 77	
1977-100D		10426	USSR	11 OCT	05 MAR 78	
1977-100E		10427	USSR	11 OCT	15 SEP 78	
1977-100F		10428	USSR	11 OCT	18 JUN 78	
1977-100G		10429	USSR	11 OCT	16 JAN 78	
1977-101A	COSMOS 959	10419	USSR	21 OCT	30 NOV 77	
1977-101B		10420	USSR	21 OCT	08 NOV 77	
1977-101C		10421	USSR	21 OCT	05 NOV 77	
1977-102A	ISEE 1	10422	US	22 OCT	26 SEP 87	
1977-102B	ISEE 2	10423	ESA	22 OCT	26 SEP 87	
1977-102C		10424	US	22 OCT	23 OCT 78	
1977-103A	COSMOS 960	10430	USSR	25 OCT	22 OCT 80	
1977-103B		10431	USSR	25 OCT	30 NOV 80	
1977-103C		10432	USSR	25 OCT	24 JUN 78	
1977-104A	COSMOS 961	10434	USSR	26 OCT	26 OCT 77	
1977-104B		10435	USSR	26 OCT	27 OCT 77	
1977-105B		10456	USSR	28 OCT	18 NOV 77	
1977-105C		10458	USSR	28 OCT	26 NOV 77	
1977-105D		10460	USSR	28 OCT	20 NOV 77	
1977-108C		10950	US	23 NOV	12 AUG 90	
1977-110A	COSMOS 964	10498	USSR	04 DEC	17 DEC 77	
1977-110B		10499	USSR	04 DEC	09 DEC 77	
1977-110C		10500	USSR	04 DEC	06 JAN 78	
1977-110D		10522	USSR	04 DEC	20 DEC 77	
1977-110E		10523	USSR	04 DEC	20 DEC 77	
1977-111A	COSMOS 965	10501	USSR	08 DEC	16 DEC 79	
1977-111AA		10910	USSR	08 DEC	04 DEC 78	

International Designation	Satellite Name	Number	Source	Launch	Decay	Note
1977-111AB		10923	USSR	08 DEC	24 AUG 78	
1977-111AC		10924	USSR	08 DEC	17 AUG 78	
1977-111B		10503	USSR	08 DEC	14 NOV 79	BQ
1977-111C		10505	USSR	08 DEC	15 APR 78	
1977-111D		10682	USSR	08 DEC	17 APR 78	
1977-111E		10683	USSR	08 DEC	01 MAY 78	
1977-111F		10741	USSR	08 DEC	15 MAY 78	
1977-111G		10742	USSR	08 DEC	18 JUL 78	
1977-111H		10780	USSR	08 DEC	24 MAY 78	
1977-111J		10782	USSR	08 DEC	18 JUL 78	
1977-111K		10789	USSR	08 DEC	06 JUN 78	
1977-111L		10795	USSR	08 DEC	01 JUN 78	
1977-111M		10797	USSR	08 DEC	18 JUL 78	
1977-111N		10807	USSR	08 DEC	23 MAY 78	
1977-111P		10809	USSR	08 DEC	20 JUN 78	
1977-111Q		10814	USSR	08 DEC	18 JUN 78	
1977-111R		10815	USSR	08 DEC	08 JUL 79	
1977-111S		10816	USSR	08 DEC	06 JUL 78	
1977-111T		10817	USSR	08 DEC	20 JUN 78	
1977-111U		10858	USSR	08 DEC	30 JUL 78	
1977-111V		10859	USSR	08 DEC	26 JUL 78	
1977-111W		10891	USSR	08 DEC	19 JUL 78	
1977-111X		10892	USSR	08 DEC	25 JUL 78	
1977-111Y		10908	USSR	08 DEC	04 AUG 78	
1977-111Z		10909	USSR	08 DEC	04 DEC 78	
1977-113A	SOYUZ 26	10506	USSR	10 DEC	16 JAN 78	E
1977-113B		10507	USSR	10 DEC	12 DEC 77	
1977-115A	COSMOS 966	10510	USSR	12 DEC	24 DEC 77	
1977-115B		10511	USSR	12 DEC	18 DEC 77	
1977-115C		10534	USSR	12 DEC	29 DEC 77	
1977-115D		10538	USSR	12 DEC	07 JAN 78	
1977-118D		11072	US	15 DEC	09 OCT 81	
1977-120A	COSMOS 969	10527	USSR	20 DEC	03 JAN 78	
1977-120B		10530	USSR	20 DEC	26 DEC 77	
1977-120C		10549	USSR	20 DEC	05 JAN 78	
1977-120D		10550	USSR	20 DEC	04 JAN 78	
1977-120E		10551	USSR	20 DEC	05 MAR 78	
1977-120F		10552	USSR	20 DEC	05 MAR 78	
1977-121B		10532	USSR	21 DEC	25 DEC 77	
1977-121BN		17731	USSR	21 DEC	15 OCT 91	
1977-121BT		19393	USSR	21 DEC	15 SEP 91	
1977-121BU		19514	USSR	21 DEC	02 FEB 89	
1977-123C		10546	USSR	27 DEC	31 DEC 77	
1977-124A	COSMOS 973	10540	USSR	27 DEC	09 JAN 78	
1977-124B		10542	USSR	27 DEC	05 JAN 78	
1977-124C		10543	USSR	27 DEC	31 DEC 77	
1977-124D		10558	USSR	27 DEC	25 JAN 78	
1977-124E		10559	USSR	27 DEC	11 JAN 78	

1978 LAUNCHES

International Designation	Satellite Name	Number	Source	Launch	Decay	Note
1978-001A	COSMOS 974	10554	USSR	06 JAN	19 JAN 78	
1978-001B		10555	USSR	06 JAN	11 JAN 78	
1978-001C		10556	USSR	06 JAN	25 JAN 78	
1978-001D		10601	USSR	06 JAN	26 JAN 78	
1978-001E		10602	USSR	06 JAN	21 JAN 78	
1978-003A	SOYUZ 27	10560	USSR	10 JAN	16 MAR 78	E
1978-003B		10583	USSR	10 JAN	13 JAN 78	
1978-006A	COSMOS 984	10592	USSR	13 JAN	26 JAN 78	
1978-006B		10593	USSR	13 JAN	22 JAN 78	
1978-008A	PROGRESS 1	10603	USSR	20 JAN	08 FEB 78	
1978-008B		10604	USSR	20 JAN	23 JAN 78	
1978-009A	MOLNIYA 3-9	10605	USSR	24 JAN	24 APR 90	
1978-009B		10606	USSR	24 JAN	12 MAR 78	
1978-009C		10608	USSR	24 JAN	03 MAR 78	
1978-009D		10610	USSR	24 JAN	08 MAR 78	
1978-009E		10802	USSR	24 JAN	09 AUG 90	
1978-010A	COSMOS 986	10607	USSR	24 JAN	07 FEB 78	
1978-010B		10609	USSR	24 JAN	28 JAN 78	
1978-011A	MAO 8	10611	PRC	26 JAN	07 FEB 78	
1978-011B		10612	PRC	26 JAN	06 FEB 78	
1978-012B		10638	US	26 JAN	21 FEB 78	
1978-013A	COSMOS 987	10639	USSR	31 JAN	14 FEB 78	
1978-013B		10640	USSR	31 JAN	04 FEB 78	
1978-013C		10641	USSR	31 JAN	05 FEB 78	
1978-013D		10670	USSR	31 JAN	17 FEB 78	
1978-013E		10671	USSR	31 JAN	20 FEB 78	
1978-014B		10665	JAPAN	04 FEB	12 FEB 79	
1978-014G		18816	JAPAN	04 FEB	10 NOV 88	
1978-014H		22008	JAPAN	04 FEB	02 AUG 92	
1978-015A	COSMOS 988	10666	USSR	08 FEB	20 FEB 78	
1978-015B		10667	USSR	08 FEB	17 FEB 78	
1978-015C		10668	USSR	08 FEB	05 MAR 78	
1978-015D		10678	USSR	08 FEB	22 FEB 78	
1978-015E		10679	USSR	08 FEB	02 MAR 78	
1978-015F		10680	USSR	08 FEB	11 MAR 78	
1978-015G		10681	USSR	08 FEB	24 FEB 78	
1978-016B		10724	US	09 FEB	31 AUG 82	
1978-016D		13998	US	09 FEB	21 JUN 89	
1978-017A	COSMOS 989	10672	USSR	14 FEB	28 FEB 78	
1978-017B		10673	USSR	14 FEB	17 FEB 78	
1978-017C		10690	USSR	14 FEB	01 MAR 78	
1978-017D		10691	USSR	14 FEB	28 FEB 78	
1978-023A	SOYUZ 28	10694	USSR	02 MAR	10 MAR 78	E
1978-023B		10695	USSR	02 MAR	05 MAR 78	
1978-024A	MOLNIYA 1-39	10696	USSR	02 MAR	09 MAR 92	
1978-024B		10697	USSR	02 MAR	10 APR 78.	
1978-024C		10698	USSR	02 MAR	09 APR 78	
1978-024D		10803	USSR	02 MAR	11 DEC 92	
1978-025A	COSMOS 992	10699	USSR	04 MAR	17 MAR 78	
1978-025B		10700	USSR	04 MAR	10 MAR 78	
1978-025C		10701	USSR	04 MAR	05 MAR 78	
1978-026AC		12187	US	05 MAR	12 FEB 88	
1978-026AD		12188	US	05 MAR	13 MAR 93	
1978-026AG		12191	US	05 MAR	28 FEB 89	
1978-026AH		12192	US	05 MAR	22 MAR 89	
1978-026AN		12197	US	05 MAR	05 FEB 82	
1978-026AQ		12199	US	05 MAR	20 FEB 90	
1978-026AT		12202	US	05 MAR	12 NOV 83	
1978-026AW		12205	US	05 MAR	21 JUL 89	
1978-026AY		12207	US	05 MAR	16 JUL 84	
1978-026BB		12210	US	05 MAR	11 FEB 90	
1978-026BL		12219	US	05 MAR	10 AUG 89	
1978-026BM		12220	US	05 MAR	13 MAR 91	
1978-026BR		12224	US	05 MAR	31 DEC 92	
1978-026BV		12228	US	05 MAR	05 OCT 90	
1978-026BW		12229	US	05 MAR	05 MAY 81	
1978-026BY		12231	US	05 MAR	19 MAR 93	
1978-026CA		12233	US	05 MAR	12 FEB 88	
1978-026CC		12235	US	05 MAR	08 FEB 89	
1978-026CF		12238	US	05 MAR	15 FEB 82	
1978-026CG		12239	US	05 MAR	05 FEB 82	
1978-026CH		12240	US	05 MAR	16 APR 84	
1978-026CJ		12241	US	05 MAR	11 APR 82	
1978-026CM		12244	US	05 MAR	18 JUN 89	
1978-026CN		12245	US	05 MAR	15 JUL 89	
1978-026CQ		12247	US	05 MAR	02 JUN 88	
1978-026CR		12248	US	05 MAR	17 AUG 81	
1978-026CS		12249	US	05 MAR	16 APR 89	
1978-026CU		12251	US	05 MAR	18 FEB 81	
1978-026CV		12252	US	05 MAR	02 JAN 82	
1978-026CW		12253	US	05 MAR	13 DEC 81	
1978-026CX		12254	US	05 MAR	20 MAR 89	
1978-026CY		12255	US	05 MAR	01 JUL 81	
1978-026DB		12258	US	05 MAR	01 JUL 81	
1978-026DC		12259	US	05 MAR	15 MAR 89	
1978-026DD		12260	US	05 MAR	29 MAY 91	
1978-026DE		12261	US	05 MAR	27 SEP 91	
1978-026DH		12264	US	05 MAR	10 DEC 81	
1978-026DQ		12271	US	05 MAR	18 NOV 81	
1978-026DR		12272	US	05 MAR	07 AUG 89	
1978-026DU		12275	US	05 MAR	27 DEC 81	
1978-026DV		12276	US	05 MAR	20 APR 87	
1978-026DW		12277	US	05 MAR	17 AUG 89	
1978-026EA		12281	US	05 MAR	07 NOV 82	
1978-026EB		12282	US	05 MAR	29 APR 82	
1978-026EF		12286	US	05 MAR	14 OCI 81	
1978-026EJ		12289	US	05 MAR	10 NOV 91	
1978-026ES		12374	US	05 MAR	27 DEC 89	
1978-026FF		13568	US	05 MAR	17 NOV 90	
1978-026FT		13580	US	05 MAR	06 MAR 83	
1978-026FX		14457	US	05 MAR	26 DEC 91	
1978-026GC		14462	US	05 MAR	05 MAY 88	
1978-026GY		17766	US	05 MAR	06 SEP 91	
1978-026GZ		17767	US	05 MAR	12 AUG 89	
1978-026HC		18301	US	05 MAR	02 OCT 88	
1978-026HG		18679	US	05 MAR	22 MAR 91	
1978-026HK		18814	US	05 MAR	11 NOV 89	
1978-026HL		19237	US	05 MAR	15 JUL 89	
1978-026X		12182	US	05 MAR	28 OCT 89	
1978-027A	COSMOS 993	10725	USSR	10 MAR	23 MAR 78	
1978-027B		10726	USSR	10 MAR	14 MAR 78	
1978-027C		10727	USSR	10 MAR	12 MAR 78	
1978-027D		10739	USSR	10 MAR	24 MAR 78	
1978-029A		10733	US	16 MAR	11 SEP 78	
1978-029C		10737	US	16 MAR	17 MAR 78	
1978-030A	COSMOS 995	10735	USSR	17 MAR	30 MAR 78	
1978-030B		10736	USSR	17 MAR	20 MAR 78	
1978-032A	COSMOS 997	10770	USSR	30 MAR	30 MAR 78	
1978-032B	COSMOS 998	10771	USSR	30 MAR	30 MAR 78	
1978-032C		10772	USSR	30 MAR	02 APR 78	
1978-032D		10775	USSR	30 MAR	03 APR 78	
1978-033A	COSMOS 999	10773	USSR	30 MAR	12 APR 78	
1978-033B		10774	USSR	30 MAR	03 APR 78	
1978-033C		10798	USSR	30 MAR	13 APR 78	
1978-033D		10799	USSR	30 MAR	16 APR 78	
1978-033E		10800	USSR	30 MAR	24 APR 78	
1978-036A	COSMOS 1001	10783	USSR	04 APR	15 APR 78	
1978-036B		10784	USSR	04 APR	06 APR 78	
1978-036C		10804	USSR	04 APR	01 JUN 78	
1978-036D		10805	USSR	04 APR	02 MAY 78	
1978-036E		10806	USSR	04 APR	25 APR 78	
1978-036F		10810	USSR	04 APR	23 MAY 78	
1978-037A	COSMOS 1002	10785	USSR	06 APR	19 APR 78	
1978-037B		10786	USSR	06 APR	10 APR 78	
1978-037C		10790	USSR	06 APR	24 APR 78	
1978-037D		10791	USSR	06 APR	24 APR 78	
1978-040A	COSMOS 1003	10811	USSR	20 APR	04 MAY 78	
1978-040B		10812	USSR	20 APR	23 APR 78	
1978-040C		10813	USSR	20 APR	22 APR 78	
1978-040D		10821	USSR	20 APR	04 MAY 78	
1978-040E		10822	USSR	20 APR	04 MAY 78	
1978-041A	HCMM	10818	US	26 APR	22 DEC 81	
1978-041B		10819	US	26 APR	24 NOV 80	
1978-042B		10853	US	01 MAY	21 OCT 91	
1978-042C		10854	US	01 MAY	06 MAR 92	

International Designation	Satellite Name	Number	Source	Launch	Decay	Note
1978-042D		10913	US	01 MAY	01 MAY 86	
1978-042E		10914	US	01 MAY	16 JUN 89	
1978-043A	COSMOS 1004	10846	USSR	05 MAY	18 MAY 78	
1978-043B		10847	USSR	05 MAY	09 MAY 78	
1978-043C		10848	USSR	05 MAY	09 MAY 78	
1978-043D		10849	USSR	05 MAY	09 MAY 78	
1978-043E		10850	USSR	05 MAY	14 MAY 78	
1978-043F		10851	USSR	05 MAY	08 JUN 78	
1978-043G		10852	USSR	05 MAY	16 JUN 78	
1978-046A	COSMOS 1006	10862	USSR	12 MAY	14 MAR 79	
1978-046B		10863	USSR	12 MAY	06 DEC 78	
1978-048A	COSMOS 1007	10895	USSR	16 MAY	29 MAY 78	
1978-048B		10896	USSR	16 MAY	20 MAY 78	
1978-048C		10897	USSR	16 MAY	18 MAY 78	
1978-048D		10921	USSR	16 MAY	31 MAY 78	
1978-048E		10922	USSR	16 MAY	03 JUL 78	
1978-049A	COSMOS 1008	10898	USSR	17 MAY	08 JAN 81	
1978-049B		10899	USSR	17 MAY	06 FEB 81	
1978-049C		10900	USSR	17 MAY	17 DEC 78	
1978-049D		10901	USSR	17 MAY	30 DEC 78	
1978-049E		10902	USSR	17 MAY	17 JUL 78	
1978-049F		10903	USSR	17 MAY	12 SEP 78	
1978-050A	COSMOS 1009	10904	USSR	19 MAY	19 MAY 78	
1978-050B		10905	USSR	19 MAY	05 JUN 78	
1978-050C		10906	USSR	19 MAY	18 JUL 78	
1978-050D		10907	USSR	19 MAY	23 MAY 78	
1978-052A	COSMOS 1010	10915	USSR	23 MAY	05 JUN 78	
1978-052B		10916	USSR	23 MAY	26 MAY 78	
1978-052C		10928	USSR	23 MAY	07 JUN 78	
1978-054A	COSMOS 1012	10919	USSR	25 MAY	07 JUN 78	
1978-054B		10920	USSR	25 MAY	28 MAY 78	
1978-055B		10926	USSR	02 JUN	21 JUN 78	
1978-055C		10927	USSR	02 JUN	17 JUN 78	
1978-055D		10929	USSR	02 JUN	25 JUN 78	
1978-057A	COSMOS 1021	10939	USSR	10 JUN	23 JUN 78	
1978-057B		10940	USSR	10 JUN	13 JUN 78	
1978-057C		10963	USSR	10 JUN	25 JUN 78	
1978-057D		10964	USSR	10 JUN	26 JUN 78	
1978-058C		10943	US	10 JUN	14 JUN 78	
1978-059A	COSMOS 1022	10944	USSR	12 JUN	25 JUN 78	
1978-059B		10945	USSR	12 JUN	16 JUN 78	
1978-059C		10965	USSR	12 JUN	28 JUN 78	
1978-059D		10966	USSR	12 JUN	28 JUN 78	
1978-060A		10947	US	14 JUN	23 AUG 81	
1978-060B		10948	US	14 JUN	24 JUL 78	
1978-060C		10951	US	14 JUN	18 JUN 78	
1978-060D		10957	US	14 JUN	18 JUN 78	
1978-060E		10958	US	14 JUN	18 JUN 78	
1978-060F		10959	US	14 JUN	18 JUN 78	
1978-061A	SOYUZ 29	10952	USSR	15 JUN	03 SEP 78	E
1978-061B		10956	USSR	15 JUN	18 JUN 78	
1978-062C		10955	US	16 JUN	20 AUG 83	
1978-064B		14244	US	27 JUN	30 OCT 86	
1978-064C		18617	US	27 JUN	25 MAR 88	
1978-064D		18649	US	27 JUN	01 MAR 88	
1978-064E		18996	US	27 JUN	10 AUG 88	
1978-064F		19103	US	27 JUN	11 JUN 88	
1978-065A	SOYUZ 30	10968	USSR	27 JUN	05 JUL 78	E
1978-065B		10969	USSR	27 JUN	30 JUN 78	
1978-066B		10971	USSR	28 JUN	31 JUL 78	
1978-066C		10972	USSR	28 JUN	12 JUL 78	
1978-069A	COSMOS 1026	10977	USSR	02 JUL	06 JUL 78	
1978-069B		10978	USSR	02 JUL	07 JUL 78	
1978-070A	PROGRESS 2	10979	USSR	07 JUL	04 AUG 78	
1978-070B		10980	USSR	07 JUL	10 JUL 78	
1978-071B		10982	US	14 JUL	11 DEC 78	
1978-072A	MOLNIYA 1-41	10984	USSR	14 JUL	08 FEB 92	
1978-072B		10985	USSR	14 JUL	30 AUG 78	
1978-072C		10986	USSR	14 JUL	23 AUG 78	
1978-072D		11073	USSR	14 JUL	17 MAR 92	
1978-073B		10988	USSR	18 JUL	20 JUL 78	
1978-073C		10989	USSR	18 JUL	19 JUL 78	
1978-076A	COSMOS 1028	10995	USSR	05 AUG	04 SEP 78	
1978-076B		10996	USSR	05 AUG	07 AUG 78	
1978-076C		10997	USSR	05 AUG	08 AUG 78	
1978-076D		11014	USSR	05 AUG	04 SEP 78	
1978-077A	PROGRESS 3	10999	USSR	07 AUG	23 AUG 78	
1978-077B		11000	USSR	07 AUG	09 AUG 78	
1978-078A	PIONEER VENUS MULTIPROBE BUS	11001	US	08 AUG	09 DEC 78	
1978-078B		11002	US	08 AUG	08 AUG 78	
1978-078D	PIONEER PROBE	12103	US	08 AUG	09 DEC 79	
1978-078E	PIONEER PROBE	12104	US	08 AUG	09 DEC 79	
1978-078F	PIONEER PROBE	12105	US	08 AUG	09 DEC 79	
1978-078G	PIONEER PROBE	12106	US	08 AUG	09 DEC 79	
1978-079B		11005	US	12 AUG	23 OCT 78	
1978-080B		11008	USSR	22 AUG	15 SEP 78	
1978-080C		11009	USSR	22 AUG	06 SEP 78	
1978-081A	SOYUZ 31	11010	USSR	26 AUG	02 NOV 78	E
1978-081B		11011	USSR	26 AUG	28 AUG 78	
1978-082A	COSMOS 1029	11012	USSR	29 AUG	08 SEP 78	
1978-082B		11013	USSR	29 AUG	03 SEP 78	
1978-082C		11018	USSR	29 AUG	11 SEP 78	
1978-082D		11019	USSR	29 AUG	10 SEP 78	
1978-083B		11016	USSR	06 SEP	08 OCT 78	
1978-083C		11017	USSR	06 SEP	21 SEP 78	
1978-084B		11077	USSR	09 SEP	10 SEP 78	
1978-084C		11021	USSR	09 SEP	10 SEP 78	
1978-084D	VENERA 11	12027	USSR	09 SEP	25 DEC 78	
1978-085A	COSMOS 1031	11022	USSR	09 SEP	22 SEP 78	
1978-085B		11023	USSR	09 SEP	14 SEP 78	
1978-085C		11024	USSR	09 SEP	17 APR 79	
1978-085D		11031	USSR	09 SEP	24 SEP 78	
1978-086B		11026	USSR	14 SEP	15 SEP 78	
1978-086C	VENERA 12	12028	USSR	14 SEP	21 DEC 78	
1978-088A	COSMOS 1032	11029	USSR	19 SEP	02 OCT 78	
1978-088B		11030	USSR	19 SEP	22 SEP 78	
1978-088C		11037	USSR	19 SEP	02 OCT 78	
1978-088D		11038	USSR	19 SEP	03 OCT 78	
1978-089A	COSMOS 1033	11039	USSR	03 OCT	16 OCT 78	
1978-089B		11041	USSR	03 OCT	06 OCT 78	
1978-089C		11063	USSR	03 OCT	17 OCT 78	
1978-090A	PROGRESS 4	11040	USSR	03 OCT	26 OCT 78	
1978-090B		11043	USSR	03 OCT	05 OCT 78	
1978-092A	COSMOS 1042	11052	USSR	06 OCT	19 OCT 78	
1978-092B		11053	USSR	06 OCT	10 OCT 78	
1978-092C		11070	USSR	06 OCT	21 OCT 78	
1978-092D		11071	USSR	06 OCT	21 OCT 78	
1978-093B		11078	US	07 OCT	25 DEC 91	
1978-094C		22536	USSR	10 OCT	11 MAR 93	
1978-095B		11058	USSR	13 OCT	28 OCT 78	
1978-095C		11059	USSR	13 OCT	20 OCT 78	
1978-095D		11064	USSR	13 OCT	31 OCT 78	
1978-096D		18532	US	13 OCT	20 OCT 88	
1978-096E		18786	US	13 OCT	25 FEB 89	
1978-097A	COSMOS 1044	11065	USSR	17 OCT	30 OCT 78	
1978-097B		11066	USSR	17 OCT	22 OCT 78	
1978-097C		11067	USSR	17 OCT	21 OCT 7a	
1978-097D		11068	USSR	17 OCT	21 OCT 78	
1978-097E		11069	USSR	17 OCT	21 OCT 78	
1978-098C		18605	US	24 OCT	20 OCT 88	
1978-099A	INTERCOSMOS 18	11082	USSR	24 OCT	17 MAR 81	
1978-099B		11083	USSR	24 OCT	13 MAY 81	
1978-099C	MAGION	11110	CZCH	24 OCT	11 SEP 81	
1978-100AF		19352	USSR	26 OCT	29 MAR 92	
1978-100AG		19364	USSR	26 OCT	26 JUN 92	
1978-100AJ		19578	USSR	26 OCT	23 MAY 93	
1978-100E		11177	USSR	26 OCT	02 FEB 89	
1978-101A	PROGNOZ 7	11088	USSR	30 OCT	22 OCT 80	
1978-101B		11090	USSR	30 OCT	25 NOV 78	
1978-101C		11089	USSR	30 OCT	17 NOV 78	
1978-102A	COSMOS 1046	11098	USSR	01 NOV	13 NOV 78	
1978-102B		11099	USSR	01 NOV	07 NOV 78	
1978-102C		11103	USSR	01 NOV	14 NOV 78	
1978-102D		11104	USSR	01 NOV	15 NOV 78	
1978-102E		11105	USSR	01 NOV	14 NOV 78	
1978-102F		11106	USSR	01 NOV	14 NOV 78	
1978-102G		11107	USSR	01 NOV	14 NOV 78	
1978-103A	HEAO 2	11101	US	13 NOV	25 MAR 82	
1978-103B		11102	US	13 NOV	13 MAR 79	
1978-104A	COSMOS 1047	11108	USSR	15 NOV	28 NOV 78	
1978-104B		11109	USSR	15 NOV	19 NOV 78	
1978-104C		11120	USSR	15 NOV	02 DEC 78	
1978-106B		11116	US	19 NOV	12 DEC 78	
1978-106C		11117	US	19 NOV	28 NOV 82	
1978-107A	COSMOS 1049	11118	USSR	21 NOV	04 DEC 78	
1978-107B		11119	USSR	21 NOV	26 NOV 78	
1978-107C		11125	USSR	21 NOV	10 DEC 78	
1978-107D		11126	USSR	21 NOV	08 DEC 78	
1978-107E		11127	USSR	21 NOV	11 DEC 78	
1978-108A	COSMOS 1050	11121	USSR	28 NOV	12 DEC 78	
1978-108B		11122	USSR	28 NOV	08 DEC 78	
1978-108C		11123	USSR	28 NOV	02 JAN 79	
1978-108D		11124	USSR	28 NOV	02 JAN 79	
1978-108E		11143	USSR	28 NOV	15 DEC 78	
1978-110A	COSMOS 1059	11137	USSR	07 DEC	20 DEC 78	
1978-110B		11138	USSR	07 DEC	11 DEC 78	
1978-110C		11157	USSR	07 DEC	23 DEC 78	
1978-110D		11160	USSR	07 DEC	23 DEC 78	
1978-111A	COSMOS 1060	11139	USSR	08 DEC	21 DEC 78	
1978-111B		11140	USSR	08 DEC	13 DEC 78	
1978-113C		11146	US	14 DEC	14 DEC 78	
1978-114A	COSMOS 1061	11148	USSR	14 DEC	27 DEC 78	
1978-114B		11149	USSR	14 DEC	20 DEC 78	
1978-114C		11167	USSR	14 DEC	04 JAN 79	
1978-115A	COSMOS 1062	11150	USSR	15 DEC	20 APR 81	
1978-115B		11151	USSR	15 DEC	20 JUN 81	
1978-115C		11152	USSR	15 DEC	07 MAR 79	
1978-116B		11154	US	16 DEC	23 APR 79	
1978-118B		11159	USSR	19 DEC	21 DEC 78	
1978-119A	COSMOS 1064	11161	USSR	20 DEC	12 NOV 89	
1978-119B		11162	USSR	20 DEC	15 NOV 88	
1978-120A	COSMOS 1065	11163	USSR	22 DEC	01 AUG 79	
1978-120B		11164	USSR	22 DEC	18 AUG 79	
1978-120C		11243	USSR	22 DEC	01 MAR 79	
1978-120D		11244	USSR	22 DEC	02 MAR 79	
1978-120E		11248	USSR	22 DEC	03 MAR 79	
1978-120F		11249	USSR	22 DEC	01 MAR 79	
1978-120G		11424	USSR	22 DEC	16 JUL 79	
1978-120H		11426	USSR	22 DEC	05 AUG 79	
1978-123A	COSMOS 1068	11169	USSR	26 DEC	08 JAN 79	
1978-123B		11171	USSR	26 DEC	31 DEC 78	
1978-123C		11172	USSR	26 DEC	01 JAN 79	
1978-123D		11226	USSR	26 DEC	09 JAN 79	
1978-123E		11227	USSR	26 DEC	09 JAN 79	
1978-124A	COSMOS 1069	11173	USSR	28 DEC	10 JAN 79	

International Designation	Satellite Name	Number	Source	Launch	Decay	Note
1978-124B		11174	USSR	28 DEC	05 JAN 79	
1978-124C		11175	USSR	28 DEC	31 DEC 78	
1978-124D		11176	USSR	28 DEC	16 JAN 79	
1978-124E		11228	USSR	28 DEC	11 JAN 79	
1978-124F		11237	USSR	28 DEC	22 JAN 79	
1979 LAUNCHES						
1979-001A	COSMOS 1070	11229	USSR	11 JAN	20 JAN 79	
1979-001B		11230	USSR	11 JAN	16 JAN 79	
1979-001C		11231	USSR	11 JAN	18 JAN 79	
1979-001D		11232	USSR	11 JAN	19 JAN 79	
1979-001E		11234	USSR	11 JAN	22 JAN 79	
1979-001F		11245	USSR	11 JAN	27 JAN 79	
1979-001G		11246	USSR	11 JAN	21 JAN 79	
1979-001H		11247	USSR	11 JAN	20 JAN 79	
1979-002A	COSMOS 1071	11233	USSR	13 JAN	26 JAN 79	
1979-002B		11235	USSR	13 JAN	17 JAN 79	
1979-002C		11236	USSR	13 JAN	18 JAN 79	
1979-002D		11250	USSR	13 JAN	27 JAN 79	
1979-002E		11253	USSR	13 JAN	27 JAN 79	
1979-002F		11254	USSR	13 JAN	29 JAN 79	
1979-004B		11241	USSR	18 JAN	31 JAN 79	
1979-004C		11242	USSR	18 JAN	29 JAN 79	
1979-005C		18529	USSR	25 JAN	25 AUG 88	
1979-006A	COSMOS 1073	11255	USSR	30 JAN	12 FEB 79	
1979-006B		11257	USSR	30 JAN	03 FEB 79	
1979-006C		11264	USSR	30 JAN	14 FEB 79	
1979-006D		11265	USSR	30 JAN	13 FEB 79	
1979-007B		11258	US	30 JAN	25 FEB 79	
1979-008B		11260	USSR	31 JAN	02 FEB 79	
1979-008C		11315	USSR	31 JAN	10 AUG 79	
1979-008D		11316	USSR	31 JAN	06 JUN 79	
1979-008E		11317	USSR	31 JAN	06 AUG 79	
1979-008F		11318	USSR	31 JAN	30 MAY 79	
1979-008G		11319	USSR	31 JAN	22 JUN 79	
1979-010A	COSMOS 1075	11262	USSR	08 FEB	19 OCT 81	
1979-010B		11263	USSR	08 FEB	29 AUG 80	
1979-013A	SAGE	11270	US	18 FEB	11 APR 89	
1979-013B		11271	US	18 FEB	19 SEP 81	
1979-013C		12405	US	18 FEB	14 AUG 82	
1979-014A	CORSA B	11272	JAPAN	21 FEB	15 APR 85	
1979-014B		11280	JAPAN	21 FEB	08 FEB 82	
1979-015B		11274	USSR	21 FEB	22 FEB 79	
1979-015C		11275	USSR	21 FEB	22 FEB 79	
1979-016A	COSMOS 1078	11276	USSR	22 FEB	02 MAR 79	
1979-016B		11277	USSR	22 FEB	24 FEB 79	
1979-016C		11287	USSR	22 FEB	02 MAR 79	
1979-017A	SOLWIND	11278	US	24 FEB	20 JUL 92	
1979-017AA		16072	US	24 FEB	07 SEP 88	
1979-017AB		16073	US	24 FEB	01 MAY 89	
1979-017AC		16074	US	24 FEB	12 JAN 89	
1979-017AD		16075	US	24 FEB	07 JUL 86	
1979-017AE		16076	US	24 FEB	02 DEC 86	
1979-017AF		16078	US	24 FEB	16 JUL 88	
1979-017AG		16079	US	24 FEB	08 APR 88	
1979-017AH		16080	US	24 FEB	24 SEP 88	
1979-017AJ		16081	US	24 FEB	30 NOV 88	
1979-017AK		16082	US	24 FEB	09 JUL 88	
1979-017AL		16083	US	24 FEB	27 FEB 91	
1979-017AP		16086	US	24 FEB	19 SEP 88	
1979-017AQ		16087	US	24 FEB	28 MAR 90	
1979-017AR		16130	US	24 FEB	10 OCT 85	
1979-017AS		16131	US	24 FEB	15 DEC 85	
1979-017AT		16132	US	24 FEB	24 OCT 85	
1979-017AU		16133	US	24 FEB	27 NOV 87	
1979-017AV		16147	US	24 FEB	07 JUN 87	
1979-017AW		16148	US	24 FEB	20 DEC 85	
1979-017AX		16149	US	24 FEB	19 APR 87	
1979-017AY		16150	US	24 FEB	13 OCT 87	
1979-017AZ		16151	US	24 FEB	10 MAY 88	
1979-017B		11279	US	24 FEB	11 OCT 83	
1979-017BA		16152	US	24 FEB	28 OCT 86	
1979-017BB		16153	US	24 FEB	21 JAN 86	
1979-017BC		16154	US	24 FEB	24 MAR 88	
1979-017BD		16155	US	24 FEB	19 JUN 88	
1979-017BE		16156	US	24 FEB	21 DEC 87	
1979-017BF		16157	US	24 FEB	27 JUN 88	
1979-017BG		16158	US	24 FEB	12 OCT 88	
1979-017BH		16278	US	24 FEB	13 NOV 88	
1979-017BJ		16279	US	24 FEB	15 APR 87	
1979-017BK		16280	US	24 FEB	10 MAR 89	
1979-017BL		16281	US	24 FEB	13 NOV 87	
1979-017BM		16282	US	24 FEB	28 JAN 88	
1979-017BN		16283	US	24 FEB	10 NOV 87	
1979-017BP		16284	US	24 FEB	25 MAY 88	
1979-017BQ		16298	US	24 FEB	15 FEB 86	
1979-017BR		16299	US	24 FEB	04 MAY 86	
1979-017BS		16300	US	24 FEB	12 JUL 88	
1979-017BT		16301	US	24 FEB	15 FEB 89	
1979-017BU		16302	US	24 FEB	23 APR 87	
1979-017BV		16303	US	24 FEB	04 FEB 89	
1979-017BW		16304	US	24 FEB	20 MAY 86	
1979-017BY		16309	US	24 FEB	27 DEC 89	
1979-017BZ		16310	US	24 FEB	25 AUG 89	
1979-017C		11291	US	24 FEB	29 JUL 79	
1979-017CB		16312	US	24 FEB	24 AUG 89	
1979-017CC		16313	US	24 FEB	21 DEC 89	
1979-017CD		16314	US	24 FEB	11 DEC 88	
1979-017CE		16315	US	24 FEB	03 OCT 86	
1979-017CF		16316	US	24 FEB	29 DEC 88	
1979-017CG		16317	US	24 FEB	13 MAR 89	
1979-017CH		16318	US	24 FEB	30 OCT 86	
1979-017CJ		16319	US	24 FEB	10 JUL 87	
1979-017CK		16320	US	24 FEB	24 APR 79	
1979-017CL		16321	US	24 FEB	13 APR 88	
1979-017CM		16322	US	24 FEB	17 DEC 87	
1979-017CN		16323	US	24 FEB	03 JAN 88	
1979-017CP		16324	US	24 FEB	14 MAR 87	
1979-017CQ		16325	US	24 FEB	21 NOV 86	
1979-017CR		16333	US	24 FEB	14 NOV 88	
1979-017CS		16334	US	24 FEB	06 JAN 87	
1979-017CT		16335	US	24 FEB	22 SEP 86	
1979-017CU		16336	US	24 FEB	19 JAN 89	
1979-017CV		16337	US	24 FEB	08 JUN 86	
1979-017CW		16338	US	24 FEB	03 MAR 86	
1979-017CX		16341	US	24 FEB	30 SEP 88	
1979-017CY		16342	US	24 FEB	25 MAR 86	
1979-017CZ		16343	US	24 FEB	10 SEP 91	
1979-017D		11322	US	24 FEB	27 FEB 82	
1979-017DA		16344	US	24 FEB	01 MAY 87	
1979-017DB		16345	US	24 FEB	08 JUL 87	
1979-017DC		16346	US	24 FEB	06 MAY 88	
1979-017DD		16347	US	24 FEB	06 APR 88	
1979-017DE		16348	US	24 FEB	23 NOV 88	
1979-017DF		16349	US	24 FEB	10 JUL 87	
1979-017DG		16350	US	24 FEB	26 JUL 86	
1979-017DH		16351	US	24 FEB	08 MAY 86	
1979-017DJ		16352	US	24 FEB	12 MAY 86	
1979-017DK		16360	US	24 FEB	10 OCT 87	
1979-017DL		16361	US	24 FEB	15 NOV 86	
1979-017DM		16362	US	24 FEB	25 JAN 86	
1979-017DN		16363	US	24 FEB	08 AUG 87	
1979-017DP		16364	US	24 FEB	03 OCT 88	
1979-017DQ		16365	US	24 FEB	20 MAR 86	
1979-017DR		16366	US	24 FEB	02 SEP 87	
1979-017DS		16367	US	24 FEB	12 OCT 87	
1979-017DT		16413	US	24 FEB	11 AUG 89	
1979-017DU		16414	US	24 FEB	14 OCT 88	
1979-017DV		16415	US	24 FEB	01 MAR 86	
1979-017DW		16416	US	24 FEB	11 DEC 87	
1979-017DX		16417	US	24 FEB	30 MAY 87	
1979-017DY		16418	US	24 FEB	23 JAN 89	
1979-017DZ		16419	US	24 FEB	01 JAN 86	
1979-017E		16027	US	24 FEB	02 JAN 86	
1979-017EA		16420	US	24 FEB	02 SEP 86	
1979-017EB		16421	US	24 FEB	26 AUG 86	
1979-017EC		16422	US	24 FEB	08 AUG 87	
1979-017ED		16423	US	24 FEB	19 JUN 86	
1979-017EE		16424	US	24 FEB	17 APR 86	
1979-017EF		16425	US	24 FEB	27 SEP 86	
1979-017EG		16458	US	24 FEB	12 JUN 88	
1979-017EH		16459	US	24 FEB	17 OCT 86	
1979-017EJ		16460	US	24 FEB	19 MAY 89	
1979-017EK		16461	US	24 FEB	21 APR 86	
1979-017EL		16462	US	24 FEB	28 JAN 86	
1979-017EM		16463	US	24 FEB	06 MAR 86	
1979-017EN		16464	US	24 FEB	22 JAN 86	
1979-017EP		16465	US	24 FEB	06 MAR 86	
1979-017EQ		16466	US	24 FEB	28 OCT 88	
1979-017ER		16467	US	24 FEB	30 JAN 89	
1979-017ES		16468	US	24 FEB	02 JUL 88	
1979-017ET		16469	US	24 FEB	04 FEB 86	
1979-017EU		16470	US	24 FEB	29 OCT 88	
1979-017EV		16471	US	24 FEB	22 OCT 87	
1979-017EW		16472	US	24 FEB	02 AUG 89	
1979-017EX		16473	US	24 FEB	19 MAY 86	
1979-017EY		16474	US	24 FEB	09 JUL 88	
1979-017EZ		16475	US	24 FEB	14 OCT 87	
1979-017F		16028	US	24 FEB	03 APR 88	
1979-017FA		16476	US	24 FEB	24 JAN 88	
1979-017FB		16477	US	24 FEB	29 APR 86	
1979-017FC		16478	US	24 FEB	06 NOV 89	
1979-017FD		16479	US	24 FEB	13 APR 86	
1979-017FH		16486	US	24 FEB	21 JAN 88	
1979-017FJ		16487	US	24 FEB	11 DEC 86	
1979-017FK		16488	US	24 FEB	01 FEB 89	
1979-017FL		16489	US	24 FEB	17 APR 89	
1979-017FM		16515	US	24 FEB	28 DEC 88	
1979-017FN		16516	US	24 FEB	25 AUG 87	
1979-017FP		16517	US	24 FEB	13 APR 86	
1979-017FQ		16518	US	24 FEB	12 OCT 88	
1979-017FR		16519	US	24 FEB	20 DEC 86	
1979-017FS		16520	US	24 FEB	16 MAR 88	
1979-017FT		16521	US	24 FEB	14 FEB 87	
1979-017FU		16522	US	24 FEB	21 OCT 87	
1979-017FV		16523	US	24 FEB	30 JAN 88	
1979-017FW		16524	US	24 FEB	16 OCT 86	
1979-017FX		16525	US	24 FEB	07 NOV 86	
1979-017FY		16534	US	24 FEB	20 FEB 89	
1979-017FZ		16535	US	24 FEB	13 MAR 87	
1979-017G		16029	US	24 FEB	11 NOV 89	
1979-017GA		16536	US	24 FEB	11 MAY 86	
1979-017GB		16537	US	24 FEB	29 JAN 88	
1979-017GC		16538	US	24 FEB	09 JAN 88	
1979-017GD		16539	US	24 FEB	11 APR 86	

International Designation	Satellite Name	Number	Source	Launch	Decay	Note
1979-017GE		16542	US	24 FEB	25 OCT 86	
1979-017GF		16543	US	24 FEB	20 APR 87	
1979-017GG		16549	US	24 FEB	24 DEC 88	
1979-017GH		16550	US	24 FEB	02 JAN 90	
1979-017GK		16552	US	24 FEB	18 SEP 87	
1979-017GL		16553	US	24 FEB	24 FEB 88	
1979-017GM		16554	US	24 FEB	27 MAY 87	
1979-017GN		16555	US	24 FEB	29 JUL 87	
1979-017GP		16556	US	24 FEB	12 FEB 86	
1979-017GQ		16557	US	24 FEB	20 JUN 86	
1979-017GR		16558	US	24 FEB	10 MAY 87	
1979-017GS		16559	US	24 FEB	30 APR 86	
1979-017GT		16560	US	24 FEB	24 FEB 87	
1979-017GU		16561	US	24 FEB	16 JAN 88	
1979-017GV		16562	US	24 FEB	25 JUN 86	
1979-017GW		16563	US	24 FEB	26 AUG 86	
1979-017GY		16565	US	24 FEB	11 MAY 88	
1979-017GZ		16566	US	24 FEB	10 OCT 88	
1979-017H		16030	US	24 FEB	23 MAR 89	
1979-017HA		16567	US	24 FEB	22 NOV 89	
1979-017HB		16568	US	24 FEB	01 JUN 88	
1979-017HC		16569	US	24 FEB	15 NOV 87	
1979-017HD		16570	US	24 FEB	06 DEC 92	
1979-017HE		16571	US	24 FEB	28 APR 88	
1979-017HF		16572	US	24 FEB	15 JUL 88	
1979-017HG		16573	US	24 FEB	12 DEC 87	
1979-017HH		16574	US	24 FEB	20 OCT 88	
1979-017HJ		16575	US	24 FEB	17 JUL 87	
1979-017HK		16576	US	24 FEB	08 AUG 88	
1979-017HL		16577	US	24 FEB	05 NOV 86	
1979-017HM		16578	US	24 FEB	27 MAR 86	
1979-a17HN		16579	US	24 FEB	27 MAR 86	
1979-017HP		16580	US	24 FEB	09 SEP 87	
1979-017HQ		16581	US	24 FEB	17 JAN 89	
1979-017HR		16582	US	24 FEB	15 JAN 89	
1979-017HS		16583	US	24 FEB	06 MAY 87	
1979-017HT		16584	US	24 FEB	09 SEP 88	
1979-017HU		16585	US	24 FEB	23 JUL 86	
1979-017HV		16586	US	24 FEB	10 FEB 86	
1979-017HW		16587	US	24 FEB	22 JAN 89	
1979-017HX		16588	US	24 FEB	01 JUL 88	
1979-017HY		16660	US	24 FEB	02 FEB 89	
1979-017HZ		16661	US	24 FEB	26 OCT 87	
1979-017J		16031	US	24 FEB	01 JAN 86	
1979-017JA		16662	US	24 FEB	30 OCT 86	
1979-017JB		16663	US	24 FEB	31 AUG 88	
1979-017JC		16726	US	24 FEB	20 NOV 88	
1979-017JD		16843	US	24 FEB	23 JUL 88	
1979-017JE		16877	US	24 FEB	04 MAR 88	
1979-017JG		16879	US	24 FEB	11 MAR 87	
1979-017JJ		17095	US	24 FEB	13 NOV 88	
1979-017JK		17100	US	24 FEB	22 SEP 88	
1979-017JL		17101	US	24 FEB	28 MAR 89	
1979-017JM		17165	US	24 FEB	27 JAN 89	
1979-017JP		17167	US	24 FEB	03 DEC 87	
1979-017JQ		17172	US	24 FEB	15 AUG 87	
1979-017JR		17173	US	24 FEB	06 MAY 87	
1979-017JS		17174	US	24 FEB	23 JUN 87	
1979-017JT		17770	US	24 FEB	26 FEB 89	
1979-017JU		17771	US	24 FEB	13 JUL 88	
1979-017JV		17772	US	24 FEB	21 MAR 89	
1979-017JW		17773	US	24 FEB	72 JAN 88	
1979-017JX		17774	US	24 FEB	26 APR 89	
1979-017JY		17775	US	24 FEB	06 AUG 88	
1979-017JZ		17776	US	24 FEB	25 JUN 88	
1979-017K		16032	US	24 FEB	16 AUG 87	
1979-017KA		17777	US	24 FEB	16 DEC 88	
1979-017KB		17778	US	24 FEB	16 SEP 88	
1979-017KC		17779	US	24 FEB	05 JUN 87	
1979-017KD		17780	US	24 FEB	17 JUN 88	
1979-017KE		17781	US	24 FEB	10 NOV 91	
1979-017KF		17782	US	24 FEB	21 FEB 88	
1979-017KG		17783	US	24 FEB	26 AUG 88	
1979-017KH		17784	US	24 FEB	04 JUN 88	
1979-017KJ		17822	US	24 FEB	29 APR 88	
1979-017KK		17823	US	24 FEB	12 NOV 88	
1979-017KL		17824	US	24 FEB	04 MAY 91	
1979-017KM		17825	US	24 FEB	22 OCT 91	
1979-017KN		17826	US	24 FEB	21 MAY 89	
1979-017KP		17827	US	24 FEB	07 OCT 88	
1979-017KQ		17828	US	24 FEB	15 AUG 87	
1979-017KR		17829	US	24 FEB	12 FEB 88	
1979-017KS		17830	US	24 FEB	19 SEP 88	
1979-017KT		17831	US	24 FEB	05 OCT 88	
1979-017KU		17832	US	24 FEB	13 FEB 89	
1979-017KV		18250	US	24 FEB	30 SEP 87	
1979-017KW		18251	US	24 FEB	11 SEP 87	
1979-017KX		18252	US	24 FEB	07 OCT 89	
1979-017KY		18253	US	24 FEB	29 NOV 87	
1979-017KZ		18254	US	24 FEB	28 OCT 87	
1979-017L		16033	US	24 FEB	04 APR 88	
1979-017LA		18255	US	24 FEB	23 JUN 90	
1979-017LB		18303	US	24 FEB	30 JAN 88	
1979-017LC		18304	US	24 FEB	22 JUN 89	
1979-017LD		18423	US	24 FEB	26 MAR 89	
1979-017LE		18424	US	24 FEB	21 SEP 92	
1979-017LF		18425	US	24 FEB	01 DEC 88	
1979-017LG		18464	US	24 FEB	21 MAR 90	
1979-017LH		18486	US	24 FEB	30 OCT 89	
1979-017LJ		18487	US	24 FEB	28 JAN 89	
1979-017LK		18546	US	24 FEB	13 MAR 89	
1979-017LL		18604	US	24 FEB	21 APR 89	
1979-017LM		18612	US	24 FEB	05 AUG 89	
1979-017LN		18646	US	24 FEB	19 SEP 88	
1979-017LP		18695	US	24 FEB	14 JAN 88	
1979-017LQ		18776	US	24 FEB	06 APR 89	
1979-017LR		18818	US	24 FEB	10 NOV 88	
1979-017LS		18867	US	24 FEB	20 OCT 88	
1979-017LT		18871	US	24 FEB	14 FEB 90	
1979-017LU		18873	US	24 FEB	26 APR 89	
1979-017LV		18915	US	24 FEB	10 DEC 89	
1979-017LW		19060	US	24 FEB	12 MAR 89	
1979-017LX		19061	US	24 FEB	04 NOV 88	
1979-017LY		19156	US	24 FEB	17 MAR 89	
1979-017LZ		19350	US	24 FEB	07 APR 89	
1979-017M		16034	US	24 FEB	27 JUN 86	
1979-017N		16035	US	24 FEB	31 OCT 88	
1979-017P		16036	US	24 FEB	12 MAY 90	
1979-017Q		16037	US	24 FEB	08 OCT 85	
1979-017R		16038	US	24 FEB	27 DEC 88	
1979-017S		16039	US	24 FEB	09 AUG 88	
1979-017T		16040	US	24 FEB	23 FEB 87	
1979-017U		16041	US	24 FEB	20 AUG 89	
1979-017V		16042	US	24 FEB	29 JUL 88	
1979-017W		16043	US	24 FEB	18 DEC 85	
1979-017X		16044	US	24 FEB	20 MAY 89	
1979-017Y		16045	US	24 FEB	23 MAY 89	
1979-017Z		16071	US	24 FEB	22 JUL 87	
1979-018A	SOYUZ 32	11281	USSR	25 FEB	13 JUN 79	
1979-018B		11282	USSR	25 FEB	27 FEB 79	
1979-019A	COSMOS 1079	11283	USSR	27 FEB	11 MAR 79	
1979-019B		11284	USSR	27 FEB	02 MAR 79	
1979-020C		11323	USSR	27 FEB	02 SEP 80	
1979-022A	PROGRESS 5	11292	USSR	12 MAR	05 APR 79	
1979-022B		11293	USSR	12 MAR	14 MAR 79	
1979-023A	COSMOS 1080	11294	USSR	14 MAR	28 MAR 79	
1979-023B		11295	USSR	14 MAR	16 MAR 79	
1979-023C		11310	USSR	14 MAR	27 MAR 79	
1979-023D		11311	USSR	14 MAR	27 MAR 79	
1979-023E		11312	USSR	14 MAR	28 MAR 79	
1979-025A		11305	US	16 MAR	22 SEP 79	
1979-025C		11307	US	16 MAR	18 MAR 79	
1979-008A	COSMOS 1074	11259	USSR	31 MAR	01 APR 79	
1979-027A	COSMOS 1090	11313	USSR	31 MAR	13 APR 79	
1979-027B		11314	USSR	31 MAR	05 APR 79	
1979-029A	SOYUZ 33	11324	USSR	10 APR	12 APR 79	
1979-029B		11325	USSR	10 APR	13 APR 79	
1979-031B		11329	USSR	12 APR	08 MAY 79	
1979-031C		11330	USSR	12 APR	05 MAY 79	
1979-033A	COSMOS 1094	11333	USSR	18 APR	07 NOV 79	
1979-033B		11334	USSR	18 APR	18 APR 79	
1979-033C		11367	USSR	18 APR	17 JUN 79	
1979-034A	COSMOS 1095	11335	USSR	20 APR	04 MAY 79	
1979-034B		11336	USSR	20 APR	27 APR 79	
1979-034C		11337	USSR	20 APR	01 MAY 79	
1979-034D		11351	USSR	20 APR	18 NOV 79	
1979-034E		11352	USSR	20 APR	21 JUN 79	
1979-034F		11354	USSR	20 APR	10 JUL 79	
1979-034G		11355	USSR	20 APR	01 JUN 79	
1979-035B		11344	USSR	25 APR	26 APR 79	
1979-035C		11345	USSR	25 APR	25 APR 79	
1979-035D		11552	USSR	25 APR	01 JAN 80	
1979-036A	COSMOS 1096	11346	USSR	25 APR	24 NOV 79	
1979-036B		11347	USSR	25 APR	25 APR 79	
1979-037A	COSMOS 1097	11348	USSR	27 APR	27 MAY 79	
1979-037B		11349	USSR	27 APR	30 APR 79	
1979-039A	PROGRESS 6	11356	USSR	13 MAY	09 JUN 79	
1979-039B		11357	USSR	13 MAY	15 MAY 79	
1979-040A	COSMOS 1098	11358	USSR	15 MAY	28 MAY 79	
1979-040B		11359	USSR	15 MAY	19 MAY 79	
1979-040C		11370	USSR	15 MAY	30 MAY 79	
1979-041A	COSMOS 1099	11360	USSR	17 MAY	30 MAY 79	
1979-041B		11361	USSR	17 MAY	21 MAY 79	
1979-041C		11371	USSR	17 MAY	22 JUN 79	
1979-041D		11373	USSR	17 MAY	21 JUN 79	
1979-041E		11374	USSR	17 MAY	02 JUN 79	
1979-042A	COSMOS 1100	11362	USSR	22 MAY	23 MAY 79	
1979-042B	COSMOS 1101	11363	USSR	22 MAY	23 MAY 79	
1979-042C		11364	USSR	22 MAY	26 MAY 79	
1979-042D		11365	USSR	22 MAY	16 JUN 79	
1979-042E		11366	USSR	22 MAY	25 MAY 79	
1979-043A	COSMOS 1102	11368	USSR	25 MAY	07 JUN 79	
1979-043B		11369	USSR	25 MAY	29 MAY 79	
1979-043C		11375	USSR	25 MAY	06 JUN 79	
1979-043D		11390	USSR	25 MAY	13 JUL 79	
1979-043E		11391	USSR	25 MAY	03 JUN 79	
1979-044A		11372	US	28 MAY	26 AUG 79	
1979-045A	COSMOS 1103	11376	USSR	31 MAY	14 JUN 79	
1979-045B		11377	USSR	31 MAY	24 JUN 79	
1979-045C		11380	USSR	31 MAY	05 JUN 79	
1979-045D		11381	USSR	31 MAY	08 JUN 79	
1979-045E		11402	USSR	31 MAY	27 JUL 79	
1979-045F		11406	USSR	31 MAY	09 JUL 79	
1979-045G		11407	USSR	31 MAY	15 NOV 79	
1979-047A	UK 6	11382	UK	02 JUN	23 SEP 90	
1979-047B		11383	US	02 JUN	10 APR 88	

International Designation	Satellite Name	Number	Source	Launch	Decay	Note
1979-048A	MOLNIYA 3-12	11384	USSR	05 JUN	26 SEP 92	
1979-048B		11385	USSR	05 JUN	24 JUN 79	
1979-048C		11386	USSR	05 JUN	21 JUN 79	
1979-048D		11554	USSR	05 JUN	05 AUG 90	
1979-049A	SOYUZ 34	11387	USSR	06 JUN	19 AUG 79	
1979-049B		11388	USSR	06 JUN	09 JUN 79	
1979-050E		11411	US	06 JUN	08 JUN 82	
1979-050F		11412	US	06 JUN	23 FEB 82	
1979-051A	BHASKARA	11392	INDIA	07 JUN	17 FEB 89	
1979-051B		11393	USSR	07 JUN	17 FEB 82	
1979-052A	COSMOS 1105	11394	USSR	08 JUN	21 JUN 79	
1979-052B		11395	USSR	08 JUN	12 JUN 79	
1979-052C		11396	USSR	08 JUN	08 JUN 79	
1979-052D		11409	USSR	08 JUN	28 JUL 79	
1979-053B		11398	US	10 JUN	11 JUN 79	
1979-054A	COSMOS 1106	11399	USSR	12 JUN	25 JUN 79	
1979-054B		11400	USSR	12 JUN	15 JUN 79	
1979-054C		11401	USSR	12 JUN	07 AUG 79	
1979-054D		11415	USSR	12 JUN	30 JUN 79	
1979-055A	COSMOS 1107	11404	USSR	15 JUN	29 JUN 79	
1979-055B		11405	USSR	15 JUN	20 JUN 79	
1979-055C		11420	USSR	15 JUN	02 JUL 79	
1979-055D		11428	USSR	15 JUN	03 JUL 79	
1979-056A	COSMOS 1108	11413	USSR	22 JUN	05 JUL 79	
1979-056B		11414	USSR	22 JUN	25 JUN 79	
1979-056C		11437	USSR	22 JUN	08 JUL 79	
1979-056D		11438	USSR	22 JUN	11 JUL 79	
1979-056E		11439	USSR	22 JUN	21 JUL 79	
1979-057D		22139	US	27 JUN	12 MAY 93	
1979-058B		11418	USSR	27 JUN	26 JUL 79	
1979-058C		11423	USSR	27 JUN	13 JUL 79	
1979-059A	PROGRESS 7	11421	USSR	28 JUN	20 JUL 79	
1979-059B		11422	USSR	28 JUN	30 JUN 79	
1979-061A	COSMOS 1111	11429	USSR	29 JUN	14 JUL 79	
1979-061B		11430	USSR	29 JUN	20 JUL 79	
1979-061C		11453	USSR	29 JUN	05 DEC 79	
1979-061D		11454	USSR	29 JUN	25 AUG 79	
1979-061E		11455	USSR	29 JUN	30 JUL 79	
1979-061F		11456	USSR	29 JUN	25 AUG 79	
1979-062B		11441	USSR	05 JUL	07 JUL 79	
1979-062C		11442	USSR	05 JUL	07 JUL 79	
1979-062E		14070	USSR	05 JUL	28 APR 84	
1979-063A	COSMOS 1112	11443	USSR	06 JUL	21 JAN 80	
1979-063AA		11643	USSR	06 JUL	21 DEC 79	
1979-063AB		11644	USSR	06 JUL	21 DEC 79	
1979-063B		11444	USSR	06 JUL	08 JAN 80	BR
1979-063C		11445	USSR	06 JUL	08 JUL 79	
1979-063D		11446	USSR	06 JUL	01 AUG 79	
1979-063E		11514	USSR	06 JUL	23 SEP 79	
1979-063F		11535	USSR	06 JUL	20 OCT 79	
1979-063G		11557	USSR	06 JUL	21 OCT 79	
1979-063H		11583	USSR	06 JUL	05 NOV 79	
1979-063J		11584	USSR	06 JUL	05 NOV 79	
1979-063K		11594	USSR	06 JUL	07 NOV 79	
1979-063L		11595	USSR	06 JUL	11 NOV 79	
1979-063M		11596	USSR	06 JUL	05 JAN 80	
1979-063N		11597	USSR	06 JUL	20 FEB 80	
1979-063P		11598	USSR	06 JUL	01 NOV 79	
1979-063Q		11599	USSR	06 JUL	07 DEC 79	
1979-063R		11603	USSR	06 JUL	12 NOV 79	
1979-063S		11617	USSR	06 JUL	05 DEC 79	
1979-063T		11618	USSR	06 JUL	22 NOV 79	
1979-063U		11619	USSR	06 JUL	05 DEC 79	
1979-063V		11620	USSR	06 JUL	30 NOV 79	
1979-063W		11625	USSR	06 JUL	04 DEC 79	
1979-063X		11626	USSR	06 JUL	04 DEC 79	
1979-063Y		11627	USSR	06 JUL	05 DEC 79	
1979-063Z		11628	USSR	06 JUL	07 DEC 79	
1979-064A	COSMOS 1113	11447	USSR	10 JUL	23 JUL 79	
1979-064B		11448	USSR	10 JUL	15 JUL 79	
1979-064C		11459	USSR	10 JUL	26 JUL 79	
1979-064D		11460	USSR	10 JUL	27 JUL 79	
1979-064E		11461	USSR	10 JUL	25 JUL 79	
1979-065A	COSMOS 1114	11449	USSR	11 JUL	26 DEC 81	
1979-065B		11450	USSR	11 JUL	05 FEB 82	
1979-066A	COSMOS 1115	11451	USSR	13 JUL	26 JUL 79	
1979-066B		11452	USSR	13 JUL	17 JUL 79	
1979-066C		11462	USSR	13 JUL	20 AUG 79	
1979-067A	COSMOS 1116	11457	USSR	20 JUL	11 MAR 93	
1979-068A	COSMOS 1117	11463	USSR	25 JUL	07 AUG 79	
1979-068B		11464	USSR	25 JUL	29 JUL 79	
1979-068C		11481	USSR	25 JUL	11 AUG 79	
1979-068D		11482	USSR	25 JUL	10 AUG 79	
1979-068E		11483	USSR	25 JUL	11 AUG 79	
1979-069A	COSMOS 1118	11465	USSR	27 JUL	09 AUG 79	
1979-069B		11466	USSR	27 JUL	31 JUL 79	
1979-069C		11477	USSR	27 JUL	07 AUG 79	
1979-070B		11475	USSR	31 JUL	20 AUG 79	
1979-070C		11476	USSR	31 JUL	14 AUG 79	
1979-071A	COSMOS 1119	11478	USSR	03 AUG	18 AUG 79	
1979-071B		11479	USSR	03 AUG	07 AUG 79	
1979-072B		11490	US	10 AUG	10 AUG 79	
1979-073A	COSMOS 1120	11485	USSR	11 AUG	24 AUG 79	
1979-073B		11486	USSR	11 AUG	15 AUG 79	
1979-073C		11505	USSR	11 AUG	28 AUG 79	
1979-073D		11506	USSR	11 AUG	31 AUG 79	
1979-074A	COSMOS 1121	11487	USSR	14 AUG	13 SEP 79	
1979-074B		11488	USSR	14 AUG	18 AUG 79	
1979-075A	COSMOS 1122	11491	USSR	17 AUG	30 AUG 79	
1979-075B		11492	USSR	17 AUG	20 AUG 79	
1979-075C		11498	USSR	17 AUG	27 AUG 79	
1979-075D		11499	USSR	17 AUG	25 AUG 79	
1979-075E		11507	USSR	17 AUG	27 AUG 79	
1979-076A	COSMOS 1123	11496	USSR	21 AUG	03 SEP 79	
1979-076B		11497	USSR	21 AUG	24 AUG 79	
1979-076C		11517	USSR	21 AUG	06 SEP 79	
1979-076D		11518	USSR	21 AUG	06 SEP 79	
1979-076E		11519	USSR	21 AUG	07 SEP 79	
1979-077B		11512	USSR	28 AUG	28 SEP 79	
1979-077C		11513	USSR	28 AUG	11 SEP 79	
1979-079A	COSMOS 1126	11515	USSR	31 AUG	14 SEP 79	
1979-079B		11516	USSR	31 AUG	09 SEP 79	
1979-079C		11524	USSR	31 AUG	26 OCT 79	
1979-079D		11525	USSR	31 AUG	16 FEB 80	
1979-079E		11526	USSR	31 AUG	14 FEB 80	
1979-079F		11527	USSR	31 AUG	16 SEP 79	
1979-079G		11528	USSR	31 AUG	17 OCT 79	
1979-080A	COSMOS 1127	11520	USSR	05 SEP	18 SEP 79	
1979-080B		11521	USSR	05 SEP	10 SEP 79	
1979-080C		11522	USSR	05 SEP	06 SEP 79	
1979-080D		11531	USSR	05 SEP	30 SEP 79	
1979-081A	COSMOS 1128	11529	USSR	14 SEP	27 SEP 79	
1979-081B		11530	USSR	14 SEP	17 SEP 79	
1979-081C		11547	USSR	14 SEP	27 SEP 79	
1979-082A	HEAO 3	11532	US	20 SEP	07 DEC 81	
1979-082B		11533	US	20 SEP	13 NOV 79	
1979-083A	COSMOS 1129	11536	USSR	25 SEP	14 OCT 79	
1979-083B		11537	USSR	25 SEP	06 OCT 79	
1979-083C		11579	USSR	25 SEP	31 OCT 79	
1979-083D		11580	USSR	25 SEP	18 OCT 79	
1979-083E		11582	USSR	25 SEP	16 MAR 80	
1979-085A	COSMOS 1138	11548	USSR	28 SEP	12 OCT 79	
1979-085B		11549	USSR	28 SEP	05 OCT 79	
1979-085C		11575	USSR	28 SEP	02 NOV 79	
1979-085D		11576	USSR	28 SEP	13 OCT 79	
1979-085E		11577	USSR	28 SEP	12 NOV 79	
1979-085F		11578	USSR	28 SEP	26 FEB 80	
1979-086B		11559	US	01 OCT	07 OCT 79	
1979-087B		11562	USSR	03 OCT	05 OCT 79	
1979-088A	COSMOS 1139	11564	USSR	05 OCT	18 OCT 79	
1979-088B		11565	USSR	05 OCT	11 OCT 79	
1979-088C		11566	USSR	05 OCT	07 OCT 79	
1979-090D		11588	USSR	16 OCT	07 SEP 82	
1979-091B		11590	USSR	20 OCT	11 NOV 79	
1979-091C		11591	USSR	20 OCT	10 NOV 79	
1979-092A	COSMOS 1142	11592	USSR	22 OCT	04 NOV 79	
1979-092B		11593	USSR	22 OCT	28 OCT 79	
1979-092C		11613	USSR	22 OCT	25 MAR 80	
1979-092D		11614	USSR	22 OCT	01 DEC 79	
1979-092E		11615	USSR	22 OCT	21 NOV 79	
1979-092F		11616	USSR	22 OCT	09 MAR 80	
1979-094A	MAGSAT	11604	US	30 OCT	11 JUN 80	
1979-094B		11606	US	30 OCT	27 JAN 80	
1979-094C		11607	US	30 OCT	22 FEB 80	
1979-096A	INTERCOSMOS 20	11609	USSR	01 NOV	05 MAR 81	
1979-096B		11610	USSR	01 NOV	26 MAR 81	
1979-097A	COSMOS 1144	11611	USSR	02 NOV	04 DEC 79	
1979-097B		11612	USSR	02 NOV	05 NOV 79	
1979-098D		11624	US	21 NOV	21 NOV 79	
1979-100A	COSMOS 1146	11632	USSR	05 DEC	25 NOV 81	
1979-100B		11633	USSR	05 DEC	31 JAN 81	
1979-101B		11636	US	07 DEC	09 APR 80	
1979-101C		11637	US	07 DEC	27 MAR 80	
1979-102A	COSMOS 1147	11638	USSR	12 DEC	26 DEC 79	
1979-102B		11639	USSR	12 DEC	19 DEC 79	
1979-102C		11646	USSR	12 DEC	15 JAN 80	
1979-102D		11647	USSR	12 DEC	09 MAR 80	
1979-103A	SOYUZ T-1	11640	USSR	16 DEC	25 MAR 80	
1979-103B		11641	USSR	16 DEC	18 DEC 79	
1979-103C		11642	USSR	16 DEC	18 DEC 79	
1979-104A	ARIANE V1	11645	ESA	24 DEC	27 NOV 89	
1979-104B		11659	ESA	24 DEC	14 NOV 82	
1979-105B		11651	USSR	28 DEC	30 DEC 79	
1979-105C		11660	USSR	28 DEC	15 APR 82	
1979-105D		11661	USSR	28 DEC	03 DEC 84	
1979-106A	COSMOS 1148	11649	USSR	28 DEC	10 JAN 80	
1979-106B		11650	USSR	28 DEC	31 DEC 79	
1979-106C		11654	USSR	28 DEC	10 JAN 80	
1979-106D		11655	USSR	28 DEC	15 JAN 80	
1979-106E		11657	USSR	28 DEC	15 JAN 80	
1979-106F		11658	USSR	28 DEC	11 JAN 80	

1980 LAUNCHES

International Designation	Satellite Name	Number	Source	Launch	Decay	Note
1980-001A	COSMOS 1149	11652	USSR	09 JAN	23 JAN 80	
1980-001B		11653	USSR	09 JAN	17 JAN 80	
1980-001C		11656	USSR	09 JAN	19 JAN 80	
1980-001D		11673	USSR	09 JAN	11 JUN 80	
1980-001E		11674	USSR	09 JAN	18 FEB 80	
1980-001F		11675	USSR	09 JAN	11 JUN 80	
1980-001G		11677	USSR	09 JAN	27 JAN 80	
1980-002A	MOLNIYA 1-46	11662	USSR	11 JAN	22 OCT 92	
1980-002B		11663	USSR	11 JAN	28 JAN 80	
1980-002C		11664	USSR	11 JAN	24 JAN 80	
1980-002D		11665	USSR	11 JAN	11 JAN 80	
1980-002E		11666	USSR	11 JAN	21 JAN 80	

International Designation	Satellite Name	Number	Source	Launch	Decay	Note
1980-002F		11670	USSR	11 JAN	04 DEC 92	
1980-004B		11719	US	18 JAN	05 JAN 90	
1980-006A	COSMOS 1152	11678	USSR	24 JAN	06 FEB 80	
1980-006B		11679	USSR	24 JAN	28 JAN 80	
1980-009A	COSMOS 1155	11685	USSR	07 FEB	21 FEB 80	
1980-009B		11686	USSR	07 FEB	16 FEB 80	
1980-009C		11711	USSR	07 FEB	13 MAR 80	
1980-009D		11712	USSR	07 FEB	02 MAR 80	
1980-009E		11717	USSR	07 FEB	10 APR 80	
1980-010A		11687	US	07 FEB	30 OCT 82	
1980-010B		11688	US	07 FEB	13 MAR 80	
1980-010C		11689	US	07 FEB	08 FEB 80	
1980-012K		11725	USSR	11 FEB	05 MAR 80	
1980-013A	COSMOS 1164	11700	USSR	12 FEB	12 JAN 81	
1980-013B		11701	USSR	12 FEB	10 MAR 80	
1980-013C		11702	USSR	12 FEB	03 APR 80	
1980-013D		11949	USSR	12 FEB	08 NOV 80	
1980-014A	SMM	11703	US	14 FEB	02 DEC 89	
1980-014B		11704	US	14 FEB	16 MAY 89	
1980-014C		20325	US	14 FEB	23 NOV 89	
1980-015A	TANSEI 4	11706	JAPAN	17 FEB	12 MAY 83	
1980-015B		11707	JAPAN	17 FEB	20 JAN 84	
1980-015C		11810	JAPAN	17 FEB	27 AUG 80	
1980-015D		11948	JAPAN	17 FEB	08 OCT 80	
1980-016B		11709	USSR	20 FEB	22 FEB 80	
1980-016C		11710	USSR	20 FEB	22 FEB 80	
1980-016E		11801	USSR	20 FEB	12 JAN 81	
1980-016F		11802	USSR	20 FEB	15 APR 82	
1980-017A	COSMOS 1165	11713	USSR	21 FEB	05 MAR 80	
1980-017B		11714	USSR	21 FEB	25 FEB 80	
1980-017C		11726	USSR	21 FEB	07 MAR 80	
1980-017D		11727	USSR	21 FEB	06 MAR 80	
1980-018B		11716	JAPAN	22 FEB	28 FEB 80	
1980-018D		13262	JAPAN	22 FEB	08 JUN 82	
1980-020A	COSMOS 1166	11722	USSR	04 MAR	18 MAR 80	
1980-020B		11723	USSR	04 MAR	13 MAR 80	
1980-020C		11724	USSR	04 MAR	10 MAR 80	
1980-020D		11737	USSR	04 MAR	04 SEP 80	
1980-020E		11738	USSR	04 MAR	28 APR 80	
1980-020F		11739	USSR	04 MAR	09 APR 80	
1980-021A	COSMOS 1167	11729	USSR	14 MAR	01 OCT 81	
1980-021B		11730	USSR	14 MAR	14 MAR 80	
1980-021C		12605	USSR	14 MAR	06 AUG 81	
1980-021D		12606	USSR	14 MAR	01 DEC 82	
1980-021E		12607	USSR	14 MAR	22 AUG 81	
1980-021F		12608	USSR	14 MAR	12 AUG 81	
1980-021G		12609	USSR	14 MAR	15 OCT 81	
1980-021H		12610	USSR	14 MAR	14 DEC 81	
1980-021J		12611	USSR	14 MAR	22 AUG 81	
1980-021R		12612	USSR	14 MAR	16 AUG 81	
1980-021L		12613	USSR	14 MAR	02 AUG 81	
1980-021M		12615	USSR	14 MAR	27 AUG 81	
1980-021N		12616	USSR	14 MAR	27 AUG 81	
1980-023A	COSMOS 1169	11741	USSR	27 MAR	03 MAR 83	
1980-023B		11742	USSR	27 MAR	04 NOV 81	
1980-024A	PROGRESS 8	11743	USSR	27 MAR	26 APR 80	
1980-024B		11744	USSR	27 MAR	29 MAR 80	
1980-025A	COSMOS 1170	11747	USSR	01 APR	12 APR 80	
1980-025B		11748	USSR	01 APR	05 APR 80	
1980-025C		11749	USSR	01 APR	03 APR 80	
1980-025D		11755	USSR	01 APR	14 APR 80	
1980-025E		11756	USSR	01 APR	13 APR 80	
1980-025F		11757	USSR	01 APR	13 APR 80	
1980-027A	SOYUZ 35	11753	USSR	09 APR	03 JUN 80	
1980-027B		11754	USSR	09 APR	11 APR 80	
1980-028B		11759	USSR	12 APR	26 APR 80	
1980-028C		11760	USSR	12 APR	08 MAY 80	
1980-028D		11761	USSR	12 APR	20 APR 80	
1980-029A	COSMOS 1173	11763	USSR	17 APR	28 APR 80	
1980-029B		11764	USSR	17 APR	21 APR 80	
1980-029C		11786	USSR	17 APR	30 APR 80	
1980-029D		11787	USSR	17 APR	30 APR 80	
1980-030AA		12356	USSR	18 APR	21 MAR 82	
1980-030AB		12357	USSR	18 APR	09 NOV 89	
1980-030AC		12358	USSR	18 APR	25 OCT 82	
1980-030AD		12359	USSR	18 APR	17 AUG 81	
1980-030AF		12361	USSR	18 APR	27 JUL 81	
1980-030AG		12362	USSR	18 APR	18 APR 82	
1980-030AH		13925	USSR	18 APR	10 MAR 91	
1980-030AJ		13926	USSR	18 APR	31 AUG 88	
1980-030AR		13927	USSR	18 APR	15 MAR 89	
1980-030AL		13928	USSR	18 APR	21 JUN 88	
1980-030AN		13930	USSR	18 APR	02 MAY 91	
1980-030AP		13931	USSR	18 APR	18 FEB 91	
1980-030AR		13945	USSR	18 APR	30 SEP 83	
1980-030AS		14920	USSR	18 APR	27 DEC 89	
1980-030AT		14921	USSR	18 APR	20 OCT 88	
1980-030AU		14922	USSR	18 APR	17 JAN 89	
1980-030AV		14923	USSR	18 APR	03 FEB 89	
1980-030AW		15759	USSR	18 APR	14 FEB 88	
1980-030B		11766	USSR	18 APR	20 APR 80	
1980-030C		11771	USSR	18 APR	30 DEC 80	
1980-030D		11772	USSR	18 APR	19 APR 81	
1980-030E		11773	USSR	18 APR	17 JAN 81	
1980-030F		11774	USSR	18 APR	17 APR 81	
1980-030G		11775	USSR	18 APR	10 AUG 89	
1980-030H		11776	USSR	18 APR	17 FEB 82	
1980-030L		11779	USSR	18 APR	10 MAR 83	
1980-030M		11780	USSR	18 APR	17 FEB 81	
1980-030P		11782	USSR	18 APR	24 JUN 84	
1980-030Q		12098	USSR	18 APR	01 DEC 82	
1980-030S		12344	USSR	18 APR	08 SEP 82	
1980-030T		12345	USSR	18 APR	09 JAN 92	
1980-030U		12346	USSR	18 APR	24 APR 82	
1980-030W		12348	USSR	18 APR	26 FEB 91	
1980-030X		12349	USSR	18 APR	15 DEC 82	
1980-030Z		12355	USSR	18 APR	03 DEC 81	
1980-031A	COSMOS 1175	11767	USSR	18 APR	28 MAY 80	
1980-031B		11768	USSR	18 APR	02 MAY 80	
1980-031C		11769	USSR	18 APR	29 SEP 80	
1980-031D		11770	USSR	18 APR	30 MAY 80	
1980-033A	PROGRESS 9	11784	USSR	27 APR	22 MAY 80	
1980-033B		11785	USSR	29 APR	29 APR 80	
1980-034B		11968	USSR	29 APR	04 OCT 80	
1980-034C		11969	USSR	29 APR	12 SEP 80	
1980-035A	COSMOS 1177	11789	USSR	29 APR	12 JUN 80	
1980-035B		11790	USSR	29 APR	02 MAY 80	
1980-035C		11840	USSR	29 APR	16 JUN 80	
1980-036A	COSMOS 1178	11793	USSR	07 MAY	22 MAY 80	
1980-036B		11794	USSR	07 MAY	14 MAY 80	
1980-036C		11795	USSR	07 MAY	12 MAY 80	
1980-036D		11805	USSR	07 MAY	07 JUL 80	
1980-036E		11806	USSR	07 MAY	16 NOV 80	
1980-036F		11807	USSR	07 MAY	22 JUN 80	
1980-037A	COSMOS 1179	11796	USSR	14 MAY	18 JUL 89	
1980-037B		11797	USSR	14 MAY	11 JAN 83	
1980-038A	COSMOS 1180	11798	USSR	15 MAY	26 MAY 80	
1980-038B		11799	USSR	15 MAY	23 MAY 80	
1980-038C		11800	USSR	15 MAY	17 MAY 80	
1980-038D		11813	USSR	15 MAY	30 MAY 8Q	
1980-038E		11814	USSR	15 MAY	31 MAY 80	
1980-038F		11815	USSR	15 MAY	31 MAY 80	
1980-040A	COSMOS 1182	11808	USSR	23 MAY	05 JUN 80	
1980-040B		11809	USSR	23 MAY	26 MAY 80	
1980-040D		11824	USSR	23 MAY	09 JUN 80	
1980-041A	SOYUZ 36	11811	USSR	26 MAY	31 JUL 80	
1980-041B		11812	USSR	26 MAY	29 MAY 80	
1980-042A	COSMOS 1183	11816	USSR	28 MAY	11 JUN 80	
1980-042B		11817	USSR	28 MAY	05 JUN 80	
1980-042C		11818	USSR	28 MAY	09 JUN 80	
1980-042D		11833	USSR	28 MAY	13 JUL 80	
1980-042E		11834	USSR	28 MAY	03 AUG 80	
1980-042F		11835	USSR	28 MAY	14 JUN 80	
1980-042G		11836	USSR	28 MAY	04 DEC 80	
1980-043A	NOAA B	11819	US	29 MAY	03 MAY 81	
1980-043B		11820	US	29 MAY	11 SEP 80	
1980-043C		11874	US	29 MAY	27 AUG 80	
1980-045A	SOYUZ T-2	11825	USSR	05 JUN	09 JUN 80	
1980-045B		11826	USSR	05 JUN	07 JUN 80	
1980-046A	COSMOS 1185	11827	USSR	06 JUN	20 JUN 80	
1980-046B		11828	USSR	06 JUN	11 JUN 80	
1980-046C		11853	USSR	06 JUN	04 JUL 80	
1980-046D		11854	USSR	06 JUN	25 JUN 80	
1980-047A	COSMOS 1186	11829	USSR	06 JUN	01 JAN 82	
1980-047AA		12509	USSR	06 JUN	14 JUN 81	
1980-047AB		12510	USSR	06 JUN	20 JUN 81	
1980-047AC		12511	USSR	06 JUN	13 JUL 81	
1980-047B		11830	USSR	06 JUN	16 DEC 81	BS
1980-047C		11831	USSR	06 JUN	17 JUN 80	
1980-047D		11912	USSR	06 JUN	24 SEP 80	
1980-047E		11913	USSR	06 JUN	14 SEP 80	
1980-047F		12407	USSR	06 JUN	03 MAY 81	
1980-047G		12408	USSR	06 JUN	03 MAY 81	
1980-047H		12412	USSR	06 JUN	01 MAY 81	
1980-047J		12413	USSR	06 JUN	03 MAY 81	
1980-047K		12414	USSR	06 JUN	03 MAY 81	
1980-047L		12415	USSR	06 JUN	03 MAY 81	
1980-047M		12416	USSR	06 JUN	03 MAY 81	
1980-047N		12422	USSR	06 JUN	03 MAY 81	
1980-047P		12453	USSR	06 JUN	28 MAY 81	
1980-047Q		12459	USSR	06 JUN	20 MAY 81	
1980-047R		12480	USSR	06 JUN	22 JUN 81	
1980-047S		12481	USSR	06 JUN	18 JUN 81	
1980-047T		12482	USSR	06 JUN	15 JUN 81	
1980-047U		12483	USSR	06 JUN	22 JUN 81	
1980-047V		12484	USSR	06 JUN	18 JUN 81	
1980-047W		12485	USSR	06 JUN	17 JUN 81	
1980-047X		12486	USSR	06 JUN	29 SEP 81	
1980-047Y		12487	USSR	06 JUN	14 JUN 81	
1980-047Z		12490	USSR	06 JUN	20 JUN 81	
1980-040C		11823	USSR	10 JUN	10 JUN 80	
1980-048A	COSMOS 1187	11837	USSR	12 JUN	26 JUN 80	
1980-048B		11838	USSR	12 JUN	17 JUN 80	
1980-048C		11839	USSR	12 JUN	17 JUN 80	
1980-048D		11865	USSR	12 JUN	29 JUN 80	
1980-048E		11866	USSR	12 JUN	18 JUL 8Q	
1980-049B		11842	USSR	14 JUN	15 JUN 80	
1980-049C		11843	USSR	14 JUN	14 JUN 80	
1980-049D		11859	USSR	14 JUN	20 APR 82	
1980-049E		11860	USSR	14 JUN	22 MAR 82	
1980-050C		11845	USSR	14 JUN	17 JUL 80	
1980-050D		11846	USSR	14 JUN	06 JUL 80	
1980-051A	METEOR 1-30	11848	USSR	18 JUN	01 MAR 92	
1980-052A		11850	US	18 JUN	06 MAR 81	
1980-052B		11851	US	18 JUN	20 JUN 80	
1980-053A	MOLNIYA 1-47	11856	USSR	21 JUN	01 APR 91	

International Designation	Satellite Name	Number	Source	Launch	Decay	Note
1980-053B		11857	USSR	21 JUN	28 JUL 80	
1980-053C		11858	USSR	21 JUN	25 JUL 80	
1980-053D		11861	USSR	21 JUN	13 MAR 91	
1980-054A	COSMOS 1189	11863	USSR	26 JUN	10 JUL 80	
1980-054B		11864	USSR	26 JUN	01 JUL 80	
1980-054C		11886	USSR	26 JUN	13 JUL 80	
1980-054D		11887	USSR	26 JUN	04 AUG 80	
1980-054E		11889	USSR	26 JUN	18 JUL 80	
1980-055A	PROGRESS 10	11867	USSR	29 JUN	19 JUL 80	
1980-055B		11868	USSR	29 JUN	02 JUL 80	
1980-057B		11872	USSR	02 JUL	02 AUG 80	
1980-057C		11873	USSR	02 JUL	31 JUL 80	
1980-059A	COSMOS 1200	11884	USSR	09 JUL	23 JUL 80	
1980-059B		11885	USSR	09 JUL	15 JUL 80	
1980-059C		11901	USSR	09 JUL	30 JUL 80	
1980-059D		11902	USSR	09 JUL	17 AUG 80	
1980-059E		11903	USSR	09 JUL	27 JUL 80	
1980-059F		11904	USSR	09 JUL	28 JUL 80	
1980-060B		11891	USSR	14 JUL	16 JUL 80	
1980-060C		11892	USSR	14 JUL	15 JUL 80	
1980-060D		11914	USSR	14 JUL	20 APR 82	
1980-060E		11923	USSR	14 JUL	02 APR 81	
1980-061A	COSMOS 1201	11894	USSR	15 JUL	28 JUL 80	
1980-061B		11895	USSR	15 JUL	18 JUL 80	
1980-061C		11910	USSR	15 JUL	31 JUL 80	
1980-061D		11911	USSR	15 JUL	02 AUG 80	
1980-062A	ROHINI 1	11899	INDIA	18 JUL	20 MAY 81	
1980-062B		11900	INDIA	18 JUL	23 JUL 81	
1980-063B		11897	USSR	18 JUL	02 AUG 80	
1980-063C		11898	USSR	18 JUL	04 AUG 80	
1980-064A	SOYUZ 37	11905	USSR	23 JUL	11 OCT 80	
1980-064B		11906	USSR	23 JUL	27 JUL 80	
1980-065A	COSMOS 1202	11907	USSR	24 JUL	07 AUG 80	
1980-065B		11908	USSR	24 JUL	29 JUL 80	
1980-065C		11920	USSR	24 JUL	23 AUG 80	
1980-065D		11921	USSR	24 JUL	10 AUG 80	
1980-065E		11922	USSR	24 JUL	11 AUG 80	
1980-066A	COSMOS 1203	11915	USSR	31 JUL	14 AUG 80	
1980-066B		11916	USSR	31 JUL	06 AUG 80	
1980-066C		11919	USSR	31 JUL	02 AUG 80	
1980-066D		11927	USSR	31 JUL	30 AUG 80	
1980-066E		11928	USSR	31 JUL	21 AUG 80	
1980-066F		11929	USSR	31 JUL	19 AUG 80	
1980-067A	COSMOS 1204	11917	USSR	31 JUL	23 FEB 81	
1980-067B		11918	USSR	31 JUL	13 FEB 81	BU
1980-067C		11960	USSR	31 JUL	11 OCT 80	
1980-067D		11961	USSR	31 JUL	14 OCT 80	
1980-067E		11987	USSR	31 JUL	31 OCT 80	
1980-067F		11988	USSR	31 JUL	24 OCT 80	
1980-067G		11989	USSR	31 JUL	24 OCT 80	
1980-067H		11992	USSR	31 JUL	26 OCT 80	
1980-067J		12010	USSR	31 JUL	03 NOV 80	
1980-067K		12011	USSR	31 JUL	03 NOV 80	
1980-067L		12014	USSR	31 JUL	04 NOV 80	
1980-067M		12015	USSR	31 JUL	10 NOV 80	
1980-067N		12036	USSR	31 JUL	27 NOV 80	
1980-067P		12037	USSR	31 JUL	18 NOV 80	
1980-067Q		12038	USSR	31 JUL	18 NOV 80	
1980-067R		12047	USSR	31 JUL	23 NOV 80	
1980-067S		12048	USSR	31 JUL	03 DEC 80	
1980-067T		12049	USSR	31 JUL	19 NOV 80	
1980-067U		12052	USSR	31 JUL	17 NOV 80	
1980-067V		12053	USSR	31 JUL	17 NOV 80	
1980-067W		12056	USSR	31 JUL	16 NOV 80	
1980-067X		12057	USSR	31 JUL	23 NOV 80	
1980-067Y		12074	USSR	31 JUL	14 DEC 80	
1980-067Z		12075	USSR	31 JUL	13 DEC 80	
1980-068A	COSMOS 1205	11924	USSR	12 AUG	26 AUG 80	
1980-068B		11925	USSR	12 AUG	17 AUG 80	
1980-068C		11943	USSR	12 AUG	10 SEP 80	
1980-068D		11944	USSR	12 AUG	29 AUG 80	
1980-070A	COSMOS 1207	11938	USSR	22 AUG	04 SEP 80	
1980-070B		11939	USSR	22 AUG	26 AUG 80	
1980-070C		11954	USSR	22 AUG	08 SEP 80	
1980-070D		11959	USSR	22 AUG	08 SEP 80	
1980-071A	COSMOS 1208	11945	USSR	26 AUG	24 SEP 80	
1980-071B		11946	USSR	26 AUG	30 AUG 80	
1980-072A	COSMOS 1209	11950	USSR	03 SEP	17 SEP 80	
1980-072B		11951	USSR	03 SEP	07 SEP 80	
1980-072C		11976	USSR	03 SEP	30 SEP 80	
1980-074B		11965	US	09 SEP	01 DEC 80	
1980-075A	SOYUZ 38	11977	USSR	18 SEP	26 SEP 80	
1980-075B		11978	USSR	18 SEP	21 SEP 80	
1980-075C		11979	USSR	18 SEP	21 SEP 80	
1980-076A	COSMOS 1210	11980	USSR	19 SEP	03 OCT 80	
1980-076B		11981	USSR	19 SEP	21 SEP 80	
1980-076C		11995	USSR	19 SEP	11 OCT 80	
1980-076D		11996	USSR	19 SEP	06 OCT 80	
1980-076E		11999	USSR	19 SEP	29 OCT 80	
1980-076F		12001	USSR	19 SEP	10 OCT 80	
1980-076G		12002	USSR	19 SEP	05 OCT 80	
1980-077A	COSMOS 1211	11982	USSR	23 SEP	04 OCT 80	
1980-077B		11983	USSR	23 SEP	26 SEP 80	
1980-078A	COSMOS 1212	11985	USSR	26 SEP	09 OCT 80	
1980-078B		11986	USSR	26 SEP	29 SEP 80	
1980-078C		12006	USSR	26 SEP	10 OCT 80	
1980-078D		12007	USSR	26 SEP	11 OCT 80	
1980-079A	PROGRESS 11	11993	USSR	28 SEP	11 DEC 80	
1980-079B		11994	USSR	28 SEP	30 SEP 80	
1980-080A	COSMOS 1213	11997	USSR	03 OCT	17 OCT 80	
1980-080B		11998	USSR	03 OCT	07 OCT 80	
1980-080C		12022	USSR	03 OCT	06 NOV 80	
1980-080D		12023	USSR	03 OCT	24 OCT 80	
1980-080E		12024	USSR	03 OCT	24 OCT 80	
1980-080F		12025	USSR	03 OCT	22 OCT 80	
1980-081B		12004	USSR	05 OCT	07 OCT 80	
1980-081C		12005	USSR	05 OCT	06 OCT 80	
1980-081D		12012	USSR	05 OCT	15 APR 82	
1980-081E		12013	USSR	05 OCT	14 JUN 82	
1980-082A	COSMOS 1214	12008	USSR	10 OCT	23 OCT 80	
1980-082B		12009	USSR	10 OCT	13 OCT 80	
1980-082C		12029	USSR	10 OCT	25 OCT 80	
1980-082D		12030	USSR	10 OCT	25 OCT 80	
1980-082E		12031	USSR	10 OCT	25 OCT 80	
1980-083A	COSMOS 1215	12016	USSR	14 OCT	12 MAY 83	
1980-083B		12017	USSR	14 OCT	27 AUG 82	
1980-083C		12018	USSR	14 OCT	10 JAN 81	
1980-083D		12026	USSR	14 OCT	18 NOV 80	
1980-084A	COSMOS 1216	12019	USSR	16 OCT	30 OCT 80	
1980-084B		12020	USSR	16 OCT	23 OCT 80	
1980-084C		12021	USSR	16 OCT	22 OCT 80	
1980-084D		12040	USSR	16 OCT	05 DEC 80	
1980-084E		12041	USSR	16 OCT	26 MAR 81	
1980-084F		12043	USSR	16 OCT	13 APR 81	
1980-084G		12044	USSR	16 OCT	02 MAR 81	
1980-084H		12045	USSR	16 OCT	21 NOV 80	
1980-085B		12033	USSR	24 OCT	17 NOV 80	
1980-085C		12034	USSR	24 OCT	07 NOV 80	
1980-086A	COSMOS 1218	12039	USSR	30 OCT	12 DEC 80	
1980-086B		12042	USSR	30 OCT	02 NOV 80	
1980-088A	COSMOS 1219	12050	USSR	31 OCT	13 NOV 80	
1980-088B		12051	USSR	31 OCT	04 NOV 80	
1980-088C		12060	USSR	31 OCT	18 NOV 80	
1980-088D		12061	USSR	31 OCT	24 NOV 80	
1980-088E		12062	USSR	31 OCT	26 NOV 80	
1980-088F		12063	USSR	31 OCT	23 NOV 80	
1980-088G		12064	USSR	31 OCT	15 NOV 80	
1980-089AA		13330	USSR	C4 NOV	12 AUG 83	
1980-089AB		13331	USSR	04 NOV	28 APR 90	
1980-089AC		13333	USSR	04 NOV	16 JAN 89	
1980-089AD		13334	USSR	04 NOV	10 APR 83	
1980-089AE		13335	USSR	04 NOV	01 DEC 83	
1980-089AF		13336	USSR	04 NOV	15 MAR 86	
1980-089AG		13337	USSR	04 NOV	30 OCT 82	
1980-089AH		13338	USSR	04 NOV	10 SEP 83	
1980-089AJ		13339	USSR	04 NOV	17 AUG 85	
1980-089AK		13340	USSR	04 NOV	06 FEB 91	
1980-089AL		13341	USSR	04 NOV	07 DEC 88	
1980-089AM		13342	USSR	04 NOV	27 APR 83	
1980-089AN		13343	USSR	04 NOV	05 FEB 89	
1980-089AP		13344	USSR	04 NOV	29 AUG 88	
1980-089AQ		13347	USSR	04 NOV	22 DEC 82	
1980-089AR		13348	USSR	04 NOV	06 OCT 82	
1980-089AS		13349	USSR	04 NOV	15 DEC 82	
1980-089AT		13350	USSR	04 NOV	26 JUL 83	
1980-089AU		13351	USSR	04 NOV	09 MAY 84	
1980-089AV		13352	USSR	04 NOV	10 OCT 87	
1980-089AW		13355	USSR	04 NOV	07 MAY 81	
1980-089AX		13356	USSR	04 NOV	16 JUL 88	
1980-089AY		13357	USSR	04 NOV	23 JAN 83	
1980-089AZ		13358	USSR	04 NOV	06 JUL 89	
1980-089B		12055	USSR	04 NOV	05 NOV 80	
1980-089BA		13359	USSR	04 NOV	02 FEB 90	
1980-089BB		13506	USSR	04 NOV	25 OCT 89	
1980-089BC		13680	USSR	04 NOV	05 JUL 88	
1980-089BD		13687	USSR	04 NOV	08 SEP 83	
1980-089BE		13688	USSR	04 NOV	03 MAR 83	
1980-089BF		14348	USSR	04 NOV	24 OCT 84	
1980-089BG		14349	USSR	04 NOV	04 MAR 84	
1980-089BH		14350	USSR	04 NOV	27 MAY 88	
1980-089BJ		14351	USSR	04 NOV	09 NOV 85	
1980-089BK		14352	USSR	04 NOV	15 APR 85	
1980-089BL		14353	USSR	04 NOV	11 JUL 88	
1980-089BM		14354	USSR	04 NOV	07 FEB 89	
1980-089BN		14355	USSR	04 NOV	07 FEB 89	
1980-089BP		14431	USSR	04 NOV	18 MAR 89	
1980-089BQ		14432	USSR	04 NOV	02 MAR 89	
1980-089BR		14433	USSR	04 NOV	27 JUN 84	
1980-089BS		14449	USSR	04 NOV	03 JUL 89	
1980-089BT		14831	USSR	04 NOV	24 APR 84	
1980-089BU		14832	USSR	04 NOV	28 SEP 85	
1980-089BV		14833	USSR	04 NOV	09 SEP 85	
1980-089BW		15763	USSR	04 NOV	19 AUG 88	
1980-089BX		17654	USSR	04 NOV	24 DEC 88	
1980-089BY		17655	USSR	04 NOV	12 AUG 89	
1980-089BZ		17656	USSR	04 NOV	21 JAN 89	
1980-089C		13308	USSR	04 NOV	19 OCT 87	
1980-089CA		17657	USSR	04 NOV	20 FEB 89	
1980-089CB		18233	USSR	04 NOV	02 FEB 89	
1980-089CC		18495	USSR	04 NOV	14 MAY 89	
1980-089CD		18610	USSR	04 NOV	18 OCT 88	
1980-089CE		19148	USSR	04 NOV	16 OCT 88	
1980-089CF		19171	USSR	04 NOV	12 JAN 89	
1980-089CG		19177	USSR	04 NOV	17 JAN 89	
1980-089D		13309	USSR	04 NOV	25 AUG 84	
1980-089E		13310	USSR	04 NOV	12 NOV 87	

International Designation	Satellite Name	Number	Source	Launch	Decay	Note
1980-089F		13311	USSR	04 NOV	21 NOV 82	
1980-089G		13312	USSR	04 NOV	05 MAR 84	
1980-089H		13313	USSR	04 NOV	07 JAN 89	
1980-089J		13314	USSR	04 NOV	12 AUG 82	
1980-089K		13315	USSR	04 NOV	04 NOV 83	
1980-089L		13316	USSR	04 NOV	05 APR 88	
1980-089M		13317	USSR	04 NOV	28 MAR 83	
1980-089N		13318	USSR	04 NOV	05 SEP 89	
1980-089P		13319	USSR	04 NOV	13 FEB 90	
1980-089Q		13320	USSR	04 NOV	03 SEP 85	
1980-089R		13321	USSR	04 NOV	28 OCT 82	
1980-089S		13322	USSR	04 NOV	19 DEC 88	
1980-089T		13323	USSR	04 NOV	17 FEB 84	
1980-089U		13324	USSR	04 NOV	28 MAR 89	
1980-089V		13325	USSR	04 NOV	25 APR 84	
1980-089W		13326	USSR	04 NOV	08 JAN 88	
1980-089X		13327	USSR	04 NOV	23 JUL 82	
1980-089Y		13328	USSR	04 NOV	11 JAN 89	
1980-089Z		13329	USSR	04 NOV	13 JUL 82	
1980-090A	COSMOS 1221	12058	USSR	12 NOV	26 NOV 80	
1980-090B		12059	USSR	12 NOV	20 NOV 80	
1980-090C		12073	USSR	12 NOV	14 APR 81	
1980-090D		12076	USSR	12 NOV	03 JAN 81	
1980-090E		12083	USSR	12 NOV	22 DEC 80	
1980-091B		12336	US	15 NOV	03 MAY 81	
1980-092B		12067	USSR	16 NOV	13 DEC 80	
1980-092C		12068	USSR	16 NOV	12 DEC 80	
1980-094A	SOYUZ T-3	12077	USSR	27 NOV	10 DEC 80	
1980-094B		12079	USSR	27 NOV	29 NOV 80	
1980-095B		12080	USSR	27 NOV	22 DEC 80	
1980-095C		12081	USSR	27 NOV	21 DEC 80	
1980-095D		12082	USSR	27 NOV	08 DEC 80	
1980-096A	COSMOS 1224	12084	USSR	01 DEC	15 DEC 80	
1980-096B		12085	USSR	01 DEC	09 DEC 80	
1980-096C		12095	USSR	01 DEC	04 MAY 81	
1980-096D		12096	USSR	01 DEC	15 MAR 81	
1980-096E		12097	USSR	01 DEC	26 JAN 81	
1980-101A	COSMOS 1227	12100	USSR	16 DEC	28 DEC 80	
1980-101B		12101	USSR	16 DEC	20 DEC 80	
1980-101C		12127	USSR	16 DEC	15 JAN 81	
1980-101D		12128	USSR	16 DEC	03 JAN 81	
1980-101E		12129	USSR	16 DEC	30 DEC 80	
1980-103B		12118	USSR	25 DEC	20 JAN 81	
1980-103C		12119	USSR	25 DEC	14 JAN 81	
1980-103D		12117	USSR	25 DEC	12 JAN 81	
1980-104B		12122	USSR	26 DEC	28 DEC 80	
1980-104C		12123	USSR	26 DEC	27 DEC 80	
1980-104D		12137	USSR	26 DEC	20 APR 82	
1980-104F		16869	USSR	26 DEC	26 OCT 86	
1980-105A	COSMOS 1236	12121	USSR	26 DEC	21 JAN 81	
1980-105B		12124	USSR	26 DEC	30 DEC 80	
1980-105C		12125	USSR	26 DEC	08 JAN 81	
1980-105D		12126	USSR	26 DEC	13 JAN 81	
1981 LAUNCHES						
1981-001A	COSMOS 1237	12130	USSR	06 JAN	20 JAN 81	
1981-001B		12131	USSR	06 JAN	15 JAN 81	
1981-001C		12132	USSR	06 JAN	12 JAN 81	
1981-001D		12145	USSR	06 JAN	05 JUN 81	
1981-001E		12146	USSR	06 JAN	25 JAN 81	
1981-001F		12147	USSR	06 JAN	03 MAR 81	
1981-001G		12148	USSR	06 JAN	10 FEB 81	
1981-002C		12135	USSR	09 JAN	27 JAN 81	
1981-002D		12136	USSR	09 JAN	22 JAN 81	
1981-004A	COSMOS 1239	12140	USSR	16 JAN	28 JAN 81	
1981-004B		12141	USSR	16 JAN	20 JAN 81	
1981-005A	COSMOS 1240	12143	USSR	20 JAN	17 FEB 81	
1981-005B		12144	USSR	20 JAN	24 JAN 81	
1981-007A	PROGRESS 12	12152	USSR	24 JAN	20 MAR 81	
1981-007B		12153	USSR	24 JAN	27 JAN 81	
1981-009B		12157	USSR	30 JAN	14 FEB 81	
1981-009C		12158	USSR	30 JAN	09 FEB 81	
1981-010A	COSMOS 1243	12160	USSR	02 FEB	02 FEB 81	
1981-010B		12161	USSR	02 FEB	04 FEB 81	
1981-011A	INTERCOSMOS 21	12162	USSR	06 FEB	07 JUL 82	
1981-011B		12163	USSR	06 FEB	22 JUN 82	
1981-012B		12296	JAPAN	11 FEB	28 FEB 81	
1981-014A	COSMOS 1245	12299	USSR	13 FEB	27 FEB 81	
1981-014B		12300	USSR	13 FEB	21 FEB 81	
1981-014C		12312	USSR	13 FEB	01 APR 81	
1981-014D		12313	USSR	13 FEB	05 MAR 81	
1981-014E		12314	USSR	13 FEB	15 MAR 81	
1981-014F		12316	USSR	13 FEB	22 JUL 81	
1981-015A	COSMOS 1246	12301	USSR	18 FEB	13 MAR 81	
1981-015B		12302	USSR	18 FEB	21 FEB 81	
1981-016B		12304	USSR	19 FEB	13 MAR 81	
1981-016C		12305	USSR	19 FEB	25 FEB 81	
1981-016D		12306	USSR	19 FEB	13 APR 81	
1981-017A	ASTRO A	12307	JAPAN	21 FEB	11 JUL 91	
1981-017B		12308	JAPAN	21 FEB	26 MAY 91	
1981-017C		12310	JAPAN	21 FEB	25 NOV 81	
1981-019A		12315	US	28 FEB	20 JUN 81	
1981-020A	COSMOS 1248	12317	USSR	05 MAR	04 APR 81	
1981-020B		12318	USSR	05 MAR	08 MAR 81	
1981-021B		12550	USSR	05 MAR	22 JUN 81	
1981-021D		12552	USSR	05 MAR	19 JUL 81	
1981-023A	SOYUZ T-4	12334	USSR	12 MAR	26 MAY 81	
1981-023B		12335	USSR	12 MAR	14 MAR 81	
1981-024A	COSMOS 1258	12337	USSR	14 MAR	14 MAR 81	
1981-024B		12338	USSR	14 MAR	15 MAR 81	
1981-025B		12340	US	16 MAR	18 MAR 81	
1981-026A	COSMOS 1259	12341	USSR	17 MAR	31 MAR 81	
1981-026B		12342	USSR	17 MAR	25 MAR 81	
1981-026C		12350	USSR	17 MAR	11 APR 81	
1981-026D		12379	USSR	17 MAR	29 APR 81	
1981-026E		12380	USSR	17 MAR	08 JUL 81	
1981-026F		12381	USSR	17 MAR	30 AUG 81	
1981-026G		12382	USSR	17 MAR	19 APR 81	
1981-027B		12352	USSR	18 MAR	18 MAR 81	
1981-027C		12353	USSR	18 MAR	20 MAR 81	
1981-027D		12396	USSR	18 MAR	15 APR 82	
1981-027E		12397	USSR	18 MAR	15 APR 82	
1981-028A	COSMOS 1260	12364	USSR	20 MAR	22 MAY 82	
1981-028AA		13202	USSR	20 MAR	14 AUG 82	
1981-028AB		13203	USSR	20 MAR	21 JUN 82	
1981-028AC		13221	USSR	20 MAR	15 JUN 82	
1981-028AD		13222	USSR	20 MAR	14 AUG 82	
1981-028AE		13223	USSR	20 MAR	09 SEP 82	
1981-028AF		13224	USSR	20 MAR	31 DEC 82	
1981-028AG		13225	USSR	20 MAR	09 AUG 82	
1981-028AH		13226	USSR	20 MAR	05 OCT 82	
1981-028AJ		13227	USSR	20 MAR	02 JUN 82	
1981-028AK		13228	USSR	20 MAR	04 NOV 82	
1981-028AL		13229	USSR	20 MAR	16 FEB 89	
1981-028AM		13230	USSR	20 MAR	17 JUN 82	
1981-028AN		13231	USSR	20 MAR	19 AUG 82	
1981-028AP		13232	USSR	20 MAR	15 OCT 82	
1981-028AQ		13233	USSR	20 MAR	08 DEC 82	
1981-028AR		13234	USSR	20 MAR	25 AUG 82	
1981-028AS		13235	USSR	20 MAR	17 OCT 82	
1981-028AT		13236	USSR	20 MAR	28 SEP 82	
1981-028AU		13442	USSR	20 MAR	20 OCT 90	
1981-028AV		13443	USSR	20 MAR	29 MAR 90	
1981-028AW		13444	USSR	20 MAR	02 JAN 83	
1981-028AX		13445	USSR	20 MAR	11 NOV 82	
1981-028AY		13620	USSR	20 MAR	30 JAN 89	
1981-028AZ		13621	USSR	20 MAR	01 MAR 89	
1981-028B		12365	USSR	20 MAR	21 MAR 81	
1981-028BA		13622	USSR	20 MAR	23 SEP 88	
1981-028BB		13623	USSR	20 MAR	12 FEB 90	
1981-028BC		13635	USSR	20 MAR	13 FEB 83	
1981-028BD		13681	USSR	20 MAR	21 JAN 92	
1981-028BF		13689	USSR	20 MAR	04 FEB 91	
1981-028BG		13690	USSR	20 MAR	18 MAR 90	
1981-028BH		13691	USSR	20 MAR	04 OCT 84	
1981-028BK		13918	USSR	20 MAR	04 JUN 89	
1981-028BL		13919	USSR	20 MAR	13 NOV 89	
1981-028BM		13920	USSR	20 MAR	11 OCT 89	
1981-028BN		13921	USSR	20 MAR	06 JAN 90	
1981-028BP		13922	USSR	20 MAR	31 DEC 90	
1981-028BQ		14220	USSR	20 MAR	16 OCT 89	
1981-028BR		14221	USSR	20 MAR	04 FEB 91	
1981-028BS		14437	USSR	20 MAR	24 JAN 89	
1981-028BT		14438	USSR	20 MAR	12 OCT 88	
1981-028BU		14924	USSR	20 MAR	19 SEP 86	
1981-028BV		17671	USSR	20 MAR	26 DEC 89	
1981-028C		13180	USSR	20 MAR	05 SEP 82	
1981-028D		13181	USSR	20 MAR	10 SEP 82	
1981-028E		13182	USSR	20 MAR	07 DEC 89	
1981-028F		13183	USSR	20 MAR	06 AUG 91	
1981-028G		13184	USSR	20 MAR	09 AUG 83	
1981-028H		13185	USSR	20 MAR	04 NOV 82	
1981-028J		13186	USSR	20 MAR	11 JAN 83	
1981-028K		13187	USSR	20 MAR	10 OCT 83	
1981-028L		13188	USSR	20 MAR	15 JUN 82	
1981-028M		13189	USSR	20 MAR	22 OCT 88	
1981-028N		13190	USSR	20 MAR	02 NOV 82	
1981-028P		13191	USSR	20 MAR	19 SEP 83	
1981-028Q		13192	USSR	20 MAR	15 JUN 85	
1981-028R		13193	USSR	20 MAR	13 JUN 82	
1981-028S		13194	USSR	20 MAR	01 SEP 82	
1981-028T		13195	USSR	20 MAR	10 SEP 82	
1981-028U		13196	USSR	20 MAR	31 MAY 82	
1981-028V		13197	USSR	20 MAR	11 DEC 84	
1981-028W		13198	USSR	20 MAR	03 OCT 82	
1981-028X		13199	USSR	20 MAR	31 AUG 82	
1981-028Y		13200	USSR	20 MAR	16 JUN 82	
1981-028Z		13201	USSR	20 MAR	15 JUN 82	
1981-029A	SOYUZ 39	12366	USSR	22 MAR	30 MAR 81	
1981-029B		12367	USSR	22 MAR	25 MAR 81	
1981-028BJ		13781	USSR	24 MAR	04 AUG 89	
1981-030A	MOLNIYA 3-15	12368	USSR	24 MAR	19 OCT 92	
1981-030B		12369	USSR	24 MAR	13 APR 81	
1981-030C		12370	USSR	24 MAR	12 APR 81	
1981-030D		12383	USSR	24 MAR	29 JUN 93	
1981-031B		12377	USSR	31 MAR	17 APR 81	
1981-031C		12378	USSR	31 MAR	15 APR 81	
1981-032A	COSMOS 1262	12385	USSR	07 APR	21 APR 81	
1981-032B		12386	USSR	07 APR	13 APR 81	
1981-032C		12387	USSR	07 APR	20 APR 81	
1981-032D		12410	USSR	07 APR	23 APR 81	
1981-032E		12411	USSR	07 APR	28 APR 81	
1981-032F		12417	USSR	07 APR	24 APR 81	
1981-034A	STS 1	12399	US	12 APR	14 APR 81	BT
1981-035A	COSMOS 1264	12400	USSR	15 APR	29 APR 81	

International Designation	Satellite Name	Number	Source	Launch	Decay	Note
1981-035B		12401	USSR	15 APR	23 APR 81	
1981-035C		12430	USSR	15 APR	24 OCT 81	
1981-035D		12431	USSR	15 APR	08 JUN 81	
1981-035E		12432	USSR	15 APR	04 MAY 81	
1981-035F		12433	USSR	15 APR	30 APR 81	
1981-036A	COSMOS 1265	12402	USSR	16 APR	28 APR 81	
1981-036B		12403	USSR	16 APR	19 APR 81	
1981-036C		12425	USSR	16 APR	15 MAY 81	
1981-036D		12426	USSR	16 APR	01 MAY 81	
1981-036F		12428	USSR	16 APR	19 MAR 82	
1981-037B		12429	USSR	21 APR	20 MAY 81	
1981-037C		12434	USSR	21 APR	01 MAY 81	
1981-039A	COSMOS 1267	12419	USSR	25 APR	29 JUL 82	
1981-039B		12420	USSR	25 APR	28 APR 81	
1981-039C		12421	USSR	25 APR	28 APR 81	
1981-039D		12475	USSR	25 APR	26 MAY 81	
1981-040A	COSMOS 1268	12423	USSR	28 APR	12 MAY 81	
1981-040B		12424	USSR	28 APR	06 MAY 81	
1981-040C		12448	USSR	28 APR	20 MAY 81	
1981-040D		12449	USSR	28 APR	14 MAY 81	
1981-040E		12450	USSR	28 APR	15 MAY 81	
1981-040F		12451	USSR	28 APR	15 MAY 81	
1981-040G		12452	USSR	28 APR	14 MAY 81	
1981-042A	SOYUZ 40	12454	USSR	14 MAY	22 MAY 81	
1981-042B		12455	USSR	14 MAY	16 MAY 81	
1981-044B		12460	US	15 MAY	17 APR 82	
1981-044C		12944	US	15 MAY	08 DEC 81	
1981-045A	COSMOS 1270	12461	USSR	18 MAY	17 JUN 81	
1981-045B		12462	USSR	18 MAY	21 MAY 81	
1981-045C		12463	USSR	18 MAY	19 MAY 81	
1981-047A	COSMOS 1272	12466	USSR	21 MAY	04 JUN 81	
1981-047B		12467	USSR	21 MAY	31 MAY 81	
1981-047C		12468	USSR	21 MAY	23 MAY 81	
1981-047D		12498	USSR	21 MAY	24 NOV 81	
1981-047E		12499	USSR	21 MAY	09 JUN 81	
1981-047F		12505	USSR	21 MAY	08 JUL 81	
1981-047G		12506	USSR	21 MAY	06 JUL 81	
1981-047H		12507	USSR	21 MAY	05 AUG 81	
1981-028BW		18542	USSR	22 MAY	12 DEC 88	
1981-048A	COSMOS 1273	12469	USSR	22 MAY	04 JUN 81	
1981-048B		12470	USSR	22 MAY	25 MAY 81	
1981-048C		12500	USSR	22 MAY	05 JUN 81	
1981-048D		12501	USSR	22 MAY	11 JUN 81	
1981-048E		12502	USSR	22 MAY	06 JUN 81	
1981-048F		12503	USSR	22 MAY	06 JUN 81	
1981-049B		12473	US	22 MAY	08 SEP 81	
1981-049C		12494	US	22 MAY	15 APR 82	
1981-051A	ROHINI 2	12491	INDIA	31 MAY	08 JUN 81	
1981-051B		12492	INDIA	31 MAY	06 JUN 81	
1981-052A	COSMOS 1274	12495	USSR	03 JUN	03 JUL 81	
1981-052B		12496	USSR	03 JUN	07 JUN 81	
1981-052C		12579	USSR	03 JUN	08 JUL 81	
1981-053AH		12689	USSR	04 JUN	04 MAR 91	
1981-053BN		12721	USSR	04 JUN	28 NOV 89	
1981-053BZ		12732	USSR	04 JUN	03 AUG 82	
1981-053C8		12734	USSR	04 JUN	04 NOV 91	
1981-053CF		12738	USSR	04 JUN	12 SEP 89	
1981-053CX		12754	USSR	04 JUN	03 AUG 83	
1981-053DE		12761	USSR	04 JUN	16 MAR 90	
1981-053DF		12762	USSR	04 JUN	04 NOV 81	
1981-053DJ		12765	USSR	04 JUN	12 APR 88	
1981-053DT		12774	USSR	04 JUN	07 OCT 91	
1981-053EA		12914	USSR	04 JUN	29 OCT 81	
1981-053EF		12951	USSR	04 JUN	19 JUN 92	
1981-053EL		12965	USSR	04 JUN	01 FEB 91	
1981-053EN		13015	USSR	04 JUN	24 MAR 91	
1981-053FH		13461	USSR	04 JUN	21 MAR 90	
1981-053FK		13463	USSR	04 JUN	16 MAR 90	
1981-053GS		13760	USSR	04 JUN	09 NOV 91	
1981-053HV		14359	USSR	04 JUN	07 JUL 89	
1981-053HY		14362	USSR	04 JUN	13 MAR 92	
1981-053HZ		14363	USSR	04 JUN	20 AUG 89	
1981-053JB		14399	USSR	04 JUN	03 FEB 92	
1981-053JC		14400	USSR	04 JUN	14 MAR 90	
1981-053JJ		14439	USSR	04 JUN	14 JAN 93	
1981-053JS		14626	USSR	04 JUN	10 MAY 85	
1981-053JU		14628	USSR	04 JUN	25 JAN 88	
1981-053JW		14630	USSR	04 JUN	06 NOV 88	
1981-053LV		18499	USSR	04 JUN	27 APR 92	
1981-053LX		18516	USSR	04 JUN	26 FEB 89	
1981-053MD		18638	USSR	04 JUN	11 SEP 89	
1981-054B		12513	USSR	09 JUN	27 JUN 81	
1981-054C		12514	USSR	09 JUN	23 JUN 81	
1981-054D		12515	USSR	09 JUN	22 JUN 81	
1981-055A	COSMOS 1276	12517	USSR	16 JUN	29 JUN 81	
1981-055B		12518	USSR	16 JUN	19 JUN 81	
1981-055C		12567	USSR	16 JUN	06 JUL 81	
1981-055D		12568	USSR	16 JUN	30 JUN 81	
1981-055E		12569	USSR	16 JUN	01 JUL 81	
1981-055F		12570	USSR	16 JUN	01 JUL 81	
1981-056A	COSMOS 1277	12520	USSR	17 JUN	01 JUL 81	
1981-056B		12521	USSR	17 JUN	28 JUN 81	
1981-056C		12522	USSR	17 JUN	22 JUN 81	
1981-056D		12523	USSR	17 JUN	28 JUN 81	
1981-056E		12524	USSR	17 JUN	29 JUN 81	
1981-056F		12525	USSR	17 JUN	22 JUN 81	
1981-056G		12526	USSR	17 JUN	23 JUN 81	
1981-056H		12573	USSR	17 JUN	31 JUL 81	
1981-056J		12574	USSR	17 JUN	12 DEC 81	
1981-056K		12575	USSR	17 JUN	28 AUG 81	
1981-057D		12562	ESA	19 JUN	23 JUN 92	
1981-057E		14125	ESA	19 JUN	17 JAN 91	
1981-058B		12548	USSR	19 JUN	28 JUL 81	
1981-058C		12549	USSR	19 JUN	29 JUL 81	
1981-060A	MOLNIYA 1-50	12556	USSR	24 JUN	14 DEC 91	
1981-060B		12557	USSR	24 JUN	05 AUG 81	
1981-060C		12558	USSR	24 JUN	22 JUL 81	
1981-060D		12563	USSR	24 JUN	27 MAR 92	
1981-061B		12565	USSR	25 JUN	26 JUN 81	
1981-061C		12566	USSR	25 JUN	28 JUN 81	
1981-061D		12603	USSR	26 JUN	15 APR 82	
1981-061E		12604	USSR	26 JUN	12 SEP 83	
1981-062A	COSMOS 1279	12571	USSR	01 JUL	15 JUL 81	
1981-062B		12572	USSR	01 JUL	14 JUL 81	
1981-062C		12576	USSR	01 JUL	02 JUL 81	
1981-062D		12590	USSR	01 JUL	20 DEC 81	
1981-062E		12591	USSR	01 JUL	07 SEP 81	
1981-062F		12592	USSR	01 JUL	16 JUL 81	
1981-062G		12593	USSR	01 JUL	22 AUG 81	
1981-063A	COSMOS 1280	12577	USSR	02 JUL	15 JUL 81	
1981-063B		12578	USSR	02 JUL	08 JUL 81	
1981-063C		12594	USSR	02 JUL	02 AUG 81	
1981-063D		12595	USSR	02 JUL	10 AUG 81	
1981-063E		12596	USSR	02 JUL	22 JUL 81	
1981-063F		12597	USSR	02 JUL	22 JUL 81	
1981-064A	COSMOS 1281	12583	USSR	07 JUL	21 JUL 81	
1981-064B		12584	USSR	07 JUL	17 JUL 81	
1981-064C		12600	USSR	07 JUL	20 AUG 81	
1981-064D		12601	USSR	07 JUL	06 SEP 81	
1981-064E		12602	USSR	07 JUL	09 DEC 81	
1981-065C	ISKRA	12587	USSR	10 JUL	07 OCT 81	
1981-065D		19236	USSR	10 JUL	16 APR 90	
1981-066A	COSMOS 1282	12588	USSR	15 JUL	14 AUG 81	
1981-066B		12589	USSR	15 JUL	19 JUL 81	
1981-067A	COSMOS 1283	12598	USSR	17 JUL	31 JUL 81	
1981-067B		12599	USSR	17 JUL	19 JUL 81	
1981-067C		12621	USSR	17 JUL	05 AUG 81	
1981-067D		12622	USSR	17 JUL	08 OCT 81	
1981-067E		12623	USSR	17 JUL	21 AUG 81	
1981-068A	COSMOS 1284	12614	USSR	29 JUL	12 AUG 81	
1981-068B		12617	USSR	-29 JUL	31 JUL 81	
1981-068C		12683	USSR	29 JUL	16 OCT 81	
1981-068D		12684	USSR	29 JUL	29 AUG 81	
1981-068E		12685	USSR	29 JUL	16 AUG 81	
1981-068F		12686	USSR	29 JUL	06 SEP 81	
1981-069B		12619	USSR	30 JUL	01 AUG 81	
1981-069C		12620	USSR	30 JUL	31 JUL 81	
1981-069D		12633	USSR	30 JUL	20 APR 82	
1981-069E		12634	USSR	30 JUL	15 APR 82	
1981-070B	DE 2	12625	US	03 AUG	19 FEB 83	
1981-070C		12626	US	03 AUG	23 NOV 88	
1981-070D		12628	US	03 AUG	17 MAR 82	
1981-070F		12798	US	03 AUG	03 AUG 81	
1981-070G		12945	US	03 AUG	25 JAN 82	
1981-070H		12946	US	03 AUG	16 FEB 82	
1981-071B		12629	USSR	04 AUG	26 AUG 81	
1981-071C		12630	USSR	04 AUG	01 SEP 81	
1981-072A	COSMOS 1286	12631	USSR	04 AUG	16 OCT 82	
1981-072B		12632	USSR	04 AUG	04 AUG 81	
1981-072C		13601	USSR	04 AUG	22 OCT 82	
1981-073B		12778	US	06 AUG	16 APR 83	
1981-076B		12678	JAPAN	10 AUG	24 AUG 81	
1981-076C		12810	JAPAN	10 AUG	23 NOV 90	
1981-078A	COSMOS 1296	12687	USSR	13 AUG	13 SEP 81	
1981-078B		12688	USSR	13 AUG	17 AUG 81	
1981-079A	COSMOS 1297	12716	USSR	18 AUG	30 AUG 81	
1981-079B		12717	USSR	18 AUG	25 AUG 81	
1981-079C		12718	USSR	18 AUG	19 AUG 81	
1981-079D		12794	USSR	18 AUG	05 SEP 81	
1981-079E		12795	USSR	18 AUG	31 AUG 81	
1981-079F		12796	USSR	18 AUG	02 SEP 81	
1981-079G		12797	USSR	18 AUG	31 AUG 81	
1981-080A	COSMOS 1298	12776	USSR	21 AUG	02 OCT 81	
1981-080B		12777	USSR	21 AUG	24 AUG 81	
1981-081B		12784	USSR	24 AUG	17 NOV 81	
1981-081C		12807	USSR	24 AUG	07 SEP 81	
1981-081D		12808	USSR	24 AUG	02 FEB 89	
1981-081E		12809	USSR	24 AUG	27 SEP 81	
1981-083A	COSMOS 1301	12788	USSR	27 AUG	10 SEP 81	
1981-083B		12789	USSR	27 AUG	31 AUG 81	
1981-083C		12790	USSR	27 AUG	02 SEP 81	
1981-083D		12811	USSR	27 AUG	24 SEP 81	
1981-083E		12812	USSR	27 AUG	17 SEP 81	
1981-083F		12813	USSR	27 AUG	09 NOV 81	
1981-085A		12799	US	03 SEP	23 NOV 84	
1981-085B		12800	US	03 SEP	12 OCT 81	
1981-086A	COSMOS 1303	12801	USSR	04 SEP	18 SEP 81	
1981-086B		12802	USSR	04 SEP	12 SEP 81	
1981-086C		12805	USSR	04 SEP	05 SEP 81	
1981-086D		12806	USSR	04 SEP	05 SEP 81	
1981-086E		12839	USSR	04 SEP	09 OCT 81	
1981-086F		12840	USSR	04 SEP	25 OCT 81	
1981-086G		12841	USSR	04 SEP	06 FEB 82	
1981-088B		12819	USSR	11 SEP	05 OCT 81	
1981-088C		12820	USSR	11 SEP	03 OCT 81	
1981-088D		12821	USSR	11 SEP	13 JAN 82	

International Designation	Satellite Name	Number	Source	Launch	Decay	Note
1981-088E		12822	USSR	11 SEP	04 DEC 81	
1981-089A	COSMOS 1306	12828	USSR	14 SEP	16 JUL 82	
1981-089B		12829	USSR	14 SEP	15 SEP 81	
1981-089C		13369	USSR	14 SEP	12 DEC 82	
1981-089D		13387	USSR	14 SEP	11 OCT 82	
1981-089E		13393	USSR	14 SEP	01 MAY 83	
1981-089F		13404	USSR	14 SEP	07 MAR 89	
1981-089G		13408	USSR	14 SEP	19 AUG 82	
1981-089H		13588	USSR	14 SEP	06 DEC 82	
1981-089J		14837	USSR	14 SEP	08 APR 89	
1981-090A	COSMOS 1307	12830	USSR	15 SEP	28 SEP 81	
1981-090B		12831	USSR	15 SEP	23 SEP 81	
1981-090C		12832	USSR	15 SEP	16 SEP 81	
1981-090D		12873	USSR	15 SEP	04 JAN 82	
1981-090E		12875	USSR	15 SEP	30 SEP 81	
1981-090F		12876	USSR	15 SEP	04 NOV 81	
1981-090G		12877	USSR	15 SEP	04 FEB 82	
1981-090H		12878	USSR	15 SEP	23 OCT 81	
1981-092A	COSMOS 1309	12837	USSR	18 SEP	01 OCT 81	
1981-092B		12838	USSR	18 SEP	21 SEP 81	
1981-093A	PRC 9A	12842	PRC	19 SEP	26 SEP 81	
1981-093B	PRC 9B	12843	PRC	19 SEP	06 OCT 82	
1981-093C		12844	PRC	19 SEP	22 NOV 81	
1981-093D	PRC 9C	12845	PRC	19 SEP	17 AUG 82	
1981-093E		12846	PRC	19 SEP	10 APR 83	
1981-093F		12847	PRC	19 SEP	25 NOV 81	
1981-093G		12884	PRC	19 SEP	22 MAY 82	
1981-093H		12885	PRC	19 SEP	20 MAY 82	
1981-093J		12890	PRC	19 SEP	20 OCT 81	
1981-093K		12958	PRC	19 SEP	16 NOV 81	
1981-095A	COSMOS 1310	12852	USSR	23 SEP	03 APR 89	
1981-095B		12853	USSR	23 SEP	17 SEP 83	
1981-096B		12874	US	24 SEP	15 APR 82	
1981-097A	COSMOS 1311	12871	USSR	28 SEP	28 AUG 83	
1981-097AA		13831	USSR	28 SEP	28 FEB 83	
1981-097AB		13832	USSR	28 SEP	28 FEB 83	
1981-097B		12872	USSR	28 SEP	16 FEB 83	
1981-097C		13054	USSR	28 SEP	27 FEB 82	BV
1981-097D		13055	USSR	28 SEP	26 FEB 82	
1981-097E		13667	USSR	28 SEP	05 DEC 82	
1981-097F		13668	USSR	28 SEP	05 DEC 82	
1981-097G		13694	USSR	28 SEP	17 DEC 82	
1981-097H		13695	USSR	28 SEP	16 DEC 82	
1981-097J		13723	USSR	28 SEP	22 DEC 82	
1981-097K		13724	USSR	28 SEP	23 DEC 82	
1981-097L		13729	USSR	28 SEP	24 DEC 82	
1981-097M		13731	USSR	28 SEP	18 DEC 82	
1981-097N		13732	USSR	28 SEP	17 DEC 82	
1981-097P		13733	USSR	28 SEP	17 DEC 82	
1981-097Q		13734	USSR	28 SEP	16 DEC 82	
1981-097R		13735	USSR	28 SEP	31 DEC 82	
1981-097S		13743	USSR	28 SEP	31 DEC 82	
1981-097T		13744	USSR	28 SEP	31 DEC 82	
1981-097U		13820	USSR	28 SEP	28 FEB 83	
1981-097V		13821	USSR	28 SEP	28 FEB 83	
1981-097W		13822	USSR	28 SEP	28 FEB 83	
1981-097X		13823	USSR	28 SEP	28 FEB 83	
1981-097Y		13824	USSR	28 SEP	28 FEB 83	
1981-097Z		13828	USSR	28 SEP	28 FEB 83	
1981-099A	COSMOS 1313	12881	USSR	01 OCT	15 OCT 81	
1981-099B		12882	USSR	01 OCT	05 OCT 81	
1981-099C		12883	USSR	01 OCT	02 OCT 81	
1981-099D		12910	USSR	01 OCT	16 OCT 81	
1981-099E		12911	USSR	01 OCT	16 OCT 81	
1981-099F		12912	USSR	01 OCT	26 OCT 81	
1981-100A	SME	12887	US	06 OCT	05 MAR 91	
1981-100B	UOSAT	12888	UK	06 OCT	13 OCT 89	
1981-100D		12891	US	06 OCT	14 NOV 81	
1981-100E		13405	US	06 OCT	19 OCT 82	
1981-101A	COSMOS 1314	12895	USSR	09 OCT	22 OCT 81	
1981-101B		12896	USSR	09 OCT	11 OCT 81	
1981-101C		12900	USSR	09 OCT	09 NOV 81	
1981-101D		12921	USSR	09 OCT	25 OCT 81	
1981-101E		12922	USSR	09 OCT	25 OCT 81	
1981-101F		12923	USSR	09 OCT	25 OCT 81	
1981-102B		12898	USSR	09 OCT	11 OCT 81	
1981-102C		12899	USSR	09 OCT	10 OCT 81	
1981-102D		12901	USSR	09 OCT	01 JUL 82	
1981-102E		12902	USSR	09 OCT	13 MAY 82	
1981-104A	COSMOS 1316	12905	USSR	15 OCT	29 OCT 81	
1981-104B		12913	USSR	15 OCT	23 OCT 81	
1981-104C		12924	USSR	15 OCT	08 NOV 81	
1981-104D		12925	USSR	15 OCT	31 OCT 81	
1981-104E		12926	USSR	15 OCT	31 OCT 81	
1981-105B		12916	USSR	17 OCT	04 NOV 81	
1981-105C		12917	USSR	17 OCT	06 NOV 81	
1981-105D		12918	USSR	17 OCT	08 NOV 81	
1981-106B		12928	USSR	30 OCT	31 OCT 81	
1981-106C		12929	USSR	30 OCT	31 OCT 81	
1981-106D		15599	USSR	30 OCT	28 FEB 82	
1981-107B		12931	US	31 OCT	02 NOV 81	
1981-108B		12934	USSR	31 OCT	23 NOV 81	
1981-108C		12935	USSR	31 OCT	14 NOV 81	
1981-109A	COSMOS 1318	12936	USSR	03 NOV	04 DEC 81	
1981-109B		12937	USSR	03 NOV	06 NOV 81	
1981-110B		12939	USSR	04 NOV	05 NOV 81	
1981-110C		12941	USSR	04 NOV	05 NOV 81	
1981-110D		15600	USSR	04 NOV	05 MAR 82	
1981-111A	STS 2	12953	US	12 NOV	14 NOV 81	BT
1981-112A	COSMOS 1319	12954	USSR	13 NOV	27 NOV 81	
1981-112B		12955	USSR	13 NOV	22 NOV 81	
1981-112C		12956	USSR	13 NOV	16 NOV 81	
1981-112D		12957	USSR	13 NOV	13 NOV 81	
1981-112E		12970	USSR	13 NOV	13 DEC 81	
1981-112F		12971	USSR	13 NOV	15 APR 82	
1981-112G		12972	USSR	13 NOV	02 JAN 82	
1981-112H		12973	USSR	13 NOV	27 DEC 81	
1981-112J		12974	USSR	13 NOV	08 DEC 81	
1981-113B		12960	USSR	17 NOV	27 NOV 81	
1981-113C		12961	USSR	17 NOV	29 NOV 81	
1981-114B		13098	US	20 NOV	26 FEB 89	
1981-115A	BHASKARA 2	12968	INDIA	20 NOV	30 NOV 91	
1981-115B		12969	USSR	20 NOV	08 OCT 89	
1981-118A	COSMOS 1329	12989	USSR	04 DEC	18 DEC 81	
1981-118B		12990	USSR	04 DEC	08 DEC 81	
1981-118C		12991	USSR	04 DEC	10 DEC 81	
1981-118D		13004	USSR	04 DEC	05 JAN 82	
1981-118E		13005	USSR	04 DEC	20 DEC 81	
1981-118F		13006	USSR	04 DEC	22 DEC 81	
1981-121A	COSMOS 1330	13008	USSR	19 DEC	19 JAN 82	
1981-121B		13009	USSR	19 DEC	23 DEC 81	
1981-122C		13025	ESA	20 DEC	21 NOV 88	
1981-123B		13013	USSR	23 DEC	16 JAN 82	
1981-123C		13014	USSR	23 DEC	10 JAN 82	

1982 LAUNCHES

International Designation	Satellite Name	Number	Source	Launch	Decay	Note
1982-002A	COSMOS 1332	13031	USSR	12 JAN	25 JAN 82	
1982-002B		13032	USSR	12 JAN	16 JAN 82	
1982-004B		13088	US	16 JAN	02 NOV 88	
1982-005A	COSMOS 1334	13036	USSR	20 JAN	03 FEB 82	
1982-005B		13037	USSR	20 JAN	24 JAN 82	
1982-005C		13038	USSR	20 JAN	22 JAN 82	
1982-005D		13039	USSR	20 JAN	20 JAN 82	
1982-005E		13048	USSR	20 JAN	04 FEB 82	
1982-005F		13049	USSR	20 JAN	06 FEB 82	
1982-005G		13050	USSR	20 JAN	17 FEB 82	
1982-005H		13051	USSR	20 JAN	06 FEB 82	
1982-005J		13052	USSR	20 JAN	05 FEB 82	
1982-006A		13040	US	21 JAN	23 MAY 82	
1982-006B		13041	US	21 JAN	22 JAN 82	
1982-007A	COSMOS 1335	13042	USSR	29 JAN	05 APR 87	
1982-007AA		14068	USSR	29 JAN	09 JUN 83	
1982-007B		13043	USSR	29 JAN	24 NOV 85	
1982-007C		13044	USSR	29 JAN	30 JAN 82	BW
1982-007D		13140	USSR	29 JAN	15 JUN 82	
1982-007E		13141	USSR	29 JAN	11 JUN 82	
1982-007F		13246	USSR	29 JAN	17 AUG 82	
1982-007G		13247	USSR	29 JAN	14 AUG 82	
1982-007H		13248	USSR	29 JAN	11 AUG 82	
1982-007J		13249	USSR	29 JAN	01 AUG 82	
1982-007K		13250	USSR	29 JAN	29 JUL 82	
1982-007L		13251	USSR	29 JAN	26 JUL 82	
1982-007M		13252	USSR	29 JAN	24 JUL 82	
1982-007N		13273	USSR	29 JAN	07 JUL 82	
1982-007P		13419	USSR	29 JAN	05 OCT 82	
1982-007Q		13420	USSR	29 JAN	05 OCT 82	
1982-007R		13421	USSR	29 JAN	29 SEP 82	
1982-007S		13422	USSR	29 JAN	28 SEP 82	
1982-007T		13429	USSR	29 JAN	04 OCT 82	
1982-007U		13430	USSR	29 JAN	04 OCT 82	
1982-007V		14052	USSR	29 JAN	11 JUN 83	
1982-007W		14053	USSR	29 JAN	10 JUN 83	
1982-007X		14054	USSR	29 JAN	11 JUN 83	
1982-007Y		14055	USSR	29 JAN	09 JUN 83	
1982-007Z		14056	USSR	29 JAN	09 JUN 83	
1982-008B		13046	USSR	29 JAN	02 FEB 82	
1982-008C		13047	USSR	29 JAN	31 JAN 82	
1982-008A	COSMOS 1336	13045	USSR	30 JAN	26 FEB 82	
1982-009B		13057	USSR	05 FEB	07 FEB 82	
1982-009C		13058	USSR	05 FEB	06 FEB 82	
1982-009D		13059	USSR	05 FEB	07 SEP 91	
1982-009E		13060	USSR	05 FEB	26 NOV 84	
1982-010A	COSMOS 1337	13061	USSR	11 FEB	25 JUL 82	
1982-010B		13062	USSR	11 FEB	11 FEB 82	
1982-011A	COSMOS 1338	13063	USSR	16 FEB	02 MAR 82	
1982-011B		13064	USSR	16 FEB	24 FEB 82	
1982-011C		13076	USSR	16 FEB	21 MAR 82	
1982-011D		13077	USSR	16 FEB	27 SEP 82	
1S82-011E		13078	USSR	16 FEB	06 APR 82	
1982-011F		13079	USSR	16 FEB	19 MAR 82	
1982-014B		13136	US	26 FEB	31 AUG 89	
1982-015B		13071	USSR	26 FEB	08 MAR 82	
1982-015C		13072	USSR	26 FEB	13 MAR 82	
1982-016B		13081	USSR	03 MAR	27 MAR 82	
1982-016C		13082	USSR	03 MAR	16 MAR 82	
1982-017B		13087	US	05 MAR	06 MAY 83	
1982-018A	COSMOS 1342	13084	USSR	05 MAR	19 MAR 82	
1982-018B		13085	USSR	05 MAR	09 MAR 82	
1982-018C		13100	USSR	05 MAR	21 MAR 82	
1982-018D		13101	USSR	05 MAR	22 MAR 82	
1982-018E		13102	USSR	05 MAR	01 APR 82	
1982-020B		13093	USSR	15 MAR	16 MAR 82	
1982-020C		13094	USSR	15 MAR	16 MAR 82	
1982-020D		13095	USSR	15 MAR	16 MAR 82	
1982-020E		13099	USSR	15 MAR	14 SEP 83	

International Designation	Satellite Name	Number	Source	Launch	Decay	Note
1982-021A	COSMOS 1343	13096	USSR	17 MAR	31 MAR 82	
1982-021B		13097	USSR	17 MAR	21 MAR 82	
1982-021C		13115	USSR	17 MAR	05 APR 82	
1982-021D		13116	USSR	17 MAR	01 APR 82	
1982-021E		13117	USSR	17 MAR	01 APR 82	
1982-022A	STS 3	13106	US	22 MAR	30 MAR 82	BT
1982-023A	MOLNIYA 3-18	13107	USSR	24 MAR	23 JUN 92	
1982-023B		13108	USSR	24 MAR	12 APR 82	
1982-023C		13109	USSR	24 MAR	19 APR 82	
1982-023D		13112	USSR	24 MAR	17 SEP 92	
1982-026A	COSMOS 1345	13118	USSR	31 MAR	27 SEP 89	
1982-026B		13119	USSR	31 MAR	10 AUG 89	
1982-028A	COSMOS 1347	13122	USSR	02 APR	21 MAY 82	
1982-028B		13123	USSR	02 APR	05 APR 82	
1982-029B		13125	USSR	07 APR	05 MAY 82	
1982-029C		13126	USSR	07 APR	27 APR 82	
1982-031B		13137	US	10 APR	18 APR 84	
1982-032A	COSMOS 1350	13134	USSR	15 APR	16 MAY 82	
1982-032B		13135	USSR	15 APR	18 APR 82	
1982-033A	SALYUT 7	13138	USSR	19 APR	07 FEB 91	BX
1982-033AA		13541	USSR	19 APR	01 OCT 82	
1982-033AB		13560	USSR	19 APR	04 OCT 82	
1982-033AC		13596	USSR	19 APR	28 OCT 82	
1982-033AE		13664	USSR	19 APR	28 DEC 82	
1982-033AF		13700	USSR	19 APR	26 MAR 83	
1982-033AG		14386	USSR	19 APR	16 NOV 83	
1982-033AH		14387	USSR	19 APR	19 FEB 84	
1982-033AJ		14388	USSR	19 APR	01 NOV 83	
1982-033AK		14392	USSR	19 APR	15 MAR 84	
1982-033AL		14393	USSR	19 APR	29 OCT 83	
1982-033AM		14485	USSR	19 APR	17 NOV 83	
1982-033AN		14486	USSR	19 APR	18 NOV 83	
1982-033AP		14487	USSR	19 APR	17 NOV 83	
1982-033AQ		14507	USSR	19 APR	24 DEC 83	
1982-033AR		14508	USSR	19 APR	04 DEC 83	
1982-033AS		14857	USSR	19 APR	29 MAR 84	
1982-033AT		14858	USSR	19 APR	27 MAR 84	
1982-033AU		14859	USSR	19 APR	04 APR 84	
1982-033AV		14860	USSR	19 APR	03 APR 84	
1982-033AW		14861	USSR	19 APR	11 APR 84	
1982-033AX		14862	USSR	19 APR	13 APR 84	
1982-033AY		14863	USSR	19 APR	03 APR 84	
1982-033AZ		14864	USSR	19 APR	14 APR 84	
1982-033B		13139	USSR	19 APR	24 APR 82	BX
1982-033BA		14870	USSR	19 APR	27 APR 84	
1982-033BB		14871	USSR	19 APR	15 APR 84	
1982-033BC		14946	USSR	19 APR	17 MAY 84	
1982-033BD		14949	USSR	19 APR	16 JUN 84	
1982-033BE		14957	USSR	19 APR	13 MAY 84	
1982-033BF		14958	USSR	19 APR	14 MAY 84	
1982-033BG		14960	USSR	19 APR	21 JUN 84	
1982-033BH		14975	USSR	19 APR	24 MAY 84	
1982-033BJ		14976	USSR	19 APR	26 MAY 84	
1982-033BK		14995	USSR	19 APR	01 JUN 84	
1982-033BL		15105	USSR	19 APR	07 AUG 84	
1982-033BM		15106	USSR	19 APR	18 AUG 84	
1982-033BN		15107	USSR	19 APR	14 AUG 84	
1982-033BP		15109	USSR	19 APR	16 AUG 84	
1982-033BQ		15111	USSR	19 APR	27 JUL 84	
1982-033BR		15112	USSR	19 APR	30 JUL 84	
1982-033BS		15113	USSR	19 APR	31 AUG 84	
1982-033BT		15114	USSR	19 APR	26 AUG 84	
1982-033BU		15127	USSR	19 APR	13 OCT 84	
1982-033BV		15128	USSR	19 APR	09 AUG 84	
1982-033BW		15129	USSR	19 APR	09 OCT 84	
1982-033BX		15130	USSR	19 APR	22 SEP 84	
1982-033BY		15133	USSR	19 APR	18 AUG 84	
1982-033BZ		15134	USSR	19 APR	04 AUG 84	
1982-033CA		15135	USSR	19 APR	27 SEP 84	
1982-033CB		15137	USSR	19 APR	08 OCT 84	
1982-033CC		15138	USSR	19 APR	18 AUG 84	
1982-033CD		15174	USSR	19 APR	14 SEP 84	
1982-033CE		15175	USSR	19 APR	31 AUG 84	
1982-033CF		15176	USSR	19 APR	06 DEC 84	
1982-033CG		15177	USSR	19 APR	22 SEP 84	
1982-033CH		15189	USSR	19 APR	13 OCT 84	
1982-033CJ		15190	USSR	19 APR	18 OCT 84	
1982-033CK		15192	USSR	19 APR	21 MAY 86	
1982-033CL		15195	USSR	19 APR	12 NOV 84	
1982-033CM		15196	USSR	19 APR	04 OCT 84	
1982-033CN		15207	USSR	19 APR	03 OCT 84	
1982-033CP		15208	USSR	19 APR	05 DEC 84	
1982-033CQ		15209	USSR	19 APR	25 OCT 84	
1982-033CR		15210	USSR	19 APR	15 SEP 84	
1982-033CS		15211	USSR	19 APR	06 OCT 84	
1982-033CT		15212	USSR	19 APR	21 SEP 84	
1982-033CU		15213	USSR	19 APR	08 OCT 84	
1982-033CV		15217	USSR	19 APR	08 SEP 84	
1982-033CW		15218	USSR	19 APR	19 SEP 84	
1982-033CX		15247	USSR	19 APR	13 MAY 86	
1982-033CY		15248	USSR	19 APR	12 DEC 84	
1982-033CZ		15249	USSR	19 APR	26 DEC 84	
1982-033D		13204	USSR	19 APR	20 JUN 82	
1982-033DA		15250	USSR	19 APR	12 DEC 84	
1982-033DB		15251	USSR	19 APR	16 OCT 84	
1982-033DC		15252	USSR	19 APR	08 DEC 84	
1982-033DD		15253	USSR	19 APR	25 SEP 84	
1982-033DE		15254	USSR	19 APR	10 NOV 84	
1982-033DF		15255	USSR	19 APR	01 NOV 84	
1982-033DG		15256	USSR	19 APR	08 DEC 84	
1982-033DH		15273	USSR	19 APR	03 OCT 84	
1982-033DJ		15274	USSR	19 APR	24 DEC 84	
1982-033DK		15275	USSR	19 APR	11 DEC 84	
1982-033DL		15276	USSR	19 APR	29 NOV 84	
1982-033DM		15281	USSR	19 APR	11 JAN 85	
1982-033DN		15282	USSR	19 APR	03 JAN 85	
1982-033DP		15283	USSR	19 APR	14 FEB 85	
1982-033DQ		15284	USSR	19 APR	04 DEC 84	
1982-033DR		15297	USSR	19 APR	04 OCT 84	
1982-033DS		15303	USSR	19 APR	20 OCT 84	
1982-033DT		15304	USSR	19 APR	26 SEP 84	
1982-033DU		15305	USSR	19 APR	09 DEC 85	
1982-033DV		15306	USSR	19 APR	20 DEC 84	
1982-033DW		15307	USSR	19 APR	11 JAN 85	
1982-033DX		15311	USSR	19 APR	13 OCT 84	
1982-033DY		15312	USSR	19 APR	11 JAN 85	
1982-033DZ		15313	USSR	19 APR	11 JAN 86	
1982-033E		13270	USSR	19 APR	24 JUN 82	
1982-033EA		15314	USSR	19 APR	10 JAN 85	
1982-033EB		15315	USSR	19 APR	25 NOV 84	
1982-033EC		15316	USSR	19 APR	11 FEB 85	
1982-033ED		15317	USSR	19 APR	08 DEC 84	
1982-033EE		15365	USSR	19 APR	05 DEC 84	
1982-033EF		15366	USSR	19 APR	12 JAN 85	
1982-033EG		15367	USSR	19 APR	03 MAR 85	
1982-033EH		15368	USSR	19 APR	11 FEB 85	
1982-033EJ		15401	USSR	19 APR	04 JAN 85	
1982-033EK		15420	USSR	19 APR	24 JAN 85	
1982-033EL		15524	USSR	19 APR	17 MAR 85	
1982-033EM		15525	USSR	19 APR	10 APR 85	
1982-033EN		15526	USSR	19 APR	08 DEC 85	
1982-033EP		15527	USSR	19 APR	28 MAR 85	
1982-033EQ		15532	USSR	19 APR	15 FEB 85	
1982-033ER		15533	USSR	19 APR	01 MAR 85	
1982-033ES		15933	USSR	19 APR	19 SEP 85	
1982-033ET		15934	USSR	19 APR	24 AUG 85	
1982-033EU		15937	USSR	19 APR	12 AUG 85	
1982-033EV		15939	USSR	19 APR	25 SEP 85	
1982-033EW		15970	USSR	19 APR	15 JAN 86	
1982-033EX		16046	USSR	19 APR	24 NOV 85	
1982-033EY		16047	USSR	19 APR	02 FEB 86	
1982-033EZ		16048	USSR	19 APR	01 JAN 86	
1982-033F		13274	USSR	19 APR	27 JUN 82	
1982-033FA		16049	USSR	19 APR	24 DEC 85	
1982-033FB		16050	USSR	19 APR	25 OCT 85	
1982-033FC		16053	USSR	19 APR	31 MAY 86	
1982-033FD		16062	USSR	19 APR	03 JAN 86	
1982-033FE		16063	USSR	19 APR	30 SEP 85	
1982-033FF		16091	USSR	19 APR	20 JAN 86	
1982-033FG		16092	USSR	19 APR	22 DEC 85	
1982-033FH		16093	USSR	19 APR	26 FEB 86	
1982-033FJ		16094	USSR	19 APR	11 FEB 86	
1982-033FK		16098	USSR	19 APR	03 OCT 85	
1982-033FL		16099	USSR	19 APR	30 SEP 85	
1982-033FM		16100	USSR	19 APR	18 DEC 85	
1982-033FN		16122	USSR	19 APR	03 OCT 86	
1982-033FP		16163	USSR	19 APR	09 FEB 86	
1982-033FQ		16164	USSR	19 APR	17 FEB 86	
1982-033FR		16165	USSR	19 APR	08 JAN 86	
1982-033FS		16166	USSR	19 APR	15 JAN 87	
1982-033FT		16167	USSR	19 APR	19 FEB 86	
1982-033FU		16168	USSR	19 APR	24 JAN 86	
1982-033FV		16171	USSR	19 APR	19 JAN 86	
1982-033FW		16172	USSR	19 APR	27 FEB 86	
1982-033FX		16173	USSR	19 APR	05 FEB 86	
1982-033FY		16174	USSR	19 APR	16 FEB 86	
1982-033FZ		16175	USSR	19 APR	07 FEB 86	
1982-033G		13275	USSR	19 APR	27 JUN 82	
1982-033GA		16176	USSR	19 APR	31 DEC 85	
1982-033GB		16217	USSR	19 APR	08 AUG 86	
1982-033GC		16218	USSR	19 APR	15 APR 86	
1982-033GD		16238	USSR	19 APR	27 MAY 86	
1982-033GE		16239	USSR	19 APR	03 DEC 85	
1982-033GF		16240	USSR	19 APR	16 NOV 86	
1982-033GG		16241	USSR	19 APR	22 FEB 86	
1982-033GH		16242	USSR	19 APR	09 APR 86	
1982-033GJ		16244	USSR	19 APR	09 MAR 86	
1982-033GK		16245	USSR	19 APR	05 DEC 85	
1982-033GL		16246	USSR	19 APR	15 MAR 86	
1982-033GM		16249	USSR	19 APR	09 FEB 86	
1982-033GN		16253	USSR	19 APR	03 DEC 86	
1982-033GP		16258	USSR	19 APR	18 JAN 86	
1982-033GQ		16259	USSR	19 APR	17 FEB 86	
1982-033GR		16260	USSR	19 APR	07 JUN 86	
1982-033GS		16441	USSR	19 APR	10 MAR 86	
1982-033GT		16721	USSR	19 APR	06 AUG 86	
1982-033GU		16749	USSR	19 APR	08 JUL 86	
1982-033GV		16750	USSR	19 APR	19 JUN 86	
1982-033GW		16751	USSR	19 APR	22 OCT 86	
1982-033GX		16752	USSR	19 APR	21 JUL 86	
1982-033GY		16810	USSR	19 APR	05 SEP 86	
1982-033GZ		16811	USSR	19 APR	31 AUG 86	
1982-033H		13287	USSR	19 APR	02 JUL 82	
1982-033HA		16812	USSR	19 APR	09 SEP 86	
1982-033HB		16813	USSR	19 APR	27 JAN 87	
1982-033HC		16814	USSR	19 APR	20 FEB 87	

International Designation	Satellite Name	Number	Source	Launch	Decay	Note
1982-033HD		16815	USSR	19 APR	10 OCT 86	
1982-033HE		16821	USSR	19 APR	27 AUG 86	
1982-033HF		16822	USSR	19 APR	08 OCT 86	
1982-033HG		16852	USSR	19 APR	21 NOV 86	
1982-033HH		16853	USSR	19 APR	06 SEP 86	
1982-033HJ		17109	USSR	19 APR	29 MAR 87	
1982-033J		13291	USSR	19 APR	09 JUL 82	
1982-033K		13299	USSR	19 APR	12 JUL 82	
1982-033L		13332	USSR	19 APR	11 AUG 82	
1982-033M		13398	USSR	19 APR	21 AUG 82	
1982-033N		13409	USSR	19 APR	25 AUG 82	
1982-033P		13410	USSR	19 APR	13 AUG 82	
1982-033Q		13417	USSR	19 APR	01 SEP 82	
1982-033R		13418	USSR	19 APR	24 AUG 82	
1982-033S		13423	USSR	19 APR	31 AUG 82	
1982-033T		13424	USSR	19 APR	02 SEP 82	
1982-033U		13435	USSR	19 APR	29 AUG 82	
1982-033V		13436	USSR	19 APR	07 OCT 82	
1982-033W		13439	USSR	19 APR	06 SEP 82	
1982-033X		13440	USSR	19 APR	27 SEP 82	
1982-033Y		13522	USSR	19 APR	03 OCT 82	
1982-033Z		13523	USSR	19 APR	02 OCT 82	
1982-034A	COSMOS 1351	13142	USSR	21 APR	14 MAR 83	
1982-034AA		13866	USSR	21 APR	26 FEB 83	
1982-034AB		13867	USSR	21 APR	25 FEB 83	
1982-034B		13143	USSR	21 APR	22 FEB 83	
1982-034C		13833	USSR	21 APR	27 FEB 83	
1982-034D		13836	USSR	21 APR	25 FEB 83	
1982-034E		13837	USSR	21 APR	26 FEB 83	
1982-034F		13838	USSR	21 APR	25 FEB 83	
1982-034G		13839	USSR	21 APR	25 FEB 83	
1982-034H		13840	USSR	21 APR	03 MAR 83	
1982-034J		13841	USSR	21 APR	27 FEB 83	
1982-034K		13842	USSR	21 APR	02 MAR 83	
1982-034L		13852	USSR	21 APR	26 FEB 83	
1982-034M		13853	USSR	21 APR	26 FEB 83	
1982-034N		13854	USSR	21 APR	26 FEB 83	
1982-034P		13855	USSR	21 APR	26 FEB 83	
1982-034Q		13856	USSR	21 APR	26 FEB 83	
1982-034R		13857	USSR	21 APR	05 MAR 83	
1982-034S		13858	USSR	21 APR	27 FEB 83	
1982-034T		13859	USSR	21 APR	26 FEB 83	
1982-034U		13860	USSR	21 APR	26 FEB 83	
1982-034V		13861	USSR	21 APR	26 FEB 83	
1982-034W		13862	USSR	21 APR	26 FEB 83	
1982-034X		13863	USSR	21 APR	26 FEB 83	
1982-034Y		13864	USSR	21 APR	26 FEB 83	
1982-034Z		13865	USSR	21 APR	03 MAR 83	
1982-035A	COSMOS 1352	13144	USSR	21 APR	05 MAY 82	
1982-035B		13145	USSR	21 APR	29 APR 82	
1982-035C		13155	USSR	21 APR	14 JUN 82	
1982-035D		13156	USSR	21 APR	30 NOV 82	
1982-035E		13157	USSR	21 APR	30 MAY 82	
1982-035F		13158	USSR	21 APR	02 JUL 82	
1982-036A	COSMOS 1353	13146	USSR	23 APR	06 MAY 82	
1982-036B		13147	USSR	23 APR	26 APR 82	
1982-036C		13159	USSR	23 APR	10 MAY 82	
1982-038A	COSMOS 1355	13150	USSR	29 APR	07 MAR 84	
1982-038AA		14683	USSR	29 APR	14 FEB 84	
1982-038AB		14684	USSR	29 APR	11 FEB 84	
1982-038AC		14685	USSR	29 APR	05 FEB 84	
1982-038AD		14686	USSR	29 APR	14 FEB 84	
1982-038AE		14687	USSR	29 APR	05 FEB 84	
1982-038AF		14697	USSR	29 APR	03 FEB 84	
1982-038B		13151	USSR	29 APR	29 APR 82	
1982-038C		14265	USSR	29 APR	13 SEP 83	
1982-038D		14266	USSR	29 APR	20 AUG 83	
1982-038E		14267	USSR	29 APR	20 SEP 83	
1982-038F		14268	USSR	29 APR	31 AUG 83	
1982-038G		14269	USSR	29 APR	25 AUG 83	
1982-038H		14270	USSR	29 APR	12 SEP 83	
1982-038J		14271	USSR	29 APR	14 AUG 82	
1982-038K		14272	USSR	29 APR	12 SEP 83	
1982-038L		14273	USSR	29 APR	20 AUG 83	
1982-038M		14274	USSR	29 APR	26 SEP 83	
1982-038N		14275	USSR	29 APR	27 AUG 83	
1982-038P		14276	USSR	29 APR	22 AUG 83	
1982-038Q		14279	USSR	29 APR	29 AUG 83	
1982-038R		14280	USSR	29 APR	26 NOV 83	
1982-038S		14281	USSR	29 APR	24 AUG 83	
1982-038T		14282	USSR	29 APR	20 AUG 83	
1982-038U		14285	USSR	29 APR	23 AUG 83	
1982-038V		14286	USSR	29 APR	23 AUG 83	
1982-038W		14342	USSR	29 APR	20 NOV 83	
1982-038X		14435	USSR	29 APR	12 NOV 83	
1982-038Y		14436	USSR	29 APR	23 DEC 83	
1982-038Z		14682	USSR	29 APR	13 FEB 84	
1982-041A		13170	US	11 MAY	05 DEC 82	
1982-041B		13171	US	11 MAY	13 MAY 82	
1982-042A	SOYUZ T-5	13173	USSR	13 MAY	27 AUG 82	
1982-042B		13174	USSR	13 MAY	15 MAY 82	
1982-043B		13592	USSR	14 MAY	29 SEP 82	
1982-043C		13593	USSR	14 MAY	19 OCT 82	
1982-043E		13602	USSR	14 MAY	16 OCT 82	
1982-043F		13615	USSR	14 MAY	08 OCT 82	
1982-033C	ISKRA 2	13176	USSR	17 MAY	09 JUL 82	
1982-044B		13178	USSR	17 MAY	18 MAY 82	
1982-044C		13179	USSR	17 MAY	19 MAY 82	
1982-044D		13217	USSR	17 MAY	14 SEP 83	
1982-044E		13218	USSR	17 MAY	14 SEP 83	
1982-045B		13206	USSR	20 MAY	17 JUN 82	
1982-045C		13207	USSR	20 MAY	04 JUN 82	
1982-046A	COSMOS 1368	13208	USSR	21 MAY	03 JUN 82	
1982-046B		13209	USSR	21 MAY	29 MAY 82	
1982-046C		13254	USSR	21 MAY	06 JUN 82	
1982-046D		13255	USSR	21 MAY	06 JUN 82	
1982-046E		13256	USSR	21 MAY	09 JUN 82	
1982-047A	PROGRESS 13	13210	USSR	23 MAY	06 JUN 82	
1982-047B		13211	USSR	23 MAY	25 MAY 82	
1982-047C		13212	USSR	23 MAY	25 MAY 82	
1982-048A	COSMOS 1369	13213	USSR	25 MAY	08 JUN 82	
1982-048B		13214	USSR	25 MAY	30 MAY 82	
1982-048C		13216	USSR	25 MAY	26 MAY 82	
1982-048D		13267	USSR	25 MAY	27 JUN 82	
1982-048E		13268	USSR	25 MAY	15 JUN 82	
1982-049A	COSMOS 1370	13219	USSR	28 MAY	11 JUL 82	
1982-049B		13220	USSR	28 MAY	31 MAY 82	
1982-050A	MOLNIYA 1-54	13237	USSR	28 MAY	19 NOV 92	
1982-050B		13238	USSR	28 MAY	12 JUL 82	
1982-050C		13239	USSR	28 MAY	28 JUN 82	
1982-050D		13240	USSR	28 MAY	01 JUN 82	
1982-050E		13253	USSR	28 MAY	02 APR 93	
1982-052B		13411	USSR	01 JUN	09 SEP 82	
1982-052C		13412	USSR	01 JUN	13 AUG 82	
1982-053A	COSMOS 1373	13244	USSR	02 JUN	16 JUN 82	
1982-053B		13245	USSR	02 JUN	11 JUN 82	
1982-053C		13276	USSR	02 JUN	18 JUL 82	
1982-053D		13277	USSR	02 JUN	10 AUG 82	
1982-053E		13278	USSR	02 JUN	04 NOV 82	
1982-053F		13279	USSR	02 JUN	05 AUG 82	
1982-054A	COSMOS 1374	13257	USSR	03 JUN	03 JUN 82	
1982-054B		13258	USSR	03 JUN	05 JUN 82	
1982-054C		13280	USSR	03 JUN	03 JUN 82	
1982-055BJ		19243	USSR	06 JUN	31 DEC 89	
1982-056A	COSMOS 1376	13263	USSR	08 JUN	22 JUN 82	
1982-056B		13264	USSR	08 JUN	12 JUN 82	
1982-056C		13288	USSR	08 JUN	14 JUL 82	
1982-056D		13289	USSR	08 JUN	29 JUN 82	
1982-056E		13290	USSR	08 JUN	22 JUN 82	
1982-057A	COSMOS 1377	13265	USSR	08 JUN	22 JUL 82	
1982-057B		13266	USSR	08 JUN	12 JUN 82	
1982-058B		13294	US	09 JUN	22 MAR 90	
1982-060A	COSMOS 1379	13281	USSR	18 JUN	18 JUN 82	
1982-060B		13284	USSR	18 JUN	18 JUN 82	
1982-061A	COSMOS 1380	13282	USSR	18 JUN	27 JUN 82	
1982-061B		13286	USSR	18 JUN	25 JUN 82	
1982-062A	COSMOS 1381	13283	USSR	18 JUN	01 JUL 82	
1982-062B		13285	USSR	18 JUN	28 JUN 82	
1982-062C		13305	USSR	18 JUN	11 AUG 82	
1982-062D		13306	USSR	18 JUN	10 SEP 82	
1982-062E		13307	USSR	18 JUN	13 JUL 82	
1982-062F		13363	USSR	18 JUN	24 FEB 83	
1982-063A	SOYUZ T-6	13292	USSR	24 JUN	02 JUL 82	
1982-063B		13293	USSR	24 JUN	26 JUN 82	
1982-064B		13296	USSR	25 JUN	07 AUG 82	
1982-064C		13297	USSR	25 JUN	15 JUL 82	
1982-065A	STS 4	13300	US	27 JUN	04 JUL 82	BT
1982-067A	COSMOS 1384	13303	USSR	30 JUN	30 JUL 82	
1982-067B		13304	USSR	30 JUN	05 JUL 82	
1982-068A	COSMOS 1385	13345	USSR	06 JUL	20 JUL 82	
1982-068B		13346	USSR	06 JUL	08 JUL 82	
1982-068C		13370	USSR	06 JUL	20 JUL 82	
1982-068D		13371	USSR	06 JUL	17 AUG 82	
1982-068E		13372	USSR	06 JUL	18 OCT 82	
1982-068F		13373	USSR	06 JUL	09 AUG 82	
1982-068G		13374	USSR	06 JUL	19 AUG 82	
1982-070A	PROGRESS 14	13361	USSR	10 JUL	13 AUG 82	
1982-070B		13362	USSR	10 JUL	12 JUL 82	
1982-071A	COSMOS 1387	13365	USSR	13 JUL	26 JUL 82	
1982-071B		13366	USSR	13 JUL	17 JUL 82	
1982-071C		13388	USSR	13 JUL	31 JUL 82	
1982-071D		13389	USSR	13 JUL	27 JUL 82	
1982-072B		13368	US	16 JUL	12 JUL 84	
1982-074A	MOLNIYA 1-55	13383	USSR	21 JUL	10 OCT 92	
1982-074B		13384	USSR	21 JUL	20 SEP 82	
1982-074C		13385	USSR	21 JUL	03 SEP 82	
1982-075A	COSMOS 1396	13391	USSR	27 JUL	10 AUG 82	
1982-075B		13392	USSR	27 JUL	01 AUG 82	
1982-075C		13406	USSR	27 JUL	19 AUG 82	
1982-075D		13407	USSR	27 JUL	12 AUG 82	
1982-076A	COSMOS 1397	13394	USSR	29 JUL	18 MAY 83	BY
1982-076B		13395	USSR	29 JUL	21 APR 83	
1982-076C		13981	USSR	29 JUL	22 APR 83	
1982-076D		13982	USSR	29 JUL	21 APR 83	
1982-076E		13986	USSR	29 JUL	13 APR 83	
1982-076F		13987	USSR	29 JUL	25 APR 83	
1982-076G		13988	USSR	29 JUL	18 APR 83	
1982-076H		13989	USSR	29 JUL	13 APR 83	
1982-076J		14008	USSR	29 JUL	24 APR 83	
1982-076K		14009	USSR	29 JUL	23 APR 83	
1982-076L		14010	USSR	29 JUL	24 APR 83	
1982-076M		14011	USSR	29 JUL	25 APR 83	
1982-076N		14012	USSR	29 JUL	24 APR 83	
1982-076P		14013	USSR	29 JUL	24 APR 83	
1982-076Q		14018	USSR	29 JUL	24 APR 83	
1982-076R		14019	USSR	29 JUL	24 APR 83	

International Designation	Satellite Name	Number	Source	Launch	Decay	Note
1982-076S		14020	USSR	29 JUL	26 APR 83	
1982-076T		14021	USSR	29 JUL	24 APR 83	
1982-076U		14022	USSR	29 JUL	25 APR 83	
1982-076V		14023	USSR	29 JUL	27 APR 83	
1982-076W		14024	USSR	29 JUL	27 APR 83	
1982-076X		14025	USSR	29 JUL	24 APR 83	
1982-076Y		14031	USSR	29 JUL	24 APR 83	
1982-076Z		14035	USSR	29 JUL	25 APR 83	
1982-077A	COSMOS 1398	13396	USSR	03 AUG	13 AUG 82	
1982-077B		13397	USSR	03 AUG	07 AUG 82	
1982-077C		13414	USSR	03 AUG	13 AUG 82	
1982-077D		13415	USSR	03 AUG	13 AUG 82	
1982-078A	COSMOS 1399	13399	USSR	04 AUG	16 SEP 82	
1982-078B		13400	USSR	04 AUG	08 AUG 82	
1982-078C		13401	USSR	04 AUG	08 AUG 82	
1982-080A	SOYUZ T-7	13425	USSR	19 AUG	10 DEC 82	
1982-080B		13426	USSR	19 AUG	21 AUG 82	
1982-081A	COSMOS 1401	13427	USSR	20 AUG	03 SEP 82	
1982-081B		13428	USSR	20 AUG	25 AUG 82	
1982-081C		13494	USSR	20 AUG	07 SEP 82	
1982-081D		13495	USSR	20 AUG	09 SEP 82	
1982-081E		13496	USSR	20 AUG	22 SEP 82	
1982-082B		13438	US	26 AUG	18 SEP 82	
1982-082C		13447	US	26 AUG	23 AUG 85	
1982-083B		13433	USSR	27 AUG	12 SEP 82	
1982-083C		13434	USSR	27 AUG	10 SEP 82	
1982-083D		13437	USSR	27 AUG	02 SEP 82	
1982-084A	COSMOS 1402	13441	USSR	30 AUG	23 JAN 83	
1982-084B		13747	USSR	30 AUG	30 DEC 82	
1982-084C		13748	USSR	30 AUG	07 FEB 83	
1982-085A	COSMOS 1403	13448	USSR	01 SEP	15 SEP 82	
1982-085B		13450	USSR	01 SEP	09 SEP 82	
1982-085C		13452	USSR	01 SEP	02 SEP 82	
1982-085D		13542	USSR	01 SEP	05 APR 83	
1982-085E		13543	USSR	01 SEP	30 OCT 82	
1982-085F		13545	USSR	01 SEP	14 OCT 82	
1982-085G		13551	USSR	01 SEP	26 OCT 82	
1982-086A	COSMOS 1404	13449	USSR	01 SEP	15 SEP 82	
1982-086B		13451	USSR	01 SEP	10 SEP 82	
1982-086C		13453	USSR	01 SEP	02 SEP 82	
1982-086D		13454	USSR	01 SEP	03 SEP 82	
1982-086E		13544	USSR	01 SEP	29 OCT 82	
1982-086F		13548	USSR	01 SEP	17 OCT 82	
1982-086G		13549	USSR	01 SEP	12 APR 83	
1982-086H		13550	USSR	01 SEP	30 OCT 82	
1982-088A	COSMOS 1405	13508	USSR	04 SEP	05 FEB 84	
1982-088AA		14650	USSR	04 SEP	09 JAN 84	
1982-088AB		14651	USSR	04 SEP	05 JAN 84	
1982-088AC		14652	USSR	04 SEP	27 FEB 84	
1982-088AD		14653	USSR	04 SEP	28 JAN 84	
1982-088AE		14654	USSR	04 SEP	14 JAN 84	
1982-088AF		14655	USSR	04 SEP	03 FEB 84	
1982-088AG		14656	USSR	04 SEP	19 FEB 84	
1982-088AH		14657	USSR	04 SEP	31 JAN 84	
1982-088AJ		14658	USSR	04 SEP	21 JAN 84	
1982-088B		13509	USSR	04 SEP	05 SEP 82	
1982-088C		14599	USSR	04 SEP	03 JAN 84	
1982-088D		14600	USSR	04 SEP	12 JAN 84	
1982-088E		14601	USSR	04 SEP	03 JAN 84	
1982-088F		14602	USSR	04 SEP	01 JAN 84	
1982-088G		14603	USSR	04 SEP	08 JAN 84	
1982-088H		14604	USSR	04 SEP	05 JAN 84	
1982-088J		14605	USSR	04 SEP	17 FEB 84	
1982-088K		14606	USSR	04 SEP	03 JAN 84	
1982-088L		14636	USSR	04 SEP	08 JAN 84	
1982-088M		14637	USSR	04 SEP	10 JAN 84	
1982-088N		14638	USSR	04 SEP	31 DEC 83	
1982-088P		14639	USSR	04 SEP	01 JAN 84	
1982-088Q		14640	USSR	04 SEP	28 DEC 83	
1982-088R		14641	USSR	04 SEP	01 JAN 84	
1982-088S		14642	USSR	04 SEP	30 DEC 83	
1982-088T		14643	USSR	04 SEP	04 JAN 84	
1982-088U		14644	USSR	04 SEP	30 DEC 83	
1982-088V		14645	USSR	04 SEP	31 DEC 83	
1982-088W		14646	USSR	04 SEP	13 JAN 84	
1982-088X		14647	USSR	04 SEP	06 JAN 84	
1982-088Y		14648	USSR	04 SEP	21 FEB 84	
1982-088Z		14649	USSR	04 SEP	04 JAN 84	
1982-089A	COSMOS 1406	13519	USSR	08 SEP	21 SEP 82	
1982-089B		13520	USSR	08 SEP	11 SEP 82	
1982-089C		13581	USSR	08 SEP	25 SEP 82	
1982-089D		13582	USSR	08 SEP	23 SEP 82	
1982-090A	PRC 12	13521	PRC	09 SEP	21 SEP 82	
1982-090B		13540	PRC	09 SEP	09 SEP 82	
1982-090C		13557	PRC	09 SEP	11 SEP 82	
1982-091A	COSMOS 1407	13546	USSR	15 SEP	16 OCT 82	
1982-091B		13547	USSR	15 SEP	19 SEP 82	
1982-093B		13555	USSR	16 SEP	18 SEP 82	
1982-093C		13556	USSR	16 SEP	17 SEP 82	
1982-093D		13583	USSR	16 SEP	12 OCT 83	
1982-093E		13584	USSR	16 SEP	14 SEP 83	
1982-094A	PROGRESS 15	13558	USSR	18 SEP	16 OCT 82	
1982-094B		13559	USSR	18 SEP	20 SEP 82	
1982-095B		13586	USSR	22 SEP	17 OCT 82	
1982-095C		13587	USSR	22 SEP	13 OCT 82	
1982-097B		13599	US	28 SEP	16 FEB 92	
1982-098A	COSMOS 1411	13597	USSR	30 SEP	14 OCT 82	
1982-098B		13598	USSR	30 SEP	06 OCT 82	
1982-098C		13613	USSR	30 SEP	23 OCT 82	
1982-098D		13614	USSR	30 SEP	15 OCT 82	
1982-098E		13616	USSR	30 SEP	16 OCT 82	
1982-099B		13645	USSR	02 OCT	04 DEC 82	
1982-099C		13646	USSR	02 OCT	12 NOV 82	
1982-099D		13647	USSR	02 OCT	14 NOV 82	
1982-100B		13604	USSR	12 OCT	14 OCT 82	
1982-100C		13605	USSR	12 OCT	13 OCT 82	
1982-101A	COSMOS 1416	13611	USSR	14 OCT	28 OCT 82	
1982-101B		13612	USSR	14 OCT	23 OCT 82	
1982-101C		13633	USSR	14 OCT	30 OCT 82	
1982-101D		13634	USSR	14 OCT	08 NOV 82	
1982-103B		13625	USSR	20 OCT	22 OCT 82	
1982-103C		13626	USSR	20 OCT	21 OCT 82	
1982-103D		13629	USSR	20 OCT	05 JUL 83	
1982-103F		13644	USSR	20 OCT	21 MAY 83	
1982-104A	COSMOS 1418	13627	USSR	21 OCT	30 SEP 83	
1982-104B		13628	USSR	21 OCT	30 MAR 83	
1982-105B		13632	US	28 OCT	16 DEC 88	
1982-105C		13684	US	28 OCT	09 FEB 84	
1982-106C		13639	US	30 OCT	14 SEP 83	
1982-107A	PROGRESS 16	13638	USSR	31 OCT	14 DEC 82	
1982-107B		13640	USSR	31 OCT	02 NOV 82	
1982-108A	COSMOS 1419	13641	USSR	02 NOV	16 NOV 82	
1982-108B		13642	USSR	02 NOV	06 NOV 82	
1982-108C		13654	USSR	02 NOV	21 NOV 82	
1982-108D		13655	USSR	02 NOV	17 NOV 82	
1982-108E		13656	USSR	02 NOV	17 NOV 82	
1982-108F		13657	USSR	02 NOV	16 NOV 82	
1982-109C		14227	USSR	11 NOV	19 SEP 90	
1982-110A	STS 5	13650	US	11 NOV	16 NOV 82	BT
1982-111A		13659	US	17 NOV	13 AUG 85	
1982-111B		13660	US	17 NOV	27 DEC 82	
1982-033AD	ISKRA 3	13663	USSR	18 NOV	16 DEC 82	
1982-112A	COSMOS 1421	13661	USSR	18 NOV	02 DEC 82	
1982-112B		13662	USSR	18 NOV	22 NOV 82	
1982-112C		13665	USSR	18 NOV	20 NOV 82	
1982-112D		13672	USSR	18 NOV	05 DEC 82	
1982-112E		13673	USSR	18 NOV	06 DEC 82	
1982-112F		13674	USSR	18 NOV	14 DEC 82	
1982-113B		13670	USSR	26 NOV	27 NOV 82	
1982-113C		13671	USSR	26 NOV	28 NOV 82	
1982-113D		13675	USSR	26 NOV	15 JUN 84	
1982-113E		13676	USSR	26 NOV	12 OCT 83	
1982-114A	COSMOS 1422	13677	USSR	03 DEC	17 DEC 82	
1982-114B		13678	USSR	03 DEC	07 DEC 82	
1982-114C		13679	USSR	03 DEC	04 DEC 82	
1982-114D		13727	USSR	03 DEC	29 DEC 82	
1982-114E		13728	USSR	03 DEC	20 DEC 82	
1982-114F		13730	USSR	03 DEC	18 DEC 82	
1982-115A	COSMOS 1423	13685	USSR	08 DEC	18 JAN 86	
1982-115AA		13717	USSR	08 DEC	08 MAR 84	
1982-115AB		13721	USSR	08 DEC	15 SEP 83	
1982-115AC		13722	USSR	08 DEC	26 JAN 84	
1982-115AD		13937	USSR	08 DEC	21 APR 83	
1982-115AE		13938	USSR	08 DEC	05 MAY 84	
1982-115AF		14343	USSR	08 DEC	02 APR 85	
1982-115AG		14434	USSR	08 DEC	16 NOV 85	
1982-115AH		19666	USSR	08 DEC	08 JAN 89	
1982-115AJ		19835	USSR	08 DEC	12 MAR 89	
1982-115B		13686	USSR	08 DEC	27 DEC 82	
1982-115C		13692	USSR	08 DEC	09 DEC 82	
1982-115D		13693	USSR	08 DEC	22 DEC 82	
1982-115E		13696	USSR	08 DEC	16 MAY 88	
1982-115F		13697	USSR	08 DEC	19 DEC 82	
1982-115G		13698	USSR	08 DEC	21 DEC 82	
1982-115H		13699	USSR	08 DEC	23 DEC 82	
1982-115J		13701	USSR	08 DEC	12 DEC 82	
1982-115K		13702	USSR	08 DEC	26 DEC 82	
1982-115L		13703	USSR	08 DEC	15 DEC 82	
1982-115M		13704	USSR	08 DEC	11 APR 84	
1982-115N		13705	USSR	08 DEC	26 MAY 85	
1982-115P		13706	USSR	08 DEC	08 APR 83	
1982-115Q		13707	USSR	08 DEC	07 JUN 84	
1982-115R		13708	USSR	08 DEC	23 MAR 89	
1982-115S		13709	USSR	08 DEC	01 NOV 83	
1982-115T		13710	USSR	08 DEC	29 FEB 84	
1982-115U		13711	USSR	08 DEC	02 SEP 83	
1982-115V		13712	USSR	08 DEC	06 DEC 84	
1982-115W		13713	USSR	08 DEC	11 DEC 85	
1982-115X		13714	USSR	08 DEC	17 FEB 86	
1982-115Y		13715	USSR	08 DEC	16 JUL 84	
1982-115Z		13716	USSR	08 DEC	26 MAY 84	
1982-117A	COSMOS 1424	13725	USSR	16 DEC	28 JAN 83	
1982-117B		13726	USSR	16 DEC	19 DEC 82	
1982-118B		13737	US	21 DEC	01 OCT 91	
1982-118D		13773	US	21 DEC	06 JUN 91	
1982-118E		13774	US	21 DEC	29 OCT 91	
1982-119A	COSMOS 1425	13739	USSR	23 DEC	06 JAN 83	
1982-119B		13740	USSR	23 DEC	06 JAN 83	
1982-119C		13741	USSR	23 DEC	28 DEC 82	
1982-119D		13742	USSR	23 DEC	28 DEC 82	
1982-119E		13754	USSR	23 DEC	27 OCT 83	
1982-119F		13755	USSR	23 DEC	03 MAR 83	
1982-119G		13756	USSR	23 DEC	11 FEB 83	
1982-120A	COSMOS 1426	13745	USSR	28 DEC	05 MAR 83	
1982-120B		13746	USSR	28 DEC	07 JAN 83	
1982-120C		13749	USSR	28 DEC	23 MAR 83	

International Designation	Satellite Name	Number	Source	Launch	Decay	Note
1982-120D		13752	USSR	28 DEC	07 JAN 83	
1982-120E		13873	USSR	28 DEC	08 MAR 83	
1982-121A	COSMOS 1427	13750	USSR	29 DEC	05 OCT 89	
1982-121B		13751	USSR	29 DEC	26 MAR 88	
1983 LAUNCHES						
1983-058E		17302	ESA	16 JAN	10 JUL 90	
1983-005A	COSMOS 1438	13779	USSR	27 JAN	07 FEB 83	
1983-005B		13780	USSR	27 JAN	30 JAN 83	
1983-005C		13787	USSR	27 JAN	12 FEB 83	
1983-005D		13788	USSR	27 JAN	08 FEB 83	
1983-005E		13789	USSR	27 JAN	11 FEB 83	
1983-005F		13790	USSR	27 JAN	13 FEB 83	
1983-007A	COSMOS 1439	13784	USSR	06 FEB	22 FEB 83	
1983-007B		13785	USSR	06 FEB	10 FEB 83	
1983-007C		13843	USSR	06 FEB	22 FEB 83	
1983-009A	COSMOS 1440	13793	USSR	10 FEB	24 FEB 83	
1983-009B		13794	USSR	10 FEB	15 FEB 83	
1983-009C		13847	USSR	10 FEB	15 MAR 83	
1983-009D		13848	USSR	10 FEB	02 MAR 81	
1983-011A	ASTRO-B	13829	JAPAN	20 FEB	17 DEC 88	
1983-011B		13830	JAPAN	20 FEB	19 JAN 89	
1983-011C		13846	JAPAN	20 FEB	03 MAR 83	
1983-011D		14566	JAPAN	20 FEB	31 AUG 87	
1983-011E		14567	JAPAN	20 FEB	08 OCT 84	
1983-012A	COSMOS 1442	13850	USSR	25 FEB	11 APR 83	
1983-012B		13851	USSR	25 FEB	01 MAR 83	
1983-013A	COSMOS 1443	13868	USSR	02 MAR	19 SEP 83	
1983-013B		13869	USSR	02 MAR	05 MAR 83	
1983-013C		13872	USSR	02 MAR	04 MAR 83	
1983-013D		14305	USSR	02 MAR	12 SEP 83	
1983-014A	COSMOS 1444	13870	USSR	02 MAR	16 MAR 83	
1983-014B		13871	USSR	02 MAR	11 MAR 83	
1983-014C		13891	USSR	02 MAR	16 MAR 83	
1983-014D		13892	USSR	02 MAR	20 APR 83	
1983-014E		13893	USSR	02 MAR	26 MAY 83	
1983-014F		13894	USSR	02 MAR	04 FEB 84	
1983-015B		13876	USSR	11 MAR	27 MAR 83	
1983-015C		13877	USSR	11 MAR	01 APR 83	
1983-015D		13881	USSR	11 MAR	26 MAR 83	
1983-016B		13879	USSR	12 MAR	14 MAR 83	
1983-016C		13880	USSR	12 MAR	13 MAR 83	
1983-016D		13884	USSR	12 MAR	27 OCT 86	
1983-016E		13885	USSR	12 MAR	02 APR 86	
1983-017A	COSMOS 1445	13883	USSR	15 MAR	16 MAR 83	
1983-017B		13888	USSR	15 MAR	17 MAR 83	
1983-017C		13889	USSR	15 MAR	18 MAR 83	
1983-018A	COSMOS 1446	13886	USSR	16 MAR	30 MAR 83	
1983-018B		13887	USSR	16 MAR	29 MAR 83	
1983-018C		13951	USSR	16 MAR	04 APR 83	
1983-018D		13952	USSR	16 MAR	30 MAR 83	
1983-018E		13953	USSR	16 MAR	31 MAR 83	
1983-019B		13895	USSR	16 MAR	04 APR 83	
1983-019C		13896	USSR	16 MAR	31 MAR 83	
1983-019E		13898	USSR	16 MAR	31 MAR 83	
1983-020B		13902	USSR	23 MAR	18 JUN 85	
1983-020C		13903	USSR	23 MAR	05 AUG 85	
1983-022B		13924	US	28 MAR	12 OCT 91	
1983-022C		14477	US	28 MAR	21 MAY 91	
1983-022D		16442	US	28 MAR	09 FEB 86	
1983-022E		16443	US	28 MAR	16 JAN 86	
1983-022F		16444	US	28 MAR	20 JAN 86	
1983-022G		16502	US	28 MAR	21 FEB 88	
1983-022H		16503	US	28 MAR	19 FEB 89	
1983-022J		16504	US	28 MAR	13 NOV 86	
1983-024A	COSMOS 1449	13955	USSR	31 MAR	15 APR 83	
1983-024B		13956	USSR	31 MAR	09 APR 83	
1983-024C		13962	USSR	31 MAR	01 APR 83	
1983-024D		13963	USSR	31 MAR	10 APR 83	
1983-024E		13993	USSR	31 MAR	23 APR 83	
1983-024F		13994	USSR	31 MAR	18 MAY 83	
1983-024G		13995	USSR	31 MAR	25 MAR 84	
1983-024H		13996	USSR	31 MAR	24 JUN 83	
1983-024J		13997	USSR	31 MAR	22 APR 83	
1983-025B		13965	USSR	02 APR	01 MAY 83	
1983-025C		13966	USSR	02 APR	16 APR 83	
1983-026A	STS 6	13968	US	04 APR	09 APR 83	BT
1983-027A	COSMOS 1450	13972	USSR	06 APR	30 MAY 90	
1983-027B		13973	USSR	06 APR	23 MAR 89	
1983-028B		13976	USSR	08 APR	08 APR 83	
1983-028C		13977	USSR	08 APR	10 APR 83	
1983-028D		13979	USSR	08 APR	15 MAY 84	
1983-028E		13980	USSR	08 APR	15 JUN 84	
1983-029A	COSMOS 1451	13975	USSR	08 APR	22 APR 83	
1983-029B		13978	USSR	08 APR	09 APR 83	
1983-029C		14016	USSR	08 APR	20 MAY 83	
1983-029D		14026	USSR	08 APR	25 APR 83	
1983-029E		14027	USSR	08 APR	27 APR 83	
1983-030C		13990	US	11 APR	27 OCT 86	
1983-032A		14001	US	15 APR	21 AUG 83	
1983-033A	ROHINI 3	14002	INDIA	17 APR	19 APR 90	
1983-033B		14003	INDIA	17 APR	18 MAR 89	
1983-033C		14004	INDIA	17 APR	11 MAY 87	
1983-033D		14224	INDIA	17 APR	06 JUN 88	
1983-033E		14225	INDIA	17 APR	04 JAN 89	
1983-033F		14397	INDIA	17 APR	04 FEB 84	
1983-034A	COSMOS 1453	14006	USSR	19 APR	08 MAY 89	
1983-034B		14007	USSR	19 APR	25 MAR 89	
1983-034C		14577	USSR	19 APR	25 MAR 84	
1983-034D		14578	USSR	19 APR	01 APR 84	
1983-034E		14597	USSR	19 APR	08 JUN 84	
1983-034F		14598	USSR	19 APR	01 DEC 84	
1983-034G		14730	USSR	19 APR	21 MAY 84	
1983-034H		14731	USSR	19 APR	10 APR 84	
1983-034J		14853	USSR	19 APR	30 MAY 84	
1983-034K		14854	USSR	19 APR	06 JUN 84	
1983-034N		15007	USSR	19 APR	02 NOV 84	
1983-034P		15008	USSR	19 APR	22 NOV 84	
1983-034Q		15013	USSR	19 APR	12 DEC 84	
1983-034R		15014	USSR	19 APR	03 DEC 84	
1983-034S		15015	USSR	19 APR	22 DEC 84	
1983-034T		15016	USSR	19 APR	14 NOV 84	
1983-034U		15017	USSR	19 APR	06 NOV 84	
1983-034V		15018	USSR	19 APR	24 OCT 84	
1983-034W		15019	USSR	19 APR	19 OCT 84	
1983-034X		15020	USSR	19 APR	14 OCT 84	
1983-034Y		15376	USSR	19 APR	15 JUL 85	
1983-034Z		15377	USSR	19 APR	22 MAY 85	
1983-035A	SOYUZ T-8	14014	USSR	20 APR	22 APR 83	
1983-035B		14015	USSR	20 APR	22 APR 83	
1983-035C		14029	USSR	20 APR	27 MAY 83	
1983-036A	COSMOS 1454	14017	USSR	22 APR	22 MAY 83	
1983-036B		14028	USSR	22 APR	26 APR 83	
1983-036C		14030	USSR	22 APR	23 APR 83	
1983-038B		14036	USSR	25 APR	31 MAY 83	
1983-038C		14037	USSR	25 APR	07 MAY 83	
1983-038D		14038	USSR	25 APR	17 MAY 83	
1983-038F		14042	USSR	25 APR	06 MAY 83	
1983-038G		14043	USSR	25 APR	29 APR 83	
1983-039A	COSMOS 1457	14039	USSR	26 APR	08 JUN 83	
1983-039B		14040	USSR	26 APR	30 APR 83	
1983-040A	COSMOS 1458	14044	USSR	28 APR	11 MAY 83	
1983-040B		14045	USSR	28 APR	02 MAY 83	
1983-040C		14046	USSR	28 APR	19 MAY 83	
1983-040D		14066	USSR	28 APR	12 MAY 83	
1983-040E		14067	USSR	28 APR	17 MAY 83	
1983-041D		14196	US	28 APR	16 JUN 90	
1983-043A	COSMOS 1460	14058	USSR	06 MAY	20 MAY 83	
1983-043B		14060	USSR	06 MAY	15 MAY 83	
1983-043C		14061	USSR	06 MAY	21 MAY 83	
1983-043D		14062	USSR	06 MAY	06 MAY 83	
1983-043E		14063	USSR	06 MAY	08 MAY 83	
1983-043F		14078	USSR	06 MAY	11 JUL 83	
1983-043G		14079	USSR	06 MAY	09 APR 84	
1983-043H		14080	USSR	06 MAY	08 AUG 83	
1983-043J		14082	USSR	06 MAY	26 MAY 83	
1983-043K		14083	USSR	06 MAY	27 MAY 83	
1983-043L		14247	USSR	06 MAY	17 AUG 83	
1983-044AB		15708	USSR	07 MAY	21 OCT 89	
1983-044AC		15709	USSR	07 MAY	08 MAR 89	
1983-044AD		15710	USSR	07 MAY	05 AUG 89	
1983-044AE		15711	USSR	07 MAY	15 AUG 87	
1983-044AF		15712	USSR	07 MAY	16 DEC 88	
1983-044AG		15713	USSR	07 MAY	17 AUG 91	
1983-044AH		15716	USSR	07 MAY	23 MAY 89	
1983-044AJ		15717	USSR	07 MAY	03 JUN 89	
1983-044AK		15718	USSR	07 MAY	02 SEP 90	
1983-044AL		15719	USSR	07 MAY	22 MAR 91	
1983-044AM		15720	USSR	07 MAY	20 MAR 89	
1983-044AN		15721	USSR	07 MAY	15 APR 87	
1983-044AP		15722	USSR	07 MAY	09 APR 90	
1983-044AQ		15723	USSR	07 MAY	19 FEB 89	
1983-044AR		15724	USSR	07 MAY	26 JUN 89	
1983-044AS		15725	USSR	07 MAY	06 JUN 85	
1983-044AT		15726	USSR	07 MAY	05 MAR 89	
1983-044AU		15727	USSR	07 MAY	16 MAR 89	
1983-044AV		15728	USSR	07 MAY	28 NOV 89	
1983-044AW		15729	USSR	07 MAY	18 JUL 90	
1983-044AX		15730	USSR	07 MAY	11 JUN 85	
1983-044AY		15731	USSR	07 MAY	29 APR 90	
1983-044AZ		15736	USSR	07 MAY	26 JUN 89	
1983-044B		14065	USSR	07 MAY	07 MAY 83	
1983-044BA		15737	USSR	07 MAY	14 SEP 88	
1983-044BB		15796	USSR	07 MAY	14 JUN 89	
1983-044BC		15797	USSR	07 MAY	04 FEB 89	
1983-044BD		15798	USSR	07 MAY	26 MAY 87	
1983-044BE		15799	USSR	07 MAY	15 NOV 90	
1983-044BF		15800	USSR	07 MAY	18 APR 89	
1983-044BG		15801	USSR	07 MAY	01 DEC 90	
1983-044BH		15802	USSR	07 MAY	28 FEB 88	
1983-044BJ		15803	USSR	07 MAY	21 OCT 88	
1983-044BK		15812	USSR	07 MAY	04 MAR 88	
1983-044BL		15813	USSR	07 MAY	22 OCT 85	
1983-044BM		15814	USSR	07 MAY	03 JUN 90	
1983-044BN		15815	USSR	07 MAY	29 APR 92	
1983-044BP		15816	USSR	07 MAY	10 JUN 89	
1983-044BQ		15817	USSR	07 MAY	27 APR 89	
1983-044BR		16774	USSR	07 MAY	11 APR 89	
1983-044BS		16775	USSR	07 MAY	04 MAR 89	
1983-044BT		16776	USSR	07 MAY	16 DEC 89	
1983-044BU		16777	USSR	07 MAY	12 FEB 89	
1983-044BV		16778	USSR	07 MAY	29 MAR 89	
1983-044BW		16779	USSR	07 MAY	13 JAN 89	
1983-044BX		16780	USSR	07 MAY	02 JUL 90	
1983-044BY		16781	USSR	07 MAY	10 DEC 88	

International Designation	Satellite Name	Number	Source	Launch	Decay	Note
1983-044BZ		16782	USSR	07 MAY	22 JUN 90	
1983-044C		15601	USSR	07 MAY	23 MAR 85	
1983-044CA		17010	USSR	07 MAY	22 MAR 89	
1983-044CB		17011	USSR	07 MAY	03 JAN 89	
1983-044CC		17012	USSR	07 MAY	22 JUN 89	
1983-044CD		17013	USSR	07 MAY	25 APR 90	
1983-044CE		17014	USSR	07 MAY	27 FEB 89	
1983-044CF		17015	USSR	07 MAY	22 DEC 88	
1983-044CG		17016	USSR	07 MAY	13 MAY 89	
1983-044CH		17017	USSR	07 MAY	03 AUG 89	
1983-044CJ		17091	USSR	07 MAY	09 MAR 88	
1983-044CK		17096	USSR	07 MAY	17 APR 89	
1983-044CL		17097	USSR	07 MAY	30 APR 89	
1983-044CM		17098	USSR	07 MAY	21 OCT 90	
1983-044CN		17099	USSR	07 MAY	28 MAR 90	
1983-044CP		17163	USSR	07 MAY	13 JUN 89	
1983-044CQ		17171	USSR	07 MAY	17 MAR 89	
1983-044CR		17643	USSR	07 MAY	08 APR 89	
1983-044CS		17644	USSR	07 MAY	18 DEC 88	
1983-044CT		17645	USSR	07 MAY	11 APR 89	
1983-044CU		17646	USSR	07 MAY	25 JAN 89	
1983-044CV		17647	USSR	07 MAY	14 FEB 89	
1983-044CW		17733	USSR	07 MAY	22 DEC 88	
1983-044CX		17734	USSR	07 MAY	13 SEP 89	
1983-044CY		17735	USSR	07 MAY	11 MAY 89	
1983-044CZ		17736	USSR	07 MAY	22 JAN 88	
1983-044D		15602	USSR	07 MAY	15 MAR 85	
1983-044DA		17737	USSR	07 MAY	16 NOV 88	
1983-044DB		17738	USSR	07 MAY	02 OCT 89	
1983-044DC		17739	USSR	07 MAY	13 FEB 91	
1983-044DD		17740	USSR	07 MAY	15 OCT 89	
1983-044DE		17741	USSR	07 MAY	06 MAY 89	
1983-044DF		17742	USSR	07 MAY	03 NOV 88	
1983-044DG		17743	USSR	07 MAY	02 APR 88	
1983-044DH		17744	USSR	07 MAY	22 MAR 90	
1983-044DJ		17745	USSR	07 MAY	30 AUG 89	
1983-044DK		17746	USSR	07 MAY	12 OCT 89	
1983-044DL		17747	USSR	07 MAY	10 APR 89	
1983-044DM		17748	USSR	07 MAY	01 APR 89	
1983-044DN		17749	USSR	07 MAY	17 APR 89	
1983-044DP		17750	USSR	07 MAY	04 JUL 89	
1983-044DQ		17751	USSR	07 MAY	05 JUN 89	
1983-044DR		17752	USSR	07 MAY	27 DEC 89	
1983-044DS		17753	USSR	07 MAY	14 MAY 89	
1983-044DT		17833	USSR	07 MAY	11 MAY 89	
1983-044DU		17834	USSR	07 MAY	22 MAY 89	
1983-044DV		17835	USSR	07 MAY	19 JAN 89	
1983-044DW		17836	USSR	07 MAY	10 APR 89	
1983-044DX		17837	USSR	07 MAY	29 MAR 89	
1983-044DY		18237	USSR	07 MAY	21 DEC 88	
1983-044DZ		18243	USSR	07 MAY	07 SEP 89	
1983-044E		15603	USSR	07 MAY	29 OCT 86	
1983-044EA		18244	USSR	07 MAY	13 SEP 89	
1983-044EB		18245	USSR	07 MAY	29 NOV 88	
1983-044EC		18246	USSR	07 MAY	13 FEB 89	
1983-044ED		18247	USSR	07 MAY	12 FEB 89	
1983-044EE		18248	USSR	07 MAY	21 NOV 89	
1983-044EF		18249	USSR	07 MAY	21 DEC 88	
1983-044EG		18378	USSR	07 MAY	02 JUL 89	
1983-044EH		18379	USSR	07 MAY	04 JUN 89	
1983-044EJ		18473	USSR	07 MAY	13 APR 89	
1983-044EK		18474	USSR	07 MAY	28 AUG 90	
1983-044EM		18481	USSR	07 MAY	30 DEC 88	
1983-044EN		18482	USSR	07 MAY	24 MAR 88	
1983-044EP		18483	USSR	07 MAY	23 APR 89	
1983-044EQ		18484	USSR	07 MAY	18 OCT 88	
1983-044ER		18485	USSR	07 MAY	19 JAN 89	
1983-044ES		18493	USSR	07 MAY	02 JUL 88	
1983-044ET		18520	USSR	07 MAY	16 JUN 89	
1983-044EU		18544	USSR	07 MAY	23 MAY 89	
1983-044EV		18566	USSR	07 MAY	03 APR 89	
1983-044EW		18694	USSR	07 MAY	22 JAN 88	
1983-044EX		18797	USSR	07 MAY	11 AUG 89	
1983-044EY		18812	USSR	07 MAY	12 FEB 91	
1983-044EZ		18868	USSR	07 MAY	26 DEC 88	
1983-044F		15604	USSR	07 MAY	22 APR 85	
1983-044FA		18869	USSR	07 MAY	02 NOV 88	
1983-044FB		18975	USSR	07 MAY	24 MAY 88	
1983-044FC		19101	USSR	07 MAY	04 MAY 89	
1983-044FD		19172	USSR	07 MAY	27 OCT 89	
1983-044FE		19296	USSR	07 MAY	15 MAY 89	
1983-044FF		19303	USSR	07 MAY	04 DEC 89	
1983-044FG		19305	USSR	07 MAY	18 NOV 88	
1983-044FH		19308	USSR	07 MAY	27 MAY 89	
1983-044FJ		19310	USSR	07 MAY	30 NOV 89	
1983-044FK		19361	USSR	07 MAY	07 APR 89	
1983-044FL		19426	USSR	07 MAY	05 MAR 89	
1983-044FM		19427	USSR	07 MAY	03 MAR 89	
1983-044FN		19668	USSR	07 MAY	16 JAN 90	
1983-044FP		19947	USSR	07 MAY	19 MAY 89	
1983-044G		15605	USSR	07 MAY	04 SEP 88	
1983-044H		15606	USSR	07 MAY	22 OCT 88	
1983-044J		15681	USSR	07 MAY	18 AUG 88	
1983-044K		15682	USSR	07 MAY	18 FEB 89	
1983-044L		15683	USSR	07 MAY	24 AUG 88	
1983-044M		15684	USSR	07 MAY	06 OCT 89	
1983-044N		15685	USSR-	07 MAY	16 DEC 90	
1983-044P		15686	USSR	07 MAY	22 OCT 88	
1983-044Q		15687	USSR	07 MAY	27 MAR 89	
1983-044Q		20133	USSR	07 MAY	10 JUL 89	
1983-044R		15688	USSR	07 MAY	04 JUL 87	
1983-044S		15689	USSR	07 MAY	24 JUL 87	
1983-044T		15690	USSR	07 MAY	01 FEB 89	
1983-044U		15691	USSR	07 MAY	07 APR 88	
1983-044V		15692	USSR	07 MAY	17 JAN 89	
1983-044W		15703	USSR	07 MAY	17 FEB 89	
1983-044X		15704	USSR	07 MAY	09 MAR 89	
1983-044Y		15705	USSR	07 MAY	28 APR 89	
1983-044Z		15706	USSR	07 MAY	26 MAR 89	
1983-045A	COSMOS 1462	14071	USSR	17 MAY	31 MAY 83	
1983-045B		14072	USSR	17 MAY	23 MAY 83	
1983-045C		14073	USSR	17 MAY	21 MAY 83	
1983-045D		14074	USSR	17 MAY	09 JUN 83	
1983-045E		14098	USSR	17 MAY	07 JUN 83	
1983-045F		14099	USSR	17 MAY	20 JUN 83	
1983-046A	COSMOS 1463	14075	USSR	19 MAY	24 JAN 93	
1983-046B		14076	USSR	19 MAY	07 NOV 89	
1983-047B		14081	US	19 MAY	13 FEB 88	
1983-049A	COSMOS 1465	14087	USSR	26 MAY	23 JAN 85	
1983-049B		14088	USSR	26 MAY	11 JAN 85	
1983-049C		14091	USSR	26 MAY	15 JUL 83	
1983-049D		14092	USSR	26 MAY	02 JUL 83	
1983-049E		14093	USSR	26 MAY	05 JUL 83	
1983-049F		14094	USSR	26 MAY	09 DEC 83	
1983-049G		14212	USSR	26 MAY	02 NOV 83	
1983-049H		14213	USSR	26 MAY	31 OCT 83	
1983-049J		14214	USSR	26 MAY	29 OCT 83	
1983-049K		14215	USSR	26 MAY	25 OCT 83	
1983-050A	COSMOS 1466	14089	USSR	26 MAY	06 JUL 83	
1983-050B		14090	USSR	26 MAY	30 MAY 83	
1983-050C		14097	USSR	26 MAY	27 MAY 83	
1983-051A	EXOSAT	14095	ESA	26 MAY	06 MAY 86	
1983-051C		14525	US	26 MAY	07 APR 86	
1983-051D		14526	US	26 MAY	06 MAR 89	
1983-051E		14927	US	26 MAY	21 MAR 89	
1983-052A	COSMOS 1467	14100	USSR	31 MAY	12 JUN 83	
1983-052B		14101	USSR	31 MAY	10 JUN 83	
1983-052C		14102	USSR	31 MAY	02 JUN 83	
1983-052D		14103	USSR	31 MAY	02 JUN 83	
1983-052E		14116	USSR	31 MAY	06 JUN 84	
1983-052F		14118	USSR	31 MAY	13 SEP 83	
1983-052G		14119	USSR	31 MAY	19 JUN 83	
1983-052H		14120	USSR	31 MAY	21 JUN 83	
1983-052J		14121	USSR	31 MAY	07 AUG 83	
1983-053B		14105	USSR	02 JUN	03 JUN 83	
1983-053C		14106	USSR	02 JUN	03 JUN 83	
1983-054B		14108	USSR	07 JUN	08 JUN 83	
1983-054C		14109	USSR	07 JUN	08 JUN 83	
1983-055A	COSMOS 1468	14110	USSR	07 JUN	21 JUN 83	
1983-055B		14111	USSR	07 JUN	12 JUN 83	
1983-055C		14140	USSR	07 JUN	15 JUL 83	
1983-055D		14141	USSR	07 JUN	29 JUN 83	
1983-057A	COSMOS 1469	14123	USSR	14 JUN	24 JUN 83	
1983-057B		14124	USSR	14 JUN	23 JUN 83	
1983-057C		14126	USSR	14 JUN	04 OCT 83	
1983-057D		14127	USSR	14 JUN	15 JUN 83	
1983-057E		14149	USSR	14 JUN	27 JUN 83	
1983-057F		14150	USSR	14 JUN	04 JUL 83	
1983-058D		14151	ESA	16 JUN	03 JUL 86	
1983-059A	STS 7	14132	US	18 JUN	24 JUN 83	BT
1983-059F	SPAS-01	14142	FRG	18 JUN	24 JUN 83	35
1983-060A		14137	US	20 JUN	21 MAR 84	
1983-060B		14138	US	20 JUN	22 JUN 83	
1983-062A	SOYUZ T-9	14152	USSR	27 JUN	23 NOV 83	
1983-062B		14153	USSR	27 JUN	29 JUN 83	
1983-062C		14519	USSR	27 JUN	17 MAR 84	
1983-064A	COSMOS 1471	14156	USSR	28 JUN	28 JUL 83	
1983-064B		14157	USSR	28 JUN	03 JUL 83	
1983-064C		14232	USSR	28 JUN	28 JUL 83	
1983-064D		14233	USSR	28 JUN	28 JUL 83	
1983-065B		14159	US	28 JUN	11 AUG 83	
1983-066B		14161	USSR	30 JUN	03 JUL 83	
1983-066C		14162	USSR	30 JUN	01 JUL 83	
1983-066D		14166	USSR	30 JUN	22 FEB 84	
1983-067B		14164	USSR	01 JUL	05 AUG 83	
1983-067C		14165	USSR	01 JUL	25 JUL 83	
1983-068A	COSMOS 1472	14169	USSR	05 JUL	19 JUL 83	
1983-068B		14170	USSR	05 JUL	07 JUL 83	
1983-068C		14197	USSR	05 JUL	11 DEC 83	
1983-068D		14198	USSR	05 JUL	03 SEP 83	
1983-068E		14202	USSR	05 JUL	16 AUG 83	
1983-068F		14203	USSR	05 JUL	28 AUG 83	
1983-070B		14183	USSR	08 JUL	21 AUG 83	
1983-070C		14184	USSR	08 JUL	24 JUL 83	
1983-071A	COSMOS 1482	14185	USSR	13 JUL	27 JUL 83	
1983-071B		14186	USSR	13 JUL	25 JUL 83	
1983-071C		14187	USSR	13 JUL	15 JUL 83	
1983-071D		14188	USSR	13 JUL	09 DEC 83	
1983-071E		14217	USSR	13 JUL	20 SEP 83	
1983-071F		14218	USSR	13 JUL	14 OCT 83	
1983-071G		14219	USSR	13 JUL	20 OCT 83	
1983-071H		14230	USSR	13 JUL	25 JUN 84	
1983-071J		14231	USSR	13 JUL	31 JUL 83	
1983-073B		14200	USSR	19 JUL	20 AUG 83	
1983-073C		14201	USSR	19 JUL	11 AUG 83	
1983-073D		14206	USSR	19 JUL	03 FEB 90	

International Designation	Satellite Name	Number	Source	Launch	Decay	Note
1983-074A	COSMOS 1483	14204	USSR	20 JUL	03 AUG 83	
1983-074B		14205	USSR	20 JUL	26 JUL 83	
1983-074C		14239	USSR	20 JUL	11 AUG 83	
1983-074D		14242	USSR	20 JUL	11 AUG 83	
1983-074E		14243	USSR	20 JUL	26 AUG 83	
1983-076A	COSMOS 1485	14210	USSR	26 JUL	09 AUG 83	
1983-076B		14211	USSR	26 JUL	05 AUG 83	
1983-076C		14216	USSR	26 JUL	14 SEP 83	
1983-076D		14253	USSR	26 JUL	29 SEP 83	
1983-076E		14254	USSR	26 JUL	30 OCT 83	
1983-076F		14255	USSR	26 JUL	14 AUG 84	
1983-077B		14235	US	28 JUL	28 SEP 83	
1983-080A	COSMOS 1487	14245	USSR	05 AUG	19 AUG 83	
1983-080B		14246	USSR	05 AUG	11 AUG 83	
1983-080C		14249	USSR	05 AUG	14 AUG 83	
1983-080D		14291	USSR	05 AUG	26 AUG 83	
1983-080E		14292	USSR	05 AUG	11 SEP 83	
1983-080F		14295	USSR	05 AUG	28 AUG 83	
1983-081B		14250	JAPAN	05 AUG	27 SEP 83	
1983-081C		14287	JAPAN	05 AUG	25 MAY 91	
1983-082A	COSMOS 1488	14251	USSR	09 AUG	23 AUG 83	
1983-082B		14252	USSR	09 AUG	18 AUG 83	
1983-082C		14298	USSR	09 AUG	13 NOV 83	
1983-082D		14302	USSR	09 AUG	05 SEP 84	
1983-082E		14303	USSR	09 AUG	28 AUG 83	
1983-082F		14347	USSR	09 AUG	13 NOV 83	
1983-083A	COSMOS 1489	14256	USSR	10 AUG	23 SEP 83	
1983-083B		14257	USSR	10 AUG	14 AUG 83	
1983-083C		14263	USSR	10 AUG	11 AUG 83	
1983-083D		14371	USSR	10 AUG	24 SEP 83	
1983-084D		14261	USSR	10 AUG	11 AUG 83	
1983-084E		14262	USSR	10 AUG	12 AUG 83	
1983-085A	PROGRESS 17	14283	USSR	17 AUG	18 SEP 83	
1983-085B		14284	USSR	17 AUG	20 AUG 83	
1983-086A	PRC 13	14288	PRC	19 AUG	03 SEP 83	
1983-086B		14289	PRC	19 AUG	30 AUG 83	
1983-086C		14290	PRC	19 AUG	20 AUG 83	
1983-086D		14293	PRC	19 AUG	20 AUG 83	
1983-086E		14294	PRC	19 AUG	21 AUG 83	
1983-086F		14296	PRC	19 AUG	20 AUG 83	
1983-087A	COSMOS 1493	14299	USSR	23 AUG	06 SEP 83	
1983-087B		14300	USSR	23 AUG	31 AUG 83	
1983-087C		14304	USSR	23 AUG	30 AUG 83	
1983-087D		14322	USSR	23 AUG	18 SEP 83	
1983-087E		14323	USSR	23 AUG	06 DEC 83	
1983-087F		14324	USSR	23 AUG	10 OCT 84	
1983-087G		14325	USSR	23 AUG	29 OCT 83	
1983-087H		14513	USSR	23 AUG	15 DEC 83	
1983-088B		14308	USSR	25 AUG	28 AUG 83	
1983-088C		14309	USSR	25 AUG	26 AUG 83	
1983-088D		14310	USSR	25 AUG	06 JUL 84	
1983-088E		14311	USSR	25 AUG	27 OCT 86	
1983-089A	STS 8	14312	US	30 AUG	05 SEP 83	BT
1983-090B		14314	USSR	30 AUG	12 SEP 83	
1983-090C		14315	USSR	30 AUG	19 SEP-83	
1983-091A	COSMOS 1494	14316	USSR	31 AUG	26 SEP 85	
1983-091AA		14753	USSR	31 AUG	08 APR 84	
1983-091AB		14754	USSR	31 AUG	10 APR 84	
1983-091AC		14755	USSR	31 AUG	10 APR 84	
1983-091B		14317	USSR	31 AUG	24 MAY 85	
1983-091C		14703	USSR	31 AUG	07 APR 84	
1983-091D		14704	USSR	31 AUG	06 MAY 84	
1983-091E		14705	USSR	31 AUG	05 APR 84	
1983-091F		14706	USSR	31 AUG	09 APR 84	
1983-091G		14707	USSR	31 AUG	07 APR 84	
1983-091H		14708	USSR	31 AUG	06 APR 84	
1983-091J		14709	USSR	31 AUG	08 APR 84	
1983-091K		14710	USSR	31 AUG	09 APR 84	
1983-091L		14711	USSR	31 AUG	05 APR 84	
1983-091M		14718	USSR	31 AUG	06 APR 84	
1983-091N		14719	USSR	31 AUG	08 APR 84	
1983-091P		14720	USSR	31 AUG	06 APR 84	
1983-091Q		14721	USSR	31 AUG	04 APR 84	
1983-091R		14744	USSR	31 AUG	04 APR 84	
1983-091S		14745	USSR	31 AUG	02 APR 84	
1983-091T		14746	USSR	31 AUG	04 APR 84	
1983-091U		14747	USSR	31 AUG	05 APR 84	
1983-091V		14748	USSR	31 AUG	10 APR 84	
1983-091W		14749	USSR	31 AUG	06 APR 84	
1983-091X		14750	USSR	31 AUG	05 APR 84	
1983-091Y		14751	USSR	31 AUG	06 APR 84	
1983-091Z		14752	USSR	31 AUG	09 APR 84	
1983-092A	COSMOS 1495	14320	USSR	03 SEP	16 SEP 83	
1983-092B		14321	USSR	03 SEP	06 SEP 83	
1983-092C		14337	USSR	03 SEP	19 SEP 83	
1983-092D		14338	USSR	03 SEP	22 SEP 83	
1983-093A	COSMOS 1496	14326	USSR	07 SEP	19 OCT 83	
1983-093B		14327	USSR	07 SEP	11 SEP 83	
1983-094C		14332	US	08 SEP	21 MAY 84	
1983-095A	COSMOS 1497	14330	USSR	09 SEP	23 SEP 83	
1983-095B		14331	USSR	09 SEP	19 SEP 83	
1983-095C		14367	USSR	09 SEP	14 DEC 83	
1983-095D		14368	USSR	09 SEP	08 OCT 84	
1983-095E		14370	USSR	09 SEP	14 NOV 83	
1983-095F		14562	USSR	09 SEP	27 DEC 83	
1983-096A	COSMOS 1498	14334	USSR	14 SEP	29 SEP 83	
1983-096B		14335	USSR	14 SEP	19 SEP 83	
1983-096C		14336	USSR	14 SEP	19 SEP 83	
1983-096D		14374	USSR	14 SEP	16 OCT 83	
1983-096E		14375	USSR	14 SEP	04 OCT 83	
1983-096F		14376	USSR	14 SEP	05 OCT 83	
1983-097A	COSMOS 1499	14339	USSR	17 SEP	01 OCT 83	
1983-097B		14340	USSR	17 SEP	26 SEP 83	
1983-097C		14341	USSR	17 SEP	19 SEP 83	
1983-097D		14382	USSR	17 SEP	20 DEC 83	
1983-097E		14383	USSR	17 SEP	09 OCT 83	
1983-097F		14384	USSR	17 SEP	04 NOV 84	
1983-097G		14385	USSR	17 SEP	20 NOV 83	
1983-097H		14391	USSR	17 SEP	04 JAN 84	
1983-098B		14366	US	22 SEP	30 OCT 83	
1983-098C		14369	US	22 SEP	31 DEC 86	
1983-100B		14378	USSR	30 SEP	02 OCT 83	
1983-100C		14379	USSR	30 SEP	30 SEP 83	
1983-100D		14389	USSR	30 SEP	16 JUN 84	
1983-100E		14390	USSR	30 SEP	27 APR 84	
1983-101A	COSMOS 1501	14380	USSR	30 SEP	26 MAY 89	
1983-101AA		15864	USSR	30 SEP	26 FEB 86	
1983-101AB		15865	USSR	30 SEP	25 MAR 86	
1983-101B		14381	USSR	30 SEP	30 MAR 89	
1983-101C		14732	USSR	30 SEP	27 APR 84	
1983-101D		14733	USSR	30 SEP	28 APR 84	
1983-101E		14947	USSR	30 SEP	05 SEP 84	
1983-101F		14963	USSR	30 SEP	09 SEP 84	
1983-101G		15108	USSR	30 SEP	01 MAR 85	
1983-101H		15110	USSR	30 SEP	04 FEB 85	
1983-101J		15203	USSR	30 SEP	29 JUL 86	
1983-101K		15204	USSR	30 SEP	04 SEP 86	
1983-101L		15277	USSR	30 SEP	13 OCT 86	
1983-101M		15278	USSR	30 SEP	26 SEP 86	
1983-101N		15374	USSR	30 SEP	11 JUL 85	
1983-101P		15375	USSR	30 SEP	05 JUL 85	
1983-101Q		15425	USSR	30 SEP	01 FEB 87	
1983-101R		15426	USSR	30 SEP	07 JAN 87	
1983-101S		15512	USSR	30 SEP	06 OCT 85	
1983-101T		15513	USSR	30 SEP	03 JAN 86	
1983-101U		15586	USSR	30 SEP	23 DEC 85	
1983-101V		15587	USSR	30 SEP	20 NOV 85	
1983-101W		15588	USSR	30 SEP	20 DEC 85	
1983-101X		15594	USSR	30 SEP	02 NOV 85	
1983-101Y		15862	USSR	30 SEP	20 MAR 86	
1983-101Z		15863	USSR	30 SEP	05 MAR 86	
1983-102A	COSMOS 1502	14395	USSR	05 OCT	29 AUG 85	
1983-102B		14396	USSR	05 OCT	01 JUN 84	
1983-104A	COSMOS 1504	14403	USSR	14 OCT	06 DEC 83	
1983-104B		14404	USSR	14 OCT	17 OCT 83	
1983-105B		14423	ESA	19 OCT	04 MAY 87	
1983-106A	PROGRESS 18	14422	USSR	20 OCT	16 NOV 83	
1983-106B		14424	USSR	20 OCT	22 OCT 83	
1983-106C		14488	USSR	20 OCT	20 NOV 83	
1983-106D		14489	USSR	20 OCT	22 NOV 83	
1983-107A	COSMOS 1505	14425	USSR	21 OCT	04 NOV 83	
1983-107B		14426	USSR	21 OCT	30 OCT 83	
1983-107C		14481	USSR	21 OCT	09 DEC 84	
1983-107D		14482	USSR	21 OCT	03 JAN 84	
1983-107E		14560	USSR	21 OCT	01 FEB 84	
1983-107F		14561	USSR	21 OCT	26 FEB 84	
1983-107G		14674	USSR	21 OCT	01 FEB 84	
1983-110A	COSMOS 1507	14455	USSR	29 OCT	19 AUG 87	
1983-110B		14456	USSR	29 OCT	29 OCT 83	
1983-112A	COSMOS 1509	14490	USSR	17 NOV	01 DEC 83	
1983-112B		14498	USSR	17 NOV	22 NOV 83	
1983-112C		14535	USSR	17 NOV	27 DEC 83	
1983-112D		14536	USSR	17 NOV	05 DEC 83	
1983-112E		14539	USSR	17 NOV	03 DEC 83	
1983-112F		14540	USSR	17 NOV	07 DEC 83	
1983-112G		14541	USSR	17 NOV	02 DEC 83	
1983-113B		14553	US	18 NOV	02 SEP 91	
1983-113C		14554	US	18 NOV	02 OCT 91	
1983-113D		14609	US	18 NOV	19 JUN 92	
1983-114B		14517	USSR	23 NOV	15 DEC 83	
1983-114C		14518	USSR	23 NOV	10 DEC 83	
1983-116A	STS 9	14523	US	28 NOV	08 DEC 83	BT
1983-117A	COSMOS 1511	14530	USSR	30 NOV	13 JAN 84	
1983-117B		14531	USSR	30 NOV	05 DEC 83	
1983-118B		14533	USSR	30 NOV	02 DEC 83	
1983-118C		14534	USSR	30 NOV	01 DEC 83	
1983-118D		14537	USSR	30 NOV	10 MAY 86	
1983-118E		14538	USSR	30 NOV	23 SEP 88	
1983-119A	COSMOS 1512	14542	USSR	07 DEC	21 DEC 83	
1983-119B		14543	USSR	07 DEC	17 DEC 83	
1983-119C		14544	USSR	07 DEC	09 DEC 83	
1983-119D		14545	USSR	07 DEC	09 DEC 83	
1983-119E		14573	USSR	07 DEC	16 FEB 84	
1983-119F		14574	USSR	07 DEC	30 DEC 83	
1983-119G		14575	USSR	07 DEC	28 APR 85	
1983-119H		14576	USSR	07 DEC	20 MAR 84	
1983-119J		14579	USSR	07 DEC	09 JUL 84	
1983-119K		14580	USSR	07 DEC	28 DEC 83	
1983-119L		14581	USSR	07 DEC	26 JUN 84	
1983-119M		14673	USSR	07 DEC	30 MAR 84	
1983-121A	COSMOS 1514	14549	USSR	14 DEC	19 DEC 83	
1983-121B		14550	USSR	14 DEC	19 DEC 83	
1983-123B		14571	USSR	21 DEC	14 FEB 84	
1983-123C		14572	USSR	21 DEC	04 FEB 84	
1983-124A	COSMOS 1516	14583	USSR	27 DEC	09 FEB 84	
1983-124B		14584	USSR	27 DEC	31 DEC 83	

International Designation	Satellite Name	Number	Source	Launch	Decay	Note
1983-125A	COSMOS 1517	14585	USSR	27 DEC	27 DEC 83	
1983-125B		14586	USSR	27 DEC	29 DEC 83	
1983-126B		14588	USSR	28 DEC	08 FEB 84	
1983-126C		14589	USSR	28 DEC	04 FEB 84	
1983-127D		14593	USSR	29 DEC	31 DEC 83	
1983-127E		14594	USSR	29 DEC	29 DEC 83	
1983-127L		21860	USSR	29 DEC	18 AUG 92	
1984 LAUNCHES						
1984-002A	COSMOS 1530	14622	USSR	11 JAN	25 JAN 84	
1984-002B		14623	USSR	11 JAN	25 JAN 84	
1984-002C		14661	USSR	11 JAN	04 FEB 84	
1984-002D		14662	USSR	11 JAN	11 APR 84	
1984-002E		14663	USSR	11 JAN	16 MAR 84	
1984-002F		14664	USSR	11 JAN	26 MAR 84	
1984-002G		14929	USSR	11 JAN	21 JUN 85	
1984-004A	COSMOS 1532	14634	USSR	13 JAN	26 FEB 84	
1984-004B		14635	USSR	13 JAN	18 JAN 84	
1984-005B		14660	JAPAN	23 JAN	21 NOV 84	
1984-005C		14665	JAPAN	23 JAN	30 JUN 92	
1984-006A	COSMOS 1533	14666	USSR	26 JAN	09 FEB 84	
1984-006B		14667	USSR	26 JAN	13 FEB 84	
1984-006C		14712	USSR	26 JAN	05 MAY 84	
1984-006D		14713	USSR	26 JAN	11 APR 84	
1984-006E		14714	USSR	26 JAN	11 APR 84	
1984-006F		14762	USSR	26 JAN	14 APR 84	
1984-007A	COSMOS 1534	14668	USSR	26 JAN	20 SEP 90	
1984-007B		14669	USSR	26 JAN	30 JUN 89	
1984-008B		14671	PRC	29 JAN	08 FEB 84	
1984-008C		14672	PRC	29 JAN	30 JAN 84	
1984-008D		14678	PRC	29 JAN	15 FEB 84	
1984-009B		14676	US	31 JAN	27 FEB 84	
1984-011A	STS 41B	14681	US	03 FEB	11 FEB 84	BT
1984-011B	WESTAR 6	14688	US	03 FEB	16 NOV 84	CA
1984-011J		14698	US	03 FEB	09 FEB 84	
1984-011K		14765	US	03 FEB	28 FEB 84	
1984-011L		14766	US	03 FEB	28 FEB 84	
1984-011M		14767	US	03 FEB	28 FEB 84	
1984-011N		14768	US	03 FEB	28 FEB 84	
1984-011P		14769	US	03 FEB	28 FEB 84	
1984-011Q		14770	US	03 FEB	28 FEB 84	
1984-011R		14771	US	03 FEB	28 FEB 84	
1984-011S		14772	US	03 FEB	28 FEB 84	
1984-011T		14773	US	03 FEB	28 FEB 84	
1984-011V		14970	US	03 FEB	16 JUL 87	
1984-011W		15556	US	03 FEB	24 APR 85	
1984-011X		15557	US	03 FEB	28 JUN 85	
1984-011C	IRT	14689	US	05 FEB	11 FEB 84	35
1984-011G		14695	US	05 FEB	09 FEB 84	
1984-011H		14696	US	05 FEB	12 FEB 84	
1984-012E		14761	US	05 FEB	18 OCT 84	
1984-012G		14964	US	05 FEB	18 OCT 84	
1984-012H		14969	US	05 FEB	18 OCT 84	
1984-012M		19441	US	05 FEB	25 AUG 88	
1984-011D	PALAPA B2	14692	INDON	06 FEB	16 NOV 84	CB
1984-011U		14824	US	06 FEB	03 FEB 90	
1984-011Y		17839	US	06 FEB	12 APR 89	
1984-014A	SOYUZ T-10	14701	USSR	08 FEB	11 APR 84	
1984-014B		14702	USSR	08 FEB	10 FEB 84	
1984-014C		14916	USSR	08 FEB	26 MAY 84	
1984-014D		14917	USSR	08 FEB	15 MAY 84	
1984-014E		14918	USSR	08 FEB	20 MAY 84	
1984-015A	OHZORA	14722	JAPAN	14 FEB	19 JUL 89	
1984-015B		14723	JAPAN	14 FEB	20 APR 88	
1984-015C		14724	JAPAN	14 FEB	26 DEC 88	
1984-015D		17092	JAPAN	14 FEB	01 JAN 88	
1984-015E		17162	JAPAN	14 FEB	08 SEP 87	
1984-015F		19665	JAPAN	14 FEB	11 DEC 88	
1984-016B		14726	USSR	15 FEB	17 FEB 84	
1984-016C		14727	USSR	15 FEB	15 FEB 84	
1984-016D		14742	USSR	15 FEB	10 MAR 87	
1984-016E		14743	USSR	15 FEB	27 OCT 86	
1984-017A	COSMOS 1537	14737	USSR	16 FEB	01 MAR 84	
1984-017B		14738	USSR	16 FEB	23 FEB 84	
1984-017C		14774	USSR	16 FEB	02 MAR 84	
1984-017D		14775	USSR	16 FEB	18 MAR 84	
1984-017E		14776	USSR	16 FEB	02 MAR 84	
1984-017F		14777	USSR	16 FEB	08 MAR 84	
1984-017G		14778	USSR	16 FEB	25 MAY 84	
1984-017H		14779	USSR	16 FEB	04 MAR 84	
1984-018A	PROGRESS 19	14757	USSR	21 FEB	01 APR 84	
1984-018B		14758	USSR	21 FEB	23 FEB 84	
1984-020A	COSMOS 1539	14763	USSR	28 FEB	09 APR 84	
1984-020B		14764	USSR	28 FEB	03 MAR 84	
1984-020C		14901	USSR	28 FEB	10 APR 84	
1984-021C		14782	US	01 MAR	21 JAN 92	
1984-022B		14784	USSR	02 MAR	03 MAR 84	
1984-022C		14785	USSR	02 MAR	03 MAR 84	
1984-022D		14788	USSR	02 MAR	27 OCT 86	
1984-022E		14789	USSR	02 MAR	10 FEB 90	
1984-023B		14787	ESA	05 MAR	06 APR 93	
1984-024B		14791	USSR	06 MAR	06 APR 84	
1984-024C		14792	USSR	06 MAR	31 MAR 84	
1984-025A	COSMOS 1542	14793	USSR	07 MAR	21 MAR 84	
1984-025B		14794	USSR	07 MAR	21 MAR 84	
1984-025C		14847	USSR	07 MAR	05 MAY 84	
1984-025D		14848	USSR	07 MAR	13 OCT 85	
1984-025E		14851	USSR	07 MAR	29 MAY 84	
1984-025F		14852	USSR	07 MAR	29 MAY 84	
1984-025G		14856	USSR	07 MAR	05 JUN 84	
1984-026A	COSMOS 1543	14797	USSR	10 MAR	05 APR 84	
1984-026B		14798	USSR	10 MAR	27 MAR 84	
1984-028B		14822	USSR	16 MAR	18 MAR 84	
1984-028C		14823	USSR	16 MAR	17 MAR 84	
1984-028E		14829	USSR	16 MAR	27 OCT 86	
1984-029B		14826	USSR	16 MAR	26 APR 84	
1984-029C		14827	USSR	16 MAR	20 APR 84	
1984-030A	COSMOS 1545	14849	USSR	21 MAR	05 APR 84	
1984-030B		14850	USSR	21 MAR	29 MAR 84	
1984-030C		14855	USSR	21 MAR	29 MAR 84	
1984-030D		14889	USSR	21 MAR	15 APR 84	
1984-030E		14890	USSR	21 MAR	04 MAY 84	
1984-030F		14891	USSR	21 MAR	29 MAY 84	
1984-030G		14892	USSR	21 MAR	11 APR 84	
1984-030H		14893	USSR	21 MAR	08 APR 84	
1984-030J		14895	USSR	21 MAR	10 MAY 84	
1984-030K		14896	USSR	21 MAR	25 APR 84	
1984-031B		14868	USSR	29 MAR	31 MAR 84	
1984-031C		14869	USSR	29 MAR	29 MAR 84	
1984-031E		14888	USSR	29 MAR	27 OCT 86	
1984-032A	SOYUZ T-11	14872	USSR	03 APR	02 OCT 84	
1984-032B		14873	USSR	03 APR	05 APR 84	
1984-032C		15345	USSR	03 APR	12 DEC 86	
1984-032D		15346	USSR	03 APR	05 DEC 84	
1984-033B		14885	USSR	04 APR	14 MAY 84	
1984-033C		14886	USSR	04 APR	01 MAY 84	
1984-034A	STS 41C	14897	US	06 APR	13 APR 84	BT
1984-034B	LDEF	14898	US	06 APR	20 JAN 90	35
1984-036A	COSMOS 1548	14902	USSR	10 APR	25 MAY 84	
1984-036B		14903	USSR	10 APR	13 APR 84	
1984-036C		14991	USSR	10 APR	25 MAY 84	
1984-036D		14992	USSR	10 APR	25 MAY 84	
1984-038A	PROGRESS 20	14932	USSR	15 APR	07 MAY 84	
1984-038B		14933	USSR	15 APR	17 APR 84	
1984-039A		14935	US	17 APR	13 AUG 84	
1983-034L		14936	USSR	19 APR	05 SEP 84	
1983-034M		14937	USSR	19 APR	19 AUG 84	
1984-040A	COSMOS 1549	14938	USSR	19 APR	03 MAY 84	
1984-040B		14939	USSR	19 APR	28 APR 84	
1984-040C		14952	USSR	19 APR	29 AUG 84	
1984-040D		14953	USSR	19 APR	14 MAY 86	
1984-040E		14954	USSR	19 APR	23 JUN 84	
1984-040F		14955	USSR	19 APR	07 MAY 84	
1984-040G		14956	USSR	19 APR	13 OCT 84	
1984-041B		14941	USSR	22 APR	24 APR 84	
1984-041C		14942	USSR	22 APR	22 APR 84	
1984-041E		14944	USSR	22 APR	09 DEC 90	
1984-041F		14945	USSR	22 APR	27 OCT 86	
1984-042A	PROGRESS 21	14961	USSR	07 MAY	26 MAY 84	
1984-042B		14962	USSR	07 MAY	10 MAY 84	
1984-044A	COSMOS 1551	14967	USSR	11 MAY	23 MAY 84	
1984-044B		14968	USSR	11 MAY	17 MAY 84	
1984-044C		14986	USSR	11 MAY	24 MAY 84	
1984-044D		14987	USSR	11 MAY	25 MAY 84	
1984-044E		14988	USSR	11 MAY	31 MAY 84	
1984-044F		14990	USSR	11 MAY	27 MAY 84	
1984-045A	COSMOS 1552	14971	USSR	14 MAY	03 NOV 84	
1984-045B		14972	USSR	14 MAY	19 MAY 84	
1984-045C		15379	USSR	14 MAY	04 NOV 84	
1984-045D		15380	USSR	14 MAY	05 NOV 84	
1984-045E		15381	USSR	14 MAY	05 NOV 84	
1984-047D		14980	USSR	19 MAY	21 MAY 84	
1984-047E		14981	USSR	19 MAY	20 MAY 84	
1984-048A	COSMOS 1557	14982	USSR	22 MAY	04 JUN 84	
1984-048B		14983	USSR	22 MAY	26 MAY 84	
1984-048C		15023	USSR	22 MAY	11 JUN 84	
1984-048D		15024	USSR	22 MAY	06 JUN 84	
1984-048E		15025	USSR	22 MAY	06 JUN 84	
1984-048F		15026	USSR	22 MAY	06 JUN 84	
1984-049B		14989	ESA	23 MAY	17 MAR 86	
1984-050A	COSMOS 1558	14993	USSR	25 MAY	08 JUL 84	
1984-050B		14994	USSR	25 MAY	28 MAY 84	
1984-051A	PROGRESS 22	14996	USSR	28 MAY	15 JUL 84	
1984-051B		14997	USSR	28 MAY	30 MAY 84	
1984-053A	COSMOS 1567	15009	USSR	30 MAY	03 APR 88	
1984-053B		15010	USSR	30 MAY	31 MAY 84	
1984-054A	COSMOS 1568	15011	USSR	01 JUN	14 JUN 84	
1984-054B		15012	USSR	01 JUN	11 JUN 84	
1984-054C		15021	USSR	01 JUN	06 JUN 84	
1984-054D		15022	USSR	01 JUN	06 JUN 84	
1984-054E		15041	USSR	01 JUN	08 AUG 84	
1984-054F		15042	USSR	01 JUN	01 NOV 84	
1984-054G		15043	USSR	01 JUN	22 AUG 86	
1984-054H		15044	USSR	01 JUN	16 JUN 84	
1984-054J		15045	USSR	01 JUN	20 JUN 84	
1984-054K		15048	USSR	01 JUN	04 DEC 84	
1984-054L		15049	USSR	01 JUN	10 SEP 84	
1984-054M		15050	USSR	01 JUN	13 JUL 84	
1984-054N		15062	USSR	01 JUN	22 AUG 84	
1984-055B		15028	USSR	06 JUN	28 JUL 84	
1984-055C		15029	USSR	06 JUN	10 JUL 84	
1984-057A	INTELSAT V F9	15034	US	09 JUN	24 OCT 84	
1984-057B		15035	US	09 JUN	11 JUN 84	
1984-058A	COSMOS 1571	15036	USSR	11 JUN	26 JUN 84	
1984-058B		15037	USSR	11 JUN	24 JUN 84	

International Designation	Satellite Name	Number	Source	Launch	Decay	Note
1984-058C		15038	USSR	11 JUN	16 JUN 84	
1984-058D		15065	USSR	11 JUN	20 OCT 84	
1984-058E		15066	USSR	11 JUN	25 NOV 84	
1984-058F		15067	USSR	11 JUN	17 APR 86	
1984-058G		15068	USSR	11 JUN	01 JUL 84	
1984-058H		15069	USSR	11 JUN	22 SEP 84	
1984-058J		15072	USSR	11 JUN	08 SEP 84	
1984-060A	COSMOS 1572	15046	USSR	15 JUN	29 JUN 84	
1984-060B		15047	USSR	15 JUN	22 JUN 84	
1984-060C		15087	USSR	15 JUN	30 JUL 84	
1984-060D		15088	USSR	15 JUN	09 JUL 84	
1984-060E		15089	USSR	15 JUN	13 JUL 84	
1984-061A	COSMOS 1573	15051	USSR	19 JUN	28 JUN 84	
1984-061B		15052	USSR	19 JUN	24 JUN 84	
1984-061C		15079	USSR	19 JUN	01 JUL 84	
1984-061D		15082	USSR	19 JUN	15 JUL 84	
1984-061E		15083	USSR	19 JUN	30 JUN 84	
1984-061F		15084	USSR	19 JUN	30 JUN 84	
1984-061G		15086	USSR	19 JUN	05 JUL 84	
1984-063B		15058	USSR	22 JUN	24 JUN 84	
1984-063C		15059	USSR	22 JUN	22 JUN 84	
1984-063D		15075	USSR	22 JUN	27 OCT 86	
1984-064A	COSMOS 1575	15060	USSR	22 JUN	07 JUL 84	
1984-064B		15061	USSR	22 JUN	28 JUN 84	
1984-064C		15101	USSR	22 JUN	10 AUG 84	
1984-064D		15102	USSR	22 JUN	18 JUL 84	
1984-064E		15103	USSR	22 JUN	12 JUL 84	
1984-064F		15104	USSR	22 JUN	20 JUL 84	
1984-065A		15063	US	25 JUN	18 OCT 84	
1984-065B		15064	US	25 JUN	27 JUN 84	
1984-066A	COSMOS 1576	15070	USSR	26 JUN	24 AUG 84	
1984-066B		15073	USSR	26 JUN	29 JUN 84	
1984-066C		15074	USSR	26 JUN	01 JUL 84	
1984-066D		15222	USSR	26 JUN	25 AUG 84	
1984-068A	COSMOS 1578	15080	USSR	28 JUN	10 JAN 93	
1984-068B		15081	USSR	28 JUN	03 APR 90	
1984-069B		15327	USSR	29 JUN	30 SEP 84	
1984-069C		15328	USSR	29 JUN	05 NOV 84	
1984-070A	COSMOS 1580	15090	USSR	30 JUN	13 JUL 84	
1984-070B		15091	USSR	30 JUN	25 JUL 84	
1984-070C		15092	USSR	30 JUN	06 JUL 84	
1984-070D		15093	USSR	30 JUN	05 JUL 84	
1984-070E		15094	USSR	30 JUN	06 JUL 84	
1984-070F		15115	USSR	30 JUN	15 JUL 84	
1984-070G		15116	USSR	30 JUN	01 AUG 84	
1984-070H		15117	USSR	30 JUN	16 JUL 84	
1984-070J		15118	USSR	30 JUN	18 JUL 84	
1984-071B		15096	USSR	03 JUL	31 AUG 84	
1984-071C		15097	USSR	03 JUL	10 AUG 84	
1984-073A	SOYUZ T-12	15119	USSR	17 JUL	29 JUL 84	
1984-073B		15120	USSR	17 JUL	20 JUL 84	
1984-073C		15136	USSR	17 JUL	28 AUG 85	
1984-073D		15140	USSR	17 JUL	02 AUG 84	
1984-074A	COSMOS 1582	15121	USSR	19 JUL	02 AUG 84	
1984-074B		15122	USSR	19 JUL	27 JUL 84	
1984-074C		15150	USSR	19 JUL	09 SEP 84	
1984-074D		15151	USSR	19 JUL	14 AUG 84	
1984-074E		15154	USSR	19 JUL	15 AUG 84	
1984-074F		15155	USSR	19 JUL	18 AUG 84	
1984-075A	COSMOS 1583	15123	USSR	24 JUL	08 AUG 84	
1984-075B		15124	USSR	24 JUL	07 AUG 84	
1984-075C		15125	USSR	24 JUL	01 AUG 84	
1984-075D		15126	USSR	24 JUL	27 JUL 84	
1984-075E		15169	USSR	24 JUL	01 FEB 85	
1984-075F		15170	USSR	24 JUL	19 DEC 84	
1984-075G		15173	USSR	24 JUL	30 NOV 86	
1984-075H		15178	USSR	24 JUL	26 OCT 84	
1984-075J		15179	USSR	24 JUL	25 OCT 84	
1984-075K		15180	USSR	24 JUL	07 SEP 84	
1984-076A	COSMOS 1584	15131	USSR	27 JUL	10 AUG 84	
1984-076B		15132	USSR	27 JUL	29 JUL 84	
1984-076C		15185	USSR	27 JUL	25 AUG 84	
1984-076D		15186	USSR	27 JUL	13 AUG 84	
1984-076E		15187	USSR	27 JUL	11 AUG 84	
1984-076F		15191	USSR	27 JUL	14 AUG 84	
1984-077A	COSMOS 1585	15142	USSR	31 JUL	28 SEP 84	
1984-077B		15143	USSR	31 JUL	03 AUG 84	
1984-078B		15145	USSR	01 AUG	03 AUG 84	
1984-078C		15146	USSR	01 AUG	02 AUG 84	
1984-078D		15160	USSR	01 AUG	12 AUG 87	
1984-078E		15161	USSR	01 AUG	27 OCT 86	
1984-079B		15148	USSR	02 AUG	19 SEP 84	
1984-079C		15149	USSR	02 AUG	07 SEP 84	
1984-080B		15153	JAPAN	02 AUG	11 OCT 84	
1984-080D		15162	JAPAN	02 AUG	12 AUG 84	
1984-081C		15165	ESA	04 AUG	06 APR 90	
1984-082A	COSMOS 1587	15163	USSR	06 AUG	31 AUG 84	
1984-082B		15164	USSR	06 AUG	18 AUG 84	
1984-082C		15238	USSR	06 AUG	18 NOV 84	
1984-082D		15239	USSR	06 AUG	15 SEP 84	
1984-082E		15240	USSR	06 AUG	27 DEC 84	
1984-082F		15241	USSR	06 AUG	26 JAN 85	
1984-082G		15242	USSR	06 AUG	26 NOV 86	
1984-082H		15243	USSR	06 AUG	06 MAR 85	
1984-083A	COSMOS 1588	15167	USSR	07 AUG	17 FEB 88	
1984-083AA		16664	USSR	07 AUG	11 APR 86	
1984-083AB		16665	USSR	07 AUG	20 APR 86	
1984-083AC		16666	USSR	07 AUG	11 APR 86	
1984-083AD		16672	USSR	07 AUG	31 MAY 86	
1984-083AE		16673	USSR	07 AUG	18 APR 89	
1984-083AF		16674	USSR	07 AUG	20 APR 86	
1984-083AG		16675	USSR	07 AUG	06 JUN 86	
1984-083AH		16689	USSR	07 AUG	28 MAY 86	
1984-083AJ		16690	USSR	07 AUG	02 JUL 86	
1984-083AK		16691	USSR	07 AUG	21 MAY 86	
1984-083AL		16702	USSR	07 AUG	17 JUL 86	
1984-083AM		16703	USSR	07 AUG	02 AUG 86	
1984-083AN		16704	USSR	07 AUG	30 JUN 86	
1984-083AP		16705	USSR	07 AUG	01 JUN 86	
1984-083AQ		16706	USSR	07 AUG	21 MAY 86	
1984-083AR		16707	USSR	07 AUG	13 JUN 86	
1984-083AS		16708	USSR	07 AUG	19 JUN 86	
1984-083AT		16709	USSR	07 AUG	16 JUL 86	
1984-083AU		16710	USSR	07 AUG	31 MAY 86	
1984-083AV		16711	USSR	07 AUG	23 MAY 86	
1984-083AW		16712	USSR	07 AUG	19 JUN 86	
1984-083AX		16713	USSR	07 AUG	13 MAY 86	
1984-083B		15168	USSR	07 AUG	08 AUG 84	
1984-083C		16617	USSR	07 AUG	13 JUL 86	
1984-083D		16626	USSR	07 AUG	27 MAR 86	
1984-083E		16627	USSR	07 AUG	11 OCT 86	
1984-083F		16628	USSR	07 AUG	08 MAR 86	
1984-083G		16629	USSR	07 AUG	04 APR 86	
1984-083H		16632	USSR	07 AUG	04 APR 86	
1984-083J		16633	USSR	07 AUG	30 MAY 88	
1984-083K		16634	USSR	07 AUG	29 APR 86	
1984-083L		16635	USSR	07 AUG	21 APR 86	
1984-083M		16636	USSR	07 AUG	24 APR 86	
1984-083N		16637	USSR	07 AUG	26 APR 86	
1984-083P		16638	USSR	07 AUG	17 MAY 86	
1984-083Q		16639	USSR	07 AUG	01 DEC 87	
1984-083R		16640	USSR	07 AUG	05 APR 86	
1984-083S		16641	USSR	07 AUG	31 MAR 86	
1984-083T		16642	USSR	07 AUG	04 APR 86	
1984-083U		16651	USSR	07 AUG	13 APR 86	
1984-083V		16652	USSR	07 AUG	13 MAY 86	
1984-083W		16653	USSR	07 AUG	20 APR 86	
1984-083X		16654	USSR	07 AUG	25 MAR 86	
1984-083Y		16655	USSR	07 AUG	08 JUL 86	
1984-083Z		16656	USSR	07 AUG	16 MAY 86	
1984-085B		15183	USSR	10 AUG	11 SEP 84	
1984-085C		15184	USSR	10 AUG	05 SEP 84	
1984-086A	PROGRESS 23	15193	USSR	14 AUG	28 AUG 84	
1984-086B		15194	USSR	14 AUG	17 AUG 84	
1984-087A	COSMOS 1590	15197	USSR	16 AUG	30 AUG 84	
1984-087B		15198	USSR	16 AUG	23 AUG 84	
1984-087C		15228	USSR	16 AUG	12 SEP 84	
1984-087D		15229	USSR	16 AUG	09 SEP 84	
1984-087E		15230	USSR	16 AUG	05 SEP 84	
1984-087F		15231	USSR	16 AUG	02 OCT 84	
1984-089B		15215	USSR	24 AUG	17 SEP 84	
1984-089C		15216	USSR	24 AUG	13 SEP 84	
1984-090B		15220	USSR	24 AUG	27 AUG 84	
1984-090C		15221	USSR	24 AUG	25 AUG 84	
1984-090D		15224	USSR	24 AUG	18 APR 85	
1984-090E		15225	USSR	24 AUG	27 OCT 86	
1984-092A	COSMOS 1591	15232	USSR	30 AUG	13 SEP 84	
1984-092B		15233	USSR	30 AUG	05 SEP 84	
1984-092C		15289	USSR	30 AUG	15 SEP 84	
1984-092D		15290	USSR	30 AUG	14 OCT 84	
1984-092E		15291	USSR	30 AUG	23 SEP 84	
1984-092F		15294	USSR	30 AUG	28 SEP 84	
1984-092G		15295	USSR	30 AUG	29 SEP 84	
1984-092H		15296	USSR	30 AUG	15 SEP 84	
1984-093A	STS 41D	15234	US	30 AUG	05 SEP 84	BT
1984-094A	COSMOS 1592	15257	USSR	04 SEP	18 SEP 84	
1984-094B		15258	USSR	04 SEP	15 SEP 84	
1984-094C		15298	USSR	04 SEP	26 SEP 84	
1984-094D		15299	USSR	04 SEP	24 OCT 84	
1984-094E		15300	USSR	04 SEP	02 OCT 84	
1984-094F		15301	USSR	04 SEP	23 SEP 84	
1984-094G		15302	USSR	04 SEP	23 SEP 84	
1984-095D		15262	USSR	04 SEP	06 SEP 84	
1984-095E		15263	USSR	04 SEP	05 SEP 84	
1984-096B		15268	USSR	07 SEP	25 OCT 84	
1984-096C		15269	USSR	07 SEP	09 OCT 84	
1984-098A	PRC 16	15279	PRC	12 SEP	29 SEP 84	
1984-098B		15280	PRC	12 SEP	25 SEP 84	
1984-098C		15285	PRC	12 SEP	13 SEP 84	
1984-098D		15286	PRC	12 SEP	13 SEP 84	
1984-099A	COSMOS 1597	15287	USSR	13 SEP	26 SEP 84	
1984-099B		15288	USSR	13 SEP	19 SEP 84	
1984-099C		15320	USSR	13 SEP	04 OCT 84	
1984-099D		15321	USSR	13 SEP	27 SEP 84	
1984-099E		15322	USSR	13 SEP	29 SEP 84	
1984-099F		15323	USSR	13 SEP	28 SEP 84	
1984-101B		15309	US	21 SEP	19 NOV 84	
1984-101C		15310	US	21 SEP	12 JUN 85	
1984-102A	COSMOS 1599	15318	USSR	25 SEP	20 NOV 84	
1984-102B		15319	USSR	25 SEP	27 SEP 84	
1984-103A	COSMOS 1600	15324	USSR	27 SEP	11 OCT 84	
1984-103B		15325	USSR	27 SEP	14 OCT 84	
1984-103C		15356	USSR	27 SEP	29 OCT 86	
1984-103D		15357	USSR	27 SEP	23 MAR 85	
1984-103E		15358	USSR	27 SEP	01 MAY 85	
1984-103F		15361	USSR	27 SEP	17 DEC 84	

International Designation	Satellite Name	Number	Source	Launch	Decay	Note
1984-103G		15364	USSR	27 SEP	08 JAN 85	
1984-103H		15373	USSR	27 SEP	26 OCT 84	
1984-104A	COSMOS 1601	15326	USSR	27 SEP	29 NOV 89	
1984-104AA		16069	USSR	27 SEP	08 MAY 88	
1984-104AB		16077	USSR	27 SEP	11 JUN 88	
1984-104AC		16109	USSR	27 SEP	13 APR 88	
1984-104AD		16954	USSR	27 SEP	06 AUG 88	
1984-104AE		16955	USSR	27 SEP	18 JUN 88	
1984-104AF		17161	USSR	27 SEP	15 MAY 88	
1984-104B		15329	USSR	27 SEP	09 SEP 89	
1984-104C		15564	USSR	27 SEP	14 NOV 85	
1984-104D		15565	USSR	27 SEP	11 NOV 85	
1984-104E		15566	USSR	27 SEP	15 DEC 85	
1984-104F		15567	USSR	27 SEP	20 DEC 85	
1984-104G		15607	USSR	27 SEP	22 MAR 88	
1984-104H		15608	USSR	27 SEP	03 MAR 88	
1984-104J		15980	USSR	27 SEP	25 MAY 88	
1984-104K		15981	USSR	27 SEP	20 MAR 88	
1984-104M		15984	USSR	27 SEP	04 MAY 88	
1984-104N		15985	USSR	27 SEP	03 APR 88	
1984-104P		15987	USSR	27 SEP	07 MAY 88	
1984-104Q		15988	USSR	27 SEP	20 MAY 88	
1984-104R		15989	USSR	27 SEP	17 MAR 88	
1984-104S		15990	USSR	27 SEP	21 APR 88	
1984-104T		15991	USSR	27 SEP	17 APR 88	
1984-104U		16003	USSR	27 SEP	06 MAY 88	
1984-104V		16004	USSR	27 SEP	30 MAR 88	
1984-104W		16005	USSR	27 SEP	26 MAR 88	
1984-104X		16006	USSR	27 SEP	15 APR 88	
1984-104Y		16067	USSR	27 SEP	17 MAR 88	
1984-104Z		16068	USSR	27 SEP	07 APR 88	
1984-106AA		22157	USSR	28 SEP	10 OCT 92	
1984-106AB		22158	USSR	28 SEP	20 NOV 92	
1984-106AC		22159	USSR	28 SEP	11 OCT 92	
1984-106AD		22160	USSR	28 SEP	13 OCT 92	
1984-106B		15334	USSR	28 SEP	06 JUL 85	
1984-106D		15336	USSR	28 SEP	01 OCT 84	
1984-106E		15337	USSR	28 SEP	29 SEP 84	
1984-106G		17358	USSR	28 SEP	03 JAN 93	
1984-106H		22140	USSR	28 SEP	06 DEC 92	
1984-106J		22141	USSR	28 SEP	11 OCT 92	
1984-106K		22142	USSR	28 SEP	10 OCT 92	
1984-106L		22143	USSR	28 SEP	11 OCT 92	
1984-106M		22144	USSR	28 SEP	19 OCT 92	
1984-106N		22145	USSR	28 SEP	05 NOV 92	
1984-106P		22146	USSR	28 SEP	27 MAR 93	
1984-106Q		22147	USSR	28 SEP	08 OCT 92	
1984-106R		22148	USSR	28 SEP	13 NOV 92	
1984-106S		22149	USSR	28 SEP	11 OCT 92	
1984-106T		22150	USSR	28 SEP	14 OCT 92	
1984-106U		22151	USSR	28 SEP	14 OCT 92	
1984-106V		22152	USSR	28 SEP	16 NOV 92	
1984-106W		22153	USSR	28 SEP	23 OCT 92	
1984-106X		22154	USSR	28 SEP	20 OCT 92	
1984-106Y		22155	USSR	28 SEP	22 OCT 92	
1984-106Z		22156	USSR	28 SEP	14 OCT 92	
1984-107B		15351	USSR	04 OCT	02 DEC 84	
1984-107C		15352	USSR	04 OCT	07 NOV 84	
1984-108A	STS 41G	15353	US	05 OCT	13 OCT 84	BT
1984-110B		15363	US	12 OCT	26 JUL 88	
1984-110C		15371	US	12 OCT	20 SEP 87	
1984-110D		15372	US	12 OCT	04 JUL 87	
1984-112B		15502	USSR	31 OCT	28 MAR 85	
1984-112D		15504	USSR	31 OCT	01 OCT 85	
1984-112E		15507	USSR	31 OCT	07 FEB 85	
1984-112F		15508	USSR	31 OCT	03 JUL 85	
1984-112G		15511	USSR	31 OCT	05 MAR 85	
1984-113A	STS 51A	15382	US	08 NOV	16 NOV 84	BT
1984-114D		15389	ESA	10 NOV	27 OCT 86	
1984-116A	COSMOS 1608	15393	USSR	14 NOV	17 DEC 84	
1984-116B		15394	USSR	14 NOV	17 NOV 84	
1984-116C		15397	USSR	14 NOV	16 NOV 84	
1984-117A	COSMOS 1609	15395	USSR	14 NOV	28 NOV 84	
1984-117B		15396	USSR	14 NOV	25 NOV 84	
1984-117C		15400	USSR	14 NOV	17 NOV 84	
1984-117D		15408	USSR	14 NOV	06 JUL 85	
1984-117E		15409	USSR	14 NOV	26 APR 87	
1984-117F		15410	USSR	14 NOV	01 MAY 85	
1984-117G		15411	USSR	14 NOV	01 DEC 84	
1984-117H		15412	USSR	14 NOV	02 APR 85	
1984-117J		15413	USSR	14 NOV	08 MAR 85	
1984-117K		15418	USSR	14 NOV	30 DEC 84	
1984-117L		15419	USSR	14 NOV	22 FEB 85	
1984-117M		15421	USSR	14 NOV	11 DEC 84	
1984-117N		15428	USSR	14 NOV	24 DEC 84	
1984-119A	COSMOS 1611	15403	USSR	21 NOV	11 JAN 85	
1984-119B		15404	USSR	21 NOV	25 NOV 84	
1984-119C		15405	USSR	21 NOV	22 NOV 84	
1984-120A	COSMOS 1612	15406	USSR	27 NOV	31 JAN 85	
1984-120B		15407	USSR	27 NOV	02 DEC 84	
1984-121A	COSMOS 1613	15414	USSR	29 NOV	24 DEC 84	
1984-121B		15415	USSR	29 NOV	12 DEC 84	
1984-121C		15416	USSR	29 NOV	06 DEC 84	
1984-121D		15417	USSR	29 NOV	01 DEC 84	
1984-121E		15456	USSR	29 NOV	05 MAY 87	
1984-121F		15457	USSR	29 NOV	13 APR 85	
1984-121G		15458	USSR	29 NOV	05 JUL 85	
1984-121H		15459	USSR	29 NOV	05 OCT 85	

International Designation	Satellite Name	Number	Source	Launch	Decay	Note
1984-121J		15460	USSR	29 NOV	12 MAY 85	
1984-121K		15461	USSR	29 NOV	11 JAN 85	
1984-121L		15462	USSR	29 NOV	08 JAN 85	
1984-121M		15463	USSR	29 NOV	21 JAN 85	
1984-122B		15424	US	04 DEC	23 DEC 84	
1984-123C		15441	US	12 DEC	13 NOV 92	
1984-124B		15430	USSR	14 DEC	11 JAN 85	
1984-124C		15431	USSR	14 DEC	20 JAN 85	
1984-124D		15435	USSR	14 DEC	21 JAN 85	
1984-124E		15436	USSR	14 DEC	12 JAN 85	
1984-124F		15437	USSR	14 DEC	09 JAN 85	
1984-124G		15438	USSR	14 DEC	20 DEC 84	
1984-125B		15433	USSR	15 DEC	16 DEC 84	
1984-125C		15434	USSR	15 DEC	16 DEC 84	
1984-125E	V1 DESCENT CRAFT	15858	USSR	15 DEC	10 JUN 85	
1984-125F	VEGA 1 BALLOON	15859	USSR	15 DEC	10 JUN 85	
1984-126A	COSMOS 1614	15442	USSR	19 DEC	19 DEC 84	
1984-126B		15443	USSR	19 DEC	21 DEC 84	
1984-126C		15444	USSR	19 DEC	20 DEC 84	
1984-126D		15445	USSR	19 DEC	20 DEC 84	
1984-127A	COSMOS 1615	15446	USSR	20 DEC	15 APR 90	
1984-127B		15448	USSR	20 DEC	09 APR 89	
1984-128C		15451	USSR	21 DEC	23 DEC 84	
1984-128D		15452	USSR	21 DEC	23 DEC 84	
1984-128E	V2 DESCENT CRAFT	15856	USSR	21 DEC	14 JUN 85	
1984-128F	VEGA 2 BALLOON	15857	USSR	21 DEC	14 JUN 85	
1984-129C		15455	US	22 DEC	17 JAN 85	

1985 LAUNCHES

International Designation	Satellite Name	Number	Source	Launch	Decay	Note
1985-001C		15466	JAPAN	07 JAN	22 MAY 85	
1985-002A	COSMOS 1616	15467	USSR	09 JAN	04 MAR 85	
1985-002B		15468	USSR	09 JAN	15 JAN 85	
1985-004B		15477	USSR	16 JAN	27 MAR 85	
1985-004C		15478	USSR	16 JAN	21 MAR 85	
1985-005A	COSMOS 1623	15479	USSR	16 JAN	30 JAN 85	
1985-005B		15480	USSR	16 JAN	05 FEB 85	
1985-005C		15498	USSR	16 JAN	09 NOV 85	
1985-005D		15499	USSR	16 JAN	31 MAR 87	
1985-005E		15500	USSR	16 JAN	22 JUL 85	
1985-005F		15501	USSR	16 JAN	19 MAY 85	
1985-005G		15509	USSR	16 JAN	26 MAY 85	
1985-005H		15510	USSR	16 JAN	26 FEB 85	
1985-007B		15485	USSR	18 JAN	20 JAN 85	
1985-007C		15486	USSR	18 JAN	19 JAN 85	
1985-007E		15488	USSR	18 JAN	21 SEP 85	
1985-008A	COSMOS 1625	15492	USSR	23 JAN	25 JAN 85	
1985-008B		15493	USSR	23 JAN	24 JAN 85	
1985-009C		15497	USSR	24 JAN	02 FEB 89	
1985-010A	STS 51C	15496	US	24 JAN	27 JAN 85	
1985-012A	COSMOS 1628	15514	USSR	06 FEB	20 FEB 85	
1985-012B		15515	USSR	06 FEB	21 FEB 85	
1985-012C		15568	USSR	06 FEB	12 SEP 85	
1985-012D		15569	USSR	06 FEB	14 JUL 87	
1985-012E		15570	USSR	06 FEB	31 MAY 85	
1985-012F		15571	USSR	06 FEB	28 JUL 85	
1985-012G		15572	USSR	06 FEB	09 JAN 86	
1985-012H		15573	USSR	06 FEB	26 MAR 85	
1985-012J		15579	USSR	06 FEB	14 MAR 85	
1985-012K		15580	USSR	06 FEB	18 MAR 85	
1985-012L		15591	USSR	06 FEB	11 APR 85	
1985-015D		15563	ESA	08 FEB	05 APR 86	
1985-016B		15575	USSR	21 FEB	24 FEB 85	
1985-016C		15576	USSR	21 FEB	22 FEB 85	
1985-016D		15577	USSR	21 FEB	27 OCT 86	
1985-016E		15578	USSR	21 FEB	27 OCT 86	
1985-017A	COSMOS 1630	15582	USSR	27 FEB	23 APR 85	
1985-017B		15583	USSR	27 FEB	05 MAR 85	
1985-018A	COSMOS 1631	15584	USSR	27 FEB	08 DEC 90	
1985-018B		15585	USSR	27 FEB	20 OCT 89	
1985-019A	COSMOS 1632	15589	USSR	01 MAR	15 MAR 85	
1985-019B		15590	USSR	01 MAR	04 MAR 85	
1985-019C		15609	USSR	01 MAR	26 MAR 85	
1985-019D		15610	USSR	01 MAR	17 MAR 85	
1985-019E		15611	USSR	01 MAR	24 MAR 85	
1985-019F		15612	USSR	01 MAR	20 MAR 85	
1985-021C		15613	US	13 MAR	20 SEP 92	
1985-024B		15627	USSR	22 MAR	25 MAR 85	
1985-024C		15628	USSR	22 MAR	23 MAR 85	
1985-024E		15632	USSR	22 MAR	07 AUG 86	
1985-024F		15633	USSR	22 MAR	20 JAN 86	
1985-026A	COSMOS 1643	15634	USSR	25 MAR	18 OCT 85	
1985-026B		15635	USSR	25 MAR	29 MAR 85	
1985-027A	COSMOS 1644	15636	USSR	03 APR	17 APR 85	
1985-027B		15637	USSR	03 APR	21 APR 85	
1985-027C		15638	USSR	03 APR	07 APR 85	
1985-027D		15639	USSR	03 APR	06 APR 85	
1985-027E		15640	USSR	03 APR	07 APR 85	
1985-027F		15647	USSR	03 APR	07 JAN 86	
1985-027G		15648	USSR	03 APR	15 OCT 85	
1985-027H		15649	USSR	03 APR	26 APR 85	
1985-027J		15650	USSR	03 APR	09 JUL 87	
1985-027K		15651	USSR	03 APR	26 SEP 85	
1985-027L		15652	USSR	03 APR	14 JUN 85	
1985-027M		15657	USSR	03 APR	07 AUG 85	
1985-027N		15658	USSR	03 APR	16 JUN 85	
1985-028A	STS 51D	15641	US	12 APR	19 APR 85	
1985-029A	COSMOS 1645	15645	USSR	16 APR	29 APR 85	

International Designation	Satellite Name	Number	Source	Launch	Decay	Note
1985-029B		15646	USSR	16 APR	10 MAY 85	
1985-029C		15667	USSR	16 APR	23 JUN 85	
1985-029D		15668	USSR	16 APR	13 MAY 85	
1985-029E		15669	USSR	16 APR	04 JUN 85	
1985-029F		15670	USSR	16 APR	19 MAY 85	
1985-029G		15671	USSR	16 APR	01 MAY 85	
1985-030A	COSNOS 1646	15653	USSR	18 APR	12 MAY 88	
1985-030AA		18799	USSR	18 APR	05 JUL 88	
1985-030B		15654	USSR	18 APR	19 APR 85	
1985-030C		18619	USSR	18 APR	27 DEC 87	
1985-030D		18620	USSR	18 APR	25 MAR 88	
1985-030E		18621	USSR	18 APR	06 DEC 87	
1985-030F		18622	USSR	18 APR	06 JAN 88	
1985-030G		18623	USSR	18 APR	03 JAN 88	
1985-030H		18624	USSR	18 APR	08 DEC 87	
1985-030J		18628	USSR	18 APR	15 JAN 88	
1985-030K		18629	USSR	18 APR	15 JAN 88	
1985-030L		18630	USSR	18 APR	06 JAN 88	
1985-030M		18652	USSR	18 APR	30 DEC 87	
1985-030N		18653	USSR	18 APR	20 JAN 88	
1985-030P		18654	USSR	18 APR	02 APR 88	
1985-030Q		18655	USSR	18 APR	24 AUG 88	
1985-030R		18656	USSR	18 APR	31 DEC 87	
1985-030S		18657	USSR	18 APR	22 JAN 88	
1985-030T		18658	USSR	18 APR	09 FEB 88	
1985-030U		18659	USSR	18 APR	15 JAN 88	
1985-030V		18660	USSR	18 APR	28 DEC 87	
1985-030W		18661	USSR	18 APR	12 FEB 88	
1985-030X		18662	USSR	18 APR	02 JAN 88	
1985-030Y		18663	USSR	18 APR	28 JUL 88	
1985-030Z		18664	USSR	18 APR	21 DEC 87	
1985-031A	COSMOS 1647	15655	USSR	19 APR	11 JUN 85	
1985-031B		15656	USSR	19 APR	23 APR 85	
1985-032A	COSMOS 1648	15659	USSR	25 APR	06 MAY 85	
1985-032B		15660	USSR	25 APR	27 APR 85	
1985-032C		15672	USSR	25 APR	10 MAY 85	
1985-032D		15673	USSR	25 APR	15 MAY 85	
1985-032E		15674	USSR	25 APR	21 MAY 85	
1985-032F		15675	USSR	25 APR	11 MAY 85	
1985-032G		15676	USSR	25 APR	09 MAY 85	
1985-033B		15662	USSR	26 APR	20 JUN 85	
1985-033C		15663	USSR	26 APR	28 JUN 85	
1985-034A	STS 51B	15665	US	29 APR	06 MAY 85	
1985-034B	NUSAT 1	15666	US	29 APR	15 DEC 86	35
1985-036A	COSMOS 1649	15694	USSR	15 MAY	29 MAY 85	
1985-036B		15695	USSR	15 MAY	28 MAY 85	
1985-036C		15696	USSR	15 MAY	18 MAY 85	
1985-036D		15742	USSR	15 MAY	19 MAR 86	
1985-036E		15743	USSR	15 MAY	10 AUG 85	
1985-036F		15744	USSR	15 MAY	29 JUN 85	
1985-036G		15745	USSR	15 MAY	01 OCT 85	
1985-036H		15746	USSR	15 MAY	17 DEC 85	
1985-036J		15747	USSR	15 NAY	03 JUN 85	
1985-036K		15753	USSR	15 MAY	14 JUN 85	
1985-036L		15754	USSR	15 MAY	26 OCT 87	
1985-036M		15765	USSR	15 MAY	17 JUN 85	
1985-037D		15700	USSR	17 MAY	20 MAY 85	
1985-037E		15701	USSR	17 MAY	18 MAY 85	
1985-038A	COSMOS 1653	15732	USSR	22 MAY	05 JUN 85	
1985-038B		15733	USSR	22 MAY	02 JUN 85	
1985-038C		15790	USSR	22 MAY	26 JUN 85	
1985-038D		15791	USSR	22 MAY	06 JUN 85	
1985-038E		15792	USSR	22 MAY	13 JUL 85	
1985-038F		15793	USSR	22 MAY	18 JUN 85	
1985-038G		15794	USSR	22 MAY	06 JUN 85	
1985-039A	COSMOS 1654	15734	USSR	23 MAY	07 AUG 85	
1985-039B		15735	USSR	23 MAY	29 MAY 85	
1985-039C		15851	USSR	23 MAY	14 AUG 85	
1985-039D		15853	USSR	23 MAY	27 JUN 85	
1985-039G		15893	USSR	23 MAY	23 JUN 85	
1985-039H		15894	USSR	23 MAY	23 JUN 85	
1985-039J		15895	USSR	23 MAY	23 JUN 85	
1985-039K		15896	USSR	23 MAY	24 JUN 85	
1985-039L		15897	USSR	23 MAY	24 JUN 85	
1985-039M		15898	USSR	23 MAY	25 JUN 85	
1985-039N		15899	USSR	23 MAY	25 JUN 85	
1985-039P		15900	USSR	23 MAY	25 JUN 85	
1985-039Q		15901	USSR	23 MAY	27 JUN 85	
1985-039R		15902	USSR	23 MAY	28 JUN 85	
1985-039S		15903	USSR	23 MAY	29 JUN 85	
1985-039T		15904	USSR	23 MAY	30 JUN 85	
1985-039U		15905	USSR	23 MAY	06 JUL 85	
1985-040B		15739	USSR	29 MAY	04 JUL 85	
1985-040C		15740	USSR	29 MAY	25 JUN 85	
1985-040E		15748	USSR	29 MAY	17 JUN 85	
1985-040F		15749	USSR	29 MAY	31 MAY 85	
1985-040G		15750	USSR	29 MAY	01 JUN 85	
1985-040H		15795	USSR	29 MAY	23 JUN 85	
1985-042B		15770	USSR	30 MAY	03 JUN 85	
1985-042C		15771	USSR	30 MAY	31 MAY 85	
1985-043A	SOYUZ T-13	15804	USSR	06 JUN	26 SEP 85	
1985-043B		15805	USSR	06 JUN	08 JUN 85	
1985-043C		16097	USSR	06 JUN	21 JAN 86	
1985-039E		15854	USSR	07 JUN	18 AUG 85	
1985-039F		15855	USSR	07 JUN	07 JUL 85	
1985-044A	COSMOS 1657	15806	USSR	07 JUN	21 JUN 85	
1985-044B		15807	USSR	07 JUN	11 JUN 85	
1985-044C		15843	USSR	07 JUN	29 JUN 85	
1985-044D		15844	USSR	07 JUN	03 JUL 85	
1985-044E		15848	USSR	07 JUN	12 JUL 85	
1985-044F		15849	USSR	07 JUN	24 JUN 85	
1985-044G		15852	USSR	07 JUN	18 AUG 85	
1985-045B		15809	USSR	11 JUN	24 AUG 85	
1985-045C		15810	USSR	11 JUN	30 JUL 85	
1985-046A	COSMOS 1659	15818	USSR	13 JUN	27 JUN 85	
1985-046B		15819	USSR	13 JUN	29 JUN 85	
1985-046C		15820	USSR	13 JUN	15 JUN 85	
1985-046D		15867	USSR	13 JUN	23 OCT 85	
1985-046E		15868	USSR	13 JUN	11 NOV 87	
1985-046F		15869	USSR	13 JUN	25 DEC 85	
1985-046G		15870	USSR	13 JUN	27 NOV 85	
1985-046H		15871	USSR	13 JUN	30 MAR 86	
1985-046J		15872	USSR	13 JUN	14 JUL 85	
1985-048A	STS 51G	15823	US	17 JUN	24 JUN 85	
1985-049B		15828	USSR	18 JUN	01 SEP 85	
1985-049C		15829	USSR	18 JUN	28 SEP 85	
1985-050A	COSMOS 1662	15833	USSR	19 JUN	16 NOV 89	CD
1985-050AA		17060	USSR	19 JUN	12 JUN 88	
1985-050AB		17137	USSR	19 JUN	29 OCT 87	
1985-050AC		17158	USSR	19 JUN	27 OCT 87	
1985-050AD		17818	USSR	19 JUN	25 JUL 88	
1985-050AE		17819	USSR	19 JUN	12 JUL 88	
1985-050B		15834	USSR	19 JUN	10 OCT 89	
1985-050C		15835	USSR	19 JUN	26 OCT 86	
1985-050D		16008	USSR	19 JUN	15 NOV 86	
1985-050E		16009	USSR	19 JUN	03 NOV 86	
1985-050F		16010	USSR	19 JUN	28 NOV 86	
1985-050G		16014	USSR	19 JUN	22 DEC 86	
1985-050H		16015	USSR	19 JUN	16 OCT 86	
1985-050J		16016	USSR	19 JUN	09 OCT 86	
1985-050K		16017	USSR	19 JUN	19 JAN 87	
1985-050L		16256	USSR	19 JUN	25 MAR 87	
1985-050M		16257	USSR	19 JUN	01 DEC 86	
1985-050N		16605	USSR	19 JUN	04 MAY 87	
1985-050P		16606	USSR	19 JUN	27 APR 87	
1985-050Q		16607	USSR	19 JUN	07 MAY 87	
1985-050R		16608	USSR	19 JUN	08 APR 87	
1985-050S		16956	USSR	19 JUN	07 OCT 87	
1985-050T		16957	USSR	19 JUN	15 SEP 87	
1985-050U		16958	USSR	19 JUN	09 SEP 87	
1985-050V		16959	USSR	19 JUN	26 MAY 88	
1985-050W		16960	USSR	19 JUN	09 SEP 87	
1985-050X		16980	USSR	19 JUN	12 SEP 87	
1985-050Y		16981	USSR	19 JUN	29 SEP 87	
1985-050Z		17059	USSR	19 JUN	20 JUN 88	
1985-048E	SPARTAN 1	15831	US	20 JUN	24 JUN 85	
1985-051A	PROGRESS 24	15838	USSR	21 JUN	15 JUL 85	
1985-051B		15839	USSR	21 JUN	24 JUN 85	
1985-052A	COSMOS 1663	15840	USSR	21 JUN	05 JUL 85	
1985-052B		15841	USSR	21 JUN	30 JUN 85	
1985-052C		15845	USSR	21 JUN	24 JUN 85	
1985-052D		15846	USSR	21 JUN	24 JUN 85	
1985-052E		15880	USSR	21 JUN	18 AUG 85	
1985-052F		15881	USSR	21 JUN	19 JUL 85	
1985-052G		15882	USSR	21 JUN	06 JUL 85	
1985-052H		15883	USSR	21 JUN	27 JUL 85	
1985-052J		15884	USSR	21 JUN	16 JUL 85	
1985-053A		15842	USSR	21 JUN	28 JUN 85	
1985-053B		15847	USSR	21 JUN	24 JUN 85	
1985-053C		15850	USSR	21 JUN	27 JUN 85	
1985-054A	COSMOS 1664	15860	USSR	26 JUN	05 JUL 85	
1985-054B		15861	USSR	26 JUN	10 JUL 85	
1985-054C		15866	USSR	26 JUN	30 JUN 85	
1985-054D		15885	USSR	26 JUN	16 JUL 85	
1985-054E		15886	USSR	26 JUN	20 JUL 85	
1985-054F		15887	USSR	26 JUN	07 JUL 85	
1985-054G		15888	USSR	26 JUN	08 JUL 85	
1985-057A	COSMOS 1665	15877	USSR	03 JUL	17 JUL 85	
1985-057B		15878	USSR	03 JUL	10 JUL 85	
1985-057C		15879	USSR	03 JUL	25 AUG 85	
1985-057D		15912	USSR	03 JUL	04 AUG 85	
1985-057E		15913	USSR	03 JUL	25 JUL 85	
1985-057F		15914	USSR	03 JUL	22 JUL 85	
1985-057G		15915	USSR	03 JUL	02 SEP 85	
1985-057H		15917	USSR	03 JUL	25 JUL 85	
1985-059A	COSMOS 1667	15891	USSR	10 JUL	17 JUL 85	
1985-059B		15892	USSR	10 JUL	19 JUL 85	
1985-059C		15920	USSR	10 JUL	27 JUL 85	
1985-060A	COSMOS 1668	15906	USSR	15 JUL	29 JUL 85	
1985-060B		15907	USSR	15 JUL	24 JUL 85	
1985-060C		15908	USSR	15 JUL	16 JUL 85	
1985-060D		15921	USSR	15 JUL	03 SEP 85	
1985-060E		15922	USSR	15 JUL	04 AUG 85	
1985-060F		15923	USSR	15 JUL	02 AUG 85	
1985-060G		15924	USSR	15 JUL	01 AUG 85	
1985-060H		15926	USSR	15 JUL	09 AUG 85	
1985-060J		15927	USSR	15 JUL	31 JUL 85	
1985-060K		15928	USSR	15 JUL	02 AUG 85	
1985-061B		15910	USSR	17 JUL	22 AUG 85	
1985-061C		15911	USSR	17 JUL	23 AUG 85	
1985-062A	COSMOS 1669	15918	USSR	19 JUL	30 AUG 85	
1985-062B		15919	USSR	19 JUL	23 JUL 85	
1985-063A	STS 51F	15925	US	29 JUL	06 AUG 85	
1985-063B	PDP	15929	US	29 JUL	06 AUG 85	35
1985-064B		16195	USSR	01 AUG	27 OCT 85	
1985-064C		16196	USSR	01 AUG	08 DEC 85	

International Designation	Satellite Name	Number	Source	Launch	Decay	Note
1985-064D		16213	USSR	01 AUG	22 FEB 88	
1985-064E		18506	USSR	01 AUG	28 SEP 88	
1985-065A	COSMOS 1671	15931	USSR	02 AUG	16 AUG 85	
1985-065B		15932	USSR	02 AUG	10 AUG 85	
1985-065C		15961	USSR	02 AUG	10 SEP 85	
1985-065D		15962	USSR	02 AUG	15 SEP 85	
1985-065E		15964	USSR	02 AUG	03 SEP 85	
1985-065F		15965	USSR	02 AUG	21 AUG 85	
1985-065G		15966	USSR	02 AUG	25 AUG 85	
1985-067A	COSMOS 1672	15940	USSR	07 AUG	21 AUG 85	
1985-067B		15941	USSR	07 AUG	11 AUG 85	
1985-067C		15971	USSR	07 AUG	28 AUG 85	
1985-067D		15972	USSR	07 AUG	15 OCT 85	
1985-067E		15973	USSR	07 AUG	15 SEP 85	
1985-067F		15974	USSR	07 AUG	03 SEP 85	
1985-067G		15975	USSR	07 AUG	22 AUG 85	
1985-067H		15976	USSR	07 AUG	24 AUG 85	
1985-068A	COSMOS 1673	15942	USSR	08 AUG	19 SEP 85	
1985-068B		15943	USSR	08 AUG	14 AUG 85	
1985-068C		15949	USSR	08 AUG	10 AUG 85	
1985-070B		15947	USSR	08 AUG	11 AUG 85	
1985-070C		15948	USSR	08 AUG	09 AUG 85	
1985-070D		15956	USSR	08 AUG	31 DEC 86	
1985-070E		15957	USSR	08 AUG	27 OCT 86	
1985-071B		15953	USSR	12 AUG	23 OCT 85	
1985-071C		15954	USSR	12 AUG	27 SEP 85	
1985-071E		15958	USSR	12 AUG	01 OCT 85	
1985-071F		16061	USSR	12 AUG	30 SEP 85	
1985-072A	COSMOS 1676	15959	USSR	16 AUG	14 OCT 85	
1985-072B		15960	USSR	16 AUG	22 AUG 85	
1985-073B		15968	JAPAN	18 AUG	13 NOV 85	
1985-073D		16388	JAPAN	18 AUG	22 MAY 86	
1985-074B		15978	USSR	22 AUG	04 NOV 85	
1985-074C		15979	USSR	22 AUG	22 OCT 85	
1985-075B		16192	USSR	23 AUG	14 DEC 85	
1985-075C		16193	USSR	23 AUG	27 OCT 85	
1985-075D		16219	USSR	23 AUG	28 NOV 85	
1985-076A	STS 51I	15992	US	27 AUG	03 SEP 85	
1985-076H		16013	US	27 AUG	28 FEB 86	
1985-077A	COSMOS 1678	15997	USSR	29 AUG	12 SEP 85	
1985-077B		15998	USSR	29 AUG	02 SEP 85	
1985-077C		16002	USSR	29 AUG	06 SEP 85	
1985-077D		16021	USSR	29 AUG	15 SEP 85	
1985-077E		16022	USSR	29 AUG	03 OCT 85	
1985-077F		16023	USSR	29 AUG	25 OCT 85	
1985-077G		16024	USSR	29 AUG	14 SEP 85	
1985-077H		16025	USSR	29 AUG	14 SEP 85	
1985-077J		16026	USSR	29 AUG	27 SEP 85	
1985-078A	COSMOS 1679	15999	USSR	29 AUG	18 OCT 85	
1985-078B		16000	USSR	29 AUG	04 SEP 85	
1985-080A	COSMOS 1681	16018	USSR	06 SEP	19 SEP 85	
1985-080B		16019	USSR	06 SEP	10 SEP 85	
1985-080C		16058	USSR	06 SEP	23 SEP 85	
1985-080E		16060	USSR	06 SEP	20 SEP 85	
1985-080D		16059	USSR	16 SEP	26 SEP 85	
1985-081A	SOYUZ T-14	16051	USSR	17 SEP	21 NOV 85	
1985-081B		16052	USSR	17 SEP	20 SEP 85	
1985-081C		16261	USSR	17 SEP	29 DEC 86	
1985-082A	COSMOS 1682	16054	USSR	19 SEP	17 MAY 88	
1985-082B		16055	USSR	19 SEP	19 SEP 85	
1985-082C		17248	USSR	19 SEP	27 DEC 86	
1985-082D		17249	USSR	19 SEP	29 DEC 86	
1985-082E		17250	USSR	19 SEP	13 JAN 87	
1985-082F		17251	USSR	19 SEP	13 JAN 87	
1985-082G		17275	USSR	19 SEP	01 MAR 87	
1985-082H		17277	USSR	19 SEP	19 MAY 87	
1985-082J		17278	USSR	19 SEP	11 MAR 87	
1985-082K		17279	USSR	19 SEP	11 MAY 87	
1985-082L		17280	USSR	19 SEP	27 APR 87	
1985-082N		17282	USSR	19 SEP	12 MAR 87	
1985-082P		17283	USSR	19 SEP	13 JUL 87	
1985-082Q		17284	USSR	19 SEP	07 JAN 87	
1985-082R		17285	USSR	19 SEP	01 MAR 87	
1985-082S		17286	USSR	19 SEP	07 MAR 87	
1985-082T		17287	USSR	19 SEP	08 MAR 87	
1985-082U		17288	USSR	19 SEP	21 JAN 87	
1985-082V		17289	USSR	19 SEP	19 JUL 87	
1985-082W		17840	USSR	19 SEP	11 AUG 87	
1985-082X		17841	USSR	19 SEP	19 APR 88	
1985-082Y		18280	USSR	19 SEP	31 MAY 88	
1985-082Z		18430	USSR	19 SEP	12 OCT 88	
1985-083A	COSMOS 1683	16056	USSR	19 SEP	04 OCT 85	
1985-083B		16057	USSR	19 SEP	03 OCT 85	
1985-083C		16120	USSR	19 SEP	28 MAR 86	
1985-083D		16121	USSR	19 SEP	26 DEC 87	
1985-083E		16123	USSR	19 SEP	24 DEC 85	
1985-083F		16124	USSR	19 SEP	22 MAY 86	
1985-083G		16126	USSR	19 SEP	24 OCT 85	
1985-083H		16127	USSR	19 SEP	04 MAR 86	
1985-084B		16065	USSR	24 SEP	21 NOV 85	
1985-084C		16066	USSR	24 SEP	25 OCT 85	
1985-084E		16090	USSR	24 SEP	16 NOV 85	
1985-085A	COSMOS 1685	16088	USSR	27 SEP	10 OCT 85	
1985-085B		16089	USSR	27 SEP	08 OCT 85	
1985-085C		16134	USSR	27 SEP	08 FEB 86	
1985-085D		16135	USSR	27 SEP	24 NOV 87	
1985-085E		16136	USSR	27 SEP	23 DEC 85	
1985-085F		16145	USSR	27 SEP	30 APR 86	
1985-085G		16146	USSR	27 SEP	15 APR 86	
1985-085H		16159	USSR	27 SEP	24 FEB 86	
1985-085J		16160	USSR	27 SEP	16 OCT 86	
1985-085K		16161	USSR	27 SEP	18 OCT 85	
1985-085L		16162	USSR	27 SEP	07 NOV 85	
1985-086A	COSMOS 1686	16095	USSR	27 SEP	07 FEB 91	
1985-086B		16096	USSR	27 SEP	02 OCT 85	
1985-086C		16128	USSR	27 SEP	01 FEB 86	
1985-088B		16104	USSR	30 SEP	28 NOV 85	
1985-088C		16105	USSR	30 SEP	03 NOV 85	
1985-089A	COSMOS 1688	16107	USSR	02 OCT	02 JUL 88	
1985-089B		16108	USSR	02 OCT	19 JUN 88	
1985-091B		16113	USSR	03 OCT	15 DEC 85	
1985-091C		16114	USSR	03 OCT	16 NOV 85	
1985-092A	STS 51J	16115	US	03 OCT	07 OCT 85	
1985-094H		16264	USSR	09 OCT	23 MAR 90	
1985-094J		16265	USSR	09 OCT	13 NOV 89	
1985-095A	COSMOS 1696	16169	USSR	16 OCT	30 OCT 85	
1985-095B		16170	USSR	16 OCT	23 OCT 85	
1985-095C		16227	USSR	16 OCT	07 NOV 85	
1985-095D		16228	USSR	16 OCT	01 DEC 85	
1985-095E		16232	USSR	16 OCT	07 NOV 85	
1985-095F		16233	USSR	16 OCT	07 NOV 85	
1985-095G		16234	USSR	16 OCT	03 NOV 85	
1985-096A	PRC 17	16177	PRC	21 OCT	07 NOV 85	
1985-096B		16178	PRC	21 OCT	03 NOV 85	
1985-096C		16179	PRC	21 OCT	24 OCT 85	
1985-096D		16180	PRC	21 OCT	24 OCT 85	
1985-098B		16184	USSR	22 OCT	07 JAN 86	
1985-098C		16185	USSR	22 OCT	14 DEC 85	
1985-099B		16188	USSR	23 OCT	24 MAR 86	
1985-099C		16189	USSR	23 OCT	21 MAR 86	
1985-099D		16190	USSR	23 OCT	12 FEB 86	
1985-101A	COSMOS 1699	16198	USSR	25 OCT	23 DEC 85	
1985-101B		16212	USSR	25 OCT	30 OCT 85	
1985-102B		16200	USSR	25 OCT	28 OCT 85	
1985-102C		16201	USSR	25 OCT	26 OCT 85	
1985-102E		16215	USSR	25 OCT	27 OCT 86	
1985-102F		16216	USSR	25 OCT	18 MAY 86	
1985-103B		16221	USSR	28 OCT	04 DEC 85	
1985-103C		16222	USSR	28 OCT	18 NOV 85	
1985-103E		16224	USSR	28 OCT	20 NOV 85	
1985-103F		16225	USSR	28 OCT	29 NOV 85	
1985-103G		16226	USSR	28 OCT	01 DEC 85	
1985-104A	STS 61A	16230	US	30 OCT	06 NOV 85	
1985-104B	GLOMAR	16231	US	30 OCT	26 DEC 86	35
1985-105B		16236	USSR	09 NOV	11 JAN 86	
1985-105C		16237	USSR	09 NOV	24 DEC 85	
1985-106A	COSMOS 1702	16247	USSR	13 NOV	27 NOV 85	
1985-106B		16248	USSR	13 NOV	28 NOV 85	
1985-106C		16285	USSR	13 NOV	31 MAY 86	
1985-106D		16286	USSR	13 NOV	28 JAN 88	
1985-106E		16287	USSR	13 NOV	13 DEC 85	
1985-106F		16288	USSR	13 NOV	10 AUG 86	
1985-106G		16289	USSR	13 NOV	07 APR 86	
1985-106H		16290	USSR	13 NOV	19 MAR 86	
1985-107B		16251	USSR	15 NOV	18 NOV 85	
1985-107C		16252	USSR	15 NOV	16 NOV 85	
1985-107D		16254	USSR	15 NOV	27 OCT 86	
1985-107E		16255	USSR	15 NOV	19 JUL 86	
1985-109A	STS 61B	16273	US	27 NOV	03 DEC 85	
1985-109E	OEX TARGET	16277	US	30 NOV	02 MAR 87	35
1985-111A	COSMOS 1705	16296	USSR	03 DEC	17 DEC 85	
1985-111B		16297	USSR	03 DEC	19 DEC 85	
1985-111C		16305	USSR	03 DEC	06 DEC 85	
1985-111D		16353	USSR	03 DEC	08 JAN 86	
1985-111E		16354	USSR	03 DEC	10 JAN 86	
1985-111F		16355	USSR	03 DEC	07 MAR 86	
1985-111G		16356	USSR	03 DEC	04 APR 86	
1985-111H		16357	USSR	03 DEC	13 OCT 86	
1985-111J		16358	USSR	03 DEC	22 DEC 85	
1985-111K		16359	USSR	03 DEC	12 JAN 86	
1985-111L		16370	USSR	03 DEC	02 JUL 86	
1985-111M		16371	USSR	03 DEC	24 JAN 86	
1985-111N		16372	USSR	03 DEC	14 JAN 86	
1985-111P		16373	USSR	03 DEC	15 FEB 88	
1985-111Q		16374	USSR	03 DEC	22 FEB 86	
1985-112A	COSMOS 1706	16306	USSR	11 DEC	09 FEB 86	
1985-112B		16307	USSR	11 DEC	16 DEC 85	
1985-114A		16328	US	13 DEC	11 MAY 89	
1985-114B		16329	US	13 DEC	09 AUG 87	
1985-114C		16330	US	13 DEC	11 APR 88	
1985-114D		17247	US	13 DEC	02 MAR 87	
1985-114E		17257	US	13 DEC	23 FEB 87	
1985-115A	COSMOS 1708	16331	USSR	13 DEC	27 DEC 85	
1985-115B		16332	USSR	13 DEC	17 DEC 85	
1985-115C		16340	USSR	13 DEC	22 DEC 85	
1985-115D		16426	USSR	13 DEC	30 DEC 85	
1985-115E		16427	USSR	13 DEC	29 JAN 86	
1985-115F		16428	USSR	13 DEC	28 DEC 85	
1985-115G		16431	USSR	13 DEC	04 JAN 86	
1985-115H		16432	USSR	13 DEC	15 JAN 86	
1985-115J		16433	USSR	13 DEC	07 JAN 86	
1985-117B		16394	USSR	24 DEC	30 JAN 86	
1985-117C		16395	USSR	24 DEC	21 JAN 86	
1985-117D		16400	USSR	24 DEC	16 JAN 86	
1985-117E		16401	USSR	24 DEC	20 JAN 86	
1985-117G		16410	USSR	24 DEC	19 JAN 86	

International Designation	Satellite Name	Number	Source	Launch	Decay	Note
1985-117H		16411	USSR	24 DEC	17 JAN 86	
1985-117J		16412	USSR	24 DEC	22 JAN 86	
1985-118D		16399	USSR	24 DEC	27 DEC 85	
1985-118E		16403	USSR	24 DEC	25 DEC 85	
1985-118G		16405	USSR	24 DEC	27 DEC 85	
1985-118H		16406	USSR	24 DEC	27 DEC 85	
1985-118J		16407	USSR	24 DEC	28 DEC 85	
1985-120A	COSMOS 1713	16429	USSR	27 DEC	22 JAN 86	
1985-120B		16430	USSR	27 DEC	27 JAN 86	
1985-120C		16508	USSR	27 DEC	30 JAN 86	
1985-120D		16509	USSR	27 DEC	03 FEB 86	
1985-121A	COSMOS 1714	16434	USSR	28 DEC	27 FEB 86	
1985-121B		16435	USSR	28 DEC	11 JAN 86	
1985-121C		16436	USSR	28 DEC	08 JAN 86	
1985-121D		16437	USSR	28 DEC	08 JUL 90	
1985-121E		16438	USSR	28 DEC	22 APR 90	
1985-121F		16439	USSR	28 DEC	04 APR 90	
1985-121G		16440	USSR	28 DEC	16 NOV 89	
1986 LAUNCHES						
1986-001A	COSMOS 1715	16447	USSR	08 JAN	22 JAN 86	
1986-001B		16448	USSR	08 JAN	16 JAN 86	
1986-001C		16505	USSR	08 JAN	25 JAN 86	
1986-001D		16506	USSR	08 JAN	27 JAN 86	
1986-001E		16507	USSR	08 JAN	31 JAN 86	
1986-003A	STS 61C	16481	US	12 JAN	18 JAN 86	
1986-004A	COSMOS 1724	16490	USSR	15 JAN	15 MAR 86	
1986-004B		16491	USSR	15 JAN	21 JAN 86	
1986-004C		16492	USSR	15 JAN	16 JAN 86	
1986-007B		16498	USSR	17 JAN	20 JAN 86	
1986-007C		16499	USSR	17 JAN	18 JAN 86	
1986-007D		16500	USSR	17 JAN	12 AUG 87	
1986-009A	COSMOS 1728	16512	USSR	28 JAN	11 FEB 86	
1986-009B		16513	USSR	28 JAN	06 FEB 86	
1986-009C		16514	USSR	28 JAN	31 JAN 86	
1986-009D		16595	USSR	28 JAN	26 FEB 86	
1986-009E		16596	USSR	28 JAN	18 FEB 86	
1986-009F		16599	USSR	28 JAN	15 FEB 86	
1986-011B		16529	USSR	01 FEB	10 MAR 86	
1986-011C		16530	USSR	01 FEB	24 MAR 86	
1986-011D		16531	USSR	01 FEB	31 MAR 86	
1986-011E		16532	USSR	01 FEB	13 MAR 86	
1986-012A	COSMOS 1730	16540	USSR	04 FEB	13 FEB 86	
1986-012B		16541	USSR	04 FEB	11 FEB 86	
1986-012C		16602	USSR	04 FEB	17 MAR 86	
1986-012D		16603	USSR	04 FEB	22 FEB 86	
1986-012E		16604	USSR	04 FEB	02 MAR 86	
1986-013A	COSMOS 1731	16589	USSR	07 FEB	03 OCT 86	
1986-013B		16590	USSR	07 FEB	09 FEB 86	
1986-016B		16598	JAPAN	12 FEB	19 APR 87	
1986-016D		16601	JAPAN	12 FEB	22 FEB 86	
1986-017AA		19584	USSR	19 FEB	05 NOV 88	
1986-017AB		19585	USSR	19 FEB	26 NOV 88	
1986-017AC		19586	USSR	19 FEB	08 NOV 88	
1986-017AD		19587	USSR	19 FEB	04 NOV 88	
1986-017AE		19588	USSR	19 FEB	05 NOV 88	
1986-017AF		19604	USSR	19 FEB	10 NOV 88	
1986-017AG		19605	USSR	19 FEB	06 NOV 88	
1986-017AH		19606	USSR	19 FEB	05 JAN 89	
1986-017AJ		19607	USSR	19 FEB	22 FEB 89	
1986-017AK		19618	USSR	19 FEB	07 DEC 88	
1986-017AL		19619	USSR	19 FEB	08 DEC 88	
1986-017AM		19620	USSR	19 FEB	06 NOV 88	
1986-017AN		19624	USSR	19 FEB	06 NOV 88	
1986-017AP		19632	USSR	19 FEB	23 DEC 88	
1986-017AQ		19633	USSR	19 FEB	18 DEC 88	
1986-017AR		19634	USSR	19 FEB	25 DEC 88	
1986-017AS		19635	USSR	19 FEB	03 DEC 88	
1986-017AT		19636	USSR	19 FEB	03 DEC 88	
1986-017AU		19680	USSR	19 FEB	22 DEC 88	
1986-017AV		19681	USSR	19 FEB	28 DEC 88	
1986-017AW		19682	USSR	19 FEB	23 DEC 88	
1986-017AX		19692	USSR	19 FEB	19 DEC 88	
1986-017AY		19693	USSR	19 FEB	24 DEC 88	
1986-017AZ		19700	USSR	19 FEB	19 JAN 89	
1986-017B		16610	USSR	19 FEB	24 FEB 86	
1986-017BA		19701	USSR	19 FEB	28 DEC 88	
1986-017BB		19702	USSR	19 FEB	27 JAN 89	
1986-017BC		19703	USSR	19 FEB	22 DEC 88	
1986-017BD		19704	USSR	19 FEB	28 FEB 89	
1986-017BE		19707	USSR	19 FEB	23 JAN 89	
1986-017BF		19708	USSR	19 FEB	20 JAN 89	
1986-017BG		19736	USSR	19 FEB	22 FEB 89	
1986-017BH		19737	USSR	19 FEB	01 FEB 89	
1986-017BJ		19738	USSR	19 FEB	22 JAN 89	
1986-017BK		19739	USSR	19 FEB	17 JAN 89	
1986-017BL		19740	USSR	19 FEB	03 MAR 89	
1986-017BM		19741	USSR	19 FEB	25 JAN 89	
1986-017BN		19743	USSR	19 FEB	22 MAR 89	
1986-017BP		19744	USSR	19 FEB	05 MAR 89	
1986-017BQ		19745	USSR	19 FEB	05 MAR 89	
1986-017BR		19746	USSR	19 FEB	05 MAR 89	
1986-017BS		19747	USSR	19 FEB	11 FEB 89	
1986-017BT		19816	USSR	19 FEB	06 MAR 89	
1986-017BU		20414	USSR	19 FEB	24 JAN 90	
1986-017BV		20415	USSR	19 FEB	15 MAR 90	
1986-017BW		20416	USSR	19 FEB	04 MAR 90	

International Designation	Satellite Name	Number	Source	Launch	Decay	Note
1986-017BX		20417	USSR	19 FEB	07 FEB 90	
1986-017BY	MIR	20407	USSR	19 FEB	27 JUN 90	
1986-017BZ		20418	USSR	19 FEB	04 MAR 90	
1986-017C		17820	USSR	19 FEB	07 MAY 87	
1986-017CA		20419	USSR	19 FEB	01 MAR 90	
1986-017CB		20420	USSR	19 FEB	11 FEB 90	
1986-017CD		20459	USSR	19 FEB	08 FEB 90	
1986-017CE		20460	USSR	19 FEB	08 MAR 90	
1986-017CF		20461	USSR	19 FEB	28 JAN 90	
1986-017CG		20472	USSR	19 FEB	03 FEB 90	
1986-017CH		20475	USSR	19 FEB	07 FEB 90	
1986-017CJ		20689	USSR	19 FEB	27 SEP 90	
1986-017CK		20690	USSR	19 FEB	03 AUG 90	
1986-017CL		20699	USSR	19 FEB	21 AUG 90	
1986-017CM		20700	USSR	19 FEB	21 AUG 90	
1986-017CN		20701	USSR	19 FEB	03 AUG 90	
1986-017CP		20719	USSR	19 FEB	19 SEP 90	
1986-017CQ		20917	USSR	19 FEB	31 OCT 90	
1986-017CR		20921	USSR	19 FEB	09 NOV 90	
1986-017CS		20922	USSR	19 FEB	08 NOV 90	
1986-017CT		21050	USSR	19 FEB	04 FEB 91	
1986-017CU		21051	USSR	19 FEB	05 FEB 91	
1986-017CV		21082	USSR	19 FEB	18 MAR 91	
1986-017CW		21084	USSR	19 FEB	05 MAR 91	
1986-017CX		21085	USSR	19 FEB	01 MAR 91	
1986-017CY		21086	USSR	19 FEB	24 FEB 91	
1986-017CZ		21235	USSR	19 FEB	25 AUG 91	
1986-017D		17821	USSR	19 FEB	06 MAY 87	
1986-017DA		21236	USSR	19 FEB	24 JUN 91	
1986-017DB		21237	USSR	19 FEB	15 JUN 91	
1986-017DC		21238	USSR	19 FEB	07 JUN 91	
1986-017DD		21239	USSR	19 FEB	25 JUN 91	
1986-017DE		21240	USSR	19 FEB	11 NOV 91	
1986-017DF		21241	USSR	19 FEB	09 JUN 91	
1986-017DG		21243	USSR	19 FEB	07 JUN 91	
1986-017DJ		21409	USSR	19 FEB	24 JUL 91	
1986-017DL		21411	USSR	19 FEB	24 AUG 91	
1986-017DM		21412	USSR	19 FEB	11 AUG 91	
1986-017DN		21413	USSR	19 FEB	28 AUG 91	
1986-017DP		21414	USSR	19 FEB	20 OCT 91	
1986-017DQ		21415	USSR	19 FEB	02 SEP 91	
1986-017DR		21416	USSR	19 FEB	14 AUG 91	
1986-017DS		21417	USSR	19 FEB	30 OCT 91	
1986-017DT		21421	USSR	19 FEB	08 AUG 91	
1986-017DU		21424	USSR	19 FEB	11 JUL 91	
1986-017DW		21476	USSR	19 FEB	08 SEP 91	
1986-017DX		21477	USSR	19 FEB	12 AUG 91	
1986-017DY		21478	USSR	19 FEB	19 AUG 91	
1986-017DZ		21481	USSR	19 FEB	16 DEC 91	
1986-017E		18885	USSR	19 FEB	08 MAR 88	
1986-017EA		21482	USSR	19 FEB	07 JAN 92	
1986-017EB		21483	USSR	19 FEB	13 SEP 91	
1986-017EC		21484	USSR	19 FEB	12 JUL 91	
1986-017ED		21486	USSR	19 FEB	05 AUG 91	
1986-017EE		21487	USSR	19 FEB	06 JUL 91	
1986-017EF		21488	USSR	19 FEB	11 APR 92	
1986-017EG		21489	USSR	19 FEB	23 SEP 91	
1986-017EH		21490	USSR	19 FEB	15 SEP 91	
1986-017EJ		21526	USSR	19 FEB	05 APR 92	
1986-017EK		21530	USSR	19 FEB	01 SEP 91	
1986-017EL		21531	USSR	19 FEB	09 SEP 91	
1986-017EM		21556	USSR	19 FEB	20 SEP 91	
1986-017EN		21557	USSR	19 FEB	14 AUG 91	
1986-017EP		21571	USSR	19 FEB	11 SEP 91	
1986-017EQ		21572	USSR	19 FEB	27 AUG 91	
1986-017ER		21573	USSR	19 FEB	06 SEP 91	
1986-017ES		21579	USSR	19 FEB	28 AUG 91	
1986-017ET		21595	USSR	19 FEB	07 SEP 91	
1986-017EU		21596	USSR	19 FEB	31 AUG 91	
1986-017EV		21603	USSR	19 FEB	05 OCT 91	
1986-017EW		21618	USSR	19 FEB	15 AUG 91	
1986-017EX		21619	USSR	19 FEB	18 DEC 91	
1986-017EY		21624	USSR	19 FEB	19 AUG 91	
1986-017EZ		21625	USSR	19 FEB	06 DEC 91	
1986-017F		18886	USSR	19 FEB	21 MAR 88	
1986-017FA		21626	USSR	19 FEB	03 OCT 91	
1986-017FB		21627	USSR	19 FEB	03 DEC 91	
1986-017FC		21628	USSR	19 FEB	07 OCT 91	
1986-017FD		21629	USSR	19 FEB	30 SEP 91	
1986-017FE		21649	USSR	19 FEB	29 AUG 91	
1986-017FF		21650	USSR	19 FEB	13 AUG 91	
1986-017FG		21651	USSR	19 FEB	12 AUG 91	
1986-017FH		21652	USSR	19 FEB	21 AUG 91	
1986-017FJ		21661	USSR	19 FEB	29 AUG 91	
1986-017FK		21810	USSR	19 FEB	28 FEB 92	
1986-017FL		21811	USSR	19 FEB	17 JUN 92	
1986-017FN		21830	USSR	19 FEB	01 JUL 92	
1986-017FP		21831	USSR	19 FEB	17 JUN 92	
1986-017FQ		21832	USSR	19 FEB	04 JUN 92	
1986-017FR		21879	USSR	19 FEB	25 MAR 92	
1986-017FS		21880	USSR	19 FEB	22 MAR 92	
1986-017FT		21881	USSR	19 FEB	29 MAR 92	
1986-017FU		21882	USSR	19 FEB	21 JAN 93	
1986-017FV		21883	USSR	19 FEB	25 FEB 92	
1986-017FW		21884	USSR	19 FEB	18 JAN 93	
1986-017FX		21885	USSR	19 FEB	21 APR 92	
1986-017FY		21886	USSR	19 FEB	26 MAR 92	
1986-017FZ		21887	USSR	19 FEB	29 JUN 92	

International Designation	Satellite Name	Number	Source	Launch	Decay	Note
1986-017G		18887	USSR	19 FEB	02 MAR 88	
1986-017GA		21888	USSR	19 FEB	04 APR 92	
1986-017GB		21889	USSR	19 FEB	09 JUN 92	
1986-017GC		21896	USSR	19 FEB	17 MAY 92	
1986-017GD		21901	USSR	19 FEB	10 MAY 92	
1986-017GE		22023	USSR	19 FEB	12 DEC 92	
1986-017GF		22024	USSR	19 FEB	04 MAY 93	
1986-017GG		22045	USSR	19 FEB	04 MAY 93	
1986-017GH		22046	USSR	19 FEB	19 OCT 92	
1986-017GJ		22106	USSR	19 FEB	26 FEB 93	
1986-017GK		22107	USSR	19 FEB	10 DEC 92	
1986-017GL		22111	USSR	19 FEB	23 SEP 92	
1986-017GN		22122	USSR	19 FEB	02 DEC 92	
1986-017GQ		22124	USSR	19 FEB	03 NOV 92	
1986-017GT		22129	USSR	19 FEB	17 NOV 92	
1986-017GU		22130	USSR	19 FEB	01 NOV 92	
1986-017GV		22131	USSR	19 FEB	12 NOV 92	
1986-017GW		22209	USSR	19 FEB	09 APR 93	
1986-017GX	MAK 2	22225	USSR	19 FEB	01 APR 93	
1986-017GY		22228	USSR	19 FEB	08 DEC 92	
1986-017H		18889	USSR	19 FEB	05 MAR 88	
1986-017HB		22632	USSR	19 FEB	25 APR 93	
1986-017HE		22678	USSR	19 FEB	30 JUN 93	
1986-017HG		22680	USSR	19 FEB	27 JUN 93	
1986-017J		19271	USSR	19 FEB	10 AUG 88	
1986-017K		19272	USSR	19 FEB	25 AUG 88	
1986-017M		19387	USSR	19 FEB	24 SEP 88	
1986-017N		19408	USSR	19 FEB	16 OCT 88	
1986-017P		19409	USSR	19 FEB	31 OCT 88	
1986-017Q		19442	USSR	19 FEB	05 OCT 88	
1986-017R		19533	USSR	19 FEB	02 OCT 88	
1986-017S		19536	USSR	19 FEB	09 NOV 88	
1986-017T		19538	USSR	19 FEB	17 OCT 88	
1986-017U		19539	USSR	19 FEB	01 OCT 88	
1986-017V		19540	USSR	19 FEB	23 OCT 88	
1986-017W		19568	USSR	19 FEB	22 OCT 88	
1986-017X		19569	USSR	19 FEB	22 OCT 88	
1986-017Y		19570	USSR	19 FEB	19 OCT 88	
1986-017Z		19571	USSR	19 FEB	20 OCT 88	
1988-070C		19388	USSR	19 FEB	17 AUG 88	
1986-019AB		17132	ESA	22 FEB	28 DEC 88	
1986-019AC		17133	ESA	22 FEB	04 SEP 88	
1986-019AD		17150	ESA	22 FEB	12 JUN 90	
1986-019AE		17151	ESA	22 FEB	04 MAY 89	
1986-019AK		17156	ESA	22 FEB	07 DEC 89	
1986-019AL		17157	ESA	22 F2B	16 DEC 88	
1986-019AM		17183	ESA	22 FEB	08 NOV 89	
1986-019AP		17185	ESA	22 FEB	11 MAR 89	
1986-019AQ		17186	ESA	22 FEB	28 DEC 88	
1986-019AR		17187	ESA	22 FEB	23 JAN 89	
1986-019AS		17188	ESA	22 FEB	11 NOV 88	
1986-019AU		17190	ESA	22 FEB	20 FEB 89	
1986-019AV		17193	ESA	22 FEB	22 MAR 89	
1986-019AW		17194	ESA	22 FEB	16 FEB 89	
1986-019AY		17196	ESA	22 FEB	16 JAN 89	
1986-019AZ		17197	ESA	22 FEB	21 FEB 89	
1986-019BC		17200	ESA	22 FEB	09 JUL 91	
1986-019BE		17202	ESA	22 FEB	04 MAR 89	
1986-019BH		17205	ESA	22 FEB	27 MAR 89	
1986-019BK		17207	ESA	22 FEB	14 JAN 89	
1986-019BM		17209	ESA	22 FEB	29 APR 91	
1986-019BN		17210	ESA	22 FEB	27 FEB 89	
1986-019BP		17211	ESA	22 FEB	21 APR 89	
1986-019BS		17220	ESA	22 FEB	23 SEP 88	
1986-019BT		17221	ESA	22 FEB	19 JUN 89	
1986-019BU		17222	ESA	22 FEB	25 MAY 90	
1986-019BV		17223	ESA	22 FEB	16 MAR 92	
1986-019BW		17224	ESA	22 FEB	18 MAR 89	
1986-019BX		17225	ESA	22 FEB	16 JAN 88	
1986-019BY		17226	ESA	22 FEB	09 DEC 89	
1986-019BZ		17227	ESA	22 FEB	14 FEB 89	
1986-019CA		17228	ESA	22 FEB	13 SEP 91	
1986-019CB		17229	ESA	22 FEB	03 JUN 89	
1986-019CC		17230	ESA	22 FEB	06 SEP 89	
1986-019CD		17231	ESA	22 FEB	25 SEP 88	
1986-019CE		17232	ESA	22 FEB	27 OCT 88	
1986-019CG		17234	ESA	22 FEB	10 AUG 88	
1986-019CJ		17236	ESA	22 FEB	20 AUG 89	
1986-019CK		17237	ESA	22 FEB	09 NOV 89	
1986-019CL		17238	ESA	22 FEB	21 DEC 88	
1986-019CM		17305	ESA	22 FEB	31 MAR 90	
1986-019CN		17306	ESA	22 FEB	24 SEP 88	
1986-019CP		17307	ESA	22 FEB	06 MAR 92	
1986-019CQ		17308	ESA	22 FEB	27 NOV 89	
1986-019CR		17309	ESA	22 FEB	15 JUN 89	
1986-019CS		17310	ESA	22 FEB	08 APR 89	
1986-019CU		17312	ESA	22 FEB	19 OCT 89	
1986-019CW		17314	ESA	22 FEB	06 JAN 92	
1986-019CX		17315	ESA	22 FEB	23 JAN 92	
1986-019CZ		17317	ESA	22 FEB	20 JUN 87	
1986-019DA		17318	ESA	22 FEB	28 OCT 87	
1986-019DB		17319	ESA	22 FEB	09 NOV 88	
1986-019DC		17320	ESA	22 FEB	19 FEB 89	
1986-019DD		17321	ESA	22 FEB	01 AUG 89	
1986-019DE		17322	ESA	22 FEB	26 OCT 88	
1986-019DH		17336	ESA	22 FEB	10 NOV 89	
1986-019DJ		17337	ESA	22 FEB	06 MAR 89	
1986-019DK		17338	ESA	22 FEB	27 SEP 89	
1986-019DL		17339	ESA	22 FEB	20 APR 90	
1986-019DM		17340	ESA	22 FEB	04 NOV 88	
1986-019DN		17341	ESA	22 FEB	01 JAN 91	
1986-019DQ		17343	ESA	22 FEB	27 DEC 91	
1986-019DR		17344	ESA	22 FEB	19 JAN 89	
1986-019DS		17345	ESA	22 FEB	16 APR 89	
1986-019DT		17346	ESA	22 FEB	11 MAY 89	
1986-019DU		17347	ESA	22 FEB	15 NOV 87	
1986-019DV		17348	ESA	22 FEB	12 OCT 88	
1986-019DW		17349	ESA	22 FEB	13 FEB 89	
1986-019DX		17350	ESA	22 FEB	25 FEB 90	
1986-019DY		17351	ESA	22 FEB	28 FEB 90	
1986-019EB		17354	ESA	22 FEB	17 FEB 89	
1986-019EC		17355	ESA	22 FEB	03 DEC 89	
1986-019ED		17371	ESA	22 FEB	09 JAN 89	
1986-019EE		17372	ESA	22 FEB	27 JAN 89	
1986-019EF		17373	ESA	22 FEB	01 JAN 89	
1986-019EG		17374	ESA	22 FEB	05 JAN 90	
1986-019EH		17375	ESA	22 FEB	09 DEC 88	
1986-019EJ		17376	ESA	22 FEB	24 AUG 90	
1986-019EK		17377	ESA	22 FEB	15 FEB 88	
1986-019EL		17378	ESA	22 FEB	09 APR 89	
1986-019EM		17379	ESA	22 FEB	04 DEC 90	
1986-019EN		17380	ESA	22 FEB	10 SEP 90	
1986-019EP		17381	ESA	22 FEB	05 OCT 88	
1986-019EQ		17382	ESA	22 FEB	19 DEC 89	
1986-019ER		17383	ESA	22 FEB	08 MAR 89	
1986-019ES		17384	ESA	22 FEB	16 FEB 89	
1986-019ET		17385	ESA	22 FEB	24 APR 89	
1986-019EU		17386	ESA	22 FEB	01 MAR 89	
1986-019EV		17387	ESA	22 FEB	05 MAR 89	
1986-019EW		17388	ESA	22 FEB	26 SEP 88	
1986-019EX		17389	ESA	22 FEB	05 OCT 88	
1986-019EY		17390	ESA	22 FEB	10 NOV 89	
1986-019EZ		17391	ESA	22 FEB	09 JUL 89	
1986-019F		17111	ESA	22 FEB	07 JAN 89	
1986-019FA		17392	ESA	22 FEB	13 MAY 92	
1986-019FB		17393	ESA	22 FEB	29 DEC 88	
1986-019FC		17394	ESA	22 FEB	19 DEC 88	
1986-019FD		17395	ESA	22 FEB	17 JAN 89	
1986-019FE		17396	ESA	22 FEB	16 APR 89	
1986-019FF		17397	ESA	22 FEB	14 APR 89	
1986-019FG		17421	ESA	22 FEB	10 FEB 89	
1986-019FH		17422	ESA	22 FEB	23 JUL 88	
1986-019FK		17424	ESA	22 FEB	25 NOV 89	
1986-019FL		17425	ESA	22 FEB	02 FEB 89	
1986-019FN		17427	ESA	22 FEB	26 DEC 89	
1986-019FR		17430	ESA	22 FEB	10 AUG 88	
1986-019FS		17431	ESA	22 FEB	25 MAR 89	
1986-019FT		17432	ESA	22 FEB	13 MAR 88	
1986-019FU		17433	ESA	22 FEB	20 AUG 89	
1986-019FV		17434	ESA	22 FEB	01 APR 89	
1986-019FW		17435	ESA	22 FEB	31 DEC 89	
1986-019FX		17436	ESA	22 FEB	06 FEB 89	
1986-019FY		17437	ESA	22 FEB	19 OCT 88	
1986-019G		17112	ESA	22 FEB	14 NOV 90	
1986-019GA		17439	ESA	22 FEB	10 APR 89	
1986-019GB		17440	ESA	22 FEB	29 OCT 87	
1986-019GC		17441	ESA	22 FEB	10 NOV 88	
1986-019GD		17442	ESA	22 FEB	05 DEC 89	
1986-019GE		17443	ESA	22 FEB	15 NOV 89	
1986-019GF		17444	ESA	22 FEB	03 FEB 91	
1986-019GG		17445	ESA	22 FEB	24 JUL 90	
1986-019GH		17446	ESA	22 FEB	20 OCT 89	
1986-019GJ		17447	ESA	22 FEB	20 JUL 88	
1986-019GK		17448	ESA	22 FEB	09 SEP 88	
1986-019GL		17449	ESA	22 FEB	17 JUN 89	
1986-019GM		17450	ESA	22 FEB	18 AUG 89	
1986-019GN		17451	ESA	22 FEB	19 NOV 89	
1986-019GP		17452	ESA	22 FEB	27 FEB 89	
1986-019GQ		17453	ESA	22 FEB	18 JAN 89	
1986-019GR		17454	ESA	22 FEB	12 JUN 89	
1986-019GS		17455	ESA	22 FEB	20 JUL 91	
1986-019GT		17456	ESA	22 FEB	23 DEC 88	
1986-019GU		17457	ESA	22 FEB	19 MAY 90	
1986-019GV		17458	ESA	22 FEB	29 SEP 90	
1986-019GX		17460	ESA	22 FEB	16 MAR 89	
1986-019GY		17461	ESA	22 FEB	17 NOV 90	
1986-019GZ		17462	ESA	22 FEB	15 OCT 88	
1986-019H		17113	ESA	22 FEB	26 AUG 88	
1986-019HA		17463	ESA	22 FEB	20 OCT 88	
1986-019HB		17464	ESA	22 FEB	22 APR 90	
1986-019HC		17465	ESA	22 FEB	12 JAN 90	
1986-019HE		17467	ESA	22 FEB	19 SEP 88	
1986-019HF		17468	ESA	22 FEB	07 NOV 88	
1986-019HG		17469	ESA	22 FEB	31 JAN 89	
1986-019HH		17470	ESA	22 FEB	16 OCT 88	
1986-019HK		17472	ESA	22 FEB	29 APR 91	
1986-019HL		17473	ESA	22 FEB	26 NOV 89	
1986-019HM		17474	ESA	22 FEB	26 OCT 88	
1986-019HN		17475	ESA	22 FEB	02 DEC 88	
1986-019HP		17476	ESA	22 FEB	26 AUG 89	
1986-019HQ		17477	ESA	22 FEB	13 JUL 89	
1986-019HS		17479	ESA	22 FEB	16 JUL 89	
1986-019HT		17539	ESA	22 FEB	20 JUL 89	
1986-019HU		17540	ESA	22 FEB	20 DEC 90	
1986-019HV		17541	ESA	22 FEB	19 OCT 88	
1986-019HW		17542	ESA	22 FEB	14 DEC 88	

International Designation	Satellite Name	Number	Source	Launch	Decay	Note
1986-019HY		17544	ESA	22 FEB	13 MAR 89	
1986-019HZ		17545	ESA	22 FEB	13 DEC 89	
1986-019J		17114	ESA	22 FEB	05 APR 89	
1986-019JA		17546	ESA	22 FEB	21 JUL 89	
1986-019JB		17547	ESA	22 FEB	15 DEC 89	
1986-019JC		17548	ESA	22 FEB	15 FEB 89	
1986-019JD		17549	ESA	22 FEB	23 APR 89	
1986-019JE		17550	ESA	22 FEB	09 OCT 88	
1986-019JG		17552	ESA	22 FEB	07 SEP 89	
1986-019JH		17553	ESA	22 FEB	17 MAR 90	
1986-019JJ		17554	ESA	22 FEB	22 OCT 89	
1986-019JM		17557	ESA	22 FEB	22 DEC 90	
1986-019JN		17558	ESA	22 FEB	31 AUG 88	
1986-019JP		17574	ESA	22 FEB	02 MAR 91	
1986-019JQ		17575	ESA	22 FEB	12 JAN 89	
1986-019JR		17576	ESA	22 FEB	13 NOV 88	
1986-019JS		17591	ESA	22 FEB	07 AUG 89	
1986-019JT		17592	ESA	22 FEB	28 APR 91	
1986-019JU		17593	ESA	22 FEB	15 FEB 91	
1986-019JV		17594	ESA	22 FEB	19 MAY 89	
1986-019JW		17595	ESA	22 FEB	20 JAN 90	
1986-019JX		17596	ESA	22 FEB	15 JAN 89	
1986-019JY		17597	ESA	22 FEB	26 APR 89	
1986-019JZ		17598	ESA	22 FEB	13 JUL 88	
1986-019K		17115	ESA	22 FEB	17 JAN 89	
1986-019KA		17599	ESA	22 FEB	14 MAR 89	
1986-019KB		17600	ESA	22 FEB	21 OCT 88	
1986-019KC		17601	ESA	22 FEB	26 FEB 89	
1986-019KD		17602	ESA	22 FEB	08 APR 89	
1986-019KE		17603	ESA	22 FEB	07 FEB 90	
1986-019KG		17605	ESA	22 FEB	02 FEB 89	
1986-019KH		17606	ESA	22 FEB	23 MAY 89	
1986-019KJ		17607	ESA	22 FEB	30 OCT 89	
1986-019KL		17609	ESA	22 FEB	16 FEB 89	
1986-019KM		17610	ESA	22 FEB	19 SEP 88	
1986-019KN		17672	ESA	22 FEB	25 FEB 91	
1986-019KP		17673	ESA	22 FEB	02 MAY 88	
1986-019KQ		17674	ESA	22 FEB	19 SEP 88	
1986-019KR		17675	ESA	22 FEB	02 MAR 89	
1986-019KS		17676	ESA	22 FEB	29 APR 89	
1986-019KT		17677	ESA	22 FEB	04 MAY 88	
1986-019KU		17678	ESA	22 FEB	18 NOV 88	
1986-019KV		17679	ESA	22 FEB	28 MAR 89	
1986-019KW		17680	ESA	22 FEB	18 JAN 89	
1986-019KX		17681	ESA	22 FEB	31 OCT 88	
1986-019KY		17682	ESA	22 FEB	31 MAR 89	
1986-019KZ		17683	ESA	22 FEB	25 FEB 89	
1986-019LA		17684	ESA	22 FEB	08 MAY 89	
1986-019LB		17685	ESA	22 FEB	13 AUG 89	
1986-019LC		17686	ESA	22 FEB	11 DEC 87	
1986-019LD		17687	ESA	22 FEB	14 MAR 89	
1986-019LE		17688	ESA	22 FEB	28 APR 90	
1986-019LF		17689	ESA	22 FEB	17 JAN 88	
1986-019LG		17690	ESA	22 FEB	08 JUN 89	
1986-019LH		17691	ESA	22 FEB	14 JAN 89	
1986-019LJ		17692	ESA	22 FEB	17 FEB 89	
1986-019LK		17693	ESA	22 FEB	29 DEC 88	
1986-019LL		17694	ESA	22 FEB	30 MAR 89	
1986-019LM		17695	ESA	22 FEB	05 SEP 89	
1986-019LN		17696	ESA	22 FEB	03 DEC 89	
1986-019LP		17697	ESA	22 FEB	19 MAY 89	
1986-019LQ		17698	ESA	22 FEB	11 FEB 92	
1986-019LR		17699	ESA	22 FEB	06 JUN 90	
1986-019LS		17700	ESA	22 FEB	12 NOV 88	
1986-019LT		17701	ESA	22 FEB	05 NOV 87	
1986-019LW		17704	ESA	22 FEB	18 JAN 90	
1986-019LX		17808	ESA	22 FEB	01 MAR 88	
1986-019LY		17809	ESA	22 FEB	27 AUG 89	
1986-019LZ		17810	ESA	22 FEB	19 JAN 89	
1986-019MA		17811	ESA	22 FEB	22 JAN 88	
1986-019MB		17812	ESA	22 FEB	05 JUL 88	
1986-019MC		17813	ESA	22 FEB	15 FEB 91	
1986-019MD		17814	ESA	22 FEB	30 SEP 88	
1986-019MF		17816	ESA	22 FEB	27 MAR 89	
1986-019MG		17817	ESA	22 FEB	31 AUG 88	
1986-019MH		17942	ESA	22 FEB	13 APR 89	
1986-019MJ		17943	ESA	22 FEB	23 APR 89	
1986-019MK		17944	ESA	22 FEB	11 SEP 89	
1986-019ML		17945	ESA	22 FEB	17 MAR 89	
1986-019MM		17946	ESA	22 FEB	07 DEC 87	
1986-019MN		17947	ESA	22 FEB	06 NOV 88	
1986-019MP		17948	ESA	22 FEB	29 APR 88	
1986-019MQ		17949	ESA	22 FEB	14 AUG 88	
1986-019MR		17950	ESA	22 FEB	29 JAN 90	
1986-019MS		17951	ESA	22 FEB	11 DEC 88	
1986-019MT		17952	ESA	22 FEB	07 JAN 89	
1986-019MU		17953	ESA	22 FEB	05 JAN 89	
1986-019MV		17954	ESA	22 FEB	04 MAY 88	
1986-019MW		17955	ESA	22 FEB	15 OCT 88	
1986-019MX		17956	ESA	22 FEB	11 NOV 89	
1986-019MY		17957	ESA	22 FEB	02 NOV 89	
1986-019MZ		17958	ESA	22 FEB	27 AUG 89	
1986-019NA		17959	ESA	22 FEB	24 AUG 89	
1986-019NB		17960	ESA	22 FEB	21 AUG 89	
1986-019NC		17961	ESA	22 FEB	03 JAN 89	
1986-019ND		17977	ESA	22 FEB	04 JAN 89	
1986-019NE		17978	ESA	22 FEB	09 SEP 89	
1986-019NF		17979	ESA	22 FEB	03 FEB 88	
1986-019NG		17980	ESA	22 FEB	17 MAR 89	
1986-019NJ		17982	ESA	22 FEB	16 MAR 89	
1986-019NK		17983	ESA	22 FEB	12 MAY 91	
1986-019NL		17984	ESA	22 FEB	21 APR 89	
1986-019NM		17985	ESA	22 FEB	11 OCT 87	
1986-019NN		17986	ESA	22 FEB	02 NOV 88	
1986-019NP		17987	ESA	22 FEB	02 NOV 88	
1986-019NR		17989	ESA	22 FEB	22 MAR 89	
1986-019NS		17990	ESA	22 FEB	10 OCT 88	
1986-019NT		17991	ESA	22 FEB	30 NOV 88	
1986-019NU		17992	ESA	22 FEB	03 JUL 89	
1986-019NV		17993	ESA	22 FEB	03 JUN 89	
1986-019NW		17994	ESA	22 FEB	10 JAN 89	
1986-019NX		17995	ESA	22 FEB	20 OCT 89	
1986-019NY		17996	ESA	22 FEB	26 MAR 90	
1986-019NZ		18132	ESA	22 FEB	03 FEB 88	
1986-019P		17119	ESA	22 FEB	02 JUL 90	
1986-019PA		18133	ESA	22 FEB	02 NOV 88	
1986-019PB		18134	ESA	22 FEB	12 FEB 89	
1986-019PC		18135	ESA	22 FEB	16 OCT 88	
1986-019PD		18136	ESA	22 FEB	05 OCT 88	
1986-019PE		18137	ESA	22 FEB	03 NOV 88	
1986-019PF		18138	ESA	22 FEB	14 MAR 89	
1986-019PG		18139	ESA	22 FEB	06 JAN 89	
1986-019PH		18140	ESA	22 FEB	23 OCT 88	
1986-019PJ		18141	ESA	22 FEB	26 SEP 88	
1986-019PK		18142	ESA	22 FEB	10 NOV 87	
1986-019PL		18143	ESA	22 FEB	19 JAN 89	
1986-019PM		18144	ESA	22 FEB	14 OCT 89	
1986-019PN		18145	ESA	22 FEB	15 JUN 91	
1986-019PP		18146	ESA	22 FEB	27 MAR 90	
1986-019PQ		18147	ESA	22 FEB	14 AUG 89	
1986-019PR		18148	ESA	22 FEB	12 DEC 89	
1986-019PS		18149	ESA	22 FEB	14 DEC 88	
1986-019PT		18150	ESA	22 FEB	20 FEB 90	
1986-019PU		18151	ESA	22 FEB	19 SEP 89	
1986-019PV		18164	ESA	22 FEB	26 JUN 89	
1986-019PW		18165	ESA	22 FEB	29 AUG 87	
1986-019PX		18166	ESA	22 FEB	26 FEB 89	
1986-019PY		18167	ESA	22 FEB	21 MAR 89	
1986-019PZ		18168	ESA	22 FEB	13 MAR 89	
1986-019QA		18169	ESA	22 FEB	18 APR 90	
1986-019QB		18170	ESA	22 FEB	10 NOV 87	
1986-019QC		18171	ESA	22 FEB	21 OCT 89	
1986-019QD		18172	ESA	22 FEB	12 NOV 89	
1986-019QE		18173	ESA	22 FEB	25 MAY 89	
1986-019QF		18174	ESA	22 FEB	09 AUG 89	
1986-019QG		18175	ESA	22 FEB	19 JAN 88	
1986-019QH		18176	ESA	22 FEB	27 OCT 89	
1986-019QJ		18177	ESA	22 FEB	12 OCT 89	
1986-019QK		18178	ESA	22 FEB	26 MAR 90	
1986-019QL		18179	ESA	22 FEB	14 JAN 89	
1986-019QM		18180	ESA	22 FEB	24 JUL 89	
1986-019QN		18181	ESA	22 FEB	22 JUL 89	
1986-019QP		18182	ESA	22 FEB	13 JAN 89	
1986-019QQ		18183	ESA	22 FEB	27 APR 91	
1986-019QR		18194	ESA	22 FEB	24 MAY 88	
1986-019QS		18195	ESA	22 FEB	10 DEC 88	
1986-019QU		18197	ESA	22 FEB	02 NOV 88	
1986-019QV		18198	ESA	22 FEB	01 JUL 88	
1986-019QW		18199	ESA	22 FEB	27 FEB 90	
1986-019QX		18200	ESA	22 FEB	04 OCT 88	
1986-019QY		18201	ESA	22 FEB	23 DEC 88	
1986-019QZ		18202	ESA	22 FEB	11 NOV 89	
1986-019RA		18203	ESA	22 FEB	07 MAY 89	
1986-019RB		18204	ESA	22 FEB	22 NOV 87	
1986-019RC		18205	ESA	22 FEB	24 NOV 88	
1986-019RD		18206	ESA	22 FEB	31 MAR 90	
1986-019RE		18207	ESA	22 FEB	20 MAY 89	
1986-019RG		18209	ESA	22 FEB	25 OCT 88	
1986-019RH		18210	ESA	22 FEB	09 MAR 89	
1986-019RJ		18211	ESA	22 FEB	11 OCT 89	
1986-019RL		18213	ESA	22 FEB	06 APR 89	
1986-019RP		18234	ESA	22 FEB	05 NOV 89	
1986-019RQ		18261	ESA	22 FEB	19 FEB 89	
1986-019RR		18262	ESA	22 FEB	21 MAR 89	
1986-019RS		18263	ESA	22 FEB	18 FEB 89	
1986-019RT		18264	ESA	22 FEB	14 NOV 88	
1986-019RU		18265	ESA	22 FEB	12 FEB 89	
1986-019RV		18266	ESA	22 FEB	11 DEC 88	
1986-019RW		18267	ESA	22 FEB	05 JAN 89	
1986-019RX		18289	ESA	22 FEB	02 OCT 88	
1986-019RZ		18291	ESA	22 FEB	23 OCT 88	
1986-019SA		18292	ESA	22 FEB	10 DEC 88	
1986-019SB		18293	ESA	22 FEB	15 AUG 89	
1986-019SC		18294	ESA	22 FEB	04 FEB 89	
1986-019SD		18295	ESA	22 FEB	09 SEP 89	
1986-019SE		18296	ESA	22 FEB	10 MAR 89	
1986-019SF		18297	ESA	22 FEB	26 MAR 91	
1986-019SG		18298	ESA	22 FEB	19 APR 89	
1986-019SH		18299	ESA	22 FEB	07 JAN 89	
1986-019SJ		18300	ESA	22 FEB	22 NOV 89	
1986-019SK		18404	ESA	22 FEB	05 OCT 88	
1986-019SL		18405	ESA	22 FEB	01 JUN 88	
1986-019SM		18406	ESA	22 FEB	01 JAN 89	
1986-019SN		18450	ESA	22 FEB	22 FEB 89	
1986-019SP		18451	ESA	22 FEB	29 DEC 88	
1986-019SQ		18452	ESA	22 FEB	28 APR 89	

International Designation	Satellite Name	Number	Source	Launch	Decay	Note
1986-019SS		18454	ESA	22 FEB	01 APR 91	
1986-019ST		18455	ESA	22 FEB	12 APR 89	
1986-019SU		18456	ESA	22 FEB	23 MAR 89	
1986-019SV		18457	ESA	22 FEB	05 FEB 89	
1986-019SW		18458	ESA	22 FEB	27 OCT 88	
1986-019SX		18459	ESA	22 FEB	11 OCT 89	
1986-019SY		18460	ESA	22 FEB	14 MAY 89	
1986-019TA		18462	ESA	22 FEB	31 JAN 89	
1986-019TB		18522	ESA	22 FEB	26 NOV 88	
1986-019TD		18553	ESA	22 FEB	03 JUN 89	
1986-019TE		18554	ESA	22 FEB	01 AUG 88	
1986-019TF		18555	ESA	22 FEB	19 MAR 88	
1986-019TG		18557	ESA	22 FEB	10 MAR 89	
1986-019TH		18558	ESA	22 FEB	19 MAY 89	
1986-019TJ		18559	ESA	22 FEB	20 JAN 89	
1986-019TK		18563	ESA	22 FEB	26 SEP 88	
1986-019TL		18607	ESA	22 FEB	25 MAR 89	
1986-019TM		18611	ESA	22 FEB	11 MAY 91	
1986-019TN		18614	ESA	22 FEB	04 JUL 90	
1986-019TP		18642	ESA	22 FEB	28 MAY 91	
1986-019TQ		18674	ESA	22 FEB	06 MAR 90	
1986-019TR		18676	ESA	22 FEB	16 JUL 90	
1986-019TS		18683	ESA	22 FEB	17 JAN 89	
1986-019TT		18684	ESA	22 FEB	22 MAR 90	
1986-019TU		18685	ESA	22 FEB	04 DEC 88	
1986-019TV		18774	ESA	22 FEB	02 APR 89	
1986-019TW		18778	ESA	22 FEB	19 SEP 89	
1986-019TY		18781	ESA	22 FEB	21 FEB 89	
1986-019TZ		18782	ESA	22 FEB	31 JAN 89	
1986-019UA		18787	ESA	22 FEB	18 APR 89	
1986-019UB		18813	ESA	22 FEB	02 DEC 88	
1986-019UC		18817	ESA	22 FEB	18 SEP 88	
1986-019UD		18865	ESA	22 FEB	13 NOV 89	
1986-019UE		18903	ESA	22 FEB	26 MAR 89	
1986-019UF		18905	ESA	22 FEB	04 SEP 89	
1986-019UG		18914	ESA	22 FEB	30 MAY 89	
1986-019UH		19007	ESA	22 FEB	28 OCT 88	
1986-019UJ		19009	ESA	22 FEB	13 AUG 89	
1986-019UK		19034	ESA	22 FEB	26 FEB 88	
1986-019UM		19107	ESA	22 FEB	02 MAY 91	
1986-019UP		19240	ESA	22 FEB	24 NOV 88	
1986-019UQ		19246	ESA	22 FEB	01 NOV 89	
1986-019UR		19307	ESA	22 FEB	09 MAR 90	
1986-019UT		19313	ESA	22 FEB	02 FEB 89	
1986-019UU		19315	ESA	22 FEB	06 AUG 89	
1986-019 W		19356	ESA	22 FEB	03 DEC 89	
1986-019UX		19395	ESA	22 FEB	07 JUN 89	
1986-019UZ		19477	ESA	22 FEB	04 JAN 89	
1986-019V		17126	ESA	22 FEB	16 MAR 87	
1986-019VA		19513	ESA	22 FEB	11 SEP 89	
1986-019VC		19642	ESA	22 FEB	11 APR 89	
1986-019VD		19812	ESA	22 FEB	11 FEB 91	
1986-019VE		19834	ESA	22 FEB	23 AUG 89	
1986-019VF		19855	ESA	22 FEB	16 MAR 89	
1986-019VG		19887	ESA	22 FEB	06 MAY 89	
1986-019VH		19892	ESA	22 FEB	22 JUL 89	
1986-019VJ		19966	ESA	22 FEB	06 JUL 89	
1986-019VK		19999	ESA	22 FEB	25 SEP 89	
1986-019VL		20050	ESA	22 FEB	09 APR 90	
1986-019VM		20173	ESA	22 FEB	26 JUL 90	
1986-019W		17127	ESA	22 FEB	25 JUL 87	
1986-019UL		19050	ESA	24 FEB	26 AUG 89	
1986-020A	COSMOS 1734	16618	USSR	26 FEB	26 APR 86	
1986-020B		16619	USSR	26 FEB	03 MAR 86	
1986-020C		16692	USSR	26 FEB	27 APR 86	
1986-020D		16693	USSR	26 FEB	30 APR 86	
1986-021A	COSMOS 1735	16620	USSR	27 FEB	17 NOV 88	
1986-021B		16621	USSR	27 FEB	27 FEB 86	
1986-022A	SOYUZ T-15	16643	USSR	13 MAR	16 JUL 86	
1986-022B		16644	USSR	13 MAR	16 MAR 86	
1986-023A	PROGRESS 25	16645	USSR	19 MAR	21 APR 86	
1986-023B		16646	USSR	19 MAR	22 MAR 86	
1986-024AA		16842	USSR	21 MAR	15 NOV 86	
1986-024AB		16857	USSR	21 MAR	14 NOV 86	
1986-024AC		16858	USSR	21 MAR	28 FEB 87	
1986-024AD		16859	USSR	21 MAR	16 DEC 86	
1986-024AE		16868	USSR	21 MAR	14 NOV 86	
1986-024AF		17030	USSR	21 MAR	13 NOV 86	
1986-024B		16806	USSR	21 MAR	01 SEP 86	
1986-024C		16807	USSR	21 MAR	27 JUN 86	
1986-024D		16808	USSR	21 MAR	26 JUN 86	
1986-024F		16823	USSR	21 MAR	23 OCT 86	
1986-024G		16824	USSR	21 MAR	10 AUG 87	
1986-024H		16825	USSR	21 MAR	31 AUG 86	
1986-024J		16826	USSR	21 MAR	04 DEC 86	
1986-024K		16827	USSR	21 MAR	31 OCT 86	
1986-024L		16828	USSR	21 MAR	16 DEC 86	
1986-024M		16829	USSR	21 MAR	26 JUL 87	
1986-024N		16830	USSR	21 MAR	04 JUL 86	
1986-024P		16831	USSR	21 MAR	18 MAR 87	
1986-024Q		16832	USSR	21 MAR	15 AUG 87	
1986-024R		16833	USSR	21 MAR	06 NOV 86	
1986-024S		16834	USSR	21 MAR	04 NOV 86	
1986-024T		16835	USSR	21 MAR	23 OCT 86	
1986-024U		16836	USSR	21 MAR	06 MAR 87	
1986-024V		16837	USSR	21 MAR	14 JUL 86	
1986-024W		16838	USSR	21 MAR	10 SEP 86	
1986-024X		16839	USSR	21 MAR	28 FEB 87	
1986-024Y		16840	USSR	21 MAR	24 FEB 87	
1986-024Z		16841	USSR	21 MAR	22 JAN 88	
1986-025A	COSMOS 1737	16648	USSR	25 MAR	03 DEC 86	
1986-025B		16659	USSR	25 MAR	26 MAR 86	
1986-026D		16658	ESA	28 MAR	12 AUG 87	
1986-027B		16668	USSR	04 APR	08 APR 86	
1986-027C		16669	USSR	04 APR	05 APR 86	
1986-027D		16670	USSR	04 APR	27 OCT 86	
1986-027E		16671	USSR	04 APR	27 OCT 86	
1986-028A	COSMOS 1739	16677	USSR	09 APR	07 JUN 86	
1986-028B		16678	USSR	09 APR	14 APR 86	
1986-029A	COSMOS 1740	16679	USSR	15 APR	28 APR 86	
1986-029B		16680	USSR	15 APR	27 APR 86	
1986-029C		16694	USSR	15 APR	14 DEC 86	
1986-029D		16695	USSR	15 APR	17 AUG 86	
1986-029E		16696	USSR	15 APR	18 MAR 87	
1986-029F		16697	USSR	15 APR	31 JUL 86	
1986-029G		16698	USSR	15 APR	07 OCT 86	
1986-029H		16699	USSR	15 APR	06 MAY 86	
1986-029J		16700	USSR	15 APR	11 MAY 86	
1986-029K		16701	USSR	15 APR	07 MAY 86	
1986-029L		16714	USSR	15 APR	18 APR 88	
1986-029M		16716	USSR	15 APR	08 MAY 86	
1986-031B		16684	USSR	18 APR	04 JUL 86	
1986-031C		16685	USSR	18 APR	12 JUN 86	
1986-032A	PROGRESS 26	16687	USSR	23 APR	23 JUN 86	
1986-032B		16688	USSR	23 APR	26 APR 86	
1986-033A	COSMOS 1742	16717	USSR	14 MAY	28 MAY 86	
1986-033B		16718	USSR	14 MAY	29 MAY 86	
1986-033C		16739	USSR	14 MAY	05 SEP 87	
1986-033D		16740	USSR	14 MAY	29 DEC 86	
1986-033E		16741	USSR	14 MAY	17 NOV 86	
1986-033F		16742	USSR	14 MAY	25 APR 88	
1986-033G		16743	USSR	14 MAY	28 JUN 86	
1986-033H		16744	USSR	14 MAY	31 OCT 86	
1986-041C		16793	USSR	19 MAY	22 JUN 86	
1986-035A	SOYUZ TM	16722	USSR	21 MAY	30 MAY 86	
1986-035B		16723	USSR	21 MAY	24 MAY 86	
1986-035C		16748	USSR	21 MAY	22 MAR 87	
1986-036A	COSMOS 1744	16724	USSR	21 MAY	04 JUN 86	
1986-036B		16725	USSR	21 MAY	24 JUN 86	
1986-036C		16753	USSR	21 MAY	05 JUN 86	
1986-036D		16754	USSR	21 MAY	18 AUG 86	
1986-036E		16755	USSR	21 MAY	21 JUN 86	
1986-036F		16756	USSR	21 MAY	25 JUL 86	
1986-036G		16757	USSR	21 MAY	07 JUN 86	
1986-038B		16730	USSR	24 MAY	26 MAY 86	
1986-038C		16731	USSR	24 MAY	24 MAY 86	
1986-038F		16734	USSR	24 MAY	28 DEC 86	
1986-040A	COSMOS 1746	16737	USSR	28 MAY	12 JUN 86	
1986-040B		16738	USSR	28 MAY	02 JUN 86	
1986-040C		16747	USSR	28 MAY	08 JUN 86	
1986-040D		16783	USSR	28 MAY	13 JUN 86	
1986-040E		16784	USSR	28 MAY	13 JUN 86	
1986-040F		16785	USSR	28 MAY	14 JUN 86	
1986-040G		16786	USSR	28 MAY	13 JUN 86	
1986-040H		16787	USSR	28 MAY	07 JUL 86	
1986-040J		16788	USSR	28 MAY	23 JUN 86	
1986-041A	COSMOS 1747	16745	USSR	29 MAY	12 JUN 86	
1986-041B		16746	USSR	29 MAY	25 JUN 86	
1986-041D		16794	USSR	29 MAY	25 JUN 86	
1986-041E		16795	USSR	29 MAY	14 JUN 86	
1986-041F		16796	USSR	29 MAY	15 JUN 86	
1986-043A	COSMOS 1756	16767	USSR	06 JUN	04 AUG 86	
1986-043B		16768	USSR	06 JUN	13 JUN 86	
1986-044B		16770	USSR	10 JUN	12 JUN 86	
1986-044C		16771	USSR	10 JUN	11 JUN 86	
1986-044D		16789	USSR	10 JUN	07 MAR 90	
1986-044E		16790	USSR	10 JUN	12 AUG 87	
1986-045A	COSMOS 1757	16772	USSR	11 JUN	25 JUN 86	
1986-045B		16773	USSR	11 JUN	13 JUN 86	
1986-045C		16816	USSR	11 JUN	27 JUN 86	
1986-045D		16817	USSR	11 JUN	05 JUL 86	
1986-045E		16818	USSR	11 JUN	27 JUN 86	
1986-045F		16819	USSR	11 JUN	09 JUL 86	
1986-045G		16820	USSR	11 JUN	29 JUN 86	
1986-017DV	MAK-1	21425	USSR	17 JUN	18 OCT 91	
1986-048A	COSMOS 1760	16800	USSR	19 JUN	03 JUL 86	
1986-048B		16801	USSR	19 JUN	19 JUL 86	
1986-048C		16844	USSR	19 JUN	24 MAR 88	
1986-048D		16845	USSR	19 JUN	23 MAR 87	
1986-048E		16846	USSR	19 JUN	17 MAR 87	
1986-048F		16847	USSR	19 JUN	27 JUL 86	
1986-048G		16848	USSR	19 JUN	26 DEC 86	
1986-049B		16803	USSR	19 JUN	27 SEP 86	
1986-049C		16804	USSR	19 JUN	01 OCT 86	
1986-050B		16850	USSR	05 JUL	19 SEP 86	
1986-050C		16851	USSR	05 JUL	16 AUG 86	
1986-051A	COSMOS 1762	16855	USSR	10 JUL	24 JUL 86	
1986-051B		16856	USSR	10 JUL	15 JUL 86	
1986-051C		16871	USSR	10 JUL	21 AUG 86	
1986-051D		16872	USSR	10 JUL	03 AUG 86	
1986-051E		16873	USSR	10 JUL	31 JUL 86	
1986-051F		16876	USSR	10 JUL	22 SEP 86	
1986-051G		16880	USSR	10 JUL	05 AUG 86	
1986-053A	COSMOS 1764	16861	USSR	16 JUL	11 SEP 86	
1986-053B		16862	USSR	17 JUL	26 JUL 86	
1986-054A	COSMOS 1765	16874	USSR	24 JUL	07 AUG 86	

International Designation	Satellite Name	Number	Source	Launch	Decay	Note
1986-054B		16875	USSR	24 JUL	08 AUG 86	
1986-054C		16899	USSR	24 JUL	06 NOV 86	
1986-054D		16900	USSR	24 JUL	14 MAY 87	
1986-054E		16901	USSR	24 JUL	23 SEP 86	
1986-054F		16902	USSR	24 JUL	20 MAR 87	
1986-054G		16903	USSR	24 JUL	18 SEP 86	
1986-054H		16904	USSR	24 JUL	14 SEP 86	
1986-054J		16905	USSR	24 JUL	12 MAY 88	
1986-054K		16906	USSR	24 JUL	11 SEP 86	
1986-054L		16907	USSR	24 JUL	14 SEP 86	
1986-054M		17175	USSR	24 JUL	03 FEB 87	
1986-056A	COSMOS 1767	16883	USSR	30 JUL	16 AUG 86	
1986-056B		16884	USSR	30 JUL	04 AUG 86	
1986-056C		16888	USSR	30 JUL	18 AUG 86	
1986-056D		16892	USSR	30 JUL	11 AUG 86	
1986-056E		16893	USSR	30 JUL	12 AUG 86	
1986-056F		16894	USSR	30 JUL	20 AUG 86	
1986-057B		16886	USSR	30 JUL	19 OCT 86	
1986-057C		16887	USSR	30 JUL	05 OCT 86	
1986-058A	COSMOS 1768	16890	USSR	02 AUG	16 AUG 86	
1986-058B		16891	USSR	02 AUG	07 AUG 86	
1986-058C		16911	USSR	02 AUG	19 OCT 86	
1986-058D		16912	USSR	02 AUG	16 SEP 86	
1986-058E		16913	USSR	02 AUG	24 AUG 86	
1986-058F		16914	USSR	02 AUG	17 AUG 86	
1986-058G		16915	USSR	02 AUG	29 AUG 86	
1986-058H		16916	USSR	02 AUG	31 AUG 86	
1986-059A	COSMOS 1769	16895	USSR	04 AUG	18 FEB 88	
1986-059B		16896	USSR	04 AUG	04 AUG 86	
1986-059C		18354	USSR	04 AUG	10 OCT 87	
1986-059D		18382	USSR	04 AUG	14 DEC 87	
1986-059E		18383	USSR	04 AUG	07 OCT 87	
1986-060A	COSMOS 1770	16897	USSR	06 AUG	02 FEB 87	
1986-060B		16898	USSR	06 AUG	12 AUG 86	
1986-060C		17398	USSR	06 AUG	05 FEB 87	
1986-062B		17034	USSR	20 AUG	20 OCT 86	
1986-062D		17036	USSR	20 AUG	29 NOV 86	
1986-063A	COSMOS 1772	16918	USSR	21 AUG	03 SEP 86	
1986-063B		16919	USSR	21 AUG	02 SEP 86	
1986-063C		16930	USSR	21 AUG	09 APR 87	
1986-063D		16931	USSR	21 AUG	13 MAY 88	
1986-063E		16932	USSR	21 AUG	07 JAN 87	
1986-063F		16933	USSR	21 AUG	01 JUN 87	
1986-063G		17301	USSR	21 AUG	19 MAR 87	
1986-064A	COSMOS 1773	16920	USSR	27 AUG	21 OCT 86	
1986-064B		16921	USSR	27 AUG	02 SEP 86	
1986-065B		16923	USSR	28 AUG	22 OCT 86	
1986-065C		16924	USSR	28 AUG	01 OCT 86	
1986-066A	COSMOS 1775	16926	USSR	03 SEP	17 SEP 86	
1986-066B		16927	USSR	03 SEP	25 SEP 86	
1986-066C		16970	USSR	03 SEP	12 APR 88	
1986-066D		16971	USSR	03 SEP	01 MAR 87	
1986-066E		16972	USSR	03 SEP	20 DEC 86	
1986-066F		16973	USSR	03 SEP	02 APR 87	
1986-066G		16975	USSR	03 SEP	27 SEP 86	
1986-066H		16976	USSR	03 SEP	07 OCT 86	
1986-066J		16977	USSR	03 SEP	02 OCT 86	
1986-066K		16978	USSR	03 SEP	10 OCT 86	
1986-066L		16979	USSR	03 SEP	21 DEC 86	
1986-067A	COSMOS 1776	16928	USSR	03 SEP	15 DEC 89	
1986-067AA		19410	USSR	03 SEP	20 OCT 88	
1986-067AB		19411	USSR	03 SEP	10 OCT 88	
1986-067AC		19545	USSR	03 SEP	08 NOV 88	
1986-067AD		19546	USSR	03 SEP	31 OCT 88	
1986-067AE		20272	USSR	03 SEP	05 OCT 89	
1986-067AF		20273	USSR	03 SEP	05 OCT 89	
1986-067B		16929	USSR	03 SEP	26 OCT 89	
1986-067C		17006	USSR	03 SEP	30 OCT 87	
1986-067D		17007	USSR	03 SEP	14 OCT 87	
1986-067E		17008	USSR	03 SEP	08 NOV 87	
1986-067F		17009	USSR	03 SEP	24 OCT 87	
1986-067G		17258	USSR	03 SEP	09 DEC 87	
1986-067H		17259	USSR	03 SEP	29 NOV 87	
1986-067J		17260	USSR	03 SEP	23 NOV 87	
1986-067K		17261	USSR	03 SEP	14 NOV 87	
1986-067L		17334	USSR	03 SEP	04 DEC 87	
1986-067M		17335	USSR	03 SEP	09 NOV 87	
1986-067N		17356	USSR	03 SEP	27 DEC 87	
1986-067P		17357	USSR	03 SEP	22 NOV 87	
1986-067Q		17537	USSR	03 SEP	24 NOV 87	
1986-067R		17538	USSR	03 SEP	28 DEC 87	
1986-067S		18026	USSR	03 SEP	13 JAN 88	
1986-067T		18027	USSR	03 SEP	07 FEB 88	
1986-067U		18434	USSR	03 SEP	20 SEP 88	
1986-067V		18435	USSR	03 SEP	14 SEP 88	
1986-067W		18855	USSR	03 SEP	28 MAY 88	
1986-067X		18856	USSR	03 SEP	13 JUN 88	
1986-067Y		19375	USSR	03 SEP	18 OCT 88	
1986-067Z		19376	USSR	03 SEP	01 OCT 88	
1986-068B		16935	USSR	05 SEP	12 OCT 86	
1986-068C		16936	USSR	05 SEP	06 NOV 86	
1986-069A		16937	US	05 SEP	28 SEP 86	
1986-069B		16938	US	05 SEP	25 NOV 86	
1986-069C		16940	US	05 SEP	24 SEP 86	
1986-069D		16941	US	05 SEP	14 APR 87	
1986-069E		16942	US	05 SEP	10 NOV 86	
1986-069F		16943	US	05 SEP	11 DEC 86	
1986-069G		16944	US	05 SEP	23 OCT 86	
1986-069H		16945	US	05 SEP	31 OCT 86	
1986-069J		16946	US	05 SEP	30 MAR 87	
1986-069K		16947	US	05 SEP	16 SEP 86	
1986-069L		16948	US	05 SEP	15 NOV 86	
1986-069M		16949	US	05 SEP	23 SEP 86	
1986-069N		16950	US	05 SEP	13 SEP 86	
1986-069P		16951	US	05 SEP	21 OCT 86	
1986-069Q		17019	US	05 SEP	20 MAR 87	
1986-069R		17020	US	05 SEP	29 OCT 86	
1986-069S		17021	US	05 SEP	26 OCT 86	
1986-069T		17022	US	05 SEP	31 OCT 86	
1986-071D		16964	USSR	16 SEP	16 SEP 86	
1986-071E		16965	USSR	16 SEP	16 SEP 86	
1986-071G		16984	USSR	16 SEP	29 JAN 93	
1986-071H		16985	USSR	16 SEP	26 NOV 92	
1986-072A	COSMOS 1781	16966	USSR	17 SEP	01 OCT 86	
1986-072B		16967	USSR	17 SEP	07 OCT 86	
1986-072C		16974	USSR	17 SEP	20 SEP 86	
1986-072D		16988	USSR	17 SEP	09 APR 88	
1986-072E		16989	USSR	17 SEP	20 MAY 87	
1986-072F		16990	USSR	17 SEP	19 MAY 87	
1986-072G		16991	USSR	17 SEP	29 DEC 86	
1986-072H		16992	USSR	17 SEP	13 JAN 87	
1986-072J		17049	USSR	17 SEP	01 DEC 86	
1986-073C		16983	US	17 SEP	04 AUG 91	
1986-082M		17281	USSR	19 SEP	27 MAY 87	
1986-074C		19432	USSR	30 SEP	13 MAR 89	
1986-075B		16994	USSR	03 OCT	22 NOV 86	
1986-075C		16995	USSR	03 OCT	08 NOV 86	
1986-075E		16998	USSR	03 OCT	12 NOV 86	
1986-076A	PRC 19	17001	PRC	06 OCT	23 OCT 86	
1986-076B		17002	PRC	06 OCT	20 OCT 86	
1986-076C		17005	PRC	06 OCT	08 OCT 86	
1986-0076D		17028	PRC	06 OCT	12 OCT 86	
1986-077A	COSMOS 1784	17003	USSR	06 OCT	11 NOV 86	
1986-077B		17004	USSR	06 OCT	12 OCT 86	
1986-078B		17032	USSR	15 OCT	18 NOV 86	
1986-078C		17033	USSR	15 OCT	09 DEC 86	
1986-079B		17039	USSR	20 OCT	22 DEC 86	
1986-079C		17040	USSR	20 OCT	12 DEC 86	
1986-080A	COSMOS 1786	17042	USSR	22 OCT	06 MAR 88	
1986-080B		17043	USSR	22 OCT	08 APR 88	
1986-081A	COSMOS 1787	17044	USSR	22 OCT	04 NOV 86	
1986-081B		17045	USSR	22 OCT	30 OCT 86	
1986-081C		17058	USSR	22 OCT	20 NOV 86	
1986-081D		17061	USSR	22 OCT	17 NOV 86	
1986-081E		17062	USSR	22 OCT	31 DEC 86	
1986-081F		17063	USSR	22 OCT	09 NOV 86	
1986-081G		17064	USSR	22 OCT	08 NOV 86	
1986-082B		17047	USSR	25 OCT	28 OCT 86	
1986-082C		17048	USSR	25 OCT	26 OCT 86	
1986-083A	COSMOS 1788	17050	USSR	27 OCT	21 JAN 91	
1986-083B		17051	USSR	27 OCT	05 JAN 90	
1986-084A	COSMOS 1789	17054	USSR	31 OCT	14 NOV 86	
1986-084B		17055	USSR	31 OCT	04 NOV 86	
1986-084C		17072	USSR	31 OCT	27 AUG 87	
1986-084D		17073	USSR	31 OCT	19 DEC 86	
1986-084E		17074	USSR	31 OCT	07 JAN 87	
1986-084F		17075	USSR	31 OCT	19 NOV 86	
1986-084G		17076	USSR	31 OCT	03 JAN 87	
1986-084H		17077	USSR	31 OCT	21 FEB 87	
1986-085A	COSMOS 1790	17056	USSR	04 NOV	18 NOV 86	
1986-085B		17057	USSR	04 NOV	09 NOV 86	
1986-085C		17082	USSR	04 NOV	21 NOV 86	
1986-085D		17086	USSR	04 NOV	26 NOV 86	
1986-085E		17087	USSR	04 NOV	23 NOV 86	
1986-085F		17088	USSR	04 NOV	24 NOV 86	
1986-085G		17089	USSR	04 NOV	10 DEC 86	
1986-087A	COSMOS 1792	17068	USSR	13 NOV	05 JAN 87	
1986-087B		17069	USSR	13 NOV	19 NOV 86	
1986-089B		17079	USSR	15 NOV	19 DEC 86	
1986-089C		17080	USSR	15 NOV	28 DEC 86	
1986-090B		17084	USSR	18 NOV	21 NOV 86	
1986-090C		17085	USSR	18 NOV	19 NOV 86	
1986-090E		17148	USSR	18 NOV	12 AUG 87	
1986-091B		17135	USSR	20 NOV	27 JAN 87	
1986-091C		17136	USSR	20 NOV	23 DEC 86	
1986-095A	COSMOS 1804	17179	USSR	04 DEC	18 DEC 86	
1986-095B		17180	USSR	04 DEC	31 DEC 86	
1986-095C		17243	USSR	04 DEC	04 SEP 87	
1986-095D		17244	USSR	04 DEC	13 APR 88	
1986-095E		17245	USSR	04 DEC	26 MAY 87	
1986-095F		17246	USSR	04 DEC	15 APR 87	
1986-095G		17252	USSR	04 DEC	08 APR 87	
1986-096B		17182	US	05 DEC	13 MAR 88	
1986-098B		17214	USSR	12 DEC	23 FEB 87	
1986-098C		17215	USSR	12 DEC	10 FEB 87	
1986-099A	COSMOS 1807	17217	USSR	16 DEC	23 JAN 87	
1986-099B		17218	USSR	16 DEC	22 DEC 86	
1986-102A	COSMOS 1810	17262	USSR	26 DEC	11 SEP 87	
1986-102B		17263	YSSR	26 DEC	31 DEC 86	
1986-102C		18345	USSR	26 DEC	11 SEP 87	
1986-102D		18346	USSR	26 DEC	12 SEP 87	
1986-102E		18347	USSR	26 DEC	12 SEP 87	
1986-103B		17265	USSR	26 DEC	02 FEB 87	
1986-103C		17266	USSR	26 DEC	21 JAN 87	
1986-103E		17276	USSR	26 DEC	27 JAN 86	

1987 LAUNCHES

International Designation	Satellite Name	Number	Source	Launch	Decay	Note
1987-002A	COSMOS 1811	17292	USSR	09 JAN	13 FEB 87	
1987-002B		17293	USSR	09 JAN	16 JAN 87	
1987-002C		17294	USSR	09 JAN	11 JAN 87	
1987-004A	COSMOS 1813	17297	USSR	15 JAN	13 MAR 89	
1987-004AA		17419	USSR	15 JAN	31 AUG 87	
1987-004AB		17420	USSR	15 JAN	10 DEC 87	
1987-004AC		17488	USSR	15 JAN	05 OCT 88	
1987-004AD		17489	USSR	15 JAN	22 SEP 88	
1987-004AE		17490	USSR	15 JAN	22 MAR 87	
1987-004AF		17491	USSR	15 JAN	07 OCT 87	
1987-004AG		17492	USSR	15 JAN	17 NOV 88	
1987-004AH		17493	USSR	15 JAN	03 NOV 88	
1987-004AJ		17494	USSR	15 JAN	03 SEP 88	
1987-004AK		17495	USSR	15 JAN	07 DEC 88	
1987-004AL		17496	USSR	15 JAN	14 APR 87	
1987-004AM		17497	USSR	15 JAN	27 MAR 87	
1987-004AN		17498	USSR	15 JAN	05 DEC 88	
1987-004AP		17499	USSR	15 JAN	22 APR 88	
1987-004AQ		17500	USSR	15 JAN	30 JAN 90	
1987-004AR		17501	USSR	15 JAN	25 OCT 88	
1987-004AS		17502	USSR	15 JAN	31 DEC 88	
1987-004AT		17503	USSR	15 JAN	16 SEP 87	
1987-004AU		17504	USSR	15 JAN	02 SEP 87	
1987-004AV		17505	USSR	15 JAN	27 OCT 87	
1987-004AW		17508	USSR	15 JAN	14 APR 87	
1987-004AX		17509	USSR	15 JAN	11 AUG 87	
1987-004AY		17510	USSR	15 JAN	30 JUN 87	
1987-004AZ		17511	USSR	15 JAN	20 AUG 87	
1987-004B		17298	USSR	15 JAN	01 FEB 87	
1987-004BA		17512	USSR	15 JAN	03 JUN 87	
1987-004BB		17513	USSR	15 JAN	21 JUN 87	
1987-004BC		17514	USSR	15 JAN	22 JUN 87	
1987-004BD		17515	USSR	15 JAN	17 MAY 87	
1987-004BE		17516	USSR	15 JAN	17 SEP 88	
1987-004BF		17517	USSR	15 JAN	22 NOV 87	
1987-004BG		17518	USSR	15 JAN	27 APR 87	
1987-004BH		17519	USSR	15 JAN	16 AUG 87	
1987-004BJ		17520	USSR	15 JAN	05 MAY 87	
1987-004BK		17521	USSR	15 JAN	12 JUL 88	
1987-004BL		17522	USSR	15 JAN	29 NOV 87	
1987-004BM		17756	USSR	15 JAN	18 MAR 89	
1987-004BN		17757	USSR	15 JAN	10 OCT 87	
1987-004BP		17758	USSR	15 JAN	19 OCT 87	
1987-004BQ		17759	USSR	15 JAN	13 SEP 87	
1987-004BR		17760	USSR	15 JAN	29 DEC 87	
1987-004BS		17761	USSR	15 JAN	09 JAN 88	
1987-004BT		17762	USSR	15 JAN	17 JAN 89	
1987-004BU		17852	USSR	15 JAN	14 AUG 88	
1987-004BV		17853	USSR	15 JAN	09 JUL 87	
1987-004BW		17854	USSR	15 JAN	11 OCT 87	
1987-004BX		17855	USSR	15 JAN	31 DEC 87	
1987-004BY		17856	USSR	15 JAN	13 APR 87	
1987-004BZ		17857	USSR	15 JAN	31 OCT 87	
1987-004C		17361	USSR	15 JAN	27 JUN 87	
1987-004CA		17858	USSR	15 JAN	12 SEP 87	
1987-004CB		17859	USSR	15 JAN	09 MAY 87	
1987-004CC		17860	USSR	15 JAN	15 APR 87	
1987-004CD		17861	USSR	15 JAN	11 SEP 87	
1987-004CE		17862	USSR	15 JAN	14 SEP 87	
1987-004CF		17863	USSR	15 JAN	28 AUG 87	
1987-004CG		17864	USSR	15 JAN	06 OCT 87	
1987-004CH		17865	USSR	15 JAN	21 MAY 87	
1987-004CJ		17866	USSR	15 JAN	11 JUL 87	
1987-004CK		17867	USSR	15 JAN	13 AUG 87	
1987-004CL		17868	USSR	15 JAN	24 MAY 87	
1987-004CM		17869	USSR	15 JAN	13 JUL 87	
1987-004CN		17870	USSR	15 JAN	02 JUN 87	
1987-004CP		17871	USSR	15 JAN	15 OCT 87	
1987-004CQ		17882	USSR	15 JAN	04 MAY 88	
1987-004CR		17883	USSR	15 JAN	05 DEC 87	
1987-004CS		17884	USSR	15 JAN	03 AUG 88	
1987-004CT		17885	USSR	15 JAN	26 JAN 88	
1987-004CU		17886	USSR	15 JAN	12 SEP 88	
1987-004CV		17887	USSR	15 JAN	29 DEC 87	
1987-004CW		17888	USSR	15 JAN	17 AUG 87	
1987-004CX		17889	USSR	15 JAN	13 AUG 87	
1987-004CY		17890	USSR	15 JAN	25 AUG 87	
1987-004CZ		17891	USSR	15 JAN	25 NOV 87	
1987-004D		17362	USSR	15 JAN	01 FEB 87	
1987-004DA		17892	USSR	15 JAN	02 APR 88	
1987-004DB		17893	USSR	15 JAN	14 OCT 87	
1987-004DC		17894	USSR	15 JAN	24 SEP 87	
1987-004DD		17895	USSR	15 JAN	05 OCT 87	
1987-004DE		17896	USSR	15 JAN	26 AUG 88	
1987-004DF		17897	USSR	15 JAN	11 SEP 87	
1987-004DG		17898	USSR	15 JAN	15 SEP 88	
1987-004DH		17899	USSR	15 JAN	05 NOV 88	
1987-004DJ		17900	USSR	15 JAN	25 OCT 89	
1987-004DK		17901	USSR	15 JAN	05 JUL 88	
1987-004DL		17919	USSR	15 JAN	22 JAN 89	
1987-004DM		17920	USSR	15 JAN	03 AUG 88	
1987-004DN		17921	USSR	15 JAN	19 JUL 87	
1987-004DP		17922	USSR	15 JAN	01 SEP 87	
1987-004DQ		17923	USSR	15 JAN	25 AUG 87	
1987-004DR		17924	USSR	15 JAN	14 OCT 88	
1987-004DS		17925	USSR	15 JAN	19 JUN 89	
1987-004DT		17926	USSR	15 JAN	21 JUN 87	
1987-004DU		17927	USSR	15 JAN	22 SEP 87	
1987-004DV		17928	USSR	15 JAN	11 JUN 87	
1987-004DW		17929	USSR	15 JAN	10 OCT 87	
1987-004DX		17930	USSR	15 JAN	10 JAN 88	
1987-004DY		17931	USSR	15 JAN	13 SEP 88	
1987-004DZ		17932	USSR	15 JAN	27 FEB 88	
1987-004E		17363	USSR	15 JAN	25 APR 87	
1987-004EA		17933	USSR	15 JAN	07 SEP 87	
1987-004EB		17934	USSR	15 JAN	09 SEP 87	
1987-004EC		17935	USSR	15 JAN	27 JUN 87	
1987-004ED		17936	USSR	15 JAN	28 NOV 87	
1987-004EE		17937	USSR	15 JAN	06 OCT 87	
1987-004EF		17938	USSR	15 JAN	29 NOV 87	
1987-004EF		18030	USSR	15 JAN	13 JUN 87	
1987-004EG		18029	USSR	15 JAN	18 JUN 87	
1987-004EJ		18031	USSR	15 JAN	22 SEP 87	
1987-004EK		18032	USSR	15 JAN	29 SEP 87	
1987-004EL		18033	USSR	15 JAN	31 OCT 87	
1987-004EM		18034	USSR	15 JAN	30 SEP 87	
1987-004EN		18035	USSR	15 JAN	12 JUL 87	
1987-004EP		18036	USSR	15 JAN	10 OCT 88	
1987-004EQ		18037	USSR	15 JAN	09 JAN 88	
1987-004ER		18038	USSR	15 JAN	29 OCT 87	
1987-004ES		18039	USSR	15 JAN	22 JAN 89	
1987-004ET		18040	USSR	15 JAN	15 SEP 87	
1987-004EU		18041	USSR	15 JAN	02 SEP 87	
1987-004EV		18042	USSR	15 JAN	01 NOV 88	
1987-004EW		18043	USSR	15 JAN	06 AUG 87	
1987-004EX		18044	USSR	15 JAN	31 MAR 88	
1987-004EY		18045	USSR	15 JAN	31 JUL 88	
1987-004EZ		18046	USSR	15 JAN	15 MAR 89	
1987-004F		17364	USSR	15 JAN	14 JUN 87	
1987-004FA		18047	USSR	15 JAN	23 OCT 87	
1987-004FB		18048	USSR	15 JAN	17 AUG 87	
1987-004FC		18049	USSR	15 JAN	23 OCT 88	
1987-004FD		18050	USSR	15 JAN	18 OCT 88	
1987-004FE		18051	USSR	15 JAN	24 JUL 87	
1987-004FF		18052	USSR	15 JAN	17 JUL 87	
1987-004FG		18053	USSR	15 JAN	29 JUN 87	
1987-004FH		18054	USSR	15 JAN	24 OCT 87	
1987-004FJ		18055	USSR	15 JAN	09 AUG 87	
1987-004FK		18056	USSR	15 JAN	02 OCT 87	
1987-004FL		18057	USSR	15 JAN	24 SEP 88	
1987-004FM		18058	USSR	15 JAN	19 SEP 88	
1987-004FN		18059	USSR	15 JAN	03 SEP 87	
1987-004FP		18060	USSR	15 JAN	23 JUL 87	
1987-004FQ		18061	USSR	15 JAN	23 JUL 87	
1987-004FR		18062	USSR	15 JAN	20 OCT 87	
1987-004FS		18063	USSR	15 JAN	09 SEP 87	
1987-004FT		18064	USSR	15 JAN	13 SEP 87	
1987-004FU		18065	USSR	15 JAN	11 SEP 87	
1987-004FV		18066	USSR	15 JAN	04 AUG 87	
1987-004FW		18067	USSR	15 JAN	27 MAR 88	
1987-004FX		18068	USSR	15 JAN	25 FEB 88	
1987-004FY		18069	USSR	15 JAN	29 AUG 87	
1987-004FZ		18070	USSR	15 JAN	12 FEB 88	
1987-004G		17401	USSR	15 JAN	04 JUN 87	
1987-004GA		18071	USSR	15 JAN	10 MAR 88	
1987-004GB		18072	USSR	15 JAN	06 SEP 87	
1987-004GC		18073	USSR	15 JAN	19 SEP 87	
1987-004GD		18074	USSR	15 JAN	03 OCT 87	
1987-004GE		18075	USSR	15 JAN	12 AUG 87	
1987-004GF		18076	USSR	15 JAN	27 APR 88	
1987-004GG		18077	USSR	15 JAN	19 SEP 87	
1987-004GH		18078	USSR	15 JAN	16 FEB 88	
1987-004GJ		18079	USSR	15 JAN	19 JUL 87	
1987-004GK		18080	USSR	15 JAN	28 JUN 87	
1987-004GL		18081	USSR	15 JAN	03 SEP 87	
1987-004GM		18082	USSR	15 JAN	15 JAN 88	
1987-004GN		18230	USSR	15 JAN	24 AUG 87	
1987-004GP		18235	USSR	15 JAN	25 OCT 87	
1987-004GQ		18236	USSR	15 JAN	04 NOV 87	
1987-004GS		18305	USSR	15 JAN	17 OCT 88	
1987-004GT		18514	USSR	15 JAN	21 NOV 87	
1987-004GU		18515	USSR	15 JAN	30 JAN 88	
1987-004GV		18595	USSR	15 JAN	11 FEB 89	
1987-004GW		18596	USSR	15 JAN	05 MAR 89	
1987-004GX		18609	USSR	15 JAN	29 SEP 88	
1987-004GY		18696	USSR	15 JAN	26 MAR 88	
1987-004GZ		18815	USSR	15 JAN	09 FEB 89	
1987-004H		17402	USSR	15 JAN	30 DEC 88	
1987-004HA		18909	USSR	15 JAN	01 MAY 88	
1987-004HB		19063	USSR	15 JAN	11 MAY 88	
1987-004HC		19563	USSR	15 JAN	13 JAN 89	
1987-004J		17403	USSR	15 JAN	24 FEB 87	
1987-004K		17404	USSR	15 JAN	29 DEC 88	
1987-004L		17405	USSR	15 JAN	24 AUG 88	
1987-004M		17406	USSR	15 JAN	07 NOV 88	
1987-004N		17407	USSR	15 JAN	16 NOV 88	
1987-004P		17408	USSR	15 JAN	22 JAN 89	
1987-004Q		17409	USSR	15 JAN	22 FEB 87	
1987-004R		17410	USSR	15 JAN	28 MAR 87	
1987-004S		17411	USSR	15 JAN	20 FEB 87	
1987-004T		17412	USSR	15 JAN	07 JAN 88	
1987-004U		17413	USSR	15 JAN	27 JUL 87	
1987-004V		17414	USSR	15 JAN	17 JAN 89	
1987-004W		17415	USSR	15 JAN	13 JUN 89	

International Designation	Satellite Name	Number	Source	Launch	Decay	Note
1987-004X		17416	USSR	15 JAN	02 JUN 88	
1987-004Y		17417	USSR	15 JAN	18 JAN 88	
1987-004Z		17418	USSR	15 JAN	06 MAR 87	
1987-005A	PROGRESS 27	17299	USSR	16 JAN	25 FEB 87	
1987-005B		17300	USSR	16 JAN	20 JAN 87	
1987-007A	COSMOS 1815	17326	USSR	22 JAN	15 NOV 88	
1987-007B		17327	USSR	22 JAN	06 OCT 88	
1987-008B		17329	USSR	22 JAN	01 MAR 87	
1987-008C		17330	USSR	22 JAN	25 FEB 87	
1987-010A	COSMOS 1817	17365	USSR	30 JAN	31 JAN 87	
1987-010B		17366	USSR	30 JAN	02 FEB 87	
1987-010C		17367	USSR	30 JAN	18 MAR 87	
1987-010D		17368	USSR	30 JAN	13 FEB 87	
1987-011B		17370	USSR	01 FEB	02 FEB 87	
1987-011C		17399	USSR	01 FEB	29 JUL 90	
1987-011D		17400	USSR	01 FEB	04 MAY 90	
1987-012A	ASTRO C	17480	JAPAN	05 FEB	01 NOV 91	
1987-012C		18229	JAPAN	05 FEB	25 NOV 88	
1987-012D		18238	JAPAN	05 FEB	08 OCT 89	
1987-012E		18407	JAPAN	05 FEB	11 AUG 89	
1987-012F		18408	JAPAN	05 FEB	22 NOV 88	
1987-012G		18409	JAPAN	05 FEB	11 FEB 89	
1987-012H		18523	JAPAN	05 FEB	03 FEB 88	
1987-012J		18682	JAPAN	05 FEB	09 FEB 89	
1987-012K		18927	JAPAN	05 FEB	13 APR 93	
1987-013A	SOYUZ TM-2	17482	USSR	05 FEB	30 JUL 87	
1987-013B		17483	USSR	05 FEB	08 FEB 87	
1987-013C		18242	USSR	05 FEB	16 JAN 88	
1987-014A	COSMOS 1819	17484	USSR	07 FEB	18 FEB 87	
1987-014B		17485	USSR	07 FEB	10 FEB 87	
1987-014C		17486	USSR	07 FEB	07 FEB 87	
1987-014D		17487	USSR	07 FEB	08 FEB 87	
1987-014E		17529	USSR	07 FEB	03 MAR 87	
1987-014F		17530	USSR	07 FEB	24 MAR 87	
1987-014G		17531	USSR	07 FEB	23 FEB 87	
1987-014H		17532	USSR	07 FEB	26 FEB 87	
1987-016A	COSMOS 1820	17523	USSR	14 FEB	06 MAR 87	
1987-016B		17524	USSR	14 FEB	22 FEB 87	
1987-019A	COSMOS 1822	17533	USSR	19 FEB	05 MAR 87	
1987-019B		17534	USSR	19 FEB	01 MAR 87	
1987-019C		17568	USSR	19 FEB	11 MAR 87	
1987-019D		17569	USSR	19 FEB	17 MAR 87	
1987-019E		17570	USSR	19 FEB	10 APR 87	
1987-019F		17571	USSR	19 FEB	08 MAR 87	
1987-019G		17572	USSR	19 FEB	11 MAR 87	
1987-019H		17573	USSR	19 FEB	06 MAR 87	
1987-020AB		18750	USSR	20 FEB	28 AUG 90	
1987-020AV		18968	USSR	20 FEB	23 SEP 88	
1987-020AX		18970	USSR	20 FEB	18 MAY 90	
1987-020AY		18994	USSR	20 FEB	21 MAR 89	
1987-020AZ		18995	USSR	20 FEB	24 FEB 90	
1987-020BC		19104	USSR	20 FEB	04 JAN 89	
1987-020BF		19146	USSR	20 FEB	21 JAN 89	
1987-020BH		19149	USSR	20 FEB	17 FEB 91	
1987-020BJ		19151	USSR	20 FEB	01 DEC 88	
1987-020BK		19152	USSR	20 FEB	29 MAR 90	
1987-020BL		19153	USSR	20 FEB	13 OCT 88	
1987-020BM		19155	USSR	20 FEB	10 MAR 89	
1987-020BN		19159	USSR	20 FEB	01 OCT 88	
1987-020BP		19179	USSR	20 FEB	27 DEC 88	
1987-020BQ		19180	USSR	20 FEB	14 JUN 90	
1987-020BR		19184	USSR	20 FEB	01 FEB 91	
1987-020BS		19186	USSR	20 FEB	23 APR 89	
1987-020BT		19188	USSR	20 FEB	26 JUN 89	
1987-020BU		19247	USSR	20 FEB	25 DEC 88	
1987-020BV		19249	USSR	20 FEB	14 AUG 88	
1987-020BX		19392	USSR	20 FEB	20 MAY 90	
1987-020BY		19394	USSR	20 FEB	23 FEB 90	
1987-020BZ		19437	USSR	20 FEB	18 APR 92	
1987-020CB		19452	USSR	20 FEB	23 FEB 90	
1987-020CC		19456	USSR	20 FEB	04 NOV 88	
1987-020CD		19562	USSR	20 FEB	28 OCT 88	
1987-020CG		19644	USSR	20 FEB	25 MAY 90	
1987-020CH		19645	USSR	20 FEB	01 JAN 89	
1987-020CJ		19815	USSR	20 FEB	06 AUG 89	
1987-020CK		19833	USSR	20 FEB	21 DEC 89	
1987-020CL		19836	USSR	20 FEB	26 OCT 89	
1987-020CM		19837	USSR	20 FEB	11 JUL 89	
1987-020CN		19839	USSR	20 FEB	01 APR 89	
1987-020CP		19840	USSR	20 FEB	24 NOV 89	
1987-020CQ		19854	USSR	20 FEB	23 JUL 89	
1987-020CR		19861	USSR	20 FEB	26 SEP 89	
1987-020CS		19888	USSR	20 FEB	02 AUG 89	
1987-020CT		19890	USSR	20 FEB	13 OCT 89	
1987-020CU		19955	USSR	20 FEB	17 JUL 89	
1987-020CV		19957	USSR	20 FEB	21 DEC 89	
1987-020CW		19982	USSR	20 FEB	28 MAR 91	
1987-020CX		20003	USSR	20 FEB	28 JAN 90	
1987-020CY		20004	USSR	20 FEB	10 JUL 90	
1987-020CZ		20019	USSR	20 FEB	08 SEP 89	
1987-020DA		20020	USSR	20 FEB	17 JAN 91	
1987-020DB		20092	USSR	20 FEB	13 JUN 90	
1987-020DC		20126	USSR	20 FEB	18 AUG 91	
1987-020DD		20130	USSR	20 FEB	13 AUG 90	
1987-020DE		20154	USSR	20 FEB	16 MAY 90	
1987-020DF		20155	USSR	20 FEB	13 OCT 89	
1987-020DG		20198	USSR	20 FEB	30 MAR 91	
1987-020DH		20199	USSR	20 FEB	25 AUG 90	
1987-020DJ		20201	USSR	20 FEB	17 JUL 90	
1987-020DK		20225	USSR	20 FEB	12 OCT 91	
1987-020DL		20226	USSR	20 FEB	17 FEB 91	
1987-020DN		20228	USSR	20 FEB	01 NOV 89	
1987-020DQ		20454	USSR	20 FEB	21 FEB 90	
1987-020K		18732	USSR	20 FEB	02 MAR 91	
1987-020M		18734	USSR	20 FEB	21 NOV 92	
1987-020N		18735	USSR	20 FEB	29 JUL 89	
1987-020S		18739	USSR	20 FEB	18 FEB 91	
1987-020T		18740	USSR	20 FEB	27 FEB 92	
1987-020U		18741	USSR	20 FEB	22 MAR 89	
1987-020W		18743	USSR	20 FEB	11 JAN 89	
1986-019RM		18228	ESA	22 FEB	19 JUL 89	
1987-021A	COSMOS 1824	17559	USSR	26 FEB	22 APR 87	
1987-021B		17560	USSR	26 FEB	04 MAR 87	
1987-023A	PROGRESS 28	17564	USSR	03 MAR	28 MAR 87	
1987-023B		17565	USSR	03 MAR	06 MAR 87	
1987-025A	COSMOS 1826	17577	USSR	11 MAR	25 MAR 87	
1987-025B		17578	USSR	11 MAR	25 MAR 87	
1987-025C		17579	USSR	11 MAR	17 MAR 87	
1987-025D		17580	USSR	11 MAR	16 MAR 87	
1987-025E		17581	USSR	11 MAR	17 MAR 87	
1987-025F		17785	USSR	11 MAR	30 JUN 87	
1987-025G		17786	USSR	11 MAR	30 AUG 88	
1987-025H		17787	USSR	11 MAR	08 APR 87	
1987-025J		17788	USSR	11 MAR	16 SEP 87	
1987-025K		17789	USSR	11 MAR	21 NOV 87	
1987-025L		17790	USSR	11 MAR	03 MAY 87	
1987-025M		17791	USSR	11 MAR	18 APR 87	
1987-028B		17612	USSR	19 MAR	22 MAR 87	
1987-028C		17613	USSR	19 MAR	19 MAR 87	
1987-028E		17709	USSR	19 MAR	30 NOV 91	
1987-028F		17710	USSR	19 MAR	30 SEP 87	
1987-029B		17707	US	20 MAR	23 APR 87	
1987-029C		17708	INDON	20 MAR	18 DEC 89	
1987-030B		17846	USSR	31 MAR	04 APR 87	
1987-030C		17851	USSR	31 MAR	25 AUG 88	
1987-031A	COSMOS 1834	17847	USSR	08 APR	14 OCT 88	
1987-031B		17848	USSR	08 APR	08 APR 87	
1987-032A	COSMOS 1835	17849	USSR	09 APR	04 JUN 87	
1987-032B		17850	USSR	09 APR	14 APR 87	
1987-033A	COSMOS 1836	17876	USSR	16 APR	02 DEC 87	
1987-033B		17877	USSR	16 APR	20 APR 87	
1987-033C		18028	USSR	16 APR	20 JUN 87	
1987-033D		18587	USSR	16 APR	03 DEC 87	
1987-033E		18588	USSR	16 APR	05 DEC 87	
1987-034A	PROGRESS 29	17878	USSR	21 APR	11 MAY 87	
1987-034B		17879	USSR	21 APR	24 APR 87	
1987-035A	COSMOS 1837	17880	USSR	22 APR	28 APR 87	
1987-035B		17881	USSR	22 APR	24 APR 87	
1987-035C		17914	USSR	22 APR	18 MAY 87	
1987-035D		17915	USSR	22 APR	30 APR 87	
1987-035E		17916	USSR	22 APR	02 MAY 87	
1987-035F		17917	USSR	22 APR	08 MAY 87	
1987-035G		17918	USSR	22 APR	02 MAY 87	
1987-036A	COSMOS 1838	17902	USSR	24 APR	15 MAY 91	
1987-036B	COSMOS 1839	17903	USSR	24 APR	08 MAY 91	
1987-036C	COSMOS 1840	17904	USSR	24 APR	14 SEP 91	
1987-036D		17905	USSR	24 APR	25 APR 87	
1987-036E		17906	USSR	24 APR	27 APR 87	
1987-036F		17909	USSR	24 APR	12 AUG 87	
1987-036G		17910	USSR	24 APR	12 APR 91	
1987-036H		17913	USSR	24 APR	22 NOV 91	
1987-036J		17962	USSR	24 APR	28 OCT 87	
1987-037A	COSMOS 1841	17907	USSR	24 APR	08 MAY 87	
1987-037B		17908	USSR	24 APR	22 MAY 87	
1987-037C		17964	USSR	24 APR	09 MAY 87	
1987-037D		17965	USSR	24 APR	04 JUL 87	
1987-037E		17966	USSR	24 APR	21 MAY 87	
1987-037F		17967	USSR	24 APR	24 MAY 87	
1987-037G		17968	USSR	24 APR	11 JUN 87	
1987-039A	COSMOS 1843	17940	USSR	05 MAY	19 MAY 87	
1987-039B		17941	USSR	05 MAY	13 MAY 87	
1987-039C		17963	USSR	05 MAY	07 MAY 87	
1987-039D		18001	USSR	05 MAY	28 AUG 87	
1987-039E		18002	USSR	05 MAY	11 JUL 88	
1987-039F		18003	USSR	05 MAY	06 NOV 87	
1987-039G		18022	USSR	05 MAY	22 NOV 87	
1987-040B		17970	USSR	11 MAY	14 MAY 87	
1987-040C		17971	USSR	11 MAY	12 MAY 87	
1987-041G		18687	USSR	13 MAY	06 APR 93	
1987-042A	COSMOS 1845	17975	USSR	13 MAY	27 MAY 87	
1987-0428		17976	USSR	13 MAY	01 JUN 87	
1987-042C		18013	USSR	13 MAY	25 OCT 87	
1987-042D		18014	USSR	13 MAY	16 AUG 88	
1987-042E		18015	USSR	13 MAY	24 NOV 87	
1987-042F		18016	USSR	13 MAY	22 JUN 87	
1987-042G		18019	USSR	13 MAY	04 DEC 87	
1987-042H		18020	USSR	13 MAY	14 SEP 87	
1987-042J		18023	USSR	13 MAY	20 SEP 87	
1987-044A	PROGRESS 30	17999	USSR	19 MAY	19 JUL 87	
1987-044B		18000	USSR	19 MAY	22 MAY 87	
1987-045A	COSMOS 1846	18004	USSR	21 MAY	04 JUN 87	
1987-045B		18005	USSR	21 MAY	26 MAY 87	
1987-045C		18006	USSR	21 MAY	22 MAY 87	
1987-045D		18087	USSR	21 MAY	11 JUN 87	
1987-045E		18088	USSR	21 MAY	21 JUL 87	
1987-045F		18089	USSR	21 MAY	12 SEP 87	

International Designation	Satellite Name	Number	Source	Launch	Decay	Note
1987-045G		18090	USSR	21 MAY	14 JUN 87	
1987-045H		18091	USSR	21 MAY	04 AUG 87	
1987-045J		18092	USSR	21 MAY	15 JUL 87	
1987-045K		18093	USSR	21 MAY	19 JUL 87	
1987-045L		18094	USSR	21 MAY	05 JAN 88	
1987-046A	COSMOS 1847	18011	USSR	26 MAY	22 JUL 87	
1987-046B		18012	USSR	26 MAY	30 MAY 87	
1987-047A	COSMOS 1848	18017	USSR	28 MAY	11 JUN 87	
1987-047B		18018	USSR	28 MAY	12 JUN 87	
1987-047C		18021	USSR	28 MAY	30 MAY 87	
1987-047D		18097	USSR	28 MAY	25 JUN 87	
1987-047E		18098	USSR	28 MAY	05 NOV 87	
1987-047F		18099	USSR	28 MAY	13 AUG 87	
1987-047G		18100	USSR	28 MAY	18 NOV 87	
1987-047H		18101	USSR	28 MAY	19 SEP 87	
1987-047J		18102	USSR	28 MAY	29 SEP 88	
1987-047K		18107	USSR	28 MAY	18 JUL 87	
1987-047L		18108	USSR	28 MAY	19 SEP 87	
1987-047M		18109	USSR	28 MAY	20 SEP 87	
1987-047N		18110	USSR	28 MAY	27 JUN 87	
1987-047P		18124	USSR	28 MAY	02 AUG 87	
1987-047Q		18125	USSR	28 MAY	17 JUL 87	
1987-047R		18126	USSR	28 MAY	10 JUL 87	
1987-047S		18189	USSR	28 MAY	12 JUL 87	
1987-047T		18190	USSR	28 MAY	12 JUL 87	
1987-048B		18084	USSR	04 JUN	30 JUN 87	
1987-048C		18085	USSR	04 JUN	15 AUG 87	
1987-050B		18104	USSR	12 JUN	15 AUG 87	
1987-050C		18105	USSR	12 JUN	22 JUL 87	
1987-052B		18239	USSR	18 JUN	02 AUG 87	
1987-052C		18240	USSR	18 JUN	16 SEP 87	
1987-056A	COSMOS 1863	18155	USSR	04 JUL	18 JUL 87	
1987-056B		18156	USSR	04 JUL	18 JUL 87	
1987-056D		18158	USSR	04 JUL	07 JUL 87	
1987-056E		18216	USSR	04 JUL	20 AUG 87	
1987-056F		18217	USSR	04 JUL	13 NOV 87	
1987-056G		18218	USSR	04 JUL	23 JAN 88	
1987-056H		18219	USSR	04 JUL	08 OCT 87	
1987-056J		18220	USSR	04 JUL	09 AUG 87	
1987-056K		18221	USSR	04 JUL	23 AUG 87	
1987-056L		18224	USSR	04 JUL	07 OCT 88	
1987-057C		18157	USSR	06 JUL	08 JUL 87	
1987-058A	COSMOS 1865	18162	USSR	08 JUL	14 AUG 87	
1987-058B		18163	USSR	08 JUL	16 JUL 87	
1987-059A	COSMOS 1866	18184	USSR	09 JUL	06 NOV 87	
1987-059B		18185	USSR	09 JUL	15 JUL 87	
1987-059C		18186	USSR	09 JUL	10 JUL 87	
1987-059D		18285	USSR	09 JUL	19 AUG 87	
1987-059E		18286	USSR	09 JUL	28 AUG 87	
1987-059F		18287	USSR	09 JUL	18 AUG 87	
1987-059G		18288	USSR	09 JUL	17 SEP 87	
1987-059H		18308	USSR	09 JUL	20 SEP 87	
1987-059J		18309	USSR	09 JUL	16 OCT 87	
1987-059K		18310	USSR	09 JUL	12 OCT 87	
1987-059L		18311	USSR	09 JUL	10 OCT 87	
1987-060B		18188	USSR	10 JUL	11 JUL 87	
1987-060C		18191	USSR	10 JUL	11 FEB 90	
1987-060D		18524	USSR	10 JUL	14 MAY 90	
1987-061A	COSMOS 1868	18192	USSR	14 JUL	02 MAR 89	
1987-061B		18193	USSR	14 JUL	30 AUG 88	
1987-063A	SOYUZ TM-3	18222	USSR	22 JUL	29 DEC 87	
1987-063B		18223	USSR	22 JUL	24 JUL 87	
1987-063C		18723	USSR	22 JUL	17 JUN 88	
1987-064A	COSMOS 1870	18225	USSR	25 JUL	29 JUL 89	
1987-064B		18226	USSR	25 JUL	27 JUL 87	
1987-064C		18227	USSR	25 JUL	26 JUL 87	
1987-065A	COSMOS 1871	18259	USSR	01 AUG	10 AUG 87	
1987-065B		18260	USSR	01 AUG	03 AUG 87	
1987-066A	PROGRESS 31	18283	USSR	03 AUG	23 SEP 87	
1987-066B		18284	USSR	03 AUG	07 AUG 87	
1987-067A	PRC 20	18306	PRC	05 AUG	23 AUG 87	
1987-067B		18307	PRC	05 AUG	21 AUG 87	
1987-069A	COSMOS 1872	18314	USSR	19 AUG	3Q AUG 87	
1987-069B		18315	USSR	19 AUG	26 AUG 87	
1987-069C		18321	USSR	19 AUG	10 SEP 87	
1987-069D		18322	USSR	19 AUG	01 SEP 87	
1987-069E		18323	USSR	19 AUG	31 AUG 87	
1987-069F		18324	USSR	19 AUG	07 SEP 87	
1987-069G		18325	USSR	19 AUG	01 SEP 87	
1987-070B		18317	JAPAN	27 AUG	28 AUG 87	
1987-070C		18320	JAPAN	27 AUG	12 JUN 88	
1987-071A	COSMOS 1873	18318	USSR	28 AUG	14 SEP 87	
1987-071B		18319	USSR	28 AUG	02 SEP 87	
1987-072A	COSMOS 1874	18326	USSR	03 SEP	17 SEP 87	
1987-072B		18327	USSR	03 SEP	11 SEP 87	
1987-072C		18368	USSR	03 SEP	12 OCT 87	
1987-072D		18369	USSR	03 SEP	19 SEP 87	
1987-072E		18370	USSR	03 SEP	20 SEP 87	
1987-072F		18371	USSR	03 SEP	25 SEP 87	
1987-073B		18329	USSR	04 SEP	06 SEP 87	
1987-073C		18330	USSR	04 SEP	04 SEP 87	
1987-073F		18333	USSR	04 SEP	05 JUL 88	
1987-075A	PRC 21	18341	PRC	09 SEP	04 OCT 87	
1987-075B		18342	PRC	09 SEP	24 SEP 87	
1987-075C		18372	PRC	09 SEP	23 SEP 87	
1987-075D		18373	PRC	09 SEP	20 SEP 87	
1987-076A	COSMOS 1881	18343	USSR	11 SEP	30 MAR 88	
1987-076B		18344	USSR	11 SEP	15 SEP 87	
1987-076C		19021	USSR	11 SEP	01 APR 88	
1987-076D		19035	USSR	11 SEP	02 APR 88	
1987-077A	COSMOS 1882	18348	USSR	15 SEP	06 OCT 87	
1987-077B		18349	USSR	15 SEP	17 SEP 87	
1987-077C		18389	USSR	15 SEP	17 OCT 87	
1987-077D		18390	USSR	15 SEP	12 OCT 87	
1987-077E		18391	USSR	15 SEP	10 OCT 87	
1987-077F		18392	USSR	15 SEP	05 NOV 87	
1987-077G		18393	USSR	15 SEP	11 OCT 87	
1974-089EZ		18359	USSR	16 SEP	16 SEP 87	
1987-078C		18352	ESA	16 SEP	21 JUN 91	
1987-078D		18353	ESA	16 SEP	08 OCT 88	
1987-079D		18358	USSR	16 SEP	16 SEP 87	
1987-080D		18364	US	16 SEP	02 FEB 89	
1987-081A	COSMOS 1886	18366	USSR	17 SEP	02 NOV 87	
1987-081B		18367	USSR	17 SEP	22 SEP 87	
1987-081C		18449	USSR	17 SEP	02 NOV 87	
1987-082A	PROGRESS 32	18376	USSR	23 SEP	19 NOV 87	
1987-082B		18377	USSR	23 SEP	26 SEP 87	
1987-083A	COSMOS 1887	18380	USSR	29 SEP	12 OCT 87	
1987-083B		18381	USSR	29 SEP	19 OCT 87	
1987-083C		18398	USSR	29 SEP	08 NOV 87	
1987-083D		18399	USSR	29 SEP	05 NOV 87	
1987-083E		18400	USSR	29 SEP	21 OCT 87	
1987-083F		18401	USSR	29 SEP	28 OCT 87	
1987-084B		18385	USSR	01 OCT	04 OCT 87	
1987-084C		18386	USSR	01 OCT	02 OCT 87	
1987-084E		18388	USSR	01 OCT	16 JUL 91	
1987-084F		18420	USSR	01 OCT	06 JUN 88	
1987-085A	COSMOS 1889	18394	USSR	09 OCT	23 OCT 87	
1987-085B		18395	USSR	09 OCT	25 OCT 87	
1987-085C		18436	USSR	09 OCT	28 FEB 88	
1987-085D		18437	USSR	09 OCT	31 JAN 88	
1987-085E		18438	USSR	09 OCT	30 NOV 87	
1987-085F		18439	USSR	09 OCT	23 DEC 87	
1987-085G		18440	USSR	09 OCT	10 OCT 88	
1987-085H		18567	USSR	09 OCT	27 NOV 87	
1987-086A	COSMOS 1890	18396	USSR	10 OCT	26 DEC 88	
1987-086B		18397	USSR	10 OCT	11 OCT 87	
1987-089A	COSMOS 1893	18432	USSR	22 OCT	16 DEC 87	
1987-089B		18433	USSR	22 OCT	26 OCT 87	
1987-090B		18442	US	26 OCT	27 OCT 87	
1987-091B		18444	USSR	28 OCT	31 OCT 87	
1987-091C		18445	USSR	28 OCT	29 OCT 87	
1987-091E		18447	USSR	28 OCT	09 MAR 88	
1987-092A	COSMOS 1895	18491	USSR	11 NOV	26 NOV 87	
1987-092B		18492	USSR	11 NOV	25 NOV 87	
1987-092C		18572	USSR	11 NOV	28 NOV 87	
1987-092D		18573	USSR	11 NOV	27 NOV 87	
1987-092E		18574	USSR	11 NOV	04 DEC 87	
1987-092F		18581	USSR	11 NOV	29 NOV 87	
1987-092G		18582	USSR	11 NOV	27 NOV 87	
1987-093A	COSMOS 1896	18535	USSR	14 NOV	25 DEC 87	
1987-093B		18536	USSR	14 NOV	20 NOV 87	
1987-093C		18537	USSR	14 NOV	15 NOV 87	
1987-094A	PROGRESS 33	18568	USSR	20 NOV	19 DEC 87	
1987-094B		18569	USSR	20 NOV	23 NOV 87	
1987-096B		18576	USSR	26 NOV	28 NOV 87	
1987-096C		18577	USSR	26 NOV	27 NOV 87	
1987-096E		18579	USSR	26 NOV	12 JUL 89	
1987-096F		18580	USSR	26 NOV	18 AUG 88	
1987-099A	COSMOS 1899	18625	USSR	07 DEC	21 DEC 87	
1987-099B		18626	USSR	07 DEC	13 DEC 87	
1987-099C		18627	USSR	07 DEC	08 DEC 87	
1987-099D		18705	USSR	07 DEC	15 JAN 88	
1987-099E		18706	USSR	07 DEC	23 DEC 87	
1987-099F		18707	USSR	07 DEC	22 DEC 87	
1987-099G		18708	USSR	07 DEC	27 DEC 87	
1987-100B		18632	USSR	10 DEC	13 DEC 87	
1987-100C		18633	USSR	10 DEC	11 DEC 87	
1987-100E		18635	USSR	10 DEC	07 MAY 91	
1987-100F		18636	USSR	10 DEC	05 JUL 90	
1987-101B		19553	USSR	12 DEC	01 OCT 88	
1987-101C		19589	USSR	12 DEC	03 NOV 88	
1987-102A	COSMOS 1901	18666	USSR	14 DEC	03 FEB 88	
1987-102B		18667	USSR	14 DEC	20 DEC 87	
1987-103A	COSMOS 1902	18668	USSR	15 DEC	30 DEC 88	
1987-103B		18669	USSR	15 DEC	21 SEP 88	
1987-104A	SOYUZ TM-4	18699	USSR	21 DEC	17 JUN 88	
1987-104B		18700	USSR	21 DEC	23 DEC 87	
1987-104C		19225	USSR	21 DEC	05 NOV 88	
1987-104D		19251	USSR	21 DEC	05 JUL 88	
1987-104E		19253	USSR	21 DEC	19 JUN 88	
1987-104F		19254	USSR	21 DEC	27 JUN 88	
1987-104G		19255	USSR	21 DEC	09 JUL 88	
1987-105B		18702	USSR	21 DEC	09 FEB 88	
1987-105C		18703	USSR	21 DEC	12 JAN 88	
1987-105E		18977	USSR	21 DEC	27 MAR 88	
1987-107A	COSMOS 1905	18711	USSR	25 DEC	08 JAN 88	
1987-107B		18712	USSR	25 DEC	01 JAN 88	
1987-107C		18759	USSR	25 DEC	12 JAN 88	
1987-107D		18760	USSR	25 DEC	03 FE8 88	
1987-107E		18761	USSR	25 DEC	15 JAN 88	
1987-107F		18762	USSR	25 DEC	13 JAN 88	
1987-107G		18763	USSR	25 DEC	10 JAN 88	
1987-108A	COSMOS 1906	18713	USSR	26 DEC	13 MAR 88	
1987-108AA		18888	USSR	26 DEC	07 MAR 88	
1987-108AB		18892	USSR	26 DEC	02 APR 88	

International Designation	Satellite Name	Number	Source	Launch	Decay	Note
1987-108AC		18893	USSR	26 DEC	06 APR 88	
1987-108AD		18894	USSR	26 DEC	25 MAR 88	
1987-108AE		18895	USSR	26 DEC	12 MAR 88	
1987-108AF		18896	USSR	26 DEC	28 MAY 88	
1987-108AG		18897	USSR	26 DEC	11 MAR 88	
1987-108AH		18898	USSR	26 DEC	30 MAR 88	
1987-108AJ		18899	USSR	26 DEC	06 APR 88	
1987-108AK		18900	USSR	26 DEC	12 MAR 88	
1987-108AL		18950	USSR	26 DEC	20 MAR 88	
1987-108AM		18972	USSR	26 DEC	23 MAY 88	
1987-108AN		18973	USSR	26 DEC	20 MAR 88	
1987-108AP		18974	USSR	26 DEC	12 APR 88	
1987-108B		18714	USSR	26 DEC	29 DEC 87	
1987-108C		18825	USSR	26 DEC	19 FEB 88	
1987-108D		18826	USSR	26 DEC	28 FEB 88	
1987-108E		18827	USSR	26 DEC	23 FEB 88	
1987-108F		18828	USSR	26 DEC	01 APR 88	
1987-108G		18829	USSR	26 DEC	02 MAR 88	
1987-108H		18830	USSR	26 DEC	11 FEB 88	
1987-108J		18831	USSR	26 DEC	13 FEB 88	
1987-108K		18832	USSR	26 DEC	05 MAR 88	
1987-108L		18833	USSR	26 DEC	07 APR 88	
1987-108M		18834	USSR	26 DEC	26 MAY 88	
1987-108N		18835	USSR	26 DEC	19 MAR 88	
1987-108P		18836	USSR	26 DEC	27 FEB 88	
1987-108Q		18837	USSR	26 DEC	01 JUN 88	
1987-108R		18838	USSR	26 DEC	09 FEB 88	
1987-108S		18839	USSR	26 DEC	20 MAR 88	
1987-108T		18840	USSR	26 DEC	02 MAR 88	
1987-108U		18841	USSR	26 DEC	13 FEB 88	
1987-108V		18842	USSR	26 DEC	19 FEB 88	
1987-108W		18843	USSR	26 DEC	22 FEB 88	
1987-108X		18844	USSR	26 DEC	20 MAR 88	
1987-108Y		18862	USSR	26 DEC	13 MAR 88	
1987-108Z		18863	USSR	26 DEC	28 FEB 88	
1987-108B		18716	USSR	27 DEC	30 DEC 87	
1987-108C		18717	USSR	27 DEC	28 DEC 87	
1987-108F		18722	USSR	27 DEC	26 OCT 88	
1987-110A	COSMOS 1907	18720	USSR	29 DEC	12 JAN 88	
1987-110B		18721	USSR	29 DEC	12 JAN 88	
1987-110C		18724	USSR	29 DEC	01 JAN 88	
1987-110D		18768	USSR	29 DEC	26 FEB 88	
1987-110E		18769	USSR	29 DEC	08 MAY 88	
1987-110F		18770	USSR	29 DEC	06 DEC 88	
1987-110G		18771	USSR	29 DEC	09 APR 88	
1987-110H		18772	USSR	29 DEC	30 MAR 88	
1987-110J		18808	USSR	29 DEC	28 FEB 88	
1988 LAUNCHES						
1988-003A	PROGRESS 34	18795	USSR	20 JAN	04 MAR 88	
1988-003B		18796	USSR	20 JAN	23 JAN 88	
1988-004A	COSMOS 1915	18809	USSR	26 JAN	09 FEB 88	
1988-004B		18810	USSR	26 JAN	07 FEB 88	
1988-004C		18850	USSR	26 JAN	10 APR 88	
1988-004D		18851	USSR	26 JAN	17 FEB 88	
1988-004E		18852	USSR	26 JAN	21 FEB 88	
1988-004F		18853	USSR	26 JAN	23 FEB 88	
1988-006B		18845	US	03 FEB	29 OCT 92	
1988-006C		18846	US	03 FEB	12 MAR 92	
1988-006E		18956	US	03 FEB	02 FEB 89	
1988-007A	COSMOS 1916	18823	USSR	03 FEB	29 FEB 88	
1988-007B		18824	USSR	03 FEB	09 FEB 88	
1988-008A		18847	US	08 FEB	01 MAR 88	
1988-008B		18848	US	08 FEB	02 APR 88	
1988-008C		18849	US	08 FEB	05 MAR 88	
1988-008D		18854	US	08 FEB	13 FEB 88	
1988-009A	COSMOS 1917-19	18857	USSR	17 FEB	17 FEB 88	
1988-009B		18858	USSR	17 FEB	19 FEB 88	
1988-009C		18859	USSR	17 FEB	20 FEB 88	
1988-010A	COSMOS 1920	18860	USSR	18 FEB	09 MAR 88	
1988-010B		18861	USSR	18 FEB	20 FEB 88	
1988-010C		18924	USSR	18 FEB	11 MAR 88	
1988-010D		18925	USSR	18 FEB	10 MAR 88	
1988-010E		18926	USSR	18 FEB	06 APR 88	
1988-010F		18930	USSR	18 FEB	08 APR 88	
1988-010G		18933	USSR	18 FEB	02 JUL 88	
1988-010H		18934	USSR	18 FEB	28 MAR 88	
1988-010J		18935	USSR	18 FEB	26 MAR 88	
1988-011A	COSMOS 1921	18875	USSR	19 FEB	04 MAR 88	
1988-011B		18876	USSR	19 FEB	04 MAR 88	
1988-011C		18880	USSR	19 FEB	21 FEB 88	
1988-011D		18916	USSR	19 FEB	29 MAY 88	
1988-011E		18917	USSR	19 FEB	22 JUN 88	
1988-011F		18918	USSR	19 FEB	24 NOV 88	
1988-011G		18919	USSR	19 FEB	15 APR 88	
1988-011H		18920	USSR	19 FEB	05 JUN 88	
1988-011J		18921	USSR	19 FEB	19 APR 88	
1988-012B		18878	JAPAN	19 FEB	26 FEB 88	
1988-013B		18882	USSR	26 FEB	05 APR 88	
1988-013D		18884	USSR	26 FEB	25 MAR 88	
1988-013E		19004	USSR	26 FEB	07 APR 88	
1988-014B		18923	PRC	07 MAR	20 JUL 89	
1988-015A	COSMOS 1923	18931	USSR	10 MAR	22 MAR 88	
1988-015B		18932	USSR	10 MAR	16 MAR 88	
1988-015C		18936	USSR	10 MAR	11 MAR 88	
1988-015D		18987	USSR	10 MAR	01 APR 88	
1988-015E		18988	USSR	10 MAR	24 MAR 88	
1988-015F		18989	USSR	10 MAR	23 MAR 88	
1988-015G		18990	USSR	10 MAR	25 MAR 88	
1988-015H		18991	USSR	10 MAR	27 MAR 88	
1988-017B		18947	USSR	11 MAR	07 APR 88	
1988-017C		18948	USSR	11 MAR	09 APR 88	
1988-018D		18954	ESA	11 MAR	01 FEB 89	
1988-019B		19160	USSR	14 MAR	29 JUN 88	
1988-019C		19161	USSR	14 MAR	24 MAY 88	
1988-019E		19517	USSR	14 MAR	03 NOV 88	
1988-022B		18981	USSR	17 MAR	28 APR 88	
1988-022C		18982	USSR	17 MAR	10 APR 88	
1988-022E		19187	USSR	17 MAR	06 JUL 88	
1988-024A	PROGRESS 35	18992	USSR	23 MAR	05 MAY 88	
1988-024B		18993	USSR	23 MAR	26 MAR 88	
1988-024C		19089	USSR	23 MAR	05 MAY 88	
1988-025A	COSMOS 1935	19011	USSR	24 MAR	08 APR 88	
1988-025B		19012	USSR	24 MAR	27 MAR 88	
1988-025C		19040	USSR	24 MAR	08 APR 88	
1988-026A	SM-D	19013	ITALY	25 MAR	06 DEC 88	
1988-026B		19014	US	25 MAR	16 JUN 88	
1988-027A	COSMOS 1936	19015	USSR	30 MAR	18 MAY 88	
1988-027B		19016	USSR	30 MAR	02 APR 88	
1988-027C		19022	USSR	30 MAR	31 MAR 88	
1988-027D		19129	USSR	30 MAR	18 MAY 88	
1988-027E		19130	USSR	30 MAR	18 MAY 88	
1988-028B		19018	USSR	31 MAR	02 APR 88	
1988-028C		19019	USSR	31 MAR	31 MAR 88	
1988-030A	COSMOS 1938	19041	USSR	11 APR	25 APR 88	
1988-030B		19042	USSR	11 APR	16 APR 88	
1988-030C		19065	USSR	11 APR	29 APR 88	
1988-030D		19066	USSR	11 APR	29 APR 88	
1988-030E		19067	USSR	11 APR	07 MAY 88	
1988-030F		19068	USSR	11 APR	29 APR 88	
1988-030G		19069	USSR	11 APR	30 APR 88	
1988-031A	FOTON 1	19043	USSR	14 APR	28 APR 88	
1988-031B		19044	USSR	14 APR	29 APR 88	
1988-031C		19084	USSR	14 APR	29 MAY 88	
1988-031D		19085	USSR	14 APR	05 MAY 88	
1988-031E		19086	USSR	14 APR	15 MAY 88	
1988-031F		19088	USSR	14 APR	05 MAY 88	
1988-034B		19074	USSR	26 APR	28 APR 88	
1988-034C		19075	USSR	26 APR	26 APR 88	
1988-035A	COSMOS 1941	19079	USSR	27 APR	11 MAY 88	
1988-035B		19080	USSR	27 APR	02 MAY 88	
1988-035C		19081	USSR	27 APR	27 APR 88	
1988-035D		19087	USSR	27 APR	03 MAY 88	
1988-035E		19096	USSR	27 APR	22 MAY 88	
1988-035F		19097	USSR	27 APR	14 MAY 88	
1988-035G		19098	USSR	27 APR	13 MAY 88	
1988-035H		19099	USSR	27 APR	16 MAY 88	
1988-035J		19100	USSR	27 APR	12 MAY 88	
1988-036B		19091	USSR	06 MAY	08 MAY 88	
1988-036C		19092	USSR	06 MAY	06 MAY 88	
1988-036D		19093	USSR	06 MAY	04 MAR 90	
1988-036F		19095	USSR	06 MAY	16 JUL 89	
1988-037A	COSMOS 1942	19115	USSR	12 MAY	04 JUL 88	
1988-037B		19116	USSR	12 MAY	16 MAY 88	
1988-037C		19273	USSR	12 MAY	05 JUL 88	
1988-038A	PROGRESS 36	19117	USSR	13 MAY	05 JUN 88	
1988-038B		19118	USSR	13 MAY	15 MAY 88	
1988-041A	COSMOS 1944	19123	USSR	18 MAY	23 JUN 88	
1988-041B		19124	USSR	18 MAY	24 MAY 88	
1988-042A	COSMOS 1945	19131	USSR	19 MAY	31 MAY 88	
1988-042B		19132	USSR	19 MAY	01 JUN 88	
1988-042C		19199	USSR	19 MAY	03 JUN 88	
1988-042D		19200	USSR	19 MAY	04 JUN 88	
1988-042E		19201	USSR	19 MAY	10 JUN 88	
1988-042F		19202	USSR	19 MAY	01 JUN 88	
1988-042G		19203	USSR	19 MAY	02 JUN 88	
1988-043D		19166	USSR	21 MAY	22 MAY 88	
1988-043E		19167	USSR	21 MAY	21 MAY 88	
1988-044C		19191	USSR	26 MAY	07 JUL 88	
1988-044D		19192	USSR	26 MAY	28 JUN 88	
1988-044E		19239	USSR	26 MAY	13 JUL 88	
1988-044F		19434	USSR	26 MAY	23 SEP 88	
1988-044G		19457	USSR	26 MAY	09 OCT 88	
1988-044H		19481	USSR	26 MAY	13 OCT 88	
1988-045A	COSMOS 1949	19193	USSR	28 MAY	23 APR 90	
1988-04sB		19194	USSR	28 MAY	28 MAY 88	
1988-047A	COSMOS 1951	19197	USSR	31 MAY	14 JUN 88	
1988-047B		19198	USSR	31 MAY	03 JUN 88	
1988-047C		19212	USSR	31 MAY	17 JUL 88	
1988-047D		19213	USSR	31 MAY	19 JUN 88	
1988-047E		19214	USSR	31 MAY	17 JUN 88	
1986-017L		19386	USSR	07 JUN	06 NOV 88	
1988-048A	SOYUZ TM-5	19204	USSR	07 JUN	07 SEP 88	
1988-048B		19205	USSR	07 JUN	09 JUN 88	
1988-049A	COSMOS 1952	19206	USSR	11 JUN	25 JUN 88	
1988-049B		19207	USSR	11 JUN	17 JUN 88	
1988-049C		19208	USSR	11 JUN	12 JUN 88	
1988-049D		19209	USSR	11 JUN	12 JUN 88	
1988-049E		19265	USSR	11 JUN	20 JUL 88	
1988-049F		19266	USSR	11 JUN	02 JUL 88	
1988-049G		19267	USSR	11 JUN	26 JUN 88	
1988-049H		19268	USSR	11 JUN	28 JUN 88	
1988-049J		19269	USSR	11 JUN	30 JUN 88	
1988-049K		19270	USSR	11 JUN	27 JUN 88	
1988-052B		19224	US	16 JUN	08 MAR 89	

International Designation	Satellite Name	Number	Source	Launch	Decay	Note
1988-052C		19252	US	16 JUN	18 JUN 88	
1988-052D		19298	US	16 JUN	29 OCT 88	
1988-052E		19389	US	16 JUN	15 SEP 88	
1988-054A	COSMOS 1955	19258	USSR	22 JUN	20 AUG 88	
1988-054B		19259	USSR	22 JUN	27 JUN 88	
1988-054C		19262	USSR	22 JUN	23 JUN 88	
1988-055A	COSMOS 1956	19263	USSR	23 JUN	07 JUL 88	
1988-055A		19286	USSR	23 JUN	05 AUG 88	
1988-055B		19264	USSR	23 JUN	25 JUN 88	
1988-055C		19278	USSR	23 JUN	08 NOV 88	
1988-055D		19279	USSR	23 JUN	23 AUG 88	
1988-055E		19280	USSR	23 JUN	31 AUG 88	
1988-055F		19284	USSR	23 JUN	13 JUL 88	
1988-055G		19285	USSR	23 JUN	05 AUG 88	
1989-049B		20096	USSR	27 JUN	29 JUN 89	
1988-057A	COSMOS 1957	19276	USSR	07 JUL	21 JUL 88	
1988-057B		19277	USSR	07 JUL	09 JUL 88	
1988-057C		19326	USSR	07 JUL	30 JUL 88	
1988-057D		19327	USSR	07 JUL	27 JUL 88	
1988-057E		19328	USSR	07 JUL	24 JUL 88	
1988-057F		19329	USSR	07 JUL	22 JUL 88	
1988-058C		19283	USSR	07 JUL	08 JUL 88	
1988-059C		19289	USSR	12 JUL	13 JUL 88	
1988-059D		19292	USSR	12 JUL	13 JUL 88	
1988-060A	COSMOS 1958	19320	USSR	14 JUL	21 MAR 89	
1988-060B		19321	USSR	14 JUL	07 JAN 89	
1988-061A	PROGRESS 37	19322	USSR	18 JUL	12 AUG 88	
1988-061B		19323	USSR	18 JUL	21 JUL 88	
1988-057G		19333	USSR	21 JUL	20 AUG 88	
1988-057H		19334	USSR	21 JUL	23 JUL 88	
1988-063D		19335	ESA	21 JUL	12 JUL 89	
1988-065A	COSMOS 1960	19338	USSR	28 JUL	09 APR 90	
1988-065AA		20385	USSR	28 JUL	13 JAN 90	
1988-065AB		20386	USSR	28 JUL	13 JAN 90	
1988-065AC		20387	USSR	28 JUL	16 JAN 90	
1988-065AD		20388	USSR	28 JUL	16 JAN 90	
1988-065AE		20395	USSR	28 JUL	15 JAN 90	
1988-065AF		20396	USSR	28 JUL	13 JAN 90	
1988-065B		19339	USSR	28 JUL	08 JAN 90	
1988-065C		19340	USSR	28 JUL	19 OCT 88	
1988-065D		19341	USSR	28 JUL	19 OCT 88	
1988-065E		19342	USSR	28 JUL	18 OCT 88	
1988-065F		19343	USSR	28 JUL	18 OCT 88	
1988-065G		19506	USSR	28 JUL	30 JAN 89	
1988-065H		19507	USSR	28 JUL	27 JAN 89	
1988-065J		19551	USSR	28 JUL	31 JAN 89	
1988-065K		19552	USSR	28 JUL	12 FEB 89	
1988-065L		19602	USSR	28 JUL	25 FEB 89	
1988-065M		19603	USSR	28 JUL	10 FEB 89	
1988-065N		20365	USSR	28 JUL	04 JAN 90	
1988-065P		20366	USSR	28 JUL	04 JAN 90	
1988-065Q		20375	USSR	28 JUL	10 JAN 90	
1988-065R		20376	USSR	28 JUL	06 JAN 90	
1988-065S		20377	USSR	28 JUL	10 JAN 90	
1988-065U		20379	USSR	28 JUL	07 JAN 90	
1988-065W		20381	USSR	28 JUL	09 JAN 90	
1988-065X		20382	USSR	28 JUL	10 JAN 90	
1988-065Y		20383	USSR	28 JUL	12 JAN 90	
1988-065Z		20384	USSR	28 JUL	15 JAN 90	
1988-066B		19345	USSR	01 AUG	04 AUG 88	
1988-066C		19346	USSR	01 AUG	02 AUG 88	
1988-066F		19349	USSR	01 AUG	02 JUN 89	
1988-067A	PRC 23	19368	PRC	05 AUG	13 AUG 88	
1988-067B		19369	PRC	05 AUG	19 AUG 88	
1988-067C		19370	PRC	05 AUG	07 AUG 88	
1988-067D		19371	PRC	05 AUG	07 AUG 88	
1988-067E		19381	PRC	05 AUG	27 AUG 88	
1988-067F		19382	PRC	05 AUG	17 AUG 88	
1988-067G		19383	PRC	05 AUG	19 AUG 88	
1988-067H		19474	PRC	05 AUG	26 JUN 89	
1988-068A	COSMOS 1962	19372	USSR	08 AUG	22 AUG 88	
1988-068B		19373	USSR	08 AUG	13 AUG 88	
1988-068C		19374	USSR	08 AUG	09 AUG 88	
1988-068D		19403	USSR	08 AUG	24 AUG 88	
1988-068E		19404	USSR	08 AUG	26 AUG 88	
1988-068F		19405	USSR	08 AUG	12 SEP 88	
1988-068G		19406	USSR	08 AUG	27 AUG 88	
1988-068H		19407	USSR	08 AUG	23 AUG 88	
1988-069B		19378	USSR	12 AUG	20 SEP 88	
1988-069C		19379	USSR	12 AUG	21 SEP 88	
1988-069E		19424	USSR	12 AUG	12 SEP 88	
1988-069F		19438	USSR	12 AUG	15 SEP 88	
1988-069G		19449	USSR	12 AUG	16 SEP 88	
1988-070A	COSMOS 1963	19384	USSR	16 AUG	02 OCT 88	
1988-070B		19385	USSR	16 AUG	21 AUG 88	
1988-071B		19398	USSR	18 AUG	21 AUG 88	
1988-071C		19399	USSR	18 AUG	19 AUG 88	
1989-065C		20190	USSR	22 AUG	23 AUG 89	
1988-072A	COSMOS 1964	19412	USSR	23 AUG	07 SEP 88	
1988-072B		19413	USSR	23 AUG	28 AUG 88	
1988-072C		19417	USSR	23 AUG	24 AUG 88	
1988-072D		19418	USSR	23 AUG	24 AUG 88	
1988-072E		19469	USSR	23 AUG	10 SEP 88	
1988-072F		19470	USSR	23 AUG	11 SEP 88	
1988-072G		19471	USSR	23 AUG	25 SEP 88	
1988-072H		19472	USSR	23 AUG	09 SEP 88	
1988-073A	COSMOS 1965	19414	USSR	23 AUG	22 SEP 88	
1988-073B		19415	USSR	23 AUG	25 AUG 88	
1988-073C		19416	USSR	23 AUG	23 AUG 88	
1988-073D		19523	USSR	23 AUG	04 OCT 88	
1988-073E		19524	USSR	23 AUG	18 OCT 88	
1988-073F		19525	USSR	23 AUG	13 OCT 88	
1988-073G		19526	USSR	23 AUG	02 OCT 88	
1988-075A	SOYUZ TM-6	19443	USSR	29 AUG	21 DEC 88	
1988-075B		19444	USSR	29 AUG	31 AUG 88	
1988-076B		19446	USSR	30 AUG	30 SEP 88	
1988-076C		19447	USSR	30 AUG	16 SEP 88	
1988-079A	COSMOS 1967	19462	USSR	06 SEP	15 SEP 88	
1988-079B		19463	USSR	06 SEP	15 SEP 88	
1988-079C		19465	USSR	06 SEP	08 SEP 88	
1988-079D		19466	USSR	06 SEP	08 SEP 88	
1988-079E		19492	USSR	06 SEP	16 SEP 88	
1988-079F		19493	USSR	06 SEP	16 SEP 88	
1988-079G		19494	USSR	06 SEP	23 SEP 88	
1988-079H		19497	USSR	06 SEP	18 SEP 88	
1988-079J		19498	USSR	06 SEP	18 SEP 88	
1988-079K		19499	USSR	06 SEP	16 SEP 88	
1988-081D		19487	ESA	08 SEP	03 JAN 90	
1988-082A	COSMOS 1968	19488	USSR	09 SEP	23 SEP 88	
1988-082B		19489	USSR	09 SEP	11 SEP 88	
1988-082C		19527	USSR	09 SEP	28 SEP 88	
1988-082D		19528	USSR	09 SEP	16 OCT 88	
1988-082E		19529	USSR	09 SEP	27 SEP 88	
1988-082F		19530	USSR	09 SEP	26 SEP 88	
1988-083A	PROGRESS 38	19486	USSR	09 SEP	23 NOV 88	
1988-083B		19491	USSR	09 SEP	12 SEP 88	
1988-084A	COSMOS 1969	19495	USSR	15 SEP	13 NOV 88	
1988-084B		19496	USSR	15 SEP	19 SEP 88	
1988-084C		19500	USSR	15 SEP	16 SEP 88	
1988-085D		19504	USSR	16 SEP	16 SEP 88	
1988-086B		19509	JAPAN	16 SEP	21 SEP 88	
1988-087A	HORIZON 1	19519	ISRAL	19 SEP	14 JAN 89	
1988-087B		19520	ISRAL	19 SEP	12 JAN 89	
1988-088A	COSMOS 1973	19521	USSR	22 SEP	10 OCT 88	
1988-088B		19522	USSR	22 SEP	01 OCT 88	
1988-088C		19564	USSR	22 SEP	22 NOV 88	
1988-088D		19565	USSR	22 SEP	31 MAR 89	
1988-088E		19566	USSR	22 SEP	20 NOV 88	
1988-088F		19567	USSR	22 SEP	12 OCT 88	
1988-088G		19572	USSR	22 SEP	02 NOV 88	
1988-088H		19575	USSR	22 SEP	01 NOV 88	
1988-088J		19591	USSR	22 SEP	02 FEB 89	
1988-089C		19534	US	24 SEP	18 JUN 92	
1988-090B		19542	USSR	29 SEP	29 OCT 88	
1988-090C		19543	USSR	29 SEP	22 OCT 88	
1988-091A	STS 26	19547	US	29 SEP	03 OCT 88	
1988-092B		19555	USSR	03 OCT	23 OCT 88	
1988-092C		19556	USSR	03 OCT	28 OCT 88	
1988-092E		19669	USSR	03 OCT	31 DEC 88	
1988-094A	COSMOS 1976	19582	USSR	13 OCT	27 OCT 88	
1988-094B		19583	USSR	13 OCT	21 OCT 88	
1988-094C		19614	USSR	13 OCT	14 APR 89	
1988-094D		19615	USSR	13 OCT	12 NOV 88	
1988-094E		19616	USSR	13 OCT	18 DEC 88	
1988-094F		19617	USSR	13 OCT	10 DEC 88	
1988-094G		19623	USSR	13 OCT	16 NOV 88	
1988-095B		19597	USSR	20 OCT	22 OCT 88	
1988-095C		19598	USSR	20 OCT	21 OCT 88	
1988-096B		19609	USSR	25 OCT	25 NOV 88	
1988-096C		19610	USSR	25 OCT	09 NOV 88	
1988-097A	COSMOS 1978	19612	USSR	27 OCT	10 NOV 88	
1988-097B		19613	USSR	27 OCT	03 NOV 88	
1988-097C		19627	USSR	27 OCT	12 NOV 88	
1988-097D		19628	USSR	27 OCT	12 NOV 88	
1988-097E		19629	USSR	27 OCT	14 NOV 88	
1988-097F		19630	USSR	27 OCT	13 NOV 88	
1988-097G		19631	USSR	27 OCT	23 NOV 88	
1988-100A	BURAN	19637	USSR	15 NOV	15 NOV 88	
1988-101A	COSMOS 1979	19647	USSR	18 NOV	25 DEC 89	
1988-101B		19648	USSR	18 NOV	18 NOV 88	
1988-102G		19811	USSR	23 NOV	17 OCT 90	
1988-103A	COSMOS 1981	19651	USSR	24 NOV	08 DEC 88	
1988-103B		19652	USSR	24 NOV	10 DEC 88	
1988-103C		19653	USSR	24 NOV	27 NOV 88	
1988-103D		19654	USSR	24 NOV	28 NOV 88	
1988-103E		19655	USSR	24 NOV	28 NOV 88	
1988-103F		19674	USSR	24 NOV	17 DEC 88	
1988-103G		19675	USSR	24 NOV	09 DEC 88	
1988-103H		19676	USSR	24 NOV	11 DEC 88	
1988-104A	SOYUZ TM-7	19660	USSR	26 NOV	27 APR 89	
1988-104B		19661	USSR	26 NOV	28 NOV 88	
1988-105A	COSMOS 1982	19662	USSR	30 NOV	14 DEC 88	
1988-105B		19663	USSR	30 NOV	10 DEC 88	
1988-105C		19694	USSR	30 NOV	20 JAN 89	
1988-105D		19695	USSR	30 NOV	13 JAN 89	
1988-105E		19696	USSR	30 NOV	27 APR 89	
1988-105F		19697	USSR	30 NOV	14 JAN 89	
1988-105G		19698	USSR	30 NOV	20 DEC 88	
1988-106A	STS 27	19670	US	02 DEC	06 DEC 88	
1988-107A	COSMOS 1983	19672	USSR	08 DEC	22 DEC 88	
1988-107B		19673	USSR	08 DEC	14 DEC 88	
1988-107C		19677	USSR	08 DEC	10 DEC 88	
1988-107D		19678	USSR	08 DEC	10 DEC 88	
1988-107E		19679	USSR	08 DEC	09 DEC 88	
1988-107F		19709	USSR	08 DEC	23 DEC 88	
1988-107G		19717	USSR	08 DEC	23 DEC 88	

International Designation	Satellite Name	Number	Source	Launch	Decay	Note
1988-107H		19718	USSR	08 DEC	28 DEC 88	
1988-107J		19719	USSR	08 DEC	24 DEC 88	
1988-108B		19684	USSR	10 DEC	12 DEC 88	
1988-108C		19685	USSR	10 DEC	11 DEC 88	
1988-108E		19691	USSR	10 DEC	13 NOV 91	
1988-108F		19699	USSR	10 DEC	06 SEP 89	
1988-110A	COSMOS 1984	19705	USSR	16 DEC	13 FEB 89	
1988-110B		19706	USSR	16 DEC	20 DEC 88	
1988-112B		19714	USSR	22 DEC	02 JAN 89	
1988-112C		19715	USSR	22 DEC	02 JAN 89	
1988-112E		19722	USSR	22 DEC	03 JAN 89	
1988-112F		19723	USSR	22 DEC	30 DEC 88	
1988-112G		19724	USSR	22 DEC	24 DEC 88	
1988-112H		19725	USSR	22 DEC	03 JAN 89	
1988-113A	COSMOS 1985	19720	USSR	23 DEC	04 MAY 92	
1988-113AA		20288	USSR	23 DEC	12 NOV 89	
1988-113AB		20289	USSR	23 DEC	14 NOV 89	
1988-113AC		20290	USSR	23 DEC	13 NOV 89	
1988-113AE		20342	USSR	23 DEC	01 JAN 90	
1988-113AF		20343	USSR	23 DEC	31 DEC 89	
1988-113AG		20345	USSR	23 DEC	31 DEC 89	
1988-113AH		20346	USSR	23 DEC	31 DEC 89	
1988-113AJ		20422	USSR	23 DEC	28 FEB 90	
1988-113AK		20423	USSR	23 DEC	21 FEB 90	
1988-113AL		20424	USSR	23 DEC	25 FEB 90	
1988-113AM		20425	USSR	23 DEC	25 FEB 90	
1988-113AN		20430	USSR	23 DEC	24 FEB 90	
1988-113AP		20431	USSR	23 DEC	18 FEB 90	
1988-113B		19721	USSR	23 DEC	15 MAR 89	
1988-113C		19726	USSR	23 DEC	02 MAR 89	
1988-113D		19727	USSR	23 DEC	03 MAR 89	
1988-113E		19711	USSR	23 DEC	17 MAR 89	
1988-113F		19712	USSR	23 DEC	14 MAR 89	
1988-113G		19763	USSR	23 DEC	15 MAR 89	
1988-113J		19885	USSR	23 DEC	04 MAY 89	
1988-113K		19886	USSR	23 DEC	26 APR 89	
1988-113L		19912	USSR	23 DEC	25 APR 89	
1988-113M		19943	USSR	23 DEC	15 JUN 89	
1988-113N		19944	USSR	23 DEC	16 JUN 89	
1988-113P		19958	USSR	23 DEC	17 JUN 89	
1988-113Q		19959	USSR	23 DEC	17 JUN 89	
1988-113R		20097	USSR	23 DEC	26 AUG 89	
1988-113S		20098	USSR	23 DEC	25 AUG 89	
1988-113T		20099	USSR	23 DEC	28 AUG 89	
1988-113U		20181	USSR	23 DEC	02 OCT 89	
1988-113V		20182	USSR	23 DEC	02 OCT 89	
1988-113W		20183	USSR	23 DEC	04 OCT 89	
1988-113X		20184	USSR	23 DEC	01 OCT 89	
1988-113Y		20286	USSR	23 DEC	14 NOV 89	
1988-113Z		20287	USSR	23 DEC	13 NOV 89	
1988-114A	PROGRESS 39	19728	USSR	25 DEC	07 FEB 89	
1988-114B		19729	USSR	25 DEC	27 DEC 88	
1988-115B		19731	USSR	28 DEC	21 JAN 89	
1988-115C		19732	USSR	28 DEC	22 JAN 89	
1988-115E		19742	USSR	28 DEC	16 JAN 89	
1988-115F		19748	USSR	28 DEC	25 JAN 89	
1988-116A	COSMOS 1986	19734	USSR	29 DEC	11 FEB 89	
1988-116B		19735	USSR	29 DEC	02 JAN 89	

1989 LAUNCHES

International Designation	Satellite Name	Number	Source	Launch	Decay	Note
1989-001D		19752	USSR	10 JAN	10 JAN 89	
1989-002A	COSMOS1990	19756	USSR	12 JAN	11 FEB 89	
1989-002B		19757	USSR	12 JAN	14 JAN 89	
1989-002C		19794	USSR	12 JAN	12 FEB 89	
1989-002D		19795	USSR	12 JAN	11 FEB 89	
1989-003A	COSMOS1991	19758	USSR	18 JAN	01 FEB 89	
1989-003B		19759	USSR	18 JAN	27 JAN 89	
1989-003C		19760	USSR	18 JAN	20 JAN 89	
1989-003D		19761	USSR	18 JAN	19 JAN 89	
1989-003E		19762	USSR	18 JAN	20 JAN 89	
1989-003F		19778	USSR	18 JAN	02 MAR 89	
1989-003G		19779	USSR	18 JAN	11 FEB 89	
1989-003H		19780	USSR	18 JAN	18 JUN 89	
1989-003J		19781	USSR	18 JAN	07 MAR 89	
1989-003K		19782	USSR	18 JAN	18 FEB 89	
1989-004B		19766	USSR	26 JAN	28 JAN 89	
1989-004C		19767	USSR	26 JAN	26 JAN 89	
1989-004D		19768	USSR	26 JAN	31 JUL 91	
1989-007A	COSMOS 1993	19774	USSR	28 JAN	27 MAR 89	
1989-007B		19775	USSR	28 JAN	01 FEB 89	
1989-008A	PROGRESS 40	19783	USSR	10 FEB	05 MAR 89	
1989-008B		19784	USSR	10 FEB	12 FEB 89	
1989-008C		19871	USSR	10 FEB	07 MAR 89	
1989-008D		19872	USSR	10 FEB	07 MAR 89	
1989-010A	COSMOS 2000	19792	USSR	10 FEB	03 MAR 89	
1989-010B		19793	USSR	10 FEB	12 FEB 89	
1989-010C		19864	USSR	10 FEB	30 MAY 89	
1989-010D		19865	USSR	10 FEB	29 MAR 89	
1989-010E		19866	USSR	10 FEB	09 MAR 89	
1989-010F		19867	USSR	10 FEB	24 MAR 89	
1989-010G		19873	USSR	10 FEB	15 MAR 89	
1989-011B		19797	USSR	14 FEB	11 MAR 89	
1989-011C		19798	USSR	14 FEB	03 MAR 89	
1989-011E		19956	USSR	14 FEB	30 APR 89	
1989-012A	COSMOS 2002	19800	USSR	14 FEB	15 OCT 89	
1989-012B		19801	USSR	14 FEB	27 JUL 89	
1989-012C		19805	USSR	14 FEB	03 MAR 89	
1989-012D		19806	USSR	14 FEB	10 MAR 89	
1989-012E		19821	USSR	14 FEB	11 MAR 89	
1989-012F		19825	USSR	14 FEB	08 MAR 89	
1989-012G		19832	USSR	14 FEB	10 MAR 89	
1989-012H		19846	USSR	14 FEB	09 MAR 89	
1989-012J		19847	USSR	14 FEB	11 MAR 89	
1989-012K		19848	USSR	14 FEB	11 MAR 89	
1989-012L		19849	USSR	14 FEB	06 MAR 89	
1989-012M		19850	USSR	14 FEB	10 MAR 89	
1989-013B		19803	US	14 FEB	13 MAR 89	
1989-013C		19804	US	14 FEB	06 DEC 89	
1989-014B		19808	USSR	15 FEB	01 MAR 89	
1989-014C		19809	USSR	15 FEB	09 MAR 89	
1989-015A	COSMOS 2003	19818	USSR	17 FEB	03 MAR 89	
1989-015B		19819	USSR	17 FEB	23 FEB 89	
1989-015C		19820	USSR	17 FEB	18 FEB 89	
1989-015D		19868	USSR	17 FEB	08 MAR 89	
1989-015E		19869	USSR	17 FEB	04 MAR 89	
1989-015F		19870	USSR	17 FEB	04 MAR 89	
1989-015G		19875	USSR	17 FEB	05 MAR 89	
1989-016B		19823	JAPAN	21 FEB	04 APR 89	
1989-016D		19841	JAPAN	21 FEB	28 FEB 89	
1989-016E		19842	JAPAN	21 FEB	27 FEB 89	
1989-016F		19843	JAPAN	21 FEB	25 FEB 89	
1989-016G		19844	JAPAN	21 FEB	25 FEB 89	
1989-016H		19845	JAPAN	21 FEB	25 FEB 89	
1989-016J		19889	JAPAN	21 FEB	22 MAR 89	
1989-016L		19962	JAPAN	21 FEB	01 FEB 92	
1989-019A	COSMOS 2005	19862	USSR	02 MAR	25 APR 89	
1989-019B		19863	USSR	02 MAR	08 MAR 89	
1989-020C		19877	ESA	06 MAR	21 APR 92	
1989-020D		19878	ESA	06 MAR	22 DEC 92	
1989-021A	STS 29	19882	US	13 MAR	18 MAR 89	
1989-022A	COSMOS 2006	19893	USSR	16 MAR	31 MAR 89	
1989-022B		19894	USSR	16 MAR	03 APR 89	
1989-022C		19897	USSR	16 MAR	20 MAR 89	
1989-022D		19898	USSR	16 MAR	20 MAR 89	
1989-022E		19899	USSR	16 MAR	18 MAR 89	
1989-022F		19914	USSR	16 MAR	17 APR 89	
1989-022G		19915	USSR	16 MAR	31 MAR 89	
1989-022H		19916	USSR	16 MAR	25 AUG 89	
1989-022J		19917	USSR	16 MAR	26 APR 89	
1989-022K		19918	USSR	16 MAR	12 APR 89	
1989-023A	PROGRESS 41	19895	USSR	16 MAR	25 APR 89	
1989-023B		19896	USSR	16 MAR	18 MAR 89	
1989-023C		19940	USSR	16 MAR	03 MAY 89	
1989-024A	COSMOS 2007	19900	USSR	23 MAR	22 SEP 89	
1989-024B		19901	USSR	23 MAR	25 MAR 89	
1989-026A		19911	US	24 MAR	23 JUN 92	
1989-029A	COSMOS 2017	19923	USSR	06 APR	19 APR 89	
1989-029B		19924	USSR	06 APR	16 APR 89	
1989-029C		19925	USSR	06 APR	07 APR 89:	
1989-029D		19926	USSR	06 APR	07 APR 89	
1989-029E		19927	USSR	06 APR	08 APR 89	
1989-029F		19934	USSR	06 APR	22 APR 89	
1989-029G		19935	USSR	06 APR	27 APR 89	
1989-029H		19936	USSR	06 APR	19 APR 89	
1989-029J		19937	USSR	06 APR	21 APR 89	
1989-030B		19929	USSR	14 APR	16 APR 89	
1989-030C		19930	USSR	14 APR	14 APR 89	
1989-030E		19932	USSR	14 APR	06 FEB 90	
1989-031A	COSMOS 2018	19938	USSR	20 APR	19 JUN 89	
1989-031B		19939	USSR	20 APR	24 APR 89	
1989-032A	FOTON 2	19941	USSR	26 APR	11 MAY 89	
1989-032B		19942	USSR	26 APR	08 MAY 89	
1989-032C		19978	USSR	26 APR	31 MAY 89	
1989-032D		19979	USSR	26 APR	15 MAY 89	
1989-032E		19984	USSR	26 APR	18 MAY 89	
1989-032F		19985	USSR	26 APR	15 MAY 89	
1989-033A	STS 30	19968	US	04 MAY	08 MAY 89	
1989-034A	COSMOS 2019	19972	USSR	05 MAY	18 MAY 89	
1989-034B		19973	USSR	05 MAY	11 MAY 89	
1989-034C		19974	USSR	05 MAY	06 MAY 89	
1989-034D		19975	USSR	05 MAY	07 MAY 89	
1989-034E		19988	USSR	05 MAY	20 MAY 89	
1989-034F		19989	USSR	05 MAY	19 MAY 89	
1989-036A	COSMOS 2020	19986	USSR	17 MAY	15 JUL 89	
1989-036B		19987	USSR	17 MAY	21 MAY 89	
1989-037A	COSMOS 2021	20000	USSR	24 MAY	06 JUL 89	
1989-037B		20001	USSR	24 MAY	27 MAY 89	
1989-037C		20002	USSR	24 MAY	25 MAY 89	
1989-038A	RESURS-F	20006	USSR	25 MAY	17 JUN 89	
1989-038B		20007	USSR	25 MAY	26 MAY 89	
1989-038C		20056	USSR	25 MAY	23 JUL 89	
1989-038D		20060	USSR	25 MAY	24 JUL 89	
1989-038E		20076	USSR	25 MAY	20 JUN 89	
1989-038F		20077	USSR	25 MAY	19 JUN 89	
1989-038G		20078	USSR	25 MAY	20 JUN 89	
1989-038H		20079	USSR	25 MAY	02 JUL 89	
1989-038J		20080	USSR	25 MAY	19 JUN 89	
1989-039D		20027	USSR	31 MAY	31 JAN 89	
1989-040A	COSMOS 2025	20035	USSR	01 JUN	15 JUN 89	
1989-040B		20036	USSR	01 JUN	08 JUN 89	
1989-040C		20037	USSR	01 JUN	03 JUN 89	
1989-040D		20038	USSR	01 JUN	03 JUN 89	
1989-040E		20039	USSR	01 JUN	02 JUN 89	
1989-040F		20070	USSR	01 JUN	17 JUN 89	
1989-040G		20071	USSR	01 JUN	15 JUN 89	

International Designation	Satellite Name	Number	Source	Launch	Decay	Note
1989-040H		20072	USSR	01 JUN	01 JUL 89	
1989-041D		20043	ESA	05 JUN	10 MAY 92	
1989-043B		20053	USSR	08 JUN	05 JUL 89	
1989-043C		20054	USSR	08 JUN	08 JUL 89	
1989-043E		20057	USSR	08 JUN	06 JUL 89	
1989-043F		20058	USSR	08 JUN	29 JUN 89	
1989-043G		20059	USSR	08 JUN	30 JUN 89	
1989-044B		20062	US	10 JUN	12 JUL 89	
1989-044C		20063	US	10 JUN	06 MAY 91	
1989-045A	COSMOS 2027	20064	USSR	14 JUN	14 APR 92	
1989-045B		20065	USSR	14 JUN	25 MAR 91	
1989-047A	COSMOS 2028	20073	USSR	16 JUN	06 JUL 89	
1989-047B		20074	USSR	16 JUN	21 JUN 89	
1989-047C		20075	USSR	16 JUN	17 JUN 89	
1989-047D		20111	USSR	16 JUN	13 JUL 89	
1989-047G		20114	USSR	16 JUN	10 JUL 89	
1989-048B		20084	USSR	21 JUN	23 JUN 89	
1989-048C		20085	USSR	21 JUN	22 JUN 89	
1989-048E		20093	USSR	21 JUN	30 MAR 90	
1989-049A		20095	USSR	27 JUN	11 JUL 89	
1989-049G		20121	USSR	27 JUN	14 JUL 89	
1989-049C		20117	USSR	28 JUN	14 JUL 89	
1989-049D		20118	USSR	28 JUN	15 JUL 89	
1989-049F		20120	USSR	28 JUN	18 JUL 89	
1989-051A	COSMOS 2029	20105	USSR	05 JUL	19 JUL 89	
1989-051B		20106	USSR	05 JUL	07 JUL 89	
1989-051C		20139	USSR	05 JUL	15 AUG 89	
1989-051D		20140	USSR	05 JUL	14 OCT 89	
1989-051E		20141	USSR	05 JUL	07 AUG 89	
1989-051F		20142	USSR	05 JUL	18 AUG 89	
1989-051G		20143	USSR	05 JUL	08 AUG 89	
1989-051H		20144	USSR	05 JUL	23 JUL 89	
1989-052B		20108	USSR	05 JUL	07 JUL 89	
1989-052C		20109	USSR	05 JUL	06 JUL 89	
1989-052E		20115	USSR	05 JUL	15 AUG 91	
1989-054A	COSMOS 2030	20124	USSR	12 JUL	29 JUL 89	
1989-054B		20125	USSR	12 JUL	16 JUL 89	
1989-047E		20112	USSR	16 JUL	06 JUL 89	
1989-047F		20113	USSR	16 JUL	07 JUL 89	
1989-055A	RESURS-F3	20134	USSR	18 JUL	08 AUG 89	
1989-055B		20135	USSR	18 JUL	20 JUL 89	
1989-055C		20160	USSR	18 JUL	19 SEP 89	
1989-055D		20161	USSR	18 JUL	19 SEP 89	
1989-055E		20162	USSR	18 JUL	09 AUG 89	
1989-055F		20163	USSR	18 JUL	27 AUG 89	
1989-055G		20165	USSR	18 JUL	13 AUG 89	
1989-055H		20166	USSR	18 JUL	12 AUG 89	
1989-056A	COSMOS 2031	20136	USSR	18 JUL	15 SEP 89	
1989-056B		20137	USSR	18 JUL	21 JUL 89	
1989-056C		20138	USSR	18 JUL	19 JUL 89	
1989-056D		20206	USSR	18 JUL	04 SEP 89	
1989-056E		20207	USSR	18 JUL	04 SEP 89	
1989-056F		20208	USSR	18 JUL	04 SEP 89	
1989-056G		20209	USSR	18 JUL	04 SEP 89	
1989-056H		20210	USSR	18 JUL	04 SEP 89	
1989-056J		20211	USSR	18 JUL	04 SEP 89	
1989-056K		20212	USSR	18 JUL	04 SEP 89	
1989-056L		20213	USSR	18 JUL	03 SEP 89	
1989-057A	COSMOS 2032	20145	USSR	20 JUL	03 AUG 89	
1989-057B		20146	USSR	20 JUL	22 JUL 89	
1989-057C		20156	USSR	20 JUL	10 AUG 89	
1989-057D		20157	USSR	20 JUL	04 AUG 89	
1989-057E		20158	USSR	20 JUL	06 AUG 89	
1989-057F		20159	USSR	20 JUL	05 AUG 89	
1989-058B		20148	USSR	20 JUL	24 JUL 89	
1989-058A	COSMOS 2033	20147	USSR	24 JUL	06 JAN 91	
1989-049E		20119	USSR	25 JUL	05 AUG 89	
1989-060A	COSMOS 2035	20151	USSR	02 AUG	16 AUG 89	
1989-060B		20152	USSR	02 AUG	04 AUG 89	
1989-060C		20177	USSR	02 AUG	18 AUG 89	
1989-060D		20178	USSR	02 AUG	17 AUG 89	
1989-060E		20179	USSR	02 AUG	17 AUG 89	
1989-060F		20180	USSR	02 AUG	22 AUG 89	
1989-061A	STS-28	20164	US	08 AUG	13 AUG 89	
1989-062D		20171	ESA	08 AUG	08 MAR 92	
1989-063A	RESURS-F4	20175	USSR	15 AUG	14 SEP 89	
1989-063B		20176	USSR	15 AUG	16 AUG 89	
1989-063C		20231	USSR	16 AUG	15 SEP 89	
1989-063D		20239	USSR	16 AUG	16 SEP 89	
1989-063E		20240	USSR	16 AUG	19 SEP 89	
1989-063F		20241	USSR	16 AUG	16 SEP 89	
1989-064B		20186	US	18 AUG	09 SEP 89	
1989-064C		20187	US	18 AUG	10 JUN 90	
1989-065A	COSMOS 2036	20188	USSR	22 AUG	05 SEP 89	
1989-065B		20189	USSR	22 AUG	29 AUG 89	
1989-066A	PROGRESS/M	20191	USSR	23 AUG	01 DEC 89	
1989-066B		20192	USSR	23 AUG	24 AUG 89	
1989-065D		20214	USSR	24 AUG	10 SEP 89	
1989-065E		20215	USSR	24 AUG	05 SEP 89	
1989-065F		20216	USSR	24 AUG	05 SEP 89	
1989-067B		20194	US	27 AUG	22 OCT 89	
1989-069C		20204	US	04 SEP	24 SEP 89	
1989-071A	SOYUZ TM-8	20218	USSR	05 SEP	19 FEB 90	
1989-071B		20219	USSR	05 SEP	07 SEP 89	
1989-073A	RESURS F-5	20222	USSR	06 SEP	22 SEP 89	
1989-073B		20223	USSR	06 SEP	08 SEP 89	
1989-075A	COSMOS 2044	20242	USSR	15 SEP	29 SEP 89	
1989-075B		20243	USSR	15 SEP	18 SEP 89	
1989-075C		20267	USSR	15 SEP	01 OCT 89	
1989-075D		20268	USSR	15 SEP	30 SEP 89	
1989-075E		20269	USSR	15 SEP	29 SEP 89	
1989-073C		20248	USSR	22 SEP	12 OCT 89	
1989-073D		20249	USSR	22 SEP	26 SEP 89	
1989-073E		20250	USSR	22 SEP	28 SEP 89	
1989-073F		20251	USSR	22 SEP	25 SEP 89	
1989-073G		20252	USSR	22 SEP	24 SEP 89	
1989-076A	COSMOS 2045	20244	USSR	22 SEP	02 OCT 89	
1989-076B		20245	USSR	22 SEP	28 SEP 89	
1989-076C		20246	USSR	22 SEP	23 SEP 89	
1989-076D		20247	USSR	22 SEP	23 SEP 89	
1989-076E		20274	USSR	22 SEP	04 OCT 89	
1989-076F		20275	USSR	22 SEP	05 OCT 89	
1989-076G		20276	USSR	22 SEP	09 OCT 89	
1989-076H		20277	USSR	22 SEP	04 OCT 89	
1989-076J		20278	USSR	22 SEP	03 OCT 89	
1989-077B		20254	US	25 SEP	09 MAY 91	
1989-078B		20256	USSR	27 SEP	26 OCT 89	
1989-078C		20257	USSR	27 SEP	24 OCT 89	
1989-079A	COSMOS 2046	20259	USSR	27 SEP	16 APR 91	
1989-079B		20260	USSR	27 SEP	28 SEP 89	
1989-081B		20264	USSR	28 SEP	30 SEP 89	
1989-081C		20265	USSR	28 SEP	29 SEP 89	
1989-081E		20270	USSR	28 SEP	22 JUL 90	
1989-081F		20271	USSR	28 SEP	28 SEP 91	
1989-082A	COSMOS 2047	20279	USSR	03 OCT	21 NOV 89	
1989-082B		20280	USSR	03 OCT	06 OCT 89	
1989-083A	COSMOS 2048	20292	USSR	17 OCT	26 OCT 89	
1989-083B		20293	USSR	17 OCT	22 OCT 89	
1989-083C		20294	USSR	17 OCT	18 OCT 89	
1989-083D		20295	USSR	17 OCT	18 OCT 89	
1989-083E		20296	USSR	17 OCT	18 OCT 89	
1989-083F		20310	USSR	17 OCT	30 OCT 89	
1989-083G		20311	USSR	17 OCT	29 OCT 89	
1989-083H		20312	USSR	17 OCT	27 OCT 89	
1989-083J		20313	USSR	17 OCT	28 OCT 89	
1989-083K		20314	USSR	17 OCT	27 OCT 89	
1989-084A	STS-34	20297	US	18 OCT	23 OCT 89	
1989-085C		20304	US	21 OCT	05 DEC 91	
1989-088A	COSMOS 2049	20320	USSR	17 NOV	19 JUN 90	
1989-088B		20321	USSR	17 NOV	19 NOV 89	
1989-090A	STS-33	20329	US	23 NOV	28 NOV 89	
1989-091B		20331	USSR	23 NOV	18 DEC 89	
1989-091C		20332	USSR	23 NOV	10 DEC 89	
1989-092A	COSMOS 2051	20334	USSR	24 NOV	21 JAN 91	
1989-093B		20336	USSR	26 NOV	04 DEC 89	
1989-093C		20337	USSR	26 NOV	27 NOV 89	
1989-094C		20340	USSR	28 NOV	22 DEC 89	
1989-094D		20341	USSR	28 NOV	20 DEC 89	
1989-095A	COSMOS 2052	20350	USSR	30 NOV	24 JAN 90	
1989-095B		20351	USSR	30 NOV	03 DEC 89	
1989-096B		20353	USSR	01 DEC	09 MAR 91	
1989-096D		20358	USSR	01 DEC	15 APR 91	
1989-097C		20363	US	11 DEC	09 AUG 90	
1989-098B		20368	USSR	15 DEC	17 DEC 89	
1989-098C		20369	USSR	15 DEC	16 DEC 89	
1989-098E		20371	USSR	15 DEC	04 SEP 91	
1989-098F		20372	USSR	15 DEC	06 JUN 91	
1989-099A	PROGRESS M-2	20373	USSR	20 DEC	09 FEB 90	
1989-099B		20374	USSR	20 DEC	21 DEC 89	
1988-113AD		20291	USSR	23 DEC	13 NOV 89	
1989-100AB		21064	USSR	27 DEC	12 FEB 91	
1989-100AC		21205	USSR	27 DEC	24 APR 91	
1989-100AD		21206	USSR	27 DEC	23 APR 91	
1989-100AE		21207	USSR	27 DEC	27 APR 91	
1989-100AF		21537	USSR	27 DEC	14 AUG 91	
1989-100AG		21540	USSR	27 DEC	13 AUG 91	
1989-100AH		21767	USSR	27 DEC	23 NOV 91	
1989-100AJ		21768	USSR	27 DEC	22 NOV 91	
1989-100AK		21769	USSR	27 DEC	21 NOV 91	
1989-100AL		21770	USSR	27 DEC	22 NOV 91	
1989-100AM		21771	USSR	27 DEC	22 NOV 91	
1989-100AN		21772	USSR	27 DEC	29 NOV 91	
1989-100AP		21773	USSR	27 DEC	27 NOV 91	
1989-100AQ		21774	USSR	27 DEC	06 DEC 91	
1989-100C		20397	USSR	27 DEC	12 MAR 90	
1989-100D		20398	USSR	27 DEC	28 MAR 90	
1989-100E		20408	USSR	27 DEC	09 MAR 90	
1989-100F		20467	USSR	27 DEC	31 MAR 90	
1989-100G		20468	USSR	27 DEC	26 MAR 90	
1989-100H		20515	USSR	27 DEC	24 APR 90	
1989-100J		20522	USSR	27 DEC	28 APR 90	
1989-100K		20531	USSR	27 DEC	30 APR 90	
1989-100L		20532	USSR	27 DEC	07 MAY 90	
1989-100U		20911	USSR	27 DEC	17 NOV 90	
1989-101B		20392	USSR	27 DEC	28 DEC 89	
1989-101C		20393	USSR	27 DEC	28 DEC 89	
1989-101F		20400	USSR	27 DEC	31 JUL 91	
1989-100AA		21043	USSR	31 DEC	12 FEB 91	
1989-100M		20637	USSR	31 DEC	16 AUG 90	
1989-100N		20640	USSR	31 DEC	20 AUG 90	
1989-100P		20802	USSR	31 DEC	30 OCT 90	
1989-100Q		20803	USSR	31 DEC	30 OCT 90	
1989-100R		20821	USSR	31 DEC	16 NOV 90	
1989-100S		20822	USSR	31 DEC	21 NOV 90	
1989-100T		20823	USSR	31 DEC	16 NOV 90	
1989-100V		21020	USSR	31 DEC	06 FEB 91	

International Designation	Satellite Name	Number	Source	Launch	Decay	Note
1989-100X		21022	USSR	31 DEC	07 FEB 91	
1989-100Y		21023	USSR	31 DEC	08 FEB 91	
1989-100Z		21042	USSR	31 DEC	12 FEB 91	
1990 LAUNCHES						
1990-001C		20403	US	01 JAN	09 JAN 90	
1990-001E		20405	US	01 JAN	15 JAN 90	
1990-002A	STS 32	20409	US	09 JAN	20 JAN 90	
1990-003A	COSMOS 2055	20426	USSR	17 JAN	29 JAN 90	
1990-003B		20427	USSR	17 JAN	24 JAN 90	
1990-003C		20428	USSR	17 JAN	19 JAN 90	
1990-003D		20429	USSR	17 JAN	18 JAN 90	
1990-003E		20462	USSR	17 JAN	03 FEB 90	
1990-003F		20463	USSR	17 JAN	30 JAN 90	
1990-003G		20464	USSR	17 JAN	30 JAN 90	
1990-006B		20445	USSR	23 JAN	22 FEB 90	
1990-006D		20447	USSR	23 JAN	15 FEB 90	
1990-007C		20449	JAPAN	24 JAN	29 JAN 90	
1990-009A	COSMOS 2057	20457	USSR	25 JAN	19 MAR 90	
1990-009B		20458	USSR	25 JAN	30 JAN 90	
1990-012A	COSMOS 2059	20476	USSR	06 FEB	12 NOV 90	
1990-012B		20477	USSR	06 FEB	12 JUL 90	
1990-012D		20482	USSR	06 FEB	24 MAR 90	
1990-012E		20483	USSR	06 FEB	06 MAR 90	
1990-012F		20484	USSR	06 FEB	06 MAR 90	
1990-012G		20485	USSR	06 FEB	05 MAR 90	
1990-012H		20486	USSR	06 FEB	07 MAR 90	
1990-012J		20487	USSR	06 FEB	25 FEB 90	
1990-012K		20490	USSR	06 FEB	06 MAR 90	
1990-012L		20492	USSR	06 FEB	05 MAR 90	
1990-012M		20493	USSR	06 FEB	27 FEB 90	
1990-014A	SOYUZ TM-9	20494	USSR	11 FEB	09 AUG 90	
1990-014B		20495	USSR	11 FEB	13 FEB 90	
1990-015B		20497	US	14 FEB	24 MAY 92	
1990-015C		20498	US	14 FEB	07 DEC 90	
1990-016B		20500	USSR	15 FEB	17 FEB 90	
1990-016C		20501	USSR	15 FEB	17 FEB 90	
1990-016E		20506	USSR	15 FEB	03 DEC 90	
1990-016F		20507	USSR	15 FEB	04 SEP 91	
1990-019A	STS-36	20512	US	28 FEB	04 MAR 90	
1990-020A	PROGRESS M-3	20513	USSR	28 FEB	28 APR 90	
1990-020B		20514	USSR	28 FEB	02 MAR 90	
1990-021B		20524	US	14 MAR	28 MAR 90	
1990-022A	COSMOS 2060	20525	USSR	14 MAR	01 SEP 91	
1990-022B		20526	USSR	14 MAR	15 MAR 90	
1990-024A	COSMOS 2062	20529	USSR	22 MAR	05 APR 90	
1990-024B		20530	USSR	22 MAR	23 MAR 90	
1990-024C		20542	USSR	22 MAR	09 APR 90	
1990-024D		20543	USSR	22 MAR	14 APR 90	
1990-024E		20544	USSR	22 MAR	08 APR 90	
1990-024F		20545	USSR	22 MAR	07 APR 90	
1990-024G		20548	USSR	22 MAR	08 APR 90	
1990-025B		20534	US	26 MAR	04 APR 90	
1990-026B		20537	USSR	27 MAR	22 APR 90	
1990-026C		20538	USSR	27 MAR	10 APR 90	
1990-027A	OFEQ 2	20540	ISRAL	03 APR	09 JUL 90	
1990-027B		20541	ISRAL	03 APR	20 JUN 90	
1990-032A	FOTON 3	20566	USSR	11 APR	27 APR 90	
1990-032B		20567	USSR	11 APR	21 APR 90	
1990-032C		20588	USSR	11 APR	16 MAY 90	
1990-032D		20589	USSR	11 APR	01 MAY 90	
1990-032E		20590	USSR	11 APR	07 MAY 90	
1990-033A	COSMOS 2072	20568	USSR	13 APR	21 NOV 90	
1990-033B		20569	USSR	13 APR	16 APR 90	
1990-035A	COSMOS 2073	20573	USSR	17 APR	28 APR 90	
1990-035B		20574	USSR	17 APR	18 APR 90	
1990-035C		20591	USSR	17 APR	29 APR 90	
1990-035D		20592	USSR	17 APR	29 APR 90	
1990-035E		20593	USSR	17 APR	29 APR 90	
1990-035F		20594	USSR	17 APR	01 MAY 90	
1990-035G		20595	USSR	17 APR	13 MAY 90	
1990-035H		20601	USSR	17 APR	02 MAY 90	
1990-037A	STS-31	20579	US	24 APR	29 APR 90	
1990-038A	COSMOS 2075	20581	USSR	25 APR	20 FEB 92	
1990-038B		20582	USSR	25 APR	30 DEC 91	
1990-038C		20587	USSR	25 APR	04 JUL 90	
1990-038D		20746	USSR	25 APR	02 OCT 90	
1990-038E		20747	USSR	25 APR	28 SEP 90	
1990-038F		20748	USSR	25 APR	20 SEP 90	
1990-038G		20749	USSR	25 APR	20 SEP 90	
1990-038H		20750	USSR	25 APR	22 SEP 90	
1990-038J		20751	USSR	25 APR	25 SEP 90	
1990-038K		21864	USSR	25 APR	08 FEB 92	
1990-038L		21865	USSR	25 APR	07 FEB 92	
1990-038M		21866	USSR	25 APR	06 FEB 92	
1990-038N		21869	USSR	25 APR	13 FEB 92	
1990-038P		21870	USSR	25 APR	11 FEB 92	
1990-038Q		21871	USSR	25 APR	13 FEB 92	
1990-038R		21872	USSR	25 APR	12 FEB 92	
1990-039B		20584	USSR	26 APR	26 MAY 90	
1990-039C		20585	USSR	26 APR	23 MAY 90	
1990-040B		20597	USSR	28 APR	28 MAY 90	
1990-040C		20598	USSR	28 APR	13 MAY 90	
1990-040E		20600	USSR	28 APR	03 MAY 90	
1990-041A	PROGRESS-42	20602	USSR	05 MAY	27 MAY 90	
1990-041B		20603	USSR	05 MAY	07 MAY 90	
1990-042A	COSMOS 2077	20604	USSR	07 MAY	04 JUL 90	
1990-042B		20605	USSR	07 MAY	12 MAY 90	
1990-042C		20606	USSR	07 MAY	08 MAY 90	
1990-043G		20613	US	09 MAY	01 MAR 92	
1990-043J		20634	US	09 MAY	02 SEP 92	
1990-044A	COSMOS 2078	20615	USSR	15 MAY	28 JUN 90	
1990-044B		20616	USSR	15 MAY	19 MAY 90	
1990-044C		20617	USSR	15 MAY	16 MAY 90	
1990-045D		20622	USSR	19 MAY	19 MAY 90	
1990-047A	RESURS-F	20632	USSR	29 MAY	14 JUN 90	
1990-047B		20633	USSR	29 MAY	31 MAY 90	
1990-047C		20653	USSR	29 MAY	17 JUN 90	
1990-047D		20654	USSR	29 MAY	07 JUL 90	
1990-047E		20655	USSR	29 MAY	16 JUN 90	
1990-047F		20656	USSR	29 MAY	20 JUN 90	
1990-048B		20636	USSR	31 MAY	11 JUN 90	
1990-049B		20639	US	01 JUN	15 JAN 92	
1990-051B		20644	US	12 JUN	12 JUL 90	
1990-051C		20645	US	12 JUN	26 JUL 91	
1990-052B		20647	USSR	13 JUN	03 JUL 90	
1990-052C		20648	USSR	13 JUN	28 JUN 90	
1990-053A	COSMOS 2083	20657	USSR	19 JUN	03 JUL 90	
1990-053B		20658	USSR	19 JUN	21 JUN 90	
1990-053C		20675	USSR	19 JUN	19 JUL 90	
1990-053D		20676	USSR	19 JUN	30 SEP 90	
1990-053E		20677	USSR	19 JUN	29 JUL 90	
1990-053F		20678	USSR	19 JUN	20 JUL 90	
1990-053G		20679	USSR	19 JUN	09 AUG 90	
1990-054B		20660	USSR	20 JUN	22 JUN 90	
1990-054C		20661	USSR	20 JUN	21 JUN 90	
1990-054F		20711	USSR	20 JUN	02 JAN 91	
1990-055B		20664	USSR	21 JUN	21 JUL 90	
1990-055C		20665	USSR	21 JUN	20 JUL 90	
1990-056B		20668	US	23 JUN	26 JUN 90	
1990-058A	GAMMA	20683	USSR	11 JUL	28 FE8 92	
1990-058B		20684	US	11 JUL	12 JUL 90	
1990-059A	BADR-A	20685	PAKI	16 JUL	08 DEC 90	
1990-059B		20686	PRC	16 JUL	09 OCT 90	
1990-060A	RESURS-F7	20687	USSR	17 JUL	16 AUG 90	
1990-060B		20688	USSR	17 JUL	19 JUL 90	
1990-060C		20756	USSR	17 JUL	19 AUG 90	
1990-060D		20757	USSR	17 JUL	18 AUG 90	
1990-060E		20758	USSR	17 JUL	22 AUG 90	
1990-060F		20761	USSR	17 JUL	18 AUG 90	
1990-061B		20694	USSR	18 JUL	21 JUL 90	
1990-061C		20695	USSR	18 JUL	20 JUL 90	
1990-061E		20697	USSR	18 JUL	24 AUG 91	
1990-062A	COSMOS 2086	20702	USSR	20 JUL	03 AUG 90	
1990-062B		20703	USSR	20 JUL	22 JUL 90	
1990-062C		20727	USSR	20 JUL	08 AUG 90	
1990-062D		20728	USSR	20 JUL	16 AUG 90	
1990-062E		20729	USSR	20 JUL	04 AUG 90	
1990-062F		20730	USSR	20 JUL	06 AUG 90	
1990-062G		20731	USSR	20 JUL	04 AUG 90	
1990-064B		20708	USSR	25 JUL	22 AUG 90	
1990-064C		20709	USSR	25 JUL	08 AUG 90	
1990-064E		20714	USSR	25 JUL	14 AUG 90	
1990-064F		20715	USSR	25 JUL	08 AUG 90	
1990-064G		20716	USSR	25 JUL	05 AUG 90	
1990-067A	SOYUZ TM-10	20722	USSR	01 AUG	10 DEC 90	
1990-067B		20723	USSR	01 AUG	02 AUG 90	
1990-068B		20725	US	02 AUG	15 AUG 90	
1990-068C		20726	US	02 AUG	18 MAR 91	
1990-069A	COSMOS 2089	20732	USSR	03 AUG	01 OCT 90	
1990-069B		20733	USSR	03 AUG	09 AUG 90	
1990-069C		20734	USSR	03 AUG	04 AUG 90	
1990-071B		20743	USSR	10 AUG	10 SEP 90	
1990-071C		20744	USSR	10 AUG	02 SEP 90	
1990-072A	PROGRESS M-4	20752	USSR	15 AUG	20 SEP 90	
1990-072B		20753	USSR	15 AUG	16 AUG 90	
1990-072C		20812	USSR	15 AUG	16 OCT 90	
1990-073A	RESURS-F8	20754	USSR	16 AUG	01 SEP 90	
1990-073B		20755	USSR	16 AUG	18 AUG 90	
1990-073C		20782	USSR	16 AUG	21 SEP 90	
1990-073D		20783	USSR	16 AUG	07 SEP 90	
1990-073E		20784	USSR	16 AUG	05 SEP 90	
1990-073F		20785	USSR	16 AUG	03 SEP 90	
1990-073G		20786	USSR	16 AUG	04 SEP 90	
1990-073H		20787	USSR	16 AUG	03 SEP 90	
1990-075A	COSMOS 2096	20765	USSR	23 AUG	30 AUG 92	
1990-075B		20766	USSR	23 AUG	24 AUG 90	
1990-076B		20768	USSR	28 AUG	26 SEP 90	
1990-076C		20769	USSR	28 AUG	11 SEP 90	
1990-077B		20772	JAPAN	28 AUG	05 SEP 90	
1990-077C		20773	JAPAN	28 AUG	16 FEB 91	
1990-079D		20781	ESA	30 AUG	28 SEP 91	
1990-080A	COSMOS 2099	20779	USSR	31 AUG	14 SEP 90	
1990-080B		20780	USSR	31 AUG	01 SEP 90	
1990-080C		20796	USSR	31 AUG	10 SEP 90	
1990-080D		20806	USSR	31 AUG	23 SEP 90	
1990-080E		20807	USSR	31 AUG	16 SEP 90	
1990-080F		20808	USSR	31 AUG	15 SEP 90	
1990-080G		20809	USSR	31 AUG	17 SEP 90	
1990-080H		20810	USSR	31 AUG	16 SEP 90	
1990-081AB		20863	PRC	03 SEP	10 JAN 91	
1990-081B	PRC 31	20789	PRC	03 SEP	11 MAR 91	
1990-081BH		20897	PRC	03 SEP	21 JAN 93	
1990-081BU		20908	PRC	03 SEP	29 NOV 90	
1990-081BZ		20916	PRC	03 SEP	02 JAN 91	

International Designation	Satellite Name	Number	Source	Launch	Decay	Note
1990-081C	PRC 32	20790	PRC	03 SEP	24 JUL 91	
1990-082A	RESURS-F9	20794	USSR	07 SEP	21 SEP 90	
1990-082B		20795	USSR	07 SEP	09 SEP 90	
1990-082C		20817	USSR	07 SEP	13 OCT 90	
1990-082D		20818	USSR	07 SEP	23 SEP 90	
1990-082E		20819	USSR	07 SEP	25 SEP 90	
1990-082F		20820	USSR	07 SEP	25 SEP 90	
1990-084B		20814	USSR	20 SEP	03 OCT 90	
1990-084C		20815	USSR	20 SEP	01 OCT 90	
1990-085A		20824	USSR	27 SEP	28 NOV 90	
1990-085B		20825	USSR	27 SEP	28 SEP 90	
1990-087A	COSMOS 2101	20828	USSR	01 OCT	30 NOV 90	
1990-087B		20829	USSR	01 OCT	04 OCT 90	
1990-087C		21003	USSR	01 OCT	08 DEC 90	
1990-087D		21004	USSR	01 OCT	06 DEC 90	
1990-087E		21005	USSR	01 OCT	10 DEC 90	
1990-088B		20831	US	01 OCT	10 OCT 90	
1990-088C		20832	US	01 OCT	18 JUN 91	
1990-089A	PRC 33	20838	PRC	05 OCT	23 OCT 90	
1990-089B		20839	PRC	05 OCT	15 OCT 90	
1990-089C		20840	PRC	05 OCT	07 OCT 90	
1990-090A	STS-31	20841	US	06 OCT	10 OCT 90	
1990-091D		20875	ESA	12 OCT	02 OCT 91	
1990-092A	COSMOS 2102	20909	USSR	16 OCT	12 DEC 90	
1990-092B		20910	USSR	16 OCT	20 OCT 90	
1990-093C		20920	US	30 OCT	03 MAR 91	
1990-094B		20924	USSR	03 NOV	04 NOV 90	
1990-094C		20925	USSR	03 NOV	05 NOV 90	
1990-094F		20928	USSR	03 NOV	02 MAY 91	
1990-095B		20930	US	13 NOV	13 NOV 90	
1990-096A	COSMOS 2103	20933	USSR	14 NOV	03 APR 91	
1990-096B		20934	USSR	14 NOV	14 NOV 90	
1990-097A	STS-38	20935	US	15 NOV	20 NOV 90	
1990-098A	COSMOS 2104	20936	USSR	16 NOV	04 DEC 90	
1990-098B		20937	USSR	16 NOV	05 DEC 90	
1990-098C		20938	USSR	16 NOV	19 NOV 90	
1990-098D		20939	USSR	16 NOV	20 NOV 90	
1990-098E		20940	USSR	16 NOV	19 NOV 90	
1990-098F		20956	USSR	16 NOV	07 DEC 90	
1990-098G		20994	USSR	16 NOV	10 DEC 90	
1990-098H		20995	USSR	16 NOV	05 DEC 90	
1990-098J		20996	USSR	16 NOV	14 DEC 90	
1990-098K		21002	USSR	16 NOV	06 DEC 90	
1990-099B		20942	USSR	20 NOV	23 DEC 90	
1990-099C		20943	USSR	20 NOV	07 DEC 90	
1990-100D		20948	ESA	20 NOV	06 NOV 91	
1990-101B		20950	USSR	23 NOV	26 DEC 90	
1990-101C		20951	USSR	23 NOV	12 DEC 90	
1990-102B		20954	USSR	23 NOV	25 NOV 90	
1990-102C		20955	USSR	23 NOV	24 NOV 90	
1990-102E		20957	USSR	23 NOV	01 JUL 91	
1990-102F		20958	USSR	23 NOV	02 MAY 91	
1990-103C		20961	US	26 NOV	30 SEP 92	
1990-104AA		21994	USSR	28 NOV	07 OCT 92	
1990-104AB		21995	USSR	28 NOV	23 SEP 92	
1990-104AC		21996	USSR	28 NOV	22 SEP 92	
1990-104AD		22001	USSR	28 NOV	02 OCT 92	
1990-104AE		22002	USSR	28 NOV	06 OCT 92	
1990-104AF		22003	USSR	28 NOV	06 OCT 92	
1990-104AG		22486	USSR	28 NOV	05 APR 93	
1990-104C		20975	USSR	28 NOV	21 FEB 91	
1990-104D		20976	USSR	28 NOV	23 FEB 91	
1990-104E		20977	USSR	28 NOV	06 MAR 91	
1990-104F		21070	USSR	28 NOV	10 MAR 91	
1990-104H		21081	USSR	28 NOV	10 MAR 91	
1990-104J		21083	USSR	28 NOV	11 NOV 90	
1990-104K		21258	USSR	28 NOV	15 JUL 91	
1990-104L		21259	USSR	28 NOV	10 JUL 91	
1990-104M		21260	USSR	28 NOV	10 JUL 91	
1990-104N		21261	USSR	28 NOV	10 JUL 91	
1990-104P		21264	USSR	28 NOV	21 JUL 91	
1990-104Q		21265	USSR	28 NOV	14 JUL 91	
1990-104R		21266	USSR	28 NOV	20 JUL 91	
1990-104S		21306	USSR	28 NOV	21 JUL 91	
1990-104T		21307	USSR	28 NOV	22 JUL 91	
1990-104U		21308	USSR	28 NOV	15 JUL 91	
1990-104V		21309	USSR	28 NOV	15 JUL 91	
1990-104W		21310	USSR	28 NOV	19 MAY 91	
1990-104X		21991	USSR	28 NOV	07 OCT 92	
1990-104Y		21992	USSR	28 NOV	17 SEP 92	
1990-104Z		21993	USSR	28 NOV	17 SEP 92	
1990-105AC		21127	US	01 DEC	08 APR 91	
1990-105AD		21215	US	01 DEC	08 APR 91	
1990-105C		20983	US	01 DEC	06 FEB 91	
1990-105D		20984	US	01 DEC	04 AUG 92	
1990-105E		20987	US	01 DEC	23 FEB 91	
1990-105F		20988	US	01 DEC	02 JUN 91	
1990-105G		20989	US	01 DEC	16 FEB 91	
1990-105H		20990	US	01 DEC	10 JAN 91	
1990-105J		20991	US	01 DEC	20 SEP 91	
1990-105K		20992	US	01 DEC	02 MAR 91	
1990-105L		20993	US	01 DEC	02 SEP 91	
1990-105N		20999	US	01 DEC	02 FEB 91	
1990-105P		20997	US	01 DEC	05 JAN 91	
1990-105Q		21071	US	01 DEC	09 APR 91	
1990-105R		21072	US	01 DEC	18 MAR 91	
1990-105T		21074	US	01 DEC	09 MAR 91	
1990-105U		21075	US	01 DEC	09 FEB 91	
1990-105V		21076	US	01 DEC	21 FEB 91	
1990-105W		21077	US	01 DEC	19 MAY 91	
1990-105X		21078	US	01 DEC	12 APR 91	
1990-105Y		21079	US	01 DEC	12 APR 91	
1990-106A	STS-35	20980	US	02 DEC	11 DEC 90	
1990-107A	SOTUZ TM 11	20981	USSR	02 DEC	26 MAY 91	
1990-107B		20982	USSR	02 DEC	04 DEC 90	
1990-108A	COSMOS 2107	20985	USSR	04 DEC	05 APR 92	
1990-108B		20986	USSR	04 DEC	04 DEC 90	
1990-109A	COSMOS 2108	21000	USSR	04 DEC	28 JAN 91	
1990-109B		21001	USSR	04 DEC	09 DEC 90	
1990-110D		21009	USSR	08 DEC	08 DEC 90	
1990-110E		21010	USSR	08 DEC	09 DEC 90	
1990-112B		21017	USSR	20 DEC	22 DEC 90	
1990-112C		21018	USSR	20 DEC	21 DEC 90	
1990-112E		21024	USSR	21 DEC	31 JUL 91	
1990-113A	COSMOS 2113	21026	USSR	21 DEC	11 JUN 91	
1990-113B		21027	USSR	21 DEC	22 DEC 90	
1990-115A	COSMOS 2120	21035	USSR	26 DEC	17 JAN 91	
1990-115B		21036	USSR	26 DEC	28 DEC 90	
1990-115C		21037	USSR	26 DEC	28 DEC 90	
1990-115D		21061	USSR	26 DEC	18 JAN 91	
1990-115E		21062	USSR	26 DEC	19 JAN 91	
1990-115F		21063	USSR	26 DEC	19 JAN 91	
1990-115G		21068	USSR	26 DEC	21 JAN 91	
1989-100W		21021	USSR	27 DEC	21 DEC 90	
1990-116B		21039	USSR	27 DEC	29 DEC 90	
1990-116C		21040	USSR	27 DEC	28 DEC 90	
1990-116E		21044	USSR	27 DEC	21 JUL 91	
1991 LAUNCHES						
1991-002A	PROGRESS M6	21053	USSR	14 JAN	15 MAR 91	
1991-002B		21054	USSR	14 JAN	15 JAN 91	
1991-004A	COSMOS 2121	21059	USSR	17 JAN	10 FEB 91	
1991-004B		21060	USSR	17 JAN	19 JAN 91	
1991-004C		21095	USSR	17 JAN	21 FEB 91	
1991-004D		21096	USSR	17 JAN	12 FEB 91	
1991-004E		21097	USSR	17 JAN	12 FEB 91	
1991-004F		21098	USSR	17 JAN	14 FEB 91	
1991-004G		21099	USSR	17 JAN	11 FEB 91	
1991-005A	COSMOS 2122	21065	USSR	18 JAN	28 MAR 93	
1991-005B		21066	USSR	18 JAN	19 JAN 91	
1991-005C		21067	USSR	18 JAN	30 MAY 91	
1991-005D		22299	USSR	18 JAN	02 JAN 93	
1991-008A	COSMOS 2124	21092	USSR	07 FEB	07 APR 91	
1991-008B		21093	USSR	07 FEB	11 FEB 91	
1991-008C		21094	USSR	07 FEB	08 FEB 91	
1991-010B		21112	USSR	14 FEB	16 FEB 91	
1991-010C		21113	USSR	14 FEB	14 FEB 91	
1991-011A	COSMOS 2134	21116	USSR	15 FEB	01 APR 91	
1991-011B		21117	USSR	15 FEB	19 FEB 91	
1991-011C		21123	USSR	15 FEB	16 FEB 91	
1991-012B		21119	USSR	15 FEB	26 FEB 91	
1991-012C		21120	USSR	15 FEB	22 FEB 91	
1991-014B		21133	USSR	28 FEB	02 MAR 91	
1991-014C		21134	USSR	28 FEB	28 FEB 91	
1991-014E		21201	USSR	28 FEB	10 SEP 91	
1991-014F		21202	USSR	28 FEB	10 AUG 91	
1991-016A	COSMOS 2136	21143	USSR	06 MAR	20 MAR 91	
1991-016B		21144	USSR	06 MAR	17 MAR 91	
1991-016C		21145	USSR	06 MAR	08 MAR 91	
1991-016D		21146	USSR	06 MAR	09 MAR 91	
1991-016E		21192	USSR	06 MAR	22 MAR 91	
1991-016F		21193	USSR	06 MAR	24 MAR 91	
1991-016G		21194	USSR	06 MAR	21 MAR 91	
1991-016H		21195	USSR	06 MAR	22 MAR 91	
1991-020A	PROGRESS M-7	21188	USSR	19 MAR	07 MAY 91	
1991-020B		21189	USSR	19 MAR	20 MAR 91	
1991-022B		21197	USSR	22 MAR	02 APR 91	
1991-022C		21198	USSR	22 MAR	28 MAR 91	
1991-022E		21200	USSR	22 MAR	29 MAR 91	
1991-023A	COSMOS 2138	21203	USSR	26 MAR	24 MAY 91	
1991-023B		21204	USSR	26 MAR	29 MAR 91	
1991-024A	ALMAZ-1	21213	USSR	30 MAR	17 OCT 92	
1991-024B		21214	USSR	30 MAR	02 APR 91	
1991-025D		21219	USSR	04 APR	04 APR 91	
1991-027A		21224	US	05 APR	11 APR 91	
1991-031A	STS-39	21242	US	28 APR	06 MAY 91	
1991-031B	IBSS	21244	US	28 APR	06 MAY 91	
1991-031D		21245	US	28 APR	14 MAY 91	
1991-031E		21246	US	28 APR	12 MAY 91	
1991-031F		21247	US	28 APR	13 MAY 91	
1991-034A	SOYUZ TM-12	21311	USSR	18 MAY	10 OCT 91	
1991-034B		21312	USSR	18 MAY	19 MAY 91	
1991-035A	RESURS F-10	21313	USSR	21 MAY	20 JUN 91	
1991-035B		21314	USSR	21 MAY	23 MAY 91	
1991-035D		21480	USSR	21 MAY	20 JUN 91	
1991-035E		21485	USSR	21 MAY	23 JUN 91	
1991-036A	COSMOS 2149	21315	USSR	24 MAY	04 JUL 91	
1991-036B		21316	USSR	24 MAY	27 MAY 91	
1991-038A	PROGRESS M-8	21395	USSR	30 MAY	16 AUG 91	
1991-038B		21396	USSR	30 MAY	31 MAY 91	
1991-040A	STS-40	21399	US	05 JUN	14 JUN 91	
1991-043B		21427	USSR	18 JUN	02 JUL 91	
1991-043C		21428	USSR	18 JUN	27 JUN 91	
1991-044A	RESURS F-11	21524	USSR	28 JUN	21 JUL 91	
1991-044B		21525	USSR	28 JUN	30 JUN 91	

International Designation	Satellite Name	Number	Source	Launch	Decay	Note
1991-044C		21604	USSR	28 JUN	21 JUL 91	
1991-044D		21605	USSR	28 JUN	10 AUG 91	
1991-044E		21606	USSR	28 JUN	23 JUL 91	
1991-044F		21607	USSR	28 JUN	25 JUL 91	
1991-044G		21608	USSR	28 JUN	28 JUL 91	
1991-044H		21609	USSR	28 JUN	23 JUL 91	
1991-046B		21534	USSR	02 JUL	04 JUL 91	
1991-046C		21535	USSR	02 JUL	03 JUL 91	
1991-046F		21541	USSR	02 JUL	26 FEB 92	
1991-047B	LOSAT-X	21553	US	04 JUL	15 NOV 91	
1991-047C		21554	US	04 JUL	29 OCT 91	
1991-048A	COSMOS 2152	21558	USSR	09 JUL	23 JUL 91	
1991-048B		21559	USSR	09 JUL	10 JUL 91	
1991-048C		21613	USSR	09 JUL	27 JUL 91	
1991-048D		21614	USSR	09 JUL	31 JUL 91	
1991-048E		21615	USSR	09 JUL	25 JUL 91	
1991-048F		21616	USSR	09 JUL	24 JUL 91	
1991-048G		21617	USSR	09 JUL	26 JUL 91	
1991-049A	COSMOS 2153	21560	USSR	10 JUL	13 MAR 92	
1991-049B		21561	USSR	10 JUL	13 JUL 91	
1991-051A	MICROSAT 1	21580	US	17 JUL	23 JAN 92	
1991-051B	MICROSAT 2	21581	US	17 JUL	23 JAN 92	
1991-051C	MICROSAT 3	21582	US	17 JUL	24 JAN 92	
1991-051D	MICROSAT 4	21583	US	17 JUL	23 JAN 92	
1991-051E	MICROSAT 5	21584	US	17 JUL	24 JAN 92	
1991-051F	MICROSAT 6	21585	US	17 JUL	25 JAN 92	
1991-051G	MICROSAT 7	21586	US	17 JUL	23 JAN 92	
1991-051H		21587	US	17 JUL	08 NOV 91	
1991-052A	RESURS F-12	21611	USSR	23 JUL	08 AUG 91	
1991-052B		21612	USSR	23 JUL	25 JUL 91	
1991-052C		21643	USSR	23 JUL	12 AUG 91	
1991-052D		21644	USSR	23 JUL	10 AUG 91	
1991-052E		21645	USSR	23 JUL	28 AUG 91	
1991-052F		21646	USSR	23 JUL	15 AUG 91	
1991-052G		21647	USSR	23 JUL	11 AUG 91	
1991-053B		21631	USSR	01 AUG	02 SEP 91	
1991-053C		21632	USSR	01 AUG	03 SEP 91	
1991-054A	STS-43	21638	US	02 AUG	11 AUG 91	
1991-057A	PROGRESS M-9	21662	USSR	20 AUG	30 SEP 91	
1991-057B		21663	USSR	20 AUG	22 AUG 91	
1991-058A	RESURS F-13	21664	USSR	21 AUG	20 SEP 91	
1991-058B		21665	USSR	21 AUG	23 AUG 91	
1991-058C		21715	USSR	21 AUG	21 SEP 91	
1991-058D		21716	USSR	21 AUG	21 SEP 91	
1991-058E		21717	USSR	21 AUG	22 SEP 91	
1991-060B		21669	JAPAN	25 AUG	07 SEP 91	
1991-060C		21670	JAPAN	25 AUG	01 MAY 92	
1991-062C		21696	JAPAN	30 AUG	27 JUN 93	
1991-062E		21698	JAPAN	30 AUG	11 MAY 93	
1991-062G		21786	JAPAN	30 AUG	09 JUN 93	
1991-063A	STS-48	21700	US	12 SEP	18 SEP 91	
1991-064C		21704	USSR	13 SEP	14 SEP 91	
1991-064D		21705	USSR	13 SEP	15 SEP 91	
1991-064E		21739	USSR	13 SEP	09 APR 92	
1991-064F		21740	USSR	13 SEP	09 MAY 92	
1991-065B		21707	USSR	17 SEP	30 SEP 91	
1991-065C		21708	USSR	17 SEP	25 SEP 91	
1991-065E		21710	USSR	17 SEP	27 SEP 91	
1991-065F		21718	USSR	17 SEP	27 SEP 91	
1991-065G		21719	USSR	17 SEP	25 SEP 91	
1991-065H		21720	USSR	17 SEP	25 SEP 91	
1991-066A	COSMOS 2156	21713	USSR	19 SEP	17 NOV 91	
1991-066B		21714	USSR	19 SEP	22 SEP 91	
1991-069A	SOYUZ TM-13	21735	USSR	02 OCT	25 MAR 92	
1991-069B		21736	USSR	02 OCT	03 OCT 91	
1991-070A	FOTON 4	21737	USSR	04 OCT	20 OCT 91	
1991-070B		21738	USSR	04 OCT	16 OCT 91	
1991-070C		21755	USSR	04 OCT	25 OCT 91	
1991-070D		21756	USSR	04 OCT	25 OCT 91	
1991-070E		21757	USSR	04 OCT	27 OCT 91	
1991-070F		21758	USSR	04 OCT	09 NOV 91	
1991-071A	COSMOS 2163	21741	USSR	09 OCT	07 DEC 91	
1991-0718		21742	USSR	09 OCT	12 OCT 91	
1991-072A	COSMOS 2164	21743	USSR	10 OCT	12 DEC 92	
1991-072B		21744	USSR	10 OCT	14 MAR 92	
1991-072C		21745	USSR	10 OCT	25 OCT 91	
1991-073A	PROGRESS M-10	21746	USSR	17 OCT	20 JAN 92	
1991-073B		21747	USSR	17 OCT	18 OCT 91	
1991-0748		21760	USSR	23 OCT	25 OCT 91	
1991-074C		21761	USSR	23 OCT	24 OCT 91	
1991-074E		21763	USSR	23 OCT	02 AUG 92	
1991-074F		21764	USSR	23 OCT	16 APR 92	
1991-078A	COSMOS 2171	21787	USSR	20 NOV	17 JAN 92	
1991-078B		21788	USSR	20 NOV	24 NOV 91	
1991-078C		21843	USSR	20 NOV	18 JAN 92	
1991-079B		21790	USSR	22 NOV	24 NOV 91	
1991-079C		21791	USSR	22 NOV	23 NOV 91	
1991-079E		21793	USSR	22 NOV	23 JUN 92	
1991-080A	STS 44	21795	US	24 NOV	01 DEC 91	
1991-085A	COSMOS 2174	21816	USSR	17 DEC	30 JAN 92	
1991-085B		21817	USSR	17 DEC	22 DEC 91	
1991-087B		21822	USSR	19 DEC	21 DEC 91	
1991-087C		21823	USSR	19 DEC	20 DEC 91	
1991-087C		21823	USSR	19 DEC	20 DEC 91	
1991-087E		21828	USSR	19 DEC	13 AUG 92	
1991-087E		21828	USSR	19 DEC	13 AUG 92	
1991-088B		21834	PRC	28 DEC	12 APR 93	
1991-088B		21834	PRC	28 DEC	12 APR 93	

1992 LAUNCHES

International Designation	Satellite Name	Number	Source	Launch	Decay	Note
1992-001A	COSMOS 2175	21844	USSR	21 JAN	20 MAR 92	
1992-001B		21845	USSR	21 JAN	24 JAN 92	
1992-002A	STS-42	21846	US	22 JAN	30 JAN 92	
1992-003B		21848	USSR	24 JAN	19 FEB 92	
1992-003C		21849	USSR	24 JAN	05 FEB 92	
1992-004A	PROGRESS M-1	21851	USSR	25 JAN	13 MAR 92	
1992-004B		21852	USSR	25 JAN	26 JAN 92	
1992-005D		21856	USSR	29 JAN	30 JAN 92	
1992-005E		21857	USSR	29 JAN	30 JAN 92	
1992-011B		21898	USSR	04 MAR	28 MAR 92	
1992-011C		21899	USSR	04 MAR	19 MAR 92	
1992-014A	SOYUZ TM-14	21908	USSR	17 MAR	10 AUG 92	
1992-014B		21909	USSR	17 MAR	19 MAR 92	
1992-015A	STS 45	21915	US	24 MAR	02 APR 92	
1992-016A	COSMOS 2182	21920	USSR	01 APR	30 MAY 92	
1992-016B		21921	USSR	01 APR	04 APR 92	
1992-017B		21923	USSR	02 APR	04 APR 92	
1992-017C		21924	USSR	02 APR	02 APR 92	
1992-017E		21926	USSR	02 APR	23 OCT 92	
1992-017F		21927	USSR	02 APR	04 NOV 92	
1992-018A	COSMOS 2183	21928	USSR	08 APR	16 FEB 93	
1992-018B		21929	USSR	08 APR	10 APR 92	
1992-022A	PROGRESS M-12	21946	USSR	19 APR	27 JUN 92	
1992-022B		21947	USSR	19 APR	21 APR 92	
1992-024A	RESURS F-14	21951	USSR	29 APR	29 MAY 92	
1992-024B		21952	USSR	29 APR	01 MAY 92	
1992-024C		21975	USSR	29 APR	30 MAY 92	
1992-025A	COSMOS 2185	21953	USSR	29 APR	11 JUN 92	
1992-025B		21954	USSR	29 APR	03 MAY 92	
1992-026A	STS 49	21963	US	07 MAY	16 MAY 92	
1992-028A	SROSS 3	21968	INDIA	20 MAY	14 JUL 92	
1992-028B		21967	INDIA	20 MAY	24 JUN 92	
1992-029A	COSMOS 2186	21973	USSR	28 MAY	24 JUL 92	
1992-029B		21974	USSR	28 MAY	03 JUN 92	
1992-031B		21988	US	07 JUN	16 OCT 92	
1992-033A	RESURS F 15	21998	USSR	23 JUN	09 JUL 92	
1992-033B		21999	USSR	23 JUN	25 JUN 92	
1992-033C		22026	USSR	23 JUN	12 JUL 92	
1992-033D		22029	USSR	23 JUN	11 JUL 92	
1992-033E		22030	USSR	23 JUN	21 JUL 92	
1992-033F		22031	USSR	23 JUN	10 JUL 92	
1992-034A	STS 50	22000	US	25 JUN	09 JUL 92	
1992-035A	PROGRESS M-13	22004	USSR	30 JUN	24 JUL 92	
1992-035B		22005	USSR	30 JUN	02 JUL 92	
1992-040B		22018	USSR	08 JUL	15 AUG 92	
1992-040C		22019	USSR	08 JUL	26 JUL 92	
1992-040E		22021	USSR	08 JUL	14 JUL 92	
1992-040F		22022	USSR	08 JUL	07 AUG 92	
1992-043B		22042	USSR	14 JUL	17 JUL 92	
1992-043C		22043	USSR	14 JUL	15 JUL 92	
1992-043E		22047	USSR	14 JUL	18 JUN 93	
1992-044B		22050	US	24 JUL	16 MAR 93	
1992-045A	COSMOS 2203	22052	USSR	24 JUL	22 SEP 92	
1992-045B		22053	USSR	24 JUL	01 AUG 92	
1992-045C		22135	USSR	24 JUL	23 SEP 92	
1992-046A	SOYUZ TM-15	22054	USSR	27 JUL	01 FEB 93	
1992-046B		22055	USSR	27 JUI	29 JUL 92	
1992-047D		22059	USSR	30 JUL	30 JUL 92	
1992-048A	COSMOS 2207	22062	USSR	30 JUL	13 AUG 92	
1992-048B		22063	USSR	30 JUL	01 AUG 92	
1992-048C		22082	USSR	30 JUL	30 SEP 92	
1992-048D		22083	USSR	30 JUL	23 AUG 92	
1992-048E		22084	USSR	30 JUL	18 AUG 92	
1992-048F		22085	USSR	30 JUL	29 AUG 92	
1992-048G		22086	USSR	30 JUL	21 AUG 92	
1992-049A	STS-46	22064	US	31 JUL	08 AUG 92	
1992-050B		22069	USSR	06 AUG	20 SEP 92	
1992-050C		22070	USSR	06 AUG	29 AUG 92	
1992-050E		22095	USSR	06 AUG	08 SEP 92	
1992-051A	PRC 35	22072	PRC	09 AUG	01 SEP 92	
1992-051B		22073	PRC	09 AUG	16 AUG 92	
1992-051C		22074	PRC	09 AUG	09 AUG 92	
1992-051D		22075	PRC	09 AUG	10 AUG 92	
1992-054B		22088	PRC	13 AUG	12 DEC 92	
1992-055A	PROGRESS M-14	22090	USSR	15 AUG	21 OCT 92	
1992-055B		22091	USSR	15 AUG	18 AUG 92	
1992-055C		22092	USSR	15 AUG	17 AUG 92	
1992-056A	RESURS F-16	22093	USSR	19 AUG	04 SEP 92	
1992-056B		22094	USSR	19 AUG	21 AUG 92	
1992-056C	PION 1	22099	USSR	19 AUG	25 SEP 92	
1992-056D	PION 2	22100	USSR	19 AUG	24 SEP 92	
1992-056E		22102	USSR	19 AUG	13 SEP 92	
1992-056F		22103	USSR	19 AUG	07 SEP 92	
1992-056G		22104	USSR	19 AUG	06 SEP 92	
1992-056H		22105	USSR	19 AUG	07 SEP 92	
1992-059B		22113	USSR	10 SEP	13 SEP 92	
1992-059C		22114	USSR	10 SEP	11 SEP 92	
1992-059E		22125	USSR	10 SEP	31 MAR 93	
1992-059F		22126	USSR	10 SEP	07 MAY 93	
1992-061A	STS 47	22120	US	12 SEP	20 SEP 92	
1992-062A	COSMOS 2210	22133	USSR	22 SEP	20 NOV 92	
1992-062B		22134	USSR	22 SEP	26 SEP 92	
1992-064B	PRC 36	22162	PRC	06 OCT	31 OCT 92	
1992-064C		22163	PRC	06 OCT	20 OCT 92	
1992-064D		22164	PRC	06 OCT	09 OCT 92	
1992-064E		22165	PRC	06 OCT	08 OCT 92	

International Designation	Satellite Name	Number	Source	Launch	Decay	Note
1992-064F		22166	PRC	06 OCT	09 OCT 92	
1992-064G		22167	PRC	06 OCT	08 OCT 92	
1992-064H		22168	PRC	06 OCT	10 OCT 92	
1992-064J		22169	PRC	06 OCT	09 OCT 92	
1992-064K		22170	PRC	06 OCT	09 OCT 92	
1992-065A	FOTON 5	22173	USSR	08 OCT	24 OCT 92	
1992-065B		22174	USSR	08 OCT	22 OCT 92	
1992-065C		22199	USSR	08 OCT	20 NOV 92	
1992-065D		22200	USSR	08 OCT	31 OCT 92	
1992-065E		22201	USSR	08 OCT	30 OCT 92	
1992-065F		22202	USSR	08 OCT	03 NOV 92	
1992-067B		22179	USSR	14 OCT	28 OCT 92	
1992-067C		22180	USSR	14 OCT	24 OCT 92	
1992-067E		22193	USSR	14 OCT	28 OCT 92	
1992-069B		22190	USSR	21 OCT	25 NOV 92	
1992-069C		22191	USSR	21 OCT	07 NOV 92	
1992-070A	STS 52	22194	US	22 OCT	01 NOV 92	
1992-070C	CTA	22214	CAN	22 OCT	01 NOV 92	
1992-071A	PROGRESS M-15	22203	USSR	27 OCT	07 FEB 93	
1992-071B		22204	USSR	27 OCT	28 OCT 92	
1992-071C		22449	USSR	27 OCT	05 FEB 93	
1992-074B		22211	USSR	30 OCT	01 NOV 92	
1992-074F		22216	USSR	30 OCT	03 JUN 93	
1992-074C		22212	USSR	30 OCT	31 OCT 92	
1992-075A	RESURS 500	22217	USSR	15 NOV	22 NOV 92	
1992-075B		22218	USSR	15 NOV	18 NOV 92	
1992-075C		22234	USSR	15 NOV	30 NOV 92	
1992-075D		22235	USSR	15 NOV	28 NOV 92	
1992-077A	COSMOS 2220	22226	USSR	20 NOV	18 JAN 93	
1992-077B		22227	USSR	20 NOV	24 NOV 92	
1992-078B		22230	US	21 NOV	09 FEB 93	
1992-081B		22239	USSR	25 NOV	28 DEC 92	
1992-081C		22240	USSR	25 NOV	14 DEC 92	
1992-081E		22242	USSR	25 NOV	15 DEC 92	
1992-081F		22243	USSR	25 NOV	29 NOV 92	
1992-081G		22244	USSR	25 NOV	09 DEC 92	
1992-082B		22246	USSR	27 NOV	30 NOV 92	
1992-082C		22247	USSR	27 NOV	28 NOV 92	
1992-085B		22256	USSR	02 DEC	14 DEC 92	
1992-085C		22257	USSR	02 DEC	11 DEC 92	
1992-085E		22262	USSR	02 DEC	15 DEC 92	
1992-086A	STS 53	22259	US	02 DEC	09 DEC 92	
1992-087B		22261	USSR	09 DEC	12 DEC 92	
1992-088B		22270	USSR	17 DEC	19 DEC 92	
1992-088C		22271	USSR	17 DEC	18 DEC 92	
1992-090B		22279	PRC	21 DEC	19 JUN 93	
1992-091A	COSMOS 2225	22280	USSR	22 DEC	18 FEB 93	
1992-091B		22281	USSR	22 DEC	26 DEC 92	
1992-091C		22520	USSR	22 FEB	23 FEB 93	
1992-091D		22525	USSR	22 DEC	25 FEB 93	
1992-091E		22526	USSR	22 DEC	22 FEB 93	
1992-091F		22527	USSR	22 DEC	27 FEB 93	
1992-091G		22529	USSR	22 DEC	24 FEB 93	
1992-093FY		22483	USSR	25 DEC	22 FEB 93	
1992-093HJ		22548	USSR	25 DEC	19 MAY 93	
1992-095A	COSMOS 2229	22300	USSR	29 DEC	10 JAN 93	
1992-095B		22301	USSR	29 DEC	16 JAN 93	
1992-095C		22302	USSR	29 DEC	11 FEB 93	
1992-095D		22303	USSR	29 DEC	18 JAN 93	
1992-095E		22304	USSR	29 DEC	29 JAN 93	
1992-095F		22305	USSR	29 DEC	18 JAN 93	
1992-095G		22306	USSR	29 DEC	12 JAN 93	
1993 LAUNCHES						
1993-002C		22311	USSR	13 JAN	05 FEB 93	
1993-003A	STS 54	22313	US	13 JAN	19 JAN 93	
1993-004A	COSMOS 2231	22317	USSR	19 JAN	25 MAR 93	
1993-004B		22318	USSR	19 JAN	23 JAN 93	
1993-005B		22320	USSR	24 JAN	26 JAN 93	
1993-006B		22322	USSR	26 JAN	09 MAR 93	
1993-006C		22323	USSR	26 JAN	14 FEB 93	
1993-006E		22325	USSR	26 JAN	28 FEB 93	
1993-010D		22515	USSR	17 FEB	18 FEB 93	
1993-010E		22516	USSR	17 FEB	18 FEB 93	
1993-012A	PROGRESS M-16	22530	USSR	21 FEB	27 MAR 93	
1993-012B		22531	USSR	21 FEB	23 FEB 93	
1993-013B		22558	USSR	25 MAR	27 MAR 93	
1993-013C		22559	USSR	25 MAR	25 MAR 93	
1993-017B	SEDS I	22582	US	30 MAR	30 MAR 93	
1993-0018B		22586	USSR	30 MAR	31 MAR 93	
1993-019B		22589	USSR	31 MAR	01 APR 93	
1993-021A	COSMOS 2240	22592	USSR	02 APR	07 JUN 93	
1993-021B		22593	USSR	02 APR	07 APR 93	
1993-022B		22595	USSR	06 APR	10 MAY 93	
1993-022C		22596	USSR	06 APR	22 APR 93	
1993-022E		22629	USSR	06 APR	29 APR 93	
1993-022F		22630	USSR	06 APR	05 MAY 93	
1993-023A	STS 56	22621	US	08 APR	17 APR 93	
1993-023B	SPARTAN 201	22623	US	08 APR	13 APR 93	
1993-025B		22634	USSR	21 APR	30 MAY 93	
1993-025C		22635	USSR	21 APR	12 MAY 93	
1993-027A	STS 55	22640	US	26 APR	06 MAY 93	
1993-028A	COSMOS 2243	22641	USSR	27 APR	06 MAY 93	
1993-028B		22642	USSR	27 APR	01 MAY 93	
1993-029B		22644	USSR	28 APR	28 APR 93	
1993-033A	RESURS F-2	22663	USSR	21 MAY	20 JUN 93	
1993-033B		22664	USSR	21 MAY	24 MAY 93	
1993-033C		22681	USSR	21 MAY	22 JUN 93	
1993-033D		22682	USSR	21 MAY	22 JUN 93	
1993-033E		22685	USSR	21 MAY	25 JUN 93	
1993-033F		22686	USSR	21 MAY	23 JUN 93	
1993-034B		22667	USSR	22 MAY	25 MAY 93	
1993-035B		22672	USSR	26 MAY	16 JUN 93	
1993-035C		22673	USSR	26 MAY	06 JUN 93	
1993-040B		22697	USSR	25 JUN	27 JUN 93	

Notes on Satellites No Longer in Orbit

6. UNIDENTIFIED DEBRIS IN ORBIT.
14. UNIDENTIFIED DEBRIS IN ORBIT.
35. DEPLOYED FROM SPACE TRANSPORTATION VEHICLE.
A. ANNOUNCED LUNAR IMPACT.
B. SUCCESSFUL RE-ENTRY AND RECOVERY.
C. ANNOUNCED SUCCESSFUL RE-ENTRY AND RECOVERY.
D. SUCCESSFUL RE-ENTRY BUT NOT RECOVERY.
E. RE-ENTRY AND RECOVERY OF MANNED SPACE VEHICLE.
F. ORBITED AND RECOVERED MANNED SPACE VEHICLE.
G. LUNAR IMPACT.
H. 24 OBJECTS LAUNCHED WITH 1962 B-IOTA 1. ALL DECAYED.
J. 98 OBJECTS LAUNCHED WITH 1964 070A. ALL HAVE DECAYED.
K. 168 OBJECTS LAUNCHED WITH 1965 012A. ALL HAVE DECAYED.
L. 150 OBJECTS LAUNCHED WITH 1965 202A-C. SEE1965 IN ORBIT.
M. 24 OBJECTS LAUNCHED WITH 1965 088A. ALL HAVE DECAYED.
N. ANNOUNCED DECAY ON VENUS.
P. 15 OBJECTS LAUNCHED WITH 1965 112A. SEE1965 IN ORBIT.
Q. UNIDENTIFIED DEBRIS IN ORBIT.
R. ANNOUNCED SPACECRAFT LANDED ON MOON.
S. 40 OBJECTS LAUNCHED WITH 1966 012 A-C. ALL HAVE DECAYED.
T. LANDED ON MOON.
U. 54 OBJECTS LAUNCHED WITH 1966 046A. ALL HAVE DECAYED.
V. 82 OBJECTS LAUNCHED WITH 1966 056A. SEE1966 IN ORBIT.
W. 35 OBJECTS LAUNCHED WITH 1966 059A. ALL HAVE DECAYED.
X. 53 OBJECTS LAUNCHED WITH 1966 088A. ALL HAVE DECAYED.
Y. 41 OBJECTS LAUNCHED WITH 1966 101A. ALL HAVE DECAYED.
Z. 15 OBJECTS LAUNCHED WITH 1966 104A. ALL HAVE DECAYED.
AB. 15 OBJECTS LAUNCHED WITH 1967 011A. SEE1967 IN ORBIT.
AC. 18 OBJECTS LAUNCHED WITH 1967 024A. ALL HAVE DECAYED.
AD. RE-ENTRY AND RECOVERY OF MANNED SPACE VEHICLE.
AE. 16 OBJECTS LAUNCHED WITH 1968 025A. ALL HAVE DECAYED.
AF. 24 OBJECTS LAUNCHED WITH 1968 117A. ALL HAVE DECAYED.
AG. MANNED SPACECRAFT ORBITED MOON AND WAS RECOVERED.
AH. 23 OBJECTS LAUNCHED WITH 1969 021A. SEE1969 IN ORBIT.
AJ. MANNED. ORBITED MOON. MODULE LANDED, RECOVERED.
AK. 38 OBJECTS LAUNCHED WITH 1969 029A. SEE1969 IN ORBIT.
AL. 27 OBJECTS LAUNCHED WITH 1969 064A. SEE 969 IN ORBIT.
AM. MANNED. LANDED MOON. RETURNED ORBIT. DECAYED MOON.
AN. MANNED. PASSED AROUND MOON. RECOVERED.
AP. ANNOUNCED LANDING MOON. RETURN TO EARTH. RECOVERY.
AQ. UNIDENTIFIED DEBRIS IN ORBIT.
AR. ANNOUNCED DECAY ON MARS.
AS. 15 OBJECTS LAUNCHED WITH 1971 074A. ALL HAVE DECAYED.
AT. 29 OBJECTS LAUNCHED WITH 1971 106A. ALL HAVE DECAYED.
AU. 26 OBJECTS LAUNCHED WITH 1973 017A. ALL HAVE DECAYED.
AV. 197 OBJECTS LAUNCHED WITH 1973 021A. ALL HAVE DECAYED.
AW. 24 OBJECTS LAUNCHED WITH 1973 027A. ALL HAVE DECAYED.
AX. 14 OBJECTS LAUNCHED WITH 1973 075A. ALL HAVE DECAYED.
AY. 20 OBJECTS LAUNCHED WITH 1974 074A. ALL HAVE DECAYED.
AZ. 51 OBJECTS LAUNCHED WITH 1974 103A. ALL HAVE DECAYED.
BA. 19 OBJECTS LAUNCHED WITH 1974 104A. ALL HAVE DECAYED.
BB. 19 OBJECTS LAUNCHED WITH 1975 044A. ALL HAVE DECAYED.
BC. SUCCESSFUL APOLLO AND SOYUZ DOCKING AND RECOVERY.
BD. LANDED ON MARS.
BE. 77 OBJECTS LAUNCHED WITH 1975 080A. ALL HAVE DECAYED.
BF. 63 OBJECTS LAUNCHED WITH 1975 102A. ALL HAVE DECAYED.
BG. 17 OBJECTS LAUNCHED WITH 1976 012A. ALL HAVE DECAYED.
BH. 25 OBJECTS LAUNCHED WITH 1976 037A. ALL HAVE DECAYED.
BJ. 41 OBJECTS LAUNCHED WITH 1976 063A. ALL HAVE DECAYED.
BK. 249 OBJECTS LAUNCHED WITH 1976 072A. ALL HAVE DECAYED.
BL. ANNOUNCED RETURN TO EARTH FROM MOON. RECOVERY.
BM. 18 OBJECTS LAUNCHED WITH 1976 124A. ALL HAVE DECAYED.
BN. 22 OBJECTS LAUNCHED WITH 1977 042A. ALL HAVE DECAYED.
BP. 106 OBJECTS LAUNCHED WITH 1977 097A. ALL HAVE DECAYED.
BQ. 27 OBJECTS LAUNCHED WITH 1977 111A. ALL HAVE DECAYED.
BR. 26 OBJECTS LAUNCHED WITH 1979 063A. ALL HAVE DECAYED.
BS. 27 OBJECTS LAUNCHED WITH 1980 047A. ALL HAVE DECAYED.
BT. EARTH ORBIT AND LANDING OF MANNED SPACECRAFT.
BU. 24 OBJECTS LAUNCHED WITH 1980 067A. ALL HAVE DECAYED.
BV. 26 OBJECTS LAUNCHED WITH 1981 097A. SEE1981 IN ORBIT.
BW. 25 OBJECTS LAUNCHED WITH 1982 007A. SEE 1982 IN ORBIT.
BX. 184 OBJECTS LAUNCHED WITH 1982 033A. SEE 1982 ORBIT.
BY. 24 OBJECTS LAUNCHED WITH 1982 076A. ALL HAVE DECAYED.
BZ. 33 OBJECTS LAUNCHED WITH 1982 088A. SEE 1982 IN ORBIT.
CA. WESTAR 6 RECOVERED BY STS 51A NOV. 14. 1984.
CB. PALAPA B2 RECOVERED BY STS 51A NOV. 12. 1984.
CC. SOFT LANDED ON VENUS.
CD. 29 OBJECTS LAUNCHED WITH 1985 050A. SEE1985 IN ORBIT.

Index

AAA

DDD

EEE

FFF

GGG

HHH

KKK

LLL

MMM

NNN

OOO

PPP

QQQ

RRR

SSS

TTT

UUU

VVV